Commission of the European Communities

Eighth E.C. Photovoltaic Solar Energy Conference

Proceedings of the International Conference, held at Florence, Italy, 9-13 May 1988

Volume I

Edited by

I. SOLOMON and B. EQUER

Ecole Polytechnique, Palaiseau, Fr

and

P. HELM

WIP, Munich, F.R.G.

KLUWER ACADEMIC PUBLISHERS

DORDRECHT / BOSTON / LONDON

Library of Congress Cataloging in Publication Data

E.C. Photovoltaic Solar Energy Conference (8th : 1988 : Florence, Italy)
Eighth E.C. Photovoltaic Solar Energy Conference : proceedings of the international conference held at Florence, Italy, 9-13 May 1988 / edited by I. Solomon and B. Equer, and P. Helm.
p. cm.
At head of title: Commission of the European Communities.
Includes bibliographies and index.
ISBN 9027728178 (set). ISBN 9027728151 (v. 1). ISBN 902772816X (v. 2)
1. Photovoltaic power generation--Congresses. 2. Solar cells--Congresses. I. Solomon, I., 1929- . II. Equer, B., 1936- III. Helm, P., 1939- . IV. Commission of the European Communities. V. Title. VI. Title: 8th E.C. Photovoltaic Solar Energy Conference.
TK2960.E26 1988
621.31'244--dc19 88-23066
CIP

ISBN 90-277-2815-1 (Vol. I)
ISBN 90-277-2816-X (Vol. II)
ISBN 90-277-2817-8 (Set)

Publication arrangements by
Commission of the European Communities
Directorate-General Telecommunications, Information Industries and Innovation, Luxembourg

EUR 11780

Published by Kluwer Academic Publishers
P.O. Box 17, 3300 AA Dordrecht, The Netherlands.

Kluwer Academic Publishers incorporates the publishing programmes of
D. Reidel, Martinus Nijhoff, Dr W. Junk and MTP Press.

Sold and distributed in the U.S.A. and Canada
by Kluwer Academic Publishers,
101 Philip Drive, Norwell, MA 02061, U.S.A.

In all other countries, sold and distributed
by Kluwer Academic Publishers Group,
P.O. Box 322, 3300 AH Dordrecht, The Netherlands.

Printed in The Netherlands

Eighth E.C.
Photovoltaic Solar Energy Conference

INTRODUCTION

The 8th European Photovoltaic Solar Energy Conference was held in the city of Florence, Italy, and was co-sponsored by the Italian Section of the International Solar Energy Society (ISES). It has been, by far, the largest conference of the series and the major event of the year for the photovoltaic community.

The present proceedings include a total of 343 presentations from 41 countries. This is almost a 50% increase over the previous Conference and has necessited a two-volume publication. The progress in *photocell* technology and in photovoltaic *systems* occupy about the same space in the presentations. The problem of storage, always underlying in photovoltaics, has been explicitely included as a topic of the Conference. For the first time it has been necessary, this year, to introduce parallel sessions which, however, have been reduced to a minimum compatible with the number of oral presentations.

The attendence at the conference has gone up beyond our expectation: 753 persons from 43 countries. The largest contingent came, as before, from the Federal Republic of Germany (187), followed by Italy (131) and France (83). We are happy to acknowledge a large participation of our collegues from the USA (56) and, for the first time, the conference has been attended by several physicists of the USSR.

This large increase, besides the attraction of Florence, is a clear indication of a new optimism in Photovoltaics. The large number of Companies (37) at the Exhibition presenting a large potential of production to a large number of interested visitors; the huge demand of the developping countries to provide electricity to regions that cannot be grid connected in the near future; the report at the Conference that solar electricity can become cost effective for grid-connected applications for a volume of production only 3 to 4 times the present production; all these facts, selected among others, suggest a promising perspective for the industrialisation of photovoltaics.

The next European Photovoltaic Conference will take place on the last week of September 1989 in Freiburg, Federal Republic of Germany.

I. Solomon,
Conference Chairman.

CONFERENCE COMMITTEE

General Chairman

Prof. I. Solomon
Ecole Polytechnique, Palaiseau
France

Technical Programme Chairmen

Dr. L. Crespo
Instituto de Energias Renovables
Madrid, Spain

Prof. R. van Overstraeten
IMEC, Leuven
Belgium

Conference Director

Dr. W. Palz
CEC, Brussels
Belgium

Conference Secretariat

Dr. G. Caratti
CEC, Brussels
Belgium

Prof. B. Equer
Ecole Polytechnique, Palaiseau
France

Dr. A. Räuber
Fraunhofer Institut ISE, Freiburg
Fed. Rep. of Germany

International Organization, Administration

Mr. P. Helm
WIP, Munich
Fed. Rep. of Germany

Mrs. H. Riegel
WIP, Munich
Fed. Rep. of Germany

Local Organization, Exhibition

Dr. G. Grassi
CEC, Brussels
Belgium

Miss A. Grassi
Florence, Italy

Dr. M.M. Maltagliati
Florence, Italy

Publications

Dr. G. Willeke
CEC, Brussels
Belgium

Posters

Dr. M.M. Maltagliati
Florence, Italy

Dr. H.W. Schock
University Stuttgart
Fed. Rep. of Germany

Public Relations

Dr. G. Grassi
CEC, Brussels
Belgium

Mrs. R. Roux
AFME, Valbonne
France

Dr. C. Mastrella
ENI, Rome
Italy

NATIONAL COMMITTEE

Chairman

Dr. C. Corvi
ENEL/ISES, Rome

Dr. G. Bianchi
ENEA, Roma

Prof. C. Boffa
CNR, Torino

Dr D. Coiante
ENEA, Roma

Dr. G. Corsini
Italsolar, Roma

Dr. G. Ghilardotti
AGIP, Milano

Dr. A. Iliceto
ENEL, Milano

Dr. M.M. Maltagliati
Galileo, Firenze

Prof. G. Martinelli
Helios Technology, Padova

Dr. C. Mastrella
ENI, Roma

Dr. C. Messana
ENEA/FARE, Roma

Dr. F. Morselli
Ministry Scientific Research, Roma

Prof. S. Pizzini
University Milan, Milano

Dr. A. Pela
Ministry of Industry, Roma

Dr. A. Previ
ENEL, Milano

Dr. A. Taschini
ENEL, Milano

Dr. R. Vigotti
ISES, Roma

Dr. P.E. Zani
Imet Control, Genova

HONORARY COMMITTEE

Chairman

On. Sen. Prof. G. Spadolini
President of Italian Senate,
Roma

On. Dr. A. Battaglia
Minister of Industry, Roma

Prof. A. Ruberti
Minister of Research, Roma

Prof. P. Fasella
General Director DG XII
CEC, Brussels

Prof. G. Ammassari
General Director - Energy Sources
Ministry of Industry, Roma

Mr. G. Bartolini
President of Tuscany Regional
Council, Firenze

Dr. M. Bogiankino
Mayor of Florence, Firenze

Dr.Ing. G. Bonnani
Director of Florence ENEL Dpt.
Firenze

Dr. A. Brasca
President of Tuscany Provincial
Council, Firenze

Prof. Ing. G.B. Clavarino
President Ansaldo, Genova

Prof. U. Colombo
President ENEA, Roma

Mr. C. Maniatopoulos
DG XVII, CEC, Brussels

Dr. M. Marcucci
Responsible for Energy, Tuscany
Regional Council, Firenze

Prof. F. Reviglio
President ENI, Roma

Prof. L. Rossi-Bernardi
President CNR, Roma

Prof. F. Scaramuzzi
Chancellor Univ. of Florence
Firenze

Prof. R. Scrimaglio
President Italsolar, Nettuno

Mr. F. Traverso
President Helios Technology
Galliera Veneta

Dr. F. Viezzoli
President ENEL, Roma

SCIENTIFIC COMMITTEE

Dr. A.M. Barnett
University Delaware, Newark
USA

Dr. G. Beer
Conphoebus, Catania
Italy

Dr. B. Berkovsky
UNESCO, Paris
France

Prof. W.H. Bloss
University Stuttgart
Fed. Rep. of Germany

Prof. F. Califano
University Rome
Italy

Dr. D.E. Carlson
Solarex, Newtown, PA
USA

J. Chadjivassiliadis
Public Power Corporation, Athens
Greece

Prof. I. Chambouleyron
Inst. of Physics, Univ. Campinas
Brazil

E. Fabre
Photowatt, Rueil-Malmaison
France

Prof. A. Frova
University Rome
Italy

Dr. R. Gislon
ENEA, Rome
Italy

Prof. A. Goetzberger
Fraunhofer Inst. ISE, Freiburg
Fed. Rep. of Germany

Dr. J. Grabmaier
Siemens AG, Munich
Fed. Rep. of Germany

Prof. M. Green
Univ. New South Wales
Australia

Prof. L.J. Guimaraes
University Nova, Lisbon
Portugal

Prof. Y. Hamakawa
University Osaka
Japan

M.S. Imamura
WIP, Munich
Fed. Rep. of Germany

W. Kaut
CEC Brussels
Belgium

Dr. L. Kazmerski
SERI, Golden, Colorado
USA

Dr. K. Krebs
CEC, JRC Ispra
Italy

Dr. Y. Kuwano
Sanyo, Osaka
Japan

Dr. F. Lasnier
Asian Inst. Techn. Bangkok
Thailand

Prof. A. Luque
Polytechn. University Madrid
Spain

Prof. R. Mertens
IMEC, Leuven
Belgium

Dr. M.B. Prince
DOE Washington
USA

Prof. J.A. Roger
University Lyon
France

Dr. K. Roy
Telefunken Electronic, Heilbronn
Fed. Rep. of Germany

Prof. E. Sirtl
Heliotronic Burghausen
Fed. Rep. of Germany

Dr. J. Stone
SERI, Golden, Colorado
USA

F.C. Treble
Farnborough
UK

Prof. G. Winterling
MBB, Munich
Fed. Rep. of Germany

Dr. J.F. Wise
USAF, Wright-Patterson, AFB
USA

Prof. G. Wrixon
Microel. Res. Centre, Cork
Ireland

C O N T E N T S

VOLUME I

SESSION 08: SI-MATERIALS

SESSION 10: CONCENTRATORS AND SYSTEMS

SESSION 11: a-Si FUNDAMENTALS

VOLUME II

POSTER SESSION 13 : THIN-FILMS CELLS (NON a-Si)

SESSION 14: APPLICATIONS AND SYSTEMS

SESSION 15: CRYSTALLINE PHOTOCELLS

SESSION 20: POLYCRYSTALLINE FILMS

SESSION 21: NATIONAL PROGRAMMES

OPENING SESSION

The General Chairman: Prof. I. Solomon

N. Mosar, Member of the EEC Commission responsible for Energy

View of the Conference Room

Panel Discussion

EXHIBITION

Equipment in front of the Congress Centre

N. Mosar, Member of the Commission, at the Exhibition

Conference Committee Dinner
(Organizer P. Helm and
Chairman Prof. I. Solomon)

Palazzo of the Gala Dinner

GALA DINNER

OPENING SESSION

<u>DISCOURS D'OUVERTURE DE M. MOSAR</u>

8EME "CONFERENCE EUROPEENNE
SUR L'ENERGIE SOLAIRE PHOTOVOLTAIQUE"
Florence, 9-13 mai 1988

Mesdames, Messieurs,

Au nom de la Commission que j'ai l'honneur de représenter ici, je voudrais souhaiter la bienvenue à tous les participants à cette conférence européenne sur l'énergie solaire photovoltaïque.

Je voudrais également saluer les organisateurs de cet événement ainsi que les représentants des media ici présents. Grâce à leur travail, les résultats de cette conférence pourront recevoir un large echo dans tous les milieux concernés.

En ma qualité de responsable de l'énergie au sein de la Commission, je ne puis que me réjouir de l'intérêt soulevé par cet événement. Il nous permettra de mettre en évidence les progrès significatifs dans le développement de cette technologie prometteuse qu'est l'énergie solaire photovoltaïque.

Au cours de cette séance inaugurale, il me paraît approprié de replacer cette forme d'énergie dans le cadre plus vaste de nos préoccupations en matière d'approvisionnement énergétique.

Il est inutile de s'appesantir ici sur le rôle vital de l'énergie dans la société technologique où nous vivons. En manquer serait inconcevable. C'est la raison pour laquelle les autorités responsables ont cherché à coordonner et à gérer les ressources existantes, tout en préparant l'avenir.

Dans la Communauté, le Conseil des Ministres a adopté, en septembre 1986, des objectifs énergétiques à l'horizon 1995 qui proposent un cadre communautaire cohérent pour la production, l'importation et l'utilisation de l'énergie.

Ces objectifs forment la pierre angulaire de la politique énergétique communautaire, dont je voudrais souligner trois caractéristiques essentielles.

Tout d'abord sa <u>dimension internationale</u> évidente. Elle touche, en effet, directement une population de 320 millions d'habitants issus de 12 Etats membres en a des conséquences induites sur d'autres pays qui sont nos fournisseurs d'énergie.

La politique énergétique communautaire présente aussi une caractéristique de <u>complémentarité</u> avec les autres politiques de la Communauté. De plus, elle joue un rôle central dans d'autres politiques telles que celles des transports, de l'industrie ou de la recherche pour ne citer que celles-là.

La dernière caractéristique que je retiendrai ici est la fléxibilité d'utilisation et la diversification des approvisionnements voulues par ces objectifs. Ils visent à rendre notre économie énergétique, la plus apte possible, à s'adapter aux évolutions souvent rapides des conditions d'approvisionnement en énergie.

Le développement de l'énergie solaire photovoltaïque peut clairement contribuer à la réalisation de nos objectifs énergétiques.

L'énergie solaire photovoltaïque est une énergie renouvelable par excellence. Il convient donc d'en poursuivre le développement. Les objectifs énergétiques communautaires prévoient, en effet, que la contribution des énergies renouvelables doit augmenter substantiellement afin qu'elles jouent, à l'avenir, un rôle significatif dans le bilan énergétique total.

Bien sûr, nous vivons une période de relative abondance énergétique qui se traduit par des prix assez modérés pour les combustibles traditionnels. Ceci n'encourage évidemment pas les agents économiques à faire appel aux autres options énergétiques existantes.

Le rôle de la Commission consiste, ici, à donner l'impulsion nécessaire pour que notre avenir énergétique ne soit pas menacé lorsque les tendances actuelles du marché viendront à se modifier.

C'est ainsi que l'un des objectifs énergétiques communautaires propose la promotion des innovations technologiques par le biais de la recherche, du développement et de la démonstration.

En matière de recherche et développement, la Communauté a adopté, en 1987, un programme-cadre, qui s'étend jusqu'en 1991, et qui fixe l'action communautaire dans ce domaine.

Ce programme a été doté de près de 6.5 milliards d'ECUs. Il comporte une ligne d'action concernant l'énergie qui permettra notamment de poursuivre la recherche sur le solaire.

Dans la perspective d'une importante exploitation future de l'énergie solaire, le photovoltaïque constitue certainement la voie royale. Même dans les conditions climatiques parfois difficiles que connaît l'Europe , l'énergie photovoltaïque offre, à terme, un potentiel considérable. Mais, aujourd'hui encore, son utilisation à grande échelle reste un défi et force est de constater que l'essentiel reste à faire.

L'Europe a relevé ce défi depuis bien des années et y consacre des budgets toujours croissants. Nous pouvons ainsi constater avec satisfaction que les budgets publics engagés cette année par la Communauté Européenne et ses Pays membres pour le photovoltaïque atteignent 100 MECUs. Ils dépassent les budgets correspondants engagés aux Etats-Unis et au Japon.

A cette vaste entreprise que constitue la promotion du recours au photovoltaïque, la Commission des Communautés Européennes participe de plusieurs façons. C'est ainsi qu'elle poursuit depuis de nombreuses années la promotion des marchés et de l'industrie photovoltaïque à travers un programme de démonstration à l'échelle industrielle. Par ailleurs, dans le cadre de sa coopération avec les pays en développement, la Communauté est résolue à soutenir un usage plus important de ces nouvelles techniques au bénéfice de l'équipement énergétique des zones enclavées de ces pays. A ce titre, je voudrais mentionner le projet d'un vaste programme de pompes à eau photovoltaïques pour le Sahel, qui est en cours de préparation dans le cadre de la Convention de Lomé.

Le développement technologique du photovoltaïque mené par la Commission va donc bien au-delà d'un simple rôle de coordination des activités transnationales. Un aspect important de sa mission est le développement d'une stratégie à long terme visant la réalisation de technologies photovoltaïques appropriées pour une utilisation massive de l'énergie solaire en Europe et outre-mer. Il va sans dire que cette volonté s'inscrit dans la perspective de la recherche d'une compétitivité économique accrue. La Commission estime cependant qu'au moins 20 à 30 années de développement technologique et industriel seront necéssaires pour y parvenir.

Le programme de R&D communautaire actuel comporte 3 axes: le développement du silicium cristallin, la recherche sur les piles en couche mince et le développement de systèmes optimisés et économiques. En raison du budget très limité, il a été nécessaire de centrer notre action sur certains thèmes pour stimuler un mouvement d'envergure. Tel a notamment été le cas du silicium amorphe pour lequel il s'avère maintenant que la stratégie volontariste de la Commission a été un succès indéniable. Il est important néanmoins de noter que, malgré la concentration des moyens sur des secteurs privilégiés, la Commission s'efforce de rester présente dans tous les autres domaines stratégiquement vitaux, même si les moyens disponibles sont modestes. Ainsi, des contrats ont été consacrés à deux technologies qui constituent de véritables goulots d'étranglement pour le silicium cristallin, à savoir la purification du silicium et la découpe en tranches.

Enfin, ce qui concerne le développement des systèmes photovoltaïques, la Commission continue de mener un programme d'excellence qui a permis, entre autres, la construction d'une première grande installation de 300kw à Pellworm dans le nord de l'Allemagne. La technologie des systèmes qui se poursuit actuellement à l'intérieur d'un vaste programme d'actions concertées européennes va conduire à une première étape de perfectionnement. Ainsi sera-t-il possible de réaliser un nombre toujours croissant de systèmes opérationnels pour lesquels nous pouvons noter actuellement une forte demande dans tout le sud de l'Europe, notamment en Italie.

En complément au programme de recherche, la Communauté a lancé en 1978 un programme de démonstration "Energie" couvrant les secteurs des économies d'énergie, celui des énergies renouvelables, la substitution des hydrocarbures et la liquéfaction et gazéification des combustibles solides.

Dans ce programme, les énergies renouvelables jouent un rôle important dont le solaire photovoltaïque est un exemple représentatif.

Entre 1978 et 1987, plus de 4 600 propositions de projets de démonstration ont été soumises dont 1 445 ont pu être selectionnés en vue de recevoir une aide financière de la Communauté qui atteint 765 millions d'ECUs. Dans le domaine du solaire photovoltaïque, 63 projets sont en cours de réalisation et représentent une puissance installée d'environ 1 megawatt.

Ces projets technologiques, en grandeur réelle, se situent à une phase intermédiaire entre, d'une part, la recherche et le développement et, d'autre part, l'application commerciale.

L'appel d'offres pour 1988 vient de se clôturer et environ 500 nouvelles propositions ont été reçues. L'appel d'offres pour l'année 1989 clôturera le programme en cours qui avait démarré en 1986.

Le programme européen, loin de faire double emploi avec ceux des pays membres de la Communauté, joue un double rôle: celui de catalyseur, en contribuant au développement de programmes nationaux de soutien à la démonstration et celui de complément utile, permettant de combler les lacunes que pourraient présenter les programmes de certains Etats membres.

Par ailleurs, la diffusion des résultats à l'échelle de la Communauté est un bénéfice non négligeable pour tous les pays membres. La conférence qui s'ouvre aujourd'hui y apportera sa contribution.

Je voudrais achever ici mon intervention en renouvelant mes voeux de réussite pour cette conférence.

DISCORSO ING. CORVI PER 8° CONFERENZA FOTOVOLTAICA

E' per me motivo di grande soddisfazione porgere il mio più caloroso saluto ai presenti, non solo come Presidente dell'International Solar Energy Society - ISES Sezione Italiana e di Presidente dell'ISES Internazionale ma anche in qualità di Presidente del Comitato Nazionale della Conferenza.

Desidero porgere alcuni ringraziamenti:

- ringrazio innanzitutto la Comunità Europea, qui rappresentata autorevolmente dal Commissario Dr. Mosar, e la Commissione delle Comunità Europee, con la quale abbiamo lavorato in perfetta armonia per la buona riuscita del Congresso;
- ringrazio particolarmente il Sottosegretario al Ministero dell'Industria On. Ravaglia che ha acconsentito a partecipare alla nostra inaugurazione, contribuendo così, con il prestigio della sua presenza, al successo della manifestazione;
- ringrazio le autorità regionali, provinciali e comunali che hanno mostrato tanta disponibilità e sensibilità all'avvenimento:
- ringrazio i nostri numerosi sponsors e tutti coloro che hanno contribuito all'organizzazione del Congresso.

Questo Congresso ha luogo in un momento particolare per la sezione italiana dell'ISES.

Come ormai la maggior parte di Voi sa, l'ISES è la principale associazione culturale scientifica operante a livello mondiale nel campo delle fonti rinnovabili di energia e del risparmio energetico. Fra i suoi membri vi sono i più prestigiosi scienziati e tecnici nel campo delle fonti rinnovabili.

L'ISES ha sede in Australia, ha soci in 90 Paesi ed è articolata in 28 Sezioni nazionali, tra le quali figura la Sezione Italiana, che per numero di soci è la seconda nel mondo e si è dimostrata tra le più attive per numero e qualità delle iniziative nel settore della promozione ed informazione sulle fonti rinnovabili. Ad essa aderiscono i principali operatori del settore, quali Enti energetici, Industrie dei settori pubblico e privato, Istituti universitari e professionisti.

Una particolare attività dell'ISES è quella di fornire una corretta informativa sui problemi delle fonti rinnovabili e del risparmio energetico.

In un momento come l'attuale, nel quale attraverso i mezzi di comunicazione di massa vengono diffuse notizie spesso contraddittorie sulle possibilità di impiego e di sviluppo delle nuove fonti, riteniamo si debba fare molta attenzione a non creare nell'opinione pubblica facili ottimismi o aspettative che non possono essere realisticamente soddisfatte, ma nemmeno indurre a considerare le fonti rinnovabili inutili e dispendiosi giocattoli per adulti.

L'ISES italiana si propone pertanto come elemento coagulante e stimolante delle forze che operano in questo settore collaborando così al conseguimento di obiettivi significativi in linea con gli orientamenti energetici deliberati dal Parlamento.

La conversione fotovoltaica dell'energia solare è una delle tecnologie più promettenti tra quelle che puntano sullo sfruttamento delle fonti rinnovabili.

per la produzione di energia elettrica.

Tuttavia le valutazioni sulle sue effettive potenzialità, in termini di producibilità e costi nello scenario energetico non sono sempre risultate convergenti: tali diverse valutazioni sono d'altra parte comprensibili dal momento che, pur basandosi su tecnologie e conoscenze in parte già disponibili, il settore del fotovoltaico ha avuto un grosso impulso solo dopo la seconda crisi petrolifera.

Come qualunque altro settore innovativo questa contingente fase di decollo è stata inevitabilmente caratterizzata da sviluppi forse troppo rapidi, valutazioni forse troppo ottimiste, assenza di standardizzazione e quindi proiezioni di mercato confuse.

La vera grande importanza di questo Congresso risiede nel fatto che, in un momento di così delicate scelte energetiche, esso ci permetterà, grazie al contributo degli esperti mondiali del settore, di rispondere ai più determinanti interrogativi sulle reali possibilità di questa fonte energetica. Alcune delle domande più significative che rivolgiamo e che troveranno risposta a partire dall'intervento introduttivo del Prof. Solomon, Chairman della Conferenza, riguardano:

- **gli aspetti tecnologici**, ed in particolare la sfida rappresentata dalla messa a punto delle tecnologie di nuovi materiali, che dovrebbero consentire efficienze maggiori dei moduli fotovoltaici, limitando così l'impegno di territorio a parità di energia prodotta;
- **gli aspetti energetici**:

a) che ruolo può giocare la conversione fotovoltaica tenendo conto delle discontinuità della fonte solare e dei conseguenti problemi di interfaccia con una rete elettrica di grandi e/o piccole dimensioni;

b) quanto riusciranno ad essere affidabili i sistemi di alimentazione di utenze isolate - un problema importante per noi ma addirittura fondamentale per i Paesi in via di sviluppo;

- **gli aspetti economici**: come e quando si pensa di arrivare alla competitività con l'energia prodotta da altre fonti, siano esse convenzionali o rinnovabili;
- **gli aspetti giuridici** e normativi che acquisteranno un peso determinante con l'estendersi delle installazioni fotovoltaiche, con conseguenti problemi per l'uso multiplo del territorio;
- **gli aspetti ambientali**, quali ad esempio l'inserimento degli impianti fotovoltaici nel territorio, per i quali oggi si rileva una crescente sensibilità da parte del pubblico.

Ed infine la domanda più ovvia ma per noi la più significativa che scaturisce anche dalle domande precedenti:

quale potrà essere l'effettivo contributo della conversione fotovoltaica al soddisfacimento della domanda di energia negli anni futuri?

Il grande successo che a livello mondiale ha registrato questo Congresso dimostra quanto siano giudicate attuali queste domande e con quanto interesse si attendano le risposte.

Vi ringrazio per l'ascolto e auguro buon lavoro a tutti i congressisti.

OPENING ADDRESS

This conference, the eighth in a series which was started by the European Commission in 1977, has become one of the largest international forum for photovoltaics. We observe this year a surprising discontinuity both in the growth of the number of participants and of the number of presented papers. Althrough we are pleased with this increase which goes beyond our expectations, it raised quite a few problems.

We had to face a selection of papers that was particulary difficult and I take this occasion to acknowledge the efficient work performed by my collegues of the Scientific Comittee. One solution used to avoid an insufferable increase in the rejection rate was to suggest to authors the grouping of serveral papers whenever possible: for example for a series of papers of the same institute or for a series of studies on the same subject, allowing more time for presentation. The results turned out to be in many cases quite satisfactory both for the organisation of the conference and the scientific quality of the contributions.

Even then, the total number of presentations, oral and posters, has reached the value of 343, representing a 50% increase over the previous Conference in Servilla. To our deep regret it was thus impossible to avoid some parallel sessions (first time in this series of Conferences), but we have tried very hard to keep them to a minimum.

Another novelty in the present Conference, is the explicit introduction of a new topic : Advanced Storage for Photovoltaic. It is quite clear that the proper design of most systems have to include the storage, economically as well as technically. The space for this subject in this Conference is still modest, but it will probably grow in the future. Personally, I feel that the research effort in electrical storage is insufficient compared to the progress in PV and that it might soon become the bottleneck for the development of photovoltaics.

One achievement of this type of meeting is the perfect integration between basic research and industrial technology. This reflects an original and quite satisfactory situation in the field of P.V. : that of a rapid incorporation of the laboratory discoveries into the industrial progress.

There is, however, a separation of subjects that appears in our conferences, but it is somewhere else. This splitting, which is quite comparable to "hardware" and "software" in computer techniques, concerns two different populations (I could almost say two "cultures"): those working in *photocell* research and those active in the study of photovoltaic *systems*. These two activities are complementary and necessary to one another, and it is a mark of success of our European Conferences to have kept the proper balance between these two fields.

Device physics, strongly based on materials research has progressed in a spectacular way this past few years. Both crystalline and thin-film photocells have improved in parallel and are, for the time being, more complementary than in competition. But although it is necessary to have good, stable, high-efficiency, inexpensive cells, it is not sufficient. A poor system design can distroy all the advantages of the best device. That makes the importance of research on systems, which is a very active field, both in theory and experiment. As always, nothing can replace field experience and the multiplication for quite a few years of full-size solar energy applications in many countries provides invaluable information for the design and improvement of photovoltaic systems.

What is the present place of photovoltaics in the field of renewable energy resources ? That is the 100,000 Lire question, but we can safely claim that the substantial scientific and technological progress that we have made in the past few years are putting us very close to a real industrial take-off. It would be too long to list here all the achievements obtained recently: you will make your own opinion from the large quantity of high-quality papers that will be presented in the coming sessions. Concerning the economics of photovoltaics, already we can acknowledge a real success in micropower applications. We must not scorn at this type of application: not only because photovoltaics applied to small portable systems is already a multi-million dollars business, but also because the saving of hundreds of millions of dry batteries is a saving of pollution, of a large amount of raw material, and finally a substantial saving of energy. In any case we are getting close to medium or large scale energy applications and there is a question that cannot be avoided: how to handle the electrical power production. Should it be centralized, with all the advantages of large-scale production; or should it be distributed as is solar energy itself? This question, which is not purely scientific but has obvious political implications will be the subject of the Panel discussion on Wednesday, which we expect to be lively, and frustratingly too short as usual.

The Exhibition has now become a part of the Conference. It gives industry the opportunity to present its latest achievements, as well as the occasion of scientific and technical contacts. However we will stand this year a serious competition. We have the great chance to be in Florence, where there are many other types of exhibitions which have some similarities with our's: they also display panels, contained in a frame, that must be placed in the proper light. Our panels also have their own beauty (in a modern style, of course) and besides they produce electricity. Let us hope that they shall last as long as the paintings in Florence !

Last but not least, I must acknowledge the participation of the large number of people that have contributed to the success of this Conference. A complete list would be much too long to be given here. I shall only mention UNESCO, for its moral as well as practical support. And above all, I wish to thank our Italien friends, in particular of the International Solar Energy Society (ISES) and its President. This meeting would not have been possible without their support and their active and efficient participation to the local organisation of the Conference.

I.Solomon
Conference Chairman

Eighth European Photovoltaic Solar Energy Conference

9–13 May 1988 – Congress Centre, Florence, Italy

FINAL PRESS RELEASE MAY 13th, 1988

Today concludes the 8th European Photovoltaic Solar Energy Conference and Exhibition organized by the Commission of the European Communities in collaboration with the Italian Section of ISES. Conference chairman was Prof. I. Solomon from the Ecole Polytechnique in Paris.

This Conference has resulted, by far, the most important event for the Solar Energy Photovoltaic community for the year.

Delegates came from all European countries, the USA, Japan, the Soviet Union; many Arab countries, India, Chile, etc. All in all, more than 750 participants (50% more than any prior conference on this subject in Europe) from 43 different countries got together to discuss the present situation and the future perspective of photovoltaic research and industrial development. Associated with the conference was a major exhibition of photovoltaic industry from Europe and overseas. It is well known that photovoltaic technology is the most promising of solar energy systems. It allows direct conversion of solar radiation to electricity. Opening the conference, Mr Nic Mosar, the European Commissioner for Energy, reaffirmed the interest of this form of renewable energy in achieving the long-term objectives of energy supply for the European Community. In order to attain this, further research, development and demonstration activities will be vital as will be the role of industry.

Among the many highlights of the conference, the following are particular worth mentionning :

1) A study report prepared under a EC contract by the Fraunhofer Institut ISI in Germany on indirect electricity cost of conventional and new electricity sources such as photovoltaics was presented.

It was possible for the first time to quantify these external effects in the social and environmental field. As a result, solar electricity will become cost effective at least five years earlier than expected hitherto.

2) Industry has given evidence that the production cost for photovoltaic solar converters will decrease below the "magic" level of 1000 $ per KW in some years to come.

3) Without waiting for the single European market in 1992, European industry is working closely together today. The Joint Research Centre in Ispra, Italy, which belongs to the Commission of the European Community is playing an increasing role to prepare industrial standards for these industry products.

4) Strong links exist between public R&D programmes of the European countries which are further stimulated by the Commission's own R,D&D programmes.

5) Eventually the Mediterranean countries of the EC such as Italy, Spain, Greece and Portugal can become the first users of photovoltaics in Europe on a significant scale and take advantage of this over all development effort going on in Europe today.

6) Larger production volumes in industry will be necessary to make photovoltaics economically competitive for grid-connected applications. Investigations in Japan which had been reported at the conference show that such volumes may be around 100 MW per year (current world production is 28 MW per year).

7) Representatives from developing countries would like to use photovoltaics in a huge concerted effort to provide electricity to the hundreds of millions of people in the world who do not have any electricity as yet.

8) Photovoltaic technology has stimulated the synergy with other high-tech areas such as microelectronics and optoelectronics. In particular amorphous silicon is developing extremely fast in these areas and has already become a 500 million $ industry today. Details were presented at the conference.

SESSION 01:
ADVANCED STORAGE

Secondary Electrochemical Cells For Solar Energy Storage

Ernst Voss
Expert of the CEC
D-6237 Liederbach, Fed. Rep. of Germany
Heidestr. 14

In the last decades a number of new electrochemical systems have been under evaluation, as there are high temperature alkali metal/chalkogen, electrolyte flow and redox systems. The basic properties as well as the present status of development of these advanced systems are presented and their prospects for a potential application in solar energy storage are discussed. For reasons of comparison experimental data obtained on a solar power station which is equipped with lead-acid batteries are reported.

Introduction

For an economic utilization of solar energy the use of energy storage systems is a necessity in order to harmonize the inevitable time shift between energy supply and demand. For photovoltaic solar energy sources which provide a direct electric current secondary electrochemical cells appear to be straight-way tailor made since they need also direct current for recharge. At present commercially available cells and batteries are used exclusively in programmes demonstrating the technical feasibility of photovoltaic solar energy use and once again the lead-acid battery is the preferred working horse for this purpose.

In the last decades a number of new electrochemical systems have been under evaluation and some of them have been developed to an advanced state. The basic properties of the most promising secondary (rechargeable) storage cells as well as their present status and their prospects for a potential application in solar energy storage will be discussed in this paper. For reasons of comparison experimental data obtained on a solar power station equipped with a lead-acid battery are reported.

Fuel cells also have been proposed for solar application in particular within the frame of a future hydrogen economy. This will be discussed in a subsequent paper by K. Ledjeff [1].

The secondary (rechargeable) electrochemical cell

A secondary (rechargeable) electrochemical (galvanic) cell is an energy storage device which stores electrical energy in form of chemical energy (charge). On demand the chemical energy can be supplied in form of electrical energy again (discharge). Charge and discharge can be repeated many times. The storage capacity (Ah) of a secondary cell depends on the amount of active materials in the electrodes and eventually in the electrolyte (Faraday's law), i.e. energy storage and power cell are forming one unit.

Contrary to an arrangement of this kind the fuel cell is a power cell, its electrodes do not contain any active materials. Fuel and oxidant, the active materials, are stored apart from the power cell (mostly as gases or liquids) and are fed into the power cell for d.c. electricity generation on demand. So-called flow cells may also be regarded in this way.

Advanced secondary (rechargeable) electrochemical (galvanic cells) for solar energy storage

The most wellknown secondary cells which are commercially available since many decades are the lead-acid cell and the nickel-cadmium cell. These cells have been developed to a high standard of reliability and in particular the lead-acid cell is already applied for solar energy storage, since it appears to be the optimum choice today from a requirement and an economic point of view. Nevertheless there has been and still is a large effort spent on the development of new secondary cells all over the world mainly for economic reasons but also because of the expectation that the new "advanced" systems may exhibit a longer service life and a higher efficiency.

The most important requirements for an electrochemical storage device which is designed for solar applications are as follows:

- capability to perform high numbers of charges and discharges independent of depth of discharge
- high Coulomb and energy efficiency
- zero or at least low maintenance
- capable to accept small charge currents
- low self-discharge
- low costs

A high energy density of electrochemical cells may be of importance if the costs of buildings are effected by space and volume requirements.

At the present status of knowledge some advanced electrochemical systems may have a chance to exceed the lead-acid cell at least in one of the requirements. These are the systems

$$Na/\beta Al_2O_3/S,$$

$$Zn/ZnBr_2/Br_2$$

and the redox systems an example of which is the cell

$$Cr^{2+}/Cr^{3+}//Fe^{3+}/Fe^{2+}.$$

The system $Na/\beta Al_2O_3/S$

The basic cell reaction of the Na/S system is given by the simplified equation (1):

$$2\,Na + 3\,S \underset{\text{charge}}{\overset{\text{discharge}}{\rightleftharpoons}} Na_2S_3. \qquad (1)$$

The electrolyte is a solid Na^+ ion conductor. The reactant at the negative electrode, sodium (Na), and the reactant at the positive electrode, sulfur (S), are both in the liquid state at the operational temperature of about 330 °C. The fundamental theoretical data of the Na/S system are as follows:

- cell voltage: 2,1 V
- energy density: 792 Wh/kg, 1196 Wh/l.

The cells are basically of cylindrical design. The solid electrolyte $\beta\ Al_2O_3$ is applied in form of one end closed tube with a wall thickness of about 1 mm. It acts simultaneously as a separator.

In Europe the Na/S system is under development in France (CGE, Marcoussis), in Germany (BBC, Mannheim) and in the United Kingdom (CSPL, Runcorn). As an example the status of development at BBC is briefly reported.

BBC has developed a multicell battery which is incorporated in a thermally isolated metal box. This is a necessity in order to minimize thermal losses due to the high operational temperature which is well above ambient temperature. The present data of this battery as well as the prospective data for the 1990ies are collected in Table 1 [2].

Table 1

Characteristic data of the BBC Na/S battery

		1987	1990ies
voltage	V	120	192
energy	KWh	23	32
power	KW	21	33
energy density	Wh/kg	90	120
	Wh/l	90	130
self-discharge			
- electrochemical	%/d	0	0
- thermal	%/d	8	6
Coulomb efficiency	%		98
energy efficiency	%		85
cycle life		400	
cost estimate		as lead-acid at maximum	

The BBC battery is primarily determined for application in electric vehicle propulsion. But it is also being tested at present in a pilot project for peak shaving purposes in Technische Werke der Stadt Stuttgart [3]. It appears that the Na/S battery may meet a number of the requirements for solar energy storage. In particular the high Coulomb and energy efficiency, the self-discharge which is low even when the thermal "self-discharge" is taken into account, as well as the zero maintenance are of high advantage in comparison to other systems. The key problems on the other hand are the limited cycle life and the inability of the battery to withstand a large number of thermal cycles. In addition costs are still an uncertainty.

The system $Zn/ZnBr_2/Br_2$

The basic cell reaction of the zinc/bromine cell is given by equation (2):

$$Zn + Br_2 \underset{\text{charge}}{\overset{\text{discharge}}{\rightleftharpoons}} ZnBr_2 \qquad (2)$$

As can be seen during discharge the electrolyte zinc bromide is formed which normally is a 3,5 molar aqueous solution. A very specific feature of this system is the application of organic compounds for complexing the elementary bromine in order to reduce self-discharge and corrosive reactions.

The fundamental theoretical data of the Zn/Br_2 system are as follows:

- cell voltage: 1,83 V
- energy density: 438 Wh/kg, 1642 Wh/l.

The cell is basically of planar design and applies bipolar electrodes made from plastic bonded carbon which easily can be assembled to form multicell stacks. In many respects the Zn/Br_2 cell stack can be regarded as a power cell in the sense it was defined above. In the discharged state both the active materials are mainly stored outside the power cell in plastic containers. During operation the active materials are transported to the electrodes and back to the containers by means of two separated electrolyte fluxes, one for each electrode.

The zinc/bromine battery is mainly developed in the USA and in Japan as well as by S.E.A. (in cooperation with the Universities of Vienna and Graz) in Austria. S.E.A. has developed units with various energy contents. As an example the characteristic data of a 4 KWh version, a very advanced model, are collected in Table 2 [4] .

The Zn/Br_2 battery appears to be very promising for solar energy storage because of the high cycle numbers of more than 2 000 which are reported to possibly be extended up to 10 000. In addition the self-discharge is low and finally the energy efficiency is comparable to that of a lead-acid battery. Apart from charging there is no need for any maintenance.

The S.E.A. 4 KW unit is made mainly from plastic material. The electrodes are prepared from a carbon/polyethylene mixture. Polyethylene is showing a remarkably higher resistance against corrosive attack of bromine than polypropylene which has been used earlier. Nevertheless it may turn out that the plastic components' life will become a problem for a long lasting maintenance free operation.

Table 2

Characteristic data of the S.E.A. 4 KWh Zn/Br_2 unit

voltage		V	48
energy		KWh	4
power		KW	2 - 3
energy density		Wh/kg	160
power density		W/kg	100 - 150
self-discharge			
- electrolyte	in cell	%/d	5 first day, later very small
- no electrolyte	in cell	%/d	zero
Coulomb efficiency		%	82
energy efficieny		%	72
cycle life		100 % DOD	2500 *)
cost estimate			at maximum equal to PbO_2/Pb

*) obtained on an earlier unit by SCANDIA Lab. USA

The system $Cr^{2+}/Cr^{3+}//Fe^{3+}/Fe^{2+}$

The redox storage system chromium/iron is working with liquid reactants only, i.e. with aqueous hydrochloric acidic solutions of chromium and iron ions which are transported to the inert or activated electrodes by electrolyte circulation in two separate loops. In principle it is a combination of a power cell with two separate tanks for the storage of the two reactants. In so far the redox cell can be regarded as a regenerative fuel cell.

The cell reaction is given by equation (3):

$$Cr^{2+} + Fe^{3+} \underset{\text{charge}}{\overset{\text{discharge}}{\rightleftharpoons}} Cr^{3+} + Fe^{2+}. \qquad (3)$$

The design of the redox cell is similar to that of the Zn/Br_2 cell. An important component is the ion exchange membrane which separates the anode from the cathode volume and has to prevent the mixing of the different chromium and iron ions and on the other hand must allow the exchange of Cl^- ions and exhibit a resistance as low as possible.

The fundmental theoretical data of this redox system are as follows:

- cell voltage: 1,2 V
- energy density: 293 Wh/kg.

The development of the Fe/Cr redox system started originally in the USA. Today there are also considerable efforts in Japan and in Germany (Siemens and Frauenhofer Institut für solare Energiesysteme). The characteristic data of this system which are collected in Table 3, have been taken from a Japanese source [5]. They are valid for a 10 KW unit.

Table 3

Characteristic data of a 10 KW Fe/Cr redox battery

voltage	V	55
energy	KWh	80
power	KW	10
energy density	Wh/kg	6,3
	Wh/l	3,3
self-discharge	%/d	0,8
Coulomb efficiency	%	96
energy efficiency	%	69
cycle life		1000
costs estimate		25 % of lead-acid

The development of large Fe/Cr redox units seems to be already in a more advanced state in Japan. However, in the German programme a membrane of high selectivity has been developed which reduces the self-discharge to almost zero and has a low resistance. In any case the redox system has not proven its long term stability in photovoltaic solar energy storage.

The system $Pb/H_2SO_4/PbO_2$ (the lead-acid cell)

While the advanced systems mentioned above still have to prove their technical and economical feasibility, the lead-acid cell is a commercial and practical reality. It has been and still is used in many photovoltaic solar energy programmes. Table 4 summaries a number of examples which are sponsored by the EC [6].

Since the lead-acid system is very well known in general, it is renounced to described its basic properties. However, the characteristic data of a lead-acid cell designed for application in solar energy storage are summarized in Table 5 [7].

Many solar energy stations have been equipped with batteries of the type described in Table 5.

Table 4

EC sponsored solar energy projects
using lead-acid batteries for energy storage

project	country	application electric power for	power unit KW
Aghia Roumeli	Greece	village	50
Isle of Fota	Ireland	milk farm	50
Isle of Giglio	Italy	cold-storage depot	45
Karo	F-Guyana	village	35
Isle of Kythnos	Greece	the isle	100
Mont Bouquet Uimes	France	TV transmission	50
Nizza Airport	France	ground signals	50
Isle of Pellworm	Germany	leisure center	300
Isle of Terschelling	Netherlands	navigator school	50
Isle of Tremiti	Italy	sea water desalination	65
Isle of Vulcano	Italy	village	80

Table 5

Characteristic data of a special lead-acid cell
(VARTA bloc)

voltage	2 V
capacity range	12,5 - 2000 Ah
Coulomb efficiency (10 h)	95 %
energy efficiency (10 h)	80 %
cycle life	
75 % DOD	1300
30 % DOD	4500
self-discharge	3 % per month at RT
charge acceptance	10^{-3} of I_{10}

As an example the solar station of the Isle of Pellworm located at the North Sea coast will be described in some detail [8] . The photovoltaic generator is rated to provide a power of 300 KW. The lead-acid battery is designed and sized that it may supply the energy demand of a large leisure center (Kur-Zentrum) for a couple of days. The battery voltage is 348 V, its capacity is 6000 Ah. It is composed of 4 x 172 cells each having a capacity of 1500 Ah at the 10 h rate. The establishment started to work in June 1983.

Conclusive remarks

In this paper three advanced rechargeable battery systems have been described which have some potential prospects to be applied in solar energy storage in larger scale in the future. At present lead-acid batteries only are used for this purpose. In small scale photovoltaic solar energy storage also nickel cadmium cells have been and are still being used. These types of cells as well as nickel hydrogen cells are much more cost effective and, therefore, probably are too expensive for a penetration of a domain which appears to be furnished almost exclusively by the lead-acid system.

References

[1] K. Ledjeff — Unconventional energy storage systems, see subsequent paper of this conference

[2] H. Birnbreier — High temperature battery development, Paper presented at the Interlaken Symposium, June 15, 1987

H. Haase — Die BBC-Hochenergiebatterie und der Elektroantrieb, Paper presented at the BMW AG technical information conference, Bonn, Dec. 8, 1987

D. Reister — Elektro- und Wasserstoff-Antrieb für PKW im Systemvergleich, Paper presented at the BMW AG technical information conference, Bonn, Dec. 8, 1987

J.W. Angelis, H. Birnbreier and H. Haase — Hochenergiebatterien für Elektrostrassenfahrzeuge, BBC Druckschrift D HB 125387 D

[3] Anonymous — Neue Batterien der Technischen Werke Stuttgart Publication in "Stuttgarter Nachrichten" of Sept. 12,1987

[4] G. Tomazic, — Die Zink-Brom-Batterie ÖZE 40,(1987),p.13

K. Kordesch — personal communication

[5] Z. Kamio, T. Hiramatsu and S. Kondo — Research and development of of a 10 KW redox flow battery 22nd IECEC, Philadelphia, (1987), p. 1056

H. Gnobloch, H. Nischick, K. Pantel, K. Ledjeff, A. Heinzel and A. Reiner — Eisen-Chrom-Redoxionen-Speicher in: Dechema Monographien Vol. 109, (1986), p. 427 ed. W.-D. Lutz and H.-G. Müller

[6] H. Wilmes — Felderfahrungen mit Batterien in photovoltaischen Systemen VARTA spezial report 3/1987

[7] VARTA broschure — H 70 325 03 86 1

[8] H. Wilmes — Energie aus Sonne und Wind, Batterien für solar-elektrische Energiespeicherung VARTA spezial report 5/1983

Unconventional energy storage systems

K. Ledjeff
Fraunhofer Institute for Solar Energy Systems
Oltmannsstr. 22, D-7800 Freiburg, FRG

Abstract

Energy storage systems consist of matching devices for charging and discharging and the energy carrier. In principle all energy carriers (mechanical, electric, thermal, chemical) can be recycled, but the process differ greatly in complexity and efficiency. Storage cycles with high energy density materials show low efficiencies and vice versa. Grid-connected PV systems can use the storage capacity of the utility network and there is no need for additional energy storage in the near future. Stand-alone PV systems require energy storage systems but up to now there is no system available for long-term storage. It may be expected, that hydrogen systems will be on the market within the next few years.

1. Introduction

In conventional households the energy is supplied continuously by the utilities and all rapid rates of changes can be accomodated by storage systems or by peak load power plants. Photovoltaic systems deliver energy discontinuously depending on the variations of the availability of the sun. Some applications can be adapted to the intermittent supply of energy. Water pump sytems may operate as long as sufficient energy is delivered and stand still during periods of low insolation.

Most of the loads require a continous supply of energy and PV systems must be equipped with energy storage units to buffer the fluctuations of load and insolation. There are numerous ways in which energy can be stored but in nearly all practical systems the lead-acid battery is the obvious choice.

The commercially available batteries are very expensive and in future the growth of the PV market will depend on the development of new and reliable energy storage systems.

2. Energy storage systems

Energy storage is achieved in various ways such as mechanical, electric, magnetic, electrochemical, chemical, and thermal energy and in combinations thereof. All methods of energy storage need more or less complicated devices for transforming or matching the energy. In Fig. 1 the principle of energy storage systems is shown. All

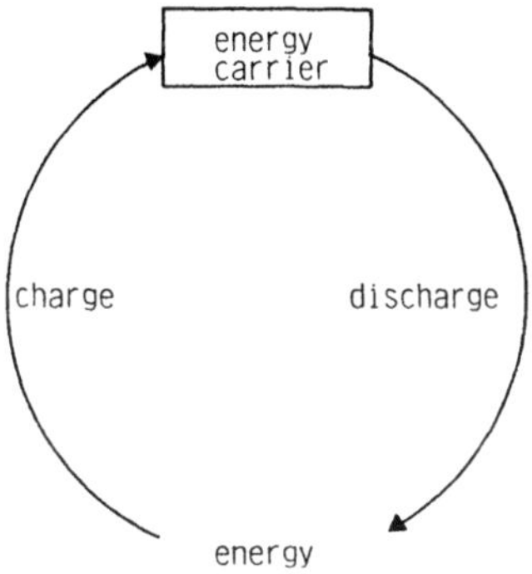

Fig. 1: Energy storage systems

the systems must consist of the energy carrier and matching systems for charging and discharging. There are many energy carriers which will fulfill the user's requirements but the technological problems caused by the matching systems reduce the number of systems that will provide the chance of realization. In principle all energy carriers can be regenerated and can be used theoretically in storage systems. The energy densitiy of various technologies are shown in Fig. 2. The energy densities of methods based on physical effects are very low, in the range of Wh/kg. Due to the higher reaction energies, chemicals have very interesting energy densities of about 500-/1500Wh/kg. The energy density of nuclear materials exceeds all the others. In recent years research on chemical energy storage systems has intensified and in Tab. 1, a few examples of chemical cycles are given./1,2/

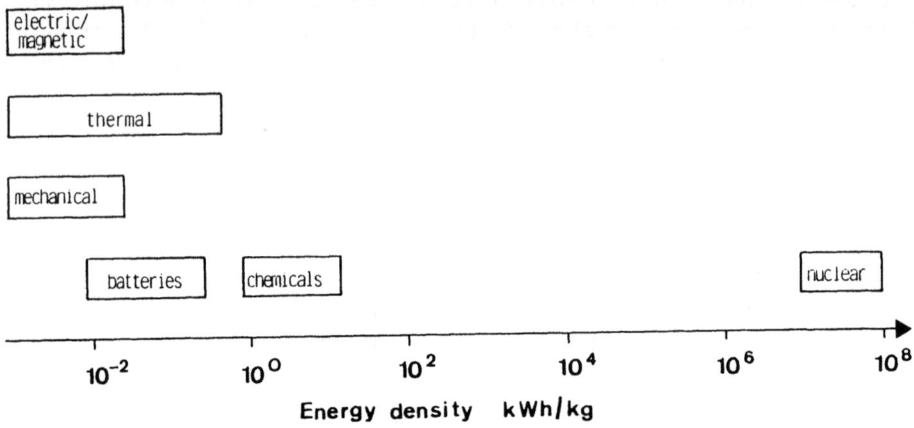

Fig. 2: Energy densities of various energy carriers

	Discharge	Charge
Hydrogen:	$H_2 + {}^1/_2\ O_2 \rightarrow H_2O$	$H_2O \rightarrow H_2 + {}^1/_2\ O_2$
Methane:	$CH_4 + 2O_2 \rightarrow CO_2 + H_2O$	$CO_2 + H_2 \rightarrow CO + H_2O$
		$CO + 3H_2 \rightarrow CH_4 + H_2O$
Methanol:	$CH_3OH + {}^3/_2O_2 \rightarrow CO_2 + 2H_2O$	$CO + 2H_2 \rightarrow CH_3OH$
Ammonia:	$NH_3 + {}^3/_4O_2 \rightarrow {}^3/_2H_2O + {}^1/_2N_2$	$N_2 + 3H_2 \rightarrow 2NH_3$
Aluminium:	$2Al + {}^3/_2O_2 \rightarrow Al_2O_3$	$Al_2O_3 \rightarrow 2Al + {}^3/_2O_2$

Tab. 1 Cycles of chemical energy carriers

The hydrogen cycle is advantageous because of the well established production and conversion technologies, the high efficiency, and the low environmental pollution level. The processes involving carbon needs more technical resources for separating CO_2, shifting and methane or methanol production. The liquid energy carriers, methanol and ammonia, have very high volumetric and gravimetric energy densities and can be substituted for gasoline. The use of metals (aluminium, calcium, magnesium, silicon) as energy carriers has limited applications in space technology for chemical rocket propellants /3/. The energy densities of metals are as high as of hydrocarbons, but the efficiencies of complete cycles are low and expensive equipment is required.

3. Grid-connected PV Systems

The public utilities are able to deliver electrical energy whenever demand arises. The load curves in Fig. 3 show the day-night and summer-winter cycles. The utili-

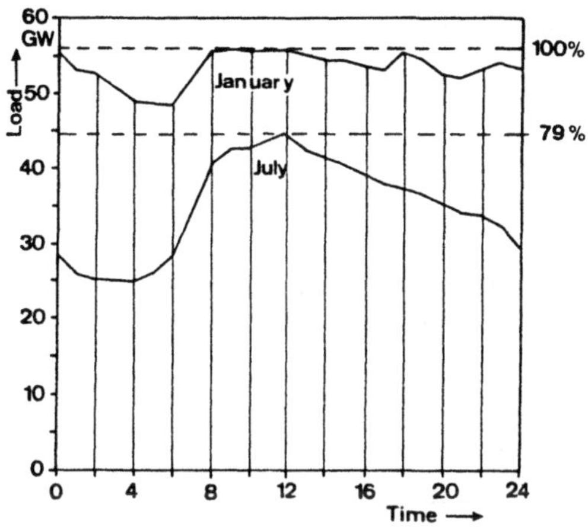

Fig. 3: Load curves (Germany 1985)

ties have organized their installations into groups according to the operation time per year. The power plants producing at the lowest costs are operated for more than 6000 h/a and constitute the base load, Fig. 4.In many countries the base load is generated in nuclear power plants. Medium load plants operate for 2000 -4000 h/a and are fuelled with coal, oil or gas. The peak load range is attractive for storage systems, because the costs of electricity production are highest and the investment in storage systems is economic. Short-term demand is satisfied within minutes by activating throttled power plants (minute reserve) and within seconds by tapping the inherently associated mechanical energy in rotating machines (spinning reserve)/4/.

Grid-connected PV systems will interact with the network and influence the load management. From the economic viewpoint it is probable that the base load will be operated under all conditions. PV will have a chance in the medium load range where the cost of fossil fuels are higher and the reduced fuel consumption brings advan-

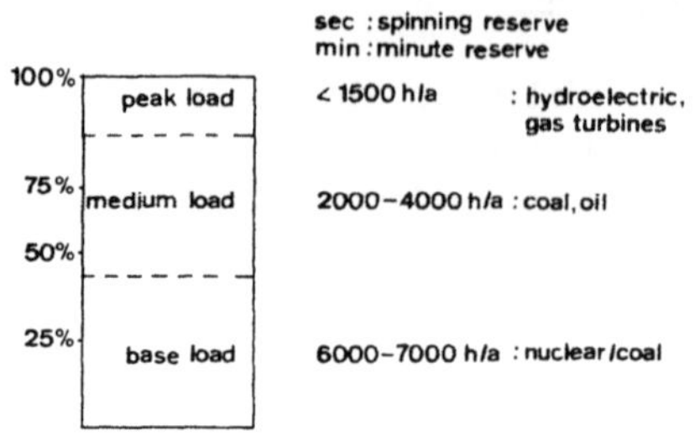

Fig. 4: Load management of utilities power plants

tages. The discontinuous electricity supply by photovoltaics requires the stand-by operation of the medium load power plants which results in higher costs due to low efficiency, Fig. 5.

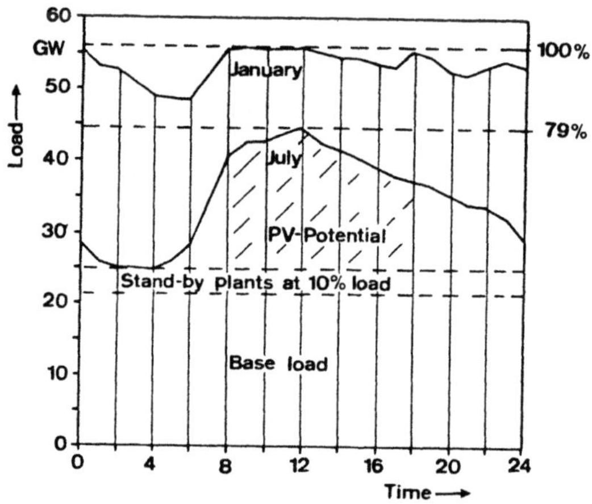

Fig. 5: Potential of grid-connected PV power plants

As a result of this estimation, one can expect a great potential for PV power plants without any need for additional energy storage systems. Increasing electricity production by PV will reduce the operating time or the medium load power plants which will be operated with a low partial load in the stand-by mode. Under these conditions power plants with high partial load efficiencies are required, which makes the use of fuel cells very attractive.

Looking far into the future there may be a demand for storage systems and candidates are:

- hydroelectric
- compressed air
- batteries
- superconducting magnets /5/
- chemicals

4. Autonomous PV Systems

Storage units in autonomous PV-systems must be selected and designed taking power and storage time into account.

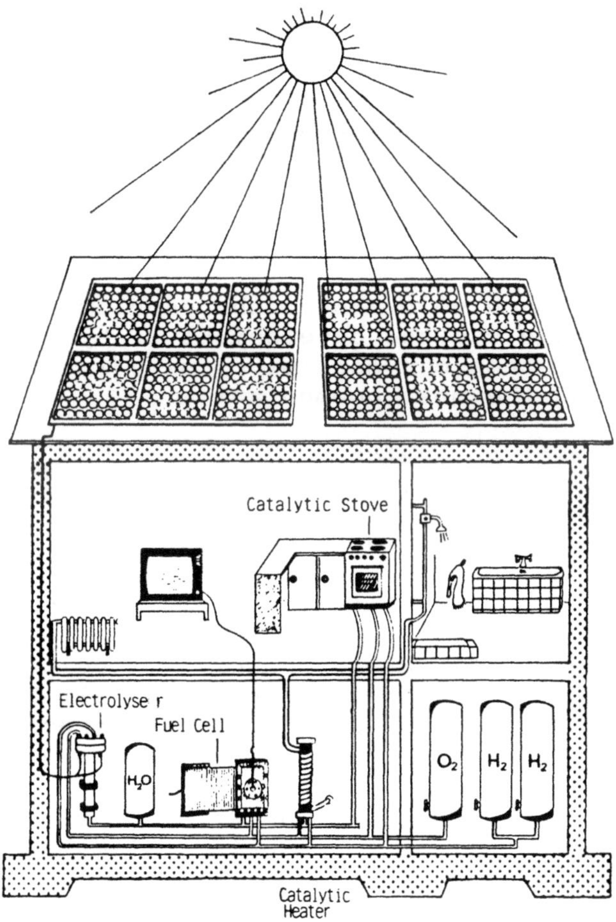

Fig. 6: Solar hydrogen oxygen house

In particular, the storage time is very important and up to now, there is no technology available for seasonal storage. Options for short-term energy storage are:

- batteries
- compressed air
- flywheels /6/
- supercapacitors

Batteries are used for low-term storage in restricted applications such as navigational aids. The disadvantages are the high costs, the low energy capacity and the need for maintenance. As a result, most of the PV systems operate without long-term storage. In future long-term storage systems on the basis of hydrogen and oxygen will be available, as they are under development now. Fig. 6 shows the design of a "solar-hydrogen-house" with H_2/O_2-electrolyser , tanks for H_2, and H_2O, fuel cell, catalytic heater, and catalytic stove. The energy storage system is sealed and no gas or water is transported from the system to the atmosphere or vice versa.

5. Hybrid Systems

The practical approach to overcome the problems related to long-term storage was the development of hybrid systems. Hybrid systems can consist of many interconnected components and the most common are:

- PV-battery-wind
- PV-battery-diesel generator
- PV-battery-thermoelectric generator /7/
- PV-battery-fuel cell/8/

Diesel generators are very cheap and available worldwide, but the low efficiency, maintenance requirements and the short lifetime are significant drawbacks.

Thermoelectric generators are very well suited to PV hybrid systems because of their silent and reliable operation. Nevertheless the use of these generators is limited to special applications because of the very low efficiency and the high investment cost.

In the future, fuel cells operating with methanol or propane will be appropriate for hybrid systems and great progress can be expected from the development of high temperature fuel cells.

References

(1) J.O.M. Bockris, Energy: The Solar Hydrogen Alternative, Australia (1975),London (1976).

(2) "Hydrogen Pilot Project-Canada", DECHEMA, Frankfurt (1978)

(3) W. Dunisch, R. Mülli, P. Kesselring, "Aluminium-A High Grade Energy Storage Medium", International Workshop on Heattransportation and Storage, Ispra, 9.-11.October 1985.

(4) R. Valentin, "Netzbelastung und Anforderungen an die Kraftwerke",VDI-Berichte **652**, Düsseldorf (1987).

(5) M. Masuda, T. Shintomi, "Conceptual Design of 5 GWH SMES", Proc. 9th Int. Conf. on Magnet Technology, MT-9, Zürich, Switzerland (1985), pp 357-359.

(6) R.C. Flanagan, "PV/Flywheel Lightstation System Design and Prototype Rotor Test", 22nd IECEC, Philadelphia ,1987, pp 1191-1196.

(7) W.C. Hall, D.J. Thornton, "Hybrid Power Thermoelectric, PV and Windenergy- A Convenient Marriage", 22nd IECEC, Philadelphia (1987), pp 1969 -1971.

(8) Proceeding of the CEC-Italien Fuel Cell Work shop(Ed. P. Zegers), 4/5 June 1987, Taormina, Italy CEC The Commission of the European Communities, Sept (1987).

FLYWHEEL ENERGY STORAGE DESIGN FOR PHOTOVOLTAIC POWERED SYSTEMS

Dr. R.C. Flanagan
Department of Mechanical Engineering
University of Ottawa
Ottawa, Canada, K1N6N5

In 1980, Canada initiated a program to design, develop and test several high energy density fibre composite rotors or flywheels. These rotors, designed for vehicular applications, stored 1.33 kWh and were rated at 105 kW and 100,000 charge/discharge cycles. In this paper, this technology and know-how are specifically directed at developing a large stationary storage system for photovoltaic powered lightstations operated by the Canadian Coast Guard, Marine Aids Division of Transport Canada. The system utilizes permanent magnet motor/generator and bearing systems together with a high performance E/XAS carbon and S2-glass composite rotor. The rotor is designed for a 20 year life and stores 8.5 kW-h at 45.7 Wh/kg. For this photovoltaic powered system, rotor or flywheel storage was shown to exhibit a number of advantages over other storage systems and further, the technical risk of development was seen to be low to nil.

INTRODUCTION

While flywheels, or momentum wheels, have been around for eons, designing a rotor for energy storage is a new technology which is just now beginning to mature. Indeed, high energy density fibre composite rotor development began in the mid to late 1970's. The Canadian program, initiated in 1980, resulted in the design, fabrication and testing of four 1.33 kW-h high performance fibre composite rotors [1]. Rotor 'FW3', shown in Figure 1, proved to be the most attractive design for several reasons [1]. Hence, the lightstation rotor presented in this paper builds on 'FW3' technology.

Energy storage rotors (ESR) should not be confused with momentum wheels (flywheels). They are a distinct and new technological component with unique inherent properties and can change the way energy/power systems are designed. Specifically, ESR's offer power/volume and power/weight ratios as much as an order of magnitude greater than any other component available and further, the higher the charge/discharge rate the more efficient is the system. On the other hand however, ESR's are not inherently energy storage devices due to the relatively large and unavoidable self-discharge rates even though specific energies up to 80 Wh/kg can be achieved in a practical design. None-the-less, in applications with high cyclic loading, turn around efficiencies usually exceed 99%.

PHOTOVOLTAIC LIGHTSTATION SYSTEM

The Canadian Coast Guard, Marine Aids Division, Transport Canada operate a number of photovoltaic powered remote lightstation systems for navigational aids. The main drawback to such a system is the very large number of lead acid batteries required to store available solar energy. This is further complicated by the low ambient winter temperatures where, in order to prevent freezing, the batteries must remain nearly fully charged thus substantially reducing their storage capacity. Thus, it is desirable to have an energy storage system which would be relatively compact, would not require large accommodations to house, would require little to no maintenance or servicing and could operate unattended, independent of climatic conditions. Hence, considerable interest exists in using a high energy density fibre composite rotor or 'flywheel' for energy storage.

For this low-power, long-term cyclic load leveling application, high performance components are required. A peak-power tracker is used to maximize energy extraction from the photovoltaic array and high efficiency solid-state DC-DC converters are used throughout. Primary power can flow directly to system load, excess power being directed through the motor drive inverter bridge to the high performance permanent-magnet motor/generator housed within the energy storage unit. Although state-of-the-art, none of the suppliers see any appreciable technical risk in system development, [2].

ENERGY STORAGE UNIT

The energy storage unit uses a single, vertically mounted, rigid shaft arrangement as shown in the general assembly (Fig. 2). The fibre composite rotor is center mounted between upper and lower magnetic bearing/suspension systems. The high speed motor/generator rotors are mounted directly on the flywheel shaft above the upper bearing. The upper and lower touch-down bearings contact only when the magnetic suspension system is not activated. Non-contact proximity probes are used to sense rotor radial (dynamic) response, rotor speed and shaft axial position. In short, the complete system acts as a single unit on a common shaft under permanent magnet levitation and electromagnet position control. The entire unit is encased in a hermetically sealed vacuum housing and operates in the upper range of the high vacuum regime (approximately 10^{-5} torr).

For the 8.5 kW-h lightstation rotor, a number of designs were investigated. The most effective combination, based on radial compatibility, assembly, cost and performance was found to be an aluminum carrier with a biannular S2-Glass/EXAS-Carbon composite ring. This triannular ring design was found to be radially compatible but does require accurate interfacial pressure control, that is, precision assembly tolerances.

Using established analytical codes, the rotor design procedure used a general optimization search from which optimal energy density geometries could be identified. Thereafter, from the design envelope, rotor geometry and component interfacial pressures were controlled/selected in a stress-analysis/manufacture/assembly iterative analytical study to establish a final design. Further, in the latter stages of design, rotor geometries were selected such that the EXAS carbon ring radial transverse stress showed the lowest safety margin thus providing for a safe failure mode should such a situation occur. The rotor specifications and stress profile are given in Table I and Figure 3 respectively.

The prototype rotor (Fig. 4) was designed to simulate the lightstation rotor and to assess the stacked ring design concept, in particular: i) radial compatibility, ii) dynamic stability, iii) structural integrity at high rotational speeds as well as iv) rotor assembly procedures. The rotor's dynamic response is displayed in Figure 5 and shows a basically "quiet" or dynamically well behaved rotor. During the tests conducted, this rotor was remarkably stable at all speeds and showed no tendency to instability if allowed to coast for long periods of time or when cycled between 20,000 and 27,000 rpm.

DISCUSSION

High energy density rotor technology has matured and large scale storage systems could be developed without significant technical risk. Rotor systems are inherently well suited to energy management of power systems, especially with high cyclic loading. However, even for the lightstation system presented, where cyclic loading was very low, the rotor system was shown to be an effective solution but did require state-of-the-art technology for the back-up systems.

ACKNOWLEDGEMENTS

1. This work was supported by the Marine Aids Division of Transport Canada under Contract No. T8080-6-2646/01SZ and in part by the Natural Sciences and Engineering Research Council of Canada under grant A1638.
2. The author recognizes the contributions of C. Aleong, F. El-Kasir, D. Flanagan, A. Laewen and P. Miller.

REFERENCES

1. R.C. Flanagan, "Design, Manufacture and Test Results for Four High Energy Density Fibre Composite Rotors", Proceedings, IECEC '86, Aug. 1986, San Diego, CA, pp. 901-907.
2. R.C. Flanagan, "Photovoltaic/Flywheel Lightstation System Design and Prototype Rotor Tests", Proceedings, IECEC '87, Philadelphia, PA, pp. 1191-1196.

Fig. 1. 'FW3' High Performance Fibre Composite Rotor |1|. 1.33 kw-h, 105 kw, 22,000 rpm, 61 cm O.D.

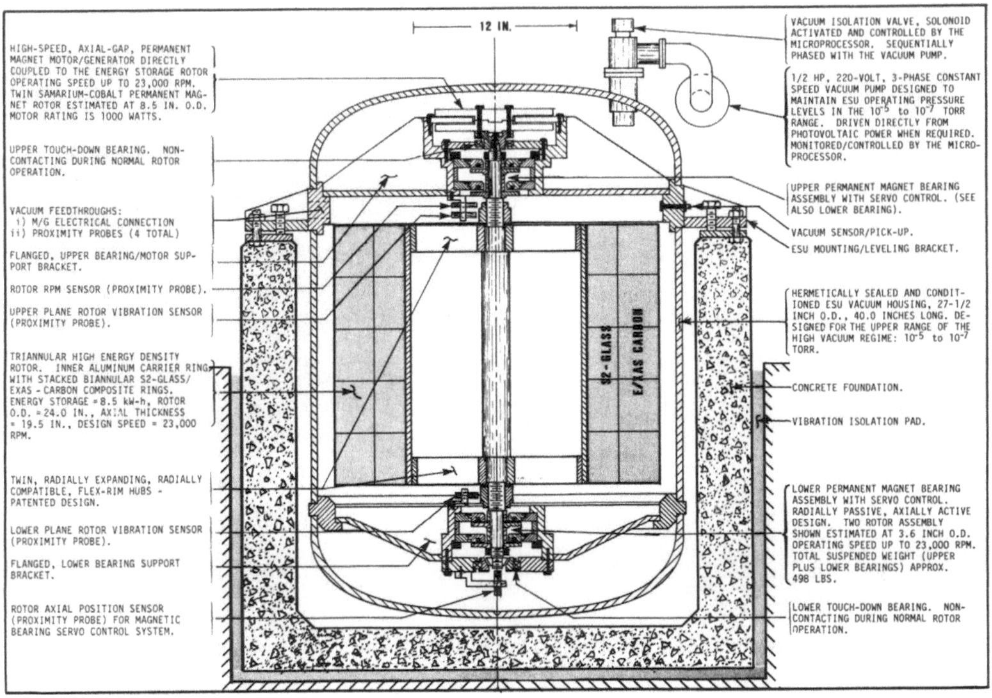

Fig. 2. General Assembly of the Flywheel Energy Storage Unit (ESU).

Table I: Lightstation rotor specifications.

ITEM	SPECIFICATION
A: DESIGN PARAMETERS	
ROTOR WEIGHT	410 lbs. (186 kg)
STORED ENERGY	8.5 kW-h
ENERGY DENSITY	20.7 Wh/lb. (45.7 Wh/kg)
DESIGN SPEED	23,000 RPM
ROTOR OUTSIDE DIAMETER	24.00 in. (60.96 cm)
ROTOR AXIAL LENGTH	19.50 in. (49.53 cm)
ASSEMBLED DIMENSIONS	
- HUB O.D.	12.60 in. (32.00 cm)
- RING 1 O.D.	13.20 in. (33.53 cm)
- RING 2 O.D.	18.00 in. (45.72 cm)
- RING 3 O.D.	24.00 in. (60.96 cm)
INTERFACIAL PRESSURES	
- RING 1/RING 2	1850 psi (12.8 MPa)
- RING 2/RING 3	2250 psi (15.5 MPa)
SWEPT VOLUME	5.1 ft.^{3} (0.14m^3)
OPERATING TEMPERATURE	-40°C to +80°C
CYCLE LIFE	20 YEARS AT 250 DEEP CYCLES/YEAR
DESIGN FAILURE MODE	RADIAL TRANSVERSE, E/XAS CARBON RING
ROTATIONAL INERTIA	7.8 lb.ft. s^2
HUB DIAMETRAL GROWTH	0.113 in. (0.287 cm)
B. ROTOR CONSTRUCTION	
ROTOR MATERIALS	
- HUB	7075-T651 ALUMINUM
- RING 1	7075-T651 ALUMINUM
- RING 2	S2-GLASS COMPOSITE
- RING 3	E/XAS-CARBON COMPOSITE
HUB DESIGN	TWIN FLEX-RIM, ANALYZED BY FINITE ELEMENT METHOD
RING DESIGN	TRIANNULAR RING STRESS ANALYSIS INCLUDING DYNAMIC LOADING, INTERFACIAL PRESSURE AND RESIDUAL STRESS
RING CONSTRUCTION	STACK OF FIVE COMPOSITE RINGS ON INNER ALUMINUM CARRIER
ROTOR ASSEMBLY	ALL COMPONENTS THERMALLY ASSEMBLED

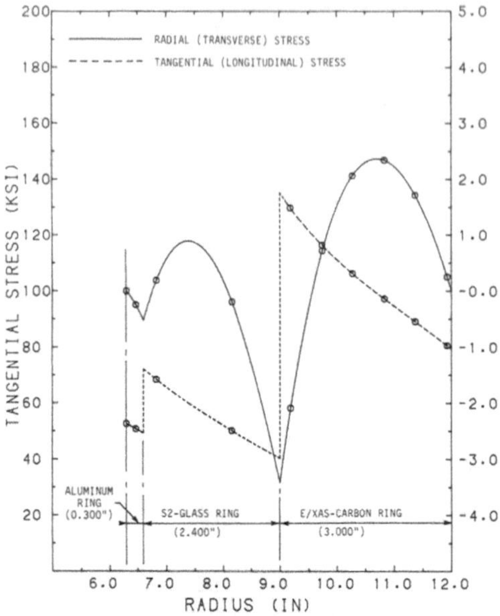

Fig. 3. Stress Profile of the Lightstation Rotor at Design Speed of 23,000 rpm.

Fig. 4. Prototype Rotor (ASR-1).

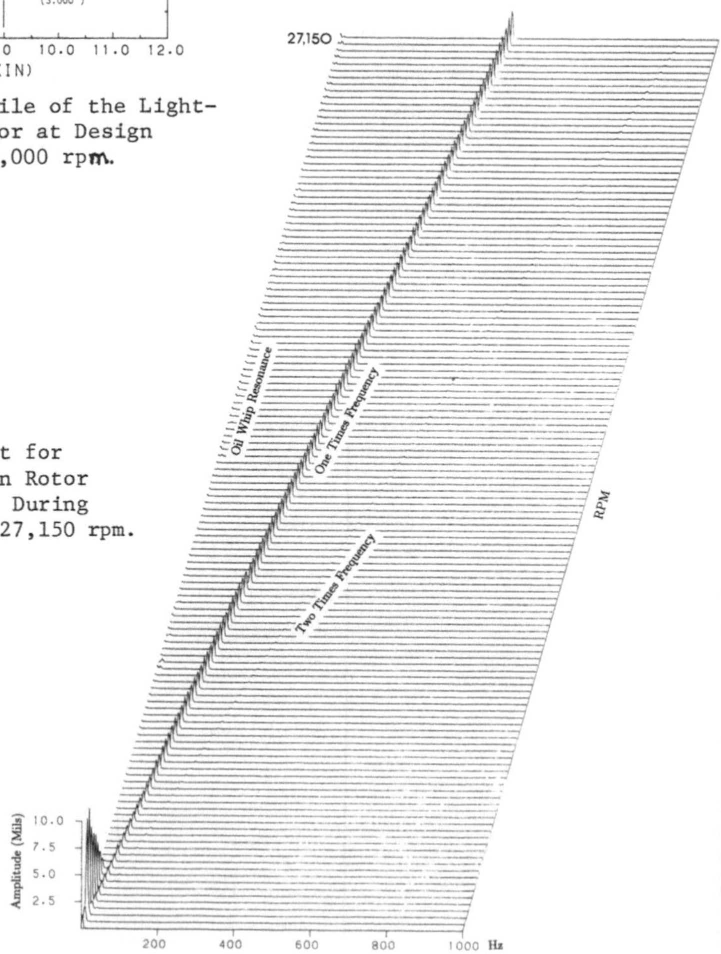

Fig. 5. Cascade Plot for Lightstation Rotor (Prototype) During Spin-Up to 27,150 rpm.

HYSOLAR 10 kW
TEST AND RESEARCH FACILITY FOR PHOTOVOLTAIC POWERED SOLAR HYDROGEN PRODUCTION

H. Steeb
A. Brinner

DFVLR, EN-TT
Pfaffenwaldring 38-40
D-7000 Stuttgart 80
Federal Republic of Germany
Tel. 0049-711-6862-618

Abstract

Within the joint Saudi Arabian - German Research, Development and Demonstration Program HYSOLAR a 10 kW test and research facility for solar hydrogen production has been designed and erected in Stuttgart. The general objective to be met using this facility is to investigate and promote the applicability of new and advanced technological concepts as well as to stimulate further R & D efforts in the field of solar hydrogen production.

The facility mainly consists of a 10 kW photovoltaic generator with variable output voltages; a test stand with several types of PV-modules; two mains-powered controllable DC supplies; a PV power conditioning system to operate in either a direct connection mode, a full DC/DC mode or a by-pass DC/DC mode; two advanced alkaline 10 kW electrolyzers and one 2 kW test electrolyzer of different technologies as well as gas handling system which includes intermediate storage, compression and gas filling station. The final installations will be completed by may 1988. Start up will be in June 1988. The paper describes the installation and presents first results.

1. Introduction

The 10 kW test and research facility represents one of six tasks of HYSOLAR. The program furthermore contains (1) a 350 kW photovoltaic-electrolytic demonstration plant in Riyadh, KSA; (2) a 2 kW photovoltaic-electrolytic research facility in Jeddah, KSA; (3) fundamental research on photoelectrochemistry, advanced alkaline electrolysis and fuel cells; (4) system studies and utilization demonstration and (5) an educational program [1,2].

The 10 kW test facility, fig. 1, has been erected to answer scientific and technical questions mainly in the following areas:

* Coupling of a photovoltaic field with the direct current load electrolyzer with respect to optimum power transfer,
* Development, investigation and comparison of different electrolyzer technologies for solar operation,
* Development and verification of feasible technical concepts for solar hydrogen plants

Therefore the test facility has been designed as a complete system of energy production, conversion and storage, suitable for specific tests, but also including all the subsystems future commercial solar-hydrogen plants might have.

2. Technical description

The test facility is composed of five main systems each with several subsystems as presented in table 1. Subsystem design data are given in table 2.

Table 1: Subsystems of the test facility

A) Energy production system
  * Photovoltaic generator
  * Power conditioning unit (PC)
  * High current supply (HCS)
  * Photovoltaic simulator (PV-S)

B) Hydrogen production system
  * Feed-water treatment unit (WT)
  * Alkaline electrolyzer 1 (E1)
  * Alkaline electrolyzer 2 (E2)
  * Alkaline electrolyzer 3 (E3)
  * Nitrogen purging system (N2)

C) Hydrogen treatment system
  * Intermediate gas storage balloon (IGS)
  * Hydrogen compressor and purification unit (HC)
  * Hydrogen filling station (GFS)
  * Hydrogen collecting pipe system (H2)
  * Oxygen collecting pipe system (O2)

D) Control System
  * Interconnection unit (ICU)
  * Safety and supervisory system (OP)
  * Compressor control unit (CC)
  * Electrolyzer 3 control unit (E3C)
  * Hydrogen safety system (H2SS)

E) Data acquisition and evaluation system
  * Data acquisition system (DAS)
  * PV-module carrier 1 (fixed angle)
  * PV-module carrier 2 (2-axis tracked)
  * Electronic load (installed in PV-S)
  * Meteorological station

2.1 Technical concept

The technical concept of the test facility is explained by means of the facility block diagram, figure 2.

Every power supply (PC, HCS, PV-S) can be connected to one of the electrolyzers (E1-E3). The connections are free-selectable by an operator with a connecting matrix (ICU). Output voltage, current and power of the rectifiers can be remote-controlled, output voltage and current of the PV generator can be adapted stepwise or continuously by the power conditioning unit (PC) to the demand of an electrolyzer. All electrolyzers can be operated in parallel. Their product gases, hydrogen and oxygen, are collected in two common gas pipe systems (H2 and O2). Oxygen is vented over roof, hydrogen is piped to the hydrogen treatment system. Here it is stored first in a gas balloon (IGS) at atmospheric pressure to decouple the discontinuous solar production from the compression. The compressor unit (HC) sucks hydrogen out of the gas balloon, purifies it and compresses the gas to 201 bar absolute pressure. The compressed hydrogen is given to a gas filling station and stored in standard gas bottles (50l, -201 bar). The electrolyzers can be purged separately with nitrogen from the common nitrogen purging system (N2).
The test facility is controlled and operated automatically by the safety and supervisory unit. Manual operation is also possible.

Selected subsystem design data are given in table 2.

Table 2: Subsystem design data

A) Photovoltaic Generator
- 10 kWp at 800 W/m², AM 1.5, ambient temp. 20°C, wind speed < 1 m/s
- Generator area: PV-module area : 183,9 m²
  active cell area : 144,0 m²
- PV-field configuration: 480 AEG modules PQ 10/10/02
  240 AEG modules PQ 10/40/02

B) Power Supplies
- HCS: max 15 kW DC, 0-15 V/0-60 V,0-1000 A/0-400 A
- PV-S: max 15 kW DC, 0-65 V, 0-300 A

C) Alkaline electrolyzers
- Electrolyzer 1 (KFA Jülich)
  10 kW (40 V, 250 A nominal)
  operating temperature 100°C, 1.05 bar
  28 cells in series, electrode areas 500 cm²
- Electrolyzer 2 (HYDROGEN SYSTEMS)
  12.9 kW (43 V, 300 A nominal)
  operating temperature 70°C, 1.05 bar
  25 cells in series, electrode areas 500 cm²
- Electrolyzer 3 (HYDROTECHNIK)
  2 kW (10 V, 200 A nominal; depending on electrode materials)
  operating temperature up to 80°C, 1.05 bar
  five cells in series, electrode areas 1000 cm²

D) Hydrogen storage
  maximum working capacity: 10 $m3_N$/h (expected: max 6.25 m^3/h)
  storage capacity: 120 $m3_N$

2.2 Facility installation

The 10 kW test facility is installed in the newly erected HYSOLAR building at the Stuttgart University campus and also houses labs for fundamental research on electrochemistry.

Figure 3 gives a sketch of part of the ground floor of the building and the facility installation. It shows from left to right the DAS evaluation room and the controlroom with the electrical facility installations, the electrolyzer room and in a extra area gas storage, compression and filling station. Besides the facility installations the building houses in its first floor three offices and a mechanical workshop for the operating force.

The Photovoltaic generator is situated south of the building on a separate steel structure. The PV-modules are mounted in five rows laying one upon another with inspection gangways between them, see fig.1.

2.3 Operating modes

A new operating system has been developed that decouples the test facility operation and supervision from a single component experiment and the data acquisition. The operator communicates with the operation program by masks and menus. Any fault operation action is blocked. The program indicates every action, subsystem status and power supply electrolyzer connection with verbal messages and a process control diagram.It is possible to select different control diagrams for indication of the complete test facility status, gas pipe system status, electrical system status and safety system status.

According to the tasks of the test facility, nine different operation modes such as test operation of single electrolyzers, permanent solar operation, permanent grid operation etc. have been developed and installed. In all modes reactions to warnings, alarms or disturbances are fixed and released automatically.

3. Start Up and first experiments

3.1 PV-field and power conditioning unit

One of the main experimental goals of the test facility is the realization and comparison of different concepts for power matching between Photovoltaic generators and electrolyzers i.e. DC loads. A very specific and unusual design of the generator was chosen to give maximum flexibility with respect to the output voltage levels.

For an optimum power matching three PV operation concepts [3] are conceivable and have been realized with a remote-controlled switch cabinet in the PV-field and a power conditioning unit with DC/DC-converters and computer control in the control room:

* Direct connection concept (A)
  The PV-field, main field (MF), is direct connected to the electrolyzer, the output voltage can be changed in three steps.
* Full DC/DC concept (B)
  The PV-field (MF) is connected to a low-setting DC/DC-converter, which decouples it from the electrolyzer. The output voltage can be adapted continuous in a wide range.
* DC/DC by-pass mode (C)
  The PV-field is subdivided into a main field (MF) and a by-pass field (BF) where the BF is connected to a high-setting DC/DC-converter. At its output both are common connected to the electrolyzer 4 power and voltage ratios between MF and BF can be selected. The common output voltage of MF and BF can be adapted continuous in a small range.

Table 3 presents MPP voltages and currents of all operation modes at 1000 W/m² and 45°C cell temperature.

Table 3: MPP voltage and currents in all PV-operation modes

No	MF voltage	BF voltage	current	power ratio (MF/BF)
A1	49.7 V		270 A	100%/0%
A2	45.5 V		288 A	100%/0%
A3	41.4 V		324 A	100%/0%
B	74.5 V		180 A	100%/0%
C1	53.8 V	20.7 V	180 A	72.2%/27.8%
C2	49.7 V	24.8 V	180 A	66.7%/33.3%
C3	45.5 V	29.0 V	180 A	61.1%/38.9%
C4	41.4 V	33.1 V	180 A	55.6%/44.4%

In all modes: output power = 13.42 kW

Figure 4 presents the PV-field characteristics in three direct connection modes at two different insolation and temperature levels. The curves have been calculated with a simulation code [4] and verified with measurements of the open circuit voltages and with different ohmic resistances. The comparison of measurements and simulation shows a misfit between them mainly resulting from a deviation of the simulated PV-field configuration and the used cell characteristic. For a better adaption of simulation the measurement data base will be extended with measurements with an electrolyzer and the verification of the cell characteristic in the next future.

The dotted lines in figures 4 show the range in which the electrolyzer characteristic can move and the PV-field operates in or near its MPP. The next experiments will show whether this range is sufficient for the solar operation with all variations of insolation, cell and electrolyzer operating temperatures over a day.

The experiments directly after installation of the PV-field have been used for the verification of its output power. It came to 12.92 kW at 1032 W/m² and 34°C. This is about 7% lower than expected but can be accepted.

3.2 Electrolysis

First functional test for solar operation have been performed with electrolyzer 3 and the high current supply HCS.

The electrolyzer has been developed in close cooperation between HYDROTECHNIK and DFVLR especially for testing new cell frame materials, electrodes and diaphragms in intermittent operation. The layout of the cell block allows the easy exchange of electrodes and diaphragms and the addition of other cells up to a number of 25. For the very first time gas separation chambers are integrated in the cell frame an a completely internal electrolyte circuit is realized in addition to an external loop.

Fig. 5 shows a cross section of the electrolyzer block and a sketch of the external electrolyte circuit. The frames of the five cells are joined in series and clenched with threaded rods through the cell frames and nuts to a filter press type arrangement.

The tests revealed, that operation is possible without the need of an electrolyte pump. The gas bubble lift in the duets from the anodic and cathodic cell compartments to the gas separation chambers is sufficient to maintain the electrolyte flow. Furthermore, the energy consumption for maintaining a passivation potential of the electrodes for corrosion prevention during night times is considerably low (0.3% of nominal power).

Fig. 6 depicts current-voltage characteristics as a function of the operating temperature T, taken with smooth Ni-electrodes and Rayton diaphragms. The strong dependency of T dearly underlines the importance of a flexible power matching for optimum power transfer from the generator as well as the need for thermal stabilization of an electrolyzer in intermittent operation. Dynamic and overload behavior of the electrolyzer is good: fast power variations as induced by insolation changes due to clouds as well as overloading by a factor of three are uncritical. Part load operation, however, gave rise to a gas purity problem; below 30% of nominal power the oxygen content in hydrogen exceeded 0.5 Vol.%. Other diaphragm/electrode combinations are tested currently to overcome this difficulty and to increase the process efficiency.

In May 88 the first system sized electrolyzer will be installed and from thereon the solar hydrogen experiments will start.

4. References

[1] H. Steeb, H. Weiss, B. Khoshaim, "HYSOLAR", Proc. 6th WHEC Vienna, July 1986

[2] H. Aba Oud, W. Grasse, "HYSOLAR 350", 8th PV Conference Florence, May 1988 (to be published)

[3] Messrs. SMA GmbH, W. Kleinkauf, W. Pigorsch, "Conception for the electrical system HYSOLAR", Final Report, SMA, Kassel 1986

[4] T. Schott, A. Siegel, "Optimization of coupled PV-electrolysis", ISES Solar World Congress, Hamburg 1987

Fig. 1: PV-field and building from south

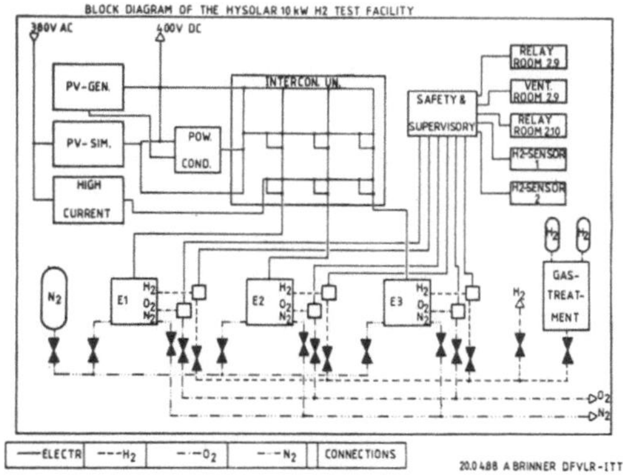

Fig. 2: Test facility block diagram

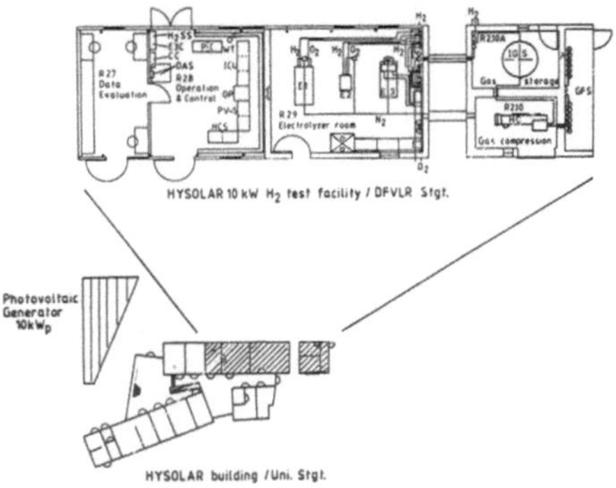

Fig. 3: Facility installation plan

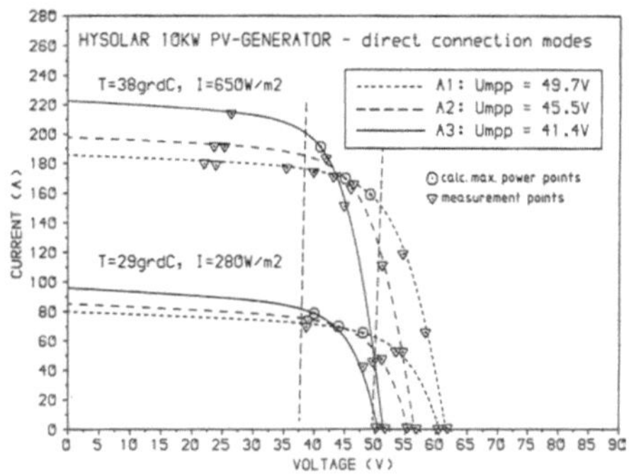

Fig. 4: PV-field characteristics in the direct connection modes

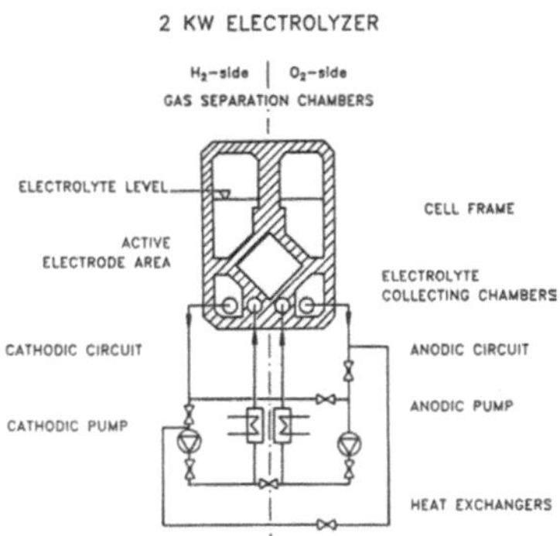

Fig. 5: Cross section of electrolyzer 3 block and electrolyte circuit

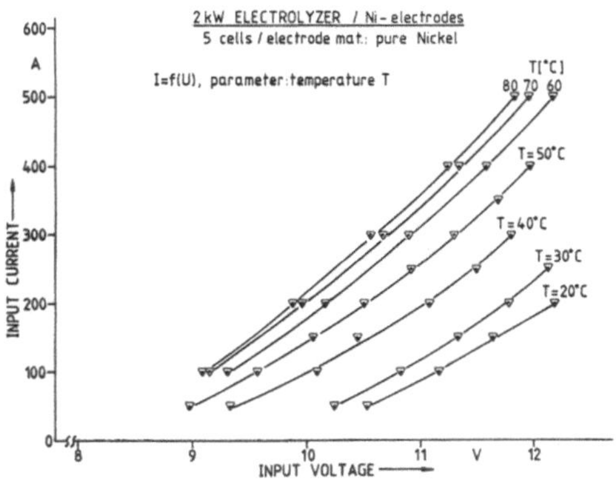

Fig. 6: I-U-characteristics of electrolyzer 3

SESSION 02:
TESTS AND TRIALS

RECENT PROGRESS IN OPERATIONAL PERFORMANCE OF STAND-ALONE AND HYBRID PHOTOVOLTAIC SYSTEMS FOR MILITARY APPLICATIONS

GARYL D. SMITH
Director of Energy Programs
Naval Weapons Center
China Lake, California 93555

ABSTRACT. The United States Department of Defense (DOD) Energy R&D Program has been involved in photovoltaic development since 1974. The Naval Weapons Center was designated lead laboratory for the DOD Photovoltaic Program in 1978. The initial effort was to coordinate the DOD participation in the U.S. Department of Energy's Federal Photovoltaic Utilization Program (FPUP) and to transfer this emerging photovoltaic technology from the research and development community to the DOD user community. The DOD has installed 300 photovoltaic power systems which range in size from 10 watts to 60 kilowatts. This experience with photovoltaic and hybrid photovoltaic power will enable the DOD Energy R&D Program to disseminate the information necessary for military activities to identify, procure and install reliable, cost-effective systems.

1. INTRODUCTION

Photovoltaic (PV) systems are the technology of choice for numerous land-based and off-shore applications within the Department of Defense. PV systems are a simple, field-proven and cost-effective power option with the potential to reduce supply vulnerability and strengthen energy security. PV systems also offer stable performance, high reliability and low maintenance characteristics. The utilization of PV for military applications is gaining significant importance as awareness of its advantages as a power supply option become more widespread.

Terrestrial photovoltaic systems have been commercially available for over 10 years. PV technology is well suited to a diverse range of military applications, particularly for remote applications. PV systems are modular by nature permitting incremental changes in generation capacity. Thousands of PV systems over the past decade attest to their field worthiness for military applications.

Within DOD, 38 different classes of applications representing over 50,000 individual power systems have been identified that could be converted to PV power. Typically, these applications require less than 5 kilowatts of peak power (kWp) and are located some distance from an electric power line (off-grid). Many of these applications have payback periods of less than one year.

In response to the Arab oil crises during the late 1970's, the Department of Energy funded the Federal Photovoltaic Utilization Program (FPUP) to test and demonstrate the value of PV systems as an alternative to conventional fuels. Under the FPUP, the DOD was authorized a total of about $5.4 million for 55 PV projects. It is important to note that the FPUP was for the most part a demonstration project. The program had many successes, identified a number of problems, and contributed to the overall understanding of systems design, application considerations and operation and maintenance requirements of PV systems under various field conditions.

FPUP helped to prove the reliability and competitiveness of PV technology in practical applications. The informantion and experience gained from the program coupled with new technological and production cost advances made by government/industry research and development have resulted in widespread acceptance of PV and the evolution of a strong, albeit small, industry base.

2. MILITARY APPLICATIONS FOR PHOTOVOLTAICS

2.1 Application Surveys

In 1986, the U.S. Navy Energy Office at China Lake, California, conducted a study of potential applications for PV. The study reviewed a variety of power-consuming applications at approximately 300 Navy facilities. An estimated 21,000 cost-effective applications falling into 38 general categories were identified. These were for remote applications ranging in size from 7 watts to 21 kilowatts of installed array power. A design and cost analysis was performed for representative systems for each of the 38 categories.

The results of this analysis (shown in Table I) demonstrate that on a projected service-wide basis, immediate utilization of PV for these 21,000 systems would result in a total annual electrical savings of approximately 40 million kWh/year or an estimated 140,000 barrels of oil equivalent/year of liquid fossil-fuel savings. Using 1986 PV systems capital and operation and maintenance (O&M) costs, the monetary savings Navy-wide would total an impressive $176 million.

The Army recently completed a study that examined the potential applications for photovoltaics at seven different Army installations in the west and southwestern United States. The study identified numerous applications where PV is now being used or where there is a near-term potential. These applications are summarized in Table II.

These studies, analyses, and examples of applications illustrate that the breadth of potential applications is extensive, and new uses are being identified every day. But many generic military applications can be cost-effectively implemented now. These applications generally fit one or more of the following characteristics:

- small load demand -- generally less than 5 kilowatts
- stand-alone -- independent of electric utility grid
- remote -- far from supply/maintenance depots or difficult to access.

These characteristics represent the types of demonstrations funded in the FPUP program as well as those applications identified by the Army and Navy as viable PV-powered alternatives to conventionally powered applications.

2.2 Categories of Photovoltaic Applications

There are three categories of applications that have the most near-term promise for the employment of PV power systems and include:

- Remote Power for
  - Lighting
  - Security Systems
  - Range Instrumentation
  - Aid to Navigation

- Communications
- Vehicle Battery Charging

2.2.1. Remote Power - Lighting and Security Systems. Lighting is among the most promising applications for PV. This can take many different forms and sizes--from a small wattage security lamp used only a few hours per week to a group of runway lights used every night of the year. Remote power applications include: munitions magazines; fuel storage areas; aircraft parking areas; small arms ranges; and vehicle parking lots.

Frequently, the combination of PV and high-efficiency, low-voltage lights makes a safe, low-cost package. These load applications are particularly well-suited to PV. Pay back periods can be as short as a few months depending on the characteristics of the individual systems.

2.2.2. Remote Power - Range Instrumentation. Various kinds of instrumenttation--from radar and telemetry to air sampling and environmental monitoring equipment--typically require only small amounts of power to operate, store and/or transmit data. This instrumentation is frequently located in remote areas and may need to be moved from one test site to another, and is an ideal application for PV. The loads can range from 7 watts to 240 watts. Payback periods are sometimes measured in months.

2.2.3. Remote Power - Navigation Aids. The United States Coast Guard has converted thousands of its buoys and shore markers to PV. Fifty eight percent were under 10 watts and all have been 35 watts or under. Similar applications exist where DOD could benefit from the experience of the Coast Guard, including hill and obstruction markers for aircraft safety. Significant savings can be realized from these applications. The Coast Guard regards the prime benefit to be decreased O&M costs resulting from much less frequent servicing requirements.

2.2.4. Communications. PV is particularly well matched for communications repeater stations. Frequently, these stations are at remote sites such as mountain tops where fuel resupply vehicles or maintenance teams have difficult access. Additionally, communications companies are using PV more and more at repeater stations that are not particularly remote. Typical sizes range from 120 to 270 watts for small stations to about 4,200 watts for larger systems. Generally, smaller systems have the most attractive payback periods.

2.2.5. Vehicle Battery Charging. In certain environments and operating scenarios, PV can be a method of overcoming the problem of parasitic battery discharge while the vehicle is inoperative. This application is similar to an already existing application in the civilian market where PV is used as a battery charger for recreational vehicles and boats. The Army Tank and Automotive Command is presently conducting a test program for such an application. Potential benefits include higher reliability, increased battery performance and longer lifetimes.

2.3. Hybrid Photovoltaic Power Systems

In addition to evaluating photovoltaic power systems the DOD Energy Program designed, installed, and evaluated a hybrid wind and photovoltaic power system for a Tactical Aircrew Combat Training System (TACTS). TACTS ranges are used to record and evaluate aircraft combat tactics both in real-time and on tape. Because a reliable power source was needed, a hybrid system was selected. The hybrid system was installed on a land-based tower at Bodie Island, North Carolina, and evaluated for 2 years. The system proved to be reliable and sufficient to perform the necessary function. Duplicate hybrid power systems have been installed on eight ocean towers off the coast of South Carolina and Georgia during 1988.

3. Summary

The United States Department of Defense Energy Research and Development Program has been evaluating renewable energy technologies since 1974, and has demonstrated the utility of using photovoltaic and hybrid PV systems in over 300 separate applications. Based on this extensive evaluation program the U.S. Department of Defense considers photovoltaic and hybrid power systems to be very reliable and cost-effective power sources for many applications. Military applications for power can be tactical, strategic, fixed, or mobile. Today's PV systems can supply power to a wide range of applications in a cost-effective manner and can reduce a systems vulnerability to energy interruptions. PV can increase energy security at military installations throughout the world. The DOD hopes that by building on its expertise and installing highly reliable, cost-effective photovoltaic power systems, this emerging technology will make a smooth transition from the R&D community to the user community.

TABLE I
NAVY SURVEY ON POTENTIAL PV APPLICATIONS

APPLICATION	NUMBER OF APPLICATIONS	TYPICAL CAPITAL COST-$	AVERAGE LIFE-CYCLE SAVINGS-$
Building Guard Station	150	30,000	61,237
Building Operations (Remote)	20	248,000	60,636
Cathodic Protection	1,000	2,400	56,870
Communication, Island	200	5,400	13,920
Commu. Repeater (Remote-Large)	300	25,600	70,080
Commu. Repeater (Large)	900	21,600	44,510
Commu. Repeater (Small)	900	10,800	48,144
Commu. Repeater (Remote-Small)	300	13,800	51,806
Communication Emergency	400	720	37,918
Communication Telephone	100	86,400	83,909
Lights, Small Craft	2,000	1,800	56,970
Lights, Target	100	14,400	66,643
Lights, Warning	1,500	1,440	116,766
Lights, Wind-Sock	200	7,200	2,176
Meteorological Station (Large)	200	21,600	141,074
Meteorological Station (Small)	1,500	140	11,564
Aid to Navigation	300	1,440	57,300
Housing (Remote)	100	15,000	66,164
Instrumentation, Camera	400	45,000	141,039
Instrumentation, Offshore	20	45,000	133,311
Lights, Buoy	1,500	1,000	111,079
Lights, Heliports	100	28,800	2,493
Lights, Magazine	600	2,880	55,225
Observation Tower	200	72,000	100,344
Radar, Beacon (Small)	500	3,600	93,341
Radar, Beacon (Large)	300	14,400	68,974
Radar, Target	400	96,000	43,442
Remote Bomb Scoring	400	4,800	55,053
Security, Sensor	1,500	1,440	56,165
Security, Visual	600	14,400	69,974
Security, Gate	100	6,000	36,500
Transportable Power	3,000	48,000	138,600
Water Pumping	200	180,000	72,027

TABLE II
TYPICAL LOAD AND POWER CHARACTERISTICS
OF POTENTIAL PV ARMY APPLICATIONS

APPLICATION	APPROXIMATE PV ARRAY SIZE (WATTS)
Test Range Equipment	25-160
Battery Chargers (for emergency power of water wells)	17-20
Clearance Lights (on water tanks)	770-2500
Global Positioning System Satellite Simulators	2000-2800
Mobile Firing Range	80-90
Mobile Generators	500-9000
Radio Repeaters	1400-1600
Firing Range Gun Position	100-110
Range Surveillance Video	1000-3500
Microwave Towers	1500-1700
Remote Data Acquisition	300-500
Meteorological Towers	200-220
Storage Facilities (Igloos)	300-320
Microwave Repeaters	1200-1600

THE SELECTION OF REFERENCE DEVICES FOR PHOTOVOLTAIC PERFORMANCE MEASUREMENTS

F.C. Treble
Consultant
43, Pierrefondes Avenue
Farnborough, Hants. GU14 8PA
United Kingdom

ABSTRACT. Recent field and laboratory experience in Europe has shown that the use of the traditional type of reference cell can sometimes lead to unacceptable errors in photovoltaic performance measurements. The reason is that, under certain test conditions, reflected light can result in the irradiance measured by the reference cell being significantly different from that incident on the test specimen. The International Electrotechnical Commission's Working Group on pv module standards has considered this problem at some length and has prepared a draft document on the subject for circulation to National Standards Committees. The draft describes old and new types of reference device, including reference modules, and indicates the kinds of simulator and outdoor test conditions for which they are suitable. The IEC proposals are presented and the supporting rationale explained.

1. INTRODUCTION

When measuring the performance of a solar cell, module or array for rating purposes, one needs a suitable reference device to monitor or set the irradiance, in order to relate the measurement to the reference spectral irradiance distribution. Significant errors can arise unless the reference device has been properly selected and calibrated.

A previous paper (1) described methods for the calibration of reference devices and proposed a procedure for future international standardisation. This paper deals with the equally important problem of selection.

The essential requirements for a reference device are :-

1) Its photovoltaic characteristics must be stable.

2) It must be a linear device, i.e. its short-circuit current must vary linearly with irradiance over the range of interest.

3) Its relative spectral response must match that of the cell, module or array to be tested to the extent necessary to limit spectral mismatch errors to less than ± 1%.

4) Its optical properties must be such that the irradiance measured by the reference device is always the same as that incident on the test specimen.

Recent experience in European laboratories and in the on-site measurement of the rated powers of the CEC's pv pilot plants during the acceptance test campaign in 1983-4 has shown that the single reference cell, procured as one would a radiometer and widely regarded as an all-purpose device, is not suitable for all applications, in that it does not always meet the last of these requirements.

The IEC's Working Group on pv module standards, TC82/WG2, has considered this problem at some length and has prepared a draft document on the subject, which describes five types of reference device and the applications for which they are suitable. The intention of the present paper is to explain these proposals and the rationale behind them.

2. UNPACKAGED CELLS AND SINGLE CELL PACKAGES

Unpackaged reference cells and the conventional reference cells in single cell packages, as supplied by manufacturers and agencies in the past, are suitable only for measurements in natural sunlight, when the direct beam is at or near normal incidence, or with a Class A (± 2% uniformity) simulator, such as the pulsed type, where the light source is not close to the test plane.

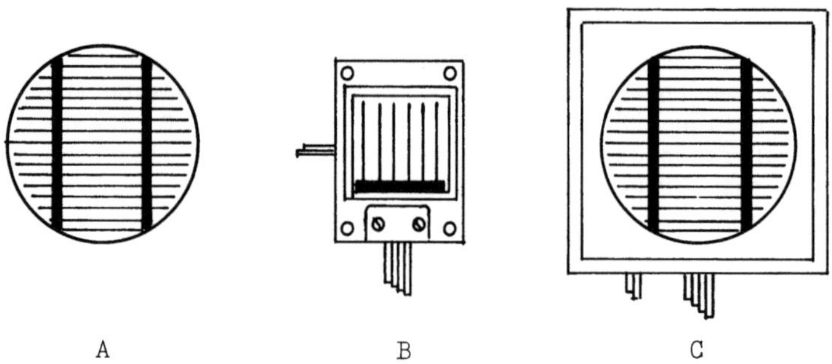

Fig. 1 Single reference cells

Unpackaged reference cells (Fig. 1, A) are mounted for use on a temperature-controlled block, with probes for making connections to the contacts. The single cell packages, B and C, are usually equipped with connecting leads and thermocouples. The type originally used for space applications, B, has a metallic housing with a flat back which can be mounted on a temperature-controlled block. The other kind, C, consists of a single cell similar to those in the modules to be tested, encapsulated and framed in a similar fashion. This type cannot easily be temperature controlled.

All single-cell packages should have a field of view of at least 160°. All surfaces within the cell's field of view should be non-reflective. The encapsulant should be electrically and optically stable. In cases where the reference cell is of a different type to those in the test modules, the window may embody a filter to achieve the necessary spectral match.

3. MULTICELL PACKAGES

When carrying out in-situ performance measurements on fixed pv arrays, it may not always be possible or convenient to work with the direct solar beam at normal incidence. In such cases, the irradiance on the modules may be enhanced by internal reflection, as illustrated in Fig. 2. The enhancement can be significant, particularly in modules with round cells and white backing.

An unpackaged reference cell or one in a single cell package will not have the same reflective properties and will therefore not be subjected to exactly the same irradiance as the modules in the array.

This phenomenon was first noticed during the acceptance tests on the CEC pilot plant at Rondulinu, Corsica, where the tilt angle of the modules is adjustable. The reference device, a single cell package of the type shown in Fig.1 C, was mounted co-planar with a module, as in Fig. 3, and the short-circuit currents of both were measured at various tilt angles between 0° and 60°, within a period of 5 minutes near local noon. Fig. 4 shows how the ratio of the short-circuit currents of the module, I_M, and the reference cell, I_R, which should remain constant, varied, in fact, by nearly 3%. So, in certain circumstances, the use of a reference cell in a single cell package could result in a measurement error of 3%.

This error can be avoided by enclosing the reference cell in a multicell package, in which the encapsulation system and the shape, size and spacing of the surrounding cells are the same as in the modules being tested. The other cells may be real ones or dummies having the same optical properties. Fig. 5 shows the minimum acceptable size.

Multicell packages may be used instead of unpackaged reference cells and

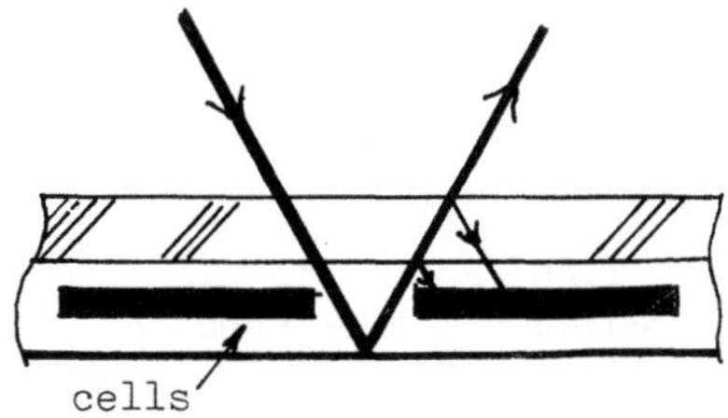

Fig. 2 Internal reflections in a module

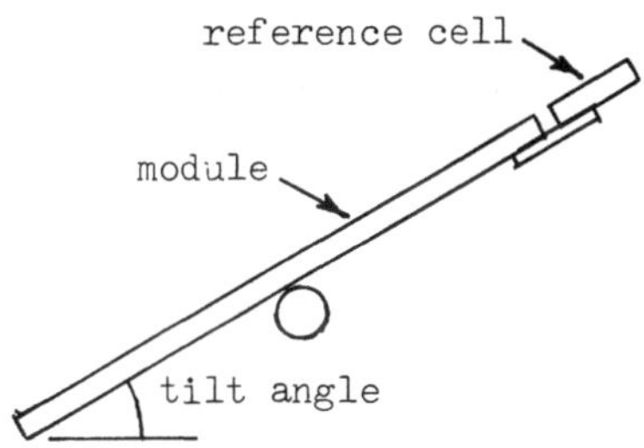

Fig. 3 Rondulinu test

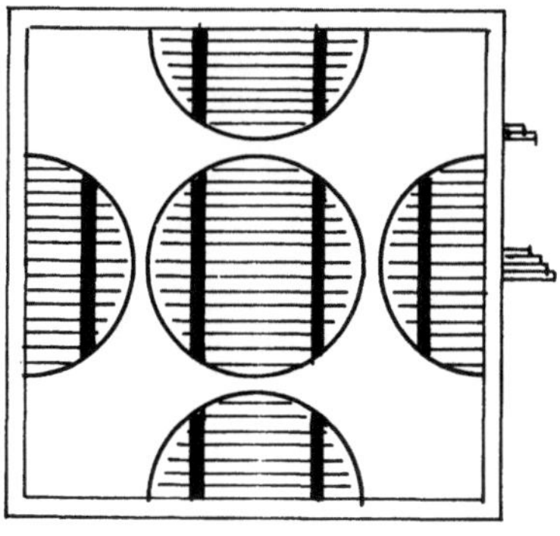

Fig. 5 Multicell package (minimum size)

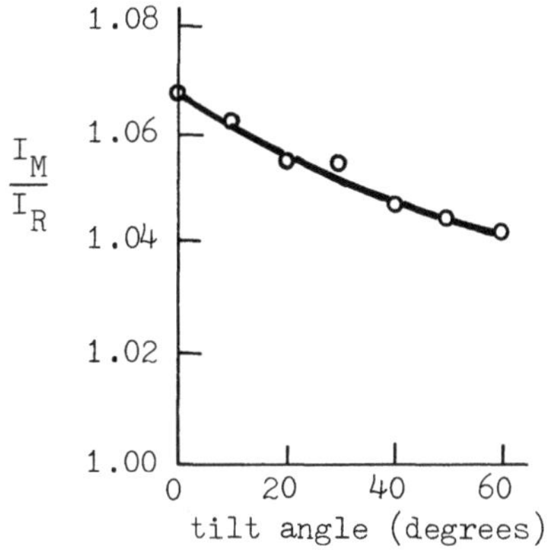

Fig. 4 Current ratio variation

those in single cell packages for the applications indicated in para. 2.

4. REFERENCE MODULES

Another problem may arise when making measurements with a simulator where the light source is close to the test plane. With such equipment, the irradiance may be enhanced by light reflected back and forth between the simulator and the module under test. Because of the restricted area of the test plane, the usual procedure with these simulators is to (a) set the irradiance with the reference device, (b) replace the reference device by the module and (c) carry out the measurement at the same setting. The problem is that, unless the reference device covers the same area as the test module and has the same reflective properties, the irradiance on the module may not be the same as that measured by the reference device.

Some members of the IEC Working Group thought that the answer to this problem was to use a reference cell in a multicell package of the same size as the module, locating the reference cell in a position where the irradiance would be the average of that over the test plane. Others argued that a reference module would give greater accuracy, particularly in cases where the multiple reflection problem was compounded by relatively poor uniformity of irradiance. It would also be more readily available.

To settle the argument, the Laboratoire Central Industries Electriques (LCIE) of Fontenay aux Roses, France, conducted a series of tests with three simulators. The first of these, S1, was a Class A pulsed simulator, with a xenon arc lamp 6m from the test plane, giving a uniformity of better than ± 0.5%. The other two, S2 and S3, comprised 77 steady-state tungsten lamps, only 0.5m from the test plane, producing uniformities of ± 3% and ± 8.5% respectively.

Two reference devices were used in turn to set the irradiance for performance measurements at 25°C on six 1.042m x 0.462m modules of 100mm square multicrystalline silicon cells. These devices were C, a reference cell in a multicell package, and M, a reference module. Both were of the same size and shape as the six test modules and they embodied the same cells and encapsulation technology. The reference cell in C was placed in the position of average irradiance, as previously determined. The cells in M were matched in terms of short-circuit current to better than ± 2%.

Fig. 6 shows how the short-circuit currents I_{SC} of the six modules, as measured with the three simulators and two reference devices, varied. There is very good agreement between the datum readings, taken with the pulsed simulator S1 and the reference module M, and those taken on the same simulator set with reference cell C. But, when the measurements were made under simulators S2 and S3, where uniformity is not so good and there is the problem of multiple reflections, the C readings change more than the M readings. In fact, with S3, the C readings are 7.5% to 9.3% lower than the datum, while the M readings, even with such poor uniformity of irradiance, are still within 2.3%.

Maximum power data from the LCIE tests show a similar result.

As a consequence of these findings, the IEC is proposing to call for a reference module to be used for setting the irradiance of simulators with a source close to the test plane, where multiple reflections are to be expected. In the case of Class B simulators, where the non-uniformity limit is ± 5%, the reference module will be required to have cells whose short-circuit currents after encapsulation are matched to better than ± 2%.

Reference modules may also be used as alternatives to single cell devices in other applications.

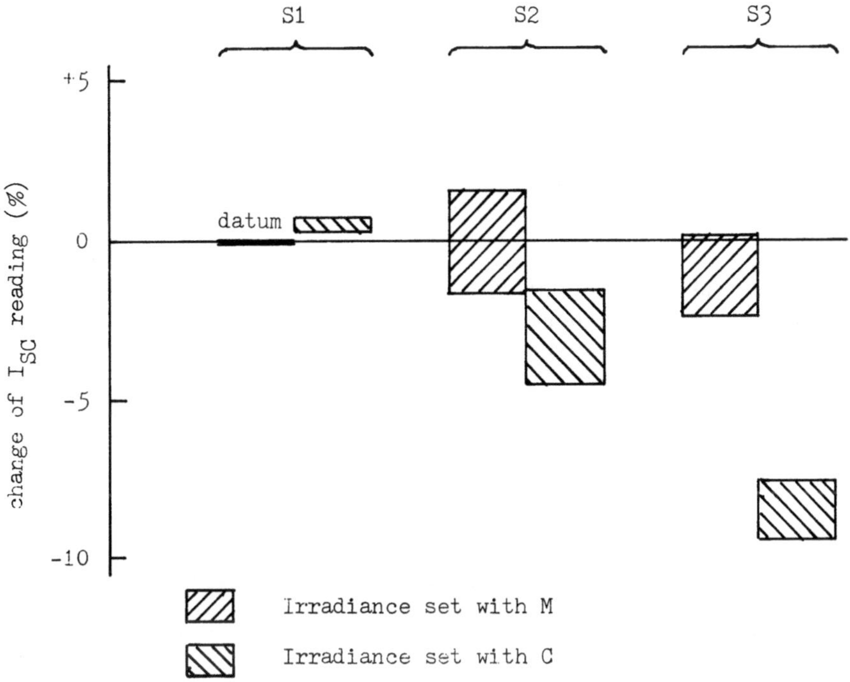

Fig. 6 L.C.I.E. results

5. CONCLUSIONS

Single reference cells, whether unpackaged or in an individual package, should be used only at normal incidence in natural sunlight or with a Class A simulator where there is no multiple reflection problem.

Reference cells in multicell packages may be used instead of individually packaged cells and also for outdoor tests at sloping incidence.

Reference modules with matched cells (± 2%) are essential for Class B simulators and may be used for all other applications.

6. ACKNOWLEDGEMENT

The author would like to thank Mr. A. Desombre of Photowatt International S.A., Caen, for arranging the LCIE test programme and supplying the data on which the analysis in this paper is based.

REFERENCE

1. Treble, F.C. "The calibration of primary terrestrial flat-plate reference devices - a proposed international standard". Seventh E.C. Photovoltaic Solar Energy Conference, Sevilla, October, 1986, pp. 89-94.

THE PHOTOVOLTAIC ADVANCED SYSTEMS TEST FACILITY*

Jay L. Chamberlin
Photovoltaic Technology Division 6221
Sandia National Laboratories
Albuquerque, New Mexico 87185, USA

ABSTRACT. Sandia National Laboratories, as part of the United States Department of Energy's (DOE) National Photovoltaics (PV) Program, operates an extensive PV testing laboratory in Albuquerque, New Mexico. The objectives of this laboratory, the Photovoltaic Advanced Systems Test Facility (PASTF), are to assess the performance of PV and balance-of-systems (BOS) hardware and to work with their developers to rectify any performance deficiencies found. The principal resources of the PASTF consist of a materials testing laboratory, a lens and lens/cell testing facility, a module testing facility, an array testing facility, a power conditioner evaluation facility, and specialized test stations for evaluation of BOS hardware. A cell characterization laboratory that is part of the larger Sandia PV effort provides evaluational support in this area. These resources are augmented by Sandia's extensive research and support facilities, principally in the areas of material characterization and environmental testing. The paper describes the resources of the PASTF along with examples of their application to the evaluation of PV hardware.

1. INTRODUCTION

The effective application of photovoltaics (PV) as a viable alternate-energy power system depends upon the ability of each component to perform, in concert with the other components, efficiently and reliably under the rigors of the use environment. While information acquired from fielded PV systems is useful in determining their performance, structured experiments conducted in a test facility generally provide more valuable data. This is particularly true when the hardware is developmental in nature, and hence not yet amenable to field applications.

Sandia National Laboratories, as part of the United States Department of Energy's (DOE) National Photovoltaics Program, operates such a laboratory, the Photovoltaic Advanced Systems Test Facility (PASTF)[1]. Over the past ten years this laboratory has provided evaluational and testing support to Sandia's Photovoltaics Projects which conduct and direct research and development programs for DOE in the areas of one-sun and concentrator cell research, concentrator collector research, systems development and evaluation, and design assistance. While most of the hardware evaluations performed by the PASTF result from these DOE-sponsored, Sandia-directed programs, some commercially available hardware is also evaluated, principally in the area of balance-of-systems (BOS) hardware. The results of these evaluations along with recommendations for hardware improvements are shared with their developers and manufacturers.

*Prepared by Sandia National Laboratories, Albuquerque, New Mexico USA for the U.S Department of Energy under Contract #DE-AC04-76DP00789.

The PASTF is located in Albuquerque, New Mexico USA, a high (1653 meters) desert location of abundant sunshine and moderate temperatures. This 35° north latitude site has an average of 174 clear days, and 108 partly cloudy days per year. The average global daily horizontal and direct normal solar radiations are 21 and 26 kJ/m^2, respectively, and the average ambient temperature is 13°C. The PASTF covers 3.5 hectares of gently sloping land situated to maximize its southern exposure. Eight buildings, housing five laboratories, a mechanical shop, an office and a storage building, are located on the site. An electronic shop, some test setups and additional storage are housed in several temporary and mobile structures also located at the site. The permanent operational staff currently consists of nine technicians. This staff is augmented by the research and developmental engineers whose hardware is under test. A complete weather station and a solar instrumentation calibration station provide resources in these critical areas.

Sandia has numerous research and testing laboratories that are also available to the PASTF. Those principally used to support the evaluation of PV hardware include the extensive material characterization laboratories and an environmental test facility. While the PASTF has both mechanical and electronics shops, these resources are frequently complemented by other Sandia shops to maximize the PASTF's PV hardware testing time. Analytical support, particularly in the areas of statistical and reliability analysis, is also available. Through the various resources of the PASTF and these associated facilities, virtually any piece of PV hardware from the component level to full systems can be evaluated. The following is a description of the PASTF resources and their typical applications.

2. THE PASTF RESOURCES

2.1 Materials Testing Laboratory

The various materials used in PV components are tested to determine the tolerance to environmental exposure in the PASTF's materials testing laboratory. This laboratory consists mainly of four high-intensity ultra-violet lamp test stands and six thermal cycling chambers along with supporting sample fabrication and analytical equipment. The lamps are used for the accelerated aging of materials exposed to solar radiation. While acceleration factors up to 85 are possible, most tests are conducted at a factor of 30. Simulated exposure times of 30,000 hours are routinely used. This is equivalent to 10 years of normal solar exposure. The laboratory also maintains an outdoor exposure area for the natural aging of specimens.

Temperature cycling is mainly used to determine the susceptibility of bonds to the stresses induced by this diurnal and cloud-produced effect. Nominal cycling temperatures of -40 to 110°C are used. Two-hundred and fifty cycles at a rate of 36 cycles per day are generally imposed on the test specimens. Tests have shown that this number of cycles is equivalent to approximately 9 years in the field. Ultrasonic and x-ray analyses, and pull tests are used to determine loss of bonding strength.

Optical materials are tested to determine any degradation of their optical properties and performance, and any changes in their dimensional stability and resiliency. Bonding agents, such as adhesives and solders, are tested for changes in their bonding strength, void fractions and electrical conductivity. One of the unique resources associated with this laboratory is a soldering facility that has the capability of bonding test specimens in selective gas environments at carefully controlled temperatures and loads. This facility is also used to solder cell assemblies for other test applications. The materials testing laboratory typically evaluates several hundred samples per year.

2.2 Lens and Lens/Cell Testing Facility

A unique resource of the PASTF is its lens and lens/cell testing facility, which permits the evaluation of the performance characteristics of these concentrator collector components. This facility consists of two precision trackers that can be adjusted to accurately follow either the path of the sun or fixed angular displacements from it. This feature allows the determination of the sensitivity of a lens or lens/cell combination to errors in tracking that might occur in use.

The lens analyzer can accommodate lenses up to 450 mm square with focal lengths up to 700 mm, and images up to 100 mm square. For analysis the image plane is raster scanned by a 0.25-mm-square silicon solar cell on an operator-selectable spacing as fine as one-hundredth of the width of the image in both directions. The flux intensities are automatically sampled and stored. Reference solar radiation data are taken by a silicon solar cell identical to that used for the image mounted in a collimator attached to the lens analyzer. During analysis, the image data are adjusted for variations in the incident solar radiation. The resulting information can be presented in a variety of graphical forms, including three-dimensional, contour, and histographic plots. Numerical data presentations include the lens' efficiency and its geometric, peak and average concentration.

The lens/cell tester can also accommodate up to 450-mm-square lenses with 700-mm focal lengths. Cells, up to 75 mm square, are soldered on thermally controlled copper paddles for testing. The lens-to-cell spacing and the position of the cell are adjustable to allow the determination of the combination's sensitivity to positional tolerances that might occur in manufacturing or use. Data is taken for the cell mounted in a collimator without the lens to determine its one-sun performance. The lens/cell combination is then measured. In both cases, current-voltage curves and the corresponding environmental data are measured. After adjusting for variations in the solar radiation, the ratio of various concentrator to one-sun measurements provides lens/cell performance information including the lens' concentration, and the lens', the cell's and the combination's efficiencies. The cell's parameters, such as short-circuit current, open-circuit voltage, maximum power and fill factor are also determined.

Tests in this area are supported by a cell characterization laboratory operated by another Sandia PV organization. This laboratory has the resources to measure a cell's current-voltage curve at concentrations up to 1000 suns, to measure its dark current-voltage and capacitance-voltage characteristics, to determine its spectral response and hemispherical reflectance, and measure its transient response to both electrical and optical excitation. The cell laboratory also has the capability to laser-scan a cell to determine its characteristics at a micro-response level.

Last year approximately 500 individual tests were conducted on lenses and lens/cell combinations.

2.3 Module Testing Facility

Sandia has developed a series of qualification tests for modules and module components that are used in DOE-sponsored programs[2]. These tests include the material and component ultra-violet exposure and thermal cycling tests already described, and the critical, with the exception of a specialized hot-spot heating test, testing of modules and submodule components. After the materials and components have passed the criteria of the qualification tests for bonding strength, electrical isolation and ultra-violet exposure, cell assemblies and receiver sections are subjected to testing which includes initial and final electrical characterizations, and thermal and thermal-humidity cycling. The thermal cycling consists of two-hundred and fifty, -40 to 110°C cycles imposed at 36 cycles per day in a nominally dry, but uncontrolled, humidity environ- ment. The thermal-humidity cycling consists of ten -40 to 85°C cycles imposed at 1 cycle per day in an environment held at 85% relative humidity throughout the cycle. Failure of electrical isolation, a loss of more than 5% of the electrical contact area or an 8% drop (relative) in conversion efficiency at maximum power results in a unit's disqualification.

After cell assembly and submodule qualifications are met, full modules are tested. These tests are similar to those used for the earlier units along with off-axis beam damage, water spray, hail damage, and hot-spot tests. For modules, the cycling limits are less severe. One-hundred and twenty -40 to 60°C thermal cycles at six per day and ten -40 to 60°C thermal-humidity cycles at one per day are imposed. The humidity levels correspond to those used for cell and receiver section testing. The same failure criteria along with the obvious water, hail, and beam survival requirements are used to judge a module's qualification.

This facility has the necessary instrumented precision trackers, computer-controlled loads, and data acquisition systems to simultaneously test up to four units. Specialized equipment, such as a water spray test stand and a hail gun, complete the qualification equipment. Diagnostic equipment, both at the PASTF and its supporting laboratories, are used in determining the underlying causes of any failure. This equipment is currently being used to test, among other things, prototype receiver sections and modules for a DOE/ENTECH/3M/City of Austin cooperative project to install a 300-kW concentrating collector system in Austin, Texas USA. During the past year, 8 complete and 8 partial module, and 4 complete and 5 partial receiver section qualifications were completed.

2.4 Array Testing Facility

Both flat plate and concentrating collector arrays with power ratings up to 30-kW are tested. Normal testing consists of a complete inspection of all of the array's components including the power generation modules, field wiring, mounting structure, and tracking subsystem if the array has one. Baseline current-voltage curves are made at the module, source circuit and full array level. The results of this inspection and baseline electrical characterization are documented and form the basis against which future evaluations are judged. The array is then resistively loaded near its maximum power point and allowed to operate in this manner between the evaluation periods. In the case of tracking systems, tracking error monitors are installed to determine the routine performance of the array in this regard. The array is periodically (nominally every three months) tested to determine any degradation in its durability or performance.

Special attention is paid to items identified in the baseline or later tests that indicate problem areas. On an infrequent basis tests are conducted to determine the extent of degradation factors. For example, the effect of array soiling is determined by measuring the array's electrical performance before and after a thorough washing.

The PASTF has a full complement of the equipment necessary to complete the testing of arrays, including computer-controlled loads to determine the array-level electrical characterization, portable curve tracers for module-level evaluations, and fixed loads for continuous operation. Arrays are routinely tested until either they fail to perform satisfactorily or until the space is required for additional experiments. Currently, two 30-kW fixed flat-plate arrays, one 16-kW single-axis tracking flat-plate array and one 10-kW two-axis tracking concentrating collector array are under test. These arrays all use crystalline or poly-crystalline silicon cells. The fixed flat-plate arrays have been under test for five years[3]; the tracking arrays for two years. Four smaller (300 to 900 W) amorphous-silicon arrays are being installed to allow evaluation of this technology.

2.5 Power Conditioner Evaluation Facility

The power conditioner evaluation facility is another unique PASTF laboratory. This facility has the resources to evaluate the performance of power electronic components and systems with power handling capabilities up to 225 kVA. Inverters, both grid-connected and stand-alone, are the most commonly tested items although other items such as battery charge controllers and array maximum power trackers are routinely tested as well. Eighteen different types of tests are typically run on an inverter under evaluation. These include conversion efficiency, operational voltage and current extremes, harmonic and waveform analysis, startup and shutdown dynamics, response to dynamic changes in both the source and the load, array and load utilization, lightning and switching surge tolerance, and various environmental evaluations. Both sinusoidal and non-sinusoidal waveforms can be analyzed.

Three redundantly instrumented test racks allow the testing of three-phase, single-phase, and stand-alone inverters. The ac measurement capabilities of these test racks are 530 V, 100 A; 270 V, 50 A; and 240 V, 20 A, respectively. The racks have comparable dc capabilities. While the data is acquired and stored by automated data acquisition systems, the racks include visual displays for test setup and monitoring. A data processing system provides a variety of hard copy and graphical outputs. Numerous instrumented ac power interfaces, including a simulated 100-kVA, three-phase utility, and instrumented variable resistive and inductive load banks of capacities up to 225 kVA are available. Various commercial loads, including a microwave oven and various types of lamps and motors, are used to determine changes in their operational performances when powered by an inverter under test. During the past year, 8 grid-connected and 12 stand-alone inverters were evaluated along with 8 other power handling devices.

2.6 Specialized Test Stands

As the need arises for hardware testing that is beyond the existing capabilities of the permanent PASTF facilities, specialized test facilities are designed, fabricated, installed, instrumented, and evaluated to satisfy the need. These test stands are generally of limited application for tests that are of short duration. However, they occasionally lead to the installation of a permanent test stand, such as was recently the case for a water pumping test facility[4]. The flexibility and experience of the PASTF staff, the capabilities of its mechanical and electronic shops, and extensive inventory of its data acquisition and control equipment, are generally sufficient to satisfy most needs in this regard. During, the past year short-term tests included such activities as determining the operational algorithms of battery charge controllers, the susceptibility of fluorescent lamps to thermal-humidity cycling, and the applicability of photovoltaics as a power source for a targeting beacon and a perimeter security system. Longer tests included verifying a computational model for the performance of batteries in photovoltaic systems, and determining the performance of photovoltaic-powered water pumping and water purification systems. Future plans include the testing of PV-powered refrigeration systems.

3. SUMMARY

The PASTF represents a unique facility for the testing of PV components, modules, arrays, subsystems and systems; BOS hardware; and loads. Design changes resulting, in part, from PASTF evaluations have led to continued improvements in and understanding of PV hardware performance, durability and reliability. The capabilities of PASTF staff and the resources of its extensive facilities have allowed successful evaluation of all commercial or developmental hardware yet presented. New facilities and capabilities are continually being added to the PASTF and it is anticipated that this facility will continue to serve an important role in the development of photovoltaics as a viable alternate-energy source.

4. ACKNOWLEDGEMENT

The author expresses his sincere appreciation to the PASTF staff for their dedicated efforts in advancing photovoltaic technology.

5. REFERENCES

[1]H. J. Gerwin, Photovoltaic Advanced Systems Test Facility: Description and Operations Plan, Sandia National Laboratories Report SAND80-1612, November 1980.

[2]R. S. Barlow and E. H. Richards, Qualification Tests for Photovoltaic Concentrator Cell Assemblies and Modules, Sandia National Laboratories Report SAND86-2743, January 1988

[3]K. J. Synder, et al, Prototype Flat-Plate Building-Block Array Fields: Two-Year Field Evaluation, Sandia National Laboratories Report SAND85-0656, February, 1986

[4]J. L. Chamberlin, An Evaluation Facility for Photovoltaic Water Pumping Systems, Proceeding of the 19th IEEE Photovoltaic Specialists Conference, New Orleans, Louisana USA, May 1987

CALIBRATION OF PRIMARY TERRESTRIAL REFERENCE CELLS WHEN COMPARED WITH PRIMARY AMO REFERENCE CELLS

K.A. Emery, C.R. Osterwald, and L.L. Kazmerski
Solar Energy Research Institute
1617 Cole Blvd., Golden, Colorado 80401 USA
R.E. Hart
NASA Lewis Research Center
21000 Brookpark Rd., Cleveland, Ohio 44135 USA

ABSTRACT. The method used at SERI to calibrate primary terrestrial reference cells should be valid for any tabular reference spectral irradiance. In this paper we report the results of a comparison between primary AMO reference cells calibrated using high altitude aircraft by NASA Lewis and primary terrestrial reference cells calibrated outdoors by SERI. The comparison showed a maximum difference of 1.5% between the primary AMO value and the short-circuit current measured in the laboratory using spectral mismatch corrections and a primary terrestrial reference cell. The comparison involved 9 different primary AMO reference cells and 8 different primary terrestrial reference cells. The agreement is within the estimated ±1% uncertainty for primary AMO and primary terrestrial reference cell calibrations and well within the estimated ±2% uncertainty for simulator based measurements at SERI.

1. INTRODUCTION

The photovoltaic conversion efficiency is normally measured with respect to a set of standard reporting conditions (SRC) given by a reference temperature, total irradiance E_{ref}, and spectral irradiance $E_{ref}(\lambda)$. The most accurate procedure for measuring the efficiency with respect to SRC is to adjust the irradiance of a solar simulator until the short-circuit current ($I_{sc}^{R,S}$) of the reference cell is equal to the calibrated short-circuit current ($I_{sc}^{R,R}$), divided by the spectral mismatch parameter (assumed to be unity for a "matched" reference cell). This paper demonstrates that primary terrestrial reference cells can be used for secondary AMO measurements with a ±2% uncertainty if proper procedures are followed.

2. PRIMARY TERRESTRIAL REFERENCE CELL CALIBRATION

The tabular method is believed to be the most accurate method for calibrating a primary terrestrial reference cell. The method involves computing the calibration number, CN_{ref}, using

$$CN_{ref} = \frac{I_{sc} \cdot \int_a^b E_{ref}(\lambda)R_{rc}(\lambda)d\lambda \cdot \int_a^b E_{sor}(\lambda)d\lambda}{E_{tot} \cdot \int_a^b E_{ref}(\lambda)d\lambda \cdot \int_a^b E_{sor}(\lambda)R_{rc}(\lambda)d\lambda} \qquad (1)$$

where E_{tot} is the total irradiance, I_{sc} is the short-circuit current of the reference cell under the incident solar spectrum $E_{sor}(\lambda)$, and $R_R(\lambda)$ is the spectral response of the reference cell [1-3]. The quantities I_{sc}, E_{tot}, and $E_{sor}(\lambda)$ are simultaneously and CN_{ref} is computed for at least 20 readings on at least 4 different days and averaged. E_{tot} is measured with a TMI MK-VI primary absolute cavity radiometer. $E_{sor}(\lambda)$ is determined using a LICOR spectroradiometer with a computer model fit to extend the wavelength range from 380-1100nm to 300-4000nm [1]. If a sufficiently accurate AM0 spectral irradiance $E_{ref}(\lambda)$, is used along with an accurate determination of the solar constant E_{ref}, then the primary terrestrial reference cell should accurately predict the AM0 $I_{sc}^{R,R}$. The estimated uncertainty (bias + precision error) in CN_{ref} is ±1% [1,3,4]. An objective of this study is to verify the accuracy of the tabular method with another method of reference cell calibration that has a proven uncertainty that is ±1% or less. Differences in CN_{ref} of 1-4% are predicted depending on the tabular AM0 spectral irradiance used [5,6]. The accepted method for calibrating a primary AM0 reference cell circumvents this problem by measuring the short-circuit current at a high altitudes so that minimal or zero corrections to $E_{ref}(\lambda)$ are required.

3. PRIMARY AM0 REFEERNECE CELL CALIBRATION

The primary AM0 reference cell calibration method used by NASA Lewis, has a proven uncertainty of less than ±1% [7,8], utilizes high altitude aircraft to measure the short-circuit current versus air mass. A Langley plot of the log of the measured I_{sc} versus air mass is extrapolated to AM0. Small (~1%) ozone correctionns are applied and the calibration is normalized to 1 astornomical unit. A Langley plot of I_{sc} versus air mass is shown in Fig. 1. It is very important that the Log I_{sc} versus air mass to AM0 extrapolation to AM0 be performed over the linear region at low air mass above the tropopause [7]. The external quantum efficiencies (electron per photon) for the 9 primary AM0 reference cells used in this study are summarized in Fig. 2.

4. COMPARISON OF PRIMARY TERRESTRIAL AND PRIMARY AM0 REFERENCE CELLS

Cost and time prevented the primary AM0 reference cells from being calibrated as primary terrestrial reference cells, so the comparison was performed in the laboratory using an "AM 1.5" global filtered solar simulator (Fig. 3). The secondary AM0 short-circuit current was computed using [1-3,9]

$$I_{sc}^{T,R} = \frac{I_{sc}^{T,S} \cdot CN_{ref} \cdot E_{ref}}{I_{sc}^{R,S} \cdot M} , \qquad (2)$$

where the first superscript T or R refers to the device under test or reference cell, respectively, the second supersript S or R refers to the source (simulator) or the reference spectrum respectively, and M is the spectral mimatch parameter. The uncertainty in the AM0 I_{sc} as determined with the spectral mismatch method has been estimated to be ±2% [1,4]. Table 1 shows the percentage difference between the primary and secondary AM0 calibrated I_{sc} for several cells using Eq. 2 and a single primary terrestrial reference cell. A value of 1367Wm^{-2} was used for E_{ref} instead of the often quoted but less accurate value of 1353Wm^{-2} [10,11]. Table 2 shows that the same secondary AM0 I_{sc} can be obtained within ±2% using Eq. 2 for a variety of primary terrestrial reference cells shown in Fig. 4.

5. Summary

The uncertainty of the tabular method for calibrating primary terrestrial reference cells has been shown to be less than ±2% and the use of Eq. 2 to obtain a secondary AM0 calibration has been shown to have an uncertainty of less than ±2% independent of the spectral mismatch error. This implies that "matched" AM0 reference cells are not required for accurate (<±2% uncertainty) AM0 measurements if spectral mismatch corrections are applied (Eq. 2,3). This study also demonstrates that primary AM0 reference cells can be used as secondary terrestrial reference cells with an uncertainty of less than ±2%.

Acknowledgements. The work at SERI was supported by the U.S Department of Energy through contract DE-AC02-83CG10093.

References

1. K. Emery and C. Osterwald, 'Solar Cell Efficiency Measurements,' Solar Cells **17**, pp.253-274 (1986).
2. C.R. Osterwald, 'Translation of Device Performance Measurements to Reference Conditions,' Solar Cells **18**, pp.269-279 (1986).
3. K.A. Emery, C.R. Osterwald, C.V. Wells, D.R. Myers, J. Burdick, T. Glatfelter, W. Czubatyj, and J. Yang, 'Methods of Measuring Solar Cell Efficiency Independent of Reference Cell or Light Source,' Proc. 18th IEEE PV Spec. Conf., pp.623-628 (1985).
4. K.A. Emery, C.R. Osterwald, and C.V. Wells, 'Uncertainty Analysis of Photovoltaic Efficiency Measurements,' Proc. 19th IEEE PV Spec. Conf., pp.153-159 (1987).
5. N.M. Pearsall, K.A. Emery, and M. Davies, 'Influence of Reference Cell and Spectra on the Measurement of Solar Cells,' Proc. 7th E.C. PV Solar Energy Conf., pp.298-303 (1986).
6. K. Emery and C. Osterwald, 'Measurement of Photovoltaic Device Current as a Function of Voltage, Temperature, Intensity and Spectrum,' Solar Cells **21**, pp.313-327 (1987).
7. H.W. Brandhorst, 'Calibration of Solar Cells Using High Altitude Aircraft,' Proc. of the International Colloquium oraganized by ECOSEC, July 6-10, 1970, Toulouse, France, pp.565-574.
8. R.E. Hart, 'Solar Cell Measurement Techniques Used at NASA Lewis Research Center,' Proc, Terrestrial Photovoltaics Measuremnt Workshop, March 19-21, 1975, NASA tech. rep. TM X-71802, pp.134-139.
9. C.H. Seaman, 'Calibration of Solar Cells by the Reference Cell Method – The Spectral Mismatch Problem,' Solar Energy **29**, pp.291-298 (1982).
10. C. Wehrli, 'Extraterrestrial Solar Spectrum,' Physical Meterological Observatory and World Radiation Center, tech. rep. No. 615, Davos-Dorf, Switzerland, July 1985.
11. R. Wilson, H. Hudson, C. Frölich, and R. Brusa, 'A Long-term Downward Trend in the Total Solar Irradiance,' Science **234**, pp.1114-1117 (1986).

Table 1. Ratio of AM0 short-circuit measured in solar simulator at SERI using terrestrial reference cell S01 and Eq. 2 to the aircraft calibrated AM0 I_{sc}. The average of these ratios are +0.33% with a standard deviation of 0.7%.

Primary AM0 reference cell Sample	Type	Secondary / Primary AM0 I_{sc} (%)
A145	Si (NASA)	-0.21
A161	Si (NASA)	-0.42
A133	Si (NASA)	0.27
30M2627	GaAS (NASA)	1.05
4593	Si (SERI)	-0.42
4606	Si (SERI)	-0.52
3	Si (SERI)	1.34
D13dd	poly Si (SERI)	1.05
2	GaAs (SERI)	0.62

Table 2. Measurement of one primary AM0 reference cell with several primary terrestrial reference cells. The poly-silicon reference cell D13dd has a primary AM0 I_{sc} of 157.8mA.

Primary terrestrial reference cell Device	type	Spectral Mismatch	Secondary AM0 $I_{sc}^{T,R}$ poly-Si cell D13dd (mA)	Secondary/ Primary AM0 I_{sc} (%)
S01	Si	1.0129	158.6	0.50
S02	Si	1.0227	157.3	-0.32
S03	Si	1.0191	157.0	-0.51
S09	KG5 filtered Si	0.9518	158.8	0.63
S10	KG5 filtered Si	0.9473	159.2	0.88
S25	GaAs	0.9824	159.4	1.00
S26	GaAs	0.9918	157.8	0.00
DSET31	poly-Si	0.9902	159.0	0.75
			mean = 158.4mA	0.4%

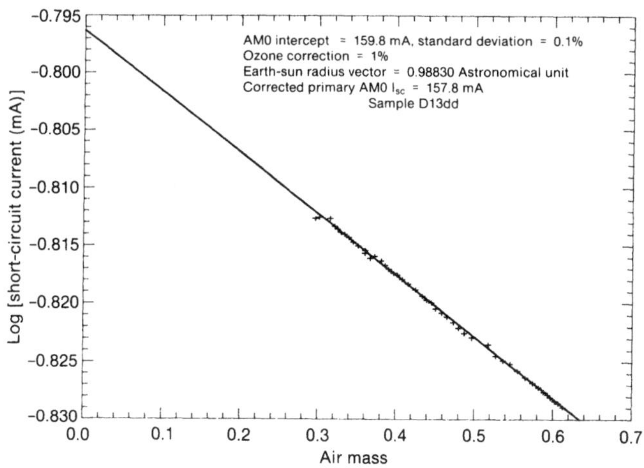

Figure 1. Langley plot for the primary AM0 calibration of the Si cell D13dd using high altitude aircraft.

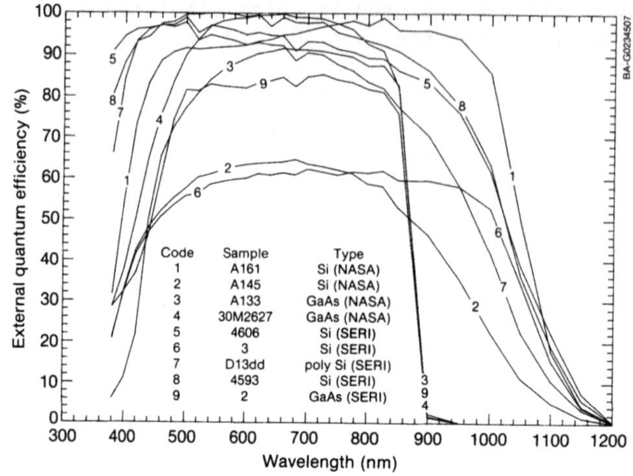

Figure 2. External quantum efficiencies for the primary AM0 reference cells used in this study.

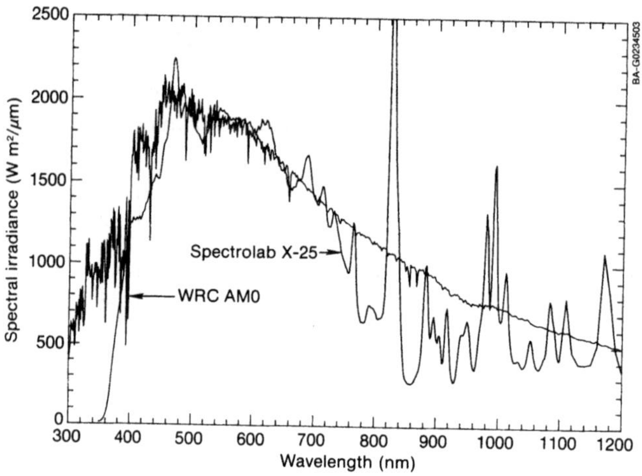

Figure 3. AM0 spectral irradiance recommended by the World Radiation Research Center compared with the Spectrolab X-25 solar simulator used in this study.

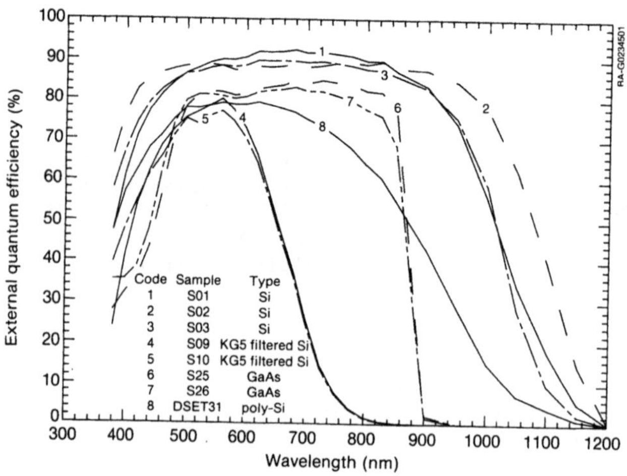

Figure 4. External quantum efficiencies for the primary terrestrial reference cells used in this study.

MONITORING THE PV DEMONSTRATION PROJECTS

by G.Blaesser

Commission of the European Communities
Joint Research Centre - Ispra Establishment - ESTI Project
21020 ISPRA (VA) Italy

SUMMARY: This report gives an overview of the monitoring data obtained so far from the PV demonstration projects which were built during the last four years with help from the EC DG XVII (New Energies Program).

1. INTRODUCTION

PV installations built within the frame of the demonstration program sponsored by DG XVII have to provide at least two years of operational data, monitored in accordance with the "Guidelines for Photovoltaic System Monitoring" /1/ developed by the European Working Group for PV System Monitoring. These guidelines distinguish between

- global monitoring: recording of daily or weekly sums of energy data to demonstrate correct plant operation
- analytical monitoring: recording of hourly sums or averages of all data listed in Table 1.

Analytical monitoring should provide a data base for testing and refining plant design and operation. It is normally required for all installations with more than 5 kWp, but in special circumstances is also provided for some of the smaller systems.

2. RESULTS FROM ANALYTICAL MONITORING

Table 2 gives use, location, nominal power (Pnom), area and "design efficiency" (DE = Pnom/area) of the projects which have provided enough data for a fair performance evaluation. Other stations have also submitted data but not yet in sufficient quantity. For each project and month we report here the following items:

hrs = total number of hours for which data were recorded (hr)
SIA/d = mean daily in-plane irradiation (kWh/sqm/d)
EAt = energy output of the PV array field (kWh)
EGt = energy supplied by an auxiliary generator (kWh)
+-ERt = energy given to (+) or taken from (-) the public grid (kWh)
PVuse = energy from PV effectively used in the loads (kWh)
Yield = 24*PVuse/hrs/Pnom (kWh/day/kWp)
PR = Performance Ratio = Yield/(SIA/d)

2.1. ANTIKYTHIRA: The Antikythira station provides PV energy to a communication link between the island of Crete and the Greek mainland; it is owned and operated by the Greek Telecommunication Organization (OTE). The PV generator reduces the fuel requirements of the 3 Diesel generators which had powered the relay link up to now. Fuel supply caused logistical problems during a large part of the year, increasing the running costs; hence this substitution appears economically attractive. The Diesel generators remain available for back-up. Reliable monitoring results are available only for the second half of

1986; they are summarized in Table 3 and show that the PV array could reduce the use of the Diesel generators to a small fraction of the total energy needed to operate this remote telecommunication relay.

2.2 KARPATHOS: The Karpathos project provides the power to pump water from a deep well for agricultural irrigation; it is owned by the Greek Development Bank. The operational data are given in Table 4. This project had considerable problems with the inverter: overheating failures caused appreciable system down-times due to lack of spare parts and the absence of maintenance personnel on site; in future installations of this kind DC pumps might be a safer choice.

2.3 FREIBURG: The Freiburg project uses PV energy and a wind turbine to operate water fountains at the Landes-Gartenschau, a public garden; its aim is to sensibilise the general public to the possibilities of PV pumping applications and also to give the owner and operator (Fraunhofer Institut) a test installation for PV systems studies. Thus this installation provides analytical data even though it is of small size (Pnom = 1.7 kW). The data given in Table 5 demonstrate that the installation and the data acquisition system generally work very well and that yields exceeding 3 kWh/day/kWp may be reached during more than half of the year even in central Europe.

2.4 OBERLIN HOUSE: The Oberlin House in Alsace is a place of international encounter, especially of youth groups. It is owned by the antroposophical Oberlin society. The PV array supplies electricity to the house (which is not connected to the public grid) with a propane-fueled motor generator as a backup. The project also serves as a test installation for the Fraunhofer Institut Freiburg and thus provides analytical monitoring. The data of Table 6 show that the contribution of the PV part to the total electricity supply was less than that of the generator; however, this system may be improved since only part of the array output is used with maximum efficiency.

2.5 ROSEMOUNT HOUSE: Rosemount House near Cork, Ireland, is a residential home; it is grid connected, and the PV installation shall show how much electric energy can be saved in a typical Irish site by adding a PV generator. The data of Table 7 cover mainly the fall and winter months; still, about 15% of the electricity used came from the PV installation.

3. PROJECTS WITH GLOBAL MONITORING

3.1 RURAL HOUSES: PV generators of different sizes (400 Wp to 1.5 kWp) were installed at 40 rural houses in Southern France: 19 houses with Pnom = 400 Wp and 4 houses with Pnom = 800 Wp used DC systems; the others (7 with 800 Wp, 8 with 1200 Wp and 2 with 1500 Wp) operated on AC. For monitoring, counters for IA, JA, JBI and JLDC are provided with DC systems, and for IA, JA, JBI, JII and ELAC in the case of AC systems. Typical results are shown in Table 8. In general, the average use of these installations is low; in fact the maximum demand occurs during the winter months and the systems are oversized for the low demand in summer. An exception is installation Nr.28 where a continuous high demand is satisfied by the PV installation as far as possible and by an auxiliary generator for the rest.

3.2 MOUNTAIN REFUGES: A number of PV generators have been installed in mountain refuges in the French Alps. They are situated between Geneva and Nice and operated by the society GTA ("Grande Traversee des Alpes"). The PV generators (of different sizes between ca 300 and 600 Wp) are the only source of electricity in these houses. Data on energy use are available for the summer seasons of 1986 and 1987 from 4 of these projects; they are summarized in Table 9. We must keep in mind that the climatic situation of Bossetan and d'Emparis is not as favorable as that of Nice or La Pare. These "sunshine" stations reach yields of 1.2 kWh/day/kWp and higher for performance ratios much below 50%. One could easily operate such stations at yields exceeding 2.5 kWh/day/kWp either by reducing the size of the array or by increasing the consumption. the other hand, the yield data of Bossetan and d'Emparis are limited by the amount of available sunshine, and little can be done to improve them.

4. CONCLUSION

The data reported here show that in many practical cases the system as a whole is still not very well optimized; in fact, while some installations are able to sustain mean yields of 2 kWh/day/kWp over the whole year, others report much lower values, even if situated in favourable climatic zones. Stand-alone systems providing only home electrification over the full year are normally designed in such a way that their mean yield is low but the long-term energy balance could be improved drastically if pumping, refrigeration or other uses of the surplus energy are included in such projects. Also it is seen that the gain in flexibility by the presence of an auxiliary generator may also visibly increase the potential of the PV part.

If the operational data from demonstration and pilot projects are translated into better design methods and improved planning of the operation of such plants the resources dedicated to PV plant monitoring are well spent and will become part of an important investment for the future of Photovoltaics.

REFERENCES

/1/ GUIDELINES FOR PHOTOVOLTAIC SYSTEM MONITORING
by G.BLAESSER and K.KREBS - ISPRA - November 1987

TABLE 1: Recorded Parameters for Analytical Monitoring

Parameter	Symbol	Units
Irradiation, global (horizontal)	IG	kWh/sqm
Irradiation, array plane	IA	kWh/sqm
Ambient temperature in the shade	TA	deg.C
Array output voltage	UA	V
Array output current (total)	JA	A
Converter output current	JC	A
Current input to battery	JBI	A
Current output from battery	JBO	A
DC line voltage (Battery voltage)	UDC	V
Current to all DC loads	JLDC	A
Inverter/rectifier DC current (+/-)	JII	A
Inverter/rectifier AC energy (+/-)	EIO	kWh
Energy to all AC loads	ELAC	kWh
Energy to(+) and from (-) grid	ER	kWh
Energy from auxiliary generator (e.g. Diesel, wind turbine etc.)	EG	kWh

TABLE 2: Demonstration Projects with Analytical Monitoring

Station	Application	Latitude	Longitude	Pnom(kW)	Area(sqm)	DE(%)
Antikythira	Telecom	35.5 N	23.2 E	20	240	8.3
Karpathos	Water-Pumping	35.3 N	27.3 E	10	138	7.2
Freiburg	Water-Pumping	48.0 N	7.5 E	1.7	22.3	7.6
Oberlin-H.	Home-Electrif.	48.2 N	7.0 E	1.56	19.8	7.9
Rosemound-H.	Home-Electrif.	51.5 N	8.0 W	5.22	68	7.7

TABLE 3: Antikythira, Jul.-Dec.86

month	Jul	Aug	Sep	Oct	Nov	Dec	total
hrs	173	329	597	661	156	420	2336
SIA/d	5.6	6.3	5.6	4.8	1.9	3.9	4.9
EAt	527	1054	1820	1798	194	993	6386
EGt	0	2	0	0	246	232	480
PVuse	372	712	915	1426	84	641	4150
Yield	2.6	2.6	1.8	2.6	0.6	1.8	2.0
PR(%)	46	41	32	54	32	46	42

TABLE 4: Karpathos, Apr.- Oct.87

month	Mar	Apr	May	Jun	Jul	Aug	Sep	Oct	total
hrs	327	146	137	330	330	144	303	332	2049
SIA/d	5.2	5.7	5.8	6.4	6.3	6.5	6.5	5.7	6.1
EAt	49	133	35	586	30	191	986	208	2218
PVuse	16	61	15	294	16	94	468	49	1013
Yield	0.1	1.0	0.3	2.1	0.1	1.5	3.7	0.4	1.1
PR(%)	2	18	5	33	2	23	57	7	25

TABLE 5: Freiburg, Sept.86 - Dec.87

month	Sep	Oct	Nov	Dec	Jan	Feb	Mar	Apr	May	Jun	Jul	Aug	Sep	Oct	Nov	Dec	Total
hrs	304	741	645	725	740	333	744	704	742	720	744	744	720	744	718	744	10812
SIA/d	5.7	3.5	3.0	1.6	1.0	2.3	3.5	4.7	3.9	5.4	5.1	4.9	4.5	2.5	1.3	1.0	3.3
EAt	90	136	111	63	54	30	153	199	173	171	205	213	200	117	71	56	2042
EGt	0	0	0	0	5	6	10	3	5	5	2	2	3	3	4	2	50
PVuse	74	102	52	77	33	34	158	173	185	172	176	167	192	120	18	53	1786
Yield	3.4	2.0	1.1	1.5	0.6	1.4	3.0	3.5	3.5	3.4	3.3	3.2	3.8	2.3	0.4	1.0	2.2
PR(%)	60	57	37	94	60	61	86	74	90	63	65	71	84	92	30	100	70

TABLE 6: Oberlin-Haus, Apr.87 - Feb.88

month	Apr	May	Jun	Jul	Aug	Sep	Oct	Nov	Dec	Jan	Feb	Total
hrs	410	725	720	744	744	698	744	719	743	744	671	7662
SIA/d	5.6	3.7	3.4	3.9	4.1	4.0	2.3	1.3	1.0	1.0	2.3	2.8
EAt	116	127	118	136	146	134	75	40	29	29	65	921
EGt	6	24	42	107	77	26	163	43	95	72	7	583
PVuse	44	71	71	0	47	108	9	18	2	0	47	370
Yield	1.7	1.5	1.5	0.0	1.0	2.4	0.2	0.4	0.0	0.0	1.1	0.9
PR(%)	30	41	44	0	24	60	9	30	0	0	48	26

TABLE 7: Rosemount House, Sept.87–Jan 88

month	Sep	Oct	Nov	Dec	Jan	Total
hrs	342	663	361	497	615	2478
SIA/d	2.6	2.6	1.2	0.9	1.1	1.7
EAt	125	236	52	33	87	533
+-ERt	-423	-869	-564	-761	-711	-3328
PVuse	110	188	34	11	45	388
Yield	1.5	1.3	0.4	0.1	0.3	0.7
PR(%)	58	50	33	11	27	36

TABLE 8: Some results from the 40 Rural Houses in Southern France

month:	jul	aug	sep	oct	nov	dec	jan	feb	mar	apr	may	jun	jul	aug
#8 (Corsica, Pnom = 400 W, Area = 4.8 sqm, DC)														
SIA/d:	6.8	7.4	7.3	4.4	2.2	3.5		3.1	3.7	3.8	6.4	6.3	6.6	
EAt:	8.9	9.8	10.1	9.6	12.8	13.8		12.7	9.8	8.8	7.1	8.0	9.1	
PVuse:	8.5	9.6	10.0	9.5	12.7	13.6		12.7	9.6	8.6	7.1	7.9	9.1	
Yield:	0.7	0.8	0.8	0.8	1.1	1.1		1.1	0.8	0.7	0.6	0.7	0.7	
PR(%):	10	11	11	18	50	32		37	22	18	9	11	11	
#1 (Provence - Cote d'Azur, Pnom = 800 W, Area = 9.6 sqm, AC)														
SIA/d:							3.4	3.1	5.1	5.5	6.0	6.6	7.2	
EAt:							26.3	19.4	26.0	16.7	15.9	18.7	20.1	
PVuse:							5.0	6.2	9.0	4.8	3.4	3.9	4.7	
Yield:							0.4	0.6	0.7	0.4	0.3	0.3	0.4	
PR(%):							12	19	13	7	5	5	6	
#28 (Provence - Cote d'Azur, Pnom = 800 W, Area = 9.6 sqm, AC)														
SIA/d:			6.1	4.6	3.4	2.8	3.4	4.3	5.3	5.3	5.6	6.9	6.7	6.5
EAt:			236	194	122	103	122	119	284	239	321	330	311	282
EGt:			31	80	87	122	113	97	59	81	52	30	58	49
PVuse:			57	39	46	30	37	33	54	36	56	47	59	43
Yield:			2.4	1.6	1.9	1.2	1.5	1.5	2.2	1.5	2.2	2.0	2.4	2.3
PR(%):			39	35	56	43	44	35	42	28	39	29	36	35

TABLE 9: Monthly Mean Yields(Wh/d/Wp) from the GTA Alpine Refuges (1986–87)

month:	Jun	Jul	Aug	Sep	Oct	Nov	Dec	Jan	Feb	Mar	Apr	May	Jun	Jul	Aug	Sep
Bossetan:	0.7	0.7	0.7	0.4	0.1	0.0	0.0	0.1	0.2	0.2	0.2	0.1	0.3	0.5	0.7	0.6
Nice:	1.8	1.2	1.5	1.2	0.6						0.4	0.1	1.1	1.7		
La Pare:			0.8	1.2										0.4	0.5	1.0
d'Emparis:			0.5	0.4									0.2	0.5	0.5	0.5

Nominal Power(W): Bossetan and Nice - 600, d'Emparis - 480, La Pare - 280

THE REFERENCE DEVICE CALIBRATION FACILITIES AT THE JRC ISPRA

Heinz A. Ossenbrink, Raymund van Steenwinkel
Commission of the European Communities
Joint Research Centre - ESTI
I-21020 Ispra (VA), Italy

ABSTRACT. This work gives a status report about the various calibration facilities developed and installed at the ESTI (European Solar Test Institute) PV laboratory of the Commission's Joint Research Centre. At present they allow calibration of radiation detectors and photovoltaic reference devices up to the size of typical modules with virtually any proposed standard calibration method. Each of these methods is tracable to a single absolute radiation standard so as deviations of the methods can be kept below ±1.5%.

1. INTRODUCTION

Solar radiation reference devices play an increasing role in the assessment of the performance of photovoltaic (PV) generators. Already in the research phase of PV devices a precise reference is needed to verify any improvements. They are of importance also at the production plant of PV devices, were the manufacturer needs the control over his efforts. At the PV application site they serve for initial approval procedures as well as for the long term monitoring task. It is obvious that any efficiency figures on these various levels of PV development are strongly related to the credibility of calibration.

The interest of PV industry and PV users in stable and precise reference devices was early recognized by standard writing committees such as ASTM and IEC which are trying to harmonize device specifications and calibration methods.

Being a test laboratory for photovoltaic devices, the problem of calibration is studied by our institute since a couple of years. Based on our experience we can contribute to current standardisation activities. Strong efforts and considerable investments into equipment has been necessary to follow the three main objectives of any calibration method:

- Make methods applicable to reference devices of any technology and size
- Use, if possible, both indoor and outdoor calibration methods
- Keep all methods tracable to an absolute radiation standard

For the calibration of PV devices we succeeded in implementing indoor and outdoor methods to be applied up to typical module sizes. Three main primary calibration tasks can be performed with our facilities:

- Outdoor calibration of thermopile sensors such as pyranometers and pyrheliometers.

- Indoor calibration of PV cells and modules
- Outdoor calibration of PV cells and modules

Using both indoor and outdoor methods for the PV device calibration, we are able to back up each calibration result and detect any systematic errors.

2. THE CALIBRATION FACILITIES

2.1. Calibration of Thermopile Sensors

In order to simulate as near as possible the conditions of later use, we calibrate these radiation sensors under natural sunlight only. The device to be calibrated is mounted on a sunfacing platform which continously tracks the sun's position. As primary reference a Practical Absolute Cavity RADiometer (PACRAD PMO-6 of CIR/Switzerland) is mounted on a separate tracking platform. The accuracy of this instrument is better than ±0.5%. The calibration of pyrheliometers is straightforward. Pyranometers with their full hemispherical viewing angle need in addition a reference instrument for the diffuse component of the total radiation. We use a precision pyranometer (Kipp & Zonen CM11), equipped with a fixed shadowing disk matching the viewing and slope angle of the PACRAD. As this instrument is also suntracking we avoid the mechanical problem of a moving shadowing disk above a horizontally mounted pyranometer. On the other hand we are limited to a total normal calibration , so cosine effects are not revealed. For this purpose however, we can fix the platform horizontally and perform a calibration against a pyranometer with known response. We limit ourself to calibration with good weather conditions, to keep the influence of the measurement of the diffuse radiation component as small as possible (amount of diffuse radiation <20% of total radiation).

The calibration error of the whole facility is to a large extent given by the radiometer error (<0.5%) and the uncertainty in the diffuse radiation measurement, which we estimate to be smaller than 0.5% on clear days.

Our calibration certificates show a conservative ± 1.5% calibration error

2.2 Indoor Calibration of PV Devices

Two indoor primary calibration methods are mainly in use at various PV laboratories. The first one uses the absolute spectral irradiance of a solar simulator to calculate together with the relative spectral response of the device its calibrated short circuit for a given standard spectrum (1). The second measures the absolute spectral responsivity of the device and calculates the calibration value directly for the spectrum needed (2). As in any case the measurement of the spectral responsivity is required, we spent our efforts for the second one aiming increased precision, tracability and repeatability. Also, we considere as difficult to measure the absolute spectral irradiance of pulsed solar simulators with the necessary accuracy.

2.2.1. Absolute Spectral Calibration of Solar Cells.

This facility basically consists of a 1 kW quartz halogen lamp, spectrally calibrated by NBS. Its chopped light passes through calibrated narrow band interference filters with a bandwidth of about 10 nm. The device to be measured receives this monochromatic light with an irradiance of between 0.1 and 1.5 W/m^2. The addition of white bias light is possible by means of halogen lamps with dichroic reflectors. The short circuit current of the test cell, which can have an area of 20 cm^2, is measured with precision shunts. The absolute spectral irradiance falling onto the cell is calculated from the absolute spectral irradiance of the standard lamp multiplied by the absolute filter transmission coefficient, which in turn is measured by a spectrophotometer regularely. The irradiance of the standard lamp can be recalibrated by a cavity radiometer; the relative spectral distribution controlled by our spectroradiometer. With this method we do not depend on any other primary standards besides the absolute radiometer.

The accuracy of the method is mainly given by the calibration and stability of the standard lamp and is estimated to be ±1.5% for a single spectral wavelength. After multiplication and integration with the standard spectrum we believe that the overall accuracy of the method is better than ±1.5%. The repeatibility over two years is better then ±0.5%.

2.2.2. Spectral Responsivity Measurements for large Areas.
This facility uses the 10000 Joule flash lamp of our commercial PASAN II sun simulator. We installed directly infront of the bulb an assembly of four interference filters (bandwidth 18 nm) tuned to the same wavelength. In the measurement plane we achieve an uniformity of ±1.5% in a circle of 1 m diameter. The monochromatic irradiance is as high as 10 W/m^2, so the presence of white bias light is often without influence to the measurement result. However, we can add bias by means of an halogen lamp array. As reference serves a silicon detector, calibrated with the method previously described. The short circuit responses of both devices are analyzed by a transient recorder. The total illumination time is about 18 ms, so temperature effects are not present as long as bias light is absent.

The overall accuracy of the method is given by the reference cell used and is estimated conservativly less than ±2 %. The repeatability has been proven to be better than ±1%.

It should be noted that the method is actually a secondary calibration method as it uses a calibrated reference detector. However, we found a superior precision compared to the usual sun simulator method, were a reference cell with spectral mismatch correction is applied (3). In particular, the calibration of amorphous cells benefits from our direct approuch.

2.3. Outdoor Primary Calibration of PV Cells and Modules.

Spectral irradiation studies performed by our laboratory showed, that the spectrum of the sunlight on clear days compares well with the AM 1.5 global spectrum proposed by the IEC and provides a better match than any of our simulators. Therefore we believe, that, by applying for PV devices the same calibration procedure as used for pyranometers one arrives at high precision and good repeatability with cost effective equipment. Also, we found during previous calibration tasks that the measurement of the solar spectrum for spectral mismatch correction can be dropped, as it introduces additional error sources which compensate the theoretically higher precision. We perform the outdoor calibration on a suntracking platform, for reasons described already previously (see 2.1). In order to keep the device under test within the temperature range of 25±2°C, we install it in temperature controlled box. This box has shutters to prevent heating by sunlight; it houses modules up to a length of 1 m. Temperature control is performed by means of an commercial air conditioning unit with thermostate, which blows air through the box. After the module stabilizes at the desired temperature, the doors open within a second, driven by pneumatic pistons. The module is exposed to the sun for about 2 seconds, during which a whole IV-curve can be measured, until doors close again to allow for cooling within a minute. These cycles repeat until a reasonable amount of data is taken. The reference instruments are, as in calibration 2.1, an PACRAD and a pyranometer with shadow disk. Until now, we have not seen any ground albedo effects, using the zenith angle of 37 degree as forseen in the calculations for the standard AM 1.5 global spectrum (4).

The accuracy of this method depends mainly of the deviation of the solar spectrum during the test time from the standard spectrum. Calculations showed, that for air masses between 1.3 and 1.8 the spectral mismatch error is below ±1% (5). Including the errors of the reference instruments we arrive at an overall accuracy of ±1.5%. This agrees well with observations found in (6).

3. RESULTS

During Round Robin campaigns we had the opportunity of confronting the different calibration methods. Various devices of different sizes and technology have been compared. The typical agreement between indoor and outdoor measurements is ±0.7%, with extrem values of about ±3%. These were often caused by optical mismatch problems, as the indoor method is a direct light technique, whereas the outdoor calibration uses the total 2πsr radiation angle. Also we noted, that both indoor and outdoor calibrations show larger (factor two) discrepancies when applied to amorphous cells. Obviously, the narrower spectral responsivity of a-Si increases the errors in calibration. No systematic difference between indoor and outdoor methods has been found until now.

4. CONCLUSIONS

We made good experiences with both indoor and outdoor methods, which are based on the same primary reference. The outdoor method does not need very sophisticated equipment, but cannot performed at any time. This is a severe drawback, if devices are calibrated for (paying) customers. The indoor method is limited only by the proper measurement techniques. Recent measurements of a tandem module indicate that transient effects on the different layers make the results at least doubtful. At present, we conclude to continue with two calibration methods.

5. REFERENCES

1.R.Shimakowa, F.Nagamine, Y.Myake,K.Fujisawa and Y.Hamakawa, Japanese Indoor Calibration Method for the Reference Solar Cell and Comparison with the Outdoor Calibration, Communication to the IEC TC82, WG2, Tokyo (1986)

2.e.g J.Metzdorf, The Differential Spectral Responsivity Method, Applied Optics, Vol.26,(9),pp.1701-1708 (1987)

3.K.A.Emery, C.R.Osterwald, T.W.Cannon and D.R.Myers, Methods for measuring Solar Cell Efficiency Independent of Reference Cell or Light Source, Proc. 18th IEEE PVSEC, Las Vegas, pp.623-628 (1985)

4.R.E.Bird and R.L.Hulstrom, Terrestrial Solar Spectral Data Sets, Solar Energy Vol.30, pp 563-573 (1983)

5.R.Shimokawa et al., Effect of Atmospheric Parameters on Solar Cell Performance under Global Irradiance, Solar Cells, Vol.19, pp.59-72 (1986)

6.F.C.Treble, The Calibration of Primary Terrestrial Flat-Plate Reference Devices - A Proposed International Standard, Proc. 7th European PVSEC, Seville, (1986)

LABORATORY AND FIELD TESTING OF AMORPHOUS SILICON MODULES AND ARRAYS

P. Ragot, R. Berre, D. Desmettre, P.Paes, D. Royer

Groupement pour l'Expérimentation des Pilotes Energétiques (GEPE)
Centre d'Etudes Nucléaires de Cadarache
13108 Saint Paul lez Durance Cedex
France

ABSTRACT. As part of the continuing photovoltaic module evaluation effort this paper presents the results obtained over the last twelve months with amorphous silicon modules of French and American manufacture. The experiments included both field tests at the Cadarache facility and laboratory tests in a controlled climatic chamber with artificial sunlight. Results are presented after one year of outdoor exposure and 800 hours at various steady-state temperatures in the climatic chamber.

INTRODUCTION

The purpose of this study was to measure performance variations in time for amorphous silicon modules and arrays under outdoor conditions. In order to discriminate among possible causes of degraded performance, however, complementary laboratory tests were also necessary.

A test procedure was defined with the following objectives:

- to measure time-dependent degradation of characteristic parameters (I_{SC}, V_{OC}, P_{max}, FF, R_S and R_{SH}).
- to perform statistical analysis of the production line module characteristics.
- to analyze and quantify the causes of degraded performance (irradiation, temperature, type of load).
- to correlate module degradation in the field and laboratory tests.

TEST FACILITY

FIELD TESTS

The facility consists of modules exposed to sunlight at a fixed 45° tilt angle, with switching provisions allowing a sequence of tests to be conducted on various series, parallel and diode configurations. The data acquisition system includes an HP 150 microcomputer, an HP 1000 minicomputer and a meteorological station.

Reference cells are used for calibration purposes. The electronic load is driven by the computer. A total of 15 voltage points are used to plot an individual I-V curve. About ten minutes are required to scan all 65 modules.

LABORATORY TESTS

In the laboratory facility the modules on a horizontal support were exposed to an illumination simulator in a controlled atmosphere chamber with temperature regulation between 10 and 50°C. The $2 \times 1\ m^2$ illumination simulator comprises 34 metal halide lamps (OSRAM HQI-E 250/NDL) with a nearly continuous spectrum covering the 300-800 nm range. Irradiance in the module plane was approximately $750\ W \cdot m^{-2}$.

METHOD

In order to address the problem of spectral variations in solar and artificial irradiation, ageing indicators were defined for characteristic parameters using the following relation (where A = I_{SC}, V_{OC}, FF, P_{max}, R_S and R_{SH}):

$$R = \frac{A_{module\ x}\ (\text{nth day})}{A_{module\ x}\ (\text{1st day})} \times \frac{A_{standard\ module}\ (\text{1st day})}{A_{standard\ module}\ (\text{nth day})}$$

The standard module was stored in darkness between each series of measurements to prevent radiation ageing. It was also checked for parameter variations in time by comparison with a primary standard using an illumination simulator.

FIELD TESTING

The current-voltage (I-V) curve was measured for each module at about noon solar time when the total irradiance was between 800 and 1000 $W \cdot m^{-2}$. The measured values were then extrapolated to the reference conditions (1000 $W \cdot m^{-2}$ at 25°C) [1]. All modules were in open circuit configuration between measurements.

The ageing tests were carried out on 65 modules from two different manufacturers ("A" and "B") as follows:

- Manufacturer "A": Ten type 1 modules, five type 2 modules and two type 3 modules
- Manufacturer "B": Forty-eight type 4 modules

LABORATORY TESTING

Temperature effects on module degradation were assessed by performing measurement in the following conditions:

- Irradiation: 700-800 $W \cdot m^{-2}$
- Air temperature: 12-32°C
- Test duration: 800 hours
- Number of modules tested: 12

For each series of measurements, four modules from 3 suppliers in France and the USA were tested under different load conditions (open circuit, short-circuit, constant load at 15 V, open circuit in darkness).

Ageing effects were investigated with 24 modules:

- Manufacturer "A": Eight type 5 modules
- Manufacturer "B": Eight type 6 modules
- Manufacturer "B": Eight type 7 modules

The module specifications are indicated in Table I.

RESULTS

FIELD TESTS

Module ageing was statistically analyzed on two representative production samples comprising ten type 1 modules and forty-eight type 4 modules. Figure 1 plots the relative variation of each mean parameter from its initial value over 344 days of exposure to sunlight with a cumulative energy of 1670 $kWh \cdot m^{-2}$; the final mean values and standard deviations are indicated in Table II.

For all the modules on the whole the short-circuit current decreased by 12 to 22%, or about the same order of magnitude as the drop in the open-circuit voltage (6-12%), while in theory the open-circuit voltage is a logarithmic function of the short-circuit current. In this respect, the combined effects of the reduced shunt resistance and the increased series resistance resulted in a lower fill factor and open-circuit voltage [2].

Significant variations in the maximum power ageing indicator were observed for module types 1, 2, 3 and 4:

Type 1: -41%	Type 3: -30%
Type 2: -29%	Type 4: -51%

LABORATORY TESTS

Variations in the characteristic parameters of three modules under continuous illumination ($G' \simeq 750\ W \cdot m^{-2}$ connected to three different loads (I_{SC}, V_{OC}, V_L) and a fourth module maintained in darkness (V_{OC}) were logged for three types of modules.

Table III indicates the parameter variations after 800 hours during which the ambient temperature in the climatic chamber was held constant at 12°C (module temperature ≃ 40°C). Little change was observed with the modules in darkness except for the shunt resistance, especially for the type 7 module. Under illumination the short-circuited modules showed the smallest variations while degradation occurred more quickly in modules connected in open circuit configuration, notably for type 7. The same phenomenon has already been reported [3].

In comparison with results obtained at an ambient temperature of 32°C several observations can be made:

- The drop in P_{max} related to the series and shunt resistance variations was lower in type 5 modules ($\bar{R}$ = 0.86 @ 32°C and 0.77 @ 12°C).
- For type 6 modules the parameter variations were practically temperature-independent except for a smaller change in the shunt resistance ($\bar{R}$ = 0.69 @ 32°C and 0.49 @ 12°C).
- With type 7 modules, P_{max} dropped more at 32°C in the open circuit and V_{Load} configurations ($\bar{R}$ = 0.60 @ 32°C and 0.76 @ 12°C), but by the same amount ($\bar{R}$ = 0.82) in open circuit.

For identical cumulative energy values, similar variations in the characteristic parameters were observed for laboratory and field tests.

CONCLUSION

Characteristic parameter variations were quantified after the first year of field testing with amorphous Si modules, revealing differences according to the type of module (up to 30-50% on the P_{max} reduction): the best results were obtained with the most recently manufactured modules (types 2 and 3).

Laboratory experiments were conducted to determine load and temperature-dependent effects on parameter variations in time. Once again, the results differed according to the type of module tested.

Similar behavior was observed for the modules in laboratory and field conditions. The method can thus be considered to provide a valid assessment of module behavior with an acceleration factor of about 4. On the whole, the results are comparable to those reported in the United States for the period covering 1985-1986 [4].

ACKNOWLEDGMENTS

The authors are grateful to the *Agence Française pour la Maîtrise de l'Energie* for providing financial support.

REFERENCES

[1] International Standard. *Procedures for Temperature and Irradiance Corrections to Measured I-V Characteristics of Crystalline Silicon Photovoltaic Devices*. International Electrotechnical Commission: IEC 891. First edition (1987).

[2] H.J. Hovel. *Semiconductors and Semimetals*. Vol 11: "Solar Cells" (R.F. Willarson and A.C. Beer). Academic Press: New York, San Francisco, London (1975).

[3] J.W. Lathrop and C.W. Davis. *Accelerated Stress Testing of Commercial a-Si:H Solar Cells*. Proceedings of the International Conference at Sevilla, Spain: October 1986, p 293.

[4] H.L. Brown, K. Firor and C. Jennings. *Thin Film Silicon Photovoltaic Module Performance Assessment*. Pacific Gas & Electric Co. Report 007.3-87.2.

Table I : Characteristics of Amorphous Si Modules Tested

Type of Module	Number of Modules	Date of Manufacture	Module Dimensions	Rated Power (W)
1	10	October 1986	1′ x 1′	3.7
2	5	January 1987	1′ x 3′	11.5
3	2	November 1986	4′ x 3′	40.4
4	48	October 1986	1′ x 1′	3.6
5	8	June 1987	1′ x 1′	3.7
6	8	September 1987	1′ x 1′	3.6
7	8	-	1′ x 1′	5.0

Table II : Relative variation of characteristic parameter values for amorphous Si modules after 344 days of sunlight exposure ($Illum = 1670\ kWh \cdot m^{-2}$)

Type of Module	$R(I_{SC})$		$R(V_{OC})$		$R(P_{max})$		R(FF)		$R(R_S)$		$R(R_{SH})$	
	aver	σ	aver	σ	aver	σ	aver	σ	aver	σ	aver	σ
1	.879	.052	.936	.030	.591	.109	.712	.078	2.697	.497	.316	.149
4	.774	.048	.877	.022	.490	.056	.721	.050	2.097	.281	.385	.097

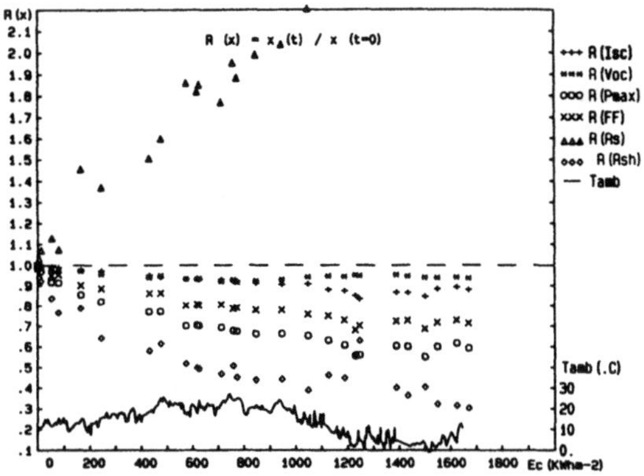

Average values for ten type 1 modules

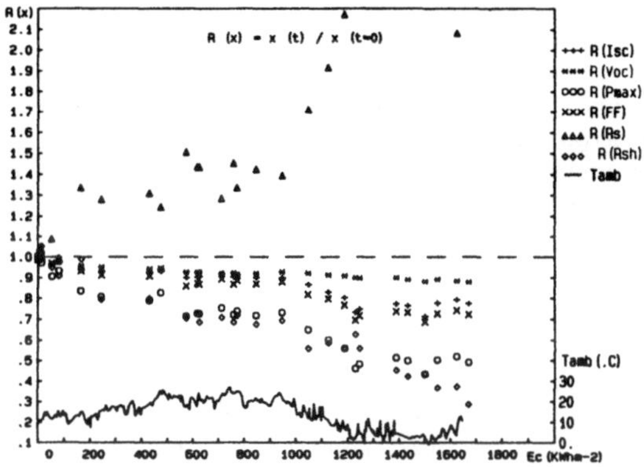

Average values for forty-eight type 4 modules

Figure 1 : Relative characteristic parameter variations
after 344 days of sunlight exposure

Table III : Relative variation of characteristic parameter values
for amorphous Si modules after 800 hours of artificial light exposure
(Illum = 600 kWh·m^{-2})
G′ ≃ 750 W.m^{-2}, T_{air} = 12°C

Type of Module	R(I_{SC})	R(V_{OC})	R(P_{max})	R(FF)	R(R_S)	R(R_{SH})	Load Conditions
5	.88	.96	.79	.94	1.13	.55	short-circuit
	.85	.98	.75	.90	1.11	.46	open circuit
	.92	.97	.77	.85	1.28	.47	VLoad = 15 V
	.98	1.02	.98	.98	.96	.82	open circ (dark)
6	.91	.95	.75	.86	1.30	.48	sc
	.87	.96	.71	.85	1.30	.43	oc
	.87	.96	.74	.88	1.20	.55	VL
	.99	1.01	.96	.98	.97	.71	oc (darkness)
7	.95	.97	.83	.90	1.07	.43	sc
	.87	.93	.71	.88	1.28	.35	oc
	.90	.96	.80	.93	1.16	.57	VL
	.97	.97	.89	.92	.98	.53	oc (darkness)

AN APPLICATION OF ON-LINE BATTERY MONITORING TO THE VULCANO PV PLANT

A.Buonarota-P.Menga
ENEL DSR - Milan

P.Ostano-V.Scarioni
CESI - Milan

ABSTRACT. The reliable knowledge of the State-Of-Charge of the battery of a PV plant can contribute to improve the system management.
Unfortunately, the technologies currently adopted to determine the battery SOC are not fully satisfactory.
The experience obtained by ENEL (the Italian Electricity Agency) on traction lead-acid batteries, operating under cyclic conditions, led to the formulation of a simple model capable of describing the relationships among the operating conditions (profile of current, temperature, etc.) and the internal SOC of the battery. This model was extended to the stationary accumulators of the Vulcano PV plant of ENEL, and checked by means of laboratory tests at CESI.
On this background, an automatic system for the on-line evaluation of the SOC of the battery has been recently set up and installed at Vulcano.
The paper presents the basis of the methodology, the layout of the system, and the preliminary experiences so far obtained.
This work is supported by a financial contribution of DG XII of the Commission of the European Communities [Contract Proposal EN3S-0135-I(A)].

INTRODUCTION

It is well recognized that the knowledge of battery SOC is necessary both for optimizing the management of the PV plant and for avoiding stresses on the battery which would result in reduction of its life.

Various methods are used for the determination of SOC in conventional applications.

The electrolyte density of the lead-acid battery is in principle related univocally to SOC. However, operations under cyclic conditions (typical of PV or traction) prevent the attainment of equilibrium in the mass of the electrolyte, thus reducing somewhat the reliability of a local measurement.

Similarly, the open circuit voltage at equilibrium is representative of SOC. On the other hand it cannot be practically used because attaining the internal state of equilibrium would require a long rest period.

Another method often employed is computing the Amperehours balance of the battery. In this case limitations arise from the fact that the charge acceptance efficiency is lower than unity, especially while approaching the threshold of gassing; this error integrates over time and leads to important inaccuracies.

The measurement of battery voltage is a very simple and widespread practice; unfortunately the voltage at the terminals is not related univocally to SOC, depending on a variety of affecting quantities (temperature, discharge current, presence of rest period, and other transient phenomena).

The difficulty to describe and forecast the behaviour and the internal state of the lead-acid battery, has stimulated in recent years the development of mathematical models as suitable tools for a better prediction of the battery response.

Many formulations proposed [1,2] are in the form of equivalent electrical circuits, which result particularly practical for the application in the electrical field. This tool can be easily oriented to the evaluation of the battery SOC.

BRIEF DESCRIPTION OF THE BATTERY MODEL

The discharge model considered here, which is described in detail in ref.[3], is based on the equivalent circuit shown in Fig.1. It consists simply of a generator with an e.m.f. E and two resistances R_Ω and R_D.
Based on experimental results it was found that:
- the quantity E is a function of extracted charge Q, temperature T and average discharge current I_m;
- the total resistance R is a function of the relative depth of discharge ρ of the battery, taken as the ratio between the extracted charge Q and the battery capacity C at average regime I_m;
- the component R_Ω takes into account the purely resistive voltage drop due to the instantaneous discharge current I;
- the component R_D represents electrolyte diffusion phenomena, whose corresponding voltage drop is evaluated on the basis of the current I_m.

To determine the values of these parameters , it is sufficient to perform simple tests, examining the response of the battery to a current step at various depths of discharge.

This procedure has been carried out on accumulators of the same type as those installed in the Vulcano PV plant; a series of laboratory tests simulating the operation of the battery in photovoltaic plants confirmed that the model represents satisfactorily the behaviour of the storage system.

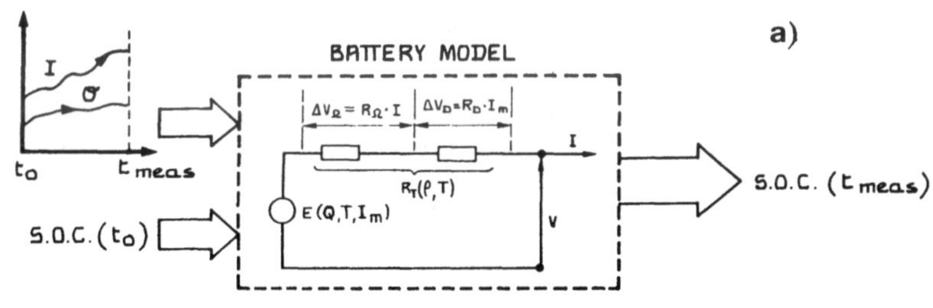

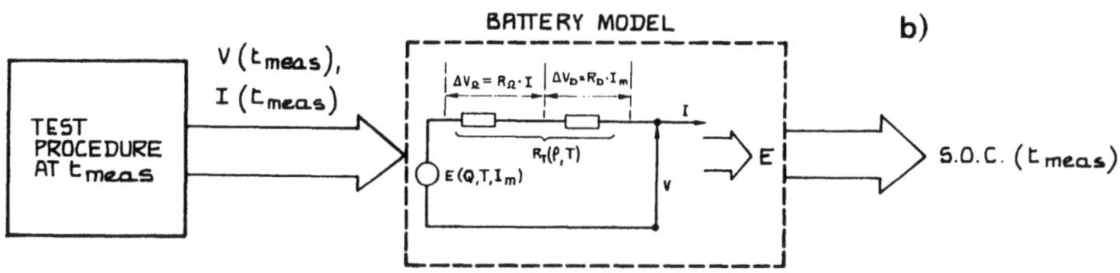

Figure 1- a) Model in open loop b) Proposed application of the model

APPLICATION OF THE BATTERY MODEL FOR THE EVALUATION OF SOC

Different approaches can be adopted for evaluating the SOC by the use of the model.

A first way, shown in Fig.1a, consists in using the model in open loop, providing it with two basic input informations: the initial charge stored into the battery at time t_0, and the profile versus time of the operating quantities, namely current and temperature. The model processes in real time these informations and evaluates the SOC.

However, the open loop operation implies the integration of the unavoidable errors of the model , thus gradually degrading the accuracy of the SOC estimate.

A possible solution [4,5] may be to close the loop through a feedback, comparing the calculated and the actual value of an appropriate output quantity used as calibrating parameter (generally the voltage at the terminals), and modifying consequently one or more internal elements of the model.

The method proposed by the authors (Fig.1b) uses the model in open loop, in such a way as to calculate the value of its internal parameters (resistances R_Ω and R_D, e.m.f. E, which are functions of SOC) through the measurement of the actual external quantities of the battery.

Specifically, it can be seen that in the model considered here, the knowledge of the output voltage V, of the instantaneous and average currents I and I_m, of the temperature T, allows the univocal determination of the SOC.

This principle has been validated through laboratory tests, and then applied to the ENEL PV plant of Vulcano.

DESCRIPTION OF THE METHOD AND OF LABORATORY TRIALS

As said, the method proposed is based on the use of a discharge model of the battery. That implies that its practicability in a PV system is limited to the discharge period, at evening or night time.

Moreover, in order to obtain a good degree of accuracy (few percent), the measurement of battery current and voltage has to be performed during a period of some minutes in steady-state conditions.

For these reasons, it is convenient to impress, on-line, one or more controlled current steps large enough to mask the load fluctuations, during night operation when the load current is generally small.

To check the reliability of this method, laboratory tests were carried out, during which operating conditions of the storage system of a PV plant were simulated.

Fig. 2 gives an example of the daily current profile, reproduced on the basis of operating data from the plant of Vulcano. For the sake of simplicity, the discharge current to the load was semplified with only two steps of constant amplitude.

The SOC is determined every night at a pre-estabilished instant of the cycle (t_{meas} in Fig.2), by applying to the battery two current steps and evaluating by the model the e.m.f. E.

Fig. 3 reports the results of the tests performed, showing the e.m.f. E versus the charge Q extracted from the battery at determined values of I_m and T, starting from full charge.

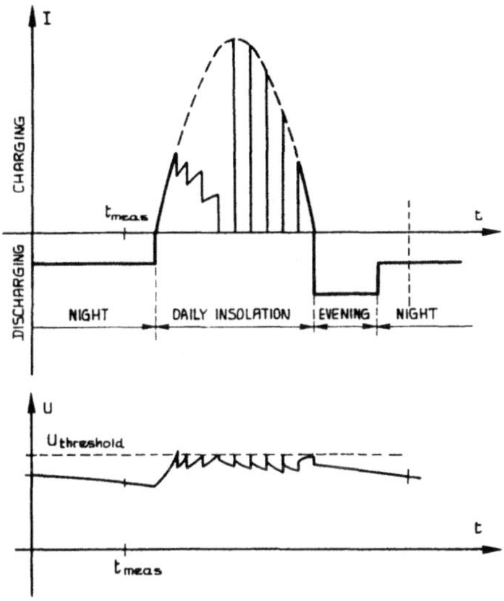

Fig. 2 - Simulation of a PV battery operation

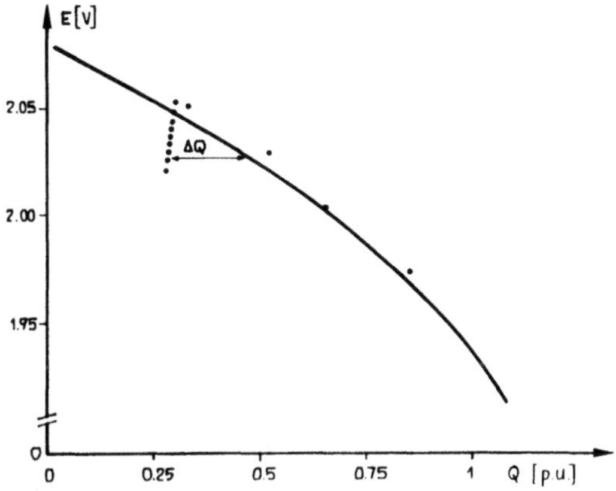

Fig.3 - Test results of SOC evaluations

It shows that in the course of a succession (1,2,..,n) of operating cycles there is a progressive difference ΔQ between the charge value calculated by means of the e.m.f. (on the continuous curve) and that calculated on the basis of the Ah balance (abscissa of each dot). As expected, the latter shows an error that increases cycle by cycle.

Various experimental checks, repeating the simulation several times for different test conditions (number of cycles, initial SOC) confirmed the reliability of the informations supplied by the method, whose error remains within 3% of battery capacity.

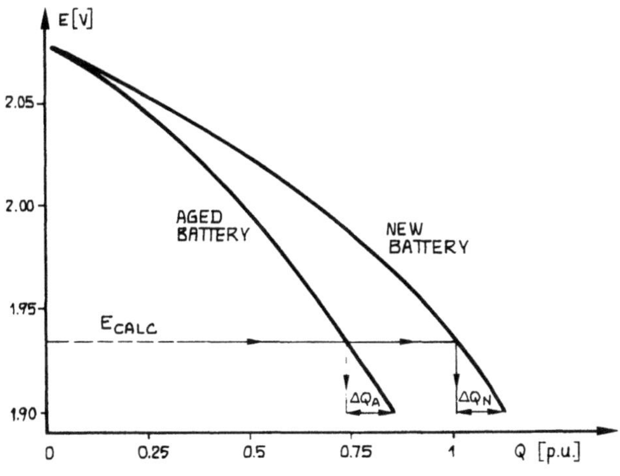

Fig. 4 - Influence of battery ageing on SOC estimate

These tests were performed on new and aged batteries. Fig. 4 shows that to a certain extent, and when the battery is sufficiently discharged,the prediction of SOC given by the method can take intrinsically into account the effect of battery ageing. In this condition, in fact, it can be seen that, although the evaluated value of the charge Q extracted from the battery is erroneous, the estimate of the residual capacity ΔQ_N results sufficiently close to the actual value ΔQ_A.

IMPLEMENTATION OF AN EXPERIMENTAL TEST SYSTEM IN THE PV PLANT

In order to evaluate the reliability of the methodology previously described, an experimental test setup has been temporarily installed in the PV plant of Vulcano.

To avoid interference with the plant and its management, a separate test hardware has been developed and installed in a small pre-fabricated building. The basic layout of the system is given in fig. 5.

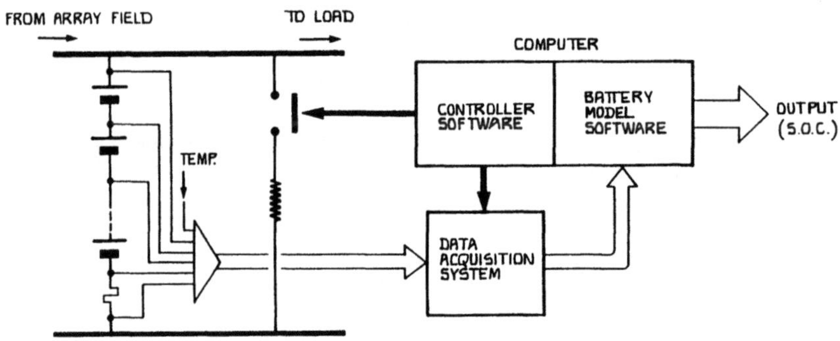

Fig. 5 - Layout of the SOC test system installed at Vulcano

A personal computer controls the whole process. Every night, at pre-established time, it closes the contactors which connect the test resistors to the battery, thus impressing current steps with appropriate intensity and duration, and records battery voltage and temperature by means of a data acquisition system. The same computer processes these data into the software that represents the battery model, and returns the SOC of the battery at that moment, as well as additional output informations.

In the period between two successive SOC determinations, the evaluation of SOC is based on the use of the battery model in open loop, as no significant error occurs over a short time.

Actually, the individual voltages of several elements of the 130-cells battery are recorded, in order to study the scattering among them and provide a diagnosis of the battery conditions. A comparison of cells with and without electrolyte agitation is also made.

The preliminary results after about two months of operation confirm that the system works correctly and returns reliable informations. The trial is continuing for a sufficient time (1 or 2 years), and will include also periodical in-field measurements of the actual charge stored into the battery, in order to verify the correctness of the SOC estimate given by the system.

Picture 1 - Pre-fabricated building hosting test hardware

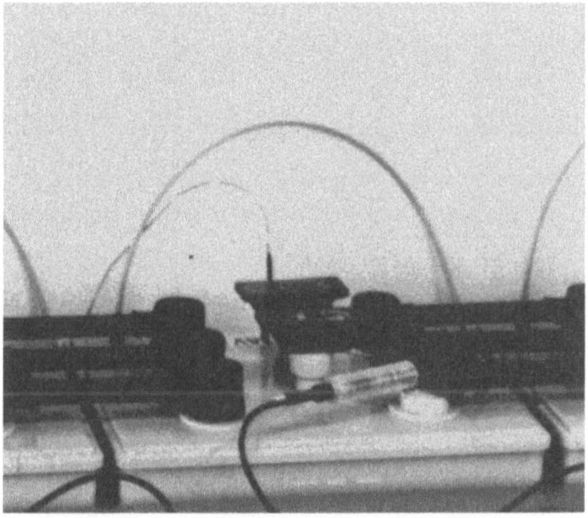

Picture 2 - Particular of test connections to cells

CONCLUSIONS

A method for good-accuracy evaluation of battery SOC, based on the use of a simple discharge model, has been developed. It has been implemented in the PV plant of Vulcano, in the form of an experimental setup which will be used for checking the long term reliability of the philosophy.

A simpler and cheaper implementation, tailored for industrial applications, will be eventually developed, after an adequate trial period.

A parallel step of the activity is the evaluation of possible modifications of the PV system management, on the basis of the accurate knowledge of battery SOC.

REFERENCES

[1] J.Meiwes, W.Schleuter: "Ein elektrisches modell zur beschreilbung des klemmenspannungsverhaltens von bleiakkumulatoren". Drive Electric, Amsterdam, October 1982

[2] N.Polese, G.Betta: "An equivalent model of batteries for electric vehicles". Drive Electric, Sorrento, October 1985

[3] R.Buccianti, R.Giglioli, P.Menga, L.Thione: "An electrical model of the lead-acid battery". EVS-7, Versailles, June 1984

[4] H.P.Schoner: "Battery state monitoring using the voltage response on drive currents". Drive Electric, Sorrento, October 1985

[5] R.Giglioli, P.Pelacchi, M.Raugi, G.Zini: "A state of charge observer for lead acid batteries". L'Energia Elettrica N. 9, 1987

REFERENCES

[illegible]
[illegible] R.Schwarz, "Sealed [illegible] using [illegible] constant [illegible] drive currents" [illegible] Electric, Toronto, October 1985.
[illegible] [illegible], [illegible], [illegible], [illegible], "A model of charge [illegible] for lead acid batteries", L'Energia Elettrica N. [illegible]

SESSION 03:
FUNDAMENTALS

NEW MATERIALS AND STRATEGIES IN PHOTOVOLTAICS

Helmut Tributsch
Hahn-Meitner-Institut
1000 Berlin 39,
Fed. Rep. of Germany

Abstract:

Several promising frontiers have not yet been explored in photovoltaic technology. One consists in identifying semiconductors with absorption constants as high as possible ($\alpha > 3x10^5 cm^{-1}$) to produce ultra-thin solar cells which could be cheaper in production and material consumption, and more versatile in handling (high flexibility and package density, low weight). Another strategy aims at an accelleration of improvement of material and device properties by in-situ quality control of electronic properties during production and the application of optimization strategies based on artificial intelligence (expert systems). Such a highly computerized approach for solar cell development will be especially promising for the preparation and production control of cascade and multi-junction systems where adjustement of a large number of material and device parameters is necessary. Expectations, progress and difficulties met with these strategies are outlined on the basis of ongoing research on FeS_2 (pyrite) as solar cell material and the installation of the first plasma deposition apparatus for a-Si:H solar cells with in-situ quality control and optimization using artificial intelligence.

Materials for ultrathin solar cells

There is an obvious technological advantage in aiming at the development of ultra-thin solar cells (less than 0.1μm). Their production would consume less time and less semiconductor material, they could be handled as flexible sheets with high package density and advantages for transportation (e.g., space applications). Most important, such solar cell sheets could directly be spread over the ground, pinned to the wall or used as tent to produce shadow. In other words, the expensive structures to hold, stabilize or orient solar cells would not have to be included into the prize of solar cells. These might be important considerations towards a future economy of solar energy. But also scientific argument favour the search for strongly light absorbing semiconductor materials for photovoltaic cells. For given typical semiconductor properties such as donor or acceptor concentrations of N_D, N_A = 10^{17} cm^{-3}, a mean bulk lifetime of $\tau = 10^{-9}$ sec, electron and hole mobilities of $\mu = 100\ cm^2\ V^{-1}\ sec^{-1}$, and an incident photon density of $I_L = 2x10^{17}\ cm^{-2}\ sec^{-1}$ an increase of the absorption constant from $a = 10^4\ cm^{-1}$ to $a = 10^6 cm^{-1}$ improves the short circuit photocurrent density through a Schottky type barrier by a factor of 2.5. Simultaneously the open circuit voltage is increased and the power output improves by a factor of 2.2. A further increase of N_D in strongly absorbing material to $N_D = 10^{18}\ cm^{-3}$ leads, for high absorption coefficients, to an additional improvement of the photovoltage /1/.

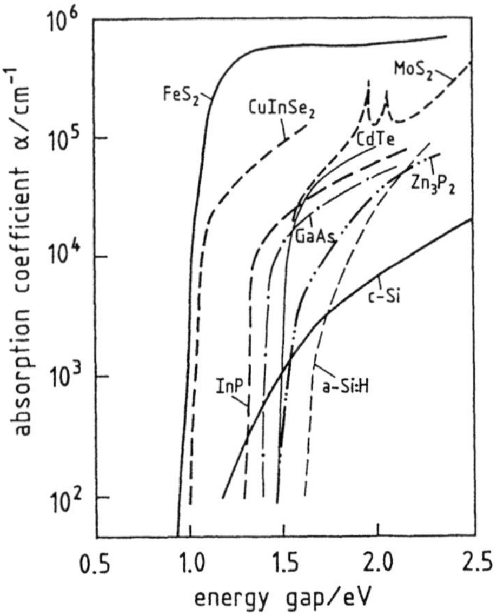

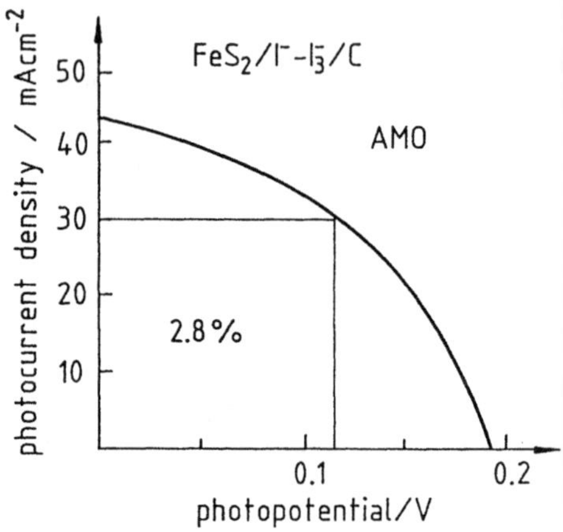

Fig.: 2. Power output characteristic of a pyrite liquid junction solar cell.

Fig.: 1. Light absorpton properties of pyrite (FeS_2) compared with crystallized and amorhous silicon and other semiconducting compounds.

After gaining some experience with transition metal chalcogenides in which transition metal d-states make up the valence and conduction band (which is of significant advantage for regenerative and fuel producing electrochemical solar cells/2/) it was decided to develope pyrite (FeS_2) as a solar energy material (approximately 100 binary and ternary compounds are known to cristallize with the pyrite structure). An absorption coefficient of $a = 6x10^5\ cm^{-1}$ for solar light, an electron mobility of $\mu = 180\ cm^2/Vs$ and a diffusion length of $L = 0.13\ \mu m$ (much bigger than the penetration depth of light) combined with a quantum efficiency of 80-90% make this stable material ($Eg = 0.95eV$) interesting for ultra-thin (a few 100 Å), extremely flexible solar cells /3/4/.

Single crystals were produced by vapor transport and from liquid phase (Te, $PbCl_2$ melts). The solid state and interfacial chemistry of this abundant iron compound turned out to be a major challenge due to the involvement of iron coordination chemistry. The science and technology of very thin photoactive films (produced by MOCVD from $Fe(CO)_5$ and S or H_2S/4/) also requires new approaches. One problem turned out to be an apparently systematic sulfur deficiency which causes FeS defects producing electronic states within the forbidden energy region. A sulfur deficiency in FeS_2 of at least 2.5% was found in all samples (powders, crystals) which we have up to now investigated /5/. These FeS defects (of the order of 10^{20} cm^{-3}) and additional impurities will have to be explored and controlled.
Partial success in eliminating or transforming interfacial FeS was obtained by selective photoelectrochemical corrosion and atomic hydrogen treatment /5/. They up to now limit the photopotential to 200mV (of theoretically possible 500 mV) so that energy conversion efficiencies of approximately 3% could not yet be exceeded (Fig.2).

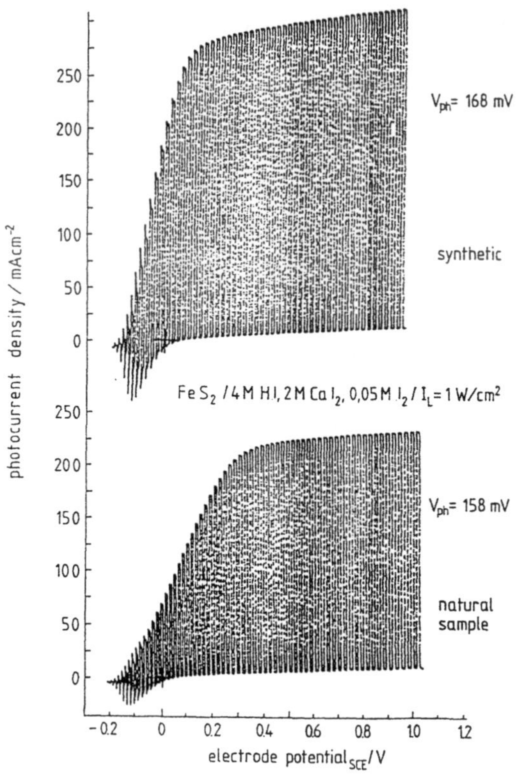

Fig.:3 . Photocurrent - voltage characteristic of FeS_2 / liquid junction (4M HI,2M CaI_2,0.05M I_2) under 1W cm^{-2} Xenon intermittant illumination. above: synthetic single crystal, below: natural

It has to be added that most effort up to now has been concentrated on liquid junction solar cells which are easy to fabricate and convenient to modify.
The most astonishing property of FeS_2 is the high quantum efficiency reached for photocurrent production with significant levels of impurities. Fig.:3 shows the photocurrent density - voltage characteristic of a pyrite/electrolyte junction illuminated intermittently with $1W/cm^2$ Xenon light (~10 solar intensities) obtained with a single crystal produced by vapor transport with addition of some ZnS /6/ (top curve) and with a natural pyrite crystal from a mine (Morgul,Turkey- bottom curve) /7/. The limiting photocurrent density reached (300 mA/cm^2) with the synthetic sample without antireflection protection or corrections for absorption losses in the electrolyte corresponds to a quantum efficiency of at least 75% (which should approach or exceed 90% after correction). It is indeed remarkable that good natural pyrite samples show quantum efficiencies only 25% smaller. The key problem with pyrite is a significant lack of knowledge on solid state and interfacial properties.
Traditional and other new materials easily produce a photovoltage equivalent to or exceeding 50% of the energy gap. $CuInS_2$, for example, which is also developed in our laboratory /8/, with its ideal energy gap of 1.52 eV, generates a photovoltage of 0.8 V, but a small quantum efficiency (20%) limits energy conversion efficiency in most samples still to typically 5-6%. This system has also the drawback of containing, with indium, a not very abundant element.

In-situ quality control for material and solar cell optimization

Even if a new cheap solar cell material is found it may take many years until a reasonably efficient thin layer solar cell is developed. The complications involved are best documented by the history of development of a-Si: solar cells. It took the effort of many groups (at present approximately 80) over a period of more than twelve years to reach a solar cell efficiency of 12-13 %. Only very few teams can perform so well in the laboratory concentrating on very small cells and might not be easily able to reproduce their results with another equipment. In fact, comercially produced a-Si:H solar cells are known to reach only significantly lower efficiencies. How long will it take to produce cascade solar cells composed of different materials, which alone can promise the approach of the high thermodynamic efficiency limits of solar energy conversion?

The development of in-situ techniques for the determination of optoelectronic properties during solar cell production is indispensable for an accelleration of development and improvement of production.We have come to the conclusion that time resolved and space resolved photoinduced microwave conductivity measurements give the most instructive information since they inform on photogenerated charge carriers which also drive the solar cell (complementary in-situ techniques like IR absorption and time resolved luminescence are equally explored). Microwave relative power absorption follows the relation

$$\Delta P(\omega,t)/P(\omega) = A(\omega)\Sigma_i \Delta n_i(t)\, \mu_i(\omega,t)$$

in which $A(\omega)$ is a frequence dependent proportionality factor, $\Delta n_i(t)$ the time dependent concentration of the photoexcited electronic species i and $\mu_i(\omega,t)$ their mobility. This contact free technique, in which a Gunn diode produces the microwave power (25.5-40 GHz) which is passed

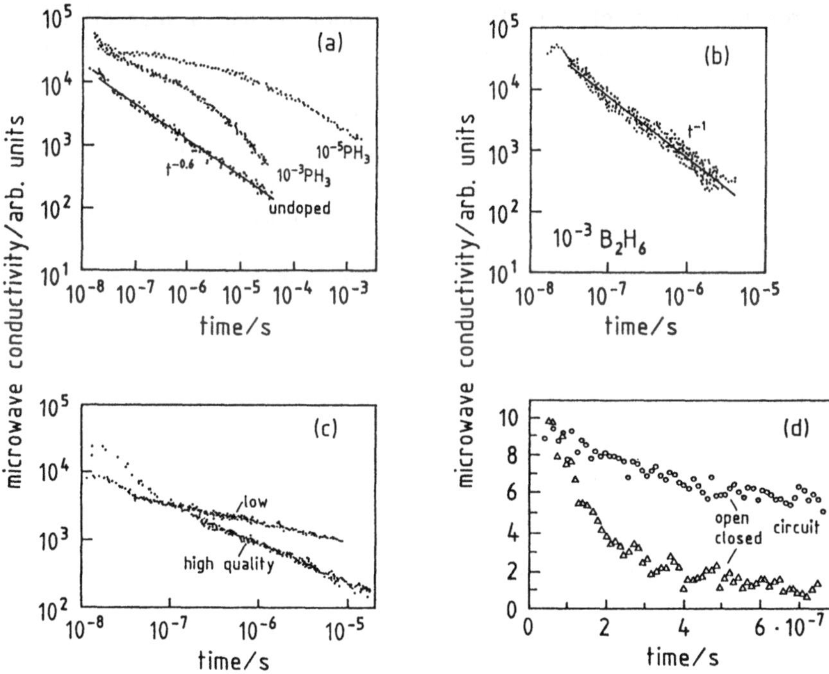

Fig.: 4. Time resolved microwave conductivity signals (Nd:YAG laser pulse, 12 ns). a) a-Si:H , undoped, n-doped; b) a-Si:H, p-doped; c) a-Si:H , intrinsic, low and high quality; d) Si:Au Schottky barrier, open and closed circuit.

through an attenuator and via a circulator which directs the microwave energy to the sample and channels the reflected wave to the detector (a thermistor) (compare Fig.: 6) detects the kinetics of light induced electronic carriers time resolved or space resolved (laser scan) /9/. Charge carrier kinetics is thus measurable in-situ during production of a thin film. During the development of these techniques valuable new insight into photoactive properties of polycrystalline materials, into the effect of surface treatment, chemisorption, catalysis and into heterojunction formation has already been obtained /10/.

For in-situ quality control and optimization it is necessary to dispose of a reliable quality signal. The magnitude of the measured time resolved microwave conductivity signal is of a certain value, but its decay kinetics is much more specific. For a high quality intrinsic a-Si:H defined by a low density of states in the mobility gap a decay according to $\Delta P(t)/P = t^{-0.6}$ is observed /10/. Inferiour quality leads to significant deviations, frequently caused be trapping of holes (Fig.: 4a). Doping by PH_3 or B_2H_6 clearly changes the recombination kinetics (Fig.: 4b and c). Also the presence of an electric field in a junction has a marked effect on the time resolved signal as shown in Fig.: 4d for a Si: Au Schottky barrier /11/.

Space resolved microwave conductivity measurements produced by laser scanning is also being developed and might later be used for the control of uniform semiconductor deposition for large area solar cells /12/. To illustrate the potential of this technique a 2 x 2mm large area (100 x 100 data points) of crystallized n-type silicon is shown in Fig.: 5 onto which a spot of a p-type polymeric semiconductor (polyfurandiylvinylene) with E_G=1.6 eV has been deposited. Microwave conductivity (above) and light reflection (below) are compared. Although the polymer film increases light reflection (below) which diminishes the absorbed light, and is itself negligibly photoconducting, the photoinduced microwave conductivity signals (the darker, the higher signals) increase where the polymer touches the silicon to form a heterojunction (Fig.: 5)/13/. The electric field produced in the interface apparently increase the carrier lifetime. In this way it may be possible to control junction formation in-situ.

Solar cell optimization using artificial intelligence

The first computer operated three-chamber plasma deposition system (with one additional preparation chamber) for p-i-n solar cells of a-Si:H which uses time resolved microwave conductivity signals for in-situ optimization of the product has been constructed in our laboratory /14/. Artificial Intelligence is being installed both for the operation of the plasma deposition equipment (in our system approximately 50 controls) and the product optimization which needs a scientific knowledge base including a wide variety of physical, solid state chemical and empirical knowledge (Fig.: 6). The use of artificial

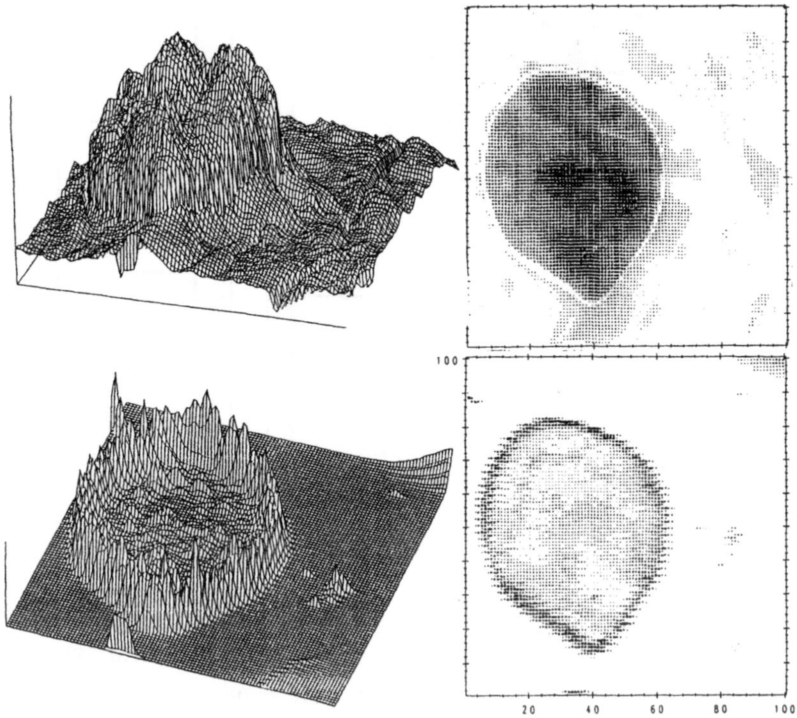

Fig.:5 Space resolved microwave conductivity (above) and light reflection (below) of n-Si (2 x 2mm) onto which a thin film spot of a polymeric p-type semiconductor was deposited (the higher signal the darker pattern).

intelligence for production control will not only have the advantage that new operations and rules can easily be included without reorganization of the computation program but it will also be convenient in guiding the apparatus through all safety and quality control procedures to the desired function.

In a successful preliminary test an optimization strategy has already been applied to optimize in a one chamber plasma deposition apparatus one parameter (the temperature) by in-situ quality control /15/. Optimization by artificial intelligence will open entirely new perspectives. The solar cell producing system will be able to learn and to deduce new empirical rules. The knowledge base of the expert system which contains scientific, technical and empirical rules for solar cell production will not remain stationary but will improve with time.

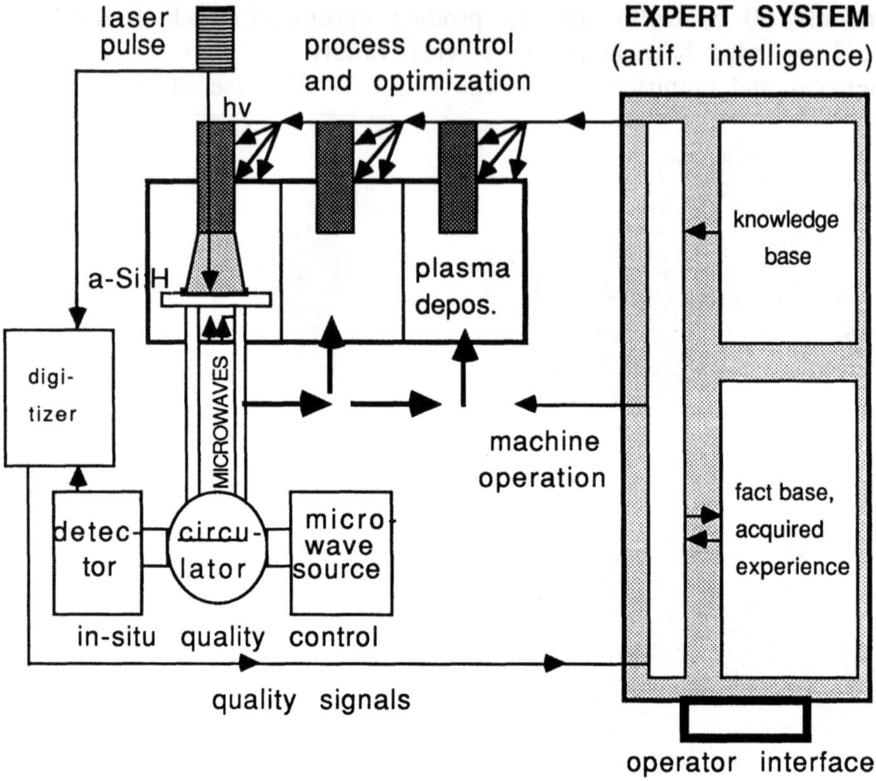

Fig.: 6. Schematic drawing explaining solar cell production using in-situ quality control (time resolved microwave conductivity measurements) and optimization with artificial intelligence.

When a new thin layer material is introduced, this intelligent production technique will be available and might guarantee a fast and straightforward optimization.

On the basis of the experience with the described pilot installation the concept of future automatic, self-optimizing research and production installations for thin layer solar may be discussed. The described approach will require further considerable effort but appears to be logic and reasonable since production of efficient multi-layer solar cells will involve a very large number of dependent production parameter which all have to be optimized and controlled to yield a high quality solar cell product.

Acknowledgement
The work was supported by BMFT grant Nr. 03E-8375-A. The author thanks for the help received from collaborators involved in the project and cited in the references.

References:

/1/ Z. Tian. , Advances in Science of China, Chemistry, Vol.**1**. (Tang Yougi, ed.) Science Press, Beijing, (1985), 107

/2/ H. Tributsch ,J. Photochem. **29** (1985)89; in " New Trends and Applications of Photocatalysis and Photoelectrochemistry for Environmental Problems" NATO Advanced Study Institute, Palermo 1987, Reidel Publ. Co. Holland (1988)

/3/ A. Ennaoui and H. Tributsch, Solar Cells **13**,(1984)197; A. Ennaoui, S. Fiechter, W. Jaegermann and H. Tributsch, J. Electrochem. Soc. **133** (1986) 97

/4/ G. Chatzitheodorou, S. Fiechter, R. Könenkamp, M. Kunst, W. Jaegermann and H. Tributsch, Mat. Res. Bull. **21** (1986)1481

/5/ N. Alonso Vante, G. Chatzitheodorou, S. Fiechter, N. Mgoduka, I. Poulios and H. Tributsch, Solar Energy Materials, submitted.

/6/ S. Fiechter, J. Mai, A. Ennaoui, W. Szacki, J. Crystal Growth, **78** (1986)438.

/7/ D. Jokisch, P. Möller and H.Tributsch in preparation.

/8/ H.J. Lewerenz, H. Goslowsky, K. D. Husemann and S. Fiechter, Nature, **321**,(1986)687

/9/ M. Kunst and H. Tributsch, Chem. Phys. Lett. **105** (1984)640;

/10/M. Kunst and A. Werner, J. Appl. Phys. **58** (1985)2236; J. Non-Cryst. Solids **97&98** (1987)611; A. Werner, M. Kunst, G. Beck, J. Lilie and H. Tributsch, Solid State Comm. **56**(1985)127

/11/M. Kunst, G.Beck and H. Tributsch, J. Electrochem. Soc. **131**(1984)954; M. Kunst and H.Wetzel, in preparation.

/12/ H. Tributsch, G. Beck and M. Kunst, European Patent Nr: 0155225; G. Beck and M. Kunst, Rev. Sci. Instrum. **57**(1986) 179

/13/ G. Schlichthörl, G. Beck, J. Lilie and H. Tributsch in preparation

/14/ G. Beck, R. Conrad, K. Emmelmann, U. Küppers, M. Kunst, H. Neitzert, A.Werner and Tributsch, H. in preparation

/15/ M. Kunst, A. Werner, G. Beck, U. Küppers, J. Lilie and H. Tributsch, Proc. Symp. on Emerging Semiconductor Technology, San Jose,USA, (1986)

NEW DIMENSIONS IN MINORITY-CARRIER LIFETIME DETERMINATION ON POLY-CRYSTALLINE SILICON SOLAR CELLS

L.A. Verhoef, J.R. Liefting, J.C. Stroom and W.C. Sinke
FOM-Institute for Atomic and Molecular Physics
Kruislaan 407, 1098 SJ Amsterdam, The Netherlands

ABSTRACT

A method has been developed for 3-dimensionally resolved minority-carrier lifetime determination in the base of poly-crystalline silicon solar cells. In this method, Light Beam Induced Current Decay (LBICD), a pulsed monochromatic light beam is employed which is focussed on- and scanned over the cell surface while monitoring the steady-state and transient short-circuit current. This yields laterally resolved lifetime determination. Varying the pulse-duration or the wavelength offers the possibility of depth-resolved lifetime determination. The minority carrier lifetime is found to be considerably smaller in proximity of a grain boundary than in the grain bulk. Computer simulations performed to fit the data could not reproduce a reduced lifetime in a region around the grain boundary, but very well fitted the steady-state current profile around the grain boundary.

INTRODUCTION

The efficiency of silicon solar cells depends strongly on the diffusion length L and recombination lifetime τ of minority carriers in the cell. In single-crystal silicon, the diffusion length and lifetime are approximately uniform throughout the base of the cell. Poly-crystalline silicon, which is often an economically more attractive material, contains high concentrations of grain- and sub-grain-boundaries, smaller physical defects and impurity clusters, which cause electrical properties and transport parameters to be position dependent. Special processes, such as hydrogen-passivation[1], preferential doping[2] and impurity-gettering[3,4], aimed at improving the average lifetime may contribute to this inhomogeneity. The performance of poly-crystalline solar cells not only depends on the spatial average of the transport parameters but also on their spatial distribution. Hence it is worthwile to develop a method for 3-dimensionally (3-D) resolved lifetime determination.

Several techniques are available for determining the minority carrier lifetime. In the Spectral Response (SR) method the steady-state short-circuit current is measured as a function of wavelength, yielding an effective diffusion length which lumps together the actual diffusion length, the back-surface recombination velocity and the back-surface reflectivity[5]. Neither lateral- nor depth-resolution is possible. Light Beam Induced Current (LBIC) involves scanning a light beam over the cell surface while simultaneously measuring the current output of the cell, enabling laterally-resolved response measurements[6,7,8]. A laterally-resolved effective minority-carrier diffusion length was obtained by varying wavelength and calculating L_{eff} locally similar to SR-measurements[9]. Photo Current Decay (PCD) employs pulsed light impinging on a short-circuited solar cell. From the current transient following the pulse, the lifetime can be determined[10]. So far PCD has been employed only using homogeneous illumination of monocrystalline silicon solar cells; the uniform lifetime is then readily derived from the data[11].

The aim of this work is to investigate the possibility of 3-D-resolved measurement of minority-carrier recombination lifetime in poly-crystalline silicon solar cells using a hybrid of PCD, SR and LBIC.

THEORY OF LIGHT BEAM INDUCED CURRENT DECAY

A pulsed laser beam is focussed on an n^+/p solar cell in short-circuit condition and the time-dependent current is monitored, see Fig. 1. At times $-T<t<0$ the light is on and minority electrons are generated in the base of the cell. The generation profile depends on the light-spot size d and the absorption coefficient α. Apart from these two parameters, the minority-electron concentration profile n(x,y,t) at t=0 depends on T and on the position(y)- and depth(x)-dependent electron recombination lifetime $\tau(x,y)$. Adjustable parameters are wavelength λ, pulse-duration T and lateral position y of the beam on the cell. The current j(t) flowing out of the base into the space charge region at x=0 is studied as a function of time. At times t>0, the current decays to zero. Except for $\tau(x,y)$, all base parameters, including electron diffusion coefficient D and the effective back-surface recombination velocity S at x=W are assumed to be position- and minority-carrier density independent. Solving this general case requires numerical methods, but in some simpler cases an analytical solution can be obtained:

First, in the simplest case under the conditions 1. steady-state $T>>\tau$, 2. homogeneous lifetime $\tau \neq \tau(x,y)$, 3. large spot size d>>L and 4. the spot is far away from the grain boundary (g.b.), a one-dimensional approach suffices and the following solution is obtained[12]:

$$j(t) = \frac{4eDF}{W^2} \sum_{i=1} j_i(\alpha', w_i)\, \tau_i\, e^{-t/\tau_i} \quad (1).$$

In this expression e is the electron charge, τ_i is the i-th mode decay constant, F is the light flux at x=0, W is the base thickness and $\alpha' = \alpha W$ is a normalized absorption coefficient. The w_i are given by the eigen value equation :

$$\frac{D}{WS} w_i = -\tan w_i \ , \text{ with } \pi(i-1/2) < w_i < \pi i \quad (2).$$

The decay constants τ_i are related to w_i and to the minority carrier lifetime τ through:

$$\frac{1}{\tau} = \frac{1}{\tau_i} - \frac{D w_i^2}{W^2} \quad (3).$$

For times $t > \tau$ the higher modes of decay have vanished in most cases (when the diffusion length is comparable to the base thickness), and the plot of log(j(t)) versus t becomes linear (pure exponential decay)[13]. The slope of the curve equals τ_1 and the extrapolation of this curve to t=0 yields the intercept $j_1(0) = (4eDF\tau_1/W^2)\, j_1(\alpha', w_1)$. From (2) and (3) it follows that $\tau = \tau_1$ in case $D\tau/W^2 << 1$.

Second, in proximity of a grain boundary and under the restriction of a small spot size d<L, lateral diffusion necessitates a two- or three-dimensional analysis. This has been evaluated analytically by Kern[14], who used a recombination velocity $S_{gb} = \infty$, yielding a result equivalent to (1). Third, the effect of pulse duration T has been taken into account by Bielle-Daspet[15]: (1) still holds, but the amplitudes $j_i(\alpha', w_i)$ become dependent on T. In case of a finite S_{gb} and uniform τ in the grain itself, $j_i(\alpha', w_i)$ become dependent on position relative to the grain boundary, whereas τ_1 remains unchanged and independent of that position.

Numerical solutions are required in more general cases. For instance, a grain boundary can be described by two models: model 1, in which the lifetime is uniform around the boundary and the boundary itself is characterized by a recombination velocity S_{gb} and model 2, in which the lifetime is reduced in a certain region around the boundary. As mentioned above, the first model is still analytically solvable, but can also be solved numerically. The second model introduces an actual position dependent lifetime $\tau(y)$, for instance induced by segregation or emission of impurities and defects. Shimokawa et.al. have found a position dependent diffusion length around a grain boundary[9]. In that case transport equations are practically insolvable and we resort to computer simulations.

Gettering processes and other processing steps might deplete or enrich certain regions of the base with impurities and defects, resulting in a depth-dependent lifetime $\tau(x)$. In principle, depth-resolved lifetime determination can be achieved by controlling the generation profile in the base of the cell, by either varying the wavelength or the pulse duration T. When the diffusion length L is large, carriers generated at large depth still contribute to the external current, so a large depth range is available. For small L only a limited region of the base can be probed. The most favorable experimental condition for depth resolution is a delta pulse in time (T -> 0; FT=constant) with variable wavelength: the initial carrier density profile n(x,y,0) is proportional to $\exp(-\alpha x)$ and diffusion only occurs during the decay; this simplifies the analysis.

EXPERIMENTAL SET-UP

The experimental set-up is shown in Fig. 2. A 5 Watt Argon-ion laser is used to pump a dye laser. By employing various dyes a wavelength range of 400-950 nm is attainable. The beam from the dye laser is chopped with an electro-optical modulator, resulting in a pulsed beam with $t_{rise} = t_{fall} = 30$ ns and adjustable pulse duration. This beam is focussed to a spot of about 20 µm onto the device under study, which is short-circuited via a fast preamplifier (DC -150 MHz). In order to reduce ringing-effects the wires are kept short and in order to eliminate RC-time effects of the cell capacitance in conjunction with the total circuit series-resistance, the samples investigated were cut down to approximately 6x6 mm^2. The preamplifier is connected to a 400 MHz digital storage scope, which is coupled to a computer. Both sampleholder and preamplifier are mounted on a translator table to enable lateral scanning.

The short-circuit current response is measured, yielding the steady-state LBIC response prior to decay and the transient for times $t>0$. Subsequently a least-squares fit is performed on the exponential tail of the current decay (aftre the light is switched off) to find τ_1.

RESULTS AND DISCUSSION

As an example of laterally resolved measurements, we first show the results of a g.b.-scan over a poly-crystalline silicon solar cell fabricated on Wacker SILSO. Pulse duration was such that steady-state was reached prior to decay .The wavelength used is 860 nm, corresponding to a penetration depth of 22 μm[16]. Thus negligible absorption takes place in the emitter and space charge region. The current decay curves for 4 different positions with respect to the grain boundary are shown in Fig. 3. For all positions pure exponential decay over two decades or more was observed. Clearly, there is a strong position dependence on the decay time up to 90 μm away from the boundary. From the decay time far away from the boundary ($y>90$ μm), we infer a diffusion length of 50 μm, using $D=25$ cm^2/s.

In Fig. 4, the decay time as function of position is shown for two g.b. regions, A and B, in combination with the steady-state level (LBIC-response) at times $t<0$. Although the LBIC signals are very similar, the decay-time profiles are different. Around boundary A, the decay time increases within 80 μm from the low value at $y=0$ to the bulk value, which is the same on both sides of the boundary. Region B apparently consists of two different grains: one with a relatively large lifetime ($y<0$) and the other with a much smaller lifetime ($y>0$). The grain boundary only influences the decay time in the one with large τ .

Computer simulations of the decay-time scans and LBIC responses as a function of position of the light spot were performed to determine the recombination velocity and lifetime in proximity of the boundary. We consider the two models for g.b. regions introduced in section 2.

First take the traditional case of L independent of y and S_{gb} affecting the current in a certain width around $y=0$. The LBIC response of both regions is readily reproduced, yielding $S_{gb}=10^4$ cm/s and $L=50$ μm, see Fig 4. The corresponding simulated decay times (the ultimate exponential mode τ_1) are independent on position y, contrary to our data. This may be due to the fact that our data were taken in a time range where the fundamental decay mode is not yet been reached; higher modes from (1) still contribute to the current[12]. However, our data clearly show pure exponential decay over more than two decades for all measurements, which suggests that indeed the fundamental mode has already been reached. So on the one hand our simulations agree with the decay time data (initial decay is much faster in proximity of the grain boundary) but on the other hand our data (two orders of magnitude decay) contradict the simulations.

Second, simulations were performed in which the diffusion length L(y) was assumed position dependent. It turns out that this description also reproduces the LBIC-profiles for both regions A and B. However, the decay time profile around the grain boundaries could not be simulated. For long times the decay always approaches the bulk value due to the finite thickness of the region of non-uniform diffusion length. The same argument as above, for the limited time range in which our data are taken, applies but also in this model no exponential decay over several decades other than the ultimate one is observed in the simulations .

The discrepancy regarding the decay time between data and simulations might be the result of a minority-carrier density dependent lifetime in proximity of the grain boundary: during the light pulse traps are filled and during the decay they are emptied. The effect of this on the current decay is not clear yet. Experiments with light bias might reveal the nature of this effect.

To study the possibilities of depth-resolved measurements, we varied the wavelength of the laser light and by doing so, the carrier generation profile. In Fig. 5 the results from scans over one grain boundary with 3 different wavelengths (860, 840 and 820 nm) are shown. The corresponding absorption depths are 22, 15and 13 μm, respectively[16]. The LBIC-signal shows the same behavior for all wavelengths, as expected from spectral response measurements: in this wavelength range the spectral response is almost independent of wavelength. The decay time τ_1 does depend on wavelength and on position. For all wavelengths τ_1 remains small on the left hand side of the grain boundary , while it increases with increasing distance on the right-hand side, as in Fig 4, but the bulk value far away from the boundary depends on wavelength. A tentative explanation is the following. For short wavelengths, penetration is small and mainly the top region of the base close to the metallurgical emitter interface is probed, where the lifetime is presumably lower.

CONCLUSIONS

A new method has been developed to perform 3-D resolved minority-carrier lifetime determinations. This method, Light Beam Induced Current Decay, employs a pulsed light beam which is focussed onto the device under study. We have been able to observe a position dependent lifetime around grain boundaries in poly-crystalline solar cell bases.

The observation of transients together with steady-state (LBIC-) signal reveals more than does LBIC alone. This is demonstrated with two measurements which showed equivalent LBIC traces, but different LBICD-response. Computer simulations were performed using two models for a grain boundary region: the traditional, where the diffusion length is constant in the grains and the boundary is represented by a recombination velocity and an alternative model, in which the diffusion length in proximity to the boundary is reduced and position-dependent. Both models reproduces the LBIC scans correctly, but are unable to match the decay time profile, indicating that other phenomena such as carrier-density dependent lifetime have to be taken into account. Experiments under a light bias possibly reveal the nature of this reduced decay time around grain boundaries.

According to our data, the wavelength of the light pulse prior to the decay did influence the decay: τ_1 was found to be smaller for light absorbed predominantly close to the metallurgical junction. This suggests the lifetime is reduced in the top region of the cell.

ACKNOWLEDGEMENTS

This work is part of the research program of the Stichting voor Fundamenteel Onderzoek der Materie (FOM) and was financially supported by the Nederlandse Organisatie voor Wetenschappelijk Onderzoek (NWO) and the Stichting Technische Wetenschappen (STW).

REFERENCES

[1]J.I. Pankove, P.J. Zanzucchi, C.W. Magee and G. Lucovsky, Appl. Phys. Lett. **46**, 421 (1985)
[2]T.H. DiStefano and J.J. Cuomo, Appl. Phys. Lett. **30**, 351 (1977)
[3]J. Culik, S. Roncin and P. Alexander, Proc. 17th IEEE Photovoltaic Solar Energy Spec. Conference 1984 (IEEE, New York), 259
[4]R.Sundaresan, D.E. Burk and J.G. Fossum, J. Appl. Phys. **55**, 1162 (1984)
[5]S.M. Sze, *Physics of Semiconductor Devices* (Wiley, New York, 1981), p. 800
[6]C.Donolato, J. Appl. Phys. **54**, 1314 (1983)
[7]J. D. Zook, Appl. Phys. Lett. **37**, 223 (1980)
[8]W.D. Sawyer, J. Appl. Phys. **59**, 2361 (1986)
[9]R. Shimokawa and Y. Hayashi, J. Appl. Phys. **59**, 2571 (1986)
[10]M. Di Giulio, S. Galassini, G. Micocci and A. Tepore, J. Appl. Phys. **52**, 7219 (1981)
[11]S.R. Dhariwal, L.S. Kothari and S.C. Jain, Solid-State Electron. **20**, 297 (1977)
[12]L.A. Verhoef, to be published
[13]A. Zondervan, L.A. Verhoef, F.A. Lindholm and A. Neugroschel, to appear in J. Appl. Phys., May 1988.
[14]R. Kern, Proc. 6th E.C. Photovolt. Solar Energy Conf. London 1985, p. 133
[15]G. Bielle-Daspet and G.D. Daspet, Solid-State Electr. **21**, 1219 (1978)
[16]M. Saritas and H.D. McKell, J. Appl. Phys. **61**, 4923 (1987)

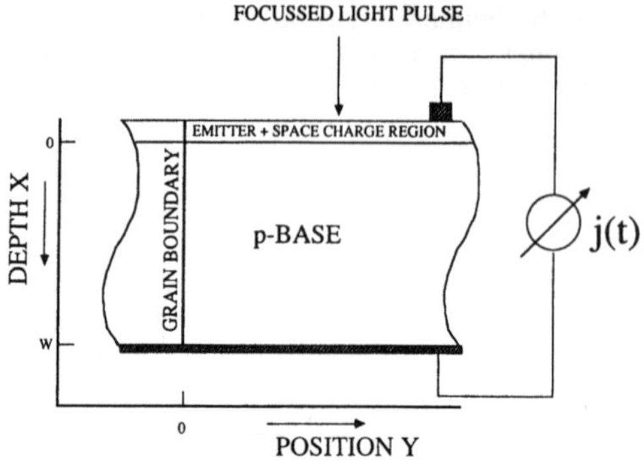

Figure 1. Schematic of the Light Beam Induced Current Decay method. A monochromatic light pulse impinges on a short-circuited solar cell, and the time-dependent current is measured. The light beam can be scanned over the cell surface.

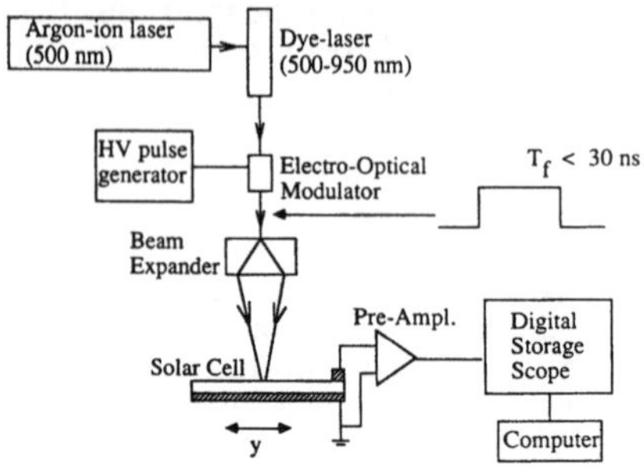

Figure 2. Experimental set-up used in LBICD. An adjustable monochromatic beam from a dye laser is chopped by an Electro-Optical Modulator and subsequently focussed onto the solar cell. The short-circuit current is amplified and measured on a digital storage scope.

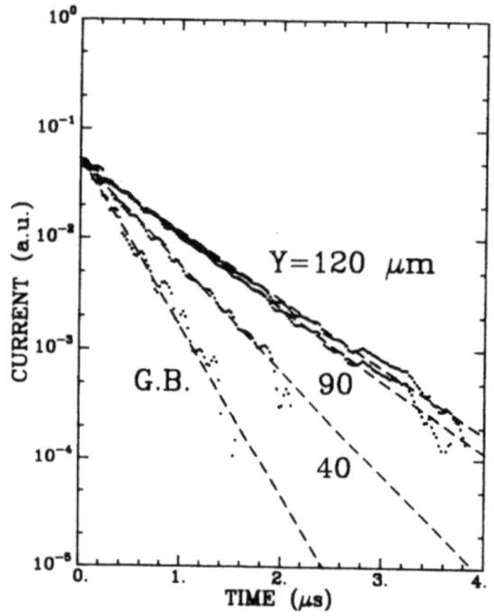

Figure 3. Current vs time for four different positions of the light spot with respect to the grain boundary. The decay is exponential over more than two decades for all measurements. t was chosen arbitrarily for each curve.

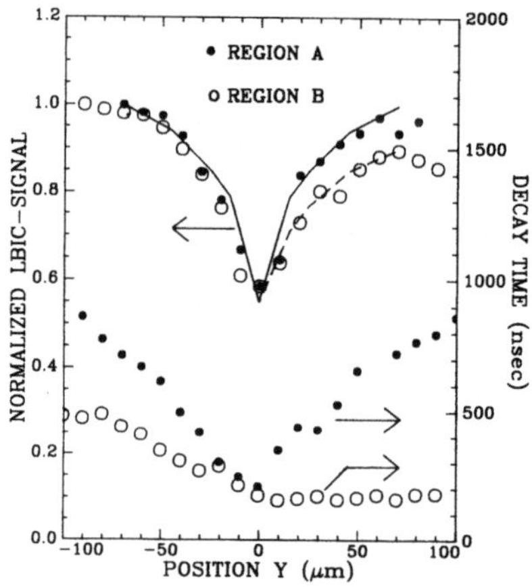

Figure 4. LBIC- and LBICD-profiles for two grain boundary regions A (closed circles) and B (open circles). The grain boundary is at y=0. The LBIC-response could be reproduced with a simulation with input $S_{gb}=10^4$ cm/s and L=50 μm (drawn lines). The simulations could not reproduce our decay time data (not shown).

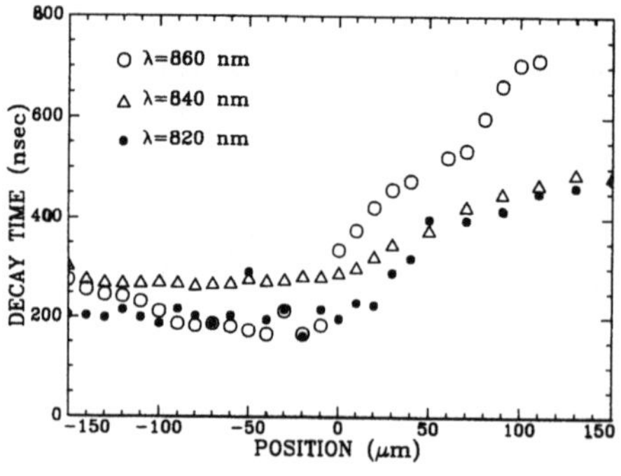

Figure 5. Decay time profiles for a grain-boundary scan with different wavelengths. Grain boundary is located at y=0.

SESSION 04;
APPLICATIONS AND SYSTEMS

MACROECONOMIC IMPACT OF PHOTOVOLTAIC ENERGY GENERATION

Olav H. Hohmeyer
Fraunhofer-Institute for Systems and Innovation Research
Breslauer Strasse 48
D - 7500 Karlsruhe
Federal Republic of Germany
Tel.: 49-721-68 09 161

ABSTRACT. Market introduction of new energy technologies based on renewable energy sources is dependent on the market price of the energy supplied. Since the competing conventional energy technologies are able to pass on to society a substantial part of their costs (external or social), renewable energy sources, which do so to a far lesser extent, are put at a systematic disadvantage. The objective of this paper is to derive the difference in the external costs of photovoltaics and conventional electricity generation and to analyse the impact of not including these costs in the market price of electricity on the competitive position and the time of market introduction of photovoltaics. The net social costs of conventional electricity generation as compared to photovoltaics are estimated to be at least in the order of 0.07 to 0.17 DM/kWh. This compares to a market price of electricity supplied to consumers in the FRG today of about 0.25 DM/kWh. It is estimated that with regard to achieving cost-effectiveness of photovoltaic systems, the inclusion of social costs in the market price of electricity would lead to a time-savings of between four and eight years.

1. INTRODUCTION

In a market economy, the basic economic problem of allocating scarce resources to competing uses is solved through the market mechanism based on the market prices of the resources. A precondition for the optimal functioning of this allocation process is that the market prices reflect all costs involved in production. If this situation is assured, the microeconomic calculations of the economic agents involved may lead to a macroeconomic optimum of the allocation process for society. If substantial costs of the production process are not reflected in the market prices because such costs are passed on to third parties not involved as consumers or producers of a product (external or social costs), the market mechanism cannot secure an optimal macroeconomic allocation. This suboptimal allocation leads to considerable losses to society. Typical costs passed on to third parties include environmental costs (e.g. forest damage), health costs, costs of intertemporal misallocation of depletable resources or R&D subsidies. Electricity production based on fossil or nuclear fuels induces substantial external costs, whereas it would appear that the use of renewable energy sources involves far fewer and lower external costs. If this is the case, then in terms of a macroeconomic optimum, there may be too little investment in technologies utilizing renewable energy sources, resulting in high costs to society.

2. THE SOCIAL COSTS OF ELECTRICITY GENERATION

The question arises as to how large the difference is between the external costs (and benefits) of photovoltaics and those of conventional electricity generation? Is this difference large enough to affect the competitive position of photovoltaics? If so, how does it affect the market introduction and diffusion of photovoltaic energy systems?

The following discussion will deal only with the direct competition between photovoltaic electricity and electricity supplied by the public grid. Special markets will not be taken into consideration.

Although it is difficult to impossible to quantify and monetarize certain external costs, particularly those in the area of health and environmental damage, this paper compares the estimated minimum net external costs of conventional electricity with those of photovoltaics. Even though full monetarization can at best remain an estimate, awareness of the minimum net external costs (the lowest possible realistic figures) cannot but help to improve an allocation process which takes into account no social costs whatsoever. The results given should be interpreted as a first systematic overview producing very crude figures which can nevertheless be used as a base for some initial corrective economic policy measures. Wherever doubt exists, assumptions have been made counter to the underlying hypothesis, namely, that the external costs of systems using renewable energy sources are considerably lower than those of systems using conventional energy. Thus, although suggested corrective measures may not be optimal, they err only insofar as they are insufficient.

The study up on which this paper is based has been conducted within the economic and administrative framework of the Federal Republic of Germany. Although the quantitative and monetary results are not directly applicable to other countries, the general approach is valid for any market-oriented economy.

Three principal areas of external effects of energy systems have been considered: environmental effects, including effects on human health, general economic effects such as changes in gross value added or employment, and subsidies paid by government agencies directly or indirectly as public provisions in kind. Public expenditure for research and development on energy technologies has been subsumed under the general heading of public subsidies. In each of these general areas, there are a variety of single external effects which are discussed in the final report of the research project (Hohmeyer, 1988).

The following effects could not be quantified and monetarized:

- the psycho-social costs of serious illnesses or deaths as well as the costs to the health care system
- the environmental effects of the production of intermediate goods used for investments in energy systems and the operation of these systems
- the environmental effects of all stages of the fuel cycles (specifically in the case of nuclear energy)
- the full cost of climatic changes
- the environmental risks of routine operation of nuclear power plants
- hidden subsidies for energy systems given under other titles.

When the quantified external effects of conventional energy systems for the production of electricity based on fossil fuels are summed up and standardized for the production of 1 kWh, gross external effects in the range of 0.04 to 0.09 DM_{82}/kWh result. For electricity generated in nuclear reactors (excluding breeder reactors), external costs in the range of 0.1 to 0.2 DM_{82}/kWh have been calculated. A weighted average for these gross external costs according to the fuel composition found in the electricity generation of the Federal Republic of Germany in 1984 is 0.05 to 0.12 DM_{82}/kWh electricity generated (see Table 1.).

Table 1.: Summary of the external effects of electricity generation based on fossil fuels, nuclear energy, and photovoltaic solar energy (1982 prices)

a) Gross external effects of electricity generated from fossil fuels (all figures are estimated minimal external costs)		
1. Environmental effects	0.0114 - 0.0609	DM_{82}/kWh_{el}
2. Depletion surcharge (1985)	0.0229	DM_{82}/kWh_{el}
3. Goods and services publicly supplied	0.0007	DM_{82}/kWh_{el}
4. Monetary subsidies (including accelerated depreciation)	0.0032	DM_{82}/kWh_{el}
5. Public R&D transfers	0.0004	DM_{82}/kWh_{el}
Total	0.0386 - 0.0881	DM_{82}/kWh_{el}
b) Gross external effects of electricity generated in nuclear reactors, excluding breeder reactors (all figures are estimated minimal external costs)		
1. Environmental effects (human health)	0.0120 - 0.1200	DM_{82}/kWh_{el}
2. Depletion surcharge (1985)	0.0591 - 0.0623	DM_{82}/kWh_{el}
3. Goods and services publicly supplied	0.0011	DM_{82}/kWh_{el}
4. Monetary subsidies	0.0014	DM_{82}/kWh_{el}
5. Public R&D transfers	0.0235	DM_{82}/kWh_{el}
Total	0.0971 - 0.2083	DM_{82}/kWh_{el}
c) Average gross external costs of the electricity generated in the FRG in 1984		
1. Costs due to electricity from fossil fuels (weighting factor 0.7444)	0.0287 - 0.0656	DM_{82}/kWh_{el}
2. Costs due to electricity from nuclear energy (weighting factor 0.2556)	0.0248 - 0.0532	DM_{82}/kWh_{el}
Total (conventional electricity)	0.0535 - 0.1188	DM_{82}/kWh_{el}
d) Solar energy (photovoltaics)		
1. Environmental effects	(-) 0.0044	DM_{82}/kWh_{el}
2. Public R&D transfers (estimate)	- 0.0052 - (-) 0.0104	DM_{82}/kWh_{el}
3. Economic net effects (not including 1982 figures)	+ 0.0240 - (+) 0.0665	DM_{82}/kWh_{el}
4. Avoided external cost of present electricity generation	+ 0.0535 - (+) 0.1188	DM_{82}/kWh_{el}
Total external benefits rounded to three digits	+ 0.068 - (+) 0.171	DM_{82}/kWh_{el}
mean	(+) 0.119	DM_{82}/kWh_{el}

The sum of external net benefits of photovoltaics lies in a range between 0.07 DM_{82}/kWh and 0.17 DM_{82}/kWh (see Table 1.). This is only a minimum estimate which may give a distorted picture of the real situation, since we have consistently down-graded the advantages of photovoltaics in all cases of doubt.

Even without the inclusion of all external effects, and with a deliberate bias against renewable energy sources, the monetarized net external effects of photovoltaic solar energy systems are of the same order of magnitude as the basic market prices of conventionally generated electricity. The handling of the external effects will have a considerable effect on the allocation process and the market introduction of photovoltaic systems in the next ten to fifteen years.

3. RESULTS

In the following, the external costs quantified above are applied to the analysis of the future competitive position and market diffusion of photovoltaics. Figure 1. illustrates the situation. In the FRG, 1982 photovoltaic prices amounted to approximately 2.7 DM_{82}/kWh, about 15 times the price of the electricity to be substituted. The inclusion of the estimated external effects shown in Figure 1. will affect the competitive situation of solar energy in about 10 to 15 years from now, if the estimated cost degression curve for photovoltaics is realistic. Based on the figures for photovoltaics given by several authors, electricity costs of 0.62 DM_{82}/kWh were estimated for the year 2000 and of 0.26 DM_{82}/kWh for the year 2050. For southern Europe, these cost figures may roughly be divided by factor 2 due to the greater amount of solar radiation per square meter and year. On the basis of production by decentralized systems, with 50 % of the electricity produced being used for private consumption and 50 % being sold to the public grid, the combined market price of the electricity to be substituted will be about 0.19 DM_{82}/kWh in 1990 in the FRG. A real price increase of 2 % annually is assumed for the electricity to be substituted.

For the photovoltaic electricity cost curve, there is an intersection with the market price curve of the electricity to be substituted termed A in Figure 1.a and 1.c. At this point of intersection, photovoltaic solar energy is competitive with the electricity to be substituted at market prices which do not include external effects. Adding the lower range of the minimum net social costs 0.07 DM/kWh to this market price curve results in a second price curve for the substituted electricity which gives the new point of cost-effectiveness for photovoltaics at B. The inclusion of these social costs shows that photovoltaics is competitive with the electricity to be substituted earlier than market prices show. Accordingly, the market introduction of photovoltaic energy systems starts earlier. This is shown in Figure 1.b. It has been assumed that a pioneer market for photovoltaic systems not oriented towards cost-effectiveness amounts to about 5 % of an assumed technically feasible potential of about 20 TWh/a in the Federal Republic of Germany, which translates into approximately 2.25 million dezentralized installations of 10 kW peak load each. After photovoltaic systems become cost-effective, market penetration should begin, reaching 95 % of the technical potential 20 years after achieving cost-effectiveness, as shown in Figure 1.b.

Figure 1.: The influence of social costs on the starting point of market penetration of decentralized photovoltaic systems and the future penetration to the year 2040 (1982 prices)
a) costs for photovoltaic electricity compared with costs for substituted electricity
b) market penetration of photovoltaic solar energy based on costs shown above
c) magnification of (a) showing the exact intersections of the cost curves

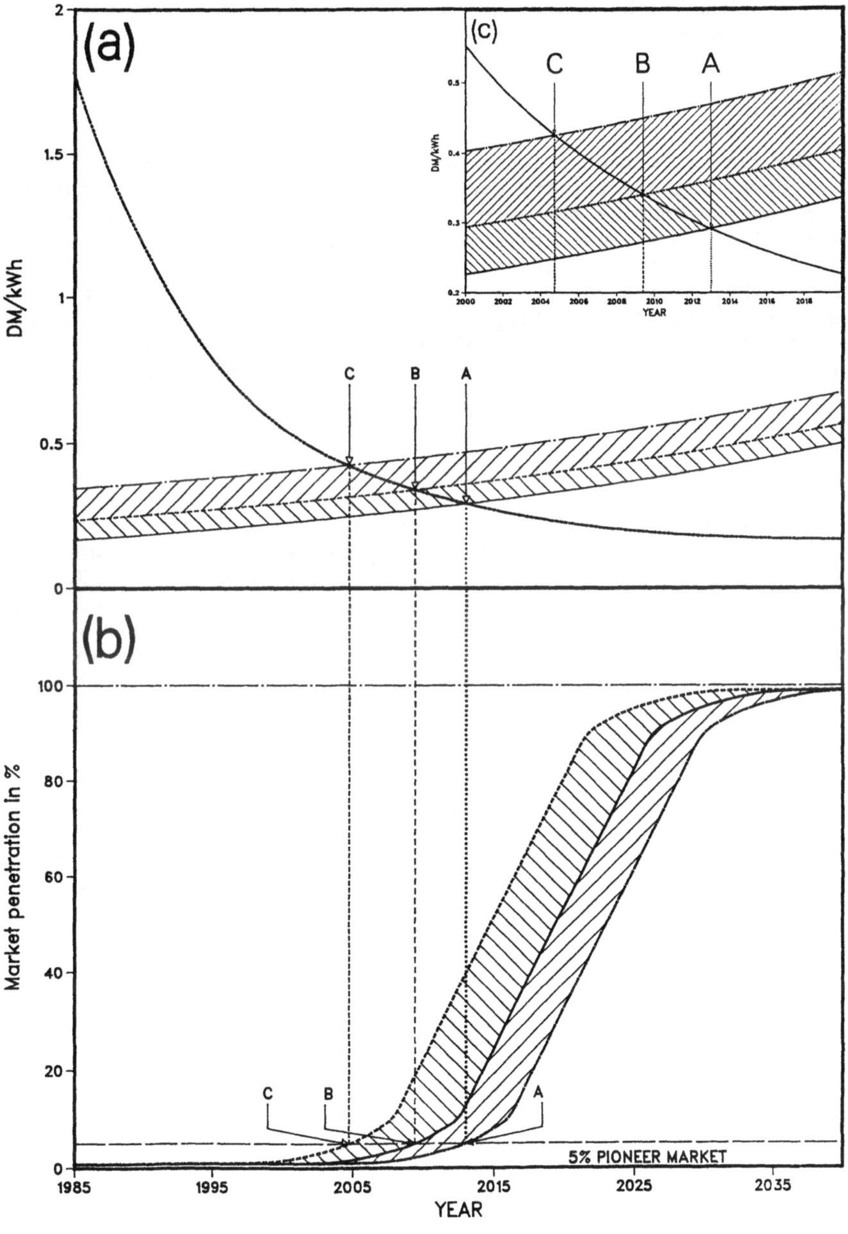

The addition of the upper range of the minimum net external effects of electricity generation (0.17 DM/kWh) to the market price of substituted electricity results in a third cost curve for the substituted electricity. This curve again gives a point of cost-effectiveness for photovoltaics as the intersection with the photovoltaic electricity cost curve C. The change in the market penetration of photovoltaic systems resulting from this altered cost situation is shown in Figure 1.b in the same way as discussed above. The inclusion of external effects in the market allocation process leads to a substantial shift in the market penetration of photovoltaic systems of about eight years. If these external effects are not included, a serious misallocation results.

4. CONCLUSION

Since the general market pricing mechanism does not work adequately in the case of competing electricity generating technologies, government has to step in and internalize the external effects of economic processes. Theoretically, this should be done by leveling a tax against all electricity generating processes causing external effects to offset the amount of these costs passed on to society. The estimated minimum costs given above could be a starting point for such a tax. But in practical politics, it will be very difficult to overcome the resistance of interested groups to such a tax. A way around the practical problem of imposing a tax would be to pay subsidies for photovoltaic systems to the extent of the social costs avoided through their operation. Since the inclusion of social costs will not make photovoltaics competitive within the next five years, a considerable amount of money should be spent on R&D grants to speed up the cost reduction of photovoltaic systems. By 1995, the internalization of external effects may start playing an important role in the market diffusion of the systems in southern Europe and shortly after the year 2000, the same will apply to countries such as the FRG. By that time, some effective policy measures for the internalization of external effects will have to be applied in order to avoid large-scale investments in the wrong electricity generating systems, a misallocation which would result in substantial losses to society.

5. ACKNOWLEDGEMENTS

This paper is based on the research project 'Social Costs of Energy' financed by the Commission of the European Communities (DG XII) under contract no. EN3S/0022/D. The final report of the project containing detailed information on the methodology employed and the results derived has just been published in English (see References).

6. REFERENCES

The paper is based on the following report, which contains all references:

Hohmeyer, Olav: The Social Costs of Energy Consumption - External Effects of Electricity Generation in the Federal Republic of Germany. Springer Verlag. Berlin, Heidelberg, New York, Paris, Tokyo 1988.

ASSESSMENT OF PHOTOVOLTAIC SYSTEMS

J.A. ROGER
Département de Physique des Matériaux (U.A. CNRS 172)
Université Claude Bernard Lyon 1
43 Bd du 11 novembre 1918
69622 Villeurbanne Cedex, France

ABSTRACT

This presentation summarizes a part of the research activities of the European Working Group on Photovoltaic Plant Monitoring. More than twenty persons from all Europe participated in these meetings instigated by the Joint Research Centre of the CEC in Ispra. The result of these discussions was the elaboration of three guidelines for assessment of photovoltaic systems. This work is now continuing in relation with a given number of Concerted Actions initiated by the CEC in Brussels. The first guidelines is related to data monitoring. The present paper addresses more directly the second guidelines concerning the analysis and the presentation of monitoring data . This paper indicates some procedures that have been retained, and the arguments that have dictated our choice. Some examples on small PV installations are given to underline the difficulties encountered when trying to generalize the scientific approach.

INTRODUCTION

The EC-funded PV demonstration or pilot projects, already installed or planned in the next coming years in Europe, will reach soon a total array power of 1 MW_p. Some projects correspond to large plants of hundreds of kW_p , but many others involve multiple installations, as for example one in France which is to supply electricity for 40 remote houses. The remarkable point is that PV systems are shifting, slowly but surely, from pilot and demonstration experience to industrial market. This is especially true in the case of small and medium size installations.

From the very beginning, PV systems appear, from desk studies, as being a cost-effective method for providing electricity in some specific applications. Long life with low maintenance, high reliabilty and environment protection are the more often heard arguments. To test these estimates, systematic and accurate monitoring is the necessary starting point, and all CEC-funded installations were provided with data acquisition systems.

The main criterion for assessing a PV system, whatever its size and purpose, seems to be its cost effectiveness - cost of the electricity supplied to the user - in comparison with alternative energy sources (windmill, diesel generator, grid extension etc..). However it appears clearly that other parameters can be as much important as cost-effectiveness, and cannot be neglected : reliability, energy availability for the users, environmental factors, social factors etc..

More than 20 experts from 12 countries participated in these working group meetings organized by the CEC Joint Research Centre of Ispra. Activities of this European WG on PV Plant Monitoring are presented in poster 7.13 by M.R. Starr and K. Krebs.

A first conclusion that emerged from discussions, was that it was urgent to develop a methodology for the assessment of PV systems. Starting from CEC pilot plant recorded data, the objective was to draw general procedures applicable to all encountered cases. The main difficulty appeared then to result from the strong differences existing between PV installations, not only in the technical design, but also in all other factors (location, type of energy used, type of service provided to the users, social factors, etc...). The logical approach was then to consider first the part of information from the PV system that could be correctly quantified.

As a consequence, three guidelines, setting out recommended procedures for monitoring, data analysis and presentation of results, have been developed :

A - PV system monitoring,

B - Analysis and presentation of monitoring data,

C - Initial and periodic tests on PV plants.

The first document has been finalized and the two others have been discussed on a draft level.

Monitoring and analysis of data indicate how well a plant is performing. Indices of performance can be derived, and the above guidelines propose unambiguous definitions of these indices and their strict conditions of validity. However the influence of other factors, such as social effects, cannot be quantified and can only be obtain through inquiries.

Evaluation of existing PV installations through data monitoring is not informative enough for a complete understanding of underlying problems. Lessons learned are as much important, and are the key factor for a correct development of new projects and the choice of PV in specific applications. Gathering of information from pilot plants and other PV site is an enormous task, initiated this year by CEC DG XII, through specialized Concerted Actions. Each part of the PV chain, from the array to the load, will be analysed from questionnaires sent to Pilot Plant Leaders and other selected experts, in order to summarize errors not to be duplicated. The objective of DG XII is to elaborate guidelines on design, operation and maintenance of PV systems.

I must stress here that these two sets of guidelines, on data monitoring and presentation of results on one side, and on lessons learned and prospects on the other side, are obviously complementary. They must be finalized as quickly as possible to provide the PV community with a powerfull tool.

In the following, a brief presentation of guidelines for the assessment of PV systems will be done with applications to some experimental cases.

1. PV SYSTEM MONITORING AS A STARTING POINT FOR ASSESSMENT

When trying to elaborate the guidelines on data monitoring, the first encountered difficulty was the elaboration of a general layout able to represent all possible PV systems whatever the size and the type of loads. We agreed on the energy flow diagram presented in figure 1 with the firm belief that some PV configurations will perhaps not be covered by this representation.

Measuring points are indicated as well as parameters to be recorded such as UA (array output voltage) and JA (array output current) as indicated as an example on figure 1. Parameters can be recorded as instantaneous values, or integrated values over a given period.

Data obtained are of interest for judging whether the design goals have been met and they should provide a base for the development of theoretical models for future design. To obtain such complete sets of data a rather detailed or *'analytical monitoring'* is necessary. In view of the corresponding cost involved in an automatic data acquisition system, this analytical monitoring will be restricted to PV installations with peak power above 5 kW.

In smaller installations a general assessment of the performance may be obtained from readings of suitable counters to determine the overall energy balance. This procedure is referred to as *'global monitoring'* . In some specific cases, anlytical monitoring can be applied for short periods to small systems in order to understand the behaviour of the installation, then revert to global monitoring.

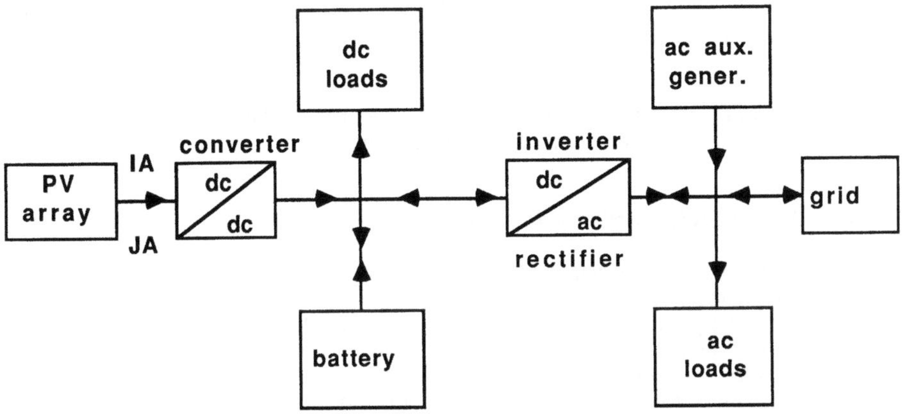

Figure1 : Schematic diagram of energy flow and measuring points [1]

Guideline A [1] presents the detailed lists of parameters that must be recorded in analytical and global monitoring respectively. A minimum set of indispensable parameters to be monitored is indicated in each case, as well as required accuracies. Recommendations for the selection of computer systems and data acquisition equipment are given and a format of records is proposed. It is not possible here to give all details of guidelines. To give just an example, table I presents the list of recorded parameters for global monitoring.

Parameter (Integrated values over monitoring steps of 1 day if possible)	Symbol	Unit
Irradiation, (global array plane)	IG	$kWh.m^{-2}$/day
Array output integrated current (total)	JA	Ah / day
Energy to all dc loads	ELDC	kWh / day
Energy to all ac loads	ELAC	kWh / day
Energy to (+) and from (-) grid	ER	kWh / day
Energy from auxiliary generator (e.g. diesel, wind turbine etc...)	EG	kWh / day
Hours load connected	TAV	hrs / day

Notes:

1. IG should be measured with a pyranometer or a reference device, and the type of apparatus clearly stated in the report. The electrical data should be determined to within +/- 2 % f.s.

2. A separate record of IG need not be made, provided it is demonstrated that the records from another recording station are representative of site conditions. However the accuracy in the data analysis is strongly reduced.

3. ELDC may be recorded as Ah if kWh recording is not appropriate.

Table I : Recorded parameters for global monitoring [1]

It is necessary to notice a very important point : footnotes in the lists of parameters give indication on the type of apparatus that should be used, and it seems necessary, for a correct treatment of collected information, to indicate with all sets of data the type of instrument used and the conditions of measurement. A suggested pro-forma for the covering note accompanying data is given in the guidelines. Log books can also provide valuable information, and a summary of sailent facts during the whole monitoring period must be added.

And this is one of the main lessons learned : records of data can be procesed correctly only if we know in what context they have been obtained.

2. DATA PROCESSING AND ANALYSIS

A second document on the analysis and the presentation of monitoring data has been discussed by our Working Group [2].

Data obtained by monitoring a PV installation need to be analysed and presented in forms suitable for several interested groups : research organisations and system designers, PV system suppliers, electricity supply utilities, users, general public and decision makers. In its final form, this information corresponds to diagrams and indices of performance that must have the same signification for everyone. It is then necessary to define unambiguous rules for the presentation of results. It is important to determine first :

- the amount of useful energy delivered by the system,
- the performance of the system as a whole,
- the performance of the principal sub-systems (obtained only in case of analytical monitoring).

The proposed document will set out standard procedures for the analysis and presentation of data. Definitions will be given for various Indices of Performance (or Figures of Merit). Some of them will be used as an example in the last part of this paper. Due to the difference in recorded data between global and analytical monitoring, two lists, giving the minimum set of energy quantities to be considered in each case, have been derived. They are defined for a given period. The main problems come then from the fact that, depending on the kind of indices of performance we want to determine, the treatment of the recorded data is not the same. For instance we do not need the same approach when comparing the output energy from a system with the meteo data (point of view of the user), than when performing the acceptance test of an array (point of view of the designer or owner who must know if the array is in conformity with data sheets).

As an illustration of treatment of data in case of global monitoring, we present below some results on a small installation in the french Alps.

In the frame of a 1985 CEC demonstration project, 22 high-mountain refuges of the Club Alpin Français (CAF) have been equiped with PV generators with powers ranging from 80 W_p to 480 W_p. All of them are between 2000 m and 3000 m altitude in the north and south of the Alpes and also in the Pyrénées. Data monitoring of such small disseminated installations was direcly bound to the budget, and for instance only three pyranometers were installed in the three typical climate regions mentionned above. However we know that micro-climate and albedo effects are very important in mountains, and extrapolation from one site to the next is made difficult.

When installation of measuring equipment occured, guidelines on monitoring were not finalized. As a consequence some parameters in the list of table I were not monitored such as 'time of load connected' or 'energy ouput from the array'. In a general way, we will see that it is difficult to strictly meet guidelines in actual practice in the case of global monitoring.

The first problem is the collection of data. Readings of counters were performed by the refuge keeper each day and written down on a piece of paper attached on the wall near the apparatus. Figure 2 presents the irradiation in the horizontal plane, recorded during the second half of august 1987. Daily values must be considered with care, since measurements were not taken at the same time each day, although asked to the refuge keeper. This is also valid for figure 4 which gives the daily consumption. The values are correct but don't correspond exactly to 24 hours. Since these refuges are used only during a short period during the year, it seams preferable to give diagrams over the whole season, and not over

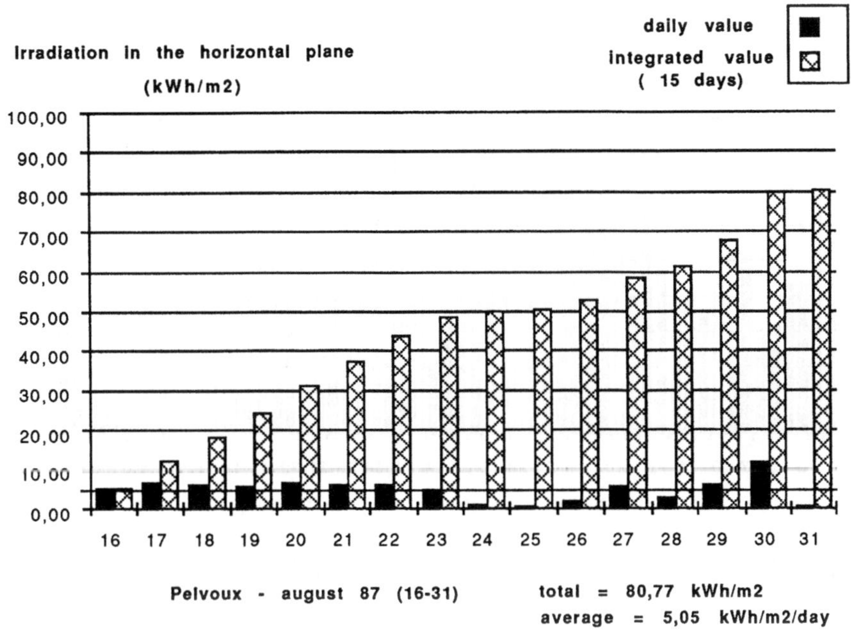

Figure 2

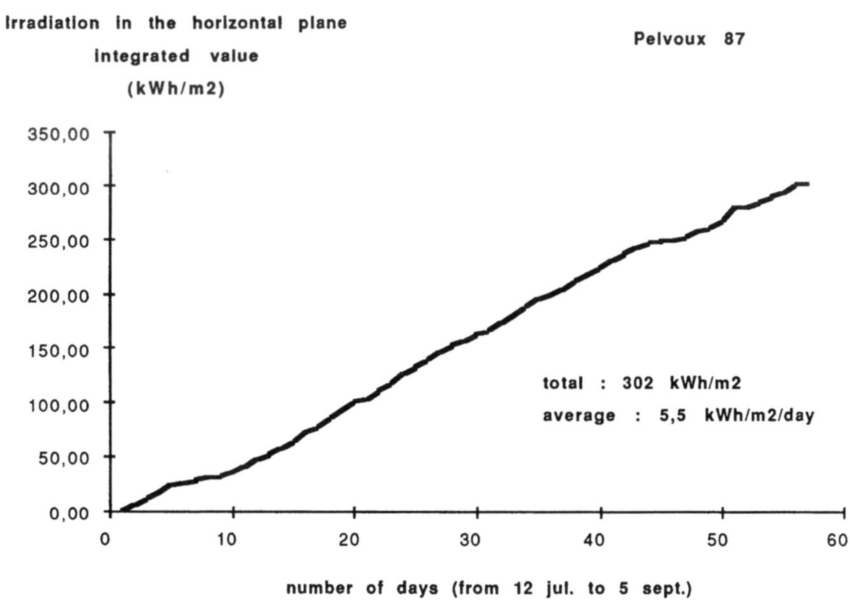

Figure 3

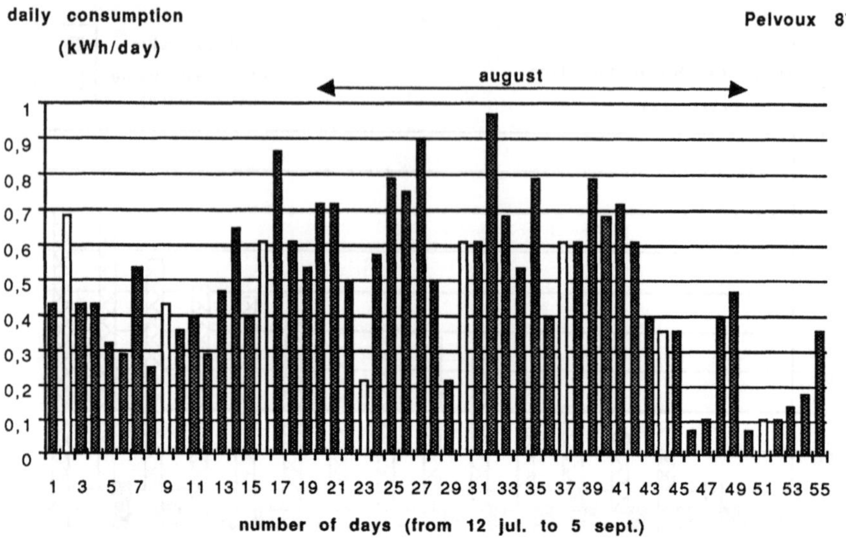

Figure 4

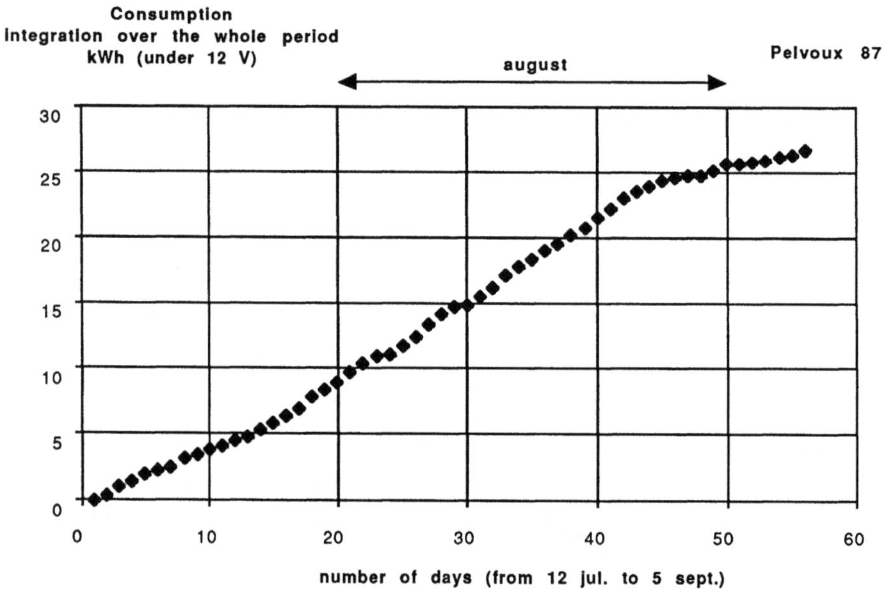

total : 26,7 kWh
average value over the season : 0,485 kWh/day

Figure 5

a month as proposed in the guidelines. This is not a very drastic divergence with the guidelines if clearly set in the presentation. Figures 3 and 5 show the integrated values, for irradiation and consumption respectively.

From these results some indices of performance may be derived. If they correspond to conditions different than stated in the guidelines, and especially if they use the same terminology, it is imperative to make them clear in the report. As an example in our case, we can define the following quantities :

overall efficiency = PV energy production to the load / irradiation measured in the horizontal plane on a surface equivalent to the surface of the modules,
= 3.7 %

It is important to note that, as indicated in the guidelines, the surface considered in the calculation includes the module frame. In our case the irradiation is measured in the horizontal plane, and extrapolation to array plane is made difficult since only daily values are available. It is then proposed to consider a in plane of the array measurement of the irradiation.

yield factor = mean daily PV energy production to the load per peak power,
= 2.42 $kWh/day/kW_p$

specific yield = mean daily PV energy production to the load per square metre of array,
= 0.2 $kWh/day/m^2$

specific area = array area for mean daily PV energy to the load of 1 kWh,
= 5 m^2

capacity factor = PV energy production to the load during a given period/ (nominal power * 24 * number of days in the period)
= 10 %

This last coefficient, derived from industrial electricity, must be used with care. Another capacity factor, defined on the day light time and not on 24 h, has been proposed. However this daylight period has to be computed.

It must be noted that all the above quantities defined to assess a given PV installation are strongly dependent on the energy transferred to the loads, e.g. the consumption.

CONCLUSION

Through the above example of global monitoring we have shown how data can be used to give diagrams and coefficients permitting a correct assessment of PV systems. In the case of analytical monitoring more information can be obtained , especially on sub-systems such as batteries, or inverters.

However assessment of PV installations may not be restricted to indices of performance. Reliability, availability and user satisfaction are as much important but rather difficult to quantify. The developed guidelines try to summarize the various procedures that could be implemented. In France a social inquiry as been conducted towards the inhabitants of the 40 solar houses, and is presented in this session of the conference.

On a general level, four factors are to be considered for the assessment of PV systems : technical factors, as in this presentation, economic factors, environmental factors, and social and institutional factors. A methodology for introducing all these factors in the assessment of PV systems has been proposed by M.R. Starr [3], who tries to give a weight coefficient to all of them. Such a procedure has now to be checked on actual installations.

References

[1] 'Guidelines for PV System Monitoring', Issue 3, CEC JRC-Ispra, Sector ESTI, I-21020 Ispra (Varese) Italy

[2] 'Guidelines for Assessment of PV Systems' draft, CEC JRC-Ispra

[3] M.R. STARR , 'Assessment of PV Plants : a Methodology', draft prepared from a report to JRC

PV RURAL ELECTRIFICATION IN THE EC ENERGY DEMONSTRATION PROGRAMME

Willi KAUT
Commission of the European Communities
Directorate-General for Energy
200 rue de la Loi, B - 1049 Brussels

Abstract

PV is very attractive and promising in rural electrification and this application plays an important role in the EC Energy Demonstration Programme.
A total of 63 PV Demonstration Projects with an installed capacity of 948 KWp are ongoing in the framework of this programme, 24 of which, representing 453 KWp, are projects in the field of rural electrification. After a short analysis of the different types of projects, the question of centralised or decentralised PV-plants and of OC or AC-plants are discussed. The importance of an adequate local distributors' and installers' infrastrucutre in PV as well as the training of future PV-users are underlined. Finally, the results of a study on the European market of rural electrification are summarised.

1. Introduction

PV is very attractive and promising in rural electrification. Consequently, this application plays an important role in the EC Energy Demonstration programme.
Under rural electrification, we include the electrification of isolated houses, groups of houses or small villages, which may serve as dwellings, agricultural exploitations, apline refuges and others. They are normally so far away from the public electric grid that the cost of grid connection by power lines is prohibitively high. Other solutions for electricity supply are a diesel generator or the use of renewable energies like wind, solar, small hydro or biogas plants.
A diesel generator pollutes the environment by exhaust-gas and noise; it needs fuel supply and much maintenance. Wind generators give no pollution, do not need fuel supply, but do need maintenance. Biogas plants need sufficient quantities of biomass and may be polluting. The best of the above-mentioned solutions seems to be small hydro, but here you need a continuous and sufficient water supply and a sufficient height. As almost all of the isolated houses without electricity are located in southern Europe with plenty of sunshine, PV solar elctricity is an ideal way of electrifying these sites. PV does not need any fuel supply, is not polluting and needs very, very little maintenance.

2. The PV-Demonstration Programme

Although the CEC Demonstration Programme in Energy started in 1978 as one means to cope with the problems caused by the oil crisis in the 70s, a real breakthrough of good PV-Demonstration projects happened only in 1983. Since then, and up to 1987, the CEC has received a total of 153 PV proposals, 73 of which have been selected with a total financial aid of 15.28 MECU.
As 10 projects have been cancelled over the years, 63 Pv-Demonstration projects are now still underway with a total installed capacity of 948.7 KWp, 24 of which, with a total installed capacity of 453 KWp are in the field of rural electrification.

	No. of projects	Inst. Capacity (KWp)	Power Range (KWp)
Single Houses	12	152	0.3 - 1.2
Small Villages	6	229	up to 35 (100)
Agricultural Exploitation	6	72	2 - 25
	--	---	
Total	24	453	

The 12 single house projects represent about 300 individual isolated houses or refuges. They are normally equipped with individual PV-installations at 12 or 24 V DC comprising some modules, a storage battery and a charge controller. As appliances some lamps, a radio or TV set and sometimes a DC-refrigerator are used. The size of these installations is between 300 and 1200 Wp with a dominance of smaller installations between 300 and 800 Wp. Plants of 1000-1200 Wp are often equipped with an inverter producing 220 VAC. In these power ranges more and more standardized systems are entering the market now, delivered as sets which can be easily installed in the do-it-yourself way.
The small village projects have one centralised PV-plant with energy distribution by line or cable. Normally, 220 VAC is produced.

3. Centralised and decentralised plants

Whilst for isolated single houses, an individual PV-plant must be used, there are two possible options given for groups of houses and small villages; either an individual plant with storage and control on each house or a central PV-plant with centralised storage and control and a distribution system to the different houses (see also end of point 2).
In the first case, the individual house owner has the possibility to manage his energy consumption according to the energy available in his own system and visible on his instruments. He also will be personally more engaged with his installation and he will keep it running and maintain it carefully. A centralised plant will certainly run at higher efficiencies. It permits a better use of the energy available in the whole system. Like in a big electric grid, a certain equalisation will occur due to different consumer habits and needs. But it also demands a high discipline and solidarity among the users. As the PV electricity is not available in unlimited quantities, instructions for the right manner of energy consumption must be established and the consumption limits set for each individual must be strictly observed. The non-observation of these limits may cause disturbances in the system and lead also to social tensions in the Community.
We have already very positive operating experience from a number of isolated individual PV-houses but still no experience from small villages. It is therefore still too early to draw final conclusions on this subject from demonstration projects. We believe, however, that in small communities and in cases of relatively low consumption, individual PV-plants on the houses are more attractive.

4. DC and AC Plants

Another point of discussion is whether in the case of individual installations at low power ranges (up to 1000-1200 Wp) DC, AC or mixed systems should be installed.

Without doubt, if only lamps, radio, TV and other DC appliances are used, DC is the most interesting solution. This system is simple, reliable, cheap and runs at good efficiencies as inverter losses are avoided. However, as a user gets more and more acquainted with electricity, his wishes and needs increase, he wants to use other appliances like flat-iron, electric tools, mixer, coffee-machine and even eventually a washing machine and dishwasher in the household. All these appliances hardly exist in DC-versions and, if so, they consume a lot of energy. Furthermore, they have rather short lifetimes as they mostly come from the camping market and are designed to be used during a 6-8 week vacation period per year but not for the whole year.
If the PV-plant is conceived as an AC-plant, then the internal consumption of the inverter (inverter losses) must be taken into account. Even if no electricity consuming equipment is switched on, the inverter consumes energy. Mixed systems allow to cope partially with this problem. The whole plant can be sub-divided into two sub-systems, a DC-system for lighting, radio etc., and an AC-system for the other appliances, which being AC-appliances should be low energy consuming. In such a system, the inverter normally will only be switched on when AC is demanded. The whole installation will be a little more complicated, but its advantages are evident.
Another possibility which might lead to simpler and cheaper DC-systems would be to encourage the industry to produce more good quality long-life DC-appliances (Ref 1.).

5. Dealers' and Installer's Infrastructure

A precondition of successful penetration of PV in rural electrification is the existence of an adequate infrastructure.
Small PV-plants in isolated sites should be installed by competent, qualified local installers, who after installing the system are available for the maintenance and also for fast intervention in case of failures of the systems. What must absolutely be avoided is that the people who live in the isolated houses, the owners and users of the PV-plants, feel lost when they run into trouble later on. They must have the feeling that in case of need, there is a specialist not too far away, who will give competent advice and rapid assistance.
Some good examples of that kind can be reported from Southern France and Spain (Ref. 2.).
The manufacturers of modules and associated PV-equipment should seriously consider the possibility of establishing a network of dealers, distributors and qualified installers. The methods employed in the field of domestic central heating boilers may provide an example.

6. Users Training

As most people living in the isolated sites have never lived with electricity, the use of PV electricity will bring fundamental changes to their way of life. In order to ensure a successful acceptance and an appropriate exploitation and use of the systems, the future owners need careful preparation and training. They must be informed about how to handle this new energy source and how to use the electric appliances and equipment. They must also learn how to adapt their consumptions to the availability of energy (hours of sunshine). They must learn that the availability of PV-electricity is limited, unlike in the case of a grid connection, and they must also learn to consume as much of the available energy as *possible in order to run the plant at a good utilisation factor. It is particularly difficult to make people understand this, especially in the case of people who are used to making economies wherever they can.*

7. Financing Aspects

In view of the continuing high cost of PV, it is quite clear that people living in isolated sites under very modest conditions normally do not have the income to afford the total financial cost of a PV-plant. Considering that it is a basic human right in today's society to have electricity available in one's dwelling, financing schemes must be found to give to all of "the less fortunate" people access to electricity at an acceptable cost.

Different approaches are to be noted in different countries. In some cases, the bodies which according to the national legislation are in charge of the electrification and more especially rural electrification (like the "Syndicats d'électrification rurale" in France) contribute to the purchase and installation of the systems and the owner of the PV-plant has to pay only 15% of the total cost. In this case, during later operation, the user has the electricity free and has only still to pay the maintenance and repair costs.

Another example comes from Italy, where the national utilitity, ENEL, finances the PV-plants, installs electricity counters in each house and the users pay the normal national tariffs. This encouraging example shows that utilities are also beginning to become interested in PV; this can also be reported from Spain.

8. Study on European Rural PV-Market

A recent study entitled "European market for local electricity production for isolated areas" shows that there are still about 1.2 Mio people occupying about 300,000 isolated houses, without electricity in the European Community. Based on different assumptions, the electrification of these houses represents a total market of more than 40 MWp in PV to be installed. This would be a total investment volume of about 735 MECU at a specific system cost of 15 ECU/Wp.

The realisation of such a tremendous programme could be an interesting challenge for the PV-industry (manufacturers and local infrastructure) as well as for utilities and other bodies in charge of electrification.

9. Conclusions

On the basis of the experience gained to date within the EC demonstration programme, small pv systems appear to be very well suited for use in rural electrification schemes in areas where houses are isolated and away from the existing grid.

More experience will come from the Programme during the next few years, but it is currently expected that the most important market will lie in relatively small systems 300-1200 Wp which are mainly used to power DC appliances. A substantial market for PV rural electrification is foreseen in the future (Ref. 3.)

REFERENCES

1. *The choice between DC and AC systems for PV Rural Electrification, W.B. Gillett, and W. Kaut, Proceedings of the 8th European Photovoltaic Solar Energy Conference, Florence, 9-13 May 1988*
2. *Proceedings of a workshop on "l'électrification photovoltaïque de 40 habitations isolées en France" held at Valbonne (France) on 22-23 October 1988.*
3. *European market for local electricity production for isolated areas, November 1987, Commission of the European Communities.*

PV - DEMONSTRATION - PROJECTS

		1983	1984	1985	1986	1987	TOTAL 1983/ 1987
Proposals received	-	29	34	27	27	36	153
Projects selected							
- number	-	10	16	18	13	16	73
- installed capacity	KWp	168.7	246	197	313	251.9	1176.6
Budget	Mio ECU	2.2	2.98	3.3	3.75	3.05	15.28
Projects cancelled							
- number	-	1	4	4	1	-	10
- installed capacity	KWp	21	149.5	26.5	26.0	-	223
Projects underway							
- number	-	9	12	14	12	16	63
- installed capacity	KWp	147.7	96.5	170.5	287	251.9	953.6

STAND-ALONE PHOTOVOLTAIC GENERATORS FOR RURAL ELECTRIFICATION IN FRANCE.

André Claverie, Patrick Coroller
French Agency for Energy Management (AFME)
Route des Lucioles, Sophia Antipolis, 06560 Valbonne, France

ABSTRACT. The specific actions undertaken by the AFME to stimulate the photovoltaic generators market in metropolitan France and her overseas territories are described. A steady socio-technical follow-up of 40 domestic stand-alone PV generators (400 to 1200 Wp) scattered in southern-France is bringing some preliminary quantitative and qualitative results. A majority of dwellers are satisfied by the rendered service but it is not easy to achieve an efficient management of this new source of electricity. In overseas territories 3500 isolated dwellings already possess a photovoltaic generator of a few hundred watts.

1. INTRODUCTION

The production of electricity by stand-alone photovoltaic (PV) generators has experienced steady expansion over the past ten years. During this period, the markets and the services rendered by photovoltaic cells have diversified and cover applications ranging from a few milliwatts to hundreds of kilowatts. Two technologies, cast multicrystalline silicon (0.18 mm-thick wafers) and thin film amorphous silicon have reached industrial stage in France.
The isolated dwellings occupy a choice position in France's strategy of market penetration. As an example, 3500 remote houses in overseas territories (Caribbean, South Pacific...) use this source of electricity. In metropolitan France, several thousand potential units are to be counted in the leisure habitat sphere.
A demonstration programme implemented by AFME, with the financial support of the Commission of the European Communities (CEC/DG 17-Energy) is exploring the different aspects relative to rural electrification by photovoltaic generators on a significant scale involving 40 permanent dwellings distributed throughout 12 departments in the south of France.

2. PERMANENT ISOLATED DWELLINGS IN METROPOLITAN FRANCE : CEC/DG17 AND AFME DEMONSTRATION PROGRAMME

The production of electricity by stand-alone photovoltaic generator presents, for dwellings not connected up to the mains network, an interesting option from technical and economic viewpoint when various problems stand in the way of classic electrification. This is particularly the case when constraints regarding environment protection or character of the relief (montainous areas in particular) make the costs of connection to the centralized mains network prohibitive.
Such a situation, rare as it is, is to be found in the South of France (Massif Central, Alps, Pyrenees...) where it is estimated that thousands of dwellings are not connected up to the Electricité de France (EDF) network.
The goal of the demonstration programme implemented by CEC/DG17 and AFME covers several aspects :
- sociological : improvement in the living conditions of users by the supply of a service superior to that given by generating diesel sets or other source (gas, oil...) ;
- technical : standardization of equipment and development of better suited domestic photovoltaic generators and particularly comparison between direct current and alternating current options ;
- administrative : adaptation of existing rural electrification procedures to renewable sources of energies ;
- economical : demonstration of the cost effectiveness of PV generator compared to grid connection on a macro-economic scale.
The cross section of the users in this programme covers a wide range of socio-economic status which is fine for the purpose of the study. Nevertheless low-income families living in rural areas are the main stream of potential PV-users.

In this programme the PV systems have been standardized by range of power :
- 400 Wp (24 V-dc, storage capacity : 8 kWh) : 19 houses;
- 800 Wp (24 V-dc, storage : 14 kWh) : 4 houses;
- 800 Wp (48 V/220V-ac, storage 17 kWh) : 7 houses ;
- 1200 Wp (48 V/220V-ac, storage 29 kWh) : 10 houses.

Both options either direct current systems or alternating current with inverter are investigated for comparison.

Crystalline silicon PV modules (consistent with french NF C 57 100 standard and CEC-ISPRA tested), 80% efficient-1kVA inverters (2 different types have been installed, their overload capability is twice the nominal power for a few minutes, sinusoidal waveform, stand-by consumption 4 W) and 350 Ah or 600 Ah lead-acid accumulators are used.

In order to prepare the dissemination of photovoltaics in the regions the systems have been set up by local electricians or small firms having experience or not in the field. They had the choice of the design of supporting structures and cabling but had to use the types of PV modules , the charge controller, the set of storage batteries , the inverters and measuring devices specified beforehand in the programme.

Each house is equipped with instrumentation devices specially developed for the programme. These devices are used for collecting the following parameters :
- Integrated input solar energy on the array plane;
- Global productivity of the PV generator ;
- Integrated energy called upon by the dweller ;
- Proportion of photovoltaic-generated power transiting through the storage battery ;
- I/P and O/P energy of inverter ;
- Frequency of use of diesel back-up set if any . A few dwellings were using one before the installation of a PV generator.

To these measurements, generalized over all the 40 dwellings, is to be added a detailed follow-up on a single house (n° 30, Gourdon in Nice's back country) . Data are sent by telephone network to the Scientific and Technical Centre for Building (CSTB) , then analysed and processed in order to improve the PV generator sizing software developed by ARMINES (both centres are based in Sophia Antipolis).

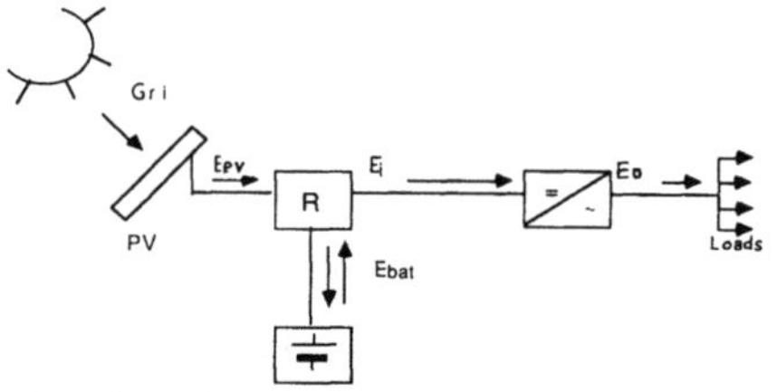

Fig. 1 : sketch of PV generator and measuring points.
Gri : global solar irradiation (Wh/m2)
S : total area of array (m2)
Epv : integrated output energy from PV array (Wh)
Ei, Eo : energy I/P & O/P inverter
Eu : used up energy (Wh)
T : duration of load use (h)

The total cost of the demonstration programme is 910 000 ECU breaking down in studies and follow-up 290 kECU and components 620 kECU.

PV modules count for 50 % of the system price, storage 12 %, power conditioning 18 to 28 %, installation 10 to 20 %.

The dwellers have paid around 20 % of the installed PV system. 40 % were brought in by CEC/DG17 and 40 % by the local section of rural electrification. Price of the installed PV system is in the range of 22-25 ECU/Wp.

A workshop held at AFME, Sophia-Antipolis, in october 1987 allowed users, installers and sponsors (CEC/DG17, local rural electrification boards and AFME) to meet and discuss their field experience. At time of commissioning and after 2 years or more of running the following has been noticed:
- modules : flawless, unchanged aspect ;
- structures : no corrosion but one structure was tilted by strong wind gust ;
- cabling : absence of protecting diodes on 3 installations (one local fitter is responsible), connection boxes not water-proof on some modules . Uncorrect cabling of electric meters on one installation ;
- charge controller : one regulator break-down on a 800Wp installation ;

- storage batteries : frost of battery due to deep discharge and bad insulation of the premises in the remote department of Lozère. Battery charging is getting difficult after deep discharge due to regulator design ;
- inverters : several failures on one type (out of two) of inverters due to lightning ;
- instrumentation : failures of a number of amperehourmeters due to over-charges and interferences caused by inverters. Pyranometer : difficulty of replacing storage batteries (6 to 12 months autonomy). Low capacity of numerical counter.

Early technical problems due to faulty equipments at the beginning of the operation have been sorted out.

2.1. Preliminary results

Users are asked to fill-in weekly data sheets after reading the various electric meters. A questionnaire (once a year) on the appreciation of the running of the generator and the problems encountered is also completed. The technical follow-up will be carried out for 2 years. The ISPRA Research Centre of the Commission of European Community is involved in the analysis of the data.

These analysis are heavy to handle. Many data have been lost due to faulty meters or careless users. Today, less than 50 % of dwellers are filling-in the data sheets.

Technical data and general information from 15 months of running have been obtained . On all the diagrams presented below (fig. 2,3,4) the mean available energy comes from solar irradiation measured at each site and considering a mean efficiency of the system of 4 % - a value confirmed by the detailed set of data acquirred in Gourdon house (see fig. 6).

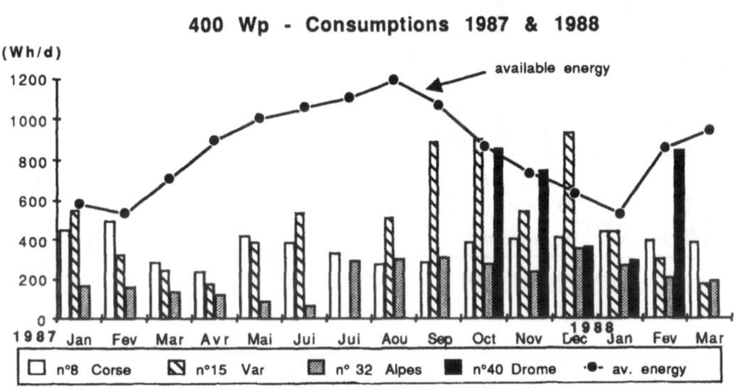

Fig. 2 : sampling of 4 typical houses equipped with a 400 Wp, 24 V-dc generator. Consumption pattern vary from a house to another . Energy used up is roughly constant during the whole year except in house n° 15 where an agricultural activity takes place. In the house n° 32 a step in consumption in July is related to the acquisition of a PV 150-litre refrigerator (250 Wh/day).The average comsumption is 300 Wh/day when 800 Wh /day is available. House n°15 in Var increased the number of equipments in september 1987, making a better use of the energy avalaible in winter. House n° 8 in Corsica is a family of 4 persons not making the best of their generator. Excess of power is available in summer season . Some dwellers use it up for pumping water for watering their garden.

Fig. 3 (next page on top) : typical house using a 800 Wp (24 V-dc) generator. House n° 23 is situated in Lozère a very cold area in Massif Central. Energy consumption is constant during the year : around 1kWh/day . This is very close to the figure that was estimated in wintertime at the design stage. Critical low threshold of the charge controller was reached a few times in winter. The user would like to have more power in winter time.

- Most of the 400 Wp (dc) and 800 Wp (dc) users are satisfied. Their electricity needs are basically lighting (around 10 lighting points) and television ; a small 300 to 400 VA inverter can be used for specific use : sewing machine , blender ... Average energy comsumption represents 50 to 80% of the available energy output of the PV system in winter and 30 to 60 % in summertime.

Fig.3 - **800 Wp-dc ; Consumptions 1987 & 1988**

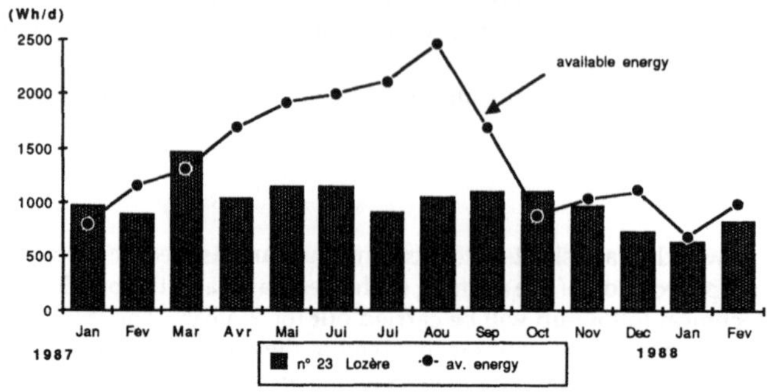

Fig. 4 - **800 Wp-ac ; Consumptions 1987 & 1988**

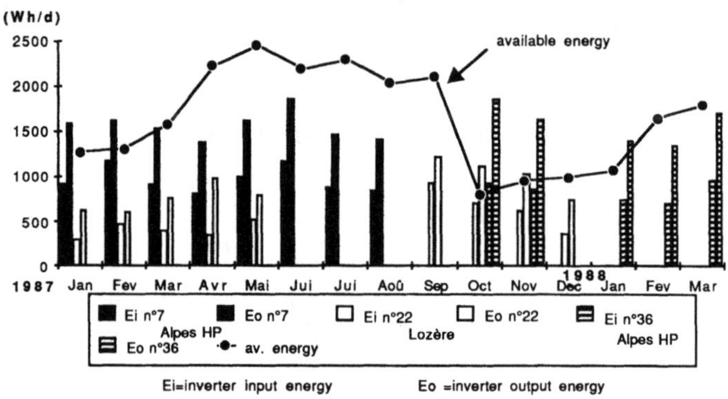

Fig. 4 : 4 installations 800 Wp (24V/220V ac) using a 1kVA inverter. Energy input (Ei) and energy output (Eo) of the inverter is plotted for each case. Poor efficiency of the inverter (50 to 60%) is shown in most of the systems. In some cases , houses n° 7 and n°36 , going through winter is difficult. Users are not really satisfied. Lighting and switching on a radio set in the morning makes the inverter working at a poor efficiency (45% at 50 VA). The system in this range of power is not optimized. If we compare 800 Wp (ac) to 800 Wp (dc) there is an overcost of 4 ECU/Wp that is to say 3200 ECU for this system and more than 30 % energy loss due to poor efficiency of the inverter.

- 800 Wp (ac) and 1200 Wp (ac) dwellers use up more energy. 60 % are satisfied. Remaining 40 % owning a deep freezer, several appliances, a washing machine etc... find it difficult to rely on their PV generator. The users are tempted to use any conventionnal 220V refrigerator or washing machine not really suited for their power limited PV generator : a source of frustration when the user compare their system to the service provided by the grid. In this range of power the users want to get more and more energy as they get acquainted with their new source of electricity.

Fig. 5 and 6 (next page) : House n° 30 in Gourdon, Alpes Maritimes. A 1280Wp-ac generator. The family runs a sheep farm. Average consumption is 2 kWh/day. The inverter efficiency is 70 to 75 %. In January and April more energy was used up than produced by the array.
Fig. 6 : Gourdon house. Mean global system efficiency (Rgi) before inverter is around 4 %. Global capacity factor (Kcgo) of the ac system is around 5 %. As a rule of thumb we can deduce that 1 Watt of permanent 24h-power is provided by a PV generator of 15 Wp in south of France.

Fig. 5 - **1280 Wp-ac ; Gourdon**

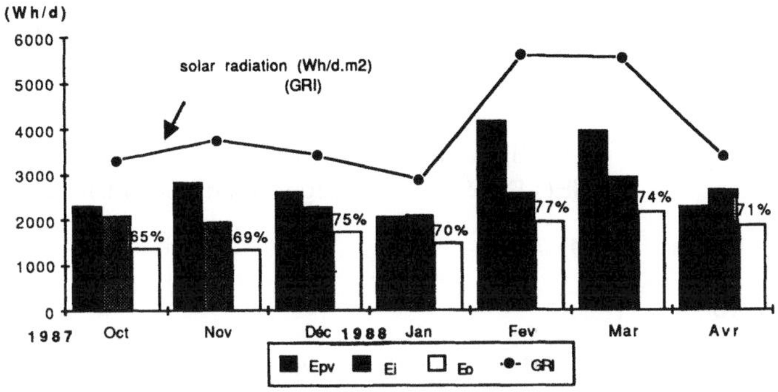

Fig. 6 - **1280 Wp - ac ; Gourdon PV generator efficiencies**

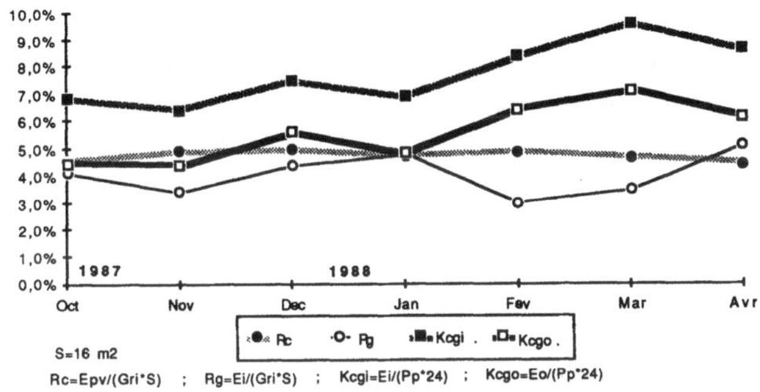

3. CONCLUSIVE REMARKS

- It has been possible to compare two options : direct current and alternating current systems.

It appears today that PV-dc systems in the range 100- 500 watts fulfill satisfactorily the basic needs such as lighting , radio, TV and small household appliances.

For PV systems in the range of 1 kWp it is suggested that basic needs should be supplied by 12/24V direct current and specific use (washing machine, vacuum cleaner ...) should be supplied through a power adapted 220V (ac) inverter. In this case 2 cabling networks are necessary in the homes.

- A thermal diesel generator as a back-up is often requested by users.
- System sizing is not easy at design stage because the actual energy needs are not well assessed by the dweller. It seems necessary to oversize the installation to allow a certain degree of improvement of user's comfort ;
- The PV system have to be carefully maintained . Help of a technician is often requested . Training is necessary and some courses are already taking place. A comprehensive operating, and maintenance manual must be available. Maintenance and handling of storage batteries must not be overlooked. Often requested is an indicator of the state of charge of the storage battery ;
- Low-income families (the main stream of potential PV users) complain that 12/24V appliances are not readily available and are expensive.

Several factors limit the large-scale dissemination of photovoltaic systems : high costs of stand-alone PV generators and lack of general information on the capabilities of the technology. The AFME photovoltaic programme as well as those of CEC/DG17 and CEC/DG12 pay particular attention to reducing these limiting factors. These actions should make it possible to reinforce the confidence of potential users in industrialized as well as in developing countries.

PV PLANTS FOR ALPINE HUTS: INSTALLATION AND OPERATING EXPERIENCE AT SEVEN ENEL PLANTS

G. Belli, A. Iliceto, A. Previ
ENEL-DSR-CREI-Cologno Monzese (MI)-Italy

SUMMARY

The problem of supplying electric power to isolated users far from the electricity distribution grid is one of general interest. Such consumers are nowadays generally supplied with electricity produced by small diesel generator-sets, and only recently have photovoltaic arrays and wind-turbines offered an alternative to the internal combustion engine.
ENEL, as a State-owned electricity utility, is interested in the development of this particular application of renewable energy sources. Enlarging a low-voltage distribution network to connect consumers whose power requirements are extremely low ($\sim$ 1000 kWh/year) may, in certain conditions, be uneconomical, both for the utility, which has to absorb most of the expense involved in construction and maintenance, and for the consumer himself. The paper reports the design criteria, the tests and the problems encountered in electrifying seven alpine huts belonging to CAI.

1. THE "CASE SPARSE" (SCATTERED DWELLINGS) PROJECT

Small photovoltaic generators (with power of a few kW) are undergoing rapid technological change that is making them both more reliable and cheaper. However, before launching into industrial applications of this new source, ENEL deemed it advisable to conduct research on its potential production and reliability and - an extremely important point - on how far it was acceptable to the public and to users. The Scattered Dwellings Project is designed to find these things out. (1) (2)

Before trying out applications with private houses, the project provides for a phase of experiments involving consumers related to the interests of the general public, such as the Alpine shelters of the "Club Alpino Italiano".

These shelters are ideal for experimental work, and especially for demonstrating the use of PV conversion: their load demand is usually below 2000 kWh/year, they are used by the public, they are isolated, and the cost of wiring them to the electricity grid would, in many cases, be prohibitive. To date, 7 shelters have been electrified. (Table I)

2. DESCRIPTION OF PLAN

The plants have been supplied by Italian manufacturers according to ENEL's specifications, except for that of the Montanaro shelter, which was entirely designed by ENEL. On-site assembly was carried out by ENEL departments that normally operate in the areas where the shelters are located.

Let us now take a look at some details of the equipment and its installation.

3. THE CONSUMER

In most cases, the units are 24 V d.c. (for lighting and refrigerators). In one case only (the Pacini shelter), the plant is 220 V a.c. In the latter case, because of the considerable distance between the building and the PV array, it was decided to convert the direct current generated by the modules to alternating current by means of a 24 V d.c./220 V a.c. inverter.

In this way, the power lost through transmission was greatly reduced. No-load inverter losses have been eliminated, since the inverter is connected up with a load-sensor.

Many of the shelters have been provided with small inverters, to enable staff to use appliances that normally have a.c. motors (slicers, drills, mixers, hair-dryers, etc.). Outside all the shelters, a special yellow light has been installed, to make it easier to spot the building at night or in fog.

4. BATTERIES

The Pacini shelter has Ni-Cd batteries, while all the others have low self-discharging Pb-lead acid batteries, made in Italy.

The decision to instal conventional Pb-acid batteries was taken after laboratory experiments to investigate the performance of these accumulators at low temperatures (down to -20°C). The test showed that, after a series of charging and discharging cycles at low temperature, conventional batteries keep up their initial performance. On the other hand the sealed batteries available on the market at the time of the experiments lost part of their initial capacity, which has been measured under standard conditions.

5. PHOTOVOLTAIC MODULES AND SUPPORT STRUCTURES

Mono- and semi-crystalline silicon modules, made in Italy, were used for the arrays. Particular care was taken over the siting of the latter, for both technical and aesthetic reasons. Installation on the roof, which is certainly preferable from the environmental point of view, was possible in two cases out of 7. (fig. 1)

6. REGULATION OF THE BATTERY CHARGE

The problem of optimal regulation of the battery charge in small photovoltaic plants remains an open question. A good regulator needs to offer a large number of advantages: - Reliability; - Simplicity of manufacture; - The ability to transfer energy from the photovoltaic array to the battery and towards the load with the maximum efficiency; - The ability to keep up battery capacity even during outages; - Low cost.

The regulators used in Alpine shelters are based on various structural principles, and the fundamental designs used are shown in fig. 2.

There are regulators in which the PV array is either completely connected or completely disconnected, or in which there is at least one level of partial utilization. A chopper, which was used in only one case, is certainly the most expensive, but is probably the one likely to ensure the longest battery life. All regulating systems use measurement of the battery's voltage for purposes of feed-back.

ENEL has decided to accept various types of regulator, in order to assess the efficiency of the various types available on the market. So far, there are no technical criteria that would justify the choice of one type rather than another. The choice generally depends on cost considerations, but the designer of PV installation must be very careful when special conditions occur.

Experimental work conducted by ENEL has shown that, in particular conditions (when the peak current of the photovoltaic array is about twice the I_{10} discharge current of the battery, and the daytime load demand is relatively low compared with peak current), the state of charge of the battery deteriorates as compared with the initial value, before finally settling down. The amount of the deterioration is dependent on the type of regulator used: the worst situation is found with a simple ON-OFF device. The situation improves if stepped regulators or a chopper are used.

7. EXPERIMENTAL RESULTS

All the plants are equipped with a system for measuring the amount of energy consumed. The meters are read periodically by shelter staff, who send the figures to ENEL. Table II shows some of the preliminary results.

8. PROBLEMS ENCOUNTERED

The technical problems encountered during the year 1986-87 were small ones, and the plants operated regularly.

However, it is worth mentioning the following points: - A relatively high fault rate in respect of inverters for fluorescent lamps (in some cases, up to 20%); - Oxidation on one of the terminal panels, resulting in interruption of the supply from a PV semi-array; - Penetration by snow of a battery-container, which led to temporary shut-down of the plant; - Faults in some of the electronic-cards of a charge regulator due to a lightning strike near the shelter; - Fault due to overvoltage in the load sensor of the inverter at the Pacini shelter; - Fault in an dedicated inverter; - Polarity inversion of a battery cell (during acceptance tests).

9. CONCLUSIONS

ENEL's experience has been successful, and it is to be hoped that CAI will consider the use of PV conversion for supplying electricity to mountain shelters in Italy. Certainly, the reaction of the public and of shelter staff has been enthusiastic. Undoubtedly, ENEL's Scattered Dwellings Project has greatly benefitted by this experience.

It is felt that the main question still outstanding is that of battery charge regulation or, more generally, that of the best way of coupling the battery to the photovoltaic array.

ACKNOWLEDGMENTS

The authors wish to thank ENEL local departments (Zona di Sondrio, Aosta, Reggio Emilia, Prato and Pistoia) for their valid assistance during the construction of the plants. A special thank is addressed to Mario Manini for his enthusiastic co-operation.

REFERENCES

(1) Previ A. and Taschini A., Photovoltaic plants as a means of supplying electricity to isolated dwellings. 5th EC Photovoltaic Solar Energy Conference, Kavouri (Athens) Oct. 1983, pp. 274-280.

(2) Belli G. and Iliceto A., The photovoltaic plants of Orbetello and Zannone. 7th EC Photovoltaic Solar Energy Conference, Sevilla Oct. 1986, pp. 245-249.

Fig. 1

a – Longoni hut
b – Montanaro resort
c – Mezzalama hut
d – Carlo Emilio resort
e – Battisti hut

TABLE I "CASE SPARSE" Project – Main features of the plants constructed by ENEL in the Alps and Apennines

Name, location and owner of the plant	Altitude (masl)	Latitude	PV ARRAY Peak power (kW)	N.° and size of the mod. (mm)	Type of cells	Tilt	E. ch. storage (kWh)	Six months' pv produc. 1/5 ÷ 30/10 (kWh)	Six months' pv produc. 1/11 ÷ 30/4 (kWh)
Mezzalama hut (AO) CAI Torino	3036	45,8°	0.35	10 (1300x350)	semicrystal.	adjustable	8.4	306	184
Carlo Emilio resort (SO) CAI Como	2140	46,2°	0.28	8 (1300x350)	semicrystal.	adjustable	6	245	147
Ponti hut (SO) CAI Milano	2559	46,2°	1.12	32 (1300x350)	semicrystal.	adjustable	24	981	590
Longoni hut (SO) CAI Seregno	2417	46,2°	0.9	24 (1041x405)	singlecrystal.	fixed 30°	26.4	728	428
Battisti hut (RE) CAI Reggio Emilia	1761	44,2°	0.7	20 (985x460)	semicrystal.	adjustable	12.6	640	401
Montanaro resort (PT) CAI Maresca	1567	44°	0.7	16 (1310x340)	singlecrystal.	fixed 60°	7.2	587	411
Pacini hut (PI) CAI Prato	1001	44°	0.8	20 (1067x445)	semicrystal.	adjustable	14	724	440
								Tilt = 30°	Tilt = 60°

TABLE II

HUT	DAILY PV PRODUCTION (SUMMER) (kWh / day)
Longoni	1.1
Battisti	1.32
Ponti	5.1
Montanaro	1.75
Pacini	2.9
Mezzalama	*
Carlo Emilio	*

* Not yet available

CHARGE CONTROL DEVICES FOR STORAGE BATTERIES INSTALLED IN SMALL STAND–ALONE PV SYSTEMS

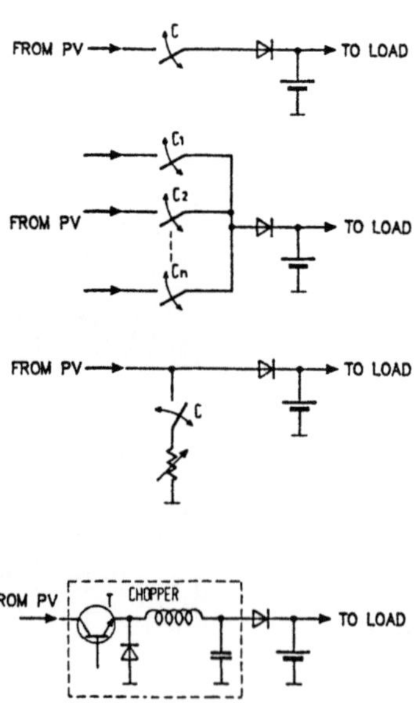

C SWITCHING DEVICE
TYPE – electromechanical
– transistor or FET
CONTROL LOGIC – ON/OFF of C following pre–established thresholds of the battery voltage (suitable delays are built–in)

T SWITCHING TRANSISTOR
CONTROL LOGIC – constant output voltage is maintained through suitable switching of T

Fig. 2

PERFORMANCE of PHOTOVOLTAIC GENERATORS of 12 REMOTE LODGINGS located in the FRENCH ALPS

by A.BONTEMPS,C.GOUBIER,C.MARQUET and J.C.SOLECKI
A.P.R.E.N. - GRENOBLE - FRANCE

Summary

Electrification, by means of photovoltaic solars cells, of 12 lodgings and huts in the mountains, regularly scattered along the Great Crossing of the French Alps, from lake Leman at Geneva to the Mediterranean, was done under contract from the Commission of European Communities and the Agence Francaise pour la Maitrise de l'Energie.

At each lodging, energy meters were used to measure the produced and consumed electrical energy. These measurements, together with detailed measurements from two installations described below, are used : (a) to verify that the present installations are correctly designed (b) to develop design criteria for future installations.

1.INTRODUCTION

Twelve remote lodgings located in the French Alps were equipped with photovoltaic generators which supply electrical energy for communication equipment such as radio emitters and also for lighting and others household appliances. The altitude of the concerned lodgings varies from 900 m to 3000 m. This electrification has several objectives :

- to use an energy source avoiding transport of fuel.
- to test solar plants in severe climatic conditions on a large scale and therefore to prepare a next stage of distribution.
- to adapt the lodging to the sociological evolution of practices of sport life in a mountain environment.
- to reduce lodging keepers' loneliness, and improve their working conditions.

At each lodging, we have been measuring by means of energy meters, the produced and consumed energy during two years. Two of the lodgings were equipped with monitoring stations in order to obtain detailed measurements of :

- solar irradiance
- PV panel temperature
- ambient temperature
- produced energy
- energy received or delivered by batteries
- consumed energy

2.TECHNICAL PRESENTATION

2.1-List of lodgings :

A list of lodgings with the number of panels, regulators and batteries is given below. In fact, 13 lodgings were equipped because some foreseen equipment were reduced at some lodgings.

Name	Commune	Dep	Altit	Pan	Reg	Bat
BOSSETAN	SAMOENS	74	1763	15	5	15
ANTERNE	SIXT	74	1807	7	3	7
LA LEISSE	TERMIGNON	73	2496	10	4	10
FOND DES FOURS	VAL D'ISERE	73	2537	7	3	7
EMPARIS	MIZOEN	38	2260	12	4	12
SOREILLER	St CHRISTOPHE	38	2780	15	5	15
CHARDONNET	NEVACHE	05	2219	7	3	7
CERVIERES	BRIANCON	05	2035	9	3	9
LA PARE	BARCELONETTE	04	1863	7	3	7
VALBELLE	SISTERON	04	917	12	4	12
LONGON	ROURE	06	1890	12	4	12
NICE	BELVEDERE	06	2232	15	5	15
SPEGGI	TENDE	06	1300	7	3	7

For local meteorological reasons, some lodgings were equipped with two orientations : some of the panels oriented to the South; the others to the East or West.

2.2 - Principle of generators :
For reliability reasons and taking account of maintenance problems, photovoltaic plants were divided into separate modules according to the figure 1

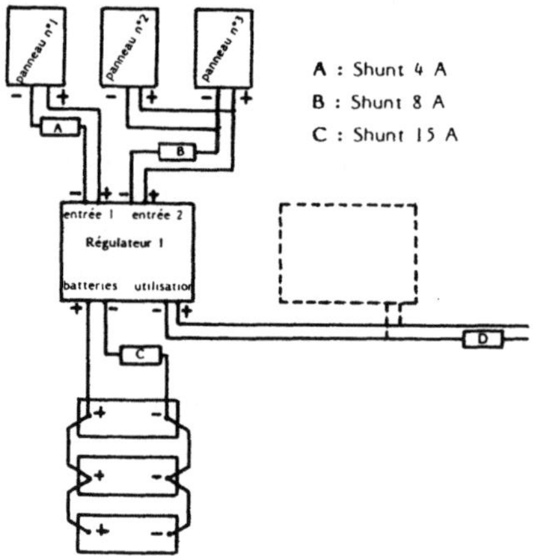

fig 1
Wiring Diagram of Panels, Regulators and Batteries

2.3 - Energy meters :

In accordance with the prescription contract, lodgings are equipped with energy meters. they have measured during 2 years : - Energy consumed by users
_ Energy displayed by panels

Displayed energy is measured on one PV panel for lodgings having one orientation and one panel of each orientation for lodgings having several orientations. The resulting displayed energy is the sum of the displayed energy of each panel multiplied by the number of panels of each orientation. Since the regulating system was sub-divided, it was not possible to have exactly the same level on each regulator. As a result, there was an uncertainty in the displayed energy.

2.4 - Monitoring station

Two lodgings - BOSSETAN and VALBELLE - were equipped with a monitoring station. A schematic lay-out is given in figure 2

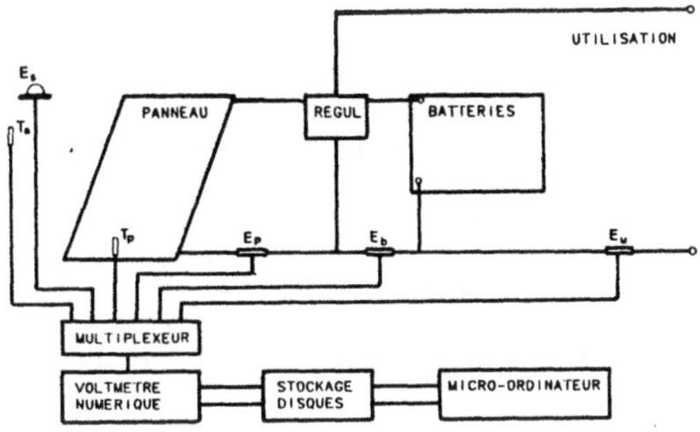

fig 2
Monitoring station

The measurement station is powered with 12 V DC delivered by the PV plant and all devices are placed in a hand-bag, thermally insulated and conditionned. Insolation measurement are made each 10 minutes in accordance with CEE recommendations. The frequency of recording other parameters is 1 minute with integration during one hour. Data are stored on discs and sent every 3 months for calculations and result analysis.

3.RESULTS and DISCUSSION :

3.1 - Meter results :

Lodging	Pan nb	Daily estim. Wh	Consumed energy 1986	Daily mean Wh	Consumed energy 1987	Daily mean Wh
BOSSETAN	15	1300	35 970	353	44 840	273
ANTERNE	7	825	19 610	228		
SOREILLER	15	1300	24 880	592		
LA PARE	7	825	5 860	293	11 350	158
NICE	15	1300	199 470	1194	60 770	921
EMPARIS	12	1050	13 270	218	20 370	251
CHARDONNET	7	825			4 540	189
LA LEISSE	10	950			23 270	423
FOND DES F.	7	825			26 330	239
CERVIERES	9	950			8 184	248
VALBELLE	12	1050			22 700	201
SPEGGI	7	825			19 010	308

These results show large differences in the way of using energy between lodging keepers. Some of them economize much energy. In all cases(NICE excepted), the consumed energy is lower than predicted but all lodgings are not yet fully equipped with electric accessories and it seems that keepers dare not use too much energy. For some lodgings, the occupation rate was less than predicted for various reasons. Comparing results measured by means of meters and monitoring stations on the two concerned lodgings (BOSSETAN and VALBELLE), we observed that energy meters do not measure low consumption due to a threshold in electro-mechanical devices. On VALBELLE, the difference between the two measurements was about 15%.

3.2 - Monitoring station results : Detailed measurements on BOSSETAN and VALBELLE show that the uncertainty due to the measurement realized with only one panel was greater than foreseen. Consequently, we have equipped the monitoring station of VALBELLE to measure produced energy on each panel and each group of 3 batteries. It's possible to see on figure 5 what happened on a typical day for panel and battery energies. For the same day, we have plotted insolation, total produced energy and total consumed energy in figure 3. In addition, the daily produced, consumed and battery energy during 72 days are shown in figure 4.

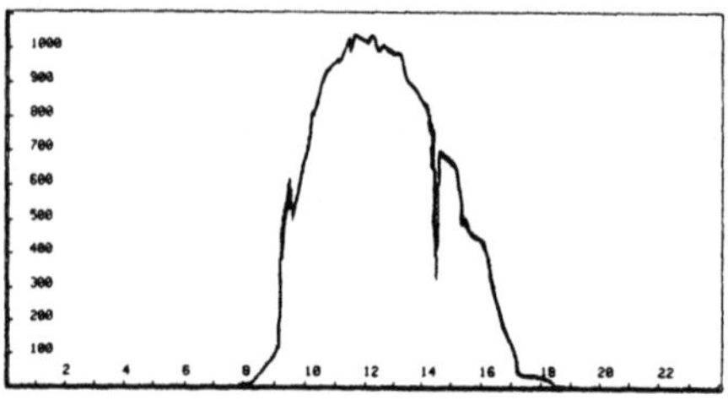

(a) Solar irradiation (W/m2)

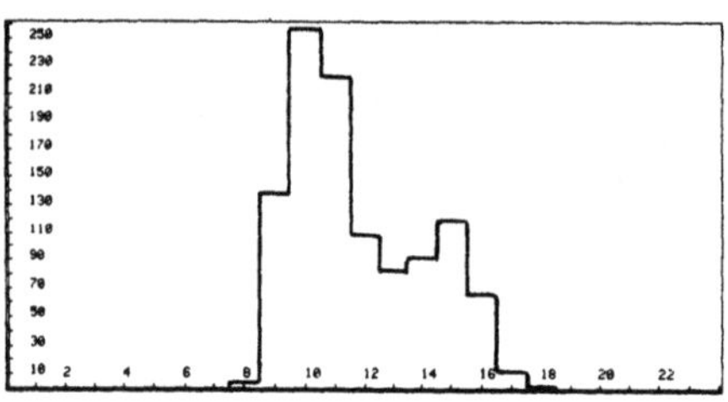

(b) Energy production (Wh)

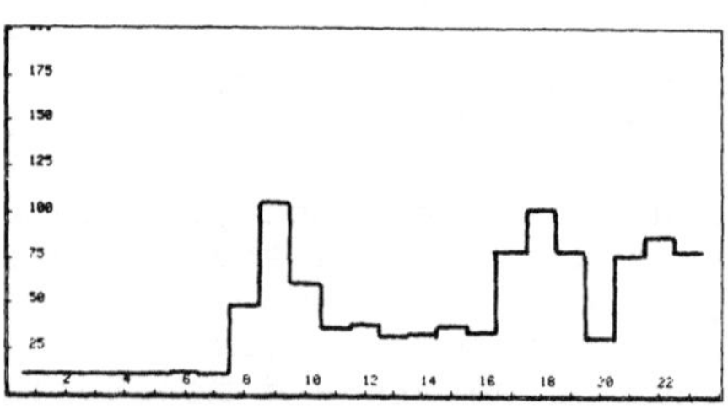

(c) Energy consumption (Wh)

fig 3

Comparison of solar produced and consumed energies.

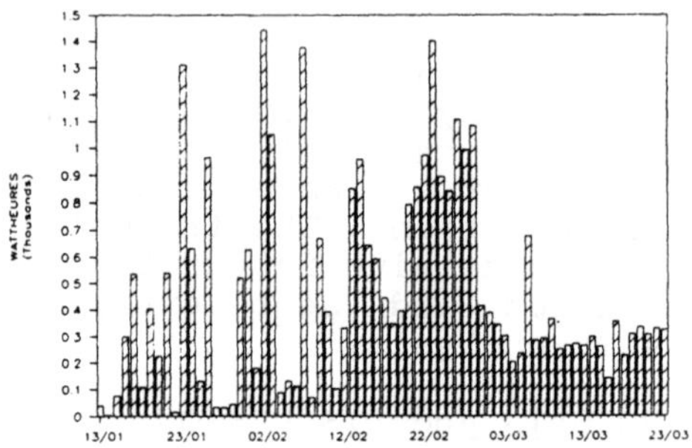

(4a)

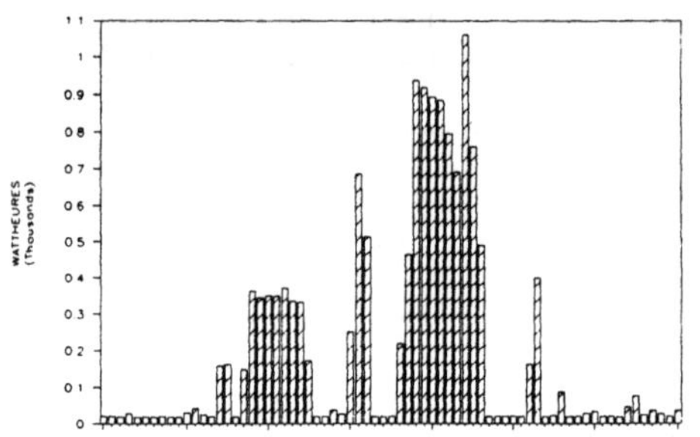

(4b)

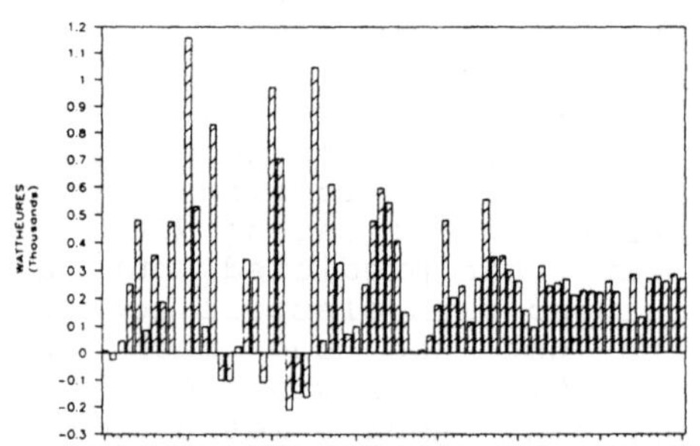

(4c)

fig 4

Daily produced (4a),consumed (4b) and stored energy (4c) during the 1rst quarter 1988

As an example of the influence of the regulator threshold, it is possible to see on the following figures (figure 5a and 5b) what happened on the same day between panels and batteries. Some panels stop quickly after a small producing period following sunrise, but the others have a sufficient production to charge all batteries. This has been observed very often on VALBELLE when there was sufficient insolation.

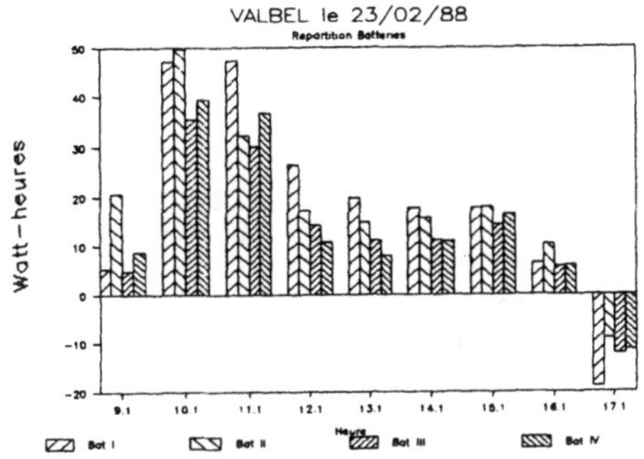

(a) Energy stored in batteries

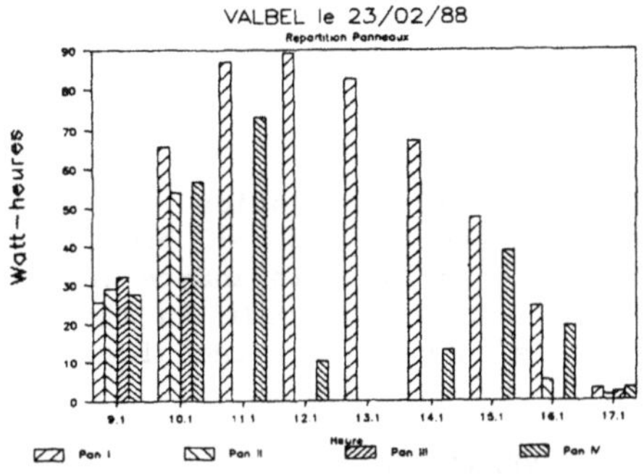

(b) Energy produced by the PV panels

fig 5

Daily stored and produced energy by the four different modules of a lodging

4.CONCLUSION

Thirteen lodgings located in the French Alps were equipped with photovoltaic generators. Produced and consumed energy was measured on each lodging with energy meters (Global measurements). Moreover, two generators were equipped with monitoring stations. From both measurements, the main conclusions are : - in some lodgings, the daily energy production was found to be the same for panels facing East and those facing South (due to the presence of afternoon clouds in mountain regions).

- it was noticed that energy meters generally used in photovoltaic systems have a threshold whith results in an energy reading about fifteen percent less.

- identical installations have shown widely different energy consumption rates which are attributed to the consumer behaviour pattern which should be taken into account in developing guidelines for energy management.

- Dividing the PV generator into separate modules allows one to obtain energy even in the case of an accidental failure and makes maintenance easier.

OPTIMIZATION OF A PHOTOVOLTAIC-DIESEL ENGINE GENERATOR HYBRID SYSTEM FOR A TELECOMMUNICATIONS MICROWAVE RELAY IN FRANCE

Marc AUBREE
FRANCE TELECOM - CNET/PAB/STC/ESA
38-40, rue du Général Leclerc
92131 ISSY-LES-MOULINEAUX
FRANCE

ABSTRACT. The combination of a PV system and a diesel engine generator allows the power supply of big remote telecom stations with an advantageous cost, with a low maintenance and a good reliability.

A concrete example in the South of France, using a data processing supervision shows the technical and economical interest of that system.

1. INTRODUCTION

During the definition of a power plant for a remote telecom station, a first step consists to verify the feasibility of a single PV system (PVS). For the big telecom stations of which the power consumption is over than 500 W, the high investment cost often make dissuasive the installation of that system ; then, a solution is to combine the PVS with a diesel engine generator (DEG).

With subsidies of the European Commission Community, the National Telecommunications Laboratory (CNET) of FRANCE TELECOM has experimented such a hybrid system on the "Gros Bessillon" microwave relay in the South East of France near Toulon.

The sizing follows the CCITT* recommandations.

2. ADVANTAGES OF THE PV/DEG HYBRID SYSTEM

The combination of the two power sources leads to a reduction of the disadvantages of each one used individually and increases the overall reliability :

- the presence of the DEG allows not to take into account only the most unfavourable solar data for the PVS sizing, but the mean yearly solar radiation.
- the presence of the PVS allows the reduction of the working time of the DEG, the fuel consumption and the number of maintenance visits.

Finally, the optimization of the hybrid system leads to the reduction of the PVS still keeping a low maintenance on the DEG.

The energy provided by the DEG is between 10 and 40 % in order to have a significant reduction of the overall costs.

Lower values will change the DEG in a simple emergency source and higher values will change the PVS in a fuel economizer.

* CCITT : "Comité Consultatif International pour le Télégraphe et le Téléphone" of the International Telecommunications Union set in GENEVA.

3. THE "GROS BESSILLON" STATION FEATURES

3.1. Telecom system and site

The telecom system is a microwave relay used for the telephone routing of several small cities and their surroundings.

The mean power consumption is 600 W/48 V.

The site is located at the altitude of 810 meters, the climate is a mediterranean type with short but hard conditions during the bad season. The access is possible by a track.

The site, frequently suffers of local thunders and lightning.

3.2. Description (fig. 1)

The hybrid system is composed of a phovoltaic generator (PVG), lightning protections (LP), a storage battery (B), a diesel engine generator (DEG), a battery control unit (BCU), a rectifier (RE), a supervision unit (US).

Owing to the energy balance, the regulation device is not necessary.

3.3. Block diagram

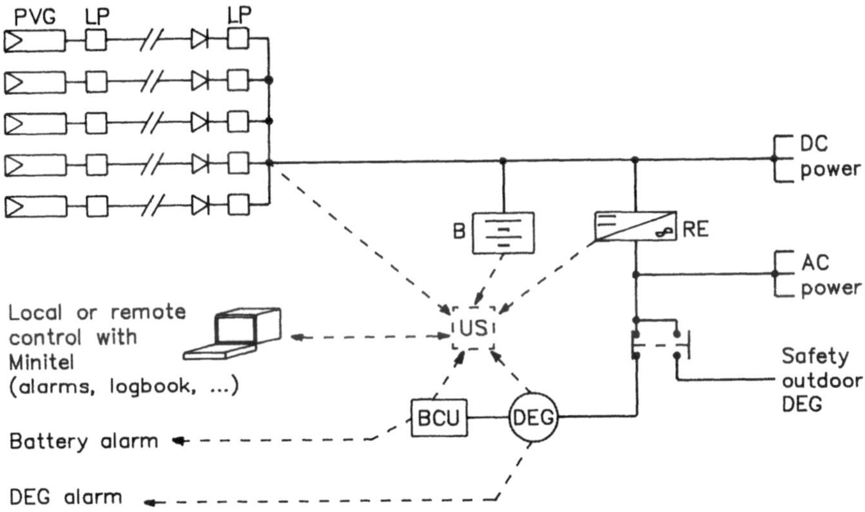

figure 1

3.4. Principle of operation

When there is sufficient solar radiation the microwave relay is supplied by the PVG and the battery.

During the night or when the weather is bad, the battery provides the energy, but only for a limited time.

If the bad period goes on, the DEG automatically starts on a voltage discharge threshold provided by the BCU. Then the DEG supplies the microwave relay through the rectifier.

The charge of the battery must be the fastest as possible (10 hours maxi) and limited to 90 % of the full charge in order to reduce the working time of the DEG.

The DEG stops with the indication of the BCU (voltage threshold or timing condition).

In case of DEG failure, an alarm is sent to the remote maintenance center. Then the battery can still provide the energy during the access time to the site (3 days maxi) of the maintenance staff.

3.5 - Using conditions of the battery (fig. 2)

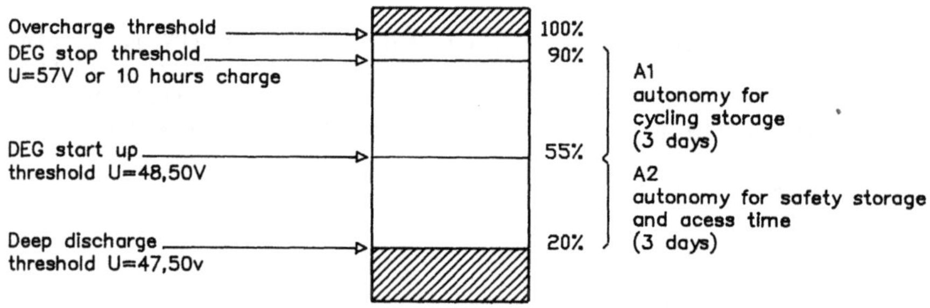

figure 2

The capacity of the battery is used as following :

A1 = an autonomy for usual charge-discharge cycles. This autonomy reduced to 3 days has a direct influence on the DEG starts up number.

A2 = an autonomy which is only conditioned by the organization of the maintenance staff and the access time to the site. This autonomy is set to 3 days (1 week-end + 1 day).

With those two conditions the start-up voltage threshold is choosen with the help of the battery manufacturer and the following data : capacity of the battery, temperature, discharge rate.

This threshold is set to 48,50 V at 25°C.

The conditions of the DEG stop are :

- end of charge voltage threshold : 57 V
- recharge time out : 10 hours

Then the GPV may complete the charge if the weather is sunny. In that case, short and not dangerous overcharge of the battery may happen.

4. SIZING AND EQUIPMENTS FEATURES

a) The system is intended to supply a permanent load of 600 W with a low unavailability (10^{-6}) and one or two maintenance visits a year.

b) The theoritical yearly percentage of energy provided by the DEG is set to about 30 %. In this case, the PVG and the battery are reduced by half, compared to a single PVS.

c) The 3 200 Wp PVG is composed of 80 multicrystalline silicon modules manufactured by PHOTOWATT company (see fig. 3). It is divided in 5 subarrays and 5 branches providing each 10 A peak current. Owing to the distance (50 meters) between the PV array and the building, lightning protections are connected on both branches extremities.

d) The solar type battery is 2 500 Ah C_{10} composed of 24 open lead acid cells from OLDHAM Company.

e) The rectifier is modular. It is a cabinet with 5 independant swiching rectifiers 20 A/57 V each, developped by HOUVENAGHEL Company.

f) The DEG is an 8 kVA SAFARI type, developped by Diesel Energie Company (see fig. 4). It integrates in a same sound proofing rack : a 350 liters fuel tank in the upper part, an air cooled engine and its alternator in the lower part, an extra large oil tank allowing 250 hours time space between oil charges, electronics control which allows 3 automatic attempts for each start-up. It is also connected to several numeric and analogic sensors which can provide data about the DEG operation. Those data are processed by the supervision unit (US) and may be displayed on a Minitel terminal.

In order to keep low maintenance and reduce the fuel transport, a 2 500 liters extra fuel tank has been connected to the integrated SAFARI tank.

g) Those DEG and rectifier are standard equipment in the french telecommunications network.

6. SUPERVISION UNIT (US)

A remote hybrid system needs a preventive maintenance tool, like for the mains connected small telecom stations.

The "Gros Bessillon" US has been developped by the CNET (fig. 5) around an 8 bit microprocessor. It is a monitoring maintenance and measurement system which provides informations (state, alarms, data) which are displayed on site or on a remote "Minitel" terminal through the telecom network.

The man-machine communication and the reading of a logbook allow the access to :

- the general synoptic of the station with refreshed measures (indoor temperature, PV current, battery voltage and current, rectifier current, telecom system current)
- dated alarms provided by the DEG (low fuel level, high oil temperature, pump failure, regulator disfunction, excess of working time, startup failure, etc...), the rectifiers (fuse failure - rectifiers failure), the battery (low voltage), the PVG (robbery), the building (irregular temperature - fire - house-breaking)
- dated events as DEG start-up and stop sequences with battery voltage, maintenances.
- statistical data : amperehours, solar energy, battery voltage are recorded and may be transmitted to a remote computer (PC).

7. CONCLUSION

The installation of amperehourmeters and the remote data transmission will give precise operation results. From now on, it can be foreseen that the operation and the low maintenance targets will be reached. (From March 1987 to March 1988 : 233 hours DEG working time, 500 liters fuel consumption).

So the PV/DEG hybrid system may be consider as an economic and reliable power source, it is also a good way for the photovoltaic development and for the power supply of several hundreds mean watts stations.

Otherwise, the remote supervision of the station is very comfortable and has a particular positive effect on the maintenance staff.

Bibliography : IUT CCITT GAS IV - Primary power sources for remote telecommunication systems
CNET - Spécification technique ST/PAB/STC/584
Station d'énergie hybride associant un générateur photovoltaïque et un groupe électrogène Diesel

fig. 3 : Gros Bessillon PV plant 3200 Wp

fig. 4 : Integrated 8 kVA DEG "SAFARI"

fig. 5 : The Supervision Unit and the "Minitel" terminal with the displayed station synoptic

SESSION 05:
Si-CELLS

EFFICIENT THIN SILICON-FILM SOLAR CELLS ON LOW COST SUBSTRATES

Allen M. Barnett
Electrical Engineering Dept.
University of Delaware
Newark, Delaware 19716 USA

and

James A. Rand, Frances A. Domian, David H. Ford,
Christopher L. Kendall, Mary Lou Rock and Robert B. Hall
AstroPower Division/Astrosystems, Inc.
30 Lovett Avenue
Newark, Delaware 19711 USA

ABSTRACT. Thin polycrystalline Silicon-Film solar cells are being developed on low cost ceramic substrates. One square centimeter solar cells with energy conversion efficiencies of 14.2% have been demonstrated. The Silicon-Film solar cell design is based on light trapping which leads to increases in the voltage, current and efficiency when compared to thick polycrystalline silicon solar cells. The light-trapping design models indicate that efficiencies in excess of 19% can be achieved.

1. INTRODUCTION

The polycrystalline Silicon-Film solar cell is designed to combine the high performance and stability inherent in crystalline silicon with the low cost potential of thin films. The design rules will be presented and performance expectations will be described. The fabrication process will be summarized and the most recent experimental results will be presented and analyzed.

2. SILICON-FILM SOLAR CELL DESIGN

The thin crystalline silicon solar cell designs with light trapping offer superior performance in comparison to thick crystalline silicon solar cells. The application of light trapping leads to increases in voltage, increases in current and accordingly increases in efficiency. The design rules which lead to these performance increases have been extended to polycrystalline thin Silicon-Film solar cells. An additional advantage of the Si-Film process is that low cost can be achieved due to the growth of thin Silicon-Film layers on inexpensive ceramic substrates.

2.1. Design rules

A set of six design rules have been developed which, when followed, will result in high performance solar cells.

1) Silicon-Film thickness: 5 to 50 microns.
2) Light trapping.
3) Planar film with single crystal grains at least twice as wide as thickness.
4) Minority-carrier diffusion length at least two times the thickness.
5) Benign grain boundaries.
6) Substrate to provide mechanical support and back plane conductor.

A drawing of the solar cell is shown in Figure 1. The drawing shows the back surface texture for light trapping. Also shown is a metallurgical barrier which prevents impurities in the substrate from contaminating the Silicon-Film. The metallurgical barrier can also serve to enhance the light reflection for the light trapping.

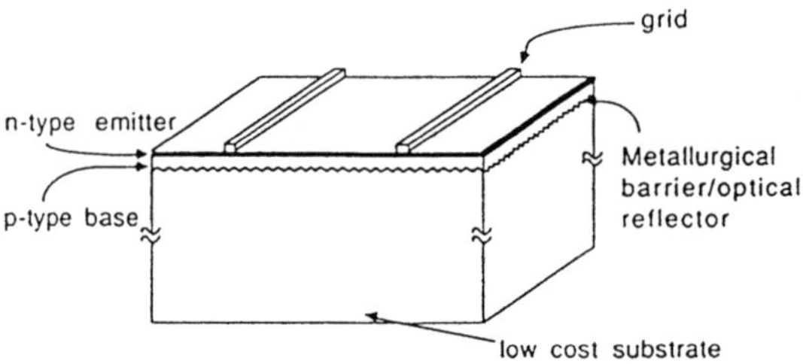

Figure 1
Silicon-Film Solar Cell.

2.2. Light Trapping Impact

Critical to this design is light trapping. Improved light absorption due to light trapping can lead to photon absorbtion increases of 37% to 26% for layers of 10 and 20 microns thick respectively [1-4]. As the effective optical thickness, Z, for a 30 micron thick solar cell with a 60 micron diffusion length is increased from Z=1 (the case with no light trapping) to Z=3, the short circuit current increases from 30.7 mA/cm^2 to 33.8 mA/cm^2. Further increases to Z=10 and Z=50 lead to Jsc of 35.9 and 37.2 mA/cm^2. This effect of increasing Z for this case is summarized in Figure 2.

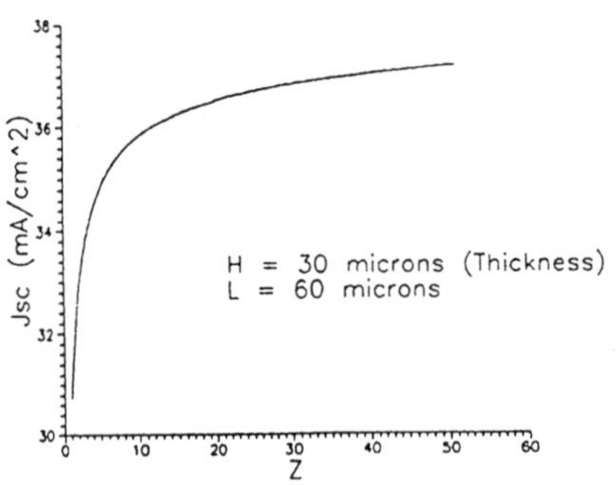

Figure 2
Current as a function of increasing Z.

2.3. Voltage Increases Due To Thickness

The requirement that diffusion lengths need only be two times the thickness allows the use of higher impurity concentrations which lead to higher voltages. In addition, the recombination volume is reduced leading to decreased Jo. Green [5] has shown that further increases in voltage can be achieved if the solar cell can be operated in high injection.

2.4. Design Predictions

The optimum thickness based on the condition that the diffusion length is equal to two times the thickness has been calculated. Bulk carrier concentrations were varied to correspond with changes in diffusion length. Lifetime values were degraded by a factor of 5 from typical single crystal values. These reduced lifetimes are used to accommodate the relatively early state of development of the Si-Film process. Front and back surface recombination of 1000 cm/sec was assumed. A reflection of 5% and a series resistance of 0.1 ohm.cm^2 was assumed. The results of this analysis are shown in Figure 3.

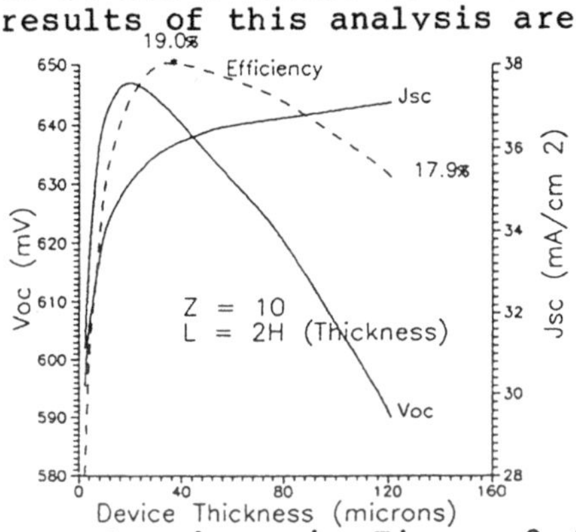

<u>Figure 3</u>
Thin silicon performance as a function of thickness. Diffusion length equals two times the thickness.

As shown in Figure 3 the optimum efficiency occurs at thicknesses between 30 to 40 microns. This design has been applied to a 30 micron thick silicon solar cell. A 60 micron diffusion length was used. Diffusion lengths of 80 microns have been observed. The design was tested for three cases i) no light trapping ii) an increase in the effective optical thickness to 10, Z = 10 and iii) an increase in the effective optical thickness to 50, Z = 50. The results are summarized in Table 1.

<u>Table 1</u>
Thin Solar Cell Parameters

	No Optical Confinement	Z=10	Z=50
Jsc	30.7	35.9	37.2
Voc	639	644	645
FF	.82	.82	.82
Eff	16.1%	19.0%	19.7%

2.5. Grain Boundary Effects

Grain boundary effects can result in reduced current due to recombination of light-generated minority carriers, reduction in voltage due to increases in Jo and shunt effects, and reductions in FF due to shunt effects (resistance and diode shunts are considered). A drawing of the structure and a lumped-parameter model are shown in Figure 4. The experimental results for the Si-Film process indicate that the primary effect of the grain boundaries is to reduce FF.

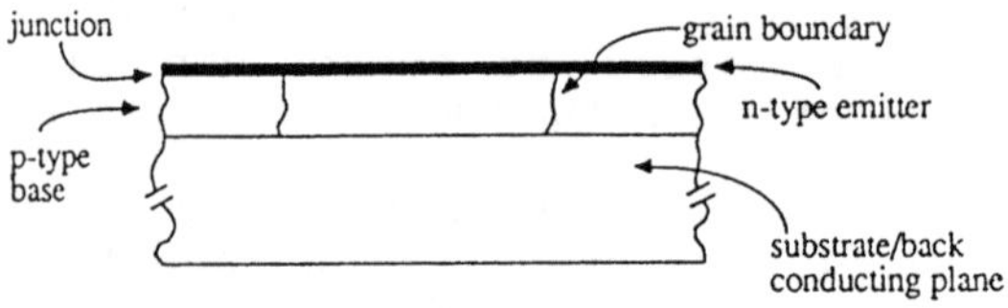

Figure 4
A) Si-Film solar cell drawing.

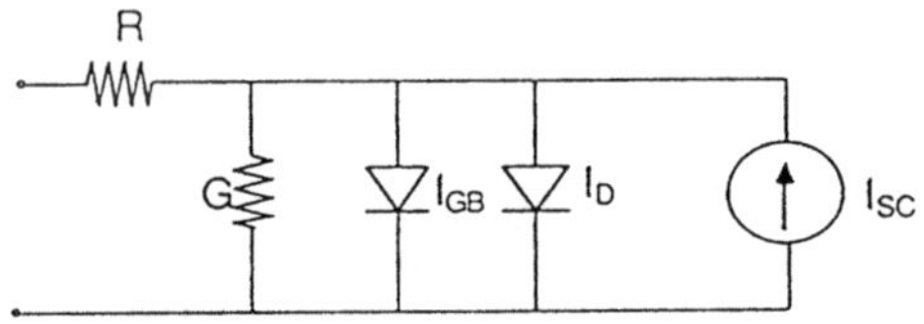

B) Equivalent circuit schematic of the lumped parameter model.

3. Si-Film Solar Cell Manufacturing and Development

The Silicon-Film substrate manufacturing process can be described by the steps of applying the metallurgical barrier/optical reflector over the ceramic substrate, deposition of the silicon (from solution) and substrate anneal to reduce thermal stress. The process is shown in Figure 5.

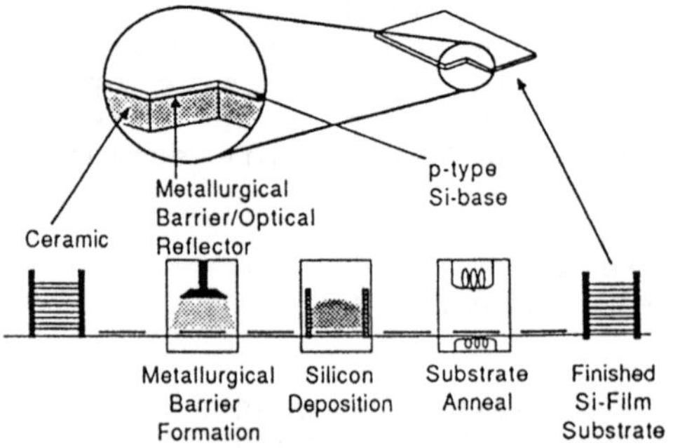

Figure 5
Si-Film substrate manufacturing process.

Upcn completion of the substrate manufacturing process, solar cells can be fabricated using standard silicon diffusion, metallization and anti-reflection coating processes.

3.1. Experimental Results

The first one cm^2 Si-Film on ceramic solar cells were fabricated in April 1987 [6] and had an efficiency of 8.4%. These solar cells used a relatively pure ceramic which reduced the requirement for the metallurgical barrier function. Accordingly, little light trapping was observed. Systematic increases first in voltage and subsequently in current and FF have led to an efficiency of 12.6% which was reported in October 1987 [7]. The highest efficiency measured to date with this relatively thick layer structure is 14.2%. The data for these solar cells is summarized in Table 2.

Table 2
Silicon-Film Development:
Progress

	4/87	6/87 (1)	8/87 (1)	3/88
Voc (mV)	533	582	600	593
Jsc (mA/cm^2)	23.4	25.0	28.4	32.4
FF (%)	67.7	70.2	74.2	74.0
Eff. (%)	8.4	10.2	12.6	14.2
Area (cm^2)	1.0	0.85	1.0	0.98

(1) Sandia measurements

The current-voltage characteristic of the best solar cell to date is shown in Figure 6.

All of these results were achieved without the metallurgical barrier. Most of the light trapping was also eliminated.

3.2. Light-Trapping

The initial light trapping structures were fabricated on highly doped single crystal substrates using a metallurgical barrier [6]. Figure 7 shows the increased long wavelength response (A) of a thin light trapping structure as compared to a thicker device on the same material (B). Thin layers have since been fabricated on ceramic substrates. The ability of the metallurgical barrier to maintain the purity of the active layer has been proven. Texturing and index of refraction mismatching at the active silicon-metallurgical barrier interface was used to enhance the light confinement trapping.

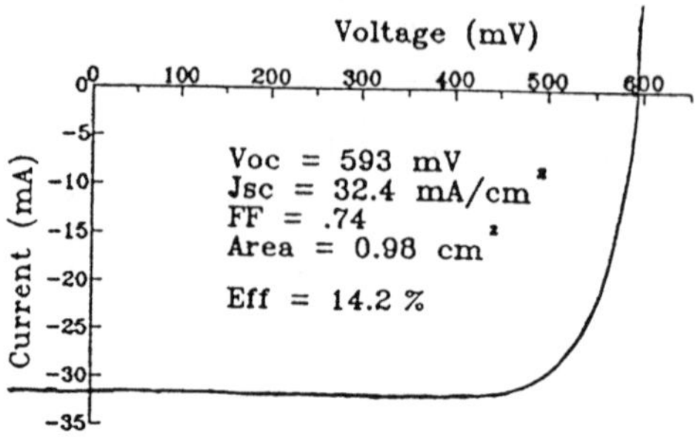

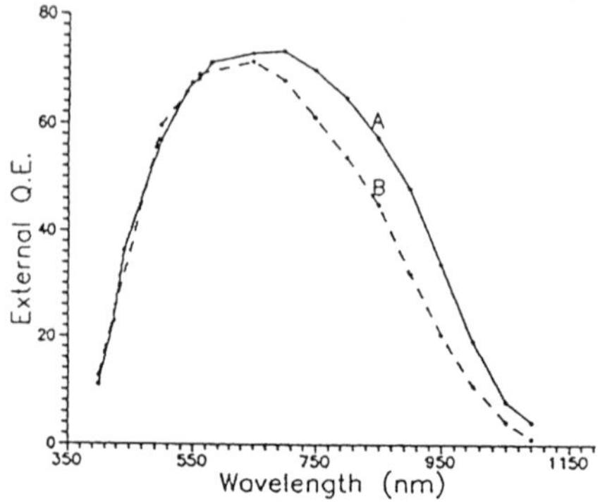

Figure 6
I-V of best Si-Film solar cell.

Figure 7
Quantum efficiency with light trapping (A) and without (B).

3.3. Large area results

Figure 8 shows several samples which indicate the present state of development of the Si-Film process in comparison with a commercially available polycrystalline substrate. 100 cm^2 ceramic substrates, and 50 cm^2 Si-Film substrates have been fabricated. Equipment and processing are presently being developed to increase the Si-Film substrate area to 100 cm^2.

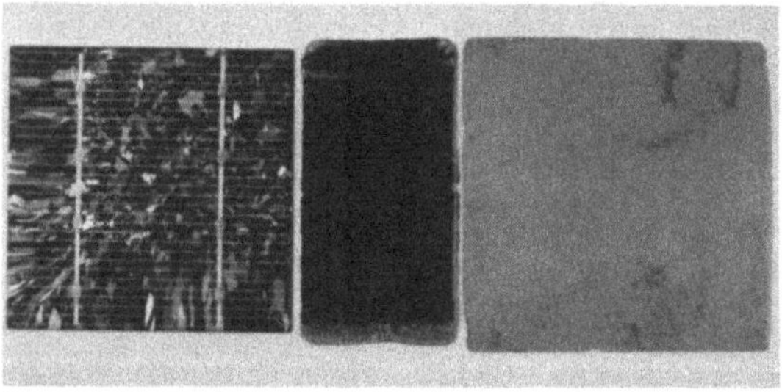

Figure 8
Right to left, a 100cm^2 solar cell, 50cm^2 silicon on ceramic, 100cm^2 ceramic.

4. SUMMARY AND CONCLUSIONS

The device analysis presented indicates that solar cell conversion efficiencies of 19% are achievable using thin (30 micron) polycrystalline Silicon-Film designs. This performance prediction exceeds that of the conventional thick (400 micron) polycrystalline silicon solar cell. Crucial to the achievement of this thin layer solar cell performance is the employment of light trapping. The Si-Film process is being developed to accommodate the design rules leading to enhanced performance using thin silicon layers, and employs a low-cost thermally matched ceramic substrate, and a metallurgical barrier.

Presently, a device configuration employing an acceptably pure, low-cost ceramic substrate and no metallurgical barrier has achieved a conversion efficiency of over 14% for a 1 cm^2 solar cell. Table 3 shows the model calculations for a 30 micron and a 100 micron thick silicon single crystal solar cell without light trapping, and the present experimental results. The existence of some light trapping is already suggested by the observed current. A current of 32.5 mA/cm^2 (compared to the observed 32.4 mA/cm^2) is predicted for a 100 micron thick (approximate thickness of experimental device) device having one normal reflection (Z=2). The V_{oc} is somewhat lower than the model calculation and is likely due to the absence of any surface passivation. The largest disparity between result and model occurs in the observed fill factor, and this is most likely due to the effects of grain boundaries.

Table 3
Thin Solar Cell Parameters
Model vs. Experimental Results

	Single Crystal Model Thin Cell H=30	Single Crystal Model Thick Cell H=100	Best Experimental Values for Each Parameter H=100
Jsc	30.7	31.6	32.4
Voc	623	617	601
FF	.82	.82	.786
Eff	15.7%	16.0%	14.2%

The Si-Film process presently under development has demonstrated 100 cm^2 ceramic substrates and 50 cm^2 device-ready substrates. The continued achievement of increased area and solar cell performance combined with the stability inherent in crystalline silicon materials, make the Si-Film on a dissimilar substrate a promising photovoltaic solar cell option.

5. ACKNOWLEDGEMENT

This work was partially supported by Sandia National Laboratories with funding from the Photovoltaic Energy Technology Division of the U.S. Department of Energy.

6. REFERENCES

[1] D. Redfield, Appl. Phys. Lett., 25, 647 (1974).
[2] A. Goetzberger, Proc. 15th IEEE Photovoltaic Specialists Conf., Orlando, Florida, 40 (1981).
[3] E. Yablonovich, Proc. 4th E.C. Photovoltaic Solar Energy Conf., (D. Reidel, London, 1982), p. 465.
[4] A.M. Barnett, R.B. Hall, D.A. Fardig and J.S. Culik, Proc. 18th IEEE Photovoltaic Specialists Conf., Las Vegas, Nevada, 1094 (1985).
[5] M.A. Green, IEEE Trans. Electron Dev., ED-31, No. 5, 671 (1984).
[6] A.M. Barnett, D.A. Fardig, R.B. Hall, J.A. Rand and D.H. Ford, Proc. 19th IEEE Photovoltaic Specialists Conf., New Orleans, Louisiana, 1266 (1987).
[7] A.M. Barnett, J.A. Rand, R.B. Hall and D.A. Fardig, Technical Digest of the International PVSEC-3, Tokyo, Japan, 101 (1987).

NEW DEVELOPMENTS IN HIGH EFFICIENCY SILICON CONCENTRATOR CELLS

R. M. Swanson
Department of Electrical Engineering
Stanford University
Stanford, California USA, 94305

Abstract

During the last several years there has been a considerable increase in the efficiency of silicon concentrator cells. Several factors have contributed to this development. New cell configurations circumvent some of the shortcomings of previous cell designs. Understanding of carrier transport and recombination in highly doped regions and dense electron-hole plasmas has improved. New computer modeling codes incorporating this understanding have enabled an accurate and predictive design process. Finally, processing techniques have improved to the point where devices are essentially free of defect related losses. This paper reviews these advances.

1 Introduction

Early concentrator cells were simply conventional flat plate cells with grids redesigned for higher current operation. Initially, there was rapid improvement as the designs were refined, culminating in the conventional n^+pp^+ or p^+nn^+ configuration. Several laboratories reported cells in the neighborhood of 20 percent in 1978[1]. There was little improvement for the following four years, however, leading many to conclude that the technology was mature and incapable of further improvements. In the meantime, gallium arsenide cells appeared in 1977 with efficiencies near 25 percent. By 1983 their efficiency had improved to 26 percent. Then, in 1985 new silicon designs emerged, eventually leading to efficiencies over 28 percent. With current understanding is felt that an efficiency of near 31 percent is obtainable following modest further improvements. Gallium arsenide cells are probably capable of comparable performance.

2 Light Trapping

Conventional p^+nn^+ devices have responsivities around 0.34 A/W (i.e., they produce 34 mA when illuminated by 100 mW/cm^2 solar radiation.) This can be improved by 23 percent to 0.42 A/W through cell design improvements and light trapping.

Effective light trapping requires that the cell have texturizing on at least one surface, to diffuse the light rays, and a back surface reflector. Upon entering the cell, incoming light rays are deflected randomly by the texturized top surface. If a photon reaches the back of the cell without being absorbed it reflects off the reflective back surface. Further, if the photon makes it back to the top, there is a high probability that it will strike the surface beyond the critical angle for total internal reflection, and hence will be reflected back into the cell. For a semiconductor with index refraction of n, it has been shown that, when the light rays are propagating in random directions, only $1/n^2$ of them will escape upon striking the top[2]. For silicon the escape probability is 8 percent. Light is thus "trapped" within the cell until it is absorbed. In actuality, the rays are not completely randomized and thus the losses are somewhat greater than this[3].

3 Electrical Design

In thinking about solar cell operation it has often been customary to visualize their operation in terms of the drift and diffusion of charge carriers. The current transport equations, along with the continuity equation, are sufficient to model cell operation; however, it is more intuitively useful to integrate the continuity equation to obtain the following description of device operation[4]:

$$I = I_{ph} - I_{rec}$$

where

$$I_{rec} = I_{bulk} + I_{surface} + I_{emitter}$$

and

$$V = \frac{kT}{q} \ln \left(\frac{p_{p\ cont} n_{n\ cont}}{n_i^2} \right) - V_{base} - V_{res}.$$

The first equation says that the terminal current is the total rate of electron-hole photo-production times q minus the new recombination current. In this case the recombination current must be defined to include that recombining in the bulk, plus that at the surfaces, plus that in the highly doped emitter regions (including the metal-semiconductor contacts). Clearly, in order to maximize performance, one must have as large a photo-current as possible (such as through light trapping) and minimize recombination as far a practical.

Recombination is generally a monotonically increasing function of the departure of the pn product (product of electron and hole densities) from its equilibrium value of n_i^2. The second equation shows, however, that large values of pn product are necessary for high terminal voltage. The voltage also includes a term due to resistance and Dember effects in the base of the cell and, of course, a term due to external resistances. Minimizing base resistance also speaks toward high values of carrier density in the base, and hence high pn product. But attaining high pn product necessitates, as seen above, minimizing recombination. Thus, much of cell design centers around carefully identifying all sources of recombination and minimizing them. Simply focusing attention on one source of recombination (such as at emitter contacts, for example) invariably results in marginal improvement because another source rapidly replaces it as the pn product increases.

As an illustration of the importance of recombination, consider the collection of photogenerated carriers in a cell with light trapping. One can define a "local quantum efficiency" as the probability that a carrier produced at some point in the cell will be collected rather than recombine. As illustrated in Figure 1, in a conventional cell

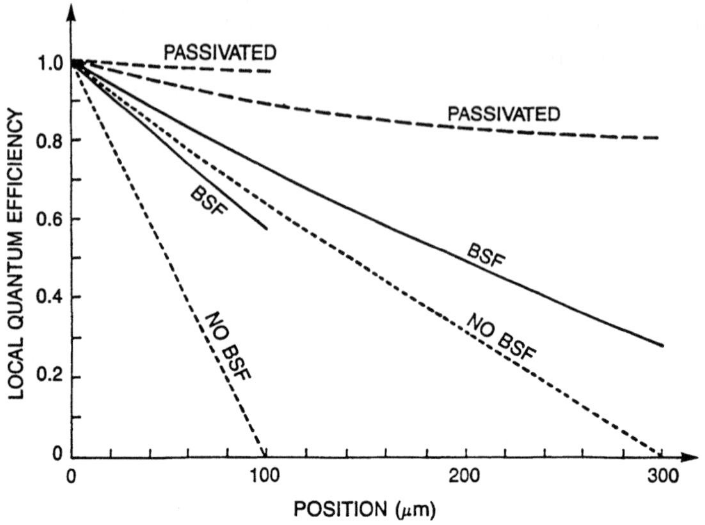

Figure 1: Calculated local quantum efficiency for a conventional n^+pp^+ cell. The base doping is 10^{17} cm^{-3}, the base lifetime 100 μsec, and the back surface field has a diffusion depth of 4 μm with a sheet resistance of 7 $\Omega/\square$. Results for both 100 and 300 μm thick cells are shown. The passivated cell has a surface recombination velocity of 20 cm/sec.

structure this value goes from essentially unity near the top of the cell to zero at the back because carriers produced at the back will recombine at the back metal contact. Such a cell will obviously not use light trapped photons very efficiently because such photons are, likely as not, absorbed deep within the cell. Usually a back surface field, in the form of a deep p^+ layer, is incorporated at the back in order to "reflect" carriers from the contact area. Figure 1 shows the result of this. There is but a slight improvement. In performing these calculations, accurate modeling of carrier transport in highly doped regions is necessary for reliable results. That back surface fields are not more effective is a result of band gap shrinkage in the doped layer. Real improvement accrues when the back surface is passivated with an oxide (putting off, for the moment, considering how electrical contact is to be made to the base). If the cell is made thin, 100 μm in the example, generation is squeezed closer to the back and the local quantum efficiency actually decreases throughout the cell due to the closer presence of the back surface — that is, unless the back is passivated. Clearly, the highest local quantum efficiency throughout occurs in thin, passivated back contact cells.

4 Limit Efficiencies

Using the preceding development it is possible to easily assess the performance of certain idealized "limit cells." First, assume that emitter recombination is reduced to negligible values, such as perhaps through a heterojunction contact (minority carrier mirror). In addition, let us neglect base and metal voltage drops and assume that the pn product is constant throughout the cell. Further, assume that defect mediated bulk recombination is made negligible through the elimination of bulk defects. This leaves radiative and Auger recombination as the only remaining possibilities.

Radiative recombination must exist thermodynamically through the principle of detailed balance (being the inverse of the desired photo-production of carriers). The volume radiative recombination rate is Bpn where $B = 1 \times 10^{-15}$ cm^3/sec. Inserting this into the modeling equations in the previous section yields a predicted efficiency of 37 percent for a 500X concentrator cell at 25°C.

Band-to-band Auger recombination, however, is also "intrinsic" to the cell material in that there is no known method for eliminating it, even conceptually. Such recombination varies as $C_n n^2 p + C_p p^2 n$ where C_n and C_p are the Auger coefficients. Using the best current values for these parameters causes the predicted limit-cell efficiency to decrease to 33 percent.

Real cells must have additional losses, such as emitter recombination and resistive voltage drops. Modeling indicates that it is reasonable to expect an actual efficiency of around 31 percent in a mature cell design.

5 Physical Parameters

It goes without saying that it is necessary to have accurate data on the transport and recombination of electrons and holes in order to properly model cell operation. Further, accurate models are needed to guide cell research. They can be used to explore new design concepts, to help interpret experimental results, and to optimize cell designs.

Often workers in the field have been limited by lack of reliable data on critical parameters. This is particularly true regarding surface recombination velocities and transport in heavily doped or highly injected regions.

5.1 Surface recombination velocity

It has been the practice to simply assume a value for surface recombination velocity, typically around 1000 cm/sec on oxide passivated surfaces. This is clearly unacceptable for modeling cells where surface recombination is an important part of the total recombination. Recent measurements have revealed that it is typically much less than this and somewhat doping dependent. Lightly doped surfaces with state-of-the-art oxide passivation layers have recombination velocities in the 2 to 5 cm/sec range[5]. As the surface doping level increases, so does the recombination velocity. For p-type surfaces, by the time the doping reaches 1×10^{17} cm^{-3} the recombination velocity is 100 cm/sec[6]. n-type surfaces have lower recombination velocity. These low values allow the consideration of cell designs which have no doped layers at the surface, or possibly very shallow doped layers.

5.2 Emitter recombination

Transport in the heavily doped regions of emitters has been the subject of considerable study and controversy. Recently, a synthesis has been achieved in which it is clear that, at least as far as modeling goes (as opposed to physical understanding), these regions can be accurately described[7,8]. One surprising result of this study is that electrons and holes as minority carriers have slightly higher mobility than when they are majority carriers at the same doping density. Additionally, band gap shrinkage in n-type material is less than usually assumed previously.

These studies have revealed that the optimum emitter under a metal contacted region is rather deep, 3 or 4 μm, with a surface concentration about 8×10^{19} cm^{-3}. Such an emitter will have a saturation current of about 5×10^{-13} A/cm^2. Under oxide passivated regions the optimum emitter is very shallow, less than 0.2 μm, with a surface concentration around 1×10^{19} cm^{-3}. Saturation currents around 1×10^{-14} A/cm^2 are possible. Of course, when the emitter must carry lateral current, as in the conventional structure, such high sheet resistance could not be used. Never-the-less, current understanding allows the modeling and design of optimum emitters for each application.

5.3 Bulk recombination

Until recently, bulk recombination was usually dominated by defect assisted processes. Today, however, a combination of float-zone grown silicon with the low process induced contamination achievable in modern semiconductor fabrication lines has resulted in defect related lifetimes well in excess of 1 ms in finished cells. This means that Auger recombination dominates in state-of-the-art cells.

For some time there has been agreement on the Auger coefficients for silicon doped in the 10^{19} to 10^{20} cm^{-3} range[9]. Recently, it has been found that these coefficients actually increase by about a factor of four when going down to doping densities of 10^{17} cm^{-3}[10]. This has the effect of decreasing the limit efficiencies slightly over previous estimates. Additionally, the best float-zone material now appears to be at the Auger lifetime limit when doped to 10^{17} cm^{-3} so no further improvement can be expected.

6 Point-Contact Cell

As an example of the direction that the considerations above have lead, a new cell, called the point-contact cell, has been under development[11]. The structure is shown in Figure 2. The base is intrinsic silicon and relies on conductivity modulation by the photogenerated and injected carriers to limit base resistance loss. The top surface is undoped, texturized and passivated by a thin SiO_2 layer. The emitters have been reduced to a small collection of alternating n and p regions on the back surface.

The purpose of the small diffused emitters is two-fold. First, emitter recombination is thereby reduced by the ratio of emitter to total surface area, and second, the reflectance of the back surface is found to be much higher in uncontacted regions. Obtaining high reflectance is important so that light trapped photons produce electron-hole pairs rather than be absorbed in the back surface. Surface recombination in regions not covered by doped emitters is limited by growing a high quality SiO_2 layer over the entire device and removing it only in the contact points.

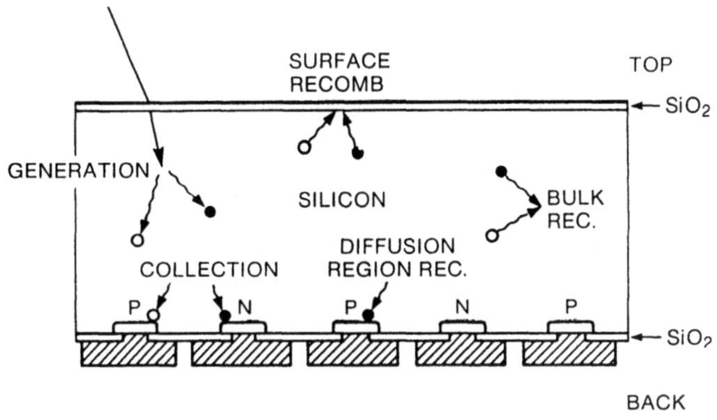

Figure 2: Cross-section of the point-contact cell illustrating various sources of recombination.

6.1 Modeling

A modeling code which handles the rather three-dimensional nature of the current flow around the contact points has been developed and used to design the cell[4]. It is found that the spacing between contact points is rather critical, the optimum being around 30 μm for 10 μm diameter contact points. This is perhaps rather non-intuitive — one might initially suppose that the contacts need only be spaced a distance less than the diffusion length, which is several millimeters in this material. Too wide a spacing results in excessive "spreading resistance" loss, however, whereas too close a spacing increases the emitter recombination excessively.

Another important finding is that the cell must be rather thin, on the order of 100 μm, to have high efficiency. This obtains because the high carrier density at the front where most of the light is absorbed is very high, resulting in considerable Auger recombination. The thinner the cell the lower the carrier density at the front and hence the less Auger. A properly designed cell is predicted to have an efficiency of around 30 percent.

6.2 Experiment

Laboratory versions of the point-contact cell have been fabricated and tested[12]. The most recent results have produced cells with an efficiency of 28.4 percent at a concentration of 100 suns. These cells are still limited by higher than desired series resistance and primitive anti-reflection coatings. Correcting these deficiencies should produce cells with efficiencies over 30 percent.

7 Conclusions

At this point, it is clear that efficiencies over 30 percent can be achieved in laboratory silicon concentrator cells. Much work remains to create a cost-effective cell fabrication process; however, this seems to be a straight-forward and doable task. This achievement would easily satisfy the DOE long-term goal for concentrator cell performance. The remaining technical hurdle at this time blocking the introduction of truly cost competitive concentrator systems is the continuing need to develop equally cost effective concentrator balance-of-system designs.

8 Acknowledgements

This work was supported by the Electric Power Research Institute under contract RP-790-2 and Sandia National Laboratories under contract 02-2338. The author is indebted to the people of the Stanford silicon solar cell research group; particularly, Ron Sinton, Dick Crane, Sharleen Beckwith, Rich King, Dave Kane, Peter Gruenbaum and J. Y. Gan.

References

[1] see for example, D. T. O'Donnell, et al, "Performanc of Silicon and Galluium Arsenide Concentrator Cells," Thirteenth IEEE Photovoltaic Specialists Conference, Washington, D.C., June 5-8, 1978, pp. 804-809.

[2] E. Yablonovich and G. D. Cody, " Intensity Inhancement in Textured Optical Sheets for Solar Cells," IEEE Transactions on Electron Devices, Vol. ED-29, No. 2, February 1982, p. 300.

[3] Martin A. Green, *High Efficiency Silicon Solar Cells*, Tans Tech Publications, 1987, p. 69.

[4] R. M. Swanson, "Point-Contact Solar Cells: Modeling and Experiment," Solar Cells, Vol. 17, 1986, pp. 85-118.

[5] W. D. Eades and R. M. Swanson, "Calculation of the surface generation and recombination velocities at the Si-SiO_2 interface," J. Appl. Phys., Vol. 58, No. 11, pp. 4267-4276, 1 December 1985.

[6] R. R. King, R. A. Sinton and R. M. Swanson, "Low Surface Recombination Velocities on Doped Silicon and their Implications for Point Contact Solar Cells," Proceedings of the Nineteenth IEEE Photovoltaic Specilists Conference, New Orleans, Louisana, May 4-8, 1987, pp. 1168-1173.

[7] J. del Alamo and R. M. Swanson, "The Physics and Modeling of Heavily Doped Emitters," IEEE Trans. Elec. Dev., Vol. ED-31, No. 12, pp. 1878-1888, Dec. 1984.

[8] R. M. Swanson and S. E. Swirhun, "Characterization of Majority and Minority Carrrier Transportr in Heavily Doped Silicon," Sandia Report SAND87-7019, November 1987.

[9] J. Dziewior and W. Schmid, " Auger Coefficients for Highly Doped and High Excited Silicon," Applied Physics Letters, Vol. 31, No. 5, 1977, pp. 1289-1294.

[10] R. A. Sinton and R. M. Swanson, "Recombination in Highly Injected Silicon," IEEE Transactions on Electron Devices, Vol. ED-34, No. 6, pp. 1380-1389, June 1987.

[11] R. M. Swanson, S. K. Beckwith, R. A. Crane, W. D. Eades, Y. H. Kwark, R. A. Sinton, and S. E. Swirhun, "Point Contact Silicon Solar Cells," IEEE Trans. on Elec. Dev., Vol. ED-31, No. 5, May 1984.

[12] R. A. Sinton and R. M. Swanson. "An Optimization Study of Si Point-Contact Solar Cells," Proceedings of the Nineteenth IEEE Photovoltaic Specilists Conference, New Orleans, Louisana, May 4-8, 1987, pp. 1201-1208.

IMPROVEMENTS IN SILICON SOLAR CELL EFFICIENCY

M.A. Green, J. Zhao, A. Wang, C.M. Chong,
S. Narayanan, A.W. Blakers, S.R. Wenham

Solar Photovoltaic Laboratory,
Joint Microelectronics Research Centre,
University of New South Wales,
Kensington, Australia, 2033.

ABSTRACT. On-going improvements in the performance of silicon solar cells are reported for a range of different cell types. Independently measured efficiencies of 21.4% and 16.6% for 4 cm^2 crystalline and polycrystalline cells are reported under the Global AM1.5 spectrum at 25°C. An efficiency of 24.1% is reported for a 12.5 mm square concentrator cell at a power output of 5W. Perhaps the most important result, however, is the reporting of an efficiency of 20.4% for a larger (12 cm^2) non-concentrating cell fabricated by sequences independently costed as being competitive with present commercial screen printing sequences.

1. INTRODUCTION

Although 20% efficiency for a laboratory cell and 18% for a commercially produced cell had long been regarded as limits for a silicon non-concentrator cell, our recent work is showing that these values are far too conservative [1].

Progress is reported in four project areas current within our Laboratory:

(i) microgrooved passivated emitter solar cells (μg PESC) [1], [2];
(ii) prismatic covered concentrator cells [1], [3];
(iii) laser grooved, buried contact solar cells [1], [4];
(iv) polycrystalline (Wacker Silso) solar cells [1], [5].

2. MICROGROOVED PESC CELLS

The structure and design features of these cells have been described elsewhere (e.g. see Figure 8.16 [1] or Figure 1 [2]). These cells are very similar to the cells which first exceeded 20% efficiency for silicon [3], although there has been considerable improvement since in top surface reflection control, bulk minority carrier lifetimes, and rear surface passivation. These improvements, particularly those in the latter area, have resulted in cell efficiency being virtually independent of cell resistivity.

Our best independently confirmed result for a 4 cm^2 cell under the global AM1.5 spectrum at 25°C has been an efficiency of 21.4% as measured by Sandia National Laboratories for a 0.2 Ωcm cell. Relevant cell parameters were V_{OC} = 669 mV, J_{SC} = 38.6 $mAcm^{-2}$, and FF = 82.9%. Sandia have measured slightly higher efficiencies for very much smaller concentrating cells under one sun illumination, namely 21.9% for 0.09 cm^2 area cells from our Laboratory and 21.7% for a 0.15 cm^2 area point contact cell from Stanford University [6].

3. PRISMATIC COVERED CONCENTRATOR CELLS

It seems that the high efficiencies under concentration demonstrated by the previous Stanford University point contact cells [6] may also be approached by much larger cells with a more conventional structure. The secret to being able to do this is to combine the previous µg PESC approach with prismatic covers which steer incoming light away from the top finger metallization [1], [7].

In this way, top contact finger shading loss can be virtually eliminated. As an example of this, Table I shows the "one sun" performance of a µg PESC cell before and after a prismatic cover is applied as measured by Sandia National Laboratories. This cell had 16 ± 1% of its top surface covered by fingers. After application of a prismatic cover by Entech [7], cell current increased by 15.5%, showing almost all shading loss has been avoided.

Cell efficiency increased by a higher figure of 18% due to additional increases in voltage output and fill factor.

Table 1: Effect of Prismatic cover on "one sun" cell performance (12.5 mm x 12.5 mm designed illumination area, 25°C)

Covered	V_{oc} (mv)	J_{sc} ($mAcm^{-2}$)	FF (%)	Efficiency (%)
No	658	32.9	76.3	16.5
Yes	664	38.0	77.1	19.5
Gain	6mV	15.5%	1%	18%

Under concentrated sunlight, the efficiency of this cell as measured by Sandia peaked at 24.1% at 120 suns concentration. Although this is the highest efficiency demonstrated to date for a concentrating cell of anywhere near comparable size, over twenty cells of markedly superior performance are presently awaiting application of covers. Efficiency in the 25-26% range would be expected from these cells.

4. LASER GROOVED, BURIED CONTACT SOLAR CELLS

Although the microgrooved PESC approach gives excellent results, processing involves too many steps for the sequence to be of interest for other than space and concentrating cells. A much more simply processed cell structure has been developed as a lower cost alternative for non-concentrating terrestrial cells. Independent costing suggests that this sequence may be more cost-effective than the screen printing sequences which form the present commercial standard in this area.

The sequence, as described elsewhere [1], [4], is based on forming grooves by laser in the top surface of the cell in combination with single top surface oxide as an antireflection coating and a surface passifier, as well as a diffusion and a metal plating mask. Cells made with this sequence have now been independently measured by Sandia National Laboratories to have passed the 20% efficiency barrier under the global AM1.5 spectrum at 25°C. The best cell measured to date was a 1.0 Ωcm cell which gave an efficiency of 20.4%.

Despite the relatively large size of this cell (12 cm^2), this is believed to be only the second silicon cell type of reasonable size (after the μg PESC described previously) to have been independently measured to have exceeded the 20% barrier under the global spectrum. Although two other groups have reported efficiency above 20% for silicon cells of reasonable size (4 cm^2 in both cases), in one case [8] independent measurement gave efficiencies below this value. In the other case [9], the measurement was under an old direct beam calibration. Measurement under the present global calibration again gave a more conservative result.

5. POLYCRYSTALLINE CELLS

Application of the PESC approach to polycrystalline silicon (Wacker SILSO) has given an efficiency of 16.6% for a 4 cm^2 cell as independently measured by Sandia National Laboratories (again under the global AM1.5 spectrum at 25°C).

Although this is believed to be the highest independently measured efficiency for any non-crystalline cell of any type (or any combination of such cells), it still falls well below the corresponding figure for crystalline silicon cells.

One reason for this is that crystallographic surface texturing cannot be used effectively for these polycrystalline cells [1]. Accordingly, a number of alternative texturing approaches are being investigated. The best have given lower reflection loss that crystallographically textured crystalline controls. Combining such texturing sequences with either the PESC or laser grooved, buried contact approach is expected to increase measured efficiency to the 17.5-18.0% range.

6. SUMMARY AND CONCLUSIONS

Table II: Summary of recent silicon solar cell results obtained at the University of New South Wales (measurements by Sandia National Laboratories under the global AM1.5 spectrum at 25°C except for the concentrator cell which was measured under direct beam AM1.5).

Cell Type	ρ (Ωcm)	Area (cm^2)	Voc (mV)	Isc (mA/cm^2)	FF (%)	Effic (%)
μg PESC	0.2	4.0	669	38.6	82.9	21.4
Laser grooved	1.0	12.0	646	39.7	79.4	20.4
Polycrystalline	1.0	4.1	605	35.2	78.2	16.6
Concentrator	0.2	1.6	783	45.4	81.1	24.1

Table II summarizes the results reported in earlier sections. It is concluded that silicon solar cell performance has not yet saturated and that further work is likely to result in near term gains. The laser grooved and concentrator cell work indicates that these efficiency improvements are likely to be able to be transferred to the field.

7. ACKNOWLEDGEMENTS

The authors would like to thank the other members of the Solar Photovoltaic Laboratory for their contributions to this work. They would also like to thank Dr. D. Helmreich and colleagues at Wacker-Chemitronic for supplying polycrystalline samples and David King, Alex Maish and colleagues at Sandia for cell characterization. This work was supported by the Australian National Energy Research Development and Demonstration Program and Sandia National Laboratories. The Joint Microelectronics Research Centre is supported by the Australian Commonwealth Special Research Centres Scheme.

8. REFERENCES

[1] M.A. Green, *High Efficiency Silicon Solar Cells*, Trans Tech Publications, Aedermannsdorf, 1987.

[2] M.A. Green, M. Taouk, A.W. Blakers, S. Narayanan, J. Zhao and P. Campbell, Appl. Phys. Lett. **49**, 194 (1986).

[3] A.W. Blakers and M.A. Green, Appl. Phys. Lett. **48**, 215 (1986).

[4] C.M. Chong, S.R. Wenham and M.A. Green, Appl. Phys. Lett. **52**, 407 (1988).

[5] S. Narayanan, S.R. Wenham and M.A. Green, Appl. Phys. Lett. **48**, 873 (1986).

[6] D.L. King and D.Arvizu, Conf. Record, 19th IEEE Photovoltaic Specialists Conf., May, 1987, pp. 43-48.

[7] M. O'Neil, Conf. Record, 18th IEEE Photovoltaic Specialists Conf., 1985, pp. 1234-9.

[8] D.B. Bickler and W. Callaghan, Conf. Record, 19th IEEE Photovoltaic Specialists Conf., May, 1987, p. 1424.

[9] T. Saitoh, T. Uematsu, Y. Kida, K. Matsukuma and K. Morita, Conf. Record, 19th IEEE Photovoltaic Specialists Conf., May, 1987, p. 1518.

OPTIMISATION OF THE BLUE-RESPONSE OF SCREEN-PRINTED SOLAR CELLS ON EPITAXIAL THIN-FILM SUBSTRATES

M. CAYMAX, M. EYCKMANS, J. NIJS, R. MERTENS ; IMEC vzw, Kapeldreef 75, B-3030 Leuven, Belgium
M. AUCOUTURIER, NGUYEN DINH HUYNH, LE HOANG THI TO, S. N. SINGH ; CNRS/LPS, 1, Place A. Briand F92195 Meudon, France.

ABSTRACT. The blue response of traditional screenprinted solar cells tends to be limited due to the typical deep junction (about 0.6 - 0.7 μm) which, according to the modelling, involves a reduction in short circuit current with approximately 10 %. In order to improve this, we have changed and optimised the firing conditions for the screenprinted P-paste in order to create a junction depth of only 0.25 μm. This, together with an optimised front side metallisation pattern involving a coverage of only 7 %, resulted in a top-value for J_{sc} of 27.8 mA/cm^2. Due to the typical thin-film-structure of epitaxial UMG-Si solar cells which involves a heavily doped substrate, creating a BSF, V_{oc} amounts up to 585 mV. This approach has resulted uptill now in efficiencies up to 11.7 %, which can still be improved by a further refinement of the junction formation step.

1. Introduction

There are two main factors that prevent a more widespread use of photovoltaic cells on earth, namely the high price and, intimately linked with this, the relatively low efficiency. A possible solution for this problem is to start from a very cheap feedstock material, i.e. UMG-Si, for making substrates, on which reasonably efficient solar cells, by means of the well known, reliable and cheap screen printing technology can be made [1]. Because the active layer is made by epitaxially growing a very pure Si-layer, it is preferable to speak here about thin-layer cells. In this paper, we will describe the optimisation of the junction formation step.

2. The influence of the junction-depth on the short circuit current

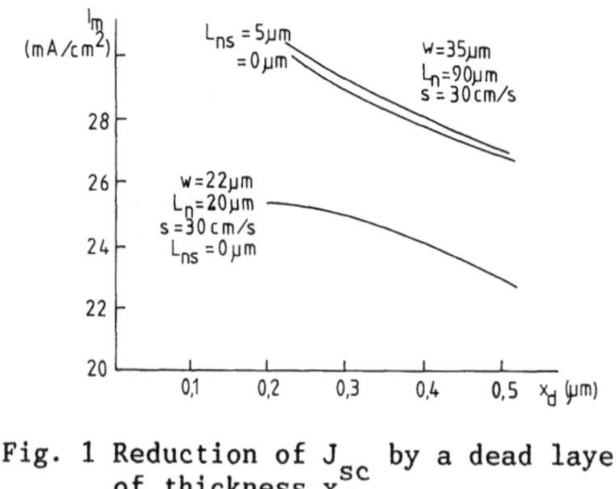

Fig. 1 Reduction of J_{sc} by a dead layer of thickness x_d

An old shortcoming of integrally screenprinted solar cells is their weak blue-response. This is due to the rather deep junction with thickness x_j including a heavily n^+-doped top-layer which constitutes a "dead layer" with thickness x_d. Any light absorbed here (the short wavelenghts !) will give rise to minority carriers that recombine before they can be collected. A quantitative approach of this problem is possible by the general model of SINGH, discussed in another paper at this conference [2]. Fig. 1 shows the influence of the dead layer thickness x_d on the

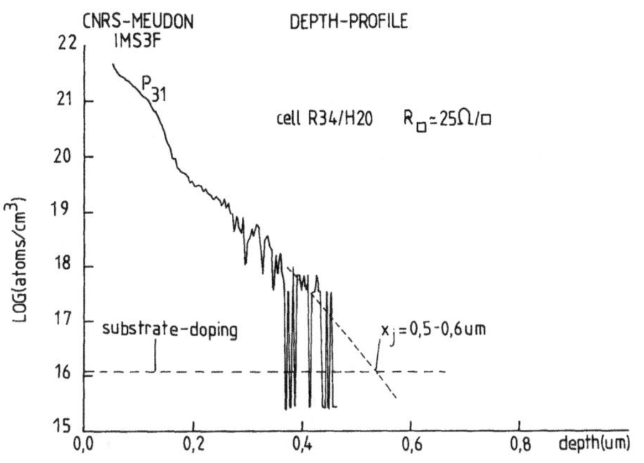

Fig. 2 SIMS-determination of junction depth

total short circuit current J_{sc}, according to this model. An increase from 0.25 to 0.5 µm for x_d involves a decrease with 3 mA/cm^2 for J_{sc}, which means a relative efficiency decrease of about 10 % ! Typical junction depths in our standard process are 0.5 - 0.6 µm as can be seen on the SIMS-measurement in Fig. 2. On the other hand, decreasing this thickness too much may give rise to other problems. First, the series resistance of the front side contact will begin to increase and will eventually lower the fill factor. Secondly, a too thin juction risks to be shortcircuited by the front-side metallisation, due to the diffusion of Ag or impurities from the metallisation material. This too can lower the fill factor.

3. Development of a thin-junction technology by screenprinting

Taking the last two problems from the previous paragraph into account, a possible approach is to make the junction in two steps as has already been proposed earlier [3] : i) a slightly doped n-part, the thickness of which does not matter too much, constituting the actual junction region, and ii) a heavily doped n^+-layer, as thin as possible, for making contact with the front metallisation and for lateral conduction of the current. While this involves a rather complicated junction formation process (with its implications on the add-on cost for cell production), we have adapted this approach to our epi-layer/UMG-Si-sandwich, without the implementation of an extra potentially expensive processing step. By growing a thin second epi-layer of n-type on the standard p-type 20 µm thick layer, the junction can be formed without a dead zone. Next, by screenprinting a standard Phosphorous-paste and firing this in appropriate conditions, a very thin superficial n^+-layer for contacting the cell can be formed. Things to be optimised in this process are the n-epi-layer thickness and the firing of the screenprinted P-paste to form the n^+-layer.

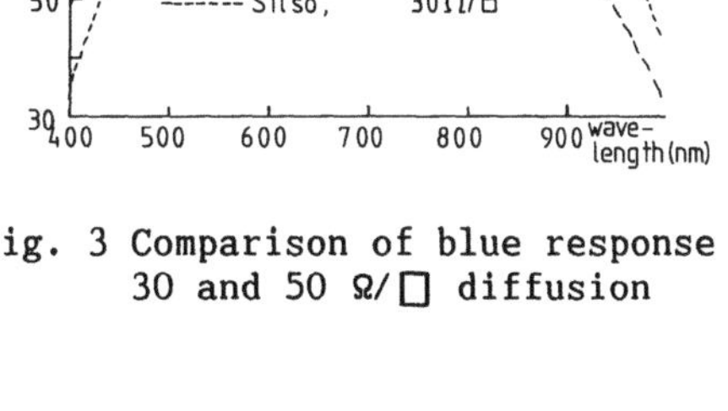

Fig. 3 Comparison of blue response of 30 and 50 Ω/□ diffusion

In order to optimize the n-epi-layer, layers with thicknesses of 1, 2 and 3 µm and with resistivities of 1 and 2.2 Ω.cm have been grown on regular SILSO-slices and on monocrystalline slices. At the same time, the firing conditions for the P-paste have been adapted to obtain two different sheet resistances, i.e. 30 and 50 Ω/□. From the completed cells, one could observe only a weak tendency for the thinner n-epi-layers to perform better. No meaningful difference could be seen between cells with either of the two doping levels in the epi-layer. Fig. 3 on the other hand, showing spectral response-measurements of SILSO-cells, one with thin n-type epi-layer and 50 Ω/□-diffusion and the other without epi-layer but with a 30 Ω/□-diffusion, indicates the improved blue-response of cells with the higher sheet resistance.

Optimisation of the firing conditions for the n^+-layer formation has been done on SILSO-wafers with a double epi-layer consisting of approx. 20 μm epi, p-type, ~ 2 Ω.cm and a 1 μm-epi of n-type, ~ 3 Ω.cm. After screenprinting the normally used Phosphorous-paste, 10 different annealing conditions have been tried out with a continuous belt furnace on batches of 3 wafers. Variable parameters are the temperature profile and the belt speed. The highest temperature used is 970°C, while the lowest one is 910°C. The belt speed varied between 12.6 and 25 cm/min. The resulting sheet resistance varied between 20 and 95 Ω/□. It has to be mentioned that various combinations of different peak temperatures and belt speeds can give rise to the same sheet resistance, but with varying peak concentrations and layer-depth. In Table I the annealing conditions and the resulting sheet resistivity of some samples are reported.
From SIMS-measurements (Fig. 4) the junction thickness x_j could be evaluated, which is also given in Table I. In order to prevent any problems with series resistance or other interferencies the cells were completed with a double metallisation pattern (double number of fingers, resulting in a metal coverage of about 21 %) and without anti-reflective coating. Because of this, the obtained currents are about 1.4 - 1.5 times lower than what is possible. Fig. 5 shows the resulting J_{sc} as a function of sheet resistivity, from which an optimal resistivity of about 40 Ω/□ can be deduced. From spectral response-measurements, the dead layer thickness x_d has been determined. This value is also given in Table I. Within the experimental error, x_d and x_j coincide quite well. One can nicely observe the decreasing dead layer depth as a function of the increasing sheet resistance.

An overall inspection of the measurement results although indicates some disadvantages of this method such as a generally lower V_{oc} than normal, probably due to a too low n-epi-layer doping. On the other hand a small experiment in which cells have been made on SILSO-slices with the optimised 40 Ω/□-diffusion but without n-type epi-layer showed that this approach not necessarily leads to the abovementioned problem

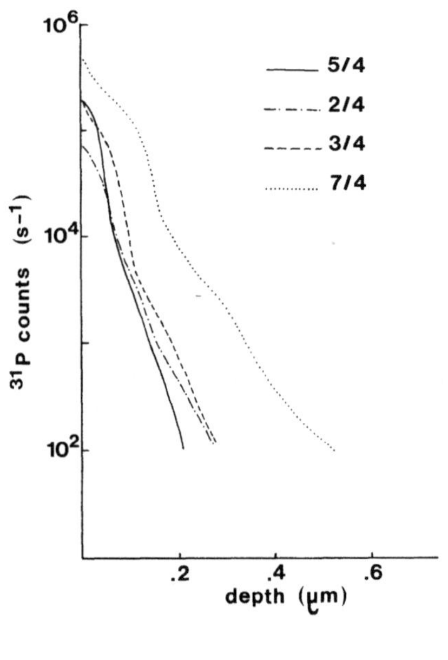

Fig. 4 P-profiles measured by SIMS

Table I : Optimisation of the junction firing conditions

Cell	Annealing conditions T_{max} (°C)	Belt speed (cm/min)	Sheet resistivity Ω/□	Junction depth x_j (μm)	Dead layer depth x_d (μm)
5/4	910	20.1	39	0.24 ± 0.05	0.19 ± 0.10
2/4	910	12.7	28	0/30 ± 0.10	0.25 ± 0.10
3/4	920	12.7	24	0.35 ± 0.10	0.42 ± 0.10
7/4	920	8.1	17	0.53 ± 0.05	0.60 ± 0.10

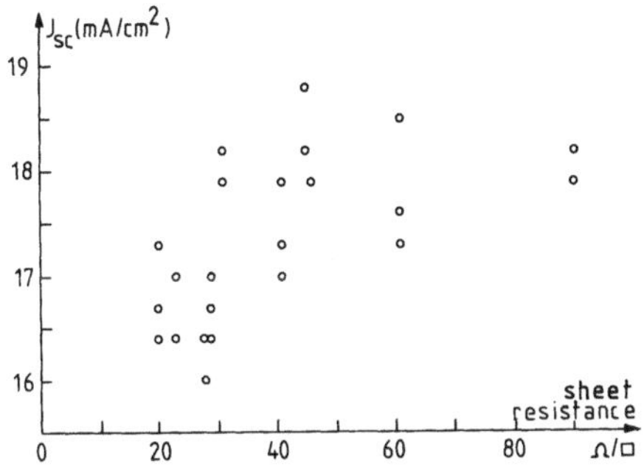

Fig. 5 Diffusion optimisation

of short-circuited junctions. As a consequence, the double epi-layer approach has been abandoned while saving the optimised junction firing conditions. A last point to be mentioned here is the optimisation of the front-side metallisation based on the measured cell parameters such as series and shunt resistance and an expected J_{sc} of about 30 mA/cm². For a cell of 2 x 2 cm², we calculated six fingers of 150 μm width and a collector with a base width of 900 μm as the optimum case. This results in a metal coverage of about 7 %.

4. Application of the screenprinted thin junction on UMG-Si substrates

The optimised parameters for the screenprinted junction-formation and the new metallisation pattern have been combined and used for making solar cells of 2 x 2 cm² on a number of UMG-Si substrates (of 5 x 5 cm² size,

Table II : Solar cells made on epitaxial UMG-Si-substrates by IMEC

Substrate	Resistivity Ω.cm	Epi-thickness μm	J_{sc} mA/cm²	V_{oc} mV	Fill factor %	η %
POLYX-F[(a)]						
651	0.02	28.2	25.9/24.5	585/572	72.4/70.2	10.9/10.0
POLYX-PC[(b)]						
P3/L2	0.035	28.3	25.6/25.2	540/530	67.6/66.7	9.3/9.0
SBB/L1	0.30	23.7	25.0/24.7	530/530	67.5/65.5	8.8/8.6
OEG/L3	0.03	28.3	27.5/27.0	560/550	71.0/69.8	10.9/10.4
IT-P[(c)]						
W279/8	2.5	28.5	27.8/27.0	550/542	73.9/71.1	11.0/10.5

Remarks
- All cells measure 1.8 x 1.8 cm² in stead of 2 x 2 cm² except IT-P-cells
- The first figure gives the top-value, the second is the average.

(a) Substrates made by Mr. FALLY of the Lab. de Marcoussis, France
(b) Made by Photowatt-Caen
(c) Made by Italsolar, Italy, by courtesy of Prof. PIZZINI ; these cells measured 2 x 2 cm²

with each 4 cells) as well as on EG-Si SILSO-slices. For unclear reasons, the resulting sheet resistance was somewhat too high and not well reproducible, i.e. between 41.4 and 50.2 Ω/□. Most of the cells have accidently been cutted to a 1.8 x 1.8 cm^2-format, which involves a 9 %-coverage instead of 7 % as previously indicated. The results have been summarised in Table II.

In Table III we tabulate the results of a number of cells on substrates of the same kind (UMG-Si-slices + epi-layer, prepared by us) but made by the Belgian company SOLPRO S.E.S. with their standard integral screenprinting production process using our set of screens. The most striking difference is their diffusion step, which gives a typical sheet resistance of 25 Ω/□ with a junction depth of 0.6 - 0.7 μm. As monitor samples 3 SILSO-slices (thus giving 12 cells) have been added. First of all, the generally good performance of cells in both series (Tables II and III) must be mentioned, with efficiencies of up to 11.7 % (measured at

Table III : Solar cells on epitaxial UMG-Si-substrates made by SOLPRO

Substrate	Resistivity Ω.cm	Epi-thickness μm	J_{sc} mA/cm^2	V_{oc} mV	Fill factor %	η %
POLYX-F						
651	0.02	28.2	26.5/26.3	585/575	76.6/75.3	11.7/11.5
686	0.06	23.7	25.5/25.3	540/530	70.8/69.5	9.5/9.3
POLYX-PC						
P3/L2	0.035	28.3	25.8/25.3	545/538	73.7/72.1	10.3/9.9
SBB/L1	0.3	23.7	25.3/24.8	540/535	75.3/72.6	10.1/9.7
OEG/L3	0.03	28.3	26.3/26.1	555/555	74.1/73.3	10.8/10.7
IT-P						
W276	0.11	28.5	26.8/26.7	570/560	75.1/72.5	11.5/10.9
W280	2.0	28.3	25.5/24.5	535/529	73.2/71.2	9.8/9.3
W281	0.5	28.5	25.3/25.0	535/534	75.9/73.0	10.3/9.8
IT-M(a)						
TW112	0.06	33.2	27.3/26.9	570/565	76.0/73.0	11.6/11.0
IT	0.03	28.5	23.8/23.2	505/492	67.4/58.6	8.1/6.8
SILSO	0.00		26.8/25.3	550/547	77.1/75.8	11.2/10.6

Remarks
- All cells measure 2 x 2 cm^2
- For substrate-manufacturers : see Table II
- The first figure is a top-value, the second gives the average.

(a) Made by ITALSOLAR, Italy, courtesy of Dr. MARGADONNA.

AM1). V_{oc} is almost equal in both series, showing no influence by the diffusion process. Nevertheless it must be remarked that V_{oc} on heavily doped UMG-Si slices reaches much higher values (up to 585 mV) than on the standard SILSO-substrates (typically 550 mV), which is due to the BSF-activity in these thin-film cells. J_{sc} is slightly better in the IMEC-cells (top value 27.8 mA/cm^2) due to the thinner dead layer. This is clear from spectral response measurements on an IMEC-cell which shows a much better blue response than a SOLPRO-cell (see Fig. 6). J_{sc} is not consistently higher in the IMEC-cells. This could be caused by the varying sheet resistance, characteristic for these first experimental trials, in

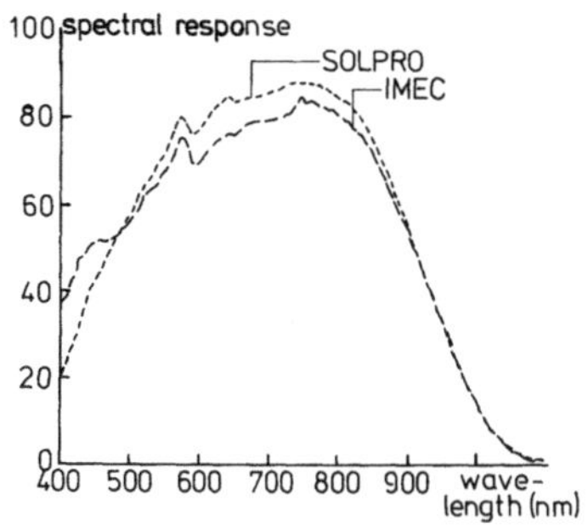

contrast with the production-environment at SOLPRO. On the other hand, the SOLPRO-cells show a very good fill factor, typically 75 %, for the complete series, whereas the IMEC-cells have some problems in this respect. Partly this has to do with probable series resistance problems accompanying our thin junctions, but partly this is also due to a problem with the printing of the fingergrid metallisation in this screenprint run (a problem that is solved at this moment), causing many interrupted fingers.

Fig. 6 Comparison of spectral response of SOLPRO and IMEC cell

5. Conclusions

After optimisation of the junction-forming-process we have succeeded in improving the short-wavelength efficiency of our cells by reducing the thickness of the dead layer. Combined with an optimised metallisation pattern, J_{sc}-values of 27.8 mA/cm^2 have been obtained. The use of heavily doped substrates, inducing an active BSF in these cells which can therefore be called thin-film cells, raised the V_{oc} up to a top-value of 585 mV. It may be remarked here that this is an important advantage which is difficult if not impossible to reproduce in standard SILSO-slices or analogues. Average efficiencies well above 10 % or even 11 % have been shown, with a top-value of 11.7 % on one of the POLYX-substrates. It may be stated here that by further optimisation of the screenprint diffusion step (especially with respect to the problem of reproducibility) and a new optimisation of the frontside metallisation (towards still finer fingers) an efficiency of 13 % is a realistic target for the near future. It is important to stress that we are using substrates made from UMG-Si feedstock material with a typical price of only 8 $/kg, that may contain thereupon a considerable amount of B, Al (the most difficult impurities for removal by cheap metallurgical treatments) and C.

Acknowledgements

We would like to express our thanks to J. FALLY, P. LAY, D. MARGADONNA and S. PIZZINI for supplying us with UMG-Si-wafers, and to L. FRISSON of SOLPRO S.E.S. for the production of a number of solar cells. We are grateful as well to P. LAERMANS for technical assistance in epi-layer growth and to A. JANSSENS for solar cell-production at IMEC. This work is financed by EEC contracts EN 3S/0074 and EN 3S/0075 and by the Belgian Ministry of Science Policy Contract CE/ENE/03.

References

1. M.CAYMAX, M.EYCKMANS, R.MERTENS, J.NIJS, M.RODOT, J.E.BOUREE, G.REVEL, J.L.PASTOL, G.SALA, J.FALLY, S.PIZZINI, 7th EPSEC, 1986, p. 800.
2. M. RODOT, S. N. SINGH, LE HOANG THI TO, V. PERRAKI, J. E. BOUREE, G. REVEL, NGUYEN DINH HUYNH, M. CAYMAX and M. EYCKMANS, this conference.
3. J. FALLY, D. GUIGNOT, L. GOEFFON ; 7th EPSEC, 1986, p. 754.

POSTER SESSION 06:
APPLICATIONS AND SYSTEMS

Experiences with a PV system for household applications*

G. Ambrosone, S. Catalanotti, A.M. Russo, G. Troise
Dipartimento di Scienze Fisiche, Università di Napoli, Facoltà di Ingegneria. Piazzale V. Tecchio 80, 80125 Napoli, Italy

U. Coscia, A. Sarno
ENEA CRIF, via Vecchio Macello, 80055 Portici (NA), Italy.

Tel: (081) 479288 - P.O.Box 83, Portici (NA), Italy

Abstract

A stand alone PV system (760 Wp) has been monitored over a long period. Features of the components and of the system are presented; experimental data of summer and winter period are compared; efficiencies, as well as surplus and deficit are discussed. Finally the values of interesting parameters are compared for typical symmer and winter days.

Introduction

We present a set of experimental data obtained in a stand alone PV system with a total peak power of 760 W. The block diagram of the system is shown in fig. 1. The PV field uses 24 modules (Ansaldo AP33D) of policrystalline cells. A standard controller (Sipre type B) is used to prevent overcharge and overdischarge of battery. The lead-acid storage battery (Tudor ATF230/21) nominal parameters are: V_o=24 V, C_o=280 Ah, I_o=28A. The load can be adjusted in a continuous way (in the range 1-50 Ω) to allow a detailed characterization of the field and battery.

The automatic monitoring of the system is performed by a computer assisted data logger that provides every 2 seconds a complete read out of data (solar flux, external and PV field temperatures; field, batteries and load voltages and currents). Field, battery and load powers are also computed, during the read out cycle; average values, of these data and powers, over a longer period (half an hour) are saved for further analysis. The experimental work has been organized according to these goals:

---

* This work has been developed under ENEA contract.

1) to obtain a complete experimental knowledge of the features and of the transfer function of the components (efficiencies can be quickly derived by the transfer functions);
2) to measure the system efficiency, surplus and deficit in different working conditions (metheorological conditions and system dimensioning).

1) Characterization of the components: field, controller and battery

a) Field

The open circuit voltage, the short circuit current and I_F, V_F (current and voltage of a working point near the maximum power) have been measured as a function of solar flux and field temperature. A three parameter fit is then used to obtain the field characteristic curves from the measured points. From this curves the maximum power P_{Fmax} (or the efficiency η_{Fmax}) have been computed. In fig. 2 is presented a fit of $P_{F\ max}$ as a function of solar flux. The spread of experimental points is mainly due to the field temperature. The statistic of data we have analysed is not yet significant to give a temperature parametrization; we only checked that the temperature dependence is correct (lower points correspond to higher temperature of the field).

b) Controller and connections

The plot of V_F (output voltage of field) as a function of V_B (input voltage of batteries or load) has shown that the voltage drop (V_F-V_B) has an initial step (controller diode) and linearly increases with increasing transferred current (ohmic drop due to overall electrical connections).

c) Battery

The battery voltage (V_B), the current (I_B) and the level of charge (Q) are measured. A typical plot is presented in fig. 3; $V_B = V_B\ (Q)$, for a constant value of I_B, exhibits a wide range of linearity. This linearity is due to the fact that the battery is used in the PV system in a "quasi stationary" and "low current" regime. In fact the solar flux, and consequently V_B and I_B, are slowly varying in time and furthermore I_B (for reasonable dimensioning of components) is a fraction of the nominal current of the battery.

The experimental transfer functions obtained for the components according to the discussed measurements are:

field

$$P_{F\ max} = \text{Area} \times \phi_{sun} \times \frac{9.26 + .00046\ \phi}{100(1+\exp\frac{40-\phi}{1000})}$$

controller and connections

$$V_B = V_F - 0.34 - 0.17\ I_F$$

battery

in linear range of charge or discharge:

$$V_{B\ Charge} = 25.6 + .0052\ Q - 2\ \exp\ (-.11\ I_B)$$

$$V_{B\ Discharge} = 21.6 + .0052\ Q + 2\ \exp\ (.11\ I_B)$$

A computer program supplies the battery transfer function in the complete range, including the non linear region.

2) Characterization of the system

We recall that $P_{F\ max}$ ($E_{F\ max}$) would be the power (the energy) delivered by the field if a maximum power tracker was used and if the system had an "infinite storage" (for instance a stand alone water pumping system with a maximum power tracker). In the system we are testing (constant load, finite storage, no maximum power tracker) the power (energy) delivered by the field, P_F (E_F), depends on the load and on the battery working point; we choose the dimensioning of the system to obtain $P_F \approx P_{F\ max}$, when the battery is in a low state of charge. With this dimensioning of the system, measurements have been performed in different periods of the year. Plots of $E_{F\ max}$ and E_F for typical winter and summer days are shown in figs. 4 and 5 respectively. The difference ($E_{F\ max} - E_F$) is the surplus: it is very noticeable in summer and practically absent in winter. The field efficiency ($\eta_F = E_F/E_{sun}$), for the same sample of winter and summer days, is plotted in fig. 6; the summer field efficiency is lower than the winter field efficiency as a consequence of the surplus.

A certain amount of the energy produced by the field is lost through the controller, the electrical connections and the charge-discharge cycle of the battery. Taking into account these losses we define the useful energy (E_u) produced by the system: $E_u = E_L + \eta_B\ \Delta E_B$ where E_L is the energy used by the load ΔE_B is the variation of the battery energy and η_B is the battery efficiency (charge or discharge efficiency for negative or positive ΔE_B respectively). The system efficiency ($\eta_s = E_u/E_{sun}$) is obviously smaller than the field efficiency, but their behaviour is substantially equivalent; numerical values for η_F and η_s are quoted in table 1.

The deficit of a system, in a certain geographical area, is strictly dependent on the number of days (n_o) during which the battery can supply energy to the load without contribution from the sun; higher is n_o, lower is the probability of deficit. For our system it is n_o=3.5, which is not high indeed; in fact in winter time the system had some deficit but no significant statistic is, at present, available.

The dependence of efficiency, surplus and deficit on the system dimensioning will be investigated in a future monitoring of the system.

Summer and winter daily mean values of E_{sun}, E_F, E_L, ΔE_B, η_F, η_s, surplus and deficit are quoted in table 1. The data used to quote these values correspond to a period of monitoring of some weeks.

Conclusions

Experimental transfer functions and efficiencies of components have been obtaione. Efficiency, surplus ad deficit of the system, in different metheorological conditions and for a fixed dimensioning, have been also measured. Once the dependence of these quantities on the dimensioning of the system will be obtained, a valuable set of experimental information will be available to dimension a stand alone PV system, and to foresee its operating conditions.

The authors would like to thank Mr. E. Coscia for his valuable help with the grafic presentation of this paper.

	winter	summer
$E_{sun,F}$	22 kWh	60 kWh
E_F	1.65 kWh	3.50 kWh
E_L	2.00 kWh	2.65 kWh
ΔE_B	- .5 kWh	.5 kWh
surplus	0 kWh	1.2 kWh
deficit	.65 kWh	0 kWh
η_F	7.5 %	5.9 %
η_S	6.8 %	5.3 %

T a b l e 1

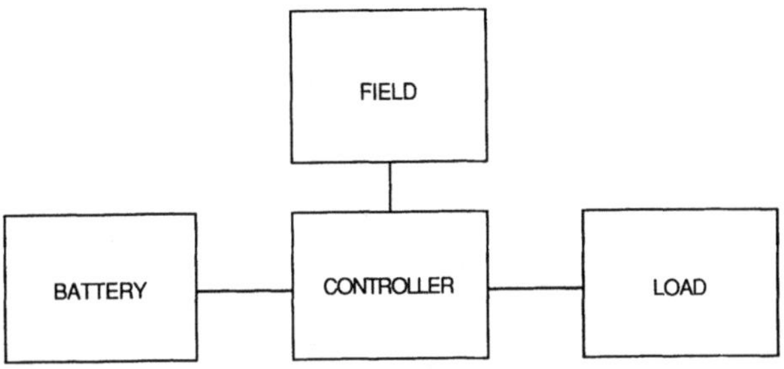

Fig. 1 - System block diagram.

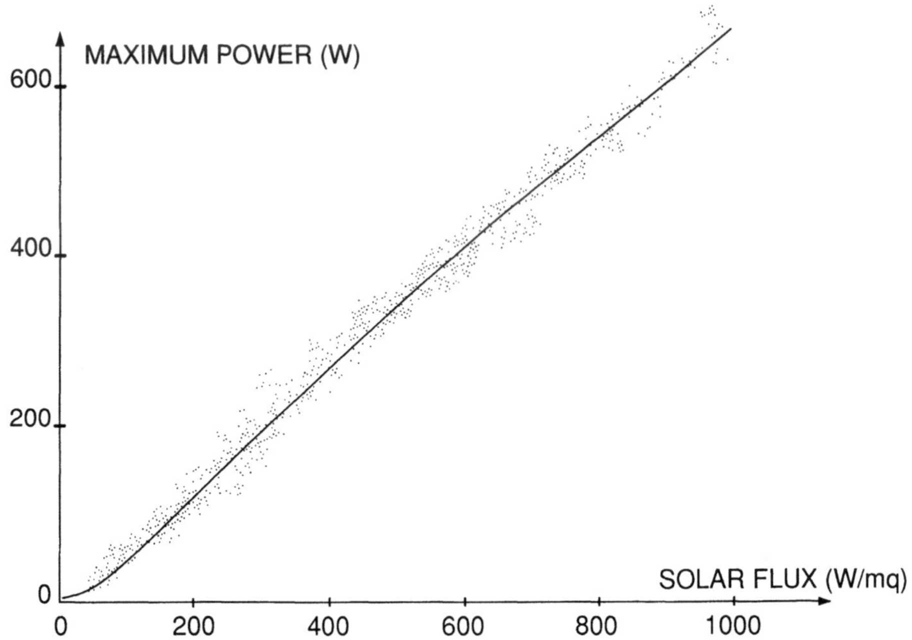

Fig.2 - Field maximum power versus solar flux

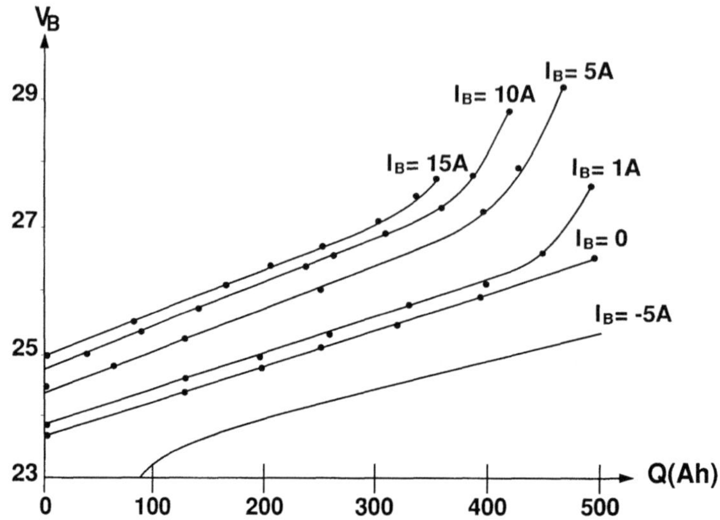

Fig.3 - Battery voltage versus level of charge (fixed value of current)

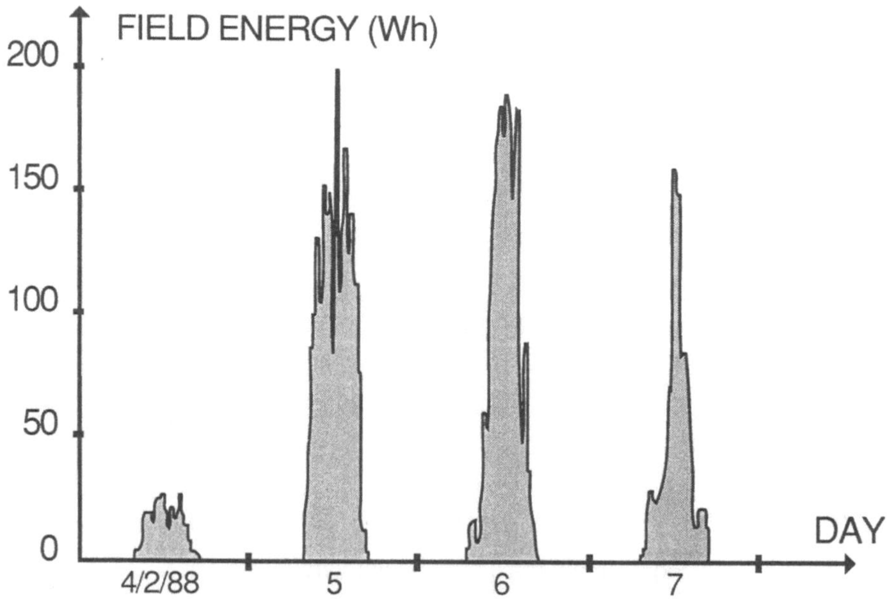

Fig.4 - $E_F max = E_F$ versus time (winter days)

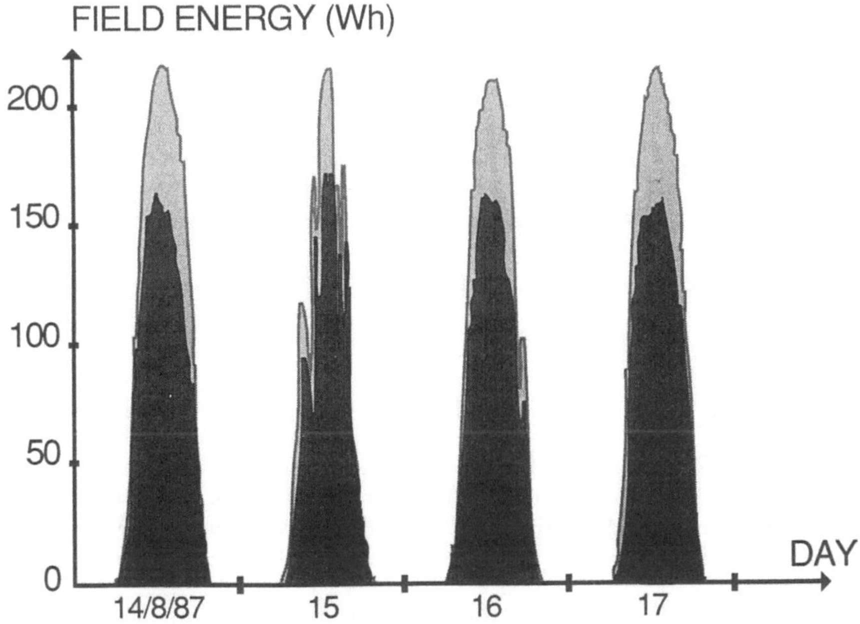

Fig.5 - E_F max (gray line), E_F (black line) versus time (summer days)

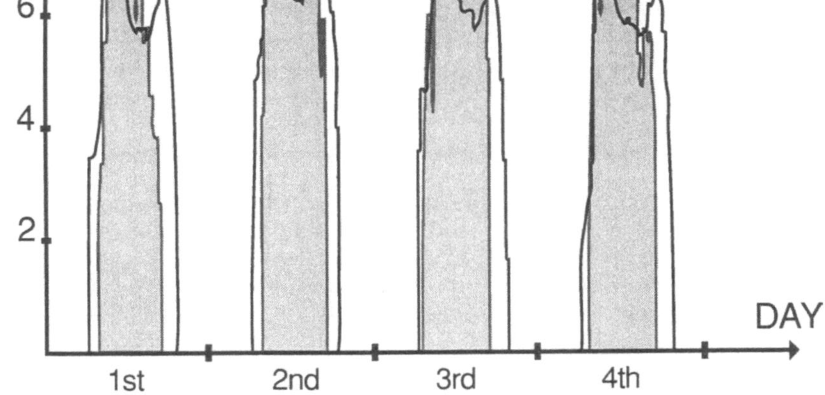

Fig.6 - Field efficiency versus time winter days (gray line), summer days (black line)

PROJECT DATA OF THE HIBRYD WIND-PHOTOVOLTAIC PLANT OF EAST OWEINAT (EGYPT)

F. Basevi - ITALSOLAR S.p.A. V. Augusto D'Andrea 6 Nettuno 00048 Italy

A. Paggi - AGIP S.p.A. V. A. Bonarelli 3 San Donato Milanese Italy

G. Santarella - S.E.S. S.p.A. V. Marino Ghetaldi 64 ROMA 00143 Italy

A. Zuffetti -AGIP S.p.A. V. A. Bonarelli 3 San Donato Milanese Italy

Summary

Italian Governament and Egyptian Governament decided to implement a Large Development Project finalized to the design, construction and operation of an agricoltural settlement, in Egypt, based on the exploitation of renewable energy sources.

The settlement is located at East Oweinat, an area of desert located 880 Km SSW of Cairo, with an exeptionally high solar insolation and a favourable wind regime.

The settlement producing sections and the related annual consumption of electrical energy are:

a) Laying-hens (future expansion) 30.000 KWh/y

b) Dairy-cows - Greenhouse - Buildings cold store, ancillary services 76.000 KWh/y

c) Irrigation 420.000 KWh/y

The wind generators, plus a 200 KW peak PV would meet the electrical demand of the settlement. An emergency generating system is garanteed by a diesel engine generators set.

Storage of energy is carried out electrically (daily storage) by lead acid batteries, and hidraulically (weekly storage) by a water basin.

1.0 INTRODUCTION

The Egyptian General Petroleum Company (GPC) and Desert Research Institute (DRI) carried out in 1982-1983 a soil and sub-soil study to define the developement potential at East Oweinat.

An estimate 1.000 billion cubic meters of good quality ground water have been discovered at shallow depths (water tables of 10-30 mt below surface), and over 3 million feddan of soil have been estimated potentially suitable for agricolture.

On this base a cooperation effort between Egyptian and Italian governaments was deemed necessary to implement a demostration project as a milestone for the development of the whole area.

The project studies and demostrates the estabilishment of a farming settlement to cultivate 166 feddan (70 hectares) based on renewable energies.

Some relevant aspects of the project are:

- the settlement is the first agricoltural and cattle-breeding farm in the western egyptian desert;
- it follows modern criteria of management and is provided of conforts for resident people;
- it is also the only agricoltural installation, built-up in the desert, completely self-sufficient from the energy point of view, making an integrated use of different sorts of renewable energies as solar energy, wind energy, biogas.

2.0 ENERGETICAL SERVICES

Energy balance of East Oweinat took into account all the available climatic data:

a) Solar radiation, temperature range, wind speed to evaluate producible energy.

b) Crops irrigation needs, and water depth, to compute the necessary energy to supply to different users:

- Irrigation pumps
- Ancillary services including biogas plant;
- Farm services (milking, milk refrigeration, etc.);
- General services of the settlement (lighting,etc.).

Energy is produced with an hybrid system composed by wind plus photovoltaic energy. Diesel generators are proveded for emergency situations.

3.0 CONTROL AND DATA ACQUISITION SYSTEM

A complete control and data acquisition system supervises the operation of the settlement.

It guarantees following functions:

- Energy surpluses sharing between elettrochemical storage and water basin (hydraulic storage);
- Solar and wind generators cut-off in case of over-production of energy;
- Priority loads selection for energy lack;

- Diesel generators automatic start-up in emergency situations.

The control system is composed by:

1) Main, central unit to process data, to display them on video and to print the relevant informations for plant management.
2) Slave, dedicated units to supervise specific plants.

4.0 PHOTOVOLTAIC PLANT

It is composed by following subsystems:

- Photovoltaic field
- Slave, dedicated control system
- Lead-acid batteries
- Power conditioning system and inverters

Photovoltaic field consists of 93 strings of series connected modules, the theoretical electric characteristics of photovoltaic array are:

- Open circuit voltage	560 Volts
- Short circuit current	640 Amps
- Voltage at MPP	380 Volts
- Current at MPP	530 Amps

All data are referred to standard conditions (AM 1.5; T=25°C).

The total peak power of photovoltaic field is 204 KWp.

Both monocristalline and policristalline modules are used.

Monocristalline modules of 72 cells 4" diameter each, and policristalline modules of 72 cells, 10cm x 10cm each, are manifactured in Italy by ITALSOLAR (ex PRAGMA) company of ENI group.

One module is composed of 6 columns of 12 cells, connected in series, and column are parallel connected inside the pannel, redundant parallel connections are provided every 4 cells connected in series. Every panel is equipped with by-bass diode.

To minimize mismatching losses, due to different characteristics of mono and polycristalline cells, strings with mono cells are composed by 36 panels in series, when strings wirh poli cells are composed by 38 panels each.

The subfield with polycristalline modules (24 strings) is a tilted system by means of a daily East/West solar traking device.

The remaining strings (69) are mounted with a fixed tilt angle of 15°.

Photovoltaic field is parallel connected with a large

batteries bank through 18 charge regulators. Each charge regulator controls 5 or 6 strings of the field.

Criteria of reliability and availability suggested the use of parallel charge regulators.

Charge regulators are realized with the use of silicon controlled rectifiers and are controlled from the dedicated control system.

The operating philosophy of the photovoltaic plant suggests the choise of a control system which has two main characteristics, such as:

- Low consumption
- High reliability (full manual back-up)

Its main function is to control the charge of the batteries by means of a two tresholds logic, the first to determine the maximum voltage allowed for the batteries, the second to set the level when the charge is reset (to avoid dangerous instability a certain difference between two levels is necessary).

The dedicated control system (slave) is connected to the main system through a serial link RS 232.

Power conditioning system is composed by the d.c. boards and the inverters section.

In the East Oweinat plant 7 inverters are used to supply a.c. energy to the loads.

Four of them are with voltage and frequency ramp start-up, to avoid reactive overloads due to the motors of the pumps.

The remaining three inverters are connected to resistive loads only.

5. WIND GENERATORS

Wind system includes the following engines:

N. 1 generator with double blade rotor

blades diameter	32 mt
hub height	26 mt (above ground)
electric generator	280 KVA (wind speed 7 mt/sec)
windward rotor	

N. 4 generators with double-blade rotor

blades diametr	8.5 mt
hub height	10 mt (above ground)
electric generator	30 KVA (wind speed 7 mt/sec)
leeward rotor	

Each engine is equipped with its own rectifier system able to transfer direct current an the common d.c. bar.

In this way wind energy will operate in parallel with

photovoltaic energy.

Dedicated control and data acquisition system have been designed to guarantee higest safety and to withstand the most severe environmental conditions.

6. DIESEL BACK-UP

Diesel back-up system is composed by two internal combustion engines, by electric generators and fuel tank.

The engines start automatically up when renewable energy production is low or in an emergency situation.

7. POWER STATION

For power station design and erection following criteria have been followed:- All the produced energy is conveyed to a common d.c. bar, at the nominal voltage of 300 Volts d.c.

Inverters to supply a.c. loads are connected to the abovesaid bar in parallel with battery bank.

For back-up purposes it is possible to by-pass inverters and send the a.c. energy generated by diesel engines directly to the loads.

Power station has been built in Italy with standard, preassembled shelters, including all electrical equipment, control boards, inverters, power centre, alternated current rectifiers and cumputerized control system.

Electrochemical storage, realized with lead-acid batteries, is placed in four modular containers with suitable ventilation, and heat removal system.

REFERENCES

1) A. PAGGI - A. ZUFFETTI Agip S.p.A. Project for the realization of a rural settlemet operating with new and renewable energy sources - October 1987.
2) EUROCONSULT - Intermediate Technology Power Ltd. - United Nations/Arab Republic of Egypt - Study of an Energy Centre at East Oweinat - December 1986.
3) A. Barra - F. Basevi - M. Farid - R. Filippa - C. Micheli and A. Zuffetti - Large Development Projects: The Egyptian-Italian Renewable Energy Settlement (EIRES), a working example.

IRAQ SOLAR REFRIGERATOR IN RURAL AREA

M.Humady, A.Maki*, S.Shawi and Y.Khader*
Scientific Research Council, Solar Energy Research Center,
P.O. Box 13026, Baghdad, Iraq.
* The light Industries Co., S.A., P.O. Box 164, Baghdad, Iraq.

Abstract

In this work a 5 cubic foot Ac refrigerator was modified to a Pv-Powered D.c refrigerator and the performance of refrigerator and power supply System has evaluated Over a period of one Year.The power supply modification was coupled to modification in the heat exchanger system where a93watt Ac compressor was replaced by a 70 watt Dc compressor.The power supply system consisted of PV modules of24 watt Peak Power and 210 A/H lead acid battery of 12 volt each coupled to an efficient charge controller.In the calculation of the power requirement the ON-OFF running time of the system was taken into considration and correlated to ambient temperature . Total Power consumption was found to be840watt per day for a summer day where ambient temperature averaged 32 °C. Under these conditions a minimum evaporating temperature of-5 C° was reached within a running time of 4 hours. Insulation material of the refeigerator consisted of 25mm layer of Polyurethane. The effects of various weather parameters on system performance was also monitored .

KEY WORDS

Refrigerator;Photovoltaic System;Compressor;Solar Intensity; Temperature;Maximum Power Point;Storage Battery

INTRODUCTION

A solar photovoltaic-based refrigerator would provide a useful facility to a community living in a rural area not covered by the national electricity power grid . A 5 cubic foot Ac powered refrigerator currently produced by local industry was modified in respect to its heat exchanger subsystem and in its energy supply system .Its Ac compressor of 1/8 Hp was changed to a Dc compressor of 1/10 Hp which is powered by solar generator system .This change eliminated the need for an invertor and allowed direct use of solar panels to meet power requirement of refrigerator.The insulation material for the refrigerator is 25mm thick polyurethane .

Sizing of the photovoltaic field

The photovoltaic system,designed for the solar refrigerator , is showm in figurel.The solar refrigerator is a variable load. However the load varies over a wide range depending on ambient temperature,with maximum power consumption in the summer and minimum in winter. In sizing the photovoltaic field it is important to find out the ambient temperature at which the refrigerator is designed to work.For this purpose the test of running and rest time was conducted at an ambient temperature of32°C in a simulation room with normal and maximum thermostat setting.The result of this test is shown in table 1.

Table 1 Test of Running and Rest time at 32 C

	cut out	cut in	cut out	cut in
EVAP.AIR TEMP. C°	-1	0	-5	-4
THERMOSTAT BULB TEMP. C°	-17	-4	-18	-4
EVAP.INLET TEMP. C°	-10	-2	-18	-6
CABINET CENTRE TEMP. C°	+5	+5	+2	+2
RUNNING TIME	12.6		19.2	
REST TIME	8.4		8.4	

The power consumption of the solar refrigerator had been found to be 70 watt at 5 C° evaporating temperature(Humady 1988).But within one hour -10 C°- evaporating temperature had been achievied and the power consumption reduced to 58 watt with thermostat set at normal position.The daily load requirement is EL=835Wh/day.To be on the safe side the daily average of solar radiation recieved on the horizontal surface is assumed to be Gsr=5kwh/m². day.
The following assumptions are made:
npv =efficiency of the photovltaic modules = 7%
nrg =efficiency of the charge regulator = 95%
nbat=efficiency of the sotrage battery block = 85%
Then the daily energy required directly from the sun is determined as (Buresch, 1983):

$$Esun = \frac{EL}{npv * nrg} = \frac{835 \text{ wh/day}}{0.07 \times 0.95} = 12.556 \quad \text{kwh/day.}$$

The area of the photovoltaic pannel(APV)which is necessary to produce the required electric energy is calculated as follows:

$$APV = \frac{Esun}{Gsr} = \frac{12.556 \text{ kwh/day}}{5 \text{ kwh/m}^2 \text{day}} = 2.5112 \text{ m}^2.$$

The peak power (PPV) corresponding to this field at the standard conditions,when the solar radiation intensity=1000 watt/m^2, and photovoltaic module temperature = 25 C^o is calculated as:

PPV = 1000(w/m^2)*2.511*0.07 = 176 watt

Long term performance tests of the photovoltaic module show that their output power may go down to 25% of its value at standard conditions when the cells temperature rises to 80°C. Therefore the actual peak power (PPVA) of the photovoltaic field is raised by one quarter of its value or:

$$PPVA = PPV + \frac{1}{4} PPV = 176 + \frac{176}{4} = 220 \text{ watt}$$

For the above purpose a single-crystalline sillicon photovoltaic module is selected.The maximum power of this module is 35 watts with a nominal voltage of 12 volt. The daily electric energy of the specified load is around 835 Wh/day. A battery block is always needed to handle the energy demands during nights and periods of very low solar radiation intensity. Lead acid storage batteries are selected for solar refrigerator.The measured amper hour (AH) efficiency of such cells is found to be86% with permitted discharge depth of 75%

$$C = \frac{835 \text{ wh/day}}{0.86 \times 0.75 \times 12} = 108 \text{ Ah/day}$$

Assuming two day with out sun,216Ah battery block of 12V is selected for the system.

Results and Discusion

The solar refrigerator has been in service for one year without interruption .No maintenance is required except cleaning the photovoltaic modules from time to time.The photovoltaic modules are tested under real weather conditions in the city of Baghdad.They were mounted on a support structure installed facing south at a till angle of 33 .

Figure 2,3 showes the effect of solar radiation and air temperature on the characteristics of these modules during the winter and summer respectively.Figure 4,showes the charge current (A H) generated from the photovoltaic modules.

On a clear day (15th of Nov.1987),the current was found to be 85 (A H).This amount of charge current is pretty high in comparism to the 35amper-hour consumption of the refrigerator during this month.This means that capacity of the battery which was initially designed to feed the refrigerator for two days without sun is sufficient to feed the refrigerator for five days without sun in winter.Also it is important to mention that 23 charge current(AH)has been generated from photovoltaic modules on a cloudy day in Dec.

CONCLUSION

The solar refrigerator has been working smoothly over the last year utilizing power derived from245watt peak power photovoltaic modules and a210AH storage battery designed for two days running of the refrigerator without sun.A charge controller was also used for protection of batteery against over-charge.the power consumption of the solar refrigerator varies widely , depending on ambient temperature .At 32°C , average for summer the power consumption,was830watt but drops down to 400 watt during the winter.
In order to minimize the heat transfer from surrounding to refrigerator the latter was insulated with25 mm polyurethane. The effect on power consumption of a200 mm thick layer of the insulator is being examined.This is expected to be effective in reducing power consumption,thus reducing the size of the P.V. system required and the cost in curred thereby.

REFERENCES

Buresch,M.(1983).System design procedures.In J.Robinson, S.Thomas(Ed.),Photovoltaic Energy System,Mcgraw Hill,New York. PP 175 - 196.

Humady,M.(1988).Proceeding Of Third Arab International Solar Energy Conferece,Solar Refrigerator,Baghdad,Iraq.chapter 3, pp 62-66.

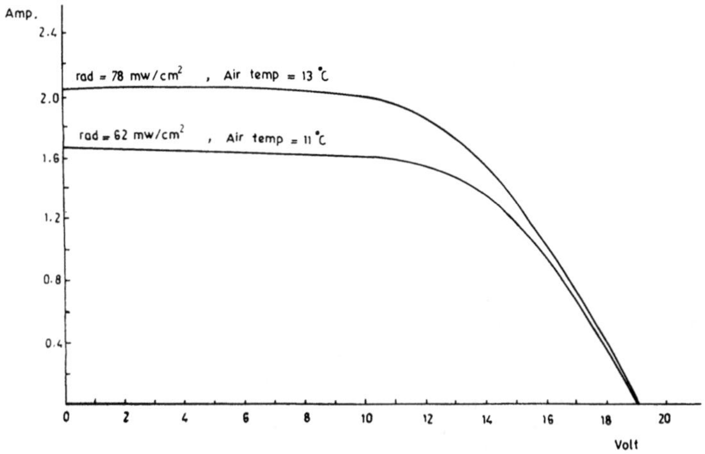

Fig. 2 The effect of solar radiation and air temperature in winter

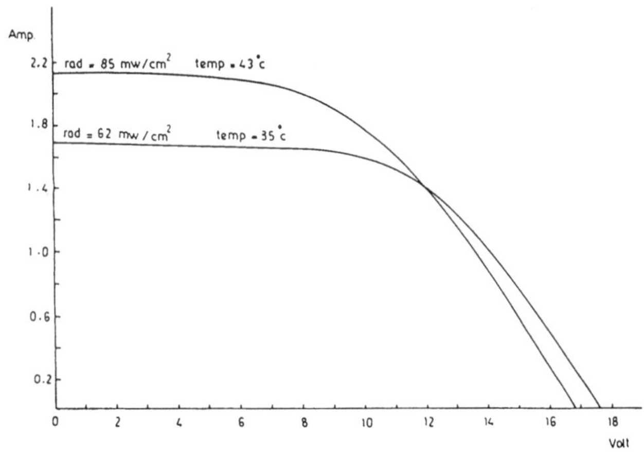

Fig. 3 The effect of solar radiation and air temperature in summer

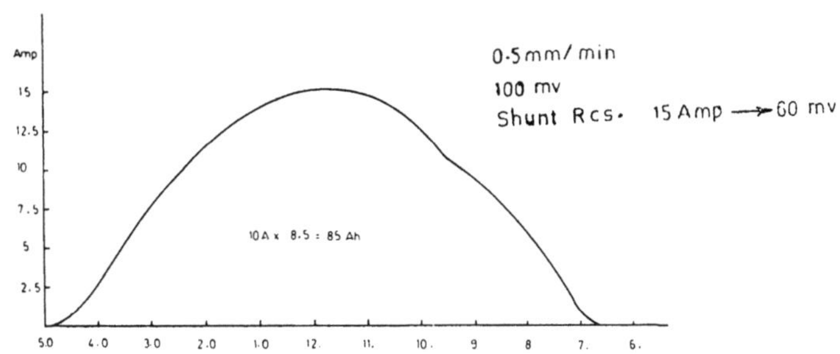

Fig. 4 Charge current (AH) generator from photovoltaic moclules at 15 NOV 1987.

THE CHOICE BETWEEN DC AND AC SYSTEMS FOR PV RURAL ELECTRIFICATION

W B Gillett & J E Bates
Sir William Halcrow & Partners
Burderop Park
Swindon SN4 0QD, UK
Tel: 44-793-812479

W Kaut
DGXVII, Commission of
the European Communities
200 Rue de la Loi,
B1049 Brussels, Belgium
Tel: 32-2-2353970

ABSTRACT. One of the main applications of PV systems is for rural electrification, where small pv generators are used to electrify rural houses, farms, hostels and mountain refuges, which are sited in remote locations far from the electricity grid. In these projects, the pv generators typically provide power for lighting and small domestic appliances such as refrigerators, televisions and kitchen equipment.

A survey of the available appliances indicates that, with the exception of dc lighting and TV's where the market is well developed, the consumer has a very limited choice of dc appliances when compared with his choice of ac appliances.

In this paper, the results of the survey are presented and the costs and performanace of typical dc and ac systems used for pv rural electrification are compared.

1. INTRODUCTION

Electricity from a pv generator may be supplied to an isolated house either as dc power or via an inverter as ac power. It is also possible to arrange for a combination of ac and dc power, in order to maximise efficiency whilst allowing a choice of conventional ac appliances for some purposes.

The relative advantages of centralised pv generators for groups of houses and individual pv generators on each house are discussed in ref 1.

When installing an individual pv generator, the options are:

(1) All dc system with dc appliances
(2) All ac system with single inverter and all ac appliances
(3) DC system with one inverter feeding a separate ac circuit
(4) DC system with several inverters to suit a range of ac appliances

The choice is influenced by the costs of the components and the suitability of the available dc and ac appliances.

2. RESULTS OF THE SURVEYS OF AVAILABLE DC APPLIANCES

2.1 General

A postal survey to more than 100 electrical appliance suppliers was carried out, first in spring 1987 and then again in spring 1988. The selected suppliers included companies involved in the PV industry, the leisure industry (boats, caravans, etc), and the conventional electrical appliance industry. The data obtained from these surveys were supplemented by those contained in a comprehensive French survey (ref 2).

The information received from companies varied greatly in quality, and only about one third of those contacted were able to supply dc appliances. Follow-up telephone calls revealed that ordinary consumers might find it quite difficult to obtain both commercial and technical information on dc appliances.

In the conventional electrical equipment industry, very few of the companies appeared to have any interest in dc appliances. In the leisure industry, companies were generally able to offer a range of dc appliances together with detailed price data. However, most were not able to provide technical data or operating efficiencies.

The pv industry response was also disappointing and suggested that few of the pv companies are yet ready to supply packaged systems for rural electrification.

In order to permit comparisons between products from different countries, all of the quoted retail prices were converted to ECU, using March 1988 rates of exchange. However, the prices quoted by different suppliers for the same product varied by up to about 20%.

2.2 Lighting

Four types of lighting were reported to be available for use with low voltage dc power (see Table I). The energy efficiencies of the different lamp types may be compared in terms of their outputs measured in lumens/watt. The costs of the bulbs and the lamps in which they are mounted are influenced by luminaire design as well as by the nominal rating of the bulb used. The cheapest and least efficient is the incandescent tungsten lamp which is commonly used for mains powered domestic lighting. The Tungsten or quartz halogen incandescent lamps often used in spot lights can be twice as efficient and their lifetime is twice that of ordinary incandescent bulbs.

Lighting Type	Watts	Lumens/Watt	Price (ECU)	
			Lamp	Bulb
Incandescent	5-100	10 - 18	9 - 20	0.4 - 3.1
Halogen	5-75	23 - 35	25 - 70	6.5 -11.5
Fluorescent	4-20	25 - 70	15 - 65	
	20-80	50 - 80	25 - 100	1.8 - 16
Sodium (low P)	18-55	100 - 130	170 - 320	11 - 50

Table 1 Available Low Voltage DC Lighting

Low voltage dc fluorescent lights employ a small, high frequency inverter and produce a light with less flicker than normal fluorescent lights running at 50 Hz. Fluorescent lighting is available in a variety of styles, including long tubes, circular tubes, and lamps which fit into an ordinary bulb holder. The range is already well developed for use in lorries, coaches, ambulances, caravans, boats, etc and is well suited for rural electrification.

Low pressure sodium lamps are extremely efficient and have a long life. However they produce a monochromatic yellow/orange light which limits their use to applications such as street lighting, where colour is not important.

2.3 Refrigerators

The available dc powered products include medical refrigerator/freezers designed for transporting and storing vaccine, portable cooler/chiller boxes, small refrigerators, refrigerator/freezers and deep freezers like those normally found in houses. All of the identified dc products run at 12 volts and/or 24 volts and several models could also run at 110V ac or 240V ac or from liquid propane gas.

Refrigerators employ either an absorption cycle or a compression cycle. Typically an absorption cycle refrigerator is rated at 70-100W, whilst compression cycle refrigerators are rated at 25-50W. However, compression cycle machines require a starting current which is usually at least 3 times their running current.

Refrigerators can be compared in terms of their average energy consumption over 24 hours with an outside ambient air temperature of 32°C. The available data on refrigerator performance and cost is shown as a function of internal capacity, in Figure 1 where it can be seen that the larger refrigerators are generally more efficient and less expensive per unit volume than smaller ones.

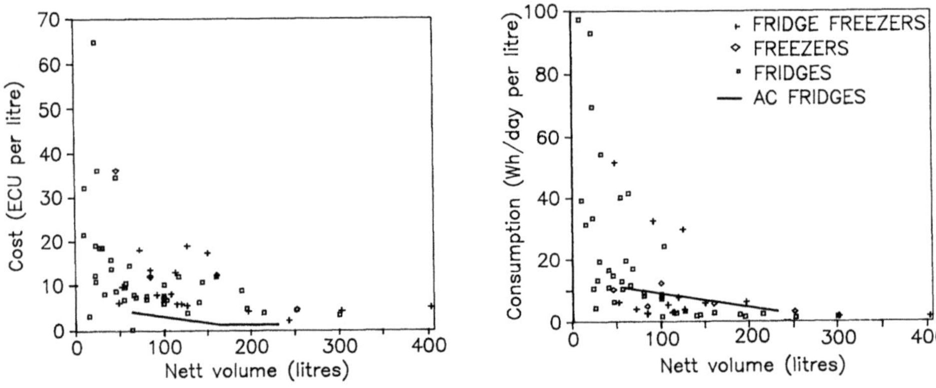

Figure 1 Energy Consumptions and Costs of dc Powered Refrigerators

The cheaper dc refrigerators are generally based on mass produced conventional models which have been converted to run with dc power. Several of the more expensive refrigerators have been specially designed with higher efficiency compressors and better standards of insulation. Prices vary between manufacturers, and there is a wide range of operating efficiencies. Many manufacturers do not publish data on energy consumption.

2.4 Televisions

Black and white televisions with 250 mm or 300 mm screens are widely available for use with 12V dc power at prices of between 70 ECU and 100 ECU.

Colour televisions with screens of between 250 mm and 375 mm are available for between 280 and 425 ECU, but usually need to be fitted with an adapter for use with a 12V dc power supply. Small black and white televisions are typically rated at 12-20W and small colour televisions at 35-50W when used with dc power.

2.5 Other dc Appliances

The survey confirmed that a wide range of other dc appliances is available. Most of these appliances are supplied by boating and caravan equipment distributors. Many portable battery powered devices are also available, such as radios, cassette players, hand held whisks, blenders and other kitchen appliances. Such appliances can be run from a low voltage dc supply or by using batteries which are recharged using a dc supply.

3. SURVEY OF AVAILABLE INVERTERS

A substantial number of small inverters was found to be available, though the majority produce a square or trapezoidal waveform rather than a sine wave. The costs of inverters increase with the quality of the waveform, and the overall efficiency, but the inverter cost per watt decreases as the inverter rating increases (see Figure 2).

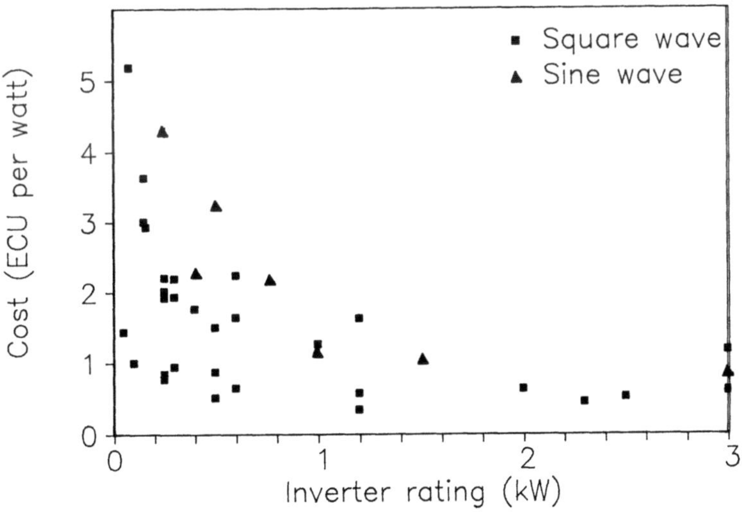

Figure 2 Costs of Available Inverters

4. TYPICAL COSTS OF STAND ALONE RURAL ELECTRIFICATION SYSTEMS

Quotations (FOB, excluding installation) were invited from ten of the leading pv systems suppliers for two alternative system specifications (A&B) to be installed in Southern Europe.

Both systems contain lights (5 @ 18W plus 10 @ 8W), a refrigerator, a small black and white television and a suitably sized pv generator. System B also has a small inverter (approx 500W) and an ac power socket.

The sizes of system offered by the suppliers varied from 318-540 Wp for system A and from 424-720 Wp for system B. The FOB prices varied substantially (see Table 2), mainly reflecting the different pv array and battery sizes proposed but also because some suppliers offered a comprehensive kit of parts whilst others offered only the main components. Clearly, some suppliers leave the installation details to the consumer, whilst others prefer to deliver a full installation package.

Price (March 1988)	System A	System B
System (ECU)	3470 - 7750	4930 - 10,035
System (ECU/Wp)	8.8 - 17.9	8.5 - 21.5
PV module (ECU/Wp)	4.1 - 6.6	4.1 - 6.6
Inverter (ECU)	-	570 - 780
Battery (ECU/kWh)	54 - 93	54 - 93

Table 2 Prices Quoted for PV Rural Electrification Systems

5 CONCLUSIONS

A DC system is the most energy efficient option for PV electrification of an isolated house provided that suitable DC appliances are available. However, DC appliances are more difficult to obtain in Europe than ac appliances, few are energy efficient and the choice is more restricted. The appliances which are sold for use in the leisure industry are usually not energy efficient and their prices vary significantly.

There is scope for more collaboration in Europe to develop cost effective packaged rural electrification systems for use in isolated houses. Since the large electrical appliance manufacturers are not interested by this market, it is necessary for the pv system suppliers to develop a range of good quality, energy efficient dc appliances and low cost inverters. To help the consumer, a European Code of Practice for PV Rural Electrification should be adopted and manufacturers should be encouraged to label their appliances, with power and standardised energy consumption ratings.

REFERENCES

(1) Kaut, W "PV Rural Electrification in the EC Energy Deomonstration Programme. Proc 8th EC PV conference,Florence 1988.

(2) Michel E, "Analyse des Equipements Electro-domestiques à Courant Continu pouvant être utilisés dans l'Habitat avec PV" Costic de Digne, Domaine de St Pierre, 0400 Digne Gaubert, France (Jan 1988). (This study was completed with financial assistance from AFME).

THE EFFECT OF SOLAR SPECTRUM VARIATIONS ON THE EFFICIENCY OF TANDEM SOLAR CELLS IN A SWEDISH CLIMATE

M. Andersson, L. Stolt, E. Wallinder, and L. Dahlgren*

Swedish Institute of Microelectronics
Box 1084, S-164 21 Kista, SWEDEN
*Swedish Meteorological and Hydrological Institute
S-601 76 Norrköping, SWEDEN

1. INTRODUCTION

Thin film tandem solar cells are being developed as one of the ways to make photovoltaic electric energy production cost effective. The tandem cells can be divided in two basic categories, two-terminal and four-terminal devices. One difference between these two cell types is how the variation of the solar spectrum affects the efficiency of the tandem cell. In the two-terminal tandem device the current generated in the high bandgap cell should be matched to the current generated in the low bandgap cell. Variations in the solar spectral distribution affects the current match and thus the cell efficiency is altered. In a four-terminal device, there is no constraint on current match and the power output from the high bandgap and low bandgap cells can be maximized independently at any solar spectrum.

The terrestrial solar spectrum vary significantly due to several reasons. Molecular scattering and scattering from aerosols, absorption by ozone, water vapor and other molecules affect the spectral distribution and the variation is dependent on the location, time of the year and time of the day. To investigate different tandem cells with respect to projected energy production it is important to know the actual variations of the solar spectrum at the location where the cells are going to be used.

In this work solar spectra have been measured at Norrköping and the data have been used to calculate energy outputs for three different tandem cells. The cell models used are based on published data from laboratory thin film cells and estimates of the performance of future thin film devices. Two-terminal and four-terminal tandem cells are compared. A two-terminal voltage-matched module is also included in the comparison.

2. SPECTRAL DATA

Spectral irradiance data have been measured since March 1986 at the Swedish Meteorological and Hydrological Institute in Norrköping. The spectra have been collected with a portable spectroradiometer (LI-COR model LI-1800) in the range 300-1100 nm at 5 nm steps. The instrument oriented towards the south and tilted 45° measures the global irradiance.

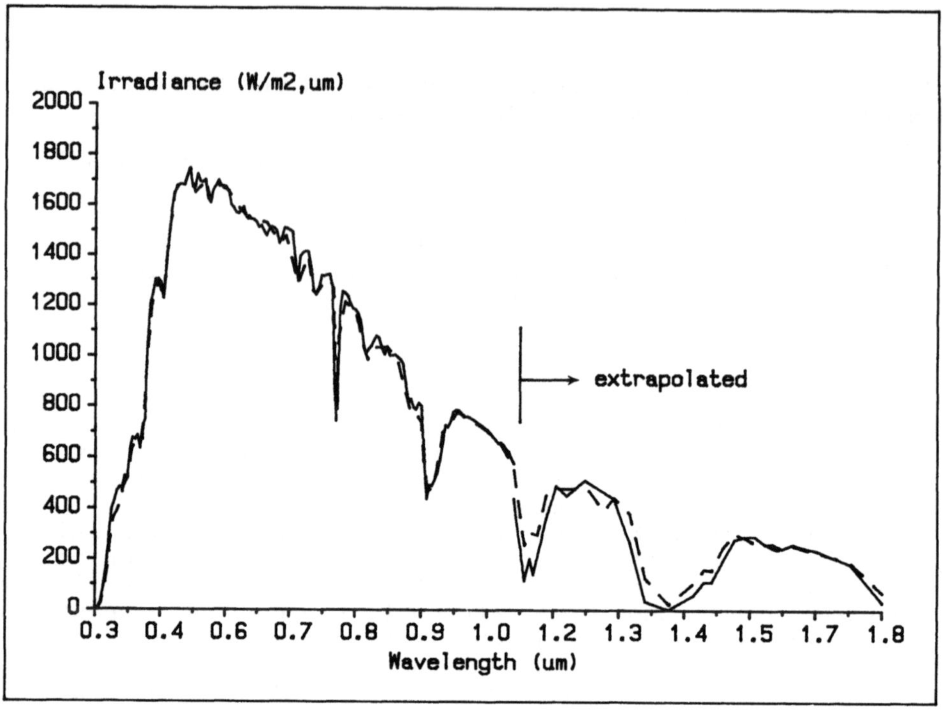

Figure 1. An example of a measured solar spectrum (300-1100 nm) extrapolated with a normalized AM 1.5 standard spectrum. Also included in the figure is a spectrum calculated by using the SERI-model fitted to the measured spectrum (dashed line).

The solar cells respond to light up to a wavelength of about 1300 nm. The measurement equipment is sensitive up to 1100 nm and to cover the missing interval we have extrapolated the measured spectra in the following simple way. A standard AM 1.5 spectrum (global irradiance, 37°) from Ref.[1] was normalized to the measured spectrum at 990 nm and inserted in the missing wavelength interval. The result for one case is shown in Fig.1 together with a spectrum calculated by using a SERI model, SPCTRAL2[2]. The SERI model is fitted to the measured spectrum and in the extrapolated part a small difference can be observed. This is due to a higher water content in the standard AM 1.5 spectrum than in the measured (and modelled) spectrum which causes stronger absorbtion around 1100 and 1400 nm. The SERI model was also fitted to a few other measured spectra and the example shown is a case where the difference in the extrapolated region is large. The SERI model is developed for clear sky conditions while the Swedish climate includes conditions with different degrees of overcast of almost equal importance (see below). For this reason we have chosen to use the AM 1.5 extrapolation in the calculations.

One problem when collecting spectra for this specific calculations (e.g. the two-terminal cell power output) is to know that the measured spectra represent true momentary spectral distribution of the irradiance. During the collection of one spectrum the instrument sweeps from 300 nm to 1100 nm in 25 seconds. When the conditions are stable the measured spectra will of course represent the momentary spectra but at unstable conditions the measured spectrum might have been distorted, for example by a cloud passing by the sun. This represents a severe problem when using automated collection of spectral data. During the collection of spectra in this work no algoritm was used to sort out distorted data. Instead a manual selection of the spectra used in the calculations was performed. This limited the number of spectra that could be used why the following approach to estimate the yearly energy production was used.

Sweden has a climate with a large fraction of diffuse irradiation which makes all levels of irradiance important. In Fig.2 the insolation year 1984 is shown as classified in different levels of irradiance. To make an estimate of the yearly energy production by tandem solar cells in this climate, 20 solar spectra covering the different levels of solar irradiance were selected. For each class of irradiance an average efficiency was calculated and by weighing these efficiencies with the yearly irradiation of the respective class, an yearly average efficiency was estimated.

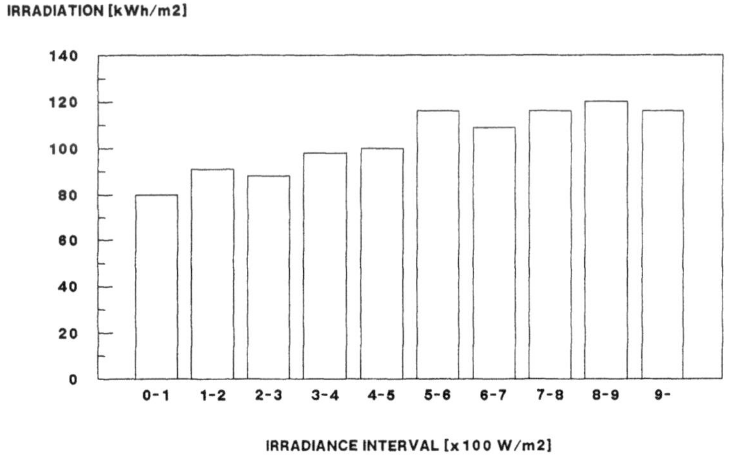

Figure 2. The irradiation classified in different levels of irradiance year 1984 on a south oriented surface tilted 45°. Location Älvkarleby, Sweden (Yearly total 1034 kWh/m^2).

3. CELL MODELLING

The tandem cells modelled in this investigation are all based on thin film concepts. All have a $CuInSe_2$ low bandgap cell (bottom cell) with a high bandgap cell (top cell) of three different types, a-Si, (ZnCd)S/$CuGaSe_2$ and ZnO/$CuGaSe_2$. The $CuInSe_2$/a-Si tandem structure is an existing device developed by ARCO Solar[3].

The $CuGaSe_2$ top cells are modelled to be fully transparent to subbandgap light, i.e. by assuming the quantum efficiency of the bottom cell to be equal to the quantum efficiency of a single junction $CuInSe_2$-cell[3] multiplied with the transmission of a $CuGaSe_2$-layer. This method gives an ideal bottom cell in the sense that no optical losses, e.g. absorbtion in the transparent back contact and reflection in boundaries between the layers of the structure, are assumed.

Fig.3 shows the quantum efficiency curves of the cells modelled. The $CuGaSe_2$-cells are of two types, one with a (ZnCd)S window layer and the other with a ZnO window layer to obtain an extended blue response. The former is modelled by using published data from experimental results of single junction cells[4,5] and the latter by using the same data with the quantum response extrapolated to shorter wavelengths by adding the quantum response of a ZnO/$CuInSe_2$ single junction cell in this interval.

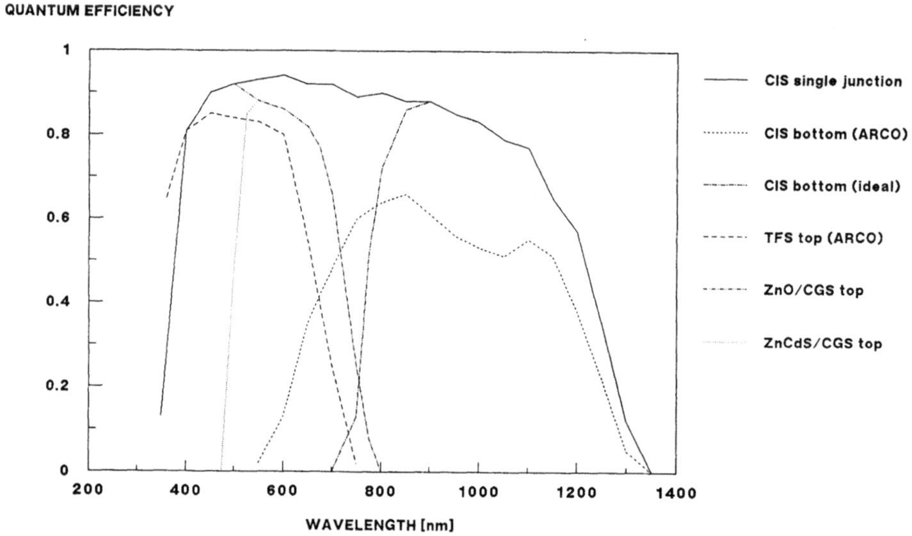

Figure 3. The quantum efficiency curves of the modelled thin film cells.

The photogenerated current I_L is calculated by multiplying the solar spectrum and the quantum efficiency at the different wavelengths and integrating over the wavelength interval.

The calculations of solar cell efficiency of the different tandem structures are made using an I-V model of the individual cells according to Eqn. 1.

$$I = I_S \cdot \left(e^{\beta \cdot (V - R_S \cdot I)} - 1 \right) - I_L \tag{1}$$

Appropriate values of the saturation current I_S, ideality factor n (β=q/nkT), series resistance R_S, have been estimated from our own cell characterization or from published I-V-data of thin film cells[5,6,7,8]. The parameter values used are presented in Table I. For comparison a single-junction $CuInSe_2$ cell is also included in the study.

Table I. The I-V-parameters of the studied cells.

Cell	I_S[A/cm²]	n	R_S[ohm]
CuInSe2 (ARCO)	$7 \cdot 10^{-7}$	1.7	1
a-Si (ARCO)	$1 \cdot 10^{-10}$	1.8	1
(ZnCd)S/CuGaSe2	$5 \cdot 10^{-12}$	1.7	1
ZnO/CuGaSe2	$5 \cdot 10^{-12}$	1.7	1
CdS/CuInSe2	$5 \cdot 10^{-7}$	1.5	1
ZnO/CuInSe2 (ARCO)	$7 \cdot 10^{-7}$	1.7	1

4. CALCULATIONS

For each combination of top and bottom cell as well as solar spectrum the current, voltage and efficiency for the individual cells and for the three different tandem configurations are numerically computed.

The tandem configurations are a four-terminal and a two-terminal device and in addition a configuration in which m bottomcells in series are connected in parallel with n topcells in series (Fig.4). The integers n and m are adjusted to have the bottomcell voltage matched to the topcell voltage in order to obtain a two-terminal tandem module without the restriction of current match as in two-terminal cell structures.

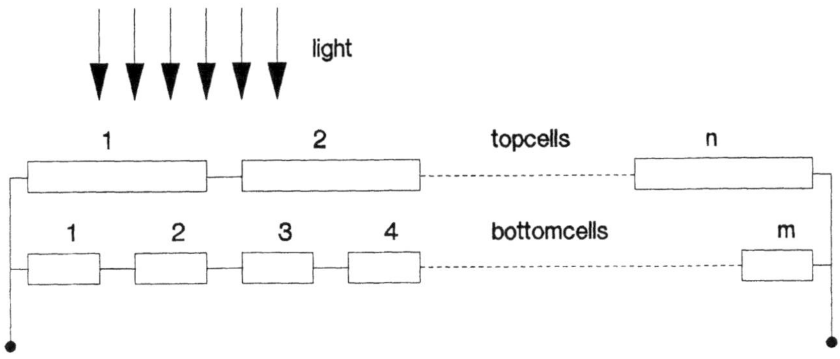

Figure 4. A voltage-matched two-terminal tandem module.

5. RESULTS

The results from the efficiency calculations of yearly averages are presented in Table II. It is clear from the table that the voltage-matched tandem configuration yields nearly the same efficiency as the four-terminal cell. The difference between the two-terminal and four-terminal configuration is about 3% in yearly energy production.

Table II. Calculated yearly average efficiency of different types of tandem solar cells.

Cell types	Tandem configuration	Yearly average efficiency [%]
a-Si (ARCO) + $CuInSe_2$ (ARCO)	4-terminal 2-terminal Voltage-matched	14.2 13.8 14.2
$(ZnCd)S/CuGaSe_2$ + $CdS/CuInSe_2$	4-terminal 2-terminal Voltage-matched	14.5 14.0 14.4
$ZnO/CuGaSe_2$ + $CdS/CuInSe_2$	4-terminal 2-terminal Voltage-matched	18.0 17.7 17.9
$ZnO/CuInSe_2$	single junction	11.9

6. CONCLUSIONS

The differences in yearly energy production in the Swedish climate between the three tandem configurations studied are small. The calculations show that on a yearly basis a two-terminal cell will produce about 3% less electric energy than a similar four-terminal cell. It was found that it is possible to obtain almost as high efficiency with a voltage matched two-terminal tandem module as with a four-terminal tandem cell.

ACKNOWLEDGEMENTS

This work has been funded by the National Energy Administration.

REFERENCES

1. Solar Cells, 15 (1985) 365
2. R. Bird and C. Riordan, SERI report TR-215-2436, Dec. 1984
3. K. Mitchell et. al., Proc. 19th IEEE Photovoltaic Specialists Conference, New Orleans 1987, p.177
4. H. W. Shock et. al., Proc. 7th E. C. Photovoltaic Solar Energy Conference, Seville 1986, p.465.
5. B. Dimmler et. al., Proc. 19th IEEE Photovoltaic Specialists Conference, New Orleans 1987, p.1454.
6. K. Mitchell et. al., Proc. 18th IEEE Photovoltaic Specialists Conference, Las Vegas 1985, p.914
7. W. Chesarek et. al., Proc. 19th IEEE Photovoltaic Specialists Conference, New Orleans 1987, p.791
8. R.R. Potter, Solar Cells, 16 (1986) 521

THE IMPLEMENTATION OF PHOTOVOLTAIC CATHODIC PROTECTION SYSTEMS

F.T.Tanasescu;N.Olariu;C.I.Popescu
ICPE
79623 Bucharest, Romania

ABSTRACT. Photovoltaic cathodic protection is the best solution for the structures subjected to corrosion, mainly in oil and gas industries. Optimum design considerations and system description are presented. The results obtained through the implementation of photovoltaic cathodic protection for oil pipes in Romania are delineated.

1. INTRODUCTION

The protection of buried metal parts subjected to corrosion is a major problem for oil and gas pipelines and wells, hazardous fluids pipelines, chemicals storage tanks, bridges and other metallic structures.
The active cathodic protection (CP) is using an electrical current to counteract the natural corrosion electrochemical currents. This current is driven from AC grid power rectifiers, or in more remote areas, one might use Diesel or natural gas driven motor generators or even thermoelectric generators.

This active cathodic protection can be hindered by the extremely high costs associated with grid power extensions or by the fuel and maintenance costs.

The development of the photovoltaic (PV) alternative has given a new reliable and economical solution to CP powering.

2. METHOD

2.1. General system description

Usually, CP systems are working in constant output current conditions (with external adjustments, if any). This approach implies real current requirements testing, soil resistivity tests and circuit resistance calculations in order to have a proper design for PV array.

However, it is implicitly assumed, that during PC systems operation, these conditions are more or less constant. This assumption is hardly met in real working conditions, when, for example, moisture content, temperature, compactness or even salt content of the soil may vary and drastically affect the cathode potential (preventing the protection).

In our opinion, the best solution was to use a two step procedure:

- a proper design based on soil tests, assuming steady state conditions (a certain working point)
- with a small cost sacrifice, to include a closed loop controller, able to self adjust the output current for changing conditions (in a range around working point).

2.2. Optimum criteria

The most important parameters to be considered for an efficient PV powered CP system are the electrical resistance due to the grounded anode and the electrical potential distribution along the pipeline.

The influence of grounded anode resistance on PV system costs was studied before (4).

In the same way, the influence of the potential distribution (and, accordingly, the length of the pipe branch to be protected by a single CP unit) could be taken into account.

The potential distribution along the pipe (figure 1) can be written as:

$$V(x) = -(WoI/2) \cdot \exp(-\gamma \cdot x) \qquad /1/$$

where: $Wo = (W/g)^{1/2}$ is the characteristic resistance, I/2 is the current in a half of the pipe branch, $\gamma = (Wg)^{1/2}$ is the decay constant, W is the resistance per length unit of pipe and g is the transfer conductance (assumed as a constant).

Consequently, the length of the pipe branch to be protected is given by:

$$L = 2/\gamma \cdot \ln(\Delta Umax/\Delta Umin) \qquad /2/$$

(refer to figure 1 for ΔUmax and ΔUmin).

Depicting, as in figure 2 the injected power per minimum injected power versus $\gamma \cdot L$ variable, one obtaines that smaller branch lengths are better (for a given γ). But, considering the PV price, a compromise can be reached for lowest total cost (figure 3), which gives:

$$L = 0.85/\gamma = 1300 \text{ meter}$$

A value of $\gamma = 5 \cdot 10^{-4}$/meter was used (from actual measurements-figure 4)

2.3. System configuration

The CP system provides the injected current into the pipe from a PV array through a charge regulator, storage battery and a closed loop current controller able to work either contnuously, or in a temporised mode.

The controller is measuring the actual voltage (Vr) of the pipe branch against a burried reference electrode $Cu/CuSO_4$ and adjusts the output current (Io), to keep Vr at the effective protection value.

Other parameters, very difficult to be considered at the very beginning, like pipe capacitance and electrochemical polarisation due to ionic conductivity of the soil, are automatically compensated in this way. Note that pipe capacitance is acting like a short term storage.

3. RESULTS

A two year testing period for a CP system composed from: 60 Watt PV array 90 Amper-hours/12 Volts DC battery, serial charge regulator and closed loop current controller, was accomplished.

In figure 4, there are presented the principal system parameters during a typical day, and in figure 5, the potential distribution along the pipe branch is depicted.

4. CONCLUSION

The economical efficiency of PV powered PC is based on the ability to reduce the voltage drops in the system by a proper design of conductors section, anode bed resistance, pipe branch length and distance between anode and pipe. A closed loop current controller is able to maintain the optimum working point, even in worse cases, avoiding the oversizing of the PV array. Depolarization time (16-18 hours for a very well insulated pipe) is very important in temporized operation, especially for a low charged battery.

5. REFERENCES

1. S.A.Jones.'Solar Power - An Alternative Energy Source for Cathodic Protection' Corrosion/80, Chicago (USA), 1980
2. C.Hemmingham.'The Intelligent Controller for Cathodic Protection' Proc.6th EC Photovoltaic Solar Energy Conference, London (UK), 1985, pp.361-365
3. G.A.Finch.'Cathodic Protection' Proc.5th EC Photovoltaic Solar Energy Conference, Athens (Greece), 1983, pp.336-340
4. C.Hemmingham.'Cost Optimisation - Solar Systems and Cathodic Protection put in Perspective' Proc.7th EC Photovoltaic Solar Energy Conference, Sevilla (Spain), 1986, pp.236-239

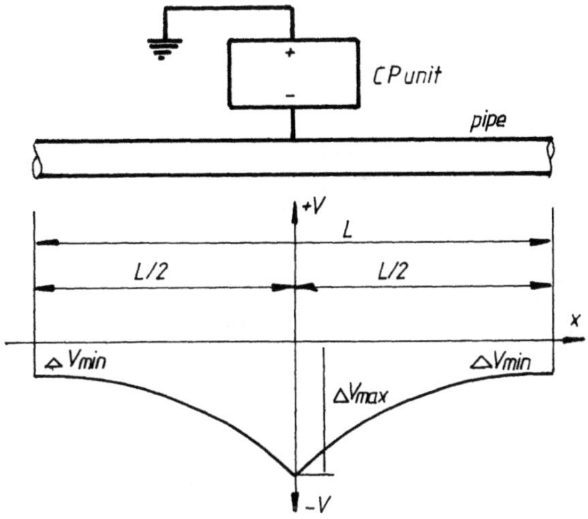

Figure 1. Pipe potential distribution

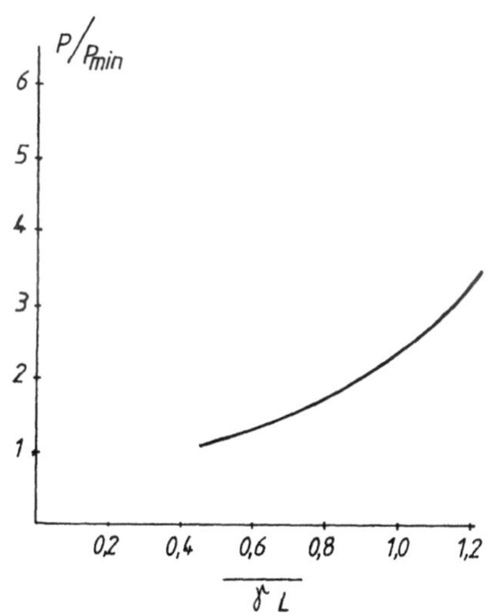

Figure 2. CP power demand
(length criteria)

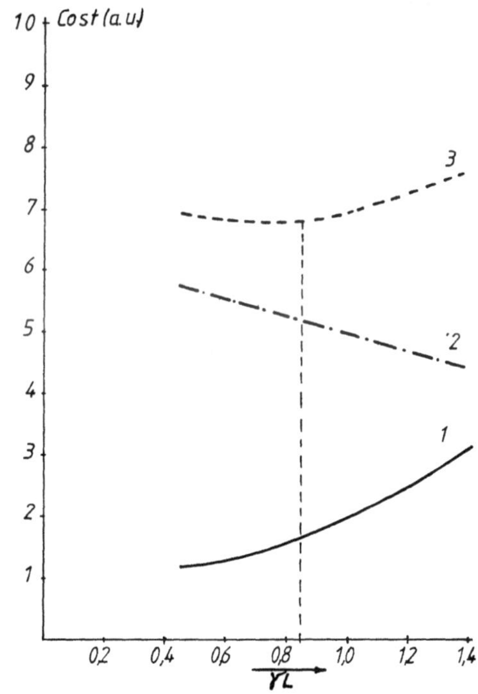

Figure 3. CP cost optimum
1:PV cost
2:anode cost
3:total cost

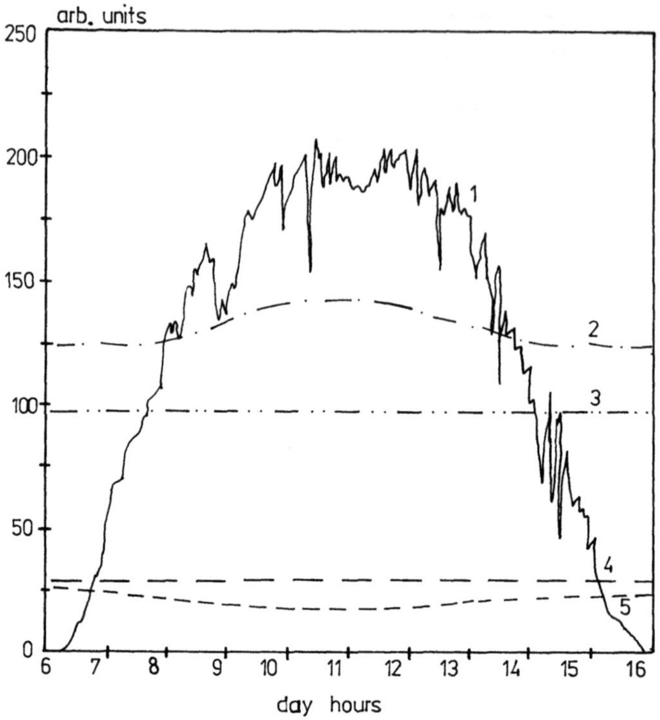

1:PV current
2:Battery voltage
3:Pipe voltage
4:Injected current
5:Battery current

Figure 4. PC system parameters

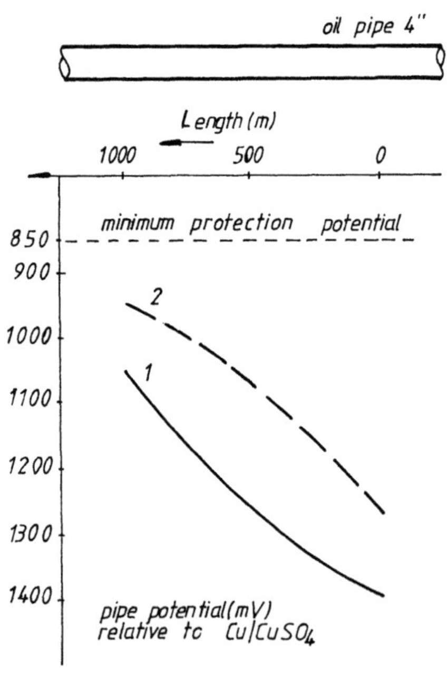

Figure 5. Pipe potential distribution against Cu/CuSO4 electrode
curve 1: sunny day
curve 2: night after cloudy day

OPERATIONAL RESULTS OF A 8.1 KwP P.V. INSTALLATION IN A SUB-SAHEL VILLAGE IN SENEGAL

E. Lorenzo, M. Montero, M.A. Egido
Instituto de Energía Solar
E.T.S.I. Telecomunicación - U.P.M.
28040 MADRID (SPAIN)

SUMMARY. Experience from 19 months of operation of a 8.1 Kwp p.v. generator in a senegalese village is explained. Technical, social and institutional aspects are separately presented. Disagreements between real energy consumptions and widely accepted suppositions are observed. The user's committment appears as a key point for the success of the project. The main problems stated at present are being derived from institutional aspects.

1. INTRODUCTION.

Noto is a subsahel senegalese village with 500 people where a 8.1 Kwp P.V. generator operates from September 1986, as the result of a cooperation project between Spain an Senegal. The generator delivers power via a 220 V-AC local grid to two water pumps, a health center, a millet mill, an ice-making machine, a recreation center (lighting, sound equipment and television) and public and private illumination (see reference (1) for details of this project).

In order to analize the performance of this installation the following parameters are recorded: radiation, ambient temperature, output from the PV array, input and output of the DC-AC inverters and consumption of each individual load. The values are manually recorded by the PV installation operator. Two different kinds of values are recorded. Instantaneous values of front and back irradiance, ambient temperature, array output current, input current to inverter, battery voltage and power delivered to load, are recorded at midday (only once a day). Daily integrated values of front and back irradiations, array output current, input current to inverter and energy delivered to each individual load are recorded, also once a day, at the early morning.

This manual adquisition data process is proving to be very efficient. Data are being regulary recorded and their analysis permits to get a complete vision of the daily system performance. An additional advantage is that it helps the constant presence and the involvement of the operator. The latter is considered of a great importance during at least the first running period of the PV system.

Despite data are still routinarily being recorded, from March-87 they are being retained by the utility committed for the plant operation (SENELEC). Because of that, they have not been analized from this data. Such a painful fact is due to problems appeared between the different senegalese Organisms involved on the project. None reasons have officially been offered at present. The answer to be continous requests of the spanish Solar Energy Institute have always been promissary but, at present, the problem remains. Consequently, the quantitative analysis of the operation plant is restricted to the first five months.

Moreover, an engineer and an anthropologist have visit Noto twice, with the main objetive to analize the user's response.

For explanation purposes will divide this paper in three different parts.

2. TECHNICAL ASPECTS.

(a) Two significant troubles have appeared at present. Both due to failures of inadequate designed components of the power conditioning system (a diode and a contactor). In both cases it was possible to repair the plant without local visits of spanish technicians. The information provided by the plant operator has permitted the power conditioning manufacturer to find the solution at factory level. This has been possible due to the high skill and positive aptitude of both, the senegalese plant operator and the spanish power conditioning manufacturer.

From such troubles three months of out or abnormal operation have been derived. We have not received notices about new failures from the last four months.

(b) Fig. 1 shows the energy flow corresponding to the first 120 days of operation. Data show a low overall energy efficiency of the inverters which stands in contradiction to the good inverter's efficiency (85%) for all loads exceeding 20% of nominal peak power. A consequently survey showed that such a low overall energy efficiency was associated to the non load operation losses derived from some troubles with the automatic switching of the inverters to accord the load condition (very often the 4000 VA inverter was switched on, while none load was operating). In February 1987 we introduced some modifications in order to solve this problem. From this date all seems to be right but the above mentioned blockage of data is obstructing the quantification of the efficiency.

(c) The accumulation of sand over the floor is causing a quick degradation of the albedo reflector (the reflecting has dropped from 0.44 to 0.26). This causes the real peak power of the PV array which is 12,4% below the initially supposed. This fact needs to be taken into account if bifacial PV panels are again employed in this particular area.

(d) Important disagreements between the real consumptions (daily water/person, etc.) and the initial hypothesis, based on the widely accepted supposition of development organizations (HWO, etc.) are observed. The current daily water consumption is about 20 l/person, far from the 40 l/person recommended by the HWO but also far from the 7 l/person consumed before PV installation. The mill yield is 15,5 kg/Kwh, about 60% of the usual claims of the european mills manufactures but very close to the figures reported by other authors[2]. The reason is that the flour for "couss-couss" (the most typical meal in this area) is significantly thinner thah the flour for bread. Consequently more energy for milling is required.

(e) The evolution of the daily energy consumption reflects the life habits adaptation to the new conditions. A clear example of the latter is the private illumination that exhibits a typical "learning shape" (see Fig. 2).

3. SOCIAL ASPECTS.

a) Social aspects have been a constant on our priorities along all the project. A social worker has been included in our team from the beginning and has participated on the design, implementation and evaluation tasks. We consider this work a key factor for the succes of the project.

b) During the installation, some of the local Authorities tried to use their social position to get more than their fair share of energy available.

c) The management of the new facilities is largely leaning upon yet existing social Organizations. For example, the women's association is committed for the management of the mill. They paid the full bill for the milling building (150.000 FCFA). The time switching for this facility reflects theirs suggestions. They charge 7 FCFA (3 pts)/Kg of grain. The cash obtained this way is covering the maintenance (tamis, etc.) the salary of a mill operator and still they have saved 50.000 CFA! (Yearly earnings in this regions are about 150.000 FCFA/family). The communal center holds the television-set, the music equipment and the ice-making machine. The management of all these facilities are committed to the case of young's association. They charge 25 FCFA (10 pts.)/l of ice. It is foreseen than the SENELEC charge all the delivered electricity at the current price (109 FCFA/Kwh). Private illumination bill will be share by the users, according to the number of lamps. The bills of milling and ice will be covered by the women's and young's organizations, respectively, and bills corresponding to water pumping and public illumination will be covered by the rural community organisation.

The starting of charging mechanism are just waiting for the signature of an agreement between the SENELEC and the MDIA (Senegalese Ministry of Artisan and Industrial Development). Such a signature is also being hidden by the same kind of problems that are blocking the analysis of recorded data.

(d) The above described way of services management not only has prevent against the possible conflicts with the existing social structure but also has contributed to reinforce it. The existence of new facilities to share (more common interests, new problems to debate, more need for meetings, etc.) is decreasing the rivalries (etnic, generational, etc.) and increasing the internal cohesion of the whole social structure.

(e) It is very difficult to analyze the effect of any particular action (i.e. installation of a PV system) on a complex social and economic structure which is already in the middle of many changes. To determine "cause-and-effect" is impossible. It is possible to isolate where certain changes are associated with certain other changes, but it is impossible to establish causality.

In particular, it is now possible to see in Noto new habits that certainly are associated to the new easiness of watering. For example, the bathing of children extremely uncommon before the PV installation, is becoming a general practice.

The farming and irrigation of "bissap" is other interesting example. The "bissap" is a well appreciated condiment for preparing drinks and meals. Before the PV system, each women directly marketed hers eventual surpluss in the neighbour market-places. Now, such surpluss is becoming more significant at their economic level and they have a delegate to carry the whole Noto surpluss to Dakar where the prices are significantly higher. The bissap yield increase is related with irrigation increase, now possible because the time and energy saved from domestic watering and milling. The implantation of a new commercialize schema has been helped by the cohesion increase of the women's association.

Other example is related to youngs. After the PV installation, they officially legalized their association and they are deploying an important activity looking for the attainment of new projects that increase the job expectancy in Noto.

4. INSTITUTIONAL ASPECTS.

The main problems stated at present in this projects are being derived from institutional aspects. The more troublesome are:

(a) The data recorded at Noto are not being translated to the other participants on the project for their analysis. This fact is seriously hidden the evaluation of the project and the possibility for future modifications.

(b) The charging for the electricity delivered is waiting for the signature of an agreement between the MDIA and the SENELEC. A first draft of this agreement was written more than one year ago. This fact is keeping the users far from the reality and more it becames a long standing custom more difficult will be to implement the absolutly needed charges.

5. CONCLUSIONS.

The main lesson learned at present in Noto can be summarized as follows:

(a) There are not major technical problems with PV rural electrification systems. However, there is still a lack of experience regarding the real user's energy demand.

(b) The consideration of social aspects is a key point for the succes of a PV rural electrification system. The users must be committed for the management of the facilities. A good way is to lay upon yet existing social Organisations.

(c) Special care must be taken in order to assure that the benefits from a PV rural electrification project reach the majority of the population. Sometimes, conflicts with the powerfuls are unavoidable.

(d) Still there is a lack of experience on legal and institutional aspects of PV rural electrification. When implementing a particular project a previous agreement between the different participants that includes unambiguous assigning of all the maintenance, evaluation and possible modifications is a key factor to assure the long-term running of the system.

6. FINAL COMMENT.

The authors want to express their wishes for a quick renewal of the transmittion data process and for the establishment of a cooperative evaluation method. They hope that the future papers about this project will include senegalese authors.

7. REFERENCES.

(1) E. Lorenzo et al. "A Photovoltaic Installation in a sub-sahel village in Senegal". Proceedings of 7th E.C. PV Solar Energy Conference. October 1986. Sevilla.

(2) J. Martz and A. Roberts. "Operational Performance of the Photovoltaic-Powered Grain Mill and Water Pump at Tangaye, Burkina Faso". NASA Technical Memorandum 86970. March 1985.

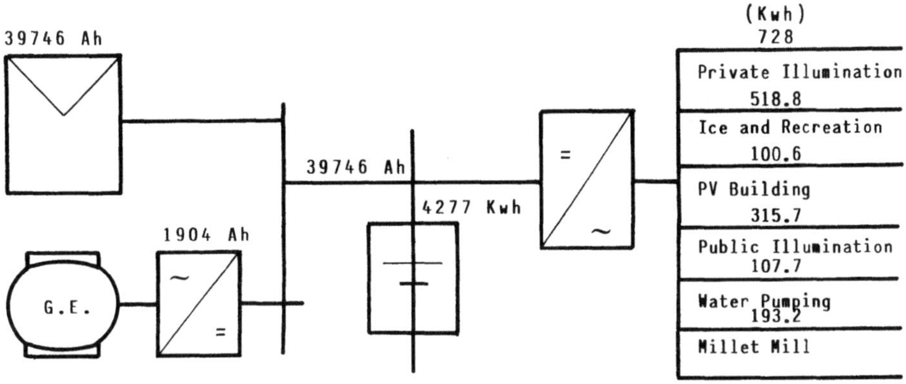

Fig.1.- Energy flow corresponding to the first 120 days.

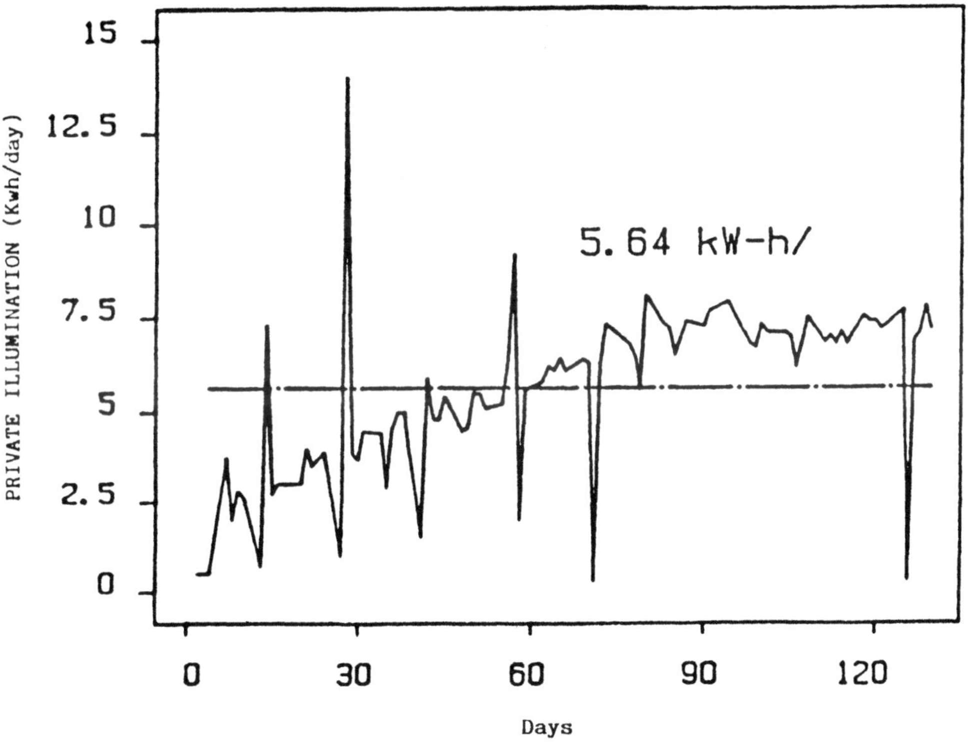

Fig.2.- Private illumination energy consumption.

PHOTOVOLTAIC RURAL ELECTRIFICATION OF 57 DWELLINGS IN LA SIERRA DE SEGURA (SPAIN)

M. Egido*, M.Erill**, R.Eyras*, A. González***, S. Gracia**, E. Lorenzo*, M. Montero*, M.Sidrach***.

* Instituto de Energía Solar
E.T.S.I. Telecomunicación
U.P.M. - 28040 MADRID (SPAIN)

** Compañía Sevillana de Electricidad
Avd. de la Borbolla, 5
41004 SEVILLA(SPAIN)

*** Instituto de Energías Renovables
CIEMAT-IER
Avd. Complutense s/n
28040 MADRID (SPAIN)

SUMMARY

This paper deals with a project, included in the framework of the DGXVII-CEE Demonstration Programme, which aim is the pv electrification of 57 dwellings located at Sierra de Segura, in the southern of Spain. The total peak power to be installed is 22 kwp. The Project includes AC/DC energy distributions and centralized and decentralized configurations regarding to the generation of Solar Photovoltaic Energy. Design and Procurement phases have been carried out. Direct costs per house will be around 1.249.280 ptas. (9.070 EECUs). Special attention is being paid to the consideration of social and institutional aspects involved.

I. INTRODUCTION

This paper describes the current state of a project which aim is the PV electrification of 57 rural dwelings located at the Sierra de Segura, an area of the southern part of Spain.

Four entities are involved: the "Compañía Sevillana de Electricidad" (CSE) that is responsible for the coordination and execution of the project, The Instituto de Energía Solar (IES)* and the Instituto de Energías Renovables (IER)** that are commited for to the technical (design, evaluation, etc.) and sociological (user's priorities and responses, etc.) aspects and Isofotón S.A. that provides the PV modules.

The project is being supported by the "Oficina para la Coordinación de la Investigación y Desarrollo Electrotécnico" (OCIDE), the CEC (DGXVII-1987 Demonstration Programme) and the Junta de Andalucía.

The starting date was september 1987. At present, the Design and Procurement phases have been completed. This paper focuses on some specific aspects of the design (different hypothesis of energy consumption, different solutions chosen, etc.), on the material costs and on main problems stated at present: (the possible concurrency of grid-extension) and the future management of the PV installations.

* From the Polytechnical University of Madrid
** From the Ministry of Industrie and Energy.

II. DESCRIPTION

The Sierra de Segura "(Jaén), is a mountaineous area of Andalucía, Spain presenting hard lacks of general infrastructure (roads, communications, etc.) and where some 1100 habitants remain grid-isolated. Because of that, a relatively important PV activity is being developped from 1982. Some 50-100 houses are currently using PV for lighting and TV. All of them are 12V-DC installations typically composed by 2 PV modules (70-90 wp) a 250 Ah battery and a charge-regulator. Such PV activity has basically been spontaneous (1) in the sens that subsidies specially devoted to PV have not exist.

A previous experimental work (2) leaded us to establish the PV energy consumption scenario summarized in table 1. The classification of the families refers to the number and age of their members. Note that this scenario excludes all heating and refrigeration applications. The reason for the latter is that butano-gas is a more economical solution than PV and that butano-gas cookers and refrigerators are already being extensively employed in this area. It's worth to pointout that this PV-butano gas combination scenario has been elaborated on the basis of and extended enquiring between grid-isolated people and that is supposed to provide the users with a similar level of confort that the grid-connected ones.

TABLE I - EXPECTED ELECTRICAL CONSUMPTION FOR RURAL DWELLINGS IN SPAIN (Wh/day)

FAMILY TYPE	EXTENDED	MEAN	REDUCED
Lighting and TV BW/color	230/414	160/288	120/216
Domestic appliances	300	150	80
Water pumping	72	54	36
TOTAL	602/786	364/492	236/332

Table I. Energy consumption scenario

The 57 houses affected by this project are located in 4 different neighbourhoods: Rio Madera - 24 grouped houses, church, bar and food-shop. Los Moralejos -12 grouped houses, all already PV powered for lighting and TV-, La Fuente de la Higuera -7 grouped houses where the proximity of the Guadalquivir river frequently originates fogs during winter which negatively affect the incident solar radiation - and Prados de la Presa - 14 dispersed houses. In the design general purposes when designing have been the comparison of several configurations in similar operational conditions and the development of standars of general validity.

As an intent of adapting the PV installations to the particularities of the four neighbourhoods we have selected four different configurations which main features are summarized in Table II.

TABLE II. FEATURES OF THE INSTALLATIONS*

VILLAGE	Nº of houses	totalpeak power(kW)	Domestic illumination and TV	Public illumination	Domestic appliances	Water pumping
Rio madera	24	9	AC - C	AC - C	AC - C	-
Prados de la Presa	14	6.5	DC - D	DC - D	AC - D	AC - D
Moralejos	12	3.5	DC - D	AC - C	AC - C	-
Fuente de la Higuera	7	3	DC - C	DC - C	AC - C	AC - C

*	Centralized	Decentralized
AC - 220V	AC - C	AC - D
DC - 24V	DC - C	DC - D

The inclusion of AC lines reflects the repeatedly expressed user's wishes for the use of standard appliances. For instance, the woman of the "Sierra de Segura" have often expressed their wishes of having standard washing machines. The introduction of special "Photovoltaic washing machines" would therefore not be recommended. All the washing machines to be powered by centralized PV installations will be time-programmed in order to avoid that the load will surpass the maximum peak power of the inverters. The use of DC for lighting try to avoid the inverter's losses associated to the low-load operational conditions. The addition of a back-up diesel generator to the PV installation of La Fuente de la Higuera try to compensate the above mentioned lowering of incident solar radiation due to fog. The 50% of the total PV modules will be of bifacial type.

The monitoring of the PV installations will be accomplished by means of an analytical system in Río Madera and of global systems in other cases, according to the recommendations of the CEC(3).

III. UP TO DATE ACTIVITIES

Collection of data field has been completed. Special attention has been payed to provide information to the users and to inquire about their wishes and priorities. For that, an anthropologist is included in the team.

The technical design has been finished, resulting on detailed specifications for the different equipment, civil work, distribution lines and indoor electrical installations. Taking into consideration that The Sierra de Segura has recently being declared Parque Nacional (Protected Natural Area) special attention has been payed to the consideration of aesthetic aspects which are mainly related to the civil works.

All the equipment manufactures and subcontractors have been selected, Table 3 shows the cost of the corresponding items. Note that the total direct costs - including the indoor electrical installations- is 1249.280 pts/house.

	Total cost (thousand pts.)	%
PV modules	34.020	48
batteries	7.302	10
Power conditioning	10.929	15
Back up.	1.000	1
Civil work	12.540	18
Electrical inst.	5.400	8
TOTAL	71.209	100
Analitical Monitoring	5.187	
Global Monitoring	3.046	

Table 3 - Cost of the different up to date acquisitions (* in thousands of pesetas).

All the PV modules have been individually tested. As an example, table 4 shows the main features of one of the modules type. The modules will be selected so that the dispersion of the current at maximum power of the individual panels of each series-asociation would be minimal.

Denomination M 40L
Nominal power 40 w
Number of modules 214

	P(w)	I_{sc}(A)	V_{oc}(V)	FF	I_{MAX}(A)	V_{MAX}(V)
Max	41.4	2.97	19.8	.76	2.73	15.95
Min	37.1	2.66	19.1	.68	2.47	14.31
Mean	39.4	2.83	19.44	.72	2.62	15.05
S.D.	0.72	0.06	0.13	.01	.05	.25

Table 4 - Measured features of a pv modules group.

4. THE USER'S APTITUDE.

Others than technical aspects are being considered of primary importance for the final succes of the project. Several members of the team are involved in the study of the PV phenomena in The "Sierra de Segura" as of 4 years ago. Because of that they are "well-known" people in this area. This fact is positively affecting the users-project relationships. Moreover, the involvement of a big utility (CSE serves electricity to 10.000.000 people) of the local Authorities and of several National Agencies defines a specially interesting framework to afford the study of social and institutional aspects related with to PV rural electrification.

Nowadays the project is being very well accepted by the great majority of the users. However two main difficulties had to be overcame in the past. On one hand, the widely extended idea of that photovoltaics is only for lighting and TV. This idea reflects the most common situation of the actual spanish pv users and, obviously, it does not help enthusiastic aptitudes. Meetings devoted to explain to the users the consumption scenario and other technical details of the project, have permit to solve this situation. On the other hand, some news about a future grid-extension affecting 30% of the houses included in this project have appeared. Some users considered the PV project as an obstacle for such a grid-extension and rejected at first. The local Authorities intervention in the sens to assure the non relation between the project and the posibility of a grid-extension solved this situation.

The main remaining project question affecting the users regards the future management and maintenance of the PV installations. All of them agree with the idea of to paying for their indoor electrical installations and for a periodic fee to assure the energy supply (like grid connected people do!). However, the implementation of such a procedure needs of a previous analysis of the legal aspects involved on it.

5. CONCLUSIONS.

The main features of a project which aim is the pv electrification of 57 dwellings in the sourthen of Spain has been presented.

Field data collection, technical design and selection of equipment manufactures and subcontrators has been completed.

The correct consideration of social and institutional factors is a key for the succes of the project.

6. REFERENCES.

1.- E. Lorenzo et al. "The rural electrification share of the spanish PV market". Proc. of the 6th E.C. Photovoltaic Solar Energy Conference. London. April. 1985.

2.- A. Krenzinger, M. Montero. "Energy consumption patterns of the rural photovoltaic market in Spain" Proceedinds of the 7th EC Photovoltaic Solar Energy Conference. Sevilla. October 1986.

3.- Guidelines for Photovoltaic System Monitoring Commission of the European Communities. JRC. Ispra. November 1987.

PV COMPONENTS FOR TRAFFIC CONTROL SYSTEM

Fabrizio Pulcinelli -Alessandro Gurioli
S.E.I.- Sistemi Energetici Integrati
Via S.Jacopo,32
50047 Prato
Italy

ABSTRACT

S.E.I. has developped a custom PV module designed to power radio-markers for Traffic Control Systems. All plastic manufacturing technology has been used. Solar cells are fixed on a printed circuit board by conductive adhesive.

Dipartimento di Elettronica of Florence has developped the production technology using a microwave oven to process the adhesive and, in a following step, to cure the silicon pottant. The first experimental system has been operating in Florence (Italy) since February 1985, while another larger system is operating in Brescia.

INTRODUCTION

Automatic traffic control system (T.C.S.) are becoming necessary for medium and large towns. Such a system controls vehicle as buses, ambulances, police cars and, anyway, every prearranged vehicle.

Tests performed in several countries and cities shows how T.C.S. allows to save about 20% of operating cost of Municipal Transport Authorities. Likewise better use of human and vehicle resources can be reached.

T.C.S. are based on:

1. Marker , a particular radio transmitter that continuously emits a binary code relating to its position in the town. Each vehicle picks-up the code of the closer marker and transmits it to the Control Unit.
2. Control Unit, that receives the codes transmitted from controlled vehicles, and allows to identify vehicles and their position on a display.

EXPERIMENTAL SYSTEM OF FIRENZE

First experimental system was performed by S.E.I in Firenze (Italy) in February 1985. The network, still working without any failure or service interruption, use fifty markers located in the main avenues connecting the town from east to west.

Each marker, based on a FM-UHF transmitter, 30mW continuous power consumption, covers about 0,02 Km^2.

The possibility to use PV modules to power the markers was considered in order to have a no maintenace system.

A not rechargeable battery had to be changed every year, with a lot of problems depending on the height of the place where the markers are positioned. S.E.I. has developped a waterproof box which contains the electronic markers.

The polycarbonate cover houses a PV module, with 18 series connected quarters of 3" mono-si cell. (Fig.1)

The cells, following a technology developped by Dipartimento di Elettronica of Università di Firenze, are bonded to a printed circuit board by silver epoxy conductive adhesive.

RTV silicon, cured by microwave oven is used for cell protection. All the cells are selected in order to charge a 6V Ni-Cd battery without any regulator.

OPERATIONAL EXPERIENCE

This experimental network allowed to collect several information about the system PV module-marker, and to arrange a computer program to design new custom modules. Main problems stands on particular kind of pollutions, that the rain can't remove, covering the module with an absorbing layer, and on the foliage, that during summertime can reduce considerably the solar radiation on the module.

Another problem is the module fixing, mainly on lighting poles or threes, generally without any possibility to optimize the sun exposition.

CONCLUSION

Using the experience of the first system S.E.I. performed some innovation in its second system in Brescia, city in the Nord of Italy, for busses control.

The system uses 350 markers. Each markers is powered by a 18 poly-si 25 x 50 mm cells module (Fig.2).

The cells are joined on a PC board by a two-components silver-epoxy adhesive. The pottant is a RTV silicon which is casted under vacuum in order to avoid air boubles.

Microwave technology has been used in curing steps of process. The curing time is 180 sec.

Automation has been introduced in inserting cells on a PCB, in adhesive dispensing and in vacuum casting RTV pottant.

Photovoltaic energy is a cost effective power for radio-markers in T.C.S. High reliability has been esperienced during the service opening new opportunities for the next future.

New technologies have been used such as conductive adhesives and microwave heating.

Fig. 1. PV Module used in Firenze system.

Fig. 2. PV Module used in Brescia system.

Fig. 3. The Radio Marker.

RAILWAYS POWERED BY SOLAR PV GENERATORS

Paolo Redi
Dipartimento d'Ingegneria Elettronica
Via Santa Marta,3
50139 Firenze
Italy

ABSTRACT.The paper investigates about the application of solar generators in powering railways,trolley-cars and trolley-buses.These vehicles uses d.c. electricity and are able to be supplied by PV generators without energy conversion.In the paper are indicated some possibilities using battery vehicles and the standard system of electrification with a contact line.

INTRODUCTION

The present paper is an investigation about the problems arising from the use of solar generators in pubblic transport supplying.Railways,trolley-cars and trolley-buses,which generally relate to d.c. electricity,can be potentially powered by PV generators.

The motor that is in use in this field is the series d.c.model.Only in the experimental phase is the asynchronous motor with a variable frequency converter.The rail vehicles and the trolley-buses for urban transport use 600V supply.The voltage is increased to 750-1500V for suburban transport.The railway d.c. voltage is 1500V in France and 3000V in several countries.The voltage value is an optimisation of different variables:the section of the contact line,the power of the vehicles,the distance between the substations,the allowable voltage drop.In 1500V railways the distance between the substations is 6-15km,in 3000V lines the distance is 12-30km depending on traffic.

The power installed is very variable:

trolley-car	600V	280kW	(Milan)
trolley-bus	600V	80kW	(Milan)
motor-coach	3000V	428kW	(Ferrovie del Gargano)
locomotive 626	3000V	2100kW	(Ferrovie dello Stato)
locomotive 633	3000V	4900kW	(Ferrovie dello Stato)

On most railways it is usual for lines which have not been electrified to be operated by diesel motor-coaches.In Germany and certain countries a tradition has grow up for branch lines service to be operated by battery motor-coaches which provide almost all the advantages of electrification without needing a contact and supply system,but at the expence of carrying a rather heavy battery.The principal limitation of any vehicle powered by stored energy is the amount of work which it is capable of performing between successive charges.This is expressed as the distance or "range" that can be traveled in its normal service between charges or "operating ratio" the amount of energy that can be stored devided by the vehicle weight.Examination of typical working roster reveals that the total energy consumed by a conventional main-line locomotive in the course of a day's working divided by its weight is upwords of 10kWh/t.Corresponding figure for a motor-coach engaged in branch-line working,which rarely exceed 300km/day,is only about 5kWh/t whereas small shunting locomotives have a daily power requirement of only 1kWh/t.Comparison with figures available from existing vehicles suggests that for the moment there is little likehood of even the smaller main-line locomotive with battery traction,values suitable for operation of motor-coaches are quite feasible with provision of a single daily charge.

The battery powered motor-coach ETA 150 of DBB (West Germany) has a weight of 47,8t,a battery of 387kWh capacity and 16t weight,giving a specific battery capacity of 24,2kWh/t,an operating ratio of 8,1kWh/t and a range of 350km.

SOLAR POWER SUPPLY

PV generators that produce d.c. electricity are potentially suitable to supply electric traction without any energy conversion.The solar modules can't in any case be placed on the roof of the vehicle.Very simple calculations can explain this conclusion.The solar generators must charge a battery that can be placed in two different positions,on the vehicle or out the vehicle. With the first solution the weight of the battery is a load which reduces the performances.The second solution requires the investiment of a contact line.

The battery on the vehicle must be charged at least every day,generally at night.The direct recharge with PV generators obviously is possible only during daily hours.That requires the provision of a simple and fast way to excange the battery .A solution could be the use of a special battery tender as employed in the past in Germany and India.The tender can be substitued at a terminus which is provided with PV arrays for charging the battery.The terminus remembers the water and coal refuelling stations for steam locomotives that are still in use in many countries.The tender could assure an autonomy lower than one day operation reducing the weight and increasing the pay-load.

Considering the operative cost,at the moment the one of battery motor-coach and light locomotive is comparable with the cost of diesel units which suffer on difficulties in maintenance and fuel availability in many developing countries.Battery trains are specially suitable for passengers transport in flat countries .PV generators could offer the power,saving precious resources.

The use of a battery out the train can considerably increase the performances.Solar generators which require a considerable area can be suitably placed along the line.The battery takes the place of the substation of present electrification system,but the particularity of solar generators allows to reduce distance between the single feeding units.Fixing the distance between the feeders in the range 5-15km,the voltage value which derives is in the range 300-1200V.It is to remenber that the voltage of batteries for railway traction in West Germany is standardised at 440V.

SYSTEM DESIGN

The PV system for these applications presents some differences in comparison to standard systems.Array voltage of 300V is at present in use in several installations.Values of 1000V are possible with few modifications of present module design.Values of 3000V require the use of insulators between the module frame and the structure,and the modification of the junction box.Module and installation cost could increase of about 20%.Cables,switches and protections are of the standard railway quality.
The batteries to be used are lead acid heavy duty traction model.The life of the batteries used by railways is usually some 5-6 years or 1600 discharge periods.Alkaline cells have also been used for traction purposes.The life is usually some 12-15 years and are more robust and better able to withstand hostile environment,overcharging and misuse.The higher investiment cost take rare the use in traction.
The high value of electricity produced by PV generators advantages the use of d.c. electronic chopper control of motors.The regenerative braking could reduce the net current consumption.The conventional form of regenerative braking would be too complicated for application to small vehicles,but the adoption of electronic control offers the possibility to provide sufficient regenerative current for this purpose.

CONCLUSION

The present paper intends to offer same suggestions about the use of PV generators to power public electric transports.The subject can be considered as an utopian project but the careful investigation on the situation of railways and public transports of many underdeveloped countries,but also in mamy other countries,opens some possibilities.There are countries that still use steam locomotives supplied by wood which is the sole resource.The power installed is very reduced.In sugar can plantations are in use locomotives with an installed power of 25kW which use the wastes otherwise utilizable for cattles feeding.
It is interesting to try to sizing a PV system for these applications.
As an example a 50km railway line with 10 motor-coach runs in a day in a flat country requires about 400kWh/day.It is able to transport about 1500 passengers/day.
In a region with an average solar irradiance of 4,5kWh/m^2 the required energy can be produced by a 120kWp solar array.The solar generators could be placed with a 10km interval using 600-750V array voltage.
Looking to a different application,in a 10km trolley-bus line with 50 runs a day the energy required is about 750kWh.The line can transport about 2500 passengers.
Using the previous insolation data the requirement for the PV array is of

225kWp.The solar generators could be placed with a 5km interval using 600V supply.

A further application of solar electric generators in standard d.c. electrification systems could be the use for voltage drop compensation between two substations.In many railway lines with old electrification system the great distance between the substations produces a large voltage drop in the contact line in some tracts of the line.A solar feeder could be used for the compensation of the voltage drop avoiding the installation of a new substation and of the high voltage power line.

A study about this application has been carried on by Sistemi Energetici Integrati s.r.l. for Ferrovie del Gargano,Italy,where the substations are distanced of 25km and at the farest terminal points the line voltage drops from the nominal value of 3000V to 2500V reducing the maximum speed of the motor-coaches.

In order to do a cost comparison with diesel and standard electric traction systems would be necessary to take in account more elements.In any case we believe that solar electric generators could be taken,also at the present,as an opportunity for applications in the transport field.

REFERENCES

1)Wilkie,G.,Modern Battery Rail-Cars. J.Inst.Locomot.Eng.,47(5)(1957)455

2)Smith,G.,Storage Batteries. Pitman,London,1964

3)Frauendorf,H., Der Akkumulatoren-Triebwagenverkehr der Deutschen Bundesbahn. Elektrische Bahnen. Vol.4, 1951

4)Andrews,H.I.,Railway Traction.Elvesier,Oxford,1986

5)Tessier,M.,Traction Electrique et Thermo-Electrique.Edition Scientifiques Riber,Paris,1978

6)Carter,F.W.,Railway Electric Traction.Edward Arnold,London,1922

7)Dover,A.T.,Electric Traction.Pitman,London,1963

8)Cotton,H.,Advanced Electrical Technology.Pitman,London,1967

9)Barwell,F.T.,Traction Research. J.Inst.Locomt.Eng.,56(")(1966-7)158

10)Barak,M., Batteries. Pitman,London,1964

11)Barak,M.,Batteries and Fuel Cells.Inst.Electr.Eng.Rev.,117(1970)1561

10 kWp-PHOTOVOLTAIC PLANT IN BERLIN

Thomas Mierke
Bernhard Voigt
EAB Energie-Anlagen Berlin GmbH
Flottwellstraße 4-5
D-1000 Berlin 30
Telefon 030/254 96-0
Telex 183634
Telefax 030/254 96 100

Rolf Hanitsch
Technical University Berlin
Institut for Electrical Machines
Einsteinufer 13-15
D-1000 Berlin 10

Klaus Bürgel
Berliner Kraft- und Licht (BEWAG)-AG
Stauffenbergstraße 26
D-1000 Berlin 30

Roland Hauk
Siemens Solar GmbH
Buchenallee 3
D-5060 Bergisch Gladbach 1

1. ABSTRACT

At present BEWAG/EAB installs a PV pilot plant with a total power of approx. 10 kWp in a residential district of Berlin (West).
The plant will be put into operation in June 1988.

The energy will be stored in a battery with a capacity of about 100 kWh and be used to cover the electric demand of a heat pump, household appliances and light of a building.
Additionally the system is connected to the public electric supply system to feed-back surplus energy.
The system will be provided with a computerized measuring and control system. The goals of the project are to investigate and demonstrate the operation of a photovoltaic plant in an urban area and to optimize the interaction of all components, i.e. the components themselves.

The project is sponsored by the Commission of the European Communities (CEC) and the Federal Ministry of Research and Technology (BMFT), FR Germany.

2. INTRODUCTION

The use of photovoltaic energy may help to solve some of our energy problems of the future.

To investigate and demonstrate the operation, BEWAG/EAB installs a 10 kWp-photovoltaic plant in Berlin (West).

3. SYSTEM DESCRIPTION

The installed peak power of at total 9,4 kWp is generated by two different modul-arrays which are mounted on two connected rack constructions. The first consists of 112 modules SM 50 (Siemens) with monocristalline structure and the second of 98 modules PQ 10/40 (AEG) with polycristalline structure.

Figure 1 shows the block diagram of the entire plant. Both arrays are connected to two separate DC/DC converters with common control. This ensures a decoupling of the different arrays.

The converters boost the generator voltage of 220 V (14 strings in series) to 280 V battery voltage.

A MPP-control automatically search for the optimum working point of the connected solar generator, which is generally temperature-dependent. It adjust their output voltage to the respective battery voltage which is determined by the battery's instantaneous state of charge. The task of the battery is to balance the fluctuations in energy supply, which is determined by the daily solar insolation, the cloudiness, and the energy demand of the consumers.

The consumers like heat pump (2,2 kW, 3-phase inverter, frequency controlled), house appliances (1-phase inverter), light devices, and charge station for an Electro Car are connected to the 280 V busbar.

If the need of the consumers is below the power output of the solar generator and the battery is fully charged, a feed-back of energy into the public grid is practiced. Indicator for a surplus of energy is the battery saturation voltage (360 V). The switch-over happens automatically.

If the battery voltage falls below the cut-off voltage (260 V) by missing solar power, the battery will be charged by the public grid.

Energy feed-back and battery charging are done by one 3-phase inverter. The reactive power is compensated for a load of 50 %.

A transformer provides a galvanic separation and an adaption of the different voltage levels.
The transformer enables an energy transfer in both directions without winding switching.

The plant operation is controlled by a personal-computer (NEC) with seriel data bus system.
Important data and signals concerning the plant conditions can be displayed and changed on-line.

To collect and evaluate the measuring data, the system will be equipped with an additional computer, provided by the BMFT. This shall ensure a standardized measuring program and data storage.

4. COMPONENTS

4.1 Solar Modules SM 50 (Siemens)

The Siemens-array consists of 112 modules. They are connected to 14 strings in series, with 8 modules per string parallel. The modules have monocristalline structure and a rated output power of 50 Wp (AM 1,5; 25 °C; 1000 W/m^2) each. The total rated output power of the array is 5,6 kWp.

4.2 Solar Modules PQ 10/40 (AEG)

The AEG-array consists of 98 modules. They are connected to 14 strings in series, with 7 modules per string parallel. The modules have poly-cristalline structure and a rated power of 38,4 Wp each. The total rated output power of the array is 3,8 kWp.

4.3 DC/DC-Converter

The DC/DC-converter is constructed as a boost-converter.
Two power modules (6 kW each) are connected together and common controlled by a 16-bit processor which generates a staggered clock rate. This leads to a low ripple. By using SIPMOS field effect power transistors the efficiency will be better than 95 %.

4.4 Battery

A battery type 4 OCSM 320 (HAGEN) is used.
The 120 hour capacity of the battery is 504 Ah, the energy 100 kWh.
The nominal voltage is 280 V. The construction is characterized by:
- High efficiency and long maintenance intervals
- Positive armor plate for a high number of cycles
- Use of a lead alloy with an antimon concentration < 3 %
- A negative grid plate with a copper expanded metal grid leads to an optimal voltage stability during discharging.

4.5 10 kW-Inverter (Simoreg)

The inverter contains two three-phase bridges. The nominal power is 10 kW. The functions are energy feed-back into the public grid and charge of the battery. The sycronization with the 50 Hz-grid happens automatically. The inverter is controlled by the 16-bit processor.

4.6 Heat Pump Inverter (Simovert)

The three-phase PWM transistor inverter is used for stepless speed control of standard induction motors. The range of frequency is 1 to 100 Hz. The nominal power is 2,2 kW. Due to a microprocessor control the adaption to the given heat pump is easy.

4.7 Monitoring and Control (Simatic)

An IBM-compatible 16-bit computer (NEC) controls the interaction of the components. The serial data transfer (current-loop, TTY) ensures trouble-free function, even over long distances.
It is possible to display the main operation parameters and to change them on-line.

4.8 Measuring System

In order to get a standardized data format the BMFT provides an unique measuring system for data logging.
The system is IBM compatible (Compaq 386) and the other components will be selected at present by the BMFT concerning both technical and economical aspects.

10 kWp-PHOTOVOLTAIC PLANT IN BERLIN

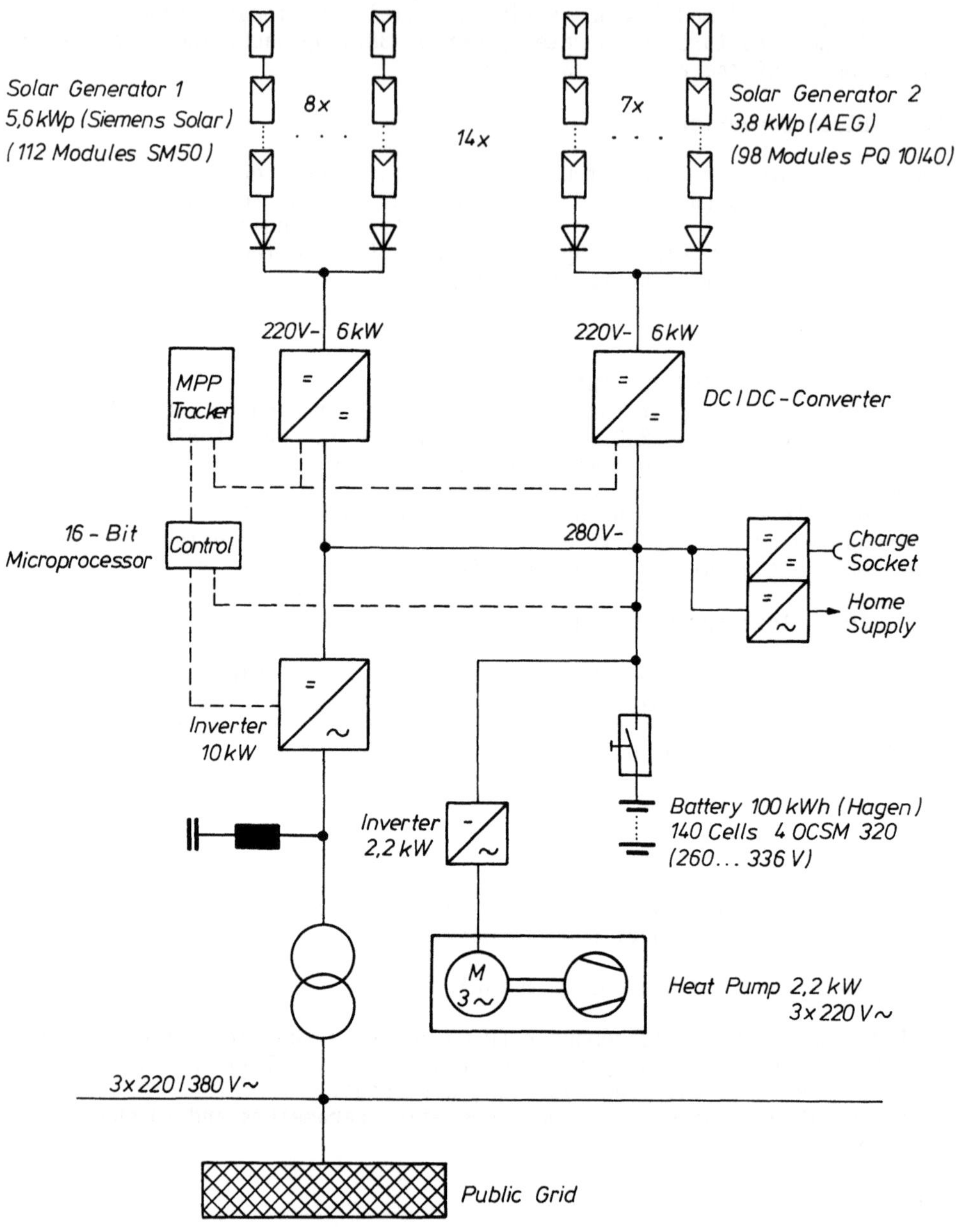

Figure 1: Blockdiagram of the Plant

Figure 2 and 3: Rack Construction During the Erection Phase, Partially Covered with PV-Modules

CALCULATION AND MEASUREMENT OF AVAILABLE SOLAR RADIATION AT DAYLIGHT

S. Aydinli
E. Krochmann
PRC Krochmann GmbH, Geneststr. 6, D - 1000 Berlin 62
J. Krochmann
Licht- und Strahlungsmeßtechnik, Geneststr. 6, D - 1000 Berlin 62
J. Reinhardt, Institut für Lichttechnik, TU Berlin
Einsteinufer 19, D - 1000 Berlin 10

ABSTRACT

The planning of voltaic solar energy installations is possible only, when the local available solar radiation can be predicted with small uncertainty. A method is developed to calculate the long time average radiant exposure on inclined surfaces. The basis of these calculations is beside the local conditions the sunshine duration and the turbidity of the atmosphere. Calculated results agree within ± 10 % with measured values.

Special radiometric equipment is developed for the measurement of solar radiation. The quality of the instruments can be characterized on the basis of the WMO recommendations and new CIE definitions for radiometers and photometers.

Data on available daylight are not sufficiently known. Therefore, the CIE proposed a "daylight measuring year" for 1991. Special photometric equipment for the measurement of the relevant daylight data is developed.

1. INTRODUCTION

Outdoor solar radiation and daylight must be measured to develop prediction methods and to examine the quality and efficiency of solar cells and solar collectors as well as to optimize the energy consumption in buildings. Radiometric data on solar radiation were measured at a larger number of meteorological stations, but the information on daylight availability is relatively poor.

2. SOLAR RADIATION ON INCLINED SURFACES

The European Committee supported a research project to develop a basic method for the prediction of global irradiance and radiant exposure on inclined surfaces at average sky conditions.

The new method was based on values of irradiance (mean hourly values), radiant exposure, sun shine duration, and turbidity, measured at different stations of the E.C. The method[1)2)], based on data for complete overcast and clear sky conditions, uses a non linear superposition according to the average sun shine duration. The correspondence to the measured data is generally < ± 10 % (fig. 1 and 2). This method can be used world wide, as the only independent variable is the sun shine duration, which is measured at many stations in the world.

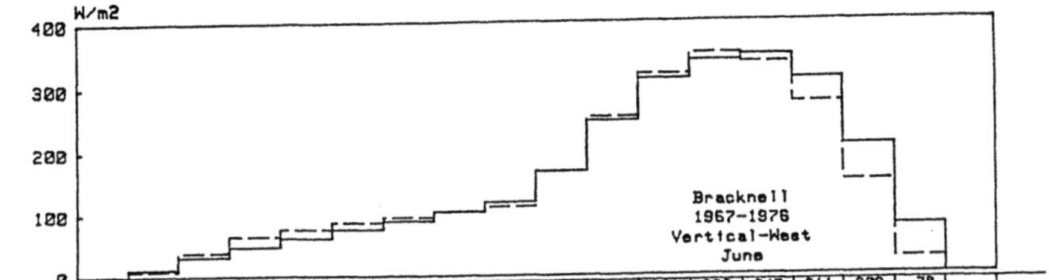

OBS.		11	31	47	61	75	88	103	118	167	247	311	339	347	311	208	78	
CAL.		8	38	64	75	86	94	103	111	167	253	318	350	336	275	150	25	
ERROR		-27%	+23%	+36%	+23%	+15%	+7%	+0%	-6%	+0%	+2%	+3%	+3%	-3%	-12%	-28%	-68%	
LOC.TIME	3- 4	4- 5	5- 6	6- 7	7- 8	8- 9	9-10	10-11	11-12	12-13	13-14	14-15	15-16	16-17	17-18	18-19	19-20	20-21 in h

Fig. 1 Comparison of the calculated and observed monthly mean hourly values of the total radiation

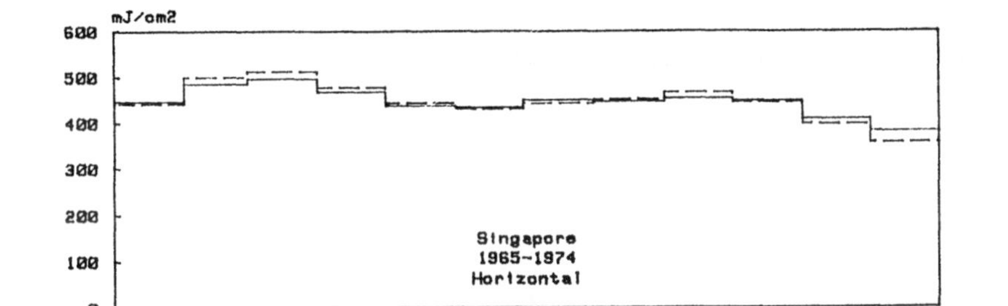

	JAN	FEB	MAR	APR	MAY	JUN	JUL	AUG	SEP	OCT	NOV	DEC	YEAR
OBS.	446.8	485.5	496.1	468.1	437.2	434.2	450.6	446.4	453.4	448.0	408.2	381.1	446.3
CAL.	441.9	499.9	512.1	477.4	443.8	430.4	443.6	451.5	466.3	444.8	397.1	356.7	448.1
ERROR	-1.1%	+3.0%	+3.2%	+2.0%	+1.5%	-0.9%	-1.6%	+1.1%	+2.9%	-0.7%	-2.7%	-6.4%	+0.4%

Fig. 2 Comparison of calculated and measured monthly mean daily values of global radiation

Using measured values of the luminous efficacy K of daylight[2)3)] (ratio of illuminance and irradiance), the same method is used for daylight calculations, too[2)].

3. SOLAR RADIATION MEASUREMENTS

Special measuring equipment for the most important solar radiation data:
- global irradiance at horizontal and 4 vertical surfaces
- diffuse (sky) and direct solar irradiance

are developed, using a solarimeter as radiometer head. The quality of these instruments can be described according to the recommendations of WMO[4)], and additionally the recommendations of the CIE on the characterization of radiometers and photometers[5)6)]. A computer controlled data acquisition system allows the nearly maintenance free measurement, storage and the fast transfer of the measured data for further calculations and evaluation.

4. INTERNATIONAL DAYLIGHT MEASURING YEAR (IDMY)

For the use of daylight - especially inside buildings - the equivalent luminous as radiant data, and additionally the sky luminance distribution must be known. The statistics of short time oscillation of daylight data and not only hourly values are important, too. Up to now, the daylight data are not sufficiently known. To encourage research on daylight availability, the CIE will start in 1991 an IDMY. In preparation of this, a symposium on "Daylight and Solar Radiation Measurement" will be held at 9/11th October, 1989, in Berlin.

Additionally, a set of daylight measuring instruments are developed for the IDMY.

5. DAYLIGHT MEASURING INSTRUMENTATION

The daylight quantities to be measured are mainly[7]:

- horizontal and 4 vertical global illuminance (photometer head, see fig. 3)
- diffuse (sky) and direct sun illuminance
- sky luminance distribution (sky scanner, see fig. 4, scanning time 40 s)

Fig. 3: Daylight photometer head for the measurement of horizontal and vertical illuminances with screening

Additional instruments are necessary for the measurement of
- reflectance ρ_{dif} (at diffuse incidence) of obstruction, ground and room surfaces (ρ_{dif} - meter, see fig. 5)
- daylight factor D (D - Meter, see fig. 6)
- special quantities (as the sky contrast coefficient)[8]

The quality of these instruments can be characterized by error values according to CIE[6] and DIN[9] and by their quality class[10].

Fig. 4: Sky scanner with data acquisition system

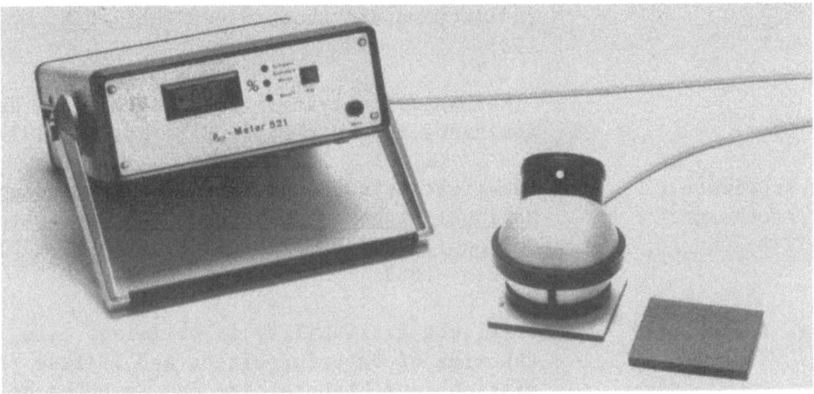

Fig. 5: ρ_{dif} - Meter with white and black reflectance standard

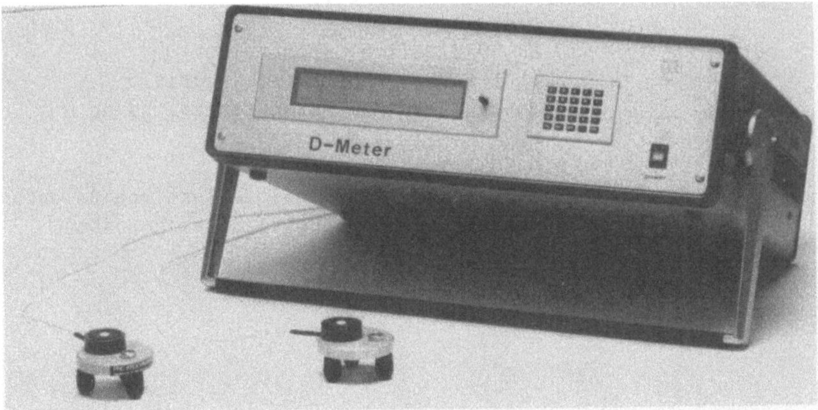

Fig. 6: D - Meter with two photometer heads

REFERENCES

1) S. Aydinli: 'Über die Berechnung der zur Verfügung stehenden Solarenergie und des Tageslichtes'. Fortschrittsberichte der VDI-Zeitung Reihe 6 No. 79, VDI-Verlag GmbH, Düsseldorf 1981

2) DIN 5034 Teil 2 (Febr. 1985). Tageslicht in Innenräumen. Grundlagen.

3) P.J. Littlefair: 'The luminous efficacy of daylight' Light. Res. & Techn., 17 (1985), No. 4, page 162-182

4) Guide to meteorological instruments and methods of observation, 5th edition WMO No. 8 World Meteorological Organization, Geneva,CH 1983

5) 'Methods of characterizing the performance of radiometers and photometers'Publ. CIE, No. 53 1983

6) 'Methods of characterizing illuminance and luminance meters'. Publ. CIE, No. 69, 1987

7) E. Krochmann, J. Krochmann: 'Daylight Measurement Equipment'Proceedings National Lighting Conference and Daylighting Colloquium 27/30 march, 1988, Cambridge CIBSE, London, 1988

8) M.R. Fontoynont: 'Daylight availability in buildings as a function of exterior direct and diffuse irradiation data' Laboratoire des Sciences de l'Habitat E.N.T.P.E. Vaulx en Velin (France) (private communication 1987)

9) DIN 5032 Teil 6 (Dec. 1985) Lichtmessung. Photometer. Begriffe, Eigenschaften und deren Kennzeichnung

10) DIN 5032 Teil 7 (Dec. 1985) Lichtmessung. Klasseneinteilung von Beleuchtungsstärke- und Leuchtdichtemeßgeräten.

HARNESSING THE MIDNIGHT SUN - PV IN ALASKA

Karin Holser
University of Alaska
Cooperative Extension Service
2221 E. Northern Lights, Suite 240
Anchorage, AK 99508-4143

ABSTRACT

Photovoltaics have been in use in Alaska since the early 1970s. One of the first installations was at the 14,000 foot level of Mt. McKinley to power a radio repeater (see Figure 1). Most of the early uses were for commercial installations. As prices have come down, more and more people living in the remote areas of Alaska have started to use photovoltaics. Since information and technical assistance is hard to find, the University of Alaska Anchorage has set up a 4-credit hour course. This course trains students in the design, installation, and use of photovoltaic systems. As the final exam students install a working system. The pictures in this presentation are of systems installed by the students.

INTRODUCTION

Three years ago the University of Alaska Anchorage started offering a course in the design and installation of photovoltaic systems. This course was developed to help Alaskans who are living off the utility grid system - which is approximately one-third of the state population. The course starts with basic electricity and follows through to the design and installation of a photovoltaic system.

The following is a sample of the course outline and its objectives.

COURSE OUTLINE

Week I	Introduction, discussion of objectives and grading criteria. History of photovoltaics. Alaskan examples.
Week II	The PV effect and today's technology - how and why it works.
Week III	Basic electricity.
Week IV	PV system components and sizing.
Week V	PV system components and sizing.
Week VI	Wire sizing.
Week VII	Safety - fusing, circuit breakers.
Week VIII	Sitting of PVs, elevation, angle, structure, shading, mounting, trackers, reflectors.
Week IX	End use, efficiencies, system layout, lighting, refrigeration, water pumping, computer entertainment equipment.
Week X	Unique Alaska systems - gen-sets, hybrid, etc.
Week XI	VACATION!!
Week XII	All day installation.
Week XIII	Interpreting, understanding, and reading meters, knowing your system. Maintenance and troubleshooting - use of hydrometers and voltmeters.
Week XIV	All day installation
Week XV	Exam Review
Week XVI	Final Exam

OBJECTIVES

1) Students will understand and be familiar with the technical concept of photovoltaics and photovoltaic systems.
2) Students will be able to give a detailed description of a stand-alone PV systems.
3) Students will know the advantages and disadvantages of PV power.
4) Students will know the parameters for a good working PV system.
5) Students will be able to design and engineer a stand-alone PV system.
6) Students will be able to install a PV system.

GRADING CRITERIA

Quiz	5 points each for total of	20 pts
Attendence		20 pts
Final Exam		20 pts
Field Installation		40 pts

The course is a full semester course offered in the evening from 7 p.m. to 10 p.m. one night a week (see Figure 2) with two weekend field installation sessions.

FIELD INSTALLATIONS

The first year the students installed one stand-alone system, a utility backup system, and redesigned and reinstalled an existing stand-alone system for a 75-year old lady who is a real pioneer. The people whose systems are being installed have to provide all the needed hardware and equipment for the project. This woman was told that in order to get the modules up high enough to receive optimum sun, it would require two large 6'x6' posts. She immediately asked if she could cut two tall spruce trees and use them. The spruce trees made great poles for mounting (see Figure 3).

As the students were doing the remounting of the modules in this particular system they discovered that the wiring being used in the house for the lights and TV were not very safe (see Figure 4), so the students rewired the house and control boxes. This system is an example of some of the do-it-yourself jobs, that can give photovoltaics and alternative energy a bad reputation. It has also helped to show the need for the class to educate people how to properly install these systems.

The next year the students installed a large stand-alone system on the hillside within the city limits of Anchorage (the largest city in Alaska). This system uses a tracker and 10 modules (see Figures 5 & 6). It does have a generator backup, which most of the systems in Alaska use, because the winter sunlight hours are not enough to keep the batteries charged and run the homes. This system was a very impressive system and even made the evening news on two of the three major TV stations in town.

This year the students will be assembling a stand-alone unit that will be used up on Mt. McKinley at the 14,000 foot level. The power will be used by the National Park Service to power 3 radios and an oximeter. This camp is a permanent part of the Park Services rescue station. The students will also be designing a system for the University. This system will be used to collect data, and provide a demonstration system for the public to view.

The course has been well attended. We have had 2 students from Latin America who have taken the class and plan to use the knowledge they have gained to encourage the use of photovoltaics in Latin America (see Figure 7). There have also been people from outside the state of Alaska attend, as well as students from different areas of the state.

This type of course needs to be developed more, so that people are taught how to install and make proper use of these systems (see Figure 8). It also helps the credibility of the industry if it is a subject that is taught in Unversities.

We are working on expanding this into a two-year degree program in alternative energy. This would include a course in Micro-Hydro, Wind Generation, Battery/Inverter systems, Efficient Home Construction, Efficient use of electricity, Passive and Active Solar Heating, Introduction to Gas and Deisel Generators, and Heat Recovery Systems.

Figure 1 Installing one of the first PV systems in Alaska on Mt. McKinley at the 14,000 foot level.

Figure 2 Classroom instruction in the theory and design of photovoltaic systems.

Figure 3 Student wiring one of the PV modules mounted on spruce poles.

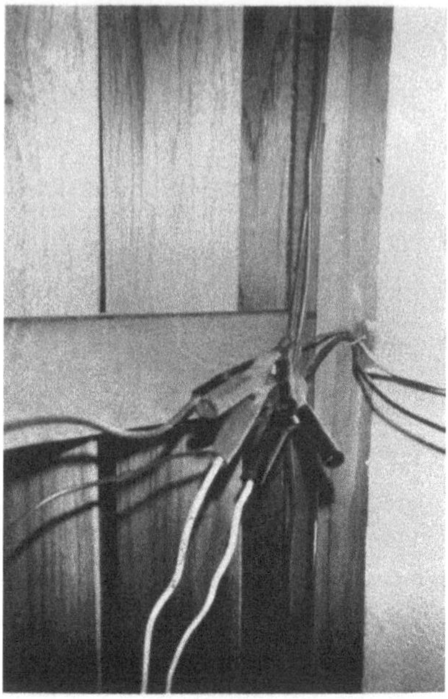

Figure 4 Hazardous wiring discovered by the students.

Figure 5 A stand-alone PV system being installed by the students on the hillside overlooking Anchorage.

Figure 6 Students learn all aspects of system installation - even trenching for a wire run.

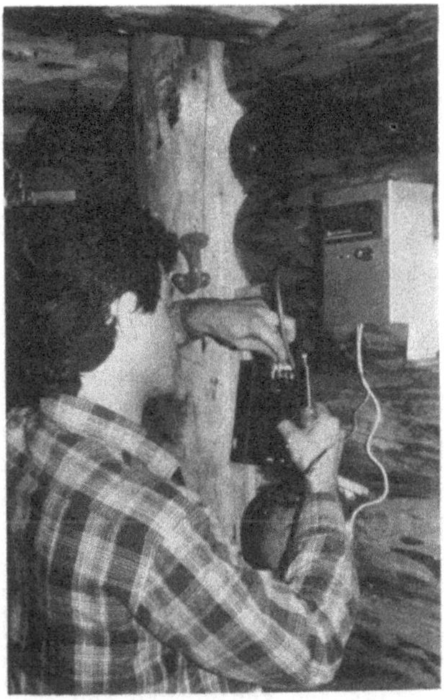

Figure 7 One of the Latin American students wiring the meter and controller.

Figure 8 Homeowner and students installing the PV array.

A NEW APPROACH TO THE ENERGY AND LOAD MANAGEMENT PROBLEM OF PV POWER SYSTEMS

Peter P. Groumpos and Kamel Y. Khouzam

Energy Research Center
Cleveland State University
Department of Electrical Engineering
Cleveland, Ohio 44115, U.S.A.

ABSTRACT

The importance of controls for the efficient operation of Photovoltaic Power Systems (PVPS) should not be overlooked. Energy and Load Management techniques when properly developed and applied to PVPS will lead to an optimal system sizing and a better utilization of the available solar energy. A new mathematical approach to this problem is presented. Loads are classified into four general categories; convenient, essential, critical, and emergency. Each category is further divided into a number of priorities. The objective is to determine, in an optimal way, the energy to be supplied to every load during the simulation period. The solution is based on the maximization of a weighted objective function which depends on the loads classification and their corresponding priorities and subject to the amount of available energy. A mathematical formulation to accomplish this is presented, where the problem is formulated as a one resource allocation problem and solved using dynamic programming techniques. A number of simulation studies illustrating the applicability of the new approach is presented. This approach will provide an efficient on-line load control sheddig in PV power systems.

1. Introduction

Despite the demonstrated versatility of photovoltaics (PV), there are limitations that have prevented their widespread use. The cost of solar cells, the timing of the PV power, the associated controls, and the space required to generate a significant amount of electricity are the most important considerations involved in the application of PV power. The importance of controls for an efficient operation of PV power systems has been addressed in a recent study [1]. Previous approaches were using mainly the state of charge (SOC) of the battery for load shedding control [2]. Cull and Eltimsahy [3] suggested the inclusion of loads with product storage for later use for better matching the insolation profile. An extension of the technique used in [3] allowed the use of a distribution of "controllable power" by the addition of multiple storage elements and sources [4]. The optimizing strategy conceived in [4] was to start with the load with the maximum efficiency at the highest power level and fulfill its energy requirement before considering the next load. The loads are classified in reference [5] into four priorities with the objective of serving the highest priority load at the expense of the less "critical" loads. In this paper, a new approach to the energy load management is presented.

The objective here is to determine, in an optimal way the energy to be supplied to each connected load in the next period of time. This is specially crucial if the energy demand is higher than the available amount of energy supply.

2. Problem Formulation

Loads are classified into four general categories as in [5]: convenient, essential, critical and emergency. The loads in each category are further sorted according to their subpriorities. Return per unit energy are assigned to each load with the highest value for the highest priority load in the emergency category. All loads are assumed controllable, even the semi-controllable or the uncontrollable loads. Table (I) gives the proposed load classification and the corresponding range of unit return value.

Table (I) Load Categories and Unit Return Values

Load	Priority	Unit Return
CONVENIENT	1	$1 \leq r_{1j} < 2$
ESSENTIAL	2	$2 \leq r_{2j} < 3$
CRITICAL	3	$3 \leq r_{3j} < 4$
EMERGENCY	4	$4 \leq r_{4j} < 5$

The solution of the addressed problem is based on the maximization of a weighted objective function which depends on the load assumed priorities subject to the amount of available energy. Specifically, it is required to solve the following optimization problem:

$$\text{Maximize: } \sum_{i=1}^{n} \sum_{j=1}^{m_i} r_{ij} E_{ij} X_{ij} \qquad (1)$$

$$\text{Subject to: } \sum_{i=1}^{n} \sum_{j=1}^{m_i} P_{ij} X_{ij} \leq E_T \qquad (2)$$

$$\text{and } \quad 0 \leq X_{ij} \leq B \qquad (3)$$

for all $i = 1$ to n and $j = 1$ to m_i

where:

i is the priority index as defined in [5]

n is 4 since four priorities are defined

j is the load index (from 1 to m_i)

m_i is the number of loads of priority i

r_{ij} is the return value associated with load j of priority i

X_{ij} is the number of units of time operating load j of priority i, which must be found by solving the optimization problem

P_{ij} is the average power required by load j of priority i in a predetermined period of time, equal E_{ij}/B

B is the total number of units of time

E_T is the total available solar energy

The return values should carefully be determined by the designer since the load priorities are dependent on the utilization of energy. For simplicity of notation we assume that the total number of loads is M where m = 1,2 to M. The supply factor SF_m is defined here by the ratio:

$$SF_m = X_m/B \tag{4}$$

Therefore the SF_m is a measure of the fraction of energy required by load m which will be actually supplied. This formulation will permit the inclusion of any variation in the load demand or priority depending upon the user's activities. It will also allow modification of the load subpriorities by keeping track of the last supply factor of the loads within each category.

3. Problem Solution

The defined mathematical formulation can be solved as one resource allocation problem [6] using Dynamic Programming (D.P.). The problem in D.P. formulation is given by:

$$f(1,y) = \max \left[R_1 X_1\right] \tag{5}$$

$$f(z,y) = \max \left[R_z X_z,\ f(z-1,\ y-E_z X_z)\right] \tag{6}$$

$$\text{for } z = 2,3,\ldots,N$$

$$0 \leqslant X_z \leqslant B$$

$$0 \leqslant y \leqslant E_T$$

where f(z,y) is the maximum total return that can be obtained by allocating y units of the available energy to loads 1 through z. So by definition y is a parameterized integer of E_T, and z is the stage number in the forward dynamic formulation. The set of recursive equations will give the final solution:

$$f(N,E_T) = \max \sum_{m=1}^{M} R_m X_m \tag{7}$$

A BASIC computer code is developed to solve the above recursive equations using Denardo [6] accelerated reaching procedure.

4. Simulation Study

Two examples are given here to demonstrate the applicability of this approach. We assume a system of four loads with fixed energy requirement. Assume that the total available energy in the next four hours is:

$$E_T = 300 + 300 \sin 30t \text{ (kWh)} \qquad (8)$$

where t = 0,1,2 and 3. In the first example we assume that the load priorities remain the same during the four hours. In the second example we assume that the load priorities are allowed to change by penalizing the unit return values of the loads by:

$$r_{ij,t+1} = r_{ij}(c - x_{ij,t}/B) \qquad (9)$$

subject to $i \leq r_{ij,t+1} < i + 1$ (10)

for $i = 1$ to 4, and $j = 1$ to m_i

where c is a penalty factor which is taken constant at 1.75. Assume that no battery is included in the system and the loads are kept in their category levels. Tables (II) and (III) give the results of the examples.

Table (II). Distribution of Energy Among the Loads for Example 1.

	Time Period E_T (kWh)			1 300	2 450	3 560	4 600
L O A D S	1.	CON	60	0	0	20	60
	2.	ESS	120	0	30	120	120
	3.	CRI	240	120	240	260	240
	4.	EME	180	180	180	180	180

Table (III). Distribution of Energy Among the Loads for Example 2.

	Time Period E_T (kWh)			1 300	2 450	3 560	4 600
L O A D S	1.	CON	60	0	0	60	60
	2.	ESS	120	0	118	80	120
	3.	CRI	240	120	152	240	240
	4.	EME	180	180	180	180	180

The results show that the loads with higher priority should be supplied before considering the less priority loads. The advantage of penalizing the loads that receive more than 75% of their demand is obvious in example 2 in which there are more than one load in the same category. Considering example 2, we notice that SF_m for load No. 2 varies in the simulation starting at 0 in the first period, then it is 0.98 in the second and 0.67 in the third period until it reaches 1 when the supply is enough. The corresponding values for load No. 3 are 0.5, 0.63, 1 and 1. Also notice that load No. 1 should get its demand in the third period inspite of the lack of available energy. This strategy allows a satisfactory load sharing for the three loads of the same category. Another penalty criteria may allow the penalty factor to be a function of the ratio of the supply to the demand.

5. Conclusion

Different approaches to the load management problem have been reviewed. Load management techniques when properly developed will lead to a better utilization of the PV energy. This paper presents a new approach to the problem. The loads are categorized into four general priorities and subpriorities are assigned to the loads. The PV energy and the loads demand should be known at least one period ahead of the simulation. The load sharing solution is based upon the load priorities. Examples illustrating the new approach are presented. Inspite of the large dimensions required the approach will lead to an efficient on-line load shedding control.

References

1. P. P. Groumpos, N. Sadati, and N. Diarra, 'The Importance of Controls for an Efficient Operation of SAPV Power Systems', Proceedings of 19th IEEE PV Specialists Conference, 1987, pp. 1246-1251.

2. P. P. Groumpos, R. C. Cull and A. F. Ratajczak, 'An Overview of Control Aspects of a Village Stand-Alone PV Power System', IEEE Transaction on Power Apparatus and Systems, Vol. PAS-103, No. 10, 1984, pp. 2845-2853.

3. R. C. Cull and A. H. Eltimshay, 'The Effects of Controls and Controllable and Storage Loads on the Performance of Stand-Alone Photovoltaic Systems', Proceedings of 15th IEEE PV Specialists Conference, 1981, pp. 621-626.

4. R. C. Cull and A. H. Eltimshay, 'Investigation of Energy Management Strategies for Photovoltaic Systems -An Analysis Technique', Proceedings of 16th IEEE PV Specialists Conference, 1982, pp. 270-275.

5. P. P. Groumpos and G. D. Papageorgiou, 'Optimum Load Management Strategies for Stand-Alone Photovoltaic Energy Power Systems', Proceedings of 17th IEEE PV Specialists Conference, 1984, pp. 647-652.

6. Eric V. Denardo, 'Dynamic Programming: Models and Applications', Prentice-Hall, Inc., Englewood Cliffs, New Jersey, 1982.

UTILIZATION OF WEATHER FORECASTS AND CLIMATIC INFORMATION FOR THE CONTROL OF RENEWABLE ENERGY SYSTEMS

Detlev Heinemann, Joachim Luther
Department of Physics, University of Oldenburg, D-2900 Oldenburg, FRG

ABSTRACT. We have systematically investigated the benefit of implementing advanced control strategies in electrical renewable energy systems (RES). Under consideration are strategies using (i) forecasts of the meteorological variables that determine the renewable energy production and (ii) climatic information obtained from meteorological time series.
In particular, the required system structures of RES in order to yield a benefit from these strategies are discussed.

1. INTRODUCTION

Besides being determined by type and size of converters and storages the performance of renewable energy systems (see Fig.1) is particularly influenced by the control strategies used and the actual control decisions taken. In a decision-theoretical sense, the current state of an energy system (characterized e.g. by storage content, load etc.) is determined by the previous state, by the actions (e.g. switchings) that are taken as a result of a decision and by events (e.g. energy input determined by meteorology) that are relevant to the outcome.
The decision-making problem is to choose the optimal action from the set of possible actions. The decision criteria in a control strategy of a RES are, e.g., minimizing the amount of auxiliary energy or minimizing surplus energy that is of no use to the system. In such a decision-making situation additional information can help to take the optimal decision (especially if some of the events are of stochastic nature). Concerning the energy input into the RES such information can be derived from medium range weather forecasts or from statistical climatic information.
In simple RES there often exist constraints that limit the allowable set of system control actions considerably; a real decision in the sense of choosing between several possible actions does not take place (e.g. turning on the Diesel-generator in cases when the load cannot be supplied by the renewable converters and the storage). In such systems there is no degree of freedom in the decision-making process. Only the existence of alternative actions leads to a situation, in which real decisions can be taken.

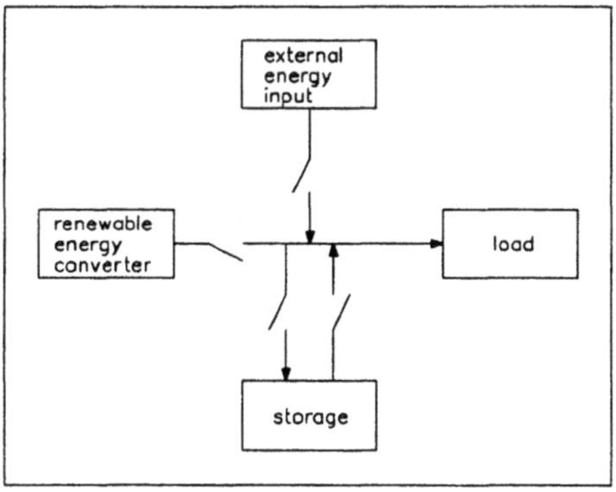

Fig.1: General structure of an electrical renewable energy system (RES).

2. REQUIRED STRUCTURE OF RENEWABLE ENERGY SYSTEMS TO BENEFIT FROM THE UTILIZATION OF ENERGY AVAILABILITY FORECASTS (EAF)

To come to an identification of possible and convenient applications of the EAF, some statements about the requirements on the structure of RES to utilize this additional information will be made. For this purpose a classification of RES has been made in such a manner that all systems belonging to one class have the same properties concerning the availability of alternative actions (same degrees of freedom). In this investigation only electrical systems with a high renewable supply are considered. System structures may be qualified with respect to the following points:

- storage: a) no storage implemented
  b) one storage medium implemented
  c) use of several storage technologies
- auxiliary energy: a) no auxiliary energy available (i.e.: the load is supplied totally by the renewable converters, the system is energy-autonomous)
  b) auxiliary energy is available (e.g. back-up-converter, grid connection)
- load: a) the load has to be supplied completely at any instant:
  $P_{storage}(t) + P_{auxiliary}(t) + P_{renewable}(t) >= P_{load}(t)$
  b) an energy amount has to be supplied in a given time interval:
  $\int_T (P_{storage}(t) + P_{auxiliary}(t) + P_{renewable}(t)) \cdot dt >= \bar{P}_{load} \cdot T$

This classification results in 12 classes of different system configurations with different sets of possible control actions.
In systems without storage devices the optimal action is determined by constraints on the control determined by the defined target states (e.g. to supply the load in any case). Such systems have no degree of freedom whether there is a back-up component or not, because no alternative actions are available. For that reason additional information (e.g. forecasts of the energy input) cannot be utilized to improve the system performance.
Regarding the availability of alternative actions the same statement is valid for systems with one storage and no auxiliary energy input. The optimal action is determined by the current state of the system, even if the specification of the load supply is less strict (type (b)).
In the case when both auxiliary energy and storage are implemented and the load is of type (a), a degree of freedom is provided under the assumption that the load cannot be supplied by the auxiliary energy converter alone. Now a decision is to be made between supplying the load with auxiliary energy or with energy from storage taking into account that the optimal action has to guarantee an adequate state of charge.
If the load does not have to be supplied at any instant (type (b)), benefit can be derived from a further degree of freedom. In addition, the decision has to be extended to the problem what part of the load shall be supplied at the moment of decision.
Finally a further method of enlarging the number of degrees of freedom is

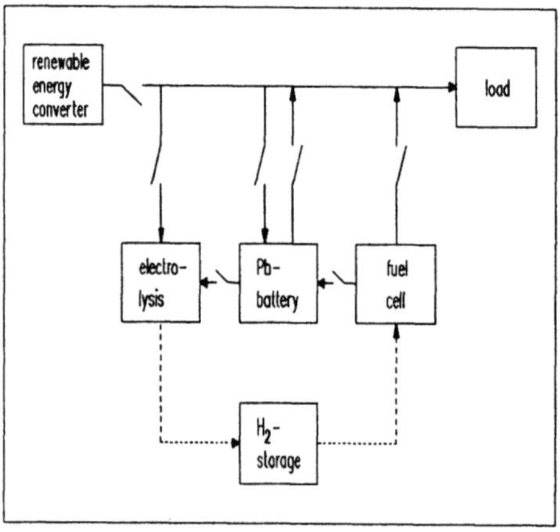

Fig.2: Structure of the investigated RES. No auxiliary energy input is available, but two storage technologies are implemented: a Pb-accumulator and a H_2-storage-system consisting of an electrolysis, a H_2-gas-storage and a fuel cell.

to implement different storage technologies in a RES. With this measure, the set of possible actions is extended considerably. However, a prerequisite for different outcomes of the decision-making process is that the storage technologies used are of different quality (efficiency, maximum conversion rate).
Even in systems without auxiliary energy input and with strict requirements of the load supply (type (a)) several alternative actions are available now, so that in the decision-making process the optimal action can be chosen from a set of possible actions. This also means that the consideration of the information from an EAF should lead to improvements in the system performance.

3. SYSTEM STRUCTURE OF THE RES UNDER INVESTIGATION

As an example, we consider an autonomous renewable energy system (no auxiliary energy input) (Fig.2) having two different storage systems: a Pb-battery and a hydrogen-storage system consisting of an electrolysis, a H_2-gas-storage and a fuel cell. The rated power of the fuel cell is assumed to cover the load. Results given below refer to storage capacities of 2 and 40 load-days for the Pb-battery and the H_2-storage, respectively. Renewable energy input is generated by a solar/wind hybrid system (equal shares of annual energy production: W/S = 50/50). In such systems the utilization of advanced control strategies can result in a reduction of system hardware necessary to yield autonomous systems.

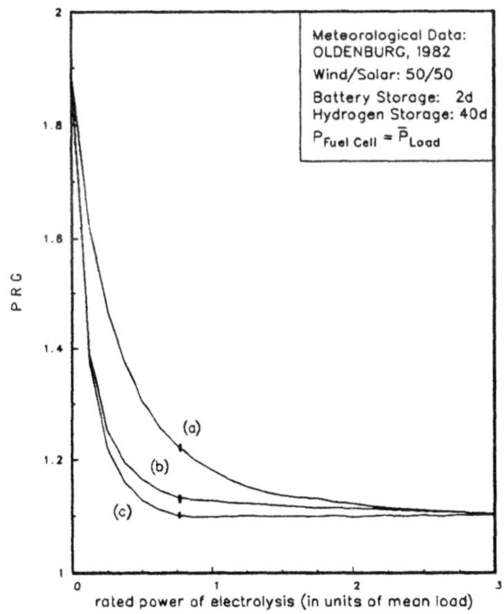

Fig.3: Required mean renewable production PRG (in units of the mean load) for system autonomy as a function of the rated power of the electrolysis. The curves (a)-(c) refer to the control strategies investigated (see text).

4. RESULTS

We have performed simulation calculations using meteorological data (hourly mean values) for the site Oldenburg (53°N latitude).
The different control strategies used are as follows:

(a) The system control makes no use of external information and the decision depends only on the state of charge of the battery. The H_2-storage is charged only when the Pb-battery is charged to its rated capacity and the renewable production exceeds the load.
(b) Considering the climatic characteristics of the site, a control threshold (lower than the rated capacity) of the Pb-battery is calculated. The electrolysis is switched on if the charge of the battery exceeds this threshold.
(c) The control strategy uses perfect meteorological forecasts of the parameters governing the renewable energy input. The decision-making problem is to distribute the energy flow between the given

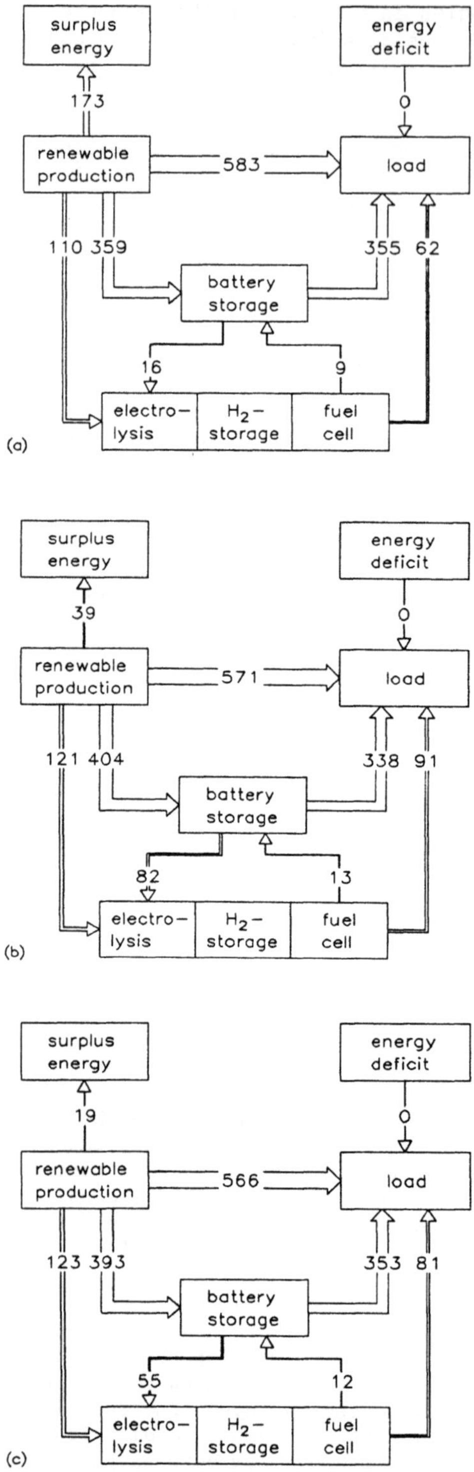

Fig.4: Annual mean energy flows between the system components for three different control strategies (see text). System configurations correspond to the marked points in Fig.3. The energy deficit of 0 characterizes an autonomous system.

storage systems in an optimal manner, so that the amount of surplus energy (that is of no use to the system) can be minimized.

The results shown in Fig.3 are obtained by utilizing these three different control strategies. The necessary mean renewable production PRG (in units of the mean load) to yield an autonomous system is plotted as a function of the rated power of the electrolysis unit (PRG is a measure of the renewable converter size). The benefit of an advanced control strategy is reflected in a smaller renewable production that is required - for a given electrolysis unit - to yield an autonomous system. The best results are obtained by utilizing the strategy including information of weather forecasts. Already considerable improvements can be made when replacing strategy (a) by strategy (b), that uses climatic information only.
For the example discussed in this paper the overall effect of these control strategies lies in a reduction of the size of the renewable converters in the range of 10 % (e.g. a reduction of the costs for PV-modules).

To give an example of the energy flows in the considered RES when using different control strategies, the flows between the system components are shown in Fig.4a-4c (system configurations as given in Fig.3 - marked points). The mean load in this example is 1000 W, the rated power of the electrolysis, 750 W. The essential changes in the energy flows when using advanced control strategies are the decreasing amount of surplus energy and the increasing energy flow

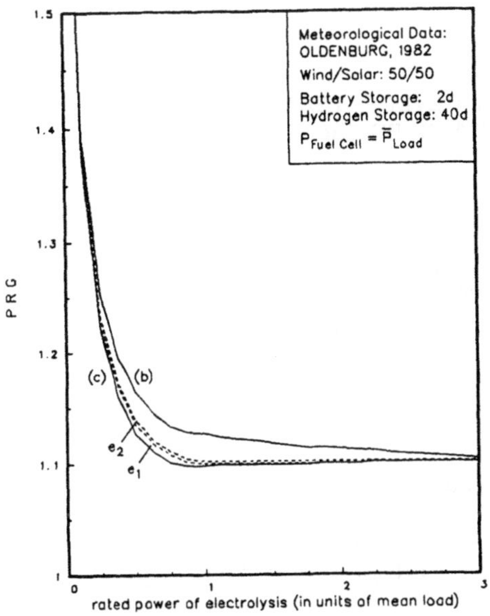

Fig.5: Required converter size (PRG) as function of the rated power of the electrolysis when using forecasts with errors according to Fig.6. The curves (b) and (c) are the same as in Fig.3.

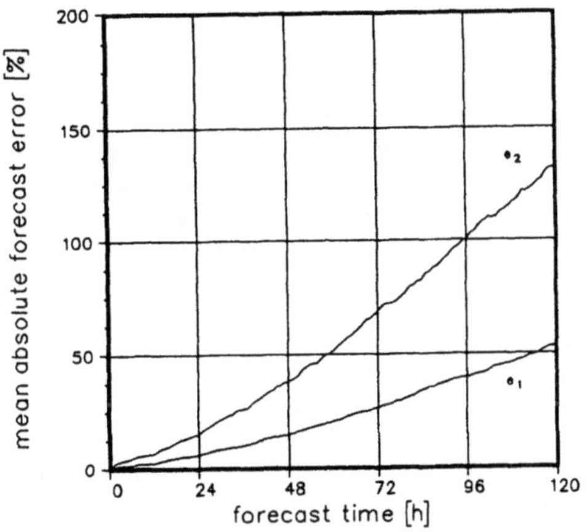

Fig.6: Mean error of synthetic forecasts as a function of time. The curves e_1 and e_2 correspond to the examples given in Fig.5.

through the H_2-storage system (the absolute value of this flow depends on the efficiencies of the converters, here assumed to be 0.7 and 0.6 for the electrolysis and the fuel cell, respectively).

Finally the sensitivity of the system performance on the forecast accuracy was a point of consideration. In Fig.5 simulation results are shown, that reflect the utilization of forecasts with defined errors instead of perfect forecasts. The synthetic forecasts are produced by multiplying the perfect forecast values by a random error term with a mean value of 1 and a standard deviation increasing with the forecast time (Fig.6). The random error terms are obtained by using the random walk method. As can be seen, considerable deviations from the perfect forecasts do not lead to a substantial increase in the required renewable energy production (at least for the given storage configuration).

5. CONCLUSIONS

Our results have shown, that the performance of renewable energy systems can be improved, when advanced control strategies are implemented. Particularly the utilization of weather forecasts can lead to reductions in installing costs for renewable converters.

ELECTRIFICATION OF THE GAVDOS ISLAND BY SOLAR PHOTOVOLTAIC POWER

J. CHADJIVASSILIADIS *) and H. KADELLA **)

*) Public Power Corporation, 10 Navarinou str., GR-106.80 Athens/Greece, Tel. (01) 36.30.845
**) Siemens Solar GmbH, Buchenallee 3, D-5060 Bergisch Gladbach 1/FRG, Tel. (0 22 04) 406-46

ABSTRACT

Gavdos is a remote island located in the Libyan Sea, south of Crete, and it is the southmost part of Europe. No electricity is existing on the island, but it offers the most favourable conditions for solar radiation and photovoltaic applications.

The purpose of the project is to demonstrate photovoltaic generators as technically suitable, economically achievable and socially well accepted as todays solution to meet the electrical energy demand of remote islands. The concept for the electrification of the island is to install a central photovoltaic power station to supply the existing small communities (Phase II) and small independent photovoltaic units for the dispersed houses (Phase I). the project is realized by the consortium PPC and Interatom/Siemens Solar GmbH within the CEC demonstration programme 1983 for the Phase II and in the frame of bilateral cooperation for the Phase I.

An architectural design was elaborated adapting the plant and the small PV units to the environment. Seasonal adjustment of the filt angle is provided in order to adapt the output of the PV generator to the demand.

Low energy consumption domestic appliances such as lighting lamps. refrigerators and TV sets, are provided for the consumers. A new type of solar inverter is developed for this project, consisting of a DC/DC-converter and a DC/AC-inverter with very high efficiency. The inverter is designed to operate in four quadrants for a possible integration of a wind converter when the demand will be increased in the future. Two transmission lines leading from the PV plant to the two communities and the relevant network have been erected to supply the consumers with AC 220/380 V, 50 Hz.

The independent PV units have been designed in three types to supply the consumers with DC 24 V.

The plant is in operation since June 87 and is ready to supply the consumers when the internal electrical installation of the houses will be completed.

1 INTRODUCTION

Gavdos Island is exactly situated at a Latitude 34° 50' N and Altitude 24° 05' E. Two communities are located on the island, Kastri and Vatsiana, and some dispersed houses. The population varies between 70 persons in winter and 90 persons in summer with additional visitors during the Summer season. A central photovoltaic plant of 20 kWp is installed between the two communities and two lines for three phase A.C. will supply the consumers, one for each community (see Fig. 1). An area of 3000 m^2 was necessary for the installation of the plant. An architectural design was elaborated adapting the plant to the environment. An old windmill is used for the installation of the electronical equipment and a small housing for the battery storage was built in a traditional way.

2 PHOTOVOLTAIC ARRAY

The rated power of the photovoltaic arrays is 20 kWp. 160 monocrystaline silicon modules type SM 144 manufactured by Siemens AG are installed in 8 arrays. The dimensions of the modules is 1020X1470 mm mounted 144 round cells of 10 cm diameter with a rated power of 130 Wp. The module frame is of anodized extruded aluminium, the module thus being capable of withstanding high surface-loading. The module is equipped with protection diodes which are located within the terminal box. 20 modules are connected in series to one branch and mounted on one supporting structure. The supporting structures are simply designed by using H section and tubes of hot galvanized steel. A good stability of the supporting is achieved by placing the middle of the module above the upper mounting tube. The construction provides an easy seasonal adjustment of the tilt angle. Lightning conductors and earthing system is installed in the generator field to protect the installation. Special devices by varistors are used also to protect the modules and the electronic equipment against lightnings.

3 POWER CONDITIONING

A storage battery consisting of 250 lead acid elements (2 V, 468 Ah each one) connected in two parallel groups (250 V, 936 Ah) is used, manufactured by TUDOR HELLAS.

A new type of solar inverter is developed for this project. In one cubicle are mounted both the DC/DC-converter and the DC/AC-inverter, (see Fig. 2).

The task of the DC/DC part of the solar inverter is the decoupling of the photovoltaic generator voltage from the battery voltage. In this way the solar generator can be utilized in an optimal way. In addition an MPP-Tracker is connected to utilize the solar field optimally. The efficiency of the DC/DC part is extremely high (96 %).

The DC/AC part is of 30 kVA rated power, three phases 230/400 V, 50 Hz. The efficiency is very high, 94 % at the rated power and above 90 % at 20 % of the rated power. The no load losses are limited to 800 W. This inverter is designed to operate in four quadrants for a possible integration of a wind converter when the demand will be increased in the future. The solar inverter is installed in an old windmill while the storage battery in a small housing.

4 PV USER APPLIANCES

The consumer will be provided by low energy consumption domestic appliances such as lighting lamps, TV sets and refrigerators. In additional special lamps will be used for street lighting:

- for the illumination, special energy saving lamps 18 W, 900 lm
- for refrigerators, 120 l with very good isolation
- TV sets, colour TV 50 W with 16" tube.

The capacity of the plant can meet the demand of the consumers during summer and winter by a seasonal adjustment of the modules tilt angle.

The transmission lines and the distribution sytem are designed to have very low losses and to protect the plant against short circuits. A superimposed control is provided to receive data and to act to the solar inverter and switching devices.

5 MONITORING SYSTEM

A monitoring system developed by Interatom according to the CEC/ISPRA guidelines is installed for the beginning of operation of the plant, for a period up to the time when the monitoring system provided by PPC will be ready for installation. This data acquisition sytem is also designed according to the ISPRA guidelines with equipment of Hewlett-Packard.

The HP 3421A Data Acquisition unit will be used with a capacity of 30 analog and/or digital indication inputs. The HP 71B Handheld computer is connected to the Data Acquisition unit via a HP-IL link. The HP-71B computer includes a quarz crystal clock and can turn on the HP3421A unit, trigger to scan a list of up to 30 inputs, instruct a mass memory to store the readings and then power down the entire system until the next time passes. The HP-9114B portable Microfloppy shall be used as mass storage media.

The active system shall be supplied by 12 V DC. The peak power consumption is estimated to be less than 200 W with an average power consumption less than 25 W.

The data scanning of the required measurements will be done on a periodic basis with a time interval of the order of about 10 seconds.

By an appropriate subroutine running in the HP-71B computer the average values of the sampled measurements will be calculated and stored in the computer's RAM every 10 minutes. The 10 minute averaged values will be transfered from the computer's RAM to the floppydisc every hour. The floppydisc will be transferred in Athens every month and from a central system a copy of a floppydisc will be produced and sent to CEC/ISPRA. The analysis of data for several useful conclusions will be done in Athens in a small central system (HP-86B).

6 DISPERSED PV-UNITS FOR REMOTE HOUSES

Besides the contral PV-plant which is designed for the electrification of the two nearest communities, six additional autonomous PV-units have been designed and erected for the dispersed houses.

There are three different types of consumers:

2 units (type A) for the shops at the harbour
3 units (type B) for the houses in Ambelos
1 unit (type C) for the school.

The units are designed for operation in 24 V DC with domestic appliances of low energy consumption. The principle diagramme and data are shown in Fig.3.

7 INSTALLATION AND COMMISSIONING

An access road to the site has been constructed and the area prepared for installation. A small housing for the batteries was built and the old windmill restored for the installation of the electronic equipment.

All the components of the plant have been transported from Germany and Athens via Crete to the island. The transportation from south Crete to the site was a very difficult task due to the lack of infrastructure and site supporting devices.

The plant and the six autonomous PV-units have been simultaneously erected during the summer season 1986 and it was completed at the end of October. Because of the difficulties to reach the island in the winter season the commissioning was foreseen for the first months of 1987. The commissioning started the first days of March 1987 but unfortunately it had to be interrupted due to the extremely hard weather conditions unusually for this region. Therefore the commissioning was continued in May 1987.

The plant and the six autonomous PV-units are now in operation since June 87 and is ready to supply the consumers when the internal electrical installation of the houses will be completed.

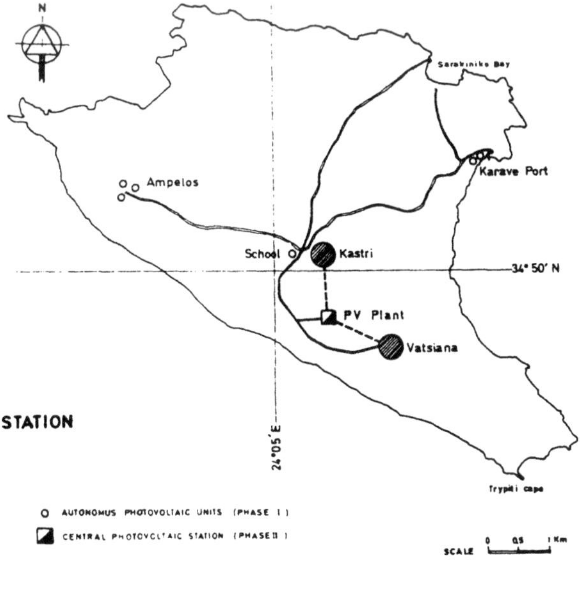

Fig. 1 GAVDOS ISLAND THE CENTRAL 20 KWp PV POWER STATION

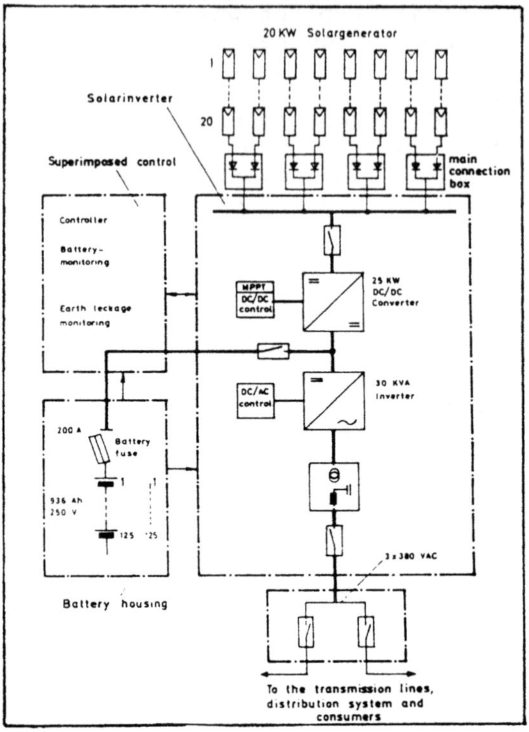

Fig. 2: Gavdos PV-Plant 20 KWp
Block Diagram

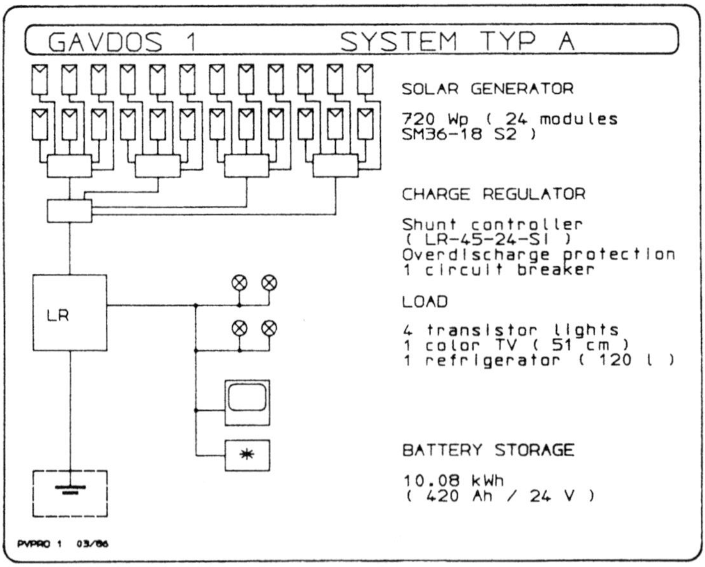

Fig. 3: Autonomous PV-Unit

CONTRACT NO SE/507/84

Solar Wind Cogeneration For Urban Residential Use

A Dichler, P R Wolfe
Solapak Ltd, Factory Three, Cock Lane
High Wycombe, Bucks HP13 7DE UK

Summary

New standards in energy efficiency for housing, office accommodation and commercial premises are being demonstrated in the UK at Milton Keynes, where a development area, the "Energy Park", has been specifically designated for schemes meeting high energy saving standards. Against this background of energy conscious design, it is appropiate to reconsider the sources of electrical energy generation in favour of "alternative" or "renewable" energies such as solar and wind power. It is particularly necessary to study the environmental impact of such "new" power sources if they are to achieve widespread use.

The Solapak scheme at Milton Keynes demonstrates how this "new" technology can be integrated into a domestic urban environment; the first mainland application of a medium sized wind turbine; unique in the nature of the non-specialist user - 9 dwellings on the Energy Park.

1.0 Introduction

The project at Milton Keynes was reported in detail at the Commission of the European Communities Second Contractors meeting where an overview of the project was given (4.1), and the Technical Scope, apparatus and operating philosophy detailed. Information regarding the monitoring scheme was also elaborated.
Figure 1 gives a block schematic indicating the system electrical configuration interrelationship between energy sources and load.

In addition to the engineering work that has been undertaken to implement the project, Solapak and Milton Keynes Development Corporation have been concerned to evaluate the impact of the scheme on the urban environment. In parallel with the engineering trials therefore, an environmental study has been commissioned to determine the social impact of the project. This is in progress and results will be reported in due course.

2.0 Implementation

From the beginning this project has set out to achieve a high degree of social and architectural integration and in the majority this has been achieved.

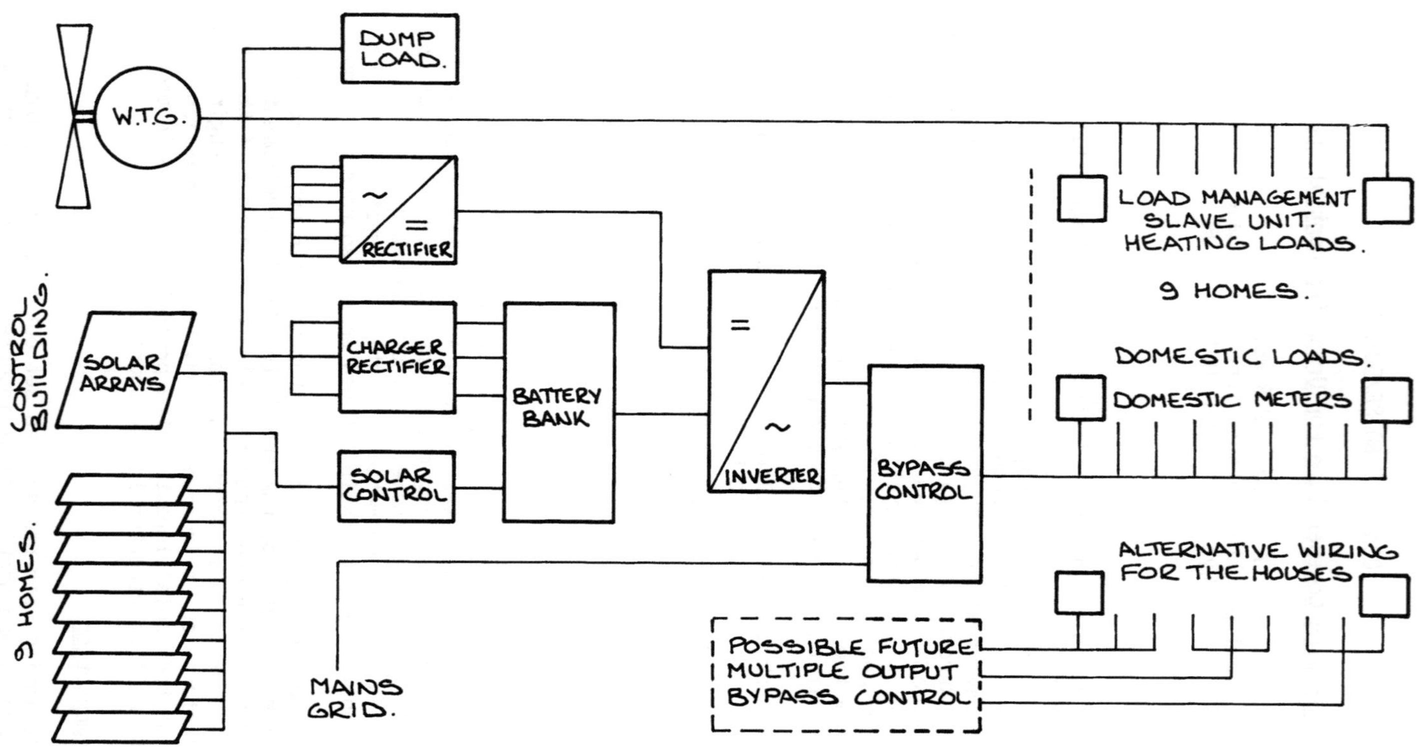

Figure 1 System Block Schematic

2.1 Solar Modules

Solar modules were specifically designed and manufactured to fit into the conservatory structure of the homes. Picture 1 shows the result which is pleasant in appearance. In addition the modules provide shading from the summer sun, which on hot days, can cause conservatories with plain glass rooves to become unusable.

Picture 1 Solar Modules Integrated Into House Architecture

2.2 Energy Management Centre

The system control and equipment shelter has been built as a garage sized property with a pitched roof upon which 2.5 kW of solar modules is mounted (Picture 2). Materials for the building are sympathetic to that of the surrounding dwellings. The result is an effectively integrated unit which blends in well with the surroundings.

Picture 2 Energy Management Centre Styled To Suit Surroundings

2.3 Wind Turbine

Attention was paid to the design of the tower. It was important to ensure that the wind turbine as an entity was a visually well balanced structure (Picture 3).

Picture 3 Wind Turbine On Tubular Tower

3.0 Preliminary Result

The scheme as an engineering project is providing valuable information pertinent to the evaluation techniques applicable to future commercial schemes of this nature. Though data collected is not yet sufficient to provide the basis for meaningful statistical analysis and prediction, the early results encourage continued evaluation.

In terms of the equipment integration into a domestic urban environment the results of the social surveys are expected to yield useful guidelines for selection of future sites.

4.0 References

(4.1) "Photovoltaic Demonstration Projects", W Gillet, J Bates, W Kaut. Proceedings of Second Contractos Meeting, Brussels, Belgium, April 1987.

PHOTOVOLTAICALLY SUPPLIED DEVICES OF LOW AND INTERMEDIATE POWER RANGE

Werner Roth, Jürgen Schmid

Fraunhofer Institut
für Solare Energiesysteme (ISE)

Oltmannsstr. 22
D-7800 Freiburg

ABSTRACT. In 1986, a project supported by the German Ministry for Research and Technology was started by ISE to systematically increase the application of photovoltaics. Over a period of five years, suitable products are being identified and appropriate photovoltaic power supplies are being developed in co-operation with industrial companies. The objectives of this project, the framework of the co-operation between industrial companies and the institute and our experience to date are discussed briefly. The specific requirements for small-scale and indoor photovoltaic systems are described in more detail.

For indoor applications it is necessary to determine the I-V characteristics of solar cells and modules under extremely low irradiation, to record the indoor irradiation resp. the indoor illumination and to select an appropriate storage battery. Examples of some innovative small power systems will be presented.

1. INTRODUCTION

Photovoltaics has the chance of making a significant contribution to an environmentally sound energy supply concept in the future. As the cost of electricity from PV generators is currently still very high compared to that from conventional power stations, this aim can however only be achieved on a long-term basis. The situation is much more favourable for electrical appliances which are not connected to the grid. Here, where it is in competition with electricity supplied by batteries or diesel generators, photovoltaics can offer a solution which is less expensive, less polluting and more easily handled. Recognition of this fact was the initial impulse for the project "Photovoltaically supplied devices of low and intermediate power range", which is supported by the German Ministry of Research and Technology, and is being carried out by ISE. This project is intended to stimulate dynamic growth in the PV market, by developing photovoltaic power supplies for electrical appliances which were previously powered by another source, as well as by developing new products, for which there is no alternative to solar electricity /1/. Achievement of this aim appears possible if products are introduced to the market which impress buyers by their utility, reliability and economy of operation.

A resulting expansion of the photovoltaic market will lead to an increase in the demand for solar cells. This in turn would allow more rational production and thus reduced solar cell prices. With the lowered costs, photovoltaics can then penetrate an increasingly attractive range of application areas.

The results show that the loads with higher priority should be supplied before considering the less priority loads. The advantage of penalizing the loads that receive more than 75% of their demand is obvious in example 2 in which there are more than one load in the same category. Considering example 2, we notice that SF_m for load No. 2 varies in the simulation starting at 0 in the first period, then it is 0.98 in the second and 0.67 in the third period until it reaches 1 when the supply is enough. The corresponding values for load No. 3 are 0.5, 0.63, 1 and 1. Also notice that load No. 1 should get its demand in the third period inspite of the lack of available energy. This strategy allows a satisfactory load sharing for the three loads of the same category. Another penalty criteria may allow the penalty factor to be a function of the ratio of the supply to the demand.

5. Conclusion

Different approaches to the load management problem have been reviewed. Load management techniques when properly developed will lead to a better utilization of the PV energy. This paper presents a new approach to the problem. The loads are categorized into four general priorities and subpriorities are assigned to the loads. The PV energy and the loads demand should be known at least one period ahead of the simulation. The load sharing solution is based upon the load priorities. Examples illustrating the new approach are presented. Inspite of the large dimensions required the approach will lead to an efficient on-line load shedding control.

References

1. P. P. Groumpos, N. Sadati, and N. Diarra, 'The Importance of Controls for an Efficient Operation of SAPV Power Systems', Proceedings of 19th IEEE PV Specialists Conference, 1987, pp. 1246-1251.

2. P. P. Groumpos, R. C. Cull and A. F. Ratajczak, 'An Overview of Control Aspects of a Village Stand-Alone PV Power System', IEEE Transaction on Power Apparatus and Systems, Vol. PAS-103, No. 10, 1984, pp. 2845-2853.

3. R. C. Cull and A. H. Eltimshay, 'The Effects of Controls and Controllable and Storage Loads on the Performance of Stand-Alone Photovoltaic Systems', Proceedings of 15th IEEE PV Specialists Conference, 1981, pp. 621-626.

4. R. C. Cull and A. H. Eltimshay, 'Investigation of Energy Management Strategies for Photovoltaic Systems -An Analysis Technique', Proceedings of 16th IEEE PV Specialists Conference, 1982, pp. 270-275.

5. P. P. Groumpos and G. D. Papageorgiou, 'Optimum Load Management Strategies for Stand-Alone Photovoltaic Energy Power Systems', Proceedings of 17th IEEE PV Specialists Conference, 1984, pp. 647-652.

6. Eric V. Denardo, 'Dynamic Programming: Models and Applications', Prentice-Hall, Inc., Englewood Cliffs, New Jersey, 1982.

4. Simulation Study

Two examples are given here to demonstrate the applicability of this approach. We assume a system of four loads with fixed energy requirement. Assume that the total available energy in the next four hours is:

$$E_T = 300 + 300 \sin 30t \text{ (kWh)} \tag{8}$$

where t = 0,1,2 and 3. In the first example we assume that the load priorities remain the same during the four hours. In the second example we assume that the load priorities are allowed to change by penalizing the unit return values of the loads by:

$$r_{ij,t+1} = r_{ij}(c - x_{ij,t}/B) \tag{9}$$

subject to $\quad i \leq r_{ij,t+1} < i + 1 \quad$ (10)

for $\quad i = 1$ to 4, and $j = 1$ to m_i

where c is a penalty factor which is taken constant at 1.75. Assume that no battery is included in the system and the loads are kept in their category levels. Tables (II) and (III) give the results of the examples.

Table (II). Distribution of Energy Among the Loads for Example 1.

	Time Period E_T (kWh)	1 300	2 450	3 560	4 600
LOADS	1. CON 60	0	0	20	60
	2. ESS 120	0	30	120	120
	3. CRI 240	120	240	260	240
	4. EME 180	180	180	180	180

Table (III). Distribution of Energy Among the Loads for Example 2.

	Time Period E_T (kWh)	1 300	2 450	3 560	4 600
LOADS	1. CON 60	0	0	60	60
	2. ESS 120	0	118	80	120
	3. CRI 240	120	152	240	240
	4. EME 180	180	180	180	180

It is therefore necessary to investigate the reliability, efficiency and lifetime of all the system components before they are taken into use. Appropriate measurement and test equipment is developed and constructed at the institute to carry out such investigations. For example, the transistorized ballasts for fluorescent lights are tested before being incorporated in photovoltaically powered lighting systems (fig.3).

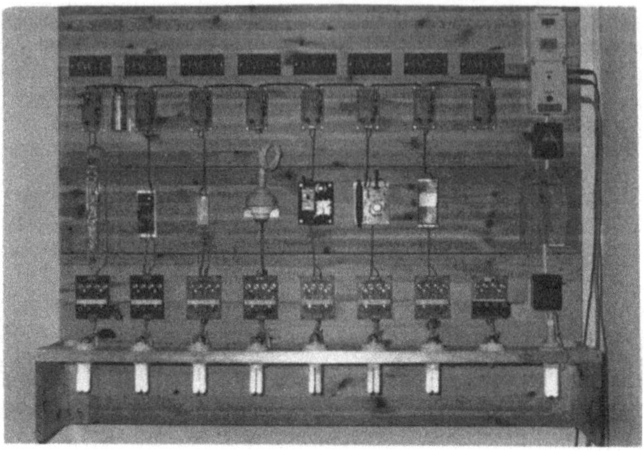

Fig.3: Equipment for testing transistorized ballasts for fluorescent lights.

Further, all the appliances developed as part of the project are thoroughly tested before they are released onto the market. This should ensure that only fully developed, functional and reliable products appear on the market.

3.4. Identification of optimal storage batteries for small-scale applications

The battery types which are most commonly used at present in photovoltaic systems are the lead acid and NiCd batteries. While the lead battery is found in large systems, the NiCd battery is preferred in small-scale applications. Its advantages are that it is maintenance-free, compact and insensitive to overcharging and extreme discharging and that it has a relatively high capacity even at low temperatures. Its main disadvantage is however the low charging efficiency for small charging currents, as are commonly encountered in small-scale photovoltaic systems. As NiCd batteries are normally charged with a constant current, there has been very little information available up to now on the charging characteristics of these batteries for very low charging currents.

A battery testing system was therefore developed and constructed to investigate the exact dependence of the charging efficiency for low charging currents. It allows more than 300 storage batteries to be tested at temperatures of -15°C, 0°C, 20°C and 40°C.

Further activities of the institute within this project are concerned with system optimization and the search for possibilities of minimizing the energy consumption of photovoltaically powered appliances.

4. EXAMPLES OF PRODUCTS DEVELOPED DURING THE PROJECT

The following examples should demonstrate the diversity and interest of photovoltaic applications in the low power range.

4.1. Solar powered ventilator

This solar powered ventilator (fig.4,5) is intended to be installed in cars, trucks, vans, caravans, boats, etc. In hot weather, it lowers the inside temperature of the vehicle without discharging the battery. During cooler weather, the solar cells charge the battery. A further advantage of this arrangement is that it prevents self-discharging of the battery when the vehicles are not used for a lengthy period.

The solar powered ventilator is completely maintenance-free, it can be operated with one hand alone and runs extremely quietly. The solar cells are well protected behind scratch-resistant, re-inforced glass 2mm thick. The solar powered ventilator will be available on the market in two sizes, with a maximum ventilation capacity of approximately $80m^3/h$ and $180m^3/h$.

Fig.4: Solar powered ventilator, top view.

Fig.5: Solar powered ventilator, bottom view.

4.2. Metal construction set with a solar generator

By using a highly efficient electric motor with additional gears, several of the modules which can be made with the construction kit were automated directly with solar electricity (fig.6).

Fig.6: Solar powered toy car.

Thus, children can already learn to handle this new type of electrical power supply while they are playing. Further, the purchase of environmentally undesirable primary batteries becomes superfluous. This product is already in commercial production too.

4.3. Solar Power Box

Figure 7 shows a transport case for a battery powered drill. Mounted on the lid of the case is a solar generator with an effective area of 6.5dm^2. Thus, the case serves as a recharging device for the battery packs. The battery packs are placed inside the case during charging. With an insolation of 1000 W/m^2, a completely discharged battery pack (energy capacity 11 Wh) can be recharged in around two hours.

Fig.7: Solar Power Box (operating prototype).

4.4. Solar powered ionic analyser for human blood tests

This is a new instrument for measuring the amounts of sodium and potassium in blood (fig.8). It can be used in routine and emergency health care at any time and anywhere, even in difficult conditions outside the clinical analytical laboratory.

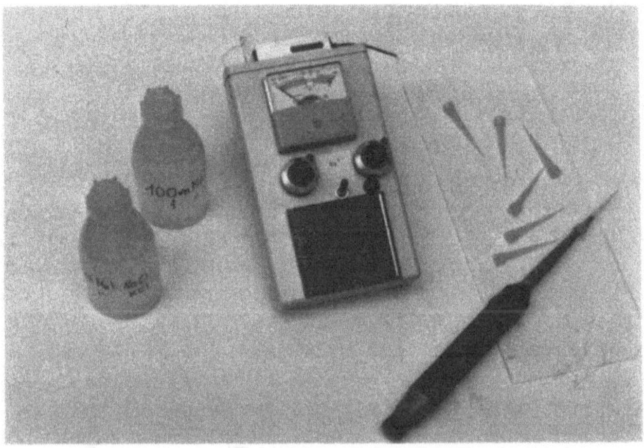

Fig.8: Solar powered ionic analyser for human blood tests.

To be able to power this instrument economically with photovoltaics, its electrical consumption had to be reduced to a fiftieth of the original value. This was possible only by using low power and low voltage integrated circuits throughout and by improving the timing controls. This device will soon be available on the market.

5. REFERENCES

/1/ J. Schmid: Small Power Photovoltaic Systems. Proc. of the 7th Photovoltaic Solar Energy Conf., Sevilla, 1986.

/2/ W. Roth: Aspekte der Auslegung photovoltaisch versorgter Indoorsysteme. Drittes nationales Symposium photovoltaischer Solarenergie, Staffelstein, 1988.

SMALL POWERBLOCK WITH PV POWER SOURCE

Miklós Pálfy
Research Institute of the Electrical Industry
Hungary, Istenhegyi út 29. 1125 Budapest

ABSTRACT. Standard type size R20 was used to equip with solar cells and battery. The rated voltage of the battery /Ni Cd/ is of 1,2 V and it has a rated capacity of 2 Ah. The rated charging current of the solar cells is 40 mAmps. For rapid charging flatplate concentrator is proposed. The measurements have shown in the case of direct radiation rate exceeding 30 % the charging of the current source can be increased advantageously by using double plates concentrator. At a direct radiation rate of 0,8. approx. a concentration factor of 2,3 was measured.

1. INTRODUCTION

The various electric and electronic equipment require an independent, autonomous power source. The most frequently used autonomous electric power sources are the primery elements and the rechargable batteries respectively. In the course of charging, in most cases the batteries must be disconnected from the equipment, they must be taken out from them and placed in the charger. At the same time in lack of an electric network the charging devices require an auxilary power source. It seems an obvious solution to apply a charging unit with solar cell, which can be mounted in unit with the battery. In this study we want to report on such a solution.

2. CONSTRUCTION OF THE CURRENT SOURCE

At the shaping, construction of the low-performance electric power source the basic consideration has been that its outer dimensions should be in conformity with the standard series introduced at the manufacture of the batteries resp. accumulators, in other words the user could apply them without any difficulty in his device. From the different standard types the choice has fallen on the type R-20.

The electric circuit and the current source are of a construction that the possibility of charging the accumulator by an external charging device has been left too.

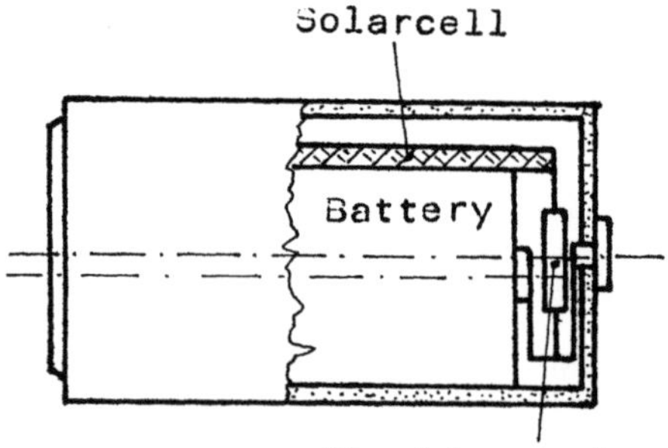

Figure 1 Construction of small powerblock with PV power source

The solar cell is composed of 5 single-crystal silicon individual cells of VKI made /1/. The individual solar cells are 10x20 mm size and connected in series their short circuit currents are I_z= 45 mA at an illumination level of 100 mA/ cm^2. The open circuit voltage of the solar cell unit is of 2,9V and at a voltage of U_T= 1,9V it supplies a current of I_T= 40 mA. The geometrical sizes of the solar cell element were practically restricted by the standard type size R20. Then, how many individual solar cell elements should be connected in series within that, has been decided by the battery being selected, of what extent the voltage drop of the blocking unit has been and at which illumination the charging is still desirable.

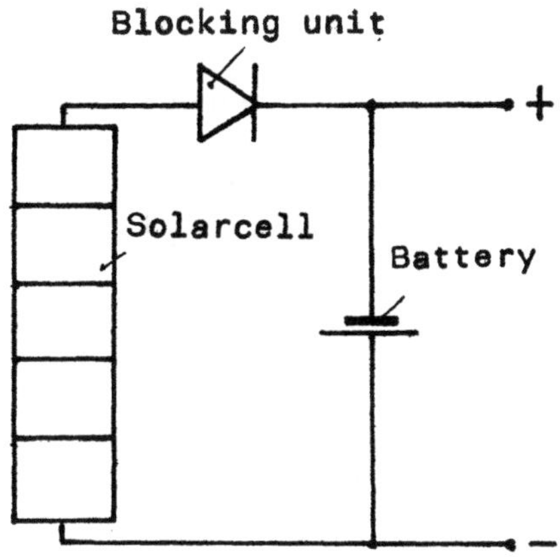

Figure 2 Schematic of powerblock

At the selection of the battery the primary consideration has been that in the given geometry the highest possible battery capacity should be built-in. In view of this, the standard NiCd battery R14 has been built-in being available. The rated voltage of the battery /NiCd/ is of 1,2 V and it has a rated capacity of 2 Ah. The possibility for the application of other batteries exists, of other batteries exists, of course, too.

The voltage drop and reverse current of the blocking unit appears anyway as a loss. By the application of a germanium diode the voltage drop could be restricted at a rated charging current of I_T= 40 mAps to 0,51 V and the value of the reverse current was also below 1 μA at a rated battery voltage of 1,2 V and at room temperature. This latter means even at a storage in dark for one month only a discharge loss of 0,0007 Ah at a maximum.

We fixed the lower threshold value of charging at a radiation level of approx. 10 mW/cm^2, under consideration that the charging current proportional to the radiation contributes only at a very small extent already to the charging-up of the battery.

We closed down hermetically the whole unit and equipped it by electrical terminals corresponding to the standard type R20. /2/, /3/

3. MEASUREMENT OF POWERBLOCK

We carried out the tests, measurements on 2 powerblocks. For the measurement of the current sources we established, on the one hand, a charging place fixed in the open air, where we oriented the current sources in southern direction with a tilt angle of 25°, on the other hand we set up the measuring arrangement under laboratory conditions, for the discharging of the current sources.

For the amperehour efficiency of charging-discharging we obtained at

powerblock 1	85 % and at
powerblock 2	79 %.

The measurements ware carried out February.

At a radiation energy of 578 mWh/cm^2 we could take out

from powerblock 1	707 Asec and
from powerblock 2	657 Asec.

4. ENHANCING OF CHARGING

For the enhancing of the charging, in case of the charging by light, the application of a concentrating means seems to be obvious. As to producibility, the most favourable is the application of flat plate concentrators. We examined in case of fixed orientation the development of the concentration rate regarding the two flat mirrors as shown in Figure 3.

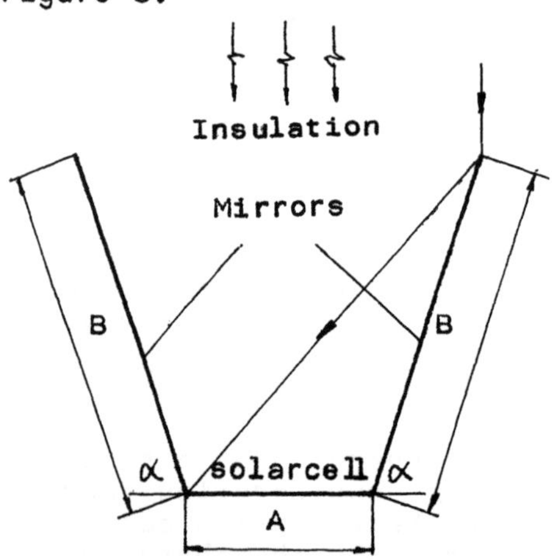

Figure 3 Flat plate concentrators

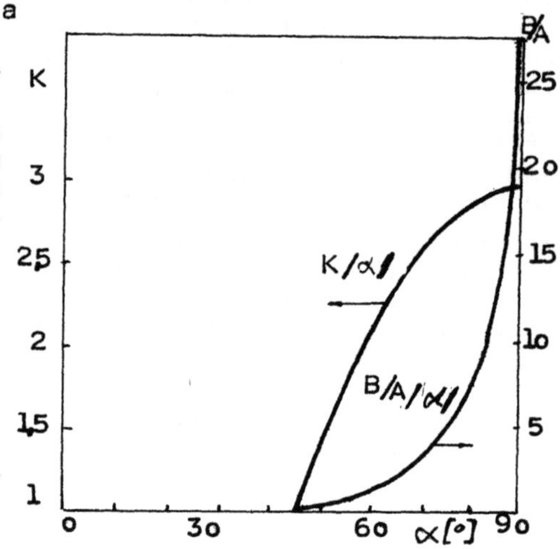

Figure 4 The extent of concentration v.s. the angle of mirrors/α/: and the ratio of the width of mirrors and the solar cell /B/A/

The extent of concentration of direct radiation /K//i.e. what the increase of direct radiation is on the solar cell of width /A/ with respect to the case without the application of the concentrator/ resulted by a simple geometrical calculation in the following:

$$K = 1 + 2\,\frac{B}{A}\cos\alpha \qquad /1/$$

wherein /B/ being the useful width of the flat plate mirrors and /α/ the angle they enclose with the solar cell.

The useful width /B/ of the flat plate mirror - i.e. that the mirror is only of a width that it should just reflect on the surface of the solar cell of width /A/ the perpendicular incident radiation - and the width /A/ of the solar cell are in the following relation to each other:

$$\frac{B}{A} = \frac{1}{\cos/90^{o} - \alpha/ \quad tg/180^{o} - 2\alpha/ - \cos\alpha} \qquad /2/$$

and then

$$K = \frac{tg\alpha . \quad tg/180^{o} - 2\alpha/ + 1}{tg\alpha . \quad tg/180^{o} - 2\alpha/ - 1} \qquad /3/$$

As far as there is only direct radiation, these relations are well applicable for the dimensioning of the mirror, it must be taken however into consideration that beside the direct radiation the indirect, or diffuse radiation is very significant at most parts of our globe. The rate of concentration depends - as it can be seen easily - to a great extent from the proportion of direct radiation and total radiation, that is $\frac{D}{T}$.

For an approximation of the problem the relations below can be taken for the extreme point of the solar cell of width /A/, resp. its middle point:

$$K_{middle} = \frac{D}{T} \frac{tg\alpha . \ tg/180^{o} - 2\alpha/ + 1}{tg\alpha . \ tg/180^{o} - 2\alpha/ - 1} +$$

$$+ /1 - \frac{D}{T}/ . /1 - \frac{1}{90} \text{ arc tg } \frac{tg\ \alpha}{tg/180^{o} - 2\alpha/ + 1} \qquad /4/$$

and

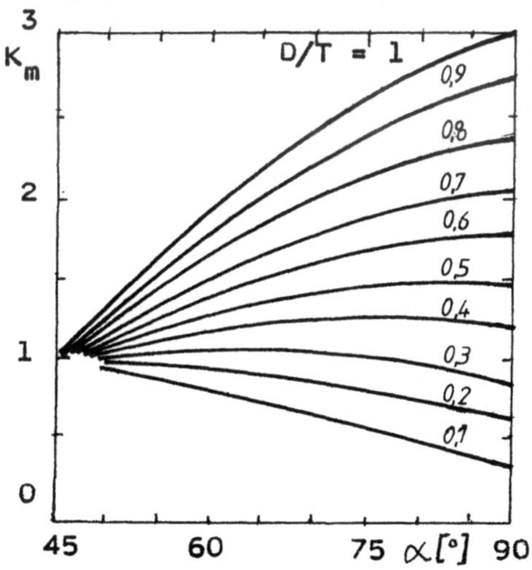

Figure 5 Extent of concentration /K_{middle}/ v.s. the angle of mirrors /α/. Parameter: the rate of direct and total radiation ($\frac{D}{T}$).

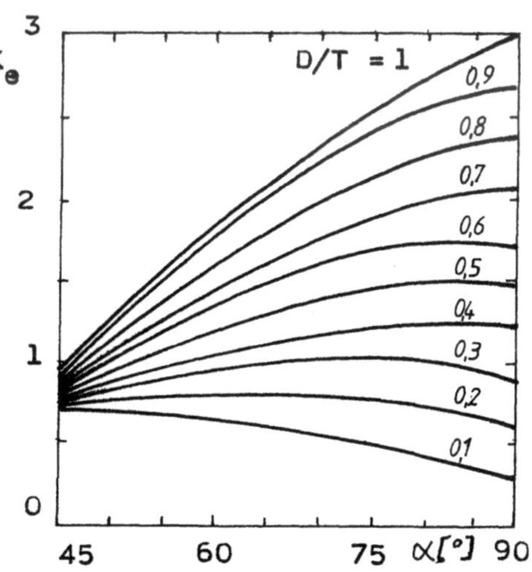

Figure 6 Extent of concentration /$K_{extreme}$/ v.s. the angle mirrors /α/. Parameter: the rate of direct and total radiation ($\frac{D}{T}$).

$$K_{extreme} = \frac{D}{T} \cdot \frac{tg\alpha \cdot tg/180^{o} - 2\alpha/ + 1}{tg\alpha \cdot tg/180^{o} - 2\alpha/ - 1} +$$

$$+ /1 - \frac{D}{T} / \cdot /1,5 - \frac{\alpha}{60} / \qquad /5/$$

In view of the above, by selecting 77° as the valus of $/\alpha/$, we prepared to the current source a simple charge-enhancing unit with two flat mirrors, of aluminium-foil surface. We selected the longitudinal size of the mirror to fourfold of the longitudinal size of the solar element. By the application of the charge-enhancing unit we carried out repeatedly the measurement as described under chapter 3 and we could take out from the powerblocks at a radiation energy of 1228 mWh/cm^2 the following amounts of charge:

from powerblock 1 3460 Asec and
from powerblock 2 3199 Asec.

Taking into account the charging-discharging efficiency determined on the basis of the measurements carried out in chapter 3, the extent of concentration K will appear with 2,29 resp. 2,3, which corresponds well to the theoretical values with a D/T of approx. 0,8.

5. CONCLUSIONS

- In summarizing the results it can be established that in small autonomous power sources the alternative charging possibility by solar cell is advantageously applicable.
- In case of direct radiation rate exceeding 30 %, the charging of the powerblock can be enhanced by the use of double flat concentrators.
- If the part of the equipment using it serving for the housing for the housing of the accumulator as a transparent lid or casing, the powerblock of solar cell can be charged also in the equipment.

REFERENCES

/1/ Miklós Pálfy: 'Electrical characteristics of solar cells'
Elektrotechnika Vol. 79, 10/1986, pp.381-389.

/2/ Ferenc Böhönyey - Miklós Pálfy:'Encapsulation of solar cells'
VKI Communications Vol. 10. 1986, pp.179-188.

/3/ Miklós Pálfy - Ferenc Böhönyey: 'Possibilities of further improvement of solar cells and some new results'
Electricity Vol. 34, 1986, 9, pp.257-260.

THE TERSCHELLING PV/WIND SYSTEM AFTER FIVE YEARS OF OPERATION

Emil W. ter Horst, Jean H. Boumans, Kornelis Blok
ECOFYS Research and Consultancy Co-operation,
Biltstraat 110, NL-3572 BJ Utrecht, The Netherlands.

SUMMARY
Within an optimization program carried out on the PV/wind system at the Terschelling Island several system components have been improved. The PV generator produces about 25 % more electricity than before. At humid weather conditions still energy leaks to the earth. The condition of the battery seems to be well, though problems are met with the battery control. The wind generator is defect and has to be replaced. Other components function satisfactory.
To find the optimum size for the new wind turbine calculations have been carried out with the simulation program SOMES. It turned out that a wind turbine of 50 kW will have the highest percentage of electricity demand cover at the lowest kWh price.
During the next two years the control strategy will be optimized for both the grid-connected and the autonomous operation mode of the system.

1 INTRODUCTION

ECOFYS is carrying out an optimization program on the PV/wind system at the Terschelling Island in the north of the Netherlands. The system has been founded in 1983 as a part of the EC PV Pilot Plant Program by HOLEC NV, the Netherlands. The energy is supplied to a Nautical College. The system consists of a solar PV generator of 50 kW_p, a windturbine with a rated power of 40 kW, a battery of 180 kWh nominal capacity and converters. The system can be operated both grid connected and autonomously. The lay-out of the system is given in figure 1.

2 STATUS OF THE SYSTEM

Improvements on the system components have been carried out from November 1987 until May 1988. The current status is given below.

2.1 The PV-generator

The PV generator consists of 2748 modules (AEG PQ/10/20/01) and has a rated power of 50 kW. There are 30 subarrays. 29 subarrays have three strings of 30 modules in serie. The subarrays are connected to the battery via a indivual Maximum Power Point Tracker (MPPT). The thirtiest subarray consists of three strings of 46 modules each. This subarray is via a diode connected to the battery.
We met the following problems:

1 CABLE CRACKS. Every year about 10-20 cable cracks do occur. These cracks can occur at any place in the cables. The cracks are easy to repair. Creeping within the cables seems to be a mayor cause of this cracking. The size of the cable pipes is also related to the problem of cable cracks. The cable pipes are too small to carry all the cables. Also the cable pipes are not enough weatherproof. The cable pipes will be replaced by other cable pipes.
2 EARTH LEAKS. Some modules have earth leaks at the place where the cables leave the module. We have located about 5 of these modules.

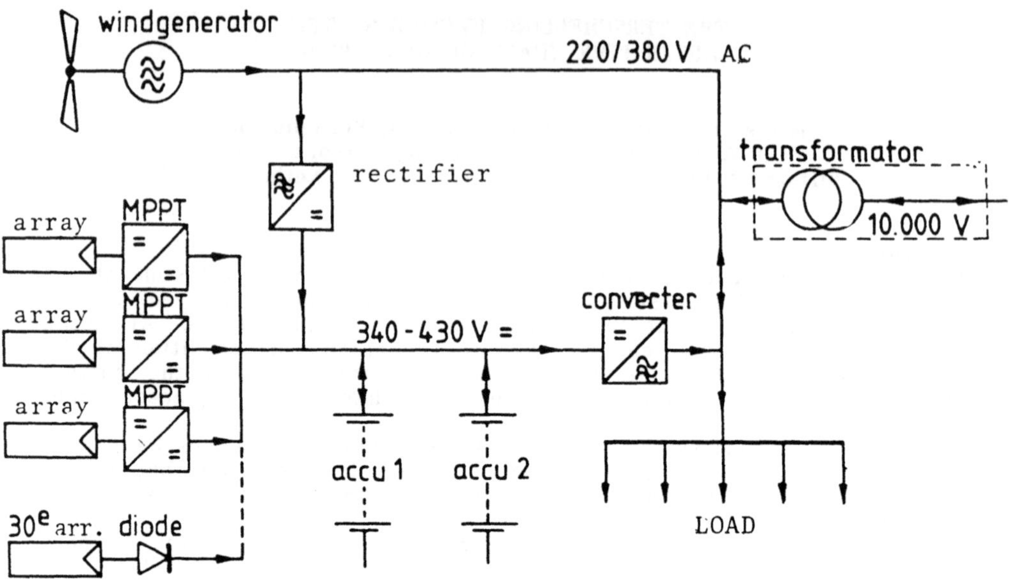

Figure 1. Lay-out of the Terschelling PV/wind system

These modules have to be replaced. We estimate that about 3-5 modules have to be replaced every year because of this problem.

During humid and low irradiation periods the production of the PV generator is much below what one expects due to diffuse earth leaks. This earth leak is increasing every year. The increasing porosity of the cable coating seems to be the cause of this problem.

3 GLASS CRACKS. Due to incorrect installation about 50 modules have glass cracks. About 20 modules do have glass cracks due to other causes, like vandalism. The glass cracks have only little direct influence on the production of these modules. One can see corrosion in the neighbourhood of older cracks, but even these modules still have a good efficiency. We intend to replace the main part of these modules, but also it is interesting to follow the production and the degradation of some the broken modules.

We have improved the PV generator on most of the aspects. The PV generator operates now in general satisfactory, which resulted in a 25 % higher energy production. At humid weather conditions a part of the productions still leaks to the earth. The improvement of the coating of the cables with porosity problems has to be further examined.

Though it has to be measured more thoroughly it can be concluded that there is almost no degradation of the modules. Failures occur due to defects in cables etc. With the new generation of modules these problems are said to be solved. Though the porosity problem still has to be solved we are optimistic about the repair the PV generator. The rated power is expected to reach 50 kW again.

2.2 The battery

The battery system at Terschelling consists of two groups of 180 Varta bloc (2305) cells. Measurements of the acid degree and the voltage have been carried out. The condition of the battery seems to be good. Problems are met with the control of the battery. We discovered during discharging

one cell being reversed (-2.4 V, while the mean voltage was 2.03 V/cell at that moment). This reversion is due to the fact that the battery system is controlled on voltage indication measured on the whole battery. This control cannot assure safe conditions for all the individual cells [1].
To improve the battery control we will install a software battery control routine into the main PLC. The state-of-charge (SOC) indication will be based on both the Shepherd model [1] and daily Ah counting. The control of the battery is based on both this SOC indication and on voltage and current measurements, performed on the total battery and on three subgroups of cells.

2.3 The wind turbine

The wind turbine at Terschelling is a so-called first generation wind turbine. Since the end of 1987 the wind turbine is defect. Major cracks of about 20 cm are discovered in two of the three rotor blades. Repair of the cracks turned out to be expensive. Because of this and because of the much higher quality of newer windturbines the installation of a new wind turbine is intended.

2.4 Other components

The other components in the system are now functioning satisfactory. In future some adaptations for blind power compensation have to be made.

3 DATA-ACQUISITION AND -MONITORING SYSTEM

A new data-acquisition system is installed based on a new PLC, which is also used for the control of the system, in communication with a MS.DOS computer (see figure 2). The PLC is used for the data-acquisition, while the micro takes care of the the data-logging and monitoring. The software is improved and it is now possible to do the first analyses of the data on site, which is of importance for a good plant management (see figure 3).
All the measurement devices are now being recalibrated.
The system functions very satisfactory.

4 OPTIMIZATION CALCULATIONS

In the next two years the control strategy of the system will be optimized both during grid-connected and during autonomous operation.
The optimization will be based on calculations by the simulationprogram SOMES of the University of Utrecht [2], and on the hour-by-hour data from the monitoring system.

We have carried out some preliminary calculations with SOMES to calculate the optimum size of the new wind turbine in this system. We searched for an optimum with a high percentage of electricity demand cover at low investment costs. Some results are summarised in table 1.
The optimum wind turbine size for the system is between 25 en 75 kW_{rated}. The calculations were made with standard prices. In searching on the wind turbine market it turned out that one 75 kW wind turbine was cheeper than the available 50 kW turbines, resulting in higher additional anual savings.

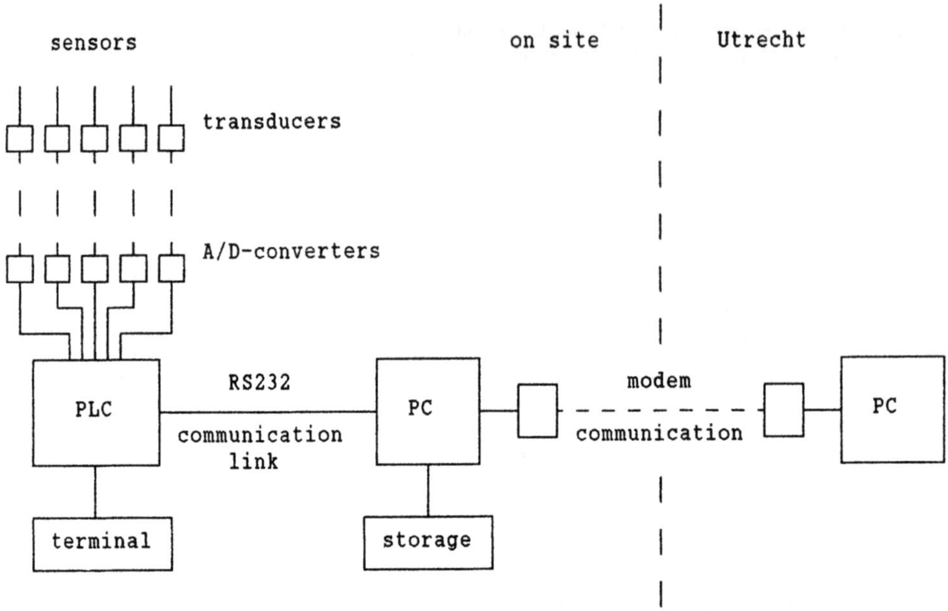

Figure 2. Lay-out of the data-monitoring system

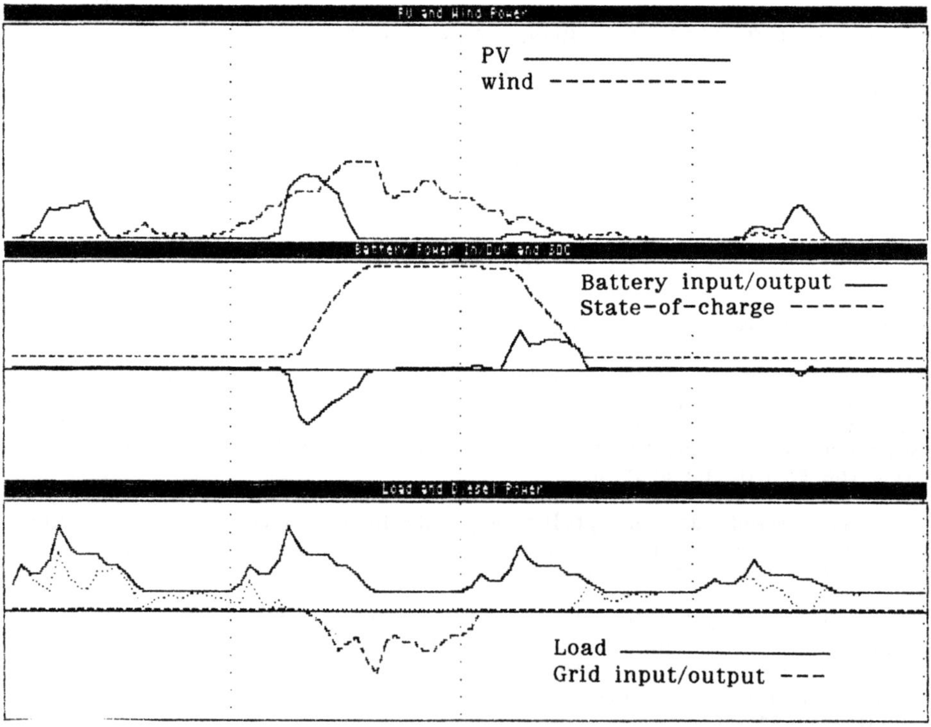

Figure 3. Example of output, e.g. for analysis of data on site

nominal wind power	energy from PV/wind		energy bought from grid		additional anual (savings - costs)
kW	kWh	%	kWh	%	in Dfl
0	35,674	21	129,330	78	-
25	70,959	43	93,902	57	2234
50	87,513	53	77,348	47	2999
75	96,354	58	68,507	42	1283
100	102,195	62	62,665	38	-1493

Table 1. Percentage of electricity demand cover and netto savings of a new wind turbine

CONCLUSIONS

The Terschelling PV/wind plant has scored in its five years of existence varying success. Though the produced energy was less than expected it can be concluded that this in principle was not due to the modules or the wind generator. We expect when the PV generator is repaired the rated power to be as at installation, though some earth leak problems still have to be solved.
The wind generator has functioned well during the last five years, but is weared out and will be replaced by a new wind turbine.
It is very important to do regular inspections on the system. We intend to do every three month a standard inspection. For the success of the system it is also important that the repair of defects repaired quickly.
The data-monitoring functions now satisfastory and online monitoring and analysis will be possible both on site and in Utrecht. It is possible to use the simulation program SOMES for the further optimization of the system.

References:

1 E.W. ter Horst, e.a.: "Battery modelling for PV applications", 8th Eur. PV Solar Energy Conf., Florence, 1988.
2 K. Blok, E.W. ter Horst: "SOMES, a Simulation and Optimization Model for Autonomous Energy Systems", University of Utrecht, June 1987.

$ The project is financially supported by European Community DG XII/B/2 and by the Netherlands Agency for Energy and Environment NOVEM

THE APPLICATION OF HYBRID a-Si:H PV AND THERMAL COLLECTORS FOR DIFFERENT USAGES

B.Lalović, Boris Kidrich Inst., Belgrade, Yugoslavia, P.O.Box 522

T.Pavlović, University of Niš, Yugoslavia

Z.Kiss and J. van Dine, Chronar Corp., Princeton, N.J., USA

ABSTRACT. An improved version of a hybrid a-Si:H PV and thermal solar collector, reported earlier, is described and its possible applications are discussed. This collector can produce about 400-600 kWh/m^2 year of thermal energy and about 70-100 kWh/m^2 year of electric energy. It is most suited in those cases where relatively large quantities of low temperature hot water are needed, such as in hotels, large apartment buildings, hospitals, colleges etc. Hybrid systems are especially attractive when electricity is used for hot water production, the investment in the thermal part being lower than in the PV part, while the contribution of thermal energy is several times higher. They use for desalination of saline water, in combination with electrodialysis, is also recommended, since the efficiency of electrodialysis increases when the feed water is preheated.

Development of a-Si:H cells on anodically oxidized aluminum substrates, aimed at construction and fabrication of simpler and cheaper compact hybrid collectors, is also reported.

1. INTRODUCTION

In a previously published paper (1) we reported about the development of a hybrid a-Si:H pv and thermal solar collector, which was constructed by using a-Si pv glass modules, bonded to a specially designed heat exchange plate, consisting of aluminum fins clad with copper tubes, through which water could flow. This hybrid unit performed well as a thermal solar collector, heating water up to 65 ^{o}C, while the electric characteristics of the pv modules showed little change. The useful peak pv power was 28 W (with a collector area of 0.8 m^2), and the thermal power could exceed 300 W.

Based on this work, an improved version of this hybrid collector, suitable for industrial production, has been developed and tested. Studies of possible applications of hybrid collectors have also been undertaken and the economics of their utilization have been estimated. Here we report about this work.

2. DESCRIPTION OF THE HYBRID COLLECTOR

The main features of the improved collector are: six glass pv modules, whose dimensions were 91.5 cm x 30.5 cm, giving a total collector area of 1.6 m^2, were used; the average module efficiency was about 5.5 percent; a tedlar foil was used as a collector top cover and a 6 cm thick polyurethane as back insulation. The cross-section of this structure is shown in Fig. 1.

The thermal efficiency of the collector was measured outdoors by using a standard procedure. The following parameters were simultaneously measured and recorded: the solar irradiance I (Wm^{-2}); the rate of water flow ($lmin^{-1}$) through the collector; the temperatures T_{in}, T_{out} (K) of water flowing in and out of the collector; the ambient temperature T_a(K). The thermal collector efficiency, as a function of the parameter $(T_{out}-T_a)/I$, is shown in Fig.2. The peak thermal power was estimated to be about 800 W. From the experience with thermal collectors of similar characteristics the average collector efficiency in practical operation can be expected to be between 30 and 40 percent. Thus in a moderate climate about 400-600 kWh/m^2 per year of thermal energy could be collected.

The I-V characteristic of the pv panel is shown in Fig.3. All six modules were connected in parallel, the voltage at peak power being about 16 V. The total peak power of the panel was about 80 W. Such a flat panel can contribute 1,5-2 kWh/ year for each watt installed or about 75-110 kWh/m^2 year.

3. APPLICATIONS OF HYBRID COLLECTORS

With the present efficiency of commercially manufactured a-Si:H modules (about 6-7 percent) the thermal energy produced in the hybrid collector is 4-6 times larger than the electric energy. This ratio is even higher in summer, since the thermal collector efficiency strongly depends on the ambient temperature. It is clear that the most suitable applications of hybrid collectors are those where relatively high consumption of hot water of low temperature (below 65 ^oC) can be expected.

In individual homes the ratio thermal/electric energy varies from country to country and from season to season, but in general the hybrid collector does not provide a well-matched energy. This is especially the case in grid-connected systems, when, because of the inverter price, the system power should exceed several kilowattts. For example, with a pv power of 8 kW, the collector area would be 100 m^2, and an average production of hot water would exceed 5000 l/day. Hence the hybrid collectors would be more suitable for a group of several residences or for larger apartment buildings.

On the other hand, hospitals, colleges, military barracks and similar institutions are good examples for the application of hybrid systems. There are many solar thermal systems consisting of several towsend collectors and, if hybrid collectors were used in them, the pv power of such systems would reach megawatt region. For example, we have proposed a centralized hybrid system of 1MW photoelectric power, which would provide the town of Dubrovnik with hot water. In summer about 20000 tourists stay in Dubrovnik daily, doubling its population. About 4000 tons of hot water is requred daily, and this could easily be supplied with the hybrid system, which could be placed on a southern slope of a nearby hill.

Because of the reduction of the balance-of-system cost the photovoltaic power in the hybrid collector is cheaper than in normal applications. In addition, its thermal part, although cheaper than the photovoltaic part, produces 4-5 times more energy. Hence the average price of the energy produced may become competitive much earlier than for the photovoltaics alone. For example, with a pv price of 4 dollars/Wp and the thermal collector price of 200 dollars/m^2 (everything included in both cases), the average price of the energy obtained (about 600 kWh/m^2year) would be less than 10 cents/kWh for a pay-back period of 10 years.

Another attractive application of hybrid systems could be desalination of saline water. J.Lundstron and G.Kvajic investigated a combined system using Si crystal pv generator and electrodialysis (ED) for potable water production (2). They proposed that preheated saline water be used, since the efficiency of ED increases with temperature. It was shown that at the temperature of 65 °C the potable water production can increase up to about 80 percent with respect to the operation at 20 °C. It is easily seen that a combined system using hybrid collectors and ED could provide energy both for preheating the saline feed water and for the operation of ED membrane stacks.

4. RECENT DEVELOPMENT OF THE HYBRID COLLECTOR

Recently we made a-Si:H cells using, as a substrate, anodicaly oxidized thin aluminum plates (3). This substrate, being unbreakable, thin, durable and lightweight, has certain advantages over glass substrate, especially for hybrid collectors. With a-Si:H cells made directly on an aluminum heat exchanger (Fig. 4), the construction and fabrication of compact hybrid absorbers can become more feasible and, of course, substiantially simpler and cheaper.

Two types of electrically insulating oxide films on top of the aluminum plates were prepared: a porous type made from phosphoric, chromic, sulfuric and oxalic acid solutions, and a homogenous type made by a hard anodization process. Resistance of all oxide films formed was in the range of 10^{12} cm, and breakdown voltages were in excess of 500 V. The n-i-p, a-Si:H films were then deposited on Al/Al_2O_3 substrate. As top electrode, indium-tin-oxide layer, deposited by Rf magnetron sputtering, was used. The best Al/Al_2O_3/Al-n-i-p-ITO solar cells had conversion efficiencies of 5.3 percent. Their I-V characteristic is shown in Fig. 5.

There is a special benefit of this a-Si:H structure for the hybrid pv-thermal collector. Since the top ITO layer has low emissivity in the infrared region, this collector has the properties of a selective absorber and its thermal radiation losses can be expected to be small.

This amorphous cell structure on aluminum substrate offers an exciting possibility of constructing and fabricating compact pv and thermal solar collectors by direct cell deposition on a suitable Al heat-exchange plate. Such an attempt is in progress.

References

1. B.Lalovic, Z.Kiss and H.Weakliam, "A Hybrid Amorphous Silicon pv and Thermal Collector", Solar Cells, 19 (1986) 131.

2. J.Lundstrom and G.Kvajic, "Solar Powered, Dual Purpose Power-Desalination PV-ED System", Alt.En. Sources V.Part B, Amsterdam, (1983) 123.

3. B.Lalovic, T.Pavlovic, J. van Dine, Z.Kiss, L.R.Shiue, T.Tonon, B.Sieb and W.Xuanzhi "a-Si:H Cells on Anodically Oxidized Al"(to be published).

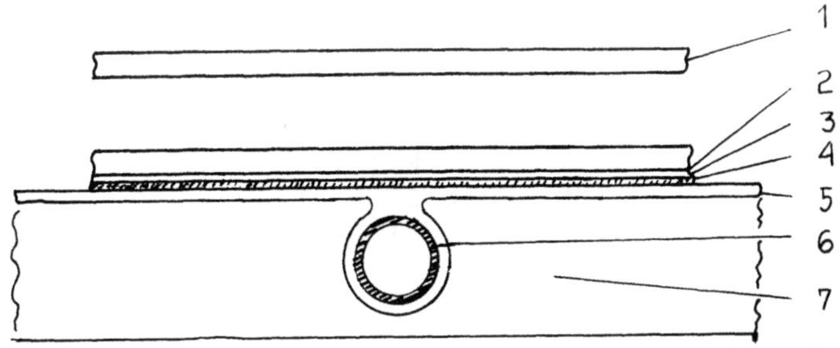

Fig.1. Cross-section of the hybrid collector: 1, glass cover; 2, pv glass module; a-Si:H layers; 4, adhesive; 5, Alfin; 6, Copper tube; 7, insulation.

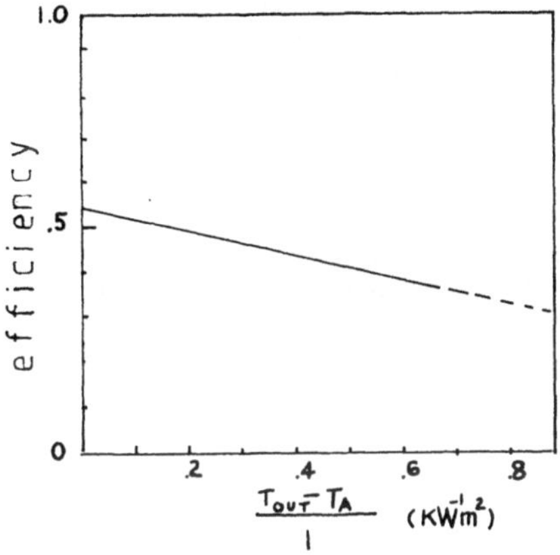

Fig.2. Thermall efficiency of the hybrid collector

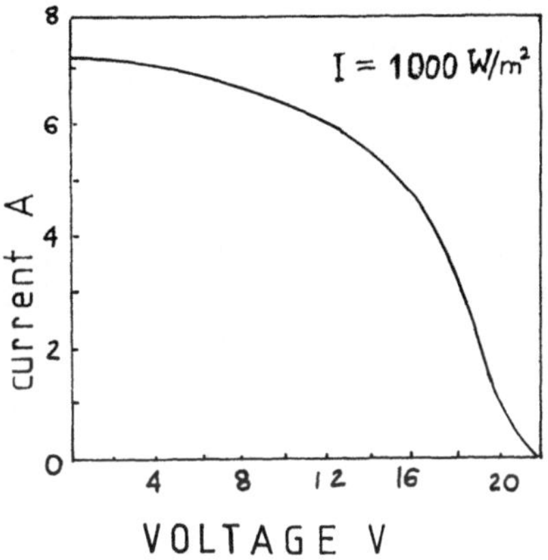

Fig.3. I-V curve of the pv panel

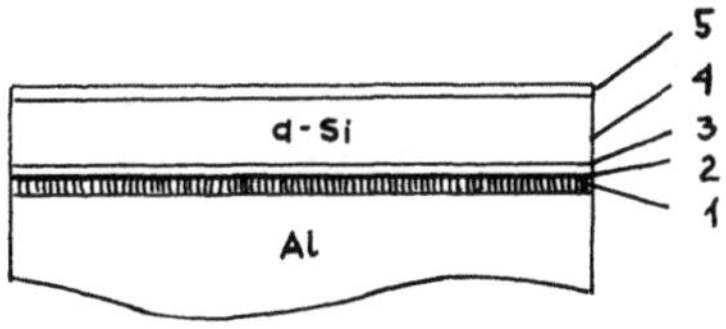

Fig.4. Structure of solar cell on Al substrate: 1, Al_2O_3, 2, Al, 3, barrier layer, 4, a-Si(n-i-p), 5, ITO

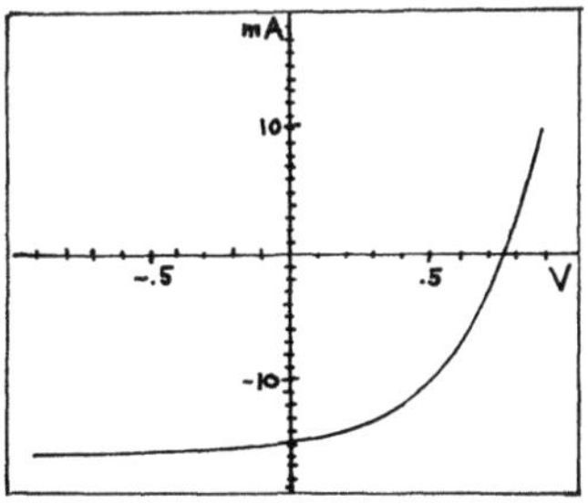

Fig.5. I-V curve for Al/Al_2O_3/Al/a-Si(n-i-p)/ITO solar cell, for solar irradiance of 1000 W/m^2.

530 SUN POINT FOCUS FRESNEL LENS STUDY WITH SECONDARY OPTIC

William Todorof
AESI
1001 West 17th Street
Costa Mesa, California 92627

By concentrating sunlight up to 500X by use of inexpensive optical lenses, concentrators have an advantage over other technologies because a small fraction of the expensive solar cell material is used to create the same energy output. The systems are more reliable than other PV options because fewer solar ells and interconnectins are necessary for concentrators than for flat-plate PV. The AESI point-focus, dual-axis array takes advantage of all the necessary features of optimum energy collection. The 640 Wp baseline array assembly uses Fresnel lenses of 530 geometric concentration to concentrate sunlight 430 times (optical) onto silicon solar cells. The complete array's turning radius is 5.2 m (17 feet) and the height is 3.4 cm (11"), contains 8 modular trays and is a complete functional entity including elevation and azimuth drive mechanisms, control electronics, and sun sensors.

In 1985 AESI developed a 280 sun concentrator with a 30 cell module. The 6.75" x 6.75" (17cm x 17cm) fresnel lens was molded into a parquet by Rochester fresnel optics (Helmut Walter) and was provided by Charles Stillwell at Sandia National Laboratories. The lens transmission efficiency was 86% and the ASEC cell efficiency was 19% at 168 suns.

During fabrication of this device, it was observed that the alignment of the cell assembly within the frame and the mounting of the components was significantly more critical than the previous 125 sun concentrator assembles. Also there was observed a lens flare problem, that resulted from using the fresnel lens parquet at 280 suns concentration.

This produced chromatic aberation (a spectral separation) at the focal point. Thus the cell lost power from wavelengths that did not impinge upon the cell.

Two months after the 280 sun concentrator had been assembled a slight corrosion was observed forming at the cell contact areas.

In view of the goal to develop a maintenance free 30 year concentrator, the presence of corrosion in a 60 day period caused concern about pre-mature cell contact failure. After a careful analysis of fabrication times, alignment procedures, number of component pieces used, potential corrosion issues, cromatic abberation and the labor intensive assembly required to achieve 280 sun concentration (using "then known" procedures), it was decided to undertake the total re-design of the system.

After an extensive series of design studies, it was decided to use two lens concentration system at 530 suns geometric concentration. We selected an existing acrylic 8.16" x 8.16" (20.7cm x 20.7cm) primary fresnel lens and had an optical glass seconcary designed to increase the concentration from 168 suns to 530 suns. The target cell size was selected to be 10mm dia (.78cm2) for use under the glass secondary lens, which provides uniform light flux density accross the cell without chromatic abberation or lens flare.

By using a 20% oversize glass secondary, we were able to tolerate tracking errors of 0.2 degrees without substancially affecting output performance.

By using an optical glass secondary top window for the cell, it was possible to design an encapsulated cell holder to eliminate the damage caused by corrosion at the cell contact interface. This encapsulated cell assembly is also removably threaded onto a heat sink which allows it to be easily replaced in the field should a cell failure ever occur. These encapsulated cell assemblys with glass secondary lens, cell and contacts are projected to cost less than $0.40 / peak watt in 2 million pieces per year production scenarios.

With a 23-26% efficient solar cell which is the projects hear term goal the peak output power at 430 x optical concentration would be approximately 8 watt at 1000 w / m2 at am 1.

At 15 mw production levels, the cost of the total system with dual axis tracking and power conditioning is less than $1.25 per peak watt of power.

At 100 mw production levels, the cost of the total system with dual axis tracking and power conditioning is less than $0.80 per peak watt.

The design goal was less than $1.00 per peak watt. Therefore we feel the project at this stage of development is successful. With an adequate funding base, the development of a 23-26 % efficient cell (at 530 x geometric) 430 suns optical concentration should be achieved in the near term future.

TRACKING ERROR (degrees) = 0

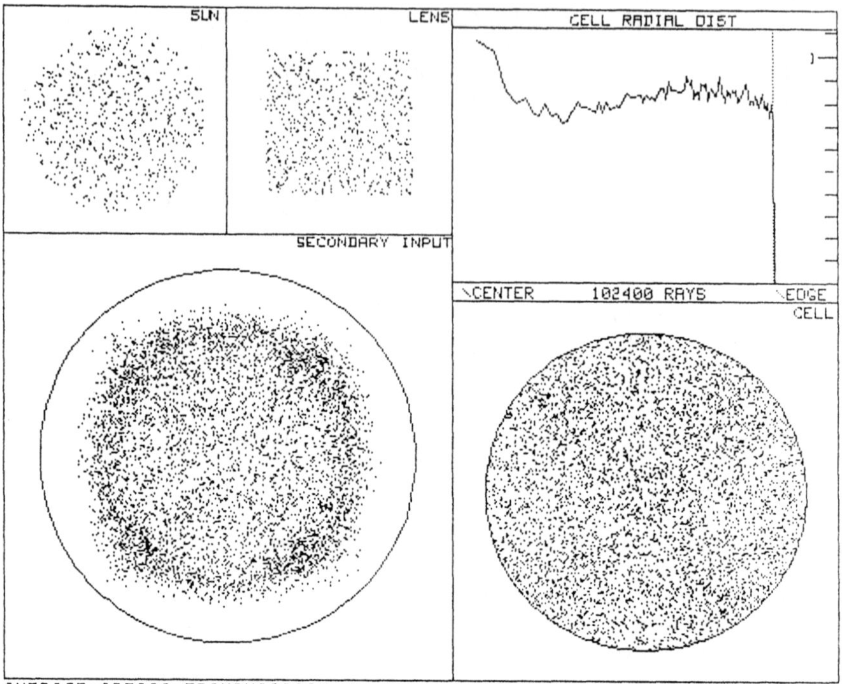

AVERAGE OPTICS TRANSMISSION	= 0.8273	+/- 0.0011
FRACTION OF RAYS LOST BY TOTAL INTERNAL REFLECTION	= 0.0000	+/- 0.0000
FRACTION OF RAYS LOST BY REFLECTION BACK UP CONE	= 0.0000	+/- 0.0000
FRACTION OF RAYS LOST BY HITTING FLASH	= 0.0000	+/- 0.0000
FRACTION OF RAYS LOST BY CONE LEAKAGE	= 0.0003	+/- 0.0001
FRACTION OF RAYS HITTING DRAFT SURFACES	= 0.0481	+/- 0.0009
FRACTION OF RAYS MISSING SECONDARY	= 0.0000	+/- 0.0000
FRACTION OF RAYS MISSING CELL	= 0.0000	+/- 0.0000
GEOMETRIC CONCENTRATION RATIO	= 593.3	SUN
PEAK FLUX CONCENTRATION RATIO	= 662 +/-23	SUNS

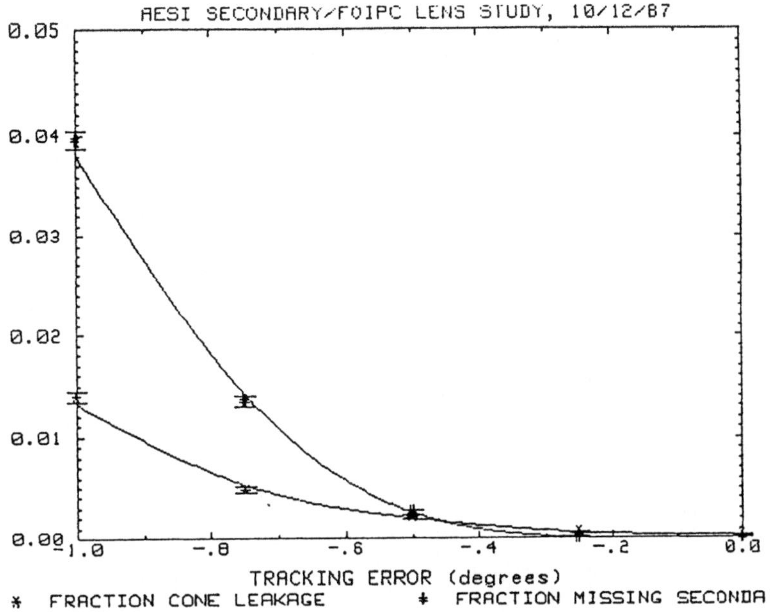
AESI SECONDARY/FOIPC LENS STUDY, 10/12/87
0.05
0.04
0.03
0.02
0.01
0.00
-1.0
-.8
-.6
-.4
-.2
0.0
TRACKING ERROR (degrees)
* FRACTION CONE LEAKAGE
* FRACTION MISSING SECONDA

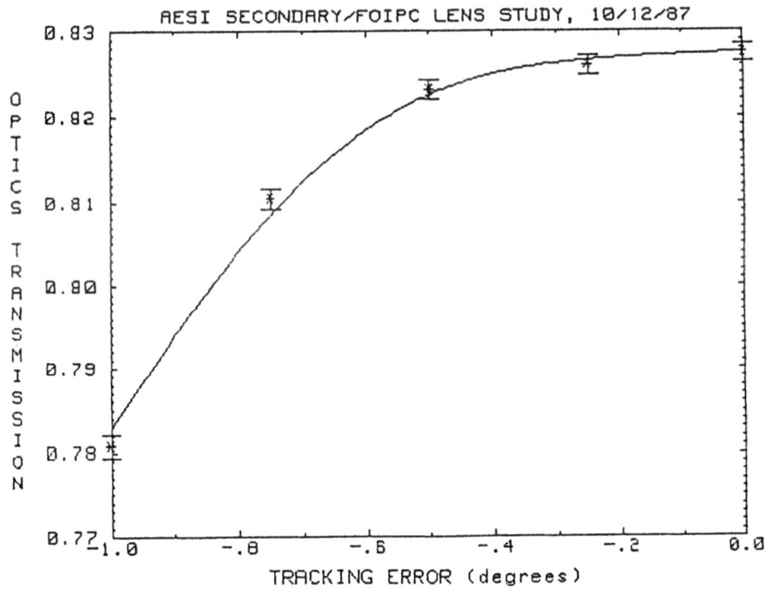
AESI SECONDARY/FOIPC LENS STUDY, 10/12/87
OPTICS TRANSMISSION
0.83
0.82
0.81
0.80
0.79
0.78
0.77
-1.0
-.8
-.6
-.4
-.2
0.0
TRACKING ERROR (degrees)

AESI SOLAR COGENERATOR

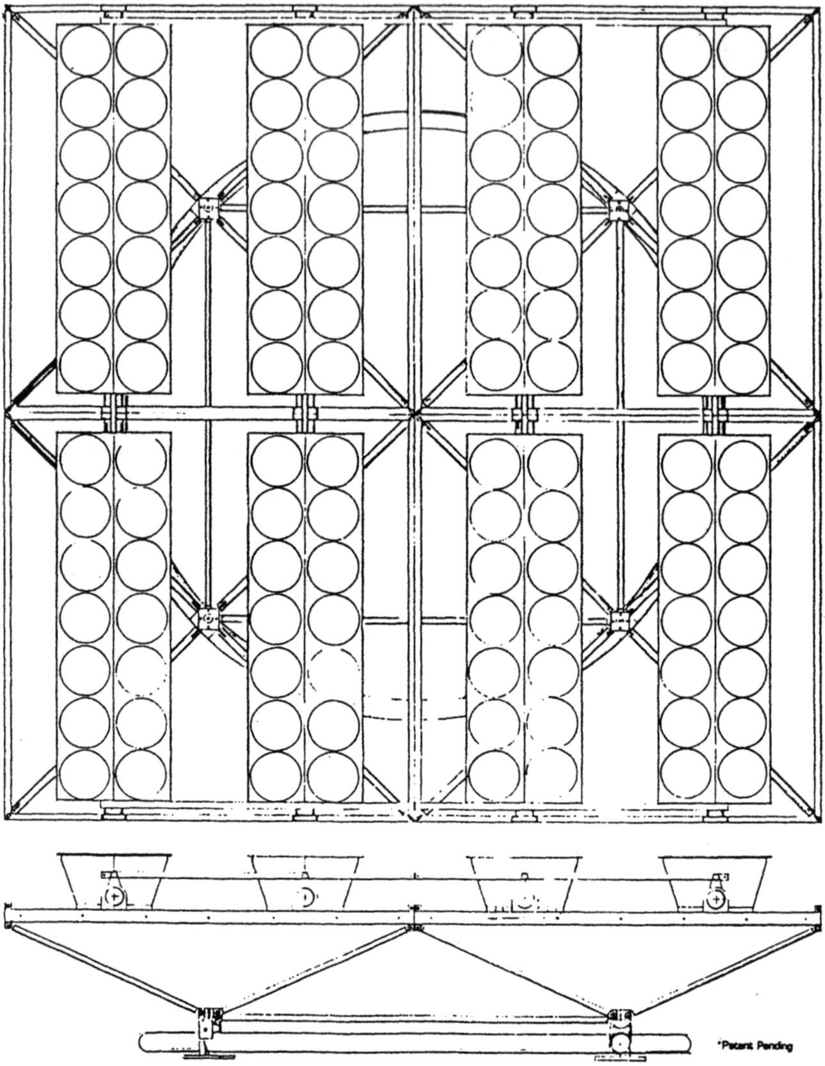

HIGH PRECISION ELECTRO-OPTICALLY CONTROLLED DUAL AXIS SUN TRACKING SYSTEM

William A. Lynch and Ziyad M. Salameh
University of Lowell
Department of Electrical Engineering
One University Avenue
Lowell, MA 01854
U.S.A.

ABSTRACT

This paper describes the design of a sun tracker which uses two electro-optic sensors and a small low cost electronic control circuit. One sensor is a four cell pyramid which is mounted on the tracker plane. The second sensor is a photo transistor beam sensor which is fixed facing south. The control circuit tracking resolution is within 0.1 degrees. This tracker does not wander on partially overcast days. It will never make multiple revolutions or face down towards the ground. Power MOSFETS are used to drive high torque DC gearbox motors.

INTRODUCTION

Sun tracking can significantly improve the efficiency of any solar array. it is highly cost effective in photovoltaic energy systems because of the high cost of photovoltaic cells and the lower cost of large aperature area concentrators such as fresnel lenses (1). Tracking flat plate photovoltaic arrays provide about 1/3 more power than fixed arrays (2). Recent advances in solar cells have resulted in silicon high concentration ratio types with up to 28% efficiency.(3)

There are three general catagories of sun trackers including passive, microprocessor, and electro-optically controlled units. Passive systems track the sun without any electronic controls or motors. These trackers are produced commercially by manufacturers such as Zomeworks (4). these trackers contain a fluid such as freon within a frame of pipes. When the array is misaligned the sun heats the freon on one side of the frame more than the other. This temperature difference causes the heated freon to evaporate. It may push a piston or may simply flow to the other side of the array and move it by gravity such as in the Zomeworks system. These trackers are simple but can only provide moderate accuracy tracking. They can get stuck in the wrong position or move in the wind. Higher torque motor driven systems do not have these problems.

Microprocessor controlled sun tracking units use mathematical formulas to predict the sun's location and therefore need not sense the sunlight. They are highly accurate, but are complex and expensive. To determine position they use stepper motors or optical encoders similar to those of industrial robots. These devices do not sense the sun, therefore they must be precisely aligned during installation. The controller must be programmed with the site latitude, longitude, and time, but should operate automatically once started. They are often used in large systems in which one controller controls many solar arrays (2). For high precision tracking, open loop microprocessor type controllers should be periodically recalibrated. Many microprocessor controlled trackers use electro-optic sensors for self calibration (5), (6). Their high component count can be a source of reliability problems (7).

In general, electro-optic sensor analog controlled trackers are simpler and less expensive than microprocessor types. One system uses four photo resistors with cylindrical shades as a sun sensor. The simple low parts count controller contains only differential amplifiers, comparitors and output components (8). When testing our first electro-optic tracker we demonstrated how it did not always track correctly. It often did not start tracking correctly when the sun was not visible or was outside the optical acceptance angle of the sun sensor. It wandered on overcast days while attempting to track bright spots in a cloudy sky. This is a waste of energy and a cause of additional wear on the mechanical components. It could detect reflected light from the ground and face down. It could also rotate through north in the morning causing multiple revolutions, resulting in twisted wires or fluid lines. The device described in this paper combines the accuracy of microprocessor types with the simplicity of electro-optically controlled units.

DESIGN AND OPERATION

This sun tracking system consists of eight subsystems as shown in the block diagram, (figure 1). The primary sensor provides information on the relative position of the sun. The fixed beam sensor determines whether a sunlight beam is present and provides information on the absolute location. The position encoder provides information on the tracker position. The track subscircuit controls the motion of the mechanical drive system and is usually controlled by the primary sensor. The shutdown subcircuit overrides primary control and disables the tracker if the beam sensor indicates no sunbeam. The search subcircuit overrides primary control and determines direction of motion if the beam sensor and position sensor indicate that the sun may be outside the optical range of the primary sensor. The output subcircuit consists of two common source complementary power MOSFET half bridge switches. The drive motors which control azimuth and zenith motion are DC gearbox servo drive motors.

The primary sensor of the tracker contains two independent differential light sensors consisting of two small photovoltaic cells wired in series. One side of the sensor is grounded. This sensor is mounted on the tracker, and moves it with the array plane. The sign of the output voltage determines which way the tracker would move. Since each solar cell operates as a current source, the sensitivity of this light sensor is very high. A small difference in cell insolation results in a large output voltage swing. Two independent sensors and a differential amplifier could have been used but the load resistance value required is low, resulting in poor sensitivity.

The primary sensor of this tracker is a four sided pyramid with a solar cell mounted on each face. It has a very wide acceptance angle of 135 degrees. A shadow box sensor shades all of the cells when aligned with the sun. It is more sensitive than the pyramid but is more complex and has a narrower acceptance angle. These sensor types are not imaging sun sensors and therefore may be affected by diffuse light resulting in tracking error. An imaging sensor does not have this problem, but its optical acceptance angle is very narrow.

The basic electro-optic tracker may move erratically or in the wrong direction if the sun is not visible to the primary sensor. The tracker always moves in a direction which equalizes the light intensity on all four primary sensor cells. A second fixed sensor is used to solve these problems. The sensor is oriented with its aperture facing south at latitude tilt. This sensor may be imaging or nonimaging, but is in all cases designed to block as much diffuse light as possible and is therefore a beam sensor. A two element beam sensor (figure 2) was used in this tracker because it is very simple, and it provides adequate position information for use with the pyramid detector. A photo-transistor is placed at the center of each half of a divided dome. The dome has sufficient aperature area to account for all sun positions and must allow some small overlap between east and west sections. This beam sensor therefore offers only moderate discrimination between diffuse and beam solar insolation. More complex beam sensors can use narrow slot aperatures or point aperatures. They can provide high sunlight beam discrimination, but require large numbers of sensors.

The sun tracker has four modes of operation which include, shutdown, search, track, and wait. The tracker is shutdown when the beam sensor indicates that no sunlight beam is present. A one bit optical encoder is located on the base of the tracker. Its output indicates whether the array is facing east or west. The outputs of the beam sensor, B1 and B2 indicate sun position in east and west respectively. When the beam sensor and encoder provide conflicting information, the search mode is triggered. When B1 and B2 are both high signals, the beam sensor indicates south and the search mode is not triggered. When the azimuth circuit is in search mode the zenith circuit is shut down. Normally the tracker is controlled by the pyramid detector. It operates in track mode when moving to find the sun, and in the wait mode if the pyramid output voltages is within the dead zone of the control circuit.

The control circuit design is shown in figure 3. This is only the azimuth circuit. The zenith circuit is similar, but it has no search subcircuit. The direction of movement is controlled by the tracking subcircuit which consists of two opamps, A4 and A5. The positive drive opamp is biased at about 75 millivolts below ground, and the negative drive opamp is biased at about 75 millivolts above ground. Each opamp has about 25 millivolts of hysteresis. The circuit uses two noninverting schmitt triggers, but the output MOSFETS Q1 and Q2 invert the opamp outputs resulting in an inverting transfer function.

The X/Y transfer plot of the tracking subcircuit contains a dead zone of about 100 millivolts centered at zero. Each side of the transfer function has a hysteresis loop of about 50 millivolts. The X/Y characteristic $Y_T = V_S(\pm R_{D2}/R_{D1} \pm R_I/R_F)$ can be adjusted by changing reference and positive feedback divider resistors. The dead zone insures that both MOSFETS can never switch on together. The hysteresis insures that no high dissipation partial on states can occur. The MOSFETS used in the output are 12 amp complementary pairs. They are low cost high power devices available in both metal and plastic packages.

The shutdown subcircuit consists of only four diodes and a resistor. When B1 and B2 are both negative, it lowers the voltage at A4 minus and A5 plus. This switches both opamps to their supply rails regardless of the pyramid sensor input. Since the MOSFETS have zero gate to source voltage they are both switched off. The search subcircuit consists of three opamps and it raises or lowers the input voltage of the tracking subcircuit to a high magnitude when the beam sensor and encoders provides opposite inputs. This forces the tracker to search in the correct direction regardless of the pyramid inputs. The search subcircuit inhibits the zenith portion of the tracking subcirucit. The tracking control unit uses superbrite LED indicators to indicate power supply status, beam sensor states and output states.

EXPERIMENTAL RESULTS

A small prototype model was used for testing the tracking system. Although it was tested with the simplest possible sensors, the tracking resolution was quite high. On a clear day, the tracker moves a few times per minute indicating approximately 0.1 degree resolution. The output duty cycle is only about 1%; therefore the power drain is low. On very hazy days the zenith tracks higher than the actual sun position because of the effects of diffuse light on the nonimaging differential pyramid light sensor. An imaging sensor would be required for use with high concentration point focus solar energy conversion systems. A nonimaging shadow box type or pyramid sensor would still be required to find the sun.

When faced away from the sun the device searches in the correct direction until the pyramid can see the sun. When a cloud passes in front of the sun, the tracker waits and does not wander. After sunset the tracker remains stationary until the next day. This tracker never makes multiple revolutions. When searching for the sun it always revolves through south. The zenith drive does not move until the search move is completed and normal tracking resumes. The solar array has no set night storage position, but will never be horizontal in high latitudes. The generally steep shutdown position limits snow buildup on the solar arrays. The tracker can be made remote controllable so the operator can adjust the storage position to minimize wind or snow loading in the event of a severe storm.

CONCLUSIONS

This tracking control unit is simple and reliable. It can be used as tested for flat plate or moderate concentration ratio systems. The device can be adapted for higher precision tracking by using imaging electro-optic sensors, and would then be applicable to any point focus solar energy system. This tracker can be applied to large scale photovoltaic power stations. Each solar array would have its own independent control unit. One fixed beam sensor could control a large group of arrays.

This sun tracker is relatively easy to install. Although the system requires initial orientation, the accuracy of the setup procedure is not critical.. All components are low cost types which are available from many suppliers. The control unit has a low component count and is a small low cost device. The cost is low enough to allow spare units to be kept on hand. In the event of a failure, the faulty unit could be immediately replaced, reducing down time to a minimum. The faulty unit could then be repaired and saved for later use.

ACKNOWLEDGEMENT

The authors would like to thank the photovoltaic program at the University of Lowell for support and encouragement.

REFERENCES

(1) Leonard Magid, 'Status and Assessment of Collector Cost Reduction Efforts', Fifteenth IEEE Photovoltaic Specialists Conference, 1981, pp. 353-355.

(2) C.F. Gay, J.W. Yerkes, J.H. Wilson, 'Performance Advantages of Two-Axis Tracking for Large Flat-Plate Photovoltaic Energy Systems,' Sixteenth IEEE Photovoltaic Specialists Conference, 1982, pp. 1368-1371.

(3) E.C. Boes, A.B. Maish, 'Advances in Concentrator Technology', Nineteenth IEEE Photovoltaic Specialists Conference, 1987, pp. 985-991.

(4) Zomeworks Coupuration, 'Passive Solar TRACK RACK,' Albuquerque, New Mexico 87125, U.S. pat No. 4275712.

(5) P.L. Swart, J.D. Vanwyk, 'Source Tracking and Power Flow Control of Terrestrial Photovoltaic Panels for Concentrated Sunlight,' Twelfth IEEE Photovoltaic Specialists Conference, 1978, pp. 701-705.

(6) B.P. Edwards, 'Computer Based Sun Following System,' Solar Energy Journal, **Vol. 21**, No. 6, 1978.

(7) A.A. Salim, F.S. Huraib, B.H. Khoshaim, N.H. Eugenio, N.R. Rao, M.S. Imamura, 'Performance Characteristics of the 350KW Concentrator Photovoltaic Array Field,' 21st Intersociety Energy Conversion Engineering Conference, 1985, pp. 1285-1291.

(8) Sode-Shinni Nmadu Rumala, 'A Shadow Method for Automatic Tracking,' Solar Energy Journal, **Vol. 37**, No. 3, 1986.

Figure 1 Tracking System Block Diagram

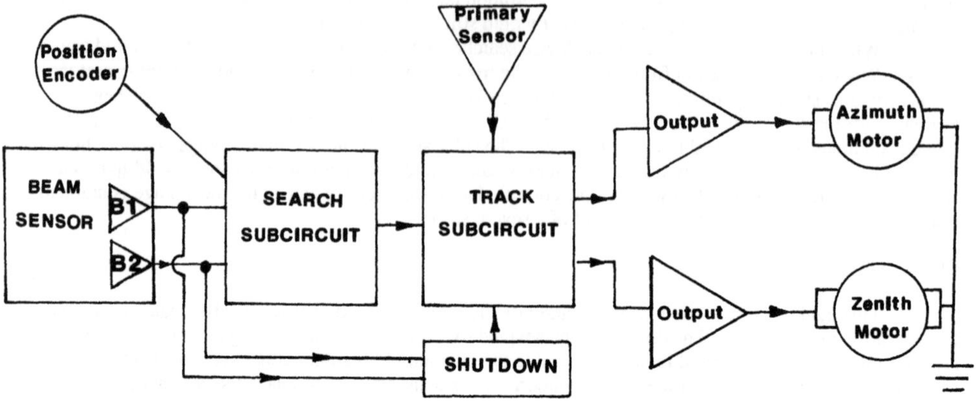

Figure 2 Fixed Beam Sensor Sunlight Detector

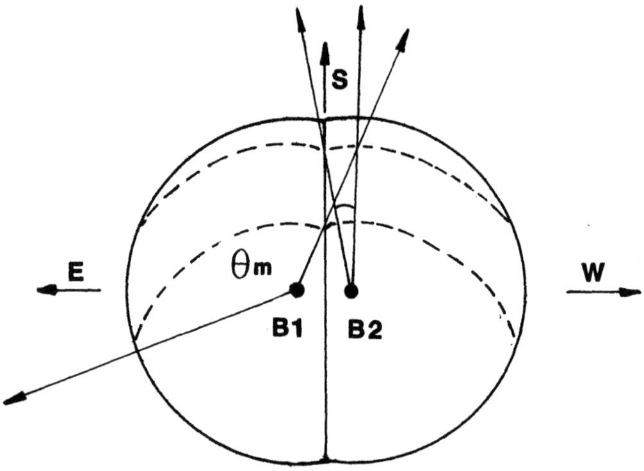

B1, B2 = phototransistor detectors located at the center of each half of the divided dome.

θm = Maximum detector acceptance angle

α = Overlap angle between East and West detectors

The dotted lines indicate the transparent operature region.

Figure 3 SUN TRACKER AZIMUTH DRIVE CONTROL CIRCUIT

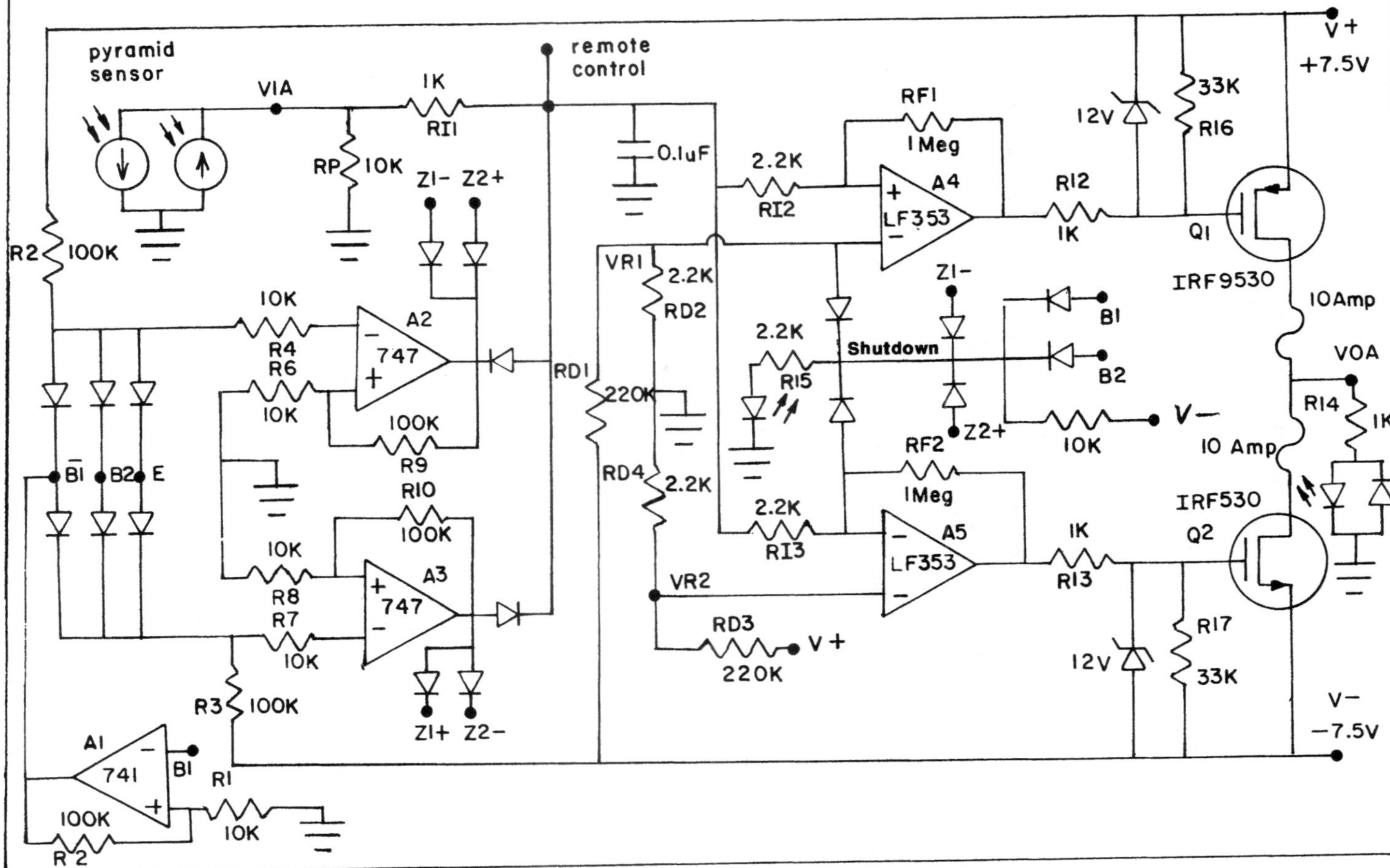

TRUNCATED COMPOUND PARABOLIC CONCENTRATORS

G. ALMONACID, A. LUQUE+

Dpto. de Electrónica y Sistemas Informáticos
Universidad de Granada. E.U. de Ingeniería Técnica Industrial
23008 JAEN (Spain)

+ Instituto de Energia Solar. Universidad Politécnica de Madrid
28040 MADRID (Spain)

ABSTRACT

The aim of this paper is the particular analysis of the truncation effects on static and quasi-static cylindrical CPCs for bifacial cells filled with diellectric, taken into account the losses of the concentrator.

By using a ray-tracing procedure we calculate the directional intercept factor, the irradiance intercept factor and the irradiance gain for these concentrators. With these results and with an analysis of the conclusions of the previous works (1,2), we find that a truncated CPC can achieve better performances than another CPC of the same height without truncation can.

INTRODUCTION

The results of the works above mentioned suggested to us to continue the analysis of the truncation Compound Parabolic Concentrators (CPCs) to make use of the advantages of truncation, i.e. a large reduction in mirrors and amount of dielectric with a small reduction in the irradiance gain for a concentrator without losses (a reduction in the irradiance gain lower than 10 % when the CPC is truncated at 50% of its total height) (1).

Two questions arise now. The first is whether a truncated CPC can achieve better irradiance gain than another CPC of the same height withot truncation. The second is the effect of the losses (reflection on mirrors and cell and oil absortion) on the preceding results.

Hence, the objetive of this work is to analyse the behaviour of several cylindrical bifacial CPCs filled with a dielectric, of the same height but with differents levels of truncation, for the case of static (non tracking) and quasi-static (with tilt adjustement - two positions in a year), taking into account the concentrator losses.

The analysis is based on the calculation of the irradiance gain, which is defined as the monofacial cell area of a losseles flat panel needed to collect the same energy as that collected by the CPC using a receiver (bifacial cell) with unit area. This is related to the cell cost (one of the main component of the concentrator cost), and the irradiance intercept factor, which is defined as the ratio of the power in the collector to the maximun power at the entry aperture. This is related to the concentrator efficiency which is another main component of the cost.

This calculation is based on a computer ray-tracing using a model of solar source which represents the yearly average radiance R(p,q) (W/m^2) of each point of the sky in Madrid (3)

ANALISIS AND DISCUSSION

A theoretical ray-tracing, made with a computer, is used to obtain the directional intercept factor of the concentrator with losses a(p,q), defined as the ratio between the power in the collector and the power inthe entry aperture when it is illuminated by light coming from a given direction. this direction is here determined by its direction cosines (p,q) with respect to the entry aperture (X,Y) axes.

With experimental measurement, made by means of a laser, on a truncated CPC prototype (1) we have proved the validity of these theoretical ray-tracings.

The irradiance gain Co, is given by the formula (4),

$$Co = I.Cg \qquad (1)$$

where Cg is the geometrical concentration of the concentrator and I is the irradiance intercept factor, defined by the equation,

$$I = R(p,q).a(p,q).dp.dq / R(p,q).dpdq \qquad (2)$$

where the integrals are extended over the source area on the p-q plane.

We made the analysis for three differents heights: 5, 10 and 20 times the cell height. We considered the concentrators filled with a dielectric of refractive index n=1.47 and exponential absortion coeficient of 0.0165/cm. The reflection cefficient of the mirror used is 0.85.

Figures 1 and 2 show the results of the irradiance intercept factor and the irradiance gain respectively vs. the meridian acceptance angle (after dielectrino filling) for the case of quasi-static cincentrators. The highest irradiance gain is produced in the three cases for a meridian acceptance angle of about ± 12º with a flat maximun. This also produces a reasonably good intercept factor of about 0.5. In the same way, for the case of static concentrators the highest irradiance gain is alwais obtained for a meridian acceptance angle of ± 24º with an intercept factor of about 0.65 (5,3).

Using the same method, we calculate the irradiance intercept factor and the irradiance gain for CPCs with a meridian acceptance angle of ± 12º after dielectric filling, with differents levels of truncation and taking into account the losses above mentioned, for the case of quasi-static concentrators. These parameter are shown in Fig.3 as a function of the level o f truncation and the final Height of the CPC. The flat maximun observed in the irradiance gain is due to the fact that the truncation decreases the average number of reflections on the mirrors (2) and the dielectric absortion is lower, because the path of the ray is shorter in the truncated concentrators.

Therefore, when we consider the losses, the result can be more favourable for the truncated concentrators.

CONCLUSIONS

The highest irradiance gain is alwais obtained, for the case of CPCs of the same height; for CPCs with a meridian acceptance angle of about 12º, for the quasi-static case, and about 24º, for the static case. That imply that a truncated concentrator can achieve better irradiance gain than another CPC of the same height without truncation (with larger acceptance angle).

For a concentrator of a given acceptance angle, truncation improve the performances of CPCs, bigger intercept factor and irradiance gain, up to certain level of truncation. The best truncation level depends on the dielectric absortion and the reflection coefficient of mirrors. For the values of the parameters mentioned above, a truncation of 30% corresponding to a height of 15.Ac (Ac is the cell height) will give the highest irradiance gain. Therefore the flat maximun allow more practical concentrators, e. g. concentrator with a height of 5.Ac with a little decrease of the irradiance gain with respect to the maxium.

REFERENCES

(1). G. Almonacid, A. Luque and A.G. Molledo. Photovoltaic static concentrator analysis. Solar Cells, 13 (1984) 163.

(2). G. Almonacid et. al. Analysis of a photovoltaic static concentrator prototype. Solar and Wind technology, Vol.4, No.2 (1987) 145.

(3). G. Almonacid. Contribución al estudio de concentradores fotovoltaicos estáticos 3D generados por diagramas 2D. Tesis Doctoral. Universidad Politécnica de Madrid. 1987.

(4). A. Luque. Theoretical Bases of photovoltaic concentrators for extended light sources. Solar Cells, 3 (1981) 355.

(5). G. Almonacid, A. Luque. Truncation effects in bifacial compound parabolic concentrators. Solar Cells, 22 (1987) 47.

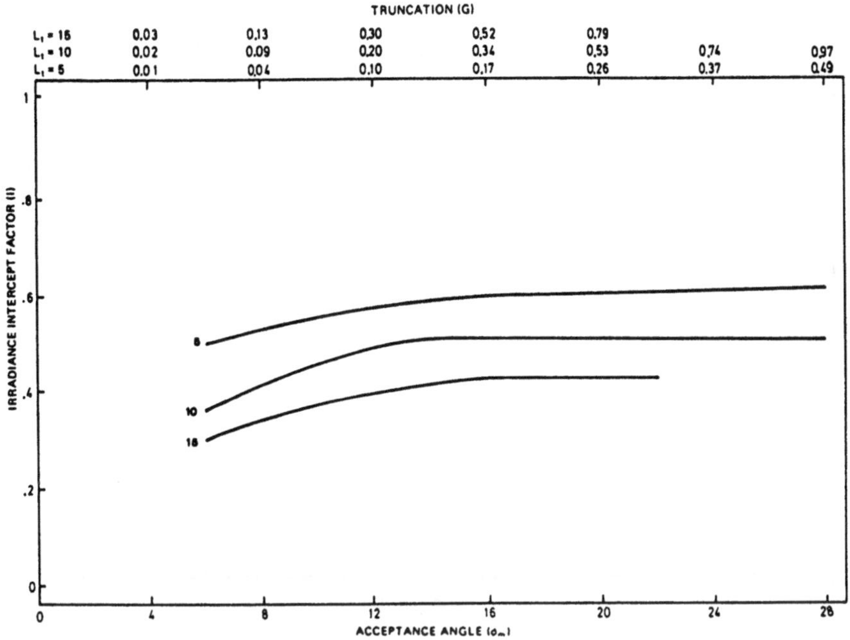

Fig. 1 Irradiance intercept factor for CPCs of the same height of 5, 10 and 20 times the cell height *vs.* the meridian acceptance angle ϕ_m after dielectric filling and the level of truncation G for each given height.

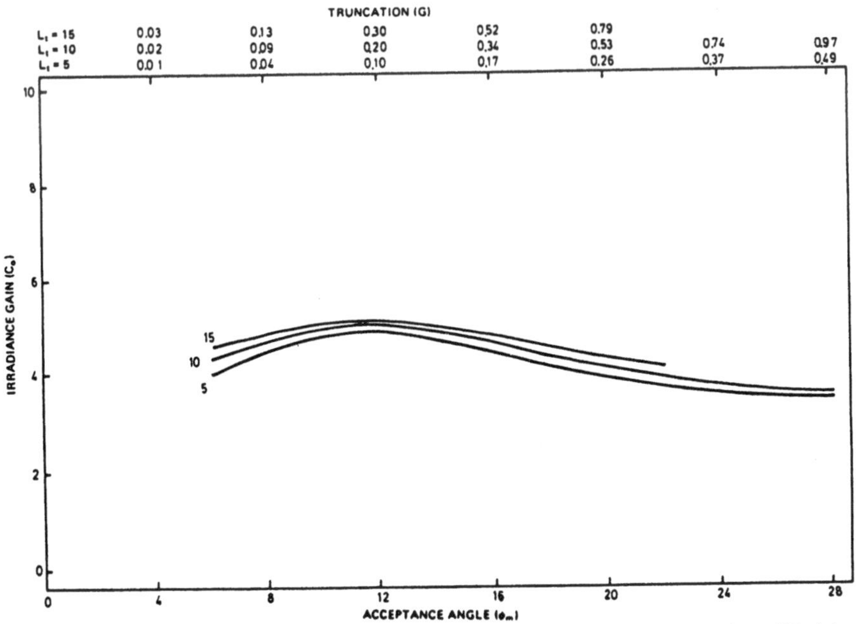

Fig. 2 Irradiance gain for CPCs of the same height of 5, 10 and 20 times the cell height *vs.* the meridian acceptance angle ϕ_m and the level of truncation G for each given height.

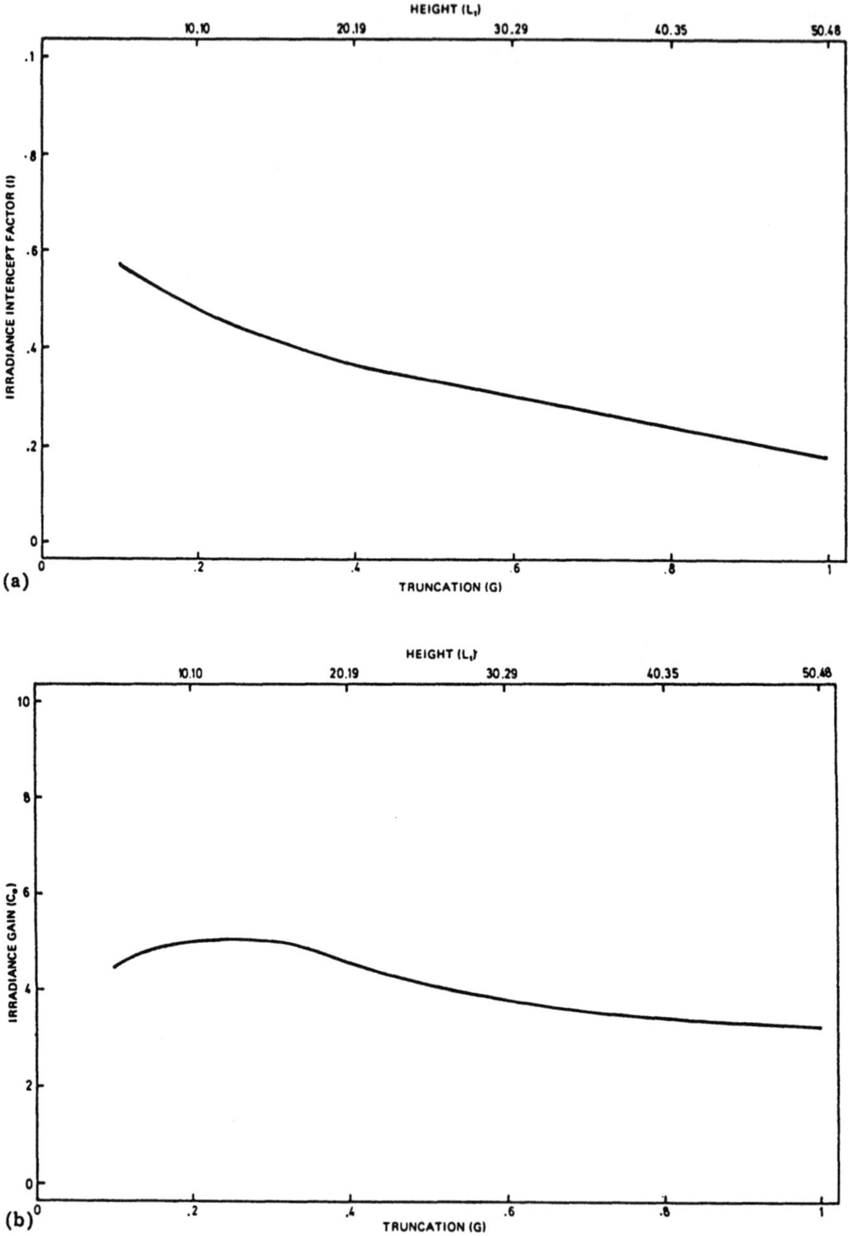

Fig. 3 (a) Irradiance intercept factor and (b) irradiance gain for truncated CPCs of meridian acceptance angle ±12° *vs.* the level of truncation G and the final height of the concentrator L_t.

RELIABILITY PREDICTION OF THE 350kW CONCENTRATING PHOTOVOLTAIC FIELD

Dr. A.R.M. Alamoud
King Saud University
E.E. Dept. College of Eng.
P.O. Box 800
Riyadh 11421
Saudi Arabia

F.S. Huraib, A.A. Salim
King Abdulaziz City for
Science and Technology
P.O. Box 6086
Riyadh 11442
Saudi Arabia

ABSTRACT. Field measurements of the 350-kW Concentrating Photovoltaic Power System (PVPS) operating since September 1981, near Riyadh, Saudi Arabia, have revealed about 1% per year degradation in the PV field power output. This is attributed to the failure of PV modules in the field. Modules are failing due to solar bond breakage at the cell to ceramic substrate and/or ceramic substrate to cell interconnect interface. The rate of failure has been almost constant at about nine half-modules per month over the past five years. The purpose of this paper is to establish certain mathematical relationships based on these factual results in order to estimate the reliability and overall life of the Photovoltaic field.

1. INTRODUCTION

In late 1981, a 350 kW photovoltaic power system (PVPS) was installed in a location 50 km northwest of the city of Riyadh, the capital of Saudi Arabia. The PVPS is made up of 160 single pedestal concentrator arrays, each containing 256 circular silicon solar cells of 2.25 inch diameter. A group of four solar cells housed togather and bypassed by a diode form a half-module. Each half-module uses a quad Fresnel lens to focus light onto four Si solar cells mounted on a heat sink. The lens serves the dual purpose of concentrating the sunlight (≈ 33 times) and providing an enclosure for the solar cells. Each array, 64 half-modules, produces about 2.2 kW of power under the standard conditions of 800 W/m^2 insolation at AM 1.5, an ambient temperature of 40°C, and wind speed less than 1 m/sec. The PV arrays are passively cooled and track the sun in two axes. The PV field is connected in such a way that every two and one-half array are connected in series to form a branch yielding the correct system operating voltage. Sixty-four branch circuits are connected in parallel to provide the 350 kW of peak power as shown in Figure 1. In addition to the 350 kW PV field, the PVPS includes a 1100 kWH lead acid batteries bank, a 300 kVA inverter, and solar weather data monitoring station. The photovoltaic power system is designed with complete automatic operation capability in standalone or cogeneration modes. Presently, the PVPS is connected to the national grid power system. Previous publications [1-3] provide additional description of various PVPS hardware along with the past performance history. The purpose of this paper is to qualitatively discuss the reliability of the photovoltaic field of the 350 kW power system.

PV Field Reliability

The PVPS reliability consists of two parts. The first one is the PV field reliability. The second is the balance of the system (BOS) reliability. The BOS reliability includes the reliability of the inverter, battery bank, and other electronic components [4]. The performance of the BOS components has been remarkable and well beyond expectation. As such the reliability of the BOS is not discusssed in this paper.

The most significant failure that has occured in the PV field is the module open circuit problem. Detailed failure analysis of two modules, by two independent laboratories has shown that the problem was due to the failure of the solder bond at the ceramic substrate to cell interconnect strap joint and cell to ceramic substrate joint [4]. A failed solar cell causes the bypass diode to be activated resulting in a loss of one half-module. Hence, the discussion that follows will be concerned only with the failure of half-modules units. Thus the complete PV field consists of 10,240 half-modules which are connected to form a network of 64 parallel branches of 160 half-modules in series.

In order to identify the defective half-modules in the PV field, bimonthly (monthly before Oct. 85) voltage tests have been conducted since June 1983. It should be noted that no tests were conducted during the year 1984. The number of defective half-modules in the entire PV field versus time is shown in Figure 2. The half-module failure data is fitted to a straight line with an R^2 value of 0.991972. Figure 2 indicates that the number of defective half-modules increases linearly with time with a constant rate of about nine half-modules per month over the past sixty months. As a result of the half-modules failure in the PV field, the power output was found to degrade about 1% per year. Laboratory tests have revealed that lenses do not contribute to the PV field output degradation. Moreover, it is believed that solar cells have not yet any contribution to the loss of output power.

Figure 3 shows a plot of the cummulative probability of failure (or failure percentage) versus time (in months) on a special Weibull probability paper. The experimental data are shown by o whereas the bold line is the best fit straight line whose equation is

$$\ell n \ell n \left[\frac{1}{1-F(t)} \right] = \beta \, \ell n \, t - \beta \, \ell n \, \eta \qquad (1)$$

where β and η are the shape parameter and characteristic life of the Weibull distribution respectively.

The values of β and η, obtained from the Weibull probability plot of Figure 3, are 1.65 and 445 months respectively. Since $\beta = 1.65$ and the failure rate is constant, the Weibull distribution approximates to the exponential distribution and an exponential reliability function results, with η given by $1/\lambda$ (the mean life) [5]. The PV field (pvf) reliability $R_{pvf}(t)$ is given by

$$R_{pvf}(t) = 1 - (1 - \prod_{i=1}^{n} R_i)^m \qquad (2)$$

where R_i is the reliability of the ith half-module, $n = 160$ is the number of half-modules per branch and $m = 64$ is the number of branches connected in parallel in the PV field. The PV field is connected to the power grid with no specific current requirement. Hence one branch is sufficient to generate the required voltage. It is clear that this system is a unit redundant one. It should be noted that failed half-modules are neither replaced nor repaired. For an n identical series units with exponential reliability function, eqn.(2) becomes

$$R_{pvf}(t) = 1 - [1 - \exp(-n\lambda t)]^m \qquad (3)$$

Eqn.(3) gives an estimate of the PV field reliability assuming no failures other than the open-circuit voltage problem take place. In other words, failures due to individual solar cell degradation are assumed to be negligible at the present time.

CONCLUSION

Field data for the 350 kW concentrating photovoltaic power system revealed that PV modules failed at constant rate of about 9 half-modules per month over the past 60 months. This results in about 1% per year degradation in the PV field power output. The reliability of the unit redundant PV field was studied. The result of a Weibull reliability plot indicates an exponential reliability function for the identical half-module units. It predicts a characteristic life of about 37 years for the PV field with no replacement of failed modules. This value does not include the effect of the solar cell degradation which is assumed to be negligible at the present time.

Finally, a mathematical relation that estimates the reliability of the PV field at a given time was found as given by eqn.(3).

ACKNOWLEDGEMENT

The authors wish to thank Dr. A.B.M. Zohrul Kabir for his helpful discussions on reliability and Mujahid Hussien for his help with experimental measurements.

REFERENCES

1. Huraib F. et al, 'Design, Installation, and Initial Performance of 350 kW Photovoltaic Power System for Saudi Arabian Villages,' Fourth E.C. Photovoltaic Solar Energy Conf., Proceedings, pp.57-66, May 1982.

2. Salim, A.A., Huraib, F.S., et al, 'Four Year Performance Summary of 350 kW Concentrating Photovoltaic Power System,' 18th IEEE Photovoltaic Specifialists Conf. Proceedings, pp.1545-1552, Oct. 1985.

3. Zakzouk, A.K.M., Alamoud, A.R.M., Khoshaim, B.H., Factors Affecting the Performance of a Photovoltaic Power System (PVPS), Int. J. Solar Energy, Vol.5, pp.67-81, 1987.

4. Khoshaim, B., et al., 'Performance of 350 KW Photovoltaic Power System for Saudi Arabian Villages After 30 months,' 17th IEEE PV Specialists Conf., Proceeding, pp.1426-1432, May 1984.

5. Shooman, M.L., Probabilistic Reliability: An Engineering Approach, McGraw Hill, 1968.

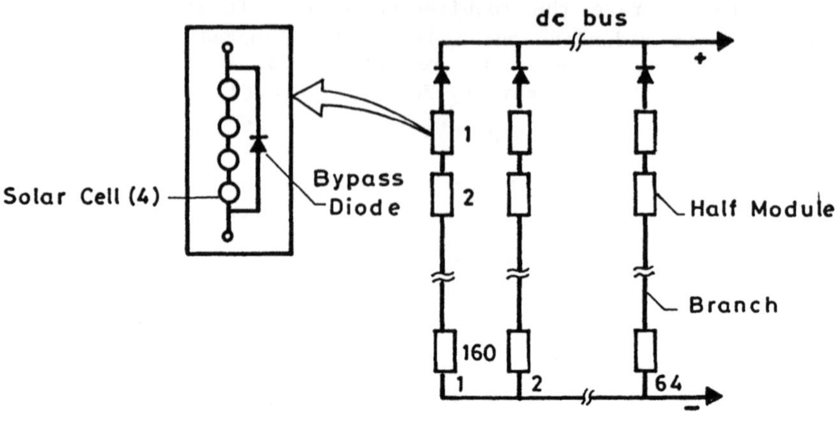

Figure 1 Simplified schematic of half-module connection in the 350 kW PV field

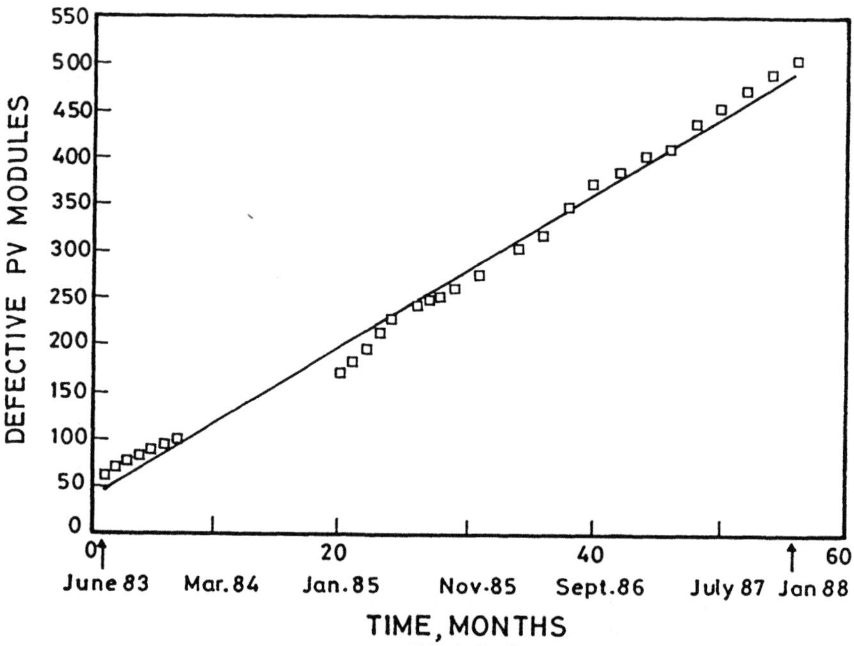

Figure 2 Number of defective modules versus time (——— best line, □ experimental points).

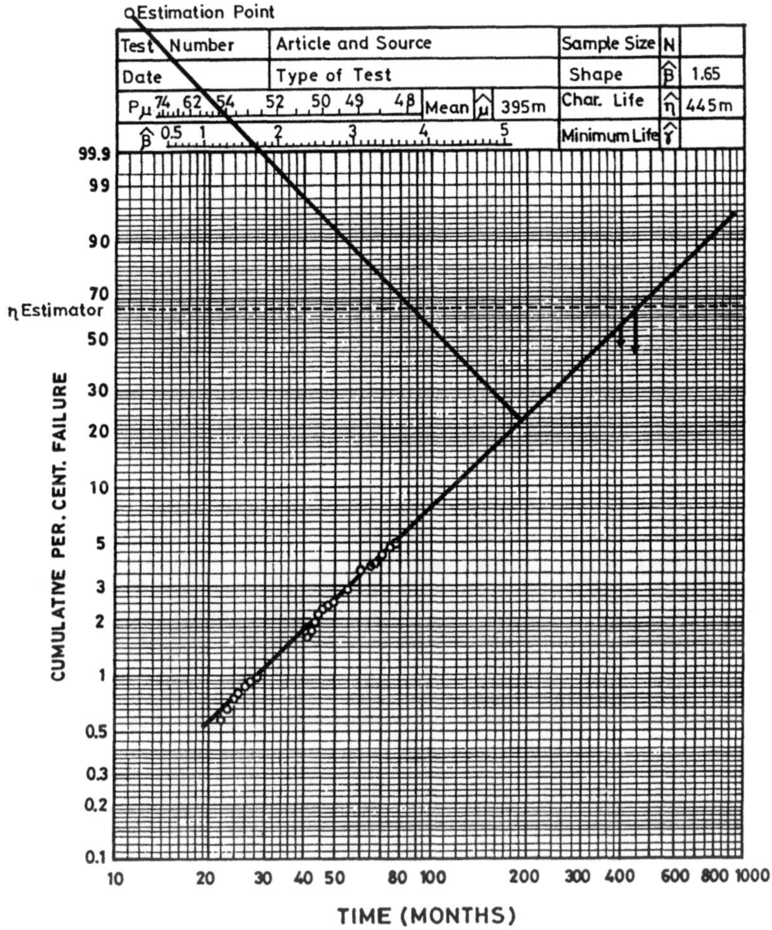

Figure 3 Cimmulative percent failure versus time plot
(——— best fit line, o experimental points)

CONCENTRATION AND LIGHT CONFINEMENT

A. Luque, G.L. Araújo and J.C. Miñano
Instituto de Energía Solar
E.T.S.I. Telecomunicación - U. Politécnica de Madrid
28040 MADRID - SPAIN

ABSTRACT. Strong light confinement, beyond the one due to the high index of refraction of the semiconductor, is achieved by either angle limited AR coating or by metal covering of the cell. Its effect in the cell efficiency is studied for cells located in concentrators or in light confining cavities.

1. INTRODUCTION.

Casting concentrated sunlight on solar cells has two purposes: one is to reduce the cell area to decrease it cost and another is to increase efficiency. Light confinement in the cell also increases the efficiency. Both techniques are analyzed in its mutual connection and also in connection with cavities in which the cell can be located to produce light confinement in the cell surroundings. The study is done for cells limited by non-radiative recombination and also for cells limited by radiative recombination that constitute a thermodynamical upper limit for the cell behaviour.

2. ABSORPTION OF LIGHT IN CELLS UNDER LIGHT CONFINEMENT.

The concept of light confinement has been sufficiently discussed in the last years[1,2]. The sunlight enters into the cell and becomes randomized by different techniques with a isotrope photon radiance (photon flux per unity of solid angle) n_{ph} which is assumed here to be monochromatic and weackly absorbed with a constant of absorption α such that w 1 (w the cell thickness). The absorption per unity of volume is n_{ph} 4 w. Let us assume that the cell has an area unity that is covered by perfectly reflecting metal on its back side and partly on its front side, leaving an uncovered surface of area f. It is also assumed that a special AR coating produces perfect transmission in an angle θ_e and zero transmission outside this angle. The rate of escaping photons is $n_{ph}\ \pi f \sin^2 \theta_e / n^2$. The rate of entering photons is $N_{ph} f$ where N_{ph} is the photon flux that is assumed to be totally casted within the angle θ_e. The balance of photons leads to the well known equation for the total photon absorption in the unity of area cell.

$$G = N_{ph} f \frac{4 \alpha w}{4 \alpha w + f(\sin^2 \theta_e / n^2)} = J_L / q \qquad (1)$$

and also represents the total carrier generation in the cell. In this equation, only the confinement by partial coverage of the front face, whose effect is given by the factor f, has not been, to our knowledge, explicitely considered in the previous literature.

3. THE RECOMBINATION TERM.

Let us consider solar cells in which the carrier movility is high enough as to guarantee an homegeneous density in the whole cell; in this case the quasi-Fermi level splitting is constant throghout the cell and takes a value qV. The recombination through the cell is also assumed to be homogeneous. In this case this recombination is proportional to the cell thickness w and are can write with generality that

$$J_r = A\ w\ \exp(V/V_T) \qquad (2)$$

where A is a constant and V_T takes the value kT/q in low injection cells 2kT/q in high injection with SRH recombination and 2kT/3q in high injection with Auger recombination.

For radiative cells the recombination produces luminiscent emission but this emission is subsequently reabsorbed an so on. The only losses are the luminiscent radiation towards the cell exterior. Light confinement enhances the reabsorption and thus reduces the recombination. The calculation of this term is classical[3,4,5]. It consists in stating two conditions. One is that in thermal equilibrium the flux of phottons emitted to the exterior equals to those entering from outside, coming from a black-body at the ambient temperature T. let n^T_{ph} be the photon radiance corresponding to this black-body with photon energy above E_g. If the temperature is such that Eg kT, we can assume that all these photons have the same monochromatic energy, slightly higher than Eg, for which the absorption ^{T} is very weack. The flux of photons entering the cell is $n^T_{ph} f \sin^2 {}^T_e$ where ^T_e is the entrance angle permitted by the angle-selective AR coating, that may be different from the one $_e$ corresponding to the sun photons, of different wavelength.

Thus the absorption of these photons is given, under light confinement by a expression similar to that in Eq. (1).

$$G^T = n^T_{ph} \sin^2\theta^T_e f \frac{4\alpha^T w}{4\alpha^T w + f \sin^2\theta^T_e / n^2} \qquad (3)$$

but in this case ^{T} is much smaller than the value of corresponding to the sun photons, of shorter wavelength, so that we can admitt that $4\alpha^T w$ is very small and

$$J_r = qG^T = qn^T_{ph}\, \pi\, 4\alpha^T w\, n^2 = A^T w \qquad (4)$$

which is also proportional to w.

The second statement to permit the calculation of the recombination is that it is proportional to the generation times exp qV/kT. So we can give for the radiative recombination an expression similar to that in Eq. (2) with $A = A^T$ and $V_T = kT/q$. Note that the value of $q\pi n^T_{ph}$ $(A/cm^2) = 407\ Eg^2 \exp(-Eg/kT)$ is well know[5]. It can be said that light confinement reduces the radiative recombination in a factor $4\alpha^T w \ll 1$. (See ref. 5).

4. OPTIMIZATION OF THE CELL THICKNESS.

Cells too thick produce almost total generation but increase the recombination. The opposite is true cells too thin. An optimal value is thus to be seek. For it the cell efficiency $\eta = (qG/Ef)\ V_{oc}$ CF is done maximum. Note that CF can be approximated as a funcion depending only on

V_{oc}/V_T. After some mathematical handling we obtain for the optimal w

$$4\alpha w = B\,\frac{f\sin^2\theta_e}{n^2}\ ;\ B = \frac{V_{oc}}{V_T}\Big/\left(1+\frac{d\ln CF}{d\ln V_{oc}}\right) \approx \frac{V_{oc}}{V_T} \qquad (5)$$

which shows that if $f\sin^2{}_e$ 1, which is the situation we refer to as weack confinement, then the condition 4 w 1 is not fulfilled so that weack absorption cannot be assumed. Although an optimal w also takes place its simple calculation is not given here: in spite of the practical interest of this weack confinement, it is not the one we are considering in here. On the contrary we are interested in the case that f sin $_e$ 1 and strong confinement is assured. For the optimized cell the G= $N_{ph}fB/(B+1)$ and thus almost total absorption of the incident photons is produced. The J-V curve is given by

$$J = qN_{ph}f\,\frac{B}{B+1} - \frac{AB}{4\alpha}\,\frac{f\sin^2\theta_e}{n^2}\exp\frac{V}{V_T} \qquad (6)$$

The efficiency of this cell is proportional to the photorresponse times V_{oc} CF ie is an increasing function of V_{oc}. We are assuming here that all the light is cast, through a concentrator, on the uncovered cell area f so that the photorresponse is independent on f. Also V_{oc} is independent on f (for the cell of optimal thickness) so that only two ways exist to increase V_{oc}: the increase of N_{ph}, ie of the concentration and the decrease of $\sin^2\theta_e$, ie. the increase of the angular confinement. The maximum of increase of N_{ph}, over the one sun value N_{ph}^{st} caused by concentration is $C\lesssim 1/\sin^2\theta_s$ where θ_s is the sun angular semidiametre and θ_e cannot be done smaller than θ_s to permit the entrance of all the sun rays.

Since from the conservation of the radiance any concentration requires a larger output angle than the input one, in a losseless ideal concentrator $C=\sin^2\theta_c/\sin^2\theta_s$ (this is actually an upper bound of C) where θ_c is the angle of rays issuing the concentrator. The optimal case is to equate θ_c and θ_e and we obtain for it.

$$J = qN_{ph}\,\frac{\sin^2\theta_c}{\sin^2\theta_s}\,f\,\frac{B}{B+1} - \frac{ABf\sin^2\theta_c}{4\alpha n^2}\exp\frac{V}{V_T} \qquad (7)$$

whose efficiency is independent on the value of θ_c. Increasing θ_c is increasing concentration and decreasing confinement. It therefore results that concentration and confinement plays the same role in the increase of efficiency. Thus, we can, in theory, adopt the value of concentration (θ_c) required by the need of reduction of cell area and yet provide sufficient confinement as to reach the efficiency upper limit. This result applies to cells based on either radiative or non-radiative recombination.

Note that the restriction of the θ_e can be technologically very difficult and that we are assuming that all the light is cast into the cell. This will require some kind of microconcentrators[2]. If the light is also cast on the metal fingers area (1-f) then some efficiency is lost.

5. USE OF EXTERNAL LIGHT ENHANCING CAVITY.

The use of light confinement in cells located in the light confining cavities suggested by Sinton and Swanson[6] merits to be studied. The light enters the cavity through a small hole of area F 1 and falls into the cavity walls where the light is reflected without absorption. The cell area is unity. A balance of energy is produced between the photons absorbed in the cell and those escaping the cavity through the hole F. The cavity walls are suposed to be totally refectant; the cell is assumed to be almost totally absorbing in an area f with the absorbance implied by Eq.(1)and a totally reflecting in the area (1-f). Under these assumptions, the cell efficiency can be maximized with W and f as independent variables.

$$G = N_{ph}F\,\frac{f}{f+F}\;\frac{4\,W}{4\,W+fF/(F+f)n^2} \tag{8}$$

For a given F and f this equation also allows for an optimization in w . The procedure is identical to that used before. It is concluded that

$$4\alpha w = \frac{B}{n^2}\;\frac{Ff}{F+f} \tag{9}$$

Note. That case in this cell the angle of light $_e$ acceptance has to be 90° because its illumination is isotropic; only area confinement can be achieved by reducing Ff/(F+f) which plays the same vole os $f\sin^2{}_e$ in Fg.(5)

The J-V equation of a thickness-optimized cell located inside the cavity is

$$J = qN_{ph}F\,\frac{f}{f+F}\;\frac{B}{B+1} - \frac{f}{f+F}\;\frac{AB}{4\alpha n^2}\,\exp\frac{V}{V_T} \tag{10}$$

We observe that V_{oc} is in this case, similar to that without cavity and $_e$=9o° but the current is reduced in a factor f/(f+F) and so is the efficiency. This reduction is slight if f≫F.

6. EXTERNAL OPTICALLY DENSE MEDIA.

Optically dense media of index of refraction n_c may exist outside the cell either because an optically dense concentrator (usually a secondary) is used or because the external cavity is filled with an optically dense medium. In this case the index of refraction in the cell confinement is to be considered relative to n_c but also the value of θ_e can be reduced because the sun angle within the medium is $\theta'_s=\sin^{-1}(\sin\theta_s/n_c)$. After these changes Eq (7) becomes

$$J= qN^{st}_{ph}\,\frac{\sin^2\theta'_c}{\sin^2\theta_s}\,n_c^2\,f\,\frac{B}{B+1} - \frac{ABf\sin^2\theta'_c}{4\,\alpha\,n^2}\,n_c^2\,\exp\frac{V}{V_T} \tag{11}$$

so that no change is actually produced in the efficiency. The prime in θ'_c is to refer to angles in the optically dense medium. Note that in an optically dense medium the irradiance concentration is $n_c^2\sin^2\theta'_c/\sin^2\theta_s$.

In the case of cavity filled with an optically dense medium then a reduction in the escape of light occurs in a factor $1/n_c^2$. If the cavity has an angle selective coating accepting rays only in an angle θ_e then this escape reduction is in $\sin^2\theta_c/n_c$. In this case F (except for its appearance as factor of N_{ph}^{st}) is to be changed into $F\sin^2\theta_c/n_c^2$ and Eq (10), if $F\sin^2\theta_c/n_c \ll 1$ and θ_e equals the angle θ_c of the rays coming from the concentrator, becomes

$$J=N_{ph}^{st}\,F\,\frac{f}{F\sin^2\theta_c/n_c^2+f}\;\frac{\sin^2\theta_c}{\sin^2\theta_s}\left(\frac{B}{B+1}\right)$$

$$-F\,\frac{f}{F\sin^2\theta c/n_c^2+f}\;\frac{AB}{4\alpha n^2}\,\sin^2\theta_c\,\exp\frac{V}{V_T} \qquad (12)$$

This equation is rather similar to Eq (7) except in the increase of the recombination current by a factor B, which is the main penalty to be paid by the increase of the light entropy before proceeding to the light absorption.

Again the formation of an angle limited coating is not easy, but in this case a secondary coupled to the cavity plays the same role. In effect, if we now couple the entry aperture with an ideal dielectric filled concentrator approximated by a CPC and we call F its entry aperture and $_{2c}$ its acceptance angle, then the cavity entry aperture is reduced to $F\sin^{2}{}_{c}/n_c^2$, ie. the interchange of photons with the exterior is reduced to the same extent as above. Eq (12) applies also to this case and is not different from Eq (10) but reveals more clearly the interrelations between concentration and light confinement.

CONCLUSIONS.

An optimal cell thickness can be found for maximum efficiency. The optimal thickness depends on the amount of light confinement. It is smaller for strong light confinement.

Strong light confinement is based on restricting the angle of entrance of the light into the cell or on covering part of its surface. The first procedure increases the efficiency in thickness optimised cells leading to the highest reachable one but the second does not because the reduction of cell thickness permitted by the increase of confinement is balanced out by the increase of cell area for a given irradiance. Increasing area confinement helps to reach the conditions for strong confinement but in this case higher efficiencies are achieved without strong confinement. In all cases it is assumed that no light is casted on the cell covered area (micro concentrator) what may be technically difficult.

Concentration is another way of producing efficiency increase. Under ideal conditions including ideal concentration on the uncovered cell area or strong light confinement, concentration is strictly equivalent to reduction of the acceptance angle at the cell. This means that the same maximum efficiency can be achieved at any level of concentration if the appropiate amount of light confinement is used.

The use of metal covered area (1-f), unless if negligeable, can only be used with the adequate microconcentrators. Reduction of angle acceptance area below $_e$ 50° is very difficult. A way to avoid these technical difficulties and those associated with the microconcentrators is the use of external light enhancing cavities that can be dielectric filled. These cavities allow, in the ideal case a higher metal coverage without using a microconcentrator. For it an increase of the costy cell area and a slight reduction of efficiency is paid.

Under strong light confinement the efficiency of thickness optimized cells with optimized metal covering present an efficiency which is independent on the entry aperture area and also on the cell metal coverage.

If dielectric filled cavities are used with selective AR coating the cell efficiency is the same as the one achieved with microconcentrators provided that $f \gg F$.

In dielectric filled cavities to use a restricted acceptance a angle coating is strictly equivalent to use an ideal secondary concentrator that restricts the acceptance angle. In the last case F is the secondary entry aperture.

In the case of dielectric secondaries of wide angular acceptance ($\sin^{-1} 1/n$) the use of the secondary is unnecesary it is sufficien to fill the cavity with the dielectric.

Against what is said in previous works we show in this one that in conditions of light confinement radiative limited cells present a recombination increasing with the cell thickness and therefore its analysis is similar to the one corresponding to nonradiative cells ie. an optimum thickness if found at which the efficiency of the thick cell often presented as a theoretical limit is overreached

REFERENCES

1) A. Yablonovitch. and D.G. Cody, IEEE Trans. Electron Devices **ED-31** (1982), 300-305.
2) P.Campbell and M.A.Green, IEEE Trans. Electron Devices **ED-33** (1986), 234-239.
3) W. Shockley and H.A.Queisser, J.Appl. Phys., **32** (1961), 510-
4) C.H.Henry, J.Appl. Phys, **51**, (1980), 4494-4500.
5) G.L.Araujo and A.Martí, Internal Report UPM/IES/LS/1688 Instituto de Energía Solar, UPM. Madrid 1988.
6) RA Sinton and R.M.Swanson, IEEE Trans. Electron Device Letters, **EDL-8**, (1987), 547-549.

AN EVALUATION OF SOME GENERIC LOW CONCENTRATION PV MODULES UTILIZING A NOVEL BIFACIAL SOLAR CELL

Ugur Ortabasi
SCI-TECH International
P.O. Box 5246
Pleasanton, CA 94566 USA

Kay Firor
Pacific Gas and Electric Company
3400 Crow Canyon Road
San Ramon, CA 94583 USA

ABSTRACT

A recent study by Pacific Gas and Electric Company and SCI-TECH International on generic concentrators for bifacial cells has led to a new design concept. A novel 20X passive, one-axis tracking concentrator which uses both direct and diffuse components of solar radiation and albedo, is now being studied.

The design involves a unique configuration which combines a linear Fresnel lens with a composite second stage consisting of an optically coupled dielectric cusp and a luminescent flat plate concentrator. The total photon flux captured is refractively guided to the light sensitive areas on both sides of the cell. A high efficiency bifacial cell made from dendritic web silicon will be used in this concentrator.

Thermal and optical features are summarized and performance predictions in terms of diffuse light and albedo contributions are discussed.

INTRODUCTION

It has been recognized that bifacial solar cells, i.e. cells with both sides sensitive to solar radiation and active, may effectively utilize total irradiance in electricity generation [1,2]. There are two basic motivations for developing such a device. One is an increased response to albedo, particularly that portion reflected from nearby grounds. Sala reported in 1983 a concentration increase of 1.6 obtained with an "albedo-collecting" static bifacial panel that used white painted floors [3]. The consideration of diffuse ground reflectance as a source of an accountable addition to the power output is justified. The second reason for using bifacial cell is the possibility to combine it easily with a non-imaging concentrator to optimize the use of diffuse light. Luque predicted in 1984 that more economical devices could be expected, if bifacial cells are integrated with modified forms of Winston-type compound parabolic concentrators filled with dielectric material [4].

A NOVEL 20X FULLY TRACKING, TOTAL RADIATION CONCENTRATOR

Previous applications of bifacial cells were mainly limited to very low concentrations (< 2X) and static concentrators. The present study involves an exploration of a fully tracking 20X Fresnel/Cusp/Luminescent (FCL) concentrator, that focuses total irradiance onto both faces of a bifacial cell.

The fundamental features of this linear FCL concentrator are shown in Figure 1. A high efficiency bifacial cell is mounted and optically coupled onto a transparent dielectric cusp, with an acceptance angle $\Theta_{max} = 90^{o}$. This cusp/bifacial cell subassembly is located in the focal area of a Fresnel lens. The concentration ratio of the Fresnel lens/bifacial fluorescent concentrator in the aperture plane of the cusp is optically and structurally coupled to the cusp. The cooling is provided via a heat sink along the center line of the cell. The unit module has a rectangular prismatic shape and embodies two parallel cell strings. The optically active luminescent plate serves also as the bottom of the module box frame.

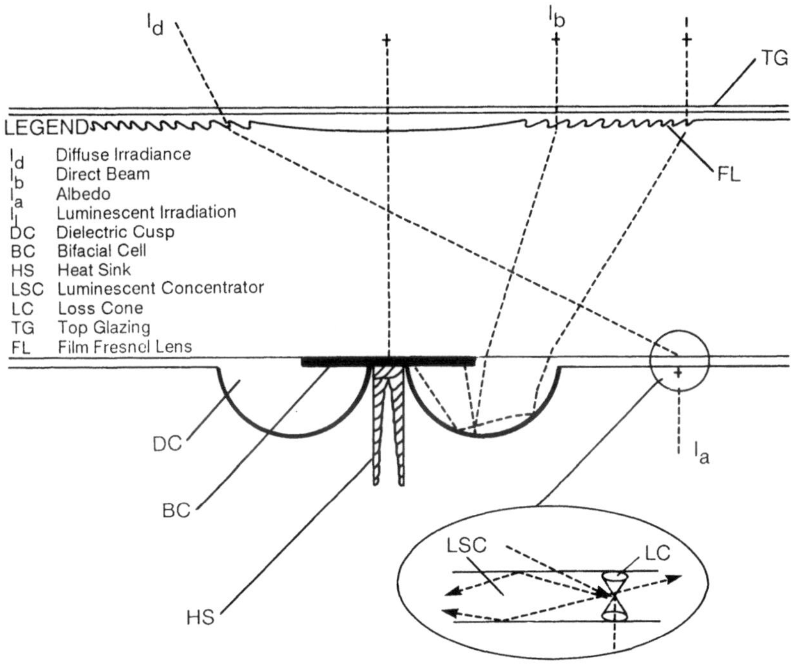

Figure 1. Principles of the FCL Concentrator

Optical Features

The flat Fresnel lens focuses the direct beam component of the solar flux on to the cusp/cell subassembly and illuminates both faces of the cell. The distribution of the flux across the cell (top and bottom) can be controlled to a certain degree by the positioning of the cell with respect to the design focal point. The optimization of the flux distribution entails compromises involving electrical cell grid design, lens losses, cell cooling and thermal extremes associated with tracking errors. A ray tracing analysis was performed for the cell both at and below the focal plane with tracking errors of 0 and 2 degrees. (See Figure 2.) It is worthwhile to note that the flux distribution on the upper face is highly uniform when the cell is placed below the focal plane. Tracking errors lead to asymmetric distributions and have to be dealt with in terms of cooling and cell design. Lens losses i.e. the number of direct rays missing the cell/cusp assembly increases as the cell moves upwards or downwards from the focal plane.

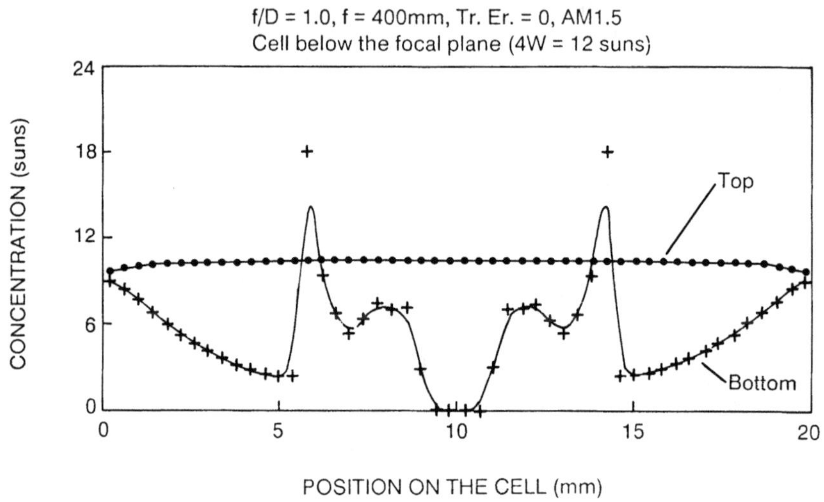

Figure 2. FCL Concentrator Intensity Distribution

Diffuse radiation transmitted through the Fresnel lens is partly absorbed in the transparent material which is appropriately doped with luminescent molecules. The dye molecules absorb strongly in certain UV and visible frequency bands of solar radiation and then emit photons in a lower frequency band. Snell's law dictates that about 75% of the emanating isotropic luminescent light is trapped as a result of total internal reflection and reaches the dielectric cusp to be further guided to the lower face of the cell. The principles and application of this type of concentrator to solar energy have been extensively investigated [5,6].

It has been projected, that if optimized silicon cells are combined with multi-dye luminescent concentrators, such a system may reach collector efficiencies ranging from 8.7% to 13.8%

Thermal Analysis

A concentration of 20X appears to be close to the practical limit of passive cooling methods for bifacial cells based on a thermal analysis. It is a convenient feature that both cell faces are identical in their electric grid patterning and the series connection is achieved simply by reversing the n^+ and p^+ faces alternately along the string. Top and bottom electrical busbars that run along the center line, serve also as a thermal heat sink. The heat removal from the cell-string to the outside of the module box is provided by a heat pipe, electrically insulated from the cell string. The circular fins attached to the condenser end are cooled passively by natural convection. The periodically distanced triangular cross bars underneath and in contact with the cell are thermally and electrically coupled to the central heat sink/busbar combination. One advantage of the heat pipe cooling is the equal temperature distribution on all cells along the string. A three dimensional finite element heat transfer program was used to find the regions of maximum temperature on the cells for normal operation and also extreme adverse conditions. The worst condition was simulated by assuming open circuitry i.e. full conversion of absorbed energy within the cell into heat, AM 1.5, no wind, 40°C ambient and 1° tracking error. The maximum temperature in the most adverse condition is about 84°C, which is 42°C higher than under normal conditions. This is acceptable as all materials utilized in the encapsulation of the bifacial cell can tolerate this temperature for several weeks without deterioration.

Bifacial Cell Design

The cell being developed for this 20X FCL concentrator is based on dendritic web silicon ribbon material approximately 100 microns thick. The bifacial cell uses the line contact principle where both junction and metal contact areas are restricted to small regions but the confined diffused regions are in the form of extended fingers.

An optimization study is being conducted to achieve a cell efficiency of 25% and an effective module efficiency of 23%. The cell features a texturized surface, light trapping, a base lifetime of about 100 microseconds and passivated surfaces.

Simulated Module Performance

To estimate the relative contribution of the diffuse radiation and albedo, a computer simulation program was developed which calculates the hourly energy output of the FCL concentrator for each radiation component separately. Pacific Gas and Electric's solar data from the Carrisa Plains, California were used. Figure 3 illustrates the simulated 1985 energy output of the FCL module.

The results show that for an average ground reflectance of 0.2, 78% of the total power output comes from direct beam irradiance, 16% from diffuse irradiance and 6% from albedo. The Carrisa Plains data represent a geographic location where the direct radiation is dominant throughout the year. In more temperate climates of the world, the FCL is expected to outperform other 20X concentrators which do not have the option of using diffuse solar radiation and albedo.

Economics

An extensive economic analysis of the FCL is now being undertaken. Final results are not available yet. However, some manufacturing cost features should be emphasized. The additional cost of a bifacial cell, cusp and luminescent concentrator is not large. The cost of manufacturing a bifacial cell varies marginally from a monofacial cell

[7,1]. Also, the FCL panel is used as a structural substrate for the cell assembly. So, the structural cost of a bifacial FCL concentrator module should not be remarkably different than a linear Fresnel lens concentrator utilizing a monofacial cell. Given the high overall efficiency of the FCL design, competitive module prices are anticipated.

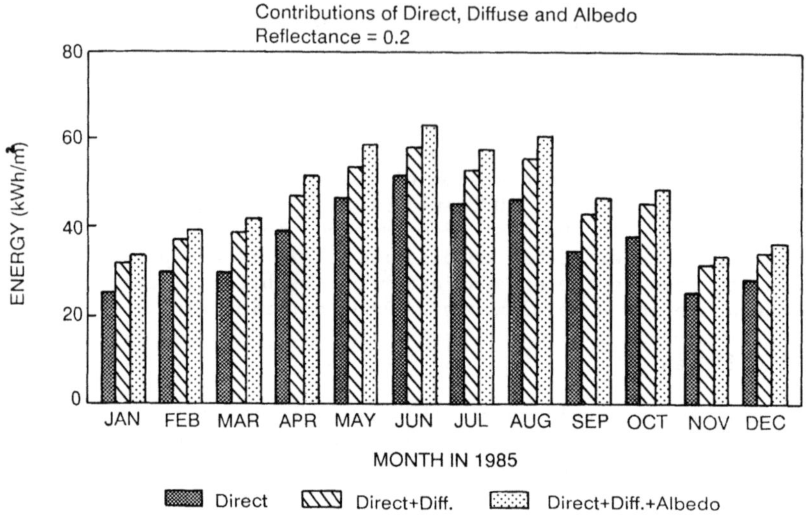

Figure 3. Simulated 1985 Energy Output of the FCL Module at Carrisa Plains, California

ACKNOWLEDGEMENT

This work is supported by the Pacific Gas and Electric Company.

REFERENCES

1. Luque, A., et al., 'Double-sided n-p-n Solar Cell for Bifacial Concentration', Solar Cells Journal, Vol.2, 1980
2. Luque, A., et al., 'New Concepts for Static Concentration of Direct and Diffuse Radiation', Proceedings from 3rd European Communities Conference on Photovoltaic Solar Energy, Cannes, France, 1980
3. Sala, G., et al., 'Albedo Collecting Photovoltaic Bifacial Panels', Proceedings from 5th Commission of the European Communities Conference on Photovoltaic Solar Energy, Athens, Greece, 1983
4. Luque, A., 'Theoretical Bases of Photovoltaic Concentrator for Extended Light Sources', Solar Cells Journal, Vol.3, 1981
5. Welford, W., et al. The Optics of Nonimaging Concentrators, Academic Press, 1978
6. Goetzberger, A., et al., 'Solar Energy Conversion with Fluorescent Collector', Journal of Applied Physics, Vol.14, 1977
7. Friedman, P., et al., Luminescent Solar Concentrator Development: Final Subcontractor Report 1, SERI/STR-211-3149, 1987

Sizing PV Concentrator and Flat Plate in Double Tracking Mode in the Presence of Energy Storage

N'To Diarra and P. P. Groumpos
Energy Research Center

Department of Electrical Engineering
Cleveland State University
Cleveland, Ohio 44115

ABSTRACT

Optimal sizing of a Photovoltaic (PV) power system, has been a challenge to engineers and system designers. The problem becomes ever more interesting and difficult when a comparison is to be made between PV concentrators and flat plates based on sizing. While sizing flat plate PV systems has attracted a lot of attention, this is not the case for PV concentrators. In this paper this problem is investigated. By proposing a suitable sizing methodology, a balancing parameter is used to size the PV concentrator and flat plate systems. The mathematical formulation takes into consideration, the site and the effect of cell temperature on sizing for both alternatives. Assumptions are made for load demand, solar insolation and environmental conditions, such that the two alternatives can be compared, in some meaningful way.

The proposed sizing methodology is used for a number of sites in the U.S.A. and around the world, and obtained results are presented. One important result is that cell temperature would affect actual system sizing.

INTRODUCTION

Tracking PV concentrator and flat plate in two dimensions (site and azimuth) results in increasing their overall efficiencies as the total solar energy collection is improved throughout the day. However, the efficiencies of such systems are affected by a number of parameters, the most important of which is the temperature of of the cells. Since the cells temperature depends on environmental parameters such as ambient temperature, insolation, and wind speed, any attempt at solving a sizing problem must take into account these parameters.

The design problem becomes even more challenging when an energy storage is desired to fulfill all requirements of a standard stand alone PV concentrator or flat plate system. In this paper we presented an improved version of the popular PV array quick sizing method.

1. PV Concentrator in Double Tracking Mode

Concentrating solar rays onto the cell produces two desirable effects:

- increasing the cell efficiency provided that a thermal control is performed on the system.

- decreasing the module cost.

Unfortunately, there are two major drawbacks:

- complexity of the system (tracking mechanism and cooling device) which would have an impact on the reliability of the systems

- the diffuse component of the insolation striking the concentrator represents an energy not available for conversion.

1.1 Sizing PV Concentrator Array

As a first order approximation, the quick sizing method which has been widely applied to the sizing of fixed tilted PV systems, is used [2]. According to that method, the area ADC required to meet daily demand DL(i) is the maximum of:

$$ADC(i) = DL(i) / \eta_{GC(i)} * [Scm(i) - Mc(i) * SScm(i)] \qquad (1)$$

where:

- $\eta_{Gc(i)}$ is the average efficiency of the concentrator for the ith month of the year

- Scm(i) in wh/m2.day is the average daily energy for the ith month of the year

- SScm in wh/m2.day, the standard deviation of the energy incident on the concentrator for the ith month.

- Mc(i) is the balancing parameter defined as

$$Mc(i) = \frac{Scm(i) - SDL(i)}{SScm(i)} \qquad (2)$$

and SDL(i) is the average daily incident energy which would meet the daily load DL(i) exactly. From equations (1) and (2) it appears clear that the new balancing parameter depends mainly on the month of the year and the concentration ratio as Scm(i) and SScm(i) is concentration ratio dependent.

Given DL(i), the sizing of the concentrator array is reduced to the determination of Scm(i), Mc(i), SScm(i) and most of all $\eta_{Gc(i)}$. Scm(i) and SScm(i) are easily computed from Collares Pereira and Rabl [1]. $\eta_{Gc(i)}$ for a given cell depends on the concentration and the environment. For simplification, we'll assume for the remainder of this paper, that $\eta_{Gc(i)} = \eta_{cD(i)}$ where $\eta_{cD(i)}$ is the efficiency on the recommended design days of the ith month under the worst environment conditions, i.e., at solar noon.

1.2 Efficiency of Concentrator Cell at Solar Noon in the Design Days

As stated earlier, $\eta_{cD(i)}$ depends on the temperature of the cell, the insolation and the concentration ratio. It can be shown that

$$\eta_{cD}(i) = \eta_c(Tc(i), Ic(i), c) =$$

$$\frac{\eta_c(Tc) * Vcm(Tc(i), Ic(i)) * Icm(Tc(i), Ic(i), c \times 1000}{Vcm(To) * Icm(To) * C * Ic(i)} \qquad (3)$$

where:

- $\eta_c(To)$ is the efficiency at one sun under NOCT.
- Vcm(To) is the voltage at maximum power under NOCT.
- Icm(To) is the current at maximum power under NOCT.
- Vcm(Tc(i), Ic(i), c) is the voltage at maximum power point under Tc(i), Ic(i) and c at solar noon on the design days of the ith month.
- Icm(Tc(i), Ic(i), c) is the current at maximum power under Tc(i), Ic(i) and c at solar noon on the design days of the ith month.
- Tc(i) is the temperature of the cell at solar noon on the design days of the ith month.
- Ic(i) is the direct insolation in w/m^2 at noon on the design days of the ith month.

Ic(i) can be computed from one of the standard insolation calculations. In a recent report [3], we have proposed a model of cell temperature which takes into account both the cell parameters and the environmental parameter and Tc(i) can be written as

$$Tc(i) = TAIR(i) + K_c{}^*A_c{}^*(1-\eta_c(To) * Ic(i) * \exp\left[-\ln(\ c-1) * \exp(-0.066 * w_s(i)\right] \qquad (4)$$

where:

- A_c is the area of the lens in m^2
- K_c is a constant which depends on the cell design
- $w_s(i)$ is the wind speed in m/s at solar noon on the design days of the ith month.

2. Sizing PV Flat Plate

Equations (1) and (2) are applicable with the subscript c changed to FP for flat plate. It is assumed for comparisons purpose, that DL(i) remain the same.

The model of cell temperature can be written as

$$T_{FP(i)} = TAIR(i) + K_{FP} * A_{FP} * \left(1-\eta_{FP}(To)\right) * I_{FP(i)} * \exp\left[-0.066w_s(i)\right] \tag{5}$$

where the different terms have the same definitions as for concentrator. It is worth noting that:

- $K_{FP} > K_c$ since no cooling device is added to flat plate.
- $I_{FP} > I_c$ as I_{FP} includes a net diffuse component
- $\eta_{FP}(To) < \eta_c(To)$.

Equation (3) is still applicable, by changing the subscript c into FP and setting c = 1.

3. Simulation Study

A computer program was developed at CSU to generate the area ADC of hypothetic concentrator and the area ADFP of flat plate required to meet a constant daily demand DL of 100 Kwh/day at Phoenix and Albuquerque in the USA, Tessalit and El Oued in the Sahara Desert. The results are shown in Tables (1) and (2) where we have determined the design month, the balancing parameters in each case.

Table i

Location / Parameters	Phoenix Latitude: 33.43	Albuquerque Latitude: 35.05
A_{DC}	76.59m^2	72m^2
D_{MC}	January	December
M_C	0.40	0.40
A_{FP}	127.32m^2	160.80m^2
D_{MFP}	December	January
M_{FP}	0.50	0.50

DMC = Design Month for Concentrator for c = 200

Table ii

Location / Parameters	Tessalit Latitude: 20.20	El Oued Latitude: 33.37
A_{DC}	$61.58m^2$	$63.54m^2$
D_{MC}	June	March
M_C	0.50	0.33
A_{FP}	$88.54m^2$	$130.58m^2$
D_{MFP}	December	December
M_{FP}	0.50	0.50

DMFP = Design Month for Flat Plate

4. Discussions and Concluding Remarks

In this paper we have presented an improved methodology to address the sizing problem of PV concentrator and flat plate operating in double tracking mode in the presence of energy storage. The balancing parameters are defined in such a way the effects of the environment are considered. One important result in this study is the fact that the design months for flat plate and concentrator are not necessarily the same, when they are designed to meet the same load demands.

Another important result is that the design month is not necessarily the one during which the lowest ratio insolation to load demand occurs. Rather, the whole thing is a combination of variable terrestrial environments. To that end, we have presented two models of cell temperature for flat plate and concentrator it remains that these models need to be checked for different cells and environment. Quantitatively, our results are obviously subject to the degree of accuracy of the proposed model. However, the major conclusions drawn are believed to be qualitatively valid.

References

1. M. Collares, Pereira and A. Rabl, "Simple Procedure for Predicting Long Term Average Performance of Noncentrating and Concentrating Solar Collectors," Argonne Laboratory No. ANL, 78-67 (1978).
2. H. L. Macomber, John B. Ruzek and Frederick A. Castello, "Photovoltaic Stand Alone Systems - Preliminary Engineering Design Handbook".
3. Peter P. Groumpos and N'To Diarra, "Technical and Economical Considerations of PV Concentrators", Technical Report ERC-R-14, Energy Research Center, Department of Electrical Engineering, Cleveland State University.

A HIGH EFFICIENCY FLUORESCENT PLANAR CONCENTRATOR STACK (FPC STACK) COMBINED WITH III-V SOLAR CELLS.

G. Kamps and K. Heidler

Fraunhofer-Institut für Solare Energiesysteme
Oltmannsstr. 22, D-7800 Freiburg, FRG

ABSTRACT. A high electrical efficiency FPC stack/III-V solar cell system with an active area of 400 cm^2 has been planned, built and characterized under outdoor conditions. The optimum configuration with the best available FPC's and GaAs solar cells was calculated using a previously described Monte-Carlo computer simulation code /1/, after measuring the spectral responses of the solar cells with an extended and improved grating monochromator set-up. The best stacks from the calculations have been built and characterized under outdoor conditions, and an overall electrical efficiency of 4.1 % at an effective concentration of 3.1 has been reached. This is the highest overall electrical efficiency reported for a FPC stack of this geometry.
The GaAs solar cells used in this paper were supplied by CISE, Milano.

1. INTRODUCTION

FPC's combine the effect of a low level concentrator and a spectrum splitter /2/. Their unique feature is the ability to concentrate diffuse light, thus eliminating the need for tracking and making low light applications highly attractive /3/. FPC's reduce the solar cell area needed compared to directly exposed generators, but as their overall efficiency is rather poor they need large areas.

One approach to increase the electrical efficiency is the FPC stack. In a recent publication /1/ we reported on the maximum overall electrical efficiency of a FPC stack achievable with the best III-V solar cells reported in the literature. For this purpose we optimized the combination, concentration and sequence of the organic dyes in a FPC stack of up to 3 single sheets with a Monte-Carlo simulation code.

In this work, we report on experimental results obtained with actually available FPC's and III-V solar cells in an optimized stack:

- In a first step we extended and improved a grating monochromator set-up to measure the spectral response of the solar cells.
- In a second step we calculated the optimum configuration with the best FPC's in the former simulations and the best solar cells using the simulation code mentioned above.
- In a third step we built and characterized the best stacks.

2. MATERIALS

2.1. Fluorescent concentrators

For the optimized stacks we used the dyes BA241 and BA339, which gave the best results in the earlier calculation /1/. Because we wanted to test the weak dependence of the efficiency on the dye concentrations, which had been predicted by the calculations, we investigated samples with different concentrations:

dyes	Maximum optical densities/sheet thicknesses		
BA241	2.49/3.22 mm	4.84/3.13 mm	7.08/3.05 mm
BA339	2.27/3.05 mm	3.48/3.04 mm	

Table 1 Parameters of the tested concentrators.

2.2. Solar cells

From CISE in Milano we obtained 10 GaAs solar cells mounted for use with FPC's. Their form and mounting are shown in the figs. 1a&b.

3. SPECTRAL MEASUREMENTS FOR SOLAR CELL CHARACTERIZATION

Because of the significant deviation of the FPC emission spectra from the solar spectrum, the spectral responses of the individual solar cells are important parameters in the calculation and optimization of the overall electrical efficiency of the stack and have to be measured.

To this purpose, an existing grating monochromator set-up (M300, Bentham) was extended and improved as given in fig. 2. It now has the following specifications:

- Solar cells up to 2x2 cm^2 can be measured.
- Spectral responses can be measured over a spectral range of 300-1200 nm in 5 nm steps, while test cell and reference cell are sequentially illuminated at each wavelength using a rotating mirror.
- The former optical system using lenses has been replaced by an achromatic imaging optical system using mirrors with more than 80% reflectance at $\lambda \geq 230$ nm.
- For the illumination, a tungsten halogen lamp is used; the additional use of a xenon arc lamp in the short wavelength range is in preparation.

For the absolute calibration of the relative spectral responses, a solar simulator has been used /4/.

4. CALCULATION OF THE BEST STACK

The figs. 3a&b present results of the spectral solar cell measurements. Using the solar cell data as well as the absorption and emission spectra of the mentioned FPC's, we calculated the overall electrical efficencies of the most promising stacks, i.e. the most promising combinations and permutations combined with the best solar cells:

	Upper position		lower position			
stack	concentrator	solar cell	concentrator	solar cell	η_{calc}[%]	$c_{opt,P}$
1	BA241/4.84	CISE No.9	BA339/2.27	CISE No.9	5.13	2.40
2	BA241/7.08	CISE No.8	BA339/2.27	CISE No.8	4.84	2.41
3	BA241/2.49	CISE No.8	BA339/2.27	CISE No.8	4.77	2.32

Table 2 Calculated overall electrical efficiencies η_{calc}[%] and optical power concentrations $c_{opt,P}$ (ratio of the output energy flux density to the input energy flux density) of the best 3 stacks. (As the best cell, No. 9, broke after the mea-

surement of the best stack, the other two were calculated and measured with cell No. 8.)

5. OUTDOOR CHARACTERIZATION

For outdoor measurements of the AM1.5 efficiency of the stacks, a sun-tracking measurement platform has been used. The individual collectors of the stacks were measured sequentially, starting by recording their I-V characteristic, then determining U_{MPP} and I_{MPP} with a MPP tracker, simultaneously measuring the solar radiation flux with a pyranometer. The diffuse radiation was also measured to yield the relationship of direct and diffuse light. The cell temperature was kept between 23° and 25°C, and the measurements were taken at a time with an AM-value near 1.5. The measurement set-up can be seen in the Figs. 4a&b.

6. RESULTS

The results of the measurements are given in table 3. The stacks are the same as calculated above.

stack	position	AM	η_{FPC}[%]	η[%]	c_{eff}	$c_{opt,P}$	η_{meas}/η_{calc}
1	upper	1.4	2.59	4.09	3.07	1.63	0.797
	lower	1.3	1.50				
2	upper	1.8-1.9	2.46	3.92	3.0	1.59	0.810
	lower	1.7-1.8	1.46				
3	upper	1.5	2.19	3.92	2.93	1.54	0.822
	lower	1.4-1.5	1.74				

Table 3
Results of the outdoor measurements. The effective concentration c_{eff} is the ratio of the I_{SC} of the solar cell when mounted on the concentrator's edge to the I_{SC} of the solar cell during exposition to direct sunlight.

7. DISCUSSION

The lower η-values compared to the calculations can be explained by the following restrictions of the programme compared to the measurements:

- The programme assumed square concentrators instead of the measured triangles with diffuse side reflectors.
- The losses by light scattering and absorption of the PMMA matrix have been neglected.
- The assumed total reflection in the sheets was not ideal.
- The programme assumed ideal optical contact of the cell with the collector's edge.
- Optical inhomogeneities in the concentrator BA241/4.84 can account for the larger deviation of stack 1 compared to the other stacks.

The measurement errors, in particular those resulting from deviation in the spectral distribution, are estimated to be appreciably lower than the effects mentioned above.

8. CONCLUSIONS

A FPC stack/III-V solar cell system with an electrical efficiency of 4.1 % at an effective concentration of 3.1 has been built and characterized. This is the highest overall electrical efficiency reported for a FPC stack of this geometry.

The comparison of simulation and experiment showed good agreement, indicating that the Monte-Carlo simulation code is a powerful tool to describe and optimize FPC systems. The deviations, which, as expected, have nearly the same value for all tested cases, can be explained by differences between the programme assumptions and the experimental set-up.

As predicted by the calculations, the overall electrical efficiency showed only a weak dependence on the dye concentration.

Further improvement can be expected, if solar cells with higher efficiencies are used. Since the GaAs cells we used were not pre-selected for high efficiencies, a further improvement of 10% can easily be obtained. Furthermore, by using AlGaAs homojunction solar cells, a better matching between emission spectrum at the concentrator's edge and solar cell bandgap can lead to improvements of about 20-30% compared to our results.

9. ACKNOWLEDGEMENTS

We gratefully acknowledge that **Dr. Carlo Flores** from **CISE**, Milano, made available the GaAs solar cells which are used and described in this work. Furthermore, we want to thank **J. Beier** and **A. Zastrow** for valuable discussions and **F. Brucker**, **J. Fusek**, **S. Kunzelmann**, **A. Schönecker**, **M. Steinert** and **H. R. Wilson** for technical assistance.

This work was supported by the Commission of the European Community under contract No. EN3S/C2/119/D.

10. REFERENCES

/1/ Kamps, G. and Heidler, K. (1987). 'Electrical efficiencies of fluorescent planar concentrator stacks (FPC stacks) combined with GaAlAs solar cells in computer simulation'. Proc. of the ISES Solar World Congress, Hamburg 1987.

/2/ Goetzberger, A., and Greubel, W. (1977). 'Solar energy conversion with fluorescent collectors'. Applied Physics, **14**, 123-139.

/3/ Heidler, K. and Kunzelmann, S. (1986). 'Solar wall clock with a fluorescent planar concentrator (FPC)'. Proc. of the 7th EC PV Sol. En. Conf., Sevilla 1986, 201-206.

/4/ Heidler, K. and Beier, J. (1988). 'Uncertainty analysis of PV efficiency measurements with a solar simulator'. Proc. of the 8th EC PV Sol. En. Conf., Florence 1988.

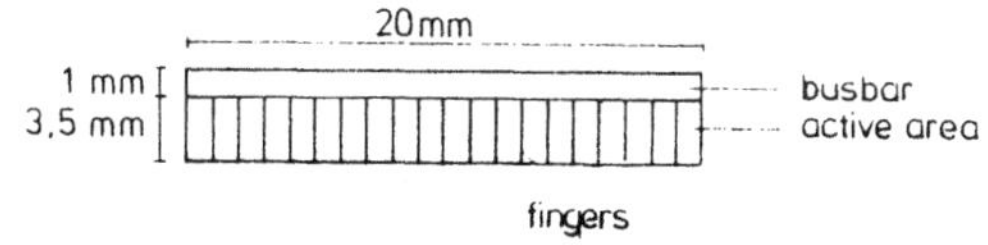

Fig. 1a Top view of a solar cell from CISE

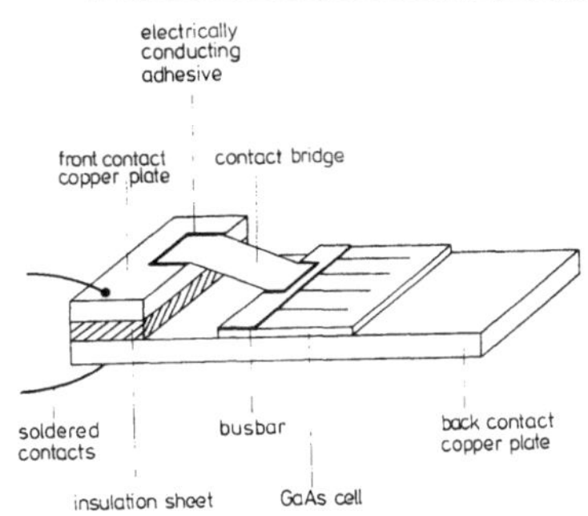

Fig. 1b Method of mounting the cell onto the FPC's edge

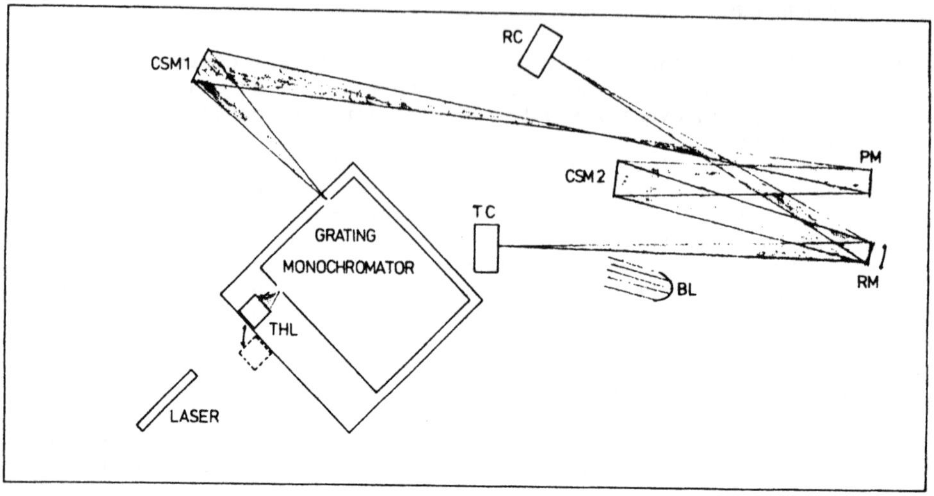

Fig. 2 Set-up for the spectral measurements
The abbreviations have the following meanings:

BL	- bias lamp	RM	- rotating mirror
CSM	- concave spherical mirror	TC	- test cell
PM	- plane mirror	THL	- tungsten halogen lamp
RC	- reference cell		

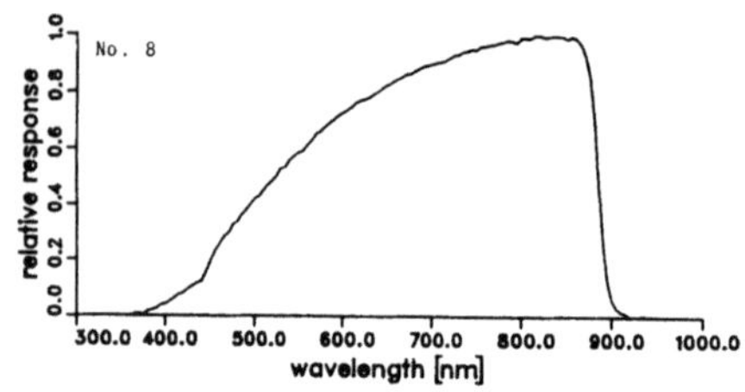

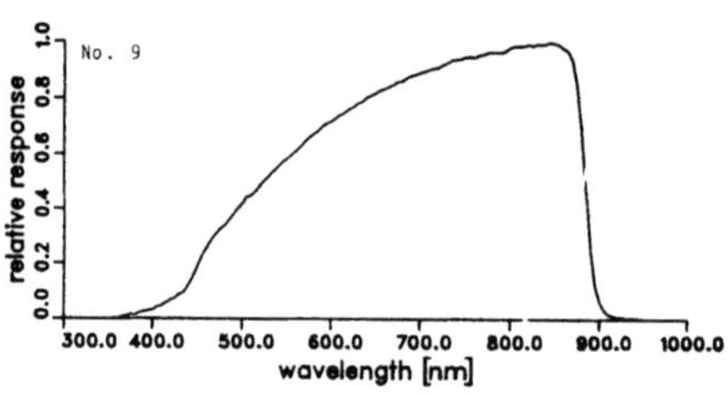

Figs. 3a&b Relative spectral responses of the CISE GaAs cells No. 8 and 9

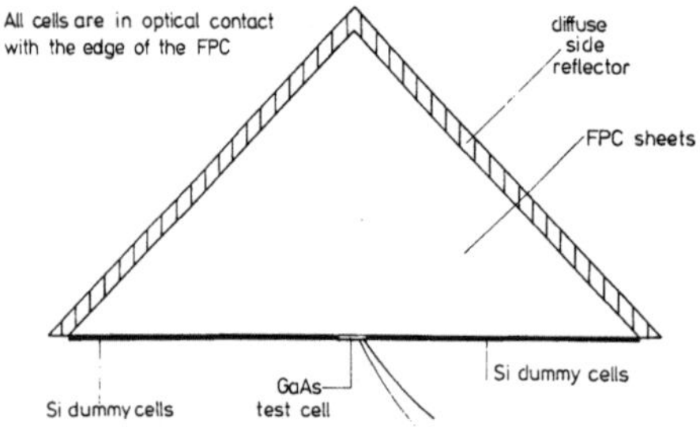

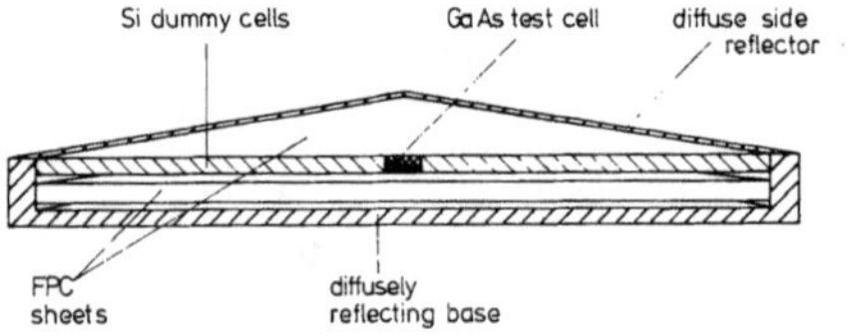

Fig. 4a Top view of the FPC stack measurement set-up

Fig. 4b Cross-section of the same device

DEVELOPMENT OF A THREE SOLAR CELL MECHANICAL STACK FOR TERRESTRIAL CONCENTRATOR SYSTEMS

Allen M. Barnett
Electrical Engineering Dept.
University of Delaware
Newark, Delaware 19716 USA

and

Sandra L. Rhoads and Gerald H. Negley
AstroPower Division/Astrosystems, Inc.
30 Lovett Avenue
Newark, Delaware 19711 USA

ABSTRACT. A mechanically stacked multijunction solar cell system comprised of a self-supporting AlGaAs top solar cell over a GaAs homojunction with a GaInAsP on InP bottom solar cell, offers an AM1.5 Direct theoretical conversion efficiency of 45.4% (38.3% with losses). The proposed structure have bandgaps of 1.93 eV, 1.43 eV, and 0.99 eV respectively. Under a concentrator configuration, the maximum AM1.5D theoretical efficiency of this stack can be further increased to 42.5% with 100X and to 45.4% with 1000X concentrating optics.

A multijunction stack requires a high performance top solar cell, a state-of-the-art middle solar cell, and a good bottom solar cell with a low bandgap. Component materials and bandgaps of the proposed design were determined through consideration of optimum efficiency predictions and technological capabilities. In this work, the demonstrated performance and expected limits of well-developed solar cells are used as the basis for predicting a practical system efficiency.

AlGaAs will be used for the top solar cell material. AstroPower has developed a technique to fabricate AlGaAs solar cells on a 100 micron thick, self-supporting, transparent AlGaAs substrate. The middle solar cell will be GaAs, and the bottom solar cell is GaInAsP lattice-matched to InP. LPE is being used to fabricate the top and bottom solar cells in our laboratory.

1. INTRODUCTION

Multijunction solar cells, mechanically-stacked or monolithic, present a major improvement in power density. One tandem approach uses the top solar cell to boost the performance of the lower bandgap solar cell. An alternative approach is based upon scavenging low energy photons from a good, relatively wide bandgap top solar cell [1]. Combining both tandem solar cell approaches leads to a three solar cell stack with efficiencies of over 40% at AM1.5D and one sun insolation.

A six-terminal wiring configuration is being proposed for the three solar cell stack (TSCS) to obtain the highest system efficiency. A two terminal wiring configuration requires the stacked solar cells to be current matched.

The key to high efficiency mechanically-stacked multijunction solar cells is the top solar cell. This top solar cell must: i) be highly efficient since one-half of the stack efficiency is the result of the top solar cell, and ii) be highly transparent to sub-bandgap photons. In this paper, the design rules for the multijunction will be presented, and the progress in the development of the top and bottom solar cell will be reviewed.

2. DESIGN

Bandgap choices of 1.93eV, 1.43eV and 0.99eV are proposed for the TSCS wired in a six-terminal configuration. Our model predicts an efficiency limit of 45.4% AM1.5D, 1X. At a concentration ratio of 100X, this bandgap combination yields a limit of 50.5%. The 100X values are obtained by assuming a 100x increase in short-circuit current values and scaling other values in accordance with the diode equation.

The model [2,3] generates maximum solar cell efficiencies for specific bandgaps. Table I illustrates the theoretical maximum efficiency predicted for the TSCS [2,3].

TABLE I
Predicted Theoretical Maximum Efficiency
for the Six-Terminal Three Solar Cell Stack
AM1.5D (75.22 mW/cm^2)

6-TERMINAL CONFIGURATION

Bandgap (eV)	Voc (volts)	Jsc (mA/cm^2)	FF	Eff[1X] (%)	Eff[100X] (%)	Eff[1000X] (%)
1.93	1.56	11.2	0.91	21.1	22.7	23.5
1.43	1.06	11.5	0.88	14.3	15.9	16.6
0.99	0.63	14.6	0.82	10.0	11.9	12.9
	Stack Efficiency =			45.4	50.5	53.0

The proposed bandgap combination for the six-terminal wiring configuration corresponds to an $Al_{.37}Ga_{.63}As$ top solar cell, a GaAs middle solar cell, and an GaInAsP bottom solar cell. AstroPower has developed a technique to fabricate AlGaAs solar cells on a transparent AlGaAs substrate. This approach utilizes the most developed wide bandgap material on a transparent substrate. The middle solar cell in the TSCS is GaAs. GaAs solar cells are approaching their practical efficiency limit; hence, they represent a mature technology in the TSCS.

GaInAsP is the material of choice for the bottom solar cell. This quaternary material offers lattice-matched alloys to InP with a tunable bandgap range of 0.75eV to 1.35eV. This material has been successfully developed for device applications in the fiber-optic and semiconductor-laser

FF=.75. In addition, we have demonstrated transparency greater than 90%.

The AlGaAs devices show good current collection. An external quantum efficiency curve is shown in Figure 2. The external quantum efficiency measurement indicates good blue response and good band-edge response. The good blue response indicates that the window layer is reducing surface recombination. The good band-edge response indicates that the diffusion length is not limiting device performance.

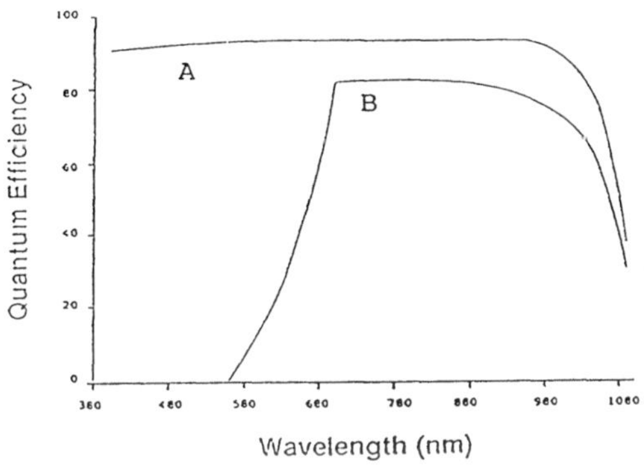

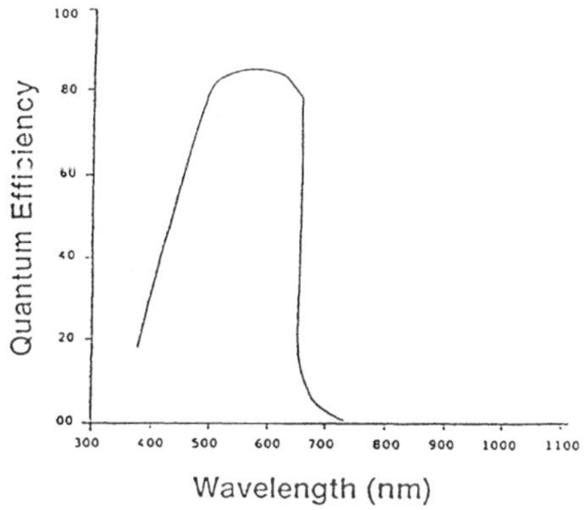

Figure 1

A) Q.E. of Silicon Solar Cell without AlGaAs filter.

B) Q.E. of Silicon Solar Cell with AlGaAs filter.

Figure 2

Representative External Q.E. Plot of an AlGaAs Top Solar Cell.

4.2 GaInAsP Bottom Solar Cell Development

GaInAsP is the material system of choice for the bottom solar cell. GaInAsP material technology has been successfully developed for device applications in the fiber-optic and semiconductor-laser field. Figure 3 is an EDS quantitative profile showing layer composition and thickness.

Homojunction 0.92eV GaInAsP solar cells have been fabricated in our laboratory. Undoped 0.92eV GaInAsP material was diffused with zinc; device areas were chemically isolated; and Au alloy contacts were applied. Open-circuit voltage values up to 390mV at AM1.5D were recorded. These results are particularly encouraging considering that open-circuit voltage values are a good indicator of the performance potential of solar cell materials, like silicon and GaAs [1].

Figure 4 shows a quantum efficiency versus wavelength plot for the 0.92eV bottom solar cell covered by an InP filter.

The sub-bandgap transparency of the filter is encouraging as is the response of the GaInAsP in the "stacked" situation.

fields. This wide technology base is directly applicable to material development for bottom solar cell designs.

3. PRACTICAL SYSTEM PERFORMANCE

The potential performance of a photovoltaic material can be predicted through the comparison of the modelled theoretical maximum performance versus the achieved performance of well-developed solar cells. A survey of the literature indicates that, in general, well-developed III-V compound solar cells achieve 96%, 91% and 96% of their theoretical limits for open-circuit voltage, short-circuit current and fill factor, respectively [1]. Using these "scale factors", one may predict the "best case" performance of a photovoltaic material including optical and electrical losses.

Assuming 100% transmission of the photons less energetic than the bandgap, the performance of each individual solar cell can be predicted. By reducing the current in the middle (bottom) solar cell by the current generated in the top (top two) solar cells, an evaluation of the stack can be done. This approach is valid since the model assumes unity quantum efficiency. Using the "scale factors", and reducing the current to simulate the TSCS, the best case prediction is given in Table II.

TABLE II
Best Case Prediction at
For The Three Solar Cell Stack
AM1.5 Direct (75.22 mW/cm^2)

6-TERMINAL CONFIGURATION

Material	Voc (volts)	Jsc (mA/cm^2)	FF	Eff[1X] (%)	Eff[100X] (%)	Eff[1000X] (%)
AlGaAs	1.50	10.2	0.88	17.9	19.2	20.7
GaAs	1.03	10.5	0.85	12.0	13.3	13.9
GaInAsP	0.60	13.3	0.79	8.4	10.0	10.8
Stack Efficiency =				38.3	42.5	45.0

4. DEVICE FABRICATION

The AlGaAs top solar cell and the GaInAsP bottom solar cell are being grown by liquid phase epitaxy (LPE).

4.1 AlGaAs Top Solar Cell Development

The rugged, self-supporting, transparent AlGaAs top solar cell can be mechanically stacked on any well developed, existing solar cell. Figure 1 shows quantum efficiency curves for a silicon solar cell with and without an AlGaAs filter. This AlGaAs filter was transparent to 91% of the photons less energetic than the bandgap of the active device layer.

AstroPower's preliminary work on this system has yielded a 10.2% (AM1.5D, 1X) AlGaAs top solar cell. The detailed characteristics were Voc=1.268 volts, Jsc=8.1 mA/cm^2, and

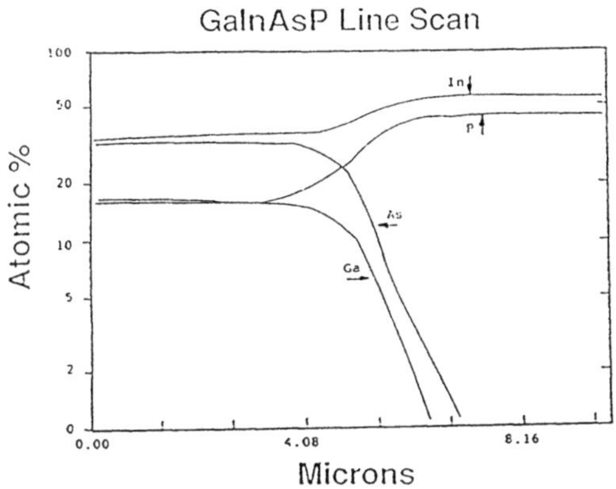

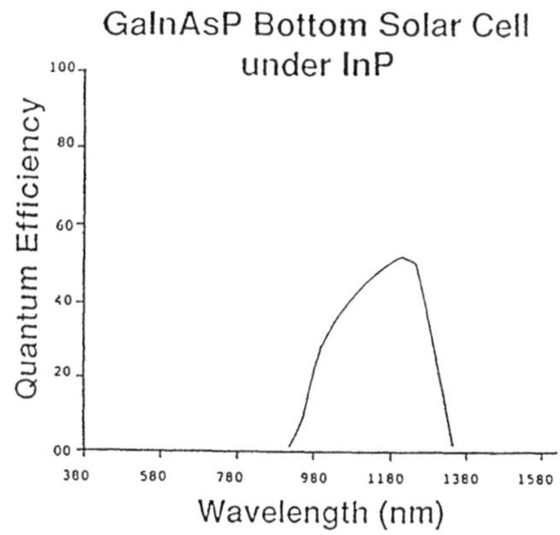

Figure 3
EDS quantitative profile of 0.92eV GaInAsP composition.

Figure 4
Quantum efficiency of 0.92eV GaInAsP solar cell under an InP filter.

5. CONCLUSION

Solar cell efficiencies greater than 30% AM1.5 Direct are realizable in the near future with a three solar cell stack (TSCS). With expected losses, a practical system efficiency of 38.3% is anticipated. These calculations are based on the 1.93eV/1.43eV/0.99eV energy bandgap combination. AlGaAs/GaAs/GaInAsP materials can be used with a six-terminal wiring configuration. Under 1000X concentrating optics, the model predicts a practical efficiency limit of 45.4%.

The self-supporting AlGaAs structure eliminates the low yield problem that others encountered when trying to remove the fragile AlGaAs from the GaAs substrate. The key to high efficiency triple stacks is in the top and middle solar cell. Both must be approaching their efficiency limit and must be highly transparent to photons less energetic than their bandgaps.

Continued progress for the AlGaAs top solar cell and the GaInAsP bottom solar cell will result in practical system efficiencies greater than 30% AM1.5D.

6. REFERENCES

[1] A.M. Barnett and J.S. Culik, "New Solar Cell Design Options", 19th IEEE Photovoltaic Specialists Conf., New Orleans, p. 931, 1987.
[2] M.E. Nell and A.M. Barnett, "The Spectral p-n Junction Model for Tandem Solar Cell Design", IEEE Trans. Elec. Dev., ED-34, 257-266, (1987).
[3] N.E. Terranova and A.M. Barnett, (Voltage Correction to Nell's Model), to be published.

Application Specific Integrated Circuits for Photovoltaic Applications

S. McCarthy and G. T. Wrixon
National Microelectronics Research Centre,
University College, Lee Maltings,
Prospect Row, Cork, Ireland.

Abstract

Application specific integrated circuits can dramatically reduce the number of chips required to implement logic functions on printed circuit boards. Reductions of chip counts of more than 70 to 1 can be achieved. This reduces the complexity of the printed circuit board, manufacturing costs, stock costs, printed circuit board size and often results in improved performance. For PV applications a reduction in chip count can result in a reduction in power demand. In particular, this could apply to low powered PV consumer products, or in the case of PV systems, in a reduction in the no-load demand of power conditioning equipment. This paper examines possible applications of ASICs to PV and it describes an ASIC designed to control the battery charging in a PV system.

1. Introduction

Most printed circuit boards currently being manufactured are collections of standard parts assembled on a custom printed circuit board. Typically, there is a high percentage of the board devoted to very small simple chips performing low level functions. An Application Specific Integrated Circuit (ASIC) is a chip that is designed to perform a specific function and it can dramatically reduce the number of chips which are required to implement the same function. Reductions in the chip count from more than 70 to 1 can be achieved. This reduces the complexity of the printed circuit board, manufacturing costs, stock costs, printed circuit board size and often results in improved circuit performance. A reduction in chip count also leads to a reduction in the power consumption of the circuit.

In photovoltaic applications a reduction in the power consumption of the load is an important design criteria. In small PV applications, such as consumer items, a small reduction in power (mW) will result in a reduction in the required solar cell size and cost. In PV systems a considerable amount of power is required to operate the electronic equipment in the system components. The most common areas are, power supplies, computer controllers, charge regulators, inverters, monitoring systems, and electronic protection systems. In the Fotavoltaic system (1), a 50kWp PV system supplying power to dairy farm on Fota Island in Cork, Ireland, the no-load demand of the system (mainly electronic equipment) is 350W continous (8.4kWh/day). A reduction in this no-load could considerably improve the overall system performance, especially during periods of low irradiation.

In the case of inverters, most inverters used in PV systems have rated efficiences greater than 90% for loads exceeding 10% rated load. However, the energy efficiency of inverters operating in PV systems rarely exceeds 70%. This is mainly due to the fact that when no load exists the PV system continually supplies the electronic boards within the inverters.

2. ASIC for Battery Control in PV Systems

An ASIC was designed and developed at the National Microelectronics Research Centre to control the battery management of a photovoltaic system. The logic used in the chip was based on the control algorithms used by the microcomputer controller in the Fotavoltaic project.

The base gate array for the ASIC was designed at the NMRC (2). It uses a 740 2-input NAND-GATE equivalent array fabricated using a 5 micron CMOS process. The chip layout was done on a Daisy system using a complete schematic capture, simulation and autorouting package. It was manufactured and tested at the NMRC.

3. Control Logic Used in ASIC

The layout of the PV array and battery is shown in figure 1. The PV array consists of 20 subarrays which are switched on and off to regulate the charge on the battery. The charging characteristic of the battery is shown in figure 2. Subarrays are connected to the battery and remain connected to it until the gassing voltage is reached (2.3V/cell). When the voltage is reached a subarray is disconnected and the current and voltage decreases. When the voltage subsequently reaches gassing voltage further subarrays are disconnected. This procedure continues until the battery is fully carged i.e. the battery is at gassing voltage and only one subarray is connected.

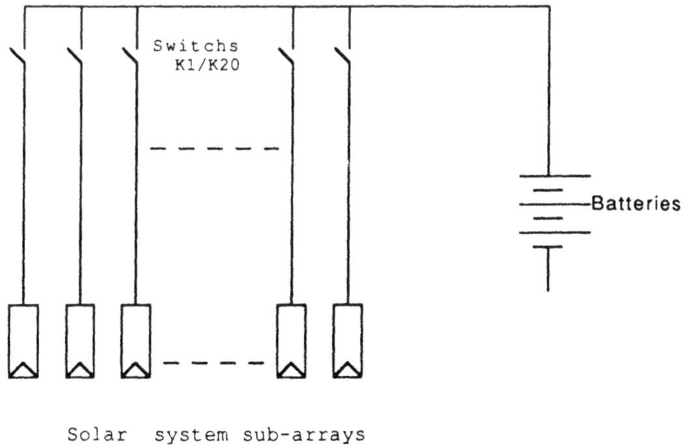

Figure 1. PV array/battery interface of PV system.

During charging the battery voltage will be maintained between two voltage levels V_L (2.16V/cell) -> V_H (2.31V/cell) and the current will be maintained between I_L ($I_{6.6}$) -> $I_H(4)$ i.e. for a 300Ah battery I_L = 45A and I_H = 60A.

For battery safety the PV array is disconnected when the battery voltage and current exceeds predefined limits.

Logic is also required to connect and disconnect the PV array for day/night conditions and for manual shut-down.

The input signal to the ASIC are:

1. Day/night signal
2. Manual start up request
3. Battery voltage < Vmax (2.4V/cell)
4. Battery current < Imax (I_4)
5. Battery voltage < V_H (2.31V/cell)
6. Battery voltage > V_L (2.16V/cell)
7. Battery current < I_H (I_4)
8. Battery current > I_L ($I_{6.6}$)

The outputs of the ASIC are signals to operate the subarray switches K1-K20.

Figure 2 is the control logic to operate one of the switches (K2). The input logic to the gates VOLTAGE IN RANGE is the logic $V_L < V < V_H$ and the CURRENT IN RANGE is $I_L < I < I_H$. VOLTAGE/CURRENT EXCESS is V > Vmax OR I > Imax.

The complete control is broken down into simple logic statements. For twenty switches this logic is duplicated 20 times.

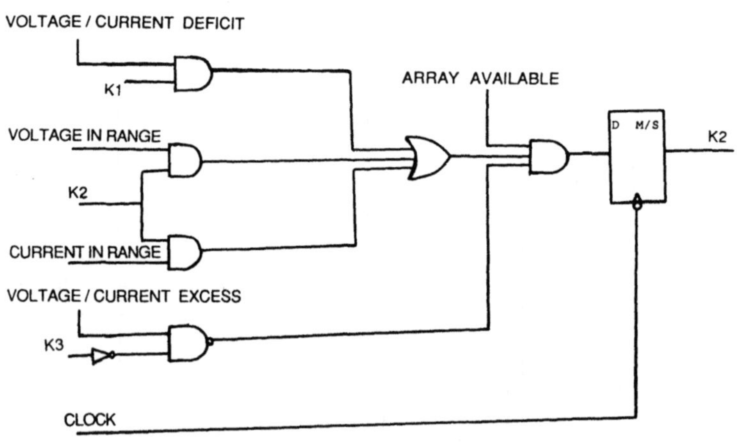

SOLUTION REQUIRES 148 GATES FOR TWENTY SWITCHS

Figure 2. Logic required to control one switch (K2).

The entire design has 8 inputs, 20 outputs and uses 148 2-input NAND-Gates. 130 dies were manufactured at the NMRC. The ASIC is currently being redeveloped using a new base gate array. When successful circuits are available they will be packaged and tested in experimental circuits.

Conclusion

Future applications of PV will depend on the reduction in the cost of solar cells, a reduction in the cost of balance-of-systems components and also in the development of low powered devices for use in PV systems. A reduction in the overall load demand may be achieved using ASICs. This paper has described how ASICs can reduce the power demand of loads and electronic circuit boards. It also described how ASIC can be used to control the battery management in PV systems and consequently reduce the power demand of the controller.

References

(1) Wrixon G. T., The Fotavoltaic Project - A 50kWp photovoltaic system to power a dairy farm on Fota Island, Cork, Ireland. Photovoltaic Power Generation, series C, Vol. 4, Reidel, Dordrecht, (153-169) 1983.

(2) Lidholm S. and A. Coffey, Manual, NMRC gate array micro cell library. National Microelectronics Research Centre, University College, Cork, Ireland.

COMPUTER SIMULATION AND OPTIMISATION OF PV SYSTEMS

S. McCarthy, G. T. Wrixon, L. Keating

National Microelectronics Research Centre,
University College, Lee Maltings,
Prospect Row, Cork,
Eire.

Abstract

The purpose of this project is to develop a computer program which can be used to simulate and optimise the performance of Photovoltaic (PV) plants. The proposed approach to the simulation represents a new departure in PV plant modelling. This approach incorporates the following advanced features:

1. A graphics interface will be available to the user for the purpose of describing the plant configuration to the computer. This will allow the user to draw the plant configuration on the computer screen, by making use of a built-in library of standard components.

2. The simulation process shall be based on the classical event queue system. A power management module shall decide on the events which should occur, and place them on the event queue. Events are then removed from the event queue, and processed by an event processor module. The processing of an event may change the status of the system, so that the power management module has to schedule further events, and so the simulation proceeds.

Adoption of this advanced approach has been made possible due to the experience already accumulated in the development of standard PV system modelling software. By making use of this experience, we are free to focus our attention on the manner in which the component models interact, rather than on the component models themselves. This will result in a more flexible program, which will allow the user to simulate the plant operation with greater precision than was ever previously possible.

Introduction

In 1987, the Directorate General DGXII of the CEC initiated a group of seven concerted action projects. The main aim of the projects was to upgrade the level of photovoltaic technology in Europe based on the experiences gained through the PV pilot plant projects commissioned in 1983. Two of these research projects are being conducted at the National Microelectronics Research Centre of Ireland, Computer Modelling and Simulation, and Battery Control and Management. This paper is a description of the former.

The Computer Modelling and Simulation project commenced officially on November 1st, 1987. The initial work consisted principally of planning and research. This involved in particular, the drawing up of a work schedule, see Fig. 1 the selection of a working group, and the compilation and distribution of a survey questionnaire on existing PV plant models. In order to determine the specific features which the model should incorporate, two principal guidelines were used:

1. The results of the questionnaire survey.

2. The specific requirements of the 3 primary model user groups, plant managers, plant designers and research institutions.

Thus, a proposal for the format and structure of the program was compiled and presented to the working group in Milan, during April. It was agreed that the program should include the following features.

1. It should be developed for use on an IBM PC or compatible, with an XT as the minimum requirement, and it should be upwardly compatible to give improved performance on superior hardware. Operating system should be MSDOS.

2. The program should incorporate optimisation techniques so that the performance of existing plants can be improved.

3. PV plant sub-component models have already been developed. Where possible, these should be evaluated in order to select the best, and incorporated into the program.

4. Attention should be focused on the interaction of the sub-component models, rather than on the models themselves. This should result in a more flexible model, capable of simulating the PV plant operation in greater detail.

5. A graphics user interface should be incorporated into the program. This would give the user complete freedom to draw a PV plant of any reasonable configuration on the computer screen, as the first step in a PV plant simulation.

6. The simulation process should be implemented by a totally unique method. (see Fig. 2). This method involves the event queue concept of classical simulation problems. Events occurring within the plant, e.g. the opening of a switch, are placed on an event queue, under the control of the power management module. Events leaving the queue are processed by an event processor which decides the consequences of a particular event. The power management module responds to these consequences, and schedules new events as necessary, and so the simulation process continues. The advantage of ths method is that the actual operation of the plant is simulated with far greater accuracy than would otherwise be possible.

Results and Conclusions

At present, the research and planning phases of the project have been completed and the software development has begun. Results to date have been derived from the analysis of the questionnaire replies and from the research studies undertaken as part of the project. They include the following:

1. PV plant computer models are used primarily in design and sizing applications. Existing plants appear to have benefited little from their availability.

2. To achieve widespread and frequent use, a modelling program must be simple to use, and its visual impact should be pleasing to the user.

3. Computer languages most favoured for existing PV Plant simulations are FORTRAN and BASIC. PASCAL and 'C' are used occasionally. Specialised simulation languages are not suited to this application, but the event queue concept used in many such languages may well be applicable.

Work on the project is on schedule. An evaluation of existing component models is currently in progress. The development of the software flowchart has commenced, and writing the computer code proper will begin in mid-June.

Figure 1

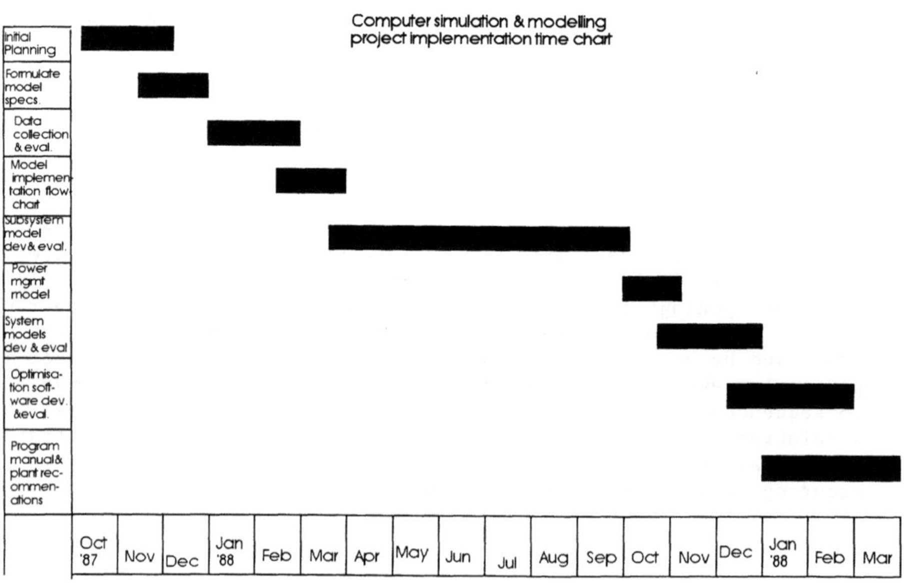

Figure 2

Generalised Flowchart Outlining The Simulation Process

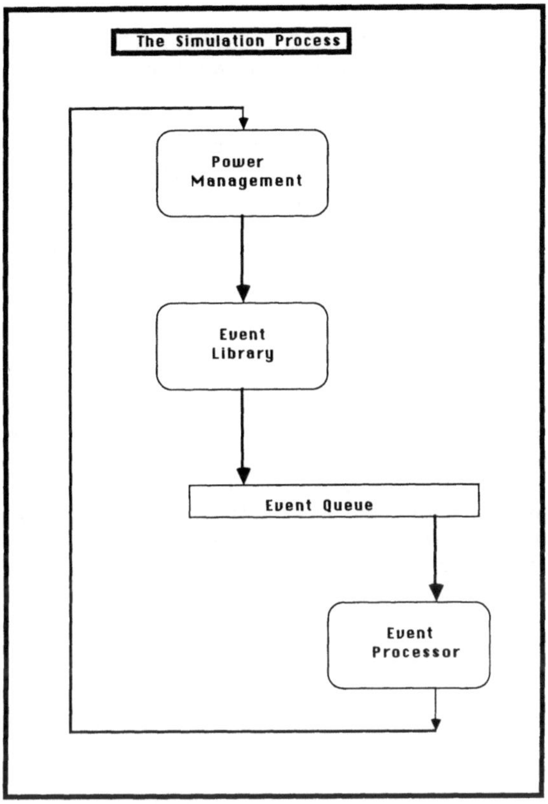

Table 1. PV Models being assessed as part of the project

Model	Name	Organisation	Country
Hermines	Mr. Didier Mayer	EMP	France
LOSADIS.PV	Dr. Denis Royer	CEN	France
PVSS	Mr. Alberto Iliceto	ENEL	Italy
SOMES	Drs. Emil Ter Horst/ Mr. Kornelius Block	ECOFYS	Netherlands
SIZ	Dr. Gert Blaesser	JRC (Ispra)	Italy
PV-FORM	Mr. Matthew Imamura	WIP	Germany

PERFORMANCE OF A BATTERY STORAGE IN WIND AND PHOTOVOLTAIC STAND ALONE ENERGY SYSTEMS

Thomas Degner, Hansjörg Gabler, Edo Wiemken
Department of Physics, University of Oldenburg
D-2900 Oldenburg, FRG

ABSTRACT. Practical experience has shown, that the operational lifetime of lead acid batteries in many solar systems is disappointingly low. Computer simulations of solar systems reveal possible unfavorable operating conditions which reduce the battery lifetime considerably. It can be demonstrated though, that an energy flow management satisfying the demands of the battery for "safe" operation has only marginal consequences on the system's energy performance.

1. INTRODUCTION

Under most circumstances, stand alone energy systems powered by wind and/or solar radiation must incorporate a battery storage to compensate for the mismatched time pattern of energy production and load. The lead acid battery built for stationary operation combines a high storage efficiency with a reasonable nominal operating life expectancy <1>.
Stimulated by early failures in the lead acid battery, which we operate at the ENERGIELABOR (a stand alone solar demonstration building at the University of Oldenburg), we started an analysis of the operating conditions of batteries in solar/wind energy systems.

2. SIMULATION OF THE OPERATING CONDITIONS OF A BATTERY STORAGE

We investigated by computer simulation the time evolution of the charge state of a battery storage in a wind and/or solar powered renewable energy system as given in figure 1.

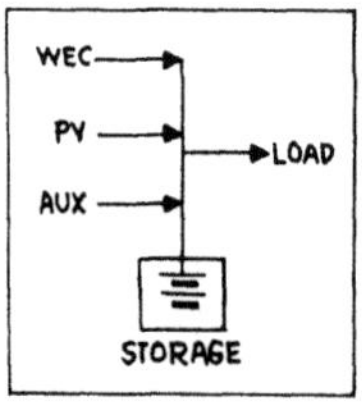

Fig. 1 Energy flow chart of a stand alone electrical supply system. Solar energy (PV), wind energy (WEC), and a conventionally fuelled motor (AUX) deliver energy to a load.

The system configuration may be characterized by the annual gross renewable production (PRG). PRG is the sum of the energy produced from renewable sources and normalized to the annual load. It is convenient to give the size of the storage (BAT) as the number of days, the load may be covered from the storage alone. The performance of the stand alone energy system is characterized by the fraction F of the load which is produced from renewable sources (F=1.0, if no motor fuel is consumed).
For the simulation the hourly values of the system states (energy production, battery charge, battery voltage etc.) are calculated and integrated over one year. Meteorological data taken in 1982 at our site in Oldenburg are input to the model.
The battery model is a modified version of the model developed by Shepherd

<2>. During charging we assume a coulombic efficiency (Ah-efficiency) of 96 %. The battery is discharged to a maximum of 80 % of its rated capacity. If further discharge is to be expected, the motor is activated producing rated power output for one hour.
As a first example of the calculations, fig. 2 gives the time evolution of the charge states and the charge histogram for a PV-system. The figure displays a characteristic pattern for solar systems in high latitudes: During a continuous period of five to six months (from October to March) the battery is never charged to its nominal capacity by the solar energy input. During the six summer months, however, a daily cycle of charging and discharging is clearly shown. The charge histogram reflects the high probability for the storage to be in a state near the upper or the lower storage limit.

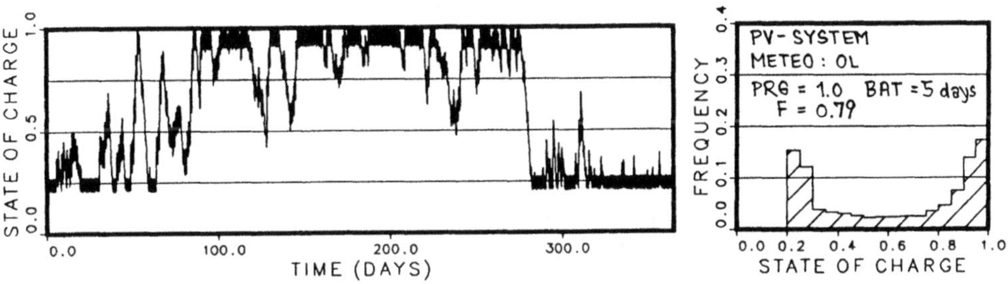

Fig. 2 Mean hourly values of the state of charge of a lead acid battery (capacity 5 load days) in a solar (PV) system at a site in northern Germany. During six months in winter the battery is never charged to full capacity by solar energy alone. The frequency distribution of the charge states (right hand side of the figure) shows, that the probability is high for the battery to be either "full" or "empty".

Fig. 3 gives calculation results for a stand alone system, combining the use of solar and wind energy. The frequency distribution is now levelled out. During two thirds of the year the charge state of the battery does not reach its upper or lower limit. The combination of the two energy sources has raised the renewable fraction F from 0.79 in the solar system to 0.90. Each of the system layouts, however, displays persistent periods of more than three months during which the battery is never charged to full capacity.

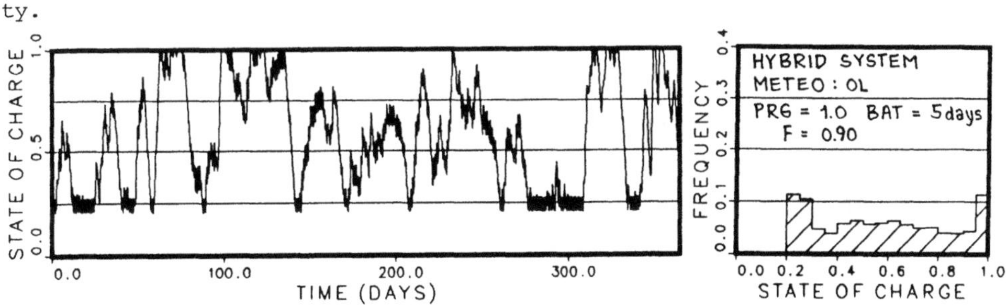

Fig. 3 Performance of a battery in a solar/wind hybrid system. The renewable energy converters are scaled to yield the same energy production as the PV-generator in fig. 2. The supplementary effect of the different time pattern of solar and wind raises the renewable fraction to 0.90. The frequency distribution of the charge states has been levelled out. Persistent periods where the battery is not charged adequately are still identified.

3. BATTERIES MAKE DEMANDS UPON THE ENERGY FLOW MANAGEMENT

We have talked to manufacturers of lead acid batteries and asked them for rules for "safe" operation of batteries in a solar system guaranteeing long operational lifetimes. To our actual knowledge the set of operating rules given below should be obeyed:

i. Avoid stratification by density differences of the electrolyte inside the battery cell. Stratification may be destroyed by "gassing", i.e. loading the battery to 120 % of its rated nominal capacity or, in order to avoid energy losses, by circulating the electrolyte during the charging of the battery.
ii. Avoid deep discharge of the battery cells, i.e. discharge of the battery to voltages below the cutoff voltage given by the manufacturers. The operation of a battery cell in deep discharge for more than a day may already be destructive.
iii. A battery should be charged to full capacity, at least every month.
iv. In charging the battery to full capacity care has to be taken, that every ampere-hour taken from the battery during the preceding discharging periods has to be replaced by 1.04 ampere-hours during recharging.

In the simulation program every Ah discharged is counted. A second counter sums up every Ah which is recharged into the storage. Every time the second counter reaches 1.04 times the content of the first counter, the battery is said to be charged to full capacity and both counters are reset to zero. Overcharging with the factor 1.04 should be sufficient to compensate for possible differences in the charge state of the positive and the negative electrode as well as for differences in the charge states among the single cells in the battery. If the "rules for safe battery operation" given above are obeyed, batteries should reach the number of possible charging/discharging cycles given by the manufacturers.

4. SIMULATION RESULTS - CONTINUED

We included the set of rules for safe battery operation into the energy load management routine of our simulation model and recalculated the system performance displayed in fig. 2. The results are given in figs. 4 and 5. Four times during the twelve month period, the battery is charged to its full capacity by the backup energy source (auxiliary). Applying the rules for safe battery operation has only marginal consequences on the system's energetic results. The renewable fraction F decreases through the implantation of the new energy management rules from 0.79 to 0.77. A large variety of calculations performed so far showed, that measures taken for safe battery operation never changed the factor F for more than 2 or 3 percents.

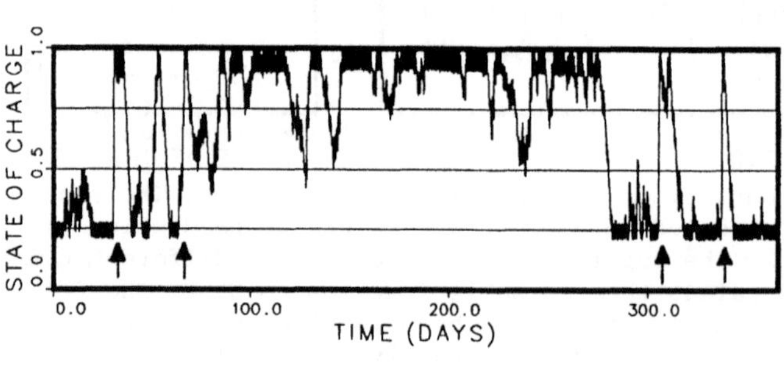

Fig. 4 The system performance displayed in fig. 2 was recalculated considering the "rules for safe battery operation". Indicated are four events where the battery is fully charged on request of the battery management routine.

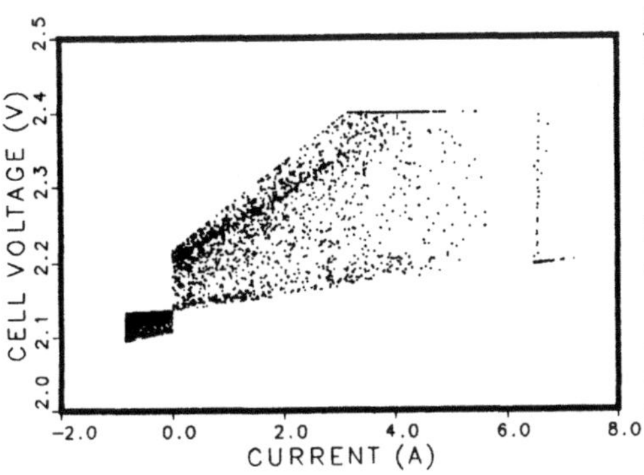

Fig. 5 Voltage/current states of a battery. All the operational states (hourly values) occupied by the battery during one year are plotted in the diagram which reflects the battery characteristics according to the battery model by Shepherd <2>. The current is scaled to a battery with 100 Ah of rated capacity. Batteries in renewable energy systems have to cope with very small currents during charging.

If renewable energy systems are not equipped with a backup energy source the "rules for safe battery operation" may only be met by a careful system design. As an example we simulated the performance of a PV-system with PRG=2.0 and BAT=5 days. Input to the calculations are radiation data from Ispra (northern Italy) (fig. 6). The calculations give a high solar fraction (F=0.98) and thus a loss of load probability of 0.02. The longest period where the battery is not charged to its rated capacity is 24 days.

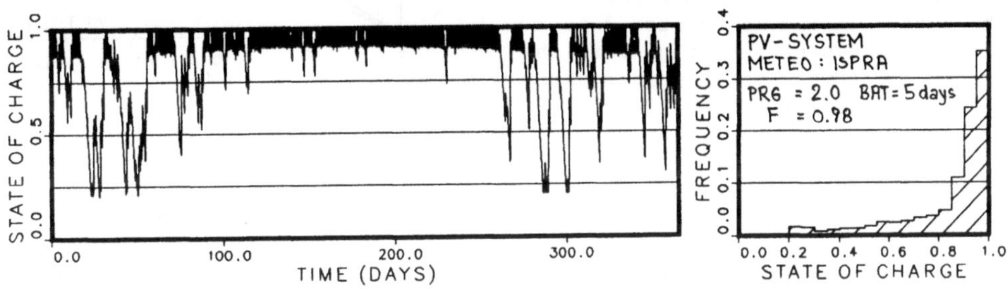

Fig. 6 Simulation of a stand alone photovoltaic system under climatic conditions of northern Italy with battery storage of "five days" and a renewable production PRG=2.0. The system displays a renewable fraction of 0.98, i.e. a low loss of load probability, if no backup system is provided. No special measures have to be taken to guarantee an adequate battery operation.

5. BATTERY LIFETIME

Under ideal operating conditions, the number of charging and discharging cycles of a lead acid battery is limited by degradation of the positive plate through shedding of the active electrode material. To a first approximation this process is independent of the cycle history, i.e. independent of charging and discharging currents or depth of discharge. In a simple model the accumulated charge taken from the battery is the limiting factor for battery operation.

Batteries recommended for use in solar energy systems exhibit considerable differences in their limiting accumulated discharge capacity. Reference <3> for example claims 600 cycles of 40 % depth of discharge which accumulates to 24000 Ah as limiting discharge capacity for a 100 Ah battery with pasted plates. For a battery with tubular ("armoured") positive plates, reference <4> claims 1500 cycles of 80 % discharge which is equivalent to 120000 Ah as limiting capacity.

We investigated the lifetime expectancy of these two solar batteries incorporated in renewable energy systems of different layouts. Results are given in figure 7.
An additional limitation to battery life is given by a slow corrosion process, which is observed as well in standby batteries not strained by cycling. It is well known that batteries with tubular plates may reach a "shelf life" of 10 to 15 years.

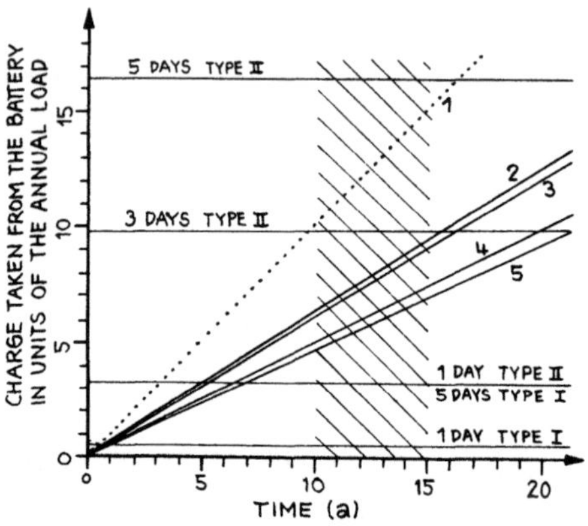

<u>Fig. 7</u> Calculation of the lifetime expectancy of lead acid batteries in renewable energy systems. The accumulated charge taken from the battery is plotted over time. In system 1 the load is completely covered from the battery, thus system 1 is the limiting case of maximum battery loading. In system 2 through 5, part of the energy production bypasses the storage, the accumulated charge taken from the battery increases much slower with time. System 2 refers to the system in fig. 6, system 3 to fig. 2 and system 5 to fig. 3 in the proceeding text. System 4 is a wind energy system with PRG=1, BAT=5. The "shelf life" expectancy is indicated by the hatched domain. The horizontal lines give the limiting accumulated charges for two different battery constructions (pasted positive plates: type I; tubular positive plates: type II) and different battery capacities (BAT=1 day, 3 days and 5 days).

We are well aware of the fact, that battery lifetime is influenced by more factors than those discussed above. The problem of an adequate design of renewable energy systems calls for a quantitative model for battery lifetime as a function of the operating parameters.

ACKNOWLEDGEMENT

We would like to thank Prof. Dr. A. Winsel and Dr. E. Voss of the VARTA-Forschungs- und Entwicklungszentrum in Kelkheim for a detailed discussion on the subject.

REFERENCES

<1> Facinelli, W.A. (1983). Modeling and Simulation of lead-acid batteries for photovoltaic systems. Ph.D. Thesis, Arizona State University.
<2> Shepherd, C.M. (1965). Design of primary and secondary cells, II. An equation describing battery discharge. <u>J. Electrochem. Soc.</u>, <u>112</u>: 657-664.
<3> Leiblein, K.H. (1987). Lead Acid Battery for small photovoltaic equipment. <u>Progress in Batteries & Solar Cells, Vol. 6</u>: 141-144.
<4> Gutzeit, K. (1987). Batteries for photovoltaic power systems. <u>Progress in Batteries & Solar Cells, Vol. 6</u>: 246-249.

DEVELOPMENT OF PHOTOVOLTAIC POWER SYSTEM USING ORTHOGONAL-CORE TYPE PUSH-PULL DC-AC CONVERTER

O. Ichinokura* and K. Murakami*
M. Maeda**, K. Mitamura**, H. Takahashi** and M. Kohno**

* Department of Electrical Engng., Tohoku University
Aramaki Aza Aoba, Sendai 980, Japan
** General R & D Center, Tohoku Electric Power CO. INC.
7-2-1, Nakayama, Sendai 980, Japan

The purpose of this study is development of a photovoltaic power system using the orthogonal-core type push-pull dc-ac converter. The converter has desirable features including simple construction and power controllability. Using the converter, a 3 kW photovoltaic power system which connects solar cells to an ac distribution line is demonstrated in this paper. The irregular power generated by solar cells is conditioned by a simple feedback control. The examination results indicate that the trial system has ease of maintenance and high reliability.

1. INTRODUCTION

Development of photovoltaic power system which connects solar cells to an ac power line has been expected because the system can utilize effectively the irregular power generated by solar cells [1]-[3].

The authors proposed a push-pull dc-ac converter for transferring the power from a dc source to an ac power line [4]. The converter is constructed with a pair of orthogonal-cores whose primary sides are excited by square-wave choppers, and has almost sinusoidal output in spite of simple construction. The transferred power is easily controlled by the phase angle of the chopper voltages.

In this paper, the authors report the orthogonal-core type push-pull dc-ac converter and its application to a photovoltaic power system which connects solar cells to ac distribution line. The phase angle of the chopper voltages is adjusted by a simple feedback control in order to transfer effectively the irregular power generated by solar cells. Good operations are obtained in the trial 3 kW photovoltaic power system.

2. ORTHOGONAL-CORE TYPE PUSH-PULL DC-AC CONVERTER

Fig. 1(a) shows a schematic diagram of the orthogonal-core and (b) the windings arrangement. N_1 and N_2 are primary and secondary windings. i_1 and i_2 are the primary and secondary currents, and the dashed curves ϕ_1 and ϕ_2 illustrate the primary and secondary fluxes.

Fig. 2 shows a circuit configuration of the dc-ac converter using the orthogonal-cores. The orthogonal-cores #1 and #2 have the same dimensions and number of turns in the windings. The primary sides of the cores are excited by square-wave transistor choppers. The secondary windings are connected in parallel through the diodes. L_{21} and L_{22} are the effective inductances of the secondary windings.

In the orthogonal-core, increase of the primary flux can produce decrease of the secondary inductance because the core reluctance increases depending on its saturation characteristic [5]. The dc-ac converter util-

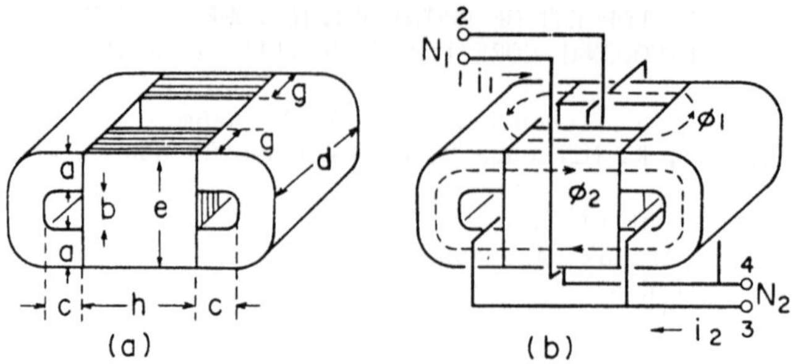

Fig. 1 Schematic diagram of the orthogonal-core. (a) Core structure. (b) Windings arrangement.

Table I Dimensions of cores and number of turns in the windings for the 3 kW system.

a=45, b=45, c=55, d=135 e=135, g=45, h=110 (mm)	N_1=264, N_2=136(turns)

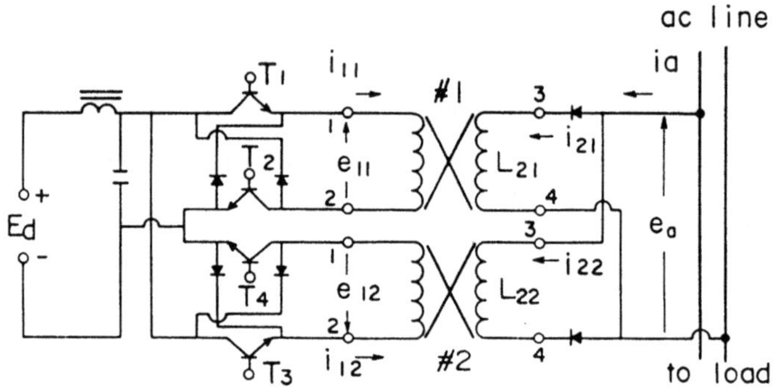

Fig. 2 Circuit configuration of the dc-ac converter.

izes the variations in the secondary inductances.

Fig. 3 depicts qualitative explanation of the converter operation. In the figure, e_a is the line voltage. e_{11} and e_{12} are the output voltages of the square-wave choppers. ϕ_{11} and ϕ_{12} are the primary fluxes of the orthogonal-cores #1 and #2, respectively. When the primary flux ϕ_{11} increases, the secondary inductance L_{21} decreases and its winding current i_{21} increases. In the same manner, the current i_{22} of the core #2 increases when the flux ϕ_{12} increases. The output current i_a is given by $i_a=i_{21}+i_{22}$.

Fig. 4 shows an example of the observed waveforms. In the figure, i_{11} and i_{12} are the primary winding currents of #1 and #2, respectively. From Fig. 3 and 4, it reveal that the orthogonal-core #2 can supply the power from the primary to the secondary in the positive half cycles of the line voltage e_a, and the core #1 in the negative half cycles. Based on this push-pull operation, the power is transferred from the dc source to the ac line and the output power has low distortion in spite of simple excitation circuit of square-wave choppers. The average power changes with the phase angle θ of chopper voltages relative to line voltage.

Fig. 5 shows the typical relations between the powers and the phase angle measured in the dc-ac converter made for 3 kW photovoltaic power sys-

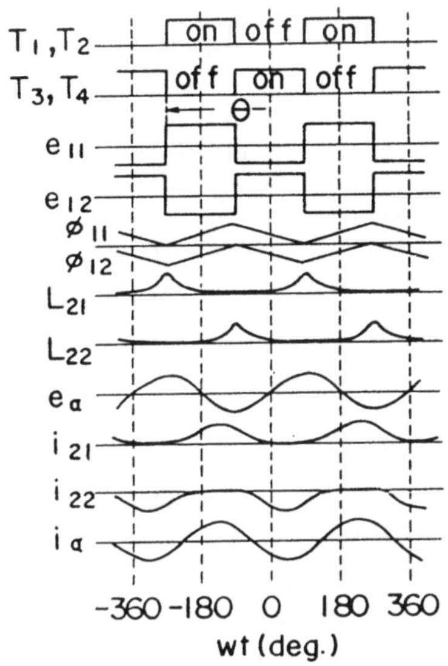

Fig. 3 Operation waveforms.

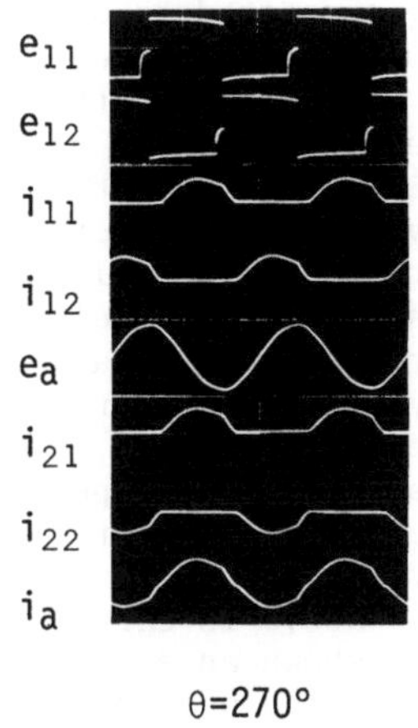

Fig. 4 Observed waveforms.

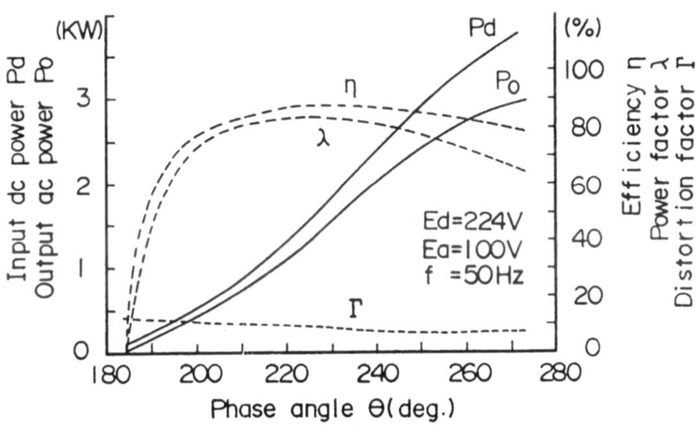

Fig. 5 Power conversion characteristics.

tem. The dimensions of the cores and number of turns in the windings are shown in Table I . The core material is silicon steel. In Fig. 5, P_d and P_o are the input and output powers. η is the conversion efficiency, λ the output power factor and Γ the distortion factor of the output current. This reveals that the transferred power is easily controlled from zero by changing the phase angle.

3. APPLICATION TO PHOTOVOLTAIC POWER SYSTEM

Using the orthogonal-core type push-pull dc-ac converter, we made a photovoltaic power system as shown in Fig. 6. The rated output voltage and power of the solar cell array are 224 V and 3 kW. The output of the dc-ac converter are connected to an commercial ac distribution line with the voltage of 100 V and the frequency of 50 Hz. The phase angle of the chopper voltages are controlled so that the array voltage maintains almost constant value by detecting only the array voltage. For this feedback control, the array power is effectively transferred to the ac line. The system starts automatically when the detected array voltage is sufficiently large and stops when it drops extremely small value.

Fig. 7 shows the control characteristic measured in the trial photovoltaic power system at actual operation. In the figure, the abscissa E_e is the intensity of the solar radiation. P_d is the array power, P_o the output power, E_d the array voltage, E_a the line voltage and θ the phase angle. This figure shows that the phase angle and powers increase when the solar radiation increases. The array voltage is constant over the wide range of the intensity of the solar radiation. This reveals that the feedback control mentioned above is carried out.

Fig. 8(a) shows an example of the operating characteristic of the trial system at fine weather. Good generation curve are obtained in the system. Fig. 8(b) shows the operating characteristic when the solar radiation changes irregularly by the clouds. This reveals that the power is effectively transferred even when the solar radiation changes vigorously. The line voltage is almost constant in spite of the drastic variation of the

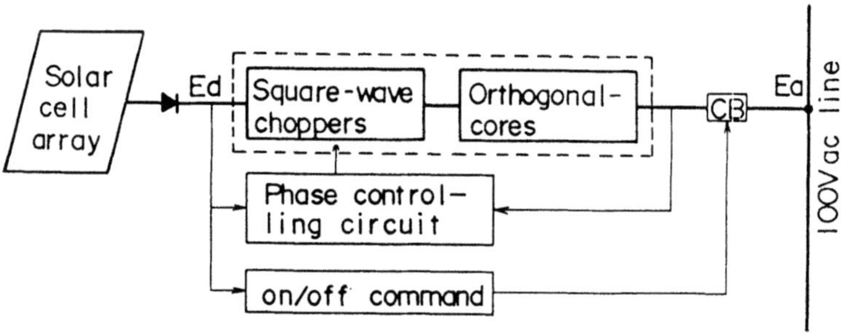

Fig. 6 Block diagram of the photovoltaic power system.

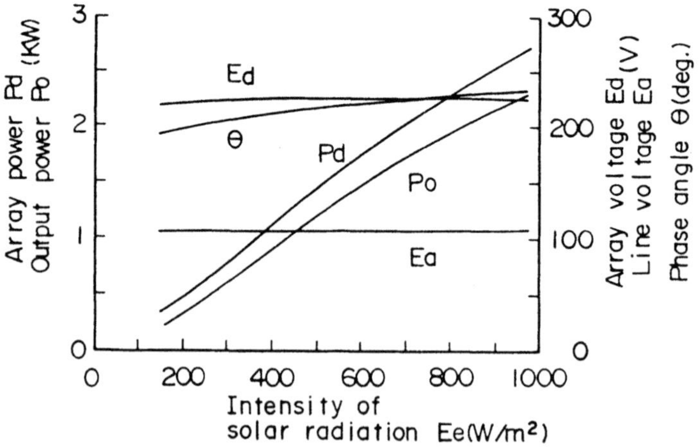

Fig. 7 Control characteristics obtained in the trial photovoltaic power system.

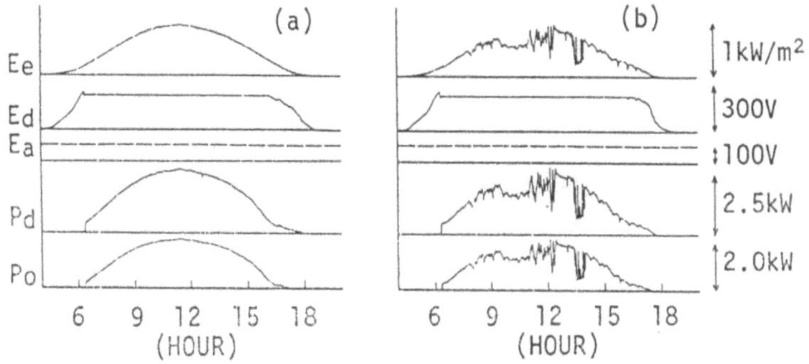

Fig. 8 Operation characteristics of the photovoltaic power system. (a) At fine weather. (b) When solar radiation is irregular.

(a) (b)

Fig. 9 General views of the experimental system. (a) Solar cell array. (b) The dc-ac converter.

transferred power.

Fig. 9 (a) shows the general view of the solar cell array and (b) the experimental apparatus of the dc-ac converter.

4. Conclusion

The application of the orthogonal-core type push-pull dc-ac converter to a photovoltaic power system is demonstrated in this paper. The system presented here is useful as a residential system because of its simple construction, maintenance free, high reliability, security and low cost.

Reference

[1] B. K. Bose, P. M. Szczesny and R. L. Steigerwald, IEEE Trans. on Industry Applications, Vol. IA-21, No. 5, p. 1182(1985)

[2] Y. Takeda, K. Takigawa, H. Kobayashi, K. Nakahara, T. Morishita, A. Kitamura and H. Matsuda, Technical Digest of the International PVSEC-3, C-IIIa-6(1987)

[3] A. Goetzberger and J. Schmid, Technical Digest of the International PVSEC-3, PL-II-2(1987)

[4] O. Ichinokura, M. Maeda, H. Takahashi and K. Murakami, Abstract book of EMMA'87, CD07(1987)

[5] O. Ichinokura and K. Murakami, IEEE Trans. on Magnetics, Vol. MAG-19, No. 5, p. 2103(1983)

A LIGHT WEIGHTED, FOUR QUADRANT, HIGH SWITCHING FREQUENCY MODULAR, PHOTOVOLTAIC DC/AC INVERTER, WITH SINUSOIDAL OUTPUT AND HIGH EFFICIENCY

S.A. Pressas and V. Makios

Laboratory of Electromagnetics, School of Engineering
University of Patras

ABSTRACT

A new High Switching Frequency DC/AC Photovoltaic inverter has been developed, based on the Sinusoidal Pulse Width Modulation - S.P.W.M. principle. The swithcing frequency is 25 KHz, the input voltage is 48 V DC and the rated power 1,5 KVA. Besides, the inverter is modular so that several units can be connected in parallel to increase the output power. The efficiency is over 85% for all loading conditions. The semiconductor switches are POWER MOSFETs and the power transformer is of a ferrite-core type. The power stage is of PUSH-PULL topology and a symmetry correction loop has been incorporated, in order to avoid core saturation. The inverter is capable of driving heavily inductive loads, becuase a path has been provided for the regenrative current to return back to the battery. Also, the inverter, due to its high switching frequency has after a minimal filtering, a purely sinusoidal output (220 V/50 Hz), with excellent output voltage regulation and frequency stability, is noiseless, light weighted, has small size and exhibits an excellent dynamic response.

INTRODUCTION

The operating principle of High Frequency Sinusoidal Pulse Width Modulation (H.F.S.P.W.M.) on which the new photovoltaic inverter developed in the Laboratory of Electromagnetics of the University of Patras is based, is a combination of two well known widely used, principles of power electronic converter operation.

a) The Low Frequency (below 2 KHz) Sinusoidal Pulse Width Modulation, used on Variable Speed Induction Motor Drives [1,2,3,4] and
b) High Frequency (over 20 KHz) Pulse Width Modulation, used on Switch Mode Power Supplies [5,6,7].

CIRCUIT DESCRIPTION

Figure 1 represents the block diagram of the inverter. The 50 Hz reference sine wave -modulating signal- is produced by a crystal controlled oscillator, while a 50 KHz triangle waveform signal generator provides the carrier signal. The rectified modulating and the carrier signal, are fed together to the control section where they are compared to each other. The comparator´s output is a high frequency pattern of pulses, which are width modulated according to the reference sinewave´s amplitude, (High Frequency Sinusoidal Pulse Width Modulation - H.F.S.P.W.M.). This Sinusoidally Pulse Width Modulated pattern of pulses, drives the power section through a totem-pole current boosting driving stage. The power semiconductor switches are Power MOSFETs connected in push-pull configuration, while the high fre-

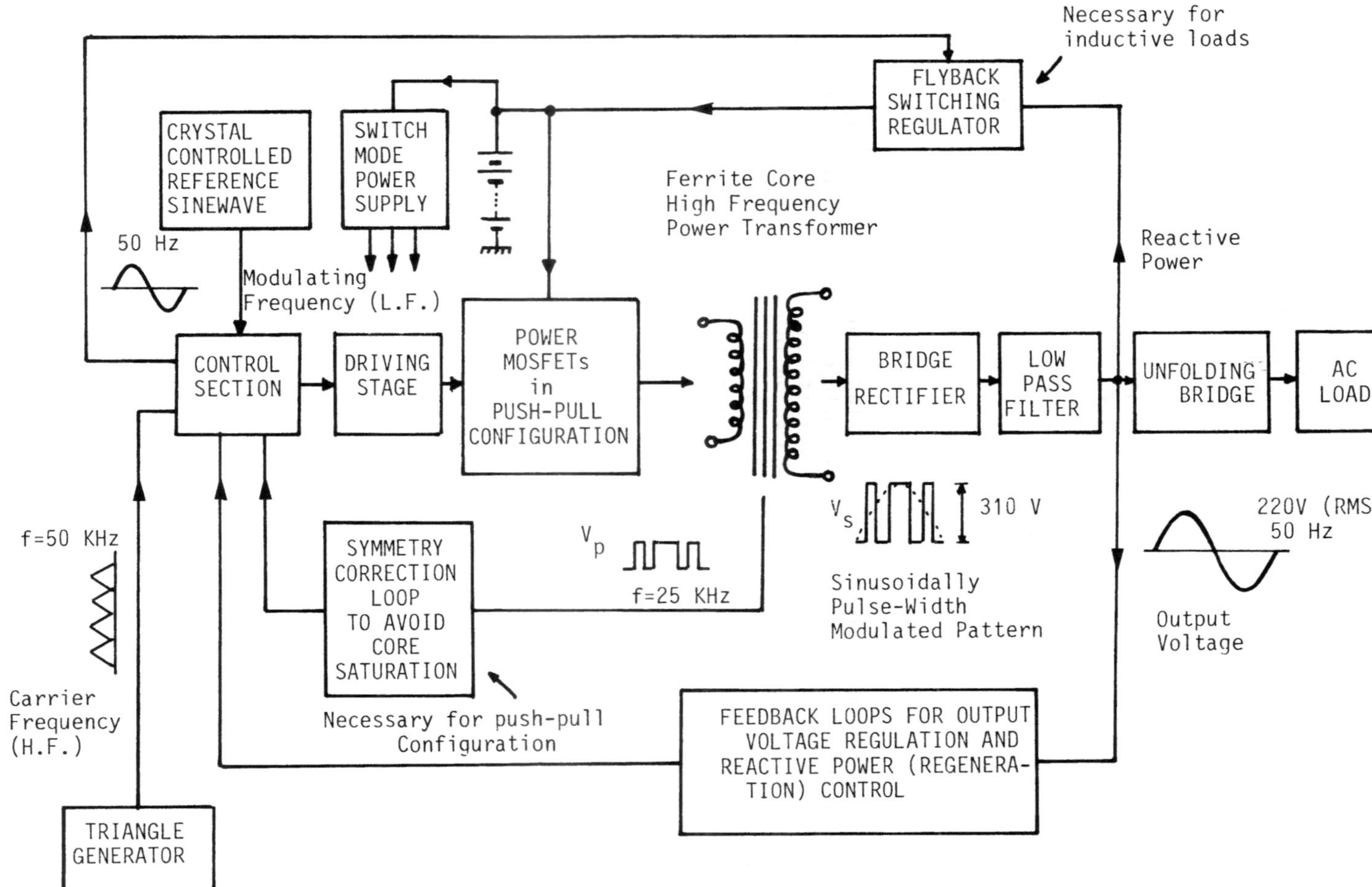

Fig.1. The Block Diagram of the Inverter.

quency power transformer is of a ferrite-core type. A symmetry correction loop is incorporated in order to avoid core saturation.

The power transformer´s secondary voltage is rectified and then filtered through a low-pass second order L-C filter and the waveform coming out from the filter is a pure 50 Hz - 220 V RMS rectified sinewave. The filter is small because its breakpoint frequency is at 5 KHz. At last, the rectified sinewave is "unfolded" by means of an unfolding bridge and so the inverter´s output is a pure 50 Hz - 220 V RMS sinewave.

Feedback loops are provided for output voltage regulation and reactive power control. Reactive power is fed back to the battery through a high frequency flyback switching regulator. Finally, a switching power supply, provides isolated low DC biasing voltages for all the signal processing and power sections.

Figures 2 and 3 present current and voltage waveforms which are indicative of the inverter´s operation.

Figures 4 and 5 present an inside view photograph of the prototype inverter.

RESULTS

A new DC/AC Photovoltaic Power Inverter with 25 KHz switching frequency has been developed. The battery voltage is 48 V, the output voltage is 220 V (RMS) - 50 Hz and the rated power is 1,5 KVA. The inverter is modular, so that several units can be connected in parallel to increase the output power. The efficiency is over 85% for all loading conditions. The inverter is capable of driving heavily inductive loads, because a path has been provided for the regenerative current to return back to the battery. Also, the inverter, due to its high switching frequency, has, after a minimal filtering, a purely sinusoidal output (220 V/50 Hz), with excellent output voltage regulation and frequency stability, is noiseless, light weighted, of a small size and exhibits an excellent dynamic response. Finally, this new inverter has current limit, short circuit portection, as well as low battery stop and high battery start for battery protection.

REFERENCES

[1] F.De Buck, P.Gistelinck and D.De Backer, "Loss-optimal PWM waveforms for variable-speed induction- motor drives", IEE Proceedings, Vol.130, Pt.B. No.5, September 1983.

[2] S.R.Bowes and J.Clare, "Steady-state performance of PWM inverter drives", IEE Proceedings, Vol.130, Pt.B. No.4, July 1983.

[3] D.A. Grant and R.Seidner, "Ratio changing in pulse-width-modulated inverters", IEE Proceedings, Vol.128, Pt.B., No.5, September 1981.

[4] Allan B.Plunkett, "Direct flux and torgue regulation in a PWM inverter-Induction Motor Drive", IEEE Transactions on Industry Applications, Vol. IA-13, No.2, March/April 1977.

[5] Herbert R.Weischedel and George R.Westerman, "A symmetry correcting pulse width modulator for power conditioning applications", IEEE Transactions on Industry Applications, Vol.IA-9, No.3, May June 1973.

[6] William M.Plivka, Alan Cocconi and Slobodan Cuk, "Use of orthogonal flux to detect impending magnetic saturation in switching converters", International PCI/MOTORCON´82 Conference, San Fransisco,California,

March 29-31, 1982.

[7] Abraham I.Pressman, "Switching and linear power supply, power converter design", Hayden Book Company Inc. Rochelle Park, New Jersey,1977.

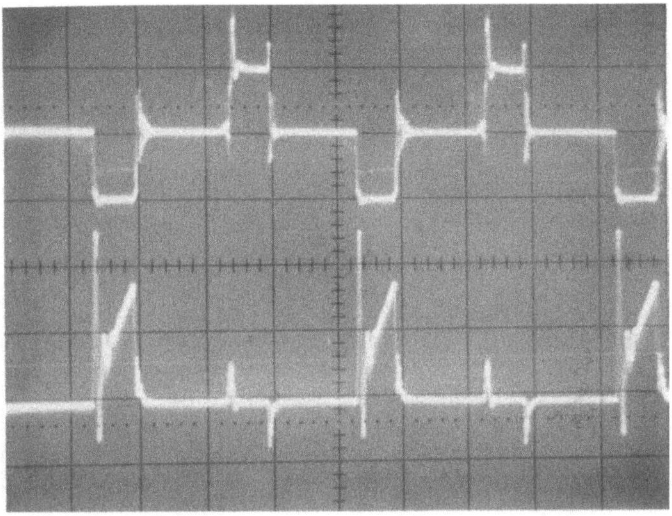

Fig.2. Power MOSFET´s Voltage and current waveforms:
top trace: V_{DS} (50 V/div), time base: 10 μs/div
bottom trace: I_D (2 A/div).

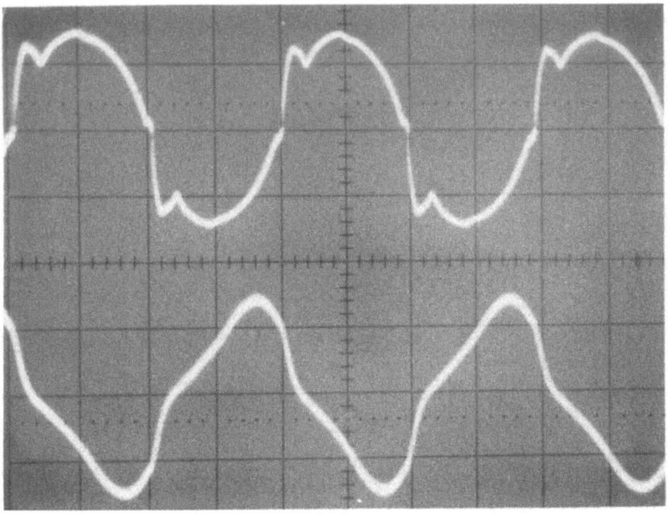

Fig.3. Induction motor´s voltage and current waveforms ($\cos\varphi = 0{,}6$)
top trace: V_L (100 V/div) time base: 5 ms/div
botton trace: I_L (2 A/div)

ACKNOWLEDGEMENTS

This work is partially supported by the European Community DGII under contract EN3S-0100-GR (T.T.).

Fig.4. The prototype inverter, Bottom view.

Fig.5. The prototype inverter, Top view.

THE PERFORMANCE OF LOADS POWERED BY A COMMON SOURCE OF SOLAR CELLS

J. Appelbaum
Faculty of Engineering
Tel Aviv University
Tel Aviv 69978, Israel

ABSTRACT: Loads of the same or different types that are in close proximity to each other may be powered either by separate solar cell sources for each one, or alternatively, by a common solar cell source for all the loads. The paper introduces a procedure for comparing the performance of the loads for these two possible setups. The criterion for the comparison is the "energy utilization efficiency". Two loads represented by an e.m.f. and an internal resistance are considered as an example. The results show that for a wide range of the load parameters, the common source system is superior. The designer of a photovoltaic system may wish to analyze the possibility of a common source/multiple load system design to achieve higher performance compared to a separate source/individual load system.

1. INTRODUCTION

In many stand-alone photovoltaic systems, the solar cell (SC) array is designed to power a specific single load, such as a warning light or a water pump. Several loads of the same or different types that are in close proximity to each other may be powered either by a separate SC source for each one, or alternatively, by a common SC source for all the loads. Loads of either the same or different types may have different I-V characteristics, and since a SC source is a finite power one and has different I-V external characteristics from the conventional power source, it can be expected that the operation of the different loads will also be different when the loads are powered either by separate sources or by a common SC source. In a common source system it is possible that the performance of all the loads is higher than the total performance of all these loads when powered by separate sources. There exists also an additional option where in certain cases, such as at low insolation, it might be advantageous to disconnect a load from the common source system in order to improve the operation of another load. In general, the designer may wish to analyze the possibility of a common source/multiple load system design to achieve higher performance compared to a separate source/individual load system. The purpose of this study is to introduce a procedure for comparing the performance between separate and common source systems. The criterion for the comparison is the "energy utilization efficiency".

Loads may be of the same or different types. It can be shown that for identical loads (the same type and size), either optimal or non-optimal, powered by identical SC arrays; and for loads of the same type but different in size that are powered by proportional sizes of SC arrays, the operation of loads are the same when powered either by separate sources or by a common SC source. There are cases, however, that systems deviate from their optimal designs to a smaller or to a larger extent. Differences in the load or array characteristics may inevitably be introduced in the system design or during system operation (e.g., shading), such that the "initially" identical system components may become non-identical. In the present study, the arrays in the separate systems are of the same sizes and in the common source system the arrays are of double size. The loads in both setups are of the same type but different in their sizes. Overall, the study deals with different solar cell systems, once powered by separate sources and the second time by a common source. Since for identical systems the performance of the loads are the same for both setups, the difference in the load performance is expected to increase as the systems become more different from each other. Two loads represented by an electro-motive-force (e.m.f.) and an internal resistance are considered as an example.

It is assumed that the operation of the loads takes place on a clear day with the variation of the insolation given by:

$$G = 100 \sin\omega \quad , \tag{1}$$

where G is the insolation in percentages of ONE SUN, and ω is the solar angle in degrees. The solar angle is related to the solar time by:

$$T = (\omega + 90)/15 \quad , \tag{2}$$

when at solar noon (T = 12:00) the solar angle is 90°. The "energy utilization efficiency" of the solar cells in the system is defined here by:

$$\eta_E = \int_{T_1}^{12:00} P(T)dT \Big/ \int_0^{12:00} P_M(T)dT \quad , \tag{3}$$

where the numerator is the input energy to the load by the SC array during the operating period between some time T_1 and noon; and the denominator is the maximum available energy that the SC array can supply. The time T_1 corresponds to some threshold operating value of the load, e.g., current, voltage or power; P is the SC array output power, and P_M is its maximum output power, both are functions of the insolation G. Because of the assumed symmetry of the insolation around noon, the calculation is performed for half a day. The η_E indicates the degree of utilization of the solar cells in the system, or in other words, the degree of matching the load to the solar cells.

The SC array used in this study is represented by the following I-V approximate equation:

$$V = -0.9\, I + \frac{1}{0.0422} \ln\left(\frac{I_{ph} - I + 0.0081}{0.0081}\right) \quad , \tag{4}$$

where for 100% insolation, I_{ph} = 13.615 A; V_{oc} = 176 V; and P_M = 1400 W. The common SC source is made up of two equal separate sources (eq. (5)) connected in parallel. The I-V equation for this array is:

$$V = -0.45\, I + \frac{1}{0.0422} \ln\left(\frac{I_{ph} - I + 0.0162}{0.0162}\right) \quad , \tag{5}$$

where for 100% insolation, I_{ph} = 27.23 A; V_{oc} = 176 V; and P_M = 2800 W.

2. Loads Represented by an E.M.F. and an Internal Resistance

Many loads, such as batteries, water electrolyzers, and other load systems containing electrical sources and internal resistances can be represented by an equivalent electro-motive-force and an internal resistance using Thevenin's Theorem. In this section we shall compare the performance of two such loads, the first time connected to separate SC sources (Eq. (4)); and the second time connected to a double-sized common SC source, (Eq. (5)), as shown in Fig. 1a and 1b, respectively.

2.1. Operation of a System of a Single Load and a Separate Source

The I-V characteristics of the load is represented by the equation:

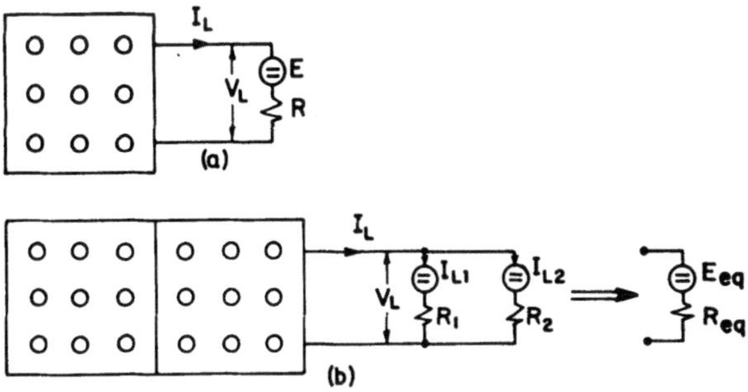

Fig. 1. Load represented by an e.m.f. and internal resistance
(a) single load and single source;
(b) combined load and common source

$$V = E + I_L R \tag{6}$$

where E is the e.m.f. and R is the internal resistance. The I-V characteristics of the SC array for different levels of insolation, in percentages of ONE SUN, computed by Eq. (5), are shown in Fig. 2. The figure also includes the maximum power line P_M and two load lines defined by

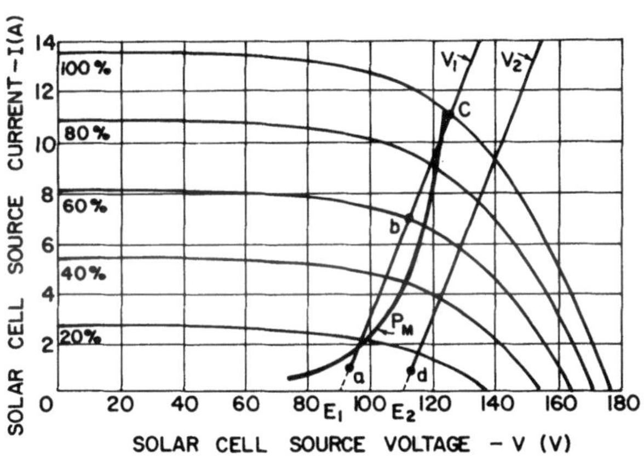

Fig. 2. I-V characteristics of single SC sources and separate loads

Eq. (7).

The equations of the two loads are:

$$\begin{aligned} V_1 &= E_1 + I_L R_1 = 90 + 3.3\, I_L \\ V_2 &= E_2 + I_L R_2 = 110 + 3.3\, I_L \end{aligned} \tag{7}$$

where 90 + 3.3 I_L is the optimal load line calculated according to the energy utilization efficiency, Eq. (3). The system operation is determined by the intersection points of the load line with the I-V characteristics of the SC source for various insolation levels, i.e., points a,b, and c.

It is assumed that the loads start to operate at insolation levels when sufficient current has been obtained, i.e., at threshold current of one ampere (see points a and d). The load currents are obtained by equating Eqs. (4) and (7) and solving I (using a computer library program), i.e.,

$$-0.9\, I_L + \frac{1}{0.0422} \ln \left[\frac{I_{ph} - I_L + 0.0081}{0.0081} \right] = 90 + 3.3\, I_L \text{ and } 110 + 3.3\, I_L \quad (8)$$

The load voltage is given by Eq. (8), and the load power is $P = V_L I_L$. The energy utilization efficiencies are calculated by Eq. (3) with the help of a computer library program based on the theory of SPLINES,

$$\eta_{s1}(V_1) = \int_{06:24}^{12:00} P(V_1)dT \Big/ \int_0^{12:00} P_M(T)dT = 99.70\ \%$$

$$\eta_{s2}(V_2) = \int_{06:34}^{12:00} P(V_2)dT \Big/ \int_0^{12:00} P_M(T)dT = 94.68\ \% \quad (9)$$

where "s" stands for separate. The average of η_{s1} and η_{s2} is:

$$\eta_s = (\eta_{s1} + \eta_{s2})/2 = 97.19\ \% \quad (10)$$

2.2. Operation of a System of Two Loads and a Common Source

The above mentioned two loads are now connected to a common source as shown in Fig. 1b. The equivalent of the parallel connection of these two loads is:

$$V_{eq} = E_{eq} + I_L R_{eq} \quad (11)$$

where

$$E_{eq} = E_1 \frac{R_2}{R_1 + R_2} + E_2 \frac{R_1}{R_1 + R_2} \quad \text{and} \quad R_{eq} = \frac{R_1 R_2}{R_1 + R_2} \quad (12)$$

The array current is obtained from the solution of Eqs. (5), (7) and (12):

$$-0.45\, I_L + \frac{1}{0.0422} \ln \left[\frac{I_{ph} - I_L + 0.0162}{0.0162} \right] = E_{eq} + I_L R_{eq} = 100 + 1.65\, I_L \quad (13)$$

The common load voltage is $V_L = 100 + 1.65\, I_L$, and the load currents are:

$$I_{L1} = (V_L - E_1)/R_1 \quad \text{and} \quad I_{L2} = (V_L - E_2)/R_2 \quad (14)$$

The common load input power is $P = V_L I_L$. The solution is again for threshold currents of one ampere. The energy utilization efficiency of the combined load in the common source system is:

$$\eta_c(V_{eq}) = \int_{06:19}^{12:00} P(V_{eq})\, dT \Big/ \int_0^{12:00} P_M(T)\, dT = 98.34\ \% \quad (15)$$

where "c" stands for common. By comparing the results of the two system setups, Eqs. (10) and (15), we see that $\eta_c > \eta_s$, i.e., the operation of the loads in the common source system is improved by 1.15% as compared to the operation in the separate source systems.

A comparison between the separate and the common source systems is also shown in Figs 3a and 3b, where the parameters of the optimal load were kept constant, i.e., $V_1 = 90 + 3.3\ I_L$, and the parameters of the second load were changed. The equation of the second load in Fig. 3a is $V_2 = E_2 + 3.3\ I_L$ where E_2 is varied in the range: $60 \le E_2 \le 130$ volts. The results show that the energy utilization efficiency of the common source system η_c is higher than that of the separate source systems η_s. For identical loads, both setups perform the same; and since the loads are optimal, the highest utilization efficiency is achieved. The equation of the second load in Fig. 3b is $V_2 = 90 + R_2 I_L$ where R_2 is varied in the range: $0.5 \le R_2 \le 10\ \Omega$. The figure shows that again the common source system η_c is superior for a wide range of internal resistances, i.e., for $R_2 \ge 1.2\ \Omega$. The highest efficiency is obtained when both loads are identical and optimal.

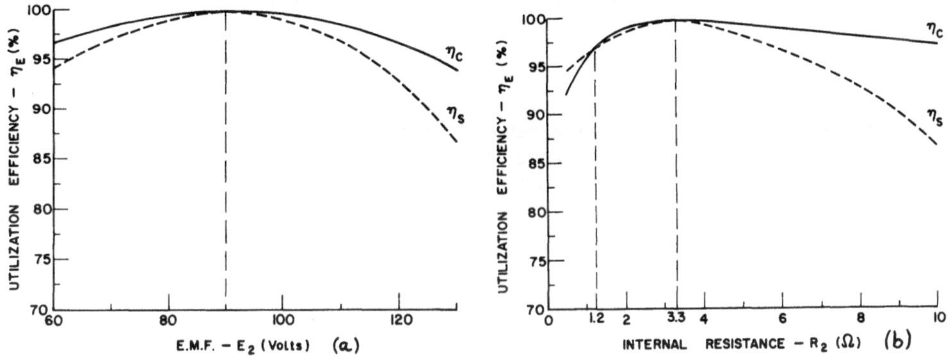

Fig. 3. Energy utilization efficiency of separate and common source systems
(a) E_1=90V, R_1=3.3Ω, E_2=variable, R_2=3.3Ω
(b) E_1=90V, R_1=3.3Ω, E_2=90V, R_2=variable

CONCLUSIONS

The paper introduces a procedure for comparing the performance of a common source/multiple load system design with a separate source/individual load system design. The criterion of the comparison is the "energy utilization efficiency". The performances of two loads represented by an e.m.f. and internal resistance were compared for separate and common source systems; the results show the preference of the common source/multiple load system.

A computer aided toolbox for PV-Hybrid System Design (PFLO1804)

H.P. Nava, R. Aringhoff, H. Hess

Flachglas Solartechnik GmbH Tel. 0221-235191
Mühlengasse 7
D-5000 Köln 1

ABSTRACT. To answer the question: "What is the best system to satisfy a required energy demand at a given location" from the technical as well as from the economic point of view, a set of useful tools has been developed.

The tools can be applied to the whole range of technically feasible systems, which can be composed by a subset of subsystems shown in fig. 1, including hybrid systems, like photovoltaics combined with wind energy converters and dieselgensets. The first step was the identification of the considered systems, for which the tools were developed. The chosen system structure is the result of the system analysis of a variety of worldwide realised projects. The toolbox consists of:

- Definition of the modular design concept and the description of the subsystems including the considered I, V and P ranges
- Data base for system components (design relevant data, cost relevant data)
- Data base for meteorological data (insolation, wind speed, temperature etc.)
- Software package for system design and energetic flow simulation
- Software package for a life cycle cost calculation

The modular design principle induces the necessary standardization of the systems. Furthermore, the tools can be very useful to identify standard systems for a specific class of applications.

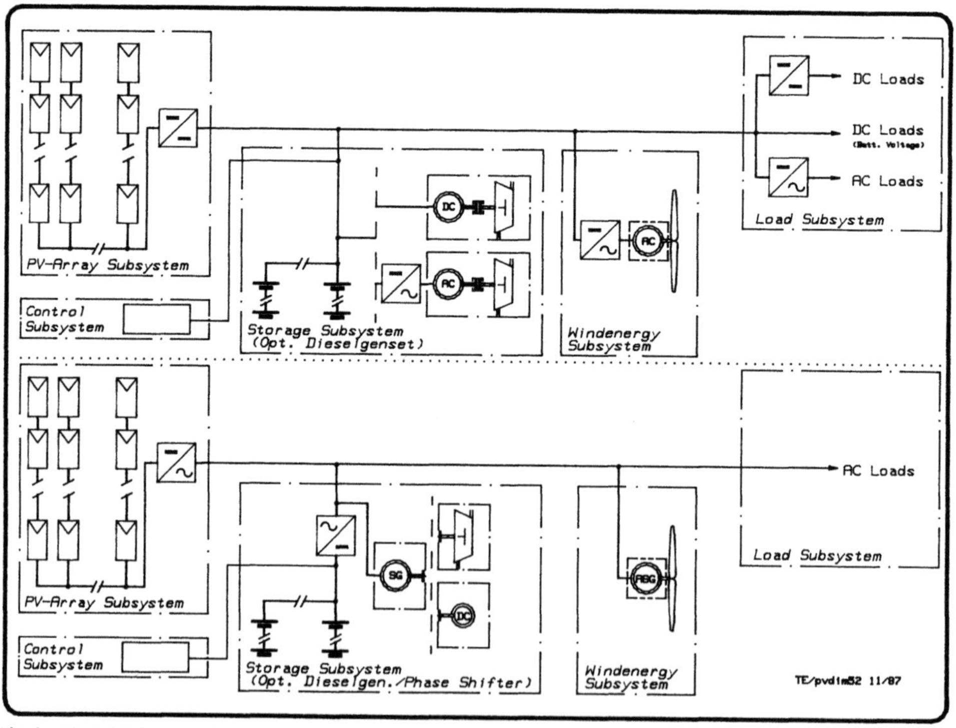

Fig.1 PV-Hybrid concepts

1. DEFINITION OF THE MODULAR DESIGN CONCEPT

From the viewpoint of software-engineering as well as from the system-engineering there is only one solution for the development of a versatile and expandable toolbox for autonomous electrical PV-Hybrid system design: modularity.

Each system which can be designed and evaluated with the computer aided toolbox is built out of all or a selected subset of: PV-Array subsystem, storage subsystem with optional Dieselgenset/rotating phaseshifter, control subsystem, windenergy subsystem and load subsystem. It is necessary to distinguish between two kinds of systems:

- System with DC-Crossbar.
  Depending on the layout of the load subsystem, these systems are able to supply any combination of DC-Loads, AC-Loads (1- and 3-phase) and loads directly connected to the crossbar (battery voltage). The inverters have to be selfcommutated.
- System with AC-Crossbar.
  3-phase AC-Loads like small 3-phase grids or load clusters are suitable for this kind of system. The inverters in the storage subsystem and in the PV-Array subsystem are grid commutated. The grid formation (control of voltage and frequency) is performed by the synchronous generator - acting as a rotating phase shifter, the inverter and the supervisory control. The choise of a synchronous generator with either a small DC start-up motor (if no backup system is desired) or a diesel engine with an overrunning clutch is mandatory.

2. DATABASE

Using a computer it should be avoided to type in any information more than once. Therefore a set of datafiles, organized as shown in fig. 2, contains most of the information which were entered when using the tools.

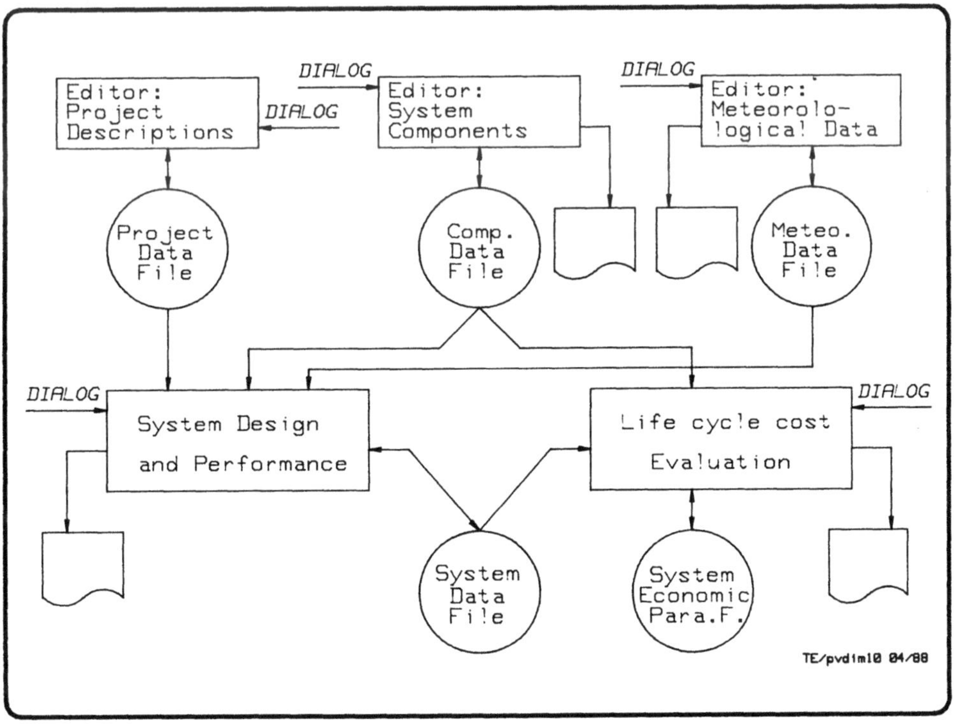

Fig.2 Data flow

2.1 Project data, system data and system cost data

Three files are keeping track of the information entered by the user or calculated during the design and evaluation process. There is one project data file for each project/investigation. The information which describes the project includes:

- Location, Latitude, Longitude etc.
- Pointer to the meteorological data file (name of the reference location)
- Restriction related to the PV-Generator (e.g limits for the tilt angle)
- Desired characteristics of the system cross bar (AC/DC, voltage)
- Loaddescription (identification, voltage, loadadaptation, etc.)
- Hypothetical daily load profiles for single loads or load clusters with a maxium resolution of one hour.

The daily load profiles can differ from month to month. For the convenience of the user a lot of possibilities for quick and easy definition of load profiles are implemented (e.g operation time for a lighthouse application: 1 hour before sunset to 1 hour after sunset)

The description of a designed system and the results from the analysis of the energy flows are stored in a project data file:

- Reference to the system data file (more than one system data file can exist for each project data file)
- System type (list of the used subsystems)
- Parameter set for subsystem description (number, type and connection of the components; parameters for the predefined operation strategies, e.g minimum operational time for dieselgenset)
- Pointer to the component data file (identification of the components used)
- Amount of the energy produced; operational time of dieselgenset; amount of the fuel used

The software package for the life cycle cost evaluation refers to the system economic parameter file which contains information on interest rate, fuelcost, price escalation rate of fuel, manpower cost per hour for O&M and subsystem specific cost calculation parameter (e.g. cost for support structure/sqm modules installed). There are default values for all these parameters but it is possible to adapt the values to the specific project requirements.

2.2 System components

This database contains all design and cost relevant data for the main components of the subsystems. Different datastructures are established for: PV-solarmodul, charge controller, PV-DC/DC-converter, battery, converter, dieselgenset/rotating phase shifter and windenergy converter. The cost information for each component includes: price ex works, design lifetime, warranty information and maintenance cost split into material (% of price/year) and personnel (hours/timeperiod).

2.3 Meteorological data

One datastructure is defined for solar data and one for wind data. Access to the data can normally be obtained by the name of the location, but it is also possible to generate a list of the stored locations sorted by their distance to given latitude and longitude.

The solar related data consist of location, latitude, longitude, altitude and monthly figures for global insolation, average Temperature, avg. min. Temperature, avg. max. Temperature and precipitation.

The wind data are stored in form of a monthly based distribution of the wind velocity. If these data are not available for a location, they can be generated from average windspeeds using the Weibull-distribution with a formfactor or the Rayleigh-distribution.

3. SOFTWAREPACKAGES

The main menue of the present version of the toolbox offers the following functions:

1. DEFINITION OF THE PROJECT
2. SYSTEM DESIGN AND PERFORMANCE
3. LIFE CYCLE COST EVALUATION
4. METEOROLOGICAL DATABASE
5. COMPONENT DATABASE
6. PRINTING OF QUESTION-SHEETS

The functions 1, 4 and 5 are used for handling of the databases described in chapter 2. Function 6 provides questionnaires for the efficient use of function 1. The functions 2 and 3 are outlined in the following.

3.1 System design and performance

This tool supports the sizing of the subsystems and predicts the output performance of the systems. After selecting the project and system file the following main steps are performed:

- Selection of the desired system configuration (PV, Wind, Diesel).
- The sizing of the subsystems (e.g. Wp for the PV-generator, kWh cap10 capacity for the storage) using the project definition file and the meteorological data base is supported by the software (see table 1) if load profiles are defined. If no load profiles are defined the sizing numbers have to be typed in by the user.
- Selection of the system components from the component data base by the user. The program checks the selected components and suggests, e.g number of modules in series and parallel connection for the Wp number provided by the sizing scheme.
- Prediction of the system output and calculation of the energy balance if load profiles are defined.
- Interactive application of the a.m. steps
- Selection of the desired documentation output. Approx. 20 different tables and figures can be produced. The user can chose between full report and short report with or without graphics or he can select the tables and figures individually.

Subsystem	Sizing as a functions of	
o Systems without dieselgenset		
PV	worst relation between insolation/load demand covered by PV	
Battery	# of autonomous days	*
Wind	desired % of load demand covered by WEC's	*
or		
PV & Battery	Loss-of-load probability /1/	*
o Systems with dieselgenset. Dieselgenset used as backup.		
PV	worst relation between insolation/load demand covered by PV	
Battery	# of autonomous hours	*
Wind	desired % of load demand covered by WEC's	*
Dieselgenset	peakload demand	
o Systems with dieselgenset. PV and/or Wind used as fuelsaver.		
Dieselgenset	average load demand	
Battery	load demand and distribution of load demand	
PV	desired % of load demand covered by PV	*
Wind	desired % of load demand covered by WEC's	*

* Parameter chooseable by the user

Tab. 1 Sizing scheme

3.2 Lifecycle cost calculation

The cost of a kWh produced - splitted in investment, O&M and fuelcost - are calculated on the basis of the information stored in the project ,system and system cost data files. The annuities for investment and O&M are calculated using the prices, lifetime, material and manpower requirements for O&M stored in the component data base plus a subsystem specific overhead. To calculate the lifecycle cost now it is only necessary to adapt the parameters stored in the system economic parameter files (see chapter 2.3)

4. EXAMPLE / REMARKS

Table 2 shows the energy balance of a PV/Wind/Diesel-Hybrid system.

Project: Lavrion System: SMA_V

INVESTIGATION: SYSTEM BALANCE (PV/WIND/DIESEL/BATTERY/LOADS)

	Available energy from PV	Available energy from wind	System energy demand	Balance PV+WIND/ demand	% Sum PV+Wind of dem.
	<-------------	(MWh/month)		------------>	
Jan	4.23	5.32	9.52	.03	100.29
Feb	5.56	7.95	8.60	4.91	157.10
Mar	5.12	1.85	9.52	-2.54	73.27
Apr	5.43	1.80	9.21	-1.98	78.49
May	4.95	1.20	9.52	-3.37	64.58
Jun	4.93	2.21	9.21	-2.07	77.49
Jul	4.99	7.80	9.52	3.27	134.40
Aug	5.11	5.47	9.52	1.05	111.07
Sep	5.24	7.18	9.21	3.20	134.76
Oct	4.76	5.86	9.52	1.10	111.52
Nov	4.01	6.03	9.21	.83	108.96
Dec	3.96	6.02	9.52	.46	104.82
SUM	58.28	58.70	112.10	4.88	104.35

	Storage subsyst. battery Input	Output	Energy from diesel	% Sum diesel of dem.	Unused excess energy	unsat. energy requ.
	<-----(MWh/month)		------>		<-(MWh/month)->	
Jan	3.55	3.54	2.84	29.87	.77	0.00
Feb	2.16	2.06	1.34	15.57	4.42	0.00
Mar	4.07	4.09	4.68	49.13	.04	0.00
Apr	3.72	3.70	4.04	43.90	.13	0.00
May	4.48	4.49	5.65	59.31	0.00	0.00
Jun	3.73	3.72	4.01	43.51	0.00	0.00
Jul	2.15	2.05	.69	7.28	2.46	0.00
Aug	2.69	2.66	1.69	17.70	1.11	0.00
Sep	2.20	2.12	1.09	11.80	2.63	0.00
Oct	3.08	3.05	2.21	23.20	1.40	0.00
Nov	3.18	3.15	2.54	27.55	1.39	0.00
Dec	3.35	3.32	2.87	30.13	1.25	0.00
SUM	38.37	37.97	33.64	30.01	15.61	0.00

BATTERY: (Refer also to the list of system components)

Setup: 59.1 (%) in April	End: 54.7 (%)	
Rated battery capacity	CAP 10	55.0 (kWh)
Worst case battery capacity	CAP 4	48.1 (kWh)
Min./max. allowed batt. cap.		20/100 (%)
Max. day/night cycle		67.1 (%)
Min. battery capacity		32.9 (%)
Max. possible battery operating period		3.8 (hrs)

Tab.2 Energy balance

Still under work is the validation of the performance model using the measurement data of a well instrumented PV/Wind- and a Wind/Diesel-System.

The software package is written in HP BASIC 5.0 and runs on HP9000 series 200/300 workstations.

REFERENCES

/1/ Chapman, R.N., "Sizing Handbook for Stand-Alone Photovoltaic/Storage Systems", Sandia National Laboratories, Albuquerque, NM 87185, April 1987

P.V. Models Vs Real data

S. LABED

Haut Commissariat à la Recherche
C.D.E.R. Route de l'Observatoire
B.P. 62 - Bouzaréah - ALGER
(ALGERIA)

ABSTRACTS : This paper is a summary of one year research period at JRC (Ispra - Italy). It consisted of a comparison between some models widely used in photovoltaics and real data collected by the european community in more than 15 PV power plants.

INTRODUCTION :

Despite the progressive and gradual cost reductions of the PV kWh during these last years, the construction of a PV power plant in order to collect data ; remains expensive. A mathematical model or a simulation program is by far less expensive to get reliable data with the condition to have the basis data (input). Models used in photovoltaics to evaluate the performance of a plant and to test the behaviour of its components under certain conditions are the same as those used in other engineering fields.
The main objective of this article is to test and compare these models with real data of E.C PV power plants [1]. For this purpose different simulation programs were written. In what follows we shall describe each model briefly and discuss it by means of graphical results.

INSOLATION MODEL :

The insolation model is the well known Liu and Jordan isotropic sky model [2,3]. For a tilted collector facing directly towards the equator the average daily tilted radiation can be expressed as follows :

$$\overline{I}t = \overline{R}.\overline{H}$$

where $\overline{H}$ is defined as the average daily insolation on a horizontal surface and $\overline{R}$ is defined as the ratio of the average daily radiation on a tilted surface to that on a horizontal surface for each month. Whith the assumption that the diffuse and reflected are isotropic ; $\overline{R}$ can be estimated by considering the beam, diffuse and reflected components.

$$\overline{R} = (1 - \overline{H}d/\overline{H}).\overline{R}b + (\overline{H}d/2\overline{H}) \ . \ [(1 + \cos\beta) + (1 - \cos\beta) \ .\rho]$$

where $\overline{H}d$ is the daily diffuse radiation averaged over a month $\overline{R}b$ is the ratio of the average beam radiation on the tilted surface to that on a horizontal surface for each month, β is the tilt angle, ρ the ground reflectance (albedo)which varies from 0.2 to 0.7.
Though for surfaces facing directly towards the equator :

$$Rb = \frac{\cos(L-\beta).\cos\delta \ \sin ws + (\pi/180).ws.\sin(L-\beta).\sin\delta}{\cos L \ . \ \cos\delta \ \sin ws + (\pi/180).ws.\sin L.\sin\delta}$$

where ws is the sunset hour angle for the horizontal surface, L is the latitude, δ is the declination of the sun and ws the sunset hour angle for the tilted surface and is given by :

$$ws = \min \left[ws, \arccos(-\tan(L - \beta) \cdot \tan\delta) \right]$$

the day by day simulation program written for this purpose compares results of the model and real data. Four input parameters are required for each station :

1. Daily global horizontal radiation (Wh/day.m^2)
2. Latitude of the site (degrees)
3. Tilt angle (degrees)
4. Albedo

CELL TEMPERATURE MODEL :

The nominal operating cell temperature of PV modules is a very important characteristic and is defined as the cell temperature of a module irradiated at 80 mW/cm^2 , 1 m/sec wind speed and 20 C ambient temperature and open circuited. Normally ; the NOCT is measured out of doors in natural sunlight. An accurate determination of NOCT is difficult because of variations in the solar irradiance, ambient temperature and wind velocity and direction [4].
A common model in photovoltaics to calculate the cell temperature is given by the following equation :

$$T_{cell} = T_{amb} + \left[(NOCT - 20)/80\right] \times I_t$$

where It is the tilted irradiance in mW/cm^2. NOCT values used in the simulation program were results of the acceptance tests done by ESTI group [5] (JRC-Ispra) on different type of modules.

MODULE EFFICIENCY MODEL :

It is well known that silicon cell efficiency is a function of cell temperature but varies very little with it. A common relation between cell efficiency and cell temperature is given by :

$$\eta_{cell} = \eta_0 \left[1 - \gamma\,(T_{cell} - T_{ref})\right]$$

where η_0 is the cell efficiency under standard conditions (1000 W/m^2 - 25 C - AM 1.5), γ is the rate of change of cell efficiency with respect to cell temperature, Tref is a reference temperature (25 C). An attempt was made here to evaluate γ for some of the P.V power plants of the E.C.

DAILY ARRAY OUTPUT :

A simulation program comprising the 3 models described above was written to compare simulated and measured daily array output. It requires 6 input parameters :

1. Daily horizontal global radiation (Wh/day.m^2)
2. Average ambient temperature (C)
3. Latitude (degrees).

4. Tilt angle (degrees)
5. Albedo
6. Array area (m^2)

and outputs the daily amount of DC power produced by the array wich is given by [6] :

$$Pd.c\ (Wh/day) = \eta cell \times It \times Area$$

BATTERIES ENERGY BALANCE :

An attempt was made here to evaluate the daily energy around batteries of EARS installation [7]. An array composed of different type of modules under outdoor tests charges the batteries during daytime. A one kilowatt DC/AC inverter supplies the data acquisition system.
The simulation program writen for this purpose include all the models described above with the input parameters plus the daily consumption. Three types of graphs are outputed for each model :

1. Daily array output energy superimposed with consumption
2. Daily input energy to batteries
3. Daily output energy from batteries

DISCUSSIONS :

Results of Liu and Jordan insolation model and measured data show good agreement (fig. 1,2). Differences were due mainly to calibration and measurement errors combined with the fact that the albedo was kept constant for the whole year ; wich is not very accurate since soiling variations were very large in some regions. A weakness of the model was observed with meteorological conditions at high level of radiation, where the tilted radiation was underestimated.
The linear relationship between insolation flux and solar cell temperature with respect to the ambient temperature shows a good correlation with measured data (fig. 3). Large differences should be attributed to the dependancy of NOCT with the type of encapsulation, wind velocity, ambient temperature combined with measurement errors in the field.
The module efficiency model seems to be a good approximation of the real module efficiency. Results (fig. 4) confirm clearly that efficiency varies very little with cell temperature. This variation ( ) has been found to be from 0.3 to 0.5 %/C. We should notice here that module efficiency was difficult to evaluate since cell temperature is not uniform in the field. This leads sometimes to large differences (mismatch losses) because module efficiency is not an accurate representation of the array efficiency.
Despite all these inconvenients the array output simulation program using the models described herein gave good results (fig. 5).
Finally the batteries energy balance simulation program gave good results (figs. 6, 7, 8) even with the assumption that the bank of batteries is large enough to absorb all the excess energy from the modules.

CONCLUSION :

Results of different models of common use in photovoltaics were tested and compared with data collected by JRC-Ispra in more than 15 stations of the european community. Most of the models were very simple but gave encouraging results. Accuracy of better than 10 % was obtained in general and sometimes less than 5 %. Large differences if any were due basically to measurement and calibration errors combined with parameters that were not taken into account in the simulation programs as wind velocity.
With this work the author wanted to test the validity of all these models described herein. Results obtained on a daily basis were acceptable and encouraging.

ACKNOWLEDGEMENTS :

This work has been supported by 'Le Haut Commissariat à la Recherche' (Algeria) and 'the International Centre of Theoretical Physics' (Trieste-Italy) and the 'Joint Research Centre' (Ispra-Italy).
The author is grateful to Dr. K.KREBS as head of the physics division and Dr. G. BLAESSER as head of the ESTI groupe for their help and support during his research period at JRC (Ispra).
Special thanks for Mr. A. TOURNIER for his valuable help and suggestions to write the simulation programs.

References

[1] G. BLAESSER, K.KREBS J.R.C. Ispra - Italy
'Summary of PV pilot plant monitoring data 1984-1985'
7th E.C P.V. solar energy conference - Sevilla-Oct. 86

[2] Benjamin Y.H.LIU and Richard C. JORDAN
'Daily insolation on surfaces toward the equator'
ASHRAE Journal Oct. 1961, p. 53-59.

[3] Benjamin Y.H.LIU and Richard C. JORDAN
'The interrelationship and characteristic distribution of direct, diffuse and total solar radiation'
Solar Energy 4,1960 - p. 1-19.

[4] John S.GRIFFITH, Mulchand S.RATHOD and Joel PASLASKI/JPL
'Some tests of flate-plate photovoltaic module cell temperatures in simulated field conditions'
I.E.E.E. conference 1981 p. 822-830.

[5] E.ROSSI, K. KREBS J.R.C. Ispra -Italy
'Survey of module tests executed by the joint research centre of the EC according to EUR specifications 101 and 501 (502)'
7 th E.C. P.V Solar energy conference - Sevilla - Oct. 86

[6] D.L. EVANS, W.A. FACINELLI, L.P. KOEHLLER
'Simulation and simplified design of P.V power systems'
I.E.E.E. conference 1981 p. 845-850.

[7] G. BLAESSER, S. KUMPF, H. OSSENBRINK, E. ROSSI J.R.C. (Ispra)
'The P.V test field EARS and the ENRA facility'
7 th E.C P.V solar energy conference - Sevilla - Oct. 86.

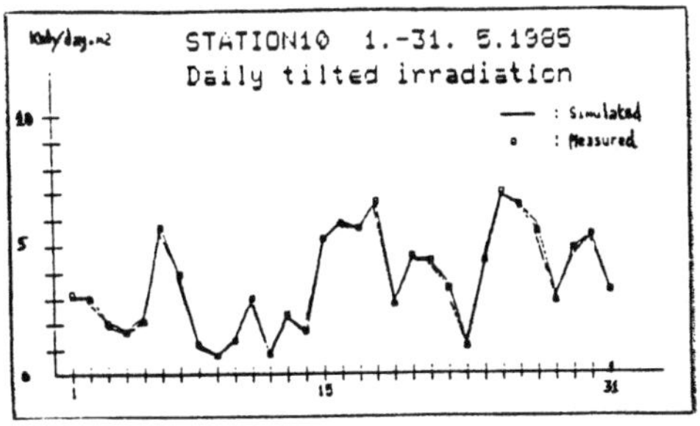

Fig. 1.

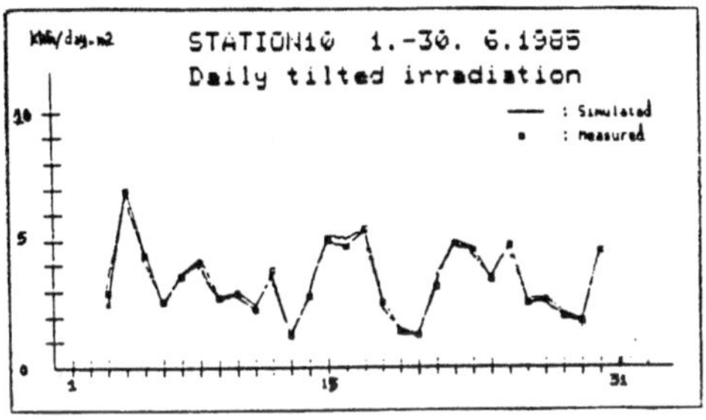

Fig. 2.

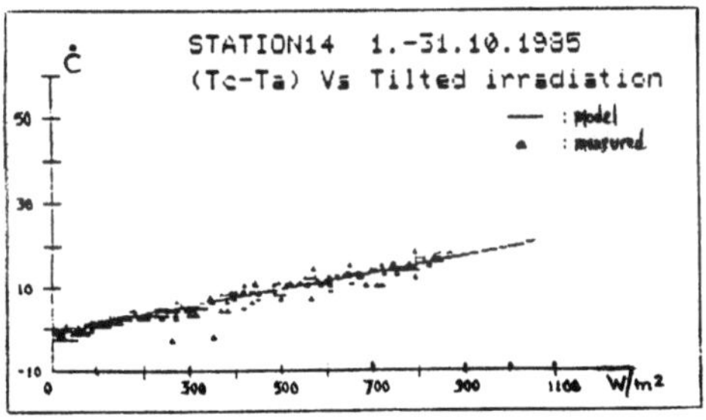

Fig. 3.

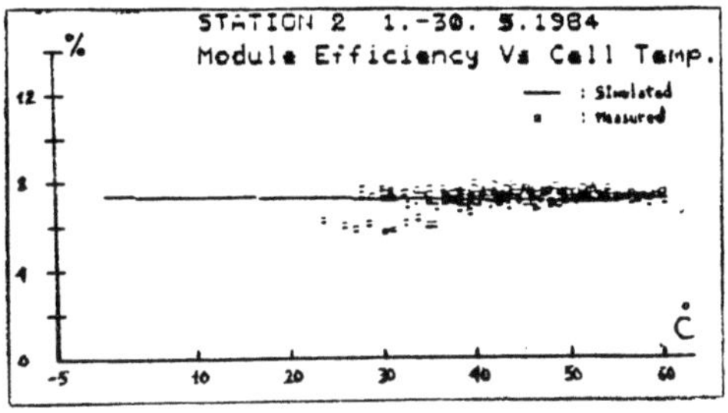

Fig. 4.

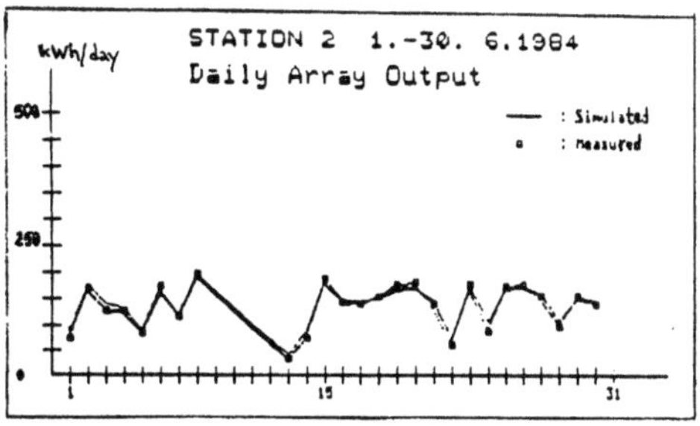

Fig. 5.

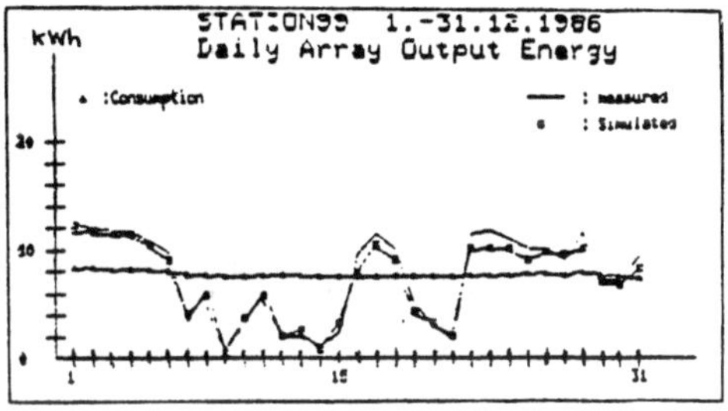

Fig. 6.

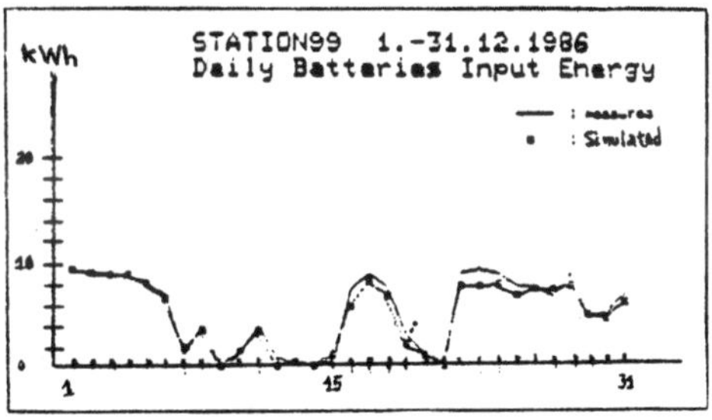

Fig. 7.

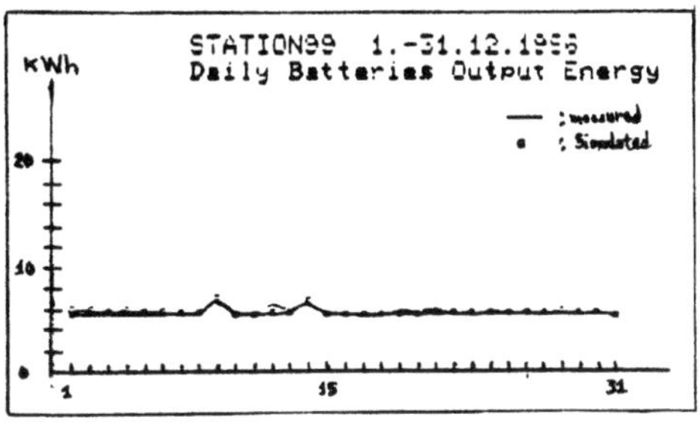

Fig. 8.

Verification of a Simulation Model for Photovoltaic Generators

T. Schott *, A. Zeidler

Deutsche Forschungs- und Versuchsanstalt
für Luft- und Raumfahrt e.V.
Institut für Technische Thermodynamik
Studiengruppe Energiesysteme
D-7000 Stuttgart 80, FRG

* WISA-Energiesysteme GmbH
D-7000 Stuttgart 1, FRG

Abstract

New experiments with the 10 kW photovoltaic generator operated in the German-Brazilian cooperation project Fernando de Noronha have been analyzed for verification of the simulation model. The grid connected operation and control characteristic of the solar inverter has been measured and adapted for stability. The thermal model proved to be sufficiently accurate for the calculation of cell operating temperatures resp. generator efficiencies. The power output of the generator is predicted with good precision - also down to dynamical time scales. Improvements of the model are possible and based on the existing model ansatz.

Keywords

PV-simulation, operating temperature model, model verification, dynamical control, frequency-power static, hybrid system, optimization.

Introduction

Concerning optimal design of photovoltaic generators, simulation programs are of practical and theoretical interest. Numerical calculations are necessary for selection of proper design parameters and for the development of operation strategies. In the context of hybrid systems, i.e. operation of a photovoltaic generator as part of an even more complex supply system (e. g. diesel generator, battery, wind converter etc.) the theoretical preanalysis of system interaction is indispensable to achieve clear operation strategies. The simulation model SIRENE (SImulation of REgenerative Network integrated Energy Systems) has been developed in parallel to experiments with a test system. The latter includes a 10 kW-photovoltaic generator with a line commutated inverter operating in a small artificial diesel-powered grid (fig. 1). This system has been described earlier in detail /1, 2/. As experimental results with this test system are now available, analysis of model verification has been started. In /2/, the basic characteristics of the solar generator in comparison with data sheet properties of single modules have been verified together with operational properties of the inverter system under stationary conditions. Presently we concentrate on the verification of system control properties, of efficiency resp. temperature model and of dynamical output simulation.

There are three types of data recorded as raw data (480 ms sample rate) and as averages over two different interval lengths (30 s resp. 10 min. average values). In the following we consider the relevance of the developed model in all three ranges of data aquisition. The results are derived from a number of experiments performed during 1987 on the test site in Natal (Brazil).

The simulation model has been described earlier /2, 5/ and only a few basic properties shall be repeated dealing with the solar generator. The procedure of modelling of a solar generator partly follows recipes as presented for instance in /4/. A basic voltage-current characteristic describes a single PV-module at defined operational conditions (standard 1000 W/m², 25 °C). In our case, this function is approximated (radiation, temperature) analytically by a so called two diode model leaving a number of parameters, which may also interpreted as physical parameters (resistances, photo current etc.), for a numerical fit to the measured module characteristic. This characteristic serves as a building block of representation of a single module. The whole generator is then given as a number of modules in parallel and in series to form a total voltage-current characteristic of proper magnitude. According to a linearized transformation rule this characteristic is then

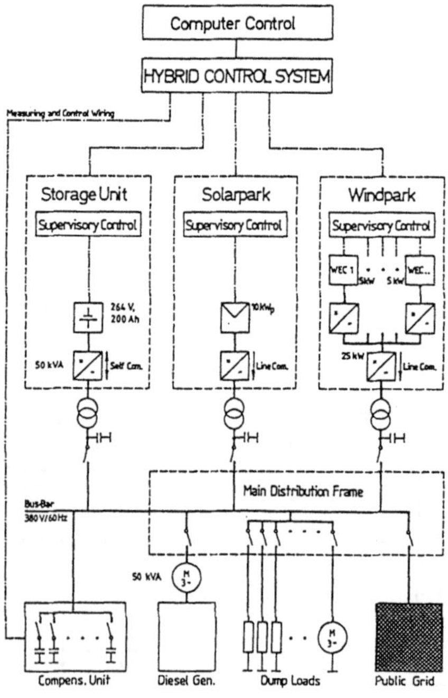

Figure 1. Blockdiagram testfield Natal

modified from standard to actual operation conditions (cf. /4/). The radiation is given as an external variable (either from a climate model itself or from measurements) and is assumed to represent the total global radiation intensity within the surface plane of the generator. The operating temperature of the module is calculated from a submodel which takes ambient air temperature wind speed and global radiation as time dependent input and a number of physical parameters as structural input. This model has been discussed in detail in /3/ and verified experimentally. Smaller differences are expected due to the different climate and due to the extended structure of the generator.

System Control

In the context of the inverter operation in a weak diesel grid, the present test system has been equipped with a sophisticated power control based on a direct characteristic response of the power output of each generating subsystem to the grid frequency. The latter is an overall system variable representing the general load situation. The range of variation of this parameter in small diesel grids usually is of the order $\pm$ 1 Hz (in this case with respect to a rated frequency of 60 Hz). The static characteristic and the dynamical response of this frequency is mainly determined by the rotating masses in the system especially of the diesel generator and is to be set by certain control parameters according to well known stability criteria (cf. /5/).

The new element here is, that the input power from solar generator (and ev. other generators) is fully accepted up to a certain upper limit of the frequency (signifying low load conditions) where further reduction of power output from the diesel is assumed to become uneconomic or unacceptable due to operational requirements. The power output of the solar generator is then limited with further increasing frequency according to a linear characteristic, inherently given in the control system by overruling the MPP-tracking process (i.e. by varying voltage). The verification of this function is shown in the measurements drawn in fig. 2. The both curves shown represent two choices of a control parameter which relates the frequency signal to the limitation setpoint for the AC-power output of the solar inverter. The control philosophy requires a relatively strong reaction of this power limitation within a small interval of the grid frequency. As the first experiment (done by imposing a certain frequency) with the original parameter choice (a $\cong$ 0,3 Hz interval) shows, a qualified control is hard to realize as there arise fading effects and influences from higher order grid parameters (wave distortion, resolution of frequency measurements etc.). By stretching the frequency interval to $\cong$ 0,8 Hz, the situation becomes much more stable (second curve/experiment).

This example shows, that power control via a frequency-power governor requires a sufficient broad frequency bandwidth during operation, a situation which is, however, quite typical for the socalled weak diesel grid. As the experiment also shows, the assumption of a linear frequency-power characteristic has been verified for the simulation (as already presumed in SIRENE).

It is, however, again necessary to analyze the dynamical response, as the frequency itself is a highly dynamical variable and the strong feed-back via power generation must be realized in a stable way not leading to oscillatory interdependence between all generators.

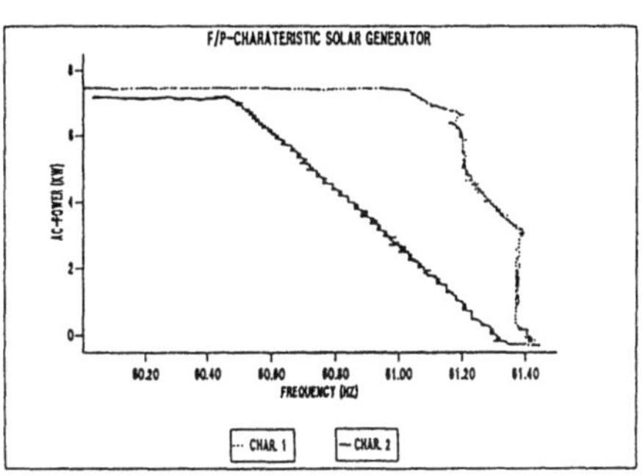

Figure 2. Frequency/power-characteristic of the solar generator (two different parameter settings)

Thermal Model

The thermal modelling of the solar generator considers as variables a cell operating temperature and an external ambient air temperature. Both variables are measured, the cell temperature being somewhat critical with respect to the exact position of the measuring sensor within the layers of the module construction (from the typical thermal masses and heat conduction properties, a temperature drop of a few degrees is expected from the inner cell temperature to the back surface temperature).
The effects of the conductive heat balance together with the convective and radiation effects (short wave and long wave spectra) have been summed up in a simplified but nonlinear model /3/ of operation temperature as a function of global radiation (Φ), ambient air temperature T_a, wind speed v_w (at generator height) and electric power output (represented as $P = \eta \cdot \Phi$ where η is the actual total efficiency for the conversion of the radiation). The following formula describes the relative temperature difference between cell and ambient air

$$T_g - T_a = [(\alpha - \eta) \cdot \Phi + (a + b \cdot T_a)]/(c + \lambda)$$

where α is the overall absorption constant of the module surface and $\lambda = \lambda_0 + \lambda_W \cdot v_W$ is a linearized convective parameter. The constants a, b, c come from a quite general linearization of the long wave radiation exchange terms and are explained in /3/. This formula has been verified through an experiment with a standard (single module for the climate of Stuttgart). As almost similar modules are taken for the 10 kW-generator in the test system, the basic constants are assumed to be the same. Minor differences are expected from the different design of an extended string of modules. The experiments under the tropical climate are then an interesting extension of this parametrization to other ranges of parameters. The value of this model lies in its simplicity by taking nonlinear effects in a proper way. It does not represent so far dynamical effects with a thermal mass distribution.

A typical ensemble of experimental results of the tests under the climate of Natal (10 min-averages of all variables recorded over different days) are depicted in fig. 3a for the temperature differences $T_g - T_a$. Neglecting the dispersion, the numerical approximation would be a linear dependence of the type

$$T_g - T_a \sim 0.023 \cdot \Phi \quad (T[K], \ \Phi[W/m^2]),$$

which is compatible with usual estimations and may be used as a good "thumb rule".

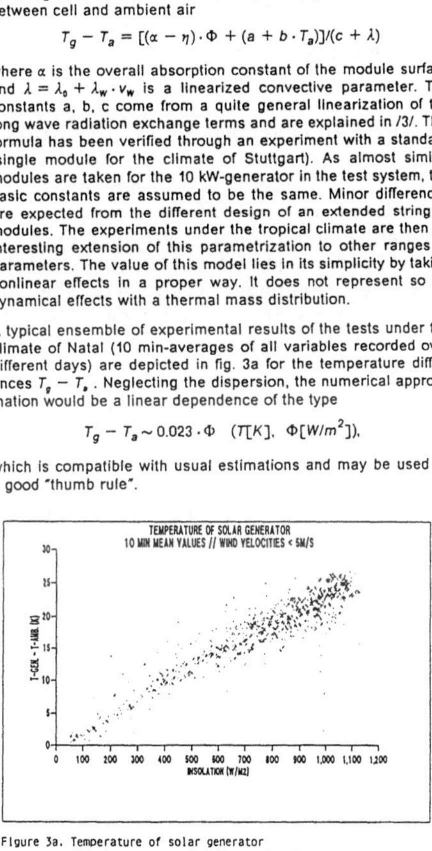

Figure 3a. Temperature of solar generator (10 min averages of different days)

Fig. 3a also reveals that the curve ends at small negative temperature differences if extended to zero radiation. This confirms that infrared emittance of the solar generator usually is higher compared to the environment leading to a surface temperature below ambient air temperature (the effect, however, is only a few degrees). A classification of this correlation with respect to wind speeds turned out to be not conclusive so far. The wind speeds are measured at a height of 1 m above the generator surface. Which is approximately 3 m above ground. Observed 10 min-averages of wind speeds range from 0 - 5 m/s and did not lead to a significant effect on the temperatures. It must, however, be stated that the wind speed usually shows a large variance within a 10 min-average (comp. fig. 3b). As the effect of convective cooling is alone. For example, with a pv price of 4 dollars/Wp and the thermal collector price of 200 dollars/m^2 (everything included in both cases), the average price of the energy obtained (about 600 kWh/m^2year) would be less than 10 cents/kWh for a pay-back period of 10 years.

Another attractive application of hybrid systems could be desalination of saline water. J.Lundstron and G.Kvajic investigated a combined system using Si crystal pv generator and electrodialysis (ED) for potable water production (2). They proposed that preheated saline water be used, since the efficiency of ED increases with temperature. It was shown that at the temperature of 65 °C the potable water production can increase up to about 80 percent with respect to the operation at 20 °C. It is easily seen that a combined system using hybrid collectors and ED could provide energy both for preheating the saline feed water and for the operation of ED membrane stacks.

4. RECENT DEVELOPMENT OF THE HYBRID COLLECTOR

Recently we made a-Si:H cells using, as a substrate, anodicaly oxidized thin aluminum plates (3). This substrate, being unbreakable, thin, durable and lightweight, has certain advantages over glass substrate, especially for hybrid collectors. With a-Si:H cells made directly on an aluminum heat exchanger (Fig. 4), the construction and fabrication of compact hybrid absorbers can become more feasible and, of course, substiantially simpler and cheaper.

Two types of electrically insulating oxide films on top of the aluminum plates were prepared: a porous type made from phosphoric, chromic, sulfuric and oxalic acid solutions, and a homogenous type made by a hard anodization process. Resistance of all oxide films formed was in the range of 10^{12} cm, and breakdown voltages were in excess of 500 V. The n-i-p, a-Si:H films were then deposited on Al/Al_2O_3 substrate. As top electrode, indium-tin-oxide layer, deposited by Rf magnetron sputtering, was used. The best Al/Al_2O_3/Al-n-i-p-ITO solar cells had conversion efficiencies of 5.3 percent. Their I-V characteristic is shown in Fig. 5.

There is a special benefit of this a-Si:H structure for the hybrid pv-thermal collector. Since the top ITO layer has low emissivity in the infrared region, this collector has the properties of a selective absorber and its thermal radiation losses can be expected to be small.

This amorphous cell structure on aluminum substrate offers an exciting possibility of constructing and fabricating compact pv and thermal solar collectors by direct cell deposition on a suitable Al heat-exchange plate. Such an attempt is in progress.

References

1. B.Lalovic, Z.Kiss and H.Weakliam, "A Hybrid Amorphous Silicon pv and Thermal Collector", Solar Cells, 19 (1986) 131.

2. J.Lundstrom and G.Kvajic, "Solar Powered, Dual Purpose Power-Desalination PV-ED System", Alt.En. Sources V.Part B, Amsterdam, (1983) 123.

3. B.Lalovic, T.Pavlovic, J. van Dine, Z.Kiss, L.R.Shiue, T.Tonon, B.Sieb and W.Xuanzhi "a-Si:H Cells on Anodically Oxidized Al"(to be published).

The measured power curve appears somewhat smoother which is due to a thermal averaging wheras the temperature model is stationary and therefore transfers all wind speed variation (which are still substantial even after 30 sec-averaging) directly to the power output calculated by the efficiency model. Incidentally, the short term deviation of measured and calculated power output (observed at t ≅ 940) is correct as there has been recorded a reduced radiation value. The fact, that the measured power has not been reducing, is interpreted as an artificial reduction of the signal due to partly shading of the sensor. The break down at t ≅ 850 is due to a short switch off of the system for battery test initiation. In general, the observed overall deviation of the simulated power from the actual measured power output of the solar generator, lies within the range of a number of smaller uncertainties inherent in the system.

As additional sources of uncertainty may be mentioned:

- sensor failure resp. calibration errors
- auxiliary power consumption (e.g. measuring devices power supply, inverter/MPP control)
- mismatching effects
- transducer errors

All these errors may add up to a error range of up to 10 % under bad circumstances. Also the modelling itself contains sources of possible systematic errors. Up to this point, the present model has not been further calibrated to the experimental results! Instead, the properties of the model were still derived from the basic properties as verified for the single module. A favourite candidate for explanation of the apparent difference in power output is, however, the possible mismatching of DC-strings. This effect (as already stated in /2/) is of the order of 5 % at full radiation level. As the DC-strings (11 in the case of the 10 kW-generator) may be operated in groups or individually switched, the mismatching is accessable to direct measurement. This experiment, however, requires rather constant ambient conditions and is not easy to perform. More attention will also be paid to this analysis in the future.

Conclusions

The present simulation model SIRENE has been validated on different scales of time resolution by a number of experiments with a 10 kW photovoltaic generator under tropical conditions. This scale ranges down to seconds where still dynamical effects from the meteorological conditions are expected and of relevance for grid control in hybrid systems. The temperature model also turned out to be sufficiently accurate but better calibration is even possible by tracing individual material and structural properties of the photovoltaic generator back to the original model.

There is, however, experimental evidence that the cell operating temperature differs substantially from the model calculation under operation at short circuit conditions. The magnitude of the effect presently is not clearly understood.

The dynamical reaction of the frequency-power inverter control is complicated and may lead to instabilities principally unavoidable. At the present stage of operational experience with the test system this is avoided by a sufficiently soft frequency-power characteristic. The subject is, however, of importance to operational optimization criteria and deserves further experiments.

References

/1/ Schott, T.; Reiniger, K.: Integrated Energy Supply Concept for Fernando de Noronha. In: Energy for Rural and Island Communities III, Proc., Inverness, 1983

/2/ Schott, T.; Zeidler, A.: PV-System Analysis. In: ISES Solar World Congress, Proc., Hamburg 1987

/3/ Schott, T.: Operation temperatures of PV-modules - a theoretical and experimental approach. In: VI CEC-PV-conference, Proc., London, 1985

/4/ Rauschenbach, H.S. : Solar Cell Array Design Handbook, London, 1980

/5/ Schott, T.; Zeidler, A.; Reiniger, K.: Hybrid System Wind/Photovoltaic/Diesel/Battery - Theoretical and Experimental Results. In: Wind-Diesel Workshop, Proc., Rutherford Appleton Lab., UK, 1987

USE OF A DETAILED ARRAY SIMULATOR FOR THE PERFORMANCE AND DIAGNOSTICS OF THE DELPHOS PHOTOVOLTAIC PLANT

L.M. Fabriani, L.G. Gentilin, M. Romanazzo
ENEA - CRE - CASACCIA, S.P. Anguillarese km 301 00100 ROMA Italy
A. Sarno
ENEA - CRIF, POB 83 80055 Portici (Napoli) Italy
P. Spirito, G. Vitale
Dept. Electronic Engineering, Univ. of Napoli, via Claudio 21, 80125 Napoli Italy

ABSTRACT

A detailed array simulator is presented, which allows to calculate the power losses in photovoltaic arrays, due to the mismatch between cells and modules. The simulator takes into account the actual I-V characteristics of the cells and modules, and the different interconnections used in large power arrays.

The simulator is used to perform the analysis of the mismatch losses and of the fault tolerance of the DELPHOS 300 KW_p Photovoltaic plant.

1. INTRODUCTION

The power losses in photovoltaic arrays, due to the mismatch between cells and modules, have been the subject of several theoretical studies /1,2/. Most of these studies, however, consider only very simple module interconnections and take into account only the short circuit current mismatch.

In recent high-power plants, due to the different technologies presently available on the market (single crystal, poly and amorphous Silicon), it is likely that different types of modules are put together to obtain the required power. As a consequence, there is a need of an array simulator that can take into account for different I-V characteristics of the cells, including detailed models of the reverse bias branch. Moreover, the array simulator must be able to handle the complicated series-parallel connections which are commonly encountered in large-size arrays.

In this paper, we report on a detailed array simulator that meets the requirements mentioned above. The simulator is based on a model of the photovoltaic cell that takes into account for the effects of series and shunt resistances, temperature effects, and for the reverse-bias breakdown voltage. The simulator can handle multiple series/parallel connections and does include the effects of bypass diodes at any interconnection level.

The model is used to analyze the performance of the 300 KW_p DELPHOS plant, which includes different cell types (single crystal and poli Silicon) and different modules, supplied by three different manifacturers.

Results show the power degradation following the failure of one or more cells, and supplies a method to identify the existance of failed modules into the array. Moreover it allows to calculate the electrical stress induced by the failure of one or more module, and supplies a guideline to select the connection of the modules that improves the fault-tolerance of the array.

2. MODEL OF THE PHOTOVOLTAIC ARRAY

The model of the photovoltaic array allows to obtain the output current voltage characteristic of the array for multiple series/parallel configurations up to the fifth level of hierarchy, and accounting for the presence of bypass diodes in series connections at any level.

The main features of the simulator are:

1 - It permits to use cells with different output characteristics, located anywhere within the array.

2 - The reverse-bias characteristics of the cells are modeled by including both the effects of shunt resistance and of avalanche multiplication, thus allowing a accurate simulation of the actual output characteristics of the cells.

3 - Different cell failures can be simulated, that yeld to either a short-circuit or to open-circuit conditions.

4 - Any part of the photovoltaic array can be partly or totally obscured.

The basic cell equations, used in the simulator are:

a - forward bias: The output current, I, is related to the voltage, V, as follows:

$$I = I_{ph} - I_{sc}(\exp(-V_{oc}/AV_T)-1)(\exp(V/AV_T)-1) \qquad (1)$$

where, I_{ph} is the actual photogenerated current, I_{sc} is the photogenerated current under AM1 radiation, V_{oc} is the AM1 open-circuit voltage, A is a diode ideality factor that is adjusted to include the effects of series and shunt resistance of the cell. Thus the factor, A, allows to simulate the cell fill factor.

If the cell is partly (or totally) obscured, the actual photogenerated current I_{ph} is changed, and the "degree of illumination" $I^* = I_{ph}/I_{sc}$ is used to define the fraction of the cell that has been obscured.

Cell failures, that resulted in open-circuit conditions are simulated by setting $I_{sc}=0$ and $A=\infty$ in the Equation above, while if the

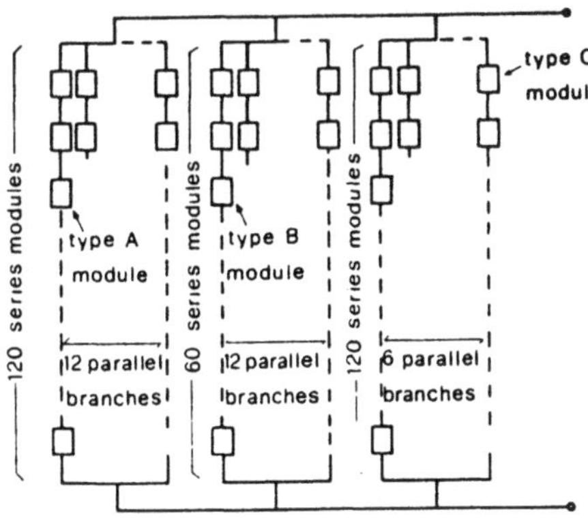

Fig.1 - The DELPHOS array

failure results in a short-circuit, the conditions are: $I_{sc}=0$, $V_{oc}=0$.

In Eq.1, the temperature effects are included both in the short-circuit current and in the thermal voltage, V_T.

b - Reverse bias: if V<0, the cell Equation is:

$$I = [I_{ph} - I_o(\exp(V/AV_T)-1)]\ M(V) - V/R_{sh}$$

where: $M(V) = 1/[1-(V/V_{br})^a]$ accounts for impact ionization, R_{sh} is the shunt resistance. A combination of these factors allows to model a large variety of experimental reverse-bias characteristics.

3. RESULTS: ANALYSIS OF THE DELPHOS PLANT

The simulator illustrated above, has been used to study the changes occurring in the output characteristics of a large array, following the failure of one or more cells. The main purpose is to identify which parameter must be monitored to detect the occurrence of a failure with the greatest sensitivity. The array structure of the DELPHOS plant has been used, whose structure is reported in Fig.1. This plant has a peak power of 300 Kw; It includes both single crystal an poli cells, coming from three different manifacturers, incapsulated into modules with

TABLE 1

Module	total series cells	parallel strings per block	series blocks per module
A (poly)	12	3	1
B (poly)	24	3	8
C (mono)	12	6	4

different number of cells and interconnections. Because of the structure of the DELPHOS array, each branch, corresponding to 120 (type A or C) or 60 (type B) series modules (according to the manifacturer) can be analyzed independently of the others. The cell connection within the modules is indicated in Table 1.

The simulation was carried out by assuming the failure of one cell within a module, yelding to a open-circuit condition at the point

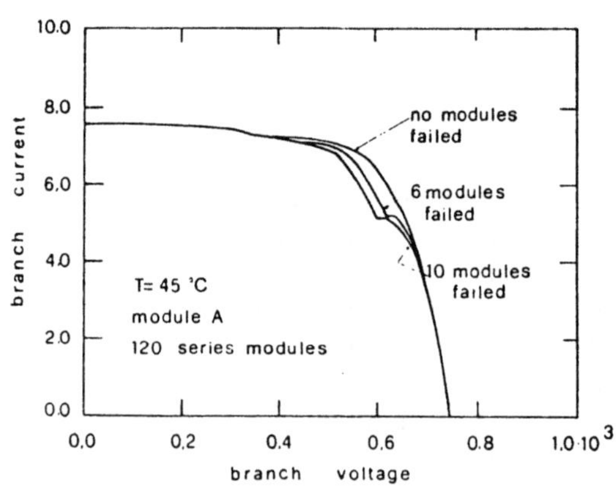

Fig.2 - The branch output curves

of failure /3/. Because of the large number of series modules, the failure of only one cell, within type A or B modules, is sufficient to put the bypass diode into operation, thus short-circuiting the whole module. In type C modules, that have been made with a larger paralleling of cells, at least two cells, within the same block, must fail, in order to bring the module reverse biased.

Fig.2 reports the modifications in the output I-V characteristics of one branch, made of type A modules, as a function of the number of failed modules. It is seen that no significant changes occur in the open-circuit voltage or in the short-circuit current, as the number of failed modules increases. Instead, significant changes occur near the maximum power point, even for a small number of failures.

Fig.3 reports the percent decrease of the main parameters of the output characteristics of the branch vs. the number of failed module, for type A module. From this plot it turns out that, by increasing the number of damaged modules, the most sensitive parameters are the branch voltage and power at the maximum power point, while the current at the maximum power point is rather insensitive to the number of failures. This is a consequence of the array structure, that consists of a large number of series modules, without cross interconnections, that makes each branch to act as a constant-current generator with respect to the failed modules.

If the branch voltage at the maximum power, V_M, is used to monitor the operation of the plant, the voltage decrease is proportional to the number of failed modules and it is possible to detect a number of failed modules less than 10% of the total modules in that branch.

If type B module is simulated, we obtain a similar dependence of the various parameters vs. module failure but, owing to the higher voltage of each type B module, the same reduction in the maximum power voltage, V_M, occurs as the number of failed modules is one-half of the previous case.

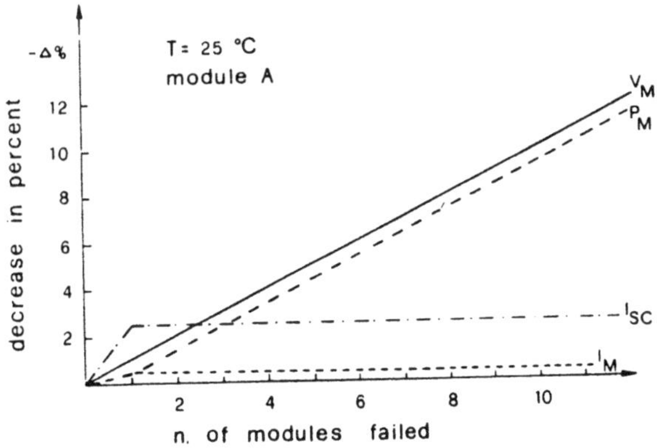

Fig.3 - Sensitivity of the branch parameters to module failure (type A modules)

The failure of type C modules, has the same effects as type A, but we have to recall that, in order to have a type C module short-circuited, at least two cell failures must occur in the same module, within the same block, a quite unlikely event.

The possibility to detect the occurrence of a module failure could be reduced, if significant temperature differences occur between different branches. Fig.4 reports the effect of temperature variations on the values of the main parameters of the output characteri stics. As seen from this Figure, referring to the voltage V_M, differences in temperature as large as 20 °C are needed, to obtain a V_M decrease comparable with that resulting from the failure of 10% of the modules. Therefore, it is concluded that temperature effects do not reduce significantly the possibility to detect the module failure.

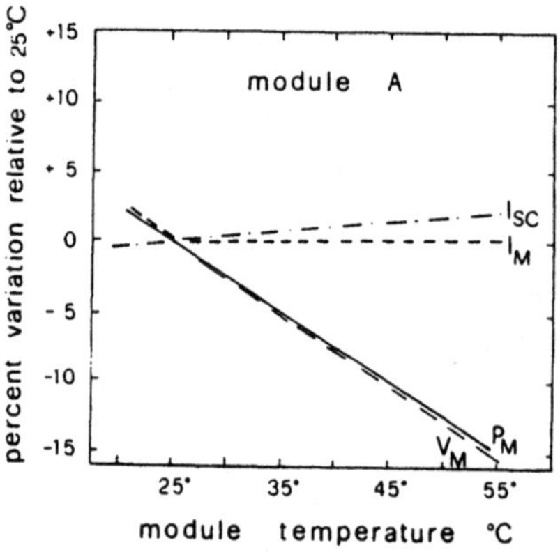

Fig.4 - Temperature dependance of branch parameters

REFERENCES

/1/ J.A.Arnett, C.C.Gonzales, Proc. 15th Phot. Spec. Conf. Orlando,1981

/2/ P.Spirito, V.Albergamo, Proc. 5th E.C. Phot. Solar Energy Conf. Athens, 1983

/3/ R.G. Ross, Proc. 4th E.C. Phot. Solar Energy Conf. Stresa, 1982, pp. 169-178

Solar Generator and Load : Automatic Control Strategy for Optimum Matching

by

S. Pace and G. dal Bon

Commission of the European Communities
Joint Research Center Ispra Establishment ESTI
21020 Ispra (Varese) Italy

Abstract

Advanced automatic control strategies, yet to be developed, may improve the efficiency to cost ratio of photovoltaic systems. The energy dissipated in the load, and the energy produced by the array, are never well-matched. An ideal photovoltaic system should dissipate exactly the amount of incident solar energy which is converted into electricity, without entering conditions of energy surplus or energy deficit. The main objective of the work presented here is to automatically adjust, in real time, the load demand to match the available electrical power. This is achieved by monitoring the battery state-of-charge through measurements of the electrolyte density, and by intervening on the system as the electrolyte density reaches certain threshold levels. In this way it becomes possible to :

- solve the difficult problem of accurately monitoring battery state-of-charge ;
- control the whole photovoltaic system ;
- ensure proper charging and discharging of the battery ;
- peak the load to match the available energy, thus improving the direct use of energy near the maximum power point.

Introduction

Stand-alone photovoltaic energy systems can be considered as small utilities in which the primary energy source varies uncontrollably. Unlike a utility in which the effort is to distribute, as far as possible, the load over the day to reduce peak demand ; a PV system requires peaking of the load to match the available energy. Consumer demand should therefore be closely matched to the profile of the generated PV energy, to improve the overall performance of the system.

The PV system test facility which we have set up at the JRC Ispra has been used for developing new, automatic, control procedures. The energy management strategy described in this article has been operating successfully for more than six months. It assumes : (1) a constant daily amount of energy consumption by the load (a pump motor), which simulates the basic needs of the consumer ; (2) periods of extra pumping, repeated as long as energy is produced, to peak the load to dissipate the available energy. The rating of the pump motor matches the array maximum power under full insolation.

Decisions for controlling the entire PV system are influenced by the battery, whose state of charge is monitored by measuring the specific gravity of the electrolyte. The battery is effectively the keystone of this system ; it is a varying load, whose size depends on its state of charge, and which affects the behaviour of the entire system.

The importance of defining control procedures for stand-alone PV systems, and of solving the problem of monitoring battery state-of-charge, have been described by several authors [1,2,3]. A recent article [4] lists various control concepts (voltage regulation, load and energy management, etc.) and their respective merits. We note that our control procedure includes all of them.

Approach and System Description

Our PV system test facility (Fig. 1) consists of a photovoltaic generator connected to a 540 Ah/24 V storage battery and a 350 W DC pump motor. An AC line, with an inverter will soon be installed. Thirty-two Arco Solar type 16-2000 modules are installed on a sun-tracking table, and series-parallel connected for a nominal total power of 1 kW at 24 V.

A computer is used for controlling the system and for monitoring its sub-units ; it also performs periodic, absolute, measurements of the battery's state of charge. A data acquisition system accepts a wide range of interdependent analog and digital inputs, and generates outputs for controlling the system, generating alarms, and actuating external devices and instruments.

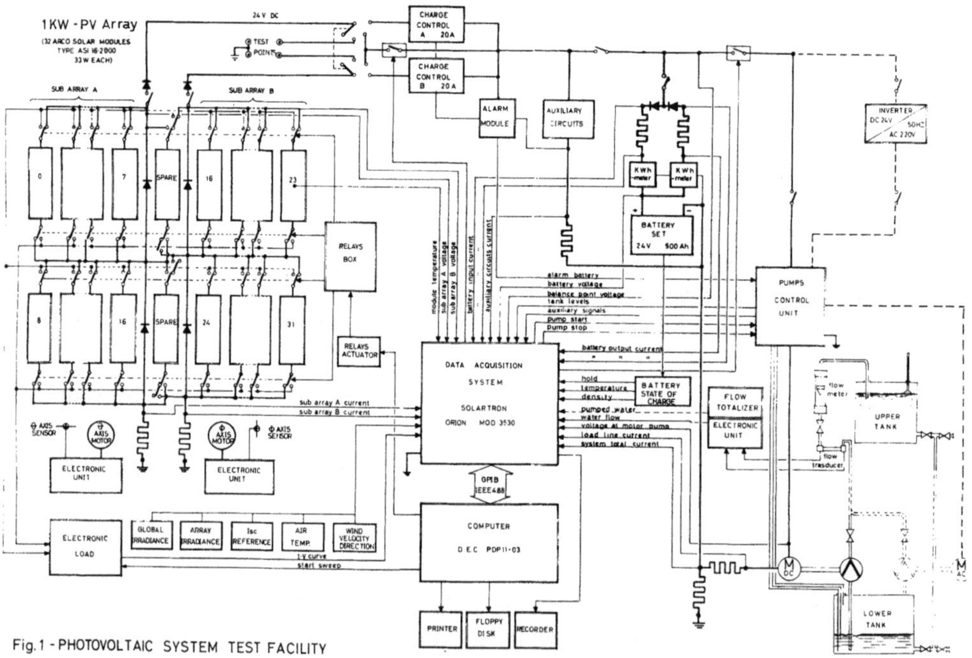

Fig. 1 - PHOTOVOLTAIC SYSTEM TEST FACILITY

An ideal PV system should dissipate exactly the amount of incident energy converted into electricity, without entering conditions of energy surplus or energy deficit. In our facility, the load demand (defined as the integral over time of the load rating) is adjusted automatically to match the available electrical power. This process involves a number of steps :

(1) The average daily energy production for each month is evaluated from the mean, over 25 years, of the monthly means of the daily sums of the global irradiation (from local meteorological records), corrected for the array conversion efficiency.

(2) The number of pumping hours per day is set as a fraction (say 20%) of those required to dissipate the daily energy production evaluated above.

(3) Energy consumption by the load is regulated as follows :

(a) The pump is switched on every day for an ordinary daily pumping period to simulate the basic needs of the consumer, dissipating a much smaller amount of energy than the foreseen daily energy production. This results in a marked tendency to keep the battery set fully charged, with the added benefits of longer battery life, and ready availability of the whole storage capacity.

(b) As the system approaches the condition of *energy surplus*, an extra-ordinary control action is performed : the ordinary daily pumping period is repeated as often as required to prevent the system from entering energy surplus, if possible. The pump is therefore actuated when, and until, there is energy production, i.e. when it is important, for better system efficiency, to peak the load and consume more power, increasing the direct use of energy, and keeping the array operating near its maximum power point.

(c) If the system does enter the energy surplus condition, the pump is kept working until the battery state of charge returns to its normal operating range. The amount of energy to be removed is periodically redefined on the basis of the past behaviour of the system. If the system shows a tendency to enter the energy surplus condition, the battery is brought more closely to its minimum state of charge, thus making subsequent entries into energy surplus less frequent, and less probable.

(d) Conversely, when *energy deficit* conditions occur, both the ordinary daily pumping period, and any extraordinary power dissipation in the load, are inhibited until the battery state of charge returns to its normal operating range. Again, the amount of energy to be recovered by the battery is calculated taking the past behaviour of the system into account ; thus the battery state of charge will be brought towards its maximum value.

Experimental Results and Discussion

The observed instantaneous correlation between electrolyte density and energy stored in the battery set is shown in Fig. 2. Density measurements are made by probes inserted into the battery cells. Charging and discharging rates affect the corresponding curves through time lags in the density measurements caused by geometrical effects and stratification of the electrolyte.

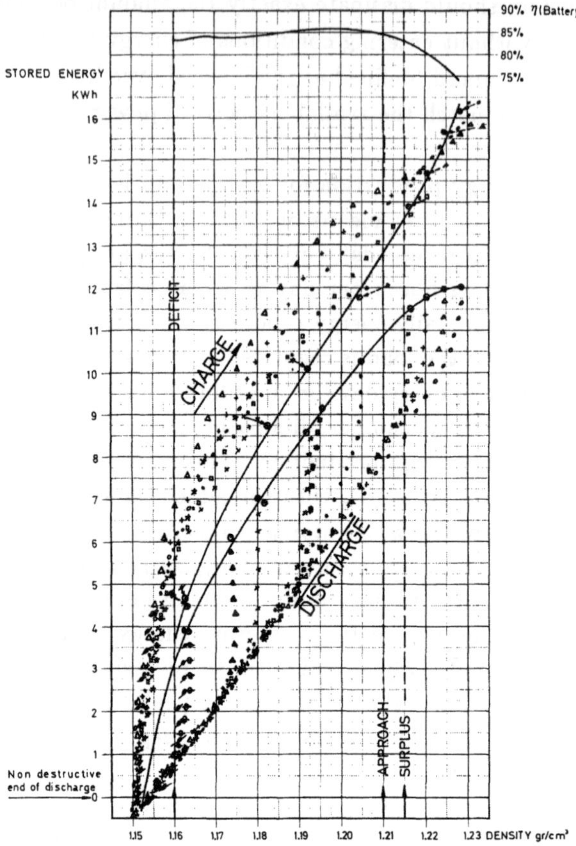

Fig. 2- Density vs. stored energy (battery)

The two continuous lines represent the charging (upper curve) and discharging (lower curve) limit curves for infinitesimal rates of charging and discharging. They were obtained during a number of complete battery cycles by inserting the correct time delay after every change in the stored energy. The lower curve is used as a reference for control purposes because it shows, for every density value, the amount of energy that can be extracted from the battery.

The battery efficiency, η, for every density value, can be obtained from the separation of the two curves, as shown in the upper part of the plot. The control thresholds (marked as *Deficit*, *Approach*, and *Surplus*) were positioned in order to force the battery to operate in the region where the η curve is flat. For density values greater than 1.215 gr/cm^3, the plot clearly shows that the two curves begin to diverge more and more, hence it is disadvantageous to operate the battery in this region, where its efficiency drops.

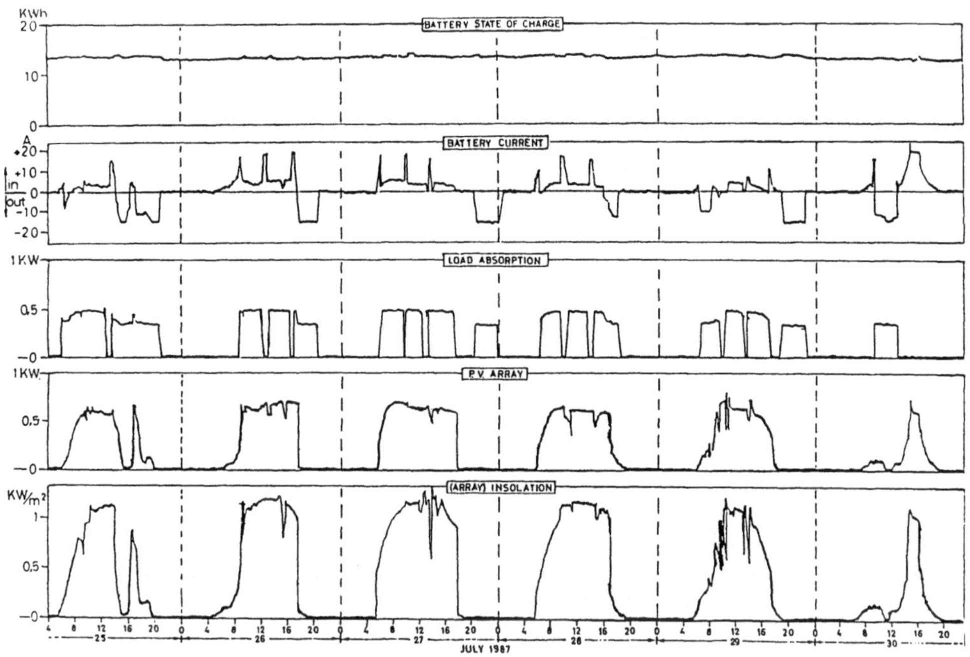

Fig.3-SYSTEM BEHAVIOUR

Fig. 3 shows the typical behaviour of the system, recorded over a period of six days in July 1987. The battery state of charge is kept high in the region where the efficiency, η, is maximum. The battery current, apart from some spikes, remains close to zero ; as it should, when most of the available energy is used directly. Finally, energy consumption by the load is well matched to peaks of energy production.

Conclusions and Future Developments

The control strategy for photovoltaic systems described here has produced satisfactory results. The density measurements are being improved, to eliminate the time lags caused by electrolyte stratification. A new, and simpler, technique for determining battery state-of-charge will be installed and tested in the near future. This technique measures the internal resistance of the battery, using very high current pulses with durations of less than a millisecond.

References

[1] R.C. Cull, A.H. Eltimsahy – Proc. IEEE PV Spec. Conf. 1982.

[2] P.P. Groumpos, R.C. Cull, A.F. Ratajaczak – IEEE Trans. on Power App. ans Sys. Vol. PAS–103, **10**, Oct. 1984.

[3] P.P. Groumpos, G.D. Papageirgiou – Proc. IEEE PV Spec. Conf. 1984.

[4] P.P. Groumpos, N. Sadati, N. To Diarra – Proc. IEEE PV Spec. Conf. 1987.

TISO SP: SIMULATION OF A PV PLANT WITH ARBITRARY CONFIGURATION

Lorenzo Castellari
Dipartimento dell'ambiente
Sezione energia e protezione dell'aria
6500 Bellinzona, Switzerland

Summary

Two sets of computer programs were developed in the frame of the project TISO. The first package, a cell level package, accepts a formalized description of a general eletric circuit composed by arbitrarily connected bipolar elements and outputs current and voltage values across every element. A photovoltaic cell or array of cells can be simulated by these programs, the cells being described by the familiar exponential five parameter model. The second package, a system level package, enables to simulate the behaviour of power plants composed by an array of identical photovoltaic modules, a power conditioning unit, a battery charge controller and a load with arbitrary profile.

1. Introduction

In the frame of the project TISO, in October 1986 we started to develop some computer programs with the aim of realizing a model for the simulation of a PV plant, to be tested with the data acquired at the TISO plant.

Our first task was to determine the electrical behaviour of the PV field, considered as composed of more elementary parts. These can be taken as the modules, the cells, or even more basic components (resistances, diodes, current and voltage sources), if a discrete-element model is taken for the modules or cells. The problem can be expressed as follows: given an electrical circuit of bipolar elements arbitrarily connected, each having a known, fixed, I-V characteristic, find the global I-V characteristic of the circuit. The circuit itself need not be bipolar, i.e. it may admit more than 2 entry point for currents.

No simplifying hypothesis is made on the connection scheme. The solution of this problem led to a first group of programs which we'll describe in section 2. below.

We turned then to the modelling of a whole PV plant, dealing with the following matters: model the way solar radiation and air temperature affect the electrical behaviour of the PV field; describe the load profile; model the electrical behaviour of an accumulator; find, at any time, the electrical working point of the system, i.e. the voltages and currents in every part of it. The model allows for the introduction of a peak power tracking device and a voltage regulator on the load. We used here a simplified model for the PV field which considers it as a multiple of a single module: the voltage and current of the whole field are obtained by multiplying the voltage and current of a single module respectively by the global number of modules in series and in parallel. This general model for a PV plant, consists of a second group of programs described in section 3.

2. **Analysis of an electrical circuit**

Both for the global circuit and for the single components, voltages are expressed as functions of currents. We want thus to determine the entry voltages of the global circuit for given entry currents. We therefore need to determine the voltage drops and the currents across every component. We can find them by requiring the 2 Kirchhoff's lemmas to hold. The analysis is formalized by considering the graph obtained by associating to every electrical component an oriented, numbered edge of the graph, and to every connection point a numbered vertex.
The analysis is peformed along the following steps:

a) acquisition of the description of the graph translated in terms of an incidence matrix A and an entry matrix B. A has 1 row per vertex and 1 column per edge of the graph, and is defined by A_{ij} being +1, −1 or 0 according to whether the j-th edge enters, leaves, or is not connected to the i-th vertex. B has 1 row per vertex, 1 column per entry and is defined in the same way as A (with entries instead of edges).

b) the I-V characteristics of all the individual components are stored as sets of parameters to be handled by appropriate subroutines;

c) by analyzing the incidence matrix a tree diagram is extracted from the graph and the edges are classified as dependent edges if they belong to the tree, as independent if instaed they close some loop;

d) if we define the vectors $\vec{C}$ and $\vec{E}$ by C_j being the current flowing in the j-th edge and E_k the current through the k-th entry, one can see that the matrix equation $A\vec{C} + B \cdot \vec{E} = 0$ rapresents the system of linear algebraic equation, one per vertex, expressing the 1st of Kichhoff's lemmas. By diagonalizing the part of A consisting of the columns corresponding to the dependent edges, we find a set of new linear equation which allow us to obtain each dependent current in terms of the independent and the entry ones. If there are no loops in the circuit (no independent currents to determine) the problem is solved since from here we can calculate immediately all of the currents;

e) if there are loops in the circuit we make use of the 2nd of Kirchhoff's lemmas to determine the independent currents. Taking these as variables and the voltage drops around the loops as functions of them (the 2nd lemma requiring these voltage drops to be zero), after assigning an arbitrary initial value to the variables, we correct them iteratively until our functions are led to zero. This can be realized by some numerical method; for us it is basically the Newton-Raphson method for many variables.

The iterative cycle is the following:

a) assign an arbitrary initial value to the variables (the independent currents); let $\vec{I}$ be the vector defined by I_m being the m-th independent current;

b) calculate the dependent currents out of these (via the linear equations of step C above);

c) calculate the voltages of the edges from the known I-V characteristics of the components, and sum them up obtaining the values of the voltages around all the loops; let $\vec{L}$ be the vector defined by L_l being the voltage drop around the l-th loop;

d) if $\vec{L} = 0$ the 2nd lemma is satisfied and the currents have just the right value; otherwise we construct the matrix D of derivatives, defined by $D_{ij} = \partial L_i / \partial I_j$; small variations $\Delta\vec{I}$ and $\Delta\vec{L}$ of our variables and functions are related by the matrix equation $\Delta\vec{L} = D \cdot \Delta\vec{I}$. According to the Newton-Raphson method we shall then take a correction for the independent currents given by $\Delta\vec{I} = -D^{-1} \cdot \vec{L}$. After correcting the value of the independent currents as $\vec{I} \rightarrow \vec{I} + \Delta\vec{I}$, we start again at (b) and repeat the cycle until we find $\vec{L} = 0$.
Once the working point of the components is found, all data about the global circuit are easily obtained.

3. Model of a PV plant

We have been following essentially the guidelines proposed in ref. (4). The input data can be divided into 4 groups:

a) choice of the configuration of the system: it contains a PV field and a load; the user can choose whether to include or not a battery, a peak power tracking device and a voltage regulator on the load;

b) qualitative description of the devices: for the modules and the battery it consists of a set of parameters to be used in ad hoc models; for the peak power tracking and the voltage regulator the efficiency curves must be given; the load is defined as a set of superimposed periodic load curves; the user fixes the length of the periods and the load profiles within each of them;

c) dimensions: number of modules globally in series and parallel; number of accumulator cells in series; capacity and initial charge of the battery; nominal power of the peak power tracking device and the voltage regulator;

d) other data: length of the simulated interval; latitude of the plant; orientation of the modules.

The output consists of the power flowing in the different components of the plant, the current and charge of the battery, given for each time interval. The working point of the system is found by the intersection of the source's I-V curve (PV field, peak power tracking, battery during discharge) with the load's I-V curve (load, voltage regulator, battery during charge). The method used to determine this point depends on the configuration of the system and may be numerical or analytical.

4. Conclusions

In the time allowed for the development of the model we could have the described models work correctly as programs, but we were not able to performe systematic testing and verification. Next developements will unify the cell- and the system-level program packages and, possibly, produce a very general and powerful simulation and analysis tool.

Acknowledgements

This project was financially supported by the Swiss Federal Office of Energy. Helpful suggestions and great support for data processing are acknowledged to Mr. G. Salvadè and Mr. C. Spinedi of the Laboratorio di Fisica Terrestre, 6952 Canobbio, Switzerland.

Bibliografy

(1) M. Camani et al: Performances of the grid Connected photovoltaic plant TISO 15, 6th EC Photovoltaic Solar Energy Conference, London, April 85

(2) P. Ceppi et al: Behaviour of the modules of the photovoltaic plant TISO 15, 7th EC Photovoltaic Solar Energy Conference, Sevilla, Spain, October 86

(3) F.P. Califano et al: La progettazione degli impianti fotovoltaici, Liguori, Napoli, 1984

(4) T.J. Lambarsky et al: Photovoltaic Transient Analysis Program User Guide, the BDM Corp., Albuquerque, N.M. Sandia lab. 1978

(5) R. Boite et J. Neirynck: Theorie des réseaux de Kirchhoff, Presses Polytechniques Romandes

REAL TIME OPTIMIZATION OF THE PHOTOVOLTAIC GENERATOR WORKING AND ENERGY TRANSFER MANAGEMENT, FROM TWO SPECIALIZED PROCESSORS

C.LISHOU - C.KRAIF - G.SOW - L.PROTIN
Laboratoire d'Energie Solaire
E.N.S.U.T BP 5085 DAKAR-FANN SENEGAL

ABSTRACT

This paper describes a research work which contributes to a project aiming to develop a wind and photovoltaic multigenerator energetics micropower generating station in which the working of each generator is optimized by a specialized processor card using the Motorola 6809 microprocessor.

The reason why this study is being carried out is explained by the important part played by renewable energies in developing countries in general, and in the Sahel area in particular (50 to 90 %) [1].

The purpose of our work is to describe an experimental device that makes it possible to optimize in real time the working of a photovoltaic generator (400w) and to carry out the management transfer of all the energy produced by a numerical control of the current into a 48 volt battery, by means of two Motorola 6809 specialized processors.

In the present version, the power optimization software developed in assembler langage, is integrated in a PROM memory of the number 1 specialized processor card.

The development system architectured around a microcomputer is designed to develop the hardware and the battery no current feedback software of the number 2 processor card.

INTRODUCTION

The aim of a micro power plant which we are about to realise at ENSUT is to simulate the functioning of a small autonomous power plant situated in a remote area (sahalian zone) and which could assure electricity supply (220v) and rural water supply.

Since solar energy is random, it is necessary to envisage a complementary energy source (e.g wind energy) to obtain a continuous functioning with a reduced electrical energy storage. After electrification, the excess power generated is used for photovoltaic water pumping [2].

We will discuss the principle of a multigenerator/multireceptor micro-power plant [3]. With this type of plant, three fondamental problems arise.

- The optimization of each renewable energy sources [4],[5]
- The coupling among each sources (fig. 1)
- The transfert in real time of all the power generated towards the receptors

The paper which we are presentig forms part of a design of micro energetics power plant. We will describe here an experimental set-up and the tests carried out in order to optimize the performance of the photovoltaic generator and to assure a real time total transfer of the energy generated towards a variable load.

SYSTEM DESCRIPTION

Since the aim of this construction is to make it simple, flexible and reliable, we have opted for a set-up that is digitally controlled by microprocessors where, a maximal number of the tasks are carried out by a software, thus simplifyant the hardware.

Our work consists essentially of two tasks. Figure 2 shows the hardware necessary for the execution of the different tasks.

Task 1 : Optimization of the performance of the photovoltaic generator by feeding-back the working voltage or tracking the extremal point. It involves a static converter and its autonomous digital control.

Task 2 : The total transfer of the energy generated by feeding-back the battery courant to a zero value. It involves a static converter and the hardware necessary for the digital control of the battery current.

The energy transfer is achieved in all the cases by the static converters. We would examine first their construction.

The static converters. We opted for static converters built around a Darlington BIP/MOS power switch which combines the advantages of a MOSFET and a bipolar transistor. The MOSFET, for its high input impedance is controlled by a unipolar voltage. This property is an advantage of the latter for interfacing with a digital control system. The reliability of the electronic set-up destined to solar installations assumes a primordial aspect. We have chosen to carry out the digital control of the static converters by a programmed logic which gives a greater flexibility of use and load matching. Before discussing digital control we will describe the acquisition of some physical quantities brought about by digital control.

The acquisition and control card. Its block diagram is shown in fig. 3. Its comprises an analog multiplexer, a sample and hold, an analog to digital converter and a PIA which serves as a parallel interface adapter between the analog to digital converter (ADC) and the bus of the microprocessor.

It also has a set of opto-couplers which work as switches in order to assure a galvanic isolation thus protecting the digital control from high voltages and induced parasitic signals.

The digital control of the static converters is assured by a fixed frequency, variable duty cycle signal. The signal is generated by the programmable timer (Motorola 6840) [5].

The specialized processor. It is built around an 8 bits microprocessor (Motorola 6809). For this reason, the compatibility, with the respect, to the software,the hardware and the control structure is perfect. Apart from the microprocessor, this acquisition card contains an UVPROM, a 2K RAM and necessary structures for exchanges between the specialized processor and the computer.

The software which is incorporated facilitates the acquisition of physical parameters (current, voltage) which are used in calculating the power and generation of control signals which allow an optimal transfer of power [6].

EVALUATION OF ENERGETICS POTENTIAL

Measurements taken over many days allowed us to evaluate practically the energetics potential of the photovoltaic generator on site at ENSUT. The resulting static characteristics I(V) and P(V) (fig.4), were plotted directly with the aid of a plotter from the bank of data (both electrical and meteorological) which we gathered days after days with the aid of our acquisition system. All the characteristics show that the optimal working zone of the photovoltaic generator is around 60 volts [7].

The negative gradient of the optimal load characteristic shows the effect of temperature on the network of these characteristics. The optimal voltage could be, under these conditions, subjected to variations of the order of 16 %. It is therefore necessary to take this into account in our control laws, for the optimization of the power generated by the photovoltaic generator.

OPTIMIZATION OF THE POWER GENERATED

Firstly, when the threashold of sunshine varies, the optimal voltage generated (Vopt) could be considered as being constant. A first approach consists of feeding back the voltage generated in a fork around Vopt (fig. 5).

Secondly, in order to take into consideration the external factors (temperature, masking, ageing) we have got a software that looks for extremal power (fig. 6).

Photograph 7 shows the response of the optimization system which was obtained, starting with a zero duty cycle, that is an open circuit generator voltage which is of the order of 75v. We could observe that the optimum is attained after three iterations which gives a convergence time of 5ms.

On the other hand, photograph 8 shows that the optimal operating point shifts easily on the vertical situated on the elbow of the characteristics (the optimal operating zone). The width of the vertical characteristic corresponds to the forks of the voltage imposed by the software for the fonctioning of the feedback by hysteresis.

ZERO CURRENT FEEDBACK OF THE BACK-UP BATTERY

The second part of our work consists once again the optimized functioning of the generator of carring out the transfert of all the energy produced towards a matched load functioning at a variable power, by controlling the current in the back-up battery. The flow chart that allows the battery to be "transparent" in relation to the energy supplied by the battery is shown in figure 9.

The control of the second static converter (fig. 2) allows us to influence the battery current Ib till it becomes zero. From that moment, the quantity of energy supplied by the generator is equal to that received by the load. Figure 10 shows the plot of the regulation of the current flowing through the battery. The variations which are positive or negative remain in the limits of the threashold imposed by the software thus confirming that the battery is functioning well as a back-up.

CONCLUSION

The aim of this work is to optimize in real time the functioning of a 400w photovoltaic generator and then assure the transfer of all the energy produised by the input sources towards a matched load.

This set-up is to be integrated into a mini multigenerator power station. We used the battery as a back-up and as an element for coupling the various generators. This functioning of the battery as a back-up by digital control of its current, constitutes a new dimension in the design of the photovoltaic systems since it tends to increase the global efficiency which is reduced when using classical storage systems.

The experimental tests have confirmed the feasibility of this device and have shown a good performance in steady state. In order to improve on the performance of the feedback in dynamic state, we are envisaging complementary studies to take into consideration not only the state variables of the system which are presently described but also the influence of the optimized sources that will be connected to it.

REFERENCES

[1] P.MATARASSO
"Projets énergétiques et projet de développement". 3ème séminaire sur l'Energie solaire - TRIESTE (Sept. 1982)

[2] B.DAKYO, S.ASTIER, J.M.ROLLAND, L.PROTIN
"Numerical control with a microcomputer of a synchronous machine fed by a photovoltaic generator by means of a thyristor inverter". E.P.E - GRENOBLE (Sept.1987)

[3] N.BEN SLAMA, M.ANNABI, J.P.REQUIER
"Power conditionning and conceptual design of photovoltaic hybrid systems". E.P.E - GRENOBLE (Sept. 1987)

[4] C.KRAIF, C.MASSELOT
"Power optimization for an aerogenerator controlled by means of a microcomputer". E.P.E - BRUSSELS (Oct. 1985)

[5] C.LISHOU
"Optimisation en temps réel du fonctionnement du générateur photovoltaïque d'une microcentrale énergétique et gestion du transfert d'énergie par processeurs spécialisés". Thèse Docteur-Ingenieur. Université de DAKAR (1987)

[6] C.KRAIF, S.BILLO, P.ROUAN, L.PROTIN
"Real time control of the power provided by two aerogenerators from a single specialized processor" I.F.A.C - ISTAMBUL (Juin 1986)

[7] C.LISHOU, G.SOW, P.ROUAN, L.PROTIN
"Control par un processeur spécialisé de la puissance fournie par un générateur photovoltaïque. Algorithme de recherche de puissances extrémales". Séminaire R.A.I.S.T - KUMASSI (Dec. 1986)

[8] C.BOISVINEAU
"Optimisation du fonctionnement d'un générateur photovoltaïque". Thèse de Doctorat de 3ème cycle .INP GRENOBLE (1981)

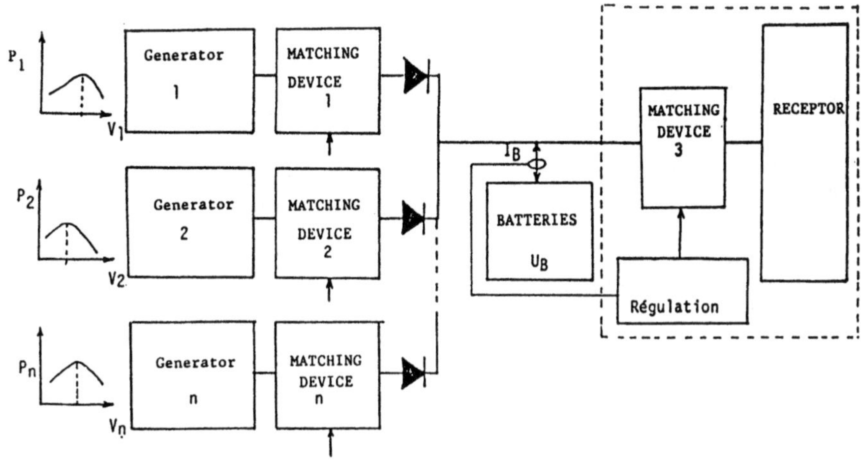

Fig 1 : MULTIGENERATOR COUPLING BY DIODES AND BACK-UP BATTERY

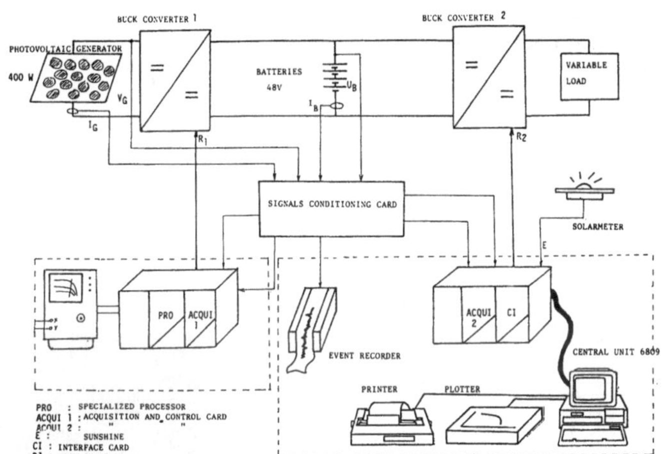

PRO : SPECIALIZED PROCESSOR
ACQUI 1 : ACQUISITION AND CONTROL CARD
ACQUI 2 : " "
E : SUNSHINE
CI : INTERFACE CARD
R1 : DUTY CYCLE 1
R2 : DUTY CYCLE 2
VG : IG GENERATOR VOLTAGE AND CURRENT
UB, IB : BATTERY " "

Fig.2 : EXPERIMENTAL SET-UP

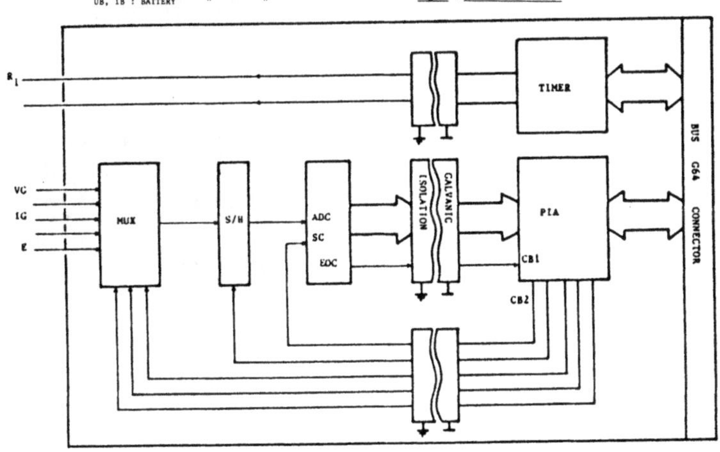

Fig. 3 : ACQUISITION AND CONTROL CARD

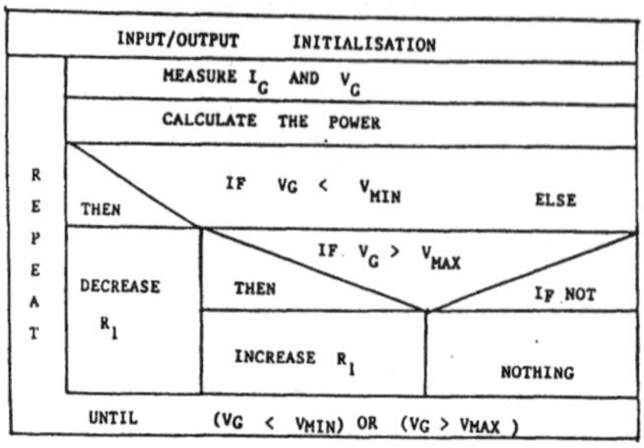

Fig.5 : FEED-BACK TO V_{OPT}

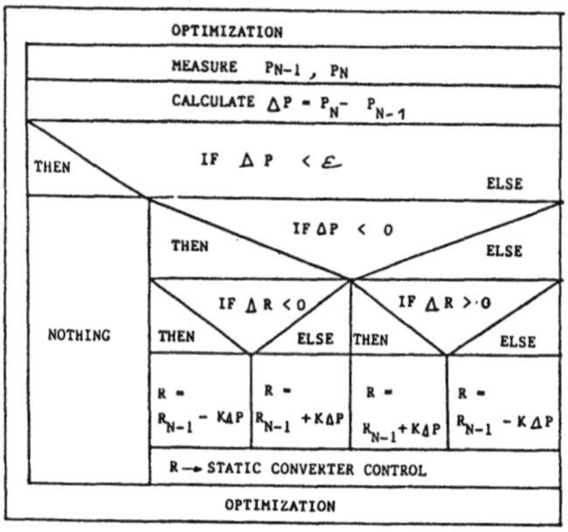

Fig. 6 : EXTREMAL TRACKING

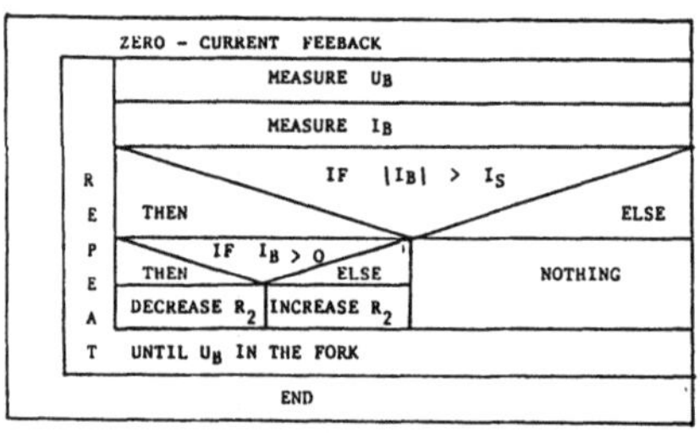

Fig.9 : FEED-BACK TO ZERO CURRENT

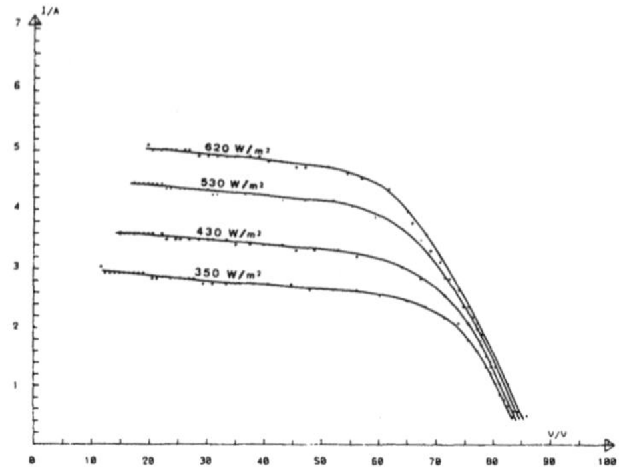

FIG.4-a THE CHARACTERISTICS I(V) OF THE GENERATOR

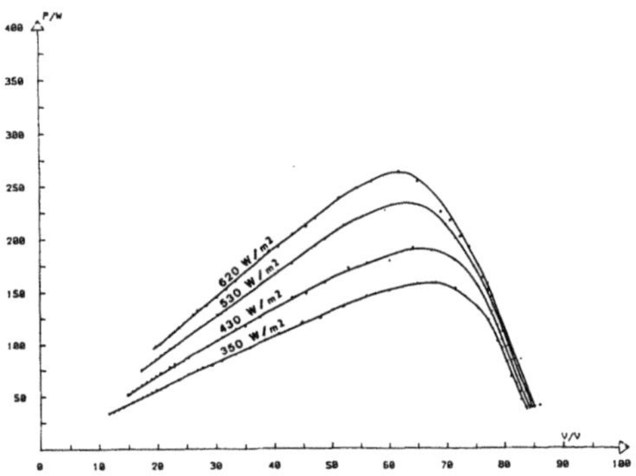

FIG.4-b THE CHARACTERISTICS P(V) OF THE GENERATOR

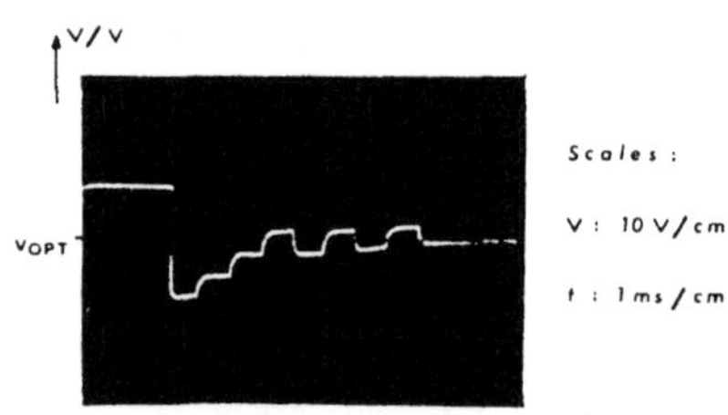

FIG. 7: CONVERGENCE OF THE GENERATOR VOLTAGE TO ITS OPTIMAL VALUE, V_{OPT}

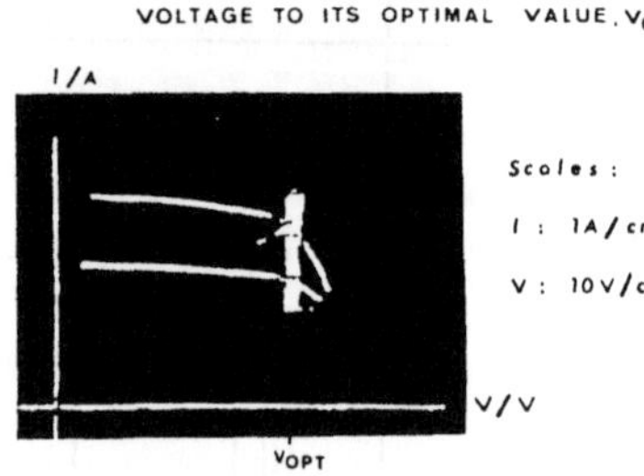

FIG.8: OPTIMIZATION OF THE GENERATOR WORKING POINT BY A SPECIALIZED PROCESSOR

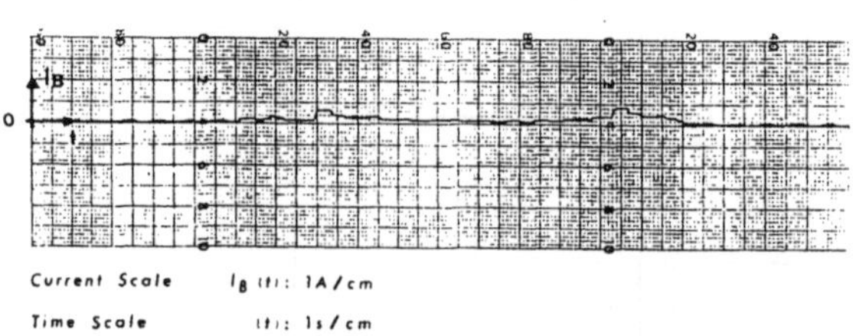

FIG.10 RECORDING OF THE BATTERIES FEEDBACK CURRENT I_B

MODULAR COMPUTER SIMULATION PACKAGE FOR PV AND SOLAR SYSTEMS

M. CASTRO, F. YEVES AND J. PEIRE

ELECTRONIC ENGINEERING DEPARTMENT / UNIVERSIDAD POLITECNICA
c/ José Gutierrez Abascal, 2. 28006 Madrid - SPAIN

ABSTRACT

This paper describes a new software computer program developed in a complete modular way in FORTRAN 77 to run in an IBM PC or compatibles, allowing to simulate Solar Systems, including PV and Solar Thermal ones. The program includes models of each block that perform the complete system, including the photovoltaic array, the DC power conditioning, the storage system and the DC-AC inverter system. The control of the system is made defining the operational control matrix which allows four different operating modes of the PV system. Annual simulation are done with three different sizes and location plants, calculating the annual average performances of the systems.

1. INTRODUCTION

The PV solar system developed and described in this paper is a flexible tool for performing non expensive time computer simulation to identify and evaluate PV applications. The program is called FVSIM and is written in FORTRAN 77 operating the computer code developed in IBM PC or compatibles computers. The program is a collection of component model subroutines linked together controlling the electric power through the system defining four different operation modes.

The structure system allows the user to model a variety of system configuration with a reduced number of input variables. The program simulates the electrical behaviour of the PV system and the evolution of the weather characteristics of the site selected. The program simulates the daily operation of the PV system allowing to select the time increment of each simulation between 5 minutes to one hour. The daily results could be integrated along the month or the year to obtain the monthly or annual characteristics of the system in evaluation.

2. PV PLANT BLOCK DIAGRAM

The PV plant block diagram is shown in the figure 1, where are included all the blocks which will compound the simulation program, (1), (2), (3). These blocks included the weather calculation one, the photovoltaic array, the DC power conditioning, the electrical energy storage, the DC-AC power converter, the DC and AC loads and the control blocks. These control blocks allow to define the four operation modes according to the possible value of each of these control blocks, which should have two differents states: connected or disconnected. The differents operating modes defined are:

Mode 0: C1 = 0, C2 = 0, PV system in no operation.
Mode 1: C1 = 1, C2 = 0, the PV array is connected directly to the load.
Mode 2, C1 = 0, C2 = 1, the energy collected in the PV array is sended completly to the electric storage.
Mode 3, C1 = 1, C2 = 1, all the plant is working depending on the input solar characteristics and the electric load ones.

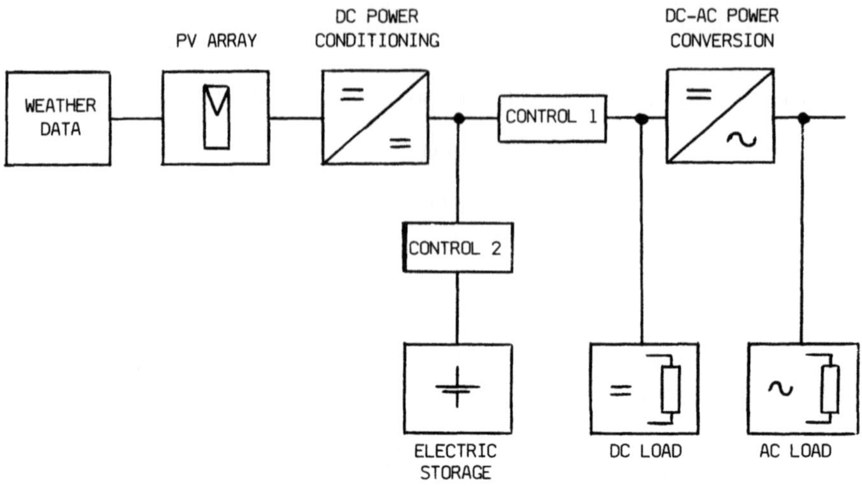

Figure 1. PV system simulation block diagram.

3. MODEL DESCRIPTION

In this section is described the simulation model selected and developed for any block of the PV system.

3.1 Weather calculation data.

The weather data is calculated every time increment selected. The variables calculated are the global and direct insolation, windspeed and ambient temperature. The insolation is calculated in a clear sky and with clouds with a time series model adjusted to some places (mainly to ALmeria, Spain). The windspeed model is a random function uniformly between the maximum and the minimum value selected. The last parameter calculated is the ambient temperature which is calculated as a sinusoidal variation between the maximum value two hours after noon and the minimum value at sunrise, (4).

3.2 PV array.

The PV array is modeled as a series-parallel connection of PV modules. Each one of these modules are modeled from its characteristic curve supplied by the manufacturer. This curve is fitted from 35 points obtaining through a three dimension least-squares adjusting program obtain a sixth degree function between output voltage and the output current and a three degree function between insolation value and the output current, (figure 2).

The user have to choose between a fixed array, a NS horizontal axis and a full tracking point focus one, having the program an account for the parasitic losses in the collector field. The model has included a new temperature model approximation based in (5) but faster in take the calculation in spite of its less accuracy. The model is fitted from (5) and the function is an exponential one as

$$Tmodule = Ki + (Kf - Ki).(1 - e^{-(1+k1.Ig).(K2+K3.Vv).Dt/Mt}) \quad (1)$$

being Tmodule the module temperature, Ki and Kf constants which value is function of the state of the module (cooling or heating from the ambient), Ig the insolation, Vv the windspeed, Dt the time increment and Mt the thermal mass of the module, (5). The square error obtained is 9.14 %.

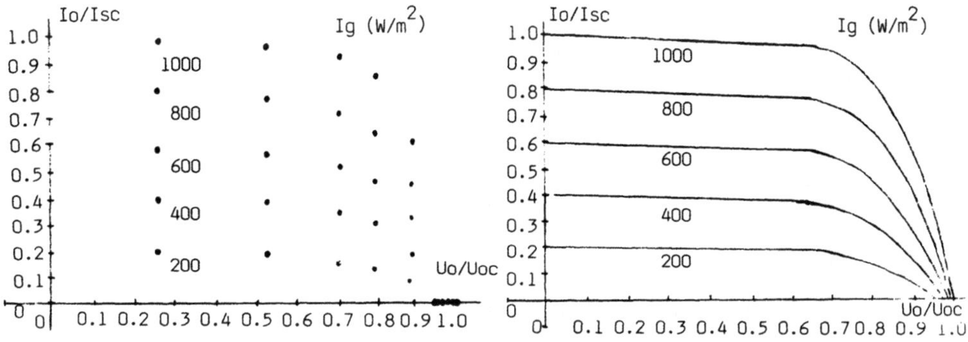

Figure 2. Operating values and curve characteristic fitted of the PV module.

3.3 DC power conditioning.
Three methods are provided for modeling the photovoltaic conversion process. These are maximum power tracking, (the DC-DC converter is based on the 3.5 section one characteristics), floating battery or voltage regulator, (6), (7), (8).

3.4 Electric battery storage.
The electric battery storage is based on the charge-discharge characteristics of an advanced lead-acid battery or a Ni-Cd one. The model selected is one tested and proved one proposed by the MONEGON, (4).

3.5 DC-AC power conversion.
This block is modeled as a single DC-AC inverter with a load dependent efficiency based on a three point adjusting better than the two point one modeled previously, (8). The function is,

$$Ncon = K1 - K2.e^{-K3.Pin/Pref} \qquad (2)$$

where are needed to adjust the model the value of Ncon (efficiency) at reference input power, Pref, the without load input losses and an intermediate value of efficiency and input power. Figure 3 shows four differents DC-AC inverter modeled with this method.

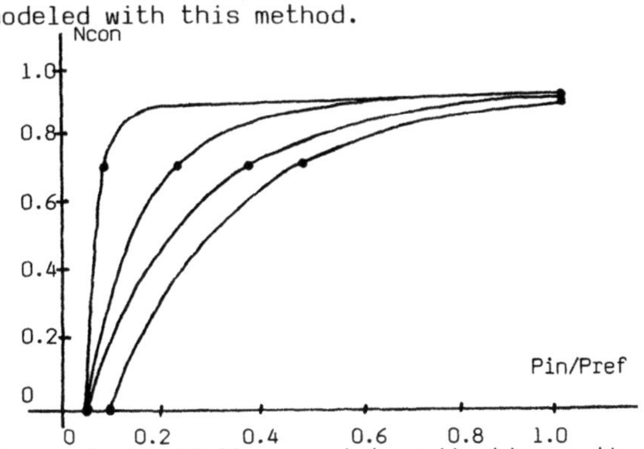

Figure 3. DC-AC three points adjusting method.

3.6 DC and AC load profile.
The user has to supply as input the DC and AC load in every simulation period to calculate the characteristics of the system and the electrical flow through it.

4. RESULTS
Several example systems are calculated and evaluated using the program FVSIM. We can remark three different simulations such as P1, system located near Madrid with an array output power of 50 kWp and a maximum energy storage of 72 kWh. Second system, P2, is a 1 MWp one without electric storage located in California near Barstow. The last system, P3, is a 20 MWp one with a maximum energy storage of 10.8 MWh located near Almeria in Spain.

Figure 4 shows a daily simulation of P1 and P3 systems, showing the I1 variable, (output collector current), I3 variable, (input/output storage current) and W7 variable, (AC output power). Table 1 shows an annual resume of the main characteristics of these three PV systems.

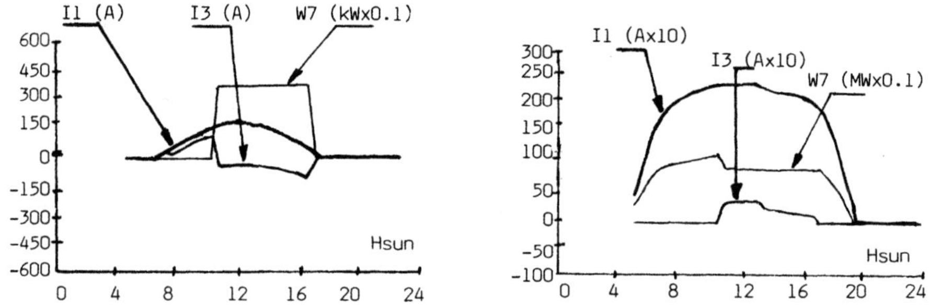

Figure 4. Daily simulation of a 50 kWp (P1) and a 20 MWp (P3) systems.

Table 1. Annual characteristics of three PV systems.

	CLEAR SKY			WITH CLOUDS		
	50 kWp	1 MWp	20 MWp	50 kWp	1 MWp	20 MWp
Collector output electric energy (MWh)	2.73	106.45	2047.73	2.18	85.16	1637.78
Net output electric energy (MWh)	2.01	84.48	1677.17	1.61	65.89	1308.19
Total electric energy stored (MWh)	0.75	0.00	991.70	0.51	0.00	713.19
Operation hours	2690.00	2620.00	2880.00	2250.00	2090.00	2290.00
Collector average efficiency (%)	8.67	10.03	11.12	8.34	9.55	10.57
DC-AC power conversion average efficiency (%)	82.00	91.00	92.00	80.50	90.50	91.00
Average PV system efficiency (%)	6.39	7.96	9.11	6.11	7.45	8.75

5. DISCUSSION

This paper had presented a new software computer program to simulate the daily operation of PV system based on a flexible definition of the characteristics of each block which composes the system. The average time to obtain one point simulation is 15 seconds running in an AT equivalent computer. Due to this speed and to the type of machines where the program is developed (IBM PC or compatibles) this program is easily installed and enough accurate to evaluate and calculate the characteristics of the system selected and designed.

6. ACKNOWLEDGEMENTS

The authors would like to express their gratitude to the Spanish-U.S. Joint Committee for the Scientific and Technologic Cooperation for sponsored this work.

7. REFERENCES

1. F. Harashima and S. Kondo. 'Optimal power flow control of solar photovoltaic energy systems'. MELECON'87. Proceedings of the International Conference held at Rome, 1987.
2. W. Palz. Photovoltaic power generation. Ed. Reidel, Dordrecht, 1983.
3. A. F. Williams. The handbook of photovoltaic applications. Building applications and system design consideration. Ed. Fairmont Press, London, 1986.
4. M. A. Egido and E. Lorenzo. 'Procesos de reducción de costes en paneles fotovoltaicos bifaciales y policristalinos. Paneles fotovoltaicos con menos de 36 células'. Internal report Solar Energy Institute. Madrid, 1986.
5. M. K. Fuentes. A simplified thermal model for flate-plate photovoltaic arrays. SAND 85-0330, Sandia National Laboratories, Albuquerque, 1985.
6. S. McCarthy and G. T. Dixon. 'The development of a computer model to optimize the performance of a 50 kWp photovoltaic system'. Sixth E.C. Photovoltaic Solar Energy Conference. Proceedings of the International Conference held at London, 1985.
7. D. F. Mennicucci. 'PVFORM - a new approach to photovoltaic systems performance modeling'. Eighteenth Photovoltaic Specialist Conference IEEE. Proceedings of the International Conference held at Las Vegas, 1985.
8. E. R. Hoover. SOLCEL-II: an improved photovoltaic system analysis program. SAND 79-1785, Sandia National Laboratories, Albuquerque, 1979.
9. J. A. Roger and C. Maguin. Photovoltaic solar panels simulation including dynamical thermal effects. Solar Energy, vol. 29, num. 3, 1982.

COMPARISON BETWEEN SIMULATION RESULTS AND ANALYTICAL SIZING METHODS FOR STAND-ALONE PHOTOVOLTAIC POWER SYSTEMS

Egido M.A.*, Lorenzo E.*, Macagnan M.**
*Instituto de Energía Solar
E.T.S.I. de Telecomunicación
Ciudad Universitaria
28040-MADRID, SPAIN

**Universidad Federal do Rio Grande do Sul
Porto Alegre, BRAZIL

ABSTRACT

The design of a stand-alone PV system often requires the determination of the loss of load probability (LLP). This paper presents a comparison between different LLP production models that are available in the litterature. The analytical and very simple Barra's model leads to the better results.

I. INTRODUCTION

The basic problem when designing a stand alone photovoltaic power system (SAPV) is to determine the PV array and battery sizes so as to obtain a specified loss-of-load probability (LLP). The most obvious method to evaluate the LLP for a given SAPV and a given location is to simulate the long-term SAPV performance using as input, the observed sequences of daily radiation data. Such sequences need to be large (10 years or more) in order to guarantee statistical consistence. This method has the inconvenience that great enough radiation secuences are available for only a very small number of locations and are clumsy to handle.

A first alternate approach is to use mathematical models, requiring only key climatological information as input, to generate statistically meaning- ful synthetic radiation secuences. For that, several models are available in the literature. The Graham's model |1| and the Aguiar's model |2| seem to be the most complete and they have been here considered.

Another simplest approach is to relay on analytical models diretly relating the array and battery sizes with the LLP. As far as we know, three different models are available in the literature: The Barra's model |3|, The Bucciarelli's model |4| and the Gordon's model |5|. All of them have been considered in this work.

This paper, analyse the validity of the five above mentioned models, by comparing their LLP predictions with the obtained from the SAPV performance simulation using the observed sequences of radiation data corresponding to three different locations of Spain: Madrid, Murcia and Salamanca.

II. MODELS DESCRIPTION

II.1. The Graham's model produces sequences of daily clearness index, K_t, using a simple autoregressive model of the form

$$X_n = \Phi_1 X_{n-1} + W_n \qquad (1)$$

Where X_n is related with K_t by the transformation function given by

$$\frac{1}{2}\left[1+\text{erf}\left(\frac{X}{2}\right)\right] = F\left[K_t, \bar{K}_t\right] \qquad (2)$$

where $F[K_t, \bar{K}_t]$ is the well know probability distribution function for K_t.

This model requires as input the average monthly radiation and the first-order autorregresive parameter Φ. If the latter is not available, an average value of 0.29 is recommended.

II.2. The Aguiar's model also produces sequences of K_t, using a library of Markov transition matrices, each corresponding to a specific internal in K_t. From the analysis of observed sequences of data in several stations, the autor's of this model propose a library wich validity is supposed to be universal. This model use as input only the overage monthly radiation.

II.3. The Barra's model assumes that the performance of a SAPV is related with the array and battery sizes throw the hyperbola curve

$$(X_m - Y_m)(1 - Y_m) = \gamma(\tau m) \qquad (3)$$

where X_m represents the array size normalized to the load in the mth month, i.e, X_m=1 means the maximum ammount of energy wich the array can deliver all along the mth month equals the load consumption during such a month. Y_m is the fraction of the load covered by the SAPV (note that Y_m =1-LLP) and τ_m is given by

$$\tau_m = \left(\frac{C}{L}\right)_m K_m d_m \qquad (4)$$

where $(C/L)_m$, K_m and d_m are, respectively, the storage capacity normalized to the load, the average clearness index and the fractional length of the mean day in the mth month. From the analysis of real sequences radiation data corresponding to several meteorological italian stations, Barra proposed the parabolic coefficient

$$\gamma = \alpha.\tau_m^{-\beta}, \quad \alpha = 2.41.10^{-2}, \quad \beta = 2.29$$

This model requires as input only the average monthly radiation.

II.4. The Bucciarelli's model lies on the approximation of the probability density function of the difference between the daily array output and the load with only two events and on the consideration of the daily storage charge/discharge process as a one-step Markov process. To apply the model, one must know the mean, the variance and the autocorrelation of the probability density of the daily solar radiation impinging onto the PV array. When the value of the variance is not available for a given location, it can be obtained from the mean according with a procedure described in reference |5|.

Regarding the autocorrelation coeficient, Bucciarelli suggest the use of an universal coefficient equal to 0.3.

II.5 The Gordon's model is similar to that of Bucciarelli using a three event representation of the difference between the daily array output and the load. Both models lead to very similar results.

III. RESULTS AND DISCUSSION

In order to analize the validity of the above described methods, we have compare their SAPV performance predictions with the obtained from a simulation based on the observed sequences of radiation data in three different locations of Spain: Madrid, Murcia and Salamanca. The period for wich data are available is 12 years. We have to mention that such period of data is not sufficient to obtain statistically meaningfull estimates of LLP for LLP values less than about: 0.001. However, we consider this validation procedure as the best we can do.

We analysed the case of a constant daily load equally distributed on the day. The array capacity is defined as the array size normalized to the load in the worse overage month. The value of LLP is calculated as the number of days in wich the SAPV dont meet the daily load requirements divided by the total number of days included on the period under analysis Fig.1 to 10 show some selected results. The curves obtained from the observed sequences data are refered as "simulation". The most significant conclusions are:

a) The Graham's model as originally proposed leads to too optimistics predictions, i.e, the predicted LLP is much more lower than the obtained from real sequences. The apparent reason is that the monthly autocorrelation of the generated radiation sequences is generally far from the corresponding to the observed sequences. As an attempt to overcame this problem, we slightly modified the model in the sens to assume that all the failures occur in the worse month and generating a synthetic radiation serie only corresponding to this month. This modification greatly improves the prediction. However, they even are significantly far from the simulated from real radiation sequences. The curve presented in Fig. 10 corresponds to the modified method.

b) The Aguiar's model leads to similar predictions to that of Graham, as Fig. 10 shows. Hence it can't be considered as satisfactory. We used the particular version of the model suggested by it's authors in order to assure that the monthly average clearness index of the syntetic sequences are equal to the used on input. Sometimes, this particular version of the model requieres a large computational time.

c) The Barra's model leads to the best predictions, of all the here considered models. It exhibits some tendence to overestimate (underestimate) the array size for low (large) storage capacities. Moreover it's very simple to use and it only requires as input the monthly average clearness index wich is generally available. We think that this method can be improved by using different parabolic coefficients for the different geographic regions.

d) The Bucciarelli's model leads to excessively optimistic predictions. Moreover, for array capacity values larger than about 1.4 it leads to the absurd of both, the array and the battery sizes have to simultaneously increase in order to keep constant the LLP value. An additional disadvantage is that some input requirements (autocorrelation and variance of the radiation on the plane of the array) are difficult to obtain.

e) The Gordon's model leads to even slightly more optimistic predictions to that of Bucciarrelli. It do not offert any additional advantage because the calculation procedure needs the same input requirements while is more time consumer. Because of that we avoided here the presentation of it's predictions.

IV. CONCLUSIONS

a) The LLP prediction models based on the generation of synthetic sequences of daily radiation lead to excesively optimistics results. A possible reason is that the monly autocorrelation coefficients of the synthetic radiation series are generally far from the corresponding to the observed radiation series. We believe that a considerable work remains to be done on this topic.

b) Between the analytical models for the LLP prediction, the Barra's one leads to the more accurate results and is the more simple to use.

REFERENCES

1. R.J.Aguiar, M.Collares-Pereira and J.P. Conde. Simple Procedure for Generating Sequences of Daily Radiation Values Using a Library of Markov Transition Matrices. Solar Energy 3,40 (1988).
2. V.A.Graham, K.G.T.Hollands, T.E. Unny. A Time Series Model for k_t With Application to Global Synthetic Weather Generation. Solar Energy 2,40 (1988).
3. L.Barra, S.Catalanotti, F.Fontana and F.Lavorante. An Analytical Method to Determine the Optimal Size of a Photovoltaic Plant. Solar Energy 6,33 (1984).
4. Louis L.Bucciarelli,Jr. The Effect of Day-to-Day Correlation in Solar Radiation on the Probability of Loss-of-Power in a Stand-Alone Photovoltaic Energy System. Solar Energy 1,36 (1986).
5. J.M.Gordon. Optimal Sizing of Stand-Alone Photovoltaic Solar Power Systems. Solar Cells 20.295-313 (1987).

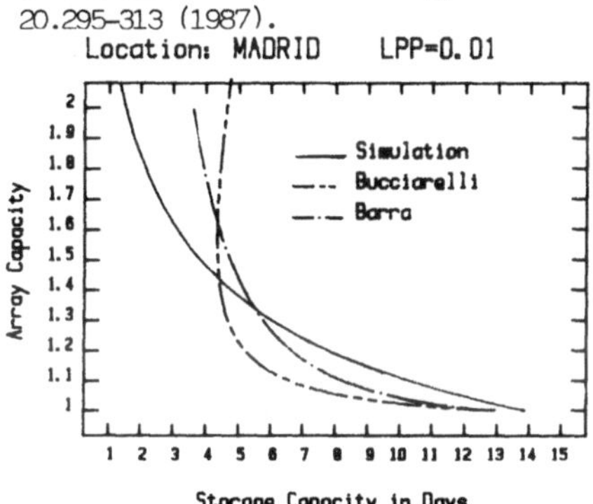

Fig. 1.- Fiability curve (Madrid, LPP=0.01)

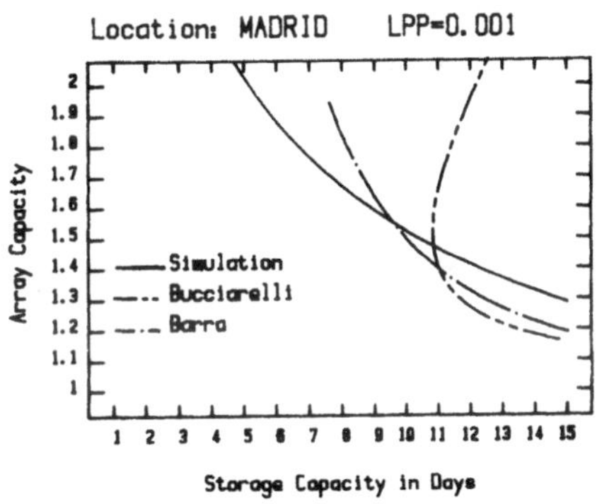

Fig. 2.- Fiability curve (Madrid, LPP=0.001)

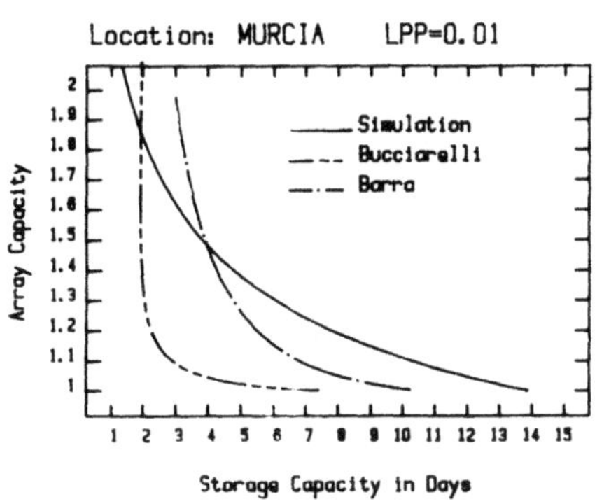

Fig. 3.- Fiability curve (Murcia, LPP=0.01)

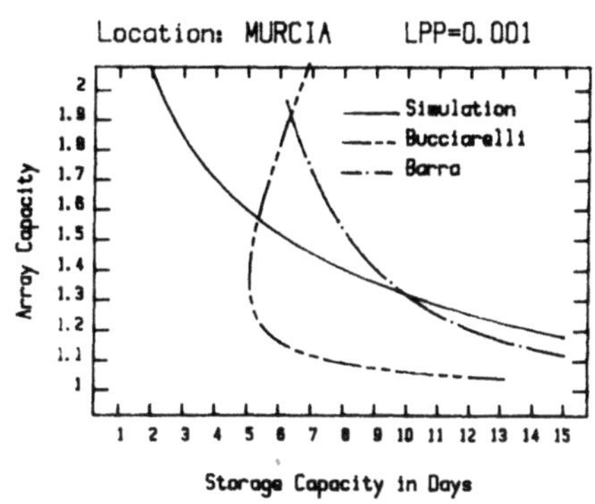

Fig. 4.- Fiability curve (Murcia, LPP=0.001)

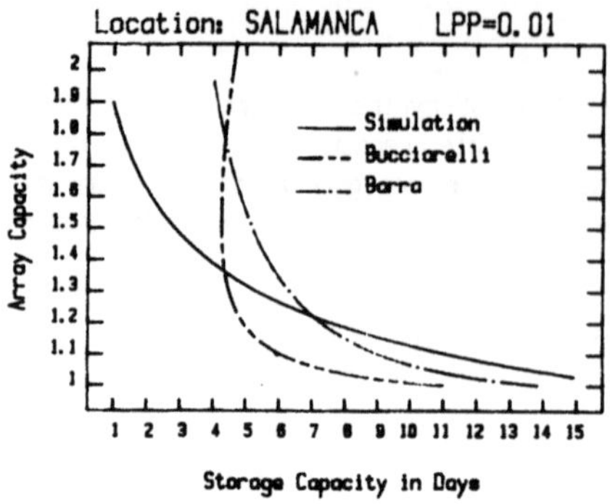

Fig. 5.- Fiability curve (Salamanca, LPP=0.01)

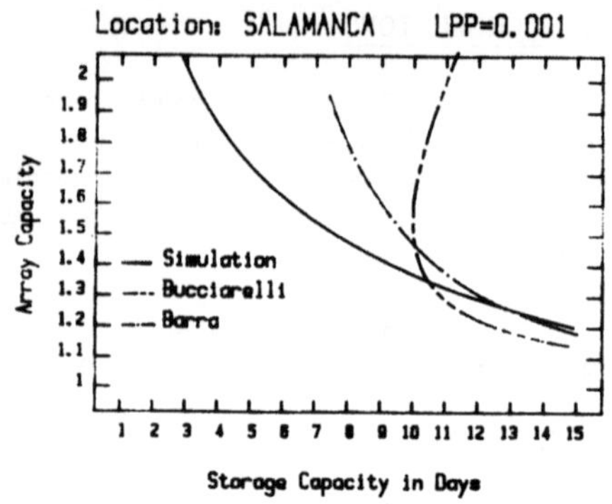

Fig. 6.- Fiability curve (Salamanca, LPP=0.001)

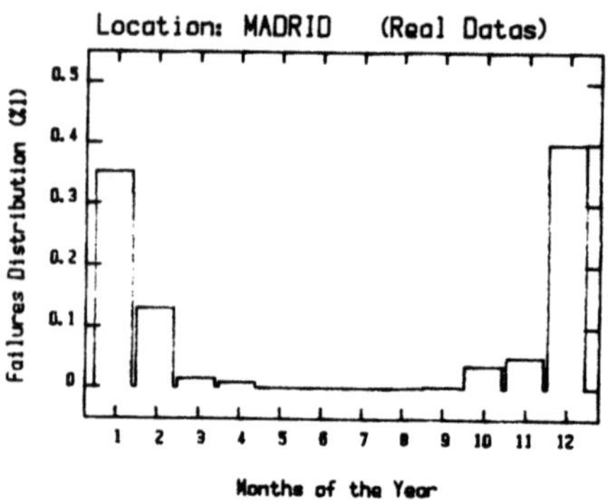

Fig. 7.- Failures distribution for Madrid using real datas.

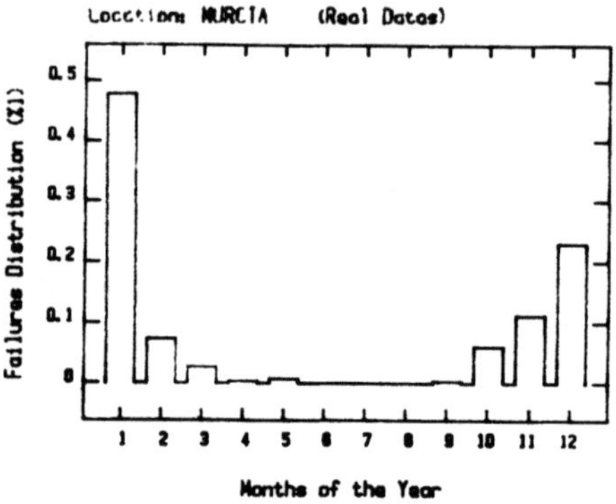

Fig. 8.- Failures distribution for Murcia using real datas.

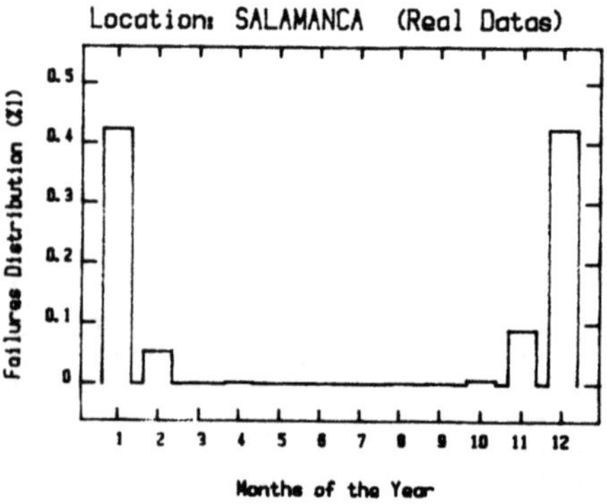

Fig. 9.- Failures distribution for Salamanca using real datas.

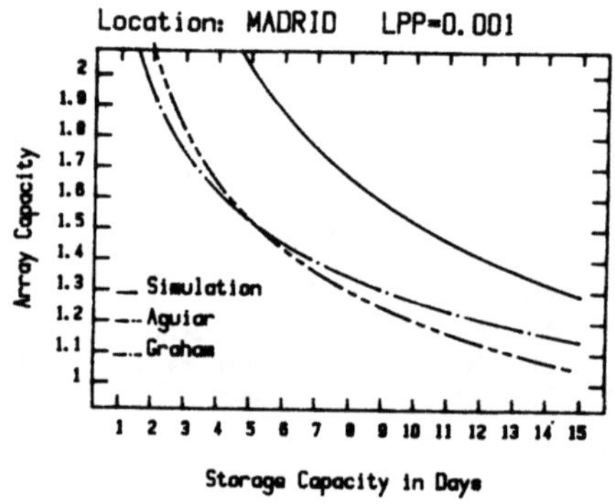

Fig. 10.- Fiability curve for sinthetic and real datas (Madrid).

ANALYTICAL TECHNIQUES FOR THE OPTIMIZATION OF PV WATER PUMPING SYSTEMS

Platon Baltas, Paul E. Russell and Christopher I. Byrnes
Department of Electrical and Computer Engineering
Arizona State University
Tempe, AZ 85287-5706
U.S.A.

ABSTRACT. System efficiency is very important when the cost of energy is high. For pv water pumping systems efficiency can be improved by the control of the input power. To control the power flow to the load while making maximum use of the array's capability requires the use of batteries. The storage requirements are small because this is short term energy but the search for the optimal power flow is time consuming and it is sometimes confined by the simulation step. The power flow that implies the maximum volume of water for a fixed energy input to the system has been derived analytically using calculus of variations techniques. A linear model for the water well and a centrifugal pump have been assumed. The analytical solution provides the optimal way of matching various size centrifugal pumps to different capacity wells. Sub-optimal ways to control the power flow are also suggested. This technique is particularly useful for low capacity wells where the drawdown varies substantially during discharge.

1. INTRODUCTION

Photovoltaics are a reliable, modular source of energy for many small scale applications, including water pumping. The high cost of photovoltaic energy justifies the use of electric energy storage and control devices that optimize system performance. Batteries and on-off switches have been previously suggested for pv water pumping systems [1], [2].

To optimize the performance one has to consider the energy availablility, the characteristics of the pumping hardware and the dynamic behavior of the water well. These are factors to be accounted for in any water pumping system design, however, photovoltaic systems have some unique characteristics that can be used to their advantage. Pv system power control hardware can be modified to turn the system on and off automatically and it is easy to interface to batteries. Design optimization consists in finding an operational mode for which a certain pumping hardware installed in a certain water well yields the maximum volume of water for a fixed pv array.

Computer simulation is a useful tool for the design of an optimal system, yet that implies extensive computational effort in some cases and the result is limited by the simulation step. Analytical methods to compute the optimal point of operation for a typical pv water pumping system are presented in this paper.

2. ASSUMPTIONS FOR THE OPTIMIZATION PROBLEM

The water well has been assumed to be well described by an n-th order ARMA model.

$$d(t) + a_1 d(t-1) + \dots + a_n d(t-n) = b_0 u(t) + \dots + b_n q(t-n) \quad (1)$$

where q(t) is the volume of water pumped during the t-T time interval and d(t) is the water well drawdown after t-T. The coefficients $a_1, \dots, a_n$,

$b_0, \ldots, b_n$ can be identified using a least squares method. This discrete time model can be easily transformed to a continuous time model using the bilinear transformation of equation (2).

$$z = \{1 + (T/2)s\} \,/\, \{1 - (T/2)s\} \qquad (2)$$

where T is the time period for the discrete model. Therefore, a continuous time transfer function can be computed from experimental data and a set of state equations for the water well can be obtained.

A centrifugal pump is assumed in the system. This type of pump is reliable, easy to start and requires little or no maintenance. The motor-pump system is assumed to be described by equation (3)

$$H = v(P) = a\,Q \qquad (3)$$

where H is the dynamic head, Q is the flow rate and v(P) is a polynomial of the input power. A second order v(P) seemed to describe accurately enough two different centrifugal pump systems tested in the laboratory.

A storage battery is also included in the system. The storage requirements are not large because the battery is not used for long term storage but rather as an energy buffer to control the power flow to the pumping system.

A first approximation to the power available to the pump can be obtained from meteorological data using a pv system simulator.

Finally, it is assumed that the system includes a check valve that does not allow for water flow from the storage tank back into the well.

A block diagram for the water well, motor-pump, check valve system is shown in Fig. 1.

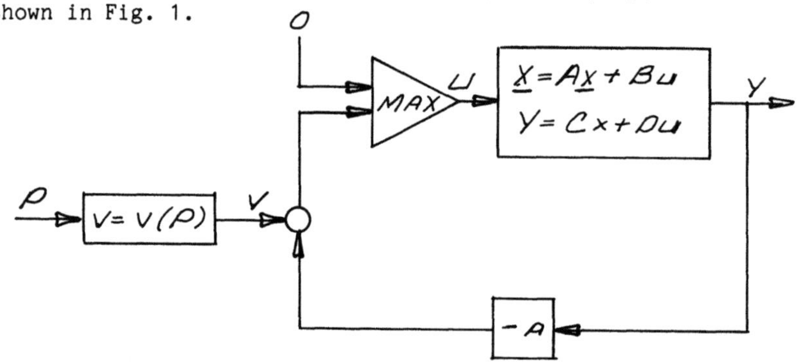

Fig. 1. System block diagram. y represents the drawdown, u is the water flow and P is the power to the system.

The problem consists in finding the maximum pumped volume during a given time period Td under the constraint of a constant input energy to the system during the same time period.

$$\frac{1}{T_d}\int_0^{T_d} P(t)\,dt = \text{constant} \qquad (4)$$

$$\frac{1}{T_d}\int_0^{T_d} u(t)\,dt = \text{maximum} \qquad (5)$$

3. SOLUTION
This problem has been solved for three different cases [3]. The solution methods have been derived from the calculus of variations.

3.1 Finite time optimization.
For this type of operation the well is pumped for a fixed finite time using a constant amount of energy. Since a storage battery is included in the system, the power can be distributed in an optimal way to obtain the maximum volume of water. A directly coupled pv pumping system does not have this ability. Considering the fixed finite time of pumping to be equal to the day length and the total energy equal to the energy obtained from the pv array, an optimal power flow to the pump can be determined and compared to the power flow in a directly coupled system.

A set of simple necessary conditions are obtained from this optimization process. The solution for a particular case is depicted in Fig. 2.

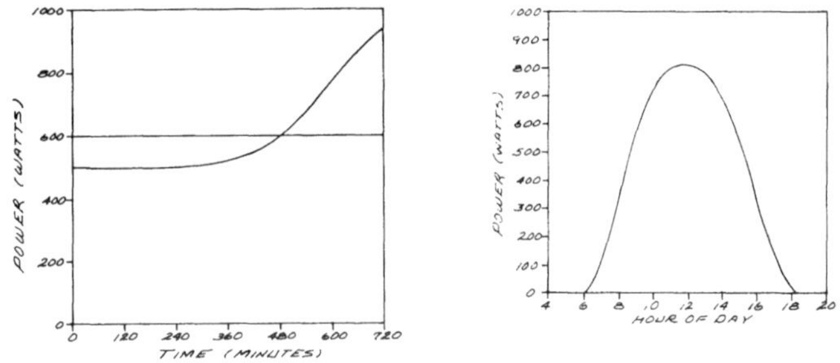

Fig. 2. a. Optimal power flow for a particular water pumping system, b. The power flow of a pv array.

This power flow is dependent on the pump and water well characteristics. It is clear that it is not close to the power flow expected in a directly coupled system.

The optimal power input does not depend on the static well level and its initial state. However, the pumped volume of water depends on these parameters. It can be concluded from Fig. 2. that it is inefficient to pump after the end of a pumping period because of the increased drawdown. If the well is left to recover overnight, then it either returns to its original state or it still "remembers" the previous day's pumping. For the low capacity wells to be pumped by photovoltaic systems, it is likely that the well will not be completely recovered. That makes it necessary to solve the optimization problem under periodic constraints.

3.2 Optimization under periodic constraints.
For this case it is assumed that the final state of the water well is the same as the original one. The set of necessary conditions imply a constant power flow as optimal. Therefore, if the system is to pump continuously it is better for the power flow to be constant. However, one can take advantage of the check valve non-linearity and operate the system in an on-off mode.

3.3 On-off optimization.
The pumping system is assumed to operate for some time and then stop to let the water well recover. The operating period could be one day (i.e.

the system operates for some time during the day and it quits at night) or only a few minutes. The water well is assumed to return to its original state at the end of the pumping period.

A set of necessary conditions that imply an optimal power flow are obtained from the Euler-Lagrange equations. These equations can be easily solved for a particular time period and duty cycle. The pumped volume for the optimal power flow as a function of the duty cycle for a constant period is plotted in Fig. 3. The optimal power flow for the maximum of this particular example is shown in Fig. 4.

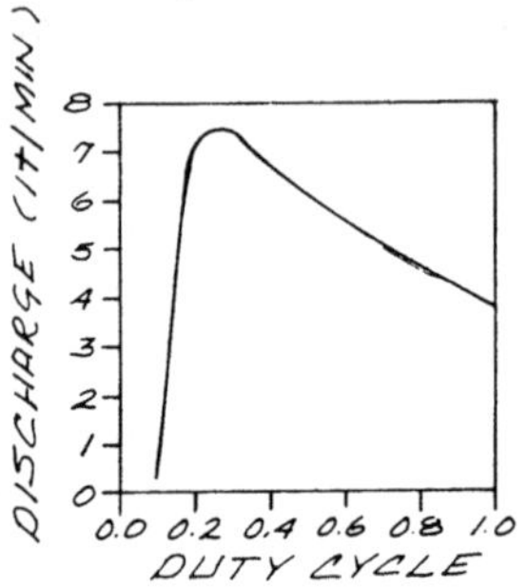

Fig. 3. Discharge as a function of the duty cycle.

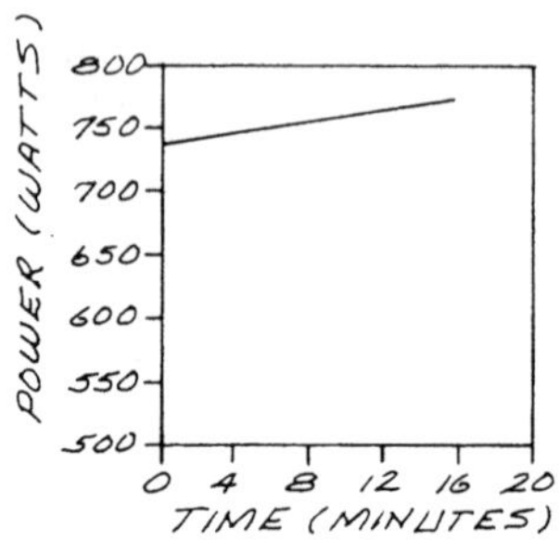

Fig. 4. Optimal power flow for maximum of Fig. 3.

The shape of these curves depends on the well and pump characteristics. The results for this particular system demonstrate that the choice of duty cycle and power flow can change substantially the performance of a water pumping system.

4. SUBOPTIMAL CONTROL

The optimal power flow of Fig. 4. is close to a linear function. As the pumping period decreases, this function approaches a constant function. For these cases the system can be controlled by a simple on-off switch to achieve a power flow very close to the optimal. If the period is small enough, the optimal power flow is determined by the characteristics of the on-off switch. However, small periods are not always desirable. Three reasons are listed here. First, this solution has not considered the transient behavior of the pump during its start up. Transient phenomena become significant if the period is small. Second, it not desirable to switch a system on and off at high frequency. Finally, some systems do not include a check valve at the bottom of the well to avoid the additional friction head. A certain volume of water is stored in the discharge pipe and returns to the well when the system is turned off. These losses are significant for small periods.

Final decision on the period T, will also depend on the well and pump characteristics.

5. DISCUSSION

This analytical solution provides the optimal way of matching various size centrifugal pumps, charactertized by equation (3), to different capacity wells, characterized by equation (1). The suggested operation implies that the well static level is not constant during pumping. One could use a high efficiency, high flow rate pump in a small capacity well to boost overall system efficiency.

Computer simulation is needed for evaluation of the array output energy, from meteorological data. Battery losses should be estimated after pumping system optimization. The adjusted available energy can be used for re-evaluation of the optimal operational mode. Note that this technique is not limited by the simulation step.

It is useful to know that the optimal performance because that performance can be used as a reference to compare sub-optimal controllers. If the reduction in hardware cost is significant, sub-optimal controllers can be used as an alternative to optimal controllers.

6. REFERENCES

[1] P. Baltas, P.E. Russell, and M. Tortoreli, "Optimum Design of Photovoltaic Water Pumping Systems," Proc. of the 7th European Photovoltaic Solar Energy Conference, Sevilla, Spain 1986, pp. 121-125

[2] P. Baltas, and P.E. Russell, "The Dynamics of Pv Water Pumping," Proc. of the 19th IEEE PV Specialists Conference, New Orleans 1987, pp. 1031-1035.

[3] P. Baltas, Optimization in Control for Photovoltaic Water Pumping Systems, Ph.D. Dissertation, Arizona State University, December 1987.

ANALYSIS AND DESIGN METHODS OF PHOTOVOLTAIC SYSTEMS

D. MAYER, P. NOLAY
ECOLE DES MINES DE PARIS
Centre d'Energétique
Rue C. Daunesse
Sophia-Antipolis
F - 06560 VALBONNE

ABSTRACT :

An analysis and a design method of photovoltaic systems are presented. The first one is based on the hourly simulation code HERMINES PV from which an accurate design method has been developed. This method allows to determine a generator/storage couple according to some technico-economical criteria and gives the complete estimation of the system performances for the selected meteorological data. The achieved solution corresponds to real marketed components.

1. INTRODUCTION

For more than ten years, the photovoltaic conversion of solar energy has known a tremendous development not only in terms of the installed power but also in terms of the range of applications. However, a great deal of uncertainties concerning the design, the management as well as the real behaviour of system from a few ten watts to several kW has to be removed. In fact, if most of the systems are working satisfactorily, the few operating analysis realized have shown, most of the time an oversizing of these installations.

To improve our knowledge on PV systems, several ways of investigation can be adopted. Full scale experimentation can give the answers but in a long or costly way. Moreover, the results cannot be generalized. Another solution consists on computer simulation which allows to treat a lot of different cases of applications, but needs a validation phasis.

It is with this aim in view that our laboratory and AFME have deciced to develop a set of methods allowing to design an installation and to predict its behaviour and its performances.

The purpose of the paper is to present such methods.

2. ANALYSIS METHOD OF PV SYSTEMS

The hourly simulation code HERMINES PV has been built for stand-alone systems such as the one presented in fig. 1. It can be easily adapted to the analysis of other types of systems, for example systems without storage (pumping, ventilation, refrigeration, ...) and systems with a standby generator.

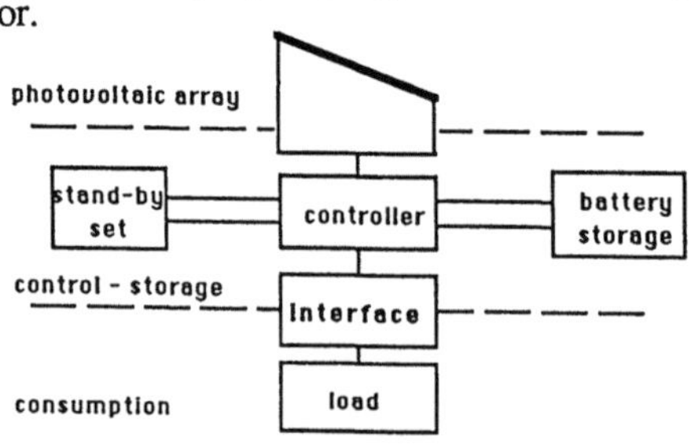

fig.1 : complete photovoltaic system

This software is composed on the one hand of models libraries related to data bases containing the real characteristics of classical components and on the other hand of routines dedicated to the system control and management. Such a modular structure allows to define a free choice architecture for different systems and different applications.

At the present time, for a same system two kind of possibilities are available :

• One to predict the performance of an installation hour by hour over a year.

• The second to simulate the behaviour of a real system from experimental data.

The flow chart of HERMINES PV is shown in fig. 2. For more precise information, refer to papers which have been already presented /1//2/.

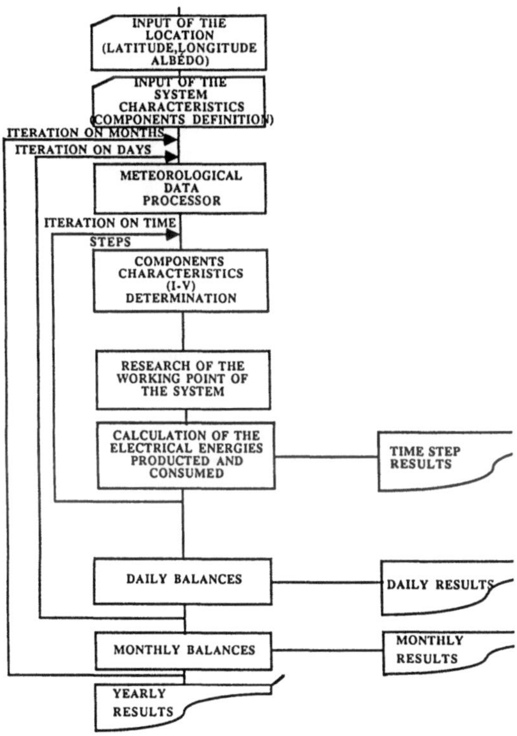

Fig. 2 : Analysis method flow chart

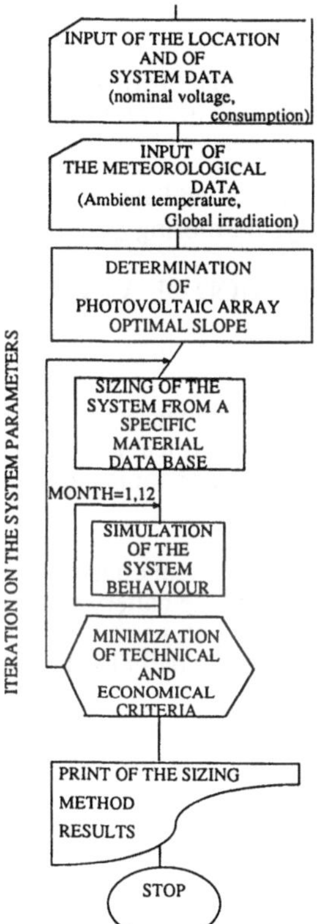

Fig. 4 : Design method flow chart

Nevertheless, to give credit to computerized results, a validation phase of the simulation tools is necessary.

To that end, an experimentation was monitored at Sophia-Antipolis, for a period of 15 days. Fig. 3 and table 1 present the comparison between calculated and measured values for the voltage and the balances of the system.

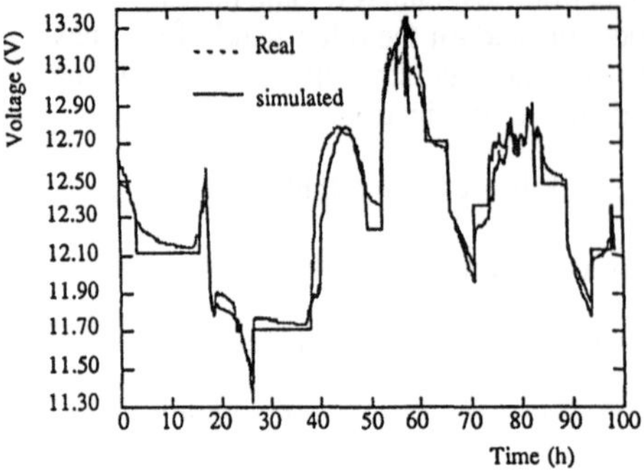

Fig. 3 : Systeme voltage

	Calculated	Measured
Battery balance (Wh)	1473	1512
Array production (Wh)	3937	3954

Table 1 : System balances

These results show that HERMINES PV is a tool able to represent the dynamic behaviour of a system with enough accuracy and to evaluate quite precisely (about a few per cent) its performances in terms of hourly, daily, or yearly balances.

Results of the same kind have been obtained on the cell temperature calculation for integrated arrays on roofing /3/.

3. DESIGN METHOD OF A PV SYSTEM

3.1 Basic principles

The main conditions requested to evaluate as accurately as possible the energy balances have been drawn from the critical analysis of the generally used design methods and from the study of the influence of some parameters on a PV system behaviour /4/. These are the following :

- Determination of the real operating point,
- Characterization of the battery state of charge on which the system operation is strongly dependent,
- Possibility to define a daily load profile allowing at least to distinguish between diurnal and nocturnal consumptions.

These are fundamental principles of the design method which is widely based on the HERMINES PV software.

3.2 Presentation of the method.

• The first phase of the method consists on determining first the optimal tilt angle from the variations of the irradience to the consumption ratio and then the month of which this ratio is the most favourable at that tilt.

• A very simplified calculation gives starting values and leads to the selection of a type of module.

• For given number of modules and battery elements, the method simulates the system operation month by month as follows :

. One day hour by hour to determine the potentiality of the system.
. Every other day by considering it as a sequence of three different periods : two nocturnal separated by a diurnal one.

• The yearly balances are analysed to calculate the new size of the array and the storage. These values are adapted according to the couples (array/storage) already tested.

• Some criteria such as the cost of installation, the required energetic behaviour and the balance of the array/storage couple are used to determine an optimal design.

The outline flowchart of the design method is shown in fig. 4.

3.3 Input and output

• Input data

* Meteorological data related to the location of the installation (monthly averages of the daily ambiant temperature and irradience),
* System voltage,
* Depth of discharge necessary to calculate the value of the load disconnection threshold,
* A percentage of the load that can be supplied by a stand-by mean, when decided,
* Monthly and daily load profiles,
* For alternative current, description of the inverter.

• Output data

Table 2 : Estimation of system performances

Load profile	: winter :1030 kWh/j	spring :	1300 kWh/j
	summer :1500 kWh/j	automn :	1300 kWh/j
Location	: Nice		
System voltage	: 24.	Tilt angle :	60
Type of module	: BPX402	Peak power :	560. Wc
Type of battery	: BAT240	Storage capacity :	240. Ah

Rough system cost : 76440.0 FF H.T.

NM	TEMP	EINC	BILBAT	EREG	ECHA	EAPP	ETOCH	RCE	RCP	REGE	REGP	CUEP	SOBA
	3C	KWh/m2.j	KAh	KWh	KWh	KWh	KWh	%	%	%	%	%	%
10	17.8	4.70	.305	53.8	40.	0.0	40.3	6.45	9.30	4.83	7.0	69.4	87.
11	13.3	3.60	.310	49.8	39.	0.0	39.0	8.05	9.54	6.30	7.5	84.4	43.
12	10.0	3.16	.407	44.7	32.	0.0	32.2	7.97	9.68	5.74	7.0	82.3	55.
1	9.2	3.36	.385	45.0	32.	0.0	32.2	7.53	9.69	5.40	6.9	77.7	63.
2	10.2	4.06	.361	41.5	29.	0.0	29.1	6.39	9.61	4.47	6.7	66.4	84.
3	12.2	4.64	.372	54.8	40.	0.0	40.3	6.66	9.55	4.90	7.0	69.8	80.
4	14.9	4.98	.361	52.8	39.	0.0	39.0	6.18	9.49	4.56	7.0	65.1	86.
5	18.6	4.89	.337	52.5	40.	0.0	40.3	6.05	9.41	4.65	7.2	64.3	80.
6	22.0	4.91	.355	57.4	45.	0.0	45.0	6.81	9.31	5.33	7.3	73.1	68
7	24.6	5.38	.424	61.3	46.	0.0	46.5	6.42	9.16	4.87	7.0	70.1	77.
8	24.2	5.40	.416	62.4	46.	0.0	46.5	6.51	9.13	4.85	6.8	71.3	79.
9	22.3	5.22	.338	52.0	39.	0.0	39.0	5.80	9.15	4.35	6.9	63.3	87
	16.6	4.53	4.37	628.	469.	0.0	469.	6.64	9.39	4.96	7.0	70.7	87

EINC : global irradience at tilt angle
BILBAT : Battery balance
EREG : Energy array production
ECHA : PV energy provided
EAPP : Energy provided by a stand-by mean
ETOCH : Energy consumed

RCE : Effective conversion efficiency
RCP : Potential conversion efficiency
REGE : Effective global energetic efficiency
REGP : Potential global energetic efficiency
CUEP : Effective to potential energy ratio
SOBA : State of charge

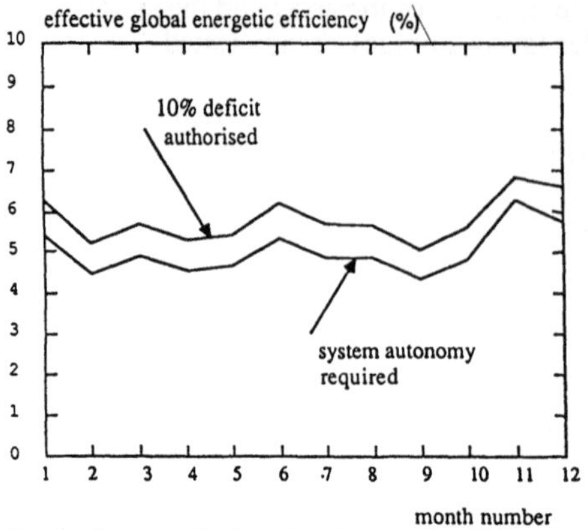

Fig. 5 : System effective global energetic efficiency

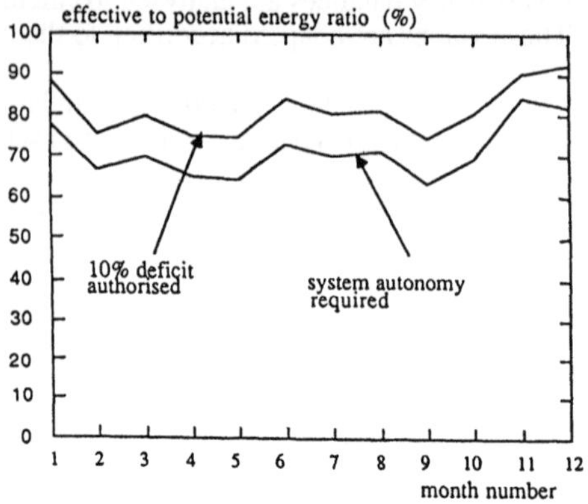

Fig. 6 : System effective to potential energy ratio

4. CONCLUSION

The use of the hourly simulation code HERMINES PV has allowed to determine which are the important parameters to be taken into account in an accurate design method.

Besides the capability to get an optimal design of a PV system, this method gives a complete estimation of the system performances for the selected meteorological data.

The accuracy of such a method depends on the climatic data available. So a data processing model by a stockastic mean in order to take into account the fluctuation of the irradience will very soon be integrated.

REFERENCES

/1/ P. Nolay, "Développement d'une méthode générale d'analyse des systèmes photovoltaïques" - Thèse de 3ème cycle Ecole des Mines de Paris, Septembre 87.
/2/ D. Mayer, P. Nolay, "Présentation du logiciel de simulation HERMINES PV" - Séminaire des contractants AFME, Novembre 87.
/3/ D. Mayer, P. Nolay, "Adaptation of the simulation code HERMINES PV to integrated PV array on roofing", ECA München, March 87.
/4/ D. Mayer, P. Nolay, "Elaboration of a new photovoltaïc system sizing method", ECA München, March 87.

SIMULATION OF RENEWABLE ELECTRICITY SUPPLY INCORPORATING A HYDROGEN STORAGE SYSTEM

Hans Georg Beyer, Joachim Luther, Jürgen Schumacher-Gröhn
Department of Physics, University of Oldenburg, D-2900 Oldenburg

ABSTRACT. We report on simulation calculations for renewable energy systems, that incorporate the following components (or a selection of these): photovoltaic array, wind energy converter, Pb-battery, hydrogen storage system (electrolyser, gas storages, fuel cell).
The aim of the investigation is the determination of system configurations that will cover a given load with a reliability close to 100 % (autonomous systems). The calculations are performed for meteorological conditions of northern Germany (Oldenburg) and northern Italy (Ispra).

1. INTRODUCTION

Due to the fluctuating time pattern of wind and solar production renewable stand alone energy systems must incorporate storage devices. Storage is essential, if a safe supply of a load has to be guaranteed without the use of auxiliary energy (autonomous systems).
In this paper we are dealing with electrical supply systems in the power range from 1 kilowatt to some 100 kilowatts. Systems of this size will typically have storage capacities from kW-hours to MW-hours.
The state of the art storage technology for renewable energy systems is lead acid batteries. This technology has the advantage of a high storage efficiency (typically > 0.8), but the cost of battery storage is too high for most long-term storage applications.
We will discuss hydrogen technologies as an alternative option for electricity storage. Hydrogen storage systems are a combination of electrolyser, gas storage and fuel cell. In contrast to batteries, the electrochemical energy conversion devices (electrolyser and fuel cell) are separated from the energy storage (gas storage). The increase of storage capacity will therefore not lead to higher expenses for the technology intensive electrochemical components. Long-term storage is within reach. H_2-storage systems will show low storage efficiencies (typically < 0.5) if realistic sizes of the electrode areas of electrolyser and fuel cell are assumed. Therefore every charge and discharge cycle of the H_2-storage systems results in high energy dissipation rates.

2. SYSTEMS UNDER INVESTIGATION AND METHOD OF APPROACH

The structure of the systems under investigation is given in figure 1. It may be characterised as follows:
- wind energy and solar energy may be combined using wind energy converters and PV-generators
- battery and H_2-storage systems may be combined.

Aim of the investigation is the determination of system configurations yielding autonomy (safe supply of the load without using auxiliary energy). This was done for different sets of meteorological input data via time step simulation with hourly time resolution.

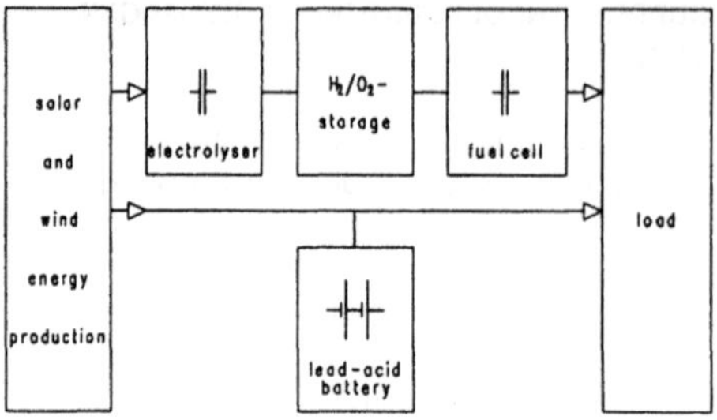

Fig.1
Configuration of the systems under investigation. Mismatch between load and renewable production is levelled out using a combined battery/hydrogen storage.

The following components are modelled in the simulation programme:
wind energy converter: Characteristics of an aerowatt FP7G generator
PV-generator : AEG crystalline cells
Pb-battery : lead acid battery for stationary use
Electrolyser : advanced Raney-Ni cells
Fuel cell : H_2/O_2 Raney-Ni cells
Gas storage : pressurized gas tanks.
The system configurations are specified by a set of 4 parameters:

- PRG gives the gross renewable production in units of mean load
- W/S gives the ratio of wind to solar production
- BAT gives the size of battery storage, characterised by the time the load is met by stored energy
- HYST gives the capacity of the hydrogen storage system, characterised by the time, the load may be supplied from stored chemical energy via the fuel cell.

It should be noted, that the parameter PRG is a function of both the absolute sizes of renewable converters and the meteorological data base. To a lower extent the operating conditions of the converters in systems with different layouts may have an influence on PRG.

3. CONSTRAINTS FOR AUTONOMOUS SYSTEMS

To reach autonomy in an energy system two conditions must be satisfied:

(1) $P_{reg} \geqq P_{dir} + (P_{load}-P_{dir})/ETA$

with: P_{reg} = mean annual power derived from renewable sources, P_{load} = mean load, P_{dir} is the power consumed directly (bypassing the storage) and ETA the mean efficiency of the combined storage system. By $(P_{load}-P_{dir})/ETA$ the energy dissipation in the storage is taken into account.

(2) The storage has to balance all the deficits between load and energy production.

4. RESULTS

We have analysed the storage configuration necessary to reach autonomy of energy systems at a site in northern Germany (Oldenburg). As an example figure 2 [1] gives the result for different converter sizes (different PRG). The calculations were performed for a wind/solar hybrid system (W/S = 50/50), the load is constant in time. Increasing the hydrogen storage in a system with fixed PRG reduces the need for battery storage. The overall storage size ("days" of battery storage plus "days" of hydrogen storage) increases with decreasing battery size. In systems with low PRG (e.g.

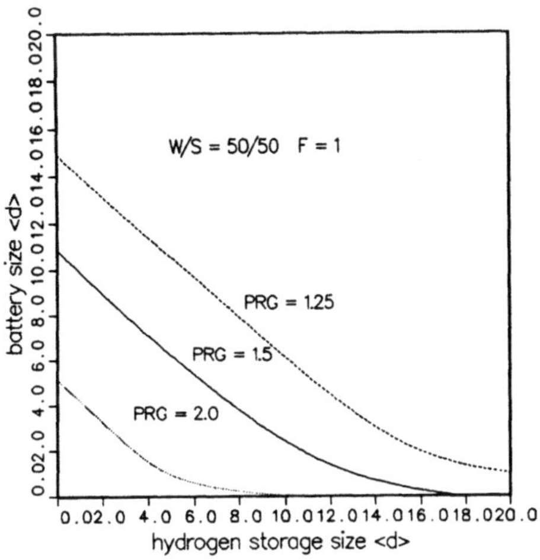

Fig.2
System configurations of autonomous systems (F=1). Combinations of storage sizes required for autonomy are shown. The curves refer to systems with a wind/solar combination W/S = 50/50. The renewable production PRG is varied. Data base: Oldenburg 1982. (Curves are smoothed)

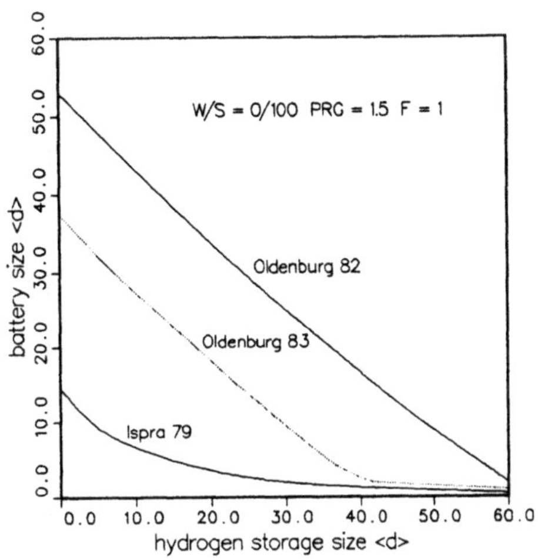

Fig.3
Same representation as Fig.2. The curves give the configurations of autonomous PV-systems (PRG = 1.5), calculated for different input data sets (Oldenburg 1982, 1983 and Ispra 1979). The axes are scaled down as compared to Fig.2.

PRG = 1.25) battery storage is indispensable. Condition 1 stated above may not be satisfied with low PRG, if the low efficiency of the hydrogen storage system dominates the storage process.
Fig.3 gives an impression of the problems associated with the use of a limited base of meteorological data. The minimal storage sizes needed for system autonomy are calculated for the meteorological conditions of two different years (data of global insolation on a plane tilted 70 deg south taken in Oldenburg 1982 and 1983, PV-system with PRG = 1.5). To correct for different energy contents in the storage at the start (January 1st) and the end of the calculation (December 31st) the calculation is continued with the identical data base until consistency is reached. Minimal storage size calculated for the 1983 data is considerably smaller than that for the 1982 data. Due to this variation of the meteorological conditions reliable values for the system layout yielding autonomy for longer periods may only be gained on the basis of the meteorological conditions of several years. Thus the relations presented here may only give a qualitative view of the system behaviour.
In Fig.3 a third curve shows the advantages of sites with better insolation conditions. The calculation is made for a PV-system at Ispra, northern Italy. (Global insolation on a plane tilted 60 deg south, 1979). It has to be noted, that the PV-converter is now scaled down, compared to the system in northern Germany to reach the same annual renewable production PRG, the total storage capacity needed, however, is considerably lower for the system in Ispra than for the system in northern Germany. This is due to the fact that seasonal changes in solar radiation are less pronounced in Ispra. Maximal battery capacity for systems without hydrogen storage system com-

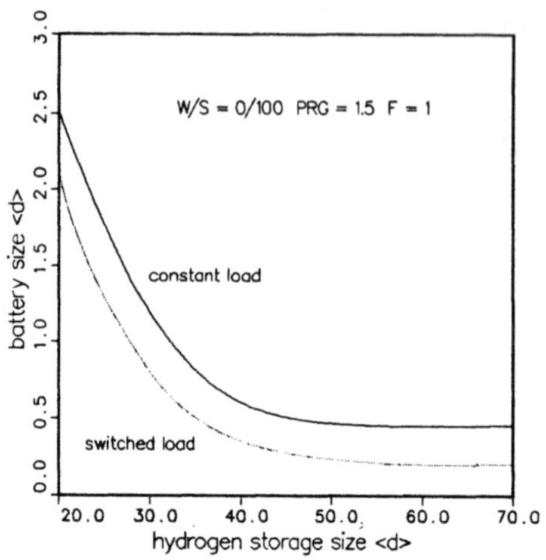

Fig.4
Configurations of autonomous PV-systems (PRG = 1.5) supplying loads with different time pattern (constant load; variable load pattern: daytime load > nighttime load - see text). Data base: Ispra 1979.

pares well with the storage needed for a hybrid (wind/solar) system in Oldenburg (14 "days" compared to 11 "days"). While for the hybrid system autonomy could be yielded using solely a hydrogen storage system, this is not possible for the PV-system at Ispra.
For the meteorological conditions of Ispra we have investigated the effect of a daily load pattern on the system performance. The mean value of the load is kept constant. During daytime (8h-18h) the load is 1.7 times the mean load. During nighttime the corresponding factor is 0.5. Figure 4 gives the effect of the load pattern on the storage configuration of autonomous systems (PRG = 1.5). The overall storage size is reduced by about 0.5 "days".

5. ANALYSES OF ENERGY FLOW

In Fig.5 the results of the analyses of the energy flow in PV-systems at Ispra (PRG = 1.5, constant load) is shown. The autonomous systems are characterised by the size of the battery. Given are the shares of energy delivered to the load via different paths (direct supply by the PV-generator, supply via the battery storage, supply via the fuel cell). 37 % of the energy consumption of the load is supplied directly by the PV-generator by-passing the storages. From condition 1 stated in chapter 3 above it may be calculated, that for autonomy the storage efficiency ETA has to be greater than 0.55 (P_{reg} = 1.5, P_{dir} = 0.37, P_{load} = 1.0). The efficiency of the H_2-storage in the computer model used is 0.48, so that the H_2-storage has to be complemented by some battery storage with good efficiencies. Even a "small" battery (BAT > 0.5d) will reduce the energy flow via the fuel cell sufficiently. A battery with a capacity equivalent to 15 days will be sufficient to reach autonomy without using a H_2-storage system.
An example for the evolution in time of the energy contents of the two storages is shown in Fig.6. The figure refers to a system with PRG = 1.5, BAT = 1d and HYST = 33d at Ispra. Given are the values of the storage contents at 6 am for every day of the year. It is remarkable that the content of the H_2-storage system is constant for nearly 3 months. With the 1-day battery the H_2-storage system will mainly provide the seasonal balance of energy.

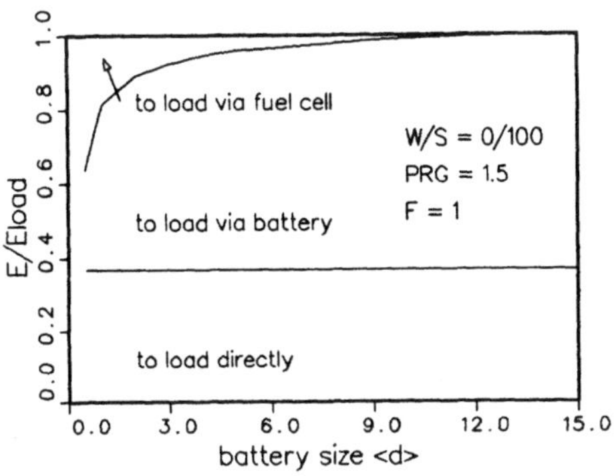

Fig.5
Normalized shares of energy supplied to the load via different paths as a function of battery size. The values are calculated for autonomous PV-systems (PRG = 1.5, data base Ispra 1979).

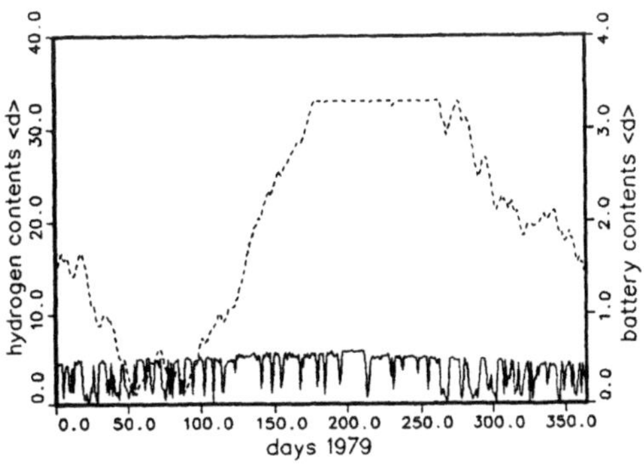

Fig.6
Annual variation of the contents of H_2-storage (dashed curve) and battery for a PV-system at Ispra (PRG = 1.5, BAT = 1d, HYST = 33d). Different scales are used for the representation of the two curves.

Acknowledgement:
This paper was supported by the Ministry of Research and Technology of the FRG (BMFT).

Reference:

[1] Beyer, H.G., J. Luther, J. Schumacher-Gröhn
Combined battery/hydrogen storage for autonomous wind/solar systems
ISES Solar World Congress, Hamburg 1987.

PHOTOVOLTAIC WATER PUMPING SYSTEM WITH PROGRESSIVE CAVITY PUMP

B. Herrmann, H. Karl, E. Kopf, G. Lehner, G. Saupe

Universität Stuttgart, Institut für Theorie der Elektrotechnik
Pfaffenwaldring 47, D-7000 Stuttgart 80, Fed. Rep. of Germany

ABSTRACT. A prototype photovoltaic pumping system capable of supplying 14 m^3 of water per day at heads of around 40 m has been developed. The system consists of a progressive cavity pump driven by a submersible ac-motor which is powered by a specially developed inverter. The PV array has a peak power of 750 W. Experiences with this system are reported. The special start-up procedure is described and some measured system characteristics are shown. The results of long term tests are presented. Furthermore, some findings of system simulations are discussed. The experiences indicate that systems using progressive cavity pumps represent an interesting alternative for photovoltaic pumping applications.

1. INTRODUCTION

In publications on photovoltaic water pumping systems the application of progressive cavity pumps is discussed controversially. For example investigations on this subject are encouraged by the World Bank /1/, whereas Matlin /2/ points out that progressive cavity pumps are unsuitable for photovoltaic applications.

Like other positive displacement pumps this type of pump offers a potentially high efficiency and a favourable part-load performance. Furthermore, it is insensitive to a certain amount of solid particles in the pumped liquid quite similar to centrifugal pumps. The most significant disadvantages of progressive cavity pumps are their high break-away torque and the possible wear of rotor and stator.

So far, experiences with photovoltaically powered progressive cavity pumps are mostly reported for systems with submersed pumps and surface mounted dc drives /e.g. 3/. In this system configuration it is relatively simple to overcome the problem of starting up the pump. However, the power transmission to the pump by means of a long vertical shaft introduces some problems.

2. SYSTEM DESCRIPTION

To combine the advantages of progressive cavity pumps with those of completely submersible motor-pump arrangements the system described in this paper was developed. It consists of a single stage pump and a water-filled asynchronous motor, both commercially available. The inverter providing ac power for the induction motor was specially developed at our institute ("ITE-inverter"). The photovoltaic array has a peak power of 750 Wp.

As mentioned above, one of the major problems related to this type of system is the stiction between the stainless steel rotor and the elastomer stator resulting in a break-away torque two to three times higher than the nominal torque. A corresponding high power is therefore temporarily required to overcome the break-away torque and to start up the system. This power cannot be delivered by the PV array and has to be provided by a storage device (electrolytic capacitor). Furthermore, a minimum speed is necessary to guarantee the lubrication of rotor and stator.

Developing the inverter for the system the pecularities mentioned above had to be taken into account. In spite of this fact the main objective of this development was to get a simple, robust and reliable device. For this reason, sensors increasing the system complexity (e.g. for the revolving speed) were eliminated. All informations needed for the system control are derived from voltage and current at the inverter input. Both signals can be measured easily. Fig. 1 shows a schematic of the ITE-inverter.

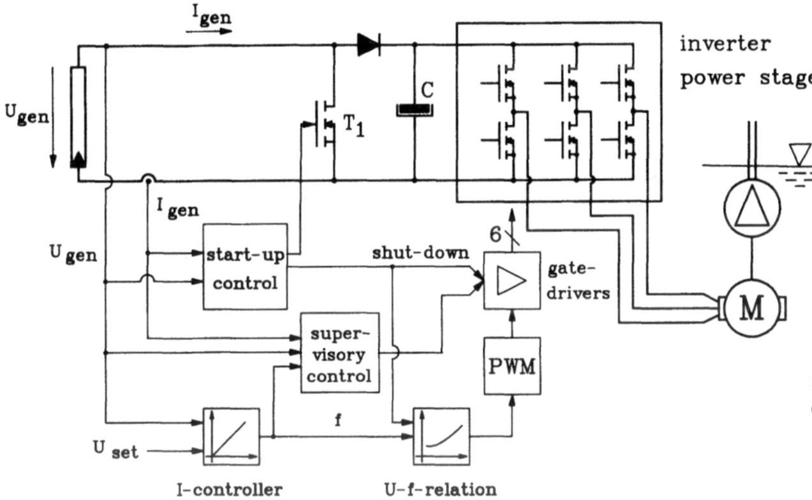

Fig.1:

Schematic of the ITE-inverter

3. START-UP PROCEDURE

The system is started up according to the following procedure: The control electronic checks the power available from the PV array by short-circuiting the solar generator temporarily. If the available power exceeds the threshold level (which depends on the minimum speed and the pumping head), the storage capacitor is charged to a voltage 20 percent above the nominal operating voltage of the PV array. Then the power stage is switched on. To overcome the break-away torque a multiple of the nominal current is impressed on the motor. This is achieved by an appropriately chosen voltage-frequency relation. The required energy is delivered by the storage capacitor. Being once in motion, the rotor is accelerated to the minimum operating speed. Fig. 2 shows the PV array voltage (resp. capacitor voltage) and the motor current during the start-up procedure.

When the minimum operating speed is reached a constant voltage control takes over the system managenment. With this control strategy the PV array voltage is kept at a fixed value, independent of variations in insolation. Thus, the PV array is always operated very close to its maximum power point /4/.

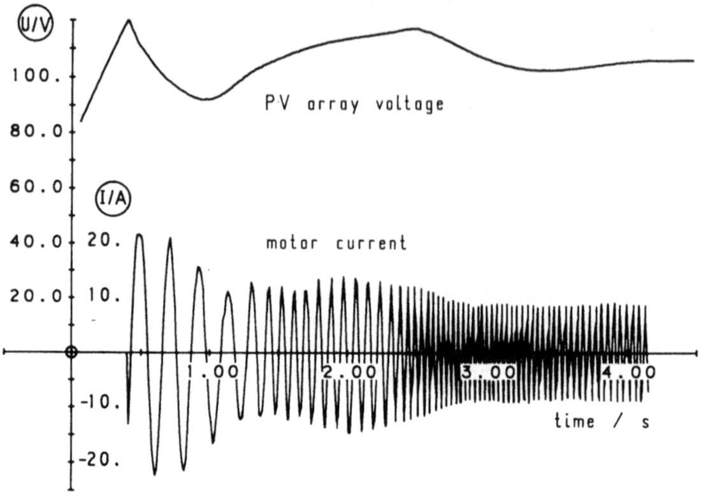

Fig. 2:

Capacitor voltage and motor current during the start-up procedure

4. SYSTEM CHARACTERISTICS

Extensive measurements were performed to characterize the developed system. Fig. 3 shows the head - flow rate characteristic of the pump with the electrical frequency as a parameter. Furthermore, the subsystem efficiency for different pumping heads is displayed in fig. 4. The expected good part-load performance is confirmed since the subsystem efficiency shows no significant decrease as a function of the input power. However, a considerable decrease in efficiency for lower pumping heads can be observed. This is a typical property of volumetric pumps.

The efficiency of the ITE-inverter reaches 90 to 96% depending on the actual input power. The subsystem efficiency in fig. 4 already includes the inverter efficiency.

Finally, fig. 5 shows the performance of the developed system as a function of the daily irradiation.

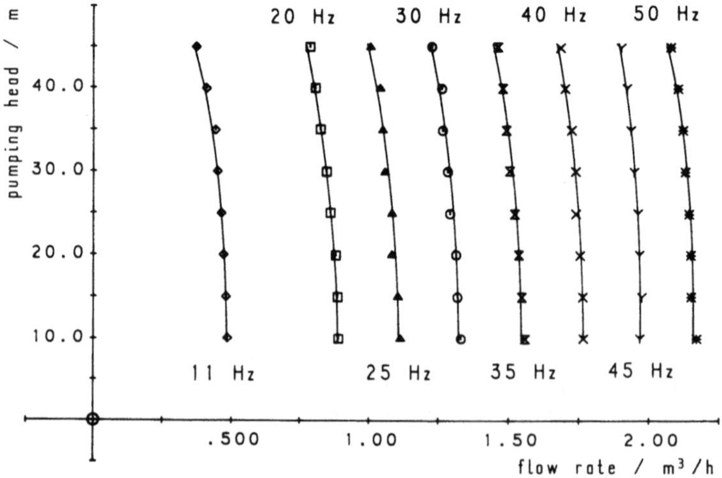

Fig 3:

Head - flow rate characteristic of the pump

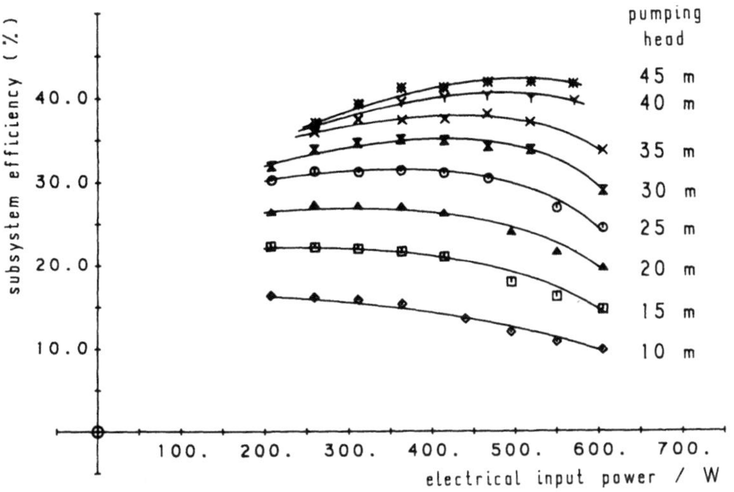

Fig.4:

Subsystem efficiency as a function of input power

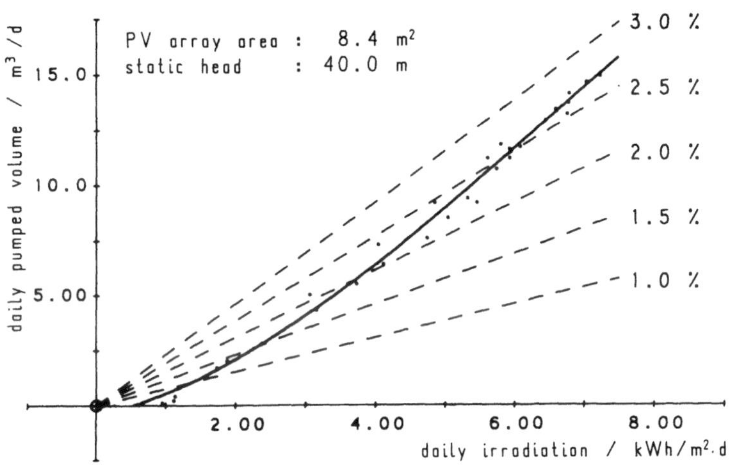

Fig. 5:

Daily pumped volume as a function of daily irradiation

5. SYSTEM SIMULATION

Besides the experimental investigations a precise model for the described system was developed and computer simulations were carried out. Among others, calculations were performed to optimize the PV array size for a given climate and pumping head. Usually the upper limit for the peak power of the PV array is determined by the maximum input power of the subsystem. However, if the inverter is able to dump a certain amount of energy at very high insolation levels, the PV array peak power may exceed the above mentioned limit. Fig 6 shows the relation between PV array size and system efficiency based on climatic data measured in 1986 (location Stuttgart). It is remarkable that the maximum system efficiency is obtained for a PV array with a peak power 24% higher than the maximum power of the subsystem. In this case about 2% of the available energy has to be dumped during times of very high insolation to protect the subsystem.

In the ITE-inverter transistor T1 (see fig. 1) being anyhow necessary for the start-up procedure can also be used to dump the surplus energy.

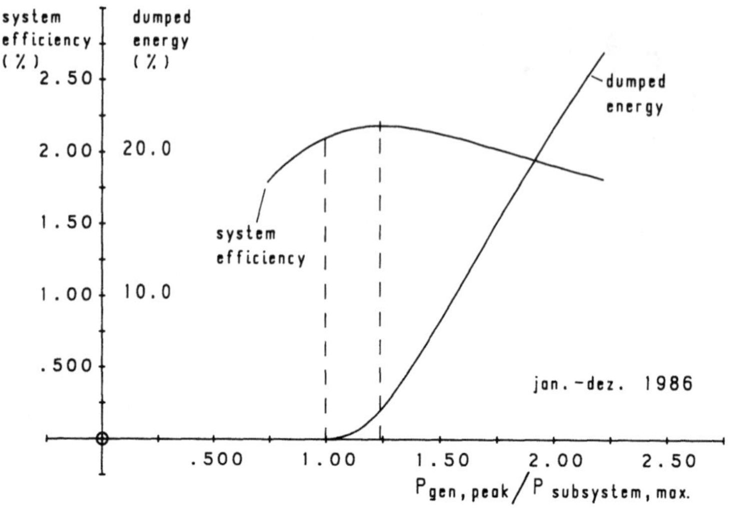

Fig. 6:

System efficiency and dumped energy as a function of PV array peak power

6. CONCLUSIONS

The reliability of the developed system with progressive cavity pump could be proven during a long term test (ca. 8 months) with the system connected to a PV array . The measured daily efficiencies are comparable to those of systems with submersible centrifugal pumps /5/. Valuating the results it should be taken into account that an off-the-shelf pump was used being the smallest one of a product family. The next larger pump already offers a significantly higher efficiency. Summarizing it can be stated that systems using progressive cavity pumps represent an interesting alternative for photovoltaic pumping applications.

7. Acknowledgement

This project is supported by the German Ministry for Research and Technology, contract no. 03E8020B1.

8. References

/1/ Sir Wiliam Halcrow and Partners: 'Small-Scale Solar-Powered Pumping Systems: The Technology, its Economics and Advancement' UNDP Project GLO/80/003, Main Report, Supp. Document 3, London, 1983

/2/ Matlin, R. W.: 'Design Optimization and Performance Charcteristics of a Photovoltaic Microirrigation System for Use in Developing Countries' Mass. Inst. of Tech., Rep. No. COO 4094-33

/3/ Pulfrey, D. L. et al.: 'A Photovoltaic-Powered System for Medium Head Pumping', Solar Energy, Vol. 38, No. 4

/4/ Herrmann, B. et al.: 'Technische Nutzung Solarer Energie, Teilprojekt Photovoltaische Systeme', Forschungsbericht T86-212 (1986), BMFT

/5/ Herrmann, B. et al.: 'Nutzung der Sonnenenergie durch photovoltaische und photothermische Energiewandlung, Teilprojekt Photovoltaische Systeme', Statusreport 1987 Photovoltaik, BMFT/KFA Jülich

DEVELOPMENT OF DC ELECTRIC MOTORS SUITABLE FOR PHOTOVOLTAIC GENERATORS. OPTIMIZATION OF A DC MOTOR-PUMP SYSTEM

A. Rosati, M. Gusso

TECHNOLOGIES 2000 s.r.l.
Via Buozzi, 16, Prato (FI) 50047, Italy

ABSTRACT. The paper illustrates the main technical features of a project concerning the development of a motor-pump unit for remote zones, having some innovative concepts, with high efficiency and high reliability.

The system is composed of a mixed flow centrifugal pump, a permanent magnet dc electric motor, with brushes, and an electronic control system. The optimization of a motor-pump unit is particularly interesting, because the problem of water pumping is very important, especially in remote zones. A good photovoltaic pumping system can have important social effects on such zones, and improve the way of living of the people living in them.

1. PHOTOVOLTAIC SOLAR PUMPS. THE STATE OF THE ART

The technical feasibility of photovoltaic solar pumps has been already demonstrated. Small outputs (up to 2 kW) are the most common and appropriate to the needs in developing countries. The storage by means of accumulators should be excluded; these reduce the total efficiency, require regular maintenance, and do not have a long lifetime. It is much cheaper to store water in tanks.

In the past years, most of the photovoltaic pumping systems were equipped with centrifugal pumps because of their favourable properties. At least for low head applications this type of pumps seems to be indeed the most suitable. The torque-speed characteristic of centrifugal pumps is such that good matching to the current-voltage characteristic of a photovoltaic array can be achieved.

Up to now, most of photovoltaic solar pumping systems have overall efficiency between 1.5 up to 4%.

The greatest percentage of investments in such systems is due, without any doubt (about 50%), to the panels.

It is, therefore, very important to try an improvement in the sub-systems.

In particular, this project tends to reach an appreciable improvement in the motor-pump system.

Our system is composed of:

a) a centrifugal pump of mixed-flow type;
b) a permanent magnet dc electric motor, of brush type;
c) an electronic control system.

The system will work without accumulators.

2. THE CENTRIFUGAL PUMP

The centrifugal pump is, without any doubt, the most common type of pump; in fact, it is employed for several applications. The progress of the development of centrifugal pumps is due to several factors:

a) their high adaptability for high-speed motors;
b) minimum of moving parts;
c) small size and low cost for the amount of liquid moved.

The pump under examination must have the following characteristics:

a) compactness, in order to be used for applications in deep wells, with small dimension boreholes;

b) high efficiency in the operating range in order to reduce the power absorbed, and hence, the photovoltaic panel surface, with an important and beneficial effect on the cost of product water.
c) high reliability, in order to guarantee a good and satisfactory running over a long time, without faults and damage.

With regard to overall dimensions, the pump external diameter must not be greater than about 10 in, in order to be placed into the wells. The pump is a multistage pump. The nominal operating conditions are:
- capacity: 3 kg/s;
- angular velocity: 3800 rpm;
- head (each stage): up to 8 m.

The high value for the angular velocity allows to obtain a high head even when the ratio between inlet and outlet diameters is low; the consequence of this fact is an appreciable reduction in the dimension of the centrifugal pump.

The design of the impeller of the centrifugal pump, under the fluid-dynamic point of view, develops in the following steps:
a) unidimensional design of the impeller;
b) bi-dimensional design of the impeller (blade-to-blade method);
c) interaction between blade-to-blade surface and meridional surface along the height of the blades (through-flow method), for eventual corrections of the results.

The unidimensional design consists of these phases:
a) determination of overall characteristics values (power, capacity, head, velocity, torque, diameter);
b) determination of geometric and kinematic quantities at entrance and discharge.

The bi-dimensional design of the impeller consists in the evaluation of the flow field inside surfaces between blades, at a certain blade height (at hub, at shroud, at camberline), in order to define the streamlines, and to design the blades avoiding high values of boundary-layer, and minimizing secondary flow losses, separations of fluid, and so on.

The numerical procedure used is based on the Finite Element Method, which is the most reliable method, in our opinion, for these fluid-dynamic calculations.

Fig. 1 shows the results of the pressure distribution on the pressure side and on the suction side from the entrance to the discharge; fig. 2 shows the flow streamlines in the R-θ plane, and fig. 3 shows the grid used for the calculation in the R-θ plane; fig. 4 shows an overall view of the flow in the R-θ plane. These figures refer to the calculation at the camberline.

The head, without considering the impeller efficiency, results 12.9 m, and the capacity results 0.31 kg/s. The impeller efficiency will be evaluated in a test bed for hydraulic machines; a mean value from 60% to 70% is, however, to be expected.

A further calculation by using the through-flow procedure has been achieved, but the results have not substantially changed.

The design of the vaned diffuser, placed downstream the impeller, has been carried out by using the blade-to-blade procedure.

The diffuser is designed in such a way that it is possible to have the flow at the inlet of the following stage, with an axial direction.

Fig. 5 shows the velocity distribution on the pressure side and on the suction side, in the vaned diffuser.

Fig. 6 shows the flow streamlines in the R-θ plane, and fig. 7 shows the grid used for the F.E.M. calculations in the R-θ plane.

3. THE PERMANENT MAGNET DC ELECTRIC MOTOR

The dc electric motor under examination is of brush type, with permanent magnets. This choice is due to the fact that, in our opinion, for applications of photovoltaic systems in remote zones, it is advantageous and more reliable to use brush motors instead of brushless motors, which show a complex control circuitry.

Two powers will be considered:

- 1 kW for a first type;
- 4.3 kW for a second type.

The aim is to obtain two important goals:

a) the highest efficiency: more than 92%;
b) the highest reliability.

The motor under examination is of "ironless type", in order to increase brush life and to decrease damping:

The possible actions for improving efficiency regard:

a) support bearings;
b) magnetic circuit;
c) motor heating.

3.1 SUPPPORT BEARINGS

The friction moment of a rolling bearing is given by a load dependent term and another load independent term. In our application very low radial loads on the bearing have to be considered.

Grease lubrication will be considered in order to avoid bearing maintenance. The only method to reduce the value of the moment of resistance (the load independent term is prevailing in comparison to the load dependent term) is the use of a high quality lubrication grease with kinematic viscosity depending little on the temperature, thus reducing power losses without compromising bearing life. A reduction of losses of 60% is expected.

3.2 MAGNETIC CIRCUIT

For the magnetic circuit, Al-Ni-Co permanent magnets with "columnar structures" have been used. The geometrical configuration of the magnetic circuit under examination is such that the working point of Al-Ni-Co permanent magnets is close to the maximum of the product BH.

The non-utilization of rare earth permanent magnets is justified by their particular geometrical configurations, which do not allow to obtain high magnetic flux concentrations.

3.3 MOTOR HEATING

The temperature inside the motor is particularly important and influences some characteristics of the motor itself. Thus, we want to maintain the temperature inside the motor at low values, in order to have a better performance. The system used is based on the "heat-pipe" principle. The heat-pipe can be considered as a closed pipe where an evaporative medium called "working fluid" is enclosed to act as a simple heat transfer element by means of latent heat through the vapor-liquid phase transition, the vapor-flow, and return-flow of the condensed liquid by means of capillary action.

The heat-pipe is composed of three principal elements: sealed container, wick and working fluid.

The application of a heat-pipe to a dc motor is very interesting, and seems to be particularly advantageous. Each motor power loss (Joule's effect, rolling and sliding friction, and so on) causes the heating of various parts of the motor, and, in particular, of the rotor. Joule's effect is surely the most significant; it is due to current conductors' resistance.

Conductors' resistance significantly increases when temperature rises, and it is therefore very important to reduce the temperature of the rotor itself in order to obtain an appreciable advantage under the efficiency point of view.

Temperature reduction also allows advantages under the reliability point of view, because the risks of damage of the insulating material of current conductor wires are lower and, furthermore, the temperature in the contact zone between commutator and brushes is lower.

This is usually the hottest part of the motor; as far as our case is concerned, we think that the temperature will oscillate around 120 °C.

The heat-pipe will be achieved using the motor shaft itself, from the zone close to the commutator to the external part of the motor with circular fins for the cooling. A schematic representation is given in fig. 8.

4. ELECTRONIC CONTROL SYSTEM

The electronic control system contains a micro-processor with EPROM memory for a stable program storage. It will use CMOS components in order to reduce the consumption to very low values.

This electronic control system will check the following parameters: dc motor feeding current, dc motor feeding voltage, pump discharge pressure, motor internal temperature, solar radiation, water level upstream the pump.

The system contains a sub-system which is used for limiting the current at the starting for a few seconds.

The current limitation can occur by means of a transistor of suitable power.

The interval between the starting and the transistor saturation must be adjusted in a simple way. The control system of the transistor must be done by the electronic control system.

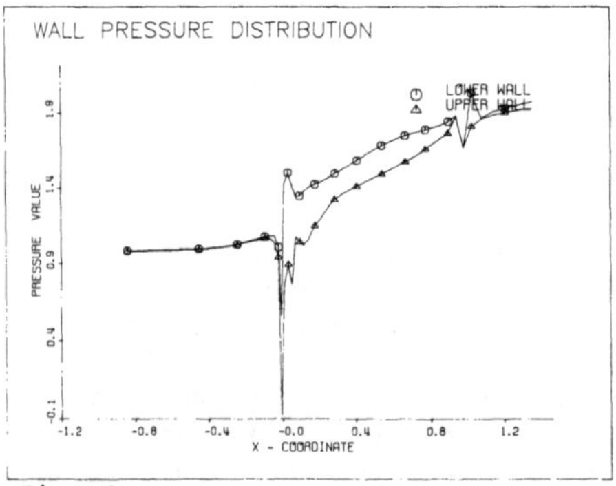

Fig. 1:
Mixed flow centrifugal pump. Pressure distribution on the pressure side and on the suction side of the impeller, at the camberline (blade-to-blade procedure).

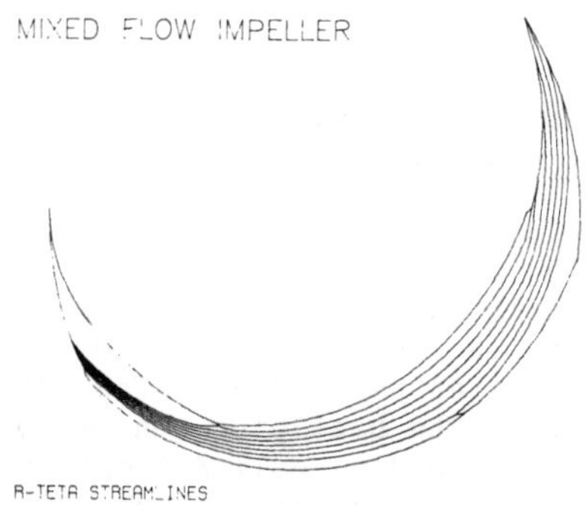

Fig. 2:
Mixed flow centrifugal pump. Flow streamlines in the impeller channel, at the chamberline.

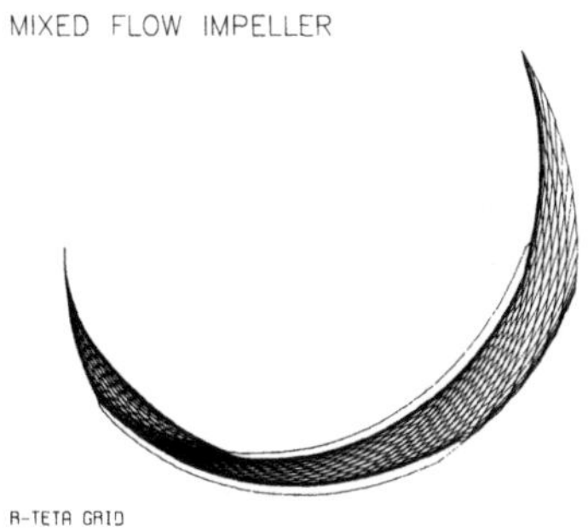

Fig. 3:
Mixed flow centrifugal pump. Grid used for the design of the impeller in the F.E.M..

Fig. 4:
Mixed flow centrifugal pump. Overall view of the flow between blades.

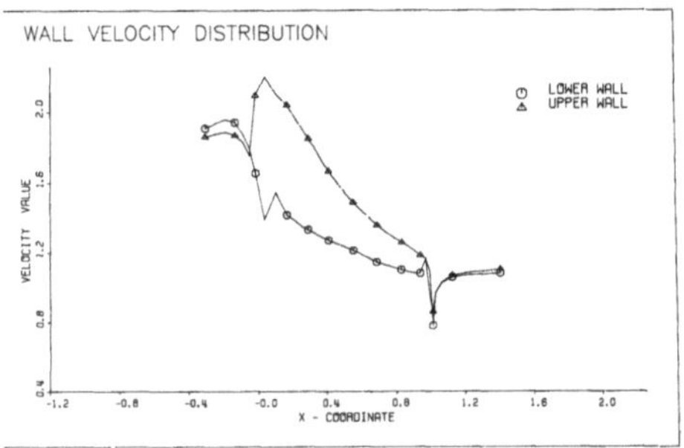

Fig. 5:
Mixed flow centrifugal pump. Velocity distribution in the vaned diffuser.

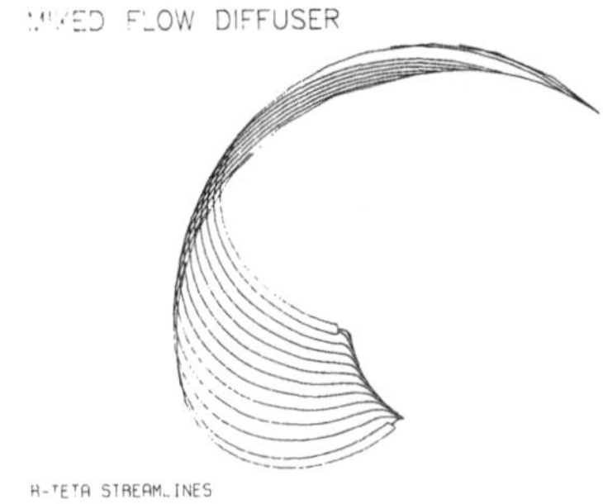

Fig. 6:
Mixed flow centrifugal pump. Flow streamlines in the vaned diffuser, at the camberline.

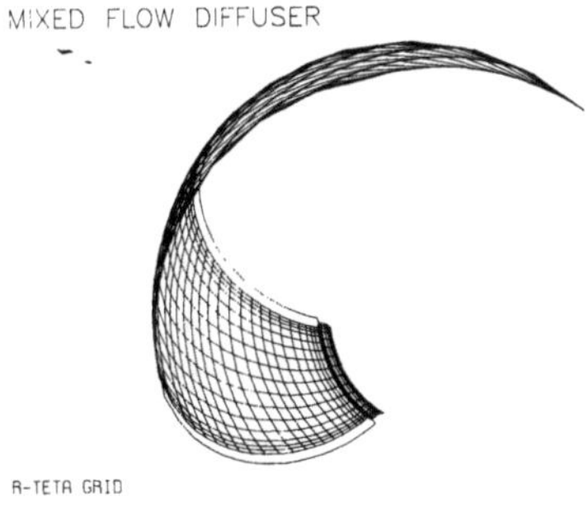

Fig. 7:
Mixed flow centrifugal pump. Grid used for the design of the vaned diffuser in the F.E.M..

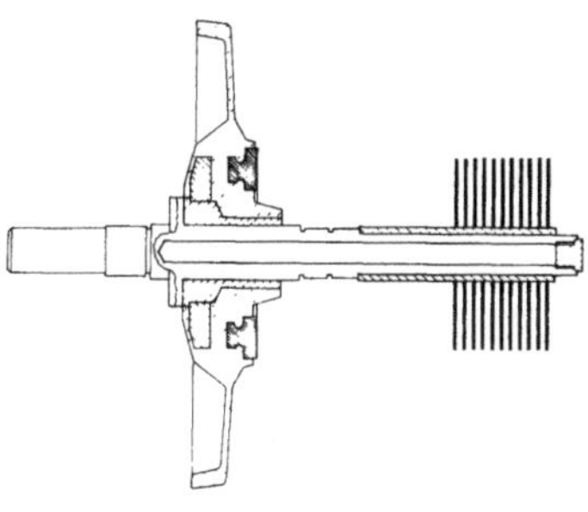

Fig. 8:
Permanent magnet dc electric motor. Cooling system of the rotor by means of "heat-pipe".

AN EVALUATION OF DUAL STEP PV PUMPING SYSTEM IN GUA GILAP, GUNUNG KIDUL - INDONESIA

Abubakar Lubis, Siswa Trihadi, Zhanggischan Udin
Directorate for Conversion and Conservation of Energy
BPP Teknologi
Jl. M.H. Thamrin no.8, 14th floor
Jakarta 10340 - INDONESIA

ABSTRACT. The dual step PV pumping system has been operating since May 1986. The power source is 6.3 kWp PV array which drives a two-stage pumping system via 2 x 3 kVA inverters and 240 - 600 Ah battery bank.
The total pumping head is 149 m (111 m for the first and 38 m for the second step) and water flow rate is around 12 m3 per day with 6 hours operation.
This paper reports an evaluation of the system.

1. INTRODUCTION

In Indonesia, especially in arid area which is typically a remote site, the problem on providing drinking water for annual requirement is still one that can not be solved yet.
One of the arid areas is Gua Gilap, Gunung Kidul.
The water source in this area is located far from habitants, as the underground river beneath a rocky hill.
To take the water from this river, the villagers have to descend the hill about 150 m. This problem occurs every year, especially in dry season (about 9 month annually).
The problem solving on this matter is through the implementation of the water pumping system which means an installation of a power source. But as the nearest power line terminated around 17 Km from the site and its accessibility,the extension of distribution line is not considered for this site. Meanwhile, the installation of on site diesel generator is also not considered as there is no skillful person for operation and maintenance of the system as well as the problem on the fuel availability.
Taking this into account a PV power supply has been installed on this site. The advantage of this system is its unattended operation and low maintenance schedule.
In addition this dual step PV pumping system is supported by NEDO (New Energy Development Organization), a Japaneese Governmental Organization which had encouraged Agency for Assessment and Application of Technology (BPP Teknologi) in executing this joint research project.

2. THE SYSTEM

The system consists of PV array (6.3 kWp), storage battery (600 Ah), control-monitor panel, inverters and pumps.
The generated DC power from PV output is converted into AC power by using two VVVF (Variable Voltage Variable Frequency) inverters. These inverters use a constant ratio between AC voltage and frequency (V/f = constant).
The system pumps up the water from the water source to the intermediate reservoir by using the submersible pump, then the centrifugal pump is used for conveying the water to the final reservoir. The system lay-out is illustrated in figure 1 and the system block diagram is illustrated in figure 2.
This system is also provided with DGDs (Data Gathering Devices) powered by separated PV array with storage battery. The DGDs collect the data every twenty minutes (20 minutes data) into the Ram Cassette.
The data gathered by DGDs in this system are PV current and voltage, cell temperature, inverters input current, water flow.
The DGD is also used for collecting Meteorological data, like irradiation, rain fall, humidity and air temperature.
In front of the control-monitor panel there are several meters for monitoring the instantaneous values such as PV current and voltage, water flow and load current; several amperehour meters (for measuring the integrated values).

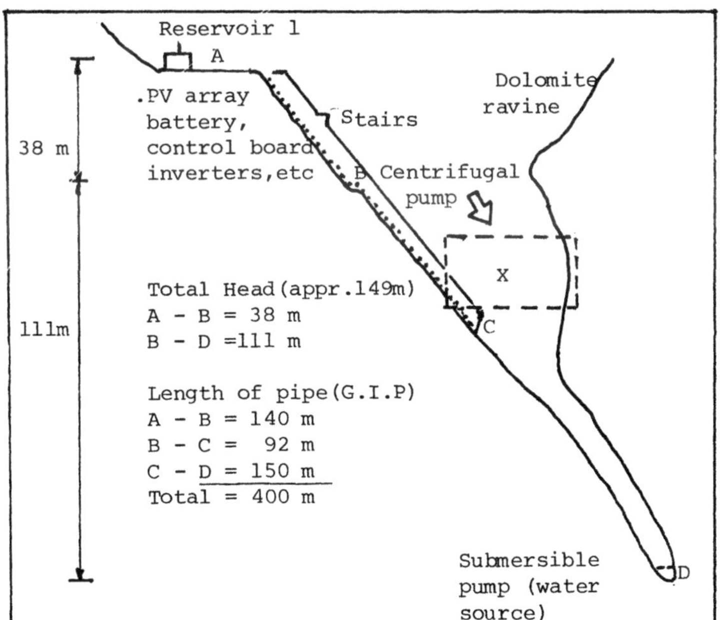

Fig.1 System lay-out

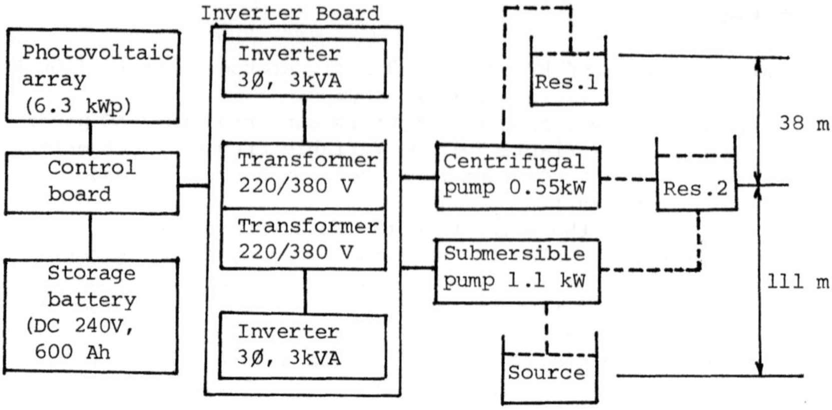

Fig.2 System block diagram

3. THE EVALUATION

In the evaluation, personal computer is used to transfer the data from ram cassette into floppy disc and process the data.
The dual step PV pumping system is considered as an integral unit for evaluation purpose. It means that the system employs a single pump operates at a total head of 149 m.
The system efficiency is calculated by using the following equations:

$$N_{system} = N_{PV} \times N_B \times N_{subsystem}$$

where :

N_{PV} = Photovoltaic efficiency is calculated as follow :

$$N_{PV} = \frac{V_p \times I_p}{Q \times S}$$

V_p, I_p = photovoltaic voltage, photovoltaic current respectively
Q = irradiation
S = effective area of PV array

N_B = Battery efficiency is calculated as follow

$$N_B = \frac{I_d \times T_d}{I_c \times T_c}$$

I_d, I_c = discharging and charging current
T_d, T_c = discharging and charging time

$N_{subsystem}$ = the efficiency of inverter 1, inverter 2, pump 1 and pump 2 assumed as an integral unit, and calculated as :

$$N_{subsystem} = \frac{Rho.G.H.Q}{V_B \times (I_{inv.1} + I_{inv.2})}$$

$Rho.G.H.Q$ = Hydraulic energy

$V_B x (I_{inv.1} + I_{inv.2})$ = Electric energy required by input inverters.

4. THE RESULTS

The results of the evaluation in this study are shown in figures 3,4 and 5.
Figure 3 shows the energy balance of the system and figure 4 shows the system efficiency where its result is nearly similar to the design point i.e 2 %.
The daily water production is at rate of 13 m3, higher than the design amount (8 m3).

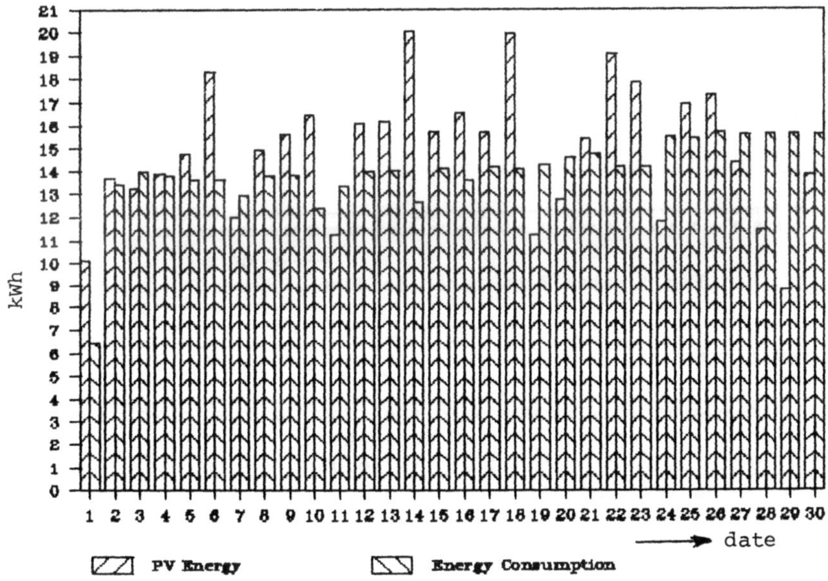

Fig.3 Energy balance (in June 1987)

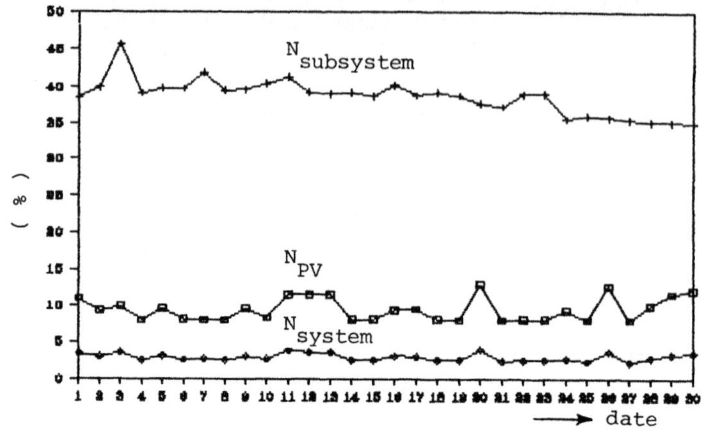

Fig.4 Efficiency of system, PV and subsystem
(in June 1987)

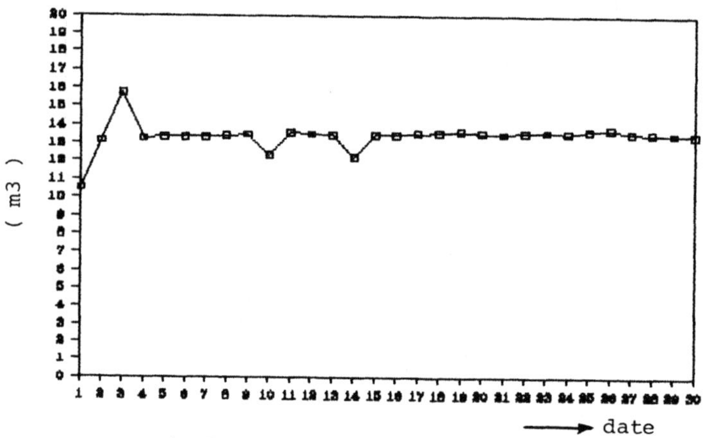

Fig.5 Water flow rate (in June 1987)

5. CONCLUSIONS

1. In the evaluation, it was used the computer system which gives the long term measurement and accuration.
2. The efficiency of photovoltaic is roughly 9 - 11 %.
3. The average system efficiency is more than 2 %.
4. The ratio of the energy consumption to the PV energy is around 85 %, meaning the energy balance fits to the optimized condition.

Motor Starting from Photovoltaic Sources

A.Y. Kheder
University of Arizona
Tucson, AZ

Paul E. Russell
Arizona State University
Tempe, AZ

ABSTRACT. At starting, electrical motors require large power and current, especially when the mechanical loads are significant. This may not be a problem in a large electrical system but it may be very severe for a limited power source such as a solar array. If a direct approach is taken, the array rating must be 5-6 times the motor rating in order to start the motor and its high inertia, high friction load. Batteries have been used to store energy and supply that energy for starting. Batteries need maintenance, and their efficiency can be a problem too.

In this study a new type of controller has been suggested and developed for use with DC motors. Computer simulation showed promising results. The controller uses the array power, which is equal to the rated power of the motor, for starting and for running conditions. Experimental results showed that the theoretical results are applicable.

1. Introduction.

The most practical use of the solar array is in the stand alone system, which basically consists of an array and the load. If the load is a small community then there should be some reserve generation or energy storage capacity to pick up any extra load so as to keep system operation under some transient disturbances. But if the load is a single pump drive in a remote location, it is difficult to justify that reserve. Array capacity is expensive, storage of electrical energy is costly too, and batteries, if used, eventually require considerable maintenance and are rather inefficient.

At starting the current in the DC motor is limited by the armature resistance only. The armature resistance is quite low and the starting current is quite high. It should be possible to use the same array for starting and for running the motor. In other words, the amount of energy used in keeping the motor running under rated condition should be made to be enough for starting the motor. To do so the system must be efficient and be able to get the maximum use of the expensive array. To implement efficiency throughout the system an efficient pump must be used. That efficient pump is often a positive displacement unit such as a jack pump. This pump is of the reciprocal type, and even though it is very efficient, it is quite hard to start because of the seals inside the pump. After sitting overnight these seals become "seated" and then it requires extra torque to "break it in" again. This extra torque must be supplied by the motor which then requires a large starting current.

The output power of the array is DC which makes it easier to use a DC motor with the array. The solution suggested works for all types of DC motors with small modification, but the analysis and the experimental part were carried out on a permanent magnet motor which is easier to study and likely to be used in water pumping.

The idea is to provide matching between the array and the load. The array rating is chosen to meet the motor rated condition, such that the array will produce the motor rated current and voltage when it operates at its maximum power point. Then the array is connected to the motor through a power conditioner circuit which assures the matching between the two units at starting. At starting the required voltage is quite low since the current is limited by the armature resistance only. What is required from the power conditioner is to work as a step down DC transformer. This can be accomplished with a power switch, transistor or thyristor, and a pulse width modulator. The output voltage of the transformer can be controlled by changing the pulse width that controls the transistor. The transistor can be switched on and off at any specified frequency.

2. The Power Conditioning Circuit.

The circuit used in the simulation is shown in Figure 1. The step down controller is to utilize the maximum power from the array and switch it to a lower voltage and higher current to start the motor and its load. This is to be accomplished by proper choice of the inductor and the capacitor. The capacitor C is to store the energy from the array and supply it to the motor as needed. The inductor L is to keep the array current near the optimum current of the array. Clearly during the charging and discharging phases of the operation,the capacitor voltage will be increasing and decreasing, changing the operating point on the i-v curve of the array. The operating point will not stay at the maximum power point. Instead it will be shifting back and forth the between points P_1 and P_2, Figure 2. The points P_1 and P_2 are chosen because they represent 90% of the maximum power available from the array. The two points correspond to the voltage V_1 and V_2. These two voltages are at ±18% from the voltage at the maximum power point [1].

At starting the induced voltage in the motor is zero so the current is limited by the armature resistance only. Supplying higher voltage means exceeding the maximum power capability of the array. Supplying smaller voltage means operating below the maximum power point. There is an optimum voltage at which there is matching between the array and the motor. To find that voltage, consider the operation as a step down transformer so that

$$V_a = V_r/X \tag{1}$$

$$I_a = I_r X \tag{2}$$

where V_r is the array voltage

I_r is the array current

V_a is the armature voltage

I_a is the armature current

X is the transformation ratio

$$V_a = E + I_a R_a$$

Substituting from equation 1 and 2 in the above equation

$$V_r/X = I_r X R_a + E$$

Solving for X and neglecting the negative solution

$$X = \frac{[E^2 + 4I_r R_a V_r]^{1/2} - E}{2I_r R_a} \qquad (3)$$

$$X \geq 1$$

X is the ratio by which the array voltage must be reduced, or divided, in order to supply the motor with the optimum voltage and keep the array operating at or near the maximum power point.

3. Choosing the Circuit Parameters

Choosing the Capacitor

During the on time the capacitor will lose some of its charge, but during the off time the capacitor will gain charge; to keep the system going, the charge lost must be equal to the charge gained. The charge lost during the on time, Q_{lost}, is given as

$$Q_{lost} = C\ \Delta V$$

where $\Delta V = V_2 - V_1$, from Figure 2.

$$Q_{lost} = \text{on time} * \text{current.}$$

$$Q_{lost} = (x-1)/x * t * I_r.$$

Equating the above two equations gives

$$C = (x-1)/x * 1/f * I_r/\Delta V \qquad (4)$$

From equation (3), it becomes clear that the larger the current of the motor, the larger the capacitor that is required. Also the larger the allowable voltage change, the smaller the capacitor. It is interesting to see that the higher the frequency of switching the smaller the capacitor requirement becomes.

Choosing the Value of the Inductor

During the operation the current will change between I_1 and I_2, see Figure 2, and the voltage between V_1 and V_2. To assure that this change will not shift the operating point away from the maximum power point we use an inductor. The inductor will be chosen such that the change in the capacitor voltage will be compensated for by the voltage drop across the inductor.

$$\Delta V = L \frac{\Delta I}{\Delta t}$$

$$L = \frac{\Delta V}{\Delta I}\, t$$

Since the t_{off} time is larger than the on time that will be chosen, that larger value requires

$$L_{off} = \frac{\Delta V}{\Delta I}\, t_{off} \quad (5)$$

4. Simulation of the Circuit and Results

Figure 1 shows the circuit that was considered in the simulation. The voltage source V represent the photovoltaic array, and the resistnace R is added to represent the slope in the characteristic of the array. there were two situations considered. The first one is that when the open circuit voltage and R where used to approximate the first part of the non-linear characteristic of Figure 2 between $V_{o.c.}$, and P_1. The second one, which represented the operating range, was considered by extending the line which passes between points P_1 and P_2 to the voltage axis. The intersection with the voltage axis gave the value of the voltage source and the slope as the resistance in series with that source.

In this study the effect of frequency on the capacitor and inductor values is very obvious, equations (4) and (5). The values of L and C are inversely proportional to the frequency and it is important to use the smaller possible values for the inductor and the capacitor. The first frequency studied was 500 Hz and the results were found and then the switching frequency was increased to 10 KHz and the effect of that on the results was found. The effect of removing the inductor from the circuit was studied too and the results were found.

The motor studied was a 1/3 H.P., rated at 90 Volts, 3.5 Amperes with an armature resistance of 3.0 ohms. Using equation (3) the transformation ratio, x, is equal to 3.0. This gives an armature current equal to 10.5 amperes and voltage equal to 22.45 volts. This set of values gives a starting torque of 3 times the rated torque. The array current remains at 3.5 Amperes and the operating point at the maximum power point. The change in the capacitor voltage is the same as the designed value. The circuit was tried experimentally and the measured values (average values) were array current = 3.75 amperes, the array voltage = 86 volts, the motor current = 10 amperes and the motor voltage = 28 volts.

5. Conclusion

It is possible to start a DC motor from a limited power source, solar array. Moreover it is possible to develop high starting torque, maybe as much as four times the rated torque. This is done by matching the load to the array all the time, by matching a changing load to the array with the help of a controller. Increasing the frequency of switching has a good effect on all system parameters. Using the inductor to keep the array current as close as possible to the rated current is effective too. For best results, a high switching frequency must be used. The higher the rated motor voltage, the larger the transformation ratio, x, and the more effective is the solution suggested.

REFERENCES

1. J. Appelbaum, J. Barry, "Performance analysis of DC motor-photovoltaic converter System-I", Solar Energy Vol. 22, pp. 439-445, Pergamon Press Ltd. 1979.

2. Z. Zinger, A. Braunstein, "Optimum operation of a combined system of a solar cell and DC motor", IEEE, Vol. PAS-100, No. 3, March 1981.

3. S.K. Datta, "Power electronics and controls", Reston

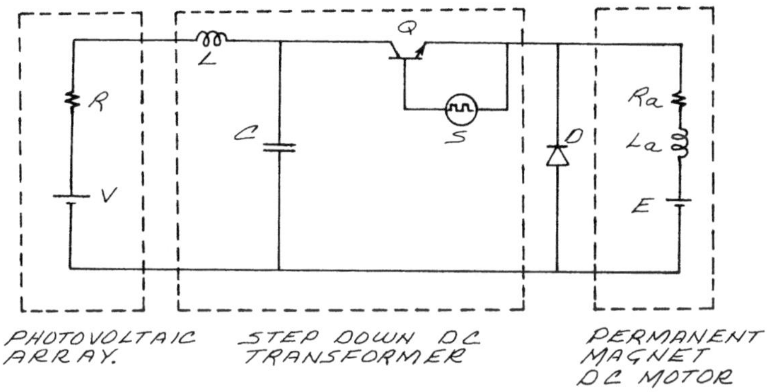

FIG 1 POWER CONDITIONER CIRCUIT.

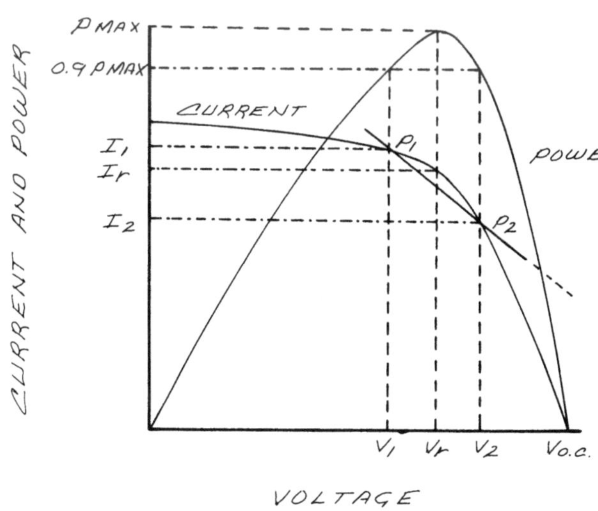

FIG 2. TYPICAL PHOTOVOLTAIC ARRAY CHARACTERISTIC.

The control of lead acid storage battery element capacity used in hybrid PV Systems .

Authors: M. PICCOLO - G. GOGLIA - G. MASCIA - D. BIONDINI.

ORGANIZATION:
MINISTERO DELLE POSTE E DELLE TELECOMUNICAZIONI
ISTITUTO SUPERORE P.T. Ufficio VI Rep. 1°
VIALE EUROPA 190
00144 ROME (ITALY)

ABSTRACT

The problem of lead acid storage batteries is of special importance when these devices are used in photovoltaic equipment which may also be fed by other energy sources (e.g. Diesel E.G.) usable - according to the state of charge of the battery - when there is need of particularly high level of power supply availability, for istance in the field of Telecommunications appliances.
In such a cases the control of electrolyte density is the best way of evaluating the state of charge of battery elements.
This paper illustrates a newly designed device which cecks either the electrolyte density or its temperature and the state of charge of the elements throungh adequate measurement instruments.

1. INTRODUCTION

A photovoltaic equipment for testing the application of photovoltaic solar energy in Telecommunications domain was implemented in 1986 by the ISPT Renewable Energy Laboratory.
Information on the performance and reliability of PV system will be obtained by studying the equipment behaviour and performing tests on it. Also useful information will be provided as for forthcoming achievements.
This equipmenent is located at the top of the P.T. Ministry building.
It is a pilot equipment designed for PV applications on TLC equipment which very often is located in difficult of access areas where connection to the network is not available.
PV generator peak power is 1320 W in 48 V Voltage.
The storage system is made up by 48 V stationary lead battery with 560 Ah C/10 capacity.
A diesel engine generator set (5 kVA, 48 V , 200 HZ) supplies power in sunless periods. A user (200 VA, cos =1,220 V, 50 Hz) is fed by a DC/AC inverter.
Fig. 1 gives an outiline of the equipment.
The graphics (Fig. 2,3,4) reproduce the results of the batteries testing

approvals.
Fig. 3 shows the discharge of one element of a 100 - hour 4.8 A constant current battery , from which the diagram of the proceeding of the residual discarge capacity according to density is obtained (curve a of Fig. 4). Storage battery charging (curve b) is also shown in the same figure during the richarging process in Fig. 2.

2. INTERVENTION OF THE DIESEL ENGINE GENERATOR.

The basic criterion supporting the design of the integrated PV system is that of limiting the capacity of the storage battery and the peak power of the PV array by fitting into the system a Diesel Engine Generator(DEG) which should both make up for expected deficiency of PV energy occuring in wintertime(15% not exceeding value over the year) and unexpected solar radiations(maximum duration time: 4days running).
The whole PV array can supply the storage battery, whose capacity is 560 Ah/10 hour, with 25A/48V maximum nominal current. It infers that recharging time of the battery by PV array will not be lower than $\frac{560}{0.90 \times 25} = 25$ hours-assumed a charge efficiency equal to 0.9- corresponding to four sunny days running. Althought the use of the DEG notably reduces the recharging time the problem, anyway, would not be solved if the intervention of the DEG were been dependent on the value of the battery voltage. In fact as the DEG two-phase charging system(60A constant current and 2.35V/el constant voltage -Fig.2-) allows to exactly specify the moment when the DEG disconnects according to the current value to battery, the right moment of DEG intervening because of battery discharge may be challengend by the presence of the high function of PV panels in connection with the battery slowly discharging, (the plotting of the charge curve in Fig.2 shows that the DEG should not work for more than twelve hours, because after this time limit, the charge increase is very slow with respect to time: 30 Ah-7%-in five hours -41%-).
It is clear that the voltage criterion can't he assumed to control the DEG intervention for uncharged storage batteries; the density criterion seems to be safer. In fact if one considers the diagram of Fig.4a, it is apparent how density behave according to the expectations and display with more accuracy the state of battery charge, not being affected by the buffer effect of PV panels.
It is possible, therefore, to perform a careful control of the battery charge by using an automatic equipment designed by an Italian manufacturer. But this criterion applies only to lead storage batteries with liquid electrolyte suitable for built-in cell.
If we compare the graphics in Fig.4(a) discharge and (b) charge process, we will see that-assumed the same residual charge value-density is lower during the charge process than during the discharge one. Obviously it depends on the fact that charging rate(through DEG) is higher than the rate of slow discharge(100 hours), the values of which are, therefore,

more riliable.
As density criterion does not apply to recharging system disconnecting current criterion has been adopted.
In consequence of what has been considered above the DEG disconnecting from the system may have an expected value of 9 A battery absorbed current. Vice versa the expected value of DEG intervention may be equal to $1148 Kg/m^3$ corresponding to 20% of residual capacity that is after eighty hours running in total absence of solar radiations.
So a remarkable margin of safety is left which ensures a good performance of the storage system for many years.
To sum up we can say that after a year of testing, density control system is proved suitable to be used in PV systems having the same features as those being tested.

3. AN OUTLINE OF DENSITY EVALUATION ASSEMBLY (DEA)

The next description takes into consideration Density Evaluation Assembly (DEA) specifically designed for density measurements as well as for internal lead storage battery electrolyte temperature. The system is made up by special measurement cells fitted into the elements to be examined and connected to a control and data acqisition set by suitable electronic circuits through connectors and flat cable.(Fig.5)
Each cell and its electronic card are fully protected against moisture and are acid-proof.Control and data collection device can "inquire" the cell picked up, making it possible to read sequentially density and temperature values relevant to the elements under inspection.
Channel selection can be made both manually and automatically.
Manual selection is performed by address keys(Fig.6). If automatic selection is chosen, channel shifting occurs sequentially with a frequency ranging from one channel every 16 seconds to anothe every 8192 seconds.
The same control box may check a hundred cells at maximum.
The indirect method of density measurement is carried out by a floating device connected to an appropriate circuit so as to make up a current specific weight transductor.
It is possible to make the current in the circuits proportional to liquid specific weight through some expedients.
As it is well know numerous small gas bubbles may be present on the inside of the acid solution of lead storage batteries; therefore an error of measurement could occur because of the gas modifying the liquid density value.
An electromagnetic bubble killer system shaking measurement elements has been introduced in order to solve this problem.
The "bubble killer" intervenes whenever channel shifting both automatic and manual is being performed so that the "inquired" cell could be freed from gas bubbles accumulated during the previous inactivity phase.
The system hold mode arrives when the "float" is under shaking phase, that

is measurement value is freezed on the display.
Bubble Killer can be performed even manually through a knob (6) no matter when.
The system is equipped with three alarms:

- density deficiency alarm
- excessive temperature alarm
- deficiency level alarm (Fig.6)

When one of the abovementioned value oversteps the preset bounds a LED on the system switchboard provides a signalling. Concurrently a relay controlling "clean" contacts from outside, steps in.
It is also possible to use the relay contacts for telealarms or ancillary devices.
Teachnical features:
Density measurements:

- Measurement interval	1100 - 1300 kg/m^3
- Accuracy and repeatability	0.7 %+1 digit
- Temperature rate	0: 50°C
- Maximum flat cable lenght	30m
- Calibration temperature	25°C
- Temperature coefficient	0.01% /°C

EXPERIMENTAL INTEGRATED ALTERNATIVE ENERGY POWER SUPPLY
I.S.P.T. PROTOTYPE (simplified block diagram)

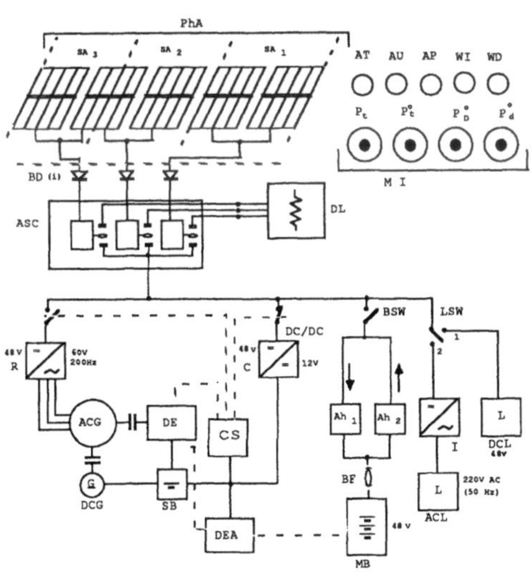

PhA	Photovoltaic Array	ACG	AC Generator
SA(i)	Subarray n. 1-2-3	DE	Diesel Engine
BD(i)	Block Diodes	CS	Control System
ASC	Autom.Switching Combiner	BF	Battery Fuse
DL	Dummy Load	DCL	DC Load
Ah(1,2)	Amper -hour 1,2	DCG	DC Generator
MI	Meteo Instruments	SB	Starting Battery
R	Rectifier	MB	Main Battery
C	Converter	I	Inverter
LSW	Load Switch	DEA	Density eval.assembly
BSW	Battery Switch	ACL	AC Load

Density Deficiency Alarm:
It can be preselected by a semifixed potentiometer located inside the control box.
Ranging field for the relay stepping in: 1150 - 1240 Kg/m^3.
The system, manufactured by an Italian Firm, is patented.

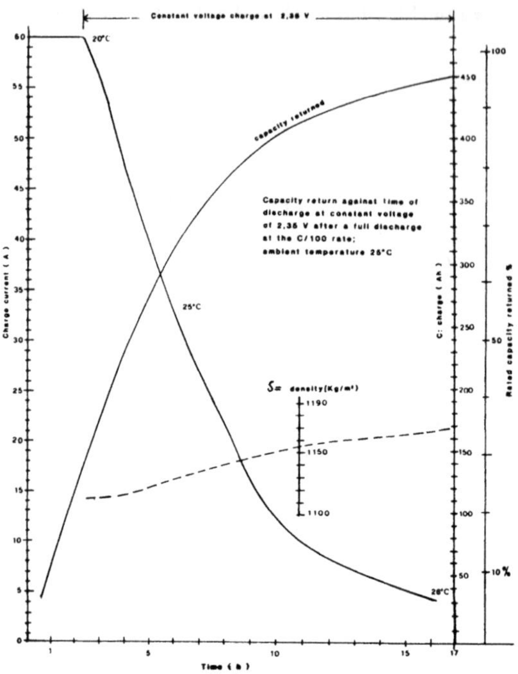

Fig. 2

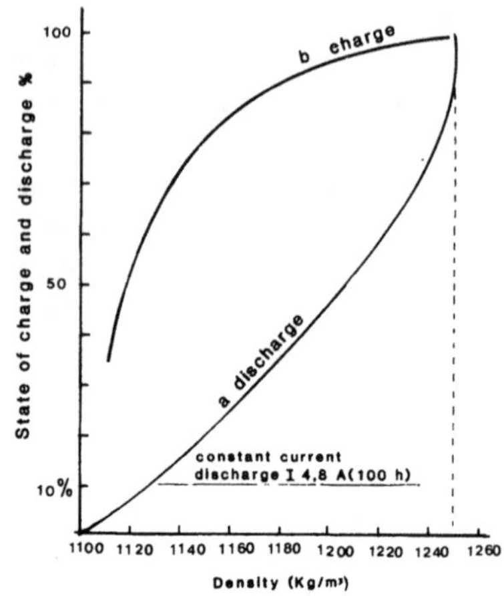

Fig. 4

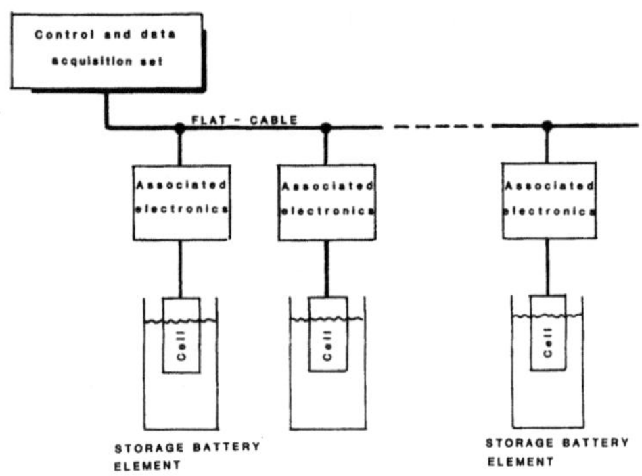

Fig. 5

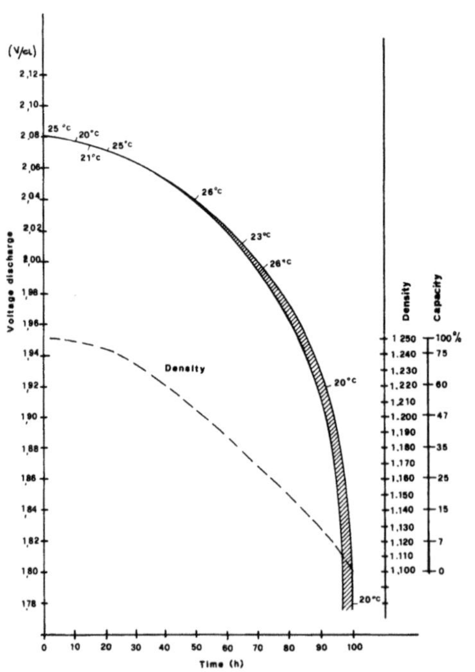

Fig. 3

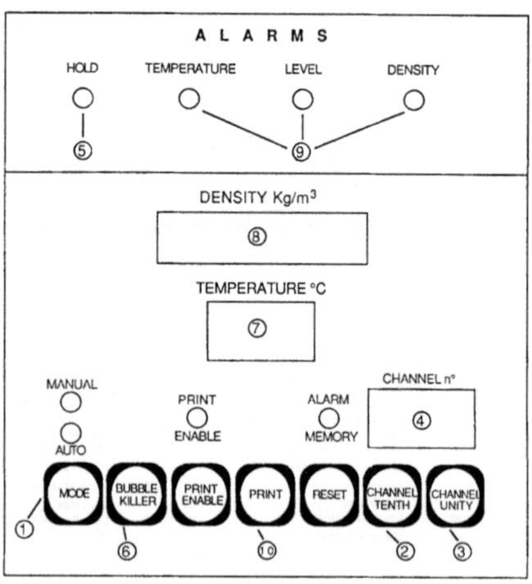

Fig. 6

Significance of Electronic Ballasts in Fluorescent Lamps for the Performance of Small-Scale Photovoltaic Lighting Systems

Klaus Haars, Winfried Müller
Interdiziplinäre Projekt Consult GmbH
Am Eisernen Schlag 31
D-6000 Frankfurt 50

Abstract

Field and laboratory tests conducted during a one-year pilot dissemination programme for Solar Home Systems (SHS's) revealed that the electronic ballasts in the lamps are the "weak link" in these small photovoltaic systems. A total of 23 ballasts were evaluated in terms of lifetime, efficiency, reliability and cost, and not a single one was found to be acceptable in all 4 of these key areas. The complexity of the ballast technology, and the demands which are placed on it in SHS applications, have been greatly underestimated. The unsatisfactory quality of the currently available models is an obstacle to dissemination of these systems on an expanded scale.

1. Introduction

In the course of a one-year testing programme financed by GTZ (Deutsche Gesellschaft für Technische Zusammenarbeit), 250 Solar Home Systems (SHS's) were installed in local smallholder households in the Altiplano area of Peru. These systems, which consisted of a 40-W_p panel, a charge regulator and a car battery, were used to provide power for fluorescent lamps and transistor radios. The main objective was to investigate the viablity of this approach as an alternative to grid-based electrification programmes in rural areas.

Figure 1: SHS installed at a farmstead near Puno

The tests demonstrated that the currently available SHS's still exhibit major technical deficiencies, primarily in the electronic components. Indeed, the 12-V DC fluorescent lamps - which are based on what is widely regarded as a very simple, proven technology - were identified as the critical "weak link" in the overall system, above all due to the poor quality of the electronic ballasts employed in them. Most DC lamp ballasts have been developed specifically for use in camping and other leisure applications and are not designed for day-in, day-out use over an extended period of time. Lifetimes of a few hundred hours and efficiencies of less than 50% - which may be adequate for such applications - are absolutely unacceptable in PV home lighting systems installed in remote areas of developing countries. Ballasts designed for transport applications are usually unsatisfactory with respect to efficiency and cost.

In an analysis of the investment and running costs of SHS's, the cost annuity figure for the lamps turned out to be almost as high as that calculated for the solar module. In other words, the poor quality of the ballasts severely undermines both the technical and the economic viability of the systems.

2. Test Results

The testing programme was supported by intensive technical research both in the laboratory and in the field. In the framework of the project, 23 electronic ballasts supplied by European and Latin American manufacturers were evaluated in terms of certain key criteria[1]. Although the tests revealed significant differences in the quality of the ballasts, even the "best" models did not meet the minimum standards in all areas, and a number of serious problems came to light. For example:

- The majority of the ballasts exhibited energy losses of more than 30%, and a value of 54% was measured for one model.

Figure 2: Lamp test stand at the project office

[1] cf. IPC: "Technischer Anhang zum Feldtest Solarer Kleinsysteme in Puno/Peru", 1987; prepared for GTZ

- With the exception of one model, the power output of all ballasts was equivalent to less than 60% of the nominal lamp power, and indeed, in some cases the value was less then 30%.
- A quarter of the ballasts would not ignite the gas discharge when the battery voltage fell below the nominal level of 12 V.
- A third of all ballasts generated an extremely uneven current flow (high distortion rate, large offset component), and two ballasts produced only one half-wave.
- 37% of all ballasts were not equipped with reverse polarity protection devices.
- A third of the ballasts caused very strong radio interference.

The results of the lifetime tests were also clearly unfavourable. With the majority of the ballasts, lamp lifetime was reduced to a few hundred hours or switching cycles. In some cases, the ballasts themselves also failed after less than 6 months of continuous operation.

3. Technical Aspects

Every fluorescent lamp must be equipped with a power conditioner, and for battery-powered operation an electronic ballast is required which transforms the electrical energy supplied by the battery (low-voltage direct current) so that it can be used to trigger and maintain the gas discharge in the tube alternating current.

Every tube has its own specific characteristics, including the nominal power (e.g. 18 W), the operating voltage (e.g. 60 V) and the tube current (e.g. 0.35 A). A specially adapted ballast must be employed in order to ensure optimal performance for a given type of tube. Use of a ballast whose nominal power differs from that of the fluorescent tube will reduce not only the ratio of the power delivered (ballast output) to the nominal lamp power, but also the electrical efficiency of the ballast. Other relevant parameters such as the peak factor are also adversely affected, which lowers both the luminous efficiency and the lamp lifetime.

Only standard fluorescent tubes are relevant for applications in the developing world. The compact lamps which have recently been developed are very expensive and are not currently available in many Third World countries, nor could they be manufactured there.

Electronic ballasts used in SHS applications must be designed to ensure long lamp **lifetimes**, must exhibit a high electrical **efficiency**, must be highly **reliable**, and - last but not least - they must be **inexpensive**.

3.1 Lifetime

The lifetime of a fluorescent lamp is in essence determined by the rate of degradation of the electrodes or lamp filaments. Thus, it is essential to utilize a reliable, high-quality ballast which will ensure that the proper lamp ignition and operating conditions are maintained. While an uneven load resulting e.g. from an offset current may considerably reduce lamp life, the decisive factor here is the ignition process, as wear and tear on the electrodes is greatest during ignition. Accordingly, the lamp lifetime should be specified not in hours but rather in terms of complete switching cycles, i.e. the total number of starts that a lamp can "withstand" before it fails.

The voltage that is required to ignite the gas discharge is considerably higher than the operating voltage, and the lifetime of the electrodes is determined by two things: the precise voltage level to which they are subjected during ignition and the temperature at which they are initially subjected to it.

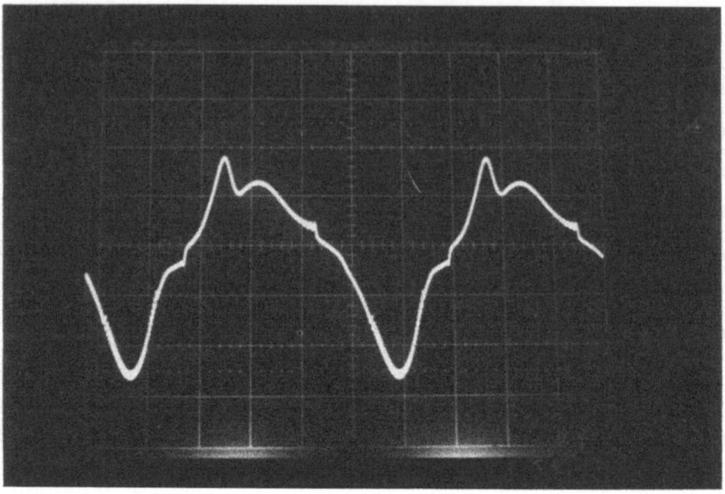

Figure 3: Voltage curve of a ballast with recognizable defects

In order to ensure trouble-free ignition even at low temperatures, the ignition voltage is usually set at a level of 600 V or higher. One way to reduce the wear and tear which this causes on the electrodes is to utilize a "voltage ramp". When such a device is used, the voltage is steadily increased, but only up to the actual point of ignition, after which it is quickly reduced to the normal operating level. However, such devices employ sophisticated electronics and thus are not only expensive (cost criterion) but also reduce the overall reliability of the ballast.

Far more significant, however, is the temperature of the filaments at ignition. Tests conducted with a recently developed ballast showed the lamp lifetime to be only about 6000 switching cycles when ignition took place under "cold start" conditions. When the same ballast was equipped with an appropriate preheating device, the figure rose by several hundreds of thousands.

3.2 Efficiency

Two types of efficiency are relevant in this context: the electrical efficiency of the ballast, i.e. the power output-input ratio, and the capacity utilization rate of the lamp, i.e. the share of the nominal lamp power which is actually supplied by the ballast.

In addition to achieving a high electrical efficiency value, the ballast should also deliver an adequate power supply to the lamp - i.e. one which is sufficient to enable it to operate at its nominal power level, and in turn achieve optimal luminous efficiency. Ballast efficiency is obviously influenced by the overall design of the circuitry of the various individual components such as the transformer, but the frequency is also an important factor here. Indeed, an increase in the frequency raises the luminous efficiency of the lamp, with the latter reaching its maximum value at about 50 kHz. However, radio interference is a problem at higher frequencies, and, as a result, the ballast frequency is normally set at about 20 - 30 kHz.

The best of the ballasts tested to date exhibits an electrical efficiency of just under 80%. Slightly higher values could probably be achieved through design improvements.

3.3 Reliability

When employed in Solar Home Systems, fluorescent lamps must operate under widely differing climatic conditions. In addition, the systems' users often know next to nothing about how the technology actually works and rarely live in areas where repair facilities or spare parts are readily available. This means that a ballast should be designed to ensure trouble-free ignition and operation even under unfavourable conditions, in particular

- ambient temperatures ranging between -10°C and +50°C,
- battery voltages of between 10 V and 15 V, and
- relative humidity levels ranging from 0% all the way up to 95%.

Moreover, it should have built-in safeguards against the specific types of technical problems and equipment damage that are most commonly encountered when users do not employ proper operating procedures, i.e.

- reverse polarity protection,
- open circuit protection,
- short circuit protection,
- radio interference suppression.

3.4 Costs

The costs of a complete lamp (fluorescent tube and fixture with built-in ballast) - between 50 and 100 DM (30 - 60 US$) - and expenditures for the replacement of worn-out tubes account for a significant share of the overall costs of the systems. Thus, is it imperative that less expensive, better quality ballasts be utilized in SHS's. This would obviously lower the investment costs of the systems, but it would also reduce the running costs by prolonging lamp lifetime. In our view, it should be feasible to produce a high-quality ballast for around 30 - 40 DM (approx. 20 US$).

4. Conclusions

The great majority of the currently available 12-V fluorescent lamps that are designed for battery (DC) operation are not suitable for use in small photovoltaic systems.

All ballasts tested so far have proved deficient with respect to key criteria such as efficiency, lifetime, and safety features.

A number of improved ballast designs are currently undergoing testing, but a truly high-quality ballast - i.e. one which would optimize the ignition process in SHS applications - is still not available on the market.

The development of a high-quality, low-cost ballast is a basic prerequisite for the dissemination of Solar Home Systems on an expanded scale in Third World countries. And while the complexity of the problems which must be addressed certainly does not call for "high tech" approaches or solutions, it is nonetheless clear that the demands placed on the ballast technology in SHS applications are usually underestimated.

5. References

Dr. C.-P. Zeitinger, W. Müller: Field Test of Solar Home Systems in Puno/Peru; GATE/GTZ 1987

PV SYSTEMS FOR RURAL AND REMOTE AREA APPLICATIONS

R. K. Jain
Semiconductors and Photovoltaics Group
Electronics Division
Bharat Heavy Electricals Ltd.
Mysore Road, P.B. No. 2606
Bangalore, INDIA-560026

Abstract

The oil crisis in the beginning of 1970's gave birth to the use of photovoltaic systems for terrestrial applications. For a fast developing country like India the use of PV energy for powering various rural and remote area needs looks very promising and cost-effective. Several PV systems have been installed and the field performance feedback have been very useful in improving the system reliability. Continuous decline in PV system prices would help in expanding the market. India represents a great market for PV systems. International cooperation in PV technology would help in improving the overall growth and socio-economic level of developing countries.

INTRODUCTION

The application of photovoltaic systems for terrestrial power opened a new market in 1973. This market began with small consumer and remote area applications and has grown strongly to a level of 29.1MW peak in 1987 (1) covering a wide variety of uses throughout the world. PV systems from few milliwatts (watches, calculators) to few millionwatts (Arco's 6.5MW plant at Carrisa Plains, California) are in operation. Thousands of photovoltaic systems have been installed as demonstration projects or pilot plants throughout the world. Steady decline in PV system prices would further open up new markets in the areas of stand-alone, residential, industrial, central power plants and would help in rapid growth of the photovoltaic industry.

It is well known that most of the developing countries are concentrated around the sunshine belt. Photovoltaic energy have the great potential to contribute to development in these countries, in particular in its rural and remote areas, which at present either have no electricity supplies or served by diesel generators. At today's price the rural and remote area applications in developing countries are economically viable using PV systems. These applications represent an important, immediate and continuously growing market. International cooperation in photovoltaic technology in the form of exchange of information/people, education, training, field testing and demonstration projects would help in the overall growth. The part aid/loan given to developing countries by international agencies could be in the form of supply and demonstration of PV systems for various applications to boost the economy and market.

For a fast developing country like India with good sunshine, the use of photovoltaic systems is quite relevant and appropriate for meeting its rural and remote area needs on priority. This would help in enhancing its growth and improving the living standard of rural people. There exists an enormous potential and market for PV systems in India. Present paper describes the developments in Indian context and the same would be certainly applicable for other developing countries of the world.

SCOPE

India is situated geographically between 8°-37° North lattitude and 68o-98o East longitude coordinates. India, with 2.2% of total world area, has nearly 16% of the world's population. It is, after China the second most populus country in the world. At present India's estimated population is about 800 million. There are about 600,000 villages in India. About 80% of population lives in villages which makes India, a land of villages. The great majority of the rural people live in small or large nucleal settlements. Areas of dispersed habitations are few, mainly in the Himalayan zone and the desert and semi-arid areas. The normal unit is the hamlet rather than the homestead, and this is true even in the hill areas where the tribals predominate. Fortunately India receives also good sunlight energy throughout the year. Figure 1 shows the distribution of mean daily global solar radiation for the year in India (2). It varies between 4.6 to 6.4 kWh/m2.day from place to place. With this background, India looks very attractive for exploiting solar energy and in particular PV energy for various rural and remote area applications. There is a great scope to serve applications such as- village lighting, pumping, refrigeration, telecommunications, and many remote area power needs. This would create a great socio-economic impact on the life of rural masses and would control drift towards urban areas.

APPLICATIONS

Due to very high cost, photovoltaic cells/arrays were used in the beginning only for space applications. Even today these systems are the most reliable with excellent power per weight ratio and would continue to play a major role in future space program. Extensive R&D throughout the world during the last fifteen years have brought PV prices significantly down to a level that several terrestrial applications (3) are economically viable. The cost-effective trend is continuing and PV market is growing strongly. More and more new applications are being created and powered by PV and it is hoped that by the year 2000 a significant proportion of the world electrical energy demand would be met by PV energy.

Apart from the many advantages PV technology offers, the modular behaviour makes it very versatile. The stand-alone and modular concepts make PV systems very suitable for rural and remote areas. Worldwide photovoltaic systems are in use for several applications ranging from few milliwatts to few millionwatts and for wide varieties of uses too numerous to mention. The priority applications would vary from country to country. For example in a remote village in a developing country few tens of watt of PV module are enough for lighting, education/entertainment through Radio/TV or save lives through medical refrigerator. This could bring tremendous happiness and change in socio-economic life of villagers. While the same small PV power is of no significant use to people in developed countries to influence their daily activities greatly or improve the standard of life. With respect to Indian scenario and priority basis following rural and remote area applications look very promising:

Street and Domestic Lighting, Water Pumping for Drinking and Micro-Irrigation, Medical Refrigeration, Community TV &Radio, Direct Reception System, Battery Charging, Village Telephone Exchange, Very Low Power TV Transmitter, Water Treatment, Grain Grinder, Small Scale Industry, Petrol Station;

Microwave Repeater Stations, Weather Monitoring Stations, Railway Signal Lighting, Wayside Railway Stations, Unmanned Railway Gates, Defense Equipments, Offshore Oil Platforms, Cathode Protection, Navigational Aids.

DEVELOPMENTS

Department of Non-Conventional Energy Sources (DNES) under Ministry of Energy, Govt. of India is coordinating the photovoltaic activities in India. Central Electronics Ltd. (CEL) at Sahibabad, Bharat Heavy Electricals Ltd. (BHEL) at Bangalore and Rajasthan Electronics & Instruments Ltd. (REIL) at Jaipur are the three PV module manufacturers, who also supply photovoltaic systems for various applications. BHEL and CEL also manufacture silicon solar cells.

PV systems supplied by Indian manufacturers have been deployed throughout the country covering Kanyakumari in south to remote places in Jammu & Kashmir in north and hilly villages of Nagaland in east to the desert areas of Rajasthan in west, including age old tribal settlements in Bastar, central India and Andaman & Nicobar islands. Indigenously designed, developed and manufactured PV systems also accompanied the expeditions to Mount Everest and Antarctica and perfomed well. India has also supplied PV systems to its neighboring countries and other developing countries of the world. Indian PV market is growing and there seems to be a good potential for its continuous growth in the future.

Large number of rural and remote area applications as mentioned in the previous section are powered by PV systems as regular supplies or as demonstration projects in India. PV operated few thousand village lighting; few hundred water pumping, community TV, railway signal lighting, battery charging, of each and few ten offshore oil platform systems have been installed. Figure 2 shows a typical PV operated village stand-alone street lighting system. Few PV systems for microwave repeater stations, weather monitoring stations, wayside railway stations, unmanned railway gates, navigational aids, medical refrigerators, village telephone exchange, VLPTs, etc have been taken up as demonstration projects to obtain feedback towards customer satisfaction and/or understanding of the system for further improvement. A 5kW PV pilot power plant is under installation by BHEL in the Lakshdweep islands which should be operational in the coming months. The experience gained on this pilot plant would be very useful for setting up similar plants. Feedback information on the field failure of village lighting, pumping and TV systems provided valuable data for subsequent redesign. Improved designs and more careful attention to quality control in manufacturing and installation have resulted in reliable systems. The performance and reliability of PV systems not only depend on the PV modules and balance of system components, but also on site characteristics. System design should incorporate this feature to guarntee trouble free operation ensuring cost-effectiveness by keeping low O&M costs.

India has a vast network of telecommunications and one of the largest railway network in the world. This alongwith some rural applications open an immediate and very big market for PV systems. For example it is estimated that 10,000 railway stations in remote areas do not have electric power, if on average 5kW PV power per station is considered, then alone this area represents a market of 50 MW of PV systems. With the continuous decline in PV system prices, it is hoped that more and more applications would be served economically. International institutions such as World Bank, U.N.; US, European and Japanese Development Aid Organizations could play a major role in improving the conditions of developing nations by financially supporting photovoltaic and integrated solar energy projects in these countries. Worldwide efforts have made the PV systems to be cost-effective and further work continues towards cost reduction so significant proportion of world electrical energy needs could be met by PV.

Acknowledgements

Author is extremely thankful to many individuals, too numerous to mention by name, from DNES, BHEL, CEL, REIL, R&D institutes in India and abroad, Railways, UNICEF, Post and Telegraph, MDL, ONGC, State Nodal Agencies, REC, Electricity Boards, customers and suppliers for having very helpful, fruitful and frank discussions during the last five years. Last but not least, I am also grateful to actual users of PV systems for their kind cooperation and giving useful feedback information which has greatly helped in system improving.

References

1. Photovoltaic Insider's Report, Ed. R. Curry, Vol. VII, No. 2, pp. 6, Feb. 1988.
2. A. Mani and S. Rangarajan, 'Solar Radiation Over India', Allied Publishers Private Ltd., 1982, pp. 47.
3. R. K. Jain, 'Silicon Photovoltaic Systems for Terrestrial Applications', Proceedings Third International Workshop on Physics of Semiconductor Devices, (Eds. S C Jain and S Radhakrishna), World Scientific Publishing Co Pte Ltd., Singapore, Nov. 27 - Dec. 2, 1985, pp. 208-222.

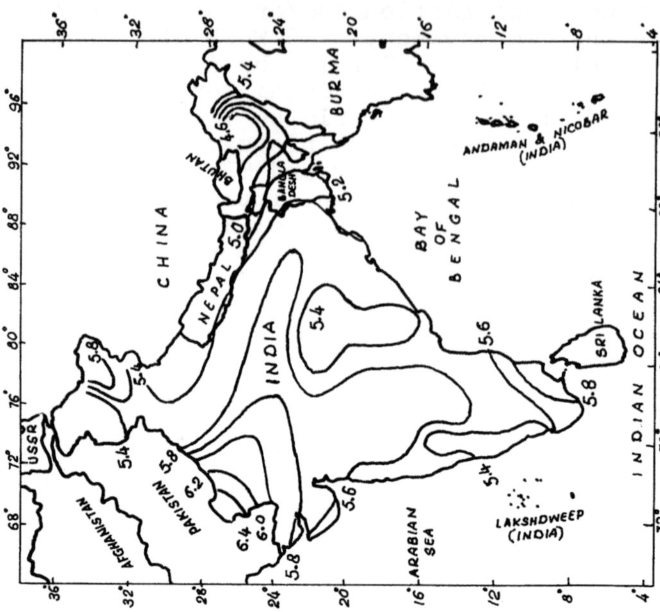

FIG. 1. DISTRIBUTION OF MEAN DAILY SOLAR RADIATION IN INDIA - ANNUAL (kWh/m². day).

Fig. 2. View of a Photovoltaic Operated Stand-Alone Village Street Lighting System.

BULK ELECTRICITY GENERATION - PROSPECTS FOR PV AND SOLAR-THERMAL SYSTEMS

H. P. Hertlein
Parkstrasse 80, D-5204 Lohmar, Fed. Rep. of Germany

ABSTRACT. Operational experience gained in the past years in field tests with PV and solarthermal systems for grid-connected electricity generation is reviewed by comparing the operating characteristics of the systems, the energy transfer characteristics, the annual energy production, and the future energy costs. While strengths and weaknesses differ, neither technology could be identified to be superior on the basis of technical performance and cost arguments. Under specific conditions, both technologies continue to hold promise for future bulk electricity generation.

1. INTRODUCTION

Over the past years, a number of system tests for grid-connected electricity generation, involving PV and solar-thermal (ST) facilities, have been carried out. Many of the tests with systems above about 25 kWe rating were engineering experiments, but several multi-year operational performance demonstrations were conducted also, mainly in the US, accumulating over 5 years experience in continuous energy production in some cases. Although frequently only indicative of first generation technology, the operating history of these field tests sheds light on the performance characteristics and differences of the respective solar technologies.

2. OPERATING CHARACTERISTICS

2.1 Experience Base

Table 1 summarizes today's experience base with larger-size PV and ST systems. So far, about 35-40 each of PV and ST facilities for grid electricity generation have been built and operated. The largest ones have been realized under commercial terms in the US Soutwest of up to 30 MWe in ST trough configurations, and up to 6.4 MWe(DC) in a flat-plate 2-axis tracking PV facility. Systems of several other configuration also were operated and tested in this region, hence having been subject to similar insolation conditions. This fact provides the basis for carrying out a system performance comparison.

2.2 General Performance Issues

Major issues and parameters indicative of similarities and differences between ST and PV power plant technologies are summarized in Table 2.

Technical viability has been demonstrated for most system concepts. Surprisingly, difficulties have been experienced more with "classical" than with solar-specific system components. Differences exist in the requirement for insolation quality and in the response capability to changes in insolation. Thermal mass and inventory is of disadvantage for ST systems in terms of energy thresholds and parasitics, but permits to incorporate energy storage and auxiliary fuel sources, and to level and shift output. Low O&M requirements and high degree of modularity are strong points of PV systems, while lack of a viable energy storage option seems the major disadvantage.

2.3 Input-Output Performance

The insolation-input to energy-output characteristics of Figs. 1 and 2 have been derived from experimental and operational data. These representations give insight into daily plant performance capabilities and are useful for analyzing performance under different insolation situations.

Energy input is the area-specific daily direct or global insolation on total mirror aperture (except for central receiver systems), or on the PV arrays. At the US Southwest test sites, direct normal insolation is about 20 % lower than global normal insolation.

ST single-unit dish/Stirling configurations achieve the most effective solar resource utilization. The high performance potential of central receiver systems is partially compensated by the need for high daily levels of direct insolation. PV systems show a characteristic temperature-induced reduction in conversion efficiency at high insolation. The performance improvement using today's PV modules (SMUD) in comparison with earlier first generation modules (LUGO) may be indicative of PV technology advances. Improvements in ST plant performance also would most likely be visible, given the opportunity to prove state-of-the-art ST technologies.

2.4 Production Performance

The year-by-year plot of output generation per kWe of installed plant capacity (Figs. 3 and 4) is the final indicator of system reliability and performance under given insolation conditions. Deviations in the slope of these curves show changes in annual insolation, operating strategy and/or system availability.

The absolute value of the MWhe/kWe ratio is sensitive to the capacity rating a plant is given; for instance, nameplate ratings of PV systems have proven to be consistently higher than the rating observed under standard operating conditions (SOC). Methods for determining SOC ratings have been established for PV systems, not yet for ST systems. The need for such rating exists, comparing expected with actual (not optimized) MWh/kW performance of ST systems (Fig. 3).

2.5 Cost Situation

Bulk electricity generation with ST and PV systems can be cost-effective today given sufficient financial incentives and favorable circumstances (insolation; peak load match; rate structure). Based on mature technologies, the levelized cost figures summarized in Tab. 3 are claimed to be achievable. No decisive advantage is obvious for either technology.

3. CONCLUSIONS

Viability of PV and ST bulk electricity generation has been establised by several multi-year field tests and commercial applications. Given adequate insolation, both technologies hold promise to generate grid electricity in the future at about equal cost. On a technical, performance or cost basis, as perceivable today, neither technology seems to command an upper edge. PV offers high reliability and little O&M needs, but ST can provide the possibility for energy storage. Facing cost disadvantages compared to conventional supplies, entrepreneurship, industrial engagement and political/financial support will be important non-technical factors on the road to maturity for both technologies.

The question which technology will proceed fastest on the learning curve towards grid-connected electricity generation will most likely be influenced more by identifying and exploiting intermediate market opportunities which offer benefits to investors/users at calculable technical and financial risk, rather than by issues of optimal performance.

4. ACKNOWLEDGEMENTS

The author is indebted to his colleagues and friends at DFVLR, SANDIA, EPRI, FhG and from many other private and industrial organisations, who provided numerous suggestions and freely shared their experience.

REFERENCES (Excerpt)

1 (1987) Today's Photovoltaic Systems: An Evaluation of their Performance Sandia Natl. Labs SAND87-2585
2 Stokes KW (1987) Photovoltaic Field Test Performance Assessment for 1986. NMSEI Final Report, EPRI Project 1607-06
3 Stokes KW (1987) Hesperia Photovoltaic Power Plant: 1985 Performance Assessment. EPRI EPRI-AP-5229
4 DeMeo EA (1987) Photovoltaics for Electric Utility Applications. Intl. PVSEC-3, Tokyo
5 (1988) Intl. 30 MWe Solar Tower Plant Phoebus, Executive Summary of Phase 1A Work. DFVLR, Cologne
6 (1987) Solar Central Receiver Technology Advancement for Electric Utility Applications - Midterm Review Documentation. Albuquerque, NM
7 Droher J (1986) Performance of the Vanguard Solar Dish-Stirling Engine Module. EPRI EPRI-AP-4608
8 Jaffe D et al (1987) The LUZ Solar Electric Generating Systems in California. ISES World Congress '87, Hamburg

Technology	Type	Approximately Installed Aperture (sqm)	Capacity (MWe)	No. of Systems inst./active
PV	Fixed Flat-plate	35,000	2.7	23/20
PV	Tracking Flat-plate	77,200	10.4	6/6
PV	Concentrators	8,100	0.8	6/5
ST	Troughs	1.220,000	160.0	14/6
ST	Dishes	50,200	7.0	15/5
ST	Central Receiver	117,000	16.0	6/1

Tab. 1: Installed Major Grid-Connected PV and ST Plant Facilities

Technology	Ops hours (h/a)	Life years	Investment ($/kWp) existing	now today	in future	Energy Cost (mills) existing	now built	in future
PV	2,080	15	12,000			1,210		
	1,500	25		8,250			439	
	1,500	25			2,500			134
ST	1,100	15	14,150			1,238		
	1,270	20		3,500			282	
	4,333	20			5,167			126
nucl.	7,000	25				<35	acceptance?	
foss.	3,750	25				35-85	fuel price?	
hydro	5,000	>40				40-100	resource?	

6.5 % interest

Tab. 3: Present and Expected Levelized Energy Costs

ISSUE	SOLAR-THERMAL TECHNOLOGIES	PHOTOVOLTAIC TECHNOLOGIES
Energy collection	- concentrate radiation by mirrors - mirrors track the sun, 1-axis for TRO, 2-axis for CRS and DISH systems	- FP systems do not concentrate - mirror or Fresnel lens CONC systems track in 1- or 2 axes
Radiation used	- need direct insolation for operation	- FP utilize total radiation - CONC need direct insolation
Land area needs demonstrated	- ~6 ha/MWe for CRS - ~3 ha/MWe for TRO - ~3 ha/MWe for DISH	- ~9 ha/MWe for F-FP - ~3.5 ha/MWe for 1-axis TR-FP ~11 ha/MWe for 2-axis TR-FP - ~14 ha/MWe for CONC
Smallest module	- ~2 MWe for TRO - ~30 MWe for CRS - ~10 kWe for single DISH	- a few kWe for all configurations
Upper module size	- ~80 MWe for TRO - ~100 MWe for CRS - ~50 kWe for single DISH	- about 5 MWe for all configurations
Operating energy threshold	- ~4 kWh/sqm.d for CRS - ~3 kWh/sqm.d for TRO - ~1.5 kWh/sqm.d for DISH	- ~0.1 kWh/sqm.d for FP - ~1.0 kWh/sqm.d for CONC
Power threshold	- about 300 W/sqm of direct insol., all configurations	- ~50 W/sqm for FP - ~150 W/sqm for CONC
Energy inventories	- thermal inventory depending on system size	- none
Response capability	- depends on system size and thermal mass	- instantaneous response to insolation
Parasitics & losses	- ~14 % of annual gross for TRO - ~36 % of annual gross for CRS	- ~6 % of gross for F-FP - ~8 % of gross for CONC
Stand-by parasitics	- yes, depending on type and size of system	- none
Soiling impact	- high reflectivity critical for performance	- soiling not critical for FP
O&M needs	- between .5 to 1.7 manyear/MWe	- ~.2 manyear/MWe
Availabilities observed	- up to 86 %/a for CRS - up to 90 %/a for TRO - up to 70 %/a for DISH	- > 95 %/a for FP - up to 97 %/a for TR-FP - up to 85 %/a for CONC
Reliability problem areas	- repetitive thermal cycling of stressed components - tracking electronics	- DC/AC inverter - ground faults and fault potection
Energy storage	- thermal storage viable for CRS and TRO	- not viable in significant amounts
Generation shifting	- possible by thermal storage or fossil hybrid source	- not a viable option
Capacity factors observed	- CRS: <24 %/mo; <12 %/a - TRO: <35 %/mo; <24 %/a - DISH: <24 %/mo; <11 %/a	- F-FP: <31 %/mo; <26 %/a - TR-FP: <46 %/mo; <35 %/a - CONC: <36 %/mo; <24 %/a

FP=Flat-plate; TR=Tracking; F=Fixed-tilt; CONC=Concentrator;
CRS=Central receiver system; TRO=Trough system

Table 2: Issues Differentiating Solar Power Plant Technologies

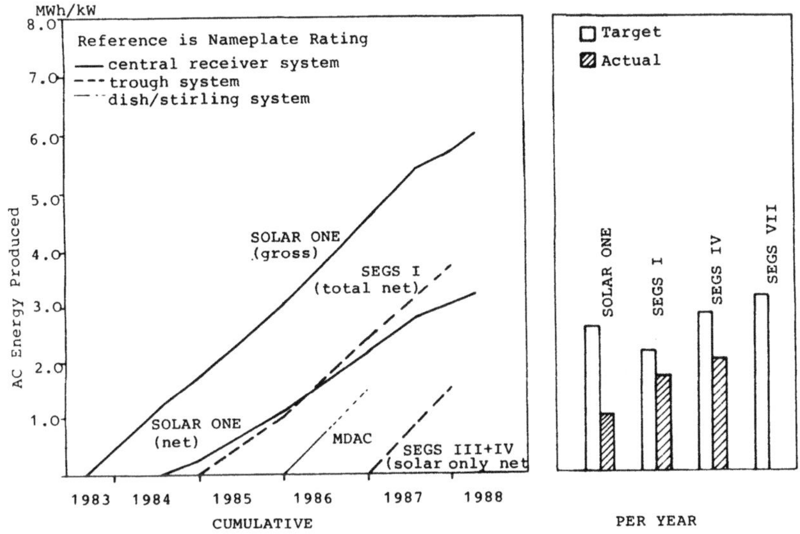

Fig. 3: Cumulative and Annual Capacity-Specific Performance of ST Plant Facilities in Continuous Production Operation

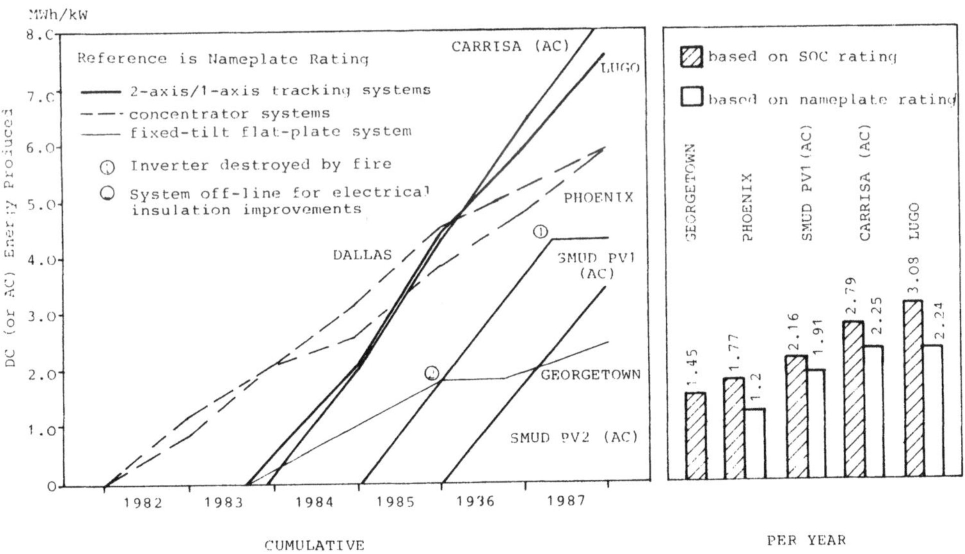

Fig. 4: Cumulative and Annual Capacity-Specific Performance of PV Plant Facilities in Continuous Production Operation

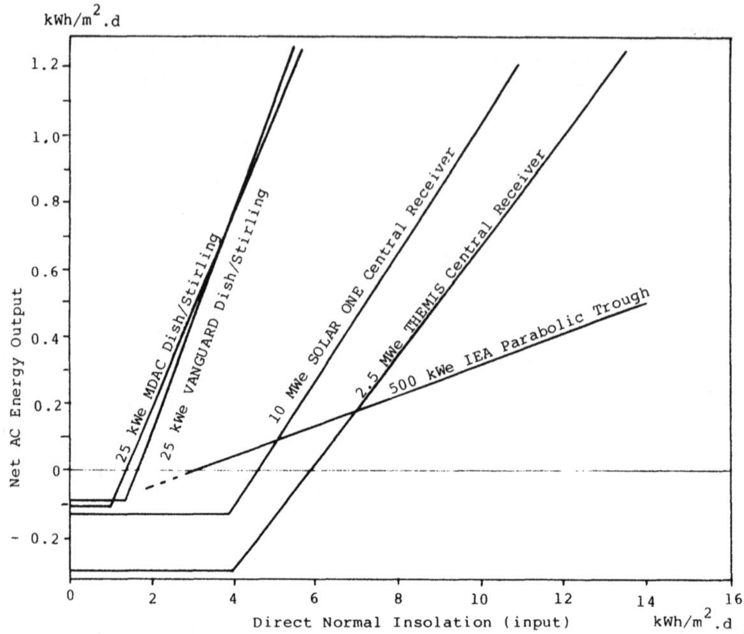

Fig. 1: Representative Energy Input-Output Performance of Trough, Central Receiver and Dish/Stirling ST Plant Facilities (derived from measured data)

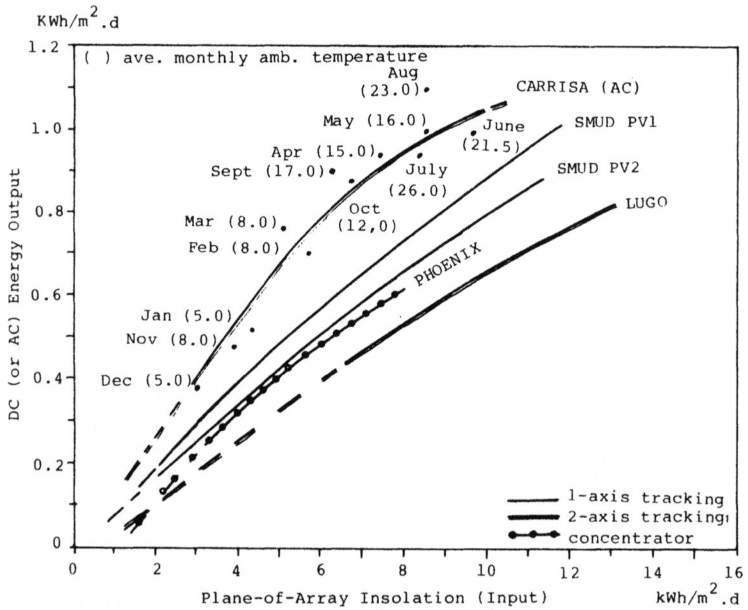

Fig. 2: Representative Energy Input-Output Performance of Tracking and Concentrator PV Plant Facilities (derived from measured data)

Power Conditioning and Control

Jürgen Schmid Herrmann Laukamp
Ralph D. von Dincklage*
Fraunhofer Institut für Solare Energiesysteme
D-7800 Freiburg, Federal Republic of Germany

May 26, 1988

Abstract

In the framework of the CEC funded *Concerted Action* the performance of of the power conditioning system of photovoltaic power plants, in particular of stand alone systems is scrutinized. Based on actual plant experience areas are to be identified where significant research and development work should be directed to in order to improve overall plant efficiency both of the present European pilot plants as well as to formulate guidelines for future plant design and plant operation.

1 Introduction

In the framework of the CEC funded *Concerted Action* the performance of photovoltaic power generating plants is investigated with the aim of increasing their overall efficiency.

The present subtask *Power Conditioning and Control* forms the interface between the photovoltaic solar cells, on one hand, and the user of the electrical power as well as the storage devices, on the other hand. 13 European photovoltaic power plants of seven member states have agreed to serve as a data base. They have accumulated tens of plant-years of operation time by now.

The first step of this subtask is to find the building blocks of *power conditioning and control* common to most plants. In the second step, performance critical components will be identified on the basis of the operational experience already gained. Subsequently, in a third step, the technical realizations of these critical components will be scrutinized. In fact, we anticipate further laboratory

*consultant to the project manager

tests und studies to be defined in the course of this third step.

In the present contribution we analyze two examples: Firstly, the potential gain of energy output of a PV plant, when some means of voltage adjustment is utilized in order to track the maximum power point and secondly the efficieny of transformer-based inverters, especially at partial load. Thirdly, we outline the development of a transformer-less inverter, presently underway at the Fraunhofer Institut at Freiburg, Federal Republic of Germany.

2 Control: Maximum Power Point Tracking

The familiar I-V characteristic of a photovoltaic cell has specific features, which require control capabilities not usually found in classical grid-connected power conditioning systems. In contrast to classical electrical generators, which are well described as either constant voltage or constant-current sources, a PV generator may present itself in either mode, depending on the actual load (the operating point) and furthermore on the temperature and irradiance.

As is well known, the short circuit current of the array is proportional to the irradiance level leaving the voltage of the maximum power point, V essentially unaffected. This maximum power point, in turn, is anti-correlated to the cell temperature, t, with a gradient of $\frac{dV}{V dt} = -0.005K^{-1}$ at the rated temperature of 25 ^oC.

The power conditioning system must adapt itself to these basic working characteristics when maximum energy is to be extracted from the PV array. The most simple mode of operation is the constant voltage mode, which is adequate if the primary variations are due to changing illumination levels. Other strategies aim at improving the energy delivered by the system by some means of adjusting the voltage to seasonal variations based on temperature sensing. More sophisticated, a continuous closed loop feedback may be employed to constantly operate near the maximum power point. This technique is called *maximum power point (MPP) tracking.*

3 Inverter efficiency

Classically designed inverters utilizes transformers. These consume some power to allow for the magnetisation, typically a few per cent of rated power.

This has an important consequence in the regime of partial load. We may

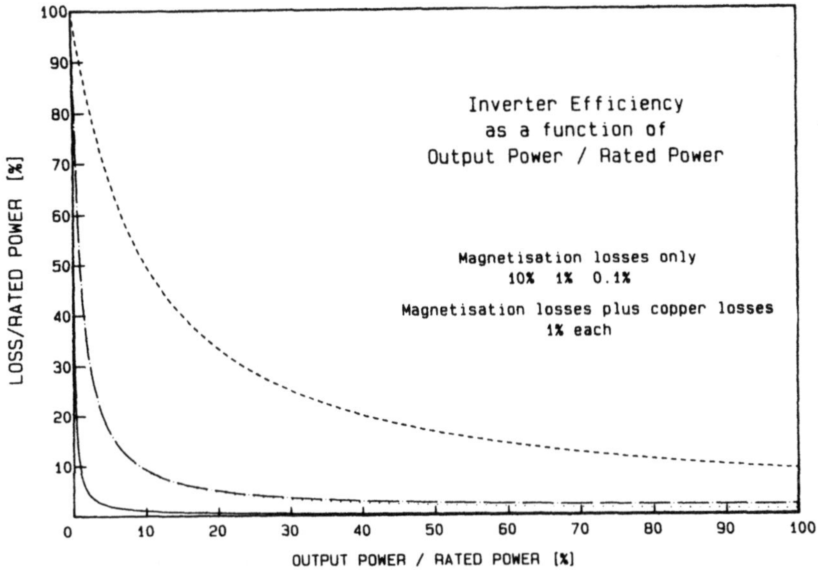

Figure 1: Inverter effiency. The ratio power loss over rated power is calculated and presented as a function of the ratio of output power over rated power. Both quantities are given in per-cent, ranging from 0 ... 100 %. The parameter is the magnetisation loss assuming three discrete values 0.1 % (dashed), 1 % (dotted), 10 % (solid). The forth curve (the dash-dotted one) includes the copper losses, set to be also 1 % . It deviates only slightly near the rated power.

emphasize it by discussing the following - extreme - example: Assume magnetisation losses amount to 1%. If the inverter is now operated at 2% of rated power, half of the power available is consumed solely in order to support the magnetic field.

This effect is shown graphically in the Figure 1. There lost power is plotted as a function of output power, both quantities taken relative to the rated power. Three curves are drawn, which differ in one parameter, namely the magnitude of the magnetisation losses, here ranging from 0.1%...10%. For comparison, one graph also shows the combined effect, i.e. when copper losses are included (here for the case of a magnetisation losses of 1% and copper losses of 1% as well. Quite expectedly, the copper losses contribute mainly near the rated power and are - by choice of parameters - equal at the rated power.

4 Transformerless Inverter

In this last section the status of a transformerless $30kW$ inverter presently being developed at the Fraunhofer Institut, Freiburg under a contract of CEC is presented.

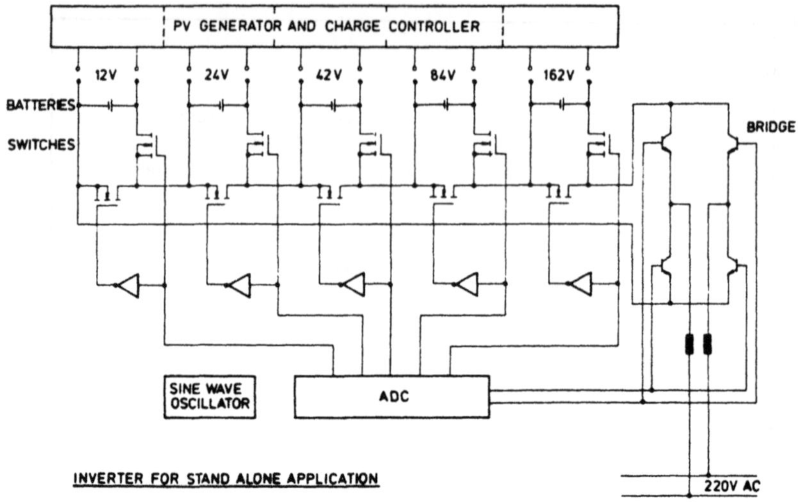

Figure 2: Circuit diagram of the transformerless inverter. Eminent features are the five discrete subfields of photovoltaic arrays with voltages approximately doubled in each step. Using a sine generator the 220 V AC is synthesized to five bit accuracy by switching the respective subfields.

It was designed specifically to comprise the following features mandatory for high-effiency PV applications: Low self-consumption implying high effiency over the full power range, low harmonic distortion ($\approx 2\%$), being able to accept and deliver reactive currents, and finally silent, compact and lightweight hardware.

The principle is shown in the diagram of figure 2 and is here based on five different PV arrays, operating at voltages e.g. 12V, 24V, 42V, 84V, 162V, which are ordered (approximately) according to the binary system. Triggered by a sine-wave generator and fast high-power switches a $220VAC$ wave can be synthesized, in the present example to five bit accuracy. The necessarily increased complexity of an array field connected to the new inverter, which needs to be subdivided into banks of arrays in such a way that they yield the above given different voltages, is more than offset by the energy gain inherent in the transformerless design.

5 Conclusion

The potential of several building blocks of a complex power conditioning subsystem have been discussed. More such issues are expected to surface in the course of the on-going *CEC funded Concerted Action.* The project is scheduled to run through spring 1989, and aims at ad-hoc plant improvements as well as the definition of longer range research and development issues.

BATTERY MANAGEMENT AND CONTROL IN PV SYSTEMS

S. McCarthy, G.T. Wrixon, M. Hill
National Microelectronics Research Centre
University College
Lee Maltings
Prospect Row
Cork
Ireland

Abstract

In May 1987 the Directorate General XII of the CEC initiated seven projects in the area of advanced Photovoltaic Plant Research. Seven concerted action groups have been formed to investigate the main problem areas in PV plant design and operation. This paper describes the work of the concerted action group on the control and management of batteries in PV systems. The main activities of this group are:

- evaluation of existing battery charge controls
- compilation of guidelines for the operation and maintenance of batteries in PV systems
- evaluation and testing of state of charge measuring equipment
- testing batteries to identify the causes of premature battery failures in PV systems.

The working group consists of representatives from the PV pilot plants, battery manufacturers and battery research institutes.

Introduction

In May 1987 the Directorate General DGXII of the CEC initiated seven projects in the area of advanced Photovoltaic Plant Research. Seven concerted action groups have been formed to investigate the main problem areas in PV systems.

The concerted action group on 'Battery Control and Management in PV Systems' was formed to identify the main problems with the batteries in the PV Pilot Plants (1) and to develop proper battery control and management procedures.

The main problems identified by the group to date are:

- inadequate maintenance procedures
- inadequate guidelines for system manager
- lack of documentation (log book) on system performance
- inadequate protection from deep discharge and overcharge
- poor charging characteristics and a lack of facilities to improve them.

The working group consists of representatives from the PV pilot plants, battery manufacturers, and battery research institutes. The main objectives of the group are to:

- evaluate existing charge/discharge procedures being used in the PV pilot plants
- compile guidelines for the operation and maintenance of batteries in the PV system
- evaluate and test existing state of charge measuring equipment
- test battery cells from the PV plants to identify the causes of premature battery cell failures.

Existing Charge Control Procedures

A variety of different charge control techniques are currently being used in the PV pilot plants.

- In Kaw and Aghia Roumeli the charge control is regulated by connecting and disconnecting the complete PV array.

- In Fota and Pellworm the battery was divided into two banks and the PV array was divided into several sub-arrays.

- In Kythnos and Terschelling DC/DC converters are used to regulate the charge.

In some systems the batteries can be recharged from the utility grid using commercially available battery chargers or using a line commutated inverter/converter.

The objective of the group is to evaluate the different techniques and to identify the advantages and disadvantages of each. The overall output would be recommendations on the proper charge control of batteries in PV systems.

Battery Operation and Maintenance Guidelines

The objective of this work is to publish a document outlining operation and maintenance procedures for batteries in PV systems. The guidelines will be used by the plant managers and system operators and will be designed to overcome the major problems experienced in the PV pilot projects.

The information contained in the guidelines is based on the experiences of the PV pilot plant managers and also the experiences of the battery experts in the working group.

The main contents of the guidelines are:
- a detailed description of battery characteristics and performance
- guidelines for routine maintenance procedures
- guidelines for quarterly and annual battery maintenance tests
- guidelines for a log-book on battery performance and test
- a check list to identify problems and to suggest correction procedures.

A draft copy of this document was presented at a recent working group meeting in Milan. The document is currently being reviewed by the experts and the first version should be available by July 1988.

Battery Modelling

The main objectives of this work are:

1. to evaluate the models which are used to calculate the state of charge in state of charge measurement devices;
2. evaluate existing battery models being used in PV system modelling;
3. Use the PV system model to evaluate the battery charge/ discharge characteristics being used in the Pilot Plants.

This work will be done in cooperation with the concerted action group on computer modelling and simulation of PV systems.

The battery models which are currently being evaluated are models by Shepard (2) and Menga (3).

Battery state of charge is an important parameter to implement a proper charge characteristic and to prevent deep-discharge and overcharge conditions. It is however a complicated function of battery type, age, history temperature and charge/discharge cycle. At present there are a number of algorithms and hardware units available for monitoring battery state of charge. One of the objectives of the concerted action group is to assess the effectiveness of both the hardware and the software of these units. This will be done by testing them under fixed laboratory conditions (Figure 1) and then testing their operation in suitable PV plants. At the recent working group meeting held in Milan the group agreed that two types of unit exist:

- an SOC unit which is used for battery management i.e. it indicates deep discharge and overcharge conditions
- an SOC unit which provides an accurate measure of residual capacity.

When evaluating the state of charge units it is important to evaluate them according to these criteria. The output of the group will be an assessment of existing units and also if necessary recommendations for the improvement of existing systems.

Battery Cell Testing

To complement the work on the above topics tests will be performed on failed cells and on a sample of operational cells from selected PV plants. The objectives of this testing will be to examine the mechanisms of capacity loss in cells in PV systems. Elements of the management and control strategy which contribute to these mechanisms can then be identified and eliminated from future charge control techniques. It is also hoped that some quantification of battery degradation can be achieved.

As well as examining these key areas of interest the group will investigate other factors contributing to poor battery performance. Two topics discussed in detail at the Milan meeting were:

- the effect of weak cells on overall battery performance
- the choice of voltage limits for simple voltage control of charge and discharge.

Conclusion

The concerted action group on battery control and management in PV systems has identified problems which are common to batteries in PV systems. The main work of the group is to analyse these problems and to provide solutions to them. The overall objectives are to improve battery performance and increase the battery life in PV systems and consequently promote PV systems as reliable sources of energy.

Fig. 1 Charge / discharge conditions for SOC unit tests.

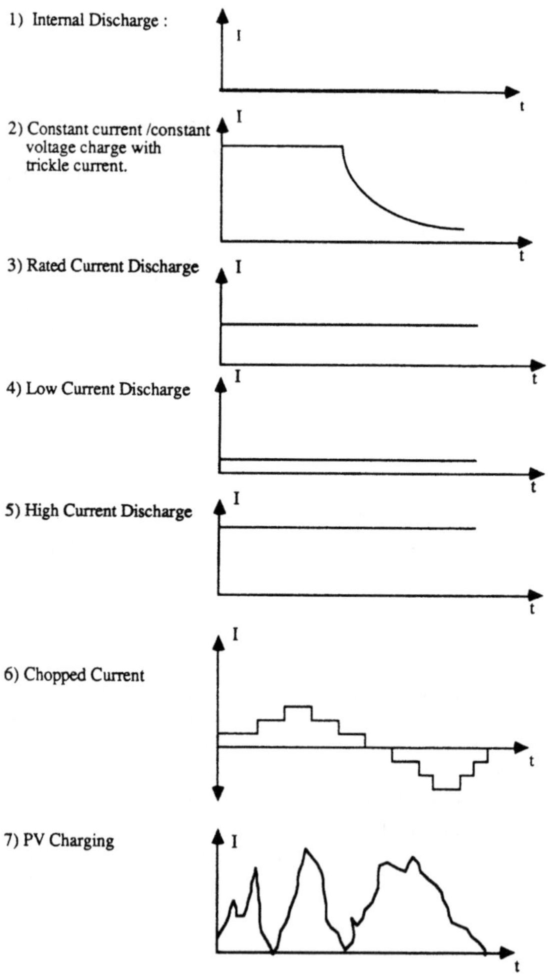

References

1. Palz W., Photovoltaic Power Generation, Series C., Vol. 4, Riedel Dordrecht.

2. Shepard C.M., U.S. Naval Research Laboratory, Washington D.C., Design of Primary and Secondary Cells. II An equation describing battery discharge, Journal of the Electrochemical Society.

3. Menga P. et al., An electrical model for discharge and recharge of lead-acid batteries, developed for idustrial applications, L'Energia Elettrica-N7-8 - 1987.

SPECIFIC ACTIONS ON THE FOTAVOLTAIC PROJECT

S. McCarthy, G. T. Wrixon, L. Keating

National Microelectronics Research Centre,
University College, Lee Maltings,
Prospect Row, Cork,
Eire.

Abstract

In 1987 the Directorate General XII of the CEC initiated an advanced photovoltaic systems research program. Part of this program involves implementing specific actions on the PV Pilot Projects. The specific actions for the Fotavoltaic project were based on the results of the computer modelling and optimization of the system i.e.

- Connection of two separate battery banks to form a single battery.
- Improve the power management of the line-commutated inverter.

This paper describes the modelling and optimisation results, the phases of the project and the work to date.

1. Introduction

In May 1987 the Directorate General XII of the CEC initiated a program on advanced photovoltaics systems research. The program was divided into two areas, viz.

Concerted actions, concentrating on areas associated with all plants e.g. battery control, inverter management, PV modules etc.

Specific actions, projects specific to individual plants.

This paper describes the specific actions for the Fotavoltaic project.(1) The Fotavoltaic project is a 50kWp PV system which is designed to supp.y electrical energy to a 250 head dairy farm on Fota Island in Cork harbour, Ireland. It is one of 15 Pilot PV Plants which were built in 1983 as part of the DGXII program on PV systems research. Throughout the operation of the plant data was monitored on an hourly basis. This data was analysed and a computer model was developed to optimise the systems performance. The results of this analysis indicated that improvements in the system performance would result if the following specific actions were implemented.

1. Connection of two separate battery banks to form a single battery.

2. Improvements to the power management of the line commutated inverter.

2. <u>Computer Model and Optimisation Results</u>

A computer model (2) was developed at the NMRC to simulate and optimise the performance of the Fotavoltaic system. The model is a collection of programs, written in Fortran, which simulate the PV system operation on an hourly basis. Its inputs are horizontal irradiation, ambient temperature, wind speed, wind direction and load demand and the results of the simulations are output in graphical and statistical format. The optimum system design was determined by using the model to analyse the systems response to simulated changes in the design criteria and by selecting the components which provide maximum performance at minimum cost. The results of the optimisation analysis of the specific actions are summarised in figures 1 and 2. The performance of the PV system is defined as the fraction of the total load energy supplied by the PV system. The optimum value is 100% i.e. total load supplied by PV.

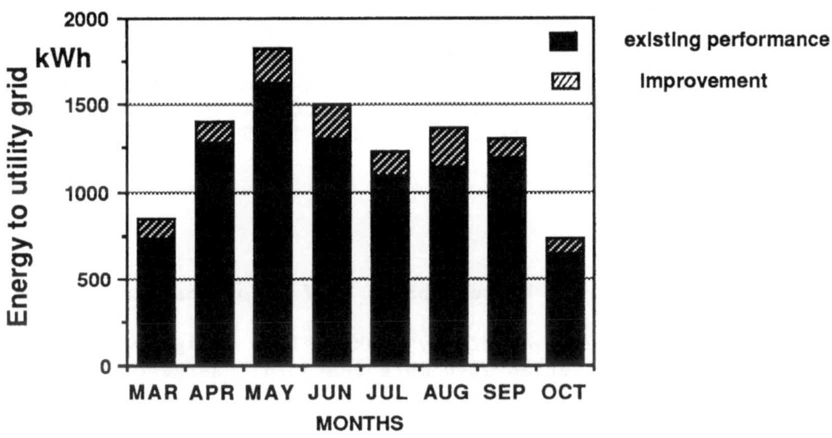

Figure 1. Performance of PV system for single and double battery operation.

Figure 1 shows that by using a single battery rather than the existing double battery the performance of the PV system could be improved during each month. In fact, using a single battery the PV system is capable of supplying 47% of the total annual load in worst case conditions. This compares with 29% using the existing double battery.

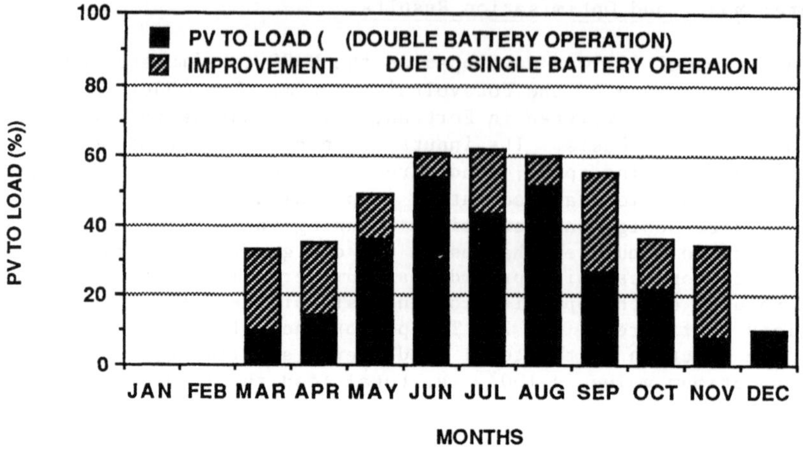

Figure 2. PV energy delivered to the utility grid for the existing peformance and for the predicted optimum performance.

Figure 2 shows the results of the simulations of improvements to the power management of the line commutated inverter. These simulations show that the output of the inverter could be increased by 10% during each month of the year. The net increase for the year would be 12.9%.

3. Specific Actions

3.1 Connection of two separate battery banks to form one battery

The existing design uses two battery banks. This design was selected because it provides both batteries with more charge/discharge cycles than is the case of a single battery. Additionally a double battery system ensures that one battery always remains at full charge in a stand-by mode. In practice it was found that double battery operation provides more charge/discharge cycles but the disadvantages are:

- the load capacity of each battery is reduced
- the control software is very complex.

In this specific action the conversion of the system from double battery operation to single battery operation can be divided into the following phases:

1. Design phase.
2. Interconnection of batteries and redesign of main DC busbar.
3. Rewrite software for battery management.
4. Recharge battery to equalise charge.
5. Test the operation of the single battery.

The time-table with the different phases is shown in table 1.

PHASE 1	Jan 88	Feb 88	Mar 88	Apr 88	May 88
Design	—				
Interconnection of batteries	—	—			
Software	—	—			
Recharge batteries		—			
Test system operation			—	—	—

3.2 Improvements to the Power Management Software

A computer model will be used to analyse the response of the system to simulated changes in the power management of the line-commutated inverter. The aim of the work is to improve the operating efficiency of the inverter and the system and to increase the amount of AC energy delivered to the utility grid. The work is divided into the following phases:

1. Definition and design of test.
2. Modelling and analysis of results.
3. Develop software for control system.
4. Test the system operation.
5. Compare results with the data predicted by the model.

Table 2 shows the time-table of the different phases.

PHASE 2	June	July	Aug	Sept	Oct	Nov	Dec
Definition	—						
Modelling		—	—				
Control Software			—				
Test System				—	—		
Analyse Results						—	
Report Results							—

4. Conclusion

The work to date is on schedule. Both batteries have been interconnected and the control software has been written. The new operation is currently being tested. The main difficulty with connecting the batteries was developing safety procedures to prevent damage to equipment or personnel. In June new software will be developed for the line-commutated inverter. In addition to improving the systems performance the results of these experiments can be used to validate the optimisation procedures used by the model. These results can in turn be applied to other PV systems as part of the concerted action group on computer modelling and simulation.

5. References

(1) Wrixon G. T., The Fotavoltaic Project - A 50kWp Photovoltaic System to Power a Dairy Farm on Fota Island, Cork, Ireland. Photovoltaic Power Generation, Series C, Vol. 4, Reidel, Dordrecht (153-169) 1983.

(2) McCarthy S., G. T. Wrixon, Optimization and Analysis of Photovoltaic Systems Using a Computer Model. Proceedings of the seventh E.C. Photovoltaic Solar Energy Conference held in Seville, September 1986.

GLOBAL ANALYSIS OF THE RONDULINU-PAOMIA PV PLANT BEHAVIOUR

A.LOUCHE, G.SIMONNOT, G.NOTTON, G.PERI
Université de Corse
Laboratoire d'hélioénergétique
Route des Sanguinaires - 20000 AJACCIO (France)

ABSTRACT. The purpose of the present paper is to show data and information on a 44 kWp PV plant runned in isolated rural site conditions.

The RONDULINU-PAOMIA pilot plant has been designed as a stand alone facility to power an hamlet ,and has been implemented since 1983 , within the solar PV R&D European Economic Community Program (CEE - DG XII).This hamlet is about 400 m height nearby the sea. Only 7 families are living there in winter due to the lack of electricity on site.The PV plant is connected to a local grid feedinghouses and street lighting. This plant presents the advantages to have a typical consumer type load and no connection to the national grid.

I. The plant (Fig 1.)

The array is divided into medium sized subarrays, each subarray being directly connected to the general power management device in the control room. It is oriented southwest ward 20 degrees from the south and it is divided into 34 rows build with 6 France-photon modules panels.

The sizing of the batteries has been made to insure a storage duration sufficient to go over the medium estimated variations of the solar energy received in winter. A seven days storage on a winter basis will give a sufficient security for most of the years. Due to the summer increase in production and consumption (some houses are only occupied during summer), this storage will give a three days reserve for summer.

The energy storage of the plant consists in lead acid batteries from Oldham. There are 84 batteries, capacity of which is 2500 Ah for a 10h discharged rate and nominal voltage is 2V.

The maximum charging and discharging bank voltage are respectively 201.6 and 155.4 V. The total storage capacity is 415 kWh at a 180 A discharge rate and 560 kWh at a 60 A discharge rate.

The inverter is a DC/AC inverter entirely built with

transistorized circuits. The three phase inverter consists in three starconnected monophase inverters and allows a perfect phase equilibrium for any type of load. A galvanic insulation between the three phases is done by an output transformer. The nominal power is 50 kVA for threes phases 220/380 V 50 Hz.

The ouput of the array consists of 17 pairs of wires arriving to the control room. After paralelling these wires the power goes to the batteries while controlled by a charge regulator featuring:

- input current from the array : 350A
- nominal voltage : 168V
- overcharge cutoff voltage : 201.6V
- overdischarge cutoff voltage : 155.4V
- output current to the inverter: 375A

When the batteries voltage is lower than 155.4V the regulation puts on a conventionnal generating set for operation. This generator is a gas engine plus an alternator fitted with a special regulation. Its nominal power output is 25 kVA and it is used to charge the batteries in case of sun radiation failure. The alternator is regulated to give a constant current at any voltage under its nominal voltage. The current is connected to DC current after passing trough a transformer and then connected directly to the battery in parallel with the solar array.

The data acquisition system consists in a static memory acquisition computer. The data are recorded every 10 seconds but for each data, only the mean value calculated over a period of 10 minutes are stored on the cartridge. The measurement system records the meteorological parameters (global horizontal, global inclined, ambient temperature, wind direction and speed), the DC electric parameters and the AC electric parameters. The total means 16 parameters every 10 minutes.

II. RESULTS

In October 1987, the Helioenergetic laboratory was elected to study and analyze the performances of this PV plant. During the period October 87/March 88 we faced many problems due to the data acquisition system because of lightning protection defects. Moreover, during the same period, only the consumption of one phase was recorded. Nevertheless, we present here the first results concerning the production of the plant. On fig 2., we plotted for January the energy produced by the solar array and the energy consummed by the users and the inverter.

As a constatation we can observe that consumption has very low value in front of production; this is due to the fact that during the period only two houses of the hamlet were occupied. The principal consumption is due to the inverter because the maximal power used by the hamlet is always lower than 10kW and consequently the inverter is operating under the worst conditions, relatively to its efficiency.

This phenomenon is more apparent on fig 3., where we plotted the output array voltage and the battery voltage versus time for two consecutive days. When the overcharge cut-off voltage has been obtained the automatism puts on a serial resistance with the array, and a weak current is maintained to prevent autodischarge from the batteries. The more important is the consumption, the more important is the time to charge the batteries. The lack of consumption is a problem for studying the efficiency of the plant; nevertheless we analyzed the performances of the solar array during the phase of charge.

The study of the solar array without serial resistance has been realized over the period mentionned above. The first results seem to be very promising; we have developped a correlation between the array output power and the insolation. As we can see on fig 4.A, there is a linear correlation between these two parameters. Moreover the coefficient corresponds to the efficiency of the array (8%), the value corresponding to the origin ordinate can be assimilated to the threshold of the array.

In an other way, we studied the array output current in the same conditions and we developped a parabolic correlation between this parameter and the irradiation (Fig 4.B). Applying this correlation with experimental data we have been able to reconstitute the output power of the array without the serial resistance. On figures 5.A and 5.B, one can observe, for two consecutive days, that for a day without overvoltage we can modelize the array output current; for a day with regulation the obtained curve, gives us the maximal output power of the array during this day.

III. CONCLUSIONS

We developped a correlation between the performances of the solar array of the Paomia pv plant and the inclined irradiation. This model allows us to caracterize a photovoltaic array without storage, i.e a photovoltaic system running the day long tracking the sun. The most important encountered problem for pv plant with storage is: how to optimize and modelize the batteries behaviour. It is a very important step in our knowledge of the plant improvment. In the same way we are studying the consumption and its modelization as function of time.

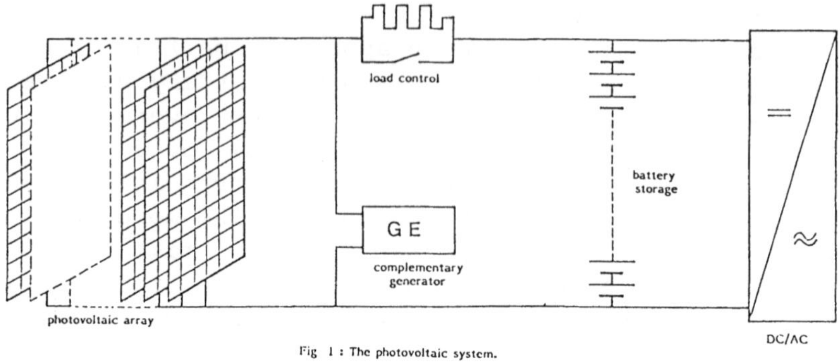

Fig 1 : The photovoltaic system.

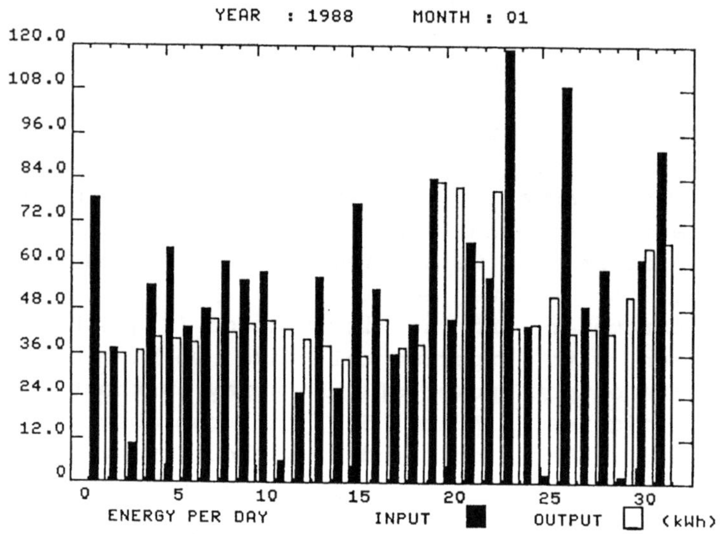

Fig 2. Energy produced by the PV array and energy consumed

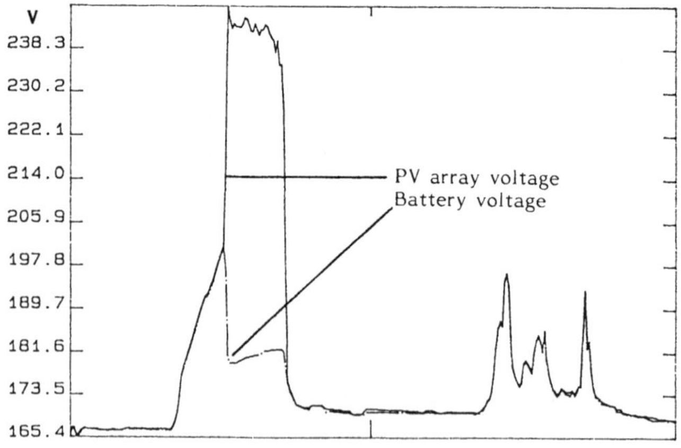

Fig 3. The output array voltage and the battery voltage

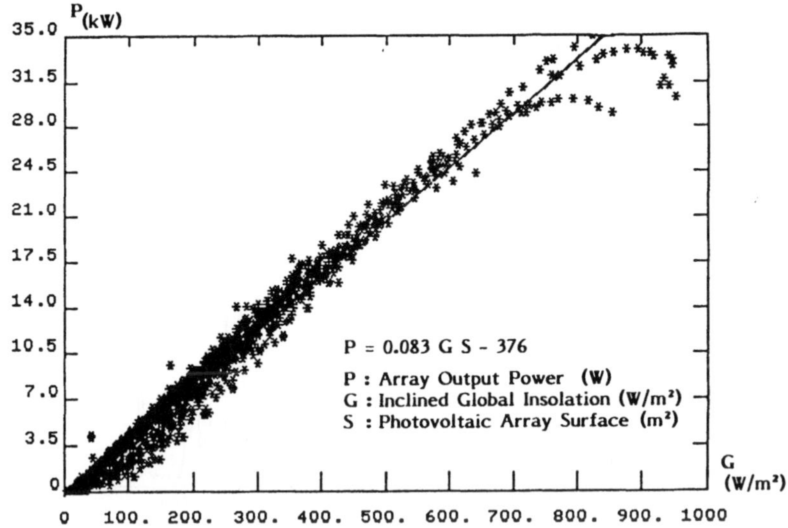

Fig 4.a LINEAR CORRELATION BETWEEN ARRAY OUTPUT POWER AND THE INSOLATION

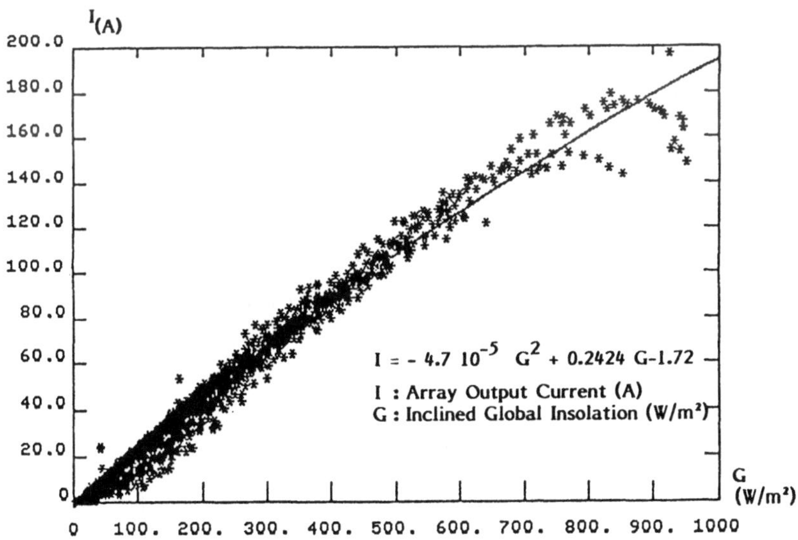

Fig 4.b PARABOLIC CORRELATION BETWEEN ARRAY OUTPUT CURRENT 'HE INSOLATION

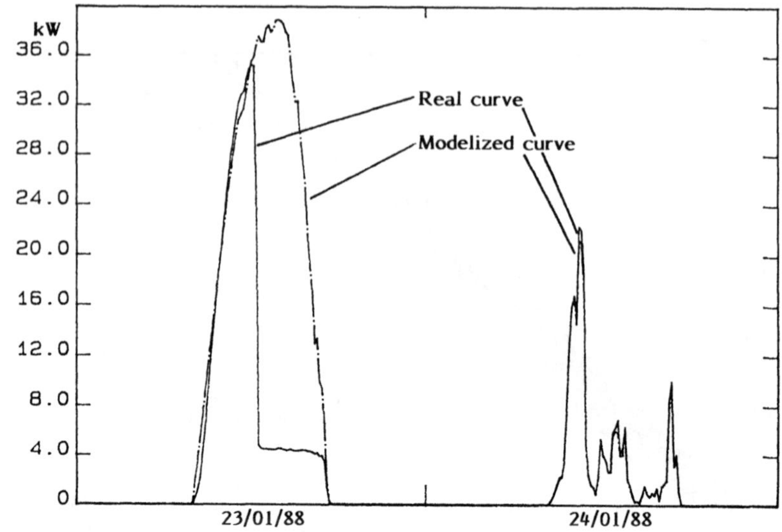

Fig 5a . Comparison between real and modelized array output power

Fig 5b. Comparison between real and modelized array output current

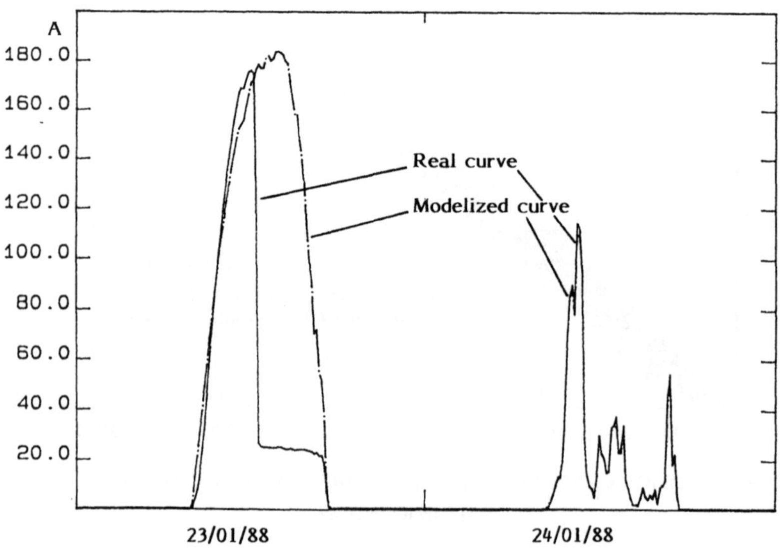

POSTER SESSION 07:
TESTS AND TRIALS

ON SITE TESTING OF LARGE PV ARRAYS USING A PORTABLE I-V CURVE PLOTTER

M. ABOU EL ELA, J.A. ROGER
Département de Physique des Matériaux (U.A. CNRS 172)
Université Claude Bernard Lyon 1
43 Bd du 11 novembre 1918
69622 Villeurbanne Cedex, France

ABSTRACT

We describe, through its block diagram, a portable equipment able to record current-voltage characteristics of PV arrays from some W_p up to 100 kW_p in some milliseconds. A fixed capacitive load is used in conjunction with a programmable sampling frequency generator. Several environmental test conditions are simultaneously recorded. A special software, used to smooth the discreet data points I-V curve and to extract the basic parameters of the array, is presented. We show that the use of a basic equation in conjunction with the determined parameters, is a possible way to extrapolate to standard test conditions. Two examples of on-site measurements are given. The first deals with the french 44 kW_p pilot plant of Rondulinu in Corsica, where the I-V characteristics of the 17 subarrays have been separately measured and the result compared with the characteristic of the total field. Information on mismatch effects is obtained. The second example refers to the 45 kW_p array of Mont Bouquet in the south of France damaged by a forest fire in 1985. We show how our apparatus has been used to check the remaining 634 modules and to sort them in order to rebuilt the PV field.

1. INTRODUCTION

Instantaneous I-V curve measurements on large PV arrays are necessary :

1) to compare actual performances to data sheets . This can be done only if a procedure of transposition to standard test conditions (STC) is available and if the equipment is able to record simultaneously other parameters such as temperature, irradiance, etc... In order to avoid irradiance corrections, the I-V curve must be recorded in some milliseconds.

2) to analyze mismatch factors by comparing the array performance to that of individual subarrays. This necessitates a control circuit in the apparatus allowing an automatic adaptation to large ranges of power.

3) to perform in-time analysis such as the study of aging effects and the detection of possible failures. Facilities for on site computations and comparison with previously recorded data must be provided.

2. PORTABLE I-V CURVE PLOTTER DESCRIPTION

The apparatus is installed in a suitcase and powered by means of two Ni/Cd batteries that can be charged either from the grid or from a PV module. The block diagram is presented on figure 1.

The plot of the I-V characteristic is obtained by sweeping the generator in some milliseconds from the short circuit point up to the the open circuit voltage by means of an electronic load.

This work has been sponsored by AFME (Agence Française pour la Maîtrise de l'Energie).

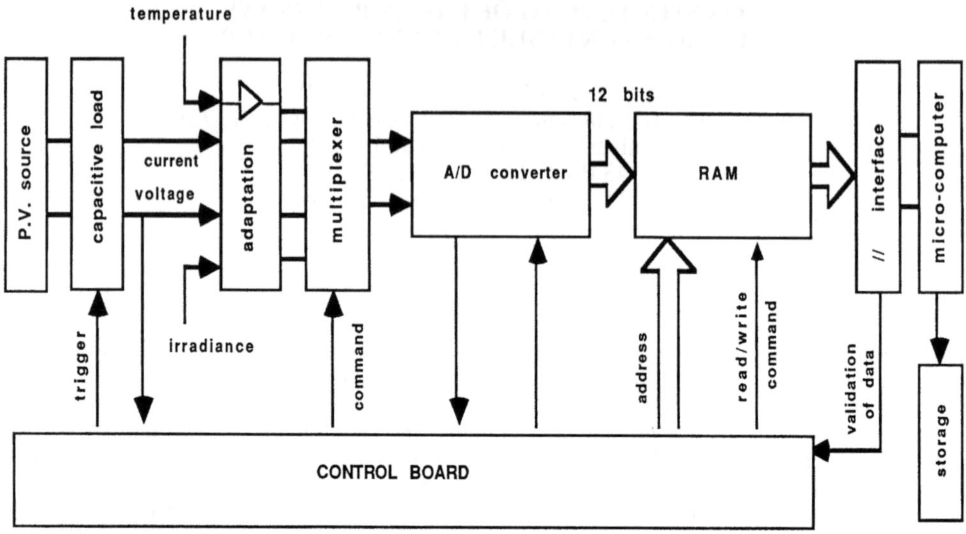

Figure 1 : General block diagram

This electronic load is a capacitive load of fixed value controlled by the trigger signal. Current and voltage data, as well as irradiance and temperature signals, are monitored through an adaptation circuit in order to match, after multiplexing, the input of a 12 bits A/D converter.

The apparatus developed some years ago by the Joint Research Centre (JRC) of CEC in Ispra [1] used a 8 bit A/D converter and worked with a fixed sampling frequency of 25 kHz. In this case such a configuration implied a variable capacitive load in order to satisfy a variable power range, up to 30 kW_p.

As shown in the diagram of figure1, a RAM (random access memory) is introduced in our apparatus between the data acquisition system and the micro-computer. This RAM enable us to work with a relatively high sampling frequency, without any restriction on the interface speed. In addition, the introduction of such an intermediate memory eliminates the necessity of a variable charging capacitor, since high values of the sampling frequency can be reached. Data are then transfered from the RAM to the micro-computer and stored. Expected short circuit current and open circuit voltage values are introduced in the computer interactive software prior to the acquisition run. The sampling frequency is then automatically adjusted, via a fully programmable frequency synthesizer, in order to record up to 252 current/voltage pairs of data point (I & V) for one characteristic. Eight channels are then left for other signals (irradiance, temperature, etc..).

In order to obtain a good distribution of data points along the I-V curve, the sampling frequency is modified inside a run, and is higher in the zone of the characteristic where the current charging the capacitance varies rapidly as shown on figure 2. This frequency switch corresponds roughly to 0,75 Voc. This frequency modification is automatically performed during acquisition via the command of a voltage controlled oscillator.

The measure of the current is not performed on a shunt, but via the evolution of the capacitor voltage with time. It is no more necessary to select various shunts for various intensities, and the measure of very high currents is made easier (figure 3). The precision of this technique has been checked by comparison with measurements performed on shunts. The maximum admissible voltage and current at inputs are 480 V and 300 A respectively.

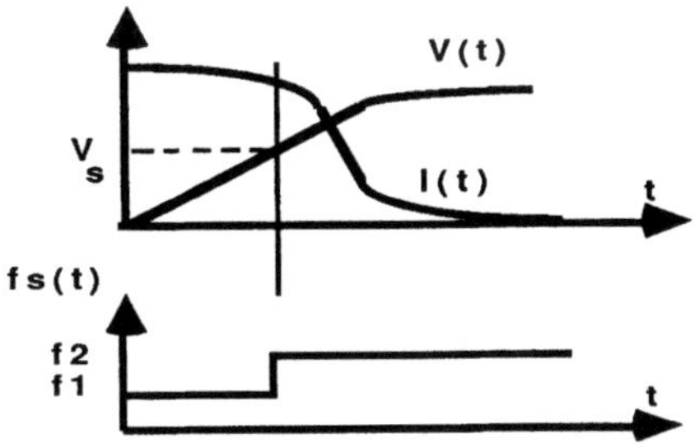

Figure 2 : Evolution of V and I as a function of time, and sampling frequency switch in a run

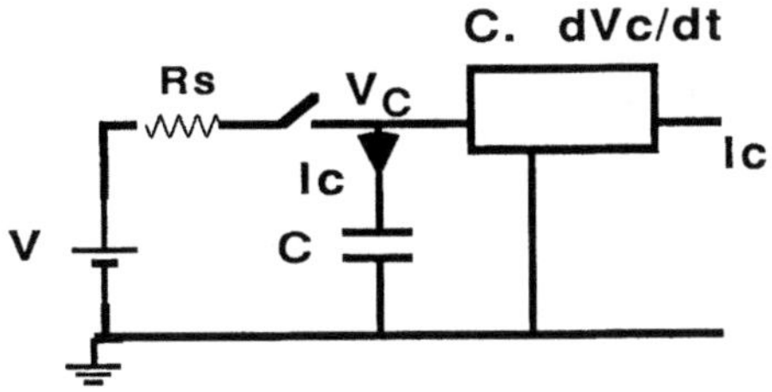

Figure 3 : Current measurement from the derivative of voltage at C terminals

3. TREATMENT OF DATA POINTS

A mean square root smoothing programme is installed in the computer and applied to recorded data points. One possibility would have been to fit an empirical equation such as given by the CEI 82 Working Group [2] :

$$I = Isc \, . \exp(A+BV+CV^2+DV^3)$$

We prefered the basic equation characterizing photovoltaic converters :

$$I = I_{sc}(\varnothing,T) - I_s(T) \left[\exp\left(\frac{e(V+R_s.I)}{nkT}\right) - 1\right] - \frac{V+R_s I}{R_{sh}}$$

This equation recalls that both the short circuit current I_{sc} and the saturation current I_s are temperature dependent, and that the ideality factor n plays a very important role. Obviously, for a large plant, measured series and shunt resistances, R_s and R_{sh}, include mismatch effects and cabling losses. The short circuit current I_{sc} is directly measured, and the other parameters are obtained by iteration procedures. The interest of such a parametric approach is that the analysis of I-V curves can give informations on defects and be used as a dignostic tool for an array, in the case of aging effects for instance. An example of a smoothed curve is given on figure 4, and the corresponding parameters are indicated.

4. EXTRAPOLATION OF ACTUAL DATA TO STANDARD TEST CONDITIONS

The recorded I-V curve gives instantaneously an information on the array performance in given site conditions. However, these data are really of interest only if we are able to convert them in order to reconstruct the I-V curve under standard test conditions (STC), which correspond to the standard spectrum AM 1.5 (S_o), a standard global irradiance of 1 $kW.m^{-2}$ (Φ_o), and a standard temperature of 25 C (T_o). This operation necessitates a reference module calibrated under S_o, Φ_o and T_o.

Two procedures have been checked, in italy and in France. The first one has been developed by JRC in Ispra and proposed as to be standardized by the IEC - WG 82 [3]. Each voltage and current value is extrapolated to STC through correction formulae, with the hypothesis that the temperature of the reference module does not differ from that of the array. A similar method has been used by the CEA Cadarache, but with different equations and the introduction of temperature measurements.

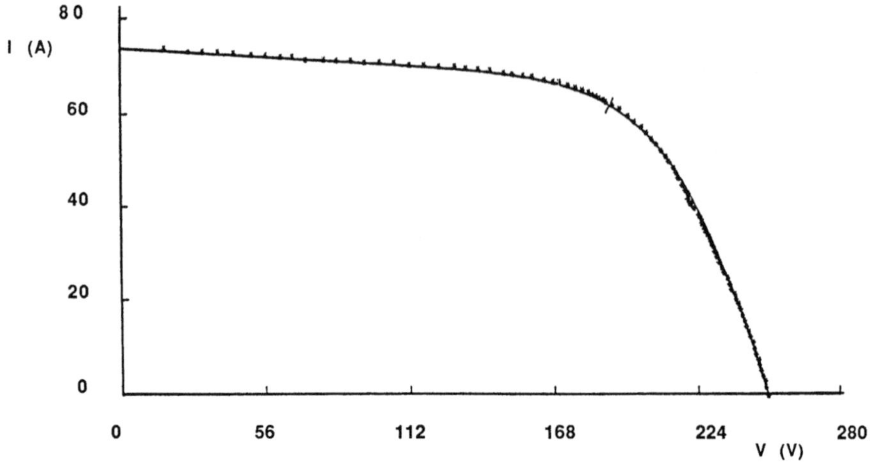

Figure 4: I-V curve recorded on the site of Rondulinu in Corsica (data points and smoothed curve) - Irradiance : 610 $W.m^{-2}$ - Module temperature : 21 C Extracted mean parameters, for this subarray of 12 modules in series and 48 strings in parallel : n = 1,97, R_s = 0,351 Ω , R_{sh} = 30,6 Ω and I_s = 697 μA

In our software we use a parametric approach based on the fondamental equation, given above for PV devices. After smoothing of the experimental curve, extrapolation to STC is obtained by taking into account the evolution of parameters with irradiance and temperature.

	Isc(A)	Voc(V)	Pmax(W)	FF	Rs(Ohm)	Rsh(Ohm)	n
Ispra	2,531	20,60	37,27	0,715	0,61	292	1,1
CEA	2,520	20,58	37,31	0,720	0,61	332	1,1
Our procedure	2,509	20,62	37,32	0,721	0,56	335	1,26

Table I : Comparison of parameter's extraction procedures in the case of a single module [4]

These three procedures have been compared on the site of Cadarache [4]. They are presented on table I as mean values for the case of 20 measurements on a module in different conditions. In the Ispra and Cadarache cases, parameters given in this table are extracted from the curve resulting from corrected data points, and the ideality factor n is imposed. In our procedure, parameters are extracted from the experimental smoothed curve, n being determined during iteration, then extrapolated to STC,

The three STC extrapolated I-V characteristics when drawn on the same graph are practically superimposed, which means that the three procedures are very close. They only differ in the evaluation of parameters such as n, Rs and Rsh. The interest of our parametric approach lies in a better determination of these last quantities due to the large number of data points recorded. Analysis of aging possible causes is then made possible.

5. PORTABLE I-V CURVE PLOTTER UTILIZATION

This apparatus has been used on the site of Paomia Rondulinu in Corsica to check the 17 blocks of this 44 kWp array. All I-V curves have been recorded at the paralleling board in the control room, and data stored on tapes. The irradiance in the plane of the modules was measured by means of a reference cell and a pyranometer, and the temperature determined from the impedance of Pt sensors attached on the lower face of modules.

The dispersion in the 17 curves extrapolated to STC has been analyzed for all parameters (Isc, Voc, Pmax, Rs, Rsh, etc...). As an example, figure 5 presents the dispersion in Pmax at STC for the 17 subarrays. The sum of Pmax is equal to 37,7 kW which gives an average value of 2,21 kW per subarray.

The I-V curve of the complete array has been recorded and extrapolated to STC, and Pmax is found to be 36,8 kW. It is then possible to evaluate losses induced by mismatch to less than 3 %. They correspond to the paralleling of subarrays. However these values remain below the rated value of 44 kWp. Other mismatch and cabling losses may also be present inside subarrays (cf number 8), and can be determined in the same way.

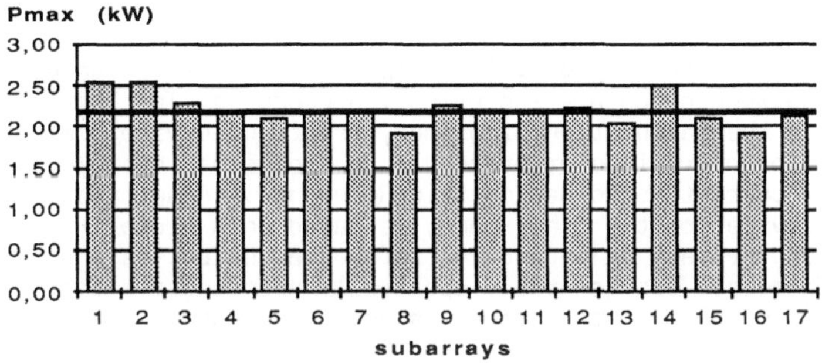

Figure 5 : Dispersion in Pmax at STC for the 17 subarrays of the Pilot Plant of Rondulinu in Corsica

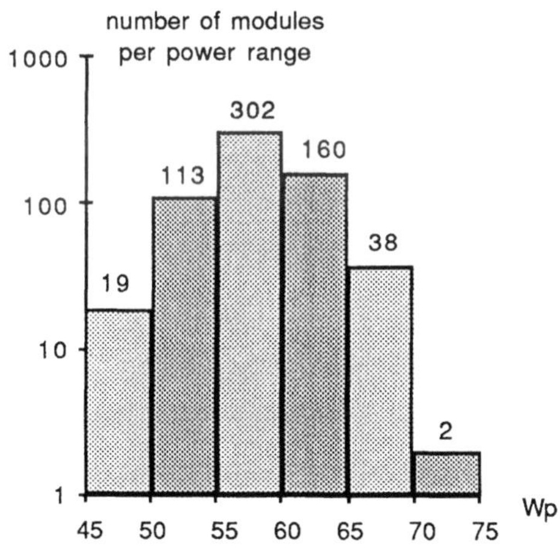

Another interest of this apparatus is the possibility for site testing of a large number of modules.This was the case when a forest fire destroyed a part of the Mont Bouquet pilot plant.In order to reconstruct the PV field it was necessary to verify each module and to sort them.

Measurements have been made on 634 modules, and the resulting histogram for Pmax is presented on figure 6, with a log. vertical scale, for the number of modules in each power range.

Figure 6 : Histogram of the 634 modules of the pilot plant of Mt Bouquet after the forest fire.

REFERENCES :

[1] G. Blaesser et al., *Proceedings of the Fifth E.C. Photovoltaic Solar Energy Conference*, Athens, 17-21 oct. 1983, p. 592 - D. Reidel Pub. Co.

[2] CEC Specification n°101, Issue 2,1980 (Doc EUR 7078 EN)

[3] G. Blaesser et al., *Proceedings of the sixth E.C.P.S.E.C.* , London, 1985, p. 481

[4] P. Pais, *Thèse de Docteur en Sciences* : 'Essais en environnement réel de champs de modules PV au silicium multicristallin et au silicium amorphe hydrogéné', Université d'Aix Marseille, 1988.

REFERENCE SOLAR CELLS IN SINGLE-CELL AND MULTI-CELL PACKAGE: DESIGN AND CALIBRATION

J. Metzdorf, T. Wittchen, B. Nawo, and H. Kaase
Physikalisch-Technische Bundesanstalt, Optics Division
Bundesallee 100, D-3300 Braunschweig, F.R.G.

ABSTRACT. Requirements for primary reference solar cells are considered and discussed in connection with necessary principles of design. As a basis for the needed primary calibration, the realization and performance of a DSR apparatus is presented which allows the application of the indoor Differential Spectral Responsivity Method to all kinds of large-area solar cells ($\lesssim$ 15 x 15 cm^2) in single-cell and multi-cell package.

1. INTRODUCTION

In the calibration of photovoltaic (pv) devices, it is generally distinguished between secondary and primary procedures. The latter are only applied to the calibration of primary reference devices based on a standard detector or radiometer traceable to a national metrological institute or to the current World Radiometric Reference. A secondary calibration of a pv device, on the other hand, is based on a properly selected and calibrated (primary or secondary) reference device. As the standard detectors/radiometers are generally spectrally nonselective thermal detectors, a large spectral mismatch between standard detector and primary reference solar cell is inevitable. Therefore, primary calibration requires: (i) an integral or direct measurement under natural or simulated solar irradiance of not less than 800 Wm^{-2} with the same spectral distribution as the standard AM1.5 spectrum; (ii) indirect, spectroradiometric measurements in the presence of proper but very simple bias irradiance as realized by the Differential Spectral Responsivity (DSR) method /1, 2/.

In this paper we firstly consider and discuss some aspects of the requirements for primary reference solar cells and secondly describe the realization and performance of a spectroradiometric primary calibration method applicable to all types of solar cells up to an area of 15 x 15 cm^2 and independent of material (quality, spectral distribution, linearity of photocurrent-irradiance characteristic).

2. REQUIREMENTS FOR REFERENCE SOLAR CELLS AND PRINCIPLES OF DESIGN

The following requirements have to be considered and weighed against each other for the design of primary reference devices: (a) spectral matching by using either proper filter glass or identical materials for solar cell, encapsulant, and window glass; (b) geometrical-optical matching realized by a multi-cell package of minimum acceptable size (see below) with the same encapsulation and package as the pv module being calibrated; (c) temperature measurement and control, where the latter is needed to maintain Standard Test Conditions (25 °C) and to determine temperature coefficients; (d) linearity where, in general, nonlinearities are connected with variations in the relative spectral responsivity; (e) stability. Both linearity and stability are desirable but may not be maintained if solar cells made of special material (e. g., amorphous silicon) shall be used. Alternatively, linear and stable FZ Si solar cells with suitable window filter glass (e. g., Schott BG18 or BG39) may be used /3/.

Fig. 1 shows an example of a 10 x 10 cm^2 solar cell in a flat single-cell package with the body made of Al because of the low thermal resistance to facilitate temperature control. A Vacon (or Kovar) plate is in-

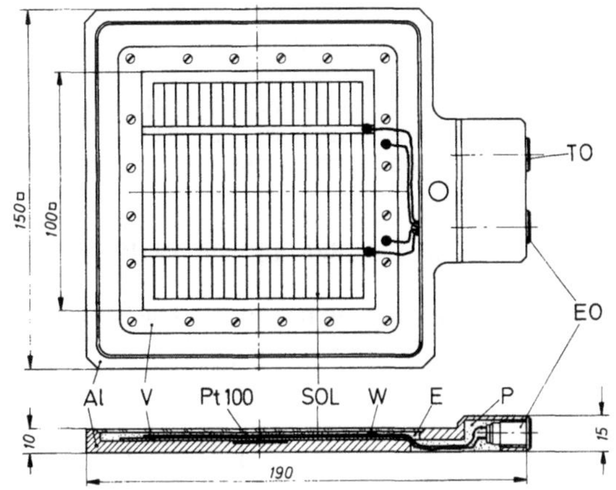

Fig. 1: Single-cell package facilitating temperature control of a 10 x 10 cm² solar cell with flat heat-conducting Al package (field of view > 160°). Thin film PT 100 resistor between Vacon plate V and Al case; SOL solar cell; W window or filter glass; E encapsulant (Silgel or EVA); P potting compound; TO thermal and EO electrical outputs (Lemosa receptacles).

Fig. 2: Two examples of multi-cell package facilitating geometrical-optical matching: 10 x 10 cm² cell with surrounding dummy cells that were bisected and remained uncut, respectively.

cluded to match the thermal expansion of the Si cell. As it is more convenient to use receptacles (Lemosa connectors with four and five pins that cannot be interchanged) instead of cables, thin-film PT 100 resistors should be used for temperature measurement instead of thermocouples in any case (see also Fig. 2). A water-cooled Al case which does not require a temperature-controlled base is generally too thick and too heavy.

Two examples of reference cells in multi-cell package shown in Fig. 2 are specially made as small modules (apart from the external dimensions identical to the pv modules being tested) to facilitate geometrical-optical matching. The reference cells are electrically connected (four-wire contact system), the neighbouring dummies (real cells without electrical connections) were bisected or remained uncut where a package with undamaged dummies, though being larger, should be favoured in practice (manufacturing more easy; no additional problems in handling and calibration). While temperature measurement of reference cells in module-like multi-cell package is straightforward with the aid of a thin-film PT 100 resistor, temperature control is difficult. Thus, in our opinion, reference solar cells in single-cell and multi-cell package are complementary primary reference devices and both versions are needed for the secondary calibration of pv modules.

3. DSR APPARATUS FOR INDOOR CALIBRATION OF PRIMARY REFERENCE SOLAR CELLS

The DSR method has already been applied to the calibration of small-area solar cells and described in /2/. As an illustration, Fig. 3 shows the block diagram of a DSR apparatus. The simple dual-beam optical arrange-

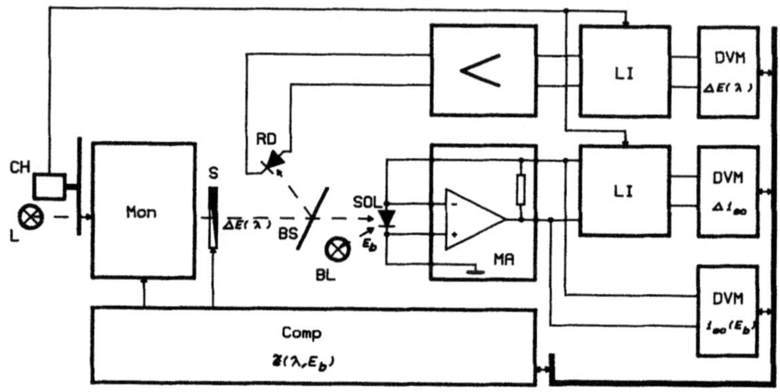

Fig. 3: Block diagram of a DSR apparatus. L lamp producing test irradiance $\Delta E(\lambda)$; CH chopper; MON monochromator; S shutter; BS beam splitter (quartz plate); RD reference or monitor detector; BL lamp producing bias irradiance E_b; SOL solar cell; MA measuring amplifier; LI lock-in amplifier; DVM digital voltmeter; Comp process computer.

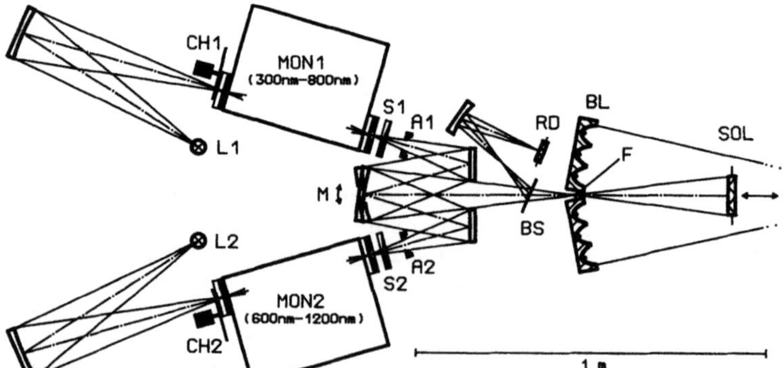

Fig. 4: Double dual-beam optical arrangement for the measurement of the relative differential spectral responsivity $\tilde{s}(\lambda, i_{sc}(E_b))$. L1 xenon arc lamp; L2 quartz halogen lamp; A1,2 apertures imaged onto the solar cell SOL; M concave mirror; F focal points of the test radiation; BL array of quartz halogen lamps with integrated parabolic multi-mirrors.

ment without any imaging optics for the measurement of the absolute differential spectral responsivity $\tilde{s}(\lambda_0, i_{sc}(E_0))$ at some discrete wavelength λ_0 in the presence of rather low but highly uniform bias irradiance E_0 has been described in /4/.

Fig. 4 shows the new double dual-beam optical arrangement for the measurement of the relative differential spectral responsivities $\tilde{s}(\lambda, i_{sc}(E_b))$ with 300 nm $\leq \lambda \leq$ 1200 nm and 0.1 $Wm^{-2} \leq E_b \leq$ 2000 Wm^{-2}. Lamps L1,2 (1000 W quartz halogen lamp, FEL type; 1000 W xenon arc lamp) and blaze wavelengths of the gratings of the double monochromators (e. g., Kratos type GM 252 D with large throughput) are properly selected according to the respective overlapping spectral ranges. The radiant power of the monochromatic test radiation and thus the mean value of the spectral irradiance in the working plane of the solar cell SOL is monitored by a reference detector behind a beam splitter BS with an additional lock-in amplifier to eliminate stray light effects (see Figs. 3 and 4). An array of 6 x 6 small long-life quartz halogen lamps (20 W to 150 W) with integrated selective reflecting parabolic multi-mirrors is used to produce the bias irradiance E_b.

The spectroradiometric requirements for the DSR apparatus due to the criteria of low uncertainty and large signal-to-noise ratio (see Fig. 5) are met: (i) the uniformities of monochromatic and bias irradiance $\Delta E(\lambda)$ and E_b over the area of the solar cell and the multi-cell package, respectively (see Fig. 6), are sufficient to carry out relative measurements /2/; (ii) high levels of irradiance up to $E_b \leq$ 2000 Wm^{-2} and $\Delta E(\lambda) \geq 10^{-5} \cdot E_b$ are produced in the whole spectral range above 320 nm (spectral bandwidth 10 nm) making possible a reasonable time (9 min) of recording the two overlapping spectra at adequate intervals of 10 nm. The small intervals (2 nm) in Fig. 5 are chosen to illustrate the low noise by increasing the number of data within the spectral bandwidth.

As the DSR calibration needs no special solar simulator or natural solar radiation, it is reproducible independent of time and place. The calibration value is based on stable and accurate detector standards that

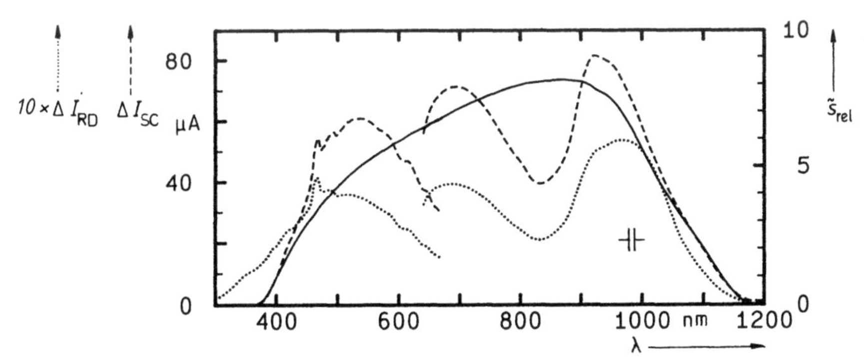

Fig. 5: Typical raw data of the ac short-circuit currents of a CZ Si solar cell (A = 100 cm²) $\Delta I_{sc}(\lambda)$ (dashed curves) and of the monitor detector $\Delta I_{RD}(\lambda)$ (dotted curves). The ratio of the photocurrents $\Delta I_{sc}/\Delta I_{RD}$ taking into account the calibration of the monitor detector is shown as a solid curve. The data measured point-by-point (2 nm intervals; 2 s integration times) are plotted as a function of wavelength without averaging to illustrate the large signal-to-noise ratio.

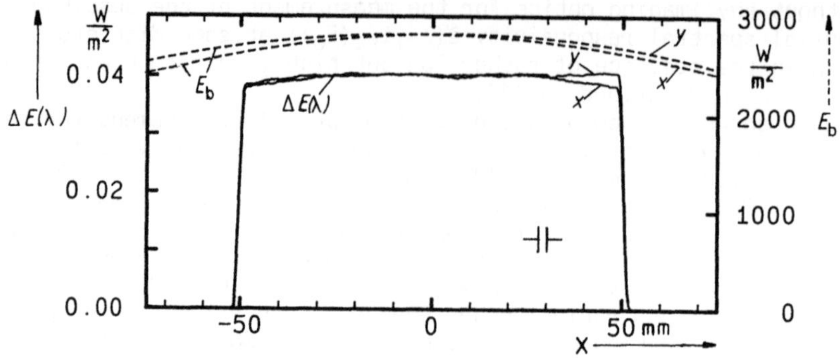

Fig. 6: Degree of uniformity of monochromatic test irradiance $\Delta E(\lambda)$ and bias irradiance E_b within the working plane (SOL in Fig. 4) as a function of horizontal displacement x and vertical displacement y (parallel to monochromator slit). Spatial resolution (2 x 2 mm^2) is limited by the area of the detector used.

are traceable to the primary radiometric standards without spectral mismatch or nonlinearity problems. The description of a simplified DSR apparatus that is easily operated and is quite suitable for installation and use in an industrial laboratory will be published separately.

ACKNOWLEDGEMENTS

The authors are indebted to AEG, Siemens, and Telefunken-electronic for specially making reference devices and to Dr. K. Möstl at PTB for calibrating the standard detectors. This work is supported by the Bundesminister für Forschung und Technologie.

REFERENCES

/1/ F.C. Treble; 'The Calibration of Primary Terrestrial Flat-Plate Reference Devices - A Proposed International Standard'. Proc. 7th EC PV Solar Energy Conf., Sevilla, 89-94 (1986).

/2/ J. Metzdorf; 'Calibration of Solar Cells 1: The Differential Spectral Responsivity Method'. Appl. Optics 26, 1701-8 (1987).

/3/ J. Metzdorf, H. Kaase; 'Photovoltaisches Referenzelement für bewertete Empfindlichkeit und hohe Bestrahlungsstärke'. In Optoelektronik in der Technik - Laser 85 Optoelektronik (Springer, Berlin 1986), pp. 250-253.

/4/ J. Metzdorf, T. Wittchen, and H. Kaase; 'Absolute Indoor Calibration of Large-Area Solar Cells'. Proc. 5th Eur. Symp. PV Generators in Space (ESA-SP-267), Scheveningen, 397-401 (1986).

Component Characterization and PV Power System Sizing Through Dynamic System Simulation

Dr. France F., Ang T.G., Wang Y.J., Hemasuk T.
Energy Technology Division
Asian Institute of Technology
GPO Box 2754 Bangkok 10501 Thailand

Abstract:

A dynamic system simulation approach to photovoltaic power system (PVPS) sizing is presented. A compact statistical solar irradiance model provides the energy input to the system. Performance characterization studies on PV modules and storage batteries are undertaken to ensure their reliable operation and to provide accurate model for simulation. Hourly simulation over a year provides informations on battery cycles and performance; loss of load frequency (LOLF) and loss of load hours (LOLH). Using a two-dimensional inverse interpolation, a family of iso-LOLH curves are generated from the simulation results. Optimum cost is achieved when the marginal rate of substitution equals the array-storage price ratio. Test-stand for array and storage battery were installed. Components characterization and system sizing have been coded to run on personal computers. Coupled with economic or financial analyses, a hybrid system may be more economical than either a PV- or conventional generator-only system.

1.0 Introduction

Every decision made during the design of a PVPS affects the cost. If the PVPS is oversized, the initial cost is increased unnecessarily. If less efficient system components are specified, maintenance and replacement costs are increased. Life-cycle cost estimates can easily double with inappropriate choices made during system design. An inexact estimate of irradiation and a faulty knowledge of the performance of system components affects system availability and cast doubts on results from any sophisticated approach to system sizing.

It is therefore profitable to have a reliable estimate of the local meteorological conditions; to be able to characterize and predict the performance of system components subjected to the installed conditions.

This paper will describe a compact random model simulation of daily totals of irradiation and hourly irradiance; the characterization of PV modules under outdoor conditions and the performance tests of storage batteries. The concept of LOLH is introduced as a reliability index to the performance of a stand-alone PVPS. Due to the substitutional possibilities between PV modules and storage batteries to meet a prescribe system reliability, economic considerations identifies a point of minimum PVPS cost for stand-alone system and an optimum energy mix in a PV/diesel hybrid system.

2.0 Irradiance Simulation

Based on detailed studies of irradiance data for Thailand [EXELL, 1981], a compact computer simulation code to generate daily total irradiation and hourly irradiance is developed. The model is expected to be applicable over the entire South-East Asia between latitudes of 25°N and 25°S. Provision is made for the separation of the solar irradiance into their direct and diffuse components and for the generation of random sequences of irradiance values with appropriate statistics.

The model constitutes the reduction of information on irradiation in Thailand to a compact form that reproduces the mean values and the random fluctuations of hourly irradiance throughout the year.

The global irradiance from a clear sky on a horizontal surface, and the direct irradiance from a clear sky on a perpendicular surface to the solar beam, are represented by empirical formulae. Allowances for variations in the intensity of solar irradiance due to variations in the distance of the sun from the earth are included. The daily solar irradiation from a clear sky consists of a Fourier series in which the arguments of the trigonometric terms are functions of the day of the year (Nd) and the Fourier coefficients are functions of latitude.

The mean daily clearness Kd is calculated by means of the truncated Fourier series

$$Kd = C0 + C1 * Cos(t - C2) + C3 * Cos(2t - C4)$$

where, $t = (Nd - 80) * 360 / 365$ degrees.

The parameters C0, C1, C2, C3 and C4 are obtain by numerical harmonic analysis of the observed mean daily clearness values for a particular location. A cumulative distribution of mean daily clearness values is modelled by a binomial distribution. Diurnal variations of irradiance are also taken into account. Finally, the hourly irradiance and the daily totals of irradiation are calculated as the product of the mean hourly or daily clearness index and the clear sky solar radiation.

3.0 PV Module Characterization

A computerized system for the characterization and performance testing of PV modules under natural sunlight and ambient conditions was designed and installed in AIT [ANG, 1987]. The system comprises two phases: first, the performance testing of PV modules through computer-controlled electronic hardware (monitoring phase); second, softwares were coded for data processing, file management and data analyses (analysis phase).

The monitoring phase consists of the PV module test-bed and software-controlled intrumentations which automate daily monitoring of module performance. The PV module test-bed is a two-axis tracking structure with power input terminals and temperature sensors to accomodate 16 modules. Standard intruments to measure tilted-global irradiance, diffuse irradiance, wind velocity and shaded ambient temperature are installed and calibrated regularly. Multiplexer and relays switch four modules for simultaneous monitoring. An N-channel 75 watt dc MOSFET device serves as the electronic variable load to each module being monitored. A ramp voltage signal drives the MOSFET to generate the I-V characteristic curve of the module.

Softwares are coded to control monitoring, measurement and data storage sequences. Provisions are made for time-interval and fixed-irradiance monitoring. Logics within the code check to ensure constant irradiance during the generation of an I-V curve.

Outdoor performance testing of PV modules requires a compilation of vast amount of data in order that module characterization can be reliable and statistically acceptable. The analysis phase consists mainly of a code named PVMOD. PVMOD version 2 handles routinary tasks such as data processing, file handling and maintenance, graphical and statistical data presentation and also provides numerical and analytical characterization of PV modules in terms of its internal physical parameters such as the shunt and series resistance, the reverse saturation current, the photo-generated current and the module ideality factor.

4.0 Storage Batteries

An electronic constant-current charge and discharge controller for storage battery is used to monitor the performance of SLI batteries under room temperature [AHMED, 1987]. Charge and discharge characteristic curves are investigated at various rates. Problems affecting self-discharge and vertical stratification of the electrolyte were simulated and measured. Charge and discharge cycle of a battery as part of a PVPS was simulated and its performance monitored to ensure proper operation when installed and to provide reliable information for system sizing and system reliability prediction.

5.0 Dynamic Sizing Of A PVPS

This section describes the process adapted in sizing PV arrays and storage batteries, which together comprise the PVPS, through hourly system simulation [WANG, 1987]. The hourly energy balance between the power system and energy consumption is controlled simultaneously by the meteorological data, the array characteristics, the charge-discharge characteristics and efficiency of the storage battery and the shape of the load profile.

The solar irradiance model of section 2.0 [EXELL, 1981] has been coded into an interactive user-friendly software [LASNIER, 1988]. This code is used to generate the meteorological input data. The ambient temperature is assumed to have a sinusoidal variation over the day throughout the year. Daily averages are also assumed to vary sinusoidally over a year.

The PVPS voltage floats with the battery voltage and the available output current from the array is a function of the incident irradiance and the cell temperature. The cell temperature is modelled by empirical formulae based on simplified theoretical considerations. The effect of current generation on cell temperature is also considered.

The array model is defined by three points from the I-V Characteristic curve [RAUSCHENBACH, 1980]: the open-circuit voltage Voc, the short-circuit current Isc, and the maximum power point Vmp and Imp. Then the array current I is expressed as a function of the array voltage V by,

$$I = Isc \{1 - C1 [Exp(V / (C2*Voc)) - 1] \}$$
$$C1 = (1 - Imp/Isc) Exp[-Vmp / (C2*Voc)]$$
$$C2 = (Vmp/Voc - 1) / Ln(1 - Imp/Isc)$$

The state of charge S(t+dt) of a storage battery can be calculated from the previous state of charge S(t) together with the other battery parameters according to the relation [ANG, 1984],

$$S(t+dt) = S(t) (1 - Ds*dt) + K1 (V*Ib - K2*Ib^2) dt$$

where Ds is the rate of self-discharge in Wh/hr; K1 is the battery charge/discharge efficiency; K2 is the battery charge/discharge apparent resistance in ohms; V is the system voltage and Ib the battery current.

The battery charge/discharge voltage and resistance are updated every dt according to the following relations,

$$R2 = S(t)/Smax$$
$$Vch = 2 + 0.148 * R2$$
$$Vdch = 1.926 + 0.124 * R2$$
$$Rch = (0.758 + 0.1309/ (1.06 - R2)) / Smax$$
$$Rdch = (0.19 + 0.1037/ (R2 - 0.14)) / Smax$$

where subscript 'ch' and 'dch' denotes charge and discharge respectively; Smax is the battery maximum state of charge.

The inverter is modelled by an efficiency value in the range of 0.85 to 0.95. A charge/discharge controller directs excess energy from the array to the storage battery; dissipating this energy when the battery is fully charged. The load is cut-off from the PVPS when the array and the storage battery can not supply enough energy to the load. This event is recorded as an LOLH.

The cumulative loss of load hours (LOLH) over a year is chosen to describe the PVPS reliability. System downtime due to component failure or maintenance are not included in the cumulative LOLH. Using a two-dimensional inverse interpolation and curve fitting, a transformed function with the LOLH and the array size as the independent variables and the storage battery as the dependent variable is achieved. A family of curves of this transformed function at various LOLH is shown in Fig. 1. These curves reveal the substitutional possibilities between arrays and storage batteries to achieve a prescribe reliablity. Along the iso-LOLH curve, the minimum PVPS cost occurs at the point where the marginal rate of substitution equals the price ratio according to the relation,

$$-d(BTYS) / d(PVS) = Pw / Pb$$

where d(BTYS)/d(PVS) is the marginal rate of substitution and Pw, Pb are the unit price of the array and storage battery respectively.

5.0 Conclusion

Hourly simulation approach to PVPS sizing based on a prescribed LOLH offers a more precise combination of array and storage battery capacity, particularly when constrained by variable load profile throughout the year. A diminishing marginal rate of reliability improvement per unit increase in array and battery capacity is evident in the figure. Energy source availabilities of 97% or higher are typically required for remote stand-alone applications. The requirement to oversize the system for availability's sake for a daily load exceeding 10 kWh may make PV-alone system uneconomical compared to a conventional generator-only system when such application are attended and at least accessible. However the PV system can be combined with the conventional generator system to obtain a hybrid system that is often more economical than either constituent alone [JONES, 1987]. If properly designed, the hybrid system can overcome the major liabilities of each system. Economic or financial analyses can identify the optimum energy mix for a hybrid system. These studies, therefore, provides a wholistic approach to optimum PVPS or PV/fossil hybrid system sizing and reliability prediction. Instrumentations and test equipments characterized system components and are also indespensable in system maintenance.

References

1. Wang Yaw-Juen (1987),"The Sizing of Stand-Alone Photovoltaic System", AIT Research Study RSPR No ET 87-8, Asian Institute of Technology, Bangkok.
2. Exell, R.H.B. (1981),"A Mathematical Model for Solar Radiation in South-East Asia (Thailand)", Solar Energy, Vol. 26, pp.161-168.
3. Ang, Tony Gan (1984),"Analysis of a Compression Refrigeration Supplied by Photovoltaic Power", AIT Thesis No ET-84-10, Asian Institute of Technology, Bangkok.
4. Rauschenbach, H. S. (1980), Solar Cell Array Design Handbook, Van Nostrand Reinhold Ltd., Toronto.
5. Ahmed Shafiq (1987),"Battery Characterization Using Auto-controlled Constant-current Battery Test-bed", AIT Research Study RSPR No ET 87-1, Asian Institute of Technology, Bangkok.
6. Ang T.G.; Hemasuk T.; Dr. Lasnier F.,"Testing of PV Modules Under Natural Condition Using the AIT PV Module Test Bed System", AIT Report, Asian Institute of Technology, Bangkok, 1987.
7. Dr. Lasnier F.; Ang T.G., Computer Package for Photovolatic Systems Sizing, Design and Financial Evaluation, AIT Software Manual, Asian Institute of Technology, Bangkok, 1988.
8. Jones G.J. and Chapman R.N.,"Photovolatic/Diesel Hybrid Systems: The Design Process", Proceedings of the 19th IEEE Photovolatic Specialists Conference, New Orleans, Louisiana, May 4-8, 1987.

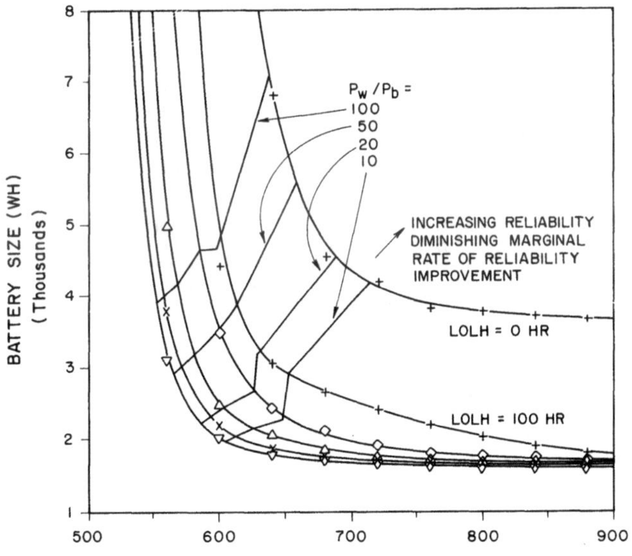

LOLH : + 0 , 100 HR , ◇ 200 HR , △ 300 HR , X 400 HR, ▽ 500 HR

Figure 1 ISO - LOLH Curves.

a) Optimum design points of four price ratio at different system reliability.

b) Diminishing marginal rate of reliability improvement.

A NEW LARGE SIZE HELIO-CLIMATIC SIMULATOR FOR COMPUTER ASSISTED LONG TERM MONITORING OF PV ARRAYS AND MODULES.

Franceschi L.,Nicoletti A.,
Lab. for Information Food and Energy (LIFE) Srl,
Via A. Guasti 9 ,00139 Rome,Italy
Barra O.A.,Bellecci C.,
University of Calabria , Dept. of Physics,
Arcavacata di Rende ,87036 Cosenza,Italy
Di Francia A.,Margadonna D.,Scrimaglio R.,
ITALSOLAR,
Via A.D'Andrea 6 ,00048 Nettuno,Italy

ABSTRACT. The need to quickly evaluate the efficiency and reliability, not only of single PV cells but also of larger systems, (modules or "strings" or "arrays") made it necessary to carry out a large size simulator. Since the most important parameters, affecting the performance of a PV system, are illumination and ambient temperature, both can be changed following controlled cycles. All the devices are remote controlled and special care has been devoted to avoid external noise. The helio-climatic chamber is described as well as the computer assisted data acquisition and control systems. The machine is presently used day by day by ITALSOLAR to test its modules, but it is available as a service for any public or private institution interested to carry out any kind of test on any type of device.

INTRODUCTION

Up to now several devices are commercially available for the evaluation of the PV cells performance (1), but the behaviour of a PV power plant is not the same as that of a single cell. In fact the series and parallel connections, the mechanical-thermal stresses, the different aging of several components when linked together, can seriously affect the overall efficiency and useful life span of the plant. Obviously not every plant will be affected by the same external agents because they depend on the installation site. For large size PV systems the same criteria generally used for the experimental evaluation of the single cell characteristics by means of solar and ambient simulators can be followed too. So a large device consisting of a completely insulated experimental chamber whose temperature can be changed, even very quickly, in a wide range was set up. Inside the chamber it is possible to lay many PV modules on a plane whose illumination can be accurately measured. The illumination is provided by means of an array of lamps, each of them is pivotable and switchable. A local computer is devoted to control all the mechanical, electrical and thermal devices, and to collect all the experimental data. A remote computer, linked to the previous one, allows to programme the experimental cycles, to control the behaviour of the simulator devices during the cycles, to analize the collected data and to store it if necessary. The italian company ANGELANTONI Spa supplied the mechanical and thermal devices and LIFE Srl provided the electronic and electrooptic systems; the simulator was installed in the outdoor experimental and evaluation tests site of the italian PV company ITALSOLAR.

EXPERIMENTAL APPARATUS

The whole experimental system is composed of an insulated chamber, its thermal apparatus, its control and data acquisition devices. The test chamber is a parallelepiped whose useful internal dimensions are 7m x 7m x 5m (h x d x w) (see fig.1).

The walls are made of prefabricated steel panels, 120 mm thick. Insulation is obtained by means of a poliurethan layer, injected inside the panels. The whole chamber is lined with Roll Bond plates linked together, in this way allowing the temperature control of the internal ambient. In fact a complex system of heaters and refrigerators is connected to the Roll Bond plates, and a ventilation system allows frequent air changes to obtain the internal thermalization. The temperature may be controlled in the range 5 - 50°C with a precision of 1°C. The rate of change is 0.2°C/min.

The main devices inside the chamber are the illumination system (see fig.2,4) and the measurements plane, facing each other.

The illumination system is composed of 36 lamps assembled into a steel chassis; the lamps are arranged in rows and columns (6 rows and 6 columns) and each one may be individually switched and oriented; besides the whole chassis may be tilted up to 45 deg from the vertical position; the total height of this system is 4 m. The lamps used are the THORN PAR 64/CSI (2), 1 kW each (for a global possible power of 36 kW); their availability on the market and their spectral distribution made them a good choice for the aim pursued. A very heavy problem is given of course by the heating of. both the chassis and the lamps when they are lighted; for this reason another independent refrigeration system is connected to the lamps chassis and allows to maintain the lamps temperature under 35°C, their maximum working temperature.

Facing the illumination system is the measurements plane, 3.6m x 1.4m (h x w), to which panels under test are bolted; the plane may be tilted up to 45 deg from the vertical position and may be rotated of 90 deg around its normal axis (see fig.4); the distance between the measurements plane and the illumination system may be chosen in the range 2m - 6m (so that it is possible to obtain radiative fluxes on the panels up to 5-6 kW/sm). The whole plane can be scanned by a radiation intensity sensor moved by means of two engines (for vertical and horizontal scanning); the movement can be continuous or discrete and after a complete scanning of the measurements plane the map of its light intensity is obtained; this kind of measurements are necessary in the regulation of radiation flux, before each measurements session.

All the control boards and devices are positioned into two rooms sideways the simulator and each function can be controlled manually even if a computer assisted control and data acquisition system is necessary to allow an easy management of the simulator.

The hardware of this contol system is divided in two parts: a) a Local Board (Slave, LB) positioned near the simulator, basically devoted to manage all the control devices of the chamber, and to collect all the experimental data; b) a Central Computer (Master, CC) positioned in the control room (see fig.3) about 30 m far from the simulator, used to programme the measurements cycles and to send to the Local Board all the corresponding executable commands; the two boards are linked together and communicate by means of an RS232-C serial interface.

The LB is assembled using ANALOG DEVICES components (3,4); it is composed of the programmable main board µMAC-5000 (64 kBytes RAM, 8088 CPU) able to preelaborate the collected data before send them to the CC, and supporting 8+8 digital inputs and outputs and 12 analog inputs for thermoresistors; a large number of different input/output boards can be connected to the main board, depending on the needs; our configuration today is composed also of one expansion board µMAC-4010 (8+8 digital inputs and outputs, 8 analog inputs for voltages and currents) and one expansion board µMAC-4040 (32+32 digital inputs and outputs). Digital inputs and outputs are mainly devoted to control the apparatus, while analog inputs are used for collecting the experimental values.

The CC, the "master unit", is a usual XT Personal Computer with 640 kBytes RAM, one 360 kB floppy disk and one 20 MB Hard Disk running under Microsoft MS DOS operative system; see fig.5 for a schematic representation of the hardware.

Each one of the two main boards is provided with a software appropriately developed for the experimental needs. In the LB, a communication software is resident in ROM, immediately acting when the LB is switched on; it allows the LB to receive direct commands and also programs to be executed immediately. The CC contains all the managing procedures so that the operator is able to decide the duration of measurements cycles, the intensity of radiation on the panels , the frequency of measurements ; the operator can also control the behaviour of each part of the plant, can store collected data, and can obtain particular elaborations of them like graphical representation of the radiation flux on the panels tested, table representation of the data, comparison between data corresponding to different cycles etc. During the initialization procedure the managing programs to be run on the LB, are sent from the mass memory of the CC to the RAM memory of the LB; in the same way information between the two computers are exchanged; the communication rate is 9600 baud.

DISCUSSION

The system so developed allows to measure the fundamental parameters necessary to evaluate the efficiency and reliability of PV plants. So temperatures in different points, output voltages and currents, radiation intensity and lifetimes of the panels tested are continuosly controlled; all the collected data can be affected by an error varying in the range 1-5%.

Up to now are possible 8 different measurement methodologies, corresponding to all the combinations of the following three couples of tests :

- Destructive test or characterizing test
- Single module test or array test
- One cycle test or multicycle test (simulating any climatic situation)

All the most significative data, converted in physical units, are also available on magnetic supports and prepared for an automatic evaluation of correlations between PV systems performance and ambient data.

At the end of each test a summary of the results is given in paper form.

REFERENCES

(1) SPIRE SUN SIMULATOR 240 A, Data Sheet, SPIRE CORPORATION, 1988

(2) THORN LAMP, PAR.64/CSI.1000 W Data Sheet, THORN, 1980

(3) Complete BASIC Programmable Stand-Alone Measurement and Control System, μMAC-5000 Data Sheet, ANALOG DEVICES, Mass. U.S.A., 1986

(4) Expander Boards and Accessories, μMAC-5000 Series Data Sheet, Mass. U.S.A., 1986

fig.1 An external view of the simulator building

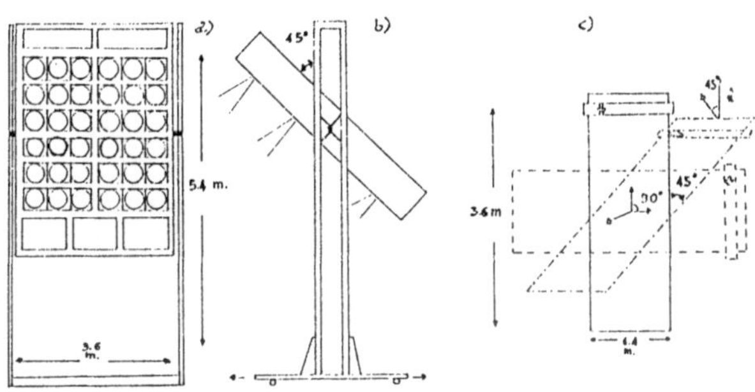

fig. 4 Lamps support: a) front view, b) lateral view, and measurements plane c).

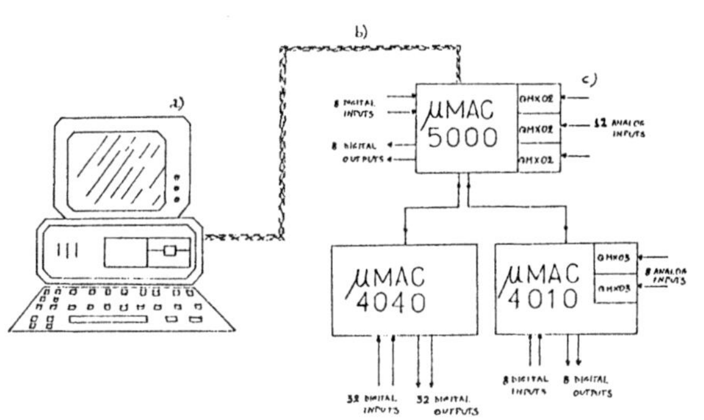

fig. 5 Schematic representation of the hardware: a) Central Computer; b)hardwired line; c)local board

fig.2 The illumination System

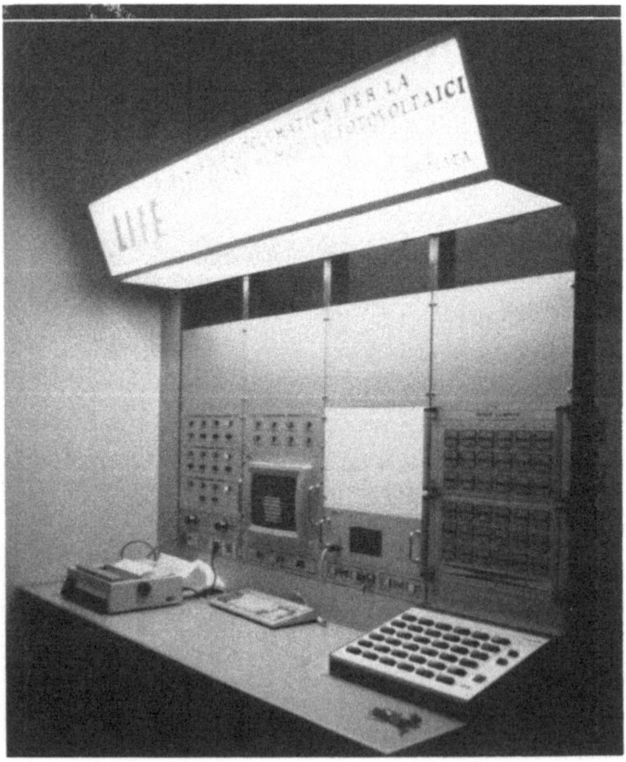

fig. 3
The control Room

DETERMINATION AND ANALYSIS OF THE PERFORMANCE AND DEGRADATION OF a-Si MODULES USING OUTDOOR, SIMULATOR AND OPEN-CIRCUIT-VOLTAGE- DECAY (OCVD) MEASUREMENTS.

F.H. Klotz*, G. Massano, A. Sarno, L. Zavarese
* Grant from E.C.

ENEA C.RI.F., POB 83, 80055 Portici (NA), Italy
Tel. (081) 47 92 88; Fax (081) 48 93 88

ABSTRACT. The purposes of this work are to verify a correct and reproducible outdoor performance evaluation for a-Si modules and to present a systematic study of the degradation effect. For this the modules are exposed to natural sunlight and periodically characterized in a solar simulator by the measurement of the I/V curve and OCVD. These procedures are applied to present state of art a-Si modules, coming from various manufacturers and realized by different technologies, for comparing their performance and degradation. Performance degradation of a-Si modules results mostly from a reduction of the Fill-Factor. Responsible for this is an increase of the serial resistance R_S and a decrease of the lifetime of the minority carriers which in turn decreases the carrier collection length. The short-circuit current and the open-circuit voltage are as well involved in the degradation, but their contribution is smaller. The OCVD technique is non destructive and very easy to carry out. For the interpretation of the OCVD transient one has to be careful because of the influences of the space charge capacitance and the shunt resistance.

1. INTRODUCTION

The outdoor performance evaluation of a-Si modules is complicated because there are no adequate stable a-Si reference cells. Also the temperature and intensity correction coefficients as for c-Si are not applicable. We present a method to compare outdoor and indoor measurements.

For scientific investigation and quality control of a-Si modules it is important to determine and analyse their performance and degradation. We want to show the influence of the light-induced degradation effect on the module performance characteristics (P_{el}, FF, V_{OC}, J_{SC}, R_S). Furthermore it is important to localize the intrinsic origin of the degradation effect. Since the FF is the most important parameter for the degradation, it is useful to analyse the minority carrier lifetime and the carrier collection length. For this we want to prove the value of the light induced OCVD technique.

And last but not least this is a comparison of the present state of art a-Si modules just now available on the market.

2. EXPERIMENTAL APPARATUS AND METHOD

The main parameters for the correct and reproducible evaluation of the module performance are: the module temperature, the global incident irradiance level and the spectral distribution of the light source. The temperature is evaluated with a thermo couple at the center of the module. The global incident irradiance level is measured with a filtered and calibrated single crystalline silicon reference cell (pseudo a-Si:H). The spectral distribution is determined with a LICOR 1800 spectroradiometer. The Air Mass is calculated with a simulation program. For the indoor measurements a Spire Pulsed Solar Simulator is used. For the outdoor measurements an electronic load is used to drive the I/V-curve.

Four types of a-Si modules (maker A,B,C and D) are exposed to natural sunlight: one in open-circuit (OC) and one in short-circuit (SC) condition; they are periodically characterized in the Solar Simulator with the measurement of the OCVD and the I/V-curve at Standard Reporting Conditions (AM 1.5, 1000 W/m^2, 25 °C). The OCVD is monitored with a Tektronix Digital Oscilloscope. For the theoretical back ground, analysis and the interpretation of the OCVD we refer to /1/. It is important to be aware that the lifetimes calculated with this model can only give an average value of the gross recombination process in the whole module.

3. RESULTS AND DISCUSSION

3.1. Comparison of indoor and outdoor performance evaluations

The performance evaluations outdoor are comparable to the simulator results (within an error of 10 %), under the following conditions. The module temperature for the outdoor measurement is adjusted by shadowing the module until 25 °C is reached. The I/V curve is driven immediately after uncovering the module. A calculation model for the Air Mass gives the daytime for AM 1.5 spectral distribution. For extrapolation of the outdoor performance of a module to 1000 W/m^2 it is necessary that the outdoor intensity is higher than 800 W/m^2.

3.2. Degradation of the I/V curve characteristics

The module types A, B, C and D have differences in bandgap, i-layer thickness, cell structure and the design of interconnection. According to this they have different initial efficiencies and degradation behavior. The results of the simulator and OCVD measurements are summarized in Table I.

In general we observed that the modules in OC-exposure degrade faster than the modules in SC-exposure. But after time specific for the module type the level of degradation becomes nearly equal. This can be explained by a material and cell structure dependent limit for the defect density which in turn determines the recombination process. For all modules the FF is the dominant degradation parameter.

As basis for the following discussion of the results we have verified the thickness of the intrinsic layer, the quantum efficiency and the optical bandgap by impedance and absolute spectral response measurements on the according solar cells. For type A, C and D cells we found approximately 500, 370 and 220 nm for the i-layer thicknesses and 1.70, 1.66 and 1.78 eV for the bandgaps.

3.2.1. <u>The Fill-Factor.</u> The change in the FF is a superposition of an increase of the serial resistance and a decrease in the carrier collection length l_c. The amount of contribution of the carrier collection length is dependent on the i-layer thickness d and his electrical field profile. For thin i-layers one can assume a constant field E ($l_c = \mu\tau E$) and the transport can be described by drift. If the i-layer is thick the field can become very low because of nonuniformities and thus the transport can also be diffusion dominated. Based on this explanation it is obvious that the strong degradation of the C type modules is dominated by this effect, because they have an i-layer thickness of about 500 nm. For Type A and D we found 370 nm and 220 nm. Type B are tandem cell modules and thus the i-layers are very thin. According to the lower thicknesses these modules are showing a smaller degradation of the FF (Fig. 1).

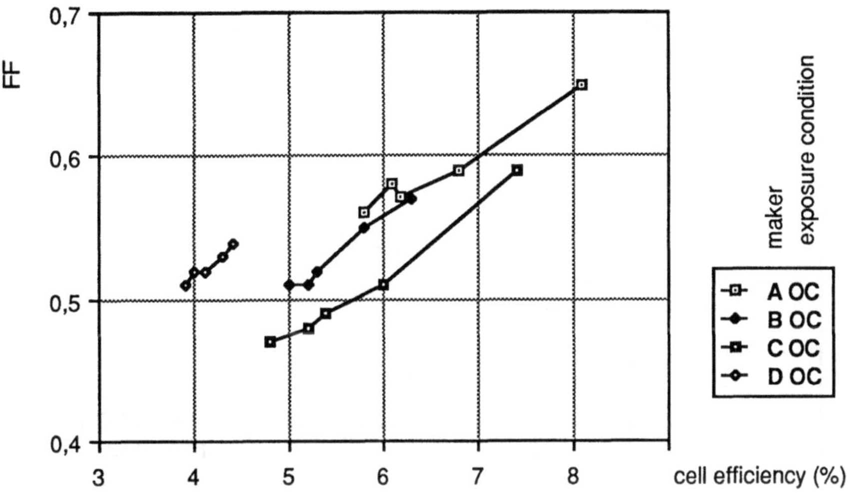

Fig. 1: The fill factor of the a-Si modules from maker A, B, C and D (in OC condition) versus their cell efficiency. Initial value and after 16, 57, 78 and 144 days of outdoor exposure.

3.2.2. The open-circuit voltage. The degradation behaviour can be expained with the correlation between the V_{oc} and the defect density. Poor quality material does not change much in the defect density and thus the V_{oc} is low and remains low (type B and D). High quality material has a high initial V_{oc} but degrades after light exposure due to an increase of defects centers as dangling bonds in the gap.

3.2.3. The short-circuit current density. The decrease of J_{sc} can be explained by an increase of the R_s and a decrease of the shunt resistance. Furthermore there is the contribution of the decrease of the minority carrier lifetime. Type C modules are showing a high initial J_{sc} due to the high absorption in the thick i-layer. For type D it is just the opposite (Fig. 2).

3.3. The OCVD technique.

This measurement technique was originally introduced for the evaluation of the minority carrier lifetime in germanium. Afterwards it was adopted for silicon devices. Because silicon has a intrinsic carrier density three orders of magnitude lower as germanium, one has to take care by the interpretation and calculation of the transient signal /2/. The space charge capacitance of the intrinsic layer can influence the decay. Furthermore a low shunt resistance of the solar cells can bend down the decay curve. In practice we have seen that the decay curve can be divided up into three linear decay regimes.

Following /1/ the first two linear decay regimes can be characterized with the lifetime τ_B due to band to band recombination and the lifetime τ_D due to defect center assisted recombination. The second decay regime is either bend up by the charging of the space charge capacitance or bend down by a low shunt resistance. Furthermore we have observed an additional third linear decay after due to the response time τ_R of the discharge of the space charge capacitance. The rusults are summarized in Table I and are showing that τ_B and τ_D are decreasing during the degradation and τ_R increases because of the increasing serial resistance.

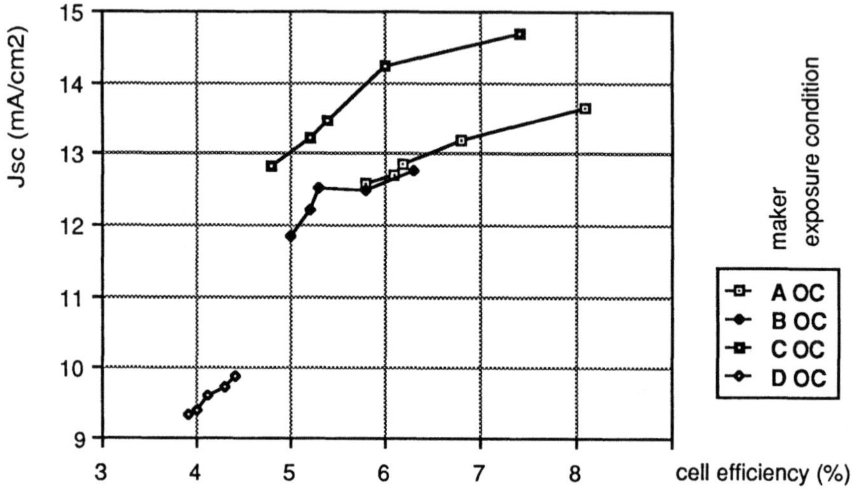

Fig. 2: The short-circuit current density of the a-Si modules from maker A, B, C and D (in OC exposure condition) versus their cell efficiency. Initial value and after 16, 57, 78 and 144 days of outdoor exposure. For comparance the values of the tandem cell module (type B) are multiplied by the factor 2.

4. CONCLUSIONS

The outdoor performance evaluations are comparable to the simulator results (within an error of 10 %), if they are performed at the same module temperature and at AM 1.5.

The dominant parameter in the degradation of a-Si modules is the decrease of the FF. This effect is very sensitive to the i-layer thickness and it is obvious that the efficiency of single cell modules is limited because of this drawback. It is possible to reduce the degradation significantly by the design of a tandem cell structure with low i-layer thicknesses. This reduces the collection losses due to a decrease of the carrier collection length. A further improvement would be the use of stacked cells with different bandgaps fitted to the spectral distribution of the sunlight.

For the analysis of the intrinsic origin of the degradation effect, the light- induced OCVD, performed with the Xenon flash of the solar simulator, is limited. The decay time of the pulse light, the space charge capacitance and the shunt resistance can influence the decay of the V_{oc}. We intend to change this method to eliminate these problems.

Table I. Results of the performance and OCVD measurements.

Maker	A		B		C		D		Exposure time (days)
Exposure Condition	OC	SC	OC	SC	OC	SC	OC	SC	
Voc*	0,954	0,982	0,983	0,983	0,984	0,983	0,998	0,987	16
	0,942	0,969	0,970	0,975	0,972	0,968	0,994	0,975	57
	0,896	0,963	0,967	0,971	0,966	0,955	0,990	0,969	78
	0,893	0,954	0,962	0,965	0,951	0,933	0,988	0,956	144
Jsc*	0,967	0,98	0,977	0,984	0,968	0,961	0,983	0,990	16
	0,942	0,949	0,981	0,961	0,917	0,912	0,971	0,971	57
	0,93	0,926	0,957	0,931	0,900	0,907	0,948	0,962	78
	0,929	0,939	0,926	0,931	0,874	0,854	0,945	0,930	144
Pel*	0,837	0,904	0,930	0,941	0,818	0,904	0,960	1,050	16
	0,767	0,822	0,850	0,894	0,728	0,790	0,918	1,010	57
	0,748	0,799	0,825	0,859	0,703	0,752	0,902	0,983	78
	0,716	0,746	0,792	0,844	0,654	0,648	0,884	0,875	144
FF*	0,908	0,938	0,965	0,982	0,864	0,965	0,981	1,078	16
	0,878	0,911	0,912	0,966	0,831	0,912	0,963	1,080	57
	0,892	0,892	0,895	0,946	0,814	0,877	0,963	1,059	78
	0,862	0,831	0,895	0,946	0,797	0,807	0,944	0,98	144
Rs*	1,355	1,133	1,116	1,034	1,155	1,144	1,129	0,850	16
	1,355	1,203	1,233	1,034	1,469	1,217	1,122	0,910	57
	1,319	1,168	1,290	1,205	1,391	1,290	0,999	1,039	78
	1,536	1,413	1,290	1,148	1,548	1,283	1,147	1,119	144
Tau B*	0,952	0,942	1,098	1,085	1,037	1	1,125	0,969	16
	0,865	0,75	1,059	1,073	1,012	1	1,104	0,992	57
	0,865	0,713	0,955	0,827	0,862	1,029	1,082	0,985	78
	0,724	0,723	0,763	0,79	0,975	0,965	1,017	0,954	144
Tau D*	0,834	0,988	0,814	0,85	0,762	1	0,906	1,024	16
	0,808	0,945	0,466	0,75	0,743	0,894	0,937	1,024	57
	0,746	0,976	0,339	0,7	0,714	0,883	0,937	0,976	78
	0,762	0,867	0,458	0,713	0,743	1,01	0,969	1,073	144
Tau R*	1,004	1,38	0,946	1,142	1,387	0,962	1,027	1,144	16
	1,127	1,444	1,263	1,362	1,452	0,913	1,171	1,595	57
	1,321	1,424	1,432	1,677	1,734	0,997	1,279	1,595	78
	1,101	1,2	1,338	1,514	1,638	0,916	1,114	1,595	144

There are four different types of modules, maker A, B, C and D. From each type one is exposed in open-circuit (OC) condition and one in short-circuit (SC) condition. All parameters are normalized to their initial values. Each column contains the parameters after 16, 57, 78 and 144 days of exposure.

5. ACKNOWLEDGEMENT

One of us (F.H.K.) is supported by the Commission of the European Communities.

5. REFERENCES

/1/ Longrigg, P., Berry, W.B., *19th IEEE PVSC*, 1488, (1987).

/2/ Lindholm F.A., et. al., *IEEE Trans. on Electron Devices* Vol. **ED-34**, No. 2, 277, (Feb. 1987).

MEASUREMENTS AND ANALYSIS OF FAILED MODULES IN A PV SYSTEM

U.Coscia,G.Massano,A.Sarno
E.N.E.A. C.RI.F.
P.O.B. 83 - 80055 PORTICI (NAPOLI) ITALY

ABSTRACT. We performed an analysis in illumination conditions to detect failed modules in a PV field consisting of parallel strings. The method is based on the estimation of the substring working voltage change when a failure occurs in the field. In this paper we apply the criterion in the case of not negligible mismatch in the current-voltage (I-V) characteristics.We have obtained on a small PV field a good agreement with the experimental results if the substring voltage variations,in absence of failures, are less than ten per cent of the substring open circuit voltage. Furthermore we suggest an alternative diagnostics to localize failed modules in dark conditions by forward bias of the strings. Finally we discuss the limits of the proposed methods.

1. INTRODUCTION.

The output power of a photovoltaic (PV) field can considerably decrease because of failed modules,therefore it is important to detect them for their replacement.

We consider the failures as modules in open circuit (o.c.) or in short circuit (s.c.) conditions.

We have suggested [1,2] a method to localize failed modules in a large PV field formed by parallel strings,in which the modules are equipped with by-pass diodes.

In this paper we apply the criterion in the case of not negligible mismatch in the (I-V) characteristics.

The proposed method [2] to carry out the diagnostics of the failures in illumination conditions (in daytime),consists in dividing a string of N modules in substrings of p modules and comparing the N/p substring voltage values.

In the case of modules having the same (I-V) characteristics, we expect [2] that the difference of voltage ΔVn (normalized to the o.c.voltage of the substring) between a substring without failures and the one containing j modules in o.c conditions is:

$$\Delta V_n = (j/N-j)\cdot(N/p)\cdot(V_n+V_d/V_{oc}) \qquad \forall\ j \ \varepsilon\ \{0,p\} \tag{1}$$

where Vd is the voltage drop of the by-pass diode corresponding to a failed module and Vn the voltage string normalized to the o.c.voltage.
Of course the formula (1) does apply in the s.c.condition by putting Vd=0.
In the case of mismatch in the module characteristics we evaluate ΔVn as follows.

We calculate the difference of voltage between the mean value of the substring voltages having high values and the one having the lowest value. Thus we normalize to the substring o.c. voltage.

We have verified [2] that for small variations in the module characteristics the relationship (1) is in good agreement with the experimental results and does not depend on the irradiance level,module temperature and string operating voltage.

In the case of large mismatch in module characteristics the analysis is more complicated and the method must be applied in a suitable way.

To clarify this point we have carried out measurements on a small PV array (about 280 Wp) formed by a string of 16 commercial modules series connected.

All modules are equipped with by-pass diodes and the array has been connected directly to a battery.

The data have been collected by a data acquisition system based on HP 310 computer and the global radiation measured by an EPPLEY precision pyranometer.

We have divided the string in 4 substrings (p=4) and measured the module mismatch effect on the substring voltages in a wide range of the irradiance level Φ (80÷1000 W/m^2) and module temperature T (25÷47 °C). Then we have analyzed at different string operating voltage the change of the substring voltages with the presence of one or two failed modules (o.c.or s.c. condition) in the same or in different substrings.

2.EXPERIMENTAL RESULTS AND DISCUSSION

Fig.1 shows,in absence of failures (j=0),the normalized maximum difference of voltage ΔVn as a function of the string normalized voltage V_n, for different irradiance levels and module temperatures.

We see ΔV_n,that for j=0 is the measure of the substring voltage mismatch, decreases for Vn ranging from 0.55 to 1 indipendently of the irradiance level.

Fig.2a shows how the normalized voltage of the substring can change in the presence of j=0,1,2 failures (o.c.modules) at $\Phi = 80$ W/m^2.

It can easily seen, for values of Vn ranging from 0 to 0.7,that ΔVn,in absence of failures is larger than the expected theoretical values calculated, for j=1, by (1). Furthermore for j=2 the experimental points are shifted with respect to the theoretical ones.This disagreement is due to the large initial voltage mismatch (ΔVn is about 0.2 for j=0).

On the other hand we obtain a good agreement between the experimental values and the calculated ones in both cases j=1 and j=2, at Φ=500 W/m^2 (see Fig.2b),in the range of Vn (from 0.45 to 0.7) where the initial voltage mismatch ΔVn assumes values lower than 0.1. We came to the same conclusions analyzing the results obtained for different irradiance levels and module temperatures.

In fig.3 we show the results in the case of a s.c. module in a substring for two irradiance levels (Φ = 260 W/m^2 and Φ = 1000 W/m^2) and in fig. 4 for two o.c. modules in two different substrings if Φ = 700 W/m^2 and Φ =1000 W/m^2. Also in these last two cases the conclusions do not change at all.

In practice the only limit of the method,for s.c. and o.c. failures as well, is due to the string operating voltage that must vary in the range

where the voltage mismatch, ΔVn ,in absence of failure, is less than ten per cent of the substring o.c.voltage.If we are not able to make measurements in the specified ranges we are forced to carry out the diagnostics in dark conditions (e.g. during the night) by forward bias the string and measuring the substring voltages Vf normalized to the voltage string Vb.
In fact,in absence of failures, the voltage drop in one substring is about p/N of the forward bias voltage (see fig.5).
On the countrary for one o.c. module,the voltage of the failed substring Vf, increases until 90 per cent of voltage Vb, because of the reverse bias of the failed module by-pass diode.
Furthermore fig.6 shows that a s.c. module causes a voltage decrease of the failed substring compared with the voltages of the unfailed ones.
Thus, by forward bias the string in the dark condition, we can easily recognize the failed substrings.
However by using the diagnostics in illumination conditions or the dark ones we are not able to localize failed modules in the case of the same number of failures in each substring. As a matter of fact,in this case, there is no difference between the substring voltage values in absence of voltage mismatch.
That's why we are forced to use a complementary effective diagnostics as for instance the periodic measurement of the open circuit voltage of the substrings.

3. CONCLUSIONS

We have proposed a method to localize failed modules in o.c. or in s.c. conditions,in the case of PV fields formed by parallel strings.
We have verified that this criterion is applicable if in absence of failures the substring voltage variations are lower than ten per cent of the substring open circuit voltage.
Furthermore we have verified that these constrains can be overcome by means of suitable dark measurements.
Finally a complementary diagnostics can be used to get further informations about an uniform distribution of the failures in the substrings.

ACKNOWLEDGEMENT

We would like to thank dr.F.Tavasso and dr.T.Contardi for useful discussions and suggestions.

REFERENCES

1) U.Coscia, A.Sarno,RT/FARE/87/9
2) U.Coscia, G.Massano, A.Sarno, Technical Digest of the International Photovoltaic Science and Engineering Conference, Nov.3-6 1987, Tokyo,Japan,pp.347-350.

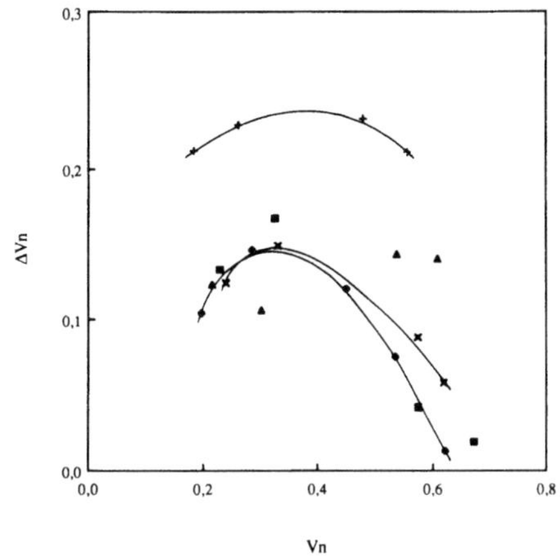

symbol	$\Phi (W/m^2)$	T(°C)
+	80	17
▲	260	23
◆	500	25
■	800	38
×	1000	46

Fig.1 - The normalized maximum voltage difference between the sub-strings, ΔVn, in absence of failures, as a function of the normalized string voltage, Vn, for different irradiance levels Φ and module temperatures T.

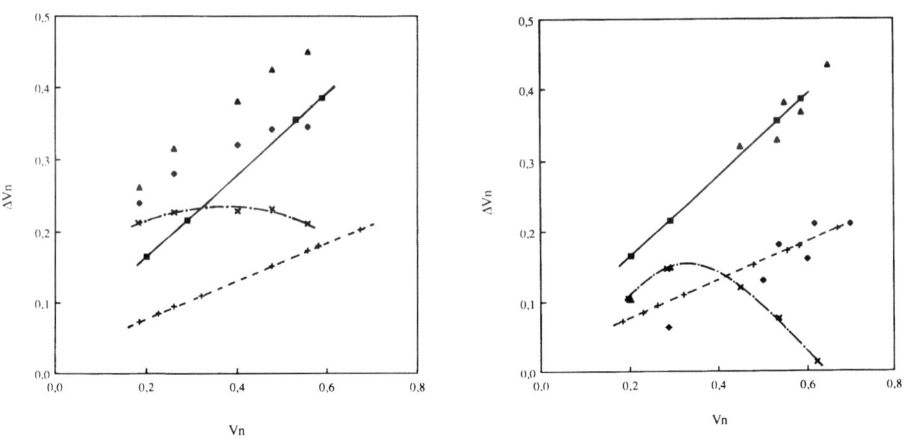

a) b)

Fig.2 - The normalized voltage difference, ΔVn, of the substring containing j=0,1,2 o.c.modules, as a function of the normalized string voltage, Vn, for two irradiance levels Φ and module temperatures T:

a) $\Phi = 80\ W/m^2$ $T = 22$ °C b) $\Phi = 500\ W/m^2$ $T = 25$ °C

Comparison between the theoretical and experimental data.

Theoretical		Experimental	
-----	j=1	×	j=0
———	j=2	◆	j=1
		▲	j=2

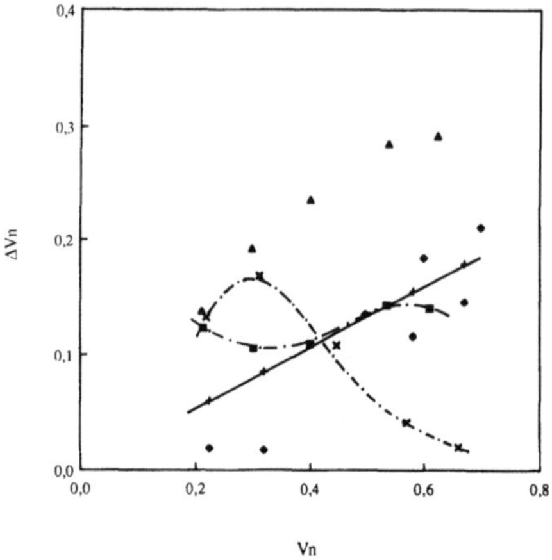

Fig.3 - The normalized voltage difference ΔVn of the substring containing j=0,1 short circuit module,as a function of the normalized string voltage Vn, for two irradiance levels Φ and module temperatures T. Experimental:
Φ=260 W/m^2 , T=23°C , (■ j=0, ▲ j=1)
Φ=800 W/m^2 , T=38°C , (✕ j=0, ◆ j=1)
Theoretical: ——— j=1

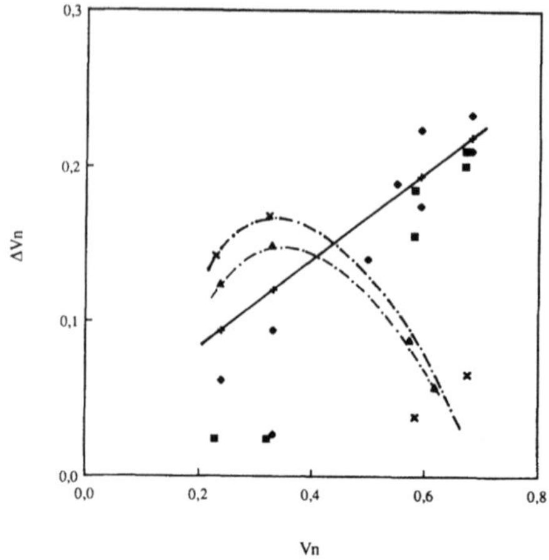

Fig.4 -The normalized voltage difference ΔVn of two different substrings,containing j=1 s.c. module each,as a function of the normalized string voltage Vn, for two irradiance levels Φ and module temperatures T. Experimental:
Φ=700 W/m^2, T=36°C, (✕ j=0, ■ j=1)
Φ=1000 W/m^2, T=46°C, (▲ j=0, ◆ j=1)
Theoretical: ——— j=1

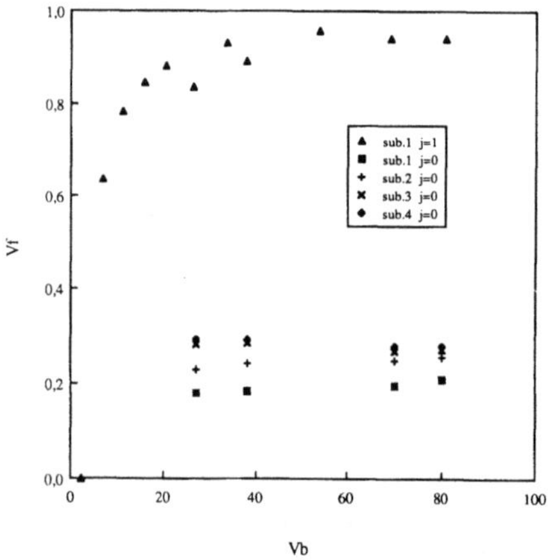

Fig.5 - The normalized voltage,Vf, of a substring (sub.1 j=1) containing one o.c.module in the dark condition, as a funtion of the string voltage Vb.Sub.1,sub.2,sub.3,sub.4 with j=0 denote the substrings before the failure.

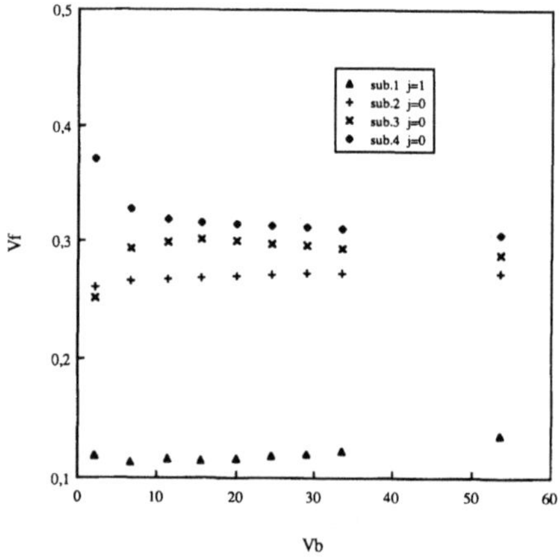

Fig.6 - The normalized voltage,Vf, for a s.c.module in substring 1(j=1) as a function of the string voltage Vb, in the dark condition.
Sub.2, sub.3, sub.4 with j=0 denote the other substrings without the failure.

ENERGY RATING OF PV MODULES BY OUTDOOR RESPONSE ANALYSIS

by

E. Gianoli–Rossi and K. Krebs

Commission of the European Communities
Joint Research Center, Ispra Establishment, Physics Division
21020 Ispra (Varese) Italy

ABSTRACT : A method for comparing PV modules, based on an energy response analysis of extended outdoor measurements, has been developed at CEC JRC Ispra. Measured data are fitted to a linear model to obtain energy output densities, energy efficiencies, and, under specific assumptions, energy costs for reference days or years. The measurement system, which is fully automatic, keeps the modules operating at maximum power. An in-plane pyranometer measures the incident irradiance. Eventual errors due to the different spectral responses of the pyranometer and of the test modules are checked by measuring irradiance values through the short-circuit currents of modules of the same type.

The measurement procedure, and the results obtained for 33 module types, are presented here. It is proposed to include this measurement as an option in the set of test procedures currently used in EUR–Specification 502.

1. INTRODUCTION

As energy conversion systems, photovoltaic devices should be rated according to their ability to convert/produce energy [1,2]. Energy production may be quantified by, e.g. the specific energy output density (or specific yield) ; the specific area needed to produce one kWh ; or a suitably defined energy efficiency. This paper describes the method for deriving such quantities from outdoor measurements, and its results for a series of modules of recent and/or current production.

The power output of a PV module is usually measured by a pulsed solar simulator under standard test conditions. Under actual outdoor conditions, power output is determined by the conversion of the varying incident irradiation, and by the temperature rise caused by illumination. The main advantage of an indoor method is the good reproducibility achievable in carefully executed measurements. The main disadvantages are that actual operating conditions are different from standard test conditions ; and that one wants to know the integrated energy response under changing conditions, and not the power output at a single instant.

2. THE ENERGY RATING (ENRA) MEASUREMENT

The method we discuss here aims to solve this problem in a simple and pragmatic way ; and to see what results are obtained with present-day modules. The method involves four steps :

1) Installation of test modules co-planar with a calibrated solar irradiation monitor, e.g. a pyranometer [1] ;

2) Recording the incident irradiation, ambient temperature, and energy output of the test modules – operating at maximum power – over a period of ten to thirty days ;
3) Reduction of measured data to a three-parameter response function using a statistical data analysis software package ;
4) Integration of the response function over a suitably chosen reference day, to obtain the energy rating. The reference day is defined by the function :

$$H(t) = \sum_{n=0}^{4} B_n t^n \ [\mathrm{W/m^2}] \ ;$$

where $B_0 =$ 2602.676742
$B_1 =$ -1399.250090
$B_2 =$ 243.511524
$B_3 =$ -15.395308
$B_4 =$ 0.319483
t = time in hours.

This function has a maximum of 900 W/m^2 at 12.00, and its integral has a value of 6.8188 kWh/m^2.d.

3. RESULTS AND DISCUSSION

Typical results from steps 2 and 3 are shown in Figs.1 and 2 (covering 33 module types, and measuring periods between July '86 and January '88). Each experimental curve shown in Fig.1 represents between 500 and 1700 energy values, obtained from 10 minute averages over 20 integration intervals of 30 seconds each.

The response functions, $\mathcal{R}(\varphi)$, shown in Fig.2, are of the type :

$$\mathcal{R}(\varphi) = A_1\varphi + A_2\varphi^2 + A_3\varphi \ln \varphi \ ;$$

where φ is the incident irradiation measured by the pyranometer.

This function was chosen, after several trials with other expressions (some explicitly containing instantaneous ambient temperature values), essentially for its shape, and not in order to relate the single A_i's to specific physical parameters of the individual modules. The fit was performed using the General Linear Model procedure of the SAS package ; the corresponding R-square values lie between 0.99 and 0.9995. We consider the $\mathcal{R}(\varphi)$ functions to be sufficient for the description of the energy behaviour of a module for most practical applications.

Fig.3 shows the minor influence of ambient temperature on the energy rating results of a specific module type which was mounted outdoors over the whole 18 month measuring period, to check the consistency of the measurement procedure. The small changes visible at high intensities are consistent with the drop in ambient temperature during the cold season. Except at low intensities, the temperature of a module depends mainly on irradiation. This explains why the output curves (Fig.1), averaged over several days, are practically only determined by intensity. On the other hand, it is interesting to note that the energy conversion efficiency itself is energy-dependent.

In Fig.4 we have plotted the ratios of the short-circuit current of reference modules to the corresponding pyranometer output. There is – apart from a greater scatter of points at small values – some indication of a spectral response mismatch. However, both effects (thermal and spectral) are negligible for the final energy rating, which involves integration over all incident energies. We therefore consider that a pyranometer is the appropriate irradiation sensor for this kind of measurement.

Using the above data and the corresponding response functions we have calculated a number of quantities which we consider to be useful for characterising the 'quality' of a photovoltaic module (see Table I). The values in the last five rows are based on the following assumptions :

Average daily incident energy $= 6.8\,\mathrm{kWh/m^2}$
Fixed charge rate $= 0.08$, $i = 5\%$, $n = 20\mathrm{y}$
Indirect cost factor $= 1.2$
Module cost $= 500\,\mathrm{ECU/m^2}$
Area-related balance-of-system cost $= 250\,\mathrm{ECU/m^2}$
Power-related balance-of-system cost $= 1500\,\mathrm{ECU/m^2.kW}$
Operational & maintenance cost $= 2.5\,\mathrm{ECU/m^2.year}$

Fig.5 shows some characteristic energy quantities ; Fig.6 gives an overview of certain cost items. Note that these values are valid for the above assumptions ; further conclusions should therefore be drawn with care. However, we believe that these results are quite useful for the assessment of modules by potential users. We therefore intend to add this procedure to the current EUR–specifications for photovoltaic module testing.

4. REFERENCES

[1] E.Gianoli-Rossi and K.Krebs, Proc. 7th EC Photovoltaic Solar Energy Conf., p.78 (1986).
[2] C.F.Gay, Proc. 4th EC Photovoltaic Solar Energy Conf., p.353 (1982).

SET ID	1/P1	1/P2	1/P3	1/P4	2/P1	2/P2	2/P3	2/P4	3/P1	3/P2	3/P3	3/P4
Module Code	WA	OI	IC	AB1	FF	GA1	OE	AB2	F	AIH	KA	AB3
Type	PWP402	BPA1215	BOS401	M53	PQ10	PSA100	BPA1235	M53	PQ10	G100	IS050	M53
Manufacturer	PW	BP	PT	AS	AEG	KYO	BP	AS	AEG	AS	ISOF	AS
TM [deg C]	25.6	25.6	25.6	25.6	27.7	27.7	27.7	27.7	21.7	21.7	21.7	21.7
Amod [m^2]	0.46	0.50	0.50	0.37	0.50	0.43	0.44	0.37	0.52	0.22	0.46	0.37
Ppower-rat [W/mod]	36.19	42.16	36.85	41.81	38.86	42.70	31.81	41.81	37.48	10.36	28.07	41.81
Penergyrat [W/mod]	29.93	36.75	29.33	34.26	32.18	36.68	29.35	34.30	27.71	7.59	26.87	34.52
ETA (STC)	7.80	8.50	7.40	11.30	7.85	10.00	7.23	11.30	7.25	4.71	6.05	11.30
ETAe (M3P,TM)	6.45	7.41	5.89	9.26	6.50	8.59	6.67	9.27	5.36	3.45	5.79	9.33
EMOD [kWh/mod.std]	0.20	0.25	0.20	0.23	0.22	0.25	0.20	0.23	0.19	0.05	0.18	0.24
YIELD [kWh/kW.std]	5.64	5.94	5.43	5.59	5.65	5.86	6.29	5.59	5.04	4.99	6.53	5.63
SYMD [kWh/m^2.std]	0.44	0.51	0.40	0.63	0.44	0.59	0.45	0.63	0.37	0.24	0.39	0.64
SEAM [m^2/kWh/std]	2.27	1.98	2.49	1.58	2.26	1.71	2.20	1.58	2.74	4.25	2.53	1.57
SAEO [kWh/m^2.y]	144.48	165.98	131.93	207.42	145.60	192.41	149.41	207.65	120.06	77.28	129.69	208.99
ACA [ECU]	33.41	35.71	35.86	26.64	35.64	30.74	31.68	26.64	37.22	15.84	33.41	26.64
ACP [ECU]	5.21	6.07	5.31	6.02	5.60	6.15	4.58	6.02	5.40	1.49	4.04	6.02
UEC [ECU/kWh]	0.59	0.52	0.65	0.44	0.59	0.46	0.57	0.44	0.71	1.05	0.64	0.43

SET ID	4/P1	4/P2	4/P3	4/P4	5/P1	5/P2	5/P3	5/P4	6/P1	6/P2	6/P3	6/P4
Module Code	V	OL	GA2	AB4	UD	UC	RB	AB5	HH	GB	JB	AB6
Type	33W	PROT	PSA100	M53	PROT	PROT	RSM5	M53	36MS	LC361	AP50	M53
Manufacturer	AUD	BP	KYO	M53	BMC	BMC	RS	AS	PRAG	KYO	AMS	AS
TM [deg C]	6.6	6.6	6.6	6.6	11.0	11.0	11.0	11.0	21.6	21.6	21.6	21.6
Amod [m^2]	0.43	0.44	0.43	0.37	0.48	0.55	0.26	0.37	0.42	0.36	0.45	0.37
Ppower-rat [W/mod]	25.47	42.68	42.70	41.81	42.21	52.46	17.02	41.81	43.04	37.59	47.28	41.81
Penergyrat [W/mod]	25.56	35.24	37.75	35.00	32.41	39.22	13.83	34.11	35.24	31.68	39.07	34.93
ETA (STC)	5.98	9.70	10.00	11.30	8.83	9.47	6.57	11.30	10.20	10.50	10.60	11.30
ETAe (M3P,TM)	6.00	8.01	8.84	9.46	6.78	7.08	5.34	9.22	8.35	8.85	8.76	9.44
EMOD [kWh/mod.std]	0.17	0.24	0.26	0.24	0.22	0.27	0.09	0.23	0.24	0.22	0.27	0.24
YIELD [kWh/kW.std]	6.84	5.63	6.03	5.71	5.24	5.10	5.54	5.56	5.58	5.75	5.64	5.70
SYMD [kWh/m^2.std]	0.41	0.55	0.60	0.65	0.46	0.48	0.36	0.63	0.57	0.60	0.60	0.64
SEAM [m^2/kWh/std]	2.44	1.83	1.66	1.55	2.16	2.07	2.75	1.59	1.76	1.66	1.67	1.55
SAEO [kWh/m^2.y]	134.40	179.42	198.01	211.90	151.87	158.59	119.61	206.53	187.04	198.24	196.22	211.45
ACA [ECU]	30.67	31.68	30.74	26.64	34.42	39.89	18.65	26.64	30.38	25.78	32.11	26.64
ACP [ECU]	3.67	6.15	6.15	6.02	6.08	7.55	2.45	6.02	6.20	5.41	6.81	6.02
UEC [ECU/kWh]	0.62	0.49	0.45	0.43	0.57	0.56	0.70	0.44	0.48	0.45	0.46	0.43

SET ID	7/P1	7/P2	7/P3	7/P4	8/P1	8/P2	8/P3	8/P4	9/P1	9/P2	9/P3	9/P4
Module Code	YB	YC	AC	AB7	SC	GC	A2I	AB8	LE	SD	HL	AB9
Type	H40	H50	M55	M53	S152	LA341	G4000	M53	SM50	MSX60	36MSC	M53
Manufacturer	HEL	HEL	AS	AS	SREX	KYO	AS	AS	SIE	SREX	PRAG	AS
TM [deg C]	25.1	25.1	25.1	25.1	25.2	25.2	25.2	25.2	9.3	9.3	9.3	9.3
Amod [m^2]	0.42	0.45	0.43	0.37	0.48	0.44	0.50	0.37	0.45	0.56	0.42	0.37
Ppower-rat [W/mod]	36.12	42.41	53.25	41.81	51.78	45.99	32.67	41.81	49.16	58.28	41.36	41.81
Penergyrat [W/mod]	30.48	36.94	44.39	34.85	38.95	39.07	25.15	34.85	41.85	50.17	34.98	36.48
ETA (STC)	8.58	9.53	12.50	11.30	10.90	10.50	6.60	11.30	10.90	10.50	9.80	11.30
ETAe (M3P,TM)	7.24	8.30	10.42	9.42	8.20	8.92	5.08	9.42	9.28	9.04	8.29	9.86
EMOD [kWh/mod.std]	0.21	0.25	0.30	0.24	0.27	0.27	0.17	0.24	0.29	0.34	0.24	0.25
YIELD [kWh/kW.std]	5.75	5.94	5.68	5.68	5.13	5.79	5.25	5.68	5.81	5.87	5.77	5.95
SYMD [kWh/m^2.std]	0.49	0.57	0.71	0.64	0.56	0.61	0.35	0.64	0.63	0.62	0.56	0.67
SEAM [m^2/kWh/std]	2.03	1.77	1.41	1.56	1.79	1.64	2.89	1.56	1.58	1.62	1.77	1.49
SAEO [kWh/m^2.y]	162.17	185.92	233.41	211.01	183.68	199.81	113.79	211.01	207.87	202.49	185.69	220.86
ACA [ECU]	30.31	32.04	30.67	26.64	34.20	31.54	35.64	26.64	32.47	39.96	30.38	26.64
ACP [ECU]	5.20	6.11	7.67	6.02	7.46	6.62	4.70	6.02	7.08	8.39	5.96	6.02
UEC [ECU/kWh]	0.54	0.47	0.40	0.43	0.49	0.45	0.74	0.43	0.43	0.44	0.48	0.41

ETAe = energy efficiency of module
SYMD = specific yield of module
SEAM = specific energy area of module
SAEO = specific annual energy output (UF=0.9)
YIELD = daily energy per peak kW of module

TAM = mean ambient daylight temperature
ACA = annual area-rel. cap. charge
ACP = annual power-rel. cap. charge
UEC = unit energy cost

TABLE I: Characteristic properties for nine sets of four module types each

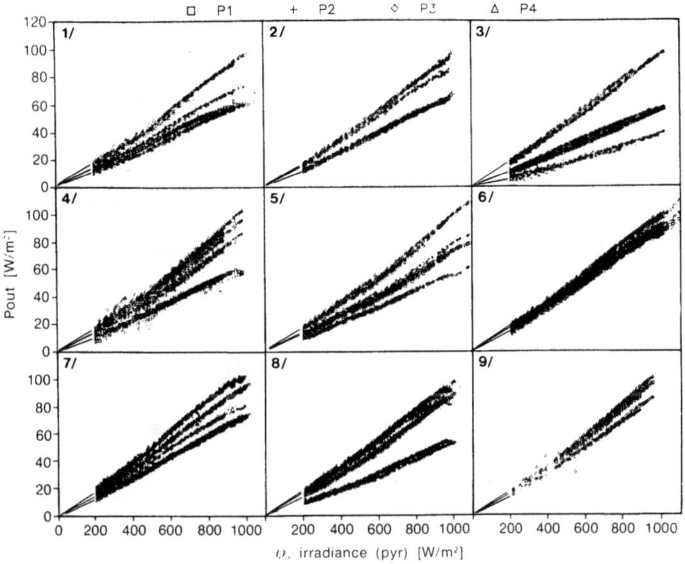

Fig. 1: Experimental data of E_{out} vsE_{in} for 33 module types (each curve consists of between 500 and 1700 single measurements)

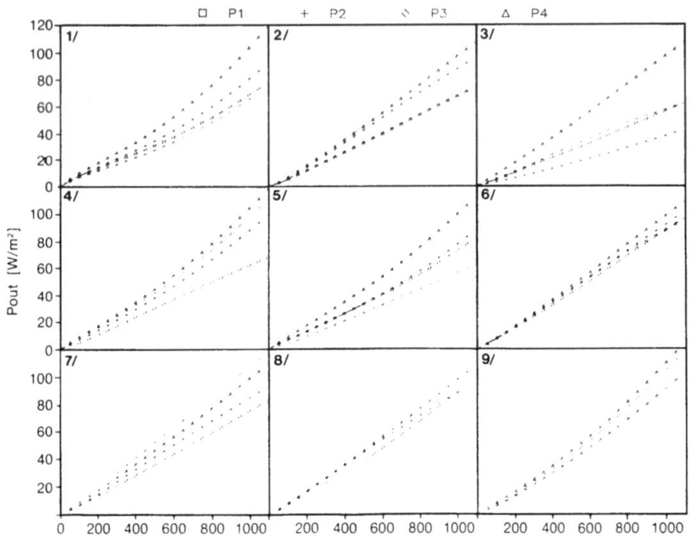

Fig. 2: Response curves obtained using the GLM procedure of the SAS package

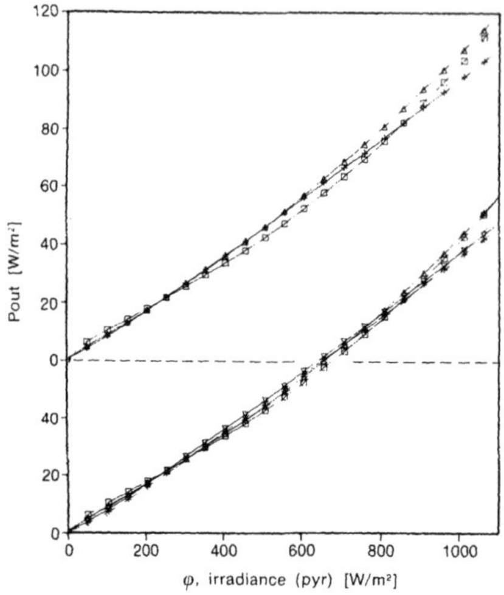

Fig. 3: Comparison of results for one type of module measured over nine periods between July '86 and January '88

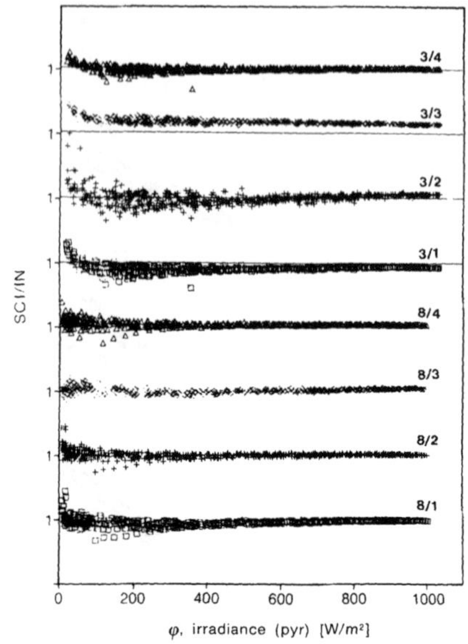

Fig. 4: Examples of the ratio between short-circuit current and pyranometer output as a function of the incident irradiation (measured by the pyranometer)

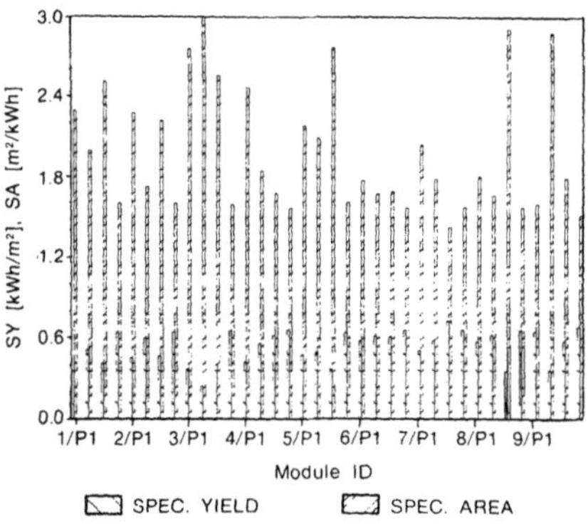

Fig. 5: Energy densities and efficiencies for 33 module types

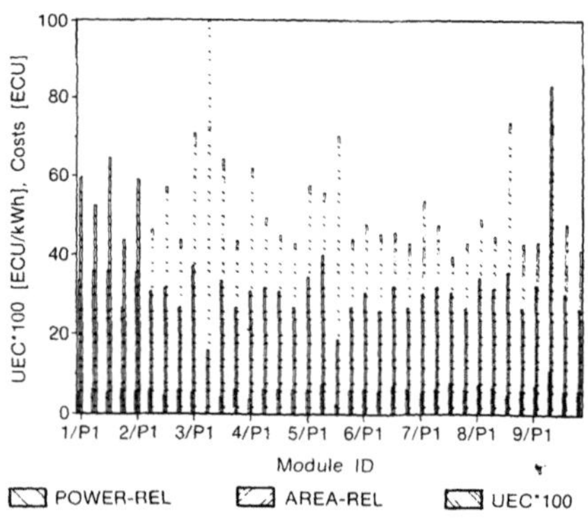

Fig. 6: Economic parameter for 33 module types

Spectroradiometer Measurements and Modelling of Solar Spectra

by

Dr. Gerhard Riesch

Commission of the European Communities
Joint Research Center, Ispra Establishment, Physics Division
21020 Ispra (Varese) Italy, Tel. **39-332-789646

Abstract : The use of a rugged, field-portable, single-channel spectroradiometer for measurements of sun and lamp spectra is described, emphasising the importance of accurate calibration of the instrument. The measured spectra are used for two purposes :

1) to compare existing, and to develop new models for the solar irradiation under different weather conditions, which serve to build up a catalogue of irradiance data for assessing the output of PV modules ;
2) to translate PV cell and module calibrations to standard reporting conditions (compensation for spectral mismatch).

Introduction

All photovoltaic devices have wavelength-dependent quantum efficiencies, i.e. different yields of charge carriers (spectral response) for different wavelengths of the incident sunlight. Spectroradiometric measurements are therefore necessary for the calibration of PV devices, and for the assessment of their output in varying sunlight.

Calibration of PV cells and modules : Exact and comparable calibrations of photovoltaic devices require a light source whose spectrum is known. These spectral irradiance data, together with the known relative spectral response of the device, allow the calibration data to be translated to standard reporting conditions (correction of *spectral mismatch*).

Assessment of output : The aim of all solar devices is to convert outdoor sunlight directly into electrical energy. To accurately predict the electrical energy obtainable at a given site, the available sunlight at the site must be known for all times and with more details than those given by the global irradiance. Usually only global irradiance (and sometimes direct irradiance) data are available ; and for most of the thermal uses of sunlight they are sufficient. PV devices, however, are particularly sensitive to the spectral composition of the incident sunlight, and hence spectral irradiance measurements are necessary for accurate estimates.

Intensity levels and spectral distributions of solar radiation vary over rather wide ranges, depending on parameters such as : geographical location and time of day (essentially parametrized by the relative air-mass) ; and atmospheric conditions (parametrized by humidity, ozone content, turbidity, and cloud cover).

Annual, complete spectral data of solar irradiance (say every 10 minutes with spectral band-widths of 5 to 10 nm) for a given site are not available. If they were available it would be impractical to store and analyse such a large amount of data.

Hence it is necessary to describe solar spectral irradiance in sufficient detail using simple models containing only a few parameters.

1. Description of the Spectroradiometer

The apparatus (see Fig.1) is a battery operated, computer controlled spectroradiometer (manufactured by GER, New York) covering the wavelength range 300 – 3000 nm, housed in two portable cases. The optical case contains the input optics ; the three gratings, driven by a stepper motor ; the order sorting filters ; and the detectors. The optical case weighs 7.3 kg (including the integrating sphere).

The electronics case contains a microcomputer ; the amplifiers ; a cassette tape recorder ; a strip chart recorder ; and two batteries (6 V, 7.5 Ah and 12 V 4.5 Ah). This case weighs 12 kg, including 3.5 kg of batteries.

Sunlight is measured on the inner surface of an integrating sphere 220 mm in diameter, coated with Kodak white reflectance coating.

The system is operated by a CMOS microcomputer, programmed to make measurements in one continuous scan through the whole wavelength range ; digitize the data ; store these raw data in CPU memory ; and output the raw data to the magnetic cassette tape recorder for future processing with a PC. Scanning the complete wavelength range from 261 to 3733 nm takes about 8 minutes in bright sunlight.

Very careful calibrations are necessary to obtain reliable data with the spectroradiometer. Four different calibration measurements are necessary to create the calibration files used for processing the data in the computer :

1. Background calibration. The background signals (noise) are obtained from dark current measurements ; these values are then subtracted from the signals of the measurement runs. Note that this procedure only accounts for electrical noise. There may be stray light in the optical pathways, giving rise to a supplementary background signal, in certain wavelength ranges, which is not detected by this background calibration.

2. Gain calibration. The gain changes (there are four, automatically switched amplification levels) are determined from measurements made at different amplifier gains, by comparing the signals before and after a gain change.

3. Wavelength calibration. A data channel number is assigned for each grating position (set by the stepper motor). Up to 1150 channels are available. Measurements are made with krypton gas discharge lamps, whose peak wavelengths are obtained from the literature. These absolute wavelength values are correlated to the channel numbers where the peaks are measured. The correlation is interpolated between these values.

4. Absolute irradiance calibration. The instrument is calibrated in absolute spectral irradiance values (e.g. $W/\mu m^2$) by measurement of a known irradiance. Such calibrations are made using a 1000 Watt (EG&G) spectral irradiance lamp (secondary standard with reported irradiance data). To increase precision, several (5 or more) calibration runs are necessary. The calibration file used for calculating the absolute values of the measured irradiance is created by averaging over these calibration runs.

2. Measurements and Analysis of Solar Spectra

Many measurements of direct and global solar spectra were made under different weather conditions. Only one typical example can be presented here. Fig.2 shows an example of the direct spectrum in the 400 – 2500 nm wavelength range, taken on a very clear, dry day at airmass 1.53.

Several models describing terrestrial solar spectra are reported in the literature [1–5]. The simple model described in [2], based mainly on [6], is the most accessible, and has been used for first comparisons with the measured data.

The calculation models use the extraterrestrial solar spectrum and formulae for extinction by the following five atmospheric attenuation effects : 1) Raleigh scattering ; 2) ozone absorption ; 3) mixed gases extinction ; 4) water vapour absorption ; 5) aerosol extinction.

A PC program has been written to plot the measured spectra on the display screen in selectable wavelength ranges (thus providing sufficient resolution), along with the calculated spectra.

The parameters of the calculation can be easily changed during the session, so that the effects of their variations on the spectra can be seen directly on the display.

The available models give satisfactory results only for atmospheric extinctions 1) to 4) above. The aerosol extinction probably comprises other absorption effects and therefore requires further study. This process characterises the envelope of the spectra and hence strongly influences the overall shape of the terrestrial spectra and the absolute value of the irradiance integral over the whole range.

The aerosol extinction is normally measured at 500 nm and the resulting parameter, $\tau_{0.5}$, is used to model the spectrum up to the infrared region (800 – 3000 nm). In this calculation, another parameter, the Ångstrom α, is used to describe the wavelength dependence of the aerosol extinction.

Preliminary analysis shows that it is probably not possible to describe aerosol extinction in the infrared region using the same τ and α values as used in the visible region. Many more spectral measurements, especially in the infrared region, and under different weather conditions, are therefore necessary to evaluate the correct modelling of the extinction effects due to aerosols.

3. Calibration of PV Cells and Translation to Reporting Conditions

PV cell or module calibrations made with light sources of any spectral composition must be translated to standard reporting conditions – for instance, to a standard AM-1.5 spectrum – to be comparable and generally usable. Translation of the calibration requires the following information :

1) the absolute spectral irradiance of the light source at the cell or module position – measureable by the IRIS spectroradiometer ;
2) the relative spectral response of the cell– obtained by independent measurements ;
3) the short circuit current.

The short circuit current must be stable during the spectral measurements, that is for about 6 minutes (or average values can be taken for slowly varying sunlight). The translation is calculated as follows.

Each of the approx. 350 IRIS measured spectral irradiance values is multiplied by the value of the relative spectral response at the corresponding wavelength. These products are then integrated over the relevant wavelength range.

The ratio of the measured short circuit current and the value of this integral is the calibration constant for the particular cell or module. The product of this calibration constant with an analogous integral – calculated for the given standard spectrum – gives the translated, normalised, and hence comparable short circuit current.

If calibrated standard cells are used for the calibration of modules, the IRIS measurements can also be used to determine the spectral mismatch correction which is applied in this alternative interpretation of the same effect.

Literature

[1] R.Guzzi et al : *Experimental validation of a spectral direct solar radiation model.* Solar Energy, **31**, pp359–363 (1983).

[2] R.Bird et al : *Simple solar spectral model for direct and diffuse irradiance on horizontal and tilted planes at the earth's surface for cloudless atmospheres.* SERI–TR–215–2436, Solar Energy Research Institute, Golden, Colorado (1984).

[3] V.E.Cachorro et al : *Comparison between various models of solar spectral irradiance and experimental data.* Appl.Optics, **24**, p3249 (1985).

[4] E.P.Shettle, R.W.Fenn : *Models for the aerosols for the lower atmosphere and the effects of humidity variations on their optical properties.* AFGL–TR–79–0214, Environ.Res.Paper No.676,94 (1979) ; Airforce Geophysics Laboratory, Bedford, Mass.

[5] F.X.Kneizys, E.P.Shettle et al : *Atmospheric transmittance/radiance computer code LOWTRAN 5.* AFGL–TR–80–0067 (1980).

[6] B.Leckner : *The spectral distribution of solar radiation at the earth's surface – Elements of a model.* Solar Energy, **20**, pp143–150 (1978).

Figure 1. The IRIS spectroradiometer

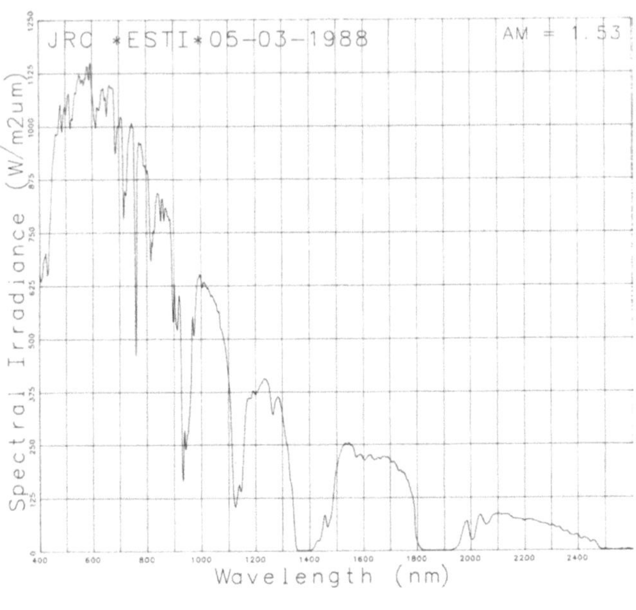

Figure 2. A typical direct irradiance solar spectrum

INTRINSIC PARAMETERS EVALUATION OF SILICON SOLAR CELLS BY A DETAILED ANALYSIS OF EXPERIMENTAL DATA WITH TWO DIFFERENT COMPUTER MODELING

R. Van Steenwinkel,
CEC-JRC 21020 ISPRA (VA), Italy, Tel 0332/780131
E. Vittone, T.A. Shamsi,
TASK R&S, via Verbano 2, ISPRA (VA), Italy, Tel 0332/781771
C. Manfredotti
Ist. Fisica Superiore, Univ. di Torino, Tel. 011/6527306
K. McAllister
University of Strathclyde, Glasgow

ABSTRACT. The objective of this work is to characterize silicon solar cells using three experimental methods and two different modeling approaches. The dark I-V characteristics, the transient behaviour and the spectral responsivity have been measured on all the samples.
A Monte Carlo and a more conventional numerical method have been used to analyze the experimental data. The lifetime and the diffusion length of the minority carriers in the base have been obtained from the opto-electronic measurements. The spectral responsivity of the devices has been successfully simulated by the Monte Carlo method in the whole wavelength range. Moreover, all the parameters involved in the dark current-voltage characteristic have been determined giving an extremely good correspondence to the experimental behaviour.

1. INTRODUCTION

The quality of n-p solar cells depends largely on the lifetime of the carriers generated by the incident light as well as on the barrier and surface properties.

In this paper we have used the spectral responsivity (S.R.) as well as the photocurrent (PCD) and photovoltage decays for the evaluation of the minority carrier lifetime in the base of three different, commercially available solar cells. The results concerning S.R. and PCD are analyzed both by conventional numerical methods [1,2,3] and by a Monte Carlo approach [4,5] which has proven to be especially effective in the analysis of the short wavelength part of the spectral response.

Moreover, a very good evaluation of the phenomenological parameters involved in the dark current-voltage characteristics was performed.

2. MATERIAL AND METHOD

2.1. Solar Cells Description

All the investigated cells are monocrystalline with an area of 4 cm^2. Cell #1 has an emitter sheet resistivity of 35-55 Ω/Sq, a junction depth of 0.4 μm, a base resistivity of 1 - 2 Ω·cm and a thickness of 370 μm with an ohmic back contact. Cells #2 and #3 have an emitter sheet resistivity of 50 Ω/Sq, a junction depth of 0.4 μm, a base resistivity of 5 Ω·cm, a thickness of 390 μm and a BSF. The front reflection of cell #1 was taken as 12 % between 400 and 1100 nm, as measured between 700 and 800 nm. For cells #2 and #3, the average front reflection was measured between 400 and 800 nm.

2.2 Experimental Set-up

The Spectral Response of the cells were measured as described in [6].

For the PVD and PCD experiments, light pulses with a duration of 10 ns and wavelength of 1060 nm were generated by a Nd-Yag laser. The laser beam was scattered uniformly over the test cell area with a set of diffusing plates. In the PVD case, the measurements were carried out using a white bias light giving a constant voltage of 0.5 Volts on the junction [7].

2.3 Analysis of the opto-electronic behaviour

The opto-electronic behaviour of the investigated cells has been simulated by the one dimensional Monte Carlo computer code "SILCE" running on a Micro VAX II at JRC of Ispra [4,5]. The results of the simulation concern both the spectral response and the photo-current transient for the whole wavelength spectra.

On the other hand, the experimental data have been analyzed using standard analytical formulas [1] and numerical methods [2,3,8]. In particular, the determination of the diffusion length was achieved with an analysis of the Spectral Response at infrared wavelengths [8], and the minority carrier lifetime in the base have been calculated from the PVD and PCD rates as described in [2,3]. In all these numerical analysis, the back recombination velocity was taken as infinite for cell #1 and as 0 for the other cells with BSF.

3. RESULTS AND DISCUSSION

3.1 Spectral Responsivity

Following the method summarized in [8], a linear regression analysis of five spectral resonse measurements between 800 and 1000 nm has been performed using the following relation :

$$(i) \qquad 1/\alpha = (F/\eta)\cdot L - L$$

where α is the optical absorption coefficient, L the diffusion length (that is obtained from the slope and from the intercept of (i)), η is the internal quantum efficiency and F is a factor whose expression can be found in [8]. The results are summarized in Table I. The experimental Spectral Response of cell #1 is drawed in Figure 1 together with the Monte Carlo and numerical simulations. For the latter, only the base contribution was considered limiting the validity of the calculations to higher wavelengths.

3.2 Photo-Voltage and Photo-Current Decay

In Table II the results of lifetime determination of the minority carrier in the base by PVD and PCD are collected. The base lifetimes have been calculated from the experimental effective ones using the relations (16a) and (16b)of [2].

The diffusion lengths corresponding to the above mentioned measured lifetimes are also included in Table I.

Figure 2 shows the PCD and PVD calculated numerically together with the PCD obtained with the Monte Carlo code for cell #1 at 1060 nm (α=10 cm^{-1}). The reflectance on the back contact was assumed to be 90 %.

3.3 Dark Current-Voltage Characteristic

The dark I-V characteristic of the three cells have been fitted with the following equation using the least square principle :

(iii) $$J_o = J_{o1} \cdot [\exp(q \cdot (V - J_o \cdot R_s)/kT) - 1] + J_{or} \cdot [\exp(q(V - J_o \cdot R_s)/nkT) - 1] + (V - J_o \cdot R_s)/R_{sh}$$

the parameters involved ($J_{o1}, J_{or}, n, R_s, R_{sh}$) are summarized in Table III.

4. CONCLUSIONS

A good agreement between the transient and the Spectral Response method has been obtained for the determination of the diffusion length if the absorption coefficients of [9] are considered. The calculation of the diffusion lengths from the Spectral Response at long wavelength has several advantages over the PVD and PCD methods at 1060 nm thanks to its weak dependence on parameters like the back surface recombination velocity S and diffusivity D, also if it is very sensitive to the absorption coefficient vs wavelength relation. As seen in Table I, there are large discrepancies between the diffusion length values obtained using relations from [9] or from [10].

The Monte Carlo approach shows real advantages with respect to numerical methods especially for the analysis of the short wavelength region of the Spectral Responsivity where the contribution of the emitter is important: the emitter properties that are difficult to handle in numerical modeling (high and non uniform doping and electrical field, variation of the lifetime and mobility with the doping concentration) are easily integrated in a Monte Carlo approach.

In the case of the simulation of the photo-current decay at long wavelengths where the contribution of the base is dominant, both methods are comparable.

Excellent fit to the forward dark current characteristic were obtained allowing a reliable determination of saturation current, recombination current, non ideality factor, series and shunt resistance of the investigated cells. The product of the non ideality factor times temperature is shown in the last column of Table III. As can be seen, this product is almost temperature independent and has a value of 800 K. The meaning of this fact is not clear yet but was already reported for high efficiency cells [8].

5. REFERENCES

[1] H.J. Hovel : Semiconductors and Semimetals Vol.11, Ac. Press (1975)

[2] B.H. Rose, H.T. Weaver : J. Appl. Phys. **54**,233 (1983)

[3] S.C. Jain and U.C. Ray : J. Appl. Phys. **54**,2079 (1983)

[4] S. Bosso, T. Shamsi, K. Krebs, C. Manfredotti : Proc. 18th PVSC (1985)

[5] E. Vittone, et al. :Proc. 7th PVSEC Conference, Sevilla (1986) p.1005

[6] R. Van Steenwinkel :Proc. 7th PVSEC Conference, Sevilla (1986) p.325

[7] A.R. Moore : RCA Review,**40**,549 (1980)

[8] M. A. Green, et al : J. Appl. Phys. **58**,4402 (1985)

[9] R.T. Swimm, K.A. Dumas : J. Appl. Phys. **53**,7502 (1982)

[10] A. Shumka : 8th PV Spec. Conf. NY p.96

[11] S.M. Sze : Fisica dei dispositivi a semiconduttori Tamburini (1973)

CELL #	T (°C)	L [2] Interc. (μm)	L [2] Slope (μm)	L [3] Interc. (μm)	L [3] Slope (μm)	L[PVD] (μm)	L[PCD] (μm)
1	20	207	199	142	142	210	170
1	40	226	216	157	153	241	-
1	60	251	237	177	170	244	-
1	80	269	254	191	183	274	-
2	23	148	135	99	94	145	140
3	24	189	158	132	113	186	170

TABLE I : Diffusion length as determined by the spectral responsivity and using the Swimm-Dumas [9] and Shumka [10] absorption coefficient vs wavelength relation. The relation between mobility and doping concentration was taken from Sze and Irvin [11]. The temperature dependence of D was assumed to be proportional to $T^{-1/2}$ corresponding to a carrier mobility dominated by acoustic phonons [11].

CELL #	T (°C)	Effective Lifetime (PVD) (μs)	Lifetime (PVD) (μs)	Effective Lifetime (PCD) (μs)	Lifetime (PCD) (μs)	Effective Lifetime (M.C.) (μs)	Lifetime (M.C.) (μs)
1	13	8.10	13.30	-	-	-	-
1	23	9.50	17.20	3.94	10.40	3.15	14.1
1	35	10.80	21.50	-	-	-	-
1	49	10.80	21.10	-	-	-	-
1	63	11.50	23.40	-	-	-	-
1	80	13.00	29.70	-	-	-	-
2	25	7.67	7.67	5.43	7.17	2.87	3.43
3	25	12.58	12.58	7.17	10.54	3.93	5.07

TABLE II : Summary of decay constants for open circuit voltage and short circuit current transients and correlated base lifetimes

CELL #	T (°C)	J_{o1} Measured (pA)	J_{or} (μA)	n	R_s (Ω)	R_{sh} (Ω)	Average Difference (%)	n·T (°K)
1	20	2.44	1.53	2.77	0.257	4074.	1.60	812.0
1	40	44.20	2.27	2.52	0.260	3675.	1.62	789.1
1	60	605.00	4.73	2.41	0.267	3647.	0.96	802.9
1	80	5037.00	8.94	2.29	0.273	2727.	0.84	808.7
2	23	2.24	1.80	2.62	0.101	3934.	1.52	775.9
3	24	1.80	0.96	2.60	0.230	3496.	1.59	772.6

TABLE III : Parametric values required to fit the experimental dark current-voltage characteristics with the equation (iii) of the text.

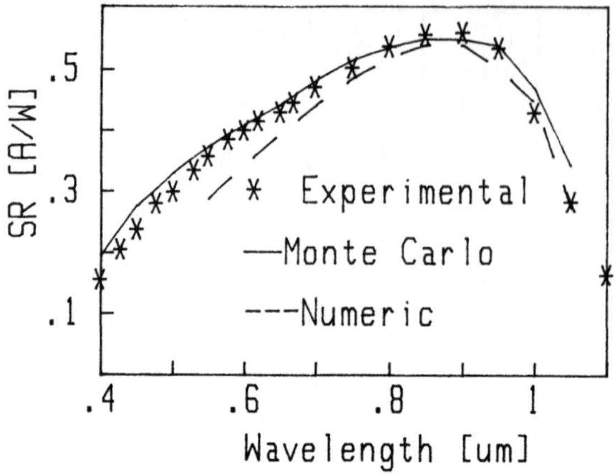

FIGURE 1
Spectral responsivity of cell # 1. The top recombination velocity has been taken as 10^4 cm/s

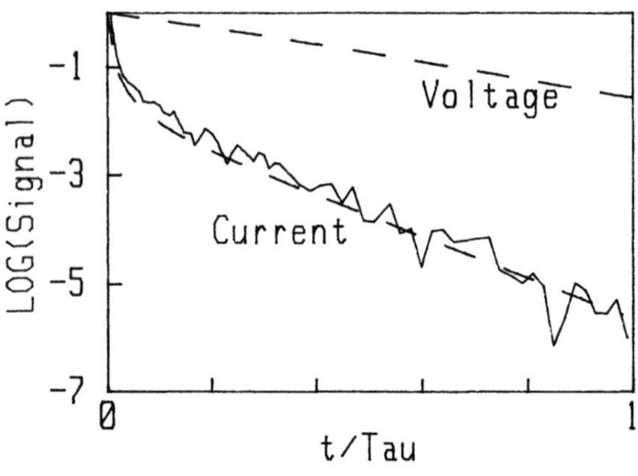

FIGURE 2
Comparison between Monte Carlo photo-current transient simulation (---) ,photo-current and photo-voltage decay (— —) as determined by analytical methods for cell # 1.

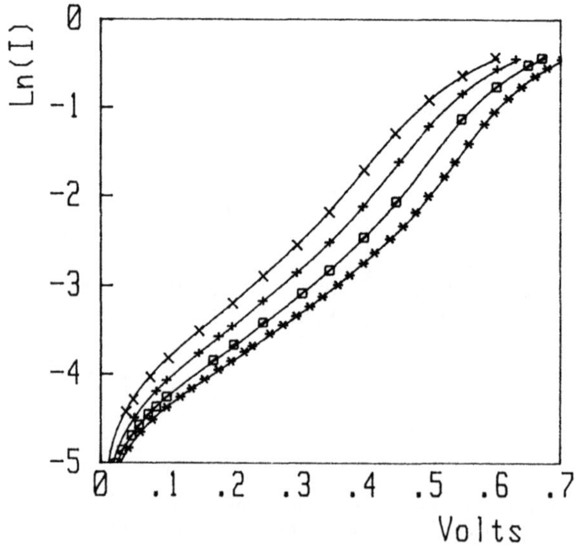

FIGURE 3
Temperature dependence of the dark current-voltage characteristics for cell # 1.
X = 80 °C, + = 60 °C,
0 = 40 °C, * = 20 °C

COMPARISON OF PV-SOLAR CELLS TECHNOLOGIES IN FLAT PLATE SOLAR MODULES

F. Chenlo, P. Román and L. Delgado
Div. Fotovoltaica. Instituto de Energías Renovables. CIEMAT.
Avda. Complutense, 22. Madrid 28040 SPAIN

Abstract.

Single crystalline and polycrystalline solar cells technologically equivalent to that used in commercially available flat plate modules tested at the IER Testing Laboratory have been systematically analyzed. Results are presented on their optical and electrical porperties, i.e. optical losses, spectral response, I-V curves under illumination, dark diode measurements, temperature coefficients.

A correlation between the different cell technologies used and the optical and the electrical parameters obtained is made, leading to some useful information on the effect of the different standard technology presently used.

1. INTRODUCTION.

Within a research program on technological improvements on commercially available flat photovoltaic modules, a search has been made to determine the effect of presently used process techniques, i.e. antireflective layer, surface treatment, metallization .., on the solar cell performance.

It is expected that if a correlation between external parameter solar cell measurements and manufacturing processes can be made, some technological improvements leading to low cost and better performance solar cells will be achieved.

2. EXPERIMENTAL.

A test sample of 11 solar cells of the type and technology described in Table I, was used.

Electrical characteristic, under illumination were made by using arc xenon solar cell siumulator, filtered to AM 1.5 global spectrum. Irradiance was set at one sun (1000 W/m^2). An electronic load was used to obtain the I-V characteristics (2,3). Temperature was controlled at 25 $\pm$1°C.

The temperature coefficients (1) were measured at 10°C intervals over a temperature range from 15°C to 75°C. To obtain temperature coefficients, a linear least-squares fit of the Jsc and Voc data for each temperature scan was performed.

Dark diode measurements (4) were performed at 25°C $\pm$ 1°C, in a voltage range between -5V to + 0.7V. Dark I-V characteristic curve is fited by a single exponential between 0.4 and 0.6 V. Thus the saturation currrent (Jo) and the ideality factor (n) are obtained. The series resistance (Rs) is calculated from the difference between the exponential equation and the dark I-V characteristic measured for voltages higher than 0.6 V.

An UV-VIS-NIR spectrophotometer with an 60 mm integrator sphere attached, was used to obtain the global reflectance measurements.

Spectral response measurements were made with a 150W Xe lamp and a high intensity 0.25 m monochromator. No bias light was used in these measurements.

CELL	CELL-TYPE	FORM	SURFACE FINISHING	METALLIZATION	TOTAL AREA cm^2	BUS-BAR AREA cm^2	TOTAL SHADOWING FACTOR (%)
1	p-Si	□	AR coating	Nickel-Tin	99.8	4.4	9.5
2	p-Si	□	Text. + AR	Aluminium	100	6.7	13.7
3	p-Si	□	AR coating	Silver	100	5.0	14.0
4	c-Si	○	Texturised	Nickel-Tin	77.9	3.5	10.8
5	c-Si	○	Texturised	Silver	77.0	4.2	15.8
6	c-Si	⯃	Texturised	Silver	92.5	4.3	11.9
7	c-Si	⯃	Text. + AR	Silver	92.5	3.8	12.3
8	c-Si	⯃	Texturised	Silver	91.9	3.8	14.0
9	c-Si	⯃	Texturised	Silver	91.9	3.5	13.6
10	c-Si	⯃	Texturised	Silver	101.7	3.5	11.5
11	c-Si	⯃	Text. + AR	Silver	101.7	4.5	10.8

Table I. Solar Cell Description.

3. RESULTS AND DISCUSSION.

3.1. Electrical and General Characteristics.

Table II presents a summary of the electrical characteristics for each solar cell.

SAMPLE	J_{sc} TOTAL (mA/cm^2)	J_{sci} Intrinsic (mA/cm^2)	V_{oc} (mV)	FF	P_{max} (w)	η (%)	$\alpha x10^{-4}$ (A/A°C)	β (mV/°Cg)	R_s (mΩ)	n	J_o (mA/cm^2)
1	29.7	32.9	571	.57	.96	9.6	1.2	-2.1	9	2.0	4.0 E - 9
2	27.8	32.2	565	.69	1.08	10.8	5.4	-2.1	5	1.5	1.9 E - 10
3	28.9	33.4	578	.69	1.16	11.6	2.9	-2.0	7.5	1.5	1.9 E - 10
4	25.4	29.4	599	.69	1.05	13.5	2.8	-2.1	8	1.6	3.5 E - 10
5	31.2	37.1	562	.61	.82	10.6	1.2	-2.2	-	2.4	3.9 E - 8
6	31.7	35.6	573	.57	.96	10.4	2.5	-2.1	9.3	2.6	6.8 E - 5
7	34.0	39.0	586	.59	1.10	11.8	-	-2.1	8	1.8	7.3 E - 10
8	30.8	39.4	595	.73	1.22	13.3	1.0	-2.1	5	1.6	3.1 E - 10
9	31.7	36.8	591	.72	1.27	13.9	1.5	-2.1	3.5	1.6	2.8 E - 10
10	31.2	35.3	581	.64	1.17	11.5	3.1	-2.0	6	2.0	2.1 E - 9
11	31.8	35.8	596	.67	1.29	12.7	4.2	-2.1	9	1.6	2.2 E - 10

Table I shows that the shadowing factor is in general, higher than data reported by manufacturers (9.5 to 15.8%).

Bus bar area is approximately 1/3 of the total shadowing area.

As it was expected, monocrystalline silicon (c-Si) cells have higher values of Isc than polycrystalline cells. Samples with AR coating and texturised surface, reach the highest short circuit current (cells 7 and 11, Table II).

c-Si cells also show larger open circuit voltages and surprisingly the best efficiency (13.9%) is obtained in a c-Si texturised cell (sample 9, Table II) but without AR coating. This is probably due to a better fill factor achieved with a well-established technological process in comparison with a more recent AR plus texturised surface process.

It is also remarkable that efficiency values of p-Si are very close to those for c-Si.

Not relevant differences in α and β between the cells measured was observed.

A correlation is establish between Rs,Jo,n and FF. Low values of FF, correspond to high values of n, Jo,and Rs. Lower efficiencies are also determined in these samples. (cells 1, 5, 6 and 10. Table II).

3.2.Reflectivity Measurements.

An estimation of the optical losses can be obtained from the solar cell reflectivity. The results are presented in Table II, where the integrated absorbance (1-R) (5) is given for each solar cell in two different wavelength ranges 350-1150 nm and 1150-2500 nm.

Large differences, up to 25% between different cell technologies are found in the optical losses. Solar cells using texturised surface plus antireflective coating show the best solar light collection. This is clearly seen in fig. 1. where the global reflectivity spectra of threee solar cells, corresponding to the same manufacturer (same AR and texturization techniques) but with differences in: i) AR layer, ii) texturized surface and iii) texturized surface plus AR layer, are presented.

As observed, the effect of combining AR layer plus surface texturization results in an increment of ~10% in the collected light relative to that collected either in AR layer or texturization surface alone.

INTEGRATED	CELL										
ABSORBANCE	1	2	3	4	5	6	7	8	9	10	11
350-1150nm	7.3	11	8.6	7.1	8.8	8.8	9.5	8.8	9.0	8.7	9.0
1150-2500nm	1.2	1.6	1.6	0.9	1.7	1.7	1.7	1.8	1.7	1.6	1.7

Table II

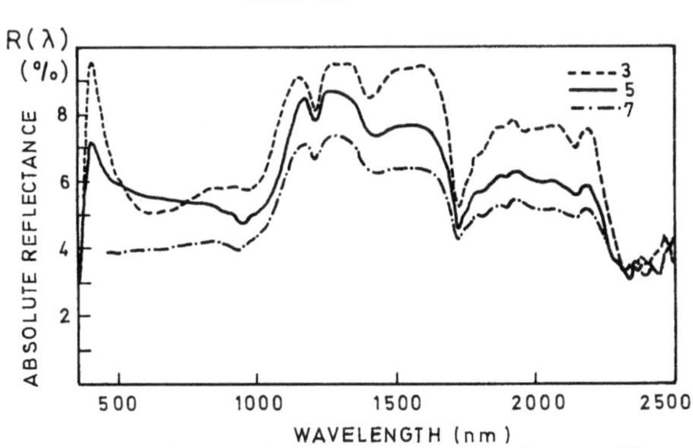

Fig. 1. Reflectivity spectra of solar cells (3) p-Si AR layer, (5) c-Si texturized surface and (7) c-Si AR layer + texturized surface.

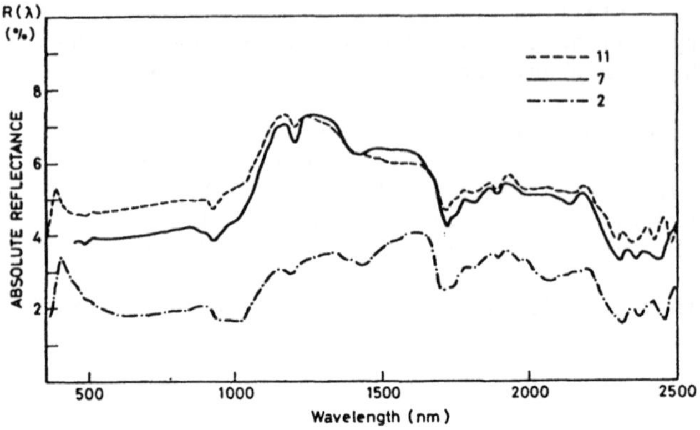

Even, with the same surface treatment, different technologies (different manufacturers) lead to different optical losses. This is presented in Fig. ., where the global reflectivity spectra of solar cells with nominally the same surface treatment, i.e., texturization plus AR layer, are presented. A difference of 5% in solar light collection can still be measured between them.

Fig. 2. Reflectivity spectra of solar cell. AR layer + texturized surface for different manufacturers.

3.3.Spectral Response.

Fig. 3. presents the relative spectral response curves corresponding to polycrystalline and c-Si cells. Measurements were performed without bias light, based on the spectral response linearity, which is supposed to be the case in the low resistivity cells here used. A shift in the spectral response of poly-Si cells towards the blue region of the solar spectrum is generally observed.

It was found that cells from the same manufacturer, i.e. with the same surface treatment show quite similar spectral response curves, thus indicating that the presence of AR layers in texturized cells has little influence if any on the spectral response curve shape.

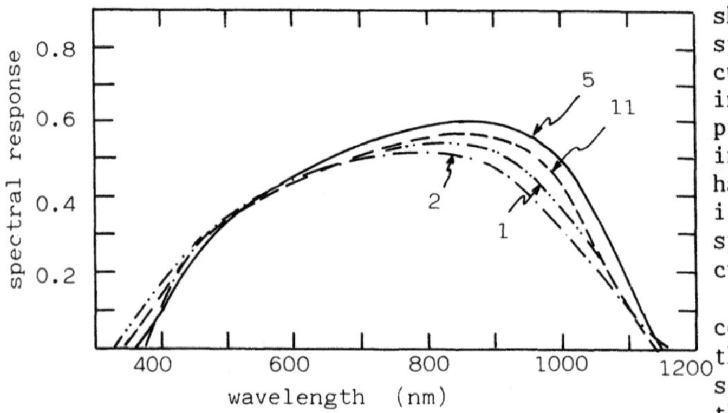

In general a good correlation between the integrated spectral response and the short circuit current was found.

Fig. 3. Spectral response of c-Si and poly-Si solar cells.

4. CONCLUSIONS.

From the results presented it is observed that solar cells with texturized surface and AR layer show larger short circuit current and solar light collection (less reflectivity). However, the larger efficiencies measured corespond to solar cells with texturized surface but without AR layer.

The prime loss of efficiency in cells with texturized surface plus AR layer, seems to be in fill factor. Therefore it can be concluded that the AR layer deposition technology on texturized cells, must be improved to avoid fill factor degradation.

Optical losses in texturized cells are primarily due to the metallization grid, rather than to reflectivity losses.

ACKNOWLEDGEMENT.

Technical assistance coming from Mr. J. Olivares and M.N.Vela are gratefully acknowledged.

REFERENCE.

(1) C.R. Osterwald, Proceeding of the 19th IEEE P.S.C. (1987) 188-193.

(2) S. Ashok, Solar Cells 14 (1985) 61-81.

(3) K.A. Emery and C.R. Osterwald, Solar Cells 19 (1986) 253-274.

(4) M.A. Hamdy and R.L. Call, Solar Cells 20 (1987) 119-127.

(5) M.B. Spitzer et. al., Solar Cells 20 (1987) 333-343.

A LOW-COST MEASUREMENT AND DATA ACQUISITION SYSTEM FOR SMALL PV POWER PLANTS

C. Brambilla, A. Iliceto
ENEL - CREL
Via A. Volta, 1
I 20093 Cologno M. MI
Italy

ABSTRACT. ENEL has developed a data monitoring system specially devoted to isolated PV plants, that provides a highly reliable and low-cost global monitoring of plant operation and performance. This D.M.S. combines several characteristics particularly suitable in isolated installations, and contains innovative dc energy measuring conuters as compared with similar equipments. This paper describes the main features of this D.M.S., shows how it is installed in small PV plants and what quantities can be acquired and calculated.

1. INTRODUCTION

Within the framework of its "Scattered Dwellings" Project, ENEL is about to instal a series of 30 photovoltaic plants, for demonstration purposes, for individual families. The project is supported by the Commission of the European Communities (DG XVII).

Each plant will be equipped with a measuring and data acquisition system GMF (Fig.1, Fig.2), which has been specially developed by ENEL in order to control the operation of the plant and to analyze energy balances.

This system combines low cost, accuracy, low consumption, environmental compatibility and reliability, all of which make it particularly suitable for data acquisition over long periods of time at small isolated plants.
Moreover, the measuring unit GMF operates on an interesting principle for d.c. energy measurement, that makes it innovative as compared with similar equipments available on the market.

2. GENERAL DESCRIPTION

The GMF consists of a main-frame (3 rack units heigh) subdivided into several circuit-card modules, which perform the following functions:

- generation of power supply voltages, needed for the various circuits, decoupled from the external power source
- d.c. energy measurement (see Fig. 3)
- signal integration (see Fig. 3)
- time-counting
- event-counting.

Each measurement is displayed by a 6-digit electromechanical counter.

In addition, an output interface, consisting of both potential-free reed contacts and voltage logic signals, can transfer the measuring pulses to an optional data-logger.

Figure 4 shows the PV plant block circuit with the measuring-points.

Connection to the circuits of the PV plant and to the data-logger are provided by fast connectors.

The main technical characteristics of the GMF are shown in Table I.

3. MEASURED AND COMPUTABLE QUANTITIES

The data recorded by this GMF, which the plant user would be asked to read every month, enable a number of interesting evaluations to be made of both plant design and plant operation.

In particular, the following quantities can be acquired or computed:

- Solar energy availability at site (E_S)
- Energy from PV array (E_{PV})
- Energy supplied to the load (E_L)
- Surplus energy ($K\,E_S - E_L$)
- Curtailed energy (estimated) $\left(\frac{E_L}{T_T - T_{LS}}\, T_{LS}\right)$
- System efficiency $\left(\frac{E_L}{k\,E_S}\,;\ \frac{E_{PV}}{k\,E_S}\,;\ \frac{E_L}{E_{PV}}\right)$
- Evaluation of system performance and life-time (long-term comparison of efficiency during corresponding periods)
- Quality of service (T_{LS}, N_{LS}, T_{AS}, N_{AS})

For further, more analytical evaluations, a supplementary programmable data-logger, with a static memory, can be connected and installed inside the GMF frame to investigate the performance of the plants in details or when the data recorded on the GMF counters show unexpedted values.

4. REMARKS

- The energy measurement performed by the GMF is not subject to additional error if the current is switched by choppers.
- Measuring and counter GMF modules can be set up in different compositions to meet any measurement purposes (e.g. as a back-up DMS in large PV plants).

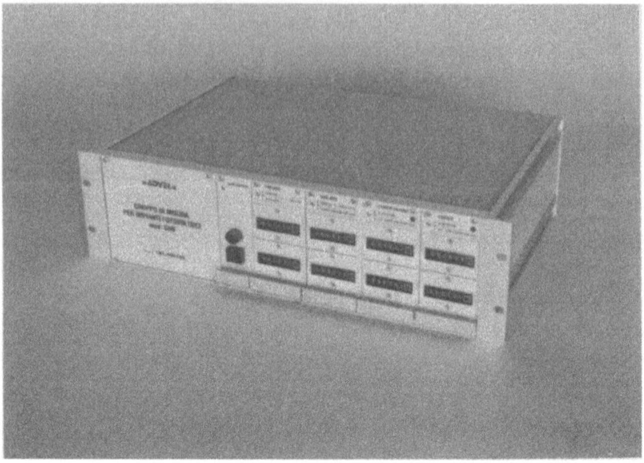

Fig. 1 - The data monitoring system GMF developed by Enel for small isolated PV plants.

Fig. 2 - Rack-monuting of the GMF at the top of the cabinet assembled for isolated PV plants.

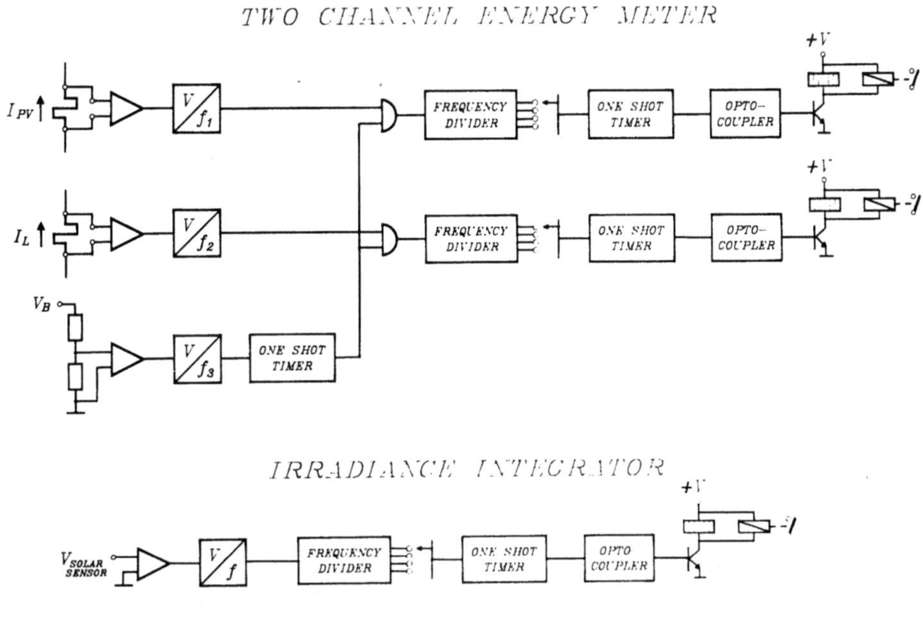

Fig. 3 - Block circuit diagram of the two channel d.c. energy meter and the irradiance integrator.

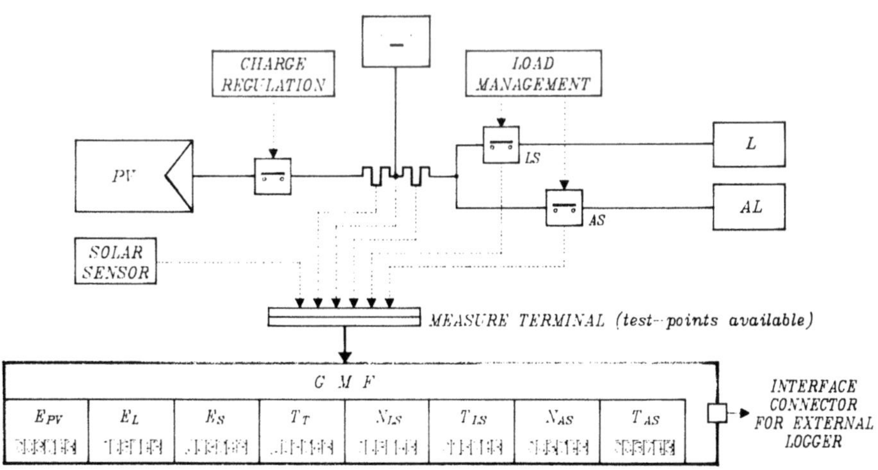

E_{PV} = energy from pv generator
E_L = energy to load
E_S = irradiation at the site
T_T = total time (system reference)
N_{LS} = n° of load disconnections
T_{LS} = total time of load disconnection
N_{AS} = n° of auxiliary load disconnections
T_{AS} = total time of auxiliary load disconnection

Fig. 4 - PV plant block circuit with measuring points.

INPUT

IDENT.	NOMINAL VALUE	
a	60 mV	(current from pv array)
b	60 mV	(current to load)
c	20 to 32 V	(battery voltage)
d	60 mV	(solar sensor)
e	ON–OFF	(load switch)
f	ON–OFF	(auxiliary load switch)

OUTPUT

$E_{PV} = \int (a \cdot c)\, dt$
$E_L = \int (b \cdot c)\, dt$
$E_S = \int (\; d \;)\, dt$
N_{DS} = "e" off–pulses counting
$T_{DS} = \int$ ("e" off–status) dt
N_{AS} = "f" off–pulses counting
$T_{AS} = \int$ ("f" off–status) dt
T_T = total operating time

output display: 6 digit electromechanical counters
resolution: – for energy measures: 10 Wh
– for time measures: 1 min
counting capability: > 10 years
repeated counter pulses from – free–potential reed contact
– logic voltage level

UNACCURACY

– for signal and energy measures: ± 1 % of measured value between 10 % and 120 % f.s.
– for time measures: 0.1 % of measured value

POWER SUPPLY

– voltage: 24 Vdc (20 to 32 Vdc)
– power consumption: 4 W max

ENVIRONMENTAL CONDITIONS

– temperature −10 °C to +55 °C
– humidity 20 % to 93 %
– metal case IP52

INSULATION LEVEL

– 1 kV d.c. between independent circuits
– 1 kV d.c. between independent circuits and earth
– 2 kV pulse 1,2/50 μs as above

EMC PERFORMANCES

electrostatic discharge test (severity level 2 IEC 801-2)
fast transient: test voltage level = 500 V

Table I – GMF TECHNICAL SPECIFICATIONS

HAILSTORMS, A REAL EXPERIENCE AT GIGLIO ISLAND

Dr. Eng. Eduardo C. L'ANGIOCOLA

TEAM Tecnologia Energia Ambiente Materiali
Via F.S. Nitti, 15 - 00191 ROME - ITALY

ABSTRACT

What happens when a real hailstorm with stones of 10 cm diameter falls on a P.V. array?
For the Giglio Island Project, the P.V. modules used had been chosen from standard production of a qualified manufacturer and had been tested and accepted by JRC (Joint Research Center of European Community) at Ispra, Italy, with routine test.
As it is well known, these modules are protected by a tempered glass cover, but during the hailstorm of September 1984 , some of them received serious damages.
The consequences of these damages are described, trying to classificate the type of cracks and holes found in the different elements. The purpose of this work is the diffusion of the data obtained from this rare experience to be used in future by project engineers, insurance companies, materials guarantee, etc; as a point of reference.
Due to the characteristics of the subject chosen, the approach is mainly descriptive with a presentation of drawings showing the damages in the array and moduls.
The description of facts and presentation of data open the way to technical and statistical hypoteses which might be studied by research laboratories and module manufacturers, contributing thus to a better knowledge of this kind of atmospheric phenomena and to the adoption of better means of protection to increase the reliability of this technology.

INTRODUCTION

The telephone call from Giglio Island announcing that a hailstorm had broken 15 modules of the P.V. array made all the managers, technicians and consultants involved shudder. I was suddenly commissioned, in my position of consultant project manager, to go and see the damages on the spot. When I arrived to the Giglio Island and saw the print of the hail-stones on the metal sheet of cars, I realized that the event was really out of the normal atmospheric phoenomena. I decided to take notes on the event and to keep these data for myself until P.V. technology should be strong enough as to prevent that these data could be used negatively against this our beautiful P.V. technology. Now I think that our child has grown enough and the information can be made public.
The first thing I did when I arrived to the plant was to check the strings of modules from the video display in the control room. I believed that from the control room I could have individalized the strings with damaged modules, but the video showed that all strings worked perfectly at the nominal current and voltage even if the damaged modules had not been still replaced. After this first checking from the control room, I went to the P.V. array to verify the damage.
At this point I must introduce some definitions so that we can classificate the visible damages.

CLASSIFICATION OF VISIBLE DAMAGES

Hole: a visible hole that allows the net passage of light through it.

First Impact: It is the impact that according to the traces left on the protective surface (i.e.: glass), can be supposed the one that produced the first damage to the module. It can either produce a hole or not; the first impact can be easily recognized due to the radial cracks that remain on the glass having the first impact as center.

Secondary impact without hole: It is the impact that has hit after the first one without causing a hole: most probably this type of impact wouldn't cause damage without a "first impact" as above defined.

Secondary impact with hole: it is the impact further a first one, causing a hole: must probably this type of impact might have caused a "first impact";

Cutting: It is the impact which has left a cut affecting the whole thickness of the module with a length ranging from 2 to 5 cm. Most probably this type of impact might have caused a "first impact" without hole;

Interruption of cell connection: It is the interruption of electrical continuity beteween two contiguous cells: it can be total or partial;

Module Characteristics:

Model 72 DG SOL , Dimensions 1308x658mm

Cells: monochristalline, 4" diameter.

Position of cells in module: 6 x 12; Connection: 6 parallels of 12 cells in series.

Nominal Power: 60 watt; V peak = 5 volts; Ipeak = 12 Amp

Tempered glass cover 3 mm thick, the glass has special no reflective surface texture on the face touching the cells that sink in transparent silicon. Back protection in white silicon.

Frame in stainless steel. Connection boxes are located at opposite shorter sides of the modules (one for each polarity).

Array Characteristics:

The whole array can be seen on Fig. 1. It consists of 24x3x12 modules plus 2 test modules, that is 866. Main angle is 35°.

Results:

The results of a minute recording of damages is presented on Table I, Figures 1 and 2.

	1	2	3	4	5	6	7	8	9	10	11	12	13	14	15
1st Impact with hole		1													
1st impact without hole	1		1	1	1	1	1	1	1	1	1	1	1*	1	1
1st impact recognizable	Y	Y	Y	Y	Y	Y	Y	Y	Y	Y	Y	N	N	Y	Y
2nd impact with hole	1	4				1		1					1	2	2
2nd impact without hole	2	5		3	4		7				6	4	6	2	4
Cutting	1				1						1		1		
Interr; of cell connection		1													
1st impact near frame	Y	Y	Y	Y	Y	Y	Y				Y	-	-		
1st impact in center area										Y		-	-		Y
1st impact in middle area								Y	Y			-	-	Y	
Total impacts	5	10	1	4	6	2	8	2	1	1	8	5	9	5	7

* Only in this case the first impact was assumed as the most possible.

DISCUSSION:

Comment on Table I

From 15 damaged modules, only in two cases it wasn't possible to recognize the first impact. In these cases the whole glass was full of cracks without a definite direction.

In only one case there was interruption of cell connection and it was not total but partial, since cells have two connection metal strips and only one was broken. Due to parallel connections of the module in no case there was interruption of current in the string.
Due to the quantity of "hole" and "no hole" impacts on module n. 2, it is my opinion that only this module was weaker than the average. In fact, it shows 10 impacts against an average of 4,93 impacts per module, and 5 of those 10 impacts are holes, against an average of 0,86 holes per damaged module, moreover this is the only module presenting a first impact with hole.
It is interesting to analyze the impacts taking into consideration the zone of the module subject to damage. Dividing the module in three areas as shown in Fig. 3, it is not difficult to realize that near the frame the elasticity of the module in the sense perpendicular to the surface (i.e.: the sense of hail grains movement) is minimum, being maximum in the center From table I, we notice that in 8 cases the first impact occurred near the frame, in three cases in the middle area and in 2 cases in the central area. Since in two cases it is not possible to recognize the first impact, the percentage on 13 modules is respectively 62%, 23% and 15% .
Impacts in the angles, that is at the most rigid point of the module, are 4 in total. Of these, 2 are holes and 1 is cutting, while only one is impact without hole.

Density of Damage

The density of damage can be quantified as 15 modules in a total of 866, this is 0,01732 or 1,73% . Should the hypotesis be accepted that module n. 2 was weaker than the average, the manufacturer deviation regarding resistance to impact is 1 in 866, this is 0,00115 or 0,115% since due to the exceptionality of the event (10 cm diameter hailgrains, see photograph° it can be considered that all modules can bear a normal hailstorm. I want to remind herethat acceptance tests are performed with ice balls of 25 mm diameter and maximum 75 mm diameter.

Conclusions

As it is well known the modules of Giglio Island Project have shown an excellent performance from the electrical point of view. After this rare experience of September 1984, the manufacturer has improved the mechanical characteristics and the thickness, introducing new materials and technologies. These new modules could bear even a "Giglio hailstorm".
As far as tests are concerned, just give a last glance to the photograph and probably the iceballs for laboratory hail-tests will grow up to 100 mm diameter.

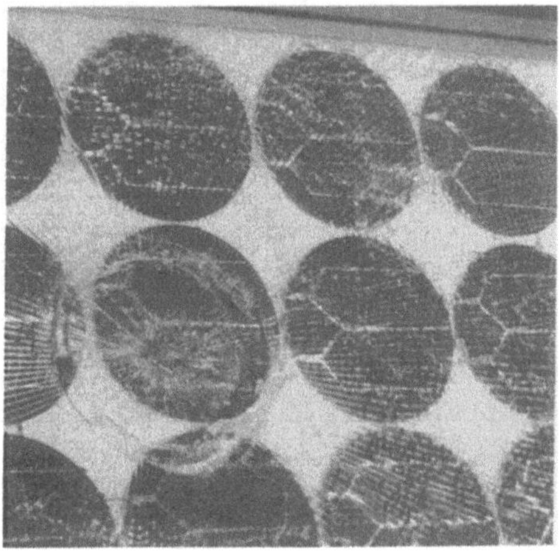

Aknowledgements

Dr. Emanuela Rossi -
JRC European Community , ISPRA

Dr. Daniele Margadonna - ITALSOLAR, NETTUNO

Eng. Silvano Loreto - ITALSOLAR, NETTUNO

Mr. Giustino Murzilli - ENEA Casaccia, ROME

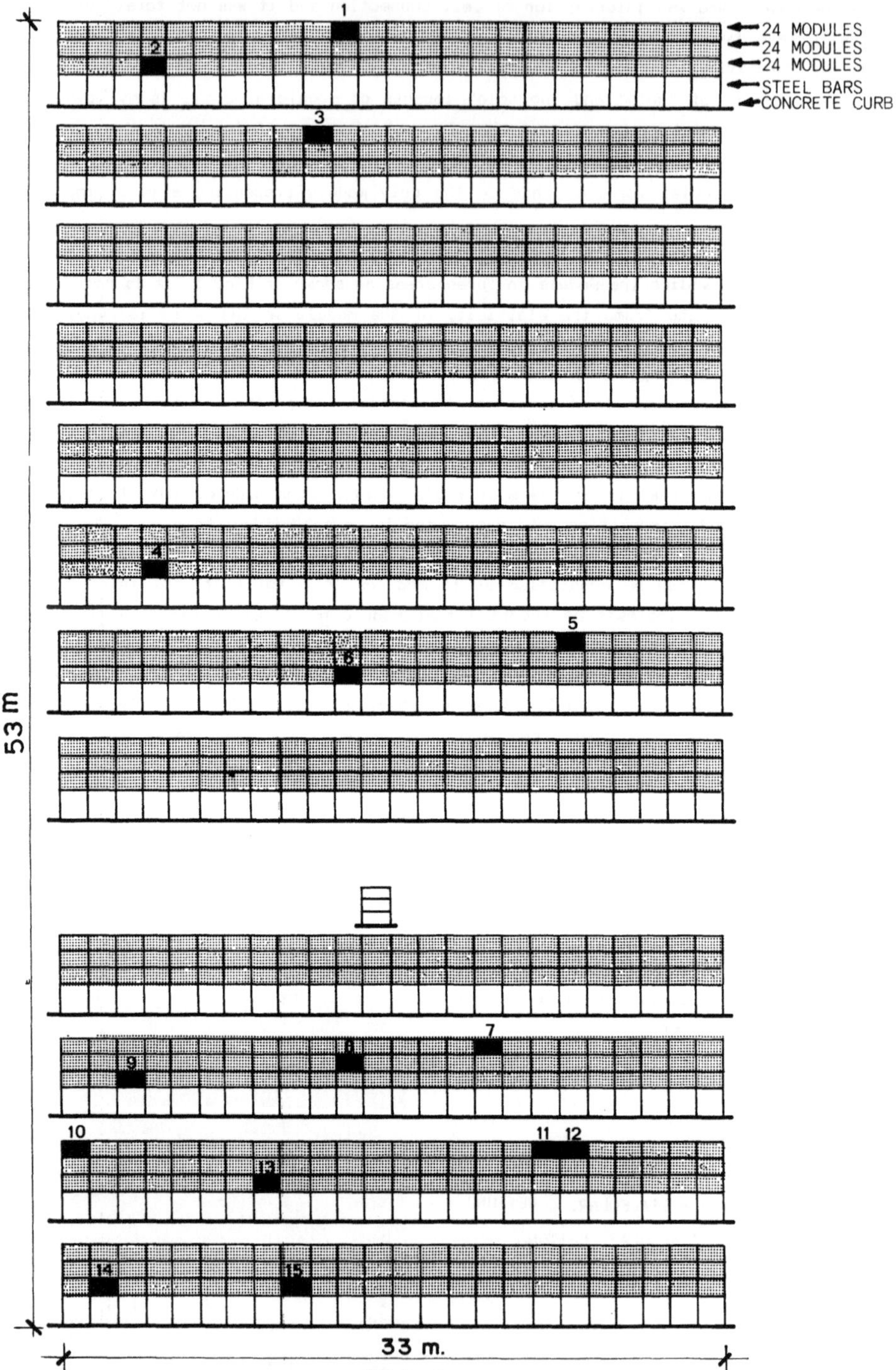
1
2
24 MODULES
24 MODULES
24 MODULES
STEEL BARS
CONCRETE CURB
3
4
5
6
53 m
7
8
9
10
11
12
13
14
15
33 m.

FIG. 1

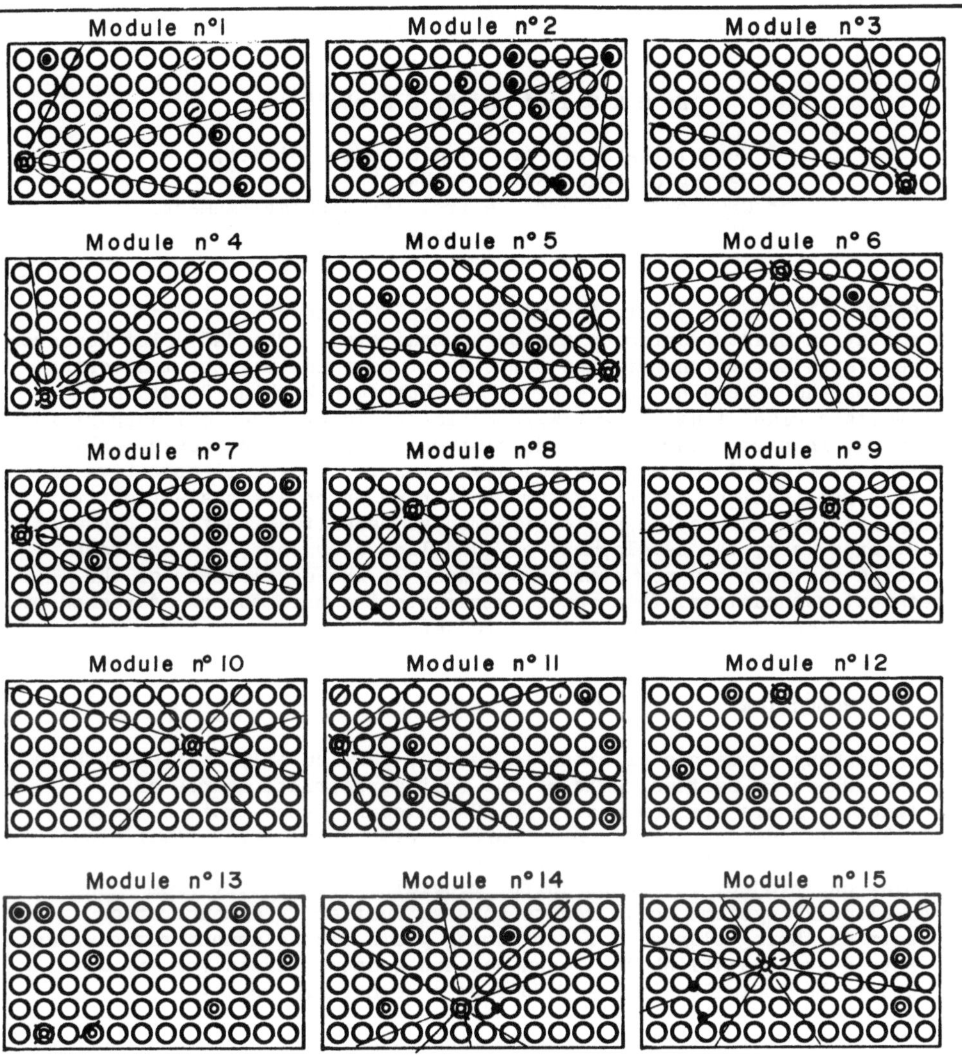

FIGURE 2

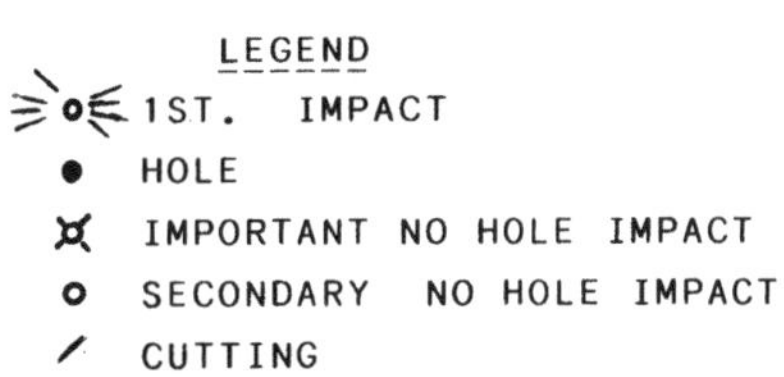

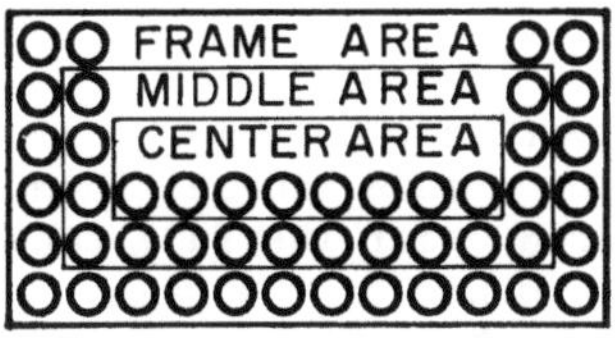

FIGURE 3

EUROPEAN WORKING GROUP ON PV PLANT MONITORING: GUIDELINES FOR MONITORING AND ASSESSMENT

Dr M R Starr* and **Dr K Krebs****

* Consulting Engineer, 23 Gloucester Street,
Faringdon, Oxon SN7 7JA, UK
** CEC Joint Research Centre, Ispra Establishment,
21020 Ispra (Varese), Italy

ABSTRACT. All the PV plants included in the EC-funded pilot and demonstration plant programmes are being monitored. The CEC Joint Research Centre, Ispra Establishment, has the task of coordinating the monitoring activities and assessing the results. The European Working Group on Photovoltaic Plant Monitoring (EWG/PPM) is providing technical support for this work, in particular for the development of guidelines covering PV plant monitoring, data analysis and assessment. The overall objective is to identify applications for PV systems in Europe which are cost-effective, taking into account social and environmental benefits. The first phase of the Group's work, the development of guideline documents, has now been completed and work has now started on Phase 2, the development of the assessment methodology. This paper summarises the main topics addressed by the EWG/PPM and the results achieved to date.

1. INTRODUCTION

There are many thousands of sites in Europe, particularly in the southern regions, where PV systems could be the most appropriate source of electricity. There are now nearly 50 EC-funded PV demonstration projects installed or under construction in Europe, with total array power about 720kWp. A number of these projects involve multiple installations; for example, one project in France is for providing electricity from stand-alone PV systems to 40 remote houses. Further PV demonstration projects are being planned for the coming years. These demonstration plants are intended to establish the technical and economic basis for similar systems to be installed in the future on a commercial scale.

The PV demonstration projects cover a wide range of applications, such as electrification of remote houses, farms and mountain refuges. There are also generators for lighthouses, radio and TV transmitters, environmental data monitoring equipment and desalination plants. Most of the projects are located in isolated sites, where grid-supplied electricity is not available.

Systematic and accurate monitoring of these demonstration plants is essential. It is not sufficient simply to install the plants and set them to work. The output of the system needs to be measured under the full range of climatic conditions. Wherever possible, the performance of the main components, such as the PV modules, the batteries and the power conditioning system, should be recorded. The actual loads imposed on the system by the user also have to be monitored, for comparison with estimates made at the design stage. It is important to know what levels of availability and reliability were actually achieved in practice. Finally, was the user satisfied with the plant or should certain aspects be changed to make it more acceptable?

In addition to social and environmental factors, an important criterion for assessing a specific PV demonstration plant will be its cost-effectiveness in comparison with alternative energy sources, such as diesel generators or grid-extension. What is the cost of electricity from the PV plant? How does this compare with the alternatives?

The answers to these questions will emerge from monitoring the performance of each PV plant and assessing the data thus obtained. All the PV plants included in the EC-funded pilot and demonstration plant programmes are being monitored to record key performance parameters and related data. The CEC Joint Research Centre, Ispra Establishment, has the task of coordinating the monitoring activities and assessing the results. The overall objective is to identify applications for PV systems in Europe which are cost-effective, taking into account social and environmental benefits.

2. THE EUROPEAN WORKING GROUP ON PV PLANT MONITORING

2.1 Objectives and Membership

The European Working Group on Photovoltaic Plant Monitoring (EWG/PPM) was established in 1985 by the JRC-Ispra to provide technical support for its work in coordinating the monitoring programme. The EWG/PPM consists of approximately 25 experts from European research organisations, industries involved in photovoltaics, and energy agencies, plus representatives from the CEC. The membership of the Group is listed in Table I.

TABLE I - MEMBERSHIP OF EWG/PPM

K Krebs, Chairman JRC-Ispra, Italy	**S McCarthy** University College, Cork, Ireland
G Blaesser, E Rossi JRC-Ispra, Italy	**A Pantoja** Hidroelectrica Espanola SA Madrid, Spain
M Andersson Institute of Microwave Technology, Stockholm, Sweden	**J L Plazy** AFME, Sophia Antipolis, Valbonne, France
J Avaritsiotis University of Athens, Greece	**P Pligoropoulos, G Karamanzanis** Public Power Corporation, Athens, Greece
P Ceppi Dipartimento Ambiente, Progetto TISO, Bellinzona, Switzerland	**J A Roger** Universite Claude Bernard, Villeurbanne, Lyon, France
M Camani Dipartimento Ambiente, Bellinzona, Switzerland	**R D W Scott** BP Solar International, Leatherhead, UK
L Delgado Instituto Energias Renovables, JEN, Madrid, Spain	**J Schmid** Fraunhofer Institut fuer Solare Energiesysteme, Freiburg, FR Germany
W B Gillett Consultant to DG XVII Halcrow, Swindon, UK	**L Selles** Renault Automation/SERI, Bois d'Arcy, France
A Iliceto, A Previ ENEL, Milan, Italy	**M R Starr** Consultant to JRC-Ispra, Faringdon, UK
M S Imamura Consultant to DG XII WIP, Muenchen, FR Germany	**E W Ter Horst** ECOFYS Research & Consultancy, Utrecht, Netherlands
H E Kadella, P Kyas Siemens Solar GmbH, Bergisch Gladbach, FR Germany	**F C Treble** Consultant, Farnborough, UK
W Kaut DG XVII, Brussels	
J Lowalt AEG AG, Wedel, FR Germany	

TABLE II - EWG/PPM SESSIONS

SESSION 1	-	ISPRA (Italy)	11-13 November 1985
SESSION 2	-	SOPHIA ANTIPOLIS (France)	10-12 March 1986
SESSION 3	-	MADRID (Spain)	22-24 October 1986
SESSION 4	-	LUGANO (Switzerland)	1-3 April 1987
SESSION 5	-	CORK (Ireland)	30 Sept - 2 Oct 1987

To date, the EWG/PPM has held five working sessions, as listed in Table II. Within the overall objective of identifying appropriate applications for PV systems in Europe, the EWG/PPM defined at an early stage its principal activities as follows:

1. The **exchange of data and experience** on PV plant monitoring in order to:
   - identify component faults and solutions to problems
   - analyse the monitoring results in the most useful ways
   - assist in the selection of appropriate monitoring equipment and instruments
   - utilise monitoring results to validate and improve system models
   - develop standards for all aspects of monitoring

2. The **dissemination of information** that will be useful to:
   - the PV industry
   - the users of PV systems
   - decision makers and the wider public.

2.2 Phased Plan of Work

The task of monitoring, data analysis and assessment may be considered in three phases. **Phase 1** covers the development of guidelines for monitoring, performance testing and analysis of data. **Phase 2** covers the development of a general methodology for the assessment of PV plants for various applications, in comparison with other energy supply technologies (wind, diesel, grid extension, etc). **Phase 3** covers the application of this methodology to actual PV demonstration plants and the subsequent dissemination of the results.

Phase 1, the development of the guidelines documents, has now been completed and work has started on Phase 2, the development of the assessment methodology. Phase 3, the application of the methodology, will start as soon as sufficient data from the first demonstration plants to be completed has been collected.

3. KEY TOPICS ADDRESSED

Comprehensive proceedings covering the matters discussed at each session of the EWG/PPM have been produced. The key topics addressed include the following:

a) **Monitoring methods and equipment**

- commercially available data acquisition systems
- when to use a Class 1 pyranometer and when a calibrated reference cell for solar irradiance measurements

- specifications for sensors and transducers
- lightning protection and surge protection devices
- frequency of sampling and error limits
- use of satellites for transfer of data from remote sites
- difficulties in monitoring batteries
- recommendations for measurements of inverters

b) **Parameters to be monitored**

- 'analytical' and 'global' monitoring
- initial and periodic tests and inspections
- requirements for Log Book to be maintained for each plant
- monitoring social factors

c) **Analysis and presentation of monitoring data**

- requirements of different interested parties (PV plant suppliers, monitoring equipment suppliers, system designers, research organisations, decision makers and the wider public)
- review of design strategies for the EC-funded PV pilot plants
- use of monitoring results for system optimisation

d) **Assessment of PV plants**

- assessment criteria
- procedure for economic analysis
- methodology for assessment

e) **Dissemination of information**

- international links
- matters to be presented in the newsletter
- guidelines documents
- proceedings of EWG/PPM sessions

4. RESULTS

The results achieved to date from Phase 1 fall into three categories:

i) Guidelines

A series of guidelines documents has been prepared under the general title **Guidelines for the Assessment of Photovoltaic Plants.**

Document A: PV System Monitoring
(third issue, 1987)
Document B: Analysis and Presentation of Monitoring Data
(first issue, 1988)
Document C: Initial and Periodic Tests on PV Plants
(first issue, 1988)

Document A sets out the requirements for both 'analytical' and 'global' monitoring, listing the parameters to be measured, the accuracy required and the sampling frequency. Standard formats for recording the data are also specified.

Document B sets out recommendations and standard procedures covering the analysis of monitored data and presentation of results. It contains sections covering initial checks for consistency, the analysis and presentation of energy data and the derivation of various Indices of Performance (i.e. Figures of Merit).

Document C gives recommendations for carrying out initial and periodic electrical performance tests on PV arrays. Details are given of the special equipment needed and the procedures to be followed for translating the results to Standard Test Conditions. A check list of points to be covered when undertaking periodic inspections of the PV modules in an array is given. Sections are also provided giving general advice on battery tests and checks and on what parameters to measure when testing power conditioning equipment.

ii) Newsletter

Four issues of the **European PV Plant Monitoring Newsletter** have been issued, dated October 1986, July 1987, December 1987 and April 1988. This is available free of charge to interested organisations on application to the JRC-Ispra.

iii) Proceedings

Comprehensive proceedings have been prepared covering the five sessions of the EWG/PPM held to date.

5. FUTURE OUTLOOK

Having completed the preparation of the first three guidelines documents, attention is now being focused on the development of the methodology for the assessment of PV plants. A preliminary draft has already been circulated to EWG/PPM members for their comments. The methodology covers the assessment of four main aspects: **technical, economic, environmental,** and **social/institutional.** An adaptible scoring system is proposed which enables the non-numerical aspects to be included in an overall assessment. The methodology is designed to facilitate the assessment of PV plants in comparison with other energy supply technologies.

Many desk studies have shown that PV systems appear to be a cost-effective method of providing electricity to remote areas. Although capital costs are high, they offer long life with low maintenance. Furthermore, they do not harm the environment. The demonstration programme means that real field experience is now being gained, enabling the true costs of supply, installation, operation and maintenance to be determined. Different design philosophies are also being investigated, to assess which approach is best for particular applications.

The European Working Group on PV Plant Monitoring is providing valuable support to the CEC in the task of monitoring the PV demonstration plants, analysing the data and assessing the results. Standard guidelines for monitoring and data analysis have been developed. A comprehensive and reliable data base is now being established. The final outcome will be that appropriate applications for PV systems will be identified and the results disseminated as widely as possible.

IEC Standards and Quality Assurance on PV Systems

C. LECLERCQ - IEC/TC 82 Chairman
Laboratoire Central des Industries Electriques
33, avenue du Général Leclerc
F-92260 FONTENAY-aux-ROSES

Abstract

The standardization in photovoltaic application is performed by the International Electrotechnical Commission (IEC). The Technical Committee 82 is in charge of this work "Solar photovoltaic energy systems". Two standards are already available and six are completed and will be published. These documents are applied on cristalline silicon solar modules. On these devices the drafts concern descriptions of environmental tests and endurance tests to be applied for qualification approval on modules. TC 82 has decided to publish description of tests when they are available and to combine in the future all the tests in one or two more complete documents. A more complex problem concerns standards on modules. The first document will be a guide and will give a general description on power generating systems and two others are under consideration, one concerning overvoltage protection and the second one measurement on PV array. The study on energy storage system and amendments on existing document to be applied on amorphous silicon module are under consideration. The final way is to use the standards for modules in the quality assurance system of the IEC.

1. Object of the IEC Work

The standardization in photovoltaic applications is performed by the International Electrotechnical Commission (IEC). This commission is constituted as a corporate association.

The object of the commission is to promote international co-operation on standardization and related matters in the fields of electrical and electronic engineering and thus to promote international understanding.

This organization is completed by a system of quality assurance which is described in the IECQ rules.

2. Organization of technical work

The technical work of the IEC shall be carried out through Technical Committees (TC) and the TC 82 is dealing with photovoltaic systems.

The TC 82 title is : Solar photovoltaic energy systems.

Any country which have a National Committee (NC) may participate to the activity of a TC.

About every 18 months TC 82 has a meeting and the drafts prepared by the secretariat of TC 82 shall be discussed after submission to the national committee for comments.

After reviews the drafts are submitted to NC for voting. If a suffisant number of approval vote is obtained the document is published as a standard.

Working groups created by the TC can help the secretariat for preparing the drafts.

In TC 82 four working groups exist : WG1 = glossary
WG2 = modules
WG3 = systems
WG4 = energy storage systems.

During a meeting the TC 82 deals with the program of work for WG and the priority of work to be done.

3. Situation of documents

3.1 Publication

Two documents are already published :

- Publication 891 edition 1987 :
  Procedures for temperature and irradiance corrections to measured I-V characteristics of cristalline silicon photovoltaic devices.
- Publication 904-1 edition 1987 :
  Measurements of photovoltaic current-voltage characteristics.

Six documents are completed and will be published :

- Requirements for reference solar cells
- Computation of spectral mismatch error introduced in the testing of a photovoltaic device
- Measurement principles for terrestrial photovoltaic solar devices with reference spectral irradiance data
- Guidance for spectral measurements of a photovoltaic (PV) device
- Solar simulation performance requirements
- Photovoltaic (PV) module twist test.

The last document is the first of a series concerning the ability of a module to be submitted to an environmental stress.

3.2 Document at voting stage

Four documents concerning modules will be submitted to NC for voting :

- Electrical leakage test on modules
- Damp heat, steady test on modules
- Salt mist corrosion test on modules
- Non-intrusive method for a hot spot endurance test on modules.

Concerning systems one document is under voting. It is submitted to National Committees; it is a guide concerning systems :

General description of Photovoltaic (PV) power generating systems.

This document is the first one concerning systems ; it gives general description of PV systems and a short description of major subsystems, components and interfaces of the PV power generating system.

For example, for an inverter the document gives a list of main characteristics (frequency, voltage level, quality of output power, etc.) and a list of parameters which describe one inverter (input conditions, output conditions, efficiency, protection equipment, etc.).

It was agreed by the TC 82 that the secretary would prepare a document on a specific subject and the document would be published as a separate standard. But at a certain stage the documents concerning modules would be published as a complete document.

This way has been used in order to obtain description of test as soon as possible. It is possible for a contract to refer to a published document in order to have a standard describing a particular test.

4. Work in progress and future of work

Some documents are under consideration by WG's.

Concerning modules

- Standard methods for the measurement of NOCT of PV solar modules
- Qualification approval test sequence.

The last document would define the sequence of test to be applied on modules for qualification. The main idea is to carry out tests in parallel on a certain number of modules in order to obtain results in few months (2 or 3). The sequence of test will include change of temperature, humidity test, mechanical test and endurance test, NOCT, hot spot, etc. . The number of modules to be submitted for qualification will be about 8 modules divided in 4 or 5 groups of tests.

Descriptions of tests would be prepared : humidity freeze, climatic sequence, marine environment test, mechanical load, robustness of terminations, etc.

These tests will complete the list of stresses to be applied on modules for qualification.

WG2 is also working on thin film device, specially amorphous silicon solar modules and the idea is to analyse admendments to existing documents on cristallin silicon modules.

Concerning systems

A draft concerning the problem of overvoltage protection for Photovoltaic (PV) power generating system is under consideration by National Committees for comments before the next meeting of TC 82.

A document relating to on-site periodic measurements on PV array is prepared by the WG3 of TC 82.

The documents concerning systems will not standardize systems but will give list of parameters describing interface problems between components of systems or sub-systems.

Concerning energy storage systems, the work is starting and there is no sufficient result for giving an idea of the way to solve the difficult problem of characterization of solar batteries for example. It is necessary to introduce some words concerning the relation with other technical committees which are working on components for instance batteries. The problem is that photovoltaic applications need specific requirements on components but the market is too small to obtain rapid results produced by a committee involved in another more important market.

5. Quality assurance system

At the begining of its work TC 82 decided to introduce modules in the IECQ system. This needs the description of tests, sequence of tests for qualification approval and for quality assurance, the definition of sampling procedures for acceptance tests and for periodic tests. The procedures of IECQ system are described in Publications QC001001 and QC001002 "Basic rules of the IEC Quality Assessment System for Electronic Components" and "Rules of Procedure of the IEC Quality Assessment System for Electronic Systems".

It is quite evident that a lot of work should be done before introducing a first module under this system. The IEC has published a Qualified Products List (QPL) QC001005 which contains devices under quality assurance defined by the system. This would facilitate commercial problems.

The application of IECQ system on a device requires a standard written in a good way and a national organization complying with rules described in basic rules. A large majority of countryshave such an organization and there will be no problem to introduce a module under the test when a full standard will be available.

6. Conclusions

The standardization on PV systems started in 1982. Many results are obtained and 2 publications are already available. An important work is still to be done. Results in the future will depend on the participation of National Committeesand experts appointed. It is quite evident that Photovoltaic is a good example of a need of standardization. This is due to the fact that in many cases the users of PV systems are not from a country were the manufacturers are located.

I would like to finish in congratulating the TC 82 secretary, my friend Mr R.J. KLEIN from IEEE (USA) and Mr WHITAKER (USA) assistant secretary for their excellent work.

ENVIRONMENTAL STRESS TESTS: ACCELERATION OF CORROSION

Alain M. Ricaud and Raymond C. Petersen
Solarex Corporation
630 Solarex Court, Frederick, Maryland 21701 USA

ABSTRACT. In a typical modern PV module (glass superstrate, cross-linked ethylene vinyl acetate encapsulant, printed silver metallization) the common proximate cause of degradation in electrical performance is corrosion of the metal grid patterns of the cells. This is revealed first in a power loss caused by an increase in the series resistance of the module and eventually in a decrease in the apparent short circuit current.

This paper is limited in scope to this particular mechanism. It proposes an Arrhenius model for the combined effects of temperature and humidity, calculates the acceleration factor using some simplifying assumptions and testing a range of activation energies, field conditions, and acceleration conditions. It is then shown that to minimize the hazards of extrapolation in stress, testing in an accelerated environment should be performed as close to the design stress as possible. General recommendations are suggested to limit relative humidity in accelerated stress tests in order to achieve meaningful results.

1. INTRODUCTION

Accelerated environmental stress tests have been designed for the purpose of assessing probable lifetimes of photovoltaic modules in service in the field. Achieving reliable correlations is difficult, because real service lifetimes of 20-30 years are expected, and this necessitates long extrapolations from accelerated test data to real world conditions. Within the photovoltaic community, the general consensus for simulation of real service lifetime is to accelerate degradation by way of high temperature storage tests, temperature cycling tests, high temperature/high humidity tests and ultraviolet irradiation tests.

The need for long extrapolation has led to two different kinds of activities in the area of accelerated testing of PV modules. On one hand, serious efforts have been made to determine quantitative relationships between accelerated test conditions and real service in the field. The most notable of this work was done at the Jet Propulsion Laboratory (JPL) of the California Institute of Technology. On the other hand, some testers have devised progressively more demanding test regimens, almost without regard for the real world relevance of their tests, ultimately reaching test conditions which cannot relate in any meaningful way to realistic service expectations.

2. AN ARRHENIUS MODEL FOR COMBINED TEMPERATURE-HUMIDITY DEGRADATION

Corrosion is a reaction between metal and water and it can be accelerated by an increase in temperature or in water concentration.

If we assume a simple corrosion reaction in which the rate is first order in water concentration (expressed by the partial pressure of water vapor in the atmosphere), then the reaction rate r is expressed by

$$r = kP \tag{1}$$

where k is the reaction rate constant and
P is the vapor pressure of water.

The rate constant k would be expected to obey the Arrhenius temperature dependence so that

$$k = A\exp(-E_a/RT) \tag{2}$$

where E_a is the activation energy
R is the gas constant (1.9872 cal/mole-deg)
A is a constant indepentent of T and
T is absolute temperature in °K.

Empirically the partial pressure of water vapor is expressed very well by

$$P = (RH/100)A'\exp(-B/T) \tag{3}$$

where P is water vapor pressure in mm of Hg
RH is relative humidity in %
$A' = 7.50 \times 10^8$ mm of Hg and $B = 5150\ (°K)^{-1}$.

Combining equations gives

$$r = AA'(RH/100)\exp(-[B+E_a/R]/T) \tag{4}$$

Comparing rates at two sets of conditions (field and test chamber)

$$r_c/r_f = (RH_c/RH_f)\exp([B+E_a/R][T_c-T_f/T_cT_f) \tag{5}$$

Where subscripts c and f refer respectively to test chamber and field.

This model can be refined by the appoximation that the module is partially exposed to a low temperature/high humidity environment (night-time) and to a high temperature/low humidity environment (day-time). This yields to the calculation of two reaction rates.

The combined reaction rate is then defined as:

$$r_f = (r_d \times t_d + r_n \times t_n)/24 \tag{6}$$

with $T_d = T_a + (NOCT - 20) \times Irr/800$
t_d = time in day conditions
t_n = time in night conditions
$RH_d = RH_a \exp B\ (1/T_d - 1/T_a)$
T_d and RH_d being respectively module daytime temperature (higher than ambient T_a) and module daytime relative humidity (lower than ambient relative humidity).

Table 1 shows how the calculated acceleration factors are sensitive to different sets of parameters:

A) Activation Energy: Acceleration factor is divided by a factor of 4 when E_a diminishes from 0.85eV (typical agreed value) to 0.50eV.

B) Nominal Operating Cell Temperature: Acceleration factor is reduced by more than a factor of 2 when the module NOCT (which depends on its construction) increases from 40°C up to 50°C. Typical module NOCT ranges between these two limits.

C) Effect of Average Irradiance in the Field: Acceleration is reduced by more than a factor of 2 when we compare a field situation of 12 hours exposure to 400 W/m^2 (4800 Wh/m^2) to a situation of 5 hours exposure to 960 W/m^2 (same accumulated energy), at same ambient temperature. The sharp temperature dependence means that time spent under lesser stress conditions contributes relatively little to the degradation process.

D) Effect of Relative Humidity of the Atmosphere: When day atmosphere relative humidity is increased from 20% to 90%, acceleration factor is reduced by more than a factor of 4. It is worth noticing that a module cannot practically reach the dew point when illuminated since the module is warmer than the surrounding atmosphere. During the night and especially at dawn, water might condense but this process is not harmful since the module temperature is relatively low at that time.

E) Acceleration Test Chamber Conditions: An environment of 90°C/95%RH provides an acceleration factor twice as high as the one obtained in a chamber at 85°C/85%RH, which could lead the experimentor to believe that he can cut the test time by a factor of 2 and expect the same failure rate. It is true, provided the failure mechanism remains the same. In fact, several cautions are necessary in employing this kind of model:

* At very high water concentration, second and higher order terms in water concentration enter the reaction kinetics picture leading to an increase in acceleration of the rate beyond that predicted by the simple model.

* Furthermore, when operating close to a dew point (95% relative humidity), very small deviations in water vapor concentration may cause water to condense; radiation from the module may lower its temperature, again causing condensation; also if the effective vapor pressure of water in contact with cell metallization is lowered (through surface tensions and capillary effects) then liquid water can condense on the metal and again a new reaction mechanism is introduced.

In Figures 2 and 3 where water vapor pressure is plotted against temperature for different values of relative humidity, we have shown in the rectangles the best performance that can be achieved with a good temperature humidity chamber in temperature accuracy and stability ($\pm$1°C) and in vapor pressure precision ($\pm$1%).

Figure 3 clearly indicates that with set points at 85°C/85%RH, the chamber works in a safe region far away from the dew point.

In Figure 2, it is obvious that the dew point can be reached at any time during the test, yielding to irrelevant mechanisms of degradation related to the condensation of water.

Moreover, according to Table 1, column E2, a module exposure to 90°C/95%RH for 2000 hours is equivalent to a lifetime of 200 years in a hot, humid climate.

Otth and Ross of JPL (Ref. 1) stated that 144 hours of exposure to 85%RH at 85°C correlates with 20 years of field exposure in Miami, Florida (a relatively hot, humid environment). These conditions match well with column D2 of Table 1. Hence, exposure times of 2000 hours are more properly regarded as torture tests than accelerated life tests.

3. GENERAL GUIDELINES FOR AN ACCELERATED TEST PLAN

The difficulty of assessing failure rates of a high reliability product through accelerate life testing is presented clearly in an excellent discussion by William Q. Meeker and Gerald J. Hahn (Ref. 2).

Long extrapolations, especially on exponential scales, invite large errors in lifetime projections, while excessive stress conditions frequently introduce new failure modes which are irrelevant under design stress (real world) conditions. For accelerated stress tests to be valid, irrelevant failure modes must be avoided. These considerations lead Meeker and Hahn to stress the importance of keeping the accelerated stress conditions as close as possible to the design stresses.

Meeker and Hahn also note the importance of selecting a physically appropriate model for the degradation or failure (implying some knowledge of the failure mechanism) and then devising an appropriate accelerated stress test. The appropriate accelerated test is specific to the failure mechanism, and not necessarily specific to the product. In the case of PV modules, this implies that an accelerated test appropriate to one type of module construction may not be appropriate to some other construction.

4. CONCLUSIONS

1) Accelerated stress conditions should be as close to the design stress as possible so as to minimize the probability of irrelevant failure modes occurring and to minimize the length of the extrapolation to real world conditions.

2) In general, some knowledge is needed of the failure mechanism in a specific product design and an accelerated stress test needs to be devised for that specific failure mechanism.

3) Four hundred hours or less of exposure to 85°C/85%RH should be a sufficient test of a PV module of the current Solarex design.

4) In general, the combination of high temperature and high relative humidity should be avoided because of the high probability of the introduction of new mechanisms of degradation related to the condensation of water. A reduction in either temperature or relative humidity will greatly reduce the probability of this occurrence.

REFERENCES

(1) Otth, D.H. and Ross, R.G., 'Assessing Photovoltaic Module Degradation and Lifetime from Long Term Environmental Tests,' Proc. of the 30th Institute of Environmental Sciences Annual Meeting, Orlando, FL, May 1-3, 1984.

(2) Meeker, W.Q. and Hahn, G.J., 'How to Plan an Accelerated Life Test - Some Practical Guidelines,' American Society for Quality Control, Basic References in Quality Control: Statistical Techniques, Vol. 10.

	A1	A2	B1	B2	C1	C2	D1	D2	E1	E2
Ea (ev)	0.50	0.85	0.85		0.85		0.85		0.85	
NoCT (°C)	45		40	50	45		45		45	
IRRAD (w/m^2)	800		800		400	960	800		800	
td (h)	6		6		12	5	6		6	
Ta day (°C)	25		25		25		25		35	
RHa (%)	50		50		50		20	90	60	
tn (h)	18		18		12	19	18		18	
Tn (°C)	15		15		15		15		15	
RHn (%)	90		90		90		90		90	
Tc (°C)	85		85		85		85		85	90
RHc (%)	85		85		85		85		85	95
Acc factor	530	2270	3530	1460	3850	1740	5160	1300	460	910
Years in field	20	20	20	20	20	20	20	20	20	20
Hours in chamber	328	77	50	120	45	101	34	135	383	192

Table 1: Calculated acceleration factors for different sets of conditions.

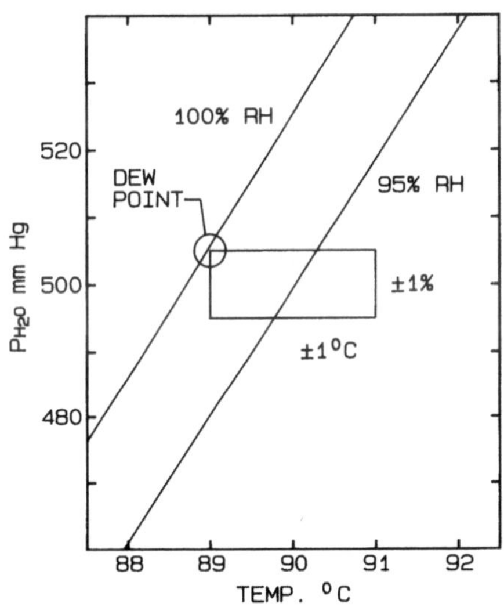

Fig. 2: Partial pressure of water vapor with relative humidity near 90 °C.

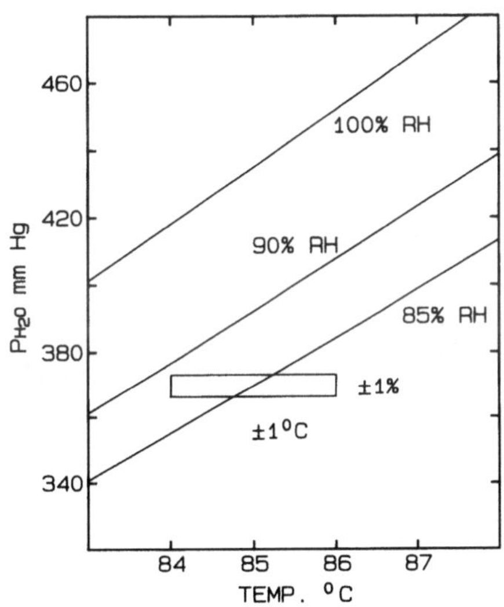

Fig. 3: Partial pressure of water vapor with relative humidity near 85 °C.

UNCERTAINTY ANALYSIS OF PV EFFICIENCY MEASUREMENTS WITH A SOLAR SIMULATOR: SPECTRAL MISMATCH, NON-UNIFORMITY AND OTHER SOURCES OF ERROR.

Klaus Heidler, Jutta Beier
Fraunhofer-Institut für Solare Energiesysteme (ISE)
Oltmannsstr. 22, D-7800 Freiburg
Federal Republic of Germany

ABSTRACT. A detailed uncertainty analysis of the solar simulator method for measuring PV efficiencies of crystalline Si solar cells leads to a 2.3 % uncertainty limit (resp. ± 4.4 % for linear addition). Comparison with the results of an international round robin indicates that many researchers underestimate their uncertainty limits.

1. INTRODUCTION

Solar cell efficiency is among the most important physical parameters in photovoltaic R & D. However, round robins revealed that the related measurement uncertainty is still rather high /1/. Keith Emery analysed uncertainty for a best and a typical case /2/. In this paper we go into more detail for the case of crystalline Si cells, area 10 x 10 cm^2 and investigate the current state of our equipment, which represents the first stage of a precision measurement facility for solar cell efficiency. Our work is done in close cooperation with the Physikalisch-Technische Bundesanstalt (PTB) which supplies us with reference cells.

The uncertainty analysis of our new equipment was so fruitful that we found it to be worth reporting, even though our set-up will now be improved to reduce some of the identified error sources. After a description of the solar simulator and the measurement procedure, we present first an analysis of the integral bias errors and then we examine the spectral bias errors. Finally, we discuss the results and draw some conclusions.

2. EFFICIENCY MEASUREMENT PROCEDURE

Our measurement routine is outlined in table I.

The reference cell is a solar cell calibrated by PTB, made of material with a similar spectral responsivity to the test cell. The location of average irradiation is determined by a scan of the test area with a detector. Contact to the test cell is made at leads, contact pads or at the outer edges of the bus bars. No action is taken to improve poor bus bar conductivity.

We use an AEG solar simulator XAT 1600 with an 1600 W Xe arc lamp and an AM 1.5 filter. A feedback cell stabi-

1 Measure spectral responsivity s(λ) of test cell (TC) and calculate spectral mismatch M

2 Put primary reference cell (RC) at location of average irradiaton in test plane

3 Adjust solar simulator so that RC reads calibration value multiplied by 1/M

4 Calibrate monitor cell MC

5 Mount TC on vacuum chuck, contact with Kelvin probes

6 Check contact resistance

7 Record I-V characteristic

Table I: Efficiency measurement procedure

lizes lamp fluctuations to less than 0.2 %. Typical non-uniformity in the 10 x 10 cm^2 test area is ± 1.8 %, measured with a 0.16 cm^2 detector.

3. UNCERTAINTY ANALYSIS

We follow the formalism adopted in /2,3,4/, where total uncertainty U is ± (B + t_{95} • S) with B for bias limit, S for precision index and t_{95} for Student's "t".

The formula for the efficiency calculation is /2/:

$$\eta = \frac{P_{max} \cdot s}{A \cdot I^{RC}} \frac{1}{M} \qquad (2)$$

P_{max} maximum power of test cell (TC) in [W]

s responsivity of reference cell (RC) in $\left[\frac{A \cdot cm^2}{W}\right]$

A total area of TC in [cm2]

I^{RC} actual I_{SC} of RC in [A]

M spectral mismatch factor

3.1 Integral bias errors

Fig. 1 lists the assumptions of our analysis, shows a graphical representation of the integral bias error distribution and gives the total bias limit for each parameter. Some of the errors need comments:

$P_{3,4}$ current of TC measured with a calibrated resistor

$P_{5,6}$ front contact BeCu, back contact Au, ΔT ≈ 7K

P_7 resolution of electronic load is 0.2 mV

P_8 (see fig. 2) TC is regarded as a parallel connection of 100 elementary cells. Non-uniformity of irradiation leads to electrical mismatch of elementary cells. Calculation for ± 10% non-uniformity.

P_9, I_{10} temperature measurement errors of TC or RC

P_{10}, I_4 half angle is 3.5°;
P_{11}, I_5 tilting of 0.5° of TC or RC
P_{12}, I_6 multiple reflections between TC/RC and last lens
P_{13}, I_7 reflections from operator or apparatus
P_{14} shadowing by probes corrected within these limits
s_1 PTB/5/;Differential Spectral Responsivity method/6/
s_2 estimate of photodiode degradation /7/
I_1 MC and RC are cleaned every 60 minutes
I_2 RC has a different reflection coefficient to TC
I_8 measurement plane is adjusted
I_9 RC is placed at location of average irradiance. Error limits give effect for a change of ± 2 cm in this position. Spatial uniformity is a function of probe area (fig. 3).

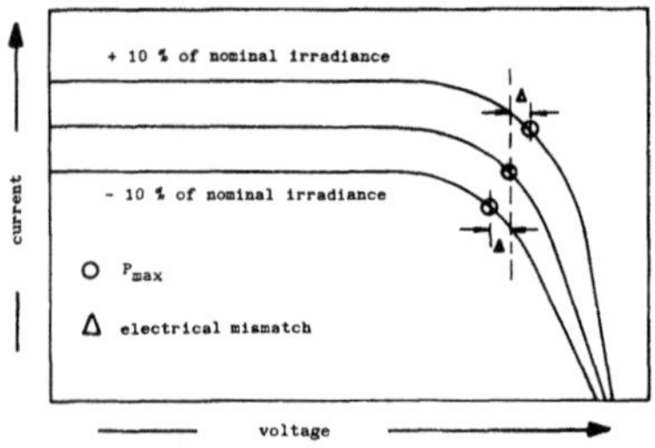

Fig 2: Electrical mismatch of elementary solar cells due to non-uniformity in the irradiance

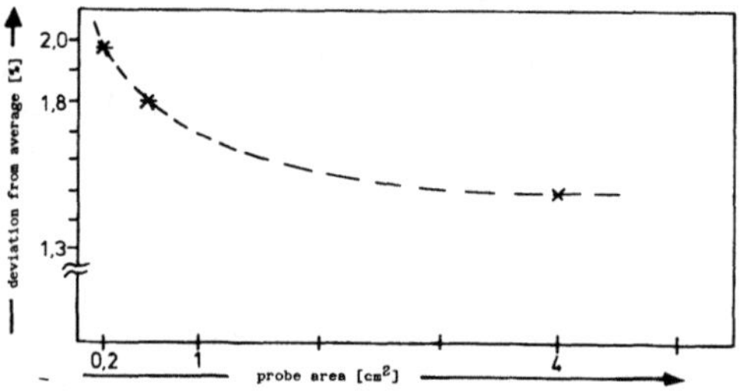

Fig 3: Maximum deviation of the irradiance from the average as a function of the detector size

3.2 Spectral bias errors

The mismatch between the spectral responsivities of RC and TC, and between the standard and the simulator spectrum is expressed by the mismatch factor M; it is defined as

$$M = \frac{\int E_{ref}(\lambda)\, s^{RC}(\lambda)\, d\lambda}{\int E_{sim}(\lambda)\, s^{RC}(\lambda)\, d\lambda} \frac{\int E_{sim}(\lambda)\, s^{TC}(\lambda)\, d\lambda}{\int E_{ref}(\lambda)\, s^{TC}(\lambda)\, d\lambda} \quad /8,9,10/,$$

where

$E_{ref}(\lambda), E_{sim}(\lambda)$: spectral irradiance of standard and simulator spectrum, resp.
$s^{RC}(\lambda), s^{TC}(\lambda)$: spectral responsivity of RC and TC,resp.

All four quantities in M appear in both the numerator and the denominator. Thus, we do not need to determine their absolute values, which would introduce large errors in M. The bias error of M can then be calculated from the bias errors of the relative values of E_{ref}, E_{sim}, s^{RC} and s^{TC} by propagation of errors.

P_{max}	+ 2000 - 2800

P_1	DVM (I)	± 200
P_2	DVM (U)	± 110
P_3	resistor	± 500
P_4	ΔT resistor	± 250
P_5	thermovoltage (I)	± 40
P_6	thermovoltage (U)	± 20
P_7	P_{max} algorithm	+ 80
P_8	non-uniformity mismatch	+ 120
P_9	ΔT test cell	± 1600
P_{10}	beam divergence	± 240
P_{11}	non-normal incidence	+ 50
P_{12}	multiple reflections	– 1000
P_{13}	stray ligth	± 500
P_{14}	shadowing	± 250

s	±11200

s_1	primary calibration	±10000
s_2	degradation	± 5000

A	±1000

A_1	area measurement	± 1000

I^{RC}	+7800 -8600

I_1	dust	± 150
I_2	reflection RC-MC	+ 60
I_3	DVM	± 800
I_4	beam divergence	+ 220
I_5	non-normal incidence	+ 50
I_6	multiple reflections	- 1000
I_7	stray light	± 500
I_8	height adjustment	± 450
I_9	spatial non-uniformity	± 7500
I_{10}	ΔT	± 400

<u>Assumptions</u>

. 10 x 10 cm^2 crystalline Si solar cell
. intrinsic inhomogeneity not more than ± 8 %
. probe connections at contact pads or at outer edge of bus bar; no correction for voltage drops on the bus bar
. nominal values I_{sc}=2000 mA, U_{oc}=500 mV, FF=0.7

Fig. 1 List of integral bias errors [ppm]

E_{ref} (λ): is assumed to have zero error.
E_{sim} (λ): we took the spectral data of the manufacturer. Our conservative estimate of the errors is based upon their information.
305 - 450 nm: ± 5 %
450 - 815 nm: ± 3 %
815 - 1200 nm: ± 5 %
s^{RC} (λ): calibration errors of PTB.
305 - 1100 nm: ± 0.6 %
1100 - 1200 nm: ± 1.5 %
s^{TC} (λ): is measured with a grating monochromator (GM).

non-linearity of the Lock-In-Amplifier (LIA):	± 1000 ppm
temperature dependence of the calibrated resistor:	+ 10 ppm/K
the LIA gain:	± 100 ppm/K
the LIA output:	+ 50 ppm/K
calibration error of the ref. cell:	± 10000 ppm

An inaccuracy of the GM of $\Delta\lambda$ = ± 0.2 nm and a shift of the band edge by ≈ ± 0.1 nm due to a temperature inaccuracy of the chuck of ± 0.3 K leads to errors at the steep flanks of s (λ); for crystalline Si:
305 - 450 nm: ± 1 %
1000 - 1200 nm: ± 1 %
All s^{TC} (λ)-errors are rounded up to
305 - 450 nm: ± 3 %
450 - 1000 nm: ± 2 %
1000 - 1200 nm: ± 3 %
With these errors for E_{sim} (λ), s^{RC} (λ) and s^{TC} (λ), a typical value for the bias error in M is

$$B_M = \pm 5400 \text{ ppm}$$

3.3 Total uncertainty limit

Table II lists bias limits (B) and precision indices (S).

	P_{max}	s	A	I^{RC}	M	η
B_i	+2000 -2800	±11200	±1000	±7800 -8600	±5400	±16000
S_i	2800	0	190	1840	300	3400

Table II: Bias limits and precision indices [ppm]

For t_{95} = 2.05 (degrees of freedom df=30) the total uncertainty limit becomes U_η = ± 2.3 % (resp. ± 4.4% for linear addition of all errors).

4. DISCUSSION

At a first glance these uncertainty limits look high, especially for the restricted assumptions. The linearly added value certainly is a conservative estimate of the known error sources. Since we never know all error sources and we want to give limits within which the true value (not the value of a calibration agency!) should fall, a conservative estimate is mandatory. Our uncertainty limits compare well with the results of the 1984/85 round robin calibration of reference cells /1/ where 8 of the internationally leading PV laboratories only agreed within ± 3 % to 4 % in the responsivity (no U_{OC}, no FF!) of crystalline Si. Therefore it seems that many researchers underestimate their uncertainty limits, most often taking reproducibility for uncertainty.

Non-uniformity is a very critical parameter and should be measured and controlled carefully. The size of the detector should correspond to the minimum dimensions of possible inhomogeneities of test cells. The CEC /11/ recommends a quarter of the TC area. For certain materials (e.g. poly-Si) this might still be too large.

5. CONCLUSIONS

A detailed uncertainty analysis reveals surprisingly many error sources and guides improvement of the equipment. The uncertainty limits of ± 2.3 % (root square sum) resp. ± 4.4 % (linear sum) compare favourably to the results of an international round robin.This indicates that researchers tend to underestimate their uncertainty limits.

6. ACKNOWLEDGEMENTS

We gratefully acknowledge the hospitality of Keith Emery and his colleagues at SERI, who passed on invaluable information about their efficiency measurements to us. Special thanks go to Jürgen Metzdorf and his group at PTB for providing us with reference cells, to Siegfried Kunzelmann and Silvia Strensch at ISE for the measurements, Armin Zastrow and other colleagues for valuable discussions and Helen Rose Wilson for correcting the manuscript. This work was supported by the German Bundesminister für Forschung und Technologie under Contract 03E-8650-B.

7. REFERENCES

/1/ H. Ossenbrink. R. Van Steenwinkel, K. Krebs, CEC preprint EUR 10613EN
/2/ K.A. Emery, C.R. Osterwald, C.V. Wells, Proc. 19th IEEE PV Spec. Conf., p. 153, New Orleans, 1987
/3/ "Measurement Uncertainty Handbook", Instrument Society of America, 1980 (ISBN 0-87664-483)
/4/ Deutsche Norm DIN 1319, Benth-Verlag, Berlin, 1983
/5/ J. Metzdorf, private communication, April 18, 1988
/6/ J. Metzdorf, Appl. Optics, 26, 9 (1987)
/7/ K.D. Stock. R. Heine, Optik 71, 4 (1985) p. 137
/8/ R. Matson, K. Emery, R. Bird, Solar Cells 11 (1984)
/9/ C.R. Osterwald, PV Insolation and Measurement Workshop, July 1-3, 1985, Vail, CO
/10/ K.A. Emery, C.R. Osterwald, Solar Cells 17 (1986).
/11/ CEC specification 101, EUR 6423 EN, JRC Ispra, 1979

HARDWARE SIMULATION OF PHOTOVOLTAIC GENERATORS

P.U. Calzolari[+], G.C. Cardinali[*], E. Faldella[+],
F. Filicori[+], G. Rulli[+]

+ Dipartimento di Elettronica, Informatica e Sistemistica
University of Bologna, Viale Risorgimento, 1
I-40136 Bologna, Italy - Tel (051) 6443001
* CNR - Istituto LAMEL, Via Castagnoli, 1
I-40126 Bologna, Italy - Tel (051) 287011

ABSTRACT. A method for the electrical simulation of Photo-Voltaic (PV) generators is proposed. It is based on the digital control, effected by means of a personal computer, of a conventional programmable power supply. Any output voltage-current characteristic defined by a mathematical model of the PV generator can be accurately simulated by means of a special-purpose control algorithm; the correct value for the voltage to be provided by the power supply is obtained, on the basis of real-time measurements of the load current, by means of an iterative approximation procedure. The simulator has been experimentally tested under several different load conditions of practical interest. In all cases good agreement has been found between the measured V-I curve and the mathematical model of the PV generator. The electrical simulator is a valuable tool in the laboratory testing and accelerated performance evaluation of PV systems.

1. INTRODUCTION

In the development of Photo-Voltaic (PV) systems designers have to deal with the difficult problem of characterizing the system performance under realistic operating conditions. For instance, it is quite important to carry out an experimental evaluation of the power matching between the generator and the load or to check the validity of the energy-management strategy adopted. In practice such measurements are difficult and time-consuming since using a PV generator in a laboratory environment is generally impractical or even impossible.

These problems can be overcome by using a hardware simulator which can be programmed in such a way as to provide the same electrical regime as the real PV generator would under given environmental conditions. To this aim, a simulator of PV generators, based on analog control circuits, has been already developed [1].

In this paper a digital approach to PV generator simulation, which does not require any special-purpose circuit to be built, is proposed; it is based on the digital control, performed by means of a Personal Computer, of a Programmable Power Supply (PPS). The computer senses, through an analog/ /digital converter, both the voltage and the current at the load and controls the power supply in order to obtain the same output voltage which would be provided by the simulated PV generator.

Accurate models residing in the PC memory allow for the simulation of any photovoltaic module on the basis of its standard experimental parameters and, therefore, of any series-parallel connected array. The interactive structure of the control program enables the user to establish the simulated environmental conditions according to different operation modes: for instance, a time dependence corresponding to an average daily irradiance curve (but obviously with a contracted time scale), or to a series of less regular events, can easily be simulated.

In the following sections the basic features of this simulator and, in particular, of its control algorithm, are described; the experimental results obtained from a prototype simulator are also presented and discussed.

2. THE HARDWARE SIMULATOR

A hardware simulator of PV generators is basically a non-linear bipole having a given voltage-current characteristic which can be defined by the non-linear equation

$$G(V, I) = 0 \tag{1}$$

For instance, if the well-known equavalent circuit in fig. 1 is adopted [2], the corresponding V-I characteristic takes the form

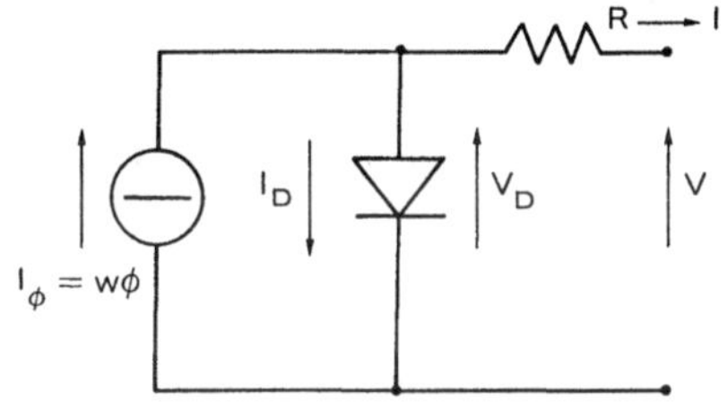

Fig. 1. A classical equivalent circuit for PV generators.

$$I_S\left[\exp\left(\frac{V + RI}{nV_T}\right) - 1\right] - I_\phi + I = 0 \tag{2}$$

where I_ϕ is the light-generated current, which is proportional to the radiation power density ϕ. Most of the model parameters and, in particular, the inverse saturation current I_S, are dependent on the cell temperature T; thus the V-I characteristic defined by eqn. 1 or 2 is dependent both on ϕ and T.

In the electrical simulation of PV generators a suitable power source, with maximum output voltage and current ratings not smaller than those of the given PV generator, is needed. To this aim a mains-fed Programmable D.C. Power Supply (PPS) (possibly one of the many commercially available units) can be conveniently used. Moreover, in order to obtain the output V-I characteristic defined by equation 1, a suitable control system for the PPS is also needed.

In most cases, in fact, the output voltage V provided by a programmable power supply is almost current-independent and only depends on the value of a controlling variable x according to a linear law

$$V = Ax .$$

Thus, for the correct simulation of a given PV generator, the value of the controlling variable x must be suitably chosen so that the output voltage V satisfies the model equation (1) for any output load. In other words, the output voltage V should satisfy both the source equation 1 and the load equation

$$I = F(V) \tag{3}$$

as shown in fig. 2.

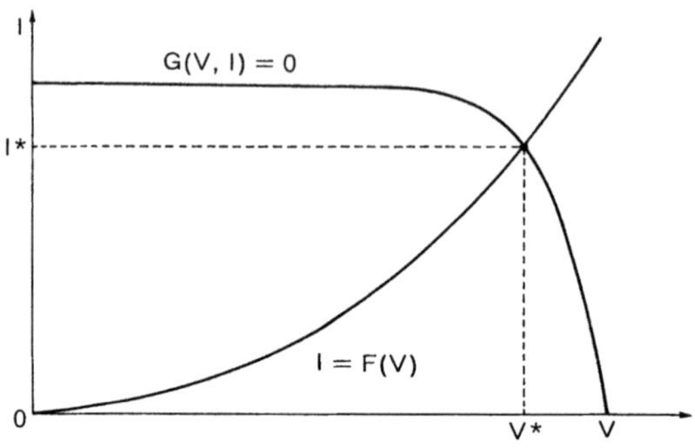

Fig. 2. Plots of the voltage-current characteristics of a PV generator (G) and the load (F); V^*, I^* is the corresponding steady-state operating point.

Since the load V-I characteristic (3) is not known in advance and may also be time-varying, a current sensor (e.g. a shunt resistor or a Hall-effect device) and a real-time feed-back control (see fig. 3) are needed in order to provide the correct value for the controlling variable x on the basis of the present value I of the current actually flowing into the load.

Such a feed-back system can be implemented analogically [1] through an operational circuit which basically reproduces at the signal level the source equivalent circuit in fig. 1. Such an analog approach, however, may be quite troublesome in practice since the open-loop gain of the feed-back controller intrinsically depends on the differential conductances both of the load and of the PV generator, which can be extremely variable in the different possible operating conditions. This may lead to poor accuracy for low open-loop gain or severe stability problems in the opposite case. Moreover, it would be quite difficult to simulate with an analog network the variations of the model parameters deriving, for instance, from temperature-related phenomena.

For such reasons our simulator is based on a digital controller consisting of a suitable control algorithm running on a general-purpose personal computer equipped with standard A/D and D/A conversion interfaces as shown in fig. 3. This approach has the advantage of allowing for the accurate simulation of PV generators without any limitation on the type of model adopted for the PVG (e.g. temperature dependence can easily be taken into account); moreover, the structure in fig. 3 can easily be assembled using commercially available equipment, with no need for special-purpose circuits to be built. This approach is, in practice, also cost-effective since a computer is needed anyway for data acquisition and processing purposes in an efficient measurement set-up.

Therefore the only special-purpose structure which is needed to implement the PVG simulator is strictly software and consists of a suitable control algorithm which is described in the following section.

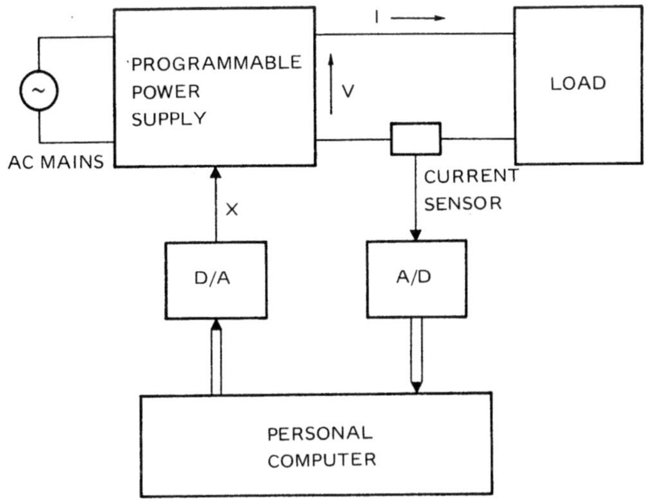

Fig. 3. Hardware simulation of PV generators based on digital feed-back control of a programmable power supply.

3. THE CONTROL ALGORITHM

When using a digital computer as the control unit for the PVG simulator, the synthesis of the control algorithm can be dealt with as the problem of solving a system of non-linear equations. In fact, in order to correctly control the PPS, the value V^* of the output voltage which satisfies both the model equation 1 and the load equation 3 must be sought for.

To this aim, a classical iterative numerical technique (e.g. the Newton-Raphson algorithm) for solving non-linear equations could be used; however it should be considered that, due to the very different features of eqns. 1 and 3, any general-purpose algorithm would be quite inefficient in our case.

The model eqn. 1 is in fact strongly non-linear, but is usually defined by a simple analytical expression (e.g. eqn. 2) which can be rapidly computed with a few arithmetical operations. The load eqn. 3, instead, is generally mildly non-linear but is not known analytically and can only be experimentally evaluated for a limited number of values of the output voltage V. In practice each evaluation of the load characteristic F is relatively time-consuming since it involves applying a given voltage V to the load through the PPS and then measuring the corresponding load current I, only after a settling time not smaller than the sum of the settling times of the PPS and the load.

Therefore, in order to make the PVG simulator controller as fast as possible, a special-purpose iterative algorithm for the solution of eqns. 1 and 3 is needed; this algorithm should require a very small number of evaluations (e.g. "measurements") of the load equation; the number of computations of the model eqn. 1, on the other hand, may necessarily be larger owing to its strong non-linearity, but is not relevant for the overall speed of the simulator since the corresponding computing times are, comparatively, much smaller.

These requirements can be met by applying separate iterative approximation procedures to the PVG model and the load equation; in particular the proposed algorithm consists of a double-loop iterative solution algorithm described by the flow-chart in fig. 4.

In the outer loop, at any k^{th} iteration the load eqn. 3 is approximated by the linear expression

$$I = F(V) \simeq I_k + (V - V_k)\,\frac{I_k - I_{k-1}}{V_k - V_{k-1}}\,, \qquad (6)$$

I_k, I_{k-1}, V_k, V_{k-1} being the load currents and voltages obtained in the last two iterations. Then a new approximated value V_{k+1} for the correct output voltage is computed by searching for the solution of eqns. 1 and 6; this problem, which is now easier since only equations defined by simple analytical expressions are involved, can be solved in the inner loop by using any classical iterative technique such as, for instance, the Newton-Raphson algorithm.

The new computed voltage V_{k+1} is then applied to the load through the PPS and, after the necessary settling time, the corresponding steady-state load current I_{k+1} is measured by the current sensor. If the inequality

$$|G(V_{k+1}, I_{k+1})| < \varepsilon \qquad (7)$$

is satisfied, a sufficiently accurate approximation of the correct output voltage has been found; otherwise more iterations of the outer loop are needed. When the accuracy test 7 is satisfied, the control loop can be stopped until a change in the simulated power density ϕ or temperature T is required or some modification in the load characteristic F occurs; in the meantime the PC can perform other auxiliary operation such as acquisition of data concerning the load operating conditions, computation of performance parameters (e.g. efficiency, output power etc.) or display of results.

Since in most practical cases the load V-I characteristic F is mildly non-linear, the convergence of the outer iterative loop is quite fast and reliable; in practice, even when starting from a rough initial estimate + for the correct output voltage, only a few iterations (e.g. 2 to 10) are necessary to achieve good simulation accuracy. Thus, taking into account that the computing times for the inner iteration (e.g. the N - R algorithm) are comparatively negligible, the average time necessary for the simulator to give the correct output voltage is generally between 2 and 10 times the sum of the settling times of the PPS and the load.

Such a response time is certainly acceptable for most testing purposes since the actual dynamics of the phenomena to be simulated are always much slower.

The above-described control algorithm also includes (see dashed line in fig. 4) the possibility of performing a sequence of simulations at different values for the power density ϕ and, possibly, also for the cell temperature T, since the values of these parameters can be assigned as given functions of time. Thus, on the basis of experimental data characterizing the daily irradiance at the sites where the PV generators should be installed, accurate accelerated tests can be carried out on solar-energy-fed systems.

+ A rough initial estimate for the output voltage, which must be assigned when starting the control algorithm, can be obtained by halving the open-circuit voltage of the PVG to be simulated.

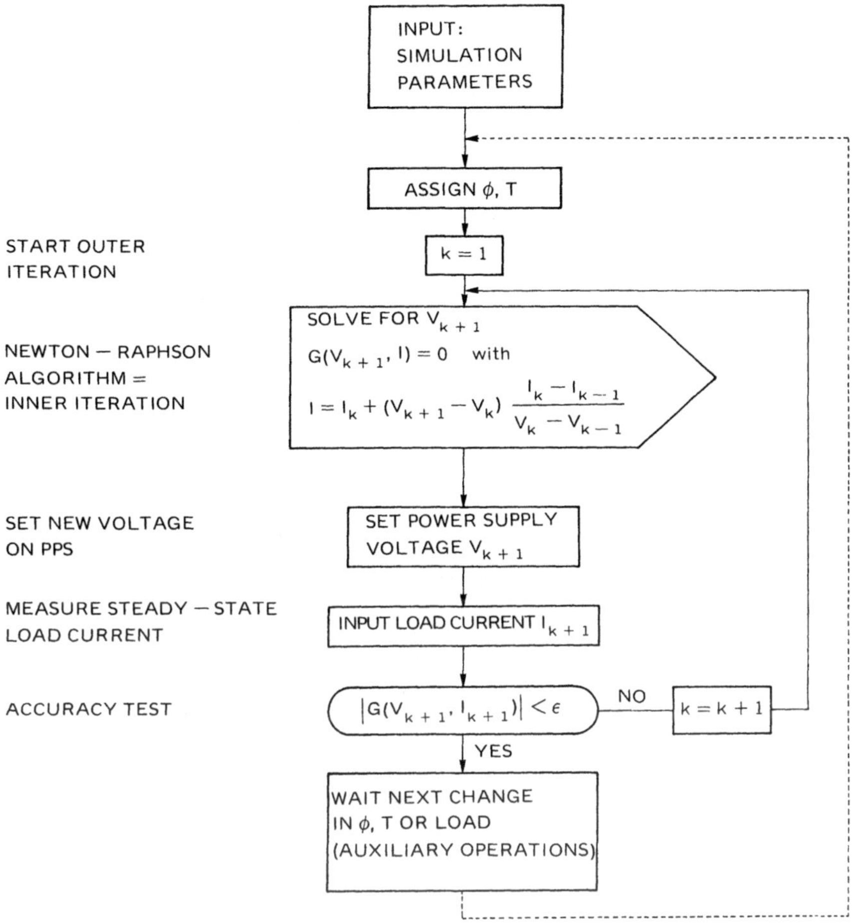

Fig. 4. Flow-chart of the digital control algorithm for the PVG simulator.

4. EXPERIMENTAL RESULTS

A prototype of the above-described simulator of PV generators was assembled by using a 150 V, 10 A programmable power supply as the power source and an IBM AT personal computer, equipped with a 12 bit analog/digital conversion interface, as the control unit. The programming of the control algorithm was carried out in FORTRAN.

The simulator was tested under several different load conditions including, for instance, a water pump driven by a d.c. motor and an a.c. operated fan driven by a transistor inverter. The convergence of the simulator to the correct output voltage was found to be fast and realiable: in most cases only a few iterations of the control algorithm (i.e. a few different trial settings of the output voltage) were sufficient to achieve good accuracy (about 1% discrepancy of the output voltage or current from the V-I curve of the given PV generator).

Some minor convergence problems were found only when simulating a very low irradiance with an electric motor as the load. The reason for this can easily be understood by considering that, with a very low power generated by the PVG, the motor is almost at a standstill and the mechanical hysteresis associated with starting friction phenomena makes the corresponding portion of the V-I curve of such a load poorly repeatable. This is clearly a special situation for which, in any case, an accurate simulation is not even necessary since the energy flow involved is practically negligible.

By using a variable load, which allowed for a gradual variation of the output current, the experimental voltage-current curve of the simulator shown in fig. 5 was obtained for a given irradiance ϕ and cell temperature T. The very small discrepancy between the measured V-I pairs and the output curve of the corresponding mathematical model confirms the validity of the simulation technique proposed.

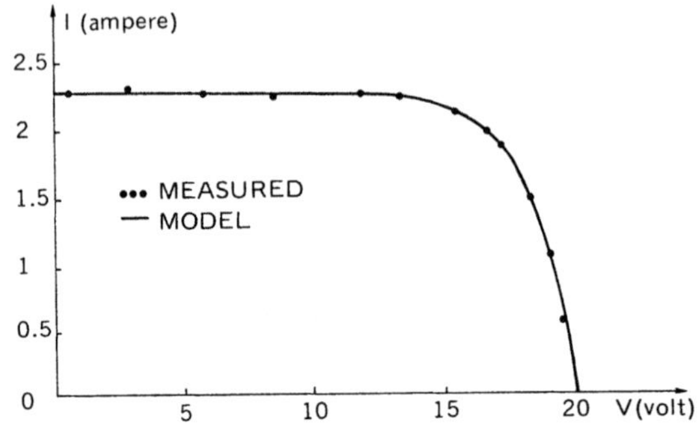

Fig. 5. Comparison between the mathematical model of a PV generator and the corresponding voltage-current curve of the electrical simulator.

5. CONCLUSIONS

For the laboratory testing of PV systems, an efficient method for the electrical simulation of PV generators has been proposed. It is based on the digital control, effected by means of a personal computer equipped with analog/digital conversion interfaces, of a conventional programmable power supply. The control algorithm consists of a special-purpose iterative technique which provides the correct value of the output voltage on the basis of real-time measurements of the load current.

A number of experimental tests carried out on a prototype simulator has demonstrated the good accuracy and robustness of the control algorithm.

Hardware simulation of PV generators is a valuable tool in the development of PV systems, since accelerated measurements of long-term performance parameters can be rapidily carried out and many different possible operating conditions can be practically tested. The simulation method proposed here has also the advantage of not requiring any special-purpose control circuit; the control function is in fact performed via software, while only conventional and commercially available hardware is needed.

REFERENCES

[1] L. Mrig, R. De Blasio, G.A. O'Sullivan and R.P.Tomko: *An advanced photo-voltaic systems simulator to demonstrate the performance of advanced photovoltaic cells and devices*, Proc. IEEE Photovoltaic Specialist Conference, pp. 199-204, 1982.

[2] P.U. Calzolari, G.C. Cardinali, E. Faldella, D. Nobili, A. Zani: *An identification technique for photovoltaic generators in operation*, Proc. MELECON '85, Madrid.

EXPERIENCE WITH A DATA ACQUISITION SYSTEM IN PAOMIA

D. MAYER, P. NOLAY
ECOLE DES MINES DE PARIS
Centre d'Energétique
Rue C. Daunesse
Sophia-Antipolis
F - 06560 VALBONNE

J. CAPILLON
ELECTRICITE DE FRANCE
Division des Etudes et des Recherches
6, quai Watier
F - 78401 CHATOU

G. PERI
UNIVERSITE DE CORSE
C N R S
Vignola
Route des Sanguinaires
F - 20000 AJACCIO

ABSTRACT

The new data acquisition system installed in PAOMIA PV plant is briefly described. The first results available for a three-month period are analysed in terms of overall efficiency. An identification method is presented for the determination of the components characteristics giving the HERMINES PV code the data required to achieve a future optimisation study of the plant performances.

1. INTRODUCTION

The analysis of the behaviour and the performances of the Paomia PV pilot plant could not be achieved in good conditions by the fact that the data acquisition system (DAS) initially installed in June 1983 only delivred adequate measurements for a three-month period.

The installation of a new DAS, based on a static memories system, has allowed to start new series of measurements since October 1987. The analysis of these data and their future treatment with the HERMINES PV simulation code will lead to an optimisation of this pilot plant operation.

2. DAS PRESENTATION

The specific innovation of the ASITEC DAS lies in the choice of an entirely static solution, for the data acquisition as well as for their storage.

A battery card allows a 36 hours off-line working without power supply.

• It is possible to acquire with the standard system 16 voltage analog channels (0 - 10 V or -10 - 10 V). An extension card allows to get 16 channels more.

• The data are stored on 512 kbytes static memory cartridges corresponding to a period of about 2 months for 16 channels (10 minutes acquisition periods).

• These data are arithmetic averages achieved from the acquired values every second or 10 seconds as you wish.

• During an experimentation it is possible to modify the acquisition period but not the number of channels.

• The data reading can be achieved directly with a micro-computer either by the mean of a RS 232 connection or in a remote place by the mean of two modems, without disturbing the acquisition.

3. THE PAOMIA PV PILOT PLANT

Only the basic characteristics of the Paomia PV pilot plant are recalled in this paper. For further information you can refer to /1/. The operating scheme of the installation is shown in fig. 1.

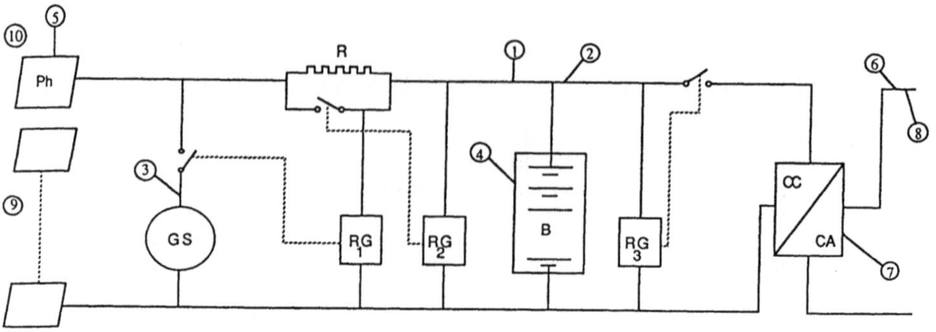

Fig. 1 : Schematic description of the Paomia plant

Ph : Photovoltaic array composed of 102 pannels in parallel and 2 pannels in series. Each pannel consists of 6 modules in series (36Wc). The maximum power delivered is 44 kW (220 A, 200 V).

GS : Emergency generating set. R : Resistance

B : Batteries composed of 84 elements of 2500 Ah. CC/CA : Inverter

RG1 : Prealarm controller for the starting of the emergency generating set when the voltage is below 160 V. It stops when the voltage reaches 185 V.

RG2 : Charge controller. When the voltage reaches 201,6 V the resistance R is connected. It is disconnected when the voltage is below 185 V.

RG3 : Discharge controller. When the voltage is below 155 V, the inverter is disconnected. The reconnection is made manually when the voltage has reached 167 V.

Location of the connected sensors.

1 : battery output current (A)
2 : battery input current (A)
3 : G.S. current (A)
4 : battery voltage (V)
5 : array output voltage (V)
6 : AC on one phase (A)
7 : A voltage on one phase (V)
8 : Active power (kW)
9 : planar irradience (W/m^2)
10 : tilted irradience (W/m^2)
11 : ambiant temperature (°C).

4. COMPONENT CHARACTERISATION

In order to simulate with a good accuracy the behaviour of the pilot plant with the HERMINES PV software, it seems quite essential to determine the real characteristics of the different components.

An identification algorithm based on the Marquardt method has been developed for the determination from experimental data of all the parameters taken into account in the components math models /1//2/.

This is achieved by minimising a norm of error between experimental and calculated values:

$J = \frac{1}{N} \sqrt{(y_{oi} - y_{im})^2}$ N : number of data

y_{im} : model output
y_{oi} : experimental data

The model structure being fixed, the algorithm convergence validates the model selection and allows to get an optimal set of parameters. Its divergence leads to change the model hypothesis.

This method has been applied to the characterisation of a part of the Paomia PV plant array. Our module model for a PV model is the following.

$$I = I_L - I_O \left[\exp\left[\frac{q}{nAkTj}(V + R_S I) - 1\right]\right] - \frac{V}{Rsh}$$

$$I_L = P_1 E (1 + P_2 (E - Eref) + P_3 (Tj - Tref))$$

$$I_O = P_4\, Tj^3\, e^{-P_{5/Tj}}$$

where :

A = number of cells in series in a module
n = ideality factor
E = irradience at the tilt angle (kW/m^2)
Tj = mean cell temperature (°C)
Rs = serial resistance
Rsh = shunt resistance
P_1 P_2 P_3 P_4 P_5 = parameters related to the junction and generated currents

In this model eight parameters (n, Rs, Rsh, Pi) cannot be accurately determined therotically. The use of the identification method gives the results reported in table 1 and fig. 2 :

	n	Rs Ω	Rsh Ω	P_1 Am^2/kW	P_2 m^2kW	P_3 1/K	P_4 A/K^3	P_5 K
PAOMIA 12 modules en série 48 en parallele	1,5	3,37	31,1	103,5	-0,2	0,0165	76	9555

Table 1 : Parameters value for parts of the PAOMIA array

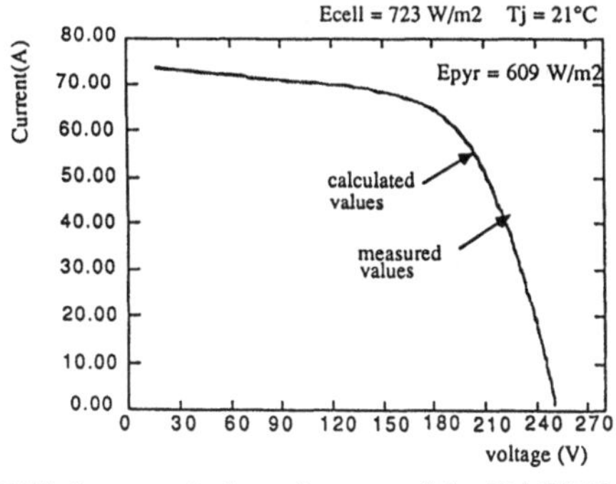

Fig. 2 : I-V characteristics of a part of the PAOMIA array

This method can also be applied to the battery model, but the data required are not so easily available.

5. FIRST RESULTS

The first experimental data have been available since October 1987. We have chosen to present the array production for a sequence in October (fig. 3) and the different results for a day which is completely representative of the global behaviour of this pilot plant. The array production, the system voltage, the consumption and the global irradience at the tilt angle are shown respectively in fig 4, 5, 6 and 7.

It seems clear that the power consumption is much too weak and the serial resistance is very soon connected limiting, by that fact , the array production.

Different kind of efficiencies defined by UTE (Union Technique de l'Electricité) for the characterisation of PV systems on site are those which are in question in the following :

- $K_{CG} = \frac{E_u}{P_o \cdot N_h}$ overall capacity ratio
- $R_C = \frac{E_{PV}}{G_{ri}}$ effective conversion efficiency
- $R_G = \frac{E_u}{G_{ri}}$ effective global energetic efficiency

	K_{CG} (%)	R_C (%)	R_G (%)
Nov. 87	3	2,7	2,4
Jan. 88	4	4,2	3,4
Global	3,6	3,5	2,9

where :

E_u : total consumed energy
E_{PV} : total energy produced
G_{ri} : global irradience at the tilt angle
P_o : peak power
N_h : number of hours in the reference period.

6. CONCLUSION

The analysis of the first results presented in this paper puts the emphasis on two major points:

- the DAS intalled now seems reliable and until now is working perfectly well.

- The adaptation of the HERMINES PV code to simulate the behaviour of the PAOMIA PV plant is available.

Now it is possible to achieve an optimisation study on this plant for which, at the present time, it seems difficult to draw any conclusions.

REFERENCES

/1/ J. CAPILLON, S. MASSAD "Point sur les centrales photovoltaïques de PAOMIA (Corse) et KAN (Guyane)" - EdF - Déc. 86.

/2/ P. NOLAY "Développement d'une méthode générale d'analyse des systèmes phoptovoltaïques" - Thèse de 3ème cycle, Sept. 87.

/3/ D.W. MARQUARDT "An algorithm for least-squares estimation of non linear parameters" J. Soc. Indust. Appl. Math. June 1963.

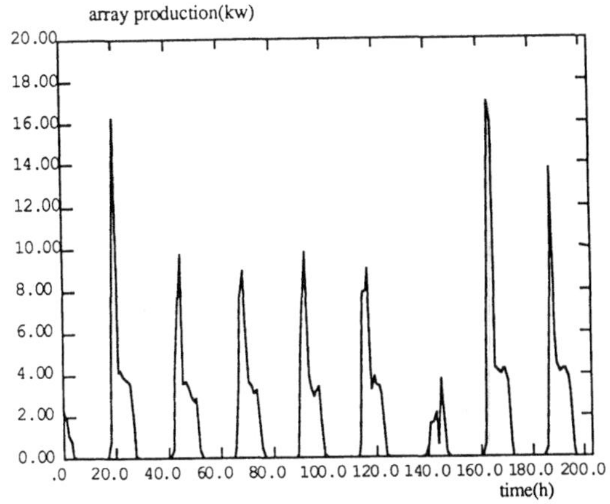

Fig. 3 : PV array production for a 8-day sequence in October

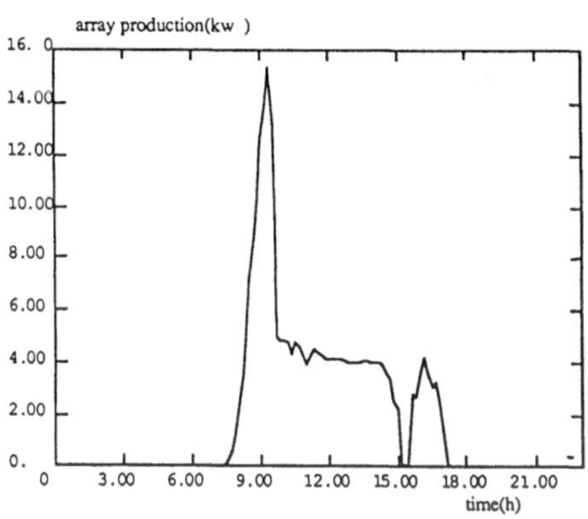

Fig. 4 : Daily array production

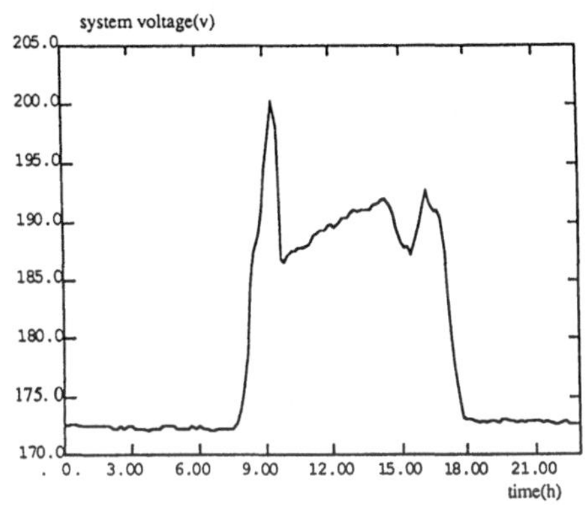

Fig. 5 : System voltage

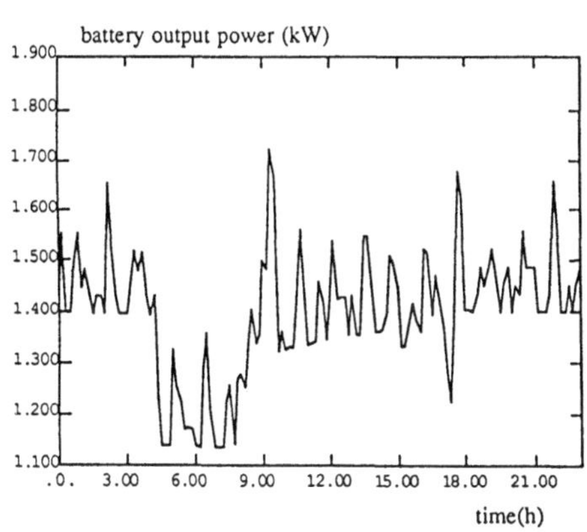

Fig. 6 : Battery daily output power

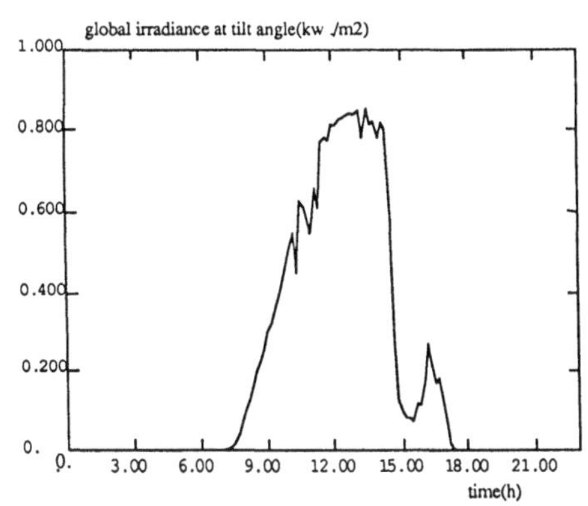

Fig. 7 : Global tilted irradience

NOVEL TEST FACILITY FOR PV PUMPING SYSTEMS

Authors: Dr. W. Bechteler, University of the Bundeswehr, Munich, *
Dr. W. Bucher, DFVLR (German Aerospace Research Establishment), Cologne ●

ABSTRACT:

For tests and development work on p.v.-pumps for arid areas the simulation of environmental conditions in the laboratory is necessary. This is performed with the help of a test equipment designed and manufactured to fulfill the requirements.
The equipment facilitates prequalification tests and comparative investigations with different pumps at well defined power conditions. Experiences with the system and results of the research work indicate that this facility is very advantageous.

For qualification tests of photovoltaic pumps and for system development sunshine conditions in Germany are very unfavourable. To make these tasks easy and not dependent on seasonal variation of solar radiation a laboratory test loop was installed, which allows testing of well pumps of up to 5 kW power in the pressure range of 15...60 m water head.

Fig. 1 shows the test loop with its hydraulic and electric components. Specific features of this test facility are:

- a pressurized tank, where the pressure level can be set to a desired value prior to starting the pump. This is essential for the start-up of pumps of the displacement type, because of the influence of the water head on the start-up-torque requirements;

- an automatic (pressure controlled) valve keeping the pressure in the preset range independently from the amount of water pumped;

- a large water reservoir, which is used as a heat sink, keeping the water temperature in the loop nearly constant, even at long-term tests;

- a computer controlled power source of 5,5 kW rated power providing the necessary DC-power from the grid. An electronic interface unit is part of the power source and gives the opportunity to influence output voltage and current by input data from a computer or by manual settings. The computer control includes a data feedback, measuring the electric output of the power source and adjusting it to the desired level;

- a set of meters and sensors for monitoring the system.

Fig. 2 is a compilation of the data sampling concept and of the sensor locations. The measured values are converted into electric signals and transferred to a data logger in preset intervals (1 s to 10 s).

* Prof. Dr. W. Bechteler, Institut für Wasserwesen, Univers. der Bundeswehr, Werner-Heisenberg-Weg 19, 8o14 Neubiberg

● Dr. W. Bucher, DFVLR; RF-ET, Linder Höhe, 5ooo Köln 9o

Besides the manual operation, which is used for efficiency measurements and long term tests, the simulation of weather conditions is the main purpose of the computer control of the power pack. An autonomous (battery powered) weather recording station, consisting of p.v. panels, a meteorological sensor equipment, and a data logger/recorder is provided to collect the necessary insolation data on site.

In Fig. 3 a flow diagram of the data transfer for power source control is plotted. Insolation data - either recorded on a standard 3,5" floppy disc or defined by a data set in a memory resident file - can be processed in a computer to calculate the power output values imposed to the power unit.

The software for the data processing and control of the operation of the power pack allows modification of parameters, e. g. the nominal output power of the DC source (panel array), fillfactor, voltage or current limits etc., but also to expand or condense the time scale (abbreviating cycling tests).

Nevertheless, the sampling rate of the feed-back circuit and the internal controls of the power pack do not allow to model quick transients (time constants less than 1 s). Dynamic response of p.v. panels was not intended to be simulated by the equipment.

Figs. 4 and 5 show some results of tests with the facility described:

In Fig. 4 two pump characteristics have been compiled: a standard centrifugal and a progressive cavity pump were put against each other. It should be noted that the efficiencies given in the figure are measured values, correlating the hydraulic output to the DC-input, thus including all losses of inverter, motor, pump and hydraulic losses in pipes and fittings.

The pumps were operated for a simulated "shunshine curve" of a fair day. Fig. 5 shows the output as a function of time. It is obvious that at part load the displacement pump is superior due to less power requirements at given water head.

Besides the comparative tests with pumps of different types tests with different motors and inverters have been performed with interesting results.

With the help of another test loop wear tests with p.v.-pumps are underway. Up to 1 % of quartz particles (medium diameter 0,2 mm) are mixed with the water circulating in a closed loop, to investigate mechanisms and results of pumping abrasive media in pumps of different types. Existing experience reported so far was found not to be satisfactory, since data available for mixtures of water and sand do not allow long-term behaviour assessment.

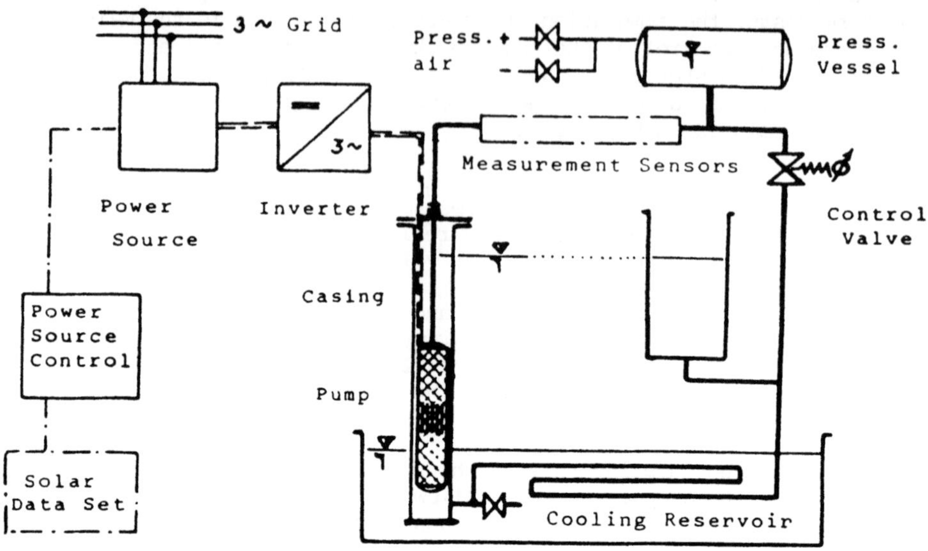

FIG. 1 : **Test facility (schematic), hydraulic test loop and electric power circuit**

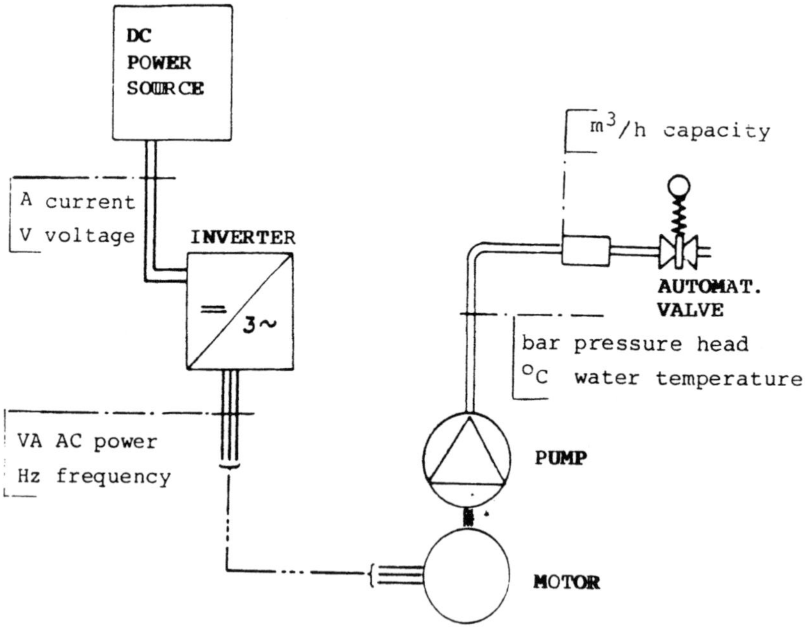

FIG. 2 : Test data sampling concept:
Indicated values are measured and recorded

FIG. 3: ⟹
Data transfer from insolation data measurement to power source control via a data processor

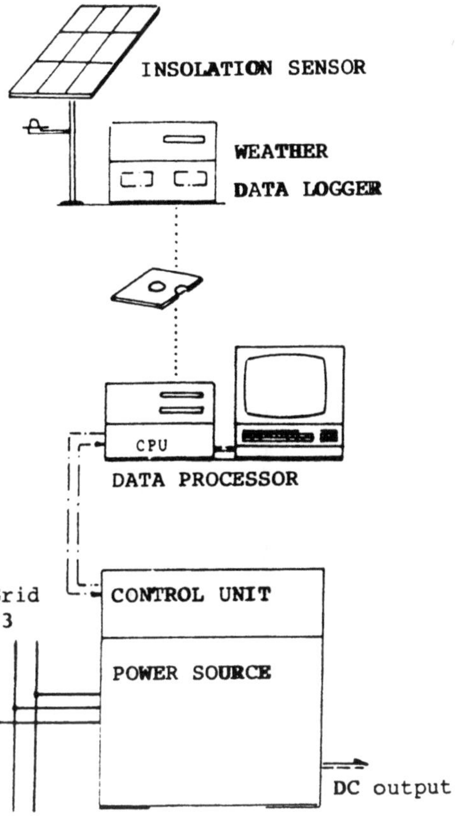

FIG. 4: ⇓
Performance charts of a centrifugal pump (left) and a progressive cavity pump (right)

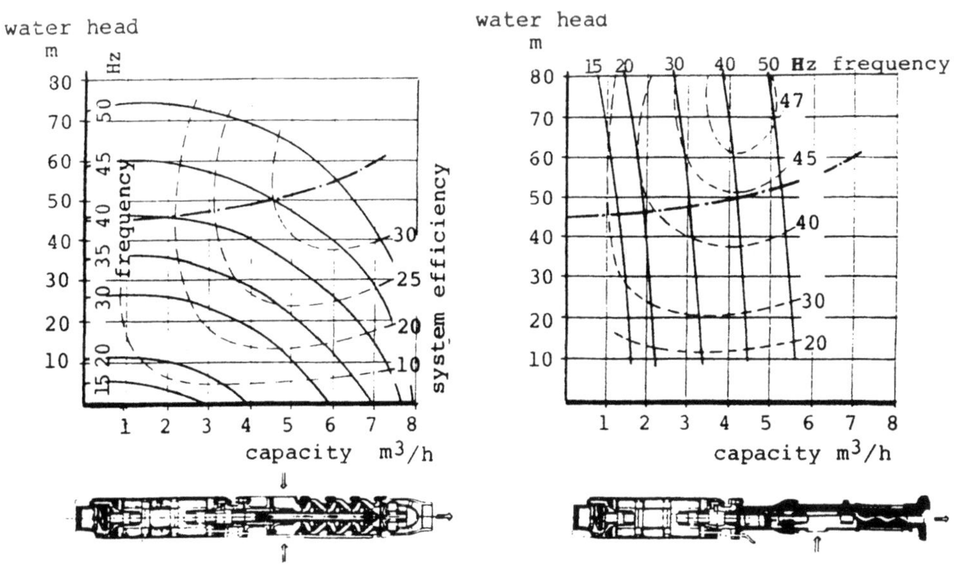

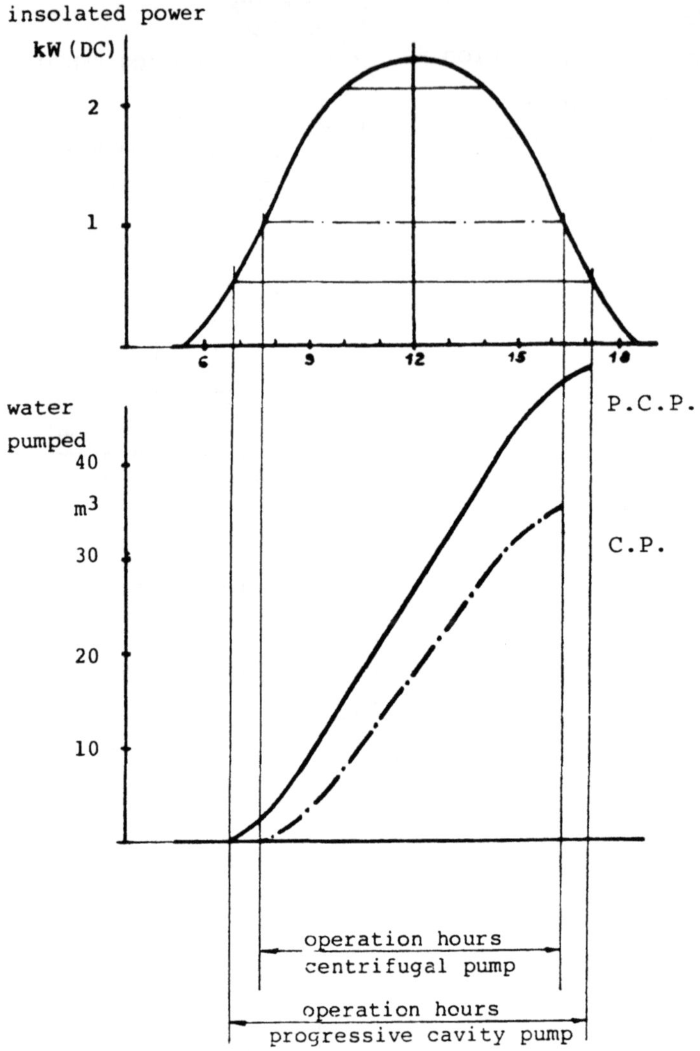

FIG. 5 : Amount of water pumped for a centrifugal (C.P.) and a progressive cavity pump (P.C.P.)

MEASUREMENT OF LOSSES IN PHOTOVOLTAIC ARRAYS ACCORDING TO THE CELL OCCULTATION PERCENTAGE

P. Ragot, D. Desmettre and P. Paes

Groupement pour l'Expérimentation des Pilotes Energétiques (GEPE)
Centre d'Etudes Nucléaires de Cadarache
13108 Saint Paul Lez Durance Cedex
France

ABSTRACT. Losses occurring when photovoltaic modules were switched into a network were quantified from field measurements on an array comprising twenty PHOTOWATT PWP 402 modules, with particular attention to the influence of the cell occultation percentage on the array productivity drop. In the test configuration (with one cell occulted in each of ten modules) occulting 50% of the cell surface area corresponded to 0.7% of the total array surface area, resulting in a loss of about 23% from the initial power level. This result shows the paramount influence of cell occultation in array performance, and the importance of maintenance practices in preventing significant PV array losses.

INTRODUCTION

As part of the French photovoltaic energy program, the *Agence Française pour la Maîtrise de l'Energie* (AFME) and the *Commissariat à l'Energie Atomique* (CEA) have set up a test facility at Cadarache to measure the performance of photovoltaic components and systems under actual outdoor conditions. One of the objectives of the program is to measure the quantitative importance of phenomena liable to occur in photovoltaic arrays in field operating conditions. We are particularly interested in cell occultation liable to occur naturally, and in evaluating the influence of the cell occultation percentage on array performance.

The study was carried out on an array of twenty PHOTOWATT PWP 402 (40 Wp) modules comprising five parallel-connected strings, each with four series-connected modules.

TEST FACILITY AND METHODOLOGY

The facility consists of modules exposed to sunlight at a fixed 45° tilt angle, with switching provisions allowing a series of tests to be conducted on various single, parallel and diode configurations. The data acquisition system includes an HP 150 microcomputer, an HP 1000 minicomputer and a meteorological station.

Reference cells are used for calibration purposes. The electronic load is driven by the computer. A total of 15 voltage points are used to plot an individual I–V curve. About four minutes are required to scan all 20 modules.

Based on measurements of individual modules, strings and string clusters, on the I–V performance characteristics at irradiation levels ranging from 800 to 1000 $W \cdot m^{-2}$ and on extrapolations from the test conditions (1000 $W \cdot m^{-2}$ at 50°C), the losses are determined for the cell occultation percentage, as well as for module connection into a network, for the presence of series diodes and for the wiring.

Ten cells (one per module) were occulted in 10% steps from 0 to 100% of the cell surface area with the following distribution: string 1 = 0 cell; string 2 = 1 cell; string 3 = 2 cells; string 4 = 3 cells; string 5 = 4 cells (Figure 1). The array power was calculated at $V_{LOAD} \approx 0.75\ V_{OC}$.

RESULTS

Figure 2 shows the I–V variation for module 15 according to the occultation percentage of one cell (τ). Some uncertainty remains on the actual τ values, due to nonlinear characteristics between 10 and 15 volts and to the fact that even after total occultation (τ = 100%) the module supplies a weak output current due to illumination of the cell edges transmitted by multiple reflections.

The most significant results are indicated in Table I. It can be seen that the available power loss is highly dependent on the load voltage, especially for low occultation values (0–10%): when τ = 10% and V_L = 14 V (≈ V_{max}) the power loss does not exceed 1.7% of the initial value. At higher τ values the power diminishes in a roughly linear manner as shown in Figure 3. The short-circuit current is relatively unaffected as a result of the parallel diode on the half-module.

A similar study of the complete generator provided the results indicated in Table II.

Once again it can be seen that the mismatch power loss is low for occultation values of less than 10%. In the arbitrary cell occultation configuration, where τ = 1.4% of the total silicon surface area, the available power (generator P_L) drops by 50% and mismatch losses represent about 15% of the total installed power rating (individual module ΣP_L).

Protection diode, cable and switching relay losses do not exceed 3% of the available power.

These results are more clearly illustrated in Figure 4, which plots the available power losses as a percentage of the installed power rating (τ = 0), and the mismatch, diode and cable losses versus the total array occultation percentage.

The effect of the protection diodes is shown in Figure 5, where the I–V characteristics for the string clusters in array M2 are plotted after extrapolation from the experimental curves obtained at 1000 $W \cdot m^{-2}$ and 50°C.

The results correspond to the selected array configuration (Figure 1) and are difficult to generalize because of the interaction between thermal and electrical effects which also depend on the applied load.

DISCUSSION

The characteristic I–V curves for photovoltaic modules, strings and string clusters (generator arrays) can be used to determine the power losses that occur when modules are connected in networks. The procedure developed by the authors for multicrystalline silicon module arrays in routine operation (i.e. functionally near the P_{max} region) has yielded the following information:

- The total electrical losses in the protection diodes and module interconnection wiring represent less than 2% of the total installed power.
- Network mismatch losses depend primarily on the dispersion of the module ratings and on the array configuration (strings of matched modules; nonreturn and/or parallel diodes).
- Generator productivity is severely penalized when the mismatch is increased by occulting some cells, even if this corresponds to a very low percentage of the total surface area. Thus, in the arbitrary configuration investigated here, occulting some 1.4% of the total silicon surface area of the array resulted in a 50% drop in available power compared with the same array when no cells were occulted.

ACKNOWLEDGMENTS

The authors are grateful to the *Agence Française pour la Maîtrise de l'Energie* for providing financial support.

Table I – PWP 402 Module Productivity versus Single Cell Occultation
(1000 $W \cdot m^{-2}$ 50°C)

Cell Occul-tation (%)	Module Occul-tation (Cell area) (%)	ISC (A)	I_L (V_L = 14 V) (A)	P_L (V_L = 14 V) (W)	$\frac{P_L(\tau)}{P_L(0)}$ (%)
0	0.00	2.75	2.51	35.20*	100.0
10	0.28	2.76	2.47	34.60	98.3
20	0.55	2.77	2.26	31.64	89.9
30	0.83	2.77	2.00	28.01	79.6
40	1.11	2.76	1.85	25.92	73.6
50	1.38	2.75	1.61	22.51	63.9
60	1.66	2.75	1.36	19.05	54.1
70	1.94	2.77	0.98	13.70	38.9
80	2.22	2.76	0.72	10.04	28.5
90	2.5	2.76	0.42	5.82	16.5
100	2.77	2.76	0.25	3.47	9.9

* (± 1.5%)

Table II – PWP 402 Generator Productivity versus Cell Occultation
(1000 $W \cdot m^{-2}$ 50°C V_L = 56 V)

Cell Occul-tation (%)	Module Occul-tation (Cell area) (%)	Individual Module ΣP_L (W)	Generator P_L w/o series diode (W)	Losses (W) Mismatch w/o series diode	Losses (W) Mismatch with series diode	Losses (W) Series Diode	Losses (W) Cables
0	0.00	686.6*	688.5*	-	5.0	5.0	6.9
10	0.14	682.7	687.8	-	4.9	4.9	6.9
20	0.28	653.8	649.1	4.7	9.4	4.7	6.1
30	0.42	619.9	594.2	25.7	30.0	4.3	5.2
40	0.55	602.9	571.1	31.8	35.8	4.0	4.6
50	0.69	569.9	528.9	40.2	43.7	3.7	4.1
60	0.83	534.7	489.7	45.0	48.4	3.4	3.5
70	0.97	479.1	423.9	55.2	58.2	3.0	2.9
80	1.11	445.3	392.4	52.9	55.6	2.7	2.6
90	1.25	404.8	357.3	47.4	49.8	2.4	2.4
100	1.39	375.9	318.9	57.0	59.1	2.1	2.2

* (± 1.5%)

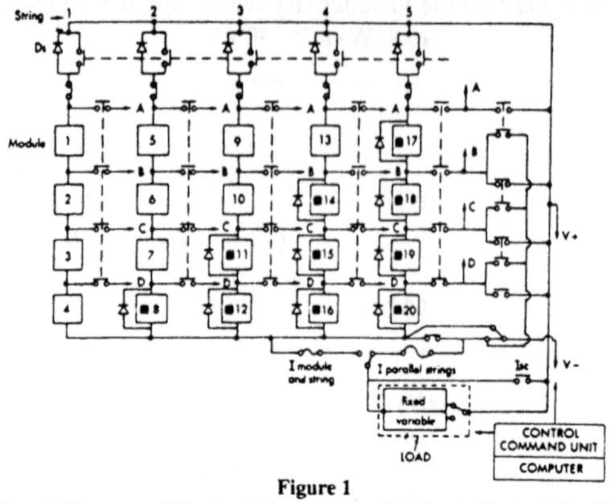

Figure 1
Circuit Diagram of Photovoltaic Array for Studies in Natural Sunlight of Individual Modules, Parallel Strings and Complete Array (1 parallel diode for 18 cells of each occulted module)

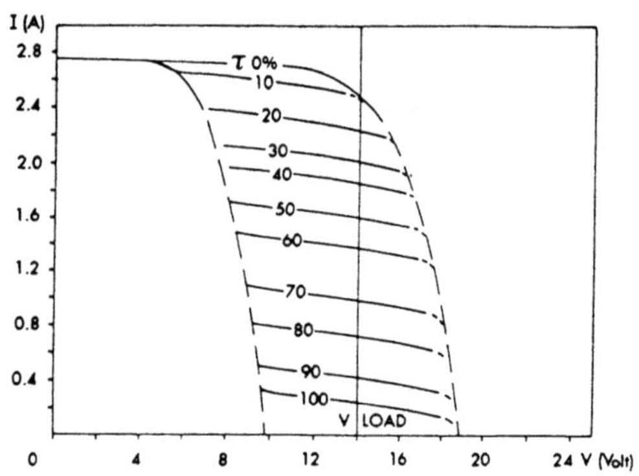

Figure 2
I-V Characteristics for Module 15 (1000 W·m^{-2}, 50°C) for Different Cell Occultation Percentages

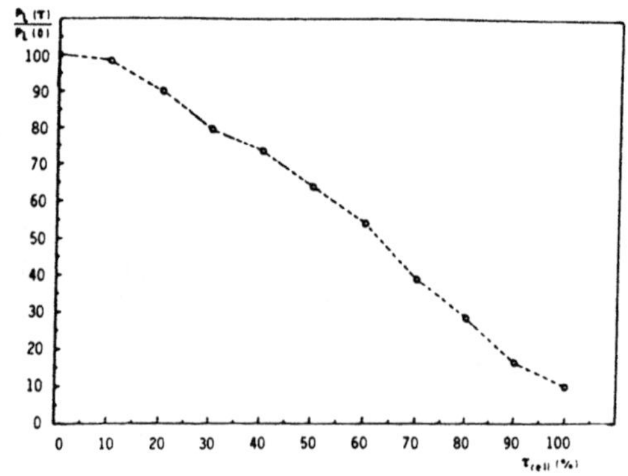

Figure 3
Relative Power Loss for a Module at $V_L \approx V_{max}$ versus Occultation Percentage for One Cell

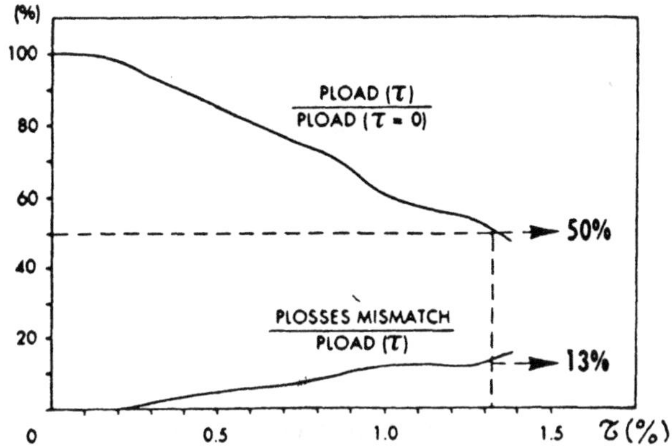

Figure 4
PWP 402 Array Power Loss versus Module Occultation Percentage

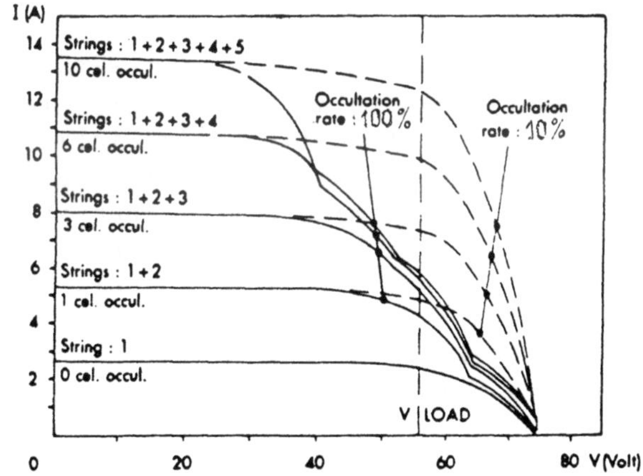

Figure 5
I-V Characteristics for Various PWP 402 Generator String Clusters (1000 W · m^{-2}, 50°C)

PHOTOVOLTAIC MODULE PERFORMANCE IN FINLAND

C. NYMAN, M. LEPPIHALME
TECHNICAL RESEARCH CENTRE OF FINLAND,
SEMICONDUCTOR LABORATORY
OTAKAARI 7 B, SF-02150 ESPOO, FINLAND

ABSTRACT. Evalution of photovoltaic module performance in Finnish climate conditions has been undertaken at two different locations in Finland. The test stations consist of modules from different manufacturers, containing cells of monocrystalline, polycrystalline and amorphous silicon. One of the monocrystalline modules works bifacially. The modules are mounted facing south at tilt angles of 45^0, 60^0, and 90^0 and the stations are located at latitudes of 60 and 67 degrees north.

1. INTRODUCTION

The purpose of this work was to study the performance and reliability of some typical solar cell modules in outdoor conditions, at two different latitudes in Finland, in order to get information for solar cell system design /1,2/. The influence of our specific climate was of special interest /3/. The test station at latitude 60^0 was constructed in 1985 on the laboratory roof in Otaniemi, close to Helsinki. The station consists of nine independent small scale units with a common measurement system. The second station is located on the top of mountain Pyhätunturi at 67^0. The station has six separate small scale units. The properties of the modules at the two stations are presented in Table I. Modules 1 and 2 of the Otaniemi station are identical but mounted at different tilt angles. The identical modules 6,7 and 8 are mounted at tilt angles of 45^0, 60^0 and 90^0, while the also identical module 9 is attached to a one axial sun tracking device.

The poly-Si modules 1 and 2 of the Pyhätunturi station are both identical. Modules 1 and 2 of the Otaniemi station and module 3 at Pyhätunturi are all identical as are both the amorphous silicon modules in Otaniemi and at Pyhätunturi (number 5).

Table I. Properties of the modules under test in Otaniemi (60^0N) and at Pyhätunturi (67^0N).

PROPERTIES OF THE MODULES UNDER TEST IN OTANIEMI 60° N

Mod. nr	type of cells	Module area(m2)	Power/Wp AM1, 25C°	Mounting
1 ■	c-Si	.28	45	45° tilt
2 ■	c-Si	.28	45	90° "
3 ■	c-Si	.28	34	60° "
4 ■	poly-Si	.40	40	60° "
5 ■	a-Si	2x.12	2x5	60° "
6 ■	c-Si	.36	43	45° "
7 ■	c-Si	.36	43	60° "
8 ■	c-Si	.36	43	90° "
9 ■	c-Si	.36	43	tracking

PROPERTIES OF THE MODULES UNDER TEST IN PYHÄTUNTURI 67° N

Mod nr	type of cells	Module area(m2)	Power/Wp AM1,25C°	Mounting
1 ■	poly-Si	.36	44	60° tilt
2 ■	poly-Si	.36	44	90° "
3 ■	c-Si	.28	45	60° "
4 ■	c-Si	.28	33,6	60° "
5 ■	a-Si	2x.12	2x5	60° "
6 ■	c-Si	.33	47	60° "

2. MODULE PERFORMANCE

2.1 OTANIEMI (60^0N)

The daily energy output in 1987 of modules 1-5 is presented in fig.1. and the daily energy output during 6/87-1/88 of modules 6-9 is presented in fig.2.

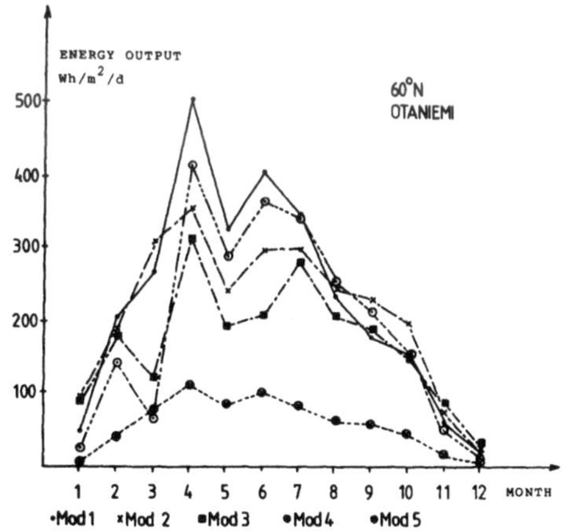

Fig.1. The daily energy output of the modules 1-5 in Otaniemi (60^0N) during 1-12/87

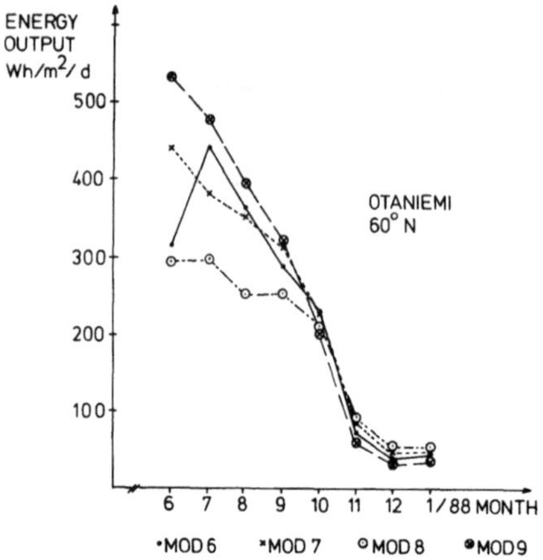

Fig. 2. The daily energy output of the modules 6-9 in Otaniemi (60^0N) during 6/87-1/88.

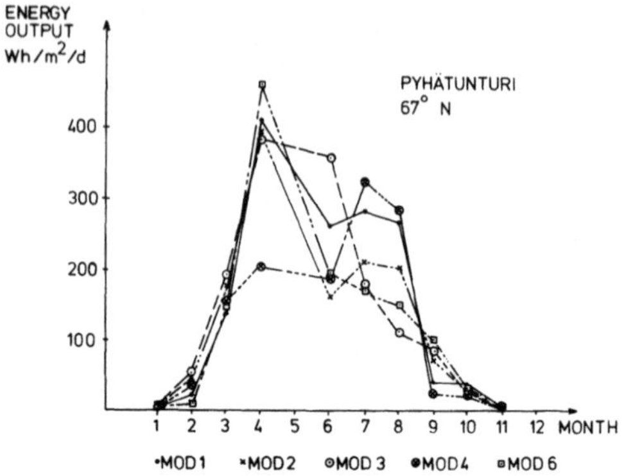

Fig. 3. The daily energy output of the modules 1-4 and 6 at Pyhätunturi (67^0N) during 10/86-9/87

In the period 1-12/87 the daily energy output of the bifacial module 1 (45^0 tilt) as average of the year is 18 % more than the module 4 (45^0 tilt) and 7 % more than the bifacial module 2 (90^0 tilt). The amorphous module 5 produces only 25 % of what of the bifacial module 1 produces.

The daily energy outputs of all modules as average of the period 6/87-3/88 show that module 9 (tracking) gives 45 % more daily energy output than module 1 (bifacial). The energy output of module 9 (tracking) is 31 % higher than that of module 8 (90^0), 13 % higher than that of module 6 (45^0) and 7 % higher than that of module 7 (60^0).

2.2 PYHÄTUNTURI (67^0 N)

The daily energy output of the modules 1-4 and 6 at Pyhätunturi is presented in fig. 3.

3. DAY-TO-DAY-CORRELATION

For optimal sizing of stand-alone photovoltaic solar power systems it is necessary to calculate the correlation coefficient. The coefficient δ, defined in ref. /4/, is the probability that a given value of energy output today is preceded by a specified value yesterday. δ=0 corresponds to mutually independent events and δ=1 represents perfect correlation (every day is identical). The correlation coefficients of radiation (45°) and the bifacial modules 1 and 2 in Otaniemi (60°) 1-12/87 are presented in fig 4.

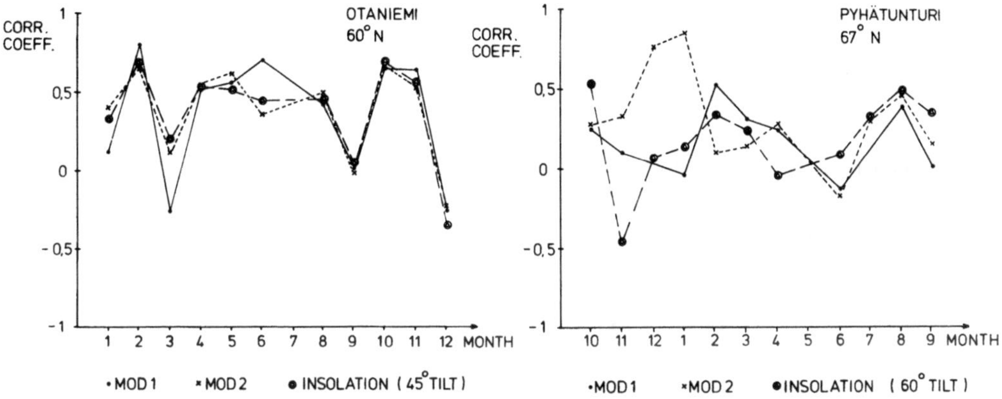

Fig. 4. Correlation coefficients of radiation and modules 1 and 2 in Otaniemi (a) and modules 1 and 2 in Pyhätunturi (b)

The correlation coefficients of radiation and the energy output of all modules in Otaniemi and Pyhätunturi was calculated. The coefficients of the available data for Otaniemi fall in the approximate range 0-0.8 with small fluctuations in summer. At Pyhätunturi the coefficient range is -0.5-0.8. The great difference between the coefficients in winter (November-February) at Pyhätunturi are due to the dark period when the insolation is zero.

4. CAPACITY FACTOR

The capacity factor of the modules and the resource capacity factors are defined in ref /5/. The annual capacity factors of Otaniemi are presented in table II and of Pyhätunturi in table III.

Table II. Average (6/87-3/88) capacity factors of modules in Otaniemi a=the whole year (24 h/d) b=measured days (24h/d).

Module:	1	2	3	4	5	6	7	8	9	INS.
a(%)	2.8	3.1	2.2	4.2	3.3	4.7	5.3	4.4	5.3	4.8
b(%)	4.2	4.3	5.1	6.4	4.1	7.2	7.6	6.2	8.1	7.0
a/b	67	72	43	66	80	65	70	71	65	68

Table III. Annual (10/86-9/87) capacity factors of modules at Pyhätunturi a=the whole year (24h/d) b=measured days 24h/d.

Module	1	2	3	4	6	INSOLATION 60^0
a(%)	1.9	2.0	1.4	1.7	1.6	4.0
b(%)	4.5	4.0	3.3	3.9	3.4	8.6
a/b	42	50	42	43	47	46

The relation a/b is the percentage.

5. CONCLUSIONS

The measured period (1/87-12/87) had much insolation in March and April and low insolation in the summer. The difference in energy output between tilt angles 45^0, 60^0, 90^0 and tracking is remarkable in summer (21 % more with tracking than 60^0 tilt in June). The tracking system does not however work in winter because of freezing. The difference between tracking and the 60^0 tilt modules is small, only 7 % more by tracking during the period 6/87-3/88. The highest capacity factor at Pyhätunturi, of measured days, is 4 % (mod 2, 50 % of days measured). In Otaniemi the corresponding factor is 8.1 % (tracking, 65 % of days measured during the period).

6. ACKNOWLEDGEMENTS

This work was supported by the Energy Department of the Ministry of Trade and Industry and done in cooperation with Neste Advanced Power Systems, Rälssitie, 01510 Vantaa, Finland

REFERENCES

1. E.Lignell, E.Siponmaa and M.Leppihalme 6th PVSEC, p.382
2. C.Nyman, Ises solar world congress. 1987
3. Measurements of solar radiation 1971-80, Finnish Meteorological Institute, Helsinki, 1982.
4. J.M. Gordon, Sol.Cells 20(1987) 245.
5. S.Hester and T.Moff, 18th IEEE PVSC p.198.

FIELD PERFORMANCE OF FLAT PLATE SOLAR CELL MODULES

Mukesh Chander and K.L. Chopra*
Thin Film Laboratory, Physics Department,
I.I.T. Hauz Khas , New Delhi-110016,INDIA.

and

J.C. Joshi
Centre for Energy Studies, I.I.T.,
Hauz Khas , New Delhi-110016, INDIA.

Summary

Flat plate modules of single crystal, polycrystalline and amorphous silicon solar cells have been studied in field conditions and a comparative study has been carried out. The measurements of electrical parameters have been done in natural sunlight. The stability of these modules have been analysed from three years output data for single crystal modules and about one year performance data of amorphous silicon modules. The degradation of crystalline silicon modules is found to be less than 7% in more than three years whereas in a- Si modules about 10% fall in efficiency has been observed in the first 50 days of exposure. It has also been observed that the rated power output was never obtained with global insolation of 1000w/m^2 and module temperature close to 28°C for all the modules. The effect of temperature on module efficiency and open circuit voltage is investigated for these modules. The effect of self shadowing (shadowing due to finite edge height of the frames) during early and late hours of the day has been compared indicating significant power loss in a-Si modules due to high cell packing factor.

INTRODUCTION

Photovoltaic flat plate modules having different kinds of solar cells and various designs of encapsulations have come commercially in the market. The basic considerations in the design of these modules are low cost ,high efficency, and stability (long life). Among thin film solar cells, amorphous silicon (hydrogenated) solar cells have shown great promise and the modules of these solar cells with module efficiency in the range of 7-8% are commercially available but the detail field study of these modules have not been done in Indian environment. It has been shown [1] that temperature has similar effect on efficiency in a-Si and crystalline Si modules but intensity has influence. The theoretical calculations have been done to show the effect of temperature on various electrical parameters of amorphous [2] and crystalline silicon [3] solar cells.

In the present work, the field performance of a-Si modules are compared with that of crystalline and poly crystalline silicon

* Present address: Director, IIT Kharagpur - 721302
West Bengal (India) .

modules. The flat plate modules have been obtained from Central Electronics Ltd. (crystalline Si), Solarex (poly and A-Si) Chronar, ARCO and ECD (a-Si).

EXPERIMENTAL DETAIL

These flat plate modules have been mounted on the fixed or tracking structures. All these modules if mounted on the fixed structure, the surface of modules is kept normal to the sun while taking instantaneous data otherwise they are mounted on the fully automatic sun tracker. The solar radiations (global and direct beam) and ambient and module glass temperatures have been measured using precision radiometers and thermistors. In order to minimise the variations in temperature and solar radiations, the data has been recorded during clear sun shine day with wind speed of about 1m/sec. The data acquisition system consists of a multiplexer which selects the module after fixed interval(10min) of time and record the the maximum power, Voc, Isc and radiations. The temperature is recorded after every 30 minutes.

The I-V curves has been plotted on X-Y plotter using maximum power point tracker or using variable load.

The effect of temperature on efficiency has been determined at constant insolation (variation < 50 w/m^2). The cell temperature can be estimated from Voc or using theoretical model developed on the basis of heat loss in encapsulation and heat generated in solar cells [4]. The degradation study of amorphous and crystalline silicon modules have been carried out. The CEL modules have been exposed to the field conditions for more than three years and performance data has been recorded every month for five to ten clear days. The Chronar module of amorphous silicon solar cells has been exposed for nearly an year and the electrical parameters have been recorded.

The typical values of electrical parameters of these modules on the clear day are given in table I.

Table I . Typical field parameters of different modules.

Module	CEL	SOLAREX	CHRONAR	ARCO
Module dimensions				
Module area(m^2)	.399	.441	0.099	0.11
Solar cells	Cz Si	poly Si	a-Si	a-Si
No. of cells	36	36	29	25
cell size(cm)	10.1 dia	10x10	30x1	29x1.18
cell area(m^2)	.288	.36	.087	.085

Contd.....

Electrical parameters:

Insolation(W/m^2)	950	950	950	1000
Pmax (W)	17.8	29.3	3.8	4.75
Isc (A)	1.8	2.7	.28	.465
Voc (V)	18.3	17.8	22.8	19.2
Fill Factor	0.54	0.61	0.59	.53
Module efficiency(%)	4.7	7.5	4.05	4.3
Module Temperature(°C)	47	46	40	48.0
Ambient Temperature(°C)	25	25	25	28.0
(dVoc/dT)/Voc*100	.39	.37	.38	.35
$(d\eta/dT)/\eta$	.48	.68	.39	--

RESULTS

The typical performance of a-Si (Chronar) module is shown in fig.1. mounted on a fixed latitude tilted mount (28.63 deg.) in the month of April. The experimental values of power are found to be comparable to that of calculated values that has been calculated from the linear equation generated from the experimental data on linear regression. The efficiency shows slight fall after midday due to increase in module temperature.

It has been found that the rated values of efficiencies of these modules are always higher than the observed values under similar conditions ($1000 w/m^2$, 28°C and 1m/s). The efficiencies of 4.5, 6.2 and 9.0% have been observed instead of 5.2, 7.0 and 9.5% for a-Si, Cz-Si and poly Si modules respectively. The power loss in these modules can be due to ohmic, mismatch, shadow and clearness index. The loss in power is about 10-15 % which is much higher than the usual ohmic and mismatch losses. The main cause is due to scatter radiations which contribute about 10% even on clear days but donot generate the proportional electric power.

The effects of temperature on efficiency and Voc are shown in fig.2 and table I. The results indicate that $(d\eta / dT)/\eta$ for a-Si module(Chronar) is about 0.39%/°C whereas in CEL module it was about 0.48%/ °C. The value of (dVoc/dT)/Voc is found to be comparable for both types of modules. But if one calculates dVoc/dT, the a-Si modules show higher values due to higher Voc.

The long term field study data has shown that crystalline silicon modules (CEL) has degraded by about 7% in three years(fig.3) in which 2-3% loss is due to reduction in photogeneration current and 1-2% loss is due increase in ohmic losses(rusting of contacts).On the other hand amorphous silicon module(Chronar) has shown about 10% loss in efficiency in about 50 days than degradation rate decreases very much and attains a uniform value of efficiency. The cause of reduction is mainly due to decrease in the value of fill factor.

The self shadowing loss [5] is due to finite edge height of the frame of the modules. At large incidence angles (>60°), it has been found that in a-Si modules(Chronar and ARCO), the power loss of about 30-40% can occur which reduce the daily energy by 2-12% in fixed tilted modules, depending upon the arrangement of cells.Figure 4 shows the power loss at different lengths of shadow for different kinds of modules. In CEL and Solarex(Poly) modules the power loss is less than 15% for shadow lengths of 3cm which require an incidence angle of greater than 80 degrees. On other hand Chronar and ARCO modules show 40-50% loss even at 1cm shadow length (about 60 degree incidence angle).

CONCLUSION

These studies indicate that the performance of flat plate modules is similar in all modules. The effect of temperature on module efficiency and open circuit voltage in a-Si modules is slightly different from that of the crystalline silicon modules. The stability results indicate that crystalline silicon modules are quiet stable in field conditions whereas amorphous silicon modules exhibit about 10% reduction in efficiency in two months time. The rated values of efficiencies are normally not achieved in field conditions for any type of module. The self shadowing loss is more prominent in a-Si modules due to their module design.

REFERENCES

1. D.L. Morel, J.P Rumburg, R.R. Gay and C.F.Gay, Field Performance of Thin film silicon modules, Proc. of 8th Biennial Congress, ISES,, Perth, P.1517, Aug. 1983.

2. Luuk Folkerts and J.M.Gordon, Amorphous silicon solar cells; Thermodynamic model for realistic performance Characteristics, Solar Cells 23, p.201,(1988).

3. H.J.Hovel, Semiconductors and Semimetals,Vol II,(Solar Cells), Academic Press, N.Y. (1975).

4. W.R. Anis,Robert P.Mertens and Roger J Van Overstraeten, Calculation of solar cell operating temperature in a flat plate PV array, 5th EC PV Solar Energy Conf., Athens, Oct 1983,p.520

5. J.C.Joshi and M.C.Bhatnagar, Self shadowing loss in PV modules, 8th Int. Conf. on Alternate Energy Sources, Miami,(1987).

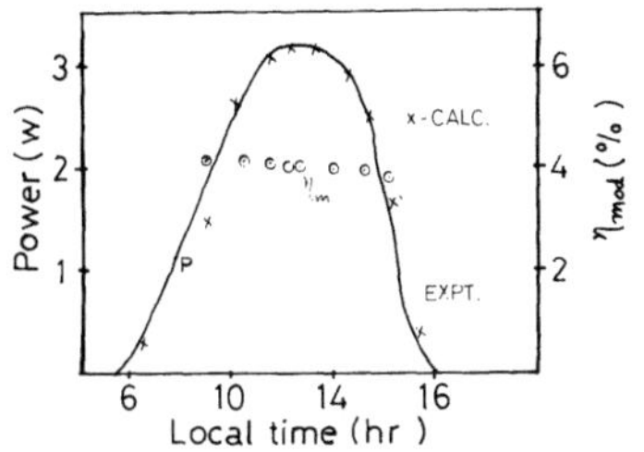

Fig. 1. Typical daily performance of a-Si (chronar) module.

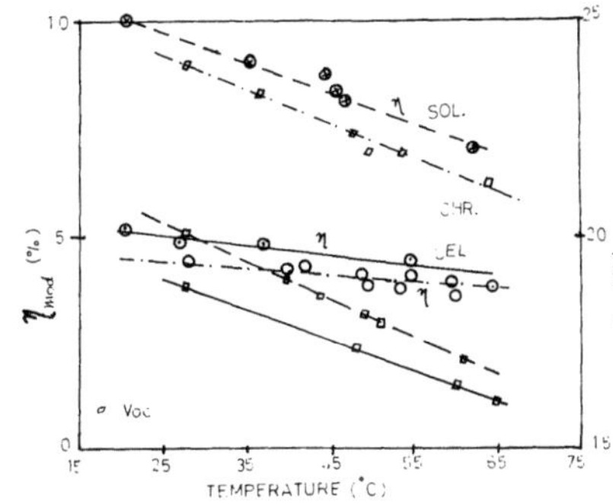

Fig. 2. Effect of temperature on V_{OC} and efficiency.

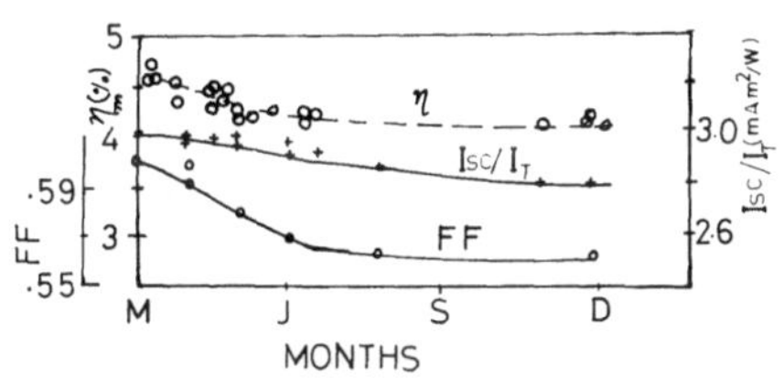

Fig. 3a. Degradation of a-Si (chronar) module in outdoor.

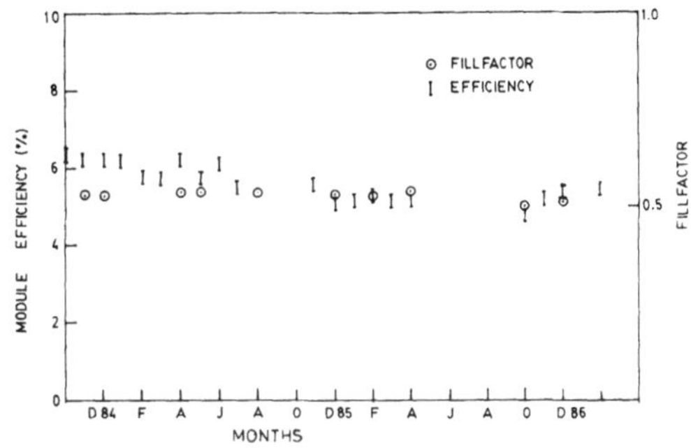

Fig. 3b. Degradation of CZ-Si(CEL) module in outdoor.

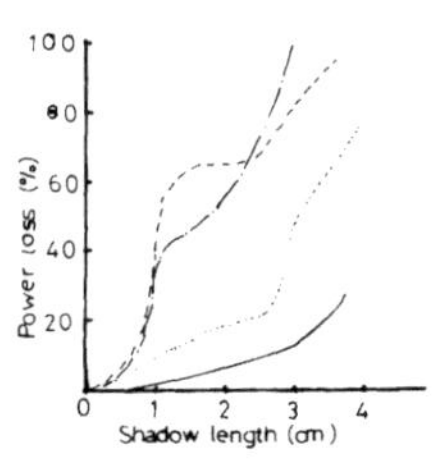

Fig. 4. Self shadow loss due to edge of frame in different modules.

TESTING TECHNIQUES AND EXPERIENCE WITH PHOTOVOLTAIC SYSTEMS

N. Locci - F. Mocci - M. Tosi
Institute of Electrotechnics, Cagliari University
Piazza D'Armi, 09123 Cagliari - Italy

ABSTRACT. The measurement systems used by the authors at the Institute of Electrotechnics of Cagliari University in experimental research on photovoltaic plants are briefly described. With a view to optimizing performance of photovoltaic systems, especially pumping plants, the investigations were directed towards the characterization of components, development of prototypes and testing of complete systems. Finally, the data acquisition and processing systems used in the tests for analysing the different design options are described.

1. INTRODUCTION

The study of photovoltaic plants necessitates suitable analytic tools for characterizing the operation of the single plant components, especially where maximizing the overall efficiency of a photovoltaic conversion chain is concerned. In this case the actual characteristics of each component must be known and the components have to be chosen and combined in such a way that the maximum possible amount of energy /1/ can be exploited. In addition suitable instruments must also be available for comparing the different design options, with both laboratory and field tests.

Drawing on past experience a test methodology was set up along with a pilot plant that proved to be of considerable help in studying plants consisting of photovoltaic generators, loads and any auxiliary circuits designed to improve the system's overall efficiency. In addition to the measuring instruments, simulation devices and computerized data acquisition systems for storing and automatically processing the measured parameters were also advantageously used in the tests.

The equipment as a whole proved useful both for comparing different design options and for developing prototypes as well as for the final testing of complete systems, with special reference to pumping plants.

With the proposed setup the following quantities can be measured:

a) electrical parameters (voltage, current, power) both in steady-state and dynamic conditions;

b) meteorological parameters (light flux density, temperature);

c) mechanical parameters (torque, speed, power, flowrate, pressure).

Tests can be performed singly or on a combination of the following plant components: photovoltaic generators, DC-DC and DC-AC convertors, loads (DC and AC motors, battery storage).

The programmable laboratory simulators allow to realize the desired characteristics for:
- photovoltaic generators (I-V, Current-Voltage function)
- mechanical loads for electrical drive (T-S, Torque-Speed function).

In this way it is possible to repeat tests in prefixed operating conditions, without being tied to contingent environmental conditions.

2. TESTING OF PV GENERATORS

Two measurement systems (options A,B) were set up for testing the photovoltaic generators. They are illustrated in the simplified block diagram of Fig. 1. For certain light flux and temperature values (measured with a solarimeter and a digital thermometer) the I-V characteristic of the generator is explored by varying the resistive load applied (option A).
The electrical characteristic can be plotted directly by means of an X-Y recorder (Philips PM 8134) while the data sampled with an HP 3421A acquisition unit are transferred to a computer (HP 71B for field tests, HP 9836 for laboratory tests) where they are stored (Mass storage unit HP 82161A), displayed and processed.
Measurements can otherwise be taken in dynamic conditions (option B), by controlling at constant switching frequency the on- and off-intervals of a static switch, so as to vary the mean current required by the load. The capacitor C, being charged and discharged during the off- and on-intervals, smooths the voltage changes thereby enabling to explore a very small portion of the I-V curve at a time. The apparent load on the PV generator varies with the on- to off-interval ratio (T_{on}/T_{off}) of the static switch, similarly to a variable ratio DC transformer.
The data acquisition system, which is completely programmable, can also be used in the field, as its components are all self-contained and are equipped with stand-by controls.

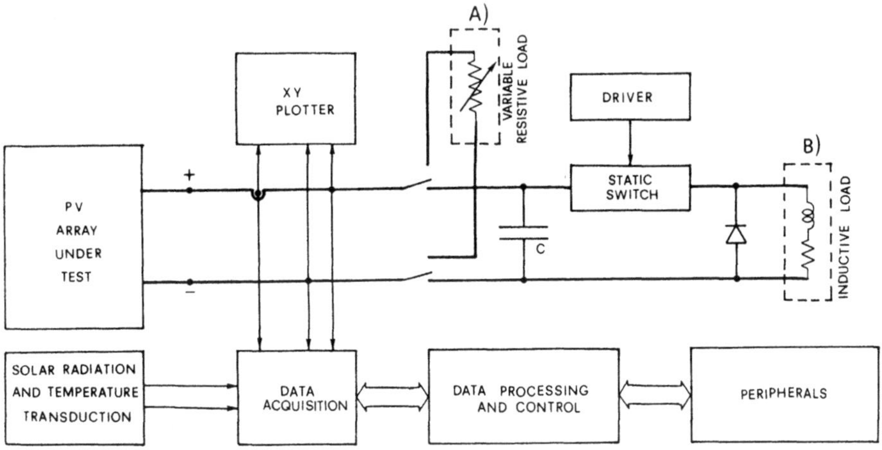

Fig. 1 - Testing of PV generators.

Laboratory tests are performed on a generator consisting of a PV array with 8 x 33 W solar modules manufactured by Ansaldo, that can be differently arranged. A programmable microprocessor-based PV simulator was also built with Voc-max = 70 V, Isc-max = 10 A and Pp-max = 500 W /2/, which allows to simulate any I-V characteristic for given meteorological parameters. The programmable device can substitute the actual PV generator reproducing cyclic meteorological conditions in virtual time. In actual practice, the daily, seasonal or yearly performance of a system can be determined with accelerated tests lasting less than two hours. This device is a precious part of the system inasmuch as its high stability and repeatability enable realistic comparisons between different design options that can be tested in strictly identical simulated meteorological conditions.

3. TESTING OF PV SYSTEMS

Laboratory tests can be performed on complete systems consisting of photovoltaic generators, conversion/matching circuits, where used, and loads (DC motors, AC motors, batteries, etc.).
The performance of a PV system can be experimentally assessed either by using the PV array in question as a power source, suitably connecting the component modules, or the simulator, programmed to represent the actual PV array intended to be used.
In order to find the optimum mechanical loads to couple to electric motors, a programmable braking system has been set up provided with all the necessary instrumentation for measuring the involved parameters. The mechanical torque, a function of speed, to be applied to the motors is

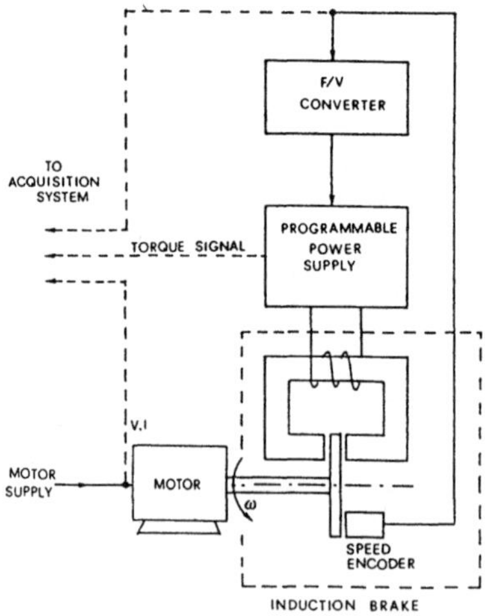

Fig. 2 - Braking system.

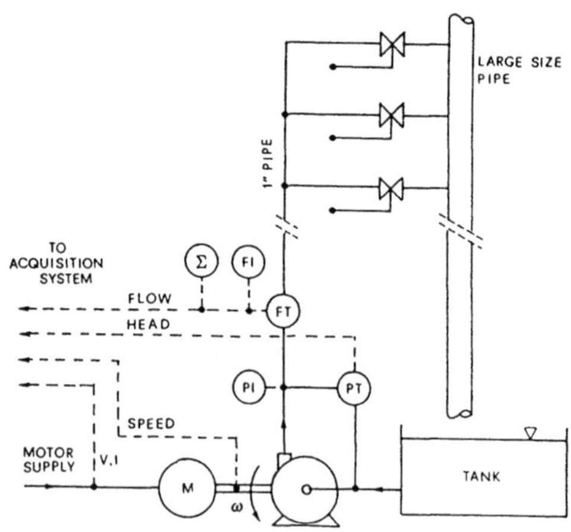

Fig. 3 - Hydraulic circuit.

provided by an induction brake, whose excitation current is controlled by an appropriate drive circuit so as to obtain the desired mechanical characteristic (Fig. 2). In this way the torque/speed characteristics typical of centrifugal pumps can be simulated and, above certain speed values, those typical of volumetric pumps, with different static heads and flowrates. The braking device makes it possible to perform, in the laboratory, all the tests on different types of drives in order to optimize the system's performance and thus compare different options (for instance with or witout DC/DC convertors for peak power tracking in the case where DC motors are employed, comparison of different types of DC motors and pumps, etc.). An application of the described apparatus is illustrated in /3/.
Furthermore, a high performance digital oscilloscope, with HPIB output interfaced with the computer for subsequent processing is employed for analysing the performance of the electronic devices used for the power tracking or monitoring particular control functions /4/.

4. PUMPING PLANTS

For the purpose of evaluating the effectiveness of pumping applications in PV systems, a closed-loop water system has been set up, suitable for tests with flowrates up to 2.5 l/s and static heads up to 12 m (Fig. 3). It consists of electric pumps, a vertical discharge pipe with diameter of 1", a gathering gravity line with large diameter, at atmospheric pressure, which conveys the water to a tank, intake for the pump. The hydraulic circuit is equipped with two turbine flowmeters with varying full-scale range, the respective convertors and integrators, and pressure meters.
Using the programmable PV simulator as generator, the volumes of water that a given plant can deliver in a certain time interval (day, month, year) for certain meteorological cycles can rapidly be determined.

5. ELECTRIC MOTORS FOR PV PUMPING PLANTS

The authors have conducted numerous theoretical and experimental investigations on both DC and AC motor /3,5,6/ supplied by photovoltaic generators for the purpose of optimizing the conversion of available electrical energy to mechanical energy, rationalizing its use.
The measurement systems have permitted the characterization, on the I-V plane, of the behaviour of different motors for a variety of conditions and new designs have been developed /1, 3/ to yield high efficiency without the need to resort to the complex circuits required for peak power tracking of the PV generators.

6. CONCLUSIONS

The testing techniques employed by the authors at the Institute of Electrotechnics of Cagliari University for photovoltaic plants are described. Given their importance, special attention has been focused on pumping plants. A pilot plant has been constructed for the global experi-

mental tests, incorporating a hydraulic circuit.
Many of the tests were made possible by the programmable PV generator simulator which permits comparative tests irrespective of meteorological conditions existing at that time and which allows to assess, in tests of short duration, the performance of the photovoltaic systems over long periods.
All the measurement data were processed implementing appropriate calculation programs.

7. REFERENCES

/1/ N. Locci, F. Mocci and M. Tosi, 'On the Optimum Design of Photovoltaic Pumping Plants', Proc. ISES Solar World Congr., Hamburg 1987.

/2/ N. Locci, F. Mocci and M. Tosi, 'A Programmable Simulator of Photovoltaic Generators', Proc. Power Electronic Specialist Conf., Vancouver 1986.

/3/ N. Locci, F. Mocci and M. Tosi, 'A Simple Electronic Control for DC Shunt Motors in Photovoltaic Pumping Plants', Proc. ISES Solar World Congr., Hamburg 1987.

/4/ N. Locci, F. Mocci and M. Tosi, 'Measurement of Instantaneous Losses in Switching Power Devices', Proc. IEEE Instr./Measur. Technology Conf., San Diego 1988.

/5/ N. Locci, F. Mocci and M. Tosi, 'Studio del motore a c.c. alimentato da generatori fotovoltaici', L'Energia Elettrica, n. 2, 1985.

/6/ N. Locci, F. Mocci and M. Tosi, 'Optimizing the operation of asynchronous motors powered by photovoltaic converters', Proc. INTERSOL 85, Montreal 1985.

SESSION 08:
SI-MATERIALS

METALLURGICAL ROUTE TO SOLAR-GRADE SILICON

J. Dietl
Heliotronic GmbH
D-8263 Burghausen, P.O. Box 1129, West-Germany

ABSTRACT. A processing route via the refinement of metallurgical-grade silicon - as studied at Wacker-Heliotronic - offers best chances of producing solar-grade silicon at costs of lowest possible levels. The refining concept is based on an optimized sequence of three metallurgical processes, the first one being alternatively replaceable by aluminothermic production of pure silicon.

1. INTRODUCTION

Competitive photovoltaic conversion of solar energy into electricity requires strong simultaneous efforts in every discipline such as material, cell and module technology. In the material sector this means the reduction of costs in the preparation of solar-grade silicon and the manufacture of square-shaped crystalline wafers. The two subtechnologies represent the basic goals of the Wacker-Heliotronic Solar Silicon Program (1).

The present paper concentrates on the development of a low-cost technology for producing solar-grade silicon as a meltstock for unconventional crystallization by ingot casting. Refining of metallurgical-grade silicon has been put in the center of the Heliotronic material concept. It is based on an optimized combination of hydrometallurgical and pyrometallurgical treatments complemented by aluminothermic processing as an alternative route. Starting from laboratory scale in 1975 the R & D activities in the meantime have reached a remarkable degree of pilot-scale character.

2. SOLAR-GRADE SPECIFICATION

Impurities in silicon affect the efficiency of solar cells in a variety of ways. The objective of the wellknown Westinghouse project (2) has been to study the effects of various elements on the performance of single crystalline solar cells. Semiconductor properties are influenced by deep-level impurities - like the transition metals - which reduce the base diffusion length. At higher concentrations of elements like carbon, interface degradation and increased junction excess current due to precipitates in the depletion region have been observed.

The term "solar-grade" characterizes a meltstock quality which is understood to be less pure - and therefore lower in cost - than electronic-grade silicon, but can nevertheless be used for photovoltaic application aiming at a comparatively high level of conversion efficiency. Due to a fairly strong correlation with the crystallization technique and the special state of cell fabrication - which at the time being is in an exciting development toward high efficiency - the specification "solar-grade" has to be defined for every particular processing sequence.

3. ROUTES TO HIGH-PURITY SILICON

The demand for low-cost silicon for terrestrial solar cells initiated the development of unconventional techniques. Schematic flow-diagrams in Fig. 1 indicate the currently most discussed processes based on the reduction of quartz. Side by side with the conventional production of electronic-grade silicon via distillation and poly-deposition of $SiHCl_3$ three metallurgical processing sequences aiming at solar-grade silicon are shown.

Two of these sequences utilize the carbothermic reduction of quartz. Whereas one route proceeds from the low-purity arc furnace product - which will be discussed later on - the approach at the right-hand side is based on an impurity-improved technique from the very beginning. This processing sequence is studied by Siemens (3) and Nippon Sheet Glass/Kawasaki Steel (4) and has some resemblance to earlier work done at Dow Corning (5) and Elkem/Exxon (6). The concept is based on an arc process running under high-purity conditions, fed with upgraded starting materials (SiO_2 and C) and completed by final purification of the reduction product.

Another technique of preparing solar-grade silicon - studied at Heliotronic as an alternative route - starts with the aluminothermic reduction of quartz sand. Whereas the Gibbs free enthalpy of the carbothermic reduction process requires temperatures above 2000 K the aluminothermic reduction is exothermic over the whole temperature range (Fig. 2). This allows the aluminothermic reduction to be conducted in a flux system at lower temperatures.

In an advanced version a flux system based on alkaline earth metal silicates is used which simultaneously serves as a solvent for the aluminum oxide formed and as a liquid-liquid extraction medium (7). In Fig. 3 the phase diagram of the system $CaO-Al_2O_3-SiO_2$ is shown. Of special interest is the marked area surrounding the pseudo-binary subsystem calcium silicate - Al_2O_3. The range up to about 60 % Al_2O_3 (corresponding to a melting temperature of 1700°C) represents the optimum field for the reduction in calcium silicate. The aluminothermically produced silicon is then purified by further metallurgical treatments, as described under chapter 4.2 and 4.3, to reach solar-grade quality.

4. REFINING OF MG-SILICON

The production of silicon in electric arc furnaces is a technically well established process. The impurities in metallurgical-grade silicon are Fe, Al, and Ca in the range of some 1000 ppm, Mn, Ti and C in the range of some 100 ppm and among the impurities below 100 ppm the doping elements B and P. These concentrations are far above the levels allowed for photovoltaic application. Nevertheless, the low cost and the industrial capacity - about 600.000 tons per year in the Western World - make the upgrading to solar-grade quality extremely attractive.

At Heliotronic, basic hydrometallurgical and pyrometallurgical refining procedures are under investigation. To define an optimum sequence of processing steps with complementary purification effects is of central importance in terms of both systematic purification and economical aspects. The central part in Fig. 1 represents an optimized refining cycle based on three metallurgical extraction processes.

4.1 Liquid-Liquid Extraction

Liquid-liquid extraction is applied to remove impurities from MG-Si in the molten state. Successful utilization requires a slag which is immiscible with silicon and has a higher chemical affinity to certain impurities than silicon. Slags composed mainly of alkaline earth metal silicates have been found to fulfill these criteria. Elements forming oxides of higher stability than SiO_2 are expected to get oxidized and transported into the slag. This has proven to be valid for elements like Ca, Mg and Al (8).

In Fig. 4 the liquid-liquid extraction process is schematically drawn. In an arc furnace intensive liquid-liquid contact between calcium silicate and MG-Si is established by forming a melt system with Si drops dispersed in the slag. After accomplishing equilibrium extraction conditions phase separation is achieved in a ladle from where prerefined P1-Si is poured into a mould.

Analytical data comparing the concentration of impurities in MG-Si and P1-Si demonstrate the purification effect by slagging in the case of Al but also of the nonmetallic contaminants B and C. Especially the potential to remove B makes the liquid-liquid extraction an important purification process.

4.2 Solid-Liquid Extraction

Solid-liquid extraction - known as acid leaching - represents the second step of the refining sequence. During solidification of P1-Si impurities with low segregation coefficients accumulate in form of precipitates mainly along grain boundaries. For effective leaching the silicon has to be milled in order to set the impurity inclusious free to become attacked by acid mixtures.

In Fig. 5 a schematic drawing of the hydrometallurgical refining process is given. Lumps of P1-Si are crushed by a sequence of jaw and roll crushers and finally pulverized in a vibration mill. Milling under water allows a hydro-cyclone to be used to classify the milled product. The purification is improved by grinding the silicon down to a particle size around 20 µm, a size well below the average grain diameter of the polycrystalline material.

Extensive experimental work has been done in optimizing type, mixture and concentration of acids (9). The leaching equipment in Fig. 5 consists mainly of a filter-extractor in which the silicon powder is leached by pumping the acid through the filter-cake followed by extensive rinsing and drying. Starting from P1-quality nearly all metal impurities can be lowered down to sub-ppma levels. Thus, hydrometallurgical processing, on the whole, represents the most effective step of the refining sequence.

4.3 Liquid-Gas Extraction

To remove impurities with unfavourable segregation coefficients which usually do not form precipitates - such as the doping element P - liquid-gas extraction is applied. Blowing reactive gases through molten silicon as well as evaporation of elemental impurities are basic procedures in this field (8). The knowledge of both the chemistry of possible high temperature reactions and vapor pressure of reaction products and elemental impurities is a prerequisite of selecting the proper experimental set-up.

In Fig. 6 a simplified view of the liquid-gas extraction process is given. HP1-Si powder is melted in a crucible and exposed - under appropriate conditions of surface to volume ratio - to gaseous reactants and/or vacuum. The conceptual approach of utilizing gaseous reactants is to form impurity compounds which can be easily removed by slagging or evaporation. To enhance the evaporation effect it is advantageous to treat the Si melt under vacuum. Application of vacuum simultaneously removes impurity elements by direct evaporation according to their vapor pressure.

Concentrations of impurities most affected by gaseous reactants and/or vacuum treatment are listed before and after liquid-gas extraction in Fig. 6. The impurity situation of the HP2-Si quality achieved is characterized by metals to the amount of 0.1 ppma, by oxygen below 10 ppma - starting from around 3000 ppma present as surface oxide - and by P below 0.05 ppma. In addition the carbon contamination is lowered to levels below 25 ppma (det. lim.).

5. CONCLUSION

The main advantage of the metallurgical route to solar-grade silicon is to produce impurity optimized silicon at lowest possible costs. At the time being solar cells of HP2-Si - crystallized by standard ingot casting and processed via conventional cell techniques - achieve conversion efficiency values up to 90 % of cells based on electronic-grade Si. To transfer this encouraging results to the goal of large-scale terrestrial photovoltaic application, both technical realization and economy of the metallurgical processing sequence have to be consolidated by final development in the pilot stage.

6. ACKNOWLEDGEMENT

The author would like to thank Prof. E. Sirtl for valuable and stimulating discussions. He is also indebted to the Bundesministerium für Forschung und Technologie for supporting this work.

7. REFERENCES

(1) E. Sirtl et al., 19th IEEE Photovolt. Spec. Conf., New Orleans, 345 (1987)

(2) J.R. Davis et al., IEEE Trans. Electron Devices, ED 27 (4), 677 (1980)

(3) H.A. Aulich et al., 4th E.C. Photovolt. Solar Energy Conf., Stresa (Reidel Publ. Comp., Dordrecht) 868 (1982)

(4) M. Ishizaki, Y. Sakaguchi, 2nd Internat. Workshop on Si for Solar Cells, Schliersee, 1987, unpubl.

(5) L.P. Hunt, V.D. Dosaj, Proc. 2nd E.C. Photovolt. Solar Energy Conf., Berlin (Reidel Publ. Comp., Dordrecht) 98 (1979)

(6) J.A. Amick et al., Proc. Symp. Mat. and New Proc. Techn. for Photovoltaics (ECS Inc., Pennington) 67 (1983)

(7) J. Dietl, C. Holm, 3rd Int. Photovolt. Sci. and Eng. Conf., Tokyo, 533 (1987)

(8) J. Dietl, Silicon Proc. for Photovolt. II, Elsevier Sci. Publ. B.V., 285 (1987)

(9) J. Dietl, Solar Cells, 10, 145 (1983)

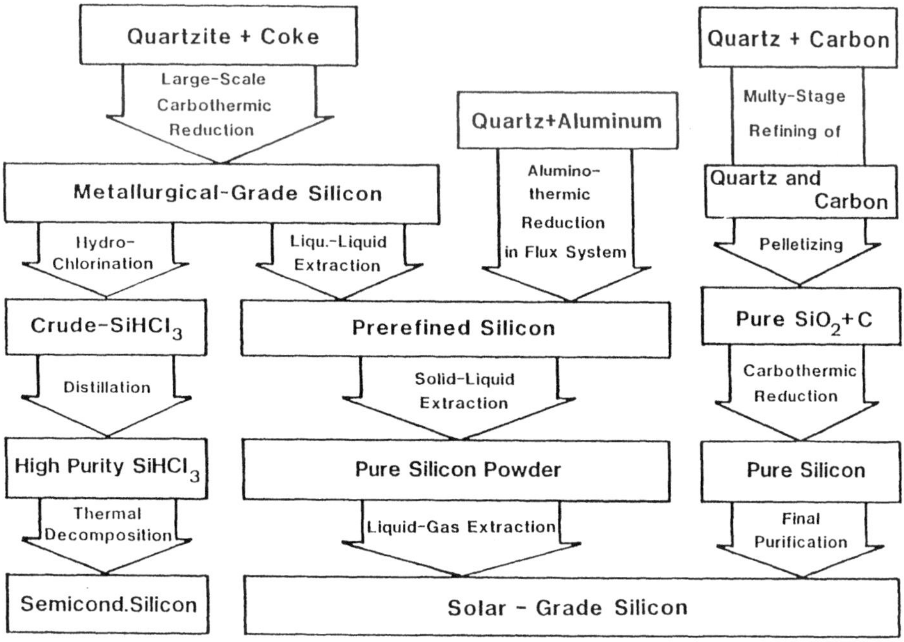

Fig.1 Conventional and unconventional routes to high-purity silicon

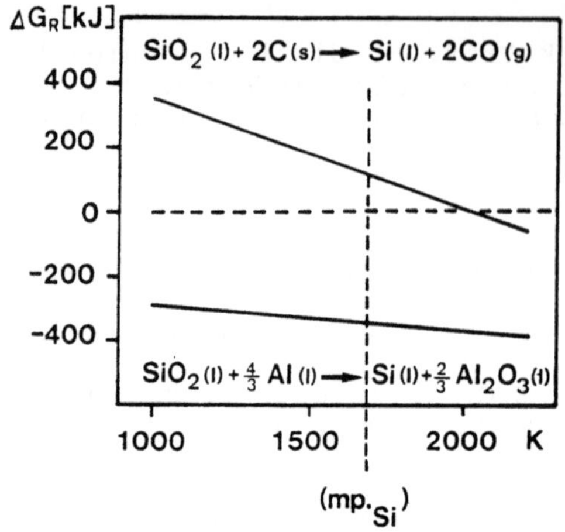

Fig.2
Gibbs free enthalpy of carbothermic and aluminothermic reduction of quartz

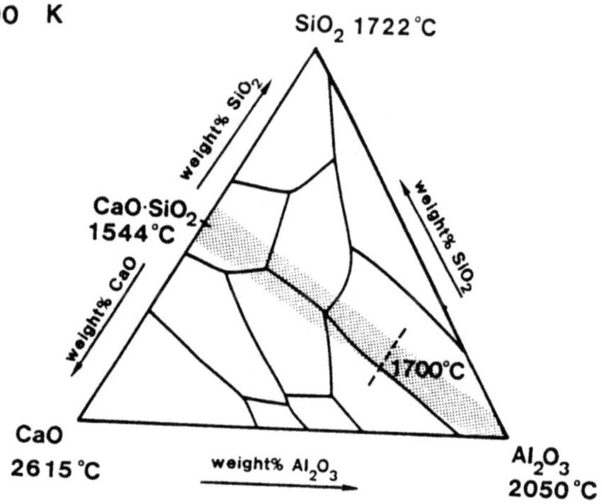

Fig.3
Phase diagram of the flux $CaO-Al_2O_3-SiO_2$ containing the subsystem $CaO \times SiO_2-Al_2O_3$ (marked area)

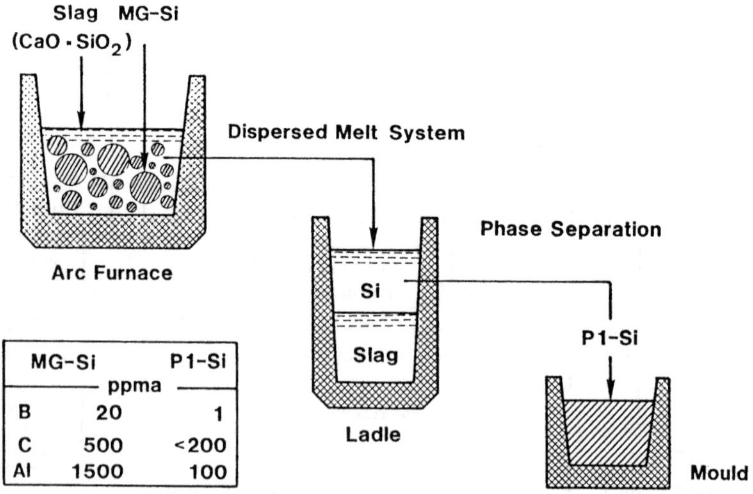

	MG-Si	P1-Si
	ppma	
B	20	1
C	500	<200
Al	1500	100

Fig.4 Schematic graph of the liquid-liquid process yielding prerefined P1-Si (purity-1-Si)

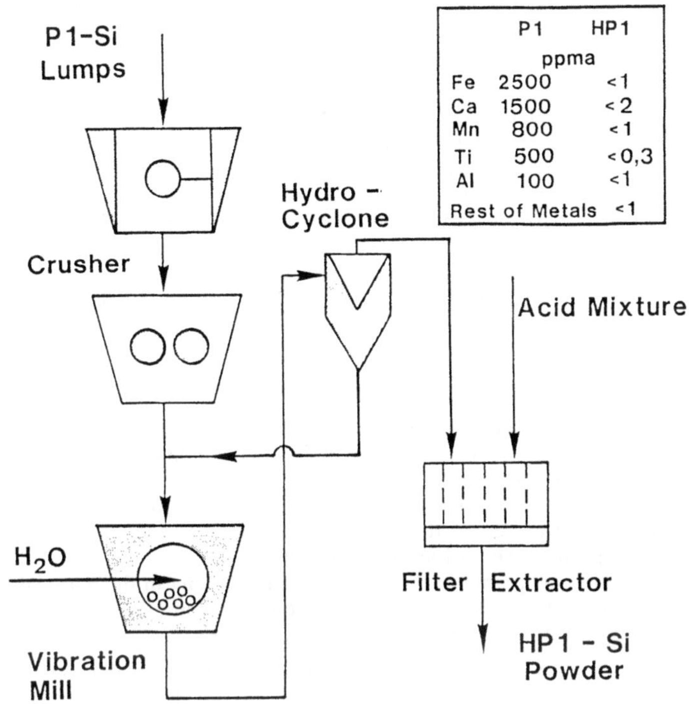

Fig.5 Flow-chart of the hydrometallurgical refining process yielding HP1-Si (high-purity-1-Si)

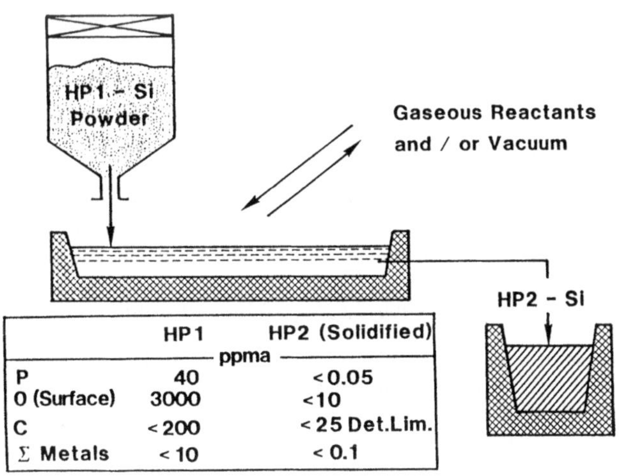

Fig.6 Liquid-gas extraction of silicon yielding HP2-quality (high-purity-2-Si)

PHOTOVOLTAIC SILICON PRODUCTION BY PLASMA TECHNIQUE. CONTROL AND REGULATION OF THE PROCESS.

D. MORVAN, F. SLOOTMAN, N. MADIGOU, J. C. ROUCHAUD*, M. FEDOROFF*, J. AMOUROUX
Univerrsité Pierre et Marie CURIE - E. N. S. C. P.
Laboratoire des Réacteurs Chimiques en Phase Plasma
11, rue Pierre et Marie Curie - 75231 PARIS CEDEX 05 FRANCE
* C. N. R. S. C. E. C. M. Vitry
Laboratoire de radioactivation
15, rue Georges Urbain - 94400 VITRY s/ SEINE -FRANCE

Abstract :

Photovoltaic silicon elaboration has been developped by plasma process using RF plasma torch. The aim of this work is to demonstrate the flexibility and the control of this tool in the way of a new concept of photovoltaic cells manufacturing.

The first part of our purpose point out the characteristics of the plasma fluid needed for purification and heat transfer. So that a mixture of argon, hydrogen, oxygen well controlled have been tested to optimize the purification mechanism.

The second part point out the extraction phenomena in the material, that means the liquid-liquid extraction between the slag and the molten silicon [1]. Then the slag composition (SiO_2, CaO, B_2O_3) is one of the main parameters of the process.

However, the evaporation mechanism at the interface plasma-slag, gives us the oportunity to control the process by using an emission spectrometer. We measure the main transport properties on the boundary layer which point out the nature of the evaporated species, the kinetics of these mechanisms and the correlation between the plasma composition and the slag properties .

At least, we have chosen the boron evaporation and the resistivity measurements of the material, to demonstrate the large correlation between the chemical properties of the plasma and the efficiency of the purification.

To conclude, photovoltaic properties and photovoltaic efficiency of solar cells have been obtained, and from metallurgical and down graded electronic material (n or p), we can reach today between 10 and 14 % of photovoltaic efficiency in standard conditions.

I - Introduction

The aim of this paper is to present the original multiphase plasma process. Starting from metallurgical grade silicon, the objective is to produce a photovoltaic one with the help of a process which takes the place of the chemical route. The promoting route of the plasma is due to the fact of the possibility to by pass all successives intermediate chemical and metallurgical steps of the classical routes (Table 1). The advantages of this way are the purification of both metallurgical grade silicon and waste material from the electronic and photovoltaic industry.

II - Experimental apparatus and technical aspect of the purification (Figure 1)

The use of a high frequency reactive plasma (4 MHz) is a good way to produce an ultra pure material (total impurity content below 1 ppm) [2]. The particular properties of the plasma (viscosity, thermal conductivity, chemical reactivity) are responsible for the way the impurities are trapped. The plasma ensures the melting of the bar, which is placed in a crucible. The purification mechanism is based upon the choice of the chemical composition of the plasma in agreement with the properties of the material. These are the reasons why we introduce in argon plasma hydrogen, in order to increase thermal conduvtivity and low quantity of oxygen, for the oxydation of the impurities, leading to the formation of a natural slag on the liquid silicon.

The bar weighs 100 grams, with a cross section area of 2.5 cm^2, measures 15 cm in length and is placed in a cooled copper crucible moving horizontally with a speed of displacement reaching 1 m/h.

In order to have a better understanding of the mass and heat transfer during the purification [3], we follow volatile atomic species [4,5,6], which leave the slag under the influence of the chemical and thermal properties of the plasma. The species undergo an electronic excitation allowing a detection by emission spectroscopy. The technical measurement consists of focusing the image of the plasma-slag interface on the entrance slit of a high resolutive spectrometer, by means of a fiber optic. The emission signal is analysed by an intensified diode array (OMA III) and is investigated by a calculator.

III - Experimental results of the purification

III.1 - Measurements of boron elimination

Silicon and boron undergo simultaneous oxidation under the influence of an oxygenated plasma. The experimental analysis of evaporation indicates that two emission lines of boron can be followed near three lines of silicon. In the case of a static fusion, that is to say the sample doesn't move horizontally under the plasma, we have demonstrate that the evaporation of boron depends on the oxygen concentration in the plasma (Figure 2).

If we follow simultaneously evaporation of boron and silicon as a function of time, with and without oxygen in the plasma, we can distinguish three steps :

- The first one consists in starting the melting of the sample with 20 cm^3/min of oxygen in the plasma. When the thermal and chemical steady state is obtained :

* Boron evaporation decreases with time.

* silicon evaporation is constant.

This is due to the evaporation of SiO and boron oxyde, under the influence of oxygen and the formation of a molten silica (SiO_2) on the surface.

- The second step consists in cutting the oxygen flow rate ; we observe that :

* Boron evaporation doesn't occur.

* Silicon evaporation is much higher than in the first step and decreases slowly with the time.

This is explained by the increase of the mass transfer of silicon, when there is no thermal barrier of silica SiO_2.

- In the third step, when we introduce oxygen again, we find that the emission intensity obtained is the same as in the first step.

Two conclusions can be presented :

The first one is that an oxygenated plasma permits to eliminate boron from silicon. The second one is that a treatment without oxygen allows us to eliminate residual oxygen content in silicon.

III.2 - Influence of synthetic slag on the purification [7].

Introduction of fluoride powder in the slag increases the elimination of impurities. In the table 2, the purification coefficients calculated (ratio of concentration of an impurity in the raw material over its final concentration) underlines that the efficiency of the slag depends on the nature of the impurity. The proportion of the synthetic slag (1-2 %) has to be considered as a function of the concentration of oxygen in the plasma, the chemical composition of the synthetic slag has to be choosen as a function of the initial concentration of the impurity in silicon, which will be probably a mixture of the fluoride used.

Table 3 supports these conclusions, showing that in the case of carbon and nitrogen, there is a good elimination of these impurities with an oxygenated plasma. Lower purification is obtained with synthetic slag, particularly in the case of CaF_2-SiO_2.

III.3 -Photovoltaic properties

Photovoltaic measurements on small photocells show :

1 - Whatever the starting material is, the resistivity increases with an oxygenated plasma treatment. This confirms the elimination of boron (Table 4). The increase of the resistivity is lower when CaF_2 and SiO_2 are present simultaneously as encapsulant or as synthetic slag. An increase of two orders of magnitude is obtained with impure metallurgical grade silicon, purified with a slag of CaF_2 and an oxygenated plasma.

2 - The plasma process can be used for purification of off grade material (Polyx with impurities as encapsulant). Furthermore, we demonstrate the possibility of inverting the type of silicon from n to p (Table 5). This way is interesting when concentration of phosphor is high and the p compensation is not possible by the classical way.

IV - Conclusion

The use of plasma technique for the production of photovoltaic silicon has to be studied as a function of the chemical composition of the starting material. The key parameters are plasma and slag compositions. The diagnostic system comprising a fiber optic and a monochromator gives us possibility to control and regulate an industrial process, which could be supplied by different kinds of raw materials : impure, such as MG Si or pure, such as waste materials, coming from the photovoltaic or electronic industry.

100 kW industrial plasmas have good energy efficiencies. We observe that the efficiency of the process is better with an increasing of the plasma power (Figure 3). The energetic consumption decreases from 75 kWh/kg in the case of a 7 kW plasma to 10 kWh/kg for a 42 kW plasma.

Acknowledgements :

The financial support of the "Agence Française pour la Maîtrise de l'Energie", the European Community, and the A. R. C. PIRSEM "Chimie des Plasmas", are greatefully acknowledged. This work is undertaken in collaboration with Photowatt Int. and "Electricité de France".

Bibliography :

[1] J. D. GILCHRIST, Extraction Metallurgy, 2nd Edition 1980, Pergamon Press.
[2] J. AMOUROUX and all, Ann. Chim. Fr., 1986, 11,pp 589-602.
[3] D. MORVAN and all, 1er Congrès National de Génie des Procédés, 21-23/09/87, Nancy-France.
[4] J. AMOUROUX and all, ISPC 8 (IUPAC), Tokyo-Japon, 31/08-04/09/87, Vol. 3, pp 1968-1873.
[5] F. SLOOTMAN and all, E. MRS Conf., 2-5/06/87, Strasbourg-France, Symp. Proc., Vol. XV, pp 393-398, Ed. Physique.
[6] F. SLOOTMAN and all, Ann. Chim. Fr., 12, pp 401-409.
[7] F. D. RICHARDSON, Physical Chemistry of Melts in Metallurgy, Vol. 1-2 (1974), Academic Press.

Table 4 : Photovoltaïc properties after plasma treatment = Influence of slag (natural : O_2 in the plasma and synthetic (fluoride on the surface)
Influence of raw material

Starting material	Nature of the slag	Measurements of small photocells (2-4 cm^2)					Photovoltaïc efficiency (%) in standard conditions with AR coating + 10 % grid shading
		Resistivity (Ω. cm)	J (mA / cm^2)	V (mV)	FF (%)	η (%)	
Impur Si MG (93 - 95 % Si) ρ = 0,04 Ω . cm	1,5 % CaF_2 + O_2 in the plasma	0,3 - 0,6	21,4	496	43	4,6	6,0
Pur Si UMG (99,5 % Si) ρ = 0,2Ω. cm	1,5 % CaF_2 + SiO_2 + O_2 in the plasma	0,6	24,75	535	53	7,0	9,3
Polyx with impurities upper part of the block ρ = 1Ω. cm	O_2 in the plasma	10	22,5	491	64	7,4	10,1
Polyx with encapsulant (CaF_2 - SiO_2) lateral part of the block - ρ = 1,17Ω. cm	+ O_2 in the plasma	1,95	29,5	533	70,8	11,0	14,3

Table 2 : Influence of slag composition on residual concentration of impurities

Impurities	INITIAL CONC. (PPM) in metallurgical grade silicon (93 - 95 % Si)	Results after plasma treatment PURIFICATION COEFFICIENT					
		without synthetic slag	with synthetic slag				
			CaF_2	MgF_2	BaF_2	LiF	NaF
Ba	117 ± 10	> 1,5 10^3	> 1,5 10^3	> 3,9 10^3	> 1,7 10^3	> 0,5 10^3	> 2,0 10^3
Fe	11957 ± 617	> 1,1 10^4	> 1,0 10^4	> 1,7 10^4	> 1,0 10^4	> 1,1 10^4	> 1,1 10^4
Ti	15900 ± 2400	1,1 10^3	2,3 10^3	> 80,0 10^3	> 11,4 10^3	0,3 10^3	> 23,0 10^3
Zr	291 ± 58	> 2,0 10^2	> 2,1 10^2	> 3,3 10^2	> 3,3 10^2	> 3,6 10^2	> 3,0 10^2

Experimental conditions :
- % weight of slag = 1,5
- number of passes = 10
- flow rate of O_2 in the pass = 30 cm^3 / min
- flow rate of H_2 = 0,30 l / min
- flow rate of Ar = 30 l / min
- power of the plasma = 8 kW

Table 3 : Influence of slag composition on residual composition of carbon and nitrogen

Impurities	Initial concentration in metallurgical grade silicon (93 - 95 % Si ; in ppm weight)	Results after plasma treatment Final concentration in ppm weight				
		without synthetic slag		with synthetic slag with O_2 in the plasma		
		without O_2 in the plasma	with O_2 in the plasma	MgF_2	BaF_2	CaF_2 - SiO_2
N	27 - 48	1,7 ± 0,3	0,3 ± 0,2 0,4 ± 0,2	4,0 ± 0,7	18 ± 2	25 ± 3 10 ± 1
C	1900	14 ± 3	2,8 ± 0,5 1,2 ± 0,3	19 ± 2	18 ± 2	69 ± 7 127 ± 12

Experimental conditions :
- % weight of slag = 1,5
- number of passes = 4
- flow rate of O_2 in the plasma = 30 cm^3 / min
- flow rate of H_2 in the plasma = 0,30 l / min
- flow rate of Ar = 30 l / min
- power of the plasma = 7 kW

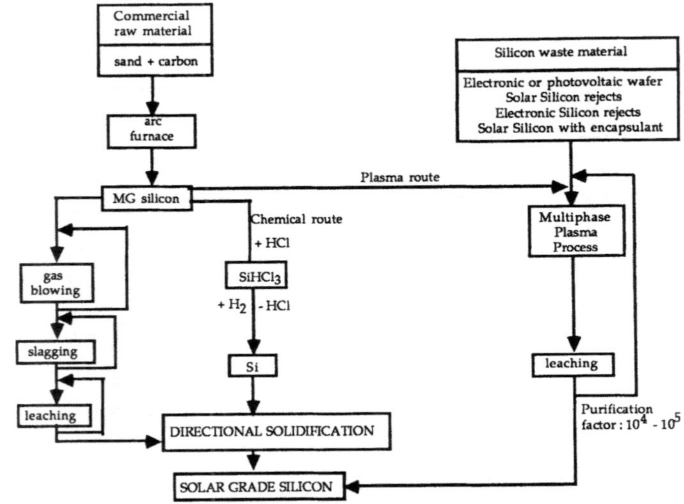

Table 1 : Schematic flow diagram for silicon elaboration

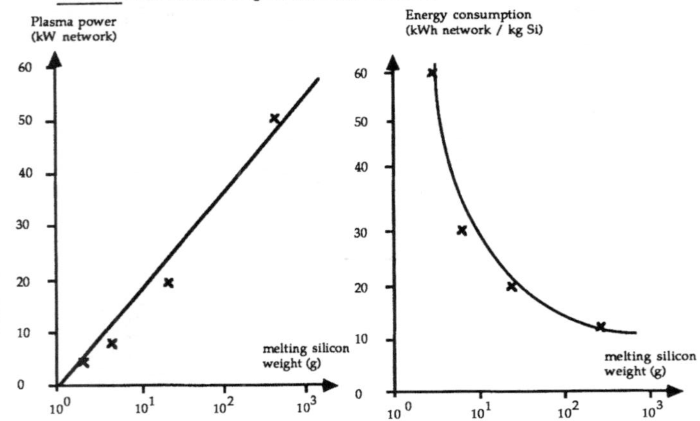

Figure 3 : Economical balance-sheet

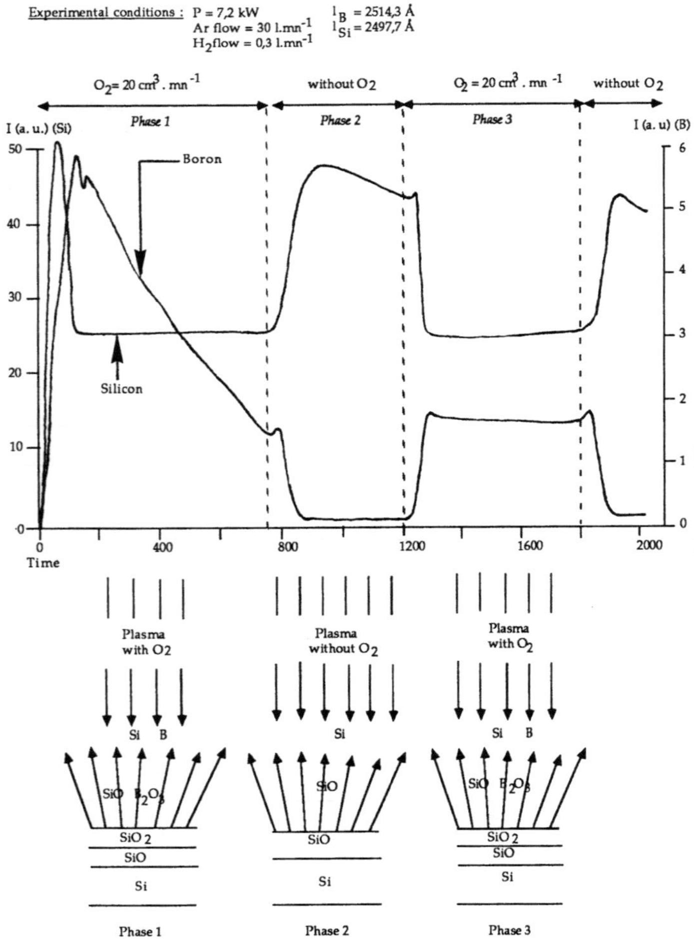

Figure 2 :Simultaneous emission curves of silicon and boron, in a silicon matrix, with or without oxygen in the plasma gas.

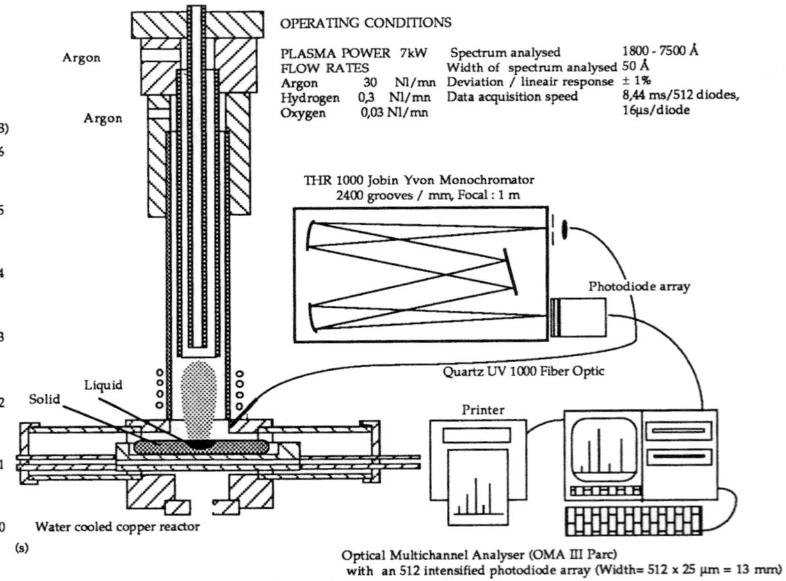

Figure 1 : Plasma reactor and emission detection system

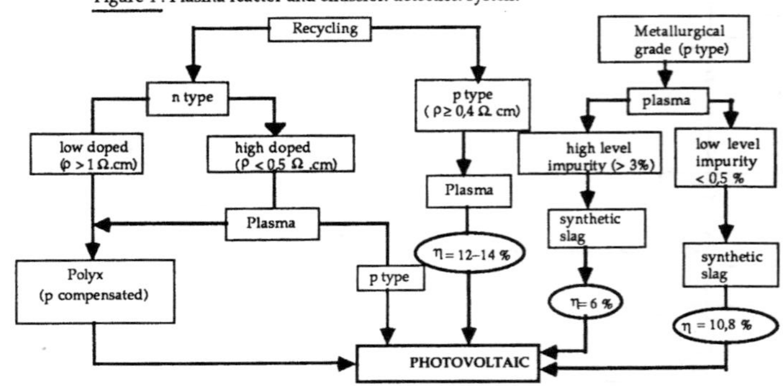

Table 5 : Schematic flow diagram for obsolete material (n, p type)

SMALL ANGLE NEUTRON SCATTERING AND ELECTRON MICROSCOPY INVESTIGATIONS OF PRECIPITATES IN POLYCRYSTALLINE SOLAR SILICON

S. Möller, Inst. of Semiconductor Technology, TU Hamburg-Harburg, 2100 Hamburg 90, FRG
H. J. Möller, Dept. of Material Science & Engineering, Case Western Reserve University, Cleveland, OH 44106, USA
T. Frisius, Research Center GKSS, 2054 Geesthacht, FRG

Abstract

The precipitation of silicon-carbide and -oxynitride particles in directionally solidified metallurgical silicon is studied. SEM and TEM in combination with EDX are applied to investigate their distribution, size and composition. In addition, it will be shown that the small angle neutron scattering is applicable to study the carbon precipitation even at the low carbon concentrations usually present in metallurgical silicon.

Introduction

Carbon is, along with oxygen, the main impurity in polycrystalline silicon, which is used for solar cell applications. The typical concentrations can easily exceed the solubility limit at the melting point (1), so that these impurities will precipitate during cooling. Although carbon and oxygen in solution are not electrically active (2), the precipitates themselves can have electrical properties (3) different from the bulk, and also induce a chain of secondary reactions. They can promote the formation of other crystal defects and can getter other impurities that are electrically active in silicon. The electrical properties of the silicon and the solar cells are thus considerably influenced, and a better control of these processes is desirable. The aim of the present study is to investigate the precipitation of carbon and other impurities in directionally solidified metallurgical silicon as a first step on the way to a correlation with electrical properties. Apart from conventional light and electron optical microscopy, the small angle neutron scattering has been applied to study the formation of carbon and other precipitates.

Experimental Procedure

Specimens of metallurgical silicon with rather high content of impurities were selected for the analysis and characterization of defects nucleating during the solidification process. The total content of all impurities was determined by wet-chemical analysis and their fraction in solution measured by Infrared-Spectroscopy (FTIR). The main impurities in all specimens were carbon, oxygen and aluminum; their concentrations are enlisted in table I. In addition, the impurities Ca, Cr, Cu, Ni, Ti, B, P, Mg, Fe and Ba were found with much lower concentrations (<3.5 ppm). The specimens have been investigated by means of Small-Angle Neutron Scattering (SANS), Light Optical Microscopy (LM) Scanning and Transmission Electron Microscopy (SEM and TEM) and Energy-Dispersive X-Ray Analysis (EDX).

In the SANS analysis, because of the low impurity concentrations, large specimens (about $9*5*5$ cm^3) had to be used. The average neutron wavelength was chosen at $7.8*10^{-1}$ nm to avoid Bragg-reflection in silicon. For the LM, SEM and EDX-investigations, the surfaces of the specimens were polished mechanically and chemically. Precipitations in the silicon matrix were imaged in the SEM by detecting the signal of mainly the backscattered

No.	C (ppm)	Al (ppm)	O (ppm)
I	140	717.6	-
II	1256	3.1	(60)
III	180.2	2.1	(39.6)

Table I: total and dissolved (in brackets) concentrations of the main impurities in directionally solidified silicon

electrons (material contrast). The chemical analysis was performed in the SEM by EDX with an ultrathin detector window, which allows the analysis of elements down to the atomic weight Z=5. Both the analyses of the precipitations and the matrix were made at different accelerating voltages of the electron beam, however, to obtain a good signal-to-noise ratio the energy of the electron beam had to be at least twice the energy of the characteristic x-rays (E_α) of the element analyzed. From FTIR, one could expect copper as the element with the highest atomic number (Z=29; E_α = 8.04 keV). Therefore, we chose U = 20 kV for the accelearting voltage to detect all elements with $5 \leqq Z \leqq 29$. A second analysis was performed at 5 kV to reduce the volume which is excited by the electron beam to emit x-rays. In that case it became possible to excite x-rays in a single precipitation only. Since none of the particles´spectra showed a signal at an energy >1.74 keV, this condition was sufficient to detect all elements present. TEM specimen were prepared by mechanical polishing and ion beam thinning. Areas of special interest were investigated in the SEM first and cut then from the bulk by an ultrasonic cutter and prepared for the TEM analysis.

Results

Large precipitates (10-20 μm) were already observed in the optical microscope in specimen III. Two different types of particles, named "A" and "B" could be distinguished (Fig. 1a). They were arranged typically in clusters or in strings, often type "A" particles show inclusions of type "B". After the specimens had been thinned down to a thickness of 3 μm, the "B" type particles appeared to be spherical and completely transparent for light; whereas type "A" remained opaque. The mean radii of the precipitates were found to be r = 11 μm (A) and r = 8 μm (B). The ratio of the volume fraction of the precipitates A : B was estimated to be 2 : 1.

The SEM images of the same particles (see left side of Fig. 1a) show a different material contrast between the precipitates of type "A" (Fig. 1b detail 1) and of type "B" (Fig. 1b detail 2). The EDX-spectra (Fig. 2) of the matrix taken at an accelerating voltage of 5 and 20 kV show the K_α -line of silicon only. In the spectrum of the "A"-particles silicon and carbon were found, whereas for "B" type precipitates nitrogen, oxygen and silicon were detected. "A" type and "B" type precipitates turned out to be electrically conductive and insulating, respectively.

For any "A" type precipitate, the intensities of the K_α -line of carbon I_C and silicon I_{Si} in the EDX-spectrum occur in the same ratio I_{Si}/I_C = 2. Since the relative peak heights of the spectrum are not directly proportional to the molar fraction of the elements (4), the spectra of the "A"-particles were compared to a pure SiC-standard. By this comparision, these particles were identified as SiC-precipitates. The EDX spectra of the "B" type particles, called Si-oxynitrides hereafter, show always two typical relations between the intensities of the k_α -lines of nitrogen I_N and oxygen I_O , either I_N / I_O= 1.2 or 1.6. Comparing these intensities to those

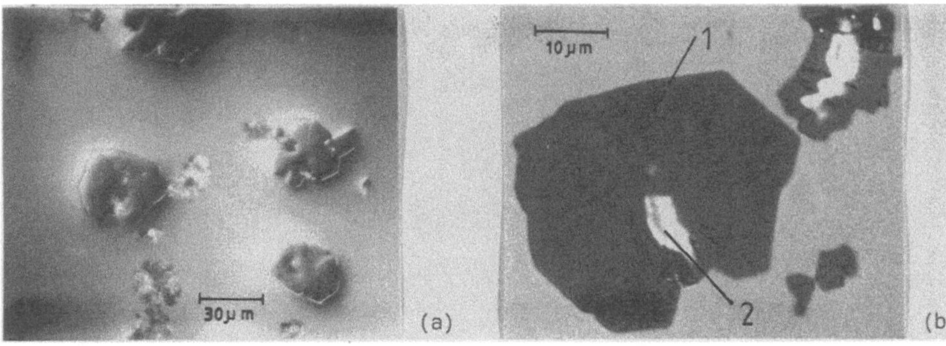

Fig. 1: LM (a) and SEM (b) micrograph of the same precipitates in metallurgical silicon

of a pure Si_3N_4 and a SiO_2 standard the composition of these Si-oxynitrides could be described by $Si_3O_2N_z$ with z = 5 or 7. For z, a mean value of 6 was assumed for further calculations because the particle size and their number with z =5 and 7 was almost equal.

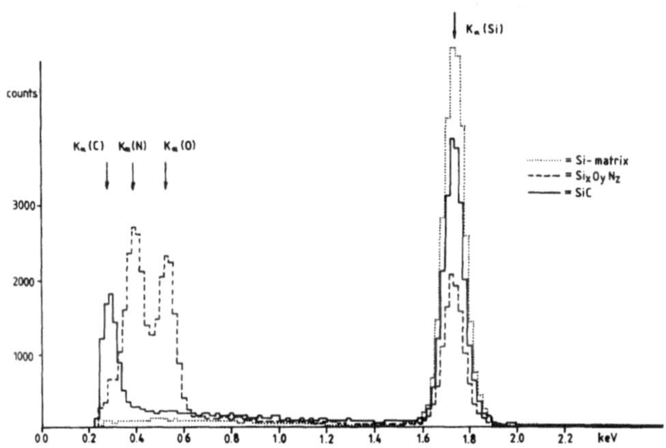

Fig 2: EDX-spectra of precipitates in metallurgical silicon

____ "A" type
---- "B" type
.... matrix

From the optical investigations the volume fraction of the precipitates $V(SiC):V(Si_3O_2N_6)$ was estimated to be 2 : 1. In specimen III the concentration of precipitated carbon corresponds to a precipitated volume of SiC of $1.7*10^{-4}$ cm^3, the volume of the precipitated Si-oxynitride is $8.6*10^{-5}$ cm^3 which is equivalent to 72 ppm of nitrogen and 24 ppm of oxygen. Apart from the large precipitates visible in the optical microscope, TEM micrographs show, in addition, a high concentration of smaller SiC precipitates (160-300 nm diameter and mean interparticle distance 200-500 nm) in the vicinity of each large Si-oxynitride particle. On the contrary, specimen I, with a comparable carbon concentration, did not contain large SiC precipitates (>2 µm). The TEM analysis revealed, however, smaller Si-oxynitride particles with a mean size of 400 nm (Fig. 3). Since the specimen volume which can be investigated by TEM is relatively small, it is difficult to observe particles of intermediate size (about 1 µm), if their density is low because they also remain undetected by optical and scanning microscopy. Therefore, it has been tried to apply SANS, which allows to investigate a large crystal volume. Fig. 4 shows the differential cross

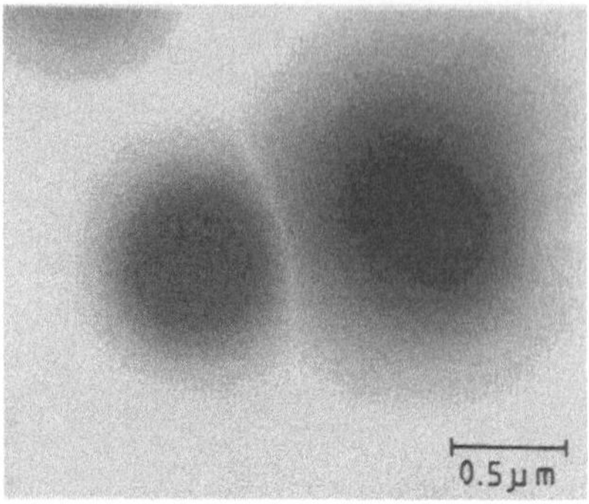

Fig. 3:
TEM picture of small $Si_3O_2N_z$ particles

section as a function of the scattering vector k for specimens I and II. For a random distribution of scattering particles, one should obtain $dS = P/k^4$ in a certain k-vector range, with

$$P = 2\pi AN (T_p - T_m)^2 \qquad \text{(I)}$$

and N number of particles per unit volume, A surface area of a single particle, $T_p = \sum_i N_{pi} b_{pi}$ with N_{pi} number of atoms of element i per unit volume of the precipitate, b_{pi} scattering length, T_m defined for the matrix respectively. The slope of the straight line obtained by plotting ln dS vs. ln k represents its k^{-4}-dependence and its intersection at k = 0 allows to determine P.

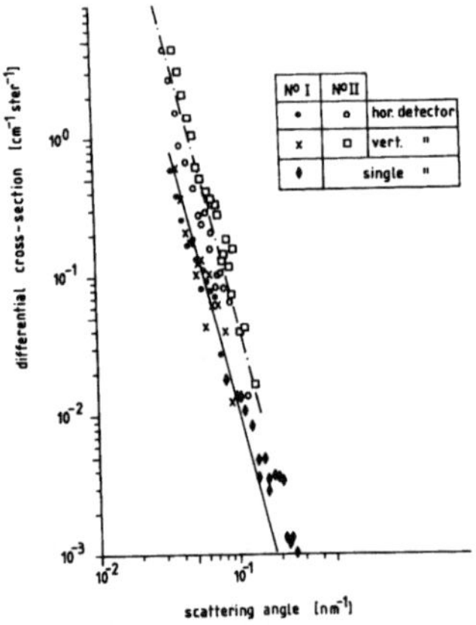

Fig. 4:
Differential cross section dS vs. scattering vector k

An interpretation of the SANS results which take into account all information obtained so far is only possible if one assumes that the main contribution to dS comes from spherical SiC particles. Since the total volume, V(SiC), of the precipitated SiC is known from chemical analysis,one can express the surface area A in eq.(I) by N, V(SiC) and r. Then, the mean size 2r, density N and particle distance d can be calculated. The scattering intensity of the smaller Si-oxynitrides particles can be neglected, however, if one assumes only these particles to contribute one obtains an upper limit for the precipitated oxygen and nitrogen concentrations (for a mean particle size of 400 nm). In table II all results for specimen I (from SANS, TEM), II (SANS) and III(LM, SEM, EDX) are presented.

No.	r(µm)	d(µm)	$N (cm^{-3})$	C_C (ppm)	C_N (ppm)	C_O (ppm)
I	1.9	60	4.3*10	130	< 9.5	< 3.2
II	3.6	53	6.8*10	1240	< 47.5	< 17.5
III	10.9	315	3.2*10	167	72	24

Table II: Mean radius r, particle density N and distance d of precipitated SiC particles. Concentration C of precipitated carbon, nitrogen and oxygen

Discussion

The results show that SiC and $Si_3O_2N_Z$ precipitate during the solidification of the silicon melt. They are distributed very inhomogeneously and preferentially in the bulk, not along grain boundaries. Since the SiC particles nucleate preferentially at or near $Si_3O_2N_Z$ precipitates, one can assume that their size and distribution control the SiC nucleation. Comparing specimen I and III which have almost the same carbon concentration, one can see that in specimen I the lower oxygen and nitrogen concentration causes a finer distribution of smaller SiC particles. Whether the $Si_3O_2N_Z$ particles are already present in the melt, or whether they nucleate during cooling, could not be determined. However, once they are formed, their interface and vicinity are preferential sites for the SiC nucleation. Although no further precipitates have been detected so far, despite the presence of a number of other impurities, one can assume that the precipitates are effective gettering sinks. The green colour of thin SiC particles, as observed in the transmisssion light microscopy, indicates, for instance, a gettering of Fe-atoms.

The different distributions and sizes of the precipitates have also an impact on the electrical resistivity of the material. Comparing the specific resistivities (measured by four-point probing and averaged over the entire surface), specimen I has a lower value (40 mΩ cm) compared to specimen III (800 m Ω cm). The experimental results have shown that the SiC particles are conducting, likely by doping with either B, Al or N, impurities which are dissolved in the bulk, whereas the $Si_3O_2N_Z$ particles turned out to be insulating. Since their volume fraction is lower in specimen I, the average resistivity should be lower as observed. Since it is desirable to correlate the spatial distribution of particles directly with local electrical parameters, like conductivity or admittance, this has not been tried yet, and will be the objective of further investigations.

The results have shown further that the SANS may be a valuable additional technique for the investigation of the density of SiC particles in silicon, provided the particles´size can be obtained by other methods. The determination of the particle size can be expected for smaller particles (< 0.1 µm), which might occur for specimens with lower oxygen or nitrogen concentrations.

References

(1) Landolt Börnstein, Zahlenwerte und Funktionen aus Naturwissenschaft und Technik, Vol. III,17 C, p.417, (1986)

(2) Kolbesen B.O., Mühlbauer A., Solid State Electronics, Vol.25, No.8, p.759, (1982)

(3) Gmelin Handbook of Inorganic Chemistry, Si Suppl., 8th Ed., Springer-Verlag Berlin, (1986)

(4) Reimer L., Rasterelektronenmikroskopie, Springer-Verlag (1977)

OPTIMIZATION OF EPR POLYSILICON SOLAR CELL CHARACTERISTICS

R. M'GHAIETH, R. GAUTHIER, P. PINARD

Laboratoire de Physique de la Matière*
Institut National des Sciences Appliquées de Lyon
20, Avenue Albert Einstein
69621 VILLEURBANNE CEDEX (FRANCE)

I. The EPR (Electron Powder Ribbon) method is used for pulling in high vacuum multicrystalline silicon ribbons by electron beam irradiation of silicon powder (3N5 or 5N) assigned to the elaboration of low cost solar cells. The elaboration of the ribbon consists of three steps : a pressing step of silicon powder (10 to 30 kN/cm^2), then a preribbon with grain size of the order of 40µm is realized by molten zone method (pulling rate ≅15 m/h) ; finally the preribbon is crystallized again (pulling rate ≅ 5 m/h). The irradiation power is between 100 and 300 watt (HV≅30 kV ; I=10mA). The electron beam is deflected at a frequency of 500-1000Hz along an axis perpendicular to the translation of the sample holder.With a thickness of the preribbon lower than 400µm, the EPR ribbon shows columnar elongated grains some millimeters wide with a thickness of 100 to 350µm, The resistivity is from 1 to 10Ω.cm corresponding to a P doping between 10E15 and 10E16 at/cm^3 (1,2).

II. The EPR technic induces the refining of the starting material as shown in Figure 1, relative to 3N5 purity grade powder and pulling rates of 10 m/h for preribbon and 4 m/h for ribbon. The different processes of refining (slagging-vacuum evaporation, zone melting) allow to obtain from a MG-HP1 powder a crystalline and flat material with HP2 or SOG purity grade (3). But, the EPR method produces as other ribbon elaboration methods (4) built in stresses which lead to a high brittleness of ribbon incompatible with elaboration process of solar cells and high density of dislocations damaging for the photovoltaic efficiency. So, we attempt to eliminate them by studying the temperature profiles during the elaboration and the repartition of thermal stresses in EPR ribbons. This last point is studied with theoretical method -simulation and calculation of the stresses- then by experimental analysis with optical method. Moreover, the highest pulling rates are estimated ; temperature profiles, stress growth and pulling rate are very strongly dependent.

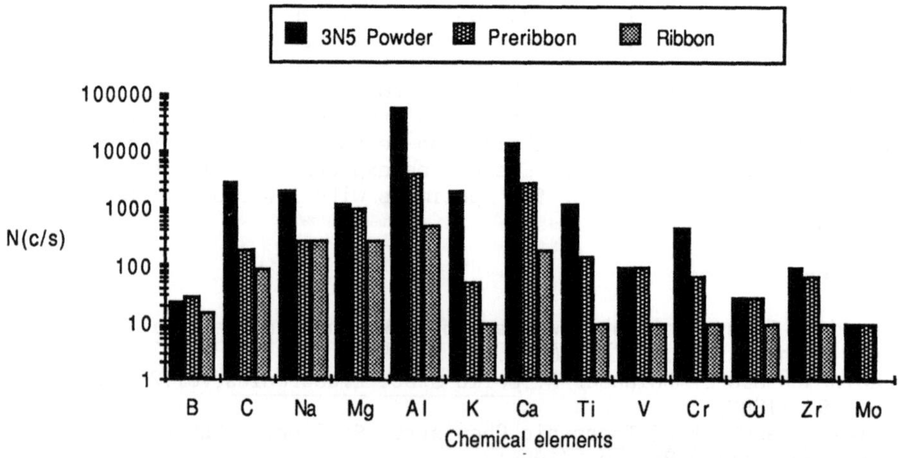

Figure 1: SIMS analysis of EPR products.

* associé au C.N.R.S.

III. The thermal stresses in the ribbon are represented by the following equations :

$$(1) \quad \sigma_{xx}=\alpha E\,(A\frac{d^2T}{dx^2})+B\frac{d^4T}{dx^4} \;;\quad \sigma_{yy}=\alpha E\,(A'\frac{d^4T}{dx^4})+B'\frac{d^6T}{dx^6} \;;\quad \sigma_{xy}=\alpha E\,(A''\frac{d^3T}{dx^3})+B''\frac{d^5T}{dx^5}$$

a, E are respectively the mean expansion coefficient ($3\ 10^{-6}.C^{-1}$) and Young modulus ($1.7\ 10^5$MPa) of silicon material. So, it is possible to calculate the profile $\sigma_{xx}(x)$ of the longitudinal stresses ; a program allows to execute the procedure for any point of a network. The process consists of two essential steps;collection of stress values by computer and processing of these data (two dimensional interpolation with triangulation and interpolation into every triangle by linear interpolation or by distance weighted least square approximation). This process loads to obtain a colour mapping of the theoretical length stresses (Figure 2).

Figure 2: Color display of calculated stresses (MPa) for EPR ribbon (Pulling rate =1m/h)

The relation (1) show that linear temperature profiles are necessary to eliminate the residual stresses in the EPR ribbons but it may be noted that a low thermal gradient is necessary to eliminate the calories of latent heat of solidification and obtain high pulling rate. So, a fore and post heating system around the molten zone was used to induce constant thermal gradients.

The experimental stress measurements are realized with computer aided infrared photoelasticimetry (PIRAO) (5). The absorption coefficient and piezo-optical coefficient of silicon are respectively of $30cm^{-1}$ and 20B at $\lambda=1.06\mu m$. These values are compatible with transmission photoelasticimetry and using of a vidicon camera ; moreover, the theoretical study of stresses has proved that the length and transverse thermal gradients only induced plane thermal stresses ($\sigma_{zz}=0$) and that the variation of these last along the thickness of the ribbon was lowest and then negligible.

IV. First, for eliminating the residual stresses, we developped a thermal treatment of the samples by the electron beam. Two electromagnetic coils B_1 and B_2 with perpendicular magnetic field induce respectively the beam deflexion along the pulling axis and along the molten zone axis ; the sample holder is moving with a speed between 4 and 15 m/h. The simultaneous injection of triangular waves in each coil induces a rectangular heating zone with size adjusted by the wave amplitude.

Moreover, a thermal treatment card used with the B_2 coil allows to act separately on the xx' deflexion amplitude during a time choised between 1 and 5 periods ; the amplitude reset during one or some period leads to the molten line yy'. The modification of scanning amplitudes during the other periods induces heating areas around the molten zone (figure 3) : so, a good choice of deflexion xx' allows to obtain a fit approach of linear profiles of temperature.

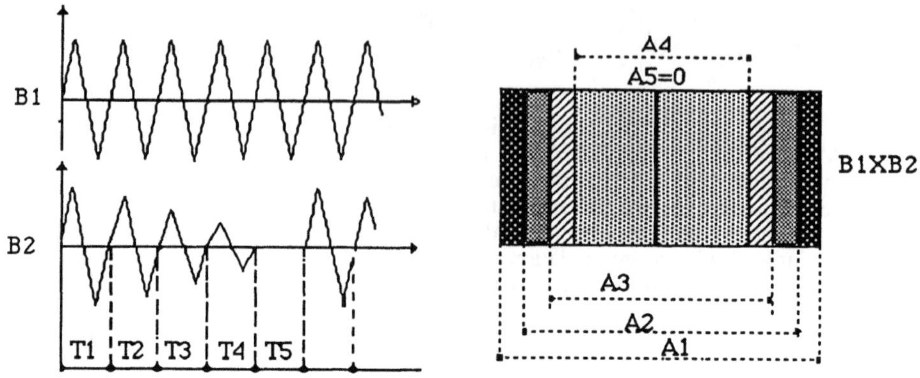

Figure 3: Principle of linearisation of temperature profile by the electron beam.

Actually, we adapt a travelling melting line on a fixed substrate, heated uniformly at 1100°C. The melting line and the heating of the sample are obtained as lastly. The Figure 4 indicates the influence of this system upon the temperature along the ribbon.

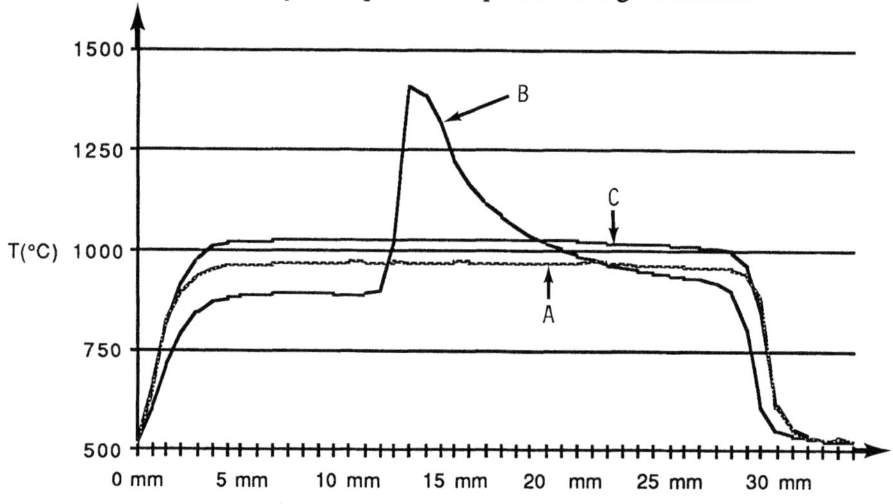

A: Temperature profile of powder heating B: Temperature profile during elaboration
C: Temperature profile of ribbon after elaboration
Figure 4: Experimental temperature profiles in EPR ribbons measured with an optical pyrometer.

It may be noted that firstly the melting line is reset at the fore end of the sample by electronic process then moved by single sweep ramp. The slope of this last can be adjusted to vary its travelling speed gradually and so to eliminate the variations of molten zonewidth due to variations of focus and to the excess of calories produced by the solidification latent heat which appear for low thermal gradients. After the molten line travelling, the electron beam is only used for sample heating and a gradual wehnelt polarization allows to set up a global and progressive cooling down of the sample ; so, a stress relaxation of the material is obtained.

V. The theoretical and experimental studies, here presented, lead to obtain good mechanical quality ; the stress values have decreased from 300 MPa to some tenth of MPa. These last values still induce a high dislocations density of the order of $10^7/cm^2$, which reduce the photovoltaic efficiency of EPR solar cells, limited to 4% without antireflection coatings.Yet, the recent results obtained with PIRAO method (Figure 5) indicate a new high decrease of stresses by using the last heating method. Moreover, the diffusion length of minority carriers, estimated by spectral energy distribution measurement with electrolytic diode and EBIC reach values of 16-20 µm; So, it may be hoped henceforth efficiencies of 7-9 %.

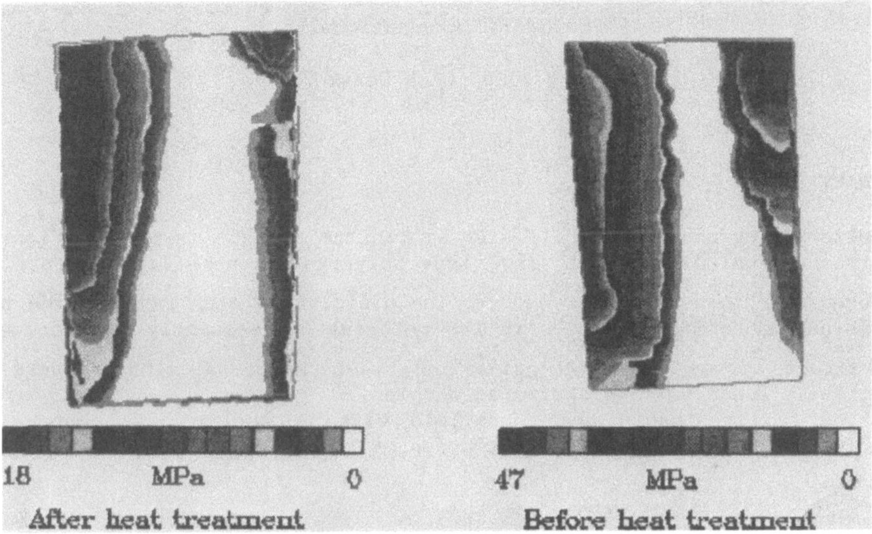

Figure 5: Color mapping of measured stresses of EPR Ribbon

ACKNOWLEDGEMENTS

We are grateful for CNRS, PIRSEM and AFME which supported this work and we thank Pr SIRTL and Dr BRAETSCH (Wacker Heliotronic), Pr MARTINUZZI (Laboratoire de photoelectricité, Université S[t]Jérome - Marseille) MM. LOUBLY and SARTI (Photowatt) for their collaboration.

References:

(1)- D.CHAMBONNET, R.GAUTHIER, R.M'GHAIETH, P.PINARD
Le ruban de silicium EPR: Elaboration et mise au point d'un dispositif de mesure des contraintes. Rev.Phys. Appl. Janv 87, 22, 557-562.
(2)- D.CHAMBONNET: Thesis, Lyon 1986.
Elaboration des rubans de silicium polycristallin EPR pour photopiles et réalisation d'un procédé de mesure des contraintes résiduelles par photoélasticimetrie infrarouge assisté par ordinateur applicable à ce type de matériau.
(3)- J.DIETL:
Hydrometallurgical purification of metallurgical grade silicon. Solar cells, 10 (1983) 145-154.
(4)- R.W. GURTLER: Nature of thermal stresses and potential for reduced thermal buckling of thin silicon ribbon grown at high speed.Journal of crystal growth, 50, pp.69-82, 1980
(5)- D.CHAMBONNET, R.GAUTHIER, R.M'GHAIETH, P.PINARD
International conference on résidual stress; Garmish-Partenkirchen (RFA) 15-17 oct 86
Resdual Stresses in Science and Technology, Vol. 1,1987, 343-349.Edit.E.Mackeruch and V.Mauk informatungesellschaft-Verlag 1987.

LOW TEMPERATURES GETTERING EFFECTS IN P-TYPE POLYCRYSTALLINE SILICON

H. POITEVIN*, C. ZURLETTO*, H. EL GHITANI*, D. SARTI** and S. MARTINUZZI*

* Laboratoire de Photoélectricité des Semi-Conducteurs
Faculté des Sciences et Techniques de Saint-Jérôme
Université d'Aix-Marseille III
F.13397 MARSEILLE Cedex 13

** PHOTOWATT International S.A.
6, rue de la Girafe - B.P. 6069
F.14002 CAEN Cedex

ABSTRACT

Pregettering by phosphorus diffusion within the surface is used to improve electron diffusion lengths L_n in P type polycrystalline silicon wafers. It is found that this treatment reduces the dislocation etch pit and the bulk recombination centre densities in the material. Consequently L_n increases. The effect of an external gettering by backside aluminum layers is drastically enhanced in pregettered samples.

The two treatments are compatible with the solar cell preparation technics and could be applied to improve or to homogenize their efficiency.

1. INTRODUCTION

The efficiency of large grained polycrystalline silicon cells is limited by recombinations within intragrain defects especially dislocations, which degrade minority carrier diffusion lengths and consequently the values of photocurrent J_{sc} and photovoltage V_{oc}. These defects have not by themselves a noticeable activity and it is the segregation of dissolved impurities like oxygen and metal atoms which constitutes the main source of recombination centres.

It is of interest to study processes, which would be compatible with the solar cells preparation technics and which can impede the segregation or can neutralize the recombination centres. External gettering effects at relatively low temperatures, applied before or after the preparation of the N^+P junctions could be candidate.

Gettering by phosphorus diffusion within the surface of silicon wafers is frequently used to improved minority carrier lifetime in the active regions of wafers in which transistors are made. It was also used to improve solar cells /1/.

A gettering effect could be also developped by a backside aluminum layer /2-6/, because during annealings at about 450°C, aluminum-silicon interface can act as a sink for metallic impurities having a high diffusivity, and this diffusivity could be enhanced by the presence of dislocation pipes.

In this paper, we have investigated the influence of a phosphorus pregettering at 850°C during 30 mn on the effective minority carrier diffusion lengths and on the distribution of dislocations in polycrystalline silicon wafers. We have also studied the effect of an additional backside gettering by aluminum, on phosphorus pregettered samples.

2. EXPERIMENTAL

Adjacent 400 and 200 μm thick wafers were cut out of the same cast silicon ingots by means of a were saw. In these wafers, the density and distribution of defects are very similar as a consequence of the columnar growth of the grains. Some wafers were used to get virgin samples, while others were submitted to phosphorus diffusion at 850°C during 30 mn, by the two faces.

Corresponding samples of 2 x 1 cm^2 were cut out of the wafers, and after etching the N^+ regions of the pregettered samples, Cr-Si semi transparent diodes (2 x 2 mm^2) were realized by the deposition of a thin chromium layer through a mask. The diodes were made at the same place in virgin and pregettered samples in order to compare the gettering effects on the same regions, i.e. on the same defects, by means of effective electron diffusion length L_n and by LBIC scan maps at λ = 940 nm.

The technics used to realize an aluminum backside gettering and the measurements methods have been described previously /5/.

After the electrical investigations, the samples were chemically etched in order to reveal dislocation etch pits in the grains and within twins and grain boundaries.

3. RESULTS

It was found that in pregettered samples, the values of L_n are neatly increased compared to those measured in virgin materials, as shown by figure 1.

The increase is the same in 400 and 200 μm thick samples, probably due to the long range effect of phosphorus gettering.

Figure 2 shows the microphotography (after etching) and the LBIC scan map of a diode made in a virgin sample. This diode contains dislocations and twins decorated by dislocations. It results an effective diffusion length of 25 μm.

In the corresponding phosphorus pregettered sample, for the diode made at the same place, the value of L_n is found to be 54 μm. The microphotography (after etching) and the LBIC scan map are given by figure 3. The etch pit density and the decoration of twins are reduced, and consequently the photocurrent in the grains are increased, while the photocurrent attenuations within twins become weak.

Virgin and pregettered samples are then submitted to an additional backside gettering at 425°C during 30 mn by an aluminum-silicon interface.

It appears that the effect of this treatment is particularly enhanced in the phosphorus pregettered samples, in which the values of L_n could be drastically increased. In the 200 μm thick samples, these increases vary in the range between 50 an 200 %, as shown by figure 4.

The aluminum gettering could be applied to mesa diode realized in solar cells /5/, in which the N^+ region is obtained by phosphorus diffusion at 850°C during 30 mn. The back contact is removed and replaced by an aluminum layer and the structure is annealed at 425°C during 1 hour.

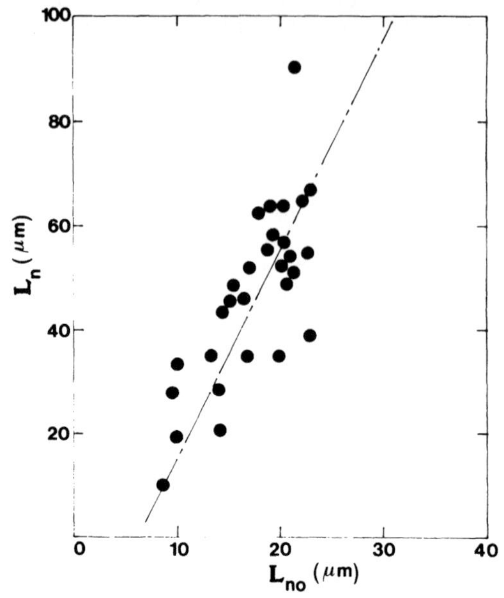

Fig.1. Effective diffusion length L_n in phosphorus gettered samples as a function of initial value L_{no}.

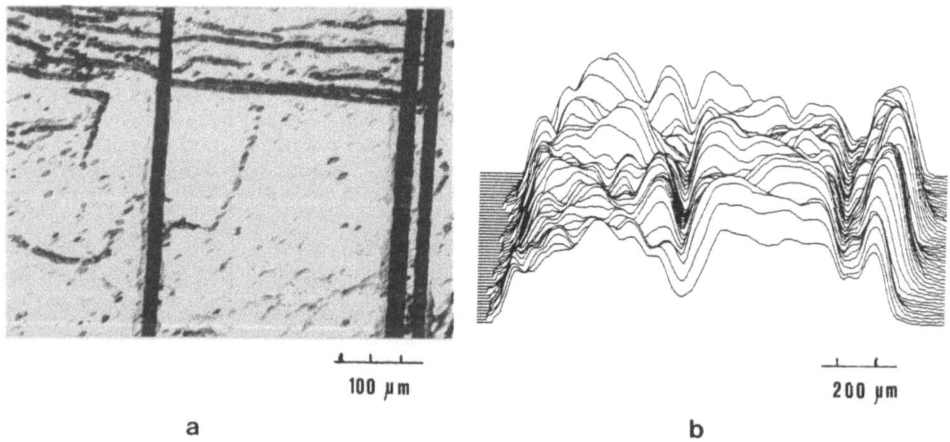

Fig.2. : (a) Microphotography and (b) corresponding LBIC scan map at λ = 940 nm (f.s. : 25 μV) of a virgin sample (etch pits formed sub grain boundaries).

Diffusion lengths values are also increased, and the experimental variations L_n/L_{no} as a function of initial value L_{no} are reported in figure 4. It is clear that the improvement of L_n observed in solar cells are in agreement with those found in wafers after that the N^+ regions have been removed.

As a consequence of the improvement of L_n, the so called saturation current J_o of the mesa diodes decreases, while V_{oc} and also J_{sc} increase.

5. CONCLUSIONS

The present work shows that a diffusion of phosphorus within the surface of polycrystalline silicon improves the electron diffusion length.

This improvement could be due to an external gettering effect and also to a modification of density and distribution of intragrain defects.

Moreover the preceeding treatment increases the sensibility of the material to an additional backside gettering by an aluminum silicon interface. As aluminum cannot diffuse significantly in silicon at 425°C, even by dislocation pipes, it is believed that the gettering by aluminum-silicon interfaces involve the participation of point defects (injection of silicon vacancies by dissolution of silicon in aluminum) and of dissolved oxygen.

The two treatments could be applied successfully to solar cells : the phosphorus pregettering could be used as a preliminar improvement of the wafers, while aluminum backside gettering could be applied after the preparation of the N^+P junction, during the fabrication of the ohmic back contact.

This work was supported by Agence Française pour la Maîtrise de l'Energie (contract 5.05.1026) and CNRS-PIRSEM (contract 8680).

REFERENCES

1 M.A. Green et al., Conf. Record 18th IEEE Photovoltaïc Spec. Conf. Las Vegas (USA) 1985, p.39.
2 W.A. Orr et al., IEEE Trans. Elect. Devices, vol.29, 8, 1151, 1982.
3 R.D. Thompson and K.N. Tu, Appl. Phys. Lett. 41, 440, 1982.
4 R. Sundaresan et al., J. Appl. Phys. 55, 1162, 1984.
5 S. Martinuzzi, M. Zehaf, H. Poitevin and G. Mathian, Conf. Record 18th IEEE Photovoltaïc Spec. Conf. Las Vegas (USA) 1985 p.1127.
6 C. Keavney, M.B. Spitzer and L.M. Geoffrey, 169th Electrochemical Society Meeting, Boston, 1986, to be published.

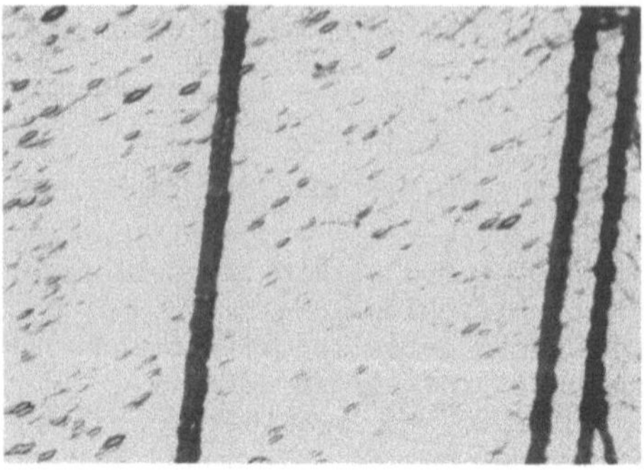

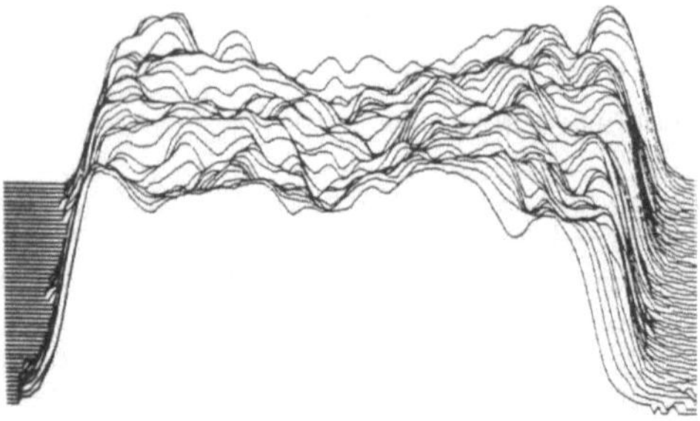

Fig.3. : (a) Microphotography and (b) corresponding LBIC scan map at λ = 940 nm of the sample of fig.2 after phosphorus gettering at 850°C during 30 mn.

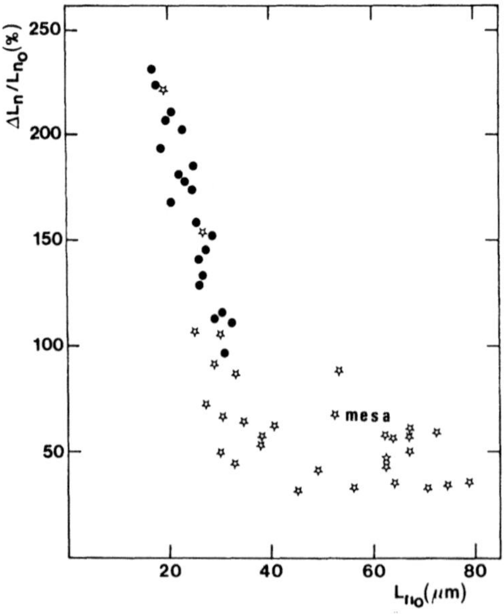

Fig.4. : Relative increase of electron diffusion length after backside gettering by Al-Si interface.

SCANNING TUNNELING MICROSCOPE AND COMPLEMENTARY MICROCHEMICAL INVESTIGATIONS OF HYDROGEN AND SHALLOW ACCEPTORS AT SILICON GRAIN BOUNDARIES

Lawrence L. Kazmerski
Solar Energy Research Institute
1617 Cole Boulevard
Golden, Colorado 80401 U.S.A.

ABSTRACT. The passivation of grain boundaries and the neutralization of impurities in polycrystalline silicon, important processes for the improvement of performance of devices fabricated from this material, are investigated using micron- to atomic level spectroscopic techniques. The incorporation of hydrogen into grain boundaries is investigated using surface analysis methods. Volume-mapping techniques are used to identify the bonding mechanisms of the hydrogen in oxygen-free and oxygen-rich intergrain regions. Interactions between shallow acceptors (B,Al,Ga and In) and hydrogen in polycrystalline Si are studied. The bonding mechanisms involved in the acceptor neutralization process at the grain boundaries are evaluated using microanalytical techniques. Volume-indexed AES and Auger difference spectroscopy data are complemented by scanning tunneling microscope spectroscopic images to confirm the direct hydrogen-silicon bonding in boron-doped grain boundaries and hydrogen-acceptor bonding in Al-doped material.

1. INTRODUCTION

A significant contribution to the recent improvements in Si solar cell performances has been the increased understanding and control of impurities in the semiconductor material and at device interfaces [1]. This applies both to purposely-placed impurities for doping and electrical passivation, and for those which inadvertently occur during the Si-production or device fabrication processes. Certainly, the ability to control such impurity species depends upon the ability to detect them, not only their *concentration levels*, but also their exact *spatial locations* within the cell structure--especially at critical interfaces--to the hundreds of Angstroms resolution. The observations that hydrogen saturates or satisfies dangling bonds have led to its use to compensate defects and amorphous material. Hydrogen has been successfully utilized to improve the electrical characteristics of polycrystalline solar cells [2-4]. Recently, it has been discovered that H can be used effectively to neutralize deep impurities and shallow acceptors in elemental and compound semiconductors [5-8]. It is the purpose of this paper to demonstrate micron to atomic level analysis techniques for the diagnostic examination of the mechanisms involved in hydrogen treatments for neutralization of electrically-active impurities in polycrystalline Si.

2. EXPERIMENTAL DETAILS

The polycrystalline Si and cells used in these investigations were produced by casting, and the properties and those of the grain boundaries have been described previously. The compositional and chemical analyses were primarily carried out using Auger electron spectroscopy *(AES)* and secondary ion mass spectrometry *(SIMS)*. Elemental and ionic spatial mapping were accomplished using volume-indexed techniques, and the instrumentation and details of this analysis have been provided previously [9]. Volume mapping provides the digital acquisition of AES or SIMS data throughout a selected microvolume, coding this information for elemental or ionic species, spatial location within the selected microvolume, and concentration. That is, this indexed information is obtained at each point in the chosen microvolume during a single depth-profile operation by control of the analyzer, with a spatial resolution of about 8000Å for SIMS and 500Å for AES. The best depth resolution is in the range 100-1000Å for each case. The software allows the operator to selectively display the information for any point, along any line, or on any plane within the analysis microvolume after a single analytical operation. The STM investigations were performed in an instrument fabricated by McAllister Technical Services, and specially modified for low-temperature and semiconductor characterizations. The software (control, data acquistion and image processing) were developed in the SERI laboratories.

3. SHALLOW IMPURITIES AT GRAIN BOUNDARIES: MICRON TO ATOMIC LEVEL ANALYSES

Measurement technologies are continually pressed to their limits and into new applications as their demonstrated contributions to their usefulness are documented. This section introduces some evolutions of the surface analytical methods with specialized applications for device investigations, and a new characterization method which is capable of the most fundamental diagnostic evaluations--those on the single-atom bonding scale. The utility of these fundamental research and diagnostic methods in photovoltaic investigations is demonstrated in the analysis of the bonding mechanisms and electrical neutralization of shallow impurities at grain boundaries in Si solar cells.

3.1. Auger Difference Spectroscopy

Previous studies have considered the incorporation of hydrogen in the defect regions of polycrystalline Si. The *electrical neutralization* of impurities in Si is of importance for the control of polycrystalline device properties. The incorporation of the hydrogen into the Si has been explained by several competing models involving the bonding of H to either Si or to the impurity itself, or the bonding via OH groups [5-8]. Theoretical calculations have supported various of these mechanisms which account for neutralization by hydrogen-acceptor pairing, and recently, some counter-proposals have suggested that the cause of passivation is *direct compensation* [9]. In any case, the chemical interaction and bonding among the Si, H, shallow acceptors and other impurity species is important for the control and modeling of the process. This paper extends the examinations of hydrogen behavior in grain boundaries to the cases in which the boundaries have concentrations of the shallow acceptors B, In, Ga and Al.

3. 2 Grain Boundaries with Shallow Acceptors

In crystalline Si, spreading resistance profiles have been used to show that there is an interaction between the *acceptor* content of the Si and the *hydrogen* introduced into the material. Treatment in atomic hydrogen increases the resistivity of p-Si, and the resistance profile suggests that the higher the acceptor background, the greater the interaction between the H and the acceptor. Both the H-Si and H-acceptor models could be used to support the H-diffusion behavior. The electrical effect of the acceptor neutralization has been reported in the EBIC studies of grain boundaries. In general, the degree of neutralization follows the bond strength trend *B<Ga<In*. The aluminum case is quite different. The response changes continuously as more H is introduced, possibly an indication that the chemistry of the introduction is different. Previously-reported spreading resistance data have also indicated anomalies for the Al-doped Si.

The *direct* observation of the hydrogen at the grain boundary requires *high spatial resolution*, such as that afforded by AES. However, hydrogen is *not directly detectable* using this surface analysis method. The effect of the chemical bonding of the H to the Si is observable in the Si-LVV Auger peak shape and position [10]. The untreated transition contains only the Si-pp bonding. The H-treated case shows the additional Si-H sp and pp bonding. The chemical effects on the AES transition are useful in determining the nature of the bonding for the shallow acceptors. In order to accomplish this, the valence AES transitions were monitored for the acceptor and Si as a function of the hydrogen processing. In each case, the elemental transitions were recorded, normalized and stored. The spectra of the hydrogenated grain boundaries were subsequently measured, and the elemental transitions were subtracted in order to produce a difference spectrum, such as that in Fig. 1. This difference spectrum for the B_{KLL} transition is essentially a straight line of zero magnitude, indicating that the B-detected (at about $10^{19}/cm^3$) is nearly elemental in chemistry. In turn, the Si_{LVV} spectrum shows the Si-H bonding. These data further support the Pankove bonding model, which predicts that the primary bonding is with the Si

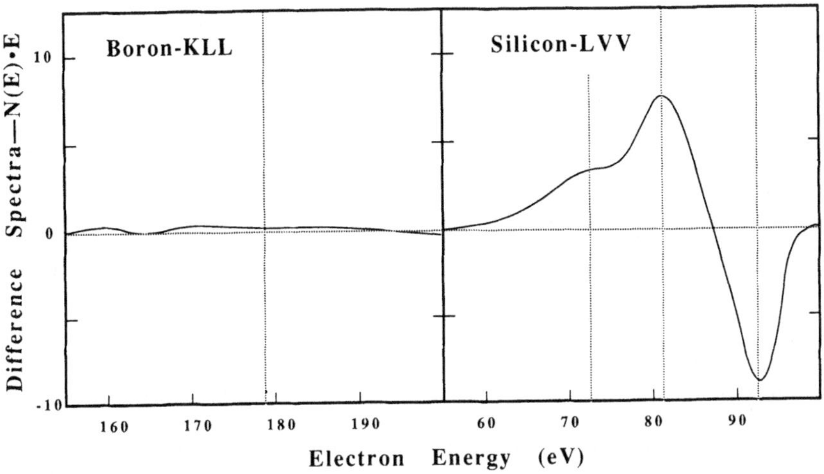

Fig. 1. *AES difference spectra for boron-rich grain boundary showing bonding of hydrogen to the silicon rather than to the boron.*

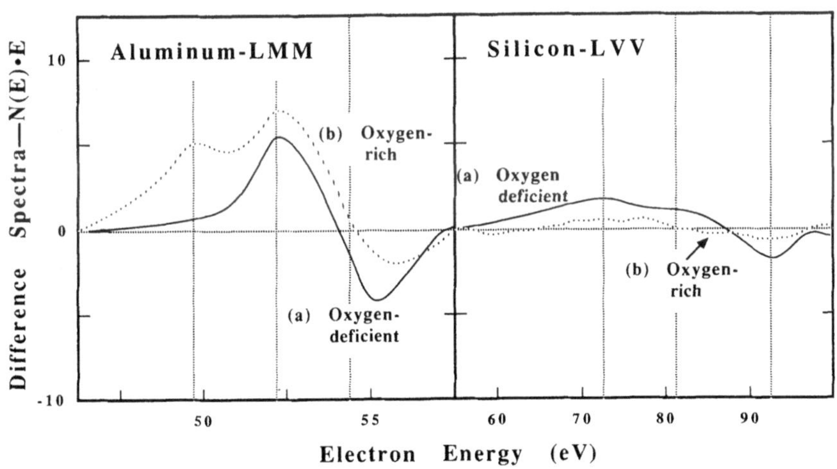

Fig. 2. *AES difference spectra for Al-rich grain boundary showing degree of Al-H bonding for oxygen-free region and Al-OH for oxygen-containing region.*

and not the B. If the hydrogen were bonding directly to the B, the B_{KLL} spectrum would be affected as shown in Fig. 1. Similar AES difference data have been reported for Ga and In (i.e., the Si-H bonding dominates during the neutralization process). Aluminum-doping is more complex--as the previously-cited electrical evidence indicated.

Figure 2 shows an apparent *direct Al-H bonding mechanism*, with a corresponding, substantially-less detected Si-H component. The resistivity for Al-doped and H-neutralized samples also has been reported to have some anomalous features. For example, the resistivity of layers has been shown to increase significantly more than for B, Ga or In-doping. Part of the explanation may have to do with alloying of the Al with the Si. However, such data also exist for samples processed well below the allowing temperature for this couple. The data of Fig. 2 indicate that the incorporation of the H may indeed be different--with the bonding directly to the Al as proposed by Sah *et al.* for such shallow acceptors. Another explanation any be that some degree of hydroxyl bonding may take place. This effect is observable in the oxygen-rich grain boundary, having approximately a $10^{19}/cm^3$ oxygen content. The AES difference spectrum, Fig. 11a, contains a measurable AlO(OH) and $Al(OH)_3$ component--species which are indistinguishable by AES or XPS but are known to exist in Si-oxides treated by H. Such a transition is absent from the oxygen-deficient ($\sim10^{17}/cm^3$ oxygen from SIMS estimates) difference spectrum of Fig. 6. Therefore, it is probable that Al-H bonding may be responsible for the anomalous results in electrical measurements. This has been further confirmed by the IR data, in which the Al-H complex is discernable for the Al-doped boundary.

3.3 Volume-Mapping Studies

The presence of impurities at the defects can *control* the chemical and diffusive properties of the grain boundary. It is important, therefore, to know something of the impurity content of not only the crystal, but also the defects when evaluating the passivation/ neutralization and material transport properties. Conventional SIMS and AES mapping has been utilized to show the distribution of impurities on grain boundary surfaces, but volume-indexing combined with the knowledge of the effects of the hydrogen on the shape and position of the Auger transitions provides a unique method of tracking the chemistry of the hydrogenation process with high spatial resolution. The initial maps were produced by mathematical deconvolution of the peaks that make up the transition, and then using that data to provide the map. However, the spatial resolution is somewhat limited, due to the time (more than 3 hrs.) required to take and process the data. The time required to sort out the bonding components using AES difference spectroscopy is reduced considerably, and a map can be produced in less than 15 min. The resolution, as indicated in Fig. 3, is increased considerably due to the minimization of data acquisition and beam exposure/ stability time [11]. Figure 3 presents a comparison of the chemical species for two cases. In the boron-doped case, the Si-H bonding clearly dominates, with no detected B-H bonding. The B is seen to be accumulated or segregated in the grain boundary region. In contrast, the Al (also accumulated in the boundary) shows far less Si-H bonding (indicated by the intensity of the mapped species), but significant acceptor-H bonding. Thus, these volume-indexed chemical maps confirm the results of the difference and IR spectroscopies, and show the penetration of the neutralizing complex down the grain boundary plane.

3.4 Scanning Tunneling Microscope Investigations

These data can be supported further by utilizing the *atomic-level imaging* capabilities of the *scanning tunneling microscope* [11]. The scanning tunneling microscope *(STM)* is an extremely high-resolution instrument, capable of structural and spectroscopic imaging of surfaces and interfaces on a scale that extends to *atomic dimensions*. This recently-developed characterization technique provides atomic imaging, information of topography, electronic and chemical structure [12]. It has lateral resolutions better than 2 Å, and can be used on metals, semiconductors and insulators. Its operation is not confined to vacuum (where the highest resolutions are producible), but can be used on gas/solid and liquid/solid environments as well. This characterization technique is still in its infancy, and its applications and strengths for electronic device and materials diagnostics is only starting to be explored.

In STM, a fine metal tip (usually an electrochemically-prepared tungsten probe) is brought within 3-10 Å of the sample surface to permit tunneling. Typically, the tunneling currents are of the order of 1 nA. Figure 4 presents a schematic representation of the STM and associated control hardware. At these extremely small distances, the wavefunctions of the atoms at the sample surface and on the fine probe overlap, and by applying a voltage difference between them (in the range 1 mV to 1 V, typically) the electrons can tunnel from the probe tip to the sample (or *vice versa*). The tunneling is actually between a single atom on the probe and the sample surface. In the STM, the probe is mounted on three orthogonal piezoelectric transducers, which allow positioning vertically as well as scanning parallel to the surface. Two modes of operation are possible. The *first* monitors changes in the tunnel current to reflect changes in the distances between the tip and surface. These are used to monitor the probe surface separation. The *second* mode is more common. The variations in tunnel current are monitored by a feedback circuit, which in turn applies a feedback voltage to the vertical control piezoelectric transducer so that a constant tunnel current is maintained while scanning. This voltage correction is measured during the horizontal raster scan and presented as a real space image of the surface corrugagtions, providing a topograph. These data can be further processed to provide three dimensional and color-coded images of the atomic arrangements.

Grain boundaries were prepared for examination in the STM by *fracture* in-vacuo to expose the region and produce a flat surface for imaging. The first grain boundary, shown in Fig. 5a, is one that is doped heavily with B. The surface is a (100)-plane. The position of the Si atoms can be easily seen in the density of states image. By properly biasing the sample during analysis, the B and H atoms can also be spectroscopically imaged, and are observed in Fig. 5b superimposed over the previous micrograph. The disruption at the grain boundary is indicated, and the atomic planes intersecting from the two grains are apparent. The relative positions of the H, Si and B suggest that the bonding of the H is to the Si rather than to the acceptor. This direct imaging gives further evidence to the Si-H bonding involved in the B-neutralization process, with the H chemically bonding to the Si, satisfying its dangling bond, leaving the B completely satisfied being bonded to three other Si atoms in the lattice. The case for Al is quite different, as indicated in Fig. 6a. In this case, a significant interact between the Al and H is observed in the STM overlays (along with some Si-H bonding). These data support the AES results that the bonding mechanisms are distinctly different for this case. For grain boundaries containing excess oxygen, hydroxyl-group bonding can also be imaged (Fig. 6b). Thus, these fundamental STM results give direct evidence for the mechanisms involved in the electrical neutralization process for shallow acceptors at Si grain boundaries.

4. SUMMARY

The microcharacterization of polycrystalline Si grain boundaries has been discussed, with spatial resolutions ranging from the multi-micrometer to the single-atom regimes. The purpose has been to provide a summary of selected and representative chemical and compositional techniques used in the analysis of photovoltaic materials, and give insight in the fundamental mechanisms controlling the electrical properties of impurities and defects in solar cells.

Acknowledgement: The author would like to acknowledge the unselfish contributions of the members of the SERI Photovoltaic Measurements and Performance Branch who continually provide their support to the photovoltaic community through measurements such as those discussed in this paper. This work was supported by the U.S. Department of Energy through contract number DE-AC02-83CH10093 with the Solar Energy Research Institute.

1. See, for example, *Proc. 19th IEEE Photovoltaic Specialist Conf.*, New Orleans (IEEE, New York; 1988).
2. C.H. Seager, D.J. Sharp, J.K.G. Panitz and J.I. Hanoka, *J. Phys.* C1 **43**, 103 (1982).
3. S. Martinuzzi, M.A. Sebbar and J. Gervais, *Appl. Phys. Lett.* **47**, 376 (1985).
4. J.A. Roger, R. Baruch, D. Bernard, J.P. David, J. Guerin, A. Mounguegi and S. Nopurredine, *Proc. Sixth Photovoltaic solar Energy Conf.*, London (D. Reidel Publ. Co., Dordrecht; 1986) pp 98-102.

5. J.I. Pankove, R.O. Wance and J.E. Berdeyheiser, *Appl Phys. Lett.* **45**, 1100 (1984). Also, J.I. Pankove, P.J. Zanazucchi and C.W. Magee, *Appl. Phys. Lett.* **46**, 787 (1985).
6. C.T. Sah, J.Y.C. Sun and J.J. Tzou, *Appl Phys. Lett.* **43**, 204 (1983). Also, C.T. Sah, J.Y.C. Sun and J.J. Tzou, *Appl. Phys. Lett.* **55**, 1525 (1984).
7. S.J. Pearton, *Phys. Rev. Lett.* 53, 855 (1984).
8. W.L. Hansen, S.J. Pearton and E.E. Haller, *Appl. Phys. Lett.* **44**, 606 (1984).
9. L.L. Kazmerski, Ref. 1, pp. ; also, *J. Vac. Sci. Technol.* A **4**, 1570 (1986).
10. L.L. Kazmerski, Solar Cells **25** (1988).
11. G. Binnig, H. Rohrer, Ch. Gerber and E. Weibel, *Phys. Rev. Lett.* **50** ,120 (1983).
12. See, for example, N. Garcia, Ed., **Scanning Tunneling Microscopy '86** (North-Holland, Amsterdam; 1987).

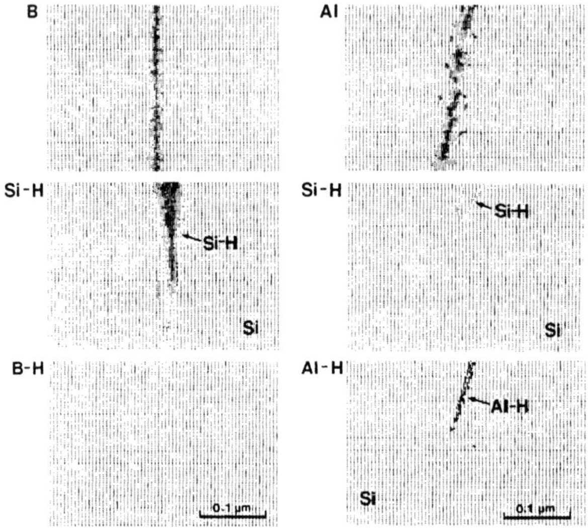

Fig. 3. *Volume-indexed AES maps of Si grain boundaries showing Si-H bonding for B-doped Si and Al-H for Al-doped Si.*

Fig. 4. *Scanning tunneling microscope (STM) configuration.*

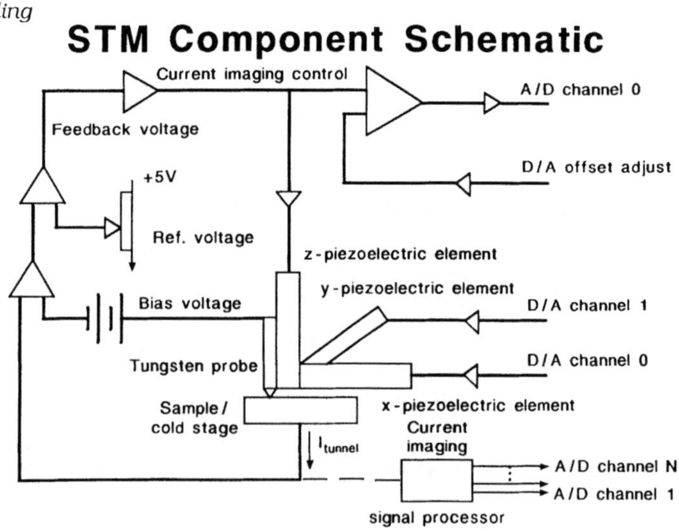

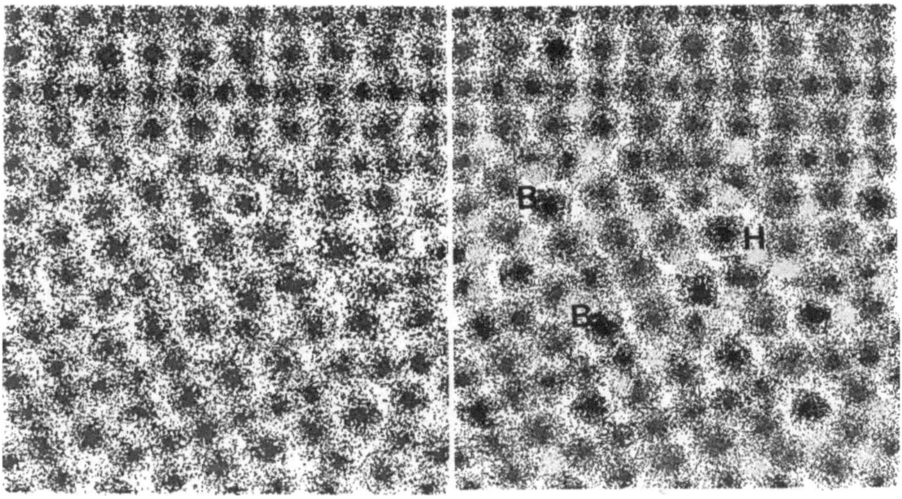

Fig. 5. *STM spectroscopic images: (a) Si grain boundary; (b) same region with overlay of boron and hydrogen atom positions under bias, showing Si-H bonding.*

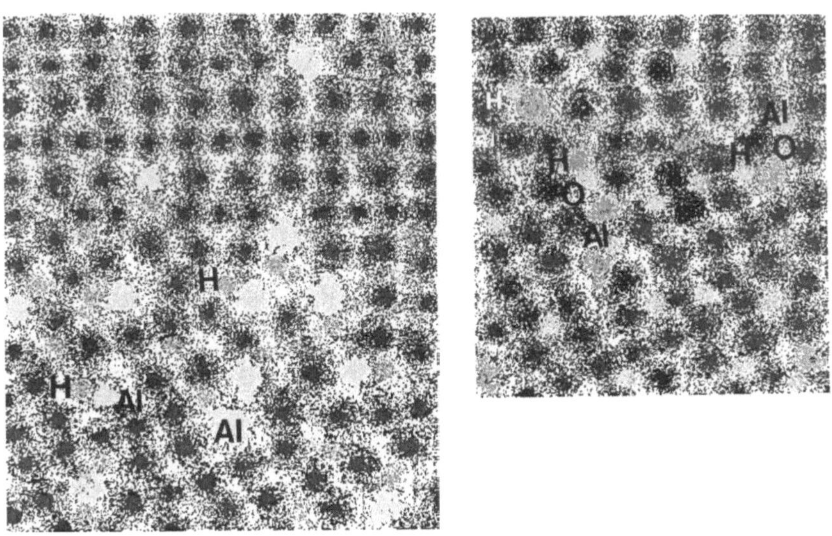

Fig. 6. *STM images of Si grain boundaries: (a) Al-doped with hydrogen (acceptor-hydrogen bonding); (b) Al-doped with hydrogen and oxygen (hydroxyl bonding).*

SESSION 09:
AMORPHOUS SI-CELLS

THE PERFORMANCE, RELIABILITY AND COST OF AMORPHOUS SILICON POWER MODULES

D. E. Carlson
Solarex Thin Film Division
826 Newtown-Yardley Road
Newtown, PA 18940 USA

ABSTRACT. Solarex has developed an in-line processing technology that has demonstrated the capability of fabricating high performance (efficiencies >8%), large area (>1000cm^2) a-Si power modules. The manufacturing process utilizes computer-controlled laser scribing to do all the patterning steps in the fabrication of these modules. All processing steps are being integrated into an automated manufacturing line with computerized controls and in-line diagnostics to insure high performance, high quality and low costs. Our present power modules consist of a thin a-Si p-i-n structure (∿0.25μm thick) deposited on textured tin oxide and sealed with either a roll-on encapsulant or EVA/Tedlar. These modules pass all JPL Block V environmental tests as well as a Solarex water immersion test, and the conversion efficiencies stabilize in the 5 to 6% range after a few months outdoors. We project that our total manufacturing costs will be about \$1.15/W$_p$ for a plant producing 10MW$_p$ of a-Si power modules (6% efficient) per year.

1. INTRODUCTION

Tremendous progress has been made in the field of amorphous silicon (a-Si) photovoltaics since the first a-Si solar cell was fabricated fourteen years ago [1]. The conversion efficiency has increased from a fraction of a percent for that first cell to values in the range of 13 to 14% for recent stacked-junction cells [2,3].

Sanyo produced the first commercial product using a-Si about eight years ago when they introduced the a-Si solar-powered calculator. Other consumer products such as solar-powered clocks, watches and hand-held games with liquid crystal displays also utilize a-Si solar cells. However, all these early products were intended mainly for indoor or low-light level use. In the last few years, several companies have introduced a-Si solar cells and modules for power generation in sunlight. In this paper, we will discuss the status of a-Si power modules with respect to their performance, reliability and cost.

2. THE MANUFACTURING PROCESS

The a-Si solar-cell structure presently used by Solarex in the manufacturing of power modules is: glass/tin oxide/p-i-n (a-Si)/Al. The glass is a common soda-lime-silicate glass that is strengthened by an ion exchange process. The p-layer is typically an a-Si:C:H alloy approximately 100Å thick. The modules consist of 30 strip cells that are automatically connected in series by three computer-controlled laser scribing steps [4]. Table 1 lists the process steps currently used by Solarex in the manufacturing of a-Si power modules (1000cm^2 in area).

The equipment used to perform the process steps listed above is computer-controlled in most cases. The glass washers, the CVD tin oxide furnace, the a-Si deposition system and the magnetron metallization system are all automated in-line systems.

We are now constructing a computer-integrated-manufacturing (CIM) line that is scheduled for completion next year. A schematic of this line is shown in Fig. 1. This CIM line will incorporate an advanced processing approach that allows us to deposit both a-Si and aluminum before performing a dual laser patterning step. These in-line laser stations utilize moving mirrors to move the laser beams and are coupled with a machine vision system for accurate registration. All these features have already been demonstrated in research systems, and the performance of large-area modules ($1000cm^2$) is comparable to that produced by the process described in Table 1. The a-Si deposition machine is a vertical, multi-substrate, multi-chamber system capable of handling substrates $>4ft^2$ in area, and this system is now undergoing engineering qualification tests.

TABLE 1

MANUFACTURING PROCESS STEPS FOR a-Si POWER MODULES

1. Clean glass
2. Deposit 500Å of SiO_2 and 6000Å of SnO_2:F (textured) by chemical vapor deposition (CVD)
3. Laser-scribe tin oxide
4. Clean substrate
5. Deposit a-Si p-i-n layers by glow discharge (2300Å i-layer)
6. Laser-scribe a-Si layers
7. Sputter-deposit aluminum (4000Å)
8. Laser-scribe aluminum
9. Test and cure segments
10. Ultrasonically bond aluminum power leads to the end segments of the modules
11. Roll-on encapsulation layer
12. Frame
13. Final test

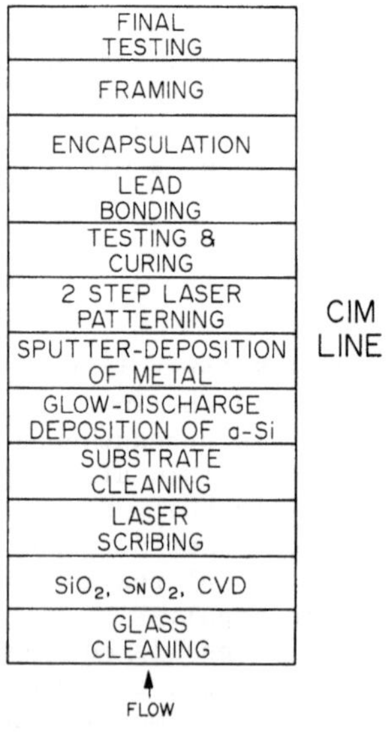

Figure 1. A schematic of the computer-integrated-manufacturing line being developed by Solarex.

3. PERFORMANCE OF a-Si POWER MODULES

At Solarex, conversion efficiencies as high as 12% have been obtained in single-junction laboratory cells ($\sim 1cm^2$) [5]. These high efficiencies are attained by using graded p/i interfaces and highly reflective back contacts of indium-tin-oxide/silver. Conversion efficiencies as high as 13.7% have been reported for stacked-junction cells using an a-Si/$CuInSe_2$ structure [2], and values of 13.0% have been reported for stacked cells with an a-Si/a-Si/a-Si:Ge structure [3].

Figure 2 shows the distribution of conversion efficiencies obtained in a production run of power modules. Conversion efficiencies in excess of 8% have been achieved for modules with i layers approximately 0.5μm thick [6]. With a cutoff at an efficiency of 5.6%, our electrical yields have averaged about 93% over the last 16 months. Many of the drop-outs are due to handling problems so we expect an even higher yield as well as a tighter distribution when our CIM line becomes operational.

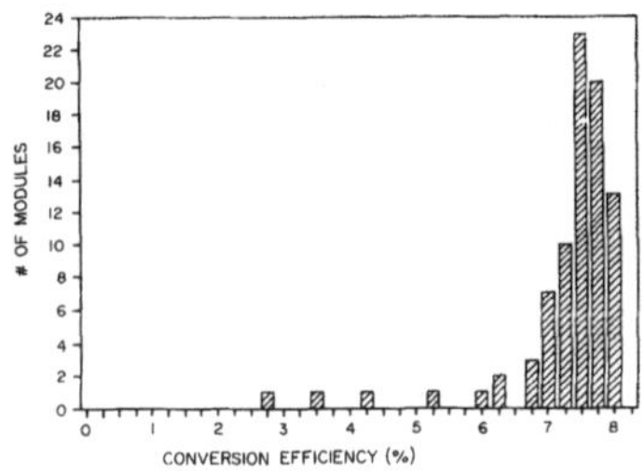

Figure 2. The distribution of conversion efficiencies obtained in a production run of large-area ($1000cm^2$) a-Si power modules.

4. RELIABILITY AND STABILITY

We periodically subject a sampling of our a-Si power modules to a battery of environmental tests in order to insure that they will function as reliable, long-term sources of electrical power. These tests include thermal cycling (50 cycles, 6 hours ea, -40°C to 90°C), humidity-freeze cycling (10 cycles, 24 hours ea; 85°C, 85% relative humidity soak with 1 hr at -40°C), hail testing (1" ice balls directed at 52 miles/hr at 5 locations over a $1000cm^2$ module), torsional testing (2cm per meter twist), high-voltage breakdown test (<50μa leakage at 1500V after spraying with water), hot water immersion testing (immersion in 40°C water for 5 days while illuminated) and humidity-soak testing (85°C 85% relative humidity for one month).

Our present power modules pass all the tests listed above. Moreover, no problems have been observed with these power modules located in outdoor test arrays. We did observe some electrochemical corrosion in earlier modules with metal frames [7], but this problem was eliminated by switching to plastic frames and improving the electrical isolation. Currently, only our 1'x4' (SA-20) modules are designed to pass the high-voltage breakdown test, but we have recently obtained breakdown voltages greater than 2000V for smaller modules (when wet) using an improved encapsulant.

All a-Si solar cells exhibit some degree of light-induced degradation depending on the impurity content [8] and microstructure [9] of the amorphous silicon and on the thickness of the i-layers [10] in single- or stacked-junction cells. Figure 3 shows light-induced degradation data for three different classifications of power modules. The degradation data labeled as A in Figure 3 are typical of that observed for single-junction a-Si power modules with thick i layers (0.6μm). The data labeled B are representative of that observed for thin i-layer (0.25μm) power modules while data labeled C are representative of a-Si/a-Si stacked-junction modules produced at Solarex in engineering runs. Data for a-Si/a-Si stacked-junction cells made in research are labeled D. All the data shown are for modules with initial efficiencies in the 6 to 8% range.

Since the light-induced degradation is reversed by high temperature annealing, the effects shown in Fig. 3 can be reduced if the modules experience periods of relatively high ambient temperature [11]. This effect in conjunction with a stacked-junction structure can lead to relatively stable a-Si power modules that exhibit less than 10% degradation after a year of outdoor operation.

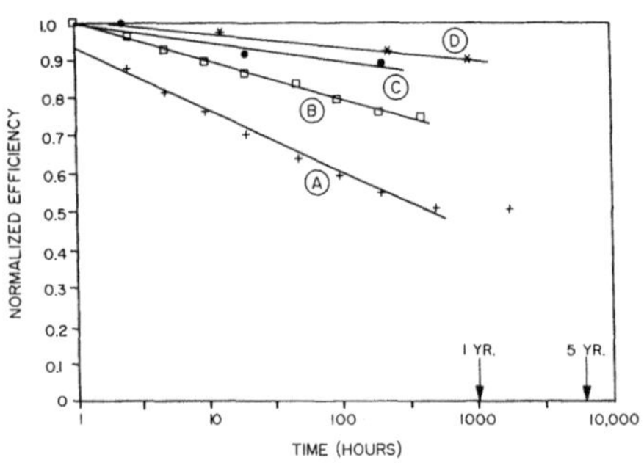

Figure 3. Normalized conversion efficiencies as a function of exposure time to 1 sun illumination for 3 classes of a-Si power modules and recent laboratory cells (see text).

5. MANUFACTURING COSTS

A number of organizations have estimated the manufacturing costs for producing a-Si power modules, and a compendium of those estimates are shown in Figure 4. All models show that the manufacturing costs are relatively high when the production volume is low (e.g., manufacturing costs are estimated to be in the range of $\$4/W_p$ to $\$8/W_p$ at production rates of $300kW_p/yr$). This is mainly due to the high capital cost of processing equipment which leads to high depreciation costs for low-volume manufacturing facilities. Moreover, most cost estimates do not include significant automation at low production rates so labor costs are relatively high.

However, as shown in Fig. 4, once production volumes exceed about $2MW_p/yr$, the manufacturing costs start to fall below $\$2/W_p$. Table 2 shows some of the key assumptions made by Solarex in estimating the manufacturing costs of a $10MW_p/yr$ facility. Many of the assumptions such as 6% efficiency, 84% line yield, single-junction p-i-n structures, etc. are similar to conditions actually experienced in our present manufacturing facility. The model assumes a high degree of automation with 3 islands of in-line automation connected by manual labor. Development of a totally integrated line would reduce the cost to somewhat less than $\$1.15/W_p$.

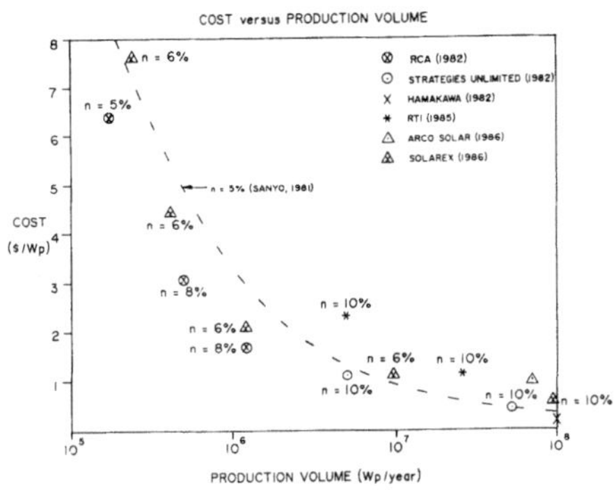

Figure 4. Estimates of manufacturing costs as a function of annual production volume. The conversion efficiencies used in the models are shown next to the data points (all costs in 1985$).

TABLE 2

COST MODEL

Assumptions

- o Production Rate: $10MW_p/yr$
- o Overall Yield: 84%
- o 5 Day Work Week, 2.5 Shifts/Day
- o Computer Integrated Manufacturing (3 sections)
- o 6% Conversion Efficiency
- o 1'x4' Modules

Summary

- o Total Manufacturing Cost: $\$1.15/W_p$
- o Direct Materials: 39.5%
  Equipment Depreciation: 35.3%
  Direct Labor: 13.5%
  Indirect Labor: 6.2%
- o 83 Direct Employees
  17 Indirect Employees
- o $20,000ft^2$ Facility
- o Capital Equipment: $16,500,000

6. CONCLUDING REMARKS

As the long-term performance of a-Si power modules improves and as the manufacturing costs fall, a-Si solar cells will gain significant market share in many power generation applications. Market penetration is occurring today in high-voltage ($\lesssim 5W_p$) applications where balance of systems costs are low. As a result of automated interconnections as described in Section 2, a-Si power modules are selling today at lower prices (in $\$/W_p$) than crystalline Si modules for high-voltage, low-power applications.

As long as the efficiency of crystalline Si solar cells exceeds that of a-Si cells, crystalline cells should retain those markets where the balance of systems costs are high.

ACKNOWLEDGEMENTS

The research reported herein was partially supported by the Solar Energy Research Institute under Subcontract No. ZB-7-06003-2.

REFERENCES

1. D.E. Carlson, U.S. Patent No. 4,064,521, December 1977.

2. K. Mitchell, R. Patter, J. Ermer, R. Wieting, C. Eberspecher, D.P. Tanner, K. Knapp and R. Gay, Conf. Record of 19th IEEE Photovoltaic Specialists Conf. (IEEE, NY) 1987, p. 13.

3. R. Ross, R. Mohr, J. Fournier and J. Yang, ibid., p. 327.

4. A.H. Firester and D.E. Carlson, RCA Engineer 28, 40 (1983).

5. D.E. Carlson, R.R. Arya, M.S. Bennett, A. Catalano, R.V. D'Aiello, C.R. Dickson, C.M. Fortmann, B. Goldstein, J. Morris, J.L. Newton and S. Wiedeman, to be published in Solar Cells.

6. D.E. Carlson, R.R. Arya, A. Catalano, R.V. D'Aiello, C.R. Dickson, J. O'Dowd and R.S. Oswald, Solar Cells 21, 15 (1987).

7. D.A. Fagnan, R.V. D'Aiello and J. Mongon, Conf. Record of 19th IEEE Photovoltaic Specialists Conf. (IEEE, NY) 1987, p. 1508.

8. D.E. Carlson, in "Tetrahedrally-Bonded Amorphous Semiconductors", eds. D. Adler & H. Fritzsche, Plenum Publishing Corp., NY 1985, p. 165.

9. C.M. Fortmann, S. Lange, M. Farley and J.G. O'Dowd, Conf. Record of 19th IEEE Photovoltaic Specialists Conf. (IEEE, NY) 1987, p. 296.

10. J.J. Hanak and V. Korsun, Conf. Record of 16th IEEE Photovoltaic Specialists Conf. (IEEE, NY) 1982, p. 1381.

11. D.E. Carlson, Proc. of the Internation Workshop on Amorphous Semiconductors, Beijing, China, ed. H. Fritzsche, D.-X. Han & C.C. Tsai (World Scientific Publishing Co., 1987) p. 193.

PROGRESS IN AMORPHOUS SILICON MATERIALS FOR HIGH PERFORMANCE a-Si SOLAR CELLS

S. Nakano, H. Tarui, T. Matsuyama, S. Okamoto, Y. Hishikawa, H. Dohjo, M. Isomura, T. Takahama, M. Nishikuni, T. Usuki, N. Nakamura, S. Tsuda, M. Ohnishi and Y. Kuwano

Functional Materials Research Center, SANYO Electric Co., Ltd.
1-18-13 Hashiridani, Hirakata City, Osaka, Japan (Phone:0720-41-1261)

Abstract

In order to improve the conversion efficiency of a-Si solar cells, $B(CH_3)_3$ was applied as a p-type doping gas for the first time. The properties of p-type a-SiC films have been improved by using $B(CH_3)_3$ and a module efficiency of 9.38% was obtained for a 10cm x 10cm integrated type submodule. Furthermore, microcrystalline SiC films have been investigated by a new plasma CVD method, called the CPM (Controlled Plasma Magnetron) method.

As for reliability, SiH_2 bond density was found to influence light induced degradation. In the cycles of light exposure and thermal annealing, an irreversible change was found in the properties of a-Si film with high SiH_2 bond density.

1. Introduction

Amorphous silicon (a-Si) solar cells have been extensively investigated as low-cost solar cells. Practical use of a-Si solar cells as an electric power source requires improvements in conversion efficiency and improvements in reliability.

In this paper, first, a high quality p-layer using a new doping gas of $B(CH_3)_3$ is reported. Second, microcrystalline SiC (μc-SiC) fabricated by a new plasma CVD method, called the CPM method, is investigated. Finally, the influence of SiH_2 bonds on light induced degradation and recovery is described.

2. High quality p-type a-SiC:H films using $B(CH_3)_3$

A new doping gas of $B(CH_3)_3$ was investigated instead of B_2H_6[1]. The conventional p-type doping gas, B_2H_6, has a B-B bond, and this B-B bond seems to remain easily in a-SiC films, resulting in high defect density. Therefore, a new doping gas of $B(CH_3)_3$ was studied.

P-type a-SiC films doped with $B(CH_3)_3$ are fabricated by both a photo-CVD method and a glow discharge (GD) method.

As for electrical properties, dark conductivity was measured for p-type a-SiC films fabricated by the photo-CVD method. The dark conductivity of a-SiC films (Eopt=2.0eV) as a function of boron concentration is shown in Fig. 1. At the same boron concentration, the dark conductivity of a-SiC films with $B(CH_3)_3$ is two orders higher than that of B_2H_6. Therefore, excellent boron doping efficiency is obtained by using the new doping gas, $B(CH_3)_3$. From the SIMS measurement, the ratio of B-B bonds/B atoms in a-SiC with $B(CH_3)_3$ is about ten times smaller than that of B_2H_6.

It is considered that the higher doping efficiency of a-SiC with $B(CH_3)_3$ is related to a decrease in defect density resulting from lower

B-B bond density.

To confirm the superiority of $B(CH_3)_3$ gas, a-SiC films fabricated by the glow discharge method have also been studied. The conventional a-SiC films fabricated by the glow discharge method with B_2H_6 have the problem of bandgap narrowing in the highly doped region. The change in the optical bandgap of a-SiC films is shown in Fig. 2 as a function of boron concentration. The optical bandgap of highly boron doped a-SiC with B_2H_6 decreases as the boron concentration increases.

In contrast, the optical bandgap of a-SiC films with $B(CH_3)_3$ is almost constant (~ 2.0eV). That is, the bandgap narrowing can be avoided by using $B(CH_3)_3$. At the same optical bandgap of 2.0eV, the photoconductivity of a-SiC film with $B(CH_3)_3$ is also improved from 1×10^{-6} $(\Omega\,cm)^{-1}$ to 2×10^{-5} $(\Omega\,cm)^{-1}$ compared to that of B_2H_6.

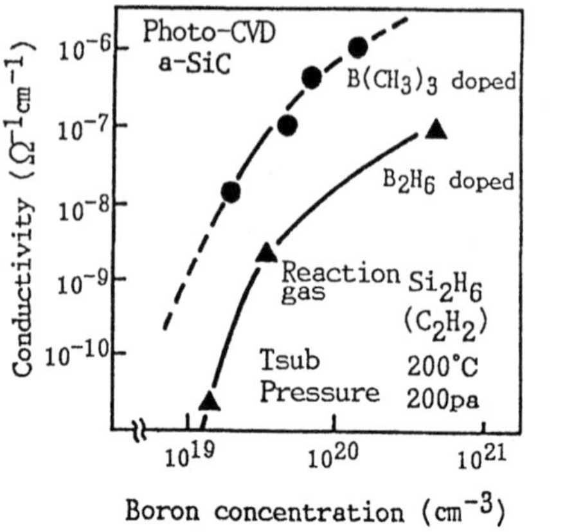

Fig. 1 Dark conductivity of p-type a-SiC films as a function of boron concentration measured by SIMS

As mentioned above, the use of $B(CH_3)_3$ instead of B_2H_6 is also effective for p-type a-SiC films. Therefore, pin a-Si solar cells whose p-layers were fabricated with $B(CH_3)_3$ were studied. Fig. 3 shows illuminated I-V characteristics of an integrated type a-Si submodule (10cm x10cm), whose p-layer was fabricated by the GD method with $B(CH_3)_3$. A module efficiency of 9.38% has been obtained.

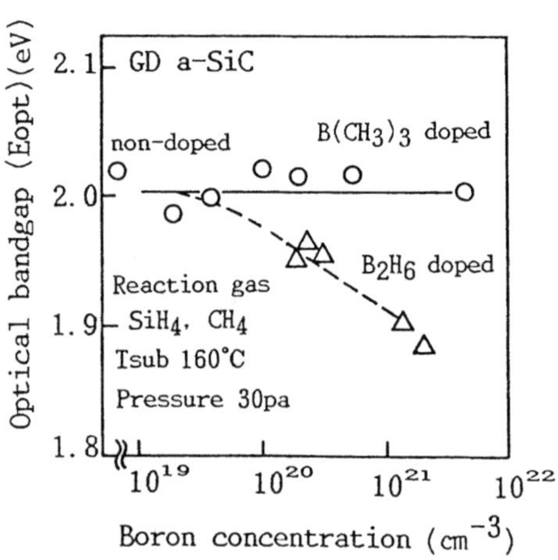

Fig. 2 Optical bandgap of p-type a-SiC films as a function of boron concentration

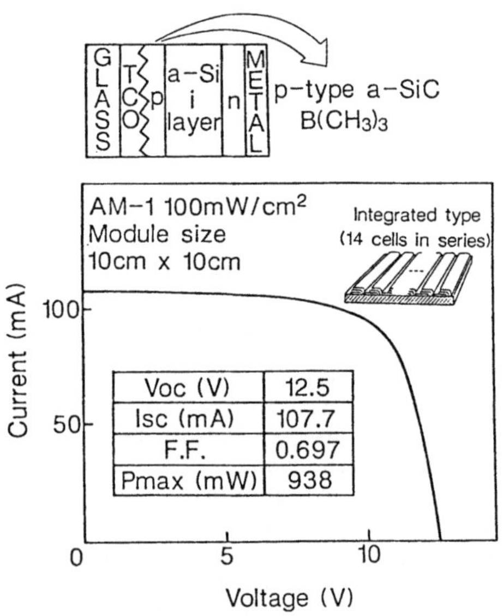

Fig. 3 Illuminated I-V characteristics of an integrated type a-Si solar cell submodule

3. Microcrystalline (μc) SiC films fabricated by the CPM method

In order to improve the conversion efficiency furthermore, especially with regard to open circuit voltage (Voc), μc-SiC films have been investigated using a new plasma CVD method, called the CPM method.

Schematic diagram of the CPM method is shown in Fig. 4. Electromagnets are positioned under the RF electrode, and plasma near the substrate is controlled by the plasma controlled electrode.

By applying a magnetic field into the plasma, high electron density plasma can be obtained. Furthermore, it is found that plasma damage in the CPM method is less than that in the conventional method[2]. From these features, the CPM method seems to be more profitable for fabricating the p-layer of a TCO/ pin/Metal type solar cell compared with the conventional GD method.

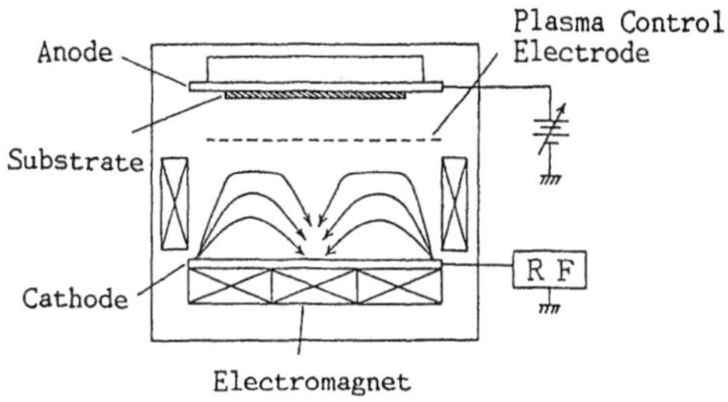

Fig. 4 Schematic diagram of the CPM method

Fig. 5 shows the conductivity and optical bandgap of undoped SiC films as a function of magnetron current. High conductivity was obtained by applying a magnetic field because the film was microcrystallized. Thus, by using the CPM method, μc-SiC films can be easily fabricated, even with low hydrogen dilution. Eopt was 2.0 ~ 2.1eV, which was larger than that of μc-Si.

Fig. 6 shows the dark conductivity of p-type μc-SiC films as a function of the optical bandgap. The dark conductivity of μc-SiC films was improved by using the CPM method, especially in the wide optical bandgap region. Voc of an a-Si solar cell whose p-layer was μc-SiC using $B(CH_3)_3$ was 0.980V. As a result, the CPM method was found to be a profitable method for fabricating a p-type μc-SiC film for a TCO/pin type solar cell.

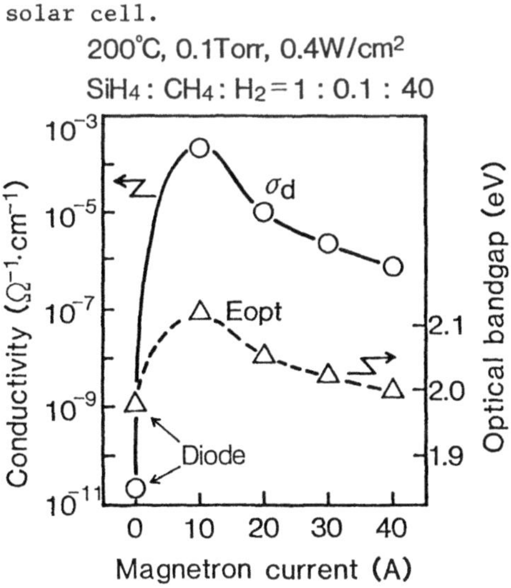

Fig. 5 Conductivity and E_{opt} of undoped SiC films as a function of magnetron current.

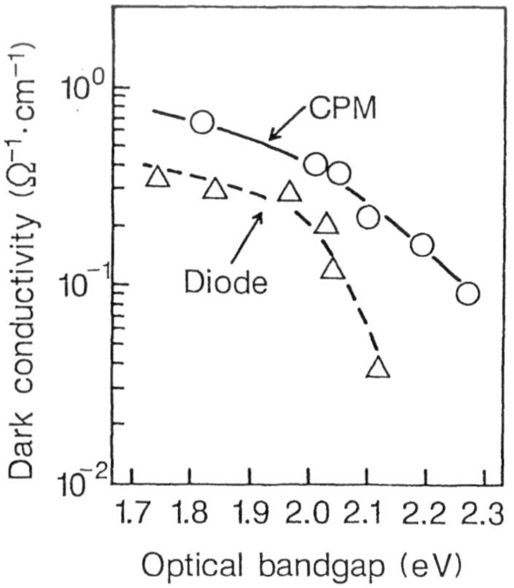

Fig. 6 Dark conductivity of p-type μc-SiC films as a function of optical bandgap

4. Influence of SiH_2 bond density on light induced degradation

To use a-Si solar cells in power systems, not only high efficiency but high reliability are required. Recently it was shown that factors associated with light-induced defects around the midgap have influence on light-induced degradation in low impurity films[3)]. Therefore, we investigated the influence of hydrogen.

Fig. 7 shows the fill factor of a-Si solar cells before and after light exposure as a function of the SiH_2 bond density in the i-layers prepared by a plasma CVD method (circles) and by a H radical CVD method (triangles). Fig. 7 indicates that reduction of SiH_2 bond density is effective for both reducing light-induced degradation and improving conversion efficiency.

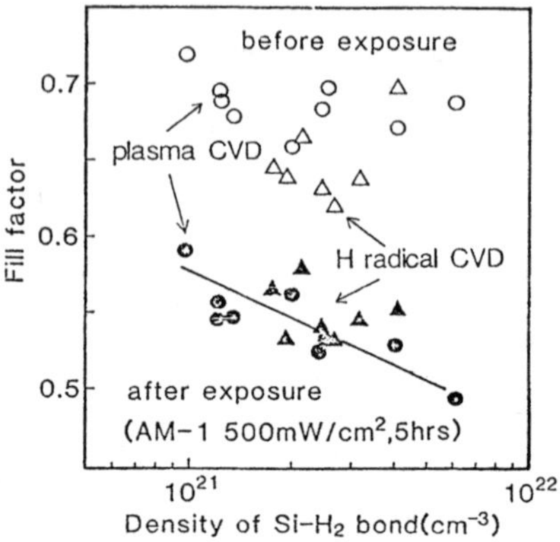

Fig. 7 Fill factor of a-Si solar cells as a function of the SiH_2 bond density

In order to clarify the role of SiH_2 bonds in the light induced effect, experiments on the recovery processes of a-Si solar cells by thermal annealing were carried out.

Experimental method is shown in the right part of Fig. 8. ESR spin density is measured after irradiation (AM-1 : 500mW/cm^2 5hrs.) and thermal annealing (180°C, 1hr.) is carried out after measurement. This experiment is repeated several times. The results are shown in the left part of Fig. 8. Although the spin density of a-Si film with low SiH_2 bond density is constant after several cycles, that of a-Si film with high SiH_2 bond density decrease gradually. Thus, after several cycles, the degradation ratio becomes smaller than that of initial state in a-Si films with high SiH_2 bond density.

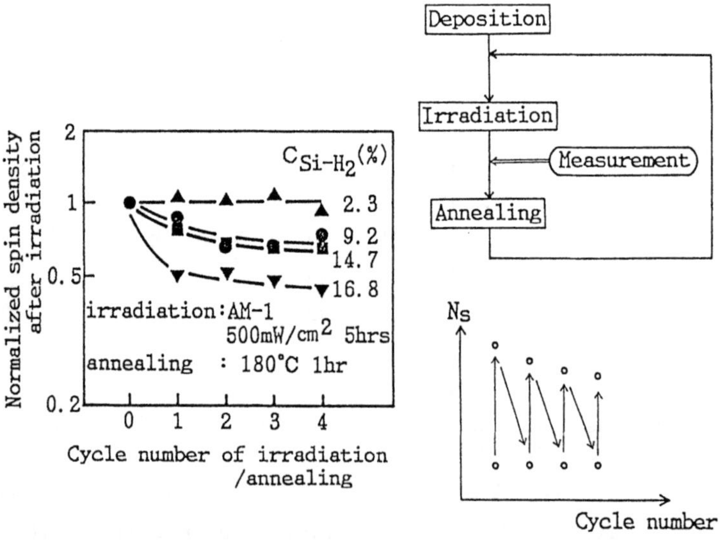

Fig. 8 The changes in spin density after irradiation as a function of the cycle number of irradiation and annealing

This behavior was not observed in the thermal annealing experiment only. The results in Fig. 8 indicate an irreversible change in the light induced effect in a-Si films with high SiH_2 bond density.

The above results will be considered as follows.

It was found that as SiH_2 bond density increased, compressive stress decreased[4]. From this result, a-Si films with high SiH_2 bond density seem to have larger structural flexibility compared to those with low SiH_2 bond density. Furthermore, some microstructural changes may be caused by hydrogen movement in light induced degradation and thermal recovery. Thus, in a-Si films with low SiH_2 bond density, the microstructure will recover to the same site as that of the initial state by thermal annealing. On the other hand, in a-Si film with high SiH_2 density, it will change to a different site from that of initial state because of their structural flexibility. As a result, an irreversible change will occur in a-Si films with high SiH_2 bond density.

5. Conclusion

Investigations are being performed for high-efficiency, high-reliability a-Si solar cells. In order to improve conversion efficiency, p-type a-SiC:H films using $B(CH_3)_3$ as a doping gas were studied for the first time. It was found that highly photoconductive p-type a-SiC:H films are obtained by using $B(CH_3)_3$. A module efficiency of 9.38% was obtained for a 10cm x 10cm integrated a-Si solar cell by using $B(CH_3)_3$.

Furthermore, μc-SiC films were investigated with a new plasma CVD method, called the CPM method which causes minimal damage to the substrate under high electron density conditions. μc-SiC films can be easily fabricated by the CPM method even with low hydrogen dilution. And V_{oc} of 0.980V was obtained for an a-Si solar cell whose p-layer was μc-SiC using $B(CH_3)_3$.

As for reliability, it was found that the SiH_2 bond plays an important role in light induced degradation. Irreversible change of a-Si films with high SiH_2 bond density was found in the light irradiation/thermal annealing cycle experiments.

Thus, we have developed a new technology to achieve high-efficiency and high-reliability a-Si solar cells.

ACKNOWLEDGEMENTS

This work is supported by NEDO as a part of the Sunshine Project under the Ministry of International Trade and Industry.

REFERENCES

1) H. Tarui et al. : Tech. Digest of International PVSEC-3 (1987) p.41
2) M. Ohnishi et al. : Jap. J. Appl. Phys. 27(1) (1988) 40
3) Y. Kuwano et al. : International Conf. Stability of Amorphous Silicon Alloy Materials and Devices, Palo Alto (1987)
4) Y. Hishikawa : J. Appl. Phys. 62 (8) (1987) 3150

LIGHT SOAKING STUDIES ON AMORPHOUS SILICON PHOTOVOLTAIC DEVICES AND MODULES

A. E. Delahoy, T. Tonon, J. A. Cambridge
M. Johnson, L. Michalski, and F. J. Kampas

CHRONAR CORP.
International Corporate Headquarters
P.O. Box 177, Princeton, NJ 08542, USA

ABSTRACT. This paper discusses and illustrates several different aspects of light soaking and annealing phenomena in a-Si:H solar cells. It is stressed that metastable defects can be created with a variable distribution of activation energies for annealing, and that the state of a light-soaked sample cannot therefore be defined in terms of a single parameter, e.g., the defect density. As a consequence, a strong distinction must be made between short term and long term soaking, and a 2½ year study demonstrates that light-soaked cells eventually reach an equilbrium state. Cells in this state cannot be annealed at temperatures lower than the soaking temperature. Analysis of fill factor versus log time plots indicates that field distortion may occur in some cells. It is also shown that a-Si:H deposition conditions can influence cell stability, and some relevant factors are identified. Finally, tandem cells are shown to be more stable than single junction cells, and outdoor data on encapsulated tandem submodules indicates that large-area a-Si:H photovoltaic panels can be fabricated having acceptable long term stability.

1. INTRODUCTION

It is the intention of this paper to draw attention to several features of light soaking and annealing phenomena in hydrogenated amorphous silicon (a-Si:H) solar cells. These features are important for understanding and predicting light soak behavior of PV panels in actual outdoor operation, and for manufacturing cells with improved stability. These features are discussed in sections 2-9 below. But first let us give a snapshot of the basic facts and models.

Illumination of a-Si:H produces metastable defects that can be removed by annealing. The bias dependence of degradation in p-i-n cells makes it clear that carrier recombination (or possibly trapping) is responsible for defect creation [1]. Indeed, we have shown that an annealed cell illuminated at 60 suns intensity, and placed under sufficient reverse bias to ensure carrier extraction, does not degrade at all [2]. Experimental data indicate that the effect of light soaking on solar cells can be well explained in terms of a reduction of the collection length (sum of electron and hole $\mu\tau E$ products) [3]. A model developed by Smith et al. [4] predicts that fill factor should decrease linearly with the log of exposure time. This model is based on Stutzmann's model for dangling bond formation [5] and Crandall's uniform field model for thin film solar cells [6]. Smith assumes that minority carrier lifetime is inversely proportional to defect density.

2. INADEQUACY OF A 1-PARAMETER DESCRIPTION OF LIGHT SOAKING

We wish now to stress that the state of a degraded sample cannot be specified by a single parameter, for example, defect density. While several authors have recognized this, others have not, and this point is important for the proper treatment of experimental data. That a

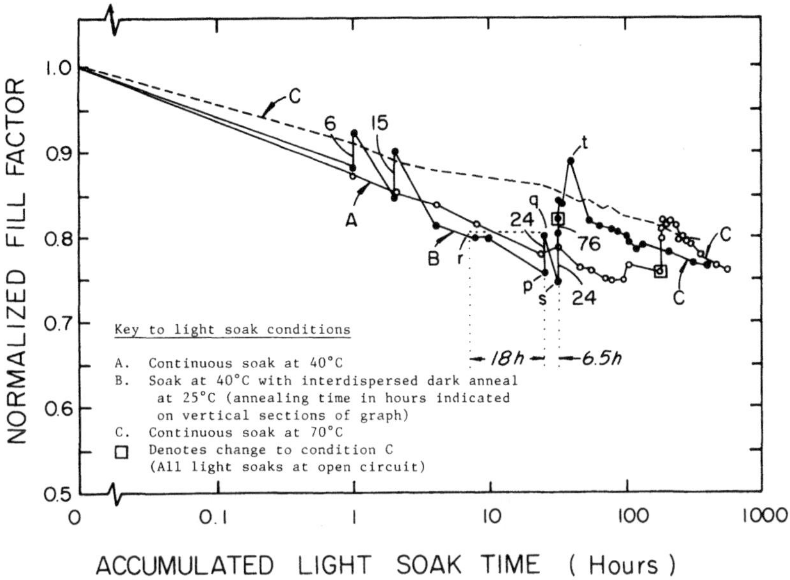

Fig. 1. Normalized fill factor versus accumulated light soak time for three cells.

1-parameter description is inadequate is clear from simple experiments involving partial annealing of degraded samples. The missing information concerns the ease with which defects can be annealed, i.e. the activation energy for defect annealing. We illustrate this by some of the data in Figure 1. This figure plots the normalized fill factor (FF) for three glass/SnO_2/p-i-n/Al solar cells as a function of accumulated light soak time for various soak and anneal sequences. We focus on the data points p, q, s and the hypothetical point r. After degradation to point p, the sample was annealed for 24 h at room temperature in the dark to point q. Now if the state of the sample at point q is fully determined by the defect density (which in turn determines FF), then point r, horizontally across from q on the initial degradation curve, must be equivalent to q, since there the cell has the same fill factor. The degradation from r to p initially took 18 hours, yet point s is reached in only 6.5 hours of soaking from point q. This contradicts the 1-parameter assumption. This, and many similar sets of data, can be interpreted only if defects are not equivalent, but are distinguished by their activation energy for annealing. In this picture, the partial anneal from p to q preferentially removes low energy defects. Furthermore, these low energy defects can be quickly restored by a short light soak. (This is corroborated by the data at very short soak times. Here substantial annealing takes place at room temperature in a short time.) There thus appears to be a correlation between the ease of creation of defects and ease of annealing, as one might expect from a simple energy barrier model.

3. SHORT-TERM AND LONG-TERM SOAKING, AND THE EQUILIBRIUM STATE

The data in Figure 1, and other data, shows that fill factor decreases as the log of soak time. At very long times, however, saturation occurs,

Table I. Efficiency and fill factor for a variety of p-i-n cells light soaked at 50°C for 2.4 years to attain an equilibrium state, and normalized fill factor for subsequent annealing at various temperatures.

		LIGHT SOAKING					ANNEALING					
		Efficiency (%) and fill factor light soak time						FF/FF₀ Anneal time (hours)				
Cell	Structure**	Annealed	3 hours	2.4 years	FF/FF_0	Stability Ranking	Anneal Temp. (°C)	0	27	75	198	363
A	S/H .43	8.21/.692	6.65/.586	5.44/.556	.80	3	80	.802	.819	.825	.841	.848
B	S .43 t	8.51/.687	6.64/.572	4.79/.521	.76	6	80	.754	.772	.782	.810	.824
C*	0.06, 0.48 t	6.58/.688	6.40/.663	5.05/.597	.87	1	60	.849	.863	.868	.872	.874
D	S .43 t	8.89/.729	6.94/.608	4.37/.529	.73	9	60	.730	.734	.738	.738	.748
E	S .65 e	9.25/.700	7.29/.594	5.15/.528	.75	7	40	.767	.769	.773	.770	.774
F	S .57 t	8.94/.665	6.91/.565	4.56/.495	.75	8	40	.753	.756	.753	.764	.753
G	S/H .43	8.20/.685	6.88/.580	5.12/.524	.77	4	25	.772	.778	.774	.789	.775
H	S/H .43	7.07/.730	5.72/.600	4.14/.516	.71	5	25	.714	.716	.715	.725	.712
I	S/H .43 i	6.82/.547	5.62/.512	4.91/.516	.94	2	25	-	-	-	-	-

* tandem cell

** S denotes 100% silane i-layer, S/H denotes silane/hydrogen mixture
number (eg .43) denotes i-layer thickness in um
t denotes presence of thin a-SiC:H transitional i-layer
e denotes 0.4 x normal silane flow rate
i denotes 0.3 x silane flow, 0.6 x hydrogen flow, 3 x normal RF power

and the fill factor asymptotically approaches a constant value. Evidence for this comes from data accumulated on a set of 23 p-i-n cells, prepared under a variety of conditions, and light soaked at Chronar for 2.4 years. The light soaking was performed by placing the cells in a circle on a rotating turntable placed under a sodium street lamp and reflector. The rotation ensured that each cell received the same light dose. A cell temperature of 50°C was maintained by a fan. Data from this experiment is shown in Table 1. The final cell efficiency was about 5%. The final normalized fill factor (after 21,000 hours of light soaking) was generally about 0.75. This latter figure is very interesting, for taken in conjunction with the data in Figure 1, it implies that a normalized fill factor of 0.75 represents a stabilized value, and that the cells had reached an equilibrium state.

An annealing study performed by keeping the above cells at 25, 40, 60 or 80°C (initiated about 1 week after cessation of light soaking) revealed the following (see Table 1). For cells annealed below the soak temperature, i.e. at 25 or 40°C, little or no recovery occurred over times as long as 363 hours. At 60°C, a slow recovery could be detected, while at 80°C a significant recovery was observed. Thus, cells brought to an equilibrium state by a sufficiently long light soak at a given temperature cannot be annealed at or below that temperature. This implies the disappearance of low energy defects through structural change, and a defect distribution probably having a sharp lower cut-off at an activation energy for annealing which is dependent on soak temperature.

4. EFFECT OF SOAK TEMPERATURE

As shown by the dashed line in Figure 1, a cell light soaked at 70°C degrades more slowly than a similar cell soaked at 40°C. It is likely that a cell brought to an equilibrium state at 70°C has a higher efficiency than one brought to an equilibrium state at 40°C. There is also evidence that the time required to reach the equilibrium state decreases with increasing soak temperature. One would further expect that the distribution of activation energies for annealing is shifted towards higher energies for light soaks performed at higher temperatures.

We have recently obtained direct evidence from DLTS measurements on p-i-n devices that the energy spectrum of defect states in the energy gap of a-Si:H is dependent on light soak temperature [7]. Figure 2 shows the normalized DLTS spectra corresponding to the "high temperature" defect band (which we believe is due to dangling bonds) for the same cell degraded by forward current injection at 27°C and at 90°C. The spectrum has a clear bulge near 60°C for the 90°C soak. Whether the depth of the defect states in the gap correlates with activation energy for annealing is an open question.

5. IS NIGHT-TIME ANNEALING IMPORTANT?

In section 2 we saw that easily annealed defects are rapidly restored by light soaking. For this reason, over many light-soak/dark-rest cycles, cell efficiency is determied principally by accumulated light soak time, regardless of the extent of interdispersed annealing. We therefore expect for panels operated in an outdoor environment, that night-time annealing, while always present to some degree, does not in itself materially increase the integrated energy output of the panel for a given time outdoors. Nevertheless, on a long time scale, slow annealing does permit a seasonal change in efficiency from winter to summer (see increases in efficiency resulting from change of light soak temperature from 40°C to 70°C in Figure 1).

6. FIELD DISTORTION IN P-I-N CELLS

For short-term soaks (and high initial FF), the relationship between drop in short-circuit current and drop in FF predicted by Crandall's uniform field model is roughly confirmed by experiment [3]. However, for long-term soaks, the drop in current is usually greater than expected. This implies excess recombination, most likely resulting from a low field region in the i-layer. A recent paper [8], and the following experiments lend support to this hypothesis.

Accelerated (30 suns) light soaks were performed on 0.1 cm^2 p-i-n cells for periods of 1, 2, 4, etc. minutes. The cells were kept near their maximum power point. The dependence of the fill factor for red light on the number of accelerated light soaks for three i-layer thicknessess is shown in Figure 3. This effectively is a plot of FF versus

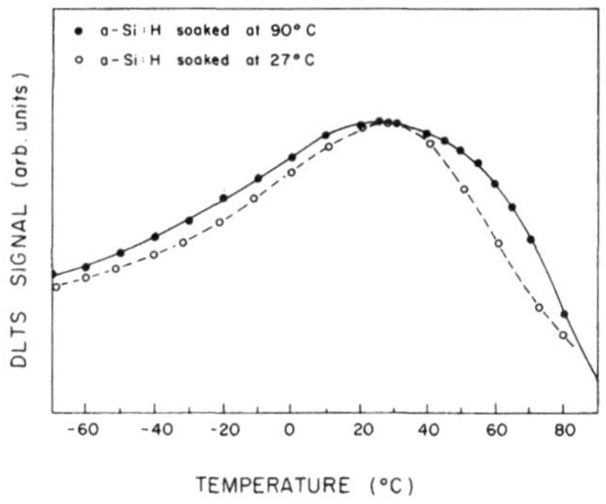

Fig. 2. Portions of DLTS spectra for a cell current-soaked at two different temperatures.

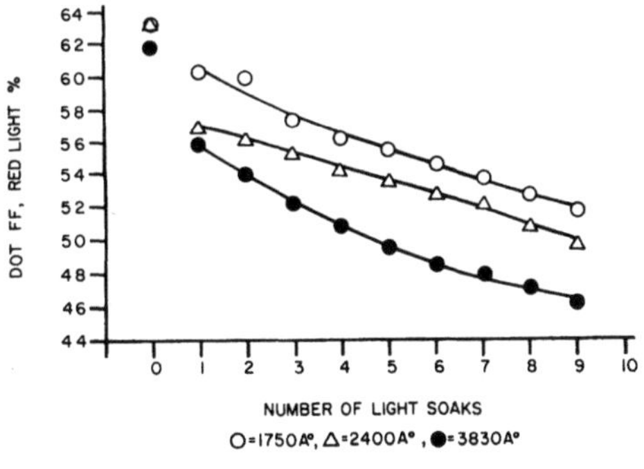

Fig. 3. Dot fill factor, measured with red light, as a function of the number of accelerated light soaks (30 suns). The length of the light soaks was 1, 2, 4, etc. minutes.

log exposure time. It is clear that the data for the thickest i-layer cell does not show a linear decrease. (The fitted lines are a quadratic fit.) Furthermore, the three sets of data exhibit an absolute drop of about .01 per light soak at long times, whereas data presented by Smith et al. [4] show a rate of decrease approximately 2½ times higher. The lower rate of decrease we found may be related to the temperature at which the accelerated light soaking was performed.

A dependence of fill factor on light soaking time similar to the 3830 A data in Fig. 3 was found by Catalano et al. [9] for devices whose i-layer had about 10^{20} cm^{-3} oxygen atoms and were prepared with 2ppm diborane in the gas phase (compensated i-layers). A linear dependence of FF on log t was obtained by these investigators for an undoped film with a higher purity i-layer. Conversely, addition of 2 ppm diborane to a low oxygen (10^{19} cm^{-3}) device showed the opposite curvature (a lower rate of initial degradation), obtained at the expense of initial efficiency. This was explained as being due to the displacement of the electric field toward the n-layer. SIMS analyses of i-layers deposited in the six-pack deposition system typically show 10^{19} - 10^{20} O cm^{-3} and 1-2 ppm B. Therefore, it is possible that the curvature in the fit to the 3830 A data in Fig. 3 can be explained by a non-uniform electric field in the i-layer.

This hypothesis is strengthened by the negative correlation we found between the violet light FF and red light FF for a large number of 4000 A cells previously subjected to 1 minute accelerated light soaking. In this connection, a non-uniform electric field which is stronger near the p-i interface will increase the violet light FF by reducing back diffusion but will reduce the red light FF by hurting hole transport near the n-layer, where holes are the minority carrier, and vice versa. Furthermore, the fact that the curvature in the fit in Fig. 3 is only significant for the thickest i-layer also supports the hypothesis that this curvature results from a non-uniform electric field. The explanation of why a non-uniform electric field results in a non-linear FF vs log t dependence probably encompasses a shift in the electric field distribution with light soaking and non-uniform spatial generation rates for defect formation.

7. INFLUENCE OF DEPOSITION CONDITIONS ON CELL STABILITY

We have performed several survey experiments in which the degradation of cells deposited under various conditions has been quantified. The cells were first annealed, and light soaking was performed at 35°C on a rotating turntable under a sodium lamp. In broadest terms, the results of one of these studies (involving 45 cells) is shown as a histogram of percent efficiency loss in Fig. 4. The distribution is relatively narrow, with half the cells exhibiting a loss of 28-34% regardless of i-layer thickness, which ranged from 0.2 to 0.65 μm. Note also that the distribution is distinctly asymmetric, with a sharp cut-off for the smaller efficiency losses and a longer tail towards higher efficiency losses.

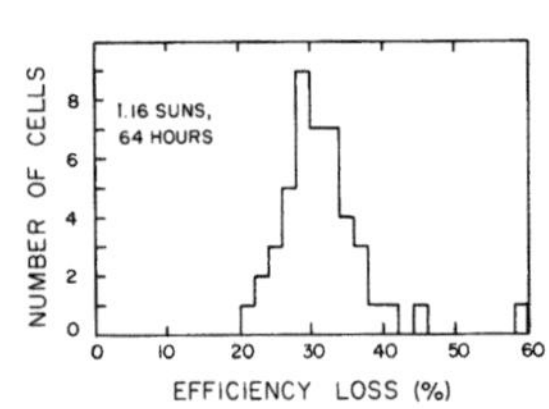

Fig. 4. Histogram of efficiency loss after light soaking for 45 differently prepared cells.

It was further concluded that, for improved stability, the cells should be a) thin, b) deposited on granular tin oxide, rather than specular, c) prepared on the anode rather than the cathode, and d) prepared from hydrogen-diluted silane rather than pure silane. It was also found that cell instability is increased at lower deposition temperatures,and that cells having an i-layer prepared from undiluted disilane are very unstable. On the other hand, substitution of a hydrogen-diluted mixture of 0.8 SiH_4 + 0.2 Si_2H_6 for pure SiH_4 did not affect cell stability, nor did substitution of deuterium for hydrogen in silane/hydrogen mixtures.

In the above study, no definite effect of a thin silicon-carbide transitional i-layer on cell stability was observed, although this conclusion should be checked using the solar spectrum for illumination. In a later study, cells of 7-8% initial efficiency with i-layers consisting entirely of a-Si:H or a-SiC:H (CH_4/SiH_4 = 0.28) were compared, with dramatic results. After a 5 day soak, cells having a silicon-carbide i-layer had lost 66% of their initial efficiency, compared to 27% for standard a-Si:H. Lastly, unusual deposition conditions often result in cells that are relatively stable, but at the expense of initial efficiency, e.g. cell 1 in Table 1.

8. SUPERIOR STABILITY OF TANDEM VERSUS SINGLE JUNCTION CELLS

The superior stability of tandem devices with respect to light soaking is expected on theoretical grounds for the following reasons. Firstly, the top i-layer of a 2-stack device is quite thin, perhaps 750 A, yet this is sufficient to absorb one half of the solar spectrum that can be usefully absorbed by a-Si:H. The high electric field in this layer not only ensures a high carrier collection efficiency, implying a low defect formation rate due to recombination, but also reduces the deleterious effect of those defects that are formed. Secondly, since the generation rate G in the bottom stack is now only ½ of its value in a single-junction cell, then, bearing in mind that the parameter controlling defect formation rate is G^2t, we see that the time scale for the generation of defects is expanded by a factor of 4.

These expectations are borne out in practice, and we see from Table 1 that the device experiencing the lowest drop in efficiency after 2.4 years sodium illumination is a tandem cell. In itself this result is not compelling, since the stability of tandem cells depends strongly on which stack is current-limiting. Additional studies performed with simulated sunlight, however, confirm that properly designed tandem cells always outperform single junction cells after extended light soaking.

9. OUTDOOR PERFORMANCE OF SUBMODULES

Stability data on four glass-glass encapsulated 930 cm^2 R&D panels operated outdoors for over 60 days are shown in Fig. 5. The panels were periodically brought inside for testing under standard 1 sun conditions. It can be seen that both single junction and tandem panels, when kept on laod, are more stable than their open-circuited counterparts. Despite having a lower initial output, the on-load tandem panel had a higher output than the on-load single junction panel at the end of the study, and had reached an almost stable condition. A longer term study conducted at SERI on a similarly encapsulated 930 cm^2 tandem panel from Chronar showed that the panel retained 83% of its initial power after 200 days of exposure (see Fig. 6). These tandem panels were not specifically designed for stability and had a 0.64 um thick i_2 layer. We hope to demonstrate even better stability using a thinner i_2 layer.

Fig. 5. (at right) Outdoor performance of single and tandem junction 930 cm^2 a-Si:H submodules.

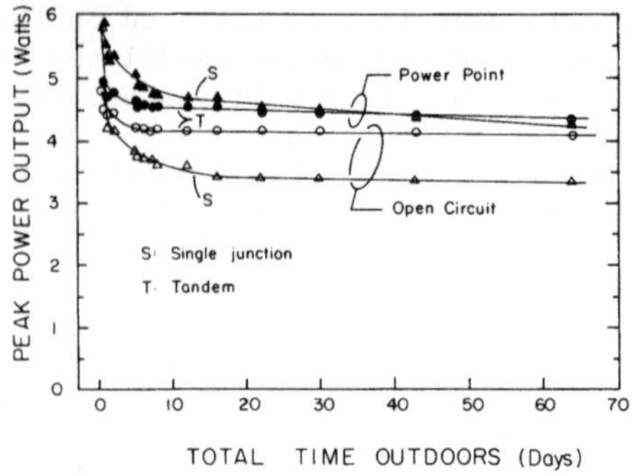

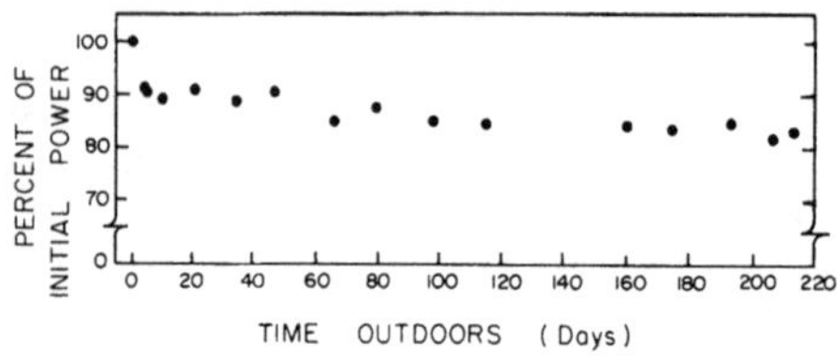

Fig. 6. (at left) Long term outdoor stability data for a 930 cm^2 tandem submodule kept on load.

ACKNOWLEDGEMENTS

This work was supported by the Solar Energy Research Institute (SERI). We thank J. Kalina for some of the measurements, and L. Mrig of SERI for kindly providing the data for Figure 6.

REFERENCES

1. D. L. Staebler, R. S. Crandall, and R. Williams, Appl. Phys. Lett. 39, 733 (1981).
2. A. E. Delahoy and T. Tonon, AIP Conf. Proc. 157, 263 (1987).
3. A. E. Delahoy et al. Semi-Annual Progress Report for the period Oct. 1984 - March 1985, SERI Subcontract No. ZB-3-03056-1. SERI/STR-211-2814 (DE85016886), available from NTIS, Virginia.
4. Z E. Smith, S. Wagner, and B. W. Faughnan, Appl. Phys. Lett. 46, 1078 (1985).
5. M. Stutzmann, W. B. Jackson, and C. C. Tsai, Appl. Phys. Lett. 45, 1075 (1984).
6. B. W. Faughnan and R. S. Crandall, Appl. Phys. Lett. 44, 537 (1984).
7. H. Schade and A. E. Delahoy, this conference.
8. R. S. Crandall, J. Kalina and A. E. Delahoy, 1988 MRS Spring Meeting, April 5-9, Reno, Nevada.
9. A. Catalano, M. Bennett, R. Arya, K. Rajan and J. Newton, 18th IEEE Photovoltaics Specialists Conference, Las Vegas, Nevada, Oct. 1985.

PERFORMANCES PRESENTED BY a-Si:C:H (DOPED AND UNDOPED) FILMS PRODUCED BY A TCDDC SYSTEM FOR PV APPLICATIONS

R. Martins, L. Guimarães, E. Fortunato, M. Vieira, N. Carvalho M. Santos and I. Ferreira.
Faculty of Science and Technology of New University of Lisbon / UNINOVA / CFMUL (INIC), Portugal.

Summary

We have investigated the optoelectronic, structure, morphology and chemical composition properties of a- or μc- Si:C:H undoped and doped films produced by a Two Consecutive Decomposition and Deposition Chamber (TCDDC) system, in the presence of controlled electromagnetic static fields. The performances of this method and its correlation with structure and transport properties presented by deposited films are discussed. The obtained results show that film's growth is controlled by plasma induced DC bias which controls gas decomposition process and the rate of impinging ions on the surface. Good quality undoped amorphous silicon films, suitable for photovoltaic (PV) applications, are produced by using power densities (dp) lower than 10 mW/cm^3. Transition from amorphous to microcrystalline phase is achieved by using hydrogen dilution ratios higher than 3% and dp > 100 mW/cm^3, whatever the kind of alloy deposited. Concerning the production of amorphous silicon carbide alloys, good correlation, between the concentration of carbon atoms in the gas phase, and those ones incorporated is achieved. Low absorption wide band gap material (WBG) with $\sigma_d \geq 10^{-1}$ $(\Omega cm)^{-1}$ and optical gap $E_{op} = 2.5$ eV have been also produced gathering attention in its application as window or back layer in the production of solar cells.

1. INTRODUCTION

Plasma assisting or inducing Chemical Vapour Deposition (P-CVD) processes have been used in numerous applications, such as to produce amorphous (a-) and/or microcrystalline (μc-) silicon (Si) films for photovoltaic (PV) and other electronic processes . However, most of the work is based on empirical recipes that the experience has proved to lead to materials with good structural and transport properties.The reason why this kind of approach is followed has to do with non-isothermal nature of the plasma and so, with the difficulty in determining and controlling the set of parameters involved in the plasma process [1].

Two main regions can be distinguished in a plasma process: one ascribed with gas (SiH_4) decomposition and other with species deposition on growing surface. The former process is controlled by gas and electron concentrations in the charged region, depending on kind of collisions (elastic or inelastic) undergone by gas molecules. The later one depends on ion flux reaching the growing surface responsible for its dehydrogenation and for the structural and transport properties of deposited films. As the deposition process is a function of the energy delivered to the plasma and on the existing relation between the residence time (t_r) and the reaction time (τ), two main regimes can be defined. One is the low energy regime where elastic collisions are dominant and discharge is used to overcome the activation energy of a kinetically hindered reaction [1,2]. Here, $t_r \ll \tau$, the system is faraway from chemical equilibrium (PCE) and deposited films are amorphous. Nevertheless, good quality a–Si films are obtained if mannomeric species reach preferentially the growing surface whilst polymerization in the gas phase is avoided [1,2]. This means that a good correlation has to be found between gas flow (at high rate), pumping speed and reactors geometry for sustaining the growing surface under bombardment of low energy ions associated to a low depletion of silane molecules in the plasma.

The other one occurs at intense discharge conditions (high energy regime) promoting strong endothermal reactions which leads to a complete change on chemistry process [1,2,3]. There, μc-Si films can be obtained if the condition $t_r \geq \tau$ is achieved, by using low flows of a high diluted carrier gas in hydrogen leading to a stable phase near PCE. This will give rise to surface dehydrogenization , passivation of the grain boundaries and improves adsorption of atoms in the matrix [1]. When impurities are present, they behave as active centres in turn of which an agglomerate of a microcrystalline phase imbibed in an amorphous tissue is obtained.

The two regimes described above are dependent on the way how plasma is produced. Thus, for instance, in diode type systems plasma is confined between electrodes and substrate is hold by one of them [1,3]. In such systems a negative dc bias is induced to cathode electrode,being inversely proportional to its area and to gas pressure used which limits control on plasma parameters and particles [1]. So, its difficult to achieve control upon induced dc bias when low values are needed, this disadvantage is overcome using high rf frequencies (f > 13.56 MHz) not only to sustain the discharge at low V_{dc} bias, but also to keep the growing surface below the threshold of structural induced surface damages.This method has been proved to lead to high quality films [1]. Another method, consists in promoting space separation between the chemistry of the plasma and that of the deposition [4]. This method (that we call TCDDC system) will allow us to reduce the number of uncontrolled parameters. In the following we shall describe this method as well as electro-optical properties of films produced in this system that can be used in PV and other related applications.

2. DESCRIPTION OF THE PREPARATION METHOD

The preparation method used is based on a two consecutive decomposition and deposition chamber system where plasma is formed in a different region of substrate location [4]. In this system, plasma ignition (determined by the appearance of a dc bias) is promoted by coupling a rf power to an inert (hydrogen) or mixed gas, fed separately from the carrier gas, as shown in figure 1.

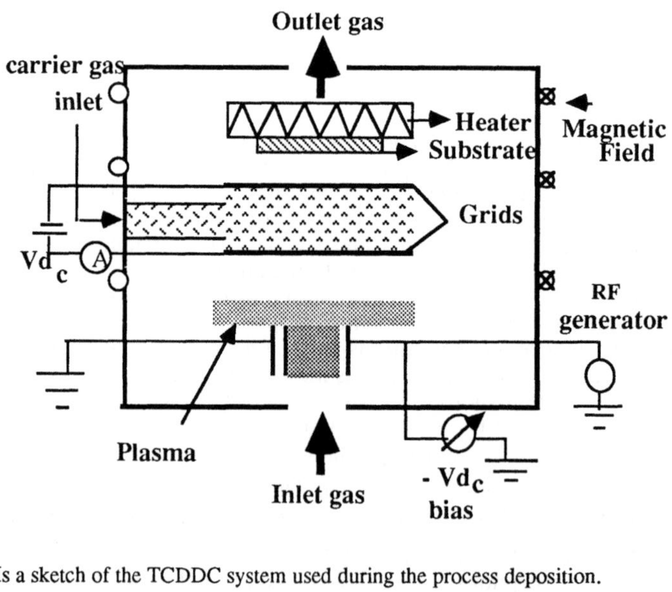

Fig.1— Is a sketch of the TCDDC system used during the process deposition.

Through this way gas phase separation is achieved even when alloys have to be deposited. For depositing good quality a-Si films, $dp \leq 10\ mWcm^{-3}$ independent of ratio between concentrations of carrier and inert gas [4] while for μc-Si films the proportion between the concentrations of carrier and inert gas has to be lower than 3:100 and $dp > 100\ mWcm^{-3}$. Control upon species formed on plasma region is achieved through two grids located between plasma and substrate holder (see figure 1).For producing high quality a-Si films at low dp, the upper grid is kept positive in relation to the substrate in order to submit the growing surface to a high flux of low energy ions. Here, the growth rate, R, is proportional to the current density detected by the upper grid, that works as a large area Langmuir probe [4]. For producing μc films (high dp) dc bias applied to the upper grid is positive or negative, depending on type of films to be produced (n- or p-type μc films, respectively). When high voltages and $t_r < 1$ s are used during the deposition process a poor quality film is obtained due to surface damages promoted by surface bombardment with high energy ions.

Plasma confinement is achieved by a magnetic field applied perpendicular to the substrate which increases the charge density within the plasma region and so it will improve the fragmentation ratio of gas molecules.

Taking into account the description given above as well as the experimentally determined R values, a simple mathematical model is used [5] to correlate R with: the density of ion species, f (proportional to the current density); the rate of silane decomposition, α; the density of atoms presented on the growing surface, F; discharge efficiency, η; plasma volume, V; and reaction constant, $\sigma\tau$ (σ is the collision cross section varying between 10^{-16} and $10^{-13}\ cm^2$ [5]). The obtained expression is:

$$R = \frac{F\alpha}{N_i} \times \frac{1}{1 + (\sigma\ \tau\ \eta\ f\ )^{-1}} \qquad (1)$$

where N_i is the number of silicon atoms per unit volume presented on the growing surface and the other letters have the meaning given above.

In this model, the reaction constant, is dependent on free energy of Gibbs (ΔG) and is expressed by:

$$\sigma\tau = 1.7 \times 10^{-22} \exp\left(-\frac{\Delta G}{RT}\right)\ [cm^2 s] \qquad (2)$$

The existing uncertainty in defining separately σ and ΔG in equation (2) is overcome by using as comparative term the residence time, t_r, given by the relation :

$$t_r = \frac{PV}{F} \qquad (3)$$

(P is the gas discharge pressure).

Thus if we ascribe $t_r > 1$s with transition between a metastable and stable phase, faraway of PCE we shall have $\sigma\ \tau\ \eta\ f << 1$ while near or at PCE, $\sigma\ \tau\ \eta\ f >> 1$. So, equation 1 can be simplified, given the following relations:

a) $R_1 \sim \frac{F}{N}\ \alpha\ (\ \sigma\ \tau\ \eta\ f\ )$ faraway of PCE $\qquad$ (4)

where f is experimentally determined by:

$$f = \frac{dp}{\xi} \times \frac{1}{e} \qquad (5)$$

b) $R_2 \sim \frac{F}{N}\,\alpha$ near or at PCE (6)

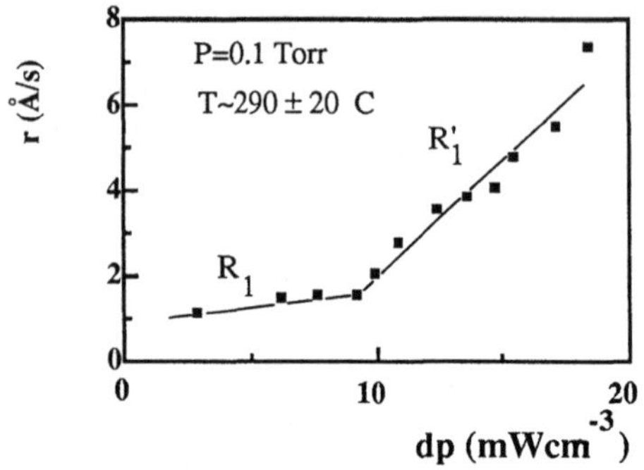

Fig.2— Shows the growth rate of undoped a–Si:H films as a function of a power density used.

Experimentally R was fitted by such expressions as shown in figure 2. The agreement obtained is quite good either a– or μc–Si films, with $10^{-5} < p\ \ 10^{-3}$ tor f in the range of 10^{15} to 10^{18} ions $cm^{-2}\ s^{-1}$ and α between 0.01 and 0.2.

3. EXPERIMENTAL DETAILS

Undoped and doped a- or μc-Si:C:H films have been produced by a TCDDC system working as described before. Typical deposition conditions are shown in table 1.

Table I - Typical deposition parameters

Gases	Gas flow (sscm)	Pressure (Torr)	Substrate temp. (°C)	Power Density ($mWcm^{-3}$)
SiH_4 CH_4 (2.5 %) PH_3/SiH_4 (5 %) B_2H_6/He H_2 He	1-15 1-15 1-2 1-2 50-400 10-20	0.1- 0.5	100-200	3-130

Electro-optical properties, microstructure, morphology and chemical composition of such films, were analysed using the apparatus and sample arrangements as described elsewhere [9]. This was done in order to see if their performances satisfy requirements concerning the production of high quality materials suitable for PV applications, namely, the production of high conductivity wide band gap materials (for decreasing photogeneration losses and improving current collection [10]) and intrinsic layers with low DOS [3]. Besides this, high resistivity ($\rho > 10^{14}\ \Omega cm$) and low absorption films (used as interlayers to sustain interdiffusion of doping species) have been produced based on undoped a-Si:C:H films using CH_4:SiH_4 mixtures in the proportion of 5:1. Figure 3 we present the change in bond configurations of IR spectra (between 950 cm^{-1} to 500 cm^{-1}) either for films produced at same dp or alloy composition.The dependence of hydrogen concentration, C_H , (inferred from IR wagging modes) and carbon contents (taken from RBS measurements) on mixture composition X, [X = $CH_4/(SiH_4+CH_4)$] are shown in figure 4.

Electro-optical performances of μc-Si:C:H n- and p- type films (using CH_4:SiH_4 mixtures in the proportion of 1:5 and a doping level of 1%), were determined by dark conductivity, σ_d, optical absorption, α, Hall effect and other structural and morphological studies. In figure 5 we show the dependence of σ_d and ΔE (activation energy obtained from dark conductivity measurements) on optical gap, Eop (inferred from Tauc's plot), while in figure 6 the dependence of α on photon energy , hυ, is shown. For comparative purposes, we show in figure 5 the corresponding behaviour of σ_d on Eop for p-type films produced by CPM and diode method [6] , whilst in figure 6 the expected behaviour of α on hυ for typical a-Si:H; μc-Si:H and c-Si films, are also shown. Hall effect measurements performed on μc-Si:C:H doped films lead us to Hall mobilities in the range between 5 $cm^2\ V^{-1}\ s^{-1}$ to 0.2 $cm^2\ V^{-1}\ s^{-1}$.

4. ANALYSIS OF THE RESULTS AND COMMENTS

From the experimental, data shown in figure 2 a kink around dp = 10mW/cm^3 is observed for the dependence of R on dp, in undoped films. This kink is ascribed with a change on electro-optical and structural properties of such films [5], limiting the obtention of good quality a-Si:H films to values of dp < 10 mW/cm^3 and so, to growth rates of the order of 3 Å/s. When μc films are produced R decreases to values of the order of 1 Å/s, depending on hydrogen gas flux used, which enhances the etching rate on the growing surface.

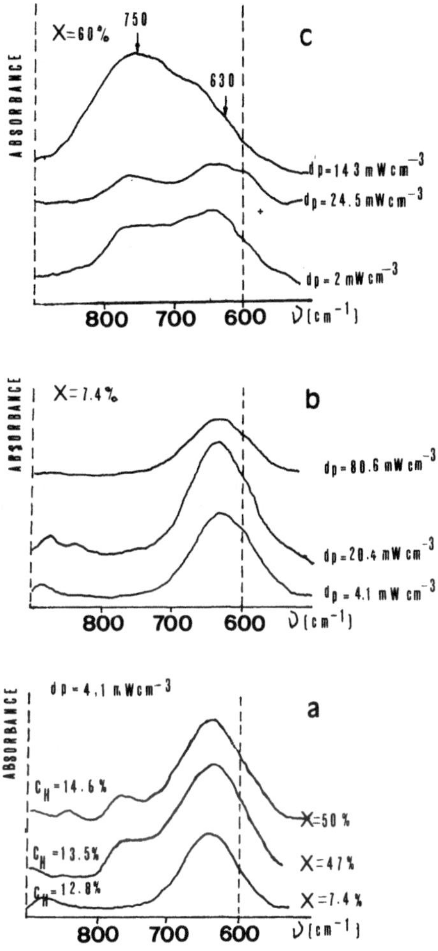

Fig.3—IR spectra of a-Si $_{1-X}$:C_X (1%B) at:a) dp=4.1mW/cm^3 and variable, X. b) X= 7.4 % and dp variable. c) X= 60% and dp variable.

Results presented in figure 3-a) show that keeping dp = 4.2 mW/cm^3 and changing X from 0.07 to 0.5 the wave number ascribed with Si:H_n wagging modes is shifted towards higher values and a new peak around 750 cm^{-1} (attributed to Si:C stretching mode [7]) starts appearing. From the obtained data we observe that the contribution of bending modes (attributed to dihydride species) are kept at low levels whatever the X value used. In figure 3-b) the behaviour of bond configuration with dp for X = 0.074 is shown. Here we observe that bond configuration do not change significantly by increasing dp and the peak around 750 cm^{-1} does not appear. The most significative effect is the increase of C_H with dp and the decrease of bending modes. In figure 3-c) we show the same dependence as before but with X = 0.6 and in the presence of a high hydrogen flux.By increasing dp a clear change in bond configuration is observed with the shift of the main peak from 630 cm^{-1} to 750 cm^{-1}. At same time, C_H decreases (from 0.19 to 0.04) when dp increases from 2.05 to 140 mW/cm^3. This change is ascribed with the increase of activated species in electrically activated states and with the transition of a- phase to μc- phases, as confirmed by normal X-Ray diffraction measurements. The observed shift of wagging modes from 630 cm^{-1} to 650 cm^{-1} and the enhancement of bending modes are attributed to incorporation of SiH_2 species in the grain boundaries of microcrystallites [7].

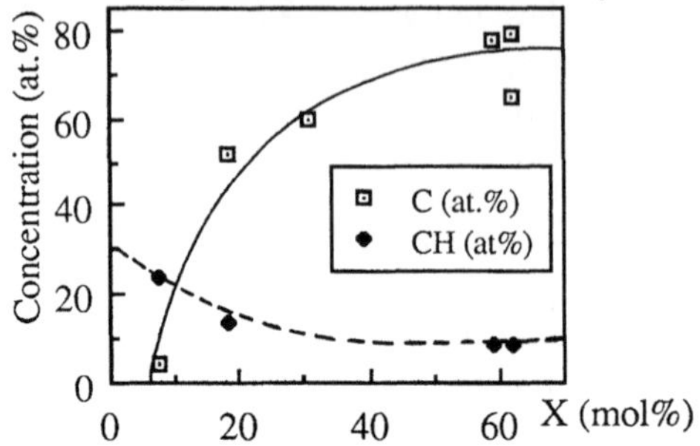

Fig.4— Dependence of C_H and C (at. %) on alloy composition X (mol%) for a–Si:C:H films.

The results in figure 4 show that carbon concentration increases as X increases whilst C_H (due to hydrogen mainly connected to silicon atoms) decreases, for dp > 60 mW/cm^3 .This also indicates a good correlation between carbon atoms in the gas phase and that ones incorporated in the film.

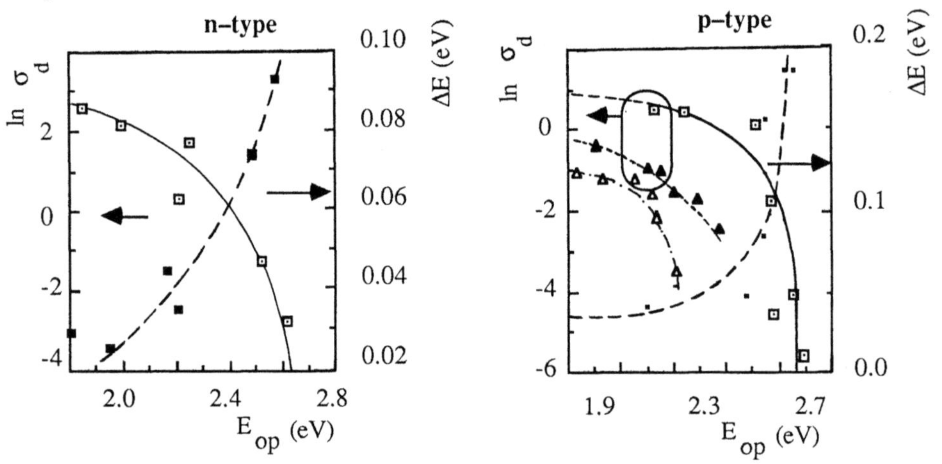

Fig.5— These plots show the dependence of dark conductivity(σ_D -▫)and activation energy (ΔE– ▪) as a function of optical gap (E_{op}) for p– and n–type wide band gap –μc films. We also show the same dependence for films produced by Diode method (▵) and by CPM method (▲), [6].

In figures 5 and 6 we show the correlation between σd and ΔE on E_{op} and of α on $h\upsilon$ for doped μc films.These results show that films produced by TCDDC method are low absorbing and present considerably high conductivities when compared with material produced either by diode or CPM method [7]. This, together with the need in getting μc material that can be used in PV applications as a window layer make this material quite promising for getting devices with high conversion efficiencies and high circuit voltages, Voc, since now band discontinuity problems related to the built potential are reduced [9]. We also would like to refer that these films present quite high mobilities when compared with normal μc films [9]. This comes in favour of our previous assumption that the number of incorporated defects on films produced by a TCDDC system is low [5].Thus the enhancement in hydrogen atoms present in this deposition process is not deleuterious concerning structural and transport properties of either a- or μc- films.

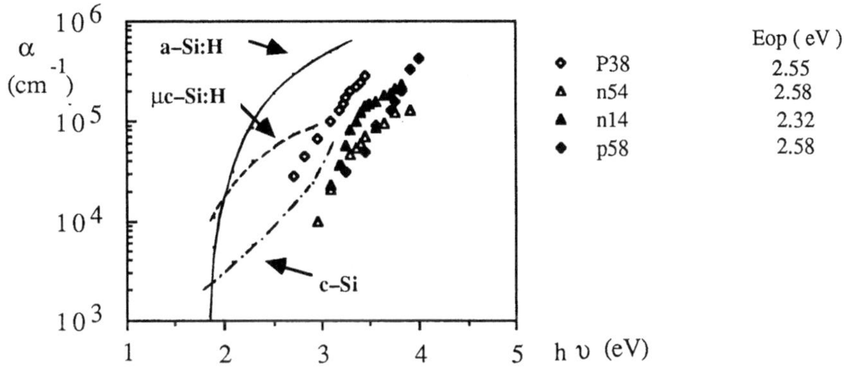

Fig.6— Dependence of absorption coefficient (α) on photon energy used for WBG–μc n– and p–type films. The corresponding dependence for a–Si (—), μc–Si (– –) and c–Si (–· –) are shown.

5. CONCLUSIONS

Overall, the experimental results show that films produced by a TCDDC system present structure and transport properties that fulfil demands for solar cells production at low costs. Good valence control was obtained in a- or μc- Si:C:H doped films by changing CH_4 concentration in the gas phase mixture. This is related to an increase in carbon incorporation (mainly in tetrahedrally bonds with silicon [7]) as observed from IR spectra shown in figure 3. Good quality undoped a-Si films are produced at low dp where the growth process is mainly ascribed with SiH precursors [5]. Concerning μc films, the results show that wide band gap/low absorption materials can be prepared by the method proposed with better performances then those ones usually reported for normal μc films. These results are advantageous for reducing photogeneration losses at the blue end of the solar spectrum (when used as front layer on pin devices) and improving current collection (when used as back layer if the corresponding metal back-contact has good reflectivity). Besides this, as ΔE on such materials is low, band discontinuities problems are overcame and so devices with high Voc are obtained [9]. AS performances in pin solar cells are (in some extension) controlled by impurities interdiffusion, in the junction regions, the use of blocking buffer interlayers able to sustain such problem are recommended. Undoped a-Si:C:H layers produced by the method described have the required performances (high resistivity, stable and low absorber) for such applications.

6. FUTURE PROSPECTS

The performances presented by films produced in a TCDDC system, make them suitable for other non–photovoltaic applications, such as light emitting diodes (LED's) and thin film transistors (TFT's).

LED's based on silicon carbide alloys and using wide band gap μc material as doped contacts, will improve brightness since band discontinuities at the interface are reduced. Besides this TFT's based on MIS (Metal Insulator Semiconductor) structures using high resistivity a-Si:C:H films as dielectric, can be produced. As a matter of fact we have already produced such devices, presenting threshold voltages below 2 V, which are correlated with the decrease on density of surface states at the interface.

ACKNOWLEDGEMENTS

The authors wish to express their thanks to Dr. G. Willeck (IMEC) for the help and contribution given on film's characterization. Thanks are also due to A.Maçarico and M. Santos for the help given during the realization of this work. This work was supported by EEC contract Nº En 3S–C1–602–P, Junta Nacional de Investigação Científica e Tecnológica (JNICT) – Project Nº 87.375 and also by Instituto Nacional de Investigação Cientifica (INIC) – Project B.

REFERENCES

[1]— S. Veprek, *Winter European Course on Amorphous silicon*, Italy (1986), *J. Phys. C: Solid State Phys*, **16** (1983) 6241-6262, *Philosophical Magazine B*, 1982 Vol. 45, nº 1, 137-145
[2]— R. Robertson et al, J. *Appl. Phys.* **59**, (1986) 3402
[3]— I. Solomon, *Winter European Course on Amorphous Silicon*, Italy (1988)
[4]— R. Martins et al, *Proc. 5th E.C. Photov. Solar En.* (1983) Conf.,778
[5]— R. Martins et al, *Journal of Non-Crystalline Solids*, **97** &**98** (1987) 1399
[6]— Y.Kuwano et al,*MRS Spring Meeting*, USA (1988)
[7]— R. Martins, *MRS Spring Meeting*, USA (1988)
[8]— M. Vieira et al, *MRS Spring Meeting*, USA (1988)
[9]— A. Madan,*MRS Spring Meeting*, USA (1988)

DEVELOPMENT AND PERFORMANCE OF 4-TERMINAL THIN FILM MODULES

D.L. Morel, K. Blaker, W. Bottenberg, L. Fabick, B. Felder, M. Gardenier, D. Kumamoto, D. Reinker and R. Rifai

ARCO Solar, Inc.
P.O. Box 2105
Chatsworth, California 91313

ABSTRACT. 4-terminal tandem hybrid thin film cells employing thin film silicon:hydrogen alloys (commonly referred to as "amorphous silicon") as the top circuit component and copper indium diselenide as the bottom circuit component are being developed in pursuit of the high efficiencies required for bulk power generation. Peak efficiencies of 15.6% in laboratory cells and 12.3% in large area modules have been achieved. Commercial quality modules have been fabricated that pass JPL Block V testing and that perform well outdoors relative to crystalline silicon.

1. INTRODUCTION

Applications of photovoltaic technology have thus far ranged from consumer electronics products to rural and to space power systems. Several large power plants have been built, but primarily as demonstration projects. The bulk power market is the ultimate target, but performance objectives must be met before the realities of this market can be accessed. Studies conducted by industry participants[1] and potential utility customers[2] have identified the required performance objectives as well as technologies that have the potential to meet those objectives. Thin film solar cells are recognized as having this potential, and efforts are under way at various institutions throughout the world to turn this potential into reality. Much of this effort has thus far been directed toward the demonstration of high efficiency, and several laboratories have reported single and multijunction device efficiencies in the 10-14% range.

In addition to demonstration of high laboratory cell efficiency, it is necessary to transfer this success to the fabrication of large area modules and to demonstrate the performance and viability of these modules under actual operating conditions. We have been pursuing the development of 4-terminal hybrid tandem (4THT) devices incorporating thin film silicon:hydrogen alloy (TFS) as the top component and copper indium diselenide (CIS) as the bottom component. We have reported the achievement of 14.2% efficiency in 4 cm^2 laboratory cells and 12.3% in 30×30 cm modules[3]. In addition, we have reported on the initial performance of this technology under actual outdoor conditions[4]. This early evaluation was conducted with 10×10 cm "mini-modules" with efficiency of about 6%. These mini-modules contained the essence of the TFS and CIS device technology, but lacked the encapsulation attributes of real modules. As early prototypes the mini-modules performed well, but the necessity of extending the studies to fully completed modules remained.

In this paper we report further progress in the development of this technology. We have fabricated commercial quality modules that have successfully completed both accelerated life testing and outdoor performance trials. We thus remain convinced of the ability of this technology to serve the bulk power market.

This work was supported in part by the California Energy Commission and the Solar Energy Research Institute.

2. MODULE EFFICIENCY

The 4-terminal hybrid tandem module (Fig. 1) is a complex stack that requires significant attention to optical matching for optimum device output. TFS devices are fabricated using processing comparable to that for our commercial modules; CIS devices use processing that is readily scalable for commercial production. Our standard laboratory test cells are 4 cm^2 in area. 30×30 cm modules consist of two individual pieces of 30×30 cm glass onto which the TFS and CIS circuits are separately deposited. The circuits are fully integrated, consisting of 50 cells in each case which are connected in series by the integration. The modules are superimposed for measurement of the output of the two circuits.

Our modeling simulations for these devices indicate expected laboratory cell efficiencies of 20-24%[5]. It is anticipated that this would result in module efficiencies of 15% and above, which is the target for bulk power implementation[1]. Table I summarizes the present status of our progress toward these objectives. As shown, we have advanced the performance of laboratory cells to 15.6%[6]. We have also achieved 12.3% efficiency in 30×30 cm modules, indicating excellent success in transferring our laboratory progress to full scale. These achievements and the extrap-

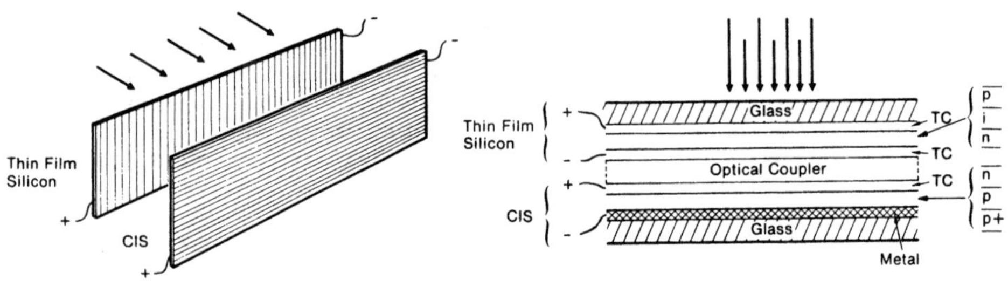

Fig. 1. Cross section and electrical configuration of tandem module.

Table I. 4-terminal tandem cell and module performance.*

4 cm^2 Cell	Eff (%)	J_{sc} (mA/cm^2)	V_{oc} (mV)	FF
Semitransparent TFS	10.3	16.4	871	0.72
Filtered CIS	5.3	17.9	432	0.68
Tandem	15.6			
Stand-Alone CIS	12.4	41.2	455	0.66

30×30 cm Module	Power (W)	Area** (cm^2)	Eff (%)	I_{sc} (mA)	V_{oc} (V)	FF
Semitransparent TFS	7.69	843	9.1	262	43.5	0.68
Filtered CIS	2.66	843	3.2	228	19.2	0.61
Tandem	10.35		12.3			
Stand-Alone CIS	7.62	844	9.0	611	21.2	0.59

*Measured at ASTM air mass 1.5, global 100 mW/cm^2, 25°C.
**Aperture area.

olations to be made from them indicate that the efficiency goals for the bulk power market will be met. It is therefore appropriate to extend these developments to the fabrication and testing of fully completed commercial quality modules.

3. MODULE FABRICATION

TFS power modules have been commercially available since 1984. Since these were initially targeted for the consumer electronics market, module design was limited in scope. As these products evolved, the intent to compete with crystalline silicon modules in the real power module arena has required the development of module packaging with long-term durability. Much of what has been learned about module construction for crystalline silicon is transferrable to thin films.

4-terminal hybrid tandem technology presents new challenges in the evolutionary development of thin film modules. The need to handle two circuits within the package complicates several aspects of module design. For example, achieving electrical isolation of the two circuits while not compromising the integrity of optical coupling between them places new demands on encapsulant materials. This challenge has been successfully met with 4THT modules that pass Jet Propulsion Laboratory (JPL) Block V testing. The most difficult criterion to pass was the Hi-pot test for leakage between the circuits and from the circuits to the frame. With respect to the latter, we have in fact extended the requirement to include wet module conditions and have passed this test as well. The need for effective electrical isolation under wet conditions is particularly important for high voltage thin film circuits in which the electrically active layers approach the frame.

The module (Fig. 2) has a metal frame and currently has two junction boxes on the back for the two circuits. The encapsulation uses advanced techniques for edge sealing and electrical isolation.

Fig. 2. Tandem module with metal frame.

4. OUTDOOR TESTING AND EVALUATION

Accelerated life testing is the first step in verifying the integrity of a new technology. It is then necessary to evaluate performance under actual operating conditions to uncover any problems that may not be revealed by accelerated testing. For tandem technology, two issues are to be addressed: (1) the standard concern with the stability and durability of the modules and (2) the need to assess the output of tandem relative to single junction technology to determine if the tandem has any operational shortcomings due to its increased complexity. To this end we have installed several of these modules on our outdoor test station and have been tracking their performance. The modules are mounted on a two-axis tracker and a complete data set is recorded every 15 minutes[7]. Reference measurements at standard conditions are taken with a large area pulsed solar simulator (LAPSS).

4.1. Stability

Table II summarizes module performance after three months of outdoor exposure during winter months. Module output dropped from an initial value of 10.4% to a "stabilized" level of 8.7% over the three-month period. All of the loss is in the TFS circuit as expected due to Staebler-Wronski effects, but is somewhat higher than typically observed. This is partially due to this having been the coldest part of the year, which exacerbates TFS degradation. We expect recovery in the upcoming warmer months to a stabilized level of greater than 85% of the initial output. The CIS circuit actually shows an increase in output over the period, which is consistent with our experience.

The module itself shows no signs of deterioration or of contributing to the TFS circuit losses. We expect this module to equilibrate at about the 9% efficiency level and to remain stable.

4.2. Daily Performance

The efficiency performance for the module and its component circuits for a typical sunny day is shown in Fig. 3; temperature and insolation profiles for the day are shown in Fig. 4. The module reaches an operating efficiency of 7-8% and remains there most of the day. The flatness of the

Table II. Module performance after three months outdoors.

	V_{oc} (volts)	I_{sc} (amps)	FF	P_{max} (watts)	Eff*
TFS Circuit					
Initial LAPSS	41.7	0.23	0.65	6.3	7.5
Final LAPSS	40.5	0.22	0.53	4.7	5.6
CIS Circuit					
Initial LAPSS	17.3	0.25	0.56	2.4	2.9
Final LAPSS	17.7	0.25	0.59	2.6	3.1
Tandem Circuit					
Initial LAPSS				8.7	10.4
Final LAPSS				7.3	8.7

*Aperture area efficiency.

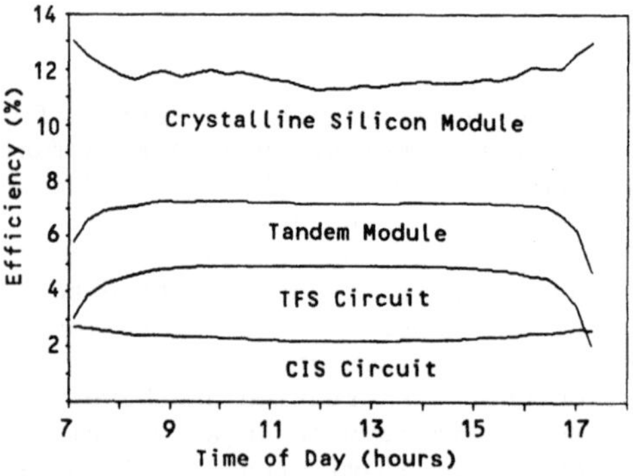

Fig. 3. Tandem efficiency over the day.

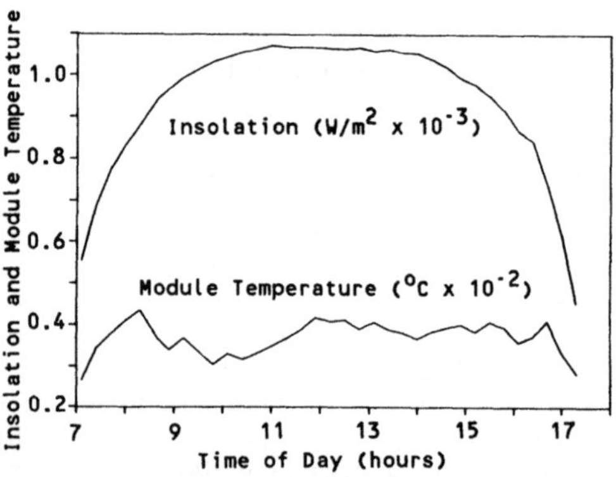

Fig. 4. Operating conditions over the day.

output over the course of the day is due to the complementary nature of the TFS and CIS component circuits. The output for a commercial crystalline silicon module which is mounted next to the tandem module is also included for comparison. Its output drops at midday due to lowering of V_{oc} with increasing temperature.

A direct comparison of the tandem to crystalline silicon module performance is given by the plot of the efficiency ratio in Fig. 5. The drop in the ratio at the ends of the day is due to factors such as lowered intensity and spectral content; however, the relative loss is small because of the low energy content of the sun's radiation in those time periods. To quantify this effect, the outputs were integrated over the day to determine the relative energy delivery of the modules. Results are summarized in Table III. The ratio of the standard conditions efficiency of tandem to crystalline silicon is .59, while that for daily energy delivery is .61. This indicates that 4THT has no shortcomings relative to crystalline silicon technology with respect to energy based rating under the conditions addressed thus far. The remaining challenge is to continue the advancement of thin film module efficiencies toward and beyond those of crystalline silicon.

5. CONCLUSIONS

With thin film tandem technology we have achieved laboratory cell and module efficiencies that are rapidly approaching those of commercial crystalline silicon. We have additionally fabricated modules with the technology that have successfully passed accelerated life testing. The modules are undergoing outdoor evaluation as well. The conclusion thus far is that the modules perform well and have no shortcomings relative to crystalline silicon in terms of energy based rating.

REFERENCES

1. G.J. Shushnar et al., *Proc. 1st Int. PV Sci. and Eng. Conf.*, Kobe, 1984, p. 331.
2. E.A. DeMeo and R.W. Taylor, *Science*, **224**, 4646(1984) 245.
3. D.L. Morel, 'Progress on High Efficiency Thin Film Solar Cells,' *Solar Cells* (to be published).
4. D.L. Morel et al., *Proc. 18th IEEE PV Spec. Conf.*, Las Vegas, 1985, p. 876.
5. K.W. Mitchell, *Proc. 1st Int. PV Sci. and Eng. Conf.*, Kobe, 1984, p. 691.
6. K.W. Mitchell, 'Copper Indium Diselenide PV Technology,' this conference.
7. L.B. Fabick et al., *Proc. 19th IEEE PV Spec. Conf.*, New Orleans, 1987, p. 182.

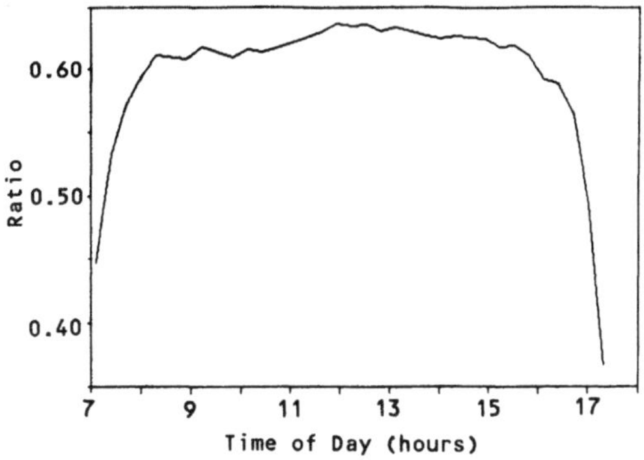

Fig. 5. Tandem/crystalline efficiency ratio.

Table III. Energy delivery of modules.

Circuit	Daily Energy (watt-hours)
TFS	43.2
CIS	21.5
Tandem	64.7
Crystalline silicon*	106.3

*Normalized to 30x30 cm.

THE TRANSITION VOLTAGE OF A-SI PIN SOLAR CELLS BEFORE AND AFTER DEGRADATION

Y.M. Sun, H. Pfleiderer and W. Krühler
Siemens Research Laboratories, D-8000 Munich 83

ABSTRACT. We examine some a-Si pin diodes serving as solar cells with different i-layer thicknesses. The observations cover the spectral response for different diode voltages and the white-light characteristics for different illumination intensities, and include the Staebler-Wronski degradation. A vanishing photocurrent defines the transition voltage. Our understanding is guided by the partition of the i-layer into 3 regions: a middle region with uniform electric field, and two boundary regions next to the pi- and in-junctions. The photocarrier recombination is concentrated in the boundary regions.

1. INTRODUCTION

Amorphous silicon (a-Si) solar cells are pin diodes deposited onto a glass substrate. The p- and n-layers establish the open-circuit voltage U_{oc}, the i-layer collects the photocurrent. The light enters through the C-doped p-layer. There are three regions within the i-layer: a middle region, where the electric field E is almost uniform, and boundary regions, where the field sharply rises towards the pi- and in-junctions. The corresponding field distribution, sketched in Fig. 1, results from numeric modeling /1/. The high-field region next to the pi-junction effectively collects photoholes, but also allows photoelectrons to recombine with holes leaking in through the pi-contact. The boundary layer next to the in-junction likewise represents a collection sink for electrons and a recombination sink for holes. The application of a forward voltage U changes the field E in the middle region. The condition E = 0 defines the flat-band voltage $U = U_f$. The appearance of a reversed field in the case $U > U_f$ (shown in Fig. 1) precedes the photocurrent reversal at the transition voltage $U = U_T$. The photocurrent I_P has (by definition) the sign of $U-U_T$. A current $I_P < 0$ is called "primary", and a current $I_P > 0$ is called "secondary". The observation of secondary photocurrents proves the existence of boundary-layer recombination /2/. We term it "surface" recombination and distinguish it from "bulk" recombination in the uniform-field region. The uniform-field model for photocurrent collection neglects the extensions of the boundary regions /3/. The variation of the i-layer thickness helps to distinguish between boundary and bulk effects. We find that our pin solar cells are essentially controlled by "surface"

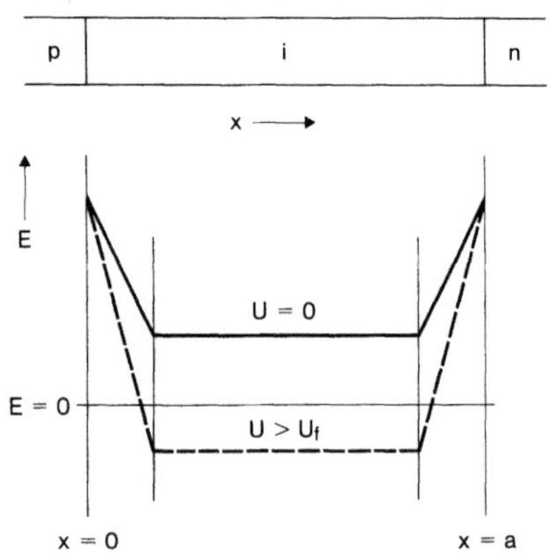

Fig.1: Electric field E, space coordinate x, applied voltage U, E = 0 for $U = U_f$

recombination and an enhancement of it causes the Staebler-Wronski (SW) degradation.

2. INTERNAL COLLECTION EFFICIENCY

The spectral response of a solar cell is expressible by the external collection efficiency Q. It represents the ratio of a photocurrent stimulated by low-intensity illumination and the photon flux onto the cell. Chopped illumination and lock-in technique yield the photocurrents for varying light wavelengths λ and diode voltages U, defining the function Q(λ,U). The SW degradation is caused by an illumination stress (with AM1 light under open circuit) of duration t. The degraded diode state is B(t). Degradation is reversible by an anneal treatment. The annealed state is A = B(0). She short-circuit efficiency Q(λ,0) of a thin diode with i-layer thickness a = 0.25 µm shows almost no SW degradation because the field E is high throughout the i-layer. But the blue response of thicker diodes with a = 0.60 µm and a = 0.68 µm degrades strongly as a consequence of increased "surface" recombination. As an example we present Q(λ,U) for the diode with a = 0.60 µm in state B(95h). Fig.2 shows the curves Q(λ) for different values of U. With increasing U the efficiency Q decreases and changes its sign. The condition Q > 0 (Q < 0) indicates primary (secondary) photocurrents. Under reverse bias U < 0 the functions Q(U) approach certain saturation values. We approximate them by $Q(U_o)$ using our lowest measuring voltage U_o = -2.5V, and define the internal collection characteristics $q(U) = -Q(U)/Q(U_o)$. The efficiencies Q and q differ in sign. The signs of I_P and q coincide. The internal efficiency q is directly derivable from I_P and solely refers to the i-layer, i.e. excludes "optical" losses. Therefore q(λ,U) is well suited to observe the SW degradation. Fig. 3 shows the functions q(U) of states A and B for a restric-

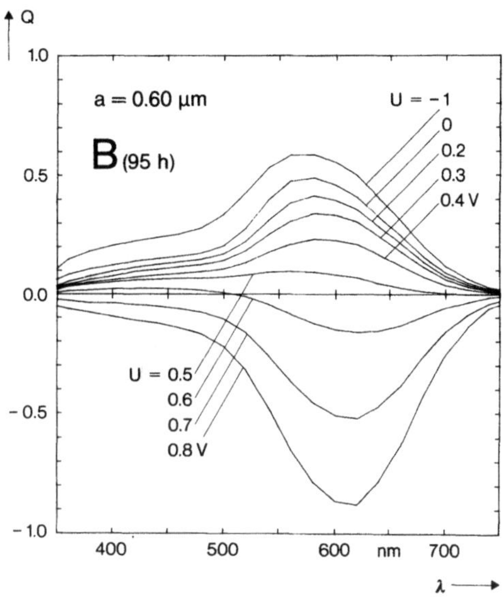

Fig.2: External collection efficiency Q versus wavelength λ for different U. Thickness of i-layer is a. Degraded state B

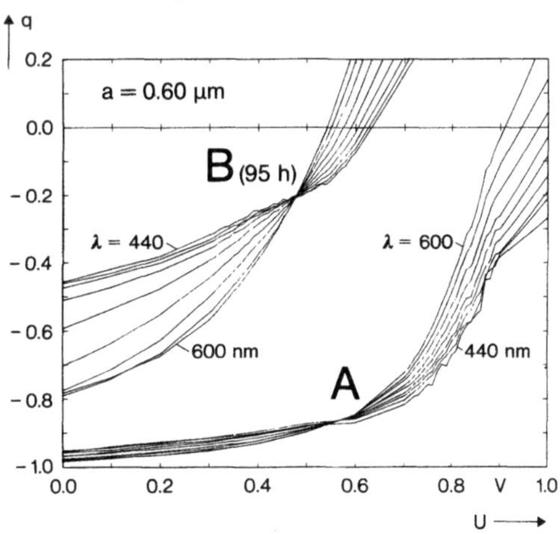

Fig.3: Internal collection efficiency q(U) for different λ. Annealed state A, degraded state B

ted λ set. The characteristics in both cases cross through one point that exhibits the flat-band voltage U_f /4/. The condition $Q = q = 0$ defines the transition voltage $U_T(\lambda)$. Carrier drift through the middle i-layer region supports primary photocurrents for $U < U_f$ but opposes to them for $U > U_f$. Carrier diffusion through the middle region supports the photocurrent within the whole primary regime, and is solely responsible of it at $U = U_f$. The drift reversal at $U = U_f$ is visible as an upward bend of the characteristics, especially in state A. With the transition from A to B the voltage U_f degrades weakly, but the efficiency $q_f = q(U_f)$ strongly. U_f depends on the built-in voltage and the space charges in the boundary regions, while q_f depends on the recombination in the boundary layers /4/.

	a [µm]	U_f [V]	λ = 600 nm U_T [V]	$-q_f$
A	0.60	0.57	0.91	0.87
	0.85	0.52	0.83	0.72
B (95 h)	0.60	0.48	0.54	0.20
	0.85	0.46	0.54	0.11

Table I: Data from q(U) for 2 diodes differing in a

The q(U) curves of the short diode with a = 0.25 µm do not cross. There seems to be no space for a uniform-field region. The thickness of a boundary region may therefore be estimated as 0.1 µm. So the Table I contains only representative data for the thick diodes. While U_f and $U_T(\lambda_o)$ with λ_o = 600 nm depend very little on a, the amount of q_f decreases with increasing a. The latter observation is compatible with negligible "bulk" recombination according to an elementary formula for q_f /4/.

3. WHITE-LIGHT CHARACTERISTICS

The AM1 characteristic is given by the current density J versus the voltage U. A sun simulator delivers the AM1 illumination. Fig. 4 shows a family of white-light characteristics for different illumination intensities Φ of a diode with a = 0.68 µm in state A. The Φ value is normalized to the AM1 intensity. Hence the parameter Φ = 1 indicates the AM1 characteristic. The other curves were obtained by darkening the illumination from the sun simulator with grey-glass filters. The intensity can be estimated through J(Φ) at U = 0. A set of functions q(U) for different λ, Fig. 3, and of J(U) for different Φ, Fig. 4, show a certain similiarity: the curves cross through one point. The functions q(Φ,U) and J(Φ,U) represent basic diode properties. The dark characteristic $J_D(U) = J(0,U)$, missing in Fig. 4, should participate in the crossing there. The white-light photocurrent is $J_P(U) = J(\Phi,U) - J(0,U)$ and $J_P = 0$

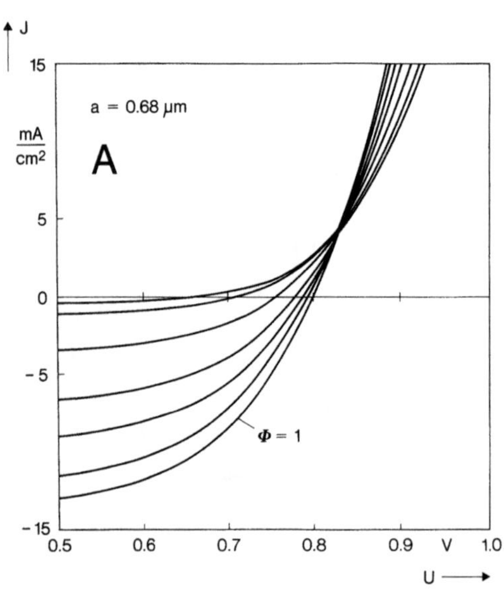

Fig.4: Current density J(U). White-light intensity Φ in fractions of AM1

defines its transition voltage U_T. If $U_T(\Phi)$ = const, then the associated characteristics cross at this voltage. That is just the case of Fig. 4. Thus an increasing photocarrier generation rate does not introduce extra space charges and leaves the "surface" recombination rate invariant. The SW degradation appears through both q and J. Most conspicuous is the degradation of the crosspoints. While degradation sharpens the crosspoint of the q(U) characteristics, it dissolves this point in the case of J(U). The degraded J(U) curves of Fig. 5 show that and now include the dark current. Fig. 6 gives an enlarged picture of the crossing area that replaces the crosspoint. The functions $J(\Phi)$ along 3 selected U values are plotted in Fig. 7. At U = 0.8 V the photocurrent J_P is secondary for $\Phi < 0.5$ and primary for $\Phi > 0.5$. Generally, the maxima of $J(\Phi)$ show a tendency towards primary currents with increasing Φ. The decrease of "surface" recombination is a possible explanation. For a special voltage, U_p = 0.823 V, the function $J(\Phi)$ develops a plateau. It corresponds to the crossing of J(U) curves within a certain Φ interval. Hence U_p may be regarded as an "effective" transition voltage. The decrease of U_p represents a special measure of the SW degradation /2/. The crossings of J(U) with $J_D(U)$, marked in Fig. 6, define the function $U_T(\Phi)$, given explicitly in Fig. 8. Three stress duration times t are considered, including the value t = 15 h valid for Figs. 5 to 7, besides t = 0, designating the annealed state A. The degradation chart of Fig. 8 shows the transition function $U_T(t,\Phi)$ decreasing with t (degradation) and increasing with Φ ("false recovery"). Fig. 9 supplements

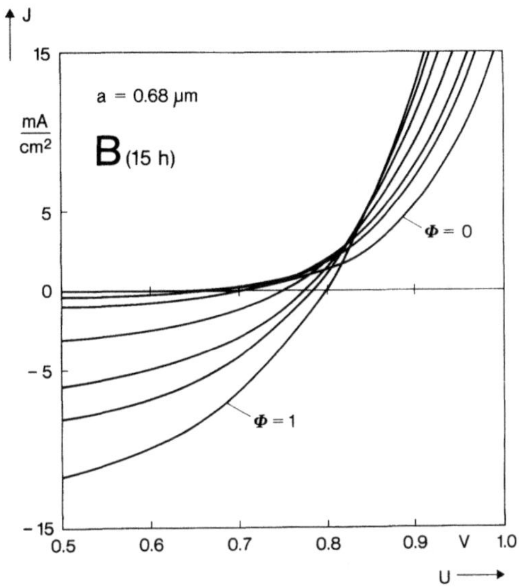

Fig.5: Like Fig.6, but state B. Φ = 0 indicates the dark characteristic

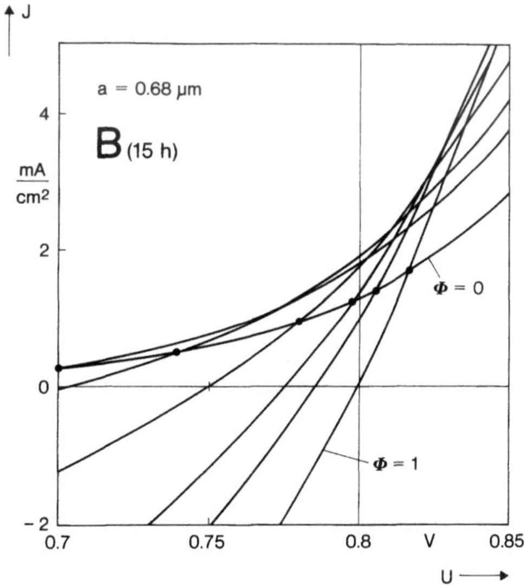

Fig.6: Part of Fig.5. The marked crossings with the dark characteristic define transition voltages U_T

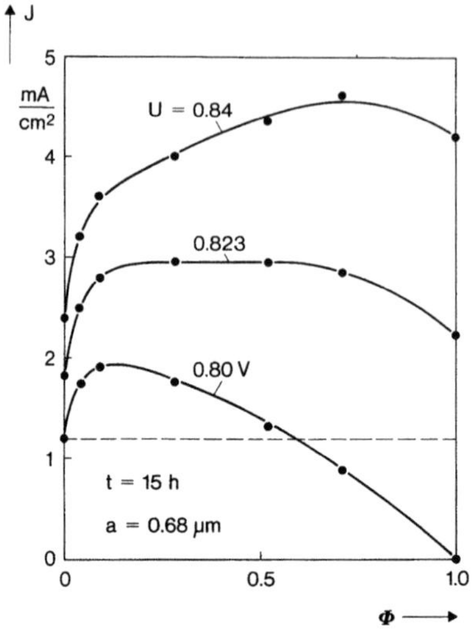

Fig.7: Functions J(Φ) for selected U values

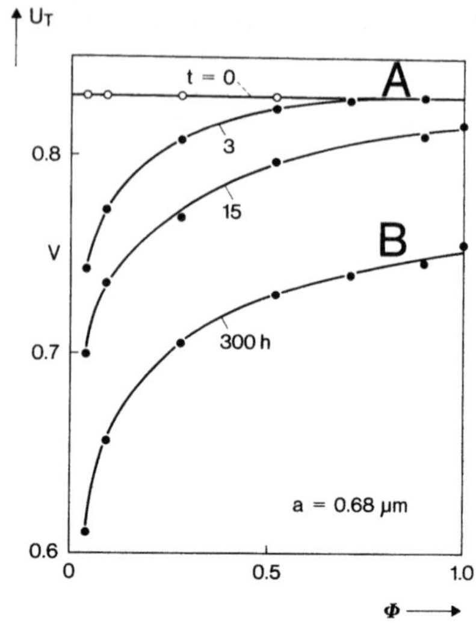

Fig.8: Transition voltage U_T versus light intensity Φ. States B(t) for different stress times t

the dependence on the i-layer thickness a. It appears that U_T is rather independent of a. Recombination and space-charge formation therefore are mainly restricted to the boundary regions of the i-layer. This remark concerns states A and B, and application of different Φ. Space-charge variations would alter the flat-band voltage U_f. The spectral response to illuminations with very low Φ shows only a weak effect of degradation on U_f according to Fig. 3 and Table I. Hence the increase of "surface" recombination should play a major part in connection with the SW degradation also for the higher illumination intensities.

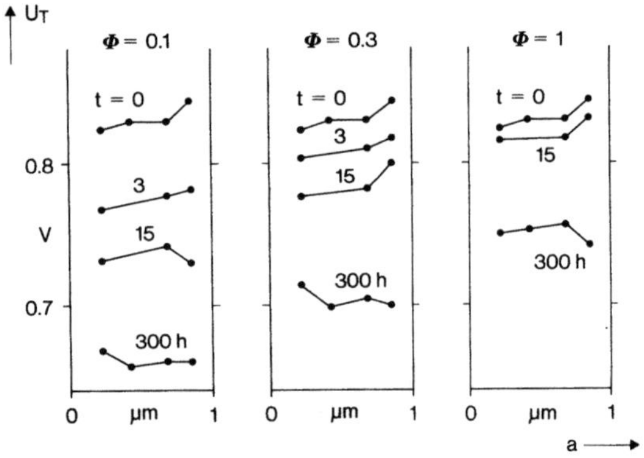

Fig.9: U_T versus i-layer thickness a. Different values of Φ and t. Measured points are marked

ACKNOWLEDGEMENT. One of the authors (S) is obliged to Siemens for a one year's research grant.

REFERENCES

/1/ M. Hack and M. Shur: J. Appl. Phys. 58 (1985) 997
/2/ W. Kusian, H. Pfleiderer and B. Bullemer: Solar Cells 22 (1987) 239
/3/ R.S. Crandall: J. Appl. Phys. 54 (1983) 7176
/4/ W. Kusian, H. Pfleiderer and B. Bullemer: Materials Research Society, 1988 Spring Meeting, Reno NV, USA

SESSION 10:
CONCENTRATORS AND SYSTEMS

PHOTOVOLTAIC CONCENTRATOR TECHNOLOGY: RECENT RESULTS*

Eldon C. Boes
Photovoltaic Technology Division 6221
Sandia National Laboratories
Albuquerque, NM 87185, USA

ABSTRACT. This paper discusses the recent results in photovoltaic concentrator technology development. The major topics presented are a detailed description of current concentrator module development efforts, a summary of concentrator component development, and a discussion of the economics of photovoltaic concentrators in light of these activities and the higher concentrator cell efficiencies that have been demonstrated.

1. INTRODUCTION

Although the applications for photovoltaics (PV) are expanding rapidly, photovoltaic power is still too expensive to compete with other power generation technologies in utility grids. Before photo-voltaics can compete in this market, and thus have an impact on the major portion of the world's electric energy economy, a significantly more cost-effective PV module must be developed. Current module prices are generally in the range of 4 to 8 US$/Wp, and prices in the range of 1 US$/Wp are probably required. Major improvements are required before flat-plate crystalline Si or any thin-film technology can achieve such prices. However, concentrator module technology probably can achieve such price goals because of the recent exciting gains in concentrator cell efficiencies.

This enhanced potential for PV concentrators has led to a renewed interest and emphasis on concentrator R&D. This paper reviews current module development efforts as well as the status of research on other concentrator components. The economic potential of concentrators is also discussed and compared with the economics of flat-plate PV module technologies.

2. CURRENT CONCENTRATOR MODULE DEVELOPMENT

There are currently more concentrator module development efforts underway than there have been for several years. Seven of the better-known efforts are listed in Table 1, and some of the particularly noteworthy aspects of these projects are described below. However, in addition to these seven projects, there are at least six more PV concentrator development projects underway in the U.S. These six projects are not discussed here because either (a) the information is proprietary, or (b) the efforts are not yet very far along.

2.1. Sandia Baseline Module

Sandia is developing this "baseline design" to demonstrate what we believe to be the best concentrator module design features for

*This work was performed at Sandia National Laboratories, supported by the U. S. Department of Energy under Contract #DE-AC04-76DP00789.

commercialization at this time. A primary project goal is achievement of a peak module efficiency of 20%. The module uses an aluminum housing which can be economically fabricated at low production volumes. The 2 x 12 matrix design means that cell temperatures will be quite uniform since every cell is adjacent to the side of the housing.

This will be the first "commercial" design to use a copper heat spreader to which the cell is directly soldered. This heat spreader is bonded to the aluminum housing with a thin electrically insulating layer. We expect the total temperature difference between the cell and the heat sink to be 4°C at an irradiance of 800 W/M^2, which implies that the cell temperature will be about 44°C above ambient temperature with a nominal wind speed of 3 m/s.

Another unique feature of this design is that it will use a concentrator cell with a parallel front metallization grid and a prism cover to eliminate grid shadowing. This approach also permits a heavier front metallization which reduces resistance losses with no optical penalty. Overall gain over a conventional design of the same cell size and concentration ratio will be approximately a 15% relative efficiency improvement. This prism cover concept has been successfully demonstrated by ENTECH for their linear Fresnel concentrators, as described below.

Martin Green of the University of New South Wales (UNSW) is supplying cells for this module. Depending on the cell efficiency achieved, we expect the peak module efficiency (at $800W/m^2$ and 27°C) will be in the range of 18 to 20%.

Although prototype fabrication is not complete and test results will not be available for several months, several firms are already adopting parts of this design for their commercial products.

2.2 Black & Veatch/Solar Kinetics/Spire Module

These three firms are cooperating in the development of the module described in Table 1. Black & Veatch is responsible for overall design and optimization; Spire will fabricate the concentrator cell assemblies; and Solar Kinetics, a company with considerable solar thermal systems experience, will fabricate the prototype modules. The first prototype modules will also use Si cells from UNSW and prism covers from ENTECH. Module testing is scheduled for later this summer.

2.3 Electric Power Research Institute (EPRI) Module

For several years, EPRI has been supporting the research and development of photovoltaic concentrator technology. It was largely through EPRI's program, along with support from Sandia, that Stanford was able to achieve 28% efficiency for Si concentrator cells.[1] At the present time, both EPRI and DOE are supporting continued development of these point-contact cells.

EPRI is also supporting both concentrator module development and array support structure design and development. The module design has been completed, and most of the components for construction of the first prototype modules have been fabricated. Preliminary tests have confirmed that the Fresnel lense parquet performs as planned and that the cell assemblies will provide the expected cell cooling. Initial electrical performance tests are expected in the near future.

2.4 ENTECH PV Concentrator Development

ENTECH is continuing to develop their linear Fresnel PV concentrator technology which was first successfully demonstrated in the 25-kW system installed at the Dallas/Ft. Worth airport. Their most recent installation is a 10-kW array of passively cooled modules at Sandia Photovoltaics Test Facility. This array has a peak efficiency above 13%.

The most important recent advances made by ENTECH in their linear Fresnel collectors are the use of prism covers and cells fabricated using 1-sun technology. They have successfully demonstrated that their prism covers effectively eliminate front grid shadowing, even with metallization fractions as high as 20 to 40%. These large metallization fractions in turn permit ENTECH to use cells fabricated with relatively low cost 1-sun technology without experiencing the large series resistance losses that would normally occur when these cells are operated at 10 to 20X (suns).

ENTECH will install a 300-kW system employing their linear Fresnel technology at a new 3M Company facility in Austin, Texas over the next year. Both 20X, 2-axis tracking arrays, and 10X, single-axis tracking arrays will be demonstrated in this system.

2.5 Intersol PV Concentrator Development

Intersol is also continuing development of their point-focus Fresnel technology. They have achieved array efficiencies above 14%, and they have several modules and arrays installed at different locations in the U.S.

2.6 Alpha Solarco's PV Concentrator Development

Alpha Solarco decided to develop PV concentrator technology a few years ago. Their first prototype array used reflective parabolic trough concentrators mounted on a pedestal, 2-axis tracking structure. They have subsequently reviewed several different PV concentrator design concepts and have begun development of a point-focus Fresnel lense module. Although not much design information is yet available from this privately funded effort, it is known that they expect to build and test prototype modules within a few months. They also have raised funding to install a complete system of several hundred kilowatts, and this is planned for completion in 1989.

2.7 Wattsun's Module Development

Wattsun Corp. is pursuing the development of a design which is quite novel in that it uses very small lenses and cells. As shown in Table 1, their initial design uses 1.2-cm-square lenses and 0.2-cm-diameter Si concentrator cells. The advantages of this approach are that the module is very thin, only 2 or 3 cm, and the lenses can be plano-convex, which eliminates the losses due to facets in Fresnel lenses. In addition, the cell assembly/receiver can probably be manufactured very economically using automated techniques from the electronics industry. Finally, the module should have relatively low material costs because it is so thin.

Wattsun has successfully fabricated several 30- x 60-cm modules with peak efficiencies above 12%. Their next efforts will include design optimization, module engineering to address durability and reliability, and design and development of a pilot production plant.

3. CONCENTRATOR CELL RESEARCH STATUS

Because overall conversion efficiency plays such an important role in determining the cost of photovoltaic power generation, the photovoltaic concentrator programs in the U.S. have always placed heavy emphasis on research and development of high efficiency concentrator cells. In the past few years, this emphasis has paid off handsomely in efficiency gains for concentrator cells. Table 2 gives a status summary for concentrator cells.

3.1 Silicon Concentrator Cells

From 1980, when the first efficiency of 20% was reported, until 1985, the best Si concentrator cell efficiencies were between 20 and 21%. Since then, dramatic Si concentrator cell efficiency gains have been made, the most notable being:

(1) An efficiency of 28% at 140X in a high-resistivity, point-contact design at Stanford University[1], and

(2) An efficiency of 25% at 110X by the University of New South Wales in low-resistivity silicon[2].

These increases resulted from a combination of improved cell designs and more advanced processing techniques and control, which permitted, for example, reduced surface recombination through better passivation layers and better light trapping.

Modelling codes such as PC1-D indicate that even higher efficiencies are possible[3], and research to achieve efficiencies above 30% is underway.

The most significant implication of these Si concentrator cell efficiencies is that they make it possible to achieve concentrator module efficiencies above 20%. As described in the section of PV Concentrator Economics, module efficiencies above 20% should lead to concentrator module prices of 1 US$/Wp and levelized energy prices of 0.12 US$/kWh. Thus, these Si concentrator cell efficiency gains represent a key step on the road to cost-effective PV power generation.

Of course, these record efficiencies were first accomplished on small experimental cells. Since then, considerable effort has been directed toward the development of batches of "module ready" cells, cells of the proper size and design to be able to be incorporated into concentrator cell assemblies for mounting in prototype modules. This transition from a few small experimental cells to batches of module-ready cells is one of the most important activities in PV concentrator technology development today.

3.2 Compound Semiconductor and Multijunction Concentrator Cells

Research is also being conducted on III-V materials and on multijunction devices for concentrator cells because these have the potential for even higher efficiencies. This area has also experienced considerable progress, with several research groups having achieved efficiencies above 25% for GaAs concentrator cells. The most important recent result is the achievement of 28% efficiency at 500X for a GaAs cell by Varian in early 1988[4].

The best efficiency reported to date for a multijunction device is 26.6% at 300X. This was achieved at Sandia using a mechanically stacked multijunction cell consisting of a GaAs cell from Varian mounted above a Si cell that was fabricated by Applied Solar Energy. The highest efficiency measured at SERI for a monolithic multijunction cell is 24% at 1-sun using an AlGaAs-on-GaAs structure in a 3-terminal configuration.

The fact that multijunction cell efficiencies are lower than the best efficiencies for single junction concentrator cells does not represent cause for pessimism regarding the ultimate potential for multijunction technology. Rather, it reflects the very different levels of maturity for these technologies. Achievement of efficiencies in the neighborhood of 35% still looks quite likely, but it will take continued research for 5 or 10 more years.

4. FRESNEL LENS RESEARCH

Nearly all of the PV concentrator module designs under development use acrylic Fresnel lenses as the solar concentrators. While acrylic lenses appear to have the efficiency and durability necessary, the most cost-effective fabrication method is not yet known. This is particularly the case for point-focus lenses. The best point-focus lenses to date have been compression-molded, but this process, in its present form, is too expensive because molding cycle times are over 30 minutes.

We are investigating two alternatives to compression-molding. The first is adaptation of 3M Company's Lensfilm capabilities to produce point-focus lenses. This company has been quite successful in fabricating linear Fresnel lenses for the ENTECH modules described earlier, and their initial attempts to make point-focus lenses were very encouraging. The other alternative is injection-molding, which is noted for being a widely used and economical process for manufacturing plastics. Initial attempts at injection molding of point-focus lenses were disappointing. The facet tips were too rounded, which lowered efficiencies to the range of 70%, well below the compression-molded lens efficiencies of about 85%. However, by using a curved-facet lens design, which reduces the number of facets by a factor of 8, and by pulling a vacuum on the mold, American Optical has achieved injection-molded lens efficiencies of 82% for single lenses. They are currently scaling up to 2 x 2 lens parquets.

It also may be possible to modify the compression-molding process to retain the good efficiency while drastically reducing the molding cycle time. Several promising ideas have been identified, but they are not yet being pursued because of a lack of adequate research funds.

5. ARRAY TRACKING CONTROL

Sandia Laboratories has designed and is developing a new microprocessor-based array tracking controller. The design incorporates initial alignment and alignment correction procedures in software. The alignment algorithms maximize array current. After alignment, the controller uses Hall effect sensors to count tracking motor turns to drive to the desired position. Similar sensors are also used to provide reference positions. Manual tracking capabilities and stow signal interfaces are included in the design. In moderate production quantities, this controller can probably be manufactured at a cost of about 250 US$, which translates into a cost of only 5 US$/m² for arrays of 50 m² area. The first controller boards have been fabricated and are currently being tested at Sandia. Additional details are available in Ref. 5.

6. THE ECONOMICS OF PHOTOVOLTAIC CONCENTRATORS

One of the most interesting issues in terrestrial photovoltaics research and development programs for the past 13 years has been the competition between the different PV collector technologies. It has always been recognized that, in order to compete with other power generation technologies, photovoltaic power system costs would have to be reduced substantially, and in particular, PV module prices would have to be reduced to at least 1 US$/peak Watt. We have seen module prices drop from about 100 US$/Wp to 5 US$/Wp. We have also seen price analyses of balance-of-systems (BOS) designs that indicate that the BOS prices can be reduced to acceptable levels for utility power generation. Thus, the remaining hurdle today for large-scale PV power generation is the module price issue that was clearly recognized at the onset of the large terrestrial PV R&D programs in 1975.

It is particularly interesting today that, after 13 years of intensive research, not one of the three general categories of PV module technologies--flat-plate crystalline Si, concentrators, and flat-plate thin films--is a clear cut winner of this competition. Even more surprising is the fact that none of these module technologies has been completely ruled out. (Of course, some subsets of these module technologies have essentially dropped out. Very few people would suggest that conventional CZ-Si, flat-plate modules or reflective parabolic trough concentrators have any hope of economic bulk power generation.)

As a framework for discussing PV module technology economics, we have plotted levelized energy prices as a function of module prices and efficiencies for fixed flat-plate modules in Fig. 1 and for concentrator modules in Fig. 2. The assumptions used in developing these plots are:

- Solar resources typical of a sunny location
- BOS costs as projected for the mid 1990' assuming that large PV power plants are being installed regularly
- Operation, maintenance, and economic factors are those of the U.S. DOE Five-Year Plan[6]
- Indirect cost factor is 50%.

It is likely that the penetration of utility markets by photovoltaic power technology will be a gradual process. We believe that the process will begin when levelized energy costs from PV power plants reach the range of 0.10 to 0.12 US$/kWh. Once utilities do begin to install PV power plants, these sales will play a major role in helping reduce PV power system costs further, which in turn will lead to additional sales. With this scenario in mind, we now discuss the three PV module technologies and what is required of them to achieve levelized energy costs of 0.12 US$/kWh.

6.1 Flat-Plate X-Si Modules

As Fig. 1 indicates, this target energy cost can be achieved with modules having peak efficiencies of 15% at costs of 165 US$/m^2. By going to peak efficiencies of 20%, the module cost can be increased to 245 US$/m^2. At present, commercial X-Si flat-plate module efficiencies are generally below 13%, and prices are generally above 400 US$/m^2. The X-Si cell efficiency gains described earlier imply that significantly higher module efficiencies are possible. However, these must be traded off against increased module costs that the higher efficiency cells may require. Module cost reductions for this technology depend almost entirely on X-Si cell cost reductions, since the cost of the solar cell represents the majority of the cost of the module, and the costs for the other module components cannot be lowered significantly. Thus, the major challenge that this technology must address in order to meet the target energy cost goal of 0.12 US$/kWh is significantly lowering the cost of X-Si cells.

6.2 Thin-Film Module Technology

About 5 years ago, both amorphous Si and polycrystalline thin-film (such as CdTe and CuInSe) technologies were expected to quickly surpass X-Si flat-plate technology by achieving good efficiencies at very low costs. However, progress in these technologies at the module level has been very slow. There are currently no polycrystalline thin-film modules on the market, and stable peak efficiencies for a-Si modules are generally below 5%. A glance at Fig. 1 tells us that substantial efficiency progress, to at least 10%, is essential for these technologies to ever supply utility power. Thus, these technologies appear to be facing the greatest challenge in reaching utility power cost goals.

6.3 Concentrator Module Technology

Figure 2 indicates that the target energy price of 0.12 US$/kWh can be achieved at 20% efficiency with a module cost of 195 US$/m^2. The Si concentrator cell efficiencies described earlier imply that concentrator module efficiencies of 20% (peak) should be achievable in the near future. Moreover, the achievement of low concentrator module costs does not depend on significantly lower solar cell costs, since concentrators require very little solar cell area. I believe that the target price of 190 US$/m^2 is achievable in moderate production quantities, with a cost breakdown such as this:

Component	US$/m^2
Concentrator cell at 300X	40
Balance of cell assembly	20
Fresnel lens	35
Housing	30
Miscellaneous	10
Labor and markup	60
Total	195

Thus, it appears that PV concentrators have a very clear path to achieving the energy price levels where utility purchases will become significant.

7. CONCLUSIONS

In the past few years, the advances in crystalline Si cell efficiencies have greatly increased the potential for both X-Si flat-plate and concentrator module technologies to achieve a target energy price of 0.12 US$/kWh. In contrast with this, progress on the efficiency and stability fronts for this film technologies has been slower than anticipated. Thus, the outcome of the competition between these three module technologies for large-scale utility markets is still unpredictable.

8. REFERENCES

[1]R. Sinton, et al, "27.5% Silicon Concentrating Cell," IEEE Electron Device Letters, Volume EDL-7, Number 10, October, 1985 p 583.

[2]M. Green, et al, "Improvements in Flat Plate and Concentrator Silicon Cell Efficiency," Proceedings of the 19th IEEE-PVSC, New Orleans, LA, May 1987.

[3]D. Rover, P. Basore, G. Thorson, "Solar Cell Modeling on Personal Computers," Proceedings of the 18th IEEE-PVSC, Las Vegas, NV, October 1985.

[4]Announced at 9th Space Photovoltaic Research and Technology Conference, Cleveland, OH, April 1988.

[5]A. Maish, "Development of an Array Tracking Controller Using a Single Chip Computer," presented at the International Workshop on Control Systems in New Energy Applications, September, 1987.

[6]Five-Year Research Plan, National Photovoltaic Program, U.S. Dept. of Energy, May 1987.

Designer	Module Description	Status
Sandia	"Sandia Baseline Module" SBM 3. Point-focus Fresnel with 2 x 12 parquet of 17-cm-square acrylic lenses, 1.25-cm-square Si cells 185X concentration, reflective secondaries, prismatic cell cover.	All components have been developed and tested. First module tests are scheduled for June 1988.
Black & Veatch	Point-focus Fresnel with 4 x 6 parquet of of 17-cm-square acrylic lenses, 1.25-cm-square Si cells, reflective secondaries.	All components are developed. Full module testing is scheduled for August 1988.
Electric Power Research Institute, Cummings Engineering	This module is designed to use the Stanford point-contact Si cell at 500X. The module uses two 4 x 6 parquets of 17.8-cm-square lenses. Module housing is sheet steel.	Several modules have been fabricated Performance testing is awaiting the delivery of mounted cells with good performance.
ENTECH	Linear Fresnel lenses are 0.9 x 3.0 m, 10X to 20X range of concentration ratios, passively (or actively) cooled, prismatic cell covers on low cost cells.	A 10-kW array of 20X modules is under test at Sandia. Peak efficiency is above 13%.
Intersol	Point-focus Fresnel with 2 x 7 parquet of 21-cm-square lenses.	Several arrays have been installed at various locations.
Alpha Solarco	Point-focus Fresnel with 4 x 4 parquet of 3M lenses at 425X. Si cells from ASEC	The first modules will be fabricated and tested this summer.
Wattsun	This "miniature" design uses a parquet of 1.2-cm-square lenses and Si cells of 2.5-mm diameter.	Several 30 x 60-cm modules have been tested with efficiencies above 12%.

Table 1. Design Descriptions for a Sample of PV Concentrator Modules Currently Being Developed.

CELL TYPE	EFFICIENCY (%) CURRENT		PROJECTED	
High Resistivity Si	100X	28		
	500X	25	300X	32
Low Resistivity Si	100X	25		
	800X	21	300X	30
"Mounted" Si	100X	24	300X	28
Low-Concentration Si	20X	19*	40X	24
GaAs	500X	28	1000X	32
Mech Stack 2 Junction	300X	27	1000X	30-35
Monolithic MJ	1X	24*	1000X	33-37

*Non-Sandia Measurement

Table 2. Status and Projections for Concentrator Cell Efficiency

FIGURE 1. FIXED FLAT PLATE ECONOMICS

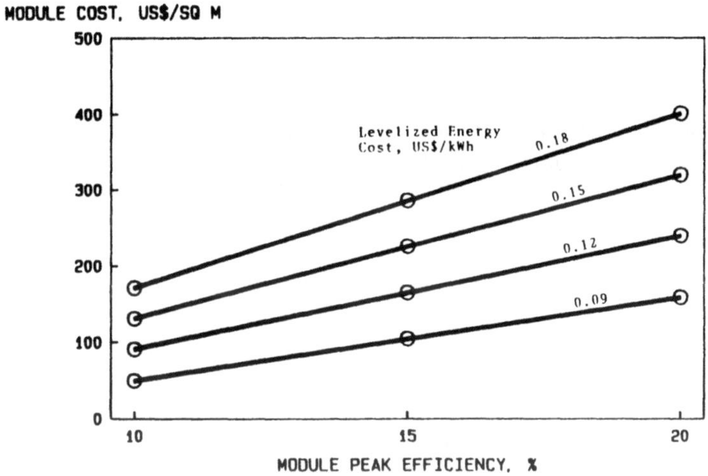

FIGURE 2. PV CONCENTRATOR ECONOMICS

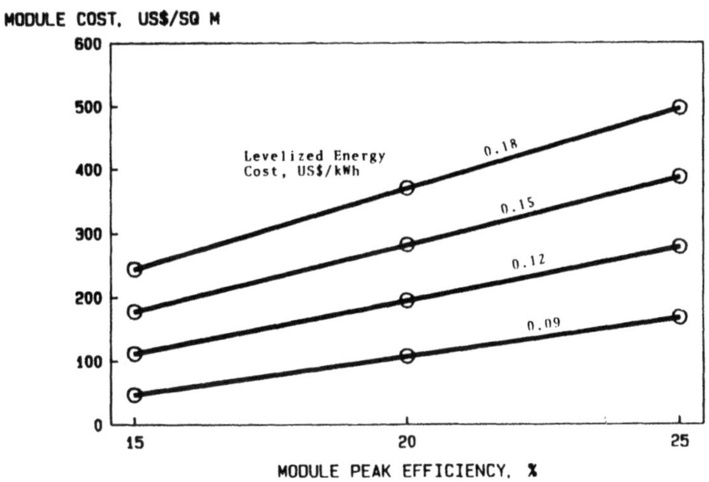

TRACKING VERSUS FIXED FLAT-PLATE ARRAYS: EXPERIMENTAL RESULTS OF ONE YEAR'S OPERATION AT THE ADRANO PILOT PLANT

S. Guastella
CONPHOEBUS S.c.r.l.
Via G. Leopardi, 60
I 95127 Catania (Italy)

A. Iliceto, V. Piazza
ENEL-CREL
Via A. Volta, 1
I 20093 Cologno MI (Italy)

ABSTRACT. To investigate into the cost/benefit analysis of sun-tracking PV systems as compared with fixed arrays, an experiment is under way at Adrano PV Plant to look into the performance of actually operating systems. This paper reports both theoretical evaluations on energy gain achievable with tracking PV plants and actual data collected over a year of continuous operation. Such data allow an analysis of financial viability, herein presented for a particular PV installation.

1. INTRODUCTION

Within the framework of the Adrano PV Project (1), which is being conducted by ENEL and Conphoebus and is supported by the Commission of the European Communities, a research programme is under way to look into the respective merits, under real operating conditions, of tracking PV systems as opposed to fixed flat plate arrays.

This research has so far made it possible to collect data on three systems, about 2.5 kWp rated, operating in parallel over a period of more than one year, which have enabled a thorough analysis to be made of the effectiveness, and led to significant evaluations on the operation of tracking systems.

This paper, after a theoretical study on the energy gain obtainable with various tracking systems, describes the plants in operation at Adrano and presents the energy data resulting from one year of actual operation of the three systems being compared.

A possible analysis of their financial viability is also presented.

2. TRACKING VERSUS FIXED PV ARRAYS: COMPUTER MODEL EVALUATIONS

Fig. 1 hereunder shows the five systems considered here, schematic diagrams of which are presented.

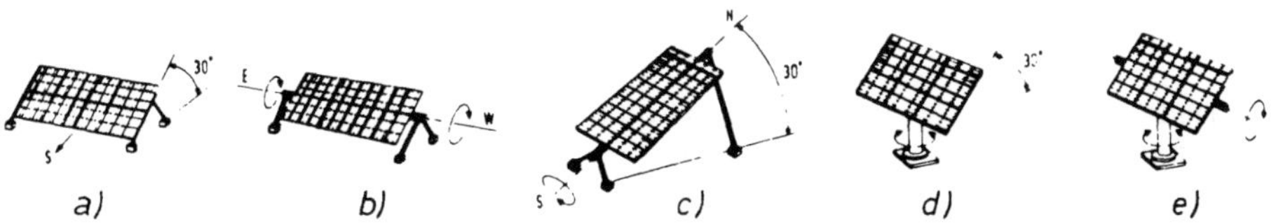

System	Azimuthal Orientation	Zenithal Orientation	Rotation Axis
a	South	30°	none
b	South	tracking	E-W horizontal
c	tracking	30°	N-S 30° tilted
d	tracking	30°	vertical
e	tracking	tracking	2 axes

Fig. 1 - The five systems considered for the computer model evaluations

The calculations that follow refer to the solar energy data recorded in Sicily over a period of 20 years from which the "average" year was selected.

The use of computer codes (2), operating on an hourly basis, made it possible to process the data available (daily global horizontal irradiation and number of daily sunny hours) and to compute the integrated values of the solar energy incident on the different fixed and tracked planes, in respect of each of the five systems considered.

These calculations produced the figures listed in Table I hereunder (one year's values), where Δ % represents the gain in annual energy collection of the tracking systems as compared with the fixed system "a".

The same calculations based on different solar data led to considerably different absolute figures, but Δ % values remained substantially unchanged.

It must be noticed that these figures, while relating to a real meteorological year, represent "solar geometry" data, and do not, therefore, take into account the technological characteristics of the devices required for the aforementioned tracking systems.

TABLE I

Irradiation available to the various tracked and fixed surfaces (yearly computed values)

System	(kWh/sqm/y)	Δ %
a	2050	0.0
b	2269	+10.7
c	2955	+44.1
d	2665	+30.0
e	3079	+50.2

3. TRACKING VERSUS FIXED PV ARRAYS: ONE YEAR'S OPERATING RESULTS AT THE ADRANO PLANT

As illustrated in the sketch in Fig. 2, the three physical systems compared (see characteristics in Table II), which are in operation at the Adrano Plant, have about the same power and are connected to the grid through three identical MPPT inverters (3), which provide perfectly symmetrical operation for the various PV generators.

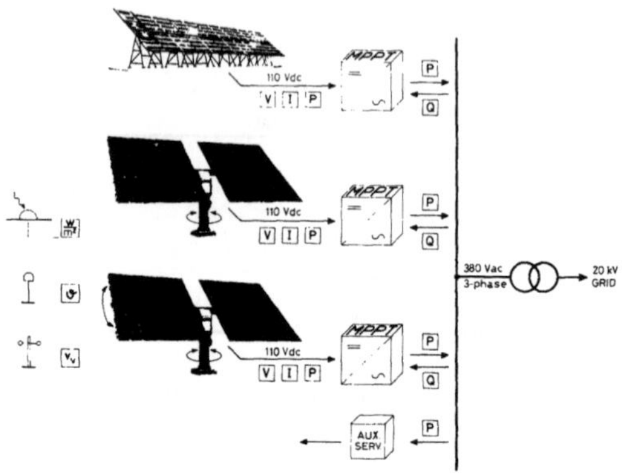

Fig. 2 - Sketch of the three systems at the Adrano PV Plant

TABLE II - Main characteristics of the three systems installed at the Adrano plant

sys.	operat.	type of cells	type of modules	n° of modules	tot.mod. area	rated voltage	rated current	module manuf.	system manuf.
a	az.South el. 30°	10x10 cm semicr.	36 cells 2 glass	88	39.5 m2	129 V	22.8 A	Ansaldo	Ansaldo
d	az.track. el. 30°	10x10 cm semicr.	20 cells 2 glass	144	37.2 m2	132 V	18.6 A	AEG	MAN
e	2 ax. track.	10x10 cm semicr.	20 cells 2 glass	144	37.2 m2	132 V	18.6 A	AEG	MAN

The two automatic tracking systems (4) have the following main functions:

- Tracking start-up with irradiation threshold equal to 200 W/sqm
- Tracking of sun with precision of +/- 5°
- Horizontal positioning under cloudy conditions for maximum collection of diffuse irradiation
- Horizontal positioning for protection against strong wind.

A high-precision data acquisition system (5) makes it possible to record all the electrical, thermal, and meteorological quantities that characterize operation of the plant (37 channels recorded on cassette;

15 mins. integration period).

After a year of operation, the data effectively available cover about 90% of total time.

Fig. 3 shows the weekly integrated energy values (52 data) for each of the three systems compared (expressed in p.u., i.e. related to 1 kWp power for each system, for comparison purposes).

Fig. 4 shows the percentage of energy gain in the tracking systems as compared with the fixed array, according to the data reported in Fig.3.

The complete figures for a whole year, are given in Table III.

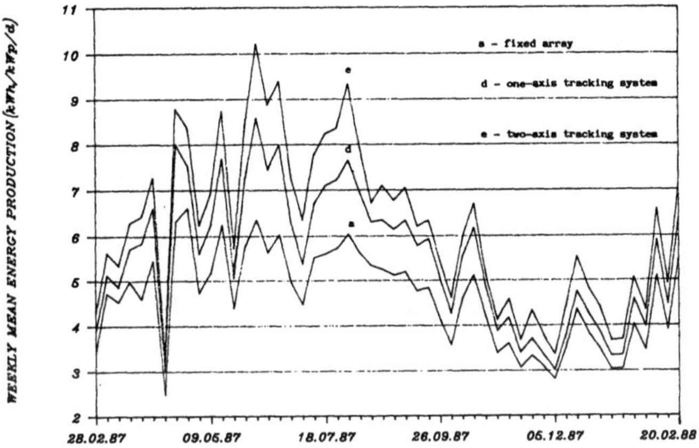

Fig. 3
Weekly mean energy production of the three systems compared

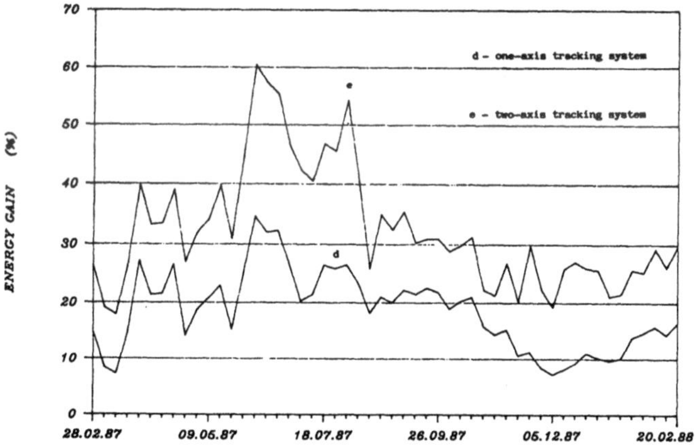

Fig. 4
Weekly energy gain in tracking systems as compared with fixed array

TABLE III - One year's energy production of the three systems in operation at Adrano Plant (from 28.2.87 to 26.2.88)

System	Energy production p.u. (kWh/kWp/y)	Difference (%)
Fixed (a)	1683	0.0
1-axis tracking (d)	2010	+19.4
2-axis tracking (e)	2257	+34.1

As will be seen, the real operating results in respect of physical systems (Table III) are considerably lower than the theoretical energy gain values for mobile surfaces (Table I), since they were determined by the operating characteristics of the plants, such as the tracking limit thresholds, the protections, the variations in performance with temperature, the dust and pollution effects, etc.

However, the theoretical values remain valid for purposes of comparing the different systems, and may be regarded as limits for ready reckoning purposes, as well as for possible technological objectives.

4. COST EVALUATION

For obvious practical reasons relating to installation and maintenance costs, it would hardly be worthwhile setting up tracking systems in photovoltaic plants for isolated users or small communities, where one of the main advantages of photovoltaic over other possible energy sources is that the plant is completely static.

In the case of large power plants, however, it might be worth assessing the financial viability of tracking rather than fixed systems.

The cost comparison, which means calculating the cost of a kWh produced with the various systems, is difficult, inasmuch as it depends on many different factors, such as latitude, meteorology (solar energy availability and direct to diffuse radiation ratio), cost and type of land, mass production of systems, cost of PV modules, possible use of reflectors and/or concentrators, O&M cost, reliability, discount rate, lifetime, etc.

This therefore means assuming a whole lot of factors that limit the validity of the conclusions arrived at in each individual case.

The following evaluations consider PV plants characterized by:

- number of generation units installed: 1000
- peak power per unit: 5 kW
- type of structures: one colum flat plate panels
- conversion efficiency of modules: 10%
- site: flat uncultivated land
- location: Sicily
- daily mean global irradiation on horizontal surface: 4.5 kWh/sqm/d
- fixed systems: 30° tilted
- tracking systems: two axis tracking
- cost of land and clearance: negligible
- lifetime: 20 years
- discount rate: 5%

Assuming all the foregoing, and given the same installed peak power, two realistic assumptions of tracking additional costs have been considered (a and b), combined with two assumptions of energy gain in tracking vs fixed systems (a1, a2 and b1, b2).

Table IV shows the results of the cost analysis performed.

TABLE IV - Tracking vs fixed PV systems: results of the cost analysis performed

		FIXED ARRAY	TRACKING SYSTEMS			
			assumption a: low additional tracking costs		assumption b: high additional tracking costs	
modules	(ECU/kWp)	4000	4000		4000	
structures & mounting	(ECU/kWp)	1000	1000		1500	
track. electr. and sensors	(ECU/kWp)	-	200		500	
mechan.& oil-hydr. devices	(ECU/kWp)	-	650		1500	
total annual capital cost	(ECU/kWp)	400	470		600	
maintenance	(ECU/kWp/y)	-	50		100	
total annual cost	(ECU/kWp/y)	400	520		700	
			assumpt. a1: high gain	assumpt. a2: low gain	assumpt. b1: high gain	assumpt. b2: low gain
tracking energy gain	(%)		+ 50	+ 30	+ 50	+ 30
cost of energy (without power conditioning)	(ECU/kWh)	0.22	0.19	0.22	0.26	0.29

Acknowledgements

The authors wish to thank MAN of Munich for their co-operation in the experimental work still in progress and for the data they kindly made available.

References

(1) Iliceto A. - The Adrano Project - Int. J. Solar Energy, 1985 Vol. 3 PP 174-190

(2) Bullo, Gasbarra, Emanuele - A detailed package of digital codes specially developed for the array design - Proc. 4th EC Photovoltaic Solar Energy Conference, Stresa, Italy, 1982

(3) Arcidiacono V., Corsi S., Lambri L. - Maximum power point tracker for photovoltaic power plants - Proc. 15th IEEE Photovoltaic Specialists Conference - S.Diego, 1982

(4) Pfeiffer - Solar guided photovoltaic generators - Proc. 6th EC Photovoltaic Solar Energy Conference, London, U.K., 1985

(5) Ghirighelli G., Russo F. - Acquisizione dati dell'ENEL - Automazione Oggi - Anno I - n° 10, November 83.

PV POWER - STUDY OF SYSTEM OPTIONS AND OPTIMIZATION

A.A. Salim, F.S. Huraib, N.N. Eugenio
King Abdulaziz City for Science and Technology
PO Box 6086, Riyadh 11442, Saudi Arabia

ABSTRACT. A multipurpose photovoltaic (PV) test system was installed in April 1987 at the Solar Village near Riyadh, Saudi Arabia to study the effects of tracking and long term accumulation of dust and dirt on the PV array energy output. The system consists of four separate branches each with the same number and type of PV modules in series and its own peak power tracking unit. The energy generated by each branch is dumped into a common load. Performance results indicate an average monthly energy gain of about 18% due to single axis and 20% due to dual axis tracking as compared to a fixed array tilted at the site latitude. Seasonal tilt angle measurement of the passive thermohydraulic two axis tracker at solar noon and its comparison with the theoretical angle revealed discrepancy in the tracker performance. Energy loss due to dust on a fixed tilted array increased to 32% in eight months and was reduced to about 2% following occasional rain. System operation todate has produced valuable information on a number of parameters including the hardware performance.

1. INTRODUCTION

A photovoltaic (PV) test system (Figures 1,2) consisting of four identical monocrystalline PV arrays - one sun tracking in two axes, one in single axis, one fixed tilted and one fixed tilted that is kept uncleaned, has been installed. Array tracking in the system is accomplished by means of solar-energy-powered totally passive thermohydraulic devices. Each array contains twelve identical PV modules in series and feeds through a maximum power point tracker to a common resistive load (Figure 3). A microprocessor controlled control and data acquisition system monitors the PV system performance as it operates continuously unattended. The data logging system continuously integrates the solar power input and array electrical power output and records them in the form of energies on a cassette tape every 15 minutes. In addition to this, input, output voltages, currents and powers are also recorded. The system is unique in design that it has several parallel arrays of identical PV modules with different orientations, which operate independent of each other, however dumping their energies into a common load. This multipurpose system has been installed to evaluate 1) the effect on energy output of PV modules (array) orientation, 2) seasonal variation in the energy gain due to tracking, 3) long term effect of accumulation of dust on the fixed tilted modules, 4) long term performance of the PV modules outdoor, 5) PV module/array power/energy efficiencies and effect of temperature on efficiencies, 6) reliability, efficiency and accuracy of the maximum power trackers, 7) reliability and accuracy of the passive thermohydraulic trackers, and 8) measurement of incident solar energy. It is also the purpose of the system to develop and validate mathematical models based on the test data obtained from the system.

2. SUMMARY OF TEST RESULTS

Results obtained from the analysis of the performance data collected for a period of one year are shown in Figures 4 through 11. Figures 4 and 5 show the total electrical energy generated during the month by each of the four branches. The gain in the energy output due to single-axis tracking as compared to the fixed tilted array (Figure 6) is about 18% per month. However, the energy gain realized due to dual-axis tracking as compared to the single-axis tracking is inaccurate except during the summer months (Figure 7). This inaccuracy was revealed by the tilt angle measurements performed on the 21st of each month at solar noon, on the dual-axis tracker (Figure 8). The array with dual-axis tracking generated on the average about 2% more energy per month as compared to the array with single-axis tracking. It would have been as high as 8% per month if the tracker had performed accurately.

Figures 9 and 10 show the effects of dust accumulation on a fixed tilted array. The monthly energy reduction in Figure 9 for the uncleaned array was obtained by comparing its performance with an identical array which was cleaned at the end of each month. The reduction in the energy output from the uncleaned array reached 32% at the end of eight months. However, the difference in the energy output improved to less than 2% in the next four months when the uncleaned array was washed by occasional rain. Figure 10 shows the adverse effect of dust accumulation on the array efficiency. The monthly solar energy input measured by the system is shown in Figure 11.

3. CONCLUSION

Major components of the PV test system which include PV modules from Dakhil-Spire, Jeddah; passive tracking systems/structures from Robbins Engineering, Ridgecrest, California; and maximum power point trackers and microprocessor controlled control/data logging system from Chronar-TriSolar, Princeton, New Jersey have performed well. The only problem encountered has been with the passive thermohydraulic tracking system on the dual-axis tracker which failed to track the sun accurately during winter months.

Test results as summarized above indicate that for increased PV energy output, single-axis tracking at the test site is definitely advantageous and sufficient. It can be optimized further by seasonal adjustment of array tilt angle due south or it can be operated at some other tilt angle to obtain a proper match between the array energy output and the load energy demand. Long term dust accumulation test on the fixed tilted array showed about 4% loss in the energy output per month. However, this loss was recovered in full following the rainy season. Based on this information and system maintenance frequency, a PV system can be sized adequately for this and similar sites.

System operation, data collection and analysis are continuing to further strengthen the above conclusions and obtain additional information. Tests are also in progress to determine the effect of array tilt angle on the accumulation of dust. Results from these tests and analyses will be reported in later publications.

Figure 1. 3-kW PV Test System

Figure 2. PV System Controls and Load

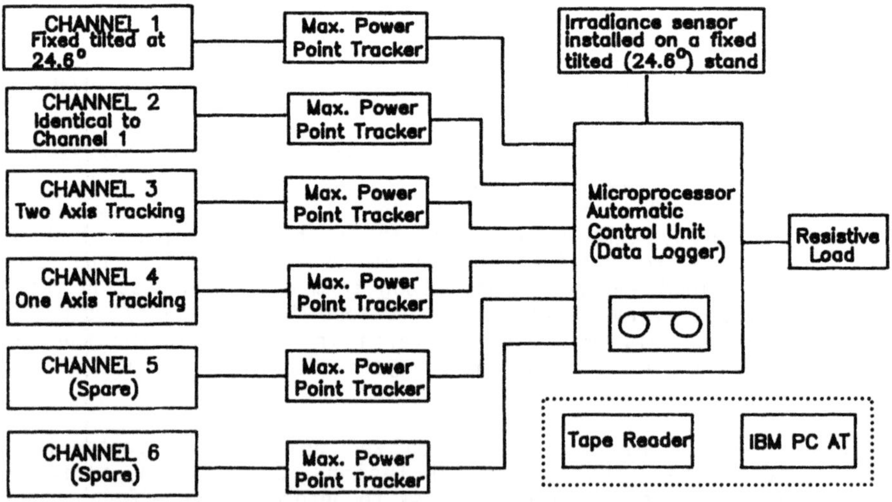

Notes: All channels contain twelve identical 40 – watt modules in series. Modules in channels 1, 3 and 4 are washed monthly. Modules in channel 2 are not washed.

Figure 3. Functional Block Diagram - PV Test System

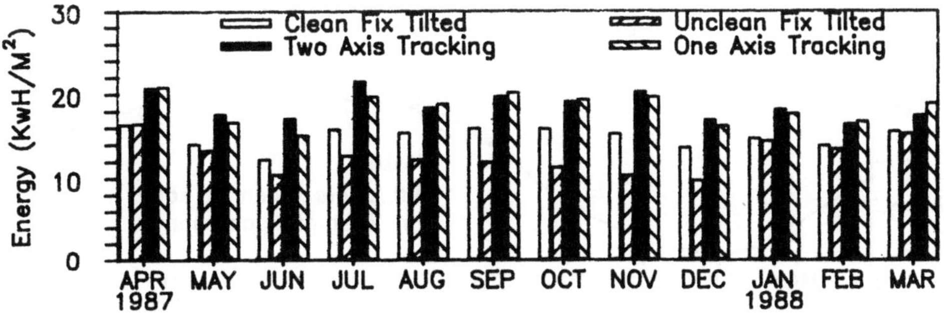

Figure 4. Energy Generated by Identical PV Arrays

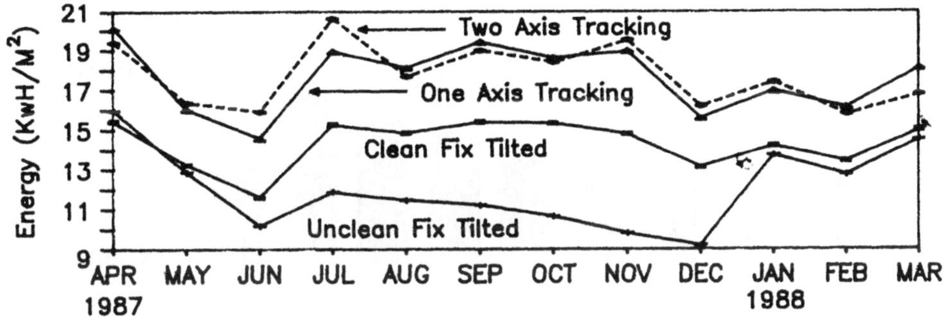

Figure 5. Output Energy Profile of Identical PV Arrays

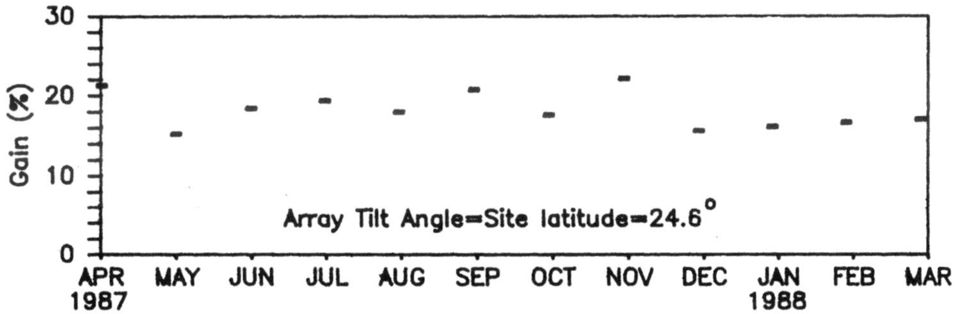

Figure 6. Energy Gain due to Single Axis Tracking as Compared to a Fixed Tilted Array

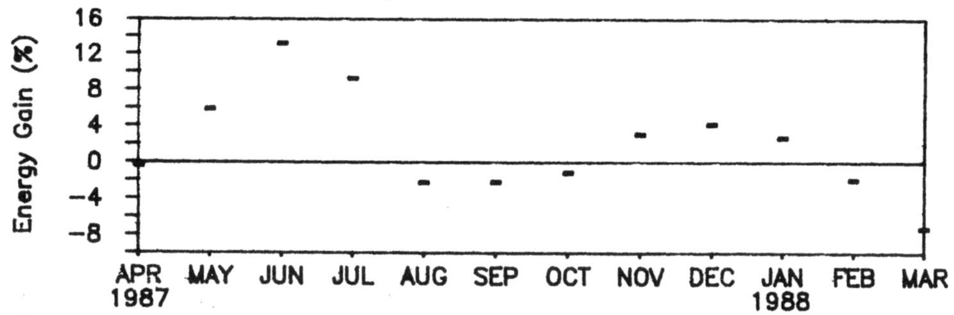

Figure 7. Energy Gain due to Dual Axis Tracking as Compared to a Single Axis Tracking Array

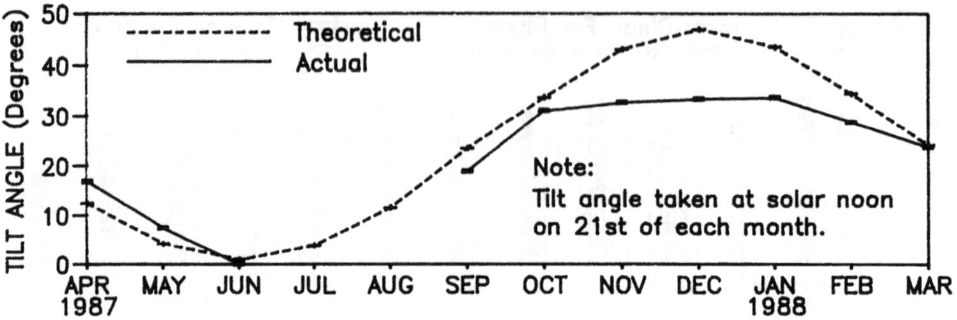

Figure 8. Performance of Thermohydraulic Tracker Installed for Elevation Tracking on a Dual Axis Tracker

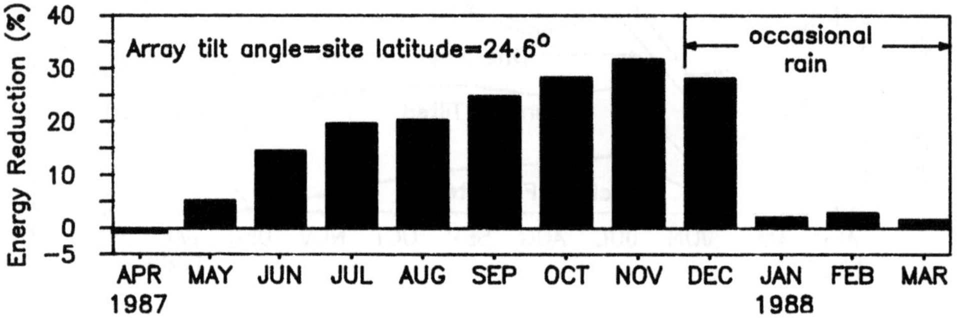

Figure 9. Effect on Energy Output of Long Term Accumulation of Dust and Dirt on a Fixed Tilted Array

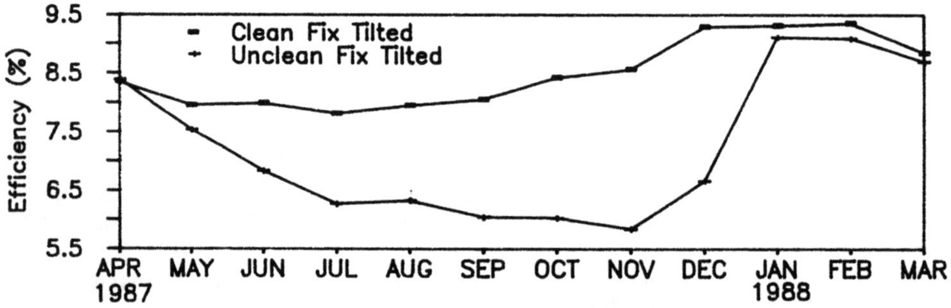

Figure 10. Effect of Dust on Array Energy Efficiency

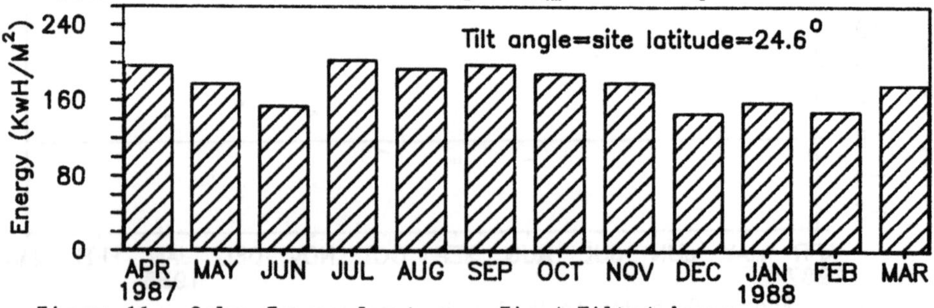

Figure 11. Solar Energy Input on a Fixed Tilted Array

POSSIBILITIES AND LIMITATIONS OF PV TECHNOLOGY IN THE PHILIPPINES

by

P.H.A. de Bakker and C. G. Zamora
Research Analysts
Philippine-German Solar Energy Project
PNOC-Energy Research and Development Center
Quezon City, Philippines

ABSTRACT

Five years of R & D in the PV technology have resulted in the introduction of some viable PV alternatives to the Philippine market. At current Watt peak prices, PV makes sense for a range of applications.

INTRODUCTION

Over the last years, the Republic of the Philippines has been regularly in the news. This Asia-Pacific archipelago is comprised of 7,000 islands where only some 2,800 are inhabited /1/. Electric power is available only in 25 of the biggest and most densely populated islands; 51% of the Philippine families still live without electricity /2/. Indications are that there exists a great demand for small decentralized power systems ranging from a few Watts to several kiloWatts. The conventional methods of power generation are generally hampered by unreliable and expensive fuel supply as well as the lack of technical know-how on engine maintenance and the scarcity of spare parts. Kerosene lamps are inconvenient and pose fire hazards, while disposable dry cell batteries are expensive and of poor quality.

The Philippine-German Solar Energy Project (PGSEP) which started in 1982 has set the stage for the nationwide commercialization of PV technology /3/. Several years of technical and economic evaluations have shed light on where and what PV applications are feasible. The project will now slowly phase out, far exceeding its original goal of only introducing the technology. All activities originated from one point, 60 km. north of Manila, in Doña Remedios Trinidad, Bulacan where a PV village power supply of 13.3 kWp and a PV field laboratory were constructed. Results of the five-year R & D on this power plant show that PV technology cannot compete with the conventional power supply in urban areas; large scale electricity production can be realized at kWh-prices that are substantially lower than a PV-produced kWh. Only for smaller capacities and in remote locations (islands) PV power generation can be competitive. (See Figure I)

1. RURAL ELECTRIFICATION

Putting up grids for small loads is quite an expensive method of rural electrification, costing approximately US$4,000 per km. grid for a simple single phase extension /4/. For islands, it is even more costly - a kilo-

----

Note: For abbreviations, kindly refer to the list at the back of this article.

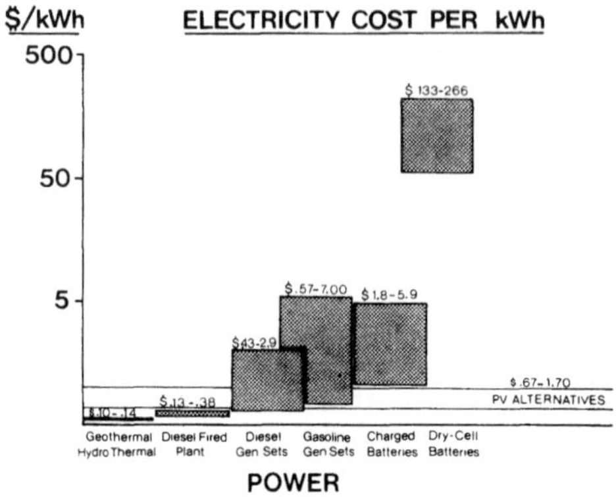

FIG 1. ELECTRICITY COST OF SOME ENERGY RELATED SOURCES

meter of submarine cable (excluding switchyards) will cost approximately US$300,000 /5/. Small diesel generator sets for reliable power supplies are another costly option because of the high purchase prices, aggravated by high fuel prices due to the poor local infrastructure. A 10 kVA diesel generator in the Philippines costs approximately US$4,800 when completely installed /6/.

Some findings of economic evaluations and comparisons:

A. Grid vs. PV

A typical village configuration of 100 households (700 people or approximately 40 kWh/day) can be supplied with competitive PV electricity of a central power plant at distances of 17 km. from the nearest existing grid (see Figure II), assuming that the existing power plants and grid can carry additional loads (which is doubtful). Figures regarding the PV power supply were derived from five years of experience with such a power plant. The average daily insolation comes close to 5 kWh/day on 1 m^2 /4/. Because of the high initial investment cost for such a central power plant, as well as for grid extensions, both options are not really considered for rural electrification.

B. Diesel Generator Sets vs. PV

Based on the kWh price, we may state that different combinations of diesel generator sets (2 x 10 kVA, 12 hr. each; or 1 x 10 kVA + 1 x 3 kVA, 12 hr. each; or 1 x 3 kVA + battery bank, 12 hr. each) all end up in the same range of kWh prices as the Central PV power plant for the same village as in case A.

Small individual 12 VDC power supplies (Solar Home Systems) and PV battery charging stations cut under these prices mainly because of absence of a distribution network /6/. (See Figure III)

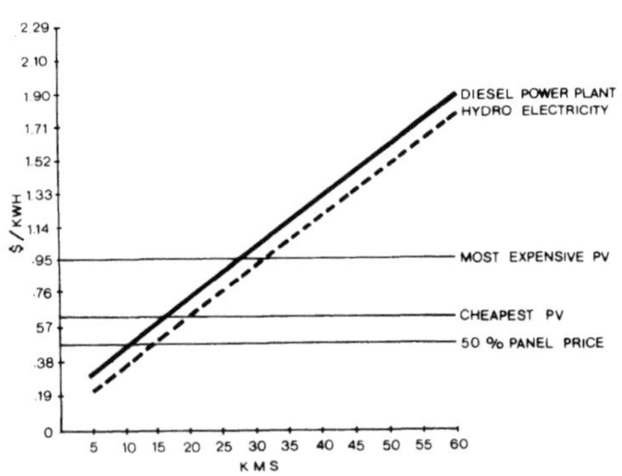

FIG. 2. $/kWh vs. DISTANCE FOR 100 HH (40 kWh/day)

C. Sea Cable vs. PV

A recent study (a combined effort of NEA/GTZ) discusses the possibility of laying 17 km. sea cable for an island of 100,000 inhabitants. It indicates that the kWh price will not be lower than US$0.70, even assuming that industrial loads such as iceplants, rice mills, etc., would also hook up. With such power prices, however, this kind of economic activities may not be expected, which will result in even higher prices for residential consumers.

D. Solar Home Systems

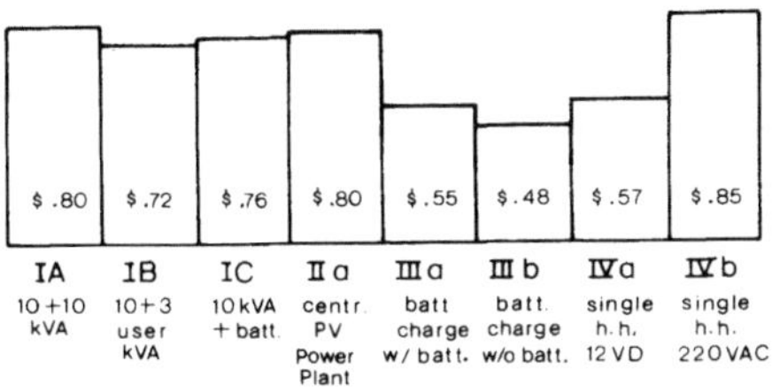

FIG. 3. Comparison of Diesel Gensets vs. PV options (all options except single hh designed for 40 kWh/day for 100 rural households.)

A small 25-30 Wp solar home system of approximately US$300 is sufficient to supply the power needs of a simple lighting fixture and a radio. At no downpayment, this system will finance itself at US$9.00/month over a three-year period.

A 50 Wp solar module will yield an average of 180 Wh/day. Such a 12 VDC power supply can power a black and white TV set, two small fluorescent tubes (20W), an incandescent bulb (10-15W) and a radio /7/. This system will cost approximately US$450 (excluding any commercial mark-up). The battery, battery control unit (BCU), voltage regulator, lamps, frames, etc., are all locally fabricated; only the solar panel is imported. At this moment, the 12 VDC ballasts and fluorescent tubes leave something to be desired. The project-designed prototypes seem more durable. At present, a reasonable lifetime can only be guaranteed for the lesser efficient 12 VDC incandescent bulbs /7/. The real monthly cost price (derived from the cost annuity) for such a system (including maintenance and anticipated replacements of several components) is less than US$5.00/month.

Many Filipino households which use kerosene and dry cell batteries currently spend between US$7.50 and US$12.50 per month for lighting and for the radio. If a financing scheme is introduced which requires a 25% downpayment and the balance as monthly installments equal to their current expenses for kerosene and batteries (i.e. US$11/month), the family will completely finance the system within three years. The system itself should well last for 20 years. If only 5% of all nonelectrified Filipino families would like and could afford such a system, still 200,000 systems can be marketed (10,000 kWp).

After having operated some 30 "Solar Homes" in Bulacan villages over the last two years, the PGSEP is currently (in complete interaction with the local industry) involved in the first real PV island electrification. A locally instituted "PV cooperative" handles all responsibilities and fee collections. Since the project implementation started only a few months ago, no conclusive data can be drawn yet. But the interest of the people is overwhelming and beyond the implementor's expectations. After this successful introduction, it will be left to Philippine enterprises to market "Solar Home Systems". Such enterprises are still scouting for the most economic PV panel on the world market.

2. INDUSTRIAL/COMMERCIAL APPLICATIONS

Mountain tops and other remote locations cannot be avoided for telecommunication relay stations. Comparative studies of current operations vs. PV alternatives for PLDT, PT&T, and RCPI indicated that PV power supplies may halve the cost for power generation for such relays even if diesel generator sets for back-up purposes are included. These findings justified two pilot projects in this field. The first reports after one year of operation are very positive. For increased reliability, the system design was based on a daily average insolation of 3.5 kWh/day on 1 m^2. The 1.8 kWp installation for the PT&T in the Zambales mountains supplied all required electricity: the back-up generator set was only started up to keep it in running condition /8/. There are 85 of such diesel-powered relay stations in the Philippines, and anywhere from two to twelve hours are spent for the delivery of fuel either by four-wheel driven vehicles or by walking, even using ladders to climb up. Thus, fuel prices may even double before reaching the relay.

Corollarily, one may also think about navigational aides: there are 66 PV-operated lighthouses in the Philippines and there is a need for more. Then there are the directional and non-directional beacons for air navigations. A pilot project at the International Airport of Manila with a runway marker proved satisfactory /9/. Since the 168 Wp PV power supply was installed not a single complaint was recorded regarding a failing power supply. There are 85 airports throughout the Philippines. After this initial positive experiences, the O.A.T. requested for a design for a PV power supply for a V.O.R. on Jomalig Island in the Pacific. There are 23 V.O.R. facilities in the country, many of them situated on remote areas.

3. SMALL ENTERPRISES

For several income-generating projects, a small PV power supply may help out in the nonelectrified areas. A PV-powered poultry incubator proved to be a simple and competitive answer to hatching problems. A 200-egg incubator will cost around US$2,150; a 600-egg incubator, approximately US$3,300 and both can be locally manufactured. Eggs can be hatched for US$0.05 - US$0.08 per live chick depending on system size and hatching success rate.

Powering a TV set, a video recorder and an amplifier by a 24 VDC was simple. Video cinemas (powered by gasoline generator sets) are spread widely throughout the Philippines. Such a PV power application (costing approximately US$1,000 including the power supply for residential power consumption) is currently being field-tested. Compared with kWh prices of gasoline generator sets, PV power supplies cost less than half (US$5.00/kWh for a gasoline generator set vs. US$1.57/kWh for a PV power supply).

Another application is the electronics repair shop. Electrified or nonelectrified, practically every family in the Philippines owns a radio or radio cassette recorder, and to a lesser extent also TV sets (operated by car batteries) penetrated the rural areas. Repair shops in remote areas use the cooking fire for soldering purposes. A US$575, 77 Wp PV power supply for soldering, drilling and testing purposes will surely draw a lot of attention. The project's initial experiences are quite positive.

PV-powered irrigation systems do not seem very promising - water can not be pumped for prices attractive to Philippine farmers for staple food production /10/. The only viable application maybe a PV irrigation system for high value cash crops with the excess electricity used for residential purposes. Other PV applications which are deemed of immediate interest are now under testing, and include PV-powered compressors for vulcanizing, shaft power for grinding, cutting, etc. in the shellcraft industry, and PV powered sewing machines. A power supply for a 1/4 Hp DC motor, four hours per day and simple lighting will cost around US$1,800 which is still beyond the reach of many rural folks.

4. COMMUNAL PROJECTS

PV refrigerators seem of some interest if used for vaccine and medicine storage. The economics of a small refrigerator/freezer becomes really clear if spoiled medicines, as a result of the unreliable direct competitor, the kerosene refrigerator, is taken into account. A 250 Wp, 150 liter refrigerator with a 30 liter freezer, as has been installed in a health clinic in the Central Visayas is actually a bit too big for its function. A 150 Wp systems would be more suitable.

For food storage and cooling drinks, no immediate market is foreseen except probably for some nonelectrified beach resorts.

Drinking water pumped by imported centrifugal pumps or by locally manufactured jackpumps can be produced for US$0.60-US$1.20 per m^3. Several of such PV drinking water systems were installed. Before, water was always fetched from afar or purchased at prices ranging from US$3.50-US$6.00 per m^3. This also explains the evident willingness of the people to pay for their much cheaper "PV water". Coastal and upland communities with small demands or limited water resources may benefit from this new technology which will find its place between handpumps and the much bigger diesel pumps.

The cheapest option for rural electrification seems to be the PV battery charging station. People bring empty lead acid batteries to the PV array (200 Wp/battery charge/average day) and have it recharged for a fee ranging from US$0.24-US$0.71 depending on the size of the system and the presence of commercial objectives /11/. The use of car batteries for TV sets, sound systems or in fishing is a common practice even if people have to travel long distances to have their batteries boost-charged (which drastically shortens the battery lifetime) for prices ranging from US$1.00 -US$2.00 per charge. PGSEP's two projects in this field were immediate successes, and because of the convenient locations of the charging station in the village, and the low-recharge prices, more people decided to buy a battery. Now, at some instances, batteries are "wait-listed". The customers themselves realize that the smooth charging of their batteries resulted in better charges and longer lifetimes /10/.

5. PHILIPPINE GOVERNMENT

The abovementioned PV applications do not cover all possibilities of this technology but the promising PV potential is clearly demonstrated. Additional research work is now being done in the field of cottage industries; e.g. small power supplies close to their source of raw materials (sea and forest products).

The Philippine Government, through its Office of Energy Affairs, after a prolonged "wait-and-see" attitude towards all renewable energy sources, seems convinced that PV may offer solutions to some of the rural and remote areas, and with several projects successfully introduced, the PV technology has emerged from a pool of renewables as a serious alternative. On a case-to-case basis, the feasibility of PV systems vs. alternatives (if any) remains to be determined, but commercialization is now within reach.

Efforts are now geared towards inviting more industries to market PV products in order to: first, supply the demand and secondly to create a healthy competition. Through testing of several PV applications and through the creation of awareness, some applications are now market-ready: they are technically proven and economically viable.

Encouraging is the recently approved incentive of the Philippine Board of Investments for a tax-free importation of renewable energy equipment/12/. To off-set the still high initial investment cost of PV power supplies, financing schemes with the Philippine Banking and Finance world are currently being worked out. Local industries are being put in place for production of some components and distribution of PV systems. Price-competitive Solar Panels for the meantime will be withdrawn from the world market.

No PV cells are yet being manufactured in the Philippines, probably because of the continued process of innovation on the production technology. Undoubtedly, the interest for PV system will increase within the next few years. And the overwhelming number of requests which the project receives daily are a sure indication of the emergence of this technology in the Philippines.

LIST OF REFERENCES

/1/ National Census and Statistics Office (NCSO), 1987.

/2/ Status of Program Implementation, NEA Annual Report, 1985.

/3/ Rainer Schroeer, 'Status of GTZ/ERDC Implemented Pilot Solar PLant', Quezon City, Feb. 1987.

/4/ P.H.A. de Bakker, PV for Rural Electrification (Part I) Economics of PV Systems vs. Grid Extension, PNOC-ERDC, Quezon City, June 1986.

/5/ Bert Bague, P.H.A. de Bakker, R. Tirones, Electrification of Burias Island, NEA/GTZ, Quezon City, April 1988.

/6/ P.H.A. de Bakker, PV for Rural Village Power Supply, PNOC-ERDC, Quezon City, 1987.

/7/ Rene Baruela, David Balleza, 'Solar Home Systems', PNOC-ERDC, Feb. 1987.

/8/ R. Arguelles, P. Boger, 'The First Photovoltaic Power Supply for a Remote Relay Station', PNOC-ERDC, Dec. 1986.

/9/ R. Arguelles, R. Barruela, 'PV Power Supply for Airport Marker at NAIA', PNOC-ERDC, Feb. 1988.

/10/ P.H.A. de Bakker, Economic Evaluations of Selected Small PV Applications, PNOC-ERDC, Quezon City, June 1986.

/11/ R. Tirones, 'The PV Battery Charging Station', PNOC-ERDC, Quezon City, 1987.

/12/ Board of Investment, Omnibus Code of 1987, Investment Priorities Plan for Energy Projects.

LIST OF ABBREVIATIONS

BCU - Battery Control Unit

GTZ - Deutsche Gesellschaft fuer Technische Zusammenarbeit (German Agency for Technical Cooperation)

NEA - National Electrification Administration

PGSEP - Philippine-German Solar Energy Project

PLDT - Philippine Long Distance and Telephone Company

PNOC-ERDC - Philippine National Oil Company-Energy Research and Development Center

PT&T - Philippine Telegraph and Telephone Company

OAT - Office of Air Transportation

OEA - Office of Energy Affairs

RCPI - Radio Communications of the Philippines, Incorporated

VOR - Very High Frequency Omni Range (A non-directional radio signal for air navigation)

SESSION 11:
a-Si FUNDAMENTALS

THE ORIGIN OF MACROSCOPIC AND MICROSCOPIC DEFECTS IN $a\text{-}Si_{1-X}Ge_X$:H ALLOYS FOR PHOTOVOLTAIC APPLICATIONS

G. Conte, L. De Angelis, D. Della Sala, R. Falesiedi, F. Galluzzi, C. Gramaccioni, G. Grillo, C. Ostrifate, C. Reita, R. Tomaciello and V. Vittori.

Eniricerche S.p.A., 00015 MONTEROTONDO, ROME (ITALY)

ABSTRACT. High-quality glow-discharge a-GeSi:H alloys obtained by using H_2 diluted gases, fluorinated gases or triode glow-discharge are currently used for the preparation of single and stacked thin film solar cells. Nevertheless, the possible further reduction of the density of dangling bonds of the Ge-rich alloys is still an open issue. We use PDS for the measurement of the density of dangling bonds and the amount of disorder of our glow-discharge alloys produced from GeH_4 + SiH_4 (both diluted 1:10 in H_2). The results are compared with some other high quality materials and with some recently published theories. We finally propose that the increasing amount of dangling bonds with the Ge fraction is an intrinsic and unavoidable effect. It is probably due to the breaking of weak Ge-Ge bonds in the amorphous network. The use of GeH_4 in the gas mixture leads also to another class of defects (1 um-thick nodules or cones). However, they turn out to be extrinsic, and can be readily eliminated with a modification of the growth procedure.

EXTRINSIC MORPHOLOGICAL DEFECTS

We found that the addition of GeH_4 to SiH_4 appreciably enhances the nucleation of characteristic cone-shaped structures, already reported for a-Si:H (1). We have established that the nucleation center of these cones resides at the film-substrate interface (Fig. 1). Microscopic inhomogeneities at the film/substrate interface can lead to the formation of such large cones. Infact the surface mobility of Ge and Si during the film formation is quite low, because of the low ratio between the growth temperature (250 °C) and the melting points of the corresponding crystalline bulk materials (2) (937 and 1410 °C for Ge and Si, respectively (3)). In this condition the shadowing effect around the substrate asperities can be effective (4) even if they have atomic-size dimensions (5). Furthermore, their spatial distribution is not constant, but is rather a replica of the holes in the stainless steel sheet that holds the substrates on the heated electrode during the deposition. Since the density of cones is larger at the center of the holes and has a similar shape, the local inhomogeneities of the glow-discharge have an important role (6). On the contrary, the possible electrical charging of the substrate is ruled out, since the films on both conducting substrates (Ta, Si, evaporated Cr on glass) and insulating ones (glass, quartz) are identical.
The way we start the discharge could lead to an enhanced nucleation of cones. Infact we rise abruptly the gas pressure

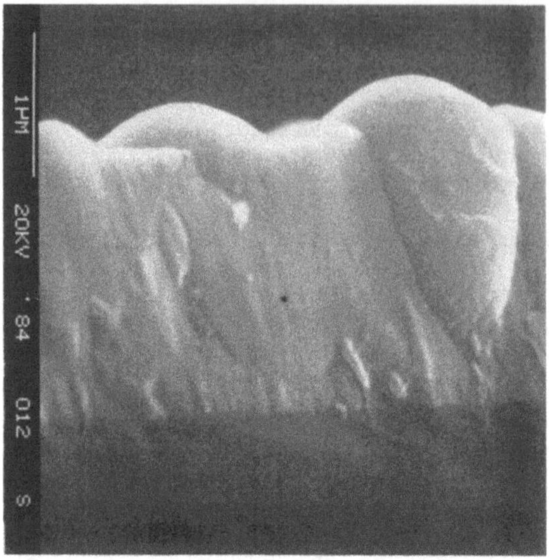

Fig. 1. SEM cross sectional view of a-GeSi:H film on cleaved crystalline silicon substrate.

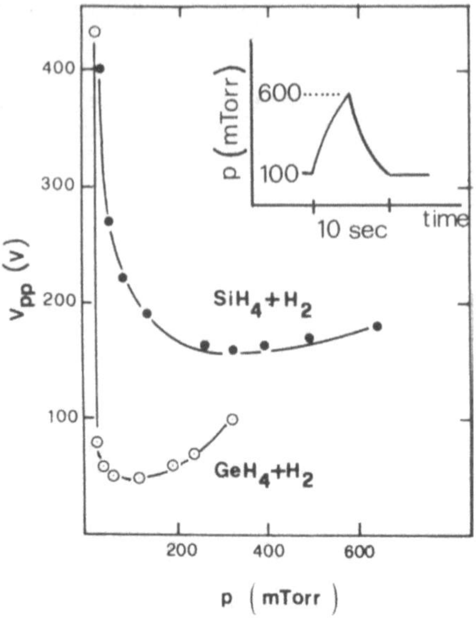

Fig.2. Paschen-like curves for our glow - discharge equipment and starting discharge procedure (insert).

from 100 to 600 mTorr while the electrodes are under bias (1200 peak-to-peak volts), in such a way that during the very early stages of the deposition the power is above the minimum value required for the plasma discharge (Fig. 2) and we work in the right branch of the Paschen-like curve. Both conditions could imply a possible polymerization of reactive species (7). Nevertheless, the IR spectra of a set of samples expressely grown at these pressures (1-2 µm-thick) do not display any appreciable enhancement of the bulk density of polymer-like groups (SiH_2, $(SiH_2)_n$).
In addition local IR transmission spectra, obtained with an IR microscope (beam diameter 0.1 mm) are identical for the regions with and without cones. Also Auger compositional measurements and X-ray transmission structural measurements are quite similar for both regions. Therefore the hydrogen content, the chemical composition and the degree of amorphization are the same within the experimental sensitivity of the above techniques ($\sim$1 at%). We have to conclude that if polymer-like material does exist, it should be restricted to a quite negligible volume fraction, e. g. to the film/substrate interface and/or cone/tissue interface (8).
We have eliminated the extrinsic morphological defects by protecting the substrates with a mechanical shutter during the stabilization time of the discharge ($\sim$5 min). Since our growth procedure consists of a preliminary cleaning discharge in pure H_2, followed by the reactive discharge in GeH_4+SiH_4+H_2, this trick has been probed for both (Tab. 1).

TABLE 1. EFFECT OF PROTECTIVE SHUTTER

	SHUTTER BEFORE H_2 DISCHARGE	SHUTTER BEFORE REACTIVE DISCHARGE	FORMATION OF DEFECTS
(I)	YES	YES	NO
(II)	NO	YES	NO
(III)	YES	NO	NO
(IV)	NO	NO	YES

It came as a surprise that the nucleation of cones can be depressed even by using the shutter before the sole cleaning discharge in H_2 (case III in Tab. 1). We argue therefore that the formation of nucleation sites for cones originates not only from an enhanced generation of microscopic particulates in the gas phase with GeH_4+SiH_4+H_2, but also because of a microdamage of the substrate during the initial stages of the cleaning discharge in H_2.

INTRINSIC DEFECTS

The total density of dangling bonds (N_s) has been determined in our alloys with the PDS technique. The calibration between the integrated sub-gap absorption and the absolute number of defects (9) has been experimentally verified in a-Ge:H with a specific EPR measurement, and has been found to be equal within 20% to that of a-Si:H (9). In Fig. 3 we show the trend of N_s and the Urbach tail width E_o as a function of the atomic

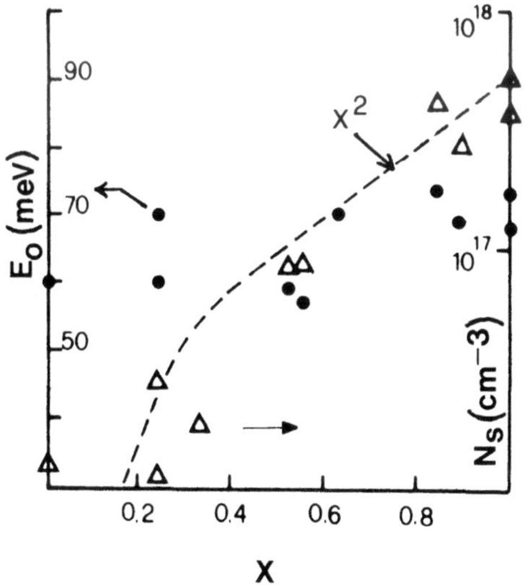

Fig.3. Urbach tail width E_o and density of dangling bonds (N_s) as a function of the alloy composition.

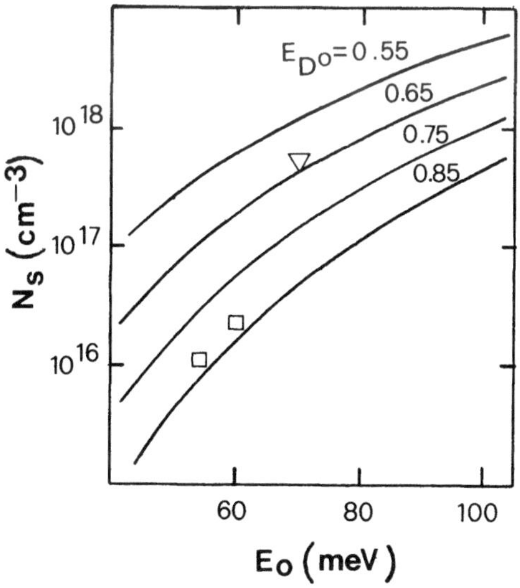

Fig. 4. Computed density of dangling bonds as a function of E_Do (eV). Symbols: experimental data for a-Si:H (□) and a-Ge:H (▽).

composition. We note that N_s is increased by a factor 50 in a-Ge:H, whereas E_o increases only by some 10%.

The slight increase of E_o from a-Si:H to a-Ge:H should be considered a well established feature for the high quality material (10,11,12). A simple model by Smith et al. in good agreement with experiment (13) states that E_o in a-Si:H corresponds to the formation energy of a strained bond. By a suitable substitution of the structural parameters of a-Si and a-Ge in their model the variation of E_o from 50 to 70 meV in Fig. 3 corresponds to an increase of the rms bond angle distribution from 8° in a-Si:H to 10° in a-Ge:H.

Also the increase of N_s of about a factor 50-100 from a-Si:H to a-Ge:H is supported from other literature data (12). However, the available literature concerning high-quality glow-discharge a-Ge:H is much less abundant, and we have made a further test. The result is that in spite of a growth rate variation from 0.9 to 4.1 Å/sec, N_s remains fixed at about $5x10^{17}$ cm^{-3} and this value appears to be a characteristic feature of high quality a-Ge:H.

The rapid rise of N_s with the increasing Ge fraction has been attributed in the past to the bad network relaxation of the Ge-rich alloy, due to the preferential attachment of H to Si rather than Ge (14). We think that also the intrinsic weakness of Ge-Ge bonds compared to Si-Si ones should play a role (the bond stregth is 39.9 and 42.2 kcal/mol, respectively (3)). A recent model for the equilibrium density of defects in a-Si:H (15) states that half the equilibrium density of defects are frozen-in at room temperature after preparation for low cooling rates, and the volume density of the neutral dangling bond states is:

$$N_D o(Si)=(1/2)\bullet(4N_{Si}\ \exp(-2\ E_D o/kT))^{(kT/(2E_o+kT))} \bullet(N_V E_o)^{(2E_o/(2E_o+kT))}$$

where N_{Si} is the volume density of Si atoms, N_V is the mobility edge value of the valence band tail density of states, $E_D o$ is the energy of the neutral dangling bond defect referred to the valence band mobility edge and T is the freeze-in temperature of defects.

The above equation can be readily extended to a-Ge:H, with the reasonable simplifications that $N_{Si} \approx N_{Ge}$ and $N_V(Si) \approx N_V(Ge)$. In this case this is actually a function of $E_D o$ and E_o. The value of $E_D o$ for a-Ge:H can be evaluated from the position of the Fermi level ($E_F-E_V=0.6$ eV in our experiments) and the correlation energy $U_{Ge}=0.1$ eV (16): $E_D o(Ge) = E_F-E_V-U_{Ge}/2 = 0.55$ eV. Furthermore, since the freeze-in temperature of defects has been determined only for a-Si:H (17,18), we assume for a-Ge:H that at slow cooling (we used 2 °C/min) the fraction of frozen-in defects and T are the same as in a-Si:H. The calculated and experimental defect densities of a-Si:H and a-Ge:H are compared in Fig. 5, which shows quite clearly that the ratio $N_s(Ge)/N_s(Si) \simeq 50$ is a real material property in agreement with the measured values of E_o.

As a final remark, we propose that the large number of dangling bonds in Ge-rich alloys originates from the breaking of Ge-Ge bonds. The purely statistical rules of random bonding hold in a-GeSi (19), without any sign of segregation of separate a-Si and a-Ge phases. Therefore the density of Ge-Ge bonds in this random alloy should increase like x^2, in fair agreement with the density of dangling bonds (Fig. 5). We conclude therefore that a roughly constant fraction of Ge-Ge bonds remains unsaturated in the alloy.

AKNOWLEDGEMENTS

This work is supported by AGIP S.p.A.

REFERENCES

1) J.C. Knights and R.A. Lujan, Appl. Phys. Lett. 35, 244 (1979).
2) J.A. Thornton, Ann. Rev. Mater. Sci. 7, 239 (1977).
3) CRC Handbook of Chemistry and Physics, 60th Ed., p. F-220 ff. (1980).
4) R.C. Ross, A.G. Johncock and A.R. Chan, J. Non-Cryst. Solids 66, 81 (1984).
5) K.H. Guenther, Applied Optics 23, 3806 (1984).
6) J.C. Knights, in 'The physics of hydrogenated amorphous silicon' Part 1, J. Joannopoulos and G. Lucovsky eds., Springer Topics in Applied Physics, Vol. 55, p. 26.
7) M.H. Brodsky, Thin Solid Films 50, 57 (1978).
8) G. Lucovsky, R.J. Nemanich and J.C. Knights, Phys. Rev. B20, 4839 (1979).
9) W.B. Jackson and N.M. Amer, Phys. Rev. B25, 5559 (1982).
10) H.G. Weller, S.M. Paasche, C.E. Nebel, F. Kessler and G.H. Bauer, Proc. 19th IEEE Photovoltaic Specialists Conf., New Orleans 1987.
11) L. Mariucci, F. Ferrazza, D. Della Sala, M. Capizzi and F. Evangelisti, J. Non- Cryst. Solids 97/98, 1075 (1987).
12) S. Aljishi, V. Chu, Z.E. Smith, D.S. Shen, J.P. Conde, D. Slobodin, J. Kolodzey and S. Wagner, J. Non-Cryst. Solids 97/98, 1023 (1987).
13) Z.E. Smith, S. Aljishi and S. Wagner, J. Non-Cryst. Solids 97/98, 775 (1987).
14) W. Paul, D.K. Paul, B. Von Roedern, J. Blake and S. Oguz, Phys. Rev. Lett. 46, 1016 (1981).
15) Z.E. Smith and S. Wagner, Phys. Rev. Lett. 59, 688 (1987).
16) M. Stutzmann, J. Stuke and H. Dersch, Phys. Status Solidi (b) 115, 141 (1983).
17) R.A. Street, J. Kakalios, C.C. Tsai and T.M. Hayes, Phys. Rev. B35, 1316 (1987).
18) Z.E. Smith and S. Wagner, Phys. Rev. B32, 5510 (1985).
19) L. Incoccia, S. Mobilio, M.G. Proietti, P. Fiorini, C. Giovannella and F. Evangelisti, Phys. Rev. B31, 1029 (1985).

GROWTH AND INVESTIGATION OF a-$Si_{1-x}Ge_x$:H ALLOYS FOR SOLAR CELLS

L.Mariucci, F.Ferrazza, F.Evangelisti and A. Frova
Department of Physics, University of Rome "La Sapienza"
Piazzale A. Moro 2, 00185 Roma, Italy

ABSTRACT. The effect of hydrogen dilution of the GeH_4+SiH_4 gas mixture on glow-discharge a-$Si_{1-x}Ge_x$ alloys was investigated. From IR data it was found that dilution brings about a decrease of the SiH_2 groups but no variation of the relative content of hydrogen attached to Si and Ge. Whereas the optical properties and the gap state density are only slightly influenced by dilution, photoconductivity but also conductivity are strongly affected. An appreciable decrease of the conductivity activation energy was found. We suggest that the dramatic increase of photoconductivity can be explained, at least partially, in terms of a photosensitization of the material due to shifting of the Fermi level toward the conduction band mobility edge.

1. INTRODUCTION

It has been reported recently that the photoconductivity of amorphous hydrogenated silicon-germanium alloys can be considerably improved in a number of different ways: GD in a triode-configuration apparatus [1], GD in a diode-configuration using SiF_4+GeF_4+H_2 [2], GD in a diode [3] or in a triode [4] configuration using SiH_4+GeH_4+H_2. All these improved alloys exhibit in general very similar physical properties.

It was suggested by Matsuda and Tanaka [1] that in order to obtain highly photosensitive alloys it is very important to have the growing surface fully passivated by hydrogen. In this condition the incoming precursor species have a large surface-diffusion length and, consequently, can find energetically favorable sites more easily. Hydrogen dilution of the gas mixture in the glow discharge favors the attainment of these growth conditions. In the present work we report on the effect of hydrogen dilution on the morphology, density of gap states and transport properties of a-$Si_{1-x}Ge_x$:H alloys for x in the range 0-0.8.

2. EXPERIMENTAL

a-$Si_{1-x}Ge_x$:H films were prepared in a standard RF Plasma-assisted Chemical Vapor Deposition apparatus with diode configuration from mixtures of SiH_4, GeH_4 and H_2.

Hydrogen dilution was varied in the range 0-90% of the total gas pressure. The resulting growth rate was found to change from $\sim$ 300 Å/min without H_2 to less than

100Å/min. The atomic concentrations of Si and Ge in the films were determined by electron microprobe analysis with an estimated uncertainty of 1 %.

The samples were characterized by IR and optical absorption, photothermal deflection spectroscopy (PDS), conductivity and photoconductivity. Conductivity and photoconductivity measurements were performed in a planar configuration; the photon flux used was 1×10^{15} photons/(s cm^2) at the wavelength corresponding to an absorption coefficient of $\sim 3 \times 10^3$ cm^{-1} in order to ensure uniform excitation inside the sample and to correct for the optical gap change with x. Whenever possible, the absorption coefficient in the low energy region was also determined by the spectral photoconductivity method [5] and compared with that obtained by PDS. A good agreement between the two measurements was found. In particular the Urbach slope showed no differences within the experimental uncertainty, while the gap state absorption as determined by PDS was a factor of 1 to 4 higher, probably due to surface state absorption [6] and/or to different transitions involved. In order to obtain N_g we have used two conversion factors: 7.9×10^{15} states/cm^2 eV for PDS [7], and 1.9×10^{16} states/cm^2 eV for photoconductivity [8].

3. RESULTS AND DISCUSSION

Features present in the IR transmittance spectra were the absorption bands due to the stretching modes af Si and Ge hydrides in the region around 2000 cm^{-1} and the corresponding wagging modes around 600 cm^{-1}. The Si-H absorption bands were dominant over the Si-H_2 ones, while the Ge-H_2 bands were not appreciable. No detectable bending mode bands were found, in agreement with the predominancy of Si-H and Ge-H bands.

Two important pieces of information on the effect of H_2 dilution were obtained by IR spectra analysis. From the intensities of the Si-H and Ge-H stretching modes normalized to the atomic content as a function of dilution, we found that the total H content does not vary appreciably, nor does the preferential attachment of H to Si with respect to Ge. A second important result [3] is that the ratio of the intensities of the Si-H and Si-H_2 stretching modes increases with dilution for all Ge concentrations, the increase being more remarkable at lower x. It is known from earlier studies [9] on a-Si:H that the presence of a large number of dihydride groups results in samples with columnar microstructures and overall poor morphology. Our data therefore suggest that dilution improves morphology by reducing dihydride related inhomogeneities.

Information on the effect of these morphology changes can be obtained by inspection of the behavior of the absorption coefficient in the pseudo-gap region. The values of the optical gap E_g, as derived by means of the Tauc plot, point out that H_2 dilution has no influence on E_g itself, consistent with the electron microprobe analysis and IR measurements which show that neither composition nor hydrogen content are changed by the variation of this growth parameter.

The behavior of the Urbach characteristic energy E_0 in the exponential region of the absorption coefficient is shown in Fig. 1 as a function of x for three H_2 dilutions. We see that alloying with a substantial amount of Ge ($x \geq 0.2$) causes an increase of E_0, which, on the other hand, remains almost constant and independent of x in the remaining composition range investigated. Hydrogen dilution somewhat reduces

E_0 values. These improved samples have a Urbach slope of ~60 meV. Similar values have been obtained by other groups [2,4] using different gas mixtures or different GD apparatus. It seems, therefore, that alloys deposited from hydrogen diluted gas mixtures exhibit the same Urbach tail regardless of GD configuration and gas species.

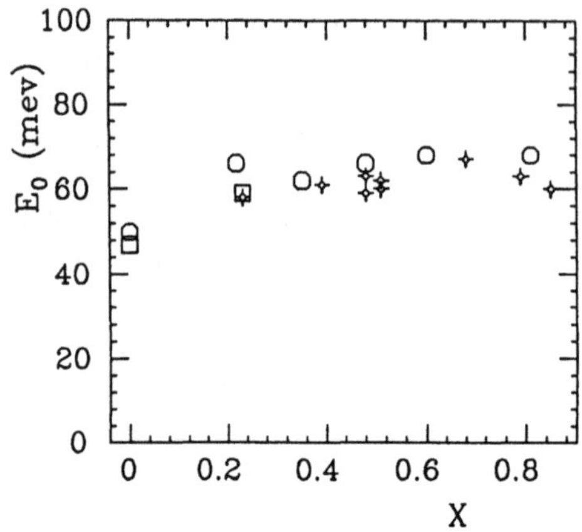

Fig.1. Urbach tail slope values vs composition for three different H_2 dilutions. Symbols, here and elsewhere:
H_2 : ○=0 □=60% ✧=80%

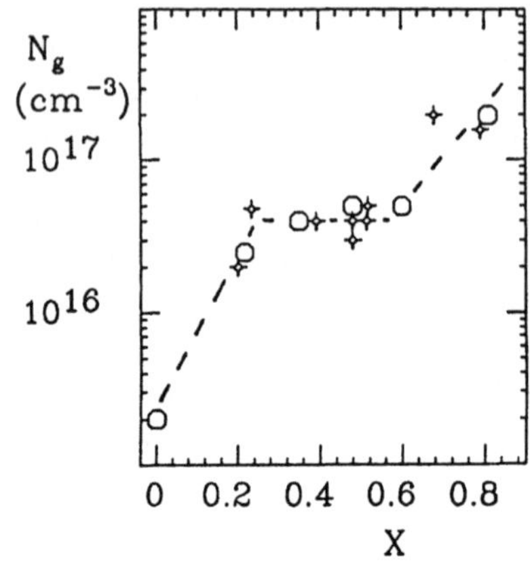

Fig.2. Dependence of the gap state density on the alloy composition for two different H_2 dilutions.

Another important physical parameter which can be extracted from the absorption data is the density of gap states. Fig. 2 shows the gap state density as a function of atomic composition for two different values of H_2 dilution. N_g increases by a factor of ~ 10 from x=0 to x=0.2 then it remains almost constant up to x=0.6 and indipendent of H_2 dilution. For x>0.6, it increases again. It should be pointed out, however, that one reason for obtaining such a large growth of the gap state density at larger x could be the lack of reoptimization of the other deposition parameters. As a matter of fact, the growing conditions used are those which, in our system, give the best materials at zero and low Ge concentrations [10]. These could be rather different from those needed at high x. The tuning of deposition parameters for with films large Ge content is presently under way in our laboratory.

It was recently reported by Stutzmann et al. [11] that optimized alloys with ~30 at.% Ge content have ~$2x10^{15}$ Si dangling bonds and ~$2x10^{16}$ Ge dangling bonds. We estimate a density of 2-$5x10^{15}$ dangling bonds in our a-Si:H samples and of 2-$6x10^{16}$ in a-$Si_{1-x}Ge_x$:H with $x \leq 0.5$; assuming that the number of Si dangling bonds does not change upon alloying, we can attribute the nearly one order of magnitude increase in gap states to Ge dangling bonds in agreement with the results of ref.[11].

As reported previously [3] the most conspicuous influence of H_2 dilution is seen in the photoelectronic properties. In Fig. 3 we show the behavior of the quantum efficiency-mobility-lifetime product $\eta\mu\tau$, as derived from steady-state photoconductivity data taken under uniform generation conditions. In Fig. 4 the dark conductivity σ_D is reported. $\eta\mu\tau$ decreases as a function of Ge content, but it is enhanced by H_2

dilution by a factor of ~ 50 on the average at all concentrations. As for σ_D, there is a general trend towards higher values for increasing x due mainly to the decrease of the mobility gap. However, for $x \leq 0.5$, H_2 dilution brings about a clear increment of the σ_D value.

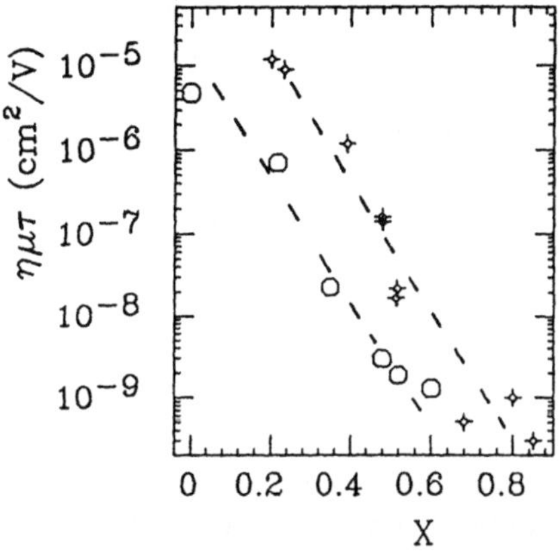

Fig.3. Decrease of the efficiency-mobility-lifetime product with increasing Ge content.

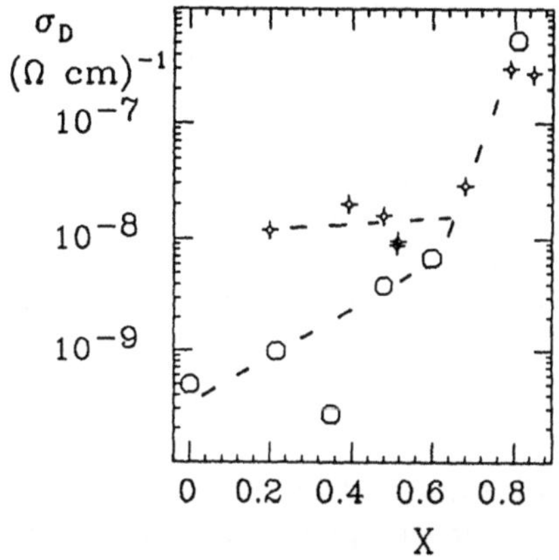

Fig.4. Change in the dark conductivity as a function of alloy composition

The simultaneous increase of photoconductivity and dark conductivity when trying to improve film quality seems to be a general characteristic of the a-$Si_{1-x}Ge_x$:H alloys [1,2,4], at least for $x \leq 0.5$-0.6.

As a consequence of the σ_D variation upon dilution, the ratio $\Delta\sigma/\sigma_D$ behaves differently from $\eta\mu\tau$. With increasing H_2 dilution, $\Delta\sigma/\sigma_D$ decreases in a-Si:H samples, stays almost constant in the alloys with lower Ge content, increases considerably at higher Ge content.

In order to ascertain the origin of conductivity and photoconductivity changes upon dilution, we have performed preliminary measurements of the conductivity activation energy on a series of samples with x=0.2 near room temperature. The data are reported in Fig. 5.

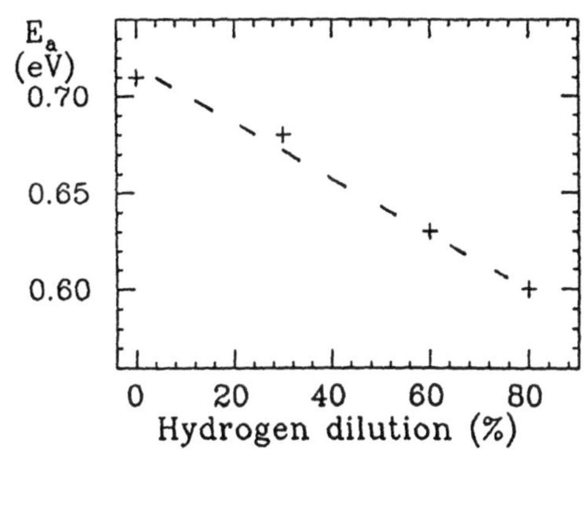

Fig.5. Dependence of the dark activation energy on H_2 dilution for fixed composition alloys (x=.2)

Identifying the activation energy with the distance of the Fermi level from the conduction-band mobility edge, we see that the conductivity change is mainly caused by a displacement of E_F toward E_c.

It is well known [12] that in a-Si:H the displacement of the Fermi level toward E_c upon doping is accompanied by a large increase in photoconductivity. Such a photosensitization is explained in terms of trapping of minority carriers in deep centers, which results in a larger lifetime for majority carriers. A similar effect is likely to be present in silicon-germanium alloys. As a consequence, our preliminary measurements of activation energy as a function of H_2 dilution show that the photoconductivity increase may be due at least partially to a photosensitization effect brought about by a displacement of E_F. However, the detailed mechanism which causes such a displacement is not presently understood. We can only speculate that H_2 dilution results in some kind of doping of the materials. Further investigations are presently under way.

5. ACKNOWLEDGEMENTS

Work supported by ENEA under ENEA/Commission EEC joint project "Low gap alloys for amorphous silicon based solar cells" (Contract EN3S - 0063-I). We wish to thank M. Vittori for electron microprobe analysis. Stimulating discussions with P. Fiorini are gratefully acknowledged.

REFERENCES

1. K. Tanaka and A. Matsuda, Mat. Res. Soc. Symp. Proc., **70** (1986) 245.
2. K.D. MacKenzie, J. Hanna, J.R. Eggert, Y.M. Li, Z.L. Sun and W. Paul, J. Non-Cryst. Solids **77-78** (1987) 881.
3. L. Mariucci, F. Ferrazza, D. della Sala, M. Capizzi, F. Evangelisti and U. Coscia, J. Non-Cryst. Solids **97-98** (1987) 1075.
4. H.C. Weller, F. Kessler, E. Lotter, C.E. Nebel. S.M. Paasche and G.H. Bauer, J. Non-Cryst. Solids **97-98** (1987) 1071.
5. F. Evangelisti, P. Fiorini, G. Fortunato, A. Frova, C.Giovannella, P. Peruzzi, J. Non-Cryst. **55** (1983) 191.
6. L. Chahed, G. Vuye, M.L. Theye, Y.M. Li, K.D. MacKenzie and W. Paul, J. Non-Cryst. Solids **97-98** (1987) 727.
7. W.B. Jackson and N.M. Amer, Phys. Rev. B**25** (1982) 5559.
8. Z.E. Smith, V. Chu, K. Shepard, S. Aljishi, D. Slobodin, J. Kolodzey, T.L. Chu and S. Wagner, Appl. Phys. Lett. **50** (1987) 1521.
9. J.C. Knights in The Physics of Hydrogenated Amorphous Silicon I, eds. J.D. Joannopoulos and G. Lucovsky (Springer-Verlag, 1984) p.5.
10. D. della Sala, M. Nsabimana, M. Capizzi, F. Evangelisti, U. Coscia and M. Vittori, Proc. 7th E.C. Photovoltaic Solar Energy Conf., eds. A. Goetzberger, W. Palz and G. Willeke (Reidel 1986) p.1199.
11. M. Stutzmann, C.C. Tsai and R.A. Street, J. Non-Cryst. Solids **97-98** (1987) 1011.
12. D.A. Anderson and W.E. Spear, Philos. Mag. **36** (1977) 695.

ELECTROREFLECTANCE AND ELECTROSTRICTION IN a-Si:H FILMS

L. HADJERIS, G. DE ROSNY, and B. EQUER
Laboratoire de Physique des Interfaces et des Couches Minces (ER258 CNRS)
Ecole Polytechnique, 91128 PALAISEAU, FRANCE

ABSTRACT. The modulation of the light reflection coefficient of an a-Si:H film, due to the application of an AC voltage between its faces, is studied as a function of wavelength, AC and DC voltages. The reflection amplitude is measured by interferometry. Evidence for electroabsorption and for electrostriction is presented.

1.INTRODUCTION

Electroreflectance in a-Si:H films has essentially been studied by two groups, Freeman et al. (1) and Nonomura et al. (2). The method consists in the detection of the modulation of the reflected intensity of a light beam when an AC electric potential is applied between two electrodes sandwiching the film, one of the electrodes being semi-transparent to allow light penetration. The effect is very small, which implies the use of a lock-in detection technique. The two groups have interpreted their observations as due to the dependence of the light absorption coefficient α with the local electric field in the film, the change of the reflectance being a consequence of the modulation of the absorption in the multireflection path between the two interfaces. The absorption coefficient variation is found to be a function of the square of the electric field. Other possible interpretations have been proposed in particular electrostriction (1,3).

If one takes as established that the effect is due to electric field dependence of the absorption coefficient, or more generally of the complex dielectric function $\varepsilon(\lambda)$, one may express its variation with the local field as : $\delta\varepsilon(\lambda) = c(\lambda).E^2(x)$
$E(x)$ is the electric field at abscissa x, $c(\lambda)$ is a complex coefficient depending on the light wavelength. In a-Si:H electroreflectance is observed in the 1.2 - 2.0 eV photon energy range (1). A detailed treatment of light propagation in non uniform media, following the calculation done by Aspnes (4), predicts that genuine electroreflection should be present as well as electroabsorption .

The ultimate aim of the experiment described in the present contribution is to seek for this genuine electroreflection, and if possible to take benefit of it to deduce the electric field profile in a-Si:H devices.

2.DESCRIPTION OF THE EXPERIMENT

As compared to the apparatus used in ref 1 and 2, the following improvements have been developped: i) not only the modulation at the excitation frequency is detected, but also the first harmonic. This is easily achieved using an EGG commercial lock-in amplifier. ii) the amplitude modulation of the reflected light is measured and not only its intensity. This is achieved by Michelson interferometry, as shown on figure 1. These improvements were necessary to exploit thoroughly the potentiality of the method, as deduced from Aspnes calculations.

3.RESULTS

Figures 2-a and 2-b show the light intensity (no interferometry) electroreflectance spectra for a 1μm thick Cr/a-Si:H/Cr sandwich, the modulation frequency is 1 kHz, there is a DC bias of 4 V applied to the structure. One observes that the modulated interference fringes are in concordance with the ones of the unmodulated reflected beam, both for the fundamental and the harmonic. This observation was already stressed by Freeman et al. (1) to emphasize the interpretation in terms of electroabsorption. Figure 2-c shows that the ratio $R_{\omega}/R_{2\omega}$ of the fundamental to the first harmonic reflection amplitudes, is remarkably wavelength independent. In figures 3-a and 3-b are shown the dependence of the ratios R_{ω}/R and $R_{2\omega}/R$ as a function of the DC bias and of the amplitude of the AC applied potential. The essential observations are i) the harmonic is DC bias independent and varies as the square of the AC voltage. ii) the fundamental varies linearly with both the DC bias and the AC voltage, as already stressed in references 1 and 2. Interpreted as done in ref. 2, the null value of the fundamental at zero bias should imply an approximately null built-in potential, which is expected from the symmetrical structure of the device.

When using the interferometer, even with a thick opaque front electrode, a modulation can be observed at all accessible wavelengths. The effect is attributed to a modulation of the optical length of the arm containing the sample, as a consequence of the film electrostriction. This interpretation is supported by the following observations:

i) when the optical length of the reference arm is varied via the piezo-electric ceramic, the unmodulated interference fringes and the modulated ones are found to be in quadrature, figures 4-a and 4-b.

ii) the modulated amplitudes, relative to the unmodulated ones, vary as $1/\lambda$, figures 4-c and 4-d, as expected from the wavelength dependence of the optical phase variation $\delta\Phi = 4\pi\delta L/\lambda$ where δL is the film thickness variation.

To study the effect with a better signal/noise ratio, an He-Ne 630 nm laser source has been substituted to the xenon arc-monochromator source. Figure 5 shows the fundamental and first harmonic modulated amplitudes, normalized to the unmodulated ones, as a function of DC bias and AC voltage. The first harmonic is DC bias independent and varies as the square of the AC amplitude whereas the fundamental varies linearely with both voltages. This is similar to the observations presented previously concerning the electroreflectance intensity with the important difference that the fundamental is not null at zero bias but rather at -18 V.

4.DISCUSSION

Let us first discuss the electrostriction which is the film contraction under the pressure generated by the forces between the charges on the metal electrodes and/or in the film itself. The observed modulation amplitudes may readily be translated in terms of modulation of the film thickness. Let us assume that the AC frequency is high enough to prevent charge transport in the film, then the electric field E(x,t) at abscissa x is the superposition of a time independent field E(x) and of a position independent, time dependent one, $(V_{AC}/L)\cos\omega t$, V_{AC} is the AC amplitude at frequency ω, and L the film thickness. From simple elastic theory, and assuming that the forces on the charges in the film are transmitted to the material, one may establish the following formulae :

$$\delta L(t) = \frac{\varepsilon^2 \varepsilon_0}{2K} \int_0^L E^2(x,t)\, dx \quad \text{with} \quad E(x,t) = E(x) + \frac{V_{AC}}{L}\cos\omega t \quad \text{, K is the film elastic constant.}$$

From this, one gets :

$$L_{\omega} = \frac{\varepsilon^2 \varepsilon_0}{K} \frac{V_{AC}}{L} \int_0^L E(x)\, dx \quad \text{and} \quad L_{2\omega} = \frac{\varepsilon^2 \varepsilon_0}{K} \frac{V_{AC}^2}{4L}$$

The integral of the field is just the built-in potential V_{bi} which depends on the DC bias. The observed AC voltage dependence of both terms, as well as the DC bias independence of the harmonic, are correctly described. The built-in potential is the sum of the zero bias value and of the DC bias under the hypothesis of simple Schottky type barriers at the metal-a-Si:H interfaces: $V_{bi}(V_{DC}) = V_{bi} + V_{DC}$, which predicts, as observed, a linear dependence of the fundamental with V_{DC}. The ratio $L_{\omega}/L_{2\omega} = 4(V_{bi} + V_{DC})/V_{AC}$ allows a direct determination of

V_{bi} from the measurements at zero bias, one finds $V_{bi} = 17.5$ V, an unrealistic value, but which corresponds exactly with the value of V_{bi} deduced from the extrapolation of the DC bias dependence of L_ω to 0. The harmonic term $L_{2\omega}$ may be used to deduce the value of the a-Si:H elastic constant. Using $\varepsilon = 12$, one gets : $K = 2\ 10^{10}$ Pa, about $1/5^{th}$ of the cristalline silicon value. So the electrostriction is adequately described by the model, with the important exception of the unrealistic value of the built-in potential. The asymetry of the response with respect to the DC bias is perhaps correlated to the stress that takes place near the interface between the film and the substrate and which could induce electric polarisation.

Let us now discuss electroreflectance. It is not a simple consequence of the film thickness variation, because the fundamental amplitudes do not disappear at a common DC bias and also because one should expect the modulated amplitudes to be null at the wavelengths where the unmodulated reflectance is stationnary, which is exactly opposite to the observation, figure 2. Although a more detailed treatment would be required, let us assume that the main contribution to the electroreflectance modulation comes from the electric field dependence of the absorption coefficient α (5), so that :

$$\delta R = A.\int \delta\alpha(x)\, dx\ , \text{ with } \delta\alpha = B.E^2(x,t) = B.(E(x) + \frac{V_{AC}}{L}\cos\omega t)^2$$

one then gets, using the same notations as for the analysis of the electrostriction :

$$\frac{R_\omega}{R} = C(\lambda).\frac{V_{AC}}{L}.(V_{bi} + V_{DC}) \quad \text{and} \quad \frac{R_{2\omega}}{R} = C(\lambda).\frac{V_{AC}^2}{4L}$$

The dependence of the modulated intensities with the DC bias and AC modulation corresponds to the observation, and the ratio of the fundamental to the harmonic is predicted, as observed, wavelength independent, moreover the ratio of the fundamental to the harmonic is : $4.(V_{bi}+V_{DC})/V_{AC}$, this leads to $V_{bi} = 0$V, corresponding nicely to the value at which $R_\omega = 0$. Here the value of V_{bi} is reasonable.

5.CONCLUSION AND FUTURE WORK

The electroreflectance data are, as stressed by Freeman et al. and Nonomura et al., a consequence of the electroabsorption, the simultaneous measurements of the fundamental and of the first harmonic modulation of the reflected intensity allows a very firm conclusion on this point. Electrostriction in a-Si:H has for the first time been studied in some detail, the interpretation is at present not completely satisfactory, some asymetry in the sample, not seen in electroreflectance has to be taken into account, it has to fit with the very simple functional dependence of electrostriction with AC and DC voltages. Genuine electroreflection, if present, will be searched using strongly absorbed wavelengths, to make this possible one has to suppress in some way the signal coming from electrostriction.

ACKNOWLEDGEMENTS

This work is supported by PIRSEM/CNRS, AFME and CEC contracts.

REFERENCES

1) E.C. Freeman, D.A. Anderson, and William Paul, Phys. Rev. B, 21, 4721 (1980)
2) S.Nonomura, H. Okamoto, and Y. Hamakawa, Appl. Phys. A, 32, 31 (1983)
3) John E. Fischer, Phys. Rev. Lett., 27, 1131 (1971)
4) D.E. Aspnes, J. Opt. Soc. Am., 63, 1380 (1973)
5) C.B. Roxlo, B. Abeles, and T. Tiedje, Phys. Rev. Lett. 52, 1994 (1984)

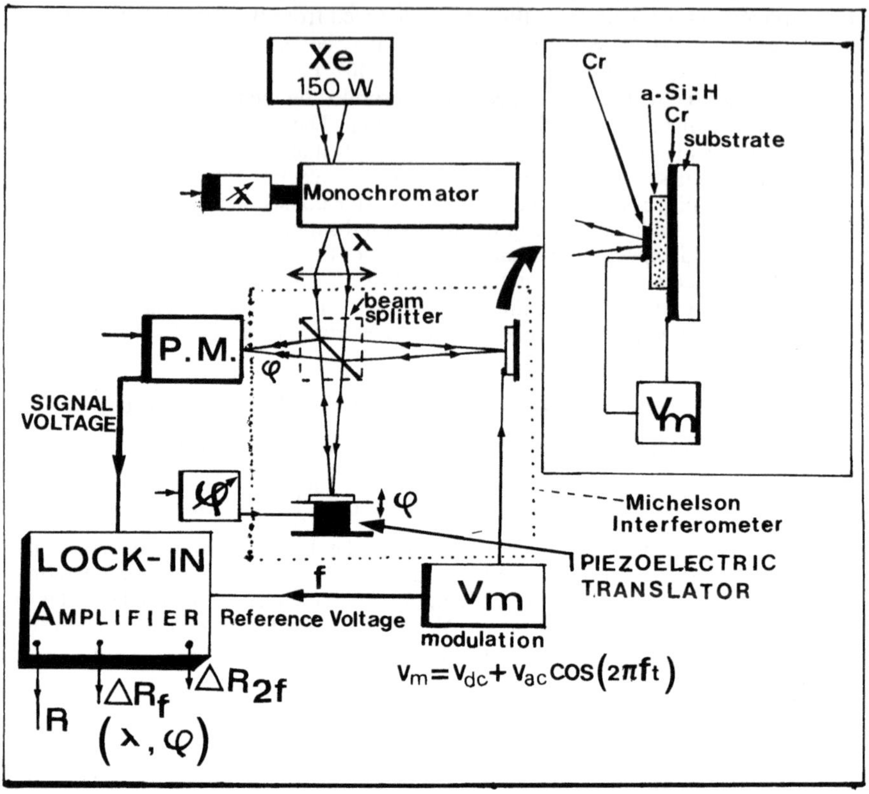

<u>Figure 1</u>: Schematic of the experimental layout. The light, issued from a Xe arc, is sent through a grating monochromator to a Michelson interferometer. The mirror of one arm is made of the a-Si:H sample under study, the path length of the other arm may be continuously varied by a piezoelectric translator. An AC voltage plus a DC bias is applied between the two faces of the sample, the front electrode is semitransparent. The light intensity modulation at fundamental and first harmonic frequencies is analysed by a lock-in amplifier.

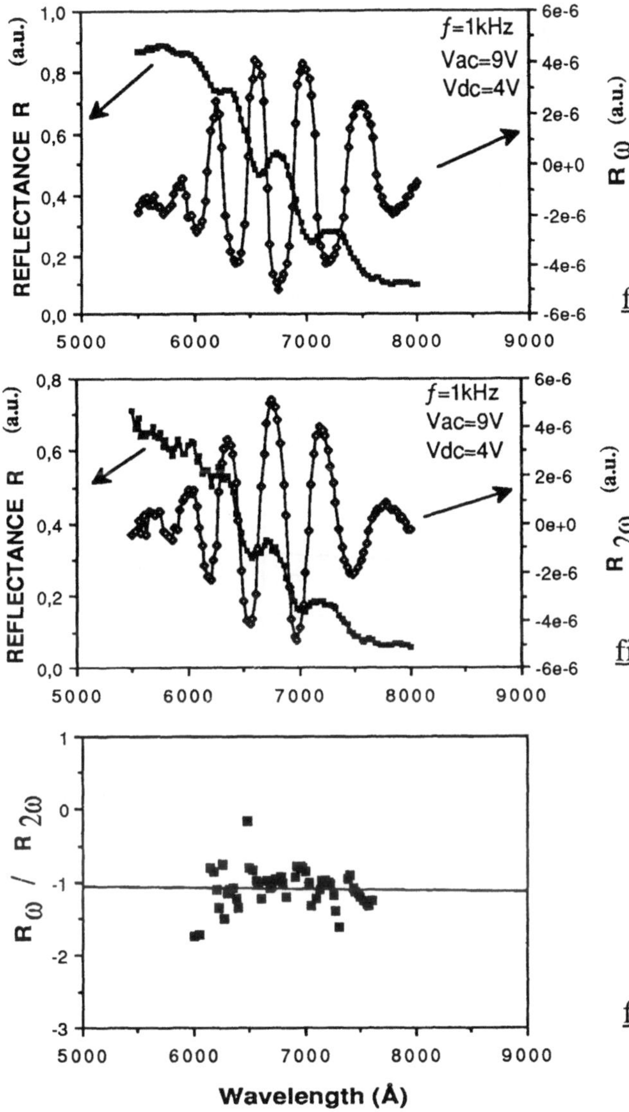

fig. 2-a

fig. 2-b

fig. 2-c

Figure 2: Wavelength dependence of the reflectance. fig. 2-a, unmodulated R and fundamental frequency R_{ω} measured simultaneously. fig 2-b, unmodulated R and first harmonic $R_{2\omega}$ measured simultanousely. fig 2-c ratio of fundamantal to first harmonic , each one normalized to its own unmodulated reflectance.
A " Cr (200Å) / i (1μm a-Si:H) / Cr (2000Å) / Corning " sample was used.

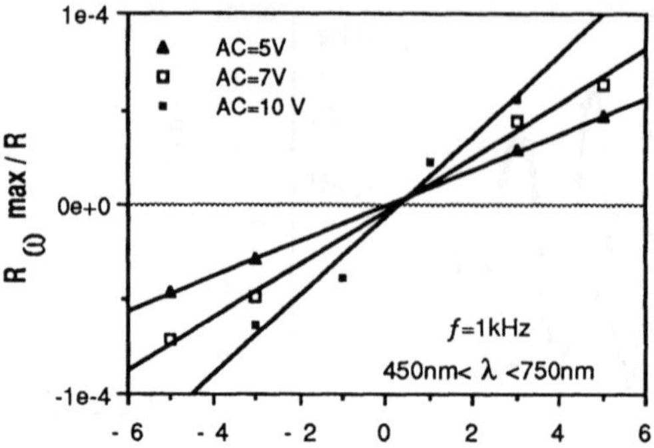

fig. 3-a

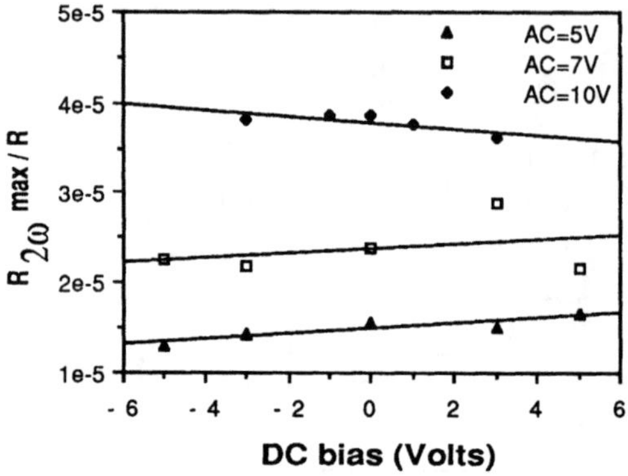

fig. 3-b

Figure 3: AC voltage and DC bias dependence of the normalized fundamental frequency (3-a) and first harmonic (3-b) reflectances.
A " Cr (200Å) / i (1µm a-Si:H) / Cr (2000Å) / Corning " sample was used.

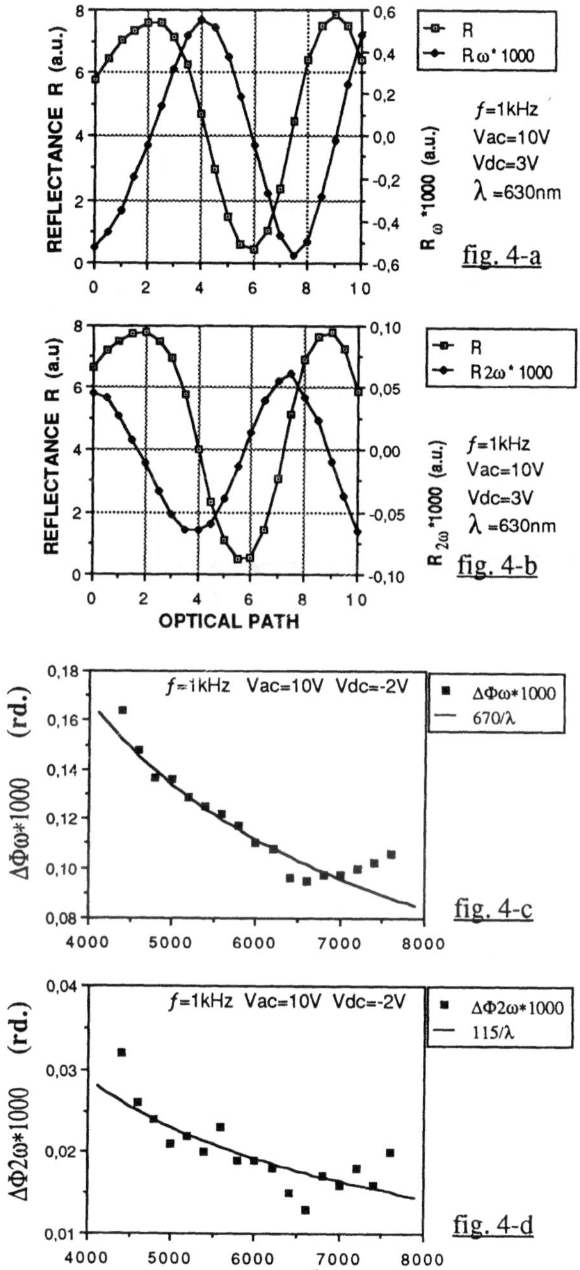

Figure 4: Fig 4-a, comparison of the unmodulated and of the fundamental frequency (ω) reflection intensities as a function of the voltage applied to the piezo translator. fig 4-b, same as fig. 4-a for the first harmonic (2ω), in both cases the quadrature condition is evidenced. Fig 4-c and 4-d, wavelength dependence of the fundamantal and first harmonic maximum intensities, expressed in term of the phase modulation of the optical path containig the sample. A " Cr (3000Å) / i (1μm a-Si:H) / Cr (2000Å) / Corning " sample was used.

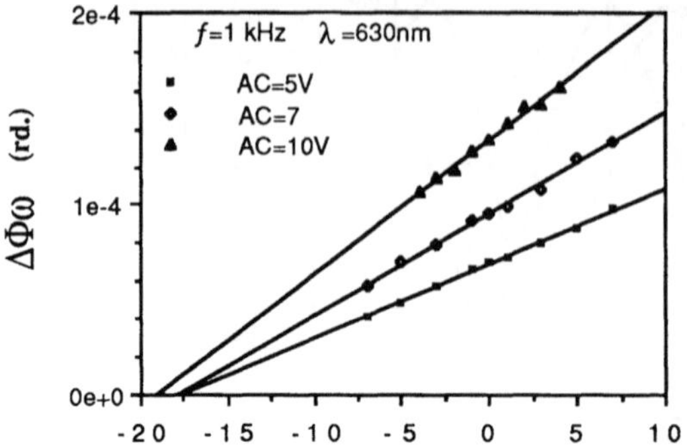

fig. 5-a

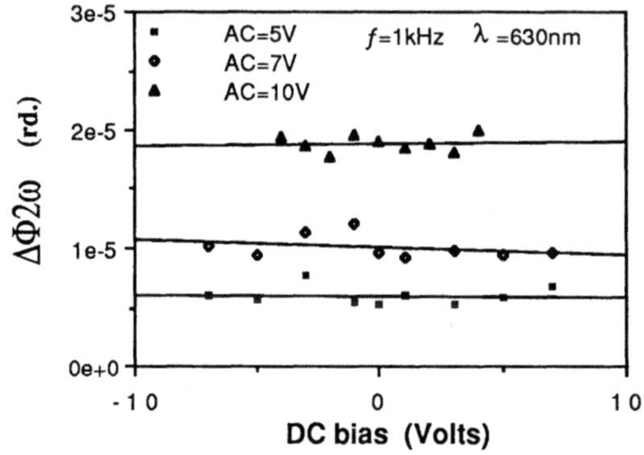

fig. 5-b

<u>**Figure 5**</u>: AC voltage and DC bias dependence of the normalized fundamental frequency (5-a) and first harmonic (5-b) maximum reflectance intensities expressed in term of the optical path containing the sample.
A " Cr (3000Å) / i (1μm a-Si:H) / Cr (2000Å) / Corning " sample was used.

XANES STUDY OF AMORPHOUS SILICON

A. Di Cicco, A. Bianconi, C. Coluzza, P. Rudolf

Dipartimento di Fisica, Università di Roma "La Sapienza", 00185 Roma, Italy.

P. Lagarde, A. M. Flank

LURE, CEA-CNRS- MEN Universite Paris Sud , Bat. 209D, Orsay, France

A.Marcelli

Laboratori Nazionali di Frascati, I.N.F.N., 00044 Frascati, Italy

Different levels of order of amorphous silicon (a-Si) samples have been measured by XANES (X-ray Absorption Near Edge Structure) analysis. We have investigated several samples of a-Si which were grown at different substrate deposition temperatures by using the ion beam sputtering technique. The multiple scattering expansion of the x-ray absorption coefficient and the spherical waves formalism were employed in the comparison between calculated and experimental results. The analysis of the amorphous silicon XANES spectra shows that increasing the substrate deposition temperature a continuous structural change and an higher level of short-range order is obtained over distances up to about 8Å. The bombardment with argon ions reduces the order in the second-neighbours coordination shell.

1.Introduction

Recent advances in data analysis of x-ray absorption spectroscopy have stimulated a great amount of work both from theoretical and from experimental point of view[1]. X-ray absorption spectroscopy is a structural technique which contains information about bond distance distribution via EXAFS (Extended X-ray Absorption Fine Structure) and disorder in bond angles via XANES (X-ray Absorption Near Edge Structure) analysis.

Many recent papers have pointed out that spherical wave formalism and multiple scattering effects are essential for the interpretation of the absorption spectra especially in its low-energy side. In particular, it has been shown how a multiple scattering signal can be extracted from an absorption spectrum by using the recently demonstrated expansion[2-5]: $\alpha(k)=\alpha_o(k)[1+\Sigma_n\chi_n(k)]$ where the absorption spectrum is described through a sum of χ_n terms due to the scattering paths of the photoelectron involving n-1 atoms around the photoabsorber. Here χ_2 is the usual EXAFS term, due to the single scattering, which is dominant at high energies of the photoelectron and probes the pair correlation function. The χ_n terms can be important in the XANES part of the spectra and probe the higher-order correlation functions.

X-ray or neutron diffraction investigation on amorphous systems furnish only the one-dimensional correlation function and cannot select between structures with different level of short-range order[6]. Therefore the analysis of XANES spectra can be an unique technique for understanding structural properties of amorphous materials .

XANES and EXAFS analysis of crystal silicon[4,5] have shown that the χ_3 signal plays in this case an important role. Disorder in bond angles in amorphous silicon determines a decrease in the intensity of the χ_3 contribution to the absorption spectrum.

In this contribution we want to present an investigation on several amorphous silicon (a-Si) samples prepared in different conditions. A connection between structural properties and preparation procedure can be established through comparison of the K-edge XANES spectra of different a-Si samples.

2.Experimental

Thin a-Si films (~ 1.5 μm) have been prepared by ion-beam sputtering: a 500 eV Ar beam is used to sputter silicon from a polycrystal target to a heated beryllium substrate. The samples have been deposited at three different SDT: room temperature (RT), 200°C and 400°C. Some of these 400°C-deposited samples were also bombarded with argon ions during the growth. X-ray absorption measurements were recorded in transmission mode at the synchrotron radiation dedicated storage ring ACO (Orsay) by using a double InSb(111) crystal monochromator and one ionization chamber as photon detector.

3.Results and discussion

X-ray absorption spectra of different a-Si films (RT-prepared, 200°C-prepared and 400°C-prepared ones) are reported in fig.1. The energy range of the measurements covers both the EXAFS and the XANES part of the spectra. In the EXAFS energy range (for wave vector values $k>4 Å^{-1}$, where $k=(1/\hbar)\sqrt{2m(E-E_o)}$ and $E_o=1836$ eV) there are no relevant differences between the spectra, on the contrary the high-resolution XANES part of the spectra is very sensitive to the preparation, i.e. to the local ordering, of the samples. These differences can be easily observed by considering the first derivatives of the spectra (derivatives' box in fig.1).

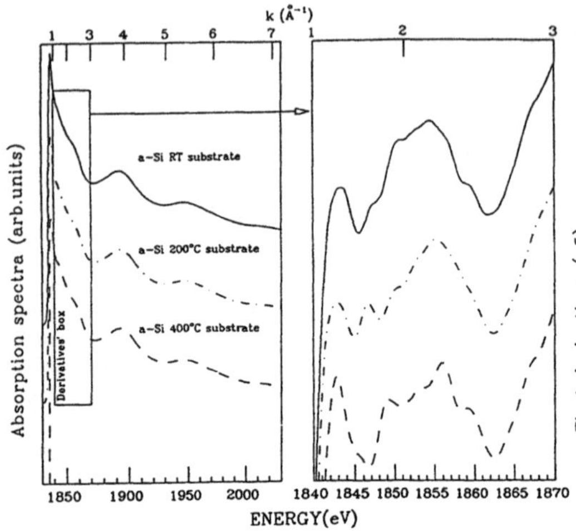

Fig.1: X-ray absorption spectra of amorphous silicon (a-Si) samples grown at different substrate deposition temperature. The panel on the left contains the normalized absorption spectra of each sample described in the text, the right one the derivatives of each spectrum drawn in the XANES energy range where there are relevant differences between the spectra.

The first step of the analysis is the evaluation of the modulating part of the absorption coefficient $\chi_{expt}(k)$, defined as:

$$\chi_{expt}(k)= [\alpha_K(k)-\alpha_o(k)]/\alpha_o(k)$$

where $\alpha_K(k)$ and $\alpha_o(k)$ are, respectively, the K-edge contribution and the atomic one to the total absorption coefficient. We have used a standard procedure for removing $\alpha_o(k)$ from the experimental spectra.

The $\chi_{expt}(k)$ signal contains information about local arrangement of the atoms around the photoabsorber which can be extracted by using the multiple scattering theory[2,7]. The region where the modulating signals $\chi_{expt}(k)$ of the different samples are significantly different is the XANES part of the spectra.

The $\chi_{expt}(k)$ signals relative to three samples of a-Si are shown in fig.2. In the EXAFS energy range we observe a regular oscillation due to the single scattering signal of the first neighbour coordination shell. A simple Fourier-analysis shows that differences in XANES spectra cannot be assigned to a change in the first

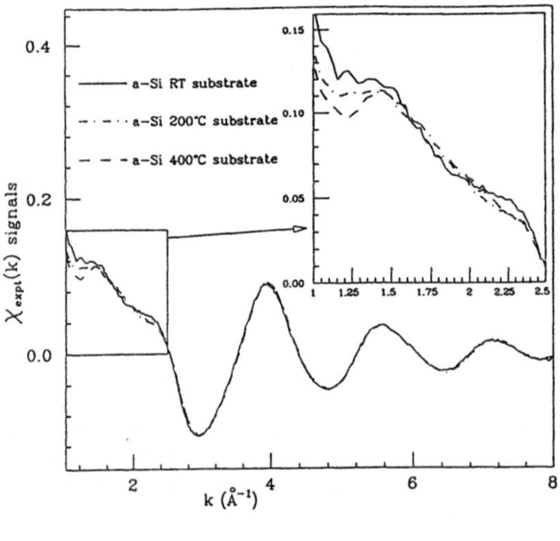

Fig.2: $\chi_{expt}(k)$ signals of a-Si samples grown at differente substrate deposition temperature. The low-energy part of the spectra has been enlarged in the panel inside the figure in order to show the different behaviour of the XANES spectra, which depend on the preparation procedures.

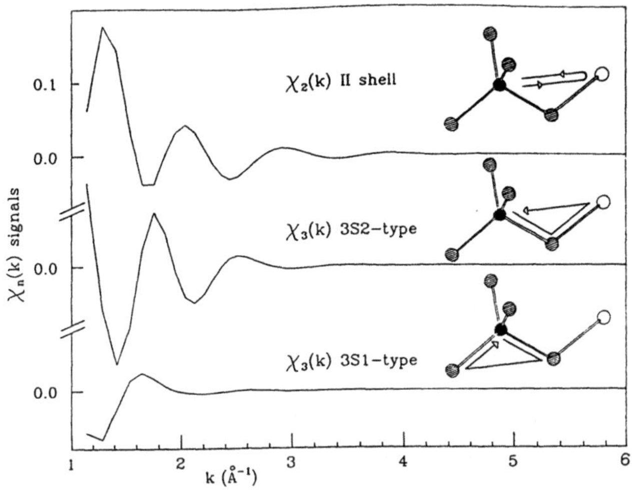

Fig.3: Calculated χ_2 and χ_3 contributions to the Si K-edge absorption coefficient for the geometry of crystal silicon. On the right the scattering pathways of the photoelectron are drawn (the photoabsorbing atom is shown as a black sphere). The signals are strongly damped by using effective Debye-Waller-like factors $[\exp(-2k^2\sigma p_n{}^2)]$ expected for the amorphous silicon case.

they can be adopted as signatures of ordering. In particular a reduction of the MSRD of each p_n contribution can explain the differences in the XANES spectra of the different a-Si samples.

In fig.3 the χ_2 and χ_3 signals are damped with reasonable values of MSRD which take into account both thermal and static disorder in amorphous silicon structure. These values were estimated starting from random-network models[11-13] and fitting procedures: $\sigma p_{2S2}{}^2$ =.09 Å^2, $\sigma p_{3S1}{}^2$ =.13 Å^2, $\sigma p_{3S2}{}^2$ =.14 Å^2. Although these relevant χ_2 and χ_3 contributions cannot explain the total high-frequency contribution to the a-Si XANES spectra they can be used as a general guide in an investigation of the transition from the amorphous to the crystalline phase.

Analysis of the Fourier-filtered difference spectra (free from low-frequency contribution) can be carried out by taking advantage of eq. (5) and of the χ_2 and χ_3 calculated signals. We have assumed a maximum level of disorder, as it is justified by the preparation and the characterization of the samples, and fixed MSRD values in the RT-prepared sample. In order to discuss the difference spectra we consider the three signal of fig. 3 which are dominant in the crystalline case by utilizing eq. 5 and limiting the series to the corresponding three paths.

A fitting procedure was employed, by using the MINUIT program of the CERN library, in order to compare experimental differences and theoretical signals.

In fig. 4 we put in comparison experimental (solid curves) and calculated signals (dot-dashed curves). We have considered in the calculation the signals of fig.3 (Fourier-filtered in the same range of the experimental differences) allowing the relative variation $\Delta\sigma p_n{}^{i2}=\sigma p_n{}^{RT2}-\sigma p_n{}^{i2}$ and $Np_n{}^i$ parameters to change in order to

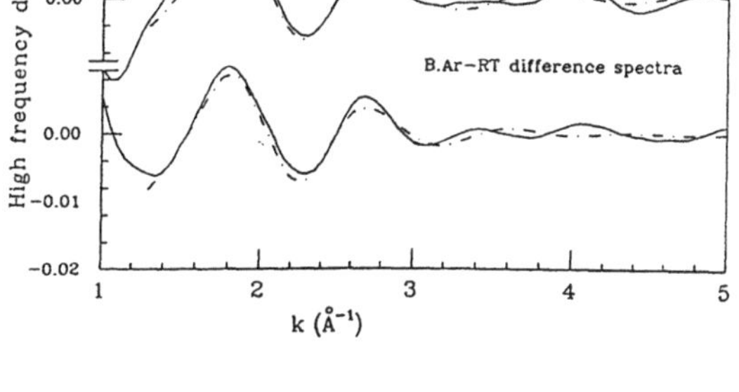

Fig.4: XANES difference spectra of the $\chi_{expt}(k)$ signals of the different a-Si samples. Solid curves are the experimental difference spectra filtered and free from low frequency contributions. The are compared with calculated results (dot-dashed curves) obtained taking into account single and double scattering contributions.

shell contribution but they are due to higher frequency signals, identified as multiple scattering and single scattering second shell terms.

Fourier-Transform technique, previously used in the a-Ge case[8], cannot be useful to analyze these energy-limited signals: in our case a subtraction procedure is more sensitive to variations of the high-frequency signals, moreover calculated multiple scattering contributions can be compared with the experimental spectra for obtaining structural information.

The multiple scattering expansion gives us a formula for analyzing XANES and EXAFS K-edge spectra[2-5]:

$$\alpha_K(k)=\alpha_0(k)[1+\sum_2^{\infty}{}_n\chi_n(k)] \quad (1) \quad ; \quad \chi_{expt}(k)=\sum_2^{\infty}{}_n\chi_n(k) \quad (2)$$

here $\chi_n(k)$ represents the contribution arising from all MS pathways p_n beginning and ending at the central atom and involving n-1 neighboring atoms.

The general expression for $\chi_n(k)$ is given by:

$$\chi_n(k)=\Sigma p_n A p_n(k)\sin[kRp_n+\phi_n(k,p_n)+2\delta_1{}^o]=\Sigma p_n A p_n(k)\sin[\phi^{eff}(k,p_n)] \quad (3)$$

where the dependence of the amplitude and phase function Ap_n and ϕ_n on the particular photoelectron path has been indicated by p_n and $\delta_1{}^o$ is the central atom phase shift. The amplitudes Ap_n's decrease with increasing order n, so that usually $\chi_2(k)$ is the dominant term in the whole energy range.

A Debye-Waller-like damping factor $\exp[-2k^2\sigma p_{2j}{}^2]$ is required to take account of thermal disorder for crystalline materials in the single scattering contribution χ_2 due to the j-th coordination shell. In amorphous systems is sometimes necessary to consider the actual asymmetric pair distribution function. For χ_n, n>2, contributions there are no simple and general expressions for introducing thermal and static disorder. A simple approach to this problem is to consider again an effective Debye-Waller-like damping factor for each particular path p_n. This point of view is supported, at least for the double scattering contribution in the case of amorphous silicon, by tentative calculations of the χ_3 contribution in this system[9]. It can be a good approximation of the actual signal in the low-energy side of the spectra ($k<4Å^{-1}$) especially for local structures with a relatively low level of disorder.

We have taken account of both inelastic and disorder effects in the calculated signals by using an effective total damping factor

$$\exp[-Rp_n/\lambda_{eff}-2k^2\sigma p_n{}^2] \quad (4)$$

for each multiple scattering path of interest in amorphous silicon.

In the rest of the paper we take the RT-prepared sample (expected to have the most disordered structure) as reference by considering the subtraction of its x-ray modulating signal from the other ones. Therefore, in terms of the multiple scattering expansion (2), we are going to consider a signal:

$$S_i(k)=\chi_{expt}{}^i(k)-\chi_{expt}{}^{RT}(k)=$$
$$=\Sigma p_n A p_n(k)\sin[\phi^{eff}(k,p_n)]\exp[-Rp_n/\lambda_{eff}]\{(Np_n{}^i)\exp[-2k^2\sigma p_n i^2]-(Np_n{}^{RT})\exp[-2k^2\sigma p_n RT^2]\} \quad (5)$$

where the index i runs over the x-ray absorption modulating signals of the a-Si samples prepared in different conditions. By writing the eq. 5 we assume that no dramatic structural changes occur in different a-Si samples in such a way that photoelectron paths of interest p_n, phase shifts, effective mean free path and average total length Rp_n remain unvaried. Mean square relative displacements (MSRDs) $\sigma p_n{}^2$ (generalized to multiple scattering geometries) and effective coordination numbers Np_n of the paths are allowed to change from the XANES spectra of the RT-prepared samples to the others. These changes characterize the gradual transition to a more ordered phase.

Previous results on crystal[5] and amorphous[4,10] silicon K-edge spectra suggest that a small cluster which comprises only two shells of neighbour atoms can be sufficient to explain the amorphous silicon XANES spectra. By considering such a 17-atoms cluster inside the crystal silicon structure three major x-ray absorption signals can be found, apart from the main first shell signal of low frequency. In fig. 3 we show these relevant signals: the single scattering χ_2 (2S2-type) contribution due to the second shell atoms (upper curve); the double scattering χ_3 (3S2-type) signal due to the shortest triangular paths involving second shell atoms (center curve); the χ_3 contribution (3S1-type) due to the shortest triangular paths findable inside the first-neighbours cluster (lower curve).

A previous paper demonstrate how these signal are detectable in the case of crystal silicon[4]. In the transition from the amorphous to the crystalline phase a raising of these contributions has to be expected. Thus

Path type p_n	$\Delta\sigma^2$ (200°C)	$\Delta\sigma^2$ (400°C)	$\Delta\sigma^2$ (B-Ar,400°C)
3S1	.04(2)	.06(2)	.06(2)
3S2	.022(5)	.030(5)	.016(5)
2Sh	.004(5)	.027(5)	.002(5)

Tab.I Parameter values obtained by best-fit calculation of the difference spectra of fig.4. We report the calculated variations $\Delta\sigma p_n i^2 = \sigma p_n RT^2 - \sigma p_n i^2$ (where i indicates 200°C 400°C and Ar bombarded a-Si samples) of the MSRDs relative to the three scattering paths p_n described in the text. The ordering tendency obtained increasing the SDT of the a-Si samples is marked by the correspondent decrease of the MSRD values.

obtain the best agreement with the experimental curve.

Coordination numbers of each path in each sample were found to be the same within the experimental errors, while MSRDs showed changes which reflect the gradual ordering of the local structure. In tab.I the calculated variations of the MRSDs of each relevant path are reported.

Looking at tab.I and fig.4 we recognize two steps in the gradual transition to the crystalline phase :

1) *in the 200°C-prepared sample an higher level of order is found inside the first-neighbour atoms coordination shell while second-shell atoms seems to be not involved in the ordering process.* In fact, we observe a remarkable change of the MSRDs relative to the 3S1-type path (completely inside the tetrahedron) and of the 3S2 one (which involves two atoms of the first shell) while the χ_2 second shell signal remains unvaried. In fig. 4 this fact can be observed by noting that the main two peaks A and B of the experimental difference spectrum correspond to the convolution of the analogous 3S1 and 3S2 peaks of the calculated signals of fig. 3.

2) *In the 400°C-prepared sample an increase of the second-shell atoms ordering is found.* The important variation of the MSRD of the χ_2 second-shell signal indicates that the ordering process now involves the second-shell atoms too. In fig. 4 the enlargement of the A peak, the slight shift of the B one and the appearence of a feature at about 3.8Å^{-1} reflect the presence of a detectable second-shell signal in the 400°C-prepared sample.

3) *The 400°C-prepared sample bombarded with Argon ions during the growth show a decreased level of local order for the second-shell atoms.* This fact can be easily seen by observing the similiraty of the Ar-Bombarded difference spectrum with the 200°C-prepared one and considering the variation of the MSRD values of tab.I.

4.Conclusions

Accurate investigation of x-ray absorption spectra of a-Si samples prepared in different conditions have shown structural differences. Comparison between calculated and experimental results has allowed to understand some features of the ordering process due to the increase of the substrate deposition temperature.

A continuous structural change is found as a function of the substrate temperature. We have shown that in the 200°C-prepared sample there is a higher level of order limited to the first neighbours while for the 400°C-prepared one the short-range ordering process is extended to the second-shell atoms.

References

1) "EXAFS and Near Edge Structure IV", *edited by P. Lagarde, D. Raoux, J. Petiau, Proc. of the international EXAFS and NES conference, Fontevraud (France), Journal de Physique , 47 Colloque C8 (1986).*
2) C.R.Natoli and M.Benfatto *in ref.1, 11 (1986)*
3) M.Benfatto, C.R.Natoli, A.Bianconi, J.Garcia, A.Marcelli, M.Fanfoni and I.Davoli, *Phys. Rev. B 34, 5774 (1986)*
4) A.Bianconi, A.Di Cicco, N.V. Pavel, M.Benfatto, A. Marcelli, C.R. Natoli, P. Pianetta, J. Woicik, *Phys. Rev. B, 36, 6426 (1987)*
5) A.Di Cicco, N.V. Pavel, A.Bianconi, *Solid State Commun. 61, 635 (1987)*
6) G. Etherington, A.C. Wright, J.T. Wenzel, J.C. Dore, J.H. Clarke, R.N. Sinclair, *Jour. Non-cryst. Sol. 48, 265 (1982)*
7) P.A.Lee, J.B.Pendry, *Phys.Rev. B, 11, 2795 (1975)*
8) F.Evangelisti, M.G.Proietti, A.Balzarotti, F.Comin, L.Incoccia and S.Mobilio, *Solid State Comm. 37, 413 (1981)*
9) A. Di Cicco, *Thesis, University of Rome (1987)*
10) A.Balerna, M.Benfatto, S.Mobilio, C.R.Natoli, A.Filipponi and F.Evangelisti *in ref.1, 63 (1986)*
11) D.E. Polk, D.S. Bodreaux, *Phys Rev. Lett. 31, 92 (1973)*
12) F.Wooten, K.Winer and D. Weaire, *Phys.Rev. Lett. 54, 1392 (1985)*
13) K. Winer *Phys. Rev. B 35, 2366 (1987).*

DENSITY OF METASTABLE STATES IN a-Si:H - SPECTROSCOPIC INVESTIGATION ON MATERIAL AND DEVICES

J.Kočka, M.Vaněček, O.Štika, Q.Dung Tring, J.Stuchlík and A.Tříska.
Institute of Physics, Czechoslovak Academy of Sciences,
Na Slovance 2, 180 40 Prague 8, Czechoslovakia

ABSTRACT

The density of metastable gap states created by light soaking has been studied directly on amorphous hydrogenated silicon solar cells. Comparison of the constant photocurrent method on forward and reverse biased 2.4 μm thick p-i-n diodes has been done in state A and B. Together with it the shift of the Fermi level of undoped middle layer has been deduced from measured temperature dependence of a.c.conductivity in a proper frequency range.

1. INTRODUCTION

To improve the efficiency of solar cells, based on amorphous hydrogenated silicon (a-Si:H), better understanding of the physics of both a-Si:H material and p-i-n solar cells is necessary. Usually the basic physical parameters like Fermi level (E_F), gap density of states (DOS) and the mobility-lifetime ($\mu\tau$) product are studied (1) on coplanar samples, prepared under conditions as similar as possible (2) to p-i-n devices and correlated with the p-i-n device performance (the efficiency for example).

The aim of this paper is to study the reversible photoinduced changes (Staebler-Wronski effect) mainly of E_F and DOS of undoped middle layer of p-i-n diode directly on the diode.

2. EXPERIMENTAL DETAILS AND RESULTS

The samples used in this work have been <u>thick p-i-n diodes</u> (see fig.1.a) with the structure : glass /ITO/ p^+-a-SiC:H (150 Å) /i-a-Si:H (2.4 μm)/ n^+-a-Si:H (400 Å) /NiCr, in which the undoped layer (DOS of deep defects in annealed state A, N(A) $\approx$ 4×10^{15} cm^{-3}) has been prepared by glow discharge decomposition under standard conditions (pure SiH_4, T_S=250°C, low RF power, growth-rate 3 Å/sec.)

The light soaking (1 hour, RT-sample air cooled, 400 mW/cm^2 of heat filtered white light) and annealing(1 hour at 165°C) have been used to obtain the state B and A respectively.

Recently we have studied on coplanar samples the influence of stress and of the light soaking (3). We have observed the " $\mu\tau$ " versus " α " duality (α - subgap optical absorption coefficient), demonstrated by strong initial decrease of $\mu\tau$ with nearly no change of " α ". This duality has been simply explained by the recharging ($D^- \rightarrow D^0$) of the silicon dangling bond (DB) related states, caused by 0.12 eV shift of the E_F towards midgap. This shows the <u>importance of the knowledge of the shift of the Fermi level.</u>

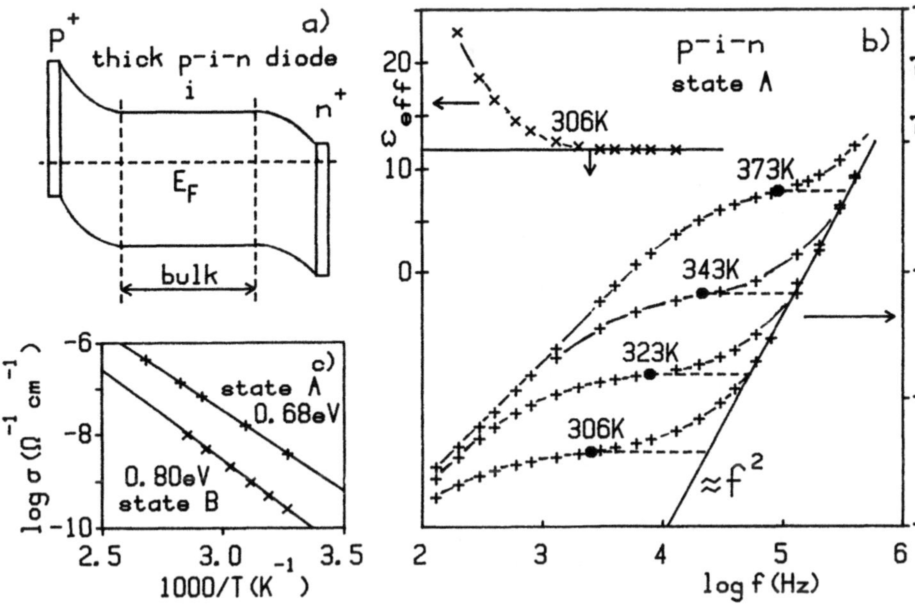

Fig.1.
a) Schematic band diagram of a thick p-i-n diode.
b) Frequency dependences of the electrical conductivity (σ) and effective permittivity (ε_{eff}) of a 2.4 μ m thick p-i-n diode in state A measured at different temperatures.
c) Temperature dependence of correct σ of undoped middle layer of p-i-n diode (deduced from a.c.conductivity in fig. 1.b) in state A and B. $E_{act} \approx E_C - E_F$ is taken as a parameter.

But how to monitor the E_F shifts in undoped middle layer of thick p-i-n cell due to the light soaking ? The only possibility of "getting inside" the p-i-n diode is to measure the a.c.conductivity at such frequencies at which the junction effects are excluded. Such experimental results are shown in fig.1.b. For each temperature we have found the frequency at which the effective permittivity ($\varepsilon_{eff} = (d/S\ \varepsilon_o) \times C_{meas}$) approaches to the correct a-Si:H value ($\varepsilon \approx 11$) and also electrical conductivity (σ) saturates. This is then the correct bulk " σ " of undoped a-Si:H. (At high frequencies $\sigma(f) \approx f^2$ is observed. This effect is related to the series resistance of ITO layer,see (4)).

We performed such measurements in the state A (see fig.1.b) and also in the state B (not shown). In fig. 1.c it is clearly demonstrated that during the light soaking of p-i-n diode the E_F of the undoped a-Si:H middle layer moves, $\approx$ 0.12eV towards midgap,similarly to coplanar samples(3). This results in a strong decrease of the p-i barrier height, the most sensitive part of the p-i-n junction. So we can say that the S-W effect is a material (bulk) effect with "device consequences".

The defect DOS is the very basic a-Si:H parameter, Si-DB in different charge states (D^o, D^+, D^-)being the most important defect. We currently use

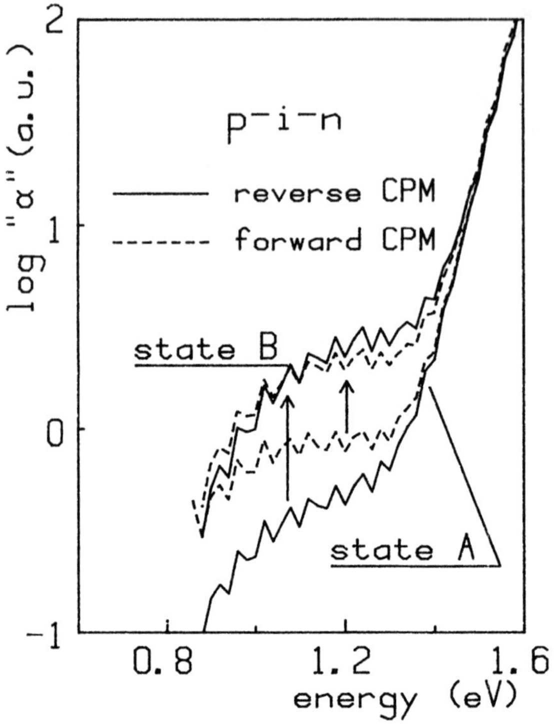

Fig.2.
The spectral dependence of CPM deduced " α " for the forward and reverse biased 2.4 μm thick p-i-n diode in state A (annealed 1 hour at 165°C) and light soaked state B (400 mW/cm^2 for 1 hour).

the deconvolution of the optical absorption coefficient " α " in the subgap region for finding of the defect DOS. For its measurement we have used the Constant Photocurrent Method (CPM - (5)), so far on coplanar samples or on forward biased Schottky diodes (6). The other technique, Photothermal Deflection Spectroscopy (PDS), successfully used (7) on coplanar samples, can hardly be used on p-i-n structure and to our knowledge the only photoconductivity technique, used on p-i-n junction in subgap region is the Modulated Photocurrent Spectroscopy(MPS),a kind of primary(8) photocurrent, measured in short circuit by Okamoto et al (9).

Important information in the above gap energy region has been recently obtained by so called DICE method (10) from the spectral dependence of the forward (secondary) photocurrent, normalized to reverse (primary) photocurrent. This, together with the fact that we have found only small phase shifts with the aid of MPS (9), stimulated us to use CPM on p-i-n junction (the same as in fig.1b) and to compare the results in forward bias (just above V_{oc} -secondary photocurrent) and reverse bias (primary photocurrent). Such results for a thick 2.4 μm p-i-n diode in the state A and B are in fig.2. Results for ≈ 0.5 μm thin p-i-n diodes with the efficiency of η ≈ 9% are more complicated and are not presented here.

3. DISCUSSION AND CONCLUSIONS

Okamoto et al (9) concluded, that instead of one contribution only, i.e. $D^o/D^+ \rightarrow$ CB (see fig.3.a) observed by us (5), there are two contributions to the subgap absorption of undoped a-Si:H : one from D^o state to CB (see fig. 3.a) and the other from VB to D^- state (D^o/D^- transition). However, there is no contradiction. Okamoto et al (9) measure the primary photocurrent (8) to which both (electron and hole) transitions contribute. We have used in CPM the secondary photocurrent, to which the predominant contribution is by the transitions creating the free majority carriers. For undoped a-Si:H (it means lightly n-type: $E_C-E_F \approx 0.7$eV) these are $D^o/D^+ \rightarrow$ CB transitions (the sign of the photocarriers can be checked by the photothermopower) and so the hole (minority) transitions could not be observed. In agreement with Okamoto et al we have found the D^o level at E_C-0.9 eV. Moreover, the position of D^- level in undoped a-Si:H, found at E_C-0.6 eV from our SCLC data (11), also agrees with that given by Okamoto et al (9).

There is one situation only where the deconvolution of secondary CPM spectra is more complicated. That is when the material is really intrinsic, with approximately equal $\mu\tau$ of electrons and holes. Then both transitions, creating free holes and free electrons, contribute even to the secondary CPM photocurrent.

On the basis of the above comments to CPM and the primary photocurrent results on Schottky diodes (12) we can explain the results in fig.2 as follows (see fig.3b) :

In the state A, where due to the position of E_F all D^o states and a substantial part of D^- states are occupied, $\mu\tau_e \gg \mu\tau_h$ and the electrons are the majority carriers. Then the "forward CPM" deduced spectrum is controlled by the $D^o/D^+ \rightarrow$ CB transitions (with D^o at $E_c - 0.9$ eV), on the other hand the "reverse CPM" spectrum is controlled by the VB $\rightarrow D^o/D^-$ transitions, with D^- at $\approx E_V + 1.2$ eV. The total number of Si-DBs estimated by deconvolution is $N_{DB}(A) \approx 4\times10^{15}$ cm^{-3}.

After light soaking (state B) $\mu\tau_e$ decreased, due to the E_F shift, $\mu\tau_e \approx \mu\tau_h$ and the number of occupied (D^o) states is equal to the number of unoccupied (D^-) states, hole and electron generation rates are comparable. In this case both carriers contribute in the "reverse" as well as in the "forward" CPM and both spectra should be similar, as observed. The total number of Si-DBs increased to $N_{DB}(B) \approx 2\times10^{16}$ cm^{-3} by light soaking.

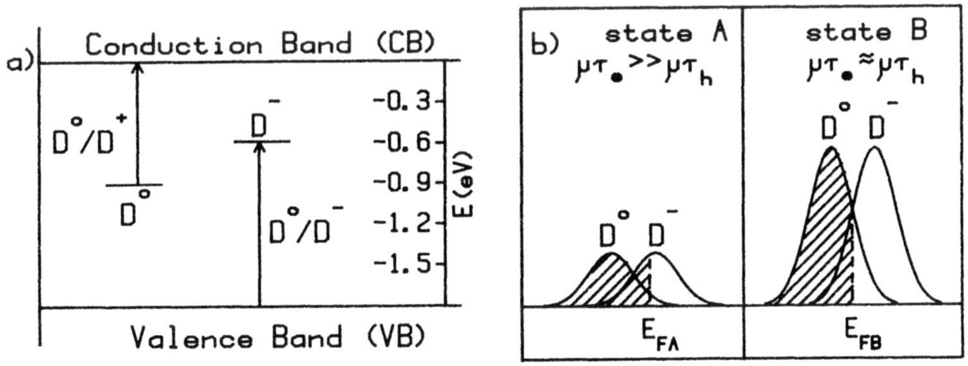

Fig.3.

a) Schematic illustration of two contributions to subgap optical absorption (Okamoto et al (9)) of undoped a-Si:H.

b) Schematic illustration of the light-induced changes of dangling bond density.

We can summarize the above results and conclude that :

a) We have for the first time demonstrated by the direct measurement of a.c. electrical conductivity on p-i-n diode that after light soaking (1 hour, 400 mW/cm^2) the Fermi level in undoped "i" layer moved of about 0.12 eV towards the midgap. The barrier height at the p-i interface as well as $\mu\tau_e$ have decreased.

b) At the same time the number of deep defects (probably Si-DBs with D^o at E_C - 0.9 eV and D^- at $\approx E_V$ + 1.2 eV) in the "i" layer of the thick p-i-n diode increased from $N_{DB}(A) \approx 4\times10^{15} cm^{-3}$ to $N_{DB}(B) \approx 2\times10^{16} cm^{-3}$.

c) The apparent controversy between our results (5,6) and results by Okamoto et al (9) has been clarified.

d) The shift of E_F can be caused by the increase of N_{DB} itself or more probably (1) by the creation of new states in the VB tail region. These could be possibly the impurity related states (O,C,N) or even Si-H related (weak) bonds.

e) We can suggest a simple rule for improvement of thick p-i-n diodes : To make the ratio of "forward" to "reverse" CPM spectrum on thick p-i-n diode at $\approx$ 1.2 eV near to unity and both spectra as low as possible.

These are the first qualitative results of CPM on a-Si:H p-i-n diodes. More detailed analysis and measurements on real solar cells (thin p-i-n diodes) have yet to be done.

4.REFERENCES

(1) M.Ohnishi, S.Tsuda, T.Takahama, M.Isomura, Y.Nakashima, N.Nakamura, S.Nakano, T.Yazaki and Y.Kuwano, Jap.J.of Appl.Phys.26,9 (1987) 1408

(2) M.S.Bennett, S.Wiedeman, J.L.Newton and K.Rajan, Proceedings of the 1987 MRS Spring Meeting (Anaheim (USA), 1987)

(3) J.Kočka, M.Vaněček, M.Nesládek, Q.Dung Tring, J.Stuchlík, O.Štika and A.Tříska, J.of Non-Cryst.Sol. 97/98 (1987) 819

(4) R.A.Street, G.Davies and A.D.Yoffe, J.of Non-Cryst.Sol.5 (1970) 276

(5) J.Kočka, M.Vaněček, F.Schauer, J.of Non-Cryst.Sol.97/98 (1987) 715

(6) M.Vaněček, A.Abrahám, O.Štika, J.Stuchlík and J.Kočka, phys.stat.sol.(a)83 (1984) 617

(7) W.B.Jackson and N.Amer, J.Phys., Paris, 42, C4 (1981) 293

(8) R.S.Crandall, in Hydrogenated Amorphous Silicon, Semiconductors and Semimetals, Volume 21 B, Academic Press (1984)

(9) H.Okamoto, M.Kida, T.Tamada and Y.Hamakawa, Phil.Mag.B 52, 6 (1985) 1115

(10) T.Takahama, M.Isomura, S.Tsuda, J.Tarni, Y.Hishikawa, N.Nakamura, Y.Nakashima, T.Matsuoka, H.Nishiwaki, M.Ohnishi, S.Nakawa and Y.Kuwano, Jap.J.of Appl.Phys.25, 10 (1986) 1538

(11) F.Schauer, J.Kočka, Phil.Mag. B 52 (1985) L25

(12) H.M.Welsh, W.Fuhs, K.H.Greeb and H.Mell, J.Phys.,Paris,42, C4 (1981) 567

DIFFUSION LENGTHS IN a-SiGe:H ALLOYS FROM OPTICAL GRATING TECHNIQUE

G.H. BAUER, C.E.NEBEL, W.H. BLOSS
Institut fuer Physikalische Elektronik, Universitaet Stuttgart,
Pfaffenwaldring 47, D-7000 Stuttgart 80, F.R. Germany

ABSTRACT. Ambipolar diffusion lengths in undoped a-SiGe:H have been analyzed by Steady State Photocarrier Grating Technique. By Ge alloying diffusion length and photoconductivity of a-SiGe:H are considerably affected. Although photoconductivity in a-SiGe:H can be improved by special deposition methods, diffusion lengths turn out to remain nearly unchanged. The comparison of L_D and σ_{ph} yields hole mobilities in a-$Si_{1-x}Ge_x$:H of $10^{-2}cm^2/Vs$ for $0 \leq x \leq 0.2$. Light soaking in a-Si:H results in noticeable changes in diffusion lengths L_D, whereas in a-SiGe:H different σ_{ph} due to different deposition methods or due to light soaking end up in changes of L_D that do not correspond to simple reduction of life time. Thus the improvement of σ_{ph} of a-SiGe:H by sophisticated deposition is by far a much more complex process than only decrease of density of midgap states.

1. INTRODUCTION

Behaviour of minority carriers is of considerable interest with respect to characterization of optoelectronic quality of materials and thus the analysis of diffusion lengths in a-Si:H and alloys has attracted high interest. Especially, changes of minority carrier behaviour in a-Si:H-alloys, which in general become inferior by adding Ge or C, are of great importance. Commonly diffusion lengths L_D in a-Si:H have been evaluated from Surface Photo Voltage measurements (SPV) /1,2,3/ which do not yield figures for L_D but rather widths of space charge regions, of Dember effect, or of band bending due to surface recombination/4/.

Recently a novel method for analysis of ambipolar L_D - Steady State Photocarrier Grating Technique (SSPG) - has been reported and been applied to a-Si:H /5,6,7/ as well as to a-SiGe:H and a-SiC:H /8/. This method bases on photoconduction vs. geometry of an illumination pattern (grating) created by interferences of two laser beams that are focused on the specimen with coplanar contacts. It has the advantage that the influence of contact effects can be eliminated and that many influences like dependence on field and temperature, on light intensity and on degree of degradation by light soaking can easily be investigated.

2. EXPERIMENTS

Amorphous hydrogenated SiGe alloys from dc-glow discharge preparation /9, 10/ have been analyzed by SSPG. For sample preparation some optimization with respect to highly photoconductive material like H_2-dilution and triode reactor has been performed /9,10/ resulting in a considerable extension of the regime of high σ_{ph} to Ge contents of $C_{Ge} \leq 35\%$ /9,10/.

The experimental setup for SSPG recently published /5-8/ is shown in fig.1. The angle δ between I_1 and I_2, and the wavelength of the light determine the grating period $\Lambda = \lambda/2\sin(\delta/2)$ /6/; the angle $\delta = 2\arctan(x_0/y)$ has been selected by the position y of the sample and the distance $2x_0$ between the two mirrors. The lateral intensity alteration consequently is

$$I=(I_1+I_2)[1+(2\sqrt{I_1I_2}/I_1+I_2)\gamma_0\cos(2\pi x/\Lambda)]$$

where $0\leq\gamma_0\leq1$ reflects departures from ideal coherence /6/. Lateral photoconductivities $\sigma_{ph,g}(I_1\oplus I_2)$ under grated illumination, $\sigma_{ph}(I_1)$, and $\sigma_{ph}(I_1+I_2)$ at I_1 and (I_1+I_2) without interferences have been monitored in one procedure. These data are plotted as a function ß vs. grating period Λ:

$$\beta(I_1,I_2,\Lambda)=[\sigma_{ph,g}(I_1\oplus I_2)-\sigma_{ph}(I_1)]/[\sigma_{ph}(I_1+I_2)-\sigma_{ph}(I_1)].$$

A numerical fit of ß vs. Λ which also includes surface recombination as well as incomplete coherence and the exponent γ_s of σ_{ph} vs. illumination /6/has been performed in order to derive diffusion lengths L_D.

2. PHOTOCONDUCTIVITY AND AMBIPOLAR DIFFUSION LENGTHS IN a-SiGe:H

In figs. 2, 3 photoconductivity σ_{ph} at monochromatic illumination and ambipolar diffusion lengths L_D of undoped a-$Si_{1-x}Ge_x$:H films vs. composition x are outlined. The photoconductivity has been monitored at the same light flux that was adjusted for the diffusion length measurement ($3x10^{17}$ photons/cm^2s). By carefully focusing optics, quality factors γ_0 in the range 0.6 to 0.9 have been achieved which limits the inaccuracy of diffusion lengths data to< ± 15%.

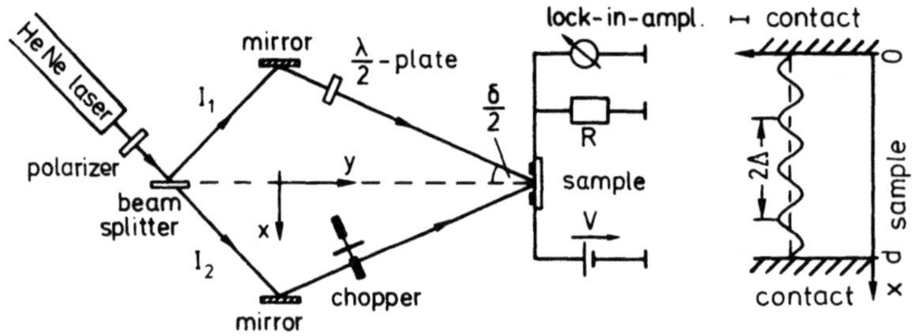

Fig.1. Experimental setup for SSPG.

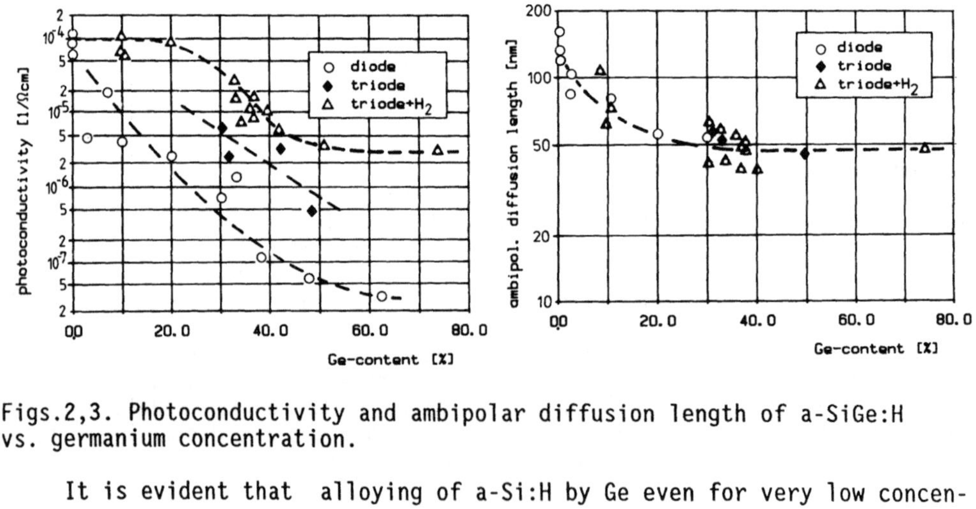

Figs.2,3. Photoconductivity and ambipolar diffusion length of a-SiGe:H vs. germanium concentration.

It is evident that alloying of a-Si:H by Ge even for very low concentrations reduces L_D considerably and the drop in our case is nearly independent of deposition method (○,◆,△), although for c_{Ge} up to about (30-40)at%, corresponding to band gaps of (1.5-1.4) eV, σ_{ph} has been improved by H_2-dilution and triode reactor. Comparable increases in σ_{ph} also accompanied by a weak shift in Fermi level to E_c has been reported by several groups applying different deposition methods (H_2-dilution of the neutral gas, triode reactor, Photo-CVD /11,12,13/), so that we assume these diffusion lenghts to some extent to be representative for those films too.

From L_D and σ_{ph} the ratio $b=\mu_n/\mu_p$ has been calculated /8/ and combined with μ_n from time-of-flight in order to derive effective hole mobilities μ_p. Fig.4 shows μ_p of a-SiGe:H deposited in a diode glow discharge without H_2 dilution. These effective hole mobilities in principal reflect the mean value of the VB tail state distribution, times its occupation, and times the energy dependent mobility $\mu_p(E)$, and since for $C_{Ge} \leq$ 20at% μ_p does not change , we can claim the VB-tail to remain nearly the same. PDS sub-band-gap-absorption outlined in fig.5 showing Urbach energies E_{0V} support the VB-tail in $C_{Ge} \leq 20\%$ to be nearly independent of Ge content. Since even for those films VB-tail in $C_{Ge} \leq 20\%$ remains nearly unchanged and as from PDS we derive that for films from triode/H_2-preparation E_{0V} doesn't increase much over $0 \leq x \leq 1$, we claim the VB-Tail also in low gap a-SiGe:H to be kept nearly unchanged, say, hole transport to be similar like in a-Si:H.

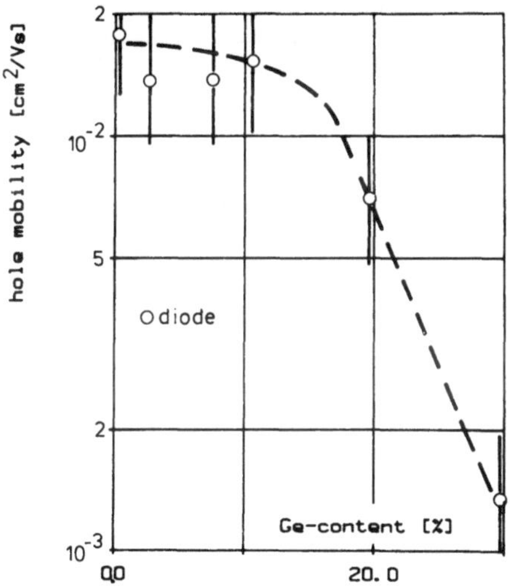

Fig.4. Hole mobility in a-SiGe:H.

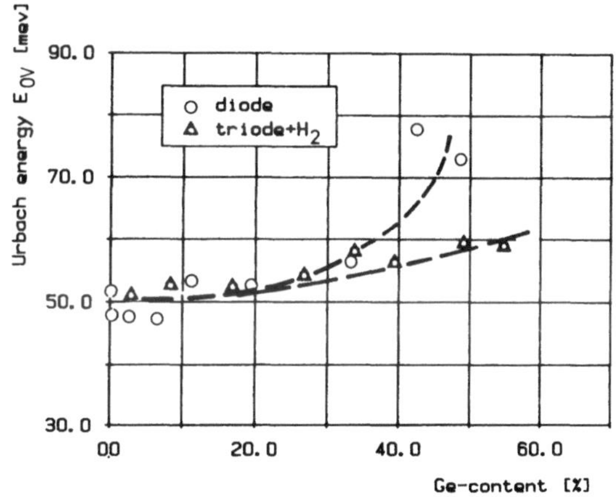

Fig.5. Urbach energy (PDS) in a-SiGe:H.

A significant difference between a-Si:H and a-SiGe:H emerges, when comparing annealed and light soaked a-Si:H with annealed and light soaked a-SiGe:H, which additionally show different σ_{ph} of the annealed states due to different preparation methods (○,△). Figs.6a,6b show ambipolar diffusion lengths L_D and photoconductivities σ_{ph} of a-Si:H in the annealed (1,4) and light-soaked state resp. (1',4', ≈50 h AM1 illum.), and those of two a-SiGe:H samples which have been grown under different conditions for high (≈10^{-4} 1/Ωcm: 3/5) or for low (≈10^{-7} 1/ Ωcm: 2/6) AM1-σ_{ph}.

In fig.6c L_D and σ_{ph} of light soaked a-SiGe:H (full lines) with high and low original σ_{ph} is outlined in comparison with their annealed data (dashed lines from fig.6b). The slopes of $\log(\sigma_{ph})$ vs. $\log(\Phi)$ are about 0.85-0.90 and correspond well with slopes in $\log(L_D)$ vs. $\log\Phi$ plot which are expected to satisfy $L_D \sim \Phi^{-(1-\gamma_s)/2}$.

Evidently, in a-Si:H and in low photoconductive a-SiGe:H by light induced generation of defects σ_{ph} and L_D decrease both attributed to an increase in density of midgap states and according to $\sigma_{ph} \sim \tau_s$ and to $L_D \sim (\tau_s)^{1/2}$. However in a-SiGe:H with high original σ_{ph} ambipolar diffusion lengths turn out to be nearly identical to those in the low σ_{ph} film. Moreover the drop in σ_{ph} by a factor of 200 after light soaking does not at all correspond to the decrease of L_D by only a factor of ≈1.5.

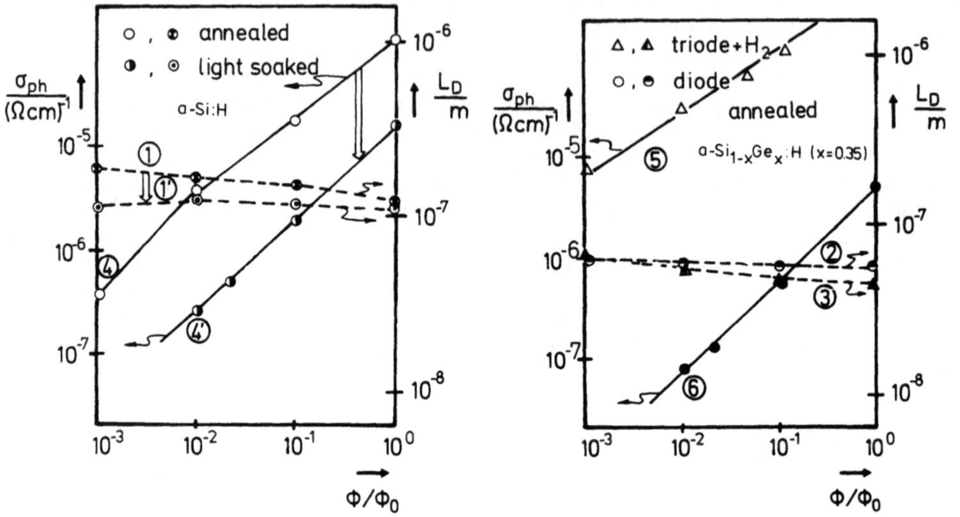

Figs.6a,6b. Photoconductivity and ambipolar diffusion lengths of a-Si:H (annealed (4,1)/light soaked(4',1')) and of annealed a-SiGe:H (C_{Ge}=0.35) from different deposition methods (diode (6,2)/triode+H_2(5,3)) versus light flux Φ.

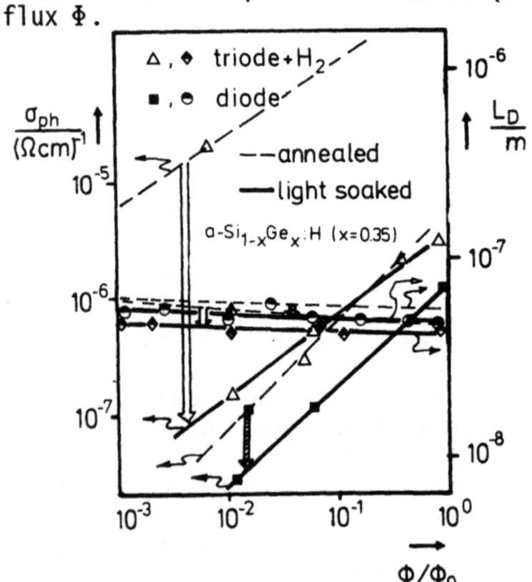

Fig.6c. Photoconductivity and ambipolar diffusion length of annealed (dashed lines from fig.6b) and light soaked (full lines) a-SiGe:H prepared by diode and triode+H_2 reactor versus light flux Φ.

3. DISCUSSION OF RESULTS

On the first view of our results on undoped films, σ_{ph} seems not to be correlated with ambipolar diffusion lengths at all, as in a-SiGe:H with similar band gap, which reads in equal Ge content, for low and for high σ_{ph}, diffusion lengths turned out to be in the same range. However in coplanar photoconductivity which approximately meets charge neutrality, σ_{ph} is additively composed of electron and hole mobility $\sigma_{ph} \sim (\mu_n + \mu_p)$, and in contrary ambipolar diffusion lengths are the product of μ_n and μ_p, $L_D \sim [(\mu_n \mu_p)/(\mu_n + \mu_p)]^{1/2}$. So in the presence of two strongly differing mobilities, the higher one determines nearly exclusively σ_{ph}, whilst the smaller one dominates L_D.

The reduction of σ_{ph} and the drop of L_D in a-Si:H and in low-σ_{ph} a-SiGe:H due to generation of additional defects can be explained quantitatively with a drop in τ; here we neither need a change in effective mobility of electrons nor of holes, which is supported by analyses of DOS indicating only minor changes of DOS in tails for light induced defect generation in a-Si:H compared with the increase in midgap density of states /14,15/.

For the explanation of the big differences in annealed state σ_{ph} (factor 100, fig.6b) of a-SiGe:H films with similar E_g showing the same L_D, we take similar μ_p (fig.4,5), calculate ratios of σ_{ph} and of L_D as fct. of τ, and get $\mu_n(\text{high } \sigma_{ph})/\mu_n(\text{low } \sigma_{ph})$ for these two a-SiGe:H films in the neighbourhood of 10^2. In addition we only need to introduce another factor of 2-4 for differences in τ in order to explain the difference in σ_{ph}. Since from PDS higher departure of DOS at midgap has been observed, the effect as to achieve such high σ_{ph} so far is not completely understood.

As the drop of σ_{ph} in highly photoconductive a-SiGe:H-films due to light soaking can not be explained either in terms of reduction of τ, we claim a considerable shift in Fermi level towards midgap over D^- and/or D^0-states (from $\sigma_{dark}(T)$: $E_C - E_F = 0.61$ eV, annealed / $E_C - E_F \geq 0.72$ eV, light soaked) that changes the recombination of electrons and holes as well.

Acknowledgement
This work has been supported by BMFT contr. no. 03E-8327-B.

References

/1/ J. Dresner et al., Appl. Phys. Lett. **38**, 998 (1981).
/2/ A.R. Moore, Appl. Phys. Lett. **40**, 403 (1982).
/3/ A.R. Moore, J. Appl. Phys. **56**, 2796 (1984).
/4/ M. Foller et al., J. Noncryst. Sol. **97/98**,567 (1987).
/5/ D. Ritter, E. Zeldov, K. Weiser, Appl. Phys. Lett. **49**, 79 (1986).
/6/ D. Ritter, K. Weiser, E. Zeldov, J. Appl. Phys. **62**, 4563 (1987).
/7/ D. Ritter et al., J. Noncryst. Sol. **97/98**, 571 (1987).
/8/ G.H. Bauer et al., Proc. MRS Spring Meeting 1988, Symp.E, in print.
/9/ H.C. Weller et al., Conf.Rec.19 IEEE PVSC, IEEE N. York(1987), p.872.
/10/ H.C. Weller et al., J. Noncryst. Sol. **97/98**, 1017 (1987).
/11/ A. Matsuda, K. Tanaka, J. Noncryst. Sol. **97/98**, 1367 (1987).
/12 V. Dalal et al., Conf.Rec.18 IEEE PVSC, IEEE N. York(1985), p.1500.
/13/ M. Konagai et al.,Proc. PVSEC-2, Adfield Co. Hong Kong(1986),p.417.
/14/ A. Glade. W. Beichler, H. Mell, J. Noncryst. Sol. **77/78**, 397 (1985).
/15/ C.E. Nebel, H.C. Weller, G.H. Bauer, Lit /8/, in print.

EXTRINSIC ORIGIN OF DEFECTS IN AMORPHOUS SILICON

David Redfield
Department of Materials Science & Engineering
Stanford University — Stanford, CA 94305 — U.S.A.

ABSTRACT

A new treatment is given for the important question as to whether defects that cause degradation in a-Si:H solar cells are intrinsic to the material or due to extrinsic factors. This treatment follows a recently completed general description of the kinetics of defect formation and anneal. These kinetic results, combined with previous experimental data, lead to the conclusions that the defects (a)are related to the built-in defects that are not light induced, and (b)both have extrinsic origin . This picture suggests a microscopic model in which 3-coordinated foreign atoms are the source of the "weak bonds" that form defects.

A central question in solar cells of hydrogenated amorphous silicon (a-Si:H) is whether light-induced degradation is caused by an inherent property of the material or by some extrinsic effects such as impurities, defects, etc. If the origin of degradation is inherent, then we cannot expect to eliminate it. On the other hand, if the origin is extrinsic, there is, in principle, the prospect of reducing the degradation. This paper addresses this question by use of the results of a recent analysis of the kinetics of generation and anneal of light-induced defects based on a newly completed rate equation whose solution has unified many observations of light-induced effects and dark-relaxation effects [1]. A summary of that analysis is given here, and it is then shown how limiting cases of interest lead to the conclusion that degradation has an extrinsic source. Thus, if means can be found to exploit this result, there is promise for improvement of cell stability. Moreover, the same analysis indicates that this kind of stability improvement may simultaneously increase the cell efficiency [2], so there could be a compound benefit.

This treatment is based on a general description of light-induced metastable defects (MSDs) that are known to cause the degradation. No atomic model of the defects is needed or used by this method; in fact this treatment leads to a model. The essential properties that describe the rates of defect generation and anneal, as well as the steady-state defect densities, are deduced from a non-specific configuration-coordinate diagram of a metastable localized center shown in Fig. 1. The metastable state corresponds to the shallower energy well, having a barrier height E_2, which is the activation energy for anneal of the MSDs, and which is about 1 eV. This metastable state has observable properties of ESR, increased carrier recombination, optical absorption, etc. The density of these MSDs, which we designate by N_M, is measurable quantitatively by use of these properties. On the other hand, the ground state, which represents latent defects, seems to have no observable properties, so the density of latent defects N_L is not directly measurable. We assume that the total density of centers $N_T \equiv N_M + N_L$ is fixed for any material, although it may vary with doping and material-preparation conditions.

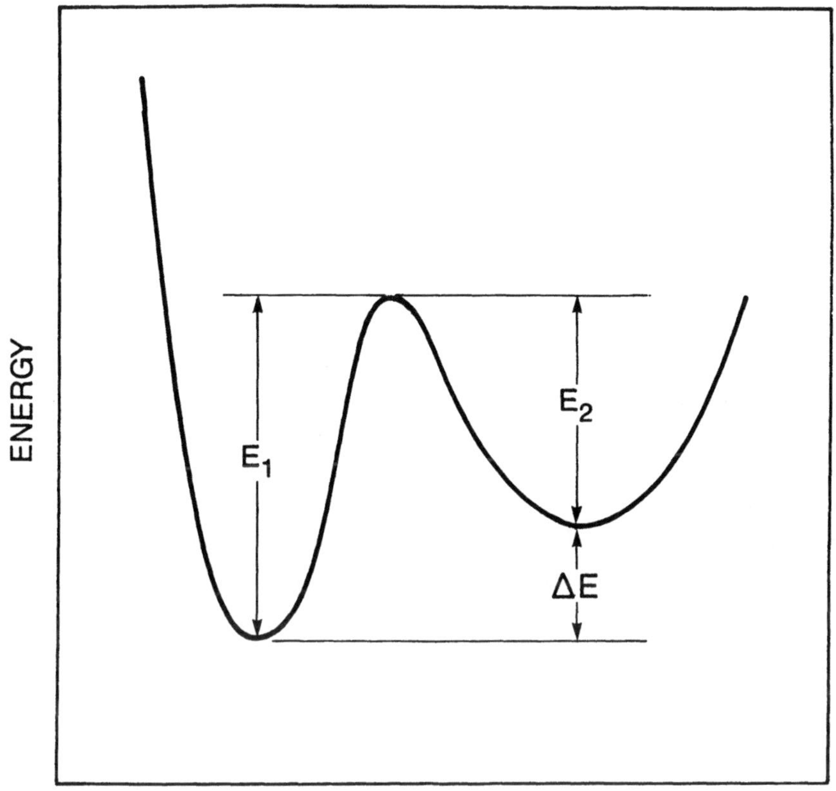

Figure 1. Configuration-coordinate diagram of localized defect centers.

Changes in N_M may be induced by light or temperature, and either of these energy sources may cause transitions over the barrier in each direction, so four processes must be considered [1]. The rate equation is taken to have a simple first-order form

$$\frac{dN_M}{dt} = C_1 N_L R - \nu_2 N_M - C_2 N_M R + \nu_1 N_L \qquad (1)$$

where R is the rate of carrier recombination and is a measure of the light intensity, C_1 and C_2 are recombination rate coefficients, ν_1 and ν_2 are thermal (vibrational) rate coefficients, with the forms $\nu_1 = \nu_{10} \exp(-E_1/kT)$ and $\nu_2 = \nu_{20} \exp(-E_2/kT)$. Note that since $E_1 > E_2$, it follows that $\nu_2 >> \nu_1$ [1].

This simple form of the rate equation has the solution

$$N_M(t) = N_0 \exp(-t/\tau) + N_{MS}[1 - \exp(-t/\tau)] \tag{2}$$

where

$$N_{MS} \equiv (C_1 R + \nu_1)\tau, \tag{3}$$

$$\tau^{-1} \equiv (C_1 + C_2)R + \nu_1 + \nu_2, \tag{4}$$

N_0 is an arbitrary initial value, and N_{MS} is the steady-state value of $N_M(t)$.

Because of the very strong temperature dependences of ν_1 and ν_2, we neglect other possible temperature dependences (which must be much weaker). It was shown in Ref. 1 that at temperatures below 350 K both ν_1 and ν_2 are negligibly small compared to the terms containing R, for light intensities near one sun. Thus this condition applies at room temperature. Note R then cancels out of N_{MS} because light-induced transitions in both directions in Fig. 1 are proportional. So in this low-temperature range, N_{MS} is independent of light intensity, whereas τ varies inversely with light intensity.

In the opposite limit of high temperatures (above ≈ 450 K), ν_1 and ν_2 are both dominant over the terms containing R, and the light intensity has no effect at all, simply because in this range, transitions between the two states are dominated by vibrational effects. Thus in both temperature limits there are simple dependences of τ and N_{MS} on the parameters of the original rate equation. These dependences are summarized in Table I for these two limiting temperature ranges. The essential feature here is that N_{MS} always scales with N_T.

TABLE I. Limiting forms of τ and N_{MS} from Eqs. 3,4 at low and high temperatures. Use has been made of the relation $\nu_2 >> \nu_1$.

	LOW TEMPERATURES	HIGH TEMPERATURES
τ	$\dfrac{1}{(C_1+C_2)R}$	$\dfrac{1}{\nu_2}$
N_{MS}	$N_T \dfrac{C_1}{C_1+C_2}$	$N_T \dfrac{\nu_1}{\nu_2}$

In this framework it is important to consider the two major effects of doping on defect densities. First, the amount of degradation increases with doping, whether by donors or acceptors [3]. Degradation at room tempreature gives the low-temperature form of N_{MS} in Table I. Second, doping increases the density of built-in defects, again for both donors and acceptors [4]. This density is determined by the deposition procedure, hence is the high-temperature form in Table I. These built-in defects were shown in Ref. 1 to be just a frozen-in fraction of the total MSDs, so the two sets are really the same. These two effects of doping, in the present framework, require that N_T increases with doping [1]. Since this total density of centers is often dominated by the number of latent defects, doping must increase the density of latent defects as well as the observed MSDs. This is the key inference, because it means that the centers that are only *potential* defects are directly related to the dopant atoms. The centers need not be in the metastable state and need not be charged to increase with doping. Thus the "weak bonds" that have been hypothesized as the latent defects are caused by the presence of dopants, hence the conclusion of extrinsic origin.

A plausible atomic model for such relations for both phosphorus donors and boron acceptors follows directly. Both P and B have preferred coordination numbers of three, so both try to establish the same kind of local environment. This environment is likely to be strained since the normal coordination of the host Si structure is four. In this picture the "weak bonds" that form the latent defects are associated with any 3-fold coordinated atoms, since they all tend to form similarly strained centers. Other foreign atoms would generally form local centers with different character, and thus may not be readily transformable to a metastable state.

There remains the need to explain MSDs in undoped a-Si:H, which controls much of solar-cell behavior. For that we hypothesize the presence of uncontrolled 3-coordinated foreign atoms, and observations require that there be $\approx 10^{17}$ cm^{-3} of these. This number can represent the sum of densities of several types of such atoms, but there are reports that boron alone is generally present at about this level in good material [5,6], even if deposited in a separated-chamber system [5]. Thus there seem to be enough of these foreign atoms, even though their source is not known. This gives a direction in which to seek better stability, and perhaps higher efficiency as well, since the built-in defects are related to the MSDs [2].

REFERENCES

1. D. Redfield, Appl. Phys. Lett. **52**, 492 (1988).
2. D. Redfield, Proc. 20th IEEE Photovoltaic Specialists Conf., 1988 (to be published).
3. A. Skumanich, N. Amer, and W. Jackson, Phys. Rev. B **31**, 2263 (1985).
4. R. Street, D. Biegelsen, and J. Knights, Phys. Rev. B **24**, 969 (1981).
5. M. Kamiyama and H. Sakai, Proc. Matl. Res. Soc. 1988 Spring Mtg.
6. A. Catalano and G. Wood, *ibid.*

POSTER SESSION 12:
THIN FILM a-Si CELLS

RECOMBINATION AND MINORITY DIFFUSION LENGHT IN UNDOPED A-Si

J. AL-HASANI *

Control & System Department University of Technology, Baghdad-IRAQ

ABSTRACT. Based on analytical approximation of our photoconductivity model of an illuminated undoped a morphous silicon, and from the defination of diffusion lenght, we derived a model representing the diffusion lenght of holes, then we analysed the diffusion lenght of holes with recombination in a-Si.

INTRODUCTION. The optimization of a-Si solar cells requires a good understanding of the factors limiting device performance. The detailed and realistic modelling of these cells is however needed in order to further increase our understanding of the device physics and hence optimize device design to increase performance.

In this paper we analysed the diffusion lenght, in a-Si on the basis of our models /1,2/.

The densities of the free and captured electrons and holes in a-Si can be determined by using Fermi- Dirac distribution /3/ which is meaningful only when no excess carriers. Thermal generations and recombinations in a-Si out of thermal equilibrium are given by /4/, the diffusion lenght, of illuminated a-Si was modelled as a function of illumination level /1/ we analysed the diffusion lenght with temperature, and the ratio of capture cross sections of electrons and holes in donor-like and acceptor-like localized centers inside the mobility gap represented by a double exponential function /5/.

THEORY. Many different models have been performed to determine the minority diffusion lenght Lp in a-Si /6-10/. It is generally believed that a primary limitation to the device performance in the small diffusion lenght Lp of the minority carriers /9/. We investigated the variation of the free carrier densities with illumination intensity in a-Si previous /2/. We shall then extend this analysis to relate the minority-carrier diffusion lenght

*Present address: Fac.of Electrotechni , Tržaška 25,Ljubljana,Yugoslavia

to illumination intensity /1/.

For the absence of electric field the material must be neutral and the generation rate equal the recombination rate. Hence

$$\rho = 0 \qquad \ldots\ldots \quad (1)$$

$$(G-R) = G_L \qquad \ldots\ldots \quad (2)$$

where ρ is the space-charge density, G-R generation-recombination rate, and G_L optical generation.

From /2/ we can obtained that

$$P = K_1 G_L^{\ B} \qquad \ldots\ldots \quad (3)$$

where K_1 is a constant, and B accounts for nonlinear variations with temperature and given by

$$B = \left(1 + \frac{KT}{wv} \left(\frac{2 - \frac{KT}{wv}}{2 - \frac{KT}{wc}} \right) \right)^{-1} \qquad \ldots\ldots \quad (4)$$

From /1/ we can obtain

$$Lp = \sqrt{K_2 Dp(P)}^{\ (1-\frac{1}{B})/2} \qquad \ldots\ldots \quad (5)$$

where K_2 is a constant, and Dp is the diffusion coefficient.

Substituting for p from equation (3). We may rewrite equation (5) as

$$Lp = \sqrt{Up \frac{KT}{q} K_2 K_1 \ (1-\frac{1}{p}) \ G_L^{\ B-1}} \qquad \ldots\ldots \quad (6)$$

or

$$Lp \propto G_L^{\ (B-1)/2} \qquad \ldots\ldots \quad (7)$$

This then gives the intensity dependence of the minority carrier diffusion lenght assuming that the hole trap quasi-Fermi level is in a distribution of donor-like states which can be assumed to vary exponentially with a characteristic energy wv in the region around E_{1D}, where

$$E_{1D} = E_i - KT\ L_n\ (R_D n + P\ /ni \qquad \ldots\ldots \quad (8)$$

and R_D is capture cross-section area ratio of donor-like states.

RESULTS AND DISCUSSION. In this section we will present the analysis results of the intensity dependence, capture cross section ratio rate mobility of hole, and temperature for the minority-carrier diffusion lenght in amorphous silicon undoped layer. Also we found the diffusion lenght with position in layer, in the bulk of the layer Lp is virtually independent of postion.

Based on a theoretical analysis given before we show that the zero field diffusion lenght is indeed intensity dependent and that this dependence can be directly related to the slope of the density of states near the valance band edge. This is to be expected as with increased intensity the hole trap quasi-Fermi level will move close to the valance band edge into a region of higher localized states density. Fig 1 show the relation of Log diffusion lenght Lp in Um and illumination level. We see that as illumination intensity is increasing Lp decreasing, which is in agreement with our analysis. In /9/ they found that the experimentally result of Wv is much greater than that used and obtained, this is because used g(E) for two exponentials whereas reality g(E) is more complex and probably is better suited to 4 or 5 exponentials. Also from Fig.1. we see the effect of increases rate of capture cross section area R for donor-like and acceptor-like of hole on the relation of Lp and intensity at the surface of layer F. Fig.2. Show the same relation with different values of mobility, and we see that as Up increases the value of Lp increases too. Fig.3. Show the relation of diffusion lenght Lp and the temperature T. We see that as T increases Lp increases too. The increase in diffusion lenght is more likely due to a decrease in g(E) near E_F and a few ev above the hole mobility edge. In other word we can say that when temperature T increases the density of localized state g(E) decreases near E_F and valance band edge.

The results obtained in this analysis agree well with other experimental published results /6,10/, but we got alonger diffusion lenght from the other, and this is due to our way to find the diffusion lenght, by interpreting as the average distance which an illuminated hole travels before recombining with an electron, which is the same if we derived a mean lifetime Tp of hole, then find the diffusion lenght Lp as

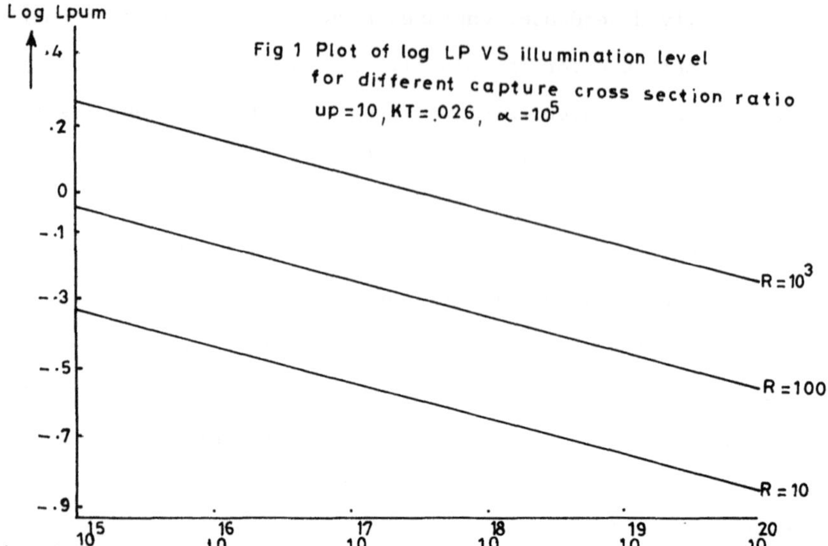

Fig 1 Plot of log LP VS illumination level for different capture cross section ratio up = 10, KT = .026, $\alpha = 10^5$

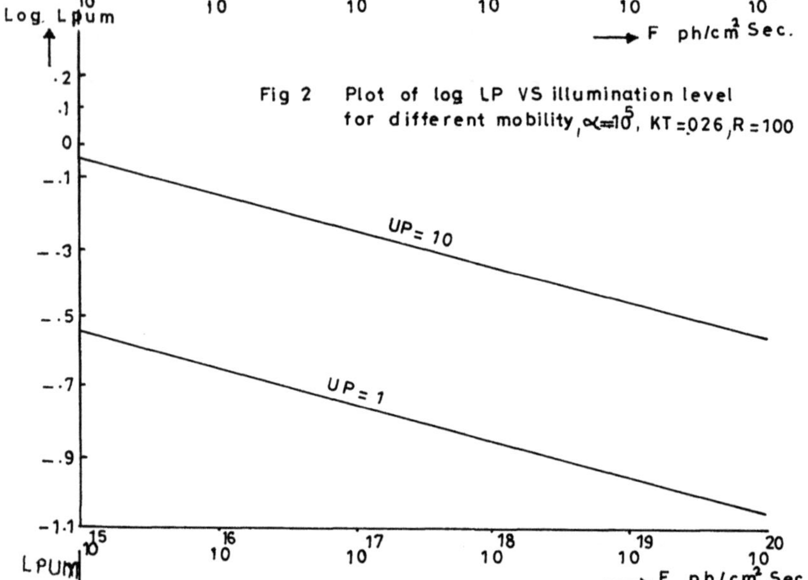

Fig 2 Plot of log LP VS illumination level for different mobility, $\alpha = 10^5$, KT = .026, R = 100

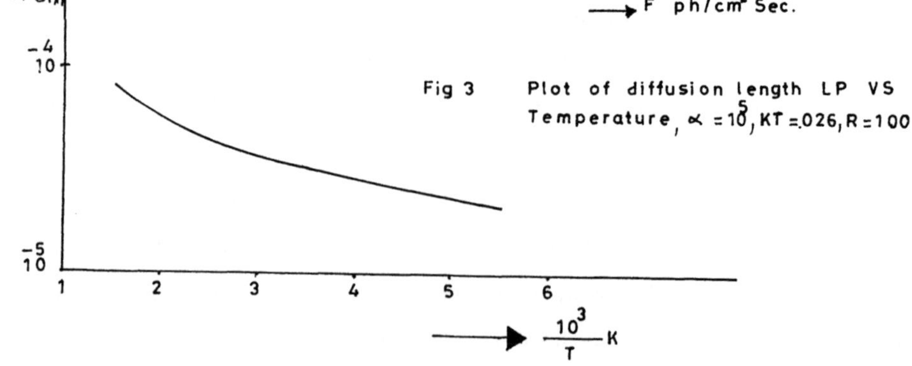

Fig 3 Plot of diffusion length LP VS Temperature, $\alpha = 10^5$, KT = .026, R = 100

$$L_p = \sqrt{D_p T_p} \qquad ..... \qquad (9)$$

while the others used the classical way to find the lifetime Tp as the ratio of free carriers to recombination rate, which is different from our result, for more details See /11/.

CONCLUSION. We concluded that the diffusion lenght which we derived by mean value is greater than the others, the diffusion lenght is indeed intensity dependent, and its value decreases with increase of F. Also the diffusion effect by capture cross-section rate R of donor-like and acceptor-like,and the mobility of carrier is effective on the value of diffusion lenght.

Acknowledgments: I would like to thank J.Furlan for helpful and valuable discussions.

REFERENCES.

1. J.AL-HASANI "It will be Published"
2. J.Furlan, J.AL-HASANI, IASTED Conf. March 3-5 Cario (1987)
3. J.Furlan, S.Amon. MIC - 86, Conf., Feb. 18-20, Innsbruck (1986)
4. J.Furlan, S.Amon. MRS Symp., April 15-18, Palo Alto (1986)
5. M.Hack, M.Shure, J.Appl., Phys., Vol.54, 5858 (1983)
6. P.Sichanugrist, M.Konagan, K.Takahashi, J.Appl.Phys.55(4),1155 (1983)
7. T. McMahon. R.Konen Kamp, IEEE, 1385 (1982)
8. B.Faughnan, A.Moare R. Crandall, Appl.Phys.Lett. 44 (6) 613 (1984)
9. M.Hack, J.McGill, W.C.Zubaty i, R.Singh, M.Shur, A.Madan, J.Appl.Phys. 53 (9) 6270 (1982).
10. M.Hack, M.Shur, J.Appl. Phys. 55 (8) 2967 (1983)
11. J.AL-HASANI, SD-87, Sept. 9-11, Topolšica (1987).

ON THE PREPARATION OF P-DOPED μc-Si:H:F FILMS AS WINDOW LAYERS FOR SOLAR CELLS

J.Cárabe, M.T. Gutierrez and L. Delgado
División Fotovoltaica. Instituto de Energías Renovables.
CIEMAT. MADRID 28040.

Abstract

The optical and photoelectronic properties of p-doped (BF_3 dopant gas) hydrogenated amorphous silicon films prepared from rf-glow discharge of silane-helium gas mixtures, are studied as a function of the rf power and the gas phase composition. The results are compared with that obtained in SiH_4+BF_3, SiH_4+Ar+BF_3 and SiH_4+CH_4+BF_3 gas mixtures. Intrinsic films produced from silane-inert gas mixtures show an increase in dark conductivity, σ_d of 3-4 orders of magnitude accompanied by a decrease in the deposition rate with increasing rf-power. At high rf-power levels, the effect of p-type doping with BF_3 results in a further increase of σ_d up to ~ 10^{-4} $(\Omega cm)^{-1}$ while the optical gap, Eg, sharply decreases with increasing BF_3 gas fraction. In contrast, in films produced from undiluted gas mixtures, the optical gap remains constant throughout the doping range. A comparison is made between the different gas mixtures used to help to optimize the relationship between σ_d and Eg.

1. INTRODUCTION.

It is known that the optoelectronic properties of the p-window layers, strongly influence the performance of p-i-n amorphous silicon solar cells (1). High dark conductivity and transparency to visible light are normally required to achieve good quality p-window layers.

In order to increase the optical band gap, silicon carbon alloys deposited from SiH_4+CH_4 gas mixtures are currently used, even if some degradation of the conductivity occurs, because of the large density of bandgap defect states introduced into the film.

More recently, the use of microcrystallyne p-layers has been promoted (2), since both dark conductivity and sunlight transparency are higher in μc-Si:H films than in a-Si:H layers. In effect, larger mobilities and difussion lengths as well as much lower absorption coefficient than that normally obtained in a-Si:H are characteristics of μc-Si. Therefore, the use of these p-type μc-Si:H films could lead to a better solar cell performance, provided that a high optical band gap can still be retained. This is important to get a high solar cell output voltage.

A widely used procedure to obtain μc-Si:H layers consists in the use of high dilution levels in hydrogen or inert gases at high rf-power levels. However degradation of the transparent conducting oxide (TCO) layer (on top of which the p-layer is deposited in p-i-n solar cells) results from the use of hydrogen rich plasmas, even at moderate substrate temperatures ~ 180°C.

On the other hand and enhancement in solar cell performance (stability and efficiency, open circuit voltage) by the use of p-doped silicon-fluorine alloys has been reported (3). The possible role of fluorine as compensating silicon dangling bonds and promoting film microcrystallization has been invoked.

As an attempt to elucidate these problems concerning optimization of p-window layers we present in this work some results on the effect of gas phase composition and rf power level on the photoelectronic properties of p-doped films prepared from SiH_4+He+BF_3. The results are compared with that previously obtained in films prepared from SiH_4+BF_3, SiH_4+CH_4+BF_3 and SiH_4+Ar+BF_3 gas mixtures.

2. EXPERIMENTAL.

Hydrogenated amorphous silicon films were deposited on 0.8 mm thickness Schott borosilicate glass in a parallel plate, two electrode (38 cm diameter, 1.5 cm apart) capacitively coupled rf system. Mass flow controllers (Tylan) were used to control the flow rate of process gases. Total gas process pressure was measured by a capacitive manometer and was kept constant at the desired process pressure by using an automatic pressure controller. Substrate temperature and process pressure were set at 250°C and 1 Torr, respectively unless otherwise indicated in the text. The reactor chamber is pumped with a rotary pump/roots blower package of 250 m^3/hr pumping speed. The chamber was evacuated at ~ 10^{-4} Torr before each deposition run.

A typical deposition run included He flow before and after plasma deposition, to minimize possible back oil streaming as well as to achiever faster substrate cooling rates. Plasma cleaning cycles (CF_4+8%O_2) are normally run, after each deposition cycle.

Optical properties, i.e. absorption coefficient, refractive index, optical thickness and optical gap were deduced from transmission and reflectivity spectra. Hydrogen content of the film was estimated from IR transmission spectra through the Si-H wagging absorption band at 630cm^{-1} (4). Conductivity measurements were made by using a coplanar two aluminium electrode, 1 mm gap configuration.

3. RESULTS AND DISCUSSION.

3.1. Silane-Helium gas mixtures.

The effect of rf-power in the film properties, i.e. deposition rate R, dark conductivity σ_d, optical gap Eg, hydrogen concentration [H] and photosensitivity σ_{AM1}/σ_d is shown in Fig. 1. a and b.

As the rf power level, P_{rf}, is increased we observe an increase in the bandgap Eg determined either as the energy at which $\alpha=10^4 cm^{-1}$ or from the straight intercept of the Tauc plot $(\alpha h\nu)^{1/2}$ vs $h\nu$. This is accompanied by a decrease in R and an increase in σ_d. At low power levels (< 50 W) σ_d and R are in the range of 10^{-9} $(\Omega cm)^{-1}$ and 2.1. A/S respectively, while at power levels of 150 W (~134 mW/cm^2) σ_d and R are in the range of 10^{-6} $(\Omega cm)^{-1}$ and 0.3 A/S respectively. This trend, i.e. increase in σ_d and decrease in R, with increasing, Prf, has been sometimes explained as indicative of film microcrystallization. However no traces of microcrystallites could be detected in our high power films by X-ray diffraction.

Hydrogen concentration and photosensitivity defined as the σ_{AM1}/σ_d ratio go through a maximum at a relatively low power density of ~50W (44mW/cm^2). This is probably related with increasing film defect density as Prf is increased. In fact, an increase in the concentration of weakly bonded hydrogen is observed as Prf is increased. This is shown in fig.2,

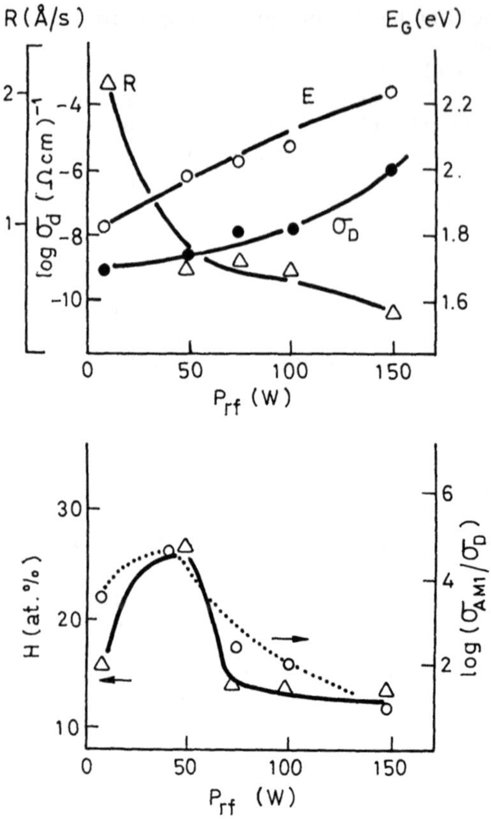

Fig. 1. Dependence of the energy gap Eg, dark conductivity σ_d, hydrogen concentration and photosensitivity σ_{AM1}/σ_d on the rf power level Prf. SiH_4/SiH_4+He=10%.

where the Si-H stretching mode absorption at 2000 cm^{-1} and the Si-H wagging mode at 640 cm^{-1} are found to decrease with increasing Prf while the amount of bonded H in dihydride mode (bands a 2079, 845 and 890 cm^{-1}) increases.

At rf power level >100W the dilution of silane in helium, results in an increase of the deposition rate, which goes through a maximum of 1.1 A/S at a silane fraction SiH_4/SiH_4+He of ~ 40%. The optical gap is moreless independent of the silane fraction at this high rf power level (Fig.3). Some plasma unstabilities are found in films prepared using mixtures of 10% silane, 90% helium at Prf>100W and low silane flow rates <10sccm.

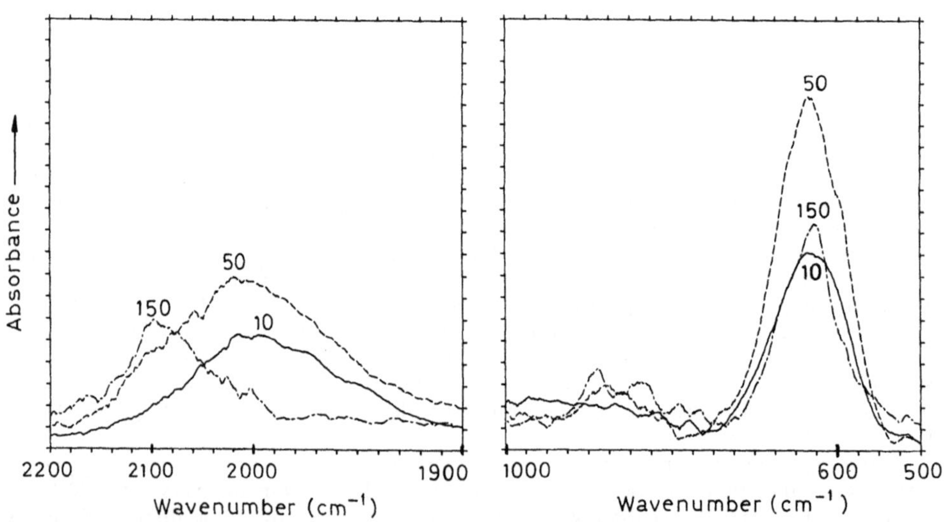

Fig. 2. IR spectra of films prepared at different power levels.

3.2.P-doped layers. Effect of BF_3 gas phase concentration.

The effect of introducing BF_3 into the gas mixture on σ_d and Eg is presented in Fig.4.

Four sets of results are presented corresponding to films prepared from gas mixtures of SiH_4+He+BF_3, SiH_4+Ar+BF_3, SiH_4+BF_3 and SiH_4+CH_4+BF_3. In each case the BF_3 flow rate was increased systematically, while the SiH_4 to He, Ar or CH_4 were held constant.

From fig. 4. it is clear that as expected in all the gas mixtures used σ_d increase with increasing BF_3/SiH_4 gas ratio.

In films prepared from silane diluted gas mixtures, the increase in σ_d is accompained by a decrease in the optical gap Eg. Therefore, values of σ_d and Eg in the range of 10^{-5}-10^{-4} $(\Omega cm)^{-1}$ and below 1.8 ev, respectively are currently obtained at BF_3/SiH_4 gas ratio >20%.

The observed decrease in Eg in diluted films contrast with the result obtained in films prepared from undilutes SiH_4 100% gas phase, where Eg,remains constant (~1.85 ev) up to a BF_3/SiH_4 gas concentration of 10%, while an increase in σ_d of ~ 4 orders of magnitude (from 10^{-10} $(\Omega cm)^{-1}$ to 10^{-6} $(\Omega cm)^{-1}$) is observed (Fig. 4).

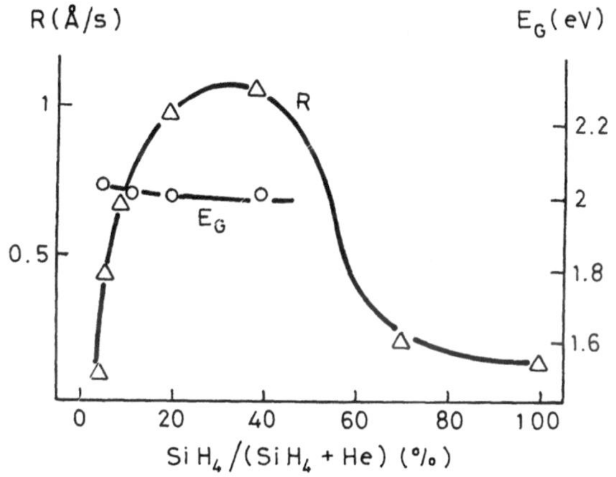

Fig. 3. Variation of R and Eg as a function of the silane fraction. Prf=150W, 1T, 250°C.

As it is also shown in fig. 4 in films prepared from SiH_4+CH_4 gas mixtures (SiH_4/CH_4=1), after an initial decrease, the optical gap remains constant at a high value 2.0 - 2.1 ev and σ_d increases, up to 10^{-6} $(\Omega cm)^{-1}$ as BF_3/SiH_4 increases.

From these results it seems that a compromise to find the optimum values of σ_d and Eg has to be established. In fig. 5. a representation of Eg, vs log (σ_d) is presented to help to find the relationship between energy gap and dar conductivity. It is clearly seen in this figure, that the degradation found in diluted band gap films, counterbalance the relatively high dark conductivities obtained. Besides, the advantage found in pure SiH_4 films, i.e. inhibition of the bandgap degradation with increasing BF_3 gas concentration has the drawback of the low σ_d ~ 10^{-6} $(\Omega cm)^{-1}$ obtained. BF_3 doped silicon-carbon films can have the potential to get the high σ_d required (>10^{-3} (Ωcm)-1)) while still retaining a high band gap (>2.0 ev). Work is now in progres in this direction.

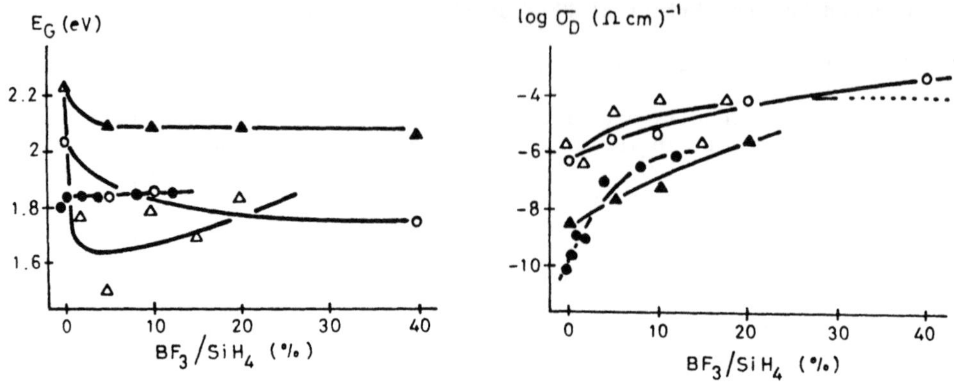

Fig. 4,. Variation of energy gap (Eg) and dark conductivity (σ_d) as a function of BF_3/SiH_4 gas phase ratio. (▲)SiH_4/CH_4=1, 250°C, 1T, 50W, (△) SiH_4/He=0.09, 1T, 250°, 150W, (O) SiH_4/Ar=0.1, 1T, 150°C, 50W. (●) SiH_4 100%, 0.75T, 150°C, 20W.

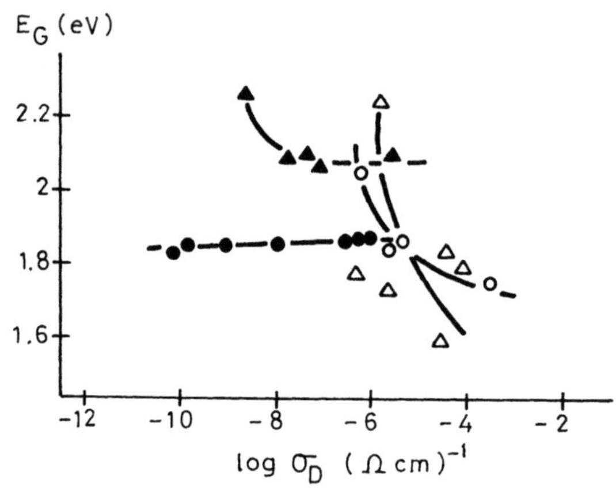

Fig. 5. Plot of Eg vs log σ_d. Symbols as in Figure 4.

REFERENCES.

1. S. Guha, J. Yang, P. Nath and M. Hack. Appli. Phys. Lett 49, (4) 1986.
2. K. Tanaka and A. Matsuda Jarect. Vol. 6. Amorphous Semiconductor Technologies & Devices (1983). Ed. Y. Hamakawa, pag. 161.
3. H. Matsumura and S. Furukawa. Jarect. Amorphous Semiconductor Technologies & Devices (1982) Ed. Y. Hawakawa, pag. 86.
4. M.T. Gutierrez, P. Román and L. Delgado, CEC Contract EN3S-0091-E. Progress Report October 1987.

IMPROVED THERMAL STABILITY OF INTERCONNECTED AMORPHOUS SILICON p-i-n SOLAR CELLS ON CONDUCTIVE TIN OXIDE SUBSTRATE

Kackadasam R. Ramaprasad
Chronar Corp., P. O. Box 177
Princeton, N.J. 08542, U.S.A.

ABSTRACT. Interconnected a-Si:H solar cell panels with SnO2 substrate and Al back contact are susceptible to degradation under thermal soaking and cycling. The formation of Al2O3 at the Al/SnO2 interface and the diffusion of Si through Al grain boundaries at the Al/n+ a-Si:H interface are responsible for module failure under thermal cycling. The latter mode of failure is also seen in integrated circuit technology where Al metal is used. This paper describes a novel method to endow greater thermal stability to both the Al/SnO2 and Al/a-Si:H interfaces. a-Si:H solar panels, after silicon scribing to expose SnO2 but before Al deposition, are dipped in a colloidal solution or a well stirred suspension of a metal hydroxide such as nickel hydroxide. After Al deposition to give interconnected cells, the panels are heated in air at 150° C. for four hours. Such panels have much greater thermal stability compared to the control as shown by thermal cycling studies at 150° C. After 180 hours at this temperature, the contact resistance of the Al/SnO2 interface increased 30 times for the control compared to only a 3-fold increase for the treated panel. The efficiency of the control decreased by about 70%, whereas the stabilized panel showed a negligible change. Further, after cycling for 632 hours at 150° C., Auger depth profiling revealed the presence of Si in Al (0.4% at. concn.) for the control panel, with no out-diffusion of Si into Al seen for the treated panel (within the Auger detectability limit).

INTRODUCTION. The successful commercialization of a-Si solar cells for power generation mandates solutions to the stability problems associated with this device. One of them is the thermal stability of the various interfaces. Interconnected solar cells on the commonly used tin oxide substrate with Al back electrode have two interfaces, viz., Al/SnO2 and Al/n+ a-Si:H, whose integrity under thermal cycling is paramount to module performance. Thermodynamically, Al/SnO2 is less stable than Al2O3/Sn [1,2], leading thus to a degradation of the former under thermal soaking, with increase in the Al/SnO2 contact resistance. The presence of humidity will have even more adverse effect [2]. The interdiffusion of Al and Si at the Al/a-Si:H interface could eventually degrade the solar cell by gradual reduction of the n+ layer thickness [1]. The problem of Si Diffusion into Al is also encountered in integrated circuit technology [3].

Various metallization schemes exist to preserve the integrity of the above two interfaces. A 100 Å Ti barrier between the Al metal and SnO2 was found to stop the Al/SnO2 degradation [2]. Blocking layers have been mentioned as endowing thermal stability to tandem cells [4]. In the context of large scale manufacture, a single metallization would be desirable, if other stabilization processes were to exist, with Al being the preferred metal because of its high reflectivity.

A novel method of stabilizing both the Al/SnO2 and Al/a-Si:H interfaces was reported earlier [5[. It consisted of treating the a-Si:H surface, after scibing it to expose the front tin oxide electrode, with a solution containing metal ions such as those of nickel. This was followed by the usual thermal evaporation of Al. The method, however novel, still involved a multi-step dip and rinse procedure which could stand further simplification from the point of view of manufacturing. This paper describes a one-step dipping procedure that, because of its simplicity, is better suited for incorporation into large scale manufacture of a-Si:H solar panels. The procedure utilizes a colloidal solution of a metal hydroxide such as that of nickel. Even a well-stirred suspension of the hydroxide in water is found to be effective.

EXPERIMENTAL. Both the control and the metal hydroxide treated samples were originally part of the same 1 ft. x 1 ft. or 1 ft. x 3 ft. panel. The size of each test sample was 4" x 12 ". Each sample will have finally 27 interconnected solar cells, with a width of 1 cm. and a length of 4.2 cm. The latter also represents the length of Al/SnO2 interface. After scribing the a-Si:H to expose the front SnO2 electrode, the sample to be stabilized is dipped for 30 seconds in a colloidal solution of nickel hydroxide. A well stirred suspension of the metal hydroxide in water was also found to be equally effective. The concentration of the nickel ion is 0.02 M, although this might not be very critical, as long as there is an uniformly dispersed gelatinous hydroxide suspension. Lower concentrations might need longer dipping times, which was not tried in this work. After the dipping step, only the back side (glass) of the sample is rinsed in de-ionized water, followed by air drying. The samples (control and the dipped one) are then subjected to Al metallization through a mask to give interconnected cells. A four hour heating in air at 150° C.completes the stabilization procedure.

Thermal stability has been monitored by heating the samples in air at 150° C. and following the photovoltaic parameters as a function of time. It is to be noted that the Al/SnO2 contact resistance reported was measured by passing a current through each cell and measuring the voltage drop and assuming that the resistance calculated is due to the Al/SnO2 interface. At the start of the thermal cycling, with low Al/SnO2 contact resistance, the constant contribution from the tin oxide sheet resistance might be even higher than the former. Hence the values seen at higher degradations of the Al/SnO2 contact are more in accord with the above method of calculation than the initial value for the resistance. Further, we are comparing the relative difference in the Al/SnO2 degradation between the control and the stabilized samples. To examine any out diffusion of Si into Al, Auger depth profiling of the Al/a-SiH interface was also done for a pair of samples.

RESULTS AND DISCUSSION. The initial heating in air at 150° C. for 4 hours after Al metallization is crucial to the present method. As shown in Fig. 1, which depicts the average values of 8 pairs of control and dipped samples from different runs, the dipped panel has a lower efficiency, initially, compared to the control. In fact, it was observed that many individual cells had low or zero Voc, indicating that the process of dipping could possibly have opened up what were potential pin holes. But the 4 hour heating produces a dramatic change, with the efficiency of the stabilized panel now being higher than that of the control at the start. The control panel itself has undergone a small reduction in its efficiency. Individual cells that showed poor performance before heating for the treated samples were found to recover remarkably after the heating. It is safe to assume that where a short existed before, this procedure has eliminated the same.

Fig. 2 is a set of graphs, wherein various normalized photovoltaic parameters have been plotted as a function of heating time at 150° C. It represents the average of 3 sample pairs. It is seen that after 180 hours (Fig. 2A), the contact resistance of the Al/SnO2 interface has increased 30 times for the control, whereas the stabilized panel shows an increase of about 3. Though there is less difference in the short circuit current deterioration between the control and the stabilized panel (Fig. 2B), the effect of the Al/SnO2 contact resistance increase on the fill factor is clearly seen from Fig. 2C. There is no change in the fill factor of the treated panel after 180 hours at 150° C., in contrast to the control panel that has suffered a loss of nearly 60% in its fill factor. This clearly demonstrates the stabilizing influence of the nickel hydroxide dip on the Al/SnO2 interface. The comparison of the efficiencies of the control and the treated panels under thermal cycling (Fig. 2D) reinforces this conclusion. The ability of the dipping procedure described here to endow thermal stability is due to nickel entity and not to any cleansing action of the method has been shown elsewhere [5]. Al diffusion into SnO2 is a precursor to the formation of Al2O3 and thus the degradation of the Al/SnO2 interface [2]. Mechanistic explanation of the success of the metal hydroxide treatment awaits further work, but it is obvious that some form of diffusion barrier is being formed at the Al/SnO2 interface.

The dipping procedure described above also acts as a diffusion barrier at the Al/n+ a-Si:H interface. Auger depth profiling of a control sample and the metal hydroxide treated one, both of which had been thermally cycled for 632 hours at 150° C., was done and the results are shown in Fig. 3. It is seen that there is Si in Al (0.4% at. concentration) for the control (Fig. 3A; note change in scale for Si). The stabilized sample had no out-diffusion of Si into Al (Fig. 3B) within the Auger detectability limit (which is less than 0.1%). We conclude that the nickel hydroxide treatment described here has a stabilizing effect on interconnected a-Si:H solar cells under thermal cycling.

ACKNOWLEDGEMENT. I thank Dr. E. Eser for useful discussions, Mr. E. Urbanski for experimental help, Mr. F. Ramos for the Auger results and Dr. L. Michalski for help with plate diagnostic measurements.

REFERENCES.

[1] E. Eser, et al, Solar Cells, 21, 25 (1987)

[2] E. Eser, F. Ramos and J. Grez, J. Appl. Phys. (to be published July, 1988)

[3] A. Paccagnella, et al, Thin Solid Films, 128, 217(1985)

[4] Y. Tawada, et al, Appl. Phys. Lett., 48, 584(1986)

[5] K. R. Ramaprasad, U. S. Patent 4,675,466

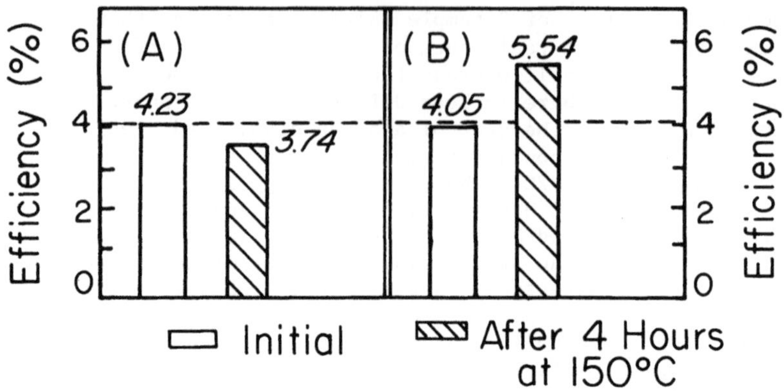

Fig. 1. Effect of initial heating(150 C; 4 hrs.) on the efficiency of a-Si:H solar panels. (A) control (B) Nickel hydroxide treated.

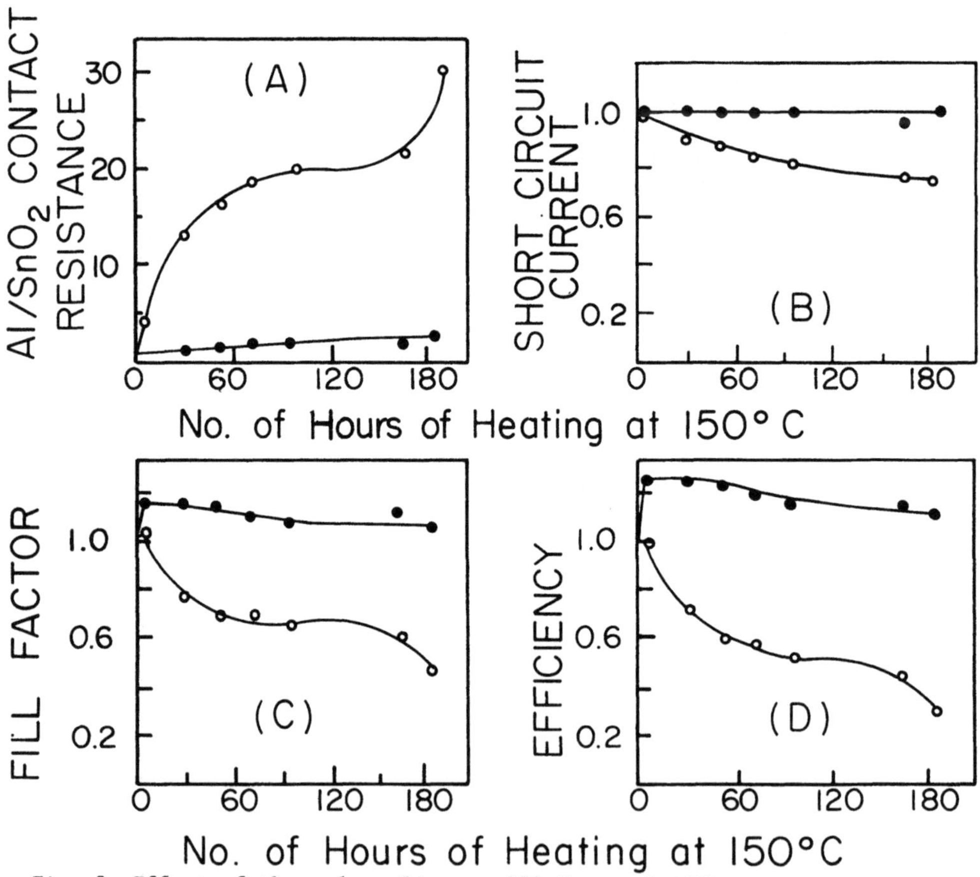

Fig. 2. Effect of thermal cycling at 150 C. on the PV parameters (normalized) of a-Si:H panels. (o) Control (●) nickel hydroxide treated

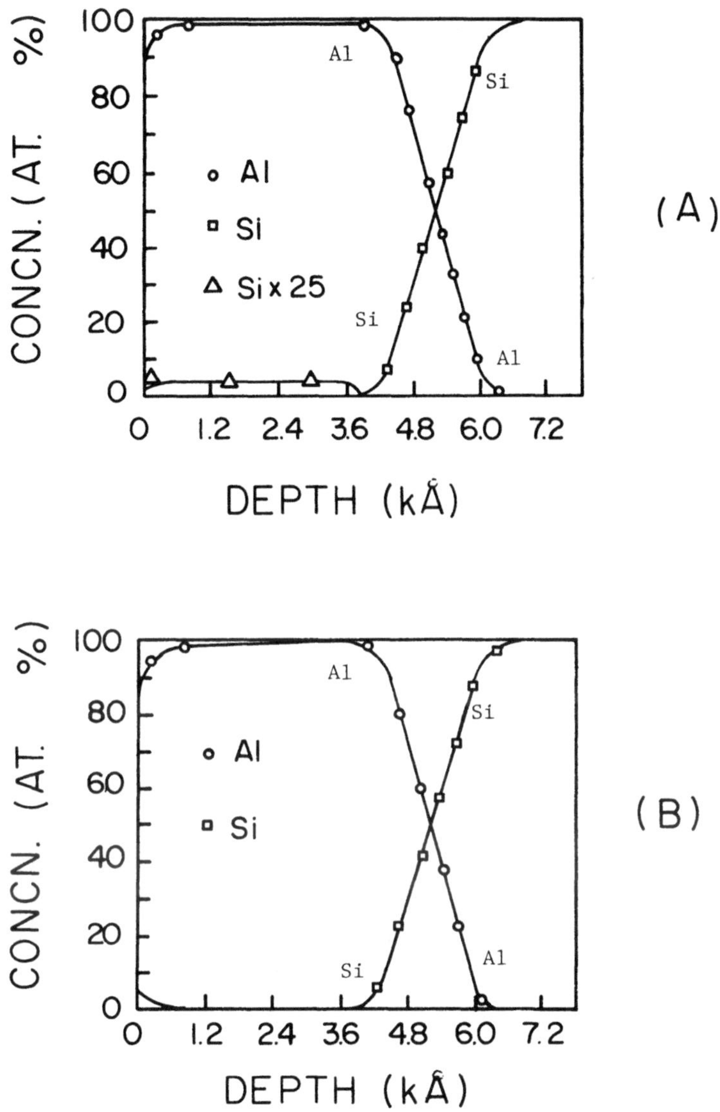

Fig. 3. Auger depth profiles of the Al/n+ a-Si:H interface after thermal cycling at 150^{0} C. for 632 hours. (A) Control panel, (B) Nickel Hydroxide treated panel (before Al deposition). Note change in scale for Si concn. in (A) at depths less than 3.6 kA

THERMALLY STIMULATED CURRENTS AND FILL FACTORS IN p-i-n AMORPHOUS SILICON SOLAR CELLS

H. Schade and A. E. Delahoy
Chronar Corporation
P. O. Box 177
Princeton, NJ 08542, USA

ABSTRACT. A correlation is established between the fill factor of completed p-i-n solar cells and the concentration and energy distribution of defect states in the i-layer. This correlation is obtained by measurements of both the fill factor and thermally stimulated currents (TSC) on the same device for a range of degradation levels. Complementary information is derived from deep level transient spectroscopy (DLTS), based on current decay after photo-excitation. The samples comprise various materials prepared by rf glow discharge deposition of a-Si:H or a-$Si_{1-x}C_x$:H, and mercury-sensitized photochemical vapor deposition of a-Si:H. Within the energy range (0.2-0.8 eV from conductive states) covered by the TSC measurement, a broad distribution of defect states is detected between 0.50 and 0.65 eV, and additionally above 0.75 eV for degraded devices. A TSC peak at about 0.25eV is ascribed to tail states. Defect concentrations and densities of state are deduced to be between $5x10^{15}$ and $5x10^{16}$ cm^{-3}, and between $1x10^{16}$ and $1x10^{17}$ $cm^{-3}eV^{-1}$ for as-deposited and degraded samples, respectively. For all samples, the fill factors and the amounts of defect charge are inversely related. For glow discharge produced samples, the collection length/ i-layer thickness ratio is found to be inversely proportional to the defect concentration, consistent with the collection length model. Experimental evidence for the effect of the energy distribution of defects on the degree of degradation is pointed out.

1. BACKGROUND AND OBJECTIVE

The degradation phenomena for both a-Si:H material and photovoltaic devices are well known. It is the purpose of this contribution to establish a direct correlation between material and device properties, deduced from the same sample. This is achieved by measurements of solar cell parameters and thermally stimulated currents (TSC) in completed devices. TSC measurements provide straightforward information on the density of states within a certain energy range of the forbidden gap. Such measurements have been frequently applied to determine trapping levels in crystalline junction devices [1], but only in a few instances to characterize a-Si:H, either by using uniform films [2,3], or Schottky barrier diodes [4].

Here we will report TSC for conventional a-Si:H p-i-n devices which is recorded as a thermally stimulated short-circuit current after photo-excitation at low temperatures. The information on the i-layer is contained in the current vs. time (temperature) curve, the so-called TSC spectrum. The integration of this curve yields the charge released from the traps, and the positions of current maxima on the temperature scale are related to energy locations of traps referred to the conductive states (either band edge).

2. EXPERIMENTS

The samples used in this study comprise conventional a-Si:H p-i-n solar cell structures, deposited either in a rf glow discharge or by mercury- sensitized photo-CVD. A sample containing a-$Si_{1-x}C_x$:H throughout the whole i-layer was also included. The device area was 0.07-0.09 cm^2. The samples were mounted in a vacuum cryostat that covered the temperature range 100-350 K. For each degradation level of the sample, i.e. as-deposited, light-soaked at about room temperature under white-light illumination at up to 13 suns for 1 h, and annealed, the fill factor was measured (at 1/100 sun to avoid the effect of series resistance), followed by the TSC run. Trap filling was accomplished by white-light excitation (1/100 sun for 10 sec) at 100 K. After this excitation the short-circuit current was measured as a function of temperature under a fairly constant heating rate (typically 0.3 deg/sec) and recorded as the TSC spectrum.

Complementary to these measurements we have also performed deep level transient spectroscopy (DLTS). Specifically, the sample is excited with a light flux of about 1×10^{13} photons/cm^2sec for 2 sec, sufficient for complete trap filling, and following this excitation, the short-circuit current difference representing the DLTS signal is measured after delays t_1=4 sec and t_2=7 sec. The cycle trap filling/current decay is repeated as a function of temperature. At any given temperature, only those states contribute to the DLTS signal for which the inverse carrier emission rate corresponds to t_1-t_2 [5].

3. RESULTS AND DISCUSSION

The information obtained by both TSC and DLTS is qualitatively quite similar, as illustrated in Fig. 1 by spectra for two different glow discharge produced samples in the as-deposited and degraded state. In both cases, a prominent low-temperature peak is reduced after degradation, while a rather broad distribution of states at higher temperatures (energies) is substantially increased.

Within the context of this study, we restrict the determination of energy levels to the crude relationship E_t= 25 kT_m, where E_t is the trap energy, and T_m the temperature of a TSC maximum. Based on this relation, the energy range of our measurements is about 0.2-0.8 eV.

The bulk of our results discussed below refers to TSC. No efforts have been made to more precisely determine energy levels and identify certain TSC peaks, except for the following assignment. The low-temperature (130 K) peak is thought to be due to tail states; the remaining distribution is related to defect states. Note that the tail state distribution is inversely related to the defect contribution, as seen in Fig. 1 for different degradation levels, and also in Fig. 2 for different materials (glow discharge vs. photo-CVD). This is explainable by a smaller shift of the quasi-Fermi level upon photo-excitation in the presence of a larger concentration of deeper defect states.

Unlike DLTS measurements, TSC yields absolute values for the charge released over a certain energy range. Table I lists defect concentrations derived from the charge released over the whole energy range assigned to defects (see Fig. 3). Also shown is the density of states, averaged over that range (0.35-0.75 eV). For both the annealed and degraded condition of the sample, these values compare well with other published results [6,7].

Since our defect concentrationswere measured on completed solar cell devices, they can be directly correlated with the cell fill factor (FF) that particularly at low illumination intensity (1/100 sun) is a good measure of the i-layer quality, without significantly being affected by series resistances. Figure 4 correlates the FF and the total charge released from defect states, showing an inverse relationship. Based on the limited number of samples, there appears to be a generic difference between glow discharge and photo-CVD produced devices. For photo-CVD, the

Table I: Defect concentrations N_s, and densities of state D(E), deduced from TSC in glow discharge produced p-i-n cells, compared to other published results

	N_s(TSC) (cm^{-3})	N_s [6] (cm^{-3})	D(E) (TSC) 0.35-0.75eV ($cm^{-3}eV^{-1}$)	D(E) [7] E_V+0.8eV ($cm^{-3}eV^{-1}$)
As-deposited	$5x10^{15}$	$4x10^{15}$	$1x10^{16}$	$1x10^{16}$
Light-soaked (10 suns, 1h)	$5x10^{16}$	$3x10^{16}$	$1x10^{17}$	$1x10^{17}$

same FF can be maintained at higher defect concentrations; however, their energy distribution is weighted towards lower energies. Due to excessive dark currents in photo-CVD samples, high-energy defects, such as those around 350 K (0.75 eV) observed in degraded glow discharge produced samples, could not be observed. Therefore, further evaluation of the relationship between FF and defect concentration is restricted to glow discharge devices.

Within the collection length model [6], the defect concentration is inversely proportional to the collection length/i-layer thickness ratio, L_c/L, while the FF, in turn, is related to L_c/L through a Hecht-type expression [8]. Replotting our experimental data according to these relationships yields Fig. 5 that to a good approximation confirms an inverse proportionality between L_c/L and the defect concentration.

It ought to be noted that the description of the degradation level (FF) by a single parameter (defect concentration) may not be fully adequate. The above-mentioned range of defect concentrations for the same FF (Fig. 4) is experimental evidence for the additional importance of the energy distribution of defects. Further evidence is the observation of a difference in the DLTS peak shape, depending on the temperature of defect generation (see Fig. 6).

ACKNOWLEDGEMENTS

We would like to thank J. Kalina for skillful performance of the measurements. We are grateful to R. S. Crandall for suggesting DLTS measurements to complement our original TSC experiments.
This research was supported by the Solar Energy Research Institute, Golden, Colorado.

REFERENCES

1. See, e.g., R. J. Basset, C. A. Hogarth, Int. J. Electronics 33 (1972) 217

2. W. Fuhs, M. Milleville, Phys. Stat. Sol. B 98 (1980) K 29

3. H. Hamdi, A. Deneuville, E. Bustarret, J. Non Cryst. Sol. 77/78 (1985) 683

4. L. Vieux-Rochaz, A. Chenevas-Paule, J. Non Cryst. Sol. 35/36 (1980) 737

5. D. V. Lang, J. D. Cohen, J. Harbison, Phys. Rev. B 25 (1982) 5285

6. Z E. Smith, S. Aljishi, D. S. Shen, V. Chu, D. Slobodin, S. Wagner, AIP Conf. Proc. 157 (1987) 171

7. J. D. Cohen, K. Mahavadi, K. Zellama, J. P. Harbison, A. E. Delahoy, Mat. Res. Symp. Proc. 70 (1986) 213

8. B. Faughnan, R. Crandall, Appl. Phys. Lett. 44 (1984) 537

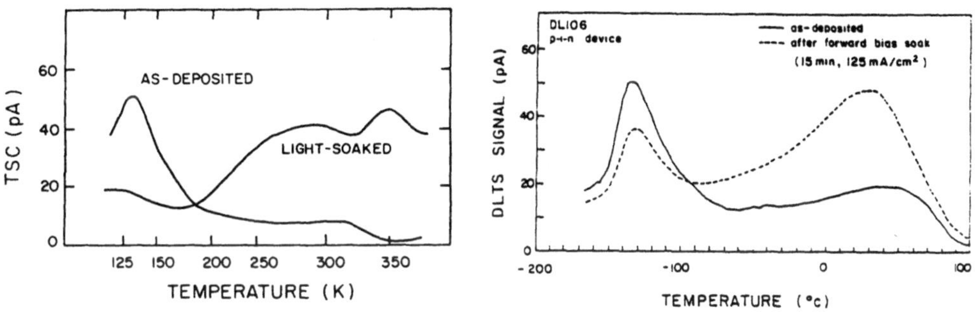

Figure 1: Comparison of TSC and DLTS spectra for as-deposited and degraded p-i-n cells prepared by glow discharge deposition

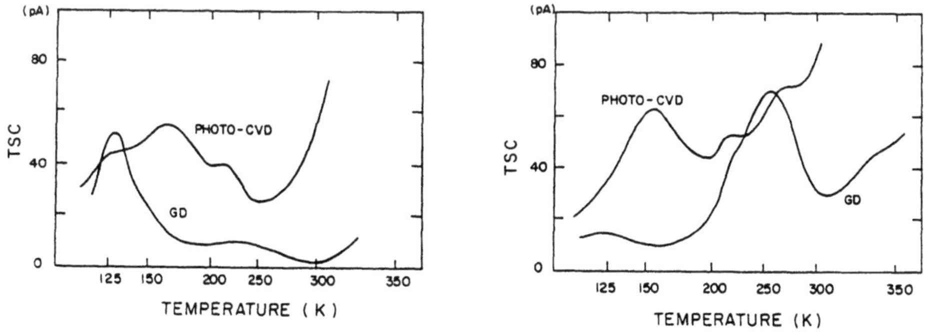

Figure 2: TSC spectra for p-i-n cells prepared by glow discharge deposition or photo-CVD, for as-deposited (left) or light-soaked (right) conditions, characterized by high (left) and low (right) fill factors, respectively

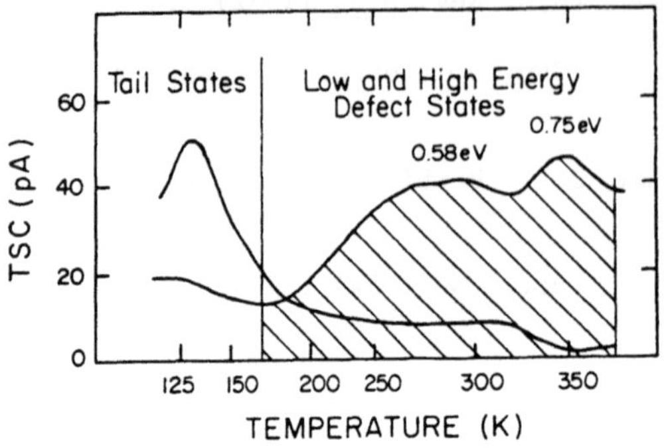

Figure 3: Evaluation of TSC spectra

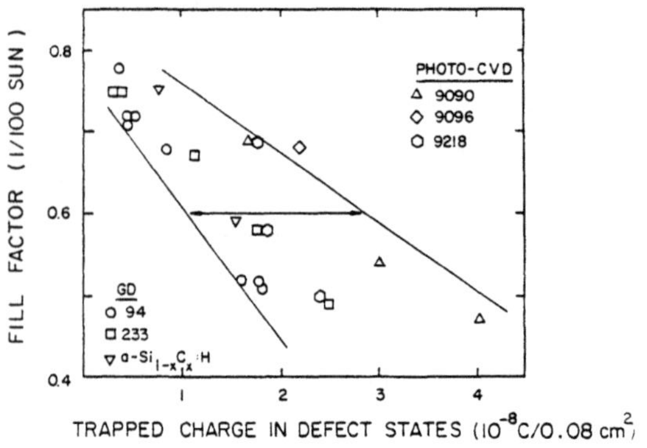

Figure 4: Correlation between the fill factor and charge released from defect states, measured on p-i-n cells deposited by glow discharge or photo-CVD

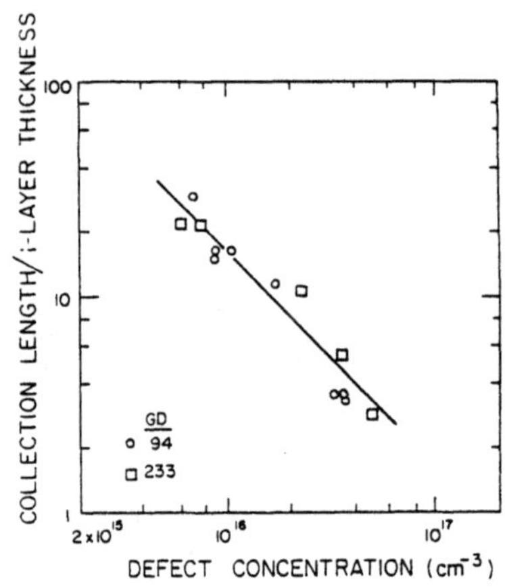

Figure 5: Inverse proportionality between collection length/ i-layer thickness ratio and defect concentration, deduced from data in Fig. 4 for glow discharge produced p-i-n cells

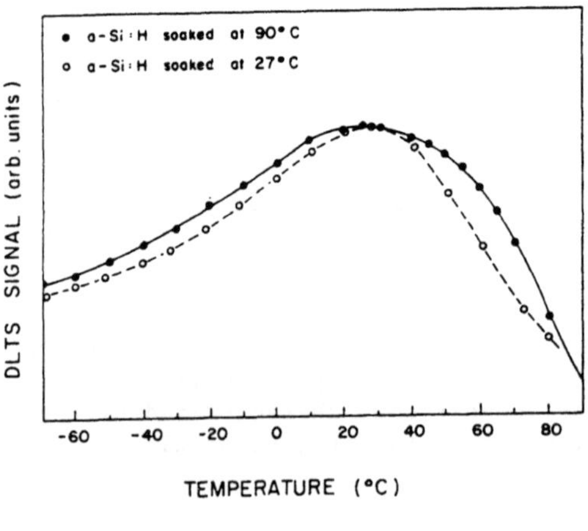

Figure 6: DLTS spectrum showing differences in shape and width of the high-temperature peak, depending on the temperature of current-soaking

A NEW MULTI-CHAMBER PECVD SYSTEM FOR THE PREPARATION OF AMORPHOUS SILICON SOLAR CELLS*

M. Van Roy, J. Smeets, P. Nagels and P. Dirven
Physics Department, S.C.K./C.E.N., B-2400 MOL (Belgium)

ABSTRACT. A continuous multi-chamber system for the preparation of amorphous silicon solar cells is presented. It consists of two load locks, three reaction and one dc magnetron sputtering chamber placed in-line. The uniformity of the deposition in the PECVD chambers is found to be strongly dependent on the gas distribution system. Preliminary results for the opto-electronic properties of material deposited in this system are presented. The intrinsic a-Si:H film is shown to be comparable in quality to material obtained elsewhere.

1. INTRODUCTION

At S.C.K./C.E.N. four single chamber plasma-enhanced CVD reactors of own construction, allowing the deposition of amorphous silicon by glow discharge decomposition, are in full operation since five years. Details about these systems can be found elsewhere (1). By careful optimization of the deposition conditions, we already achieved an efficiency of 7.8 % (cell area = 0.125 cm^2) for heterojunction solar cells of the type glass-SnO_2-p^+ a-SiC:H-i-n^+ a-Si:H-Al prepared in these systems.

A serious problem when fabricating p-i-n a-Si:H photovoltaic devices in a single chamber system, is the cross contamination of the active intrinsic layer by the boron and phosphorus dopants of the p- and n-layer. Since the p-type a-SiC:H layer is deposited prior to the photovoltaic active intrinsic layer, the walls and electrodes of the plasma chamber are covered with this layer. This acts as a source for boron and carbon incorporation in the i-layer. We already showed that the mobility-lifetime product $\mu\tau$ of the intrinsic layer and hence the cell efficiency is severely influenced by this boron inclusion in the i-layer during p-i-n deposition (2). This effect, which seems to be inherent to a single chamber system, can be avoided in a system consisting of separated chambers, where each layer is deposited in its own individual reactor.

In this paper, we report on a new multi-chamber in-line system for the preparation of amorphous silicon films and solar cells. First results on film quality are presented.

2. DESCRIPTION OF THE NEW MULTI-CHAMBER SYSTEM

Figure 1 shows a general view of the new multi-chamber system. It was constructed by Pfeiffer (FRG) in close collaboration with S.C.K./C.E.N. A schematic diagram of the system is given in Figure 2. The system consists of six separated chambers, namely a load, three plasma-enhanced CVD process, a dc magnetron sputtering and an unload chamber. Substrates to be

*Work supported by the R&D Programme "Photovoltaic Power Generation" of the Commission of the European Communities and by the Belgian National Energy R&D Programme.

coated are mounted on a U-form substrate carrier, which also acts as substrate electrode. Two 300 mm x 100 mm substrates, placed back to back, can be processed at the same time. Substrate and rf electrodes are placed vertically in order to prevent dust and flakes, that may cause pinholes in the a-Si:H film, from falling on them. The carrier is transferred from one coating chamber to another by a rack-and-pinion system. The guiding and transport system is shielded by the carrier itself to prevent the moving parts from being coated.

Figure 1. General view of the new multi-chamber plasma-enhanced CVD system

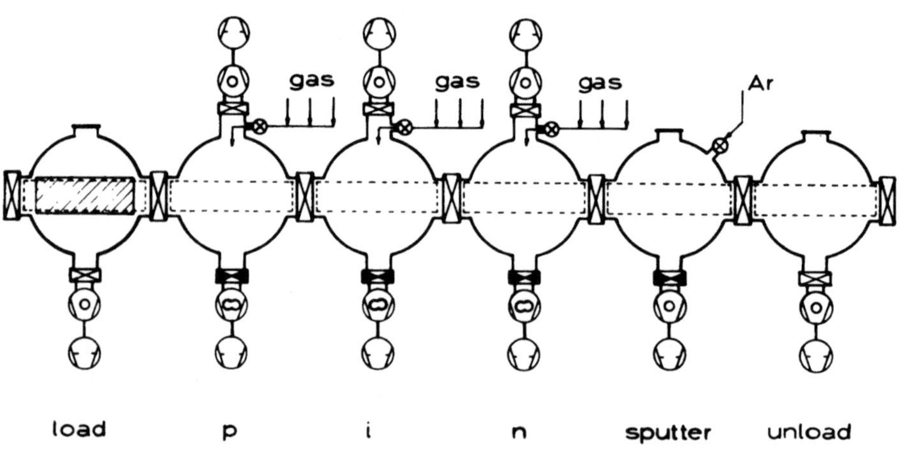

Figure 2. Schematic diagram of the new system

Each chamber is equipped with its own pumping system, consisting of a turbomolecular pump for the high vacuum and a roots pump for the process vacuum. Rotary pumps are used for backing them. Oil filtration systems are installed on the plasma chambers. Chambers are isolated by gate valves. They can be pumped down to a very low base pressure (less than 10^{-7} mbar). If one uses metal seals the system is capable of achieving 10^{-9} mbar predeposition pressure. All chambers are equipped with a radiation heater. The temperature homogeneity over the carrier surface is $\pm$ 5°C at 250°C.

The distance between the electrodes in the reaction chambers can be varied between 2 and 7 cm. The total pressure during the plasma process is measured by a MKS Baratron electronic manometer and controlled by a conductance valve. Each process chamber is equipped with three MKS mass flow controllers for automatic control of the reactant gases. The gas entry is positioned at the chamber top, the gas outlet at the bottom. The electrodes are powered by a 500 W 13.56 MHz rf generator via automatic impedance matching networks. The substrate carrier is electrically isolated and can be biased. In addition, it is possible to insert a grid electrode between the substrate and the rf electrode. This will minimize the ion bombardment of the substrates. Substrates can be transported in both directions. Therefore, the machine allows the study of multi-junction solar cells. The total system is supervised by a programmable controller. Pumping and transport can be selected manually or in automatic mode, providing a safe process run. Several substrate carriers can be controlled at the same time.

The main advantages of such a machine are obvious: better control of the impurity contamination in the intrinsic layer, higher production rate compared to a batch type process and no exposure of the reaction chambers to the open air. The direct coupling of a dc magnetron sputtering chamber to the three plasma process chambers allows us to deposit the metallic back contact (normally Al or Ag) without breaking the vacuum, so that oxygen absorption of the top a-Si:H layer can be avoided. Normally, the p-i-n coated substrates are removed from the equipment after the deposition of the n-layer and the metallic contacts are deposited in an isolated evaporation system. Following the results of Schade et al. (2), our approach is favourable for the contact resistance of the a-Si:H/metal contact and hence the cell performance.

This implies that the active area of the cell must be defined after processing. For small cells, we developed a procedure based on photolithography combined with etching for patterning the metallic contact. This pattern then eventually acts as a mask for etching the inactive a-Si:H in a CF_4+O_2 plasma. For integrated solar cells, we are studying a complete mechanical scribing process compatible with an in-line amorphous silicon/metal deposition concept (cutting only before and after deposition). This will be the ultimate step in a glass in/panel out approach.

While primarily designed for development use, this system can be used for a small production ($\pm$ 15 kW) of solar cells. In the deposition sequence of p-i-n structures, the growth of the intrinsic layer is the most time consuming process step (approximately 50 times longer then the p- or n-layer deposition). So, the intrinsic chamber will block the system's throughput. The only way to accommodate for this difference in an in-line system, is to divide the deposition process of the i-layer among several chambers. Because of the modular design of our system, additional modules can be added. In an optimized system, the throughput per hour can reach 0,36 m^2 with a cycle time of 10 minutes and 300 mm x 100 mm substrates.

3. RESULTS

3.1. Uniformity of deposition

An important requirement for attaining large area highly efficient a-Si:H solar cells is a good thickness uniformity over the total area of the substrate. Since cell efficiency is strongly thickness dependent, a non-uniform thickness distribution will severely influence the performance of monolithic, series connected a-Si:H cells. It also affects the trace quality of the patterning step of the a-Si:H (e.g. interference effects will damage the quality when using laser scribing (4)). Therefore, we first varied the operating conditions (pressure, gas flow, interelectrode distance etc.) in order to obtain a uniform deposition over the total substrate. All films were grown with an rf power of 16 W. To avoid powder formation during deposition, we found it necessary to keep the pressure sufficiently low and the gas flow sufficiently high in order to get a residence time below 5 seconds. Thicknesses were determined on each sample using Dektak profile measurements on chemically etched steps. The uniformity of deposition was found to be strongly dependent on the gas pressure and the interelectrode distance. Figure 3(a) shows the optimum thickness distribution in the process chambers without any specific gas inlet or outlet system. The film is highly non-uniform. Figure 3(b) shows the results when the gas inlet is positioned just above and the outlet just below the electrodes. Figure 3(c) shows the optimum thickness distribution when the reactant gases are fed into the discharge through a gas sprayer. These results clearly indicate that the thickness uniformity is drastically affected by the gas inlet and outlet. At present, thickness variations in each chamber are within 5 % around the mean value. Further optimization is still going on.

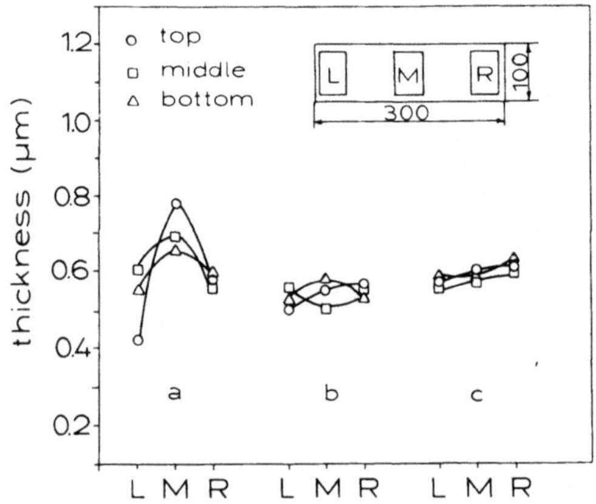

Figure 3. Thickness distribution obtained in the process chambers of the multi-chamber plasma-enhanced CVD system
a) The process chambers as received without a specific gas inlet or outlet system.
b) After installation of a new gas distribution system constructed at S.C.K./C.E.N.
c) as b) but with an extra gas sprayer in the gas inlet

3.2. Conductivity and density of states

Conductivity measurements (planar type) were performed to control the quality of individual layers. For p- and n-type doped a-Si:H layers, the results for the room temperature dark conductivity and activation energy were in agreement with earlier studies using the single chamber systems. Space Charge Limited Current measurements on Cr-n^+-i-n^+-Cr devices allow one to measure both the conductivity and the density of states (DOS) at the Fermi level of intrinsic layers avoiding contact effects (5). For a substrate temperature of 180°C, we measured a DOS of 2 x 10^{16} cm^{-3} eV^{-1} and a conductivity of 7 x 10^{-8} Ω^{-1} cm^{-1} (i-layer thickness was 1.2 μm). These results agree fairly well with the results of Den Boer (5). Further studies are being made and will be published later. They include a detailed structural characterization by IR absorption and a complete determination of the electrical and optical properties of the films.

4. CONCLUSIONS

A new multi-chamber in-line plasma-enhanced CVD system for the preparation of amorphous silicon films and solar cells was presented. It consists of a load, three reaction, a dc magnetron and an unload chamber. Two 300 mm x 100 mm glass substrates, placed back to back, can be coated in one run. The direct coupling of the dc magnetron chamber to the reaction chambers is expected to be beneficial to the quality of the semiconductor/metal interface. Ultimately, this will lead to a glass in panel out approach.

We found that the thickness uniformity was drastically affected by the gas inlet and outlet system. At the present state of optimization, thickness variations are within 5 % around the mean value. Preliminary results on the opto-electronic properties of material prepared in this system indicate that the film quality is comparable to material grown in our single chamber systems.

Further work will be aimed at further optimization of film quality and at growing highly efficient solar cells.

References

(1) P. Nagels, Photovoltaic Power Generation Vol. 1, Eds. G. Willeke and G. Grassi (1986) 23

(2) J. Smeets, M. Van Roy and P. Nagels, Proc. 7th EC Photovoltaic Solar Energy Conference, Sevilla (1986) 539

(3) H. Schade and Z.E. Smith, J. Appl. Phys. 59 (1986) 1682

(4) S. Nakano, M. Ohnishi, H. Kawada, H. Nishiwaki, S. Tsuda and Y. Kuwano, Proc. 5th EC Photovoltaic Solar Energy Conference, Athens (1983) 712

(5) W. Den Boer, J. Phys. (Paris), Colloq. C4, 42 (10, suppl.) (1981) 451

LIGHT-INDUCED METASTABLE DEFECTS IN UNDOPED a-Si:H: MIDGAP DENSITY OF STATES DETERMINATION BY CAPACITANCE, PHOTOTHERMAL DEFLECTION SPECTROSCOPY AND PHOTOCONDUCTIVITY MEASUREMENTS.

K. Zellama[+], U. Besi Vetrella, U. Coscia, P. Menna, E. Terzini, P. Roca I Cabarrocas[++], D. Mencaraglia[+++] and P.Andro[+++].
ENEA FARE/FOTO C.P. 83, 80055 PORTICI (NAPOLI), ITALY.
+ Permanent address: GPS-ENS, Université Paris 7, 2 place Jussieu 75251 Paris Cedex 05, France.
++ Laboratoire de Physique des Interfaces et des Couches Minces, Ecole Polytechnique 91128, Palaiseau Cedex, France.
+++ LGEP-SUPELEC LA au CNRS 127, Plateau du Moulon, 91190 Gif sur Yvette, France.

ABSTRACT. We have extensively investigated the light-induced defects creation, after prolonged and intense illumination in undoped hydrogenated amorphous silicon (a-Si:H) films prepared by standard r.f. glow-discharge decomposition of silane (GD) and dc sputtering triode configuration technique. We used for this study, junction capacitance method, Photothermal Deflection Spetroscopy (PDS) and Photoconductivity (PC) measurements which allow us to observe and monitor the changes in the gap density of states (DOS), due to light exposure or annealing treatments. These results are correlated with Electron Spin Resonance (ESR) and conductivity measurements. An increase in the DOS obtained from capacitance measurements was observed in going from state A (annealed state) to state B (light-soaked state). This behaviour was confirmed by PDS and PC results, while the ESR measurements did not indicate any change in the density of neutral dangling bonds (DB).The shifts in the Fermi level position, E_F, were determined by coplanar conductivity as well as sandwich configuration ac admittance. The detailed studies suggest that the predominant light-induced metastable defect is not the neutral DB, which arises from the breaking of weak Si-Si bond, as it has been proposed in the past. Moreover our results indicate that the samples already exhibited in state A different natures of defects form one sample to another.

1. INTRODUCTION

From the numerous extensive works on the reversible light-induced effects in a-Si:H, it clearly appears that prolonged and intense illumination strongly affects the electronic and optical properties of the material through the creation of metastable defects in the mobility gap, which act as recombination centres for the free carriers. Thus modifications in gap DOS accompanying this defect-creation have been observed. Also such light-induced metastable changes have important technological implications, especially for a-Si:H solar cells. A significant degradation of the efficiency of these cells due to light exposure has been reported.

Although several models have been proposed to account for the Staebler-Wronski effect (1), its physical mechanism still remains unclear. The two main models involve either the breaking of the weak Si-Si bonds (2), due to non-radiative tail-tail state recombination, or reversible changes in the charge of the hybridization states (3). However recent junction capacitance (4), optical absorption (5) and ESR (6) studies seem to disagree with the bond-breaking model (2). Finally a more recent work suggesting the breaking of weak Si-H bond (7), such as the Si-H-Si three centres bond, has been proposed.

In an attempt to have better understanding of the light-induced defects mechanism in undoped a-Si:H, we extensively studied two different kinds of samples grown either by the GD or by the sputtering technique. To monitor the changes in the gap DOS, junction Capacitance-Temperature (C-T) slope method (4,8), PDS and PC measurements have been performed. We also correlated these measurements with ESR and conductivity results.

2. EXPERIMENTAL PROCEDURES

Two series of samples (1 and 2) were grown by GD and sputtering respectively. The deposition conditions are summarized in table I. The samples of each series were co-deposited on p^+ or n^+ critalline silicon for the capacitance measurements and quartz substrates for PDS, PC, ESR and conductivity experiments. All the measurements were carried out in both state A and state B. The state A was obtained after annealing of each sample in the dark under vacuum at 200°C for 20 min.. The samples were prepared in state B by exposure at room temperature to Tungsten-Halogen unfiltred white light at an average power density of 1.2 W/cm^2 for 16 hours. For the junction capacitance, the front Schottky barrier was obtained by evaporating a semitransparent thin platinium top layer. Thus only a fraction of the incident light was absorbed into the films.

3. EXPERIMENTAL RESULTS AND DISCUSSION

As discussed previously (8), from the C-T slope method we can obtain the DOS at E_F, as well as the Fermi level position. In this technique, the bulk E_F, determined in the sandwich configuration, is insensitive to any surface charges effects.

In figures 1 and 2, we present an example of the variation of the capacitance C versus T, for three different measurement frequencies ω, obtained from the C-T experiments for sample 1 in state A and state B respectively. From the capacitance curves, we calculate the quantity $C^2(dC/dT)^{-1}$ as a function of T. The results we obtain are displayed in figure 3, corresponding to the curves presented in Fig.2. The slopes of lines of Fig.3 give the DOS at E_F. Moreover, the straight lines intercept the abcissa axis ($C^2(dC/dT)^{-1}$=O) at a temperature $T_0(\omega)$ from which, with an Arrhenius plot of ω versus the inverse of $T_0(\omega)$, we deduce the position of the bulk Fermi level.

We obtain a DOS at E_F equal to 4.5×10^{16} $cm^{-3}eV^{-1}$ and 2×10^{15} $cm^{-3}eV^{-1}$ in state A for the sputtered and the GD samples respectively. In state B, the DOS increases by a factor up to 2.5 for the sputtered samples and a factor 5 for the GD ones. The Fremi level moves towards the valence band, in going from state A to state B, from .71 to .79 eV and from .77 eV to .83 eV for the sputtered and the GD samples respectively. All these

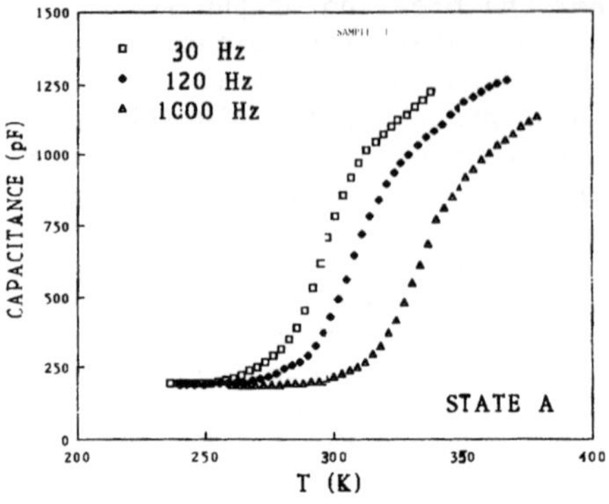

Fig. 1: Variation of C versus T of sample 1 in state A at three frequencies as indicated.

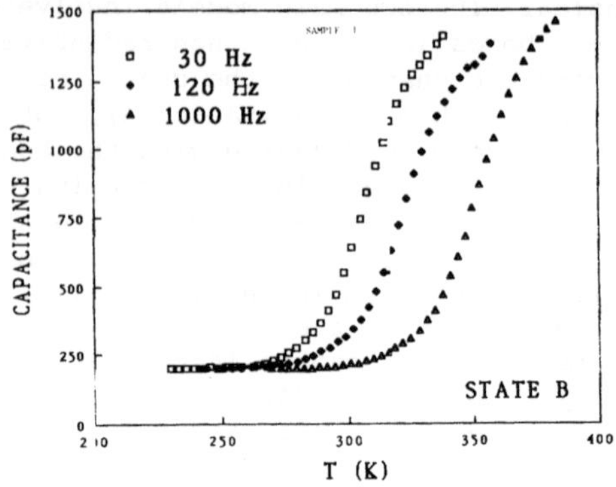

Fig. 2: Same variation in state B, for sample 1.

Fig. 3: Plot of $C^2/(dC/dT)$ versus T corresponding to the results displayed in Fig. 2, to determine the DOS at E_F and the Fermi level position.

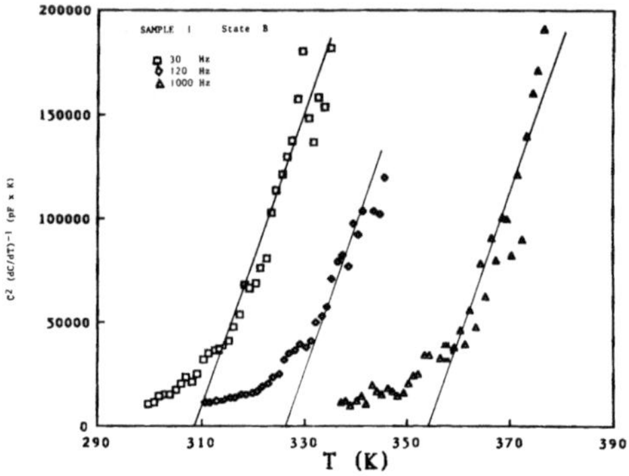

Table I. Deposition growth parameters of undoped a-Si:H films.

Sample	Deposition	T_s (°C)	RF Power (W/cm^2)	P (mTorr)	Thickness (microns)
1	Sputtered	280	---	5	1.8
2	G.D.	200	2	40	2

results are summarized in table II.

As indicated in table III, we obtain the same order of changes, between state A and state B in the DOS measured either by PDS and PC or by C-T slope method .

Table II. Junction capacitance and conductivity results(states A & B).

Sample	State (A or B)	Capacitance experiments DOS at E_F(cm-3 eV^{-1})	E_F(eV)	Conductivity E_C-E_F(eV)
1	A	4.7-5x10^{16}	.71	.71
	B	1.1-1.4x10^{17}	.79	.89
2	A	2x10^{15}	.77	.77
	B	1x10^{16}	.83	.92

Table III. PDS and PC results. For DOS from PDS we used calib. ref.10

Sample	State(A or B)	PDS DOS(cm^{-3})	PC measurements DOS(cm-3)	E_g(eV)	E_0(meV)	$\eta\mu\tau$(cm^2/V)
1	A	5x10^{16}	2.7x10^{16}	1.72	59	5.2x10^{-7}
	B	7x10^{16}	5.8x10^{16}	1.74	62	4.6x10^{-8}
2	A	3x10^{16}	5x10^{16}	1.69	72	1.9x10^{-5}
	B	7x10^{16}	2x10^{17}	1.68	90	1.0x10^{-5}

On the other hand, The ESR results indicate that, within the experimental errors, the density of neutral DB's is almost constant before (state A) and after light exposure (state B) for both sputtered (4x10^{16} cm^{-3}) and GD samples (2x10^{16} cm^{-3}), while samples studied by other groups (2) show a very large increase in the density of spins (up to 10^{18} cm-3) due to light exposure. Then, at this stage it appears that different samples can be affected in a different way by light exposure, suggesting that various kinds of defects are probably involved.

In addition the PC results indicate that the Urbach tail coefficient,E_0, measured at room temperature in both state A and state B, is almost constant (59 to 62 meV) for the sputtered samples while it increases from 72 to 90 meV for the GD ones (see table III). Moreover, the results obtained for the optical gap,E_g, do not show any significative change in E_g, in going from state A to state B, neither for the sputtered samples (E_g= 1.72 to 1.74 eV) nor for the GD ones (Eg= 1.69 to 1.68 eV), indicating that there is no significant change in the valence or the conduction band edges.

Consequently, the large change in E_0 for the GD samples suggests an increase of the disorder in this material. Such a behaviour is directly related to intrinsic properties of the material and it cannot be due to a temperature effect, as the temperature of the sample increases only from 300 to 320 K during light exposure.

On the other hand, from state A to state B, the PC results as displayed in table III, show a larger decrease in the $\eta\mu\tau$ product for the sputtered samples (5.1x10^{-7} to 4.6x10^{-8} cm^2V^{-1}) than for the GD ones (1.9x10^{-5} to 1x10^{-5} cm^2V^{-1}).

The results obtained for E_0 and the $\eta\mu\tau$ product suggest again that the light-induced defects involved in sample 1 are different from those involved in sample 2.

In an attempt to interpret more qualitatively the above results, a theoretical model is needed to correlate the DOS and the $\eta\mu\tau$ product variations in order to derive the capture cross section of the different defects existing already initially in state A or light-induced in state B. However further informations on the transport properties are needed to clearly assess which model (for instance ballistic or diffusive one (9)) is applicable for each state. But nevertheless, the PC results of table III clearly show that, in state A, samples 1 and 2 exhibit comparable values of the DOS while the $\eta\mu\tau$ product for sample 1 is two orders of magnitude lower than that of sample 2. Such a behaviour suggests that the recombination centres initially present in state A are more efficient in sample 1 than in sample 2, indicating then their different nature from one sample to another (it is noteworthy that in state A, these two samples have approximately the same optical gap and bulk Fermi level position). Then we can expect that light will affect in a different way these various kinds of defects, which has been effectively shown by our experimental results.

3. CONCLUSION

The detailed behaviour of the light-induced metastable defects in our samples shows that on one hand, as the dominant defect created after light exposure is not the neutral DB, at least another type of metastable defect must be considered; on the other hand, defects of different nature are probably involved from one sample to another. Then, to have a complete characterization of the Staebler-Wronski effect, it will be necessary to study a large variety of samples prepared in very different conditions such as those giving for example films with different hydrogen contents.

ACKNOWLEDGEMENTS

The authors wish to thank Mr. A. Mastrogiacomo and Mr. G. Fameli for their appreciable technical help.

REFERENCES

1. D.L. Staebler and C.R. Wronski, *Appl. Phys. Lett.* **31,** 292 (1977).
2. M. Stutzmann, W.B. Jackson and C.C. Tsai, *Phys. Rev.* B **32,** 23 (1985).
3. D. Adler, *Journal de Phys.,* Colloque **42,** C4, 3 (1981).
4. K.K. Mahavadi, K. Zellama, J.D. Cohen and J.P. Harbison, *Phys. Rev.* B **35,**7776 (1987), (see also refs. inside).
5. A. Skumanich, N.M. Amer and W.B. Jackson, *Phys. Rev.* B **31,** 2263 (1985).
6. C. Lee, W.D. Ohlsen, P.C. Taylor, H.S. Ullal and G.P. Ceasar, *Phys. Rev.* B **31,** 100 (1985).
7. Qin-Guo Gang and Kong Guang Lin, *Phil. Mag. Letters* **57,** 117 (1988).
8. K. Zellama, J.D. Cohen and J.P. Harbison, *Mater. Res. Soc. Sympo. Proc.,***49,** 311 (1985), (see also refs. inside).
9. R.A. Street, *Phil. Mag.* **49,** L15 (1984).
10. N.M. Amer and W.B. Jackson, *Semiconduct. and Semimet.* Vol. **21** part B, 83. Edited by J.I. Pankove, Academic Press, New York (1984).

A GROWTH STUDY OF HOMOGENEOUS-CVD (HOMOCVD) HYDROGENATED AMORPHOUS SILICON

Z.M. Qian*, A. Van Ammel, H. Michiel, J. Nijs
and R. Mertens
Interuniversitair Micro-Electronica Centrum
Kapeldreef 75, B-3030 Leuven
Belgium

Abstract

Hydrogenated amorphous silicon (α-Si:H) prepared by "homogeneous chemical vapour deposition" (HOMOCVD) has several unique properties compared with the well-known glow-discharge prepared material . In order to investigate the potential usefulness of HOMOCVD technology, we have studied and optimized the film growth conditions, such as the growth rate as a function of gas pressure, gas flow, gas temperature, and substrate temperature. The morphology and structure of the films have also been investigated by scanning electron microscopy (SEM) and transmission electron microscopy (TEM). Films grown at a deposition rate of about $150\AA/min$ show good uniformity in thickness (variation less than $\pm10\%$ over a 80mm diameter surface) and a featureless CVD-like morphology. Films grown at a high rate ($400\AA/min$) have poorer uniformity and a PVD-like structure.

1 Introduction

HOMOCVD, as an alternative deposition method to prepare α-Si:H films was first reported by Scott et al. in 1981 [1]. In HOMOCVD, silane decomposition is by pyrolysis, as in normal CVD. However, the substrate is cooled, such that enough hydrogen can be incorporated in the film in order to saturate most of the dangling bonds. The possible advantage of this deposition method is clear: whereas in the well known glow discharge technique, energetic electrons and ions bombard the surface, the mean particle energy in a CVD process is of the order of kT_{gas} only, i.e. less than 0.1eV.

This expectation is indeed confirmed by experiment. Scott and coworkers reported a low defect density and a small Staebler-Wronski effect [2] in HOMOCVD α-Si:H; the maximum conductivity is considerably higher in comparison to glow-discharge films for boron as well as phosphorus doped films [3]. Moreover, HOMOCVD films exhibit excellent photoluminescence properties [4]. A marked advantage of HOMOCVD is the possibility to vary the band gap by simply changing the substrate temperature T_s. At low T_s , more hydrogen is incorporated in the film, thus widening the gap. Band gaps up to 2.7eV may be reached.

2 Experimental

The experiments were carried out in a modified low pressure CVD (LPCVD) system. A quartz tube is mounted within a three temperature zone furnace; the cooled substrate holder is made of stainless steel and its temperature is measured by a thermocouple. Most of the experiments were carried out at $T_s = 60^0C$ by using water cooling. Only in the experiment

*on leave from Zhejiang University, Hangzhou, China

concerning the influence of the substrate temperature was compressed air used to reach a higher substrate temperature. Gallium was used as a thermal contact between the substrate holder and the substrate. The difference between the real T_s and the temperature measured by the thermocouple is estimated to be about 25^{0}C. Pure electronic grade silane was used in all cases. Our experiments were performed in the gas temperature range from 660^{0}C to 750^{0}C, with a massflow from 40 sccm to 130 sccm and an inlet pressure in the range of 470 mTorr to 590 mTorr. Films were deposited on Corning 7059 glass for the measurements of thickness, optical gap, and conductivity. The thickness of the films was measured by the stylus method (DEKTAK). Some films were deposited on silicon wafers for morphology and structural investigations by scanning electron microscope (SEM), and transmission electron microscope (TEM).

3 Results and Discussion

3.1 Growth Rate as a Function of Gas Temperature

The influence of the gas temperature on the growth rate of HOMOCVD films is given in Fig.1. In this set of experiments, the massflow and the pressure at the inlet were kept constant at 80 sccm and 486 mTorr respectively. Calculations show that,under these conditions, the incoming gas reaches the temperature of the reactor wall after a few millimeters. It is therefore safe to identify the gas temperature with the wall temperature . The activation energy deduced from Fig.1, in the temperature range from 650^{0}C to 700^{0}C, is 16 kcal/mole. The optimal temperature for obtaining the highest growth rate is about 690 ^{0} C. When the gas temperature exceeds 700 ^{0}C, the growth rate decreases rapidly. This can be explained by the extensive deposition on the hot wall of the reactor at higher temperatures. The activation energy of the homogeneous gas phase decomposition of silane is 51.2 kcal/mole [6]. We think that the discrepancy with our result is once again due to the competition of deposition of polycrystalline silicon film on the hot wall of the reactor and α-Si:H growth on the cooled substrate.

3.2 Growth Rate as a Function of Pressure

The relation between the growth rate and the pressure of silane is given in Fig.2. In these experiments, the massflow and the gas temperature were kept constant at 80 sccm and 690^0 C respectively. The gas pressure in Fig.2 is the pressure at the substrate, determined by considering the pressure drop between the gauge and the substrate. It is interesting to notice that obviously there are two regions for the HOMOCVD film growth according to the pressure of silane. For the pressure lower than 285 mTorr (region 1), the growth rate is low (around 10 Å/min) and less dependent on the pressure itself. For a pressure higher than 285 mTorr (region 2), the pressure dependence is almost exponential. It has to be mentioned that when the pressure reaches a certain value, a porous film will be formed, and the deposition fails. We call this effect "local nucleation" (LN) . LN is a phenomenon particular to HOMOCVD, which is related to the presence of a steep temperature drop in the vicinity of the substrate. It is thought that nucleation will occur locally in the narrow region of temperature decrease, at a lower pressure than homogeneous nucleation under normal LPCVD conditions. LN will be discussed in more detail elsewhere.

3.3 Growth Rate as a Function of Massflow

The massflow of silane also gives strong influence on the growth rate in our operating conditions. This is shown in Fig.3. In these experiments, the pressure at the outlet and the gas temperature were kept constant at 267 mTorr and 690 ^{0}C respectively. Qualitatively, the flow dependence is similar to the pressure dependence shown in Fig.2, with an initial plateau and a region of almost exponential increase. There is a tendency towards saturation, just as in the LPCVD case [7]. We think that this complicated behaviour is due to silane depletion on the walls of the reactor.

3.4 Growth Rate as a Function of Substrate Temperature T_s

As mentioned by Scott [1], we also found that the growth rate is not dependent on the substrate temperature T_s . The independence of the growth rate on T_s implies that the deposition process in HOMOCVD is dominated by the diffusion step, rather than the surface reaction.

3.5 The Morphology and Structure of HOMOCVD Films

Several authors reported that the optical and electrical properties of α-Si:H films and the quality of the devices are closely related to the surface morphology and the structure of the films. Tsai, Knights, et al.[8,9] found that the device-quality a-Si:H films grown by glow discharge have CVD-like morphology and structure, and that the films with PVD (physical vapour deposition)-like morphology and structure are of poor quality. CVD-like in this context stands for films which are almost featureless, even under high magnification, whereas PVD-like films show a rich morphology. This distinction was also relevant to our HOMOCVD films. The closer to the critical pressure for local nucleation the operating pressure, the more likely it is to grow a film with PVD-like morphology and columnar structure. Photographs taken by SEM and TEM demonstrate these two kinds of morphologies and structures of HOMOCVD films. Films with a CVD-like morphology and a featureless structure were grown at a pressure of 580 mTorr. Both the SEM (Fig.4a) and the TEM (Fig.4b) pictures are totally featureless. However, when the operating pressure is increased from 580 mTorr to 610 mTorr, while the other parameters remain the same , a PVD-like morphology and structure are found. The SEM picture in Fig.4c shows a rough surface, and in the cross section (Fig.4d), one sees a columnar structure. If one increases the operating pressure a little further, for instance to 620 mTorr, a yellow powder is formed on the substrate, indicative for local nucleation.

3.6 The Homogeneity of HOMOCVD Films

The homogeneity of a thin film is an important requirement for making devices. Therefore we systematically measured the thickness uniformity of the films by the stylus method (DEKTAK). For a growth rate of about 150 Å/min, the maximum thickness variations are about 10% of the thickness at the center of the holder. On the other hand, when the pressurte is increased to the vicinity of the local nucleation point, the growth rate becomes very high (about 400 Å/min), but the film is very non-uniform and of poor quality. In general, we find that good uniformity requires a rather high mass flow and a properly chosen pressure. All this holds for "successful" depositions. However, the main problem in HOMOCVD is to obtain a uniform low substrate temperature ($T_s \leq 250^0$C) in an ambient of $T_{gas} \simeq 700^0$C. We used gallium as a thermal contact layer. It sometimes occurs that the contact is not uniform, and then dark spots appear on the film. These spots are due to the locally higher T_s , which results in α-Si:H with smaller hydrogen content and smaller band gap.

3.7 The Influence of the Growth Rate on the Quality of the Film

Although this paper is mainly concerned with the dependence of the growth rate on the deposition parameters and the resulting film structure, the question naturally arises whether the quality does not degrade with the deposition rate. A detailed study will be reported separately. Here, we simply mention that measurements of dark conductivity,photoconductivity and photoluminescence indicate good quality of our material, for a growth rate of about 150 Å/min.

4 Conclusions

A systematic study on HOMOCVD film growth has been performed. It shows that our HOMOCVD films are grown with an activation energy of 16 kcal/mole and the optimal growth temperature is around 690 ^{0}C. A high massflow favors both obtaining a high growth rate and growing a uniform film. In order to obtain a reasonable growth rate with an acceptable homogeneity, a CVD-like morphology, and a featureless structure, a proper operating pressure of silane is essential. However, a limit exists in the form of the 'local nucleation' phenomenon. When the operating pressure is very close the critical pressure of local nucleation, the film not only shows poor homogeneity, but also PVD-like morphology and columnar structure with poor quality. Under certain operating parameters, CVD-like good quality HOMOCVD films with a reasonable growth rate of 150 Å/min and a good thickness uniformity have been obtained.

5 Acknowledgements

This work was performed in the frame of the Photovoltaic Programme of the Commission of the European Communities under contract number EN35-0050-B and cosponsored by the Belgian government under contract number CE/ENE/03. We thank P. Roussel and J. Vanhellemont for the SEM and TEM analyses.

References

[1] B. A. Scott, R. M. Plecenik, and E. E. Simonyi, Appl. Phys. Lett. *39*(1) (1981), 73

[2] B. A. Scott, J. A. Reimer, R.M. Plecenik, E.E. Simonyi, and W. Reuter, Appl. Phys. Lett. *40*(11) (1982) 973

[3] B. S. Meyerson, B. A. Scott, and D. J. Wolford, J. Appl. Phys. *54*(3) (1983) 1461

[4] D. J. Wolford, J. A. Reimer, and B. A. Scott, Appl. Phys. Lett. *42*(4) (1983) 369

[5] B. A. Scott, *Homogeneous Chemical Vapour Deposition*, Semiconductors and Semimetals *21 A*, ed. J. I. Pankove, Florida: Academic Press Inc. , 1984

[6] B. S. Meyerson and J. M. Jasinki, J. Appl. phys. *61*(2) (1987), 785

[7] C. H. J. Van Den Brekel and L. J. M. Bollen, J. Crystal Growth *54* (1981) 310

[8] C. C. Tsai, J. C. Knights, G. Chang, and Wacker, J. Appl. Phys. *59*(8) (1986) 2998

[9] J. C. Knights, J. Non-Crystalline. Sol. *35* - *36* (1980) 159.

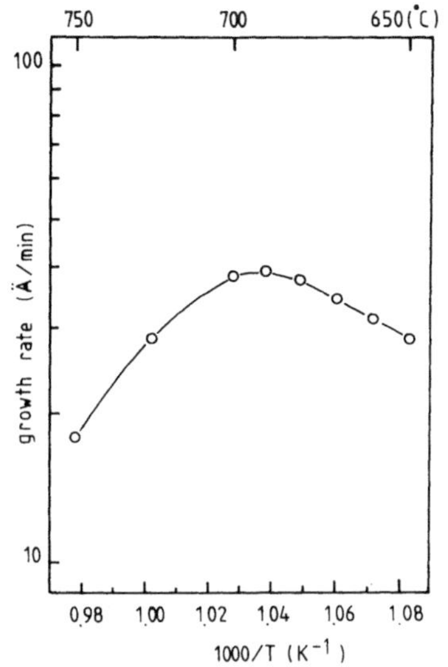

Fig.1 Dependence of the growth rate on the gas temperature. The deduced activation energy for T $\leq$ 700^{0}C is 16 kcal/mole.

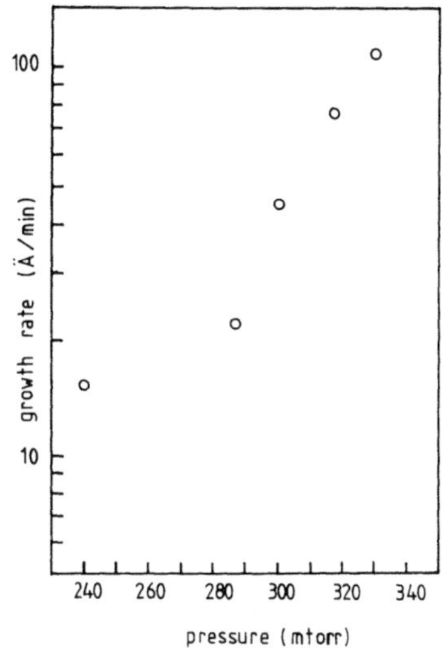

Fig.2 Dependence of the growth rate on the silane pressure. The flow is fixed at 80 sccm and T_s = 60^{0}C .

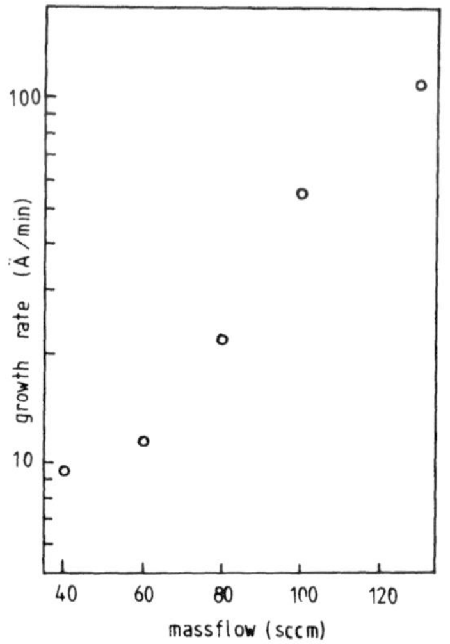

Fig.3 Dependence of the growth rate on the mass flow of silane. The pressure is fixed at 267 mTorr and T_s = 60^{0}C .

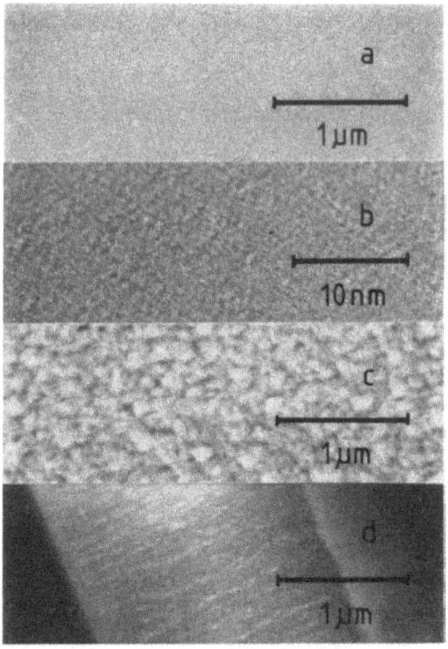

Fig.4 (a) SEM picture of featureless, CVD-like film. (b) TEM cross section of the film in (a). (c) SEM picture of PVD-like film. (d) TEM cross-section of the film in (c)

STUDY OF THE HYDROGEN INCORPORATION IN a-SiGe:H THIN FILMS THROUGH THERMAL HYDROGEN EFFUSION

G. Sardin, C. Roch, J. Andreu, J.L. Morenza, C. Godet*, A. Lloret*, J.P. Stoquert**

Departament de Fisica Aplicada i Electrònica, Universitat de Barcelona, Av. Diagonal 645, E-08028 Barcelona (Spain)

* Lab. de Physique des Interfaces et Couches Minces, Ecole Polytechnique, F-91128 Palaiseau Cedex (France)

** Centre de Recherches Nucléaires, Groupe P.H.A.S.E., F-67037 Strasbourg (France)

Abstract

Samples of a-SiGe:H have been principally analysed through thermal H effusion. The prime manifestation stands in the differentiation of the samples hydrogen content into two types, i.e. weakly and tightly bonded H (WBH and TBH). The incorporation of Ge has as side effect a strong tendency to increase the amount of weak H bonds, these being known to degrade the films electronic properties and stability. Some correlations between the WBH/(WBH+TBH) ratio and technological parameters, the Ge percent and the films thickness are presented.

Introduction

Ge alloying of a-Si:H has been thought to be a convenient way to adjust the energy gap according to the desired application, in particular for photovoltaic cells in way to optimize their efficiency. Nevertheless it has as side effect to favor the appearence of weakly bonded hydrogen (WBH) which is considered to induce poor electronic properties and low stability. Therefore H related characterizations are of fundamental interest in order to understand the H behaviour, the associated properties of such an alloy and to optimize the deposition conditions.

The variations that undergoes the silane-germane plasma from one side and the incorporation of Ge into the Si network on the other side, lead to a substantial modification of the hydrogen content as well as the proportion of the different H bonds populations in the films, i.e. the WBH/(WBH+TBH) ratio. These aspects are analysed through the H thermal effusion technique. The results are interpreted in function of some of the technological parameters such as the ratio of the germane inlet over the total gas inlet, the substrate temperature, the RF power and also of two of the films characteristics, i.e. the film Ge percent and thickness.

II. Experimental

We have studied a-SiGe:H films, 0.5 to 6.0 μm thick, grown on single crystal Si substrates, in the ARCAM reactor (LPICM, Ecole Polytechnique, Palaiseau) by means of an RF discharge in mixtures of silane, germane and hydrogen. Diverse running conditions have been effectuated such as different fluxes of the inlet gas RG=GeH4/(SiH4+GeH4) ranging from 0 to 17%, diverse RF powers (1 to 5 watts) and deposition temperatures (Ts=100,150 and 200°C), the total pressure being fixed at 90 mTorr and the fluxes ratio H2/GeH4 equal to 9.0.

The thermal effusion measurements have been effectuated by heating, up to 750°C, samples of a-SiGe:H located in a quartz tube with a static vacuum of 10^{-5} mTorr. The evolution of the H_2 partial pressure has been measured by a quadrupolar mass spectrometer, Quadruvac Q200 Leybold-Heraeus. The Rutherford Back Scattering (RBS) has been used to determine the Ge/(Si+Ge) ratio, and the Elastic Recoil Detection (ERD) to determine the H/(Si+Ge) ratio, the latter being compared to the thermal H effusion measurements.

III. Results and Discussion

The prime manifestation of the H thermal effusion stands in the differentiation of the hydrogen of the samples into two types, i.e. the weakly and the tightly bonded H (WBH and TBH). The incorporation of Ge into the network has as side effect a strong tendency to increase the amount of weak H bonds. In effect, the H effusion spectra of a-SiGe:H most often show two evolutions, one at low temperature and the other at higher temperature to which are respectively associated the WBH and TBH (fig.1). The deconvolution of the spectra allows to get the proportion between WBH and TBH. This is a quite significative quantity which in some way represents the proportion between the unusefull and usefull H. The former is unstable and related to poor electronic properties, pacifying defects in an unstable way while the second one saturates dangling bonds in a stable way. Further this quantity has the adventage over the respective absolute amount of H, not to require to know the thickness, the area and the density of the films, neither the calibration of the quadrupole, avoiding thus the experimental errors on these quantities.

The ERDA and thermal H effusion techniques give total H contents in quite good agreement. Assuming a constant atomic density ($5x10^{22}$ at.cm^{-3}) over the 0 to 40 Ge at.% range, a slight increase of the H content C(H) versus the Ge percent x(Ge), is observed at Ts=150°C (fig.2), whereas a constant or even a slightly decreasing H content is measured at Ts=200°C.

Figure 3 displays the dependence of the WBH content on the Ge incorporation x(Ge) and the deposition temperature Ts. At any value of the gas mixture composition RG, a strong increase of x(Ge) in the solid phase is observed between 100 and 150°C, followed by a saturation at higher Ts. Since the thermal decomposition of GeH4 is not significant below 300°C, this effect may be attributed to a decreasing hydrogen coverage of the growing surface. In contrast, the RF power in the range 1 to 5 W has shown little influence upon the x(Ge) and WBH contents.

The increase of the temperature Ts shows a beneficial effect whatever the proportion of the GeH4 flux, RG. At Ts=200°C almost no WBH is created by introducing GeH4, up to a typical x(Ge) of 25-30%. However, at higher Ge contents a steep increase of WBH starts to take place, whatever Ts in the 150-250°C range. Such a basic modification of the WBH incorporation, likely due to microstructures above 25-30 Ge at.%, is reflected in the sharp degradation of the electronic properties. In fig.4, this behavior is shown for two typical film thickness around 0.7 and 5.0 μm. As a consequence, one should be cautious when comparing samples with different thicknesses since they may have different microstructure.

The influence of x(Ge) on the WBH/(WBH+TBH) is dependent of the growth conditions of the films. This clearly shows the interdependence of the structural aspects with the compositional ones, e.g. for the same x(Ge), higher is Ts lower is the WBH/(WBH+TBH) ratio. It appears thus that increasing the deposition temperature conveniently reduces the proportion of WBH, nevertheless an excessive temperature will additionally reduce too much the TBH content, leading to an insufficient saturation of the dangling bonds. Hence a compromise should be reached about the selection of the deposition temperature.

One should be aware that the H based correlations have not an absolute character, but point out tendencies more or less pronounced and which are in part dependent on the specific characteristics of the reactor. In particular, different reactor geometries generate different discharge characteristics, for instance the ions bombardment, leading to structural differences, these being decisive on the H related characteristics.

Conclusion

It has been observed that alloying Si with Ge strongly favors the appearence of weakly bonded H, and that its proportion with respect to the amount of tightly bonded H increases sharply with the increment of the Ge incorporation. The WBH being relatively unstable and associated to the fall of electronic quality should thus be avoided.

The thermal H effusion spectra provides a fast access to an evaluation of the films quality since they indirectly inform about the degree of microstructure (porosity) which is known to be determinant on the electronic properties. Besides the quantitative H correlations the variety of shapes of the spectra allows qualitative differentiation of the films structure. The spectra allows also an estimation of the H solubility and of its cohesion in the host matrix, according to the degree of alloying. Thus the thermal H effusion technique proves to be a useful tool in the rough optimization of the technological parameters, the presence of WBH revealing the existence of a microstructure.

Acknowledgements: This research program on amorphous silicon is supported by the European Economic Community. The part corresponding to the University of Barcelona is additionally supported by the Instituto de Energías Renovables del CIEMAT and the CAICYT del Ministerio de Educación y Ciencia.

References

(1) Y. Yukimoto, Amorphous Semiconductor Tech. and Devices, Jarect vol.6, Y. Hamakawa ed., North-Holland Pub. Co., 1983

(2) W. Beyer, J.of Non-Crystalline Solids, 97-98, 1027 (1987)

(3) G. Sardin, J. Andreu, J.C. Delgado, J. Esteve, J.L. Morenza, 7th E.C. Photovoltaic Solar Energy Conf., D.Reidel, Dordrecht, Holland, 1986, p.577

(4) G. Sardin, J. Andreu, J.C. Delgado, J.L. Morenza, Solar Energy Mat. (in press)

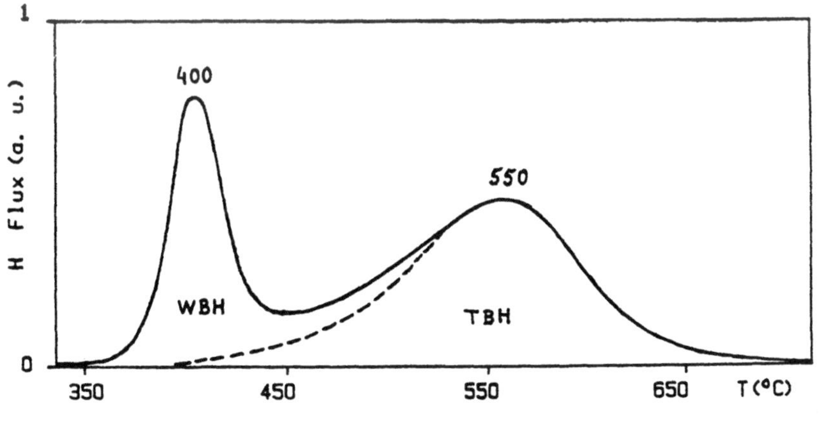

Fig. 1: Typical thermal H effusion spectrum of a-SiGe:H. Two evolutions are observed proceeding from weakly bonded H (WBH) and tightly bonded H (TBH)

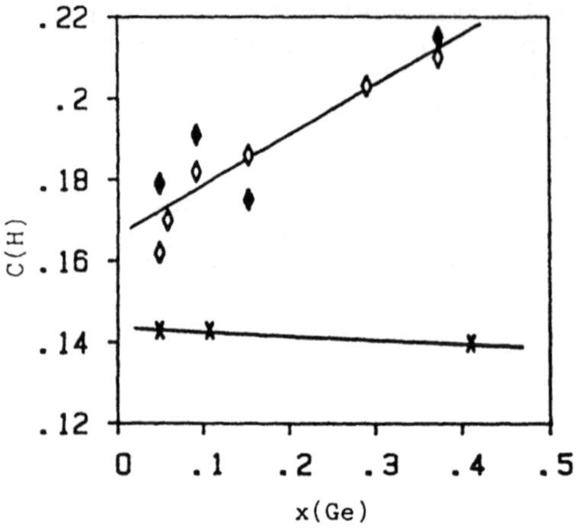

Fig. 2: H content vs Ge concentration for diverse deposition temperatures Ts (P_{rf}=1W)

◊ E.R.D. (Ts=150°C)

♦ Thermal H effusion (Ts=150°C)

X E.R.D. (Ts=200°C)

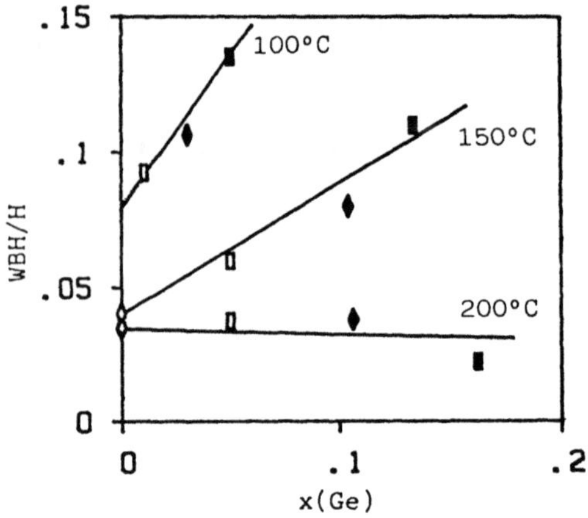

Fig. 3: WBH/H vs Ge concentration for different deposition temp.Ts.

◊ RG= 0%

▯ RG= 0.5%

♦ RG= 1.1%

▮ RG= 2.1%

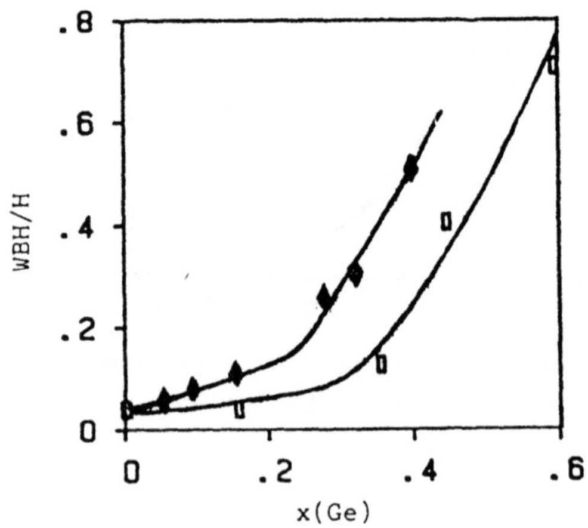

Fig. 4: WBH/H vs Ge concentration for two different ranges of film thickness. (Ts=150°C)

♦ e=0.5-1.0 µm

▯ e=4.0-6.0 µm

a-SiGe:H-BASED SOLAR CELLS: REALISTIC EVALUATION OF PERFORMANCE IN SINGLE AND MULTI-JUNCTION DEVICES.

G.Conte, D.Della Sala, F.Galluzzi, G.Grillo, C.Reita
ENIRICERCHE S.p.A., 00015 Monterotondo (Roma), Italy

ABSTRACT. We present here a realistic evaluation of single and tandem solar cells, by a model calculation wich includes a detailed treatment of the optical and transport features of the cells. Parameters desumed from our experimental data on a series of a-$Si_{1-x}Ge_x$:H alloys in the range x_{Ge} = 0 to 1 are used; various cell configurations are considered, with and without optical confinement.

1. INTRODUCTION

Amorphous hydrogenated Si-Ge alloys are very promising photovoltaic materials, owing to the "tunability" of their gap in the 1-2 eV region, which in principle allows them to be used in multi-junction stacked solar cells to get a more efficient solar spectrum absorption. The possibility of reaching high performance in these tandem or multiple devices is, however, still under investigation. In particular, the degradation of the electronic properties of the alloys with increasing Ge content has to be taken into account to choose the most promising combination of the gap values and of the thicknesses in multi-junction cells.

2. THE MODEL

<u>Cell configuration</u>.The typical configuration assumed for the tandem cells was

[1] glass/TCO/p-SiC:H/i-n-$Si_{1-x}Ge_x$:H/
/p-SiC:H/i-n-$Si_{1-x}Ge_x$:H/metal

while for the single cells obvious simplifications of this scheme were considered. The thickness of the p+ (gap 2.0 eV) layers were 10 and 5 nm and that of the n+ layers were 5 and 30 nm. The ohmic contacts were obtained by 450 nm thick fluorine-doped SnO_2 on the front and aluminum on the rear side. Both flat and textured TCO (haze rate = 10%) have been considered.

<u>Optical model</u>. For the flat cell configuration an exact Rouard recursive method (1) taking into account the optical parameters of all the layers constituting the cell was used, to obtain, for each wavelength, the total absorption and reflection and then, by calculating the light amplitudes in the cell, the individual light absorption of each layer as well as light aborption per unit length $a(z,\lambda)$ as a function of the depth z in the cell.

In textured cells, an approach similar to that used in ref. (3) was used to include the diffused light, and then $a(z,\lambda)$ was obtained step by step by calculating for each z the local absorption for the direct and the diffused light travelling in the two directions.

The charge carrier generation G(z) was calculated assuming, as usual, unit generation efficiency in the intrinsic layers and no geminate recombination:

[2] $$G(z) = e/hc \int a(z,\lambda)I(\lambda)\,\lambda\, d\lambda$$

where $I(\lambda)$ is the incident light intensity per unit area and unit wavelength. The standard AM 1.5 SERI data (4) were used for I.

The dielectric functions n, k for the amorphous layers were included in the model by assuming for each layer l a Tauc-type and a Urbach tail absorption spectrum above and below the optical gap respectively for calculating $k_l(\lambda)$, and a Cauchy expansion for $n_l(\lambda)$, using, for the relevant parameters, experimental data on a series of a-SiGe:H alloys (fig. 1). Experimental data for SnO_2 and literature data (5) for the rear metal contact were used.

Collection efficiency. A variable minority carrier model was used, in which, for each intrinsic layer, the local collection probability p(x,V) as a function of the distance from the front layer x and of the applied voltage V was calculated according to ref.(6) but the critical lenght x_c was given by

[3] $$x_c = d/(1+\mu\tau_h/\mu\tau_e)(1+\alpha\cdot d)$$

The internal electric field V_b/d was taken as uniform over the thickness of the intrinsic layer d, with

[4] $$qV_b = E_g - \Delta E_n - \Delta E_p$$

where ΔE_n, ΔE_p are the activation energies of the n+ and p+ doped layers. The mobility-lifetime values $\mu\tau\,(x_{Ge})$ were calculated assuming a lifetime inversely proportional to the density of defects:

[5] $$\mu\tau_{e,h} = C_{e,h} / N_d, \quad C_e = 2\cdot10^8/V\cdot cm \quad, C_h = 4\cdot10^7/\ V\cdot cm$$

while an exponential fit of the N_d data of fig.1 was used to represent the density of gap defects as a function of x_{Ge}:

[6] $$N_d(x_{Ge}) = 5\cdot10^{15}\cdot 10^{2.05\cdot x_{Ge}}\ cm^{-3}$$

The photocurrent is then obtained by integrating on the layer:

[7] $$I_L(V) = \int p(x,V)\cdot G(x)\cdot dx$$

Electrical model. The characteristics of the diode shunting each elementary cell in our model were assumed to depend on x_{Ge} trough the dependence of the reverse saturation current on the gap Eg and on τ (7):

[8] $$I_o = q\,N_c\cdot d/_\tau \cdot \exp(-V_{bi}/kT)$$

while the diode quality factor was held constant at a value of n = 1.5. V_{bi} was assumed to be

[9] $$V_{bi} = E_{gp} - \Delta E_p - \Delta E_n$$

where E_{gp} is the gap energy of the doped p layer of the cell.

Calculation. The numerical calculations are carried out, for a tandem cell, according to the following steps:

a) on the basis of the chosen cell configuration and materials, obtain, from the optical calculations on the complete series of layers, the two generation functions in the two intrinsic layers $G_1(x)$, $G_2(x)$.

b) separately for the two half-cells calculate the fotocurrents $I_{L1,2}(V)$ by eq. [7] for each applied V.

c) shunt each half-cell by the appropriate diode.

d) connect the two half-cells in series to obtain the final j-V curve.

3. RESULTS

Single cells. Fig. 2 shows the results of the calculation for single cells in terms of short circuit current I_{sc}, open circuit voltage V_{oc}, and efficiency η as a function of Ge content x_{Ge} in the intrinsic layer, for two series of cells deposited on a flat and on a textured substrate, and for two thicknesses of the intrinsic layer, 150 and 500 nm.

As can be seen, the sample series chosen for this simulation is not optimized for use in single a:Si:H cells due to non optimal gap value (1.85 eV) obtained for the $x_{Ge} = 0$ alloy, while the 1.7 eV alloy ($x_{Ge} = .1$) has a low mobility-lifetime value, due to the presence of Ge. If, instead, we use the combination of optimal gap and good value (as can be actually be done in real samples using a different H content) we obtain the points marked by an asterisk, which show reasonable agreement with the literature data on the best cells.

Tandem cells. Fig. 3 shows the iso-efficiency curves for tandem cells on a flat substrate in two-terminal and four-terminal configurations, as a function of the thickness of the two intrinsic layers, with gap values 1.85 eV ($x_{Ge} = 0$) and 1.5 eV (xGe = .4). A maximun of about 9.6% is observed for both configurations, which has to be compared to about 8% obtainable as a maximun in single cells on flat substrates. As expected, the maximum is much broader for the four-terminal configuration.

4. CONCLUSIONS.

A versatile model for single and multiple cells has been set up, which includes a detailed optical treatment to calculate the local charge carrier generation and collection probability, taking into account both interface and bulk recombinations. The results obtained in single cells by using experimental optical and transport data are in reasonable agreement with values obtained in real cells.

The model allows to take into accont the real behaviour of a-SiGe:H material quality in forecasting the performance of tandem cells.

ACKNOLEDGMENTS. This work was supported by AGIP S.p.A.

REFERENCES

(1) see for example O.S.Heavens, "Optical Properties of Thin Solid Films", p.63, Academic Press 1955, and F.B. Ellis, A.D.Delahoy, Solar Energy Materials 13, 109 (1986)

(2) G.Conte et al.,this conference

(3) H.Schade, Z.E.Smith, J.Appl.Phys. 57, 568 (1985)

(4) R.L.Hulstrom, SERI report No. SP-215-2784, 1985

(5) D.Y.Smith,E.Shiles,M.Inokuti in "Handbook of Optical Constants of Solids", E.D.Palik ed., p. 369, A.P. 1985.

(6) S.Nomomura, H.Okamoto, H.Kida, Y.Hamakawa, Japan. J. Appl. Phys. 21 suppl.2, 279 (1982)

(7) S.M.Sze, "Physics of Semiconductor Devices", 2nd ed., ch. 2, J.Wiley, 1981

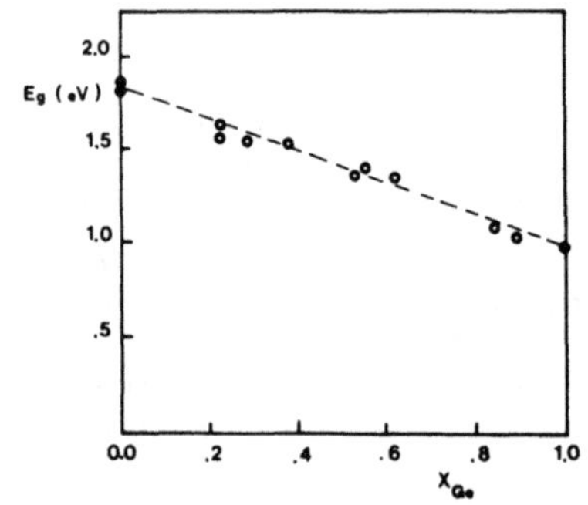

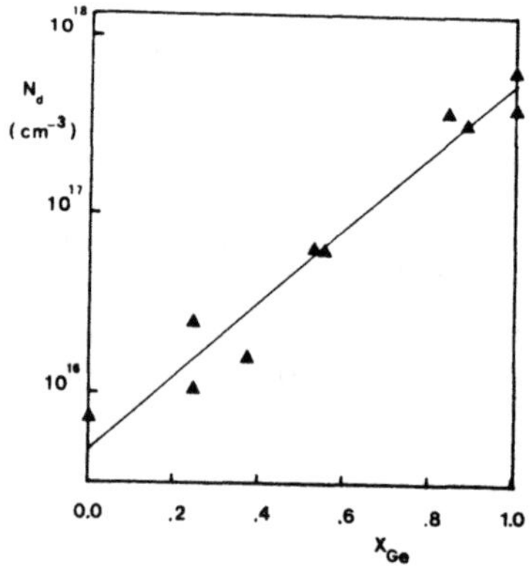

Fig. 1. Optical gap E_g and density of defects N_d from PDS measurements vs. Ge content x_{Ge} for the series of a-SiGe:H alloys considered in the calculation (from ref. 2).

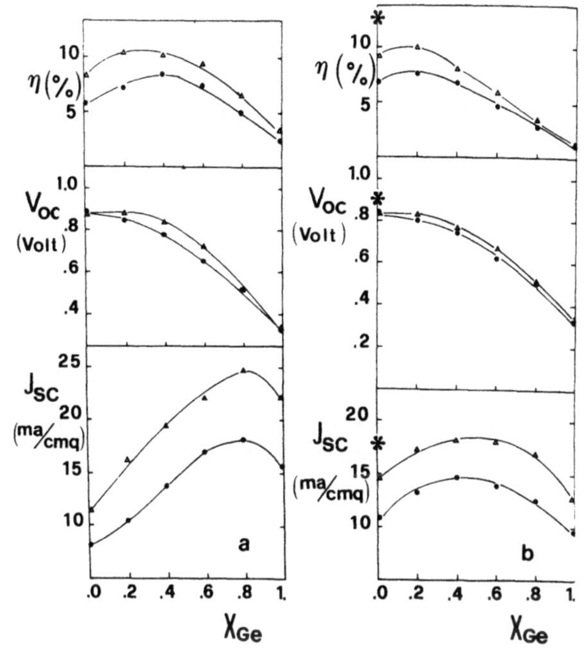

Fig. 2.

Calculated efficiency η, open circuit voltage V_{oc} and short-circuit current J_{sc} for single cells vs. atomic Ge content x_{Ge}; intrinsic layer thickness d : 150 nm (a), 500 nm (b); flat (closed circles) and textured (open triangles) substrate.

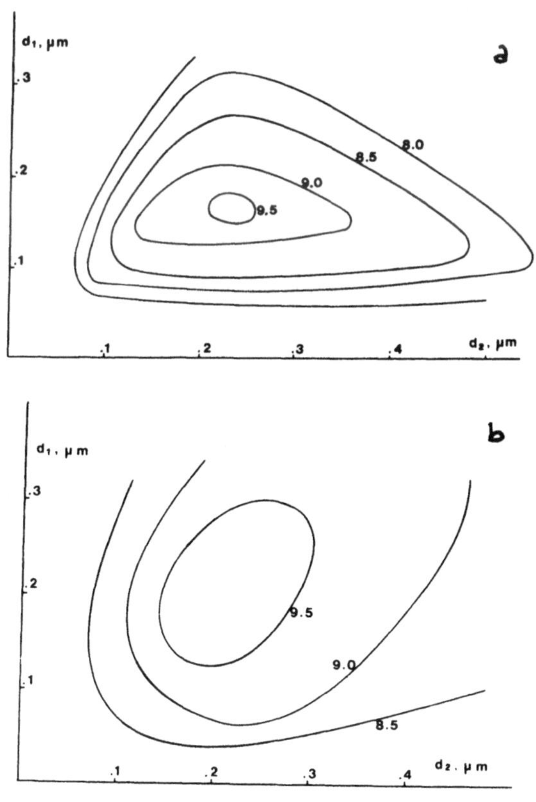

Fig. 3. Contour plots of tandem cell efficiency vs. thicknesses d_1, d_2 of the intrinsic layers; gap values E_{g1}=1.85 eV (x_{Ge}=0), E_{g2}=1.5 eV (x_{Ge}≅.4); 2-terminal (a) and 4-terminal (b) configuration; flat substrate.

CHARACTERIZATION OF a-Si:H SOLAR CELLS

J. Qiao, M. Falconieri, P. Fiorini, A. Mittiga, and F. Evangelisti
Department of Physics, University of Rome "La Sapienza"
Piazzale Aldo Moro 2, 00185 Roma, Italy

ABSTRACT

The problem of solar cell characterization is thoroughly analized. We use the variable minority carrier transport model to derive, from spectral response, the relevant parameters of devices grown with different procedures. The meaning of these parameters is studied by applying the same method to the analysis of spectral response curves simulated by a numerical model. The comparison shows that the method provides mean values for the carriers lifetimes and for the electric field in the intrinsic layer. The difficulties which arise in the interpretation of the recombination velocities at the interfaces are discussed.

INTRODUCTION

Despite the large improvement of p-i-n a-Si:H solar cells, the analysis of device performances and the determination of the actual values of the most important microscopic parameters are problems not yet solved. Numerical calculations are available to predict the behaviour of the device, but their complexity rules out any possibility to fit the experimental data for parameter determinations. Approximate models are available as proposed by Crandall [1] and Okamoto *et al.*[2]. However, since these models rely on drastic assumptions their extensive use as a characterization tool requires a careful analysis of their applicability.

In the present work, we have focused our attention on the variable minority carrier transport (VMCT) model originally proposed by Okamoto *et al.*[2]. This model is based on the regional approximation and assumes a constant electric field in the intrinsic layer of the device. The effect of the doped layers is taken into account by introducing effective recombination velocities at the interfaces. We have used this model for determining, from spectral response data, mobility-lifetime products of minority carriers, recombination velocities at interfaces and built-in potential. We have analized the conditions of its applicability and used it to fit the data from a variety of p-i-n cells prepared in different deposition conditions. Then, in order to understand the relationship between the parameters obtained by this fitting procedure and the physical properties of the constituent materials and their interfaces, we have simulated the behaviour of the experimental devices by solving numerically Poisson's and continuity equations. The generated spectral response curves have been fitted using the VMCT model. In this way the meaning of the fitting parameters was made clear.

EXPERIMENTAL AND RESULTS

The p-i-n cells investigated in this work have been prepared in a single chamber G.D. reactor, characterized by a base pressure of 10^{-6} mbar. The structure was composed

by 2000 Å of TCO, 200 Å of p-doped silicon carbon alloy with an optical gap of 2.05 eV, 4000 Å of a-Si:H, 400 Å of n-doped a-Si:H and a back contact of aluminum. We have prepared two different series of cells changing the cleaning procedures of the chamber after the p layer deposition. The first procedure (procedure (a)) consisted in simply pumping down the system to its base pressure, the second one (procedure (b)) in flowing pure silane for 15 minutes. The cells have been characterized by measuring I-V curves in dark and under AM1.5 illumination. Average efficiency was 5.5 % in the first case and 7 % in the second one. Then we have focused our attention on the spectral response $\eta(\lambda, V)$. We have measured this quantity under a positive bias of .4 Volt, a negative bias of -1 Volt and short-circuit condition, and computed the ratios $R_+ = \eta(.4)/\eta(0)$ and $R_- = \eta(-1)/\eta(0)$. Then R_+ and R_- were fitted with the VMCT model in order to determine the mobility-lifetime products for electrons and holes ($\mu_n\tau_n$ and $\mu_p\tau_p$), the recombination velocities at the p-i and i-n interfaces divided by the mobility of the relevant carrier, (S_n and S_p), and the built-in potential (V_{bi}).
A first important result was that the quality of the fit is greatly improved by the superposition of a bias light on the monochromatic beam (in particular we have used a bias light of intensity equal to a tenth of AM1.5). The reason for this can be explained as follow. The VMCT model is based on the regional approximation, that is the cell is thought of as divided in two regions, one where electrons are minority carriers, the other where holes are. The boundary x_c, between the two regions is determined by the condition $\Delta n/\tau_n = \Delta p/\tau_p$, where Δn and Δp are the excess carriers. In turns $\Delta n/\tau_n$ and $\Delta p/\tau_p$ are approximations to the recombination rate whose complete expression is:

$$R = \frac{n_0 \Delta p + p_0 \Delta n + \Delta p \Delta n}{\tau_p(n_0 + \Delta n) + \tau_n(p_0 + \Delta p)}.$$

These approximations certainly hold if $\Delta n, \Delta p \gg n_0, p_0$. However, if x_c is too close to one interface, for example the p-i one, then $p_0 \gg \Delta n, \Delta p, n_0$. As a consequence R can be approximated by $\Delta n/\tau_n$ on one side, but not by $\Delta p/\tau_p$ on the other side. Therefore x_c cannot be located too close to the p-i or to the n-i interfaces. Nevertheless this situation can happen using a weak and strongly absorbed light. In this condition the model cannot be used to fit the experimental data. The presence of a bias light moves x_c towards the middle of the cell and the model can be safely used.
A further improvement on the quality of the fit has been obtained by computing the optical generation rate inside the cell by means of an accurate optical model which uses the measured values of the optical constants of all the five layers forming the device and takes into account, by the matrix method, the multiple reflection effects.
The behaviour of R_+ and R_- for two typical cells grown with different cleaning procedures and the fits with the VMCT model are shown in Fig. 1. It is evident that the cells grown with procedure a) exhibit inferior properties in the short wavelength region. This reflects a difference in the fill factor from an average of $\sim .5$ with procedure a) to an average of $\sim .6$ with procedure b). The explanation of this fact is that the second method, as previously reported[3] and as will be discussed below, is more effective in reducing the boron contamination of the intrinsic layer.
As for the fits note that they follow the small obscillations observed in the experimental data and which originates from the interference fringes of the spectral response: this

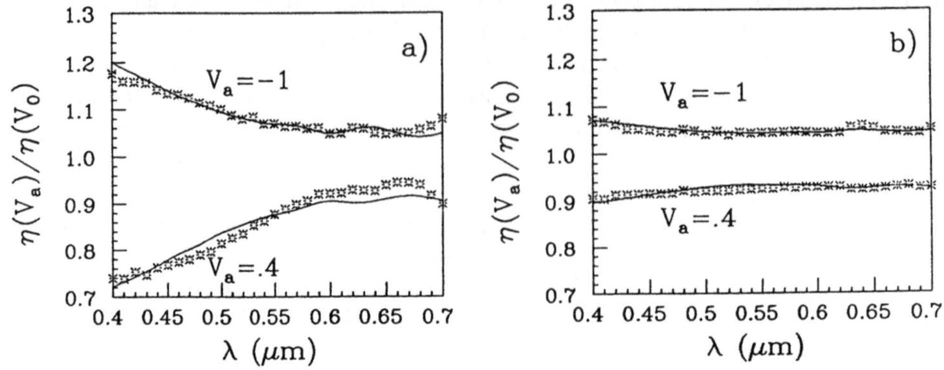

Fig.1 Ratios of spectral responce with positive and negative bias to the zero bias spectral responce.(∗) experimental, (—) fit. a) typical cell grown with procedure a). b) typical cell grown with procedure b)

indicates the correctness of our optical model. The values of the fit parameters obtained on four cells representative of the deposition procedure b) when compared to those of three cells representative of the deposition procedure a) show an appreciable ($\sim$ one order of magnitude) decrease of the p-i interface recombination velocity S_n and an equalization of the hole mobility-lifetime product $\mu_p\tau_p$ to that $\mu_n\tau_n$ for electrons. Typical values were $S_n = 0.5 - 1.0 \cdot 10^4 V/cm$ and $\mu_n\tau_n = 0.5 - 2.0 \cdot 10^{-8} cm^2/V$.

NUMERICAL SIMULATION

As mentioned previously we have simulated p-i-n solar cells by solving numerically the full set of transport equations. The numerical model we use is very similar to that presented by Hack and Shur[4], but also includes the effects due to SiC:H/Si:H heterojunction at the p-i interface. The density of states is modeled with four exponential tails of donor-like and acceptor-like states and the minimum is taken to correspond to $10^{16} cm^{-3} eV^{-1}$, a value similar to that measured in our intrinsic samples.

The effect of boron contamination present in the first series of cells is taken into account by introducing a concentration of active dopants which has a value of $4 \cdot 10^{16} cm^{-3}$ at the p-i interface and decreases exponentially in the intrinsic layer with a characteristic length of 500 Å .

Computer generated curves of R_+ and R_-, together with their fits with the VMCT model, are reported in Fig. 2 for two cells, with and without boron contamination. As can be seen they are very similar to the experimental ones of Fig. 1. Fit parameters again exibit an increase of S_n and the equalization of the $\mu\tau$ products seen in the real devices.

The relationship between fit parameters and microscopic parameters are shown in Fig. 3 and Fig. 4.

In Fig. 3 we report the electric field as a function of position together with the mean electric field determined from the VMCT model fit (V_{bi} divided by the cell thickness).It is evident that the effect of boron contamination consists mainly in reducing the electric field at the p-i interface. In Fig.4 we show the spatial dependence of electrons and holes

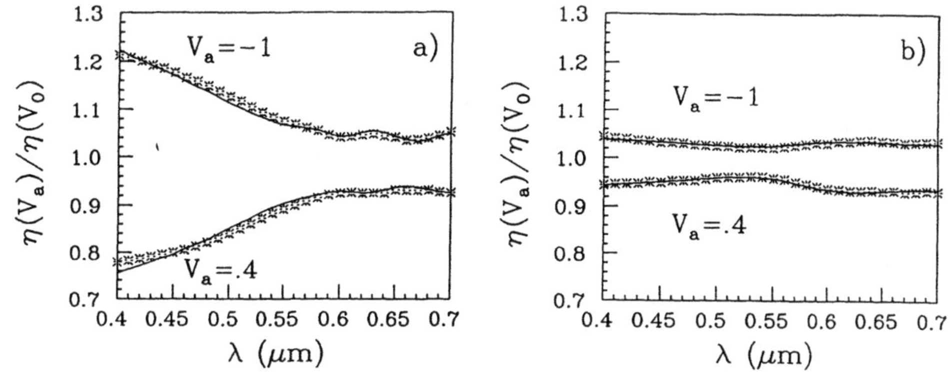

Fig.2 Ratios of spectral responce with positive and negative bias to the zero bias spectral responce.(*) numerical simulation, (—) fit. a) cell with boron contamination. b) cell without boron contamination

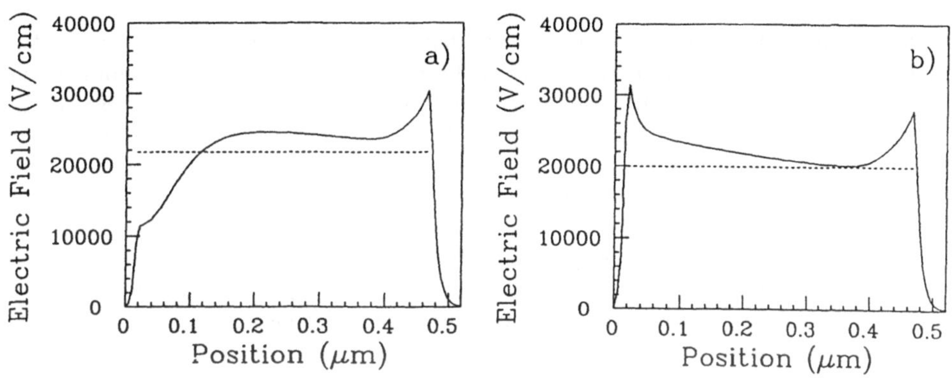

Fig.3 Electric field in the device. (—) numerical simulation. (····) from the fit. a) cell with boron contamination. b) cell without boron contamination

lifetimes, as well as the constant values of τ_n and τ_p obtained from the fit. It is possible to see that the VMCT model provides a sort of mean value of τ. The meaning of S_n is more subtle. In the VMCT model it is introduced as a boundary condition and its role is to account for the recombination at the p-i interface, in the p layer and at the p-TCO interface[5]. In reality, however, S_n behaves like a parameter which includes any effect that reduces the collection efficiency near the p-i interface. This can be clearly seen by comparing the total recombination in the intrinsic layer computed from the simulation ($\int_0^{x_c} R(x)dx$) with the approximated one obtained from the VMCT model ($\int_0^{x_c} \Delta n/\tau_n dx$ where τ_n is taken from the fit, and Δn from the simulation). The former term is larger than the latter, thus showing that the use of a constant lifetime underextimates the recombination. The missing recombination, essentially due to the decrease of τ_n close to the p layer, can be written as :

$$S_n \mu_n \Delta n(0) = \int_0^{x_c} R(x)dx - \int_0^{x_c} \frac{\Delta n(x)}{\tau_n} dx \qquad (1).$$

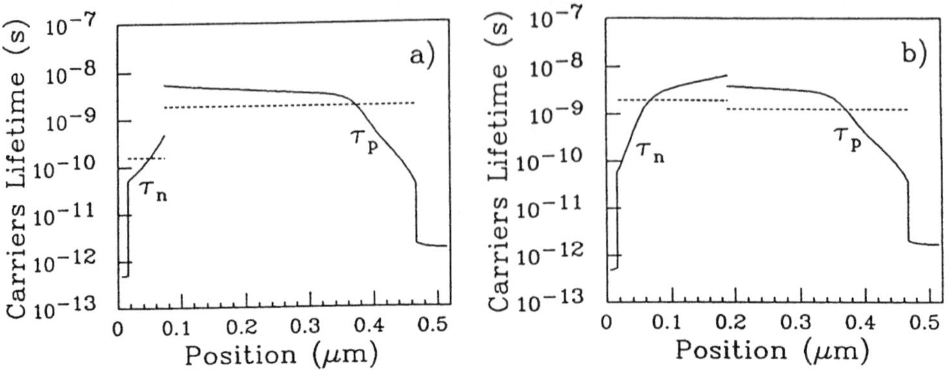

Fig.4 Lifetimes in the device. (—) numerical simulation. (· · ··) from the fit. The lifetime for each carrier is reported only in the region where it is a minority carrier. a) cell with boron contamination. b) cell without boron contamination

In the case of non contaminated cells, the above equation gives values of S_n in the range $10^3 - 10^4$ V/cm. Since $10^3 - 10^4$ is the range of values actually found in the cells grown with procedure b), we can conclude that the decrease of τ_n close to the p region gives to S_n a contribution comparable or greater than that of recombination at the p-i interface, in the p layer, and at the p-TCO interface. More complex is the analysis in the case of boron contamination. The simulation shows that, as expected, the total recombination and $\Delta n(x)$ close to the p-i interface increase (i.e. the lower electric field is less efficient in sweeping-out photogenerated carriers). Although the fit gives a lower value for τ_n, the excess recombination, given by the first term of Eq. (1), increases. This effect, essentially due to the decrease of the electric field, is taken into account in the VMCT model by increasing the value of S_n. In conclusion we believe that S_n is a phenomenological parameter which takes also into account extra effects non contained in the VMCT model and its real meaning must be established in each pratical case.

ACKNOWLEDGEMENTS

This work has been supported by PFE II (C.N.R. - ENEA) under contract N 207/86. J. Qiao acknowledges the financial support of the ICTP program for Training and Research in Italian Laboratories, Trieste, Italy. M. Falconieri and A. Mittiga acknowledge the grant from ENIRICERCHE.

REFERENCES

1. R. S. Crandall, *J. Appl. Phys.*, **54**, 7176 (1983)
2. H. Okamoto, H. Kida, S. Nonomura, and Y. Hamakawa, *Solar Cells*, **8**, 317 (1983)
3. C. R. Dickson, J. Pickens, and A. Wilczynski, *Solar Cells*, **19**, 179 (1986-87)
4. M. Hack and M. Shur, *J. Appl. Phys.*, **54**, 5858 (1983)
5. H. Okamoto, H. Kida, S. Nonomura, K. Fukumoto and Y. Hamakawa, *J. Appl. Phys.*, **54**, 3236 (1983)

INVESTIGATIONS ON THE SnO_2/a-Si:H INTERFACE

A.S. Verlinde (1), J. Smeets (2), P. Nagels (3), H.J. van Daal (1), H.H. Brongersma (1)

(1) Eindhoven University of Technology, P.O.Box 513, 5600 MB Eindhoven, The Netherlands, tel. (+)3140474281
(2) Physics Department, SCK/CEN, 2400 Mol, Belgium
(3) Rijksuniversitair Centrum, University of Antwerp, 2020 Antwerpen, Belgium

ABSTRACT. Glow-discharge deposition, a common technique for manufacturing amorphous silicon solar cells, induces degradation of the transparent conducting oxide on which the amorphous silicon is deposited. To study the extent of this degradation, amorphous silicon was glow-discharge deposited on doped tin oxide, and AES depth profiling analysis was performed on the resulting interface. The width of the interface appears to be dependent on the surface roughness of the SnO_2 layer. A least squares method was developed to determine the relative abundances of oxidized and reduced tin, and to determine the associated chemical shift. Reduction of the tin oxide substrate has been demonstrated.

1. INTRODUCTION

The most widely used method for depositing hydrogenated amorphous silicon (a-Si:H) layers for solar cell applications is the glow-discharge technique. The presence of the silane plasma, however, induces a degradation of the transparent conducting oxide (TCO) on which the a-Si:H is deposited [1]. This affects the conversion efficiency of the solar cell. The degradation of the TCO consists of two mechanisms, viz. reduction of the oxide, and diffusion of the boundary layer constituents. The present study attempts to obtain information on the distribution and chemical state of these constituents by means of Auger Electron Spectroscopy depth profiling, starting from doped tin oxide as a TCO material. The possible effect of the TCO surface roughness is hereby taken into account.

2. EXPERIMENTAL

Intrinsic layers of a-Si:H have been deposited on commercially available glass/TCO substrates (manufactured by Corning and Glaverbel) by means of the glow-discharge decomposition of silane (SiH_4). The Corning TCO is sputter deposited, the Glaverbel TCO is deposited by means of spray pyrolysis. Transparent conducting oxides have been shown to degrade under various kinds of plasma [2]. Therefore, pre-cleaning of the substrates with an argon plasma was omitted. Deposition parameters have been described in detail in previous publications [3]. As we are primarily interested in the a-Si:H/TCO interface, we did not apply any boron doping of the a-Si:H layer, nor did we incorporate carbon to increase the optical band gap. The layer thickness was kept small (approx. 50 nm) in order to avoid excessive sputter times in the depth profiling analyses.
Auger Electron Spectroscopy depth profiling is performed with a Physical Electronics Model 550 ESCA/Auger/SIMS spectrometer. A fully dedicated PDP

11/04 computer is used for data acquisition, data storage, and instrument control. The primary electron beam energy is 3 keV, the beam current 9 μA. The spot diameter of the beam is approximately 50 μm. The energy intervals that contain the Auger transitions of tin, silicon, and oxygen are scanned repetitively in differential (dN/dE) mode. The peak-to-peak value of the signals, expressed in number of counts, is a measure of the abundance of a certain element.
Simultaneous sputtering with an argon ion beam is used to investigate the depth distribution of an element. The ion beam has an energy of 3 keV and a current of 140 nA. The beam is scanned over an area of 2 x 2 mm^2. The pressure in the vacuum chamber during sputter operation is in the 10-7 mbar range. The a-Si:H layer thickness is measured before profiling. Together with the depth profile (atomic concentrations vs. sputter time), this yields a sputter rate of about 2.5 nm/min. The sputter rates of a-Si:H and tin oxide are assumed the same.
Roughness measurements are performed with a newly developed apparatus, equipped with a Philips compact disc sensor. With this experimental set-up, a lateral resolution of 1 μm, together with a depth resolution of 3 nm, can be achieved [4].

3. RESULTS

During the AES measurements we observe a change in the spectrum of the double tin MNN transition (figure 1). This is ascribed to a change in the relative abundances of reduced, i.e. metallic, and oxidized tin in the boundary layer. The two minima in the spectrum of pure metallic tin occur at energies of 429 eV and 437 eV. When replacing pure metallic by pure oxidized tin the spectrum undergoes a chemical shift of about -5 eV to yield minima at energies of 423 eV and 432 eV. Apart from this shift the spectra practically have the same form. In the presence of both oxidized and metallic tin the observed spectrum is a superposition of the spectra corresponding to these chemical states. As these spectra largely overlap, their superposition is subject to interference. This is taken to account

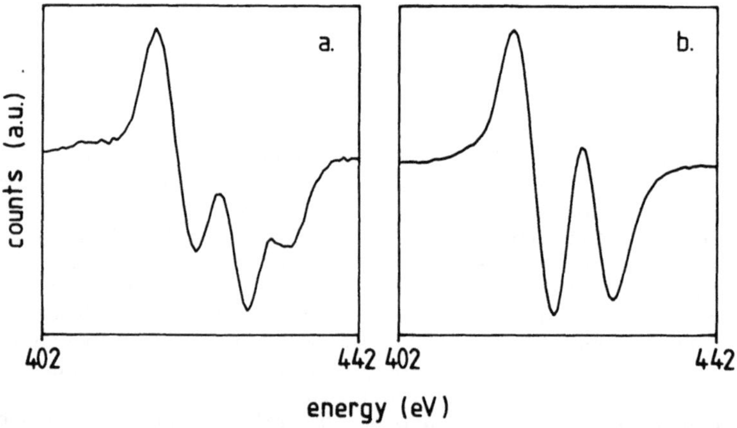

Fig.1: Change in the tin Auger spectrum; a. spectrum taken in the boundary layer; b. spectrum taken in the tin oxide bulk.

for the observed spectral change.

For each of the tin Auger spectra observed in the profile, we apply the following procedure to determine the relative abundances of oxidized and metallic tin. At the end of the profile, i.e. for large sputter times, the observed spectrum is that of pure oxidized tin. At any other position in the profile the measured spectrum is compared to calculated spectra, composed of reduced and oxidized tin spectra of chosen amplitudes and energy separation. The best fit between measured and calculated spectra, with the two amplitudes and the energy separation as parameters, is obtained by a least squares fit. The deviations between the observed spectra and the spectra calculated in this way roughly fall within the noise of the measured signal. At any position in the profile the energy separations obtained in this way lie satisfactorily in the range -4.5 ± 0.5 eV. The peak-to-peak amplitudes of the calculated spectra are a direct measure of the relative abundances of both chemical states.

In the case of Si the distinction between semiconducting, i.e. covalently bonded, and oxidized material is simple. The Auger LMM peaks occur at energies of 92 eV and 77 eV, respectively, and do not overlap. We only need to take into account that the background signal in this interval is not uniform, but increases linearly with energy (to a good approximation). Relative atomic concentrations are obtained by dividing the amplitudes of the spectra by empirically determined sensitivity factors.

An AES depth profile through the a-Si:H/SnO_2 interface, obtained by means of this procedure, is shown in figure 2. The profile given displays the atomic concentrations of metallic and oxidized tin, semiconducting and oxidized silicon, and oxygen. The maximum atomic concentration of reduced tin amounts to about 5 at%. Oxidized silicon is observed in relatively

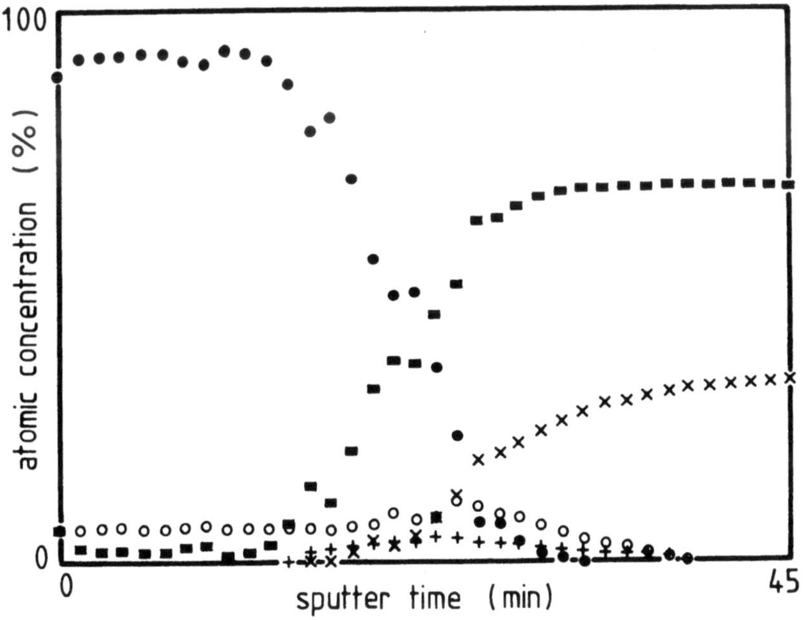

Fig.2: AES depth profile of a-Si:H/Corning TCO. Shown are the atomic concentrations of oxidized tin (×), reduced tin (+), semiconducting silicon (•), oxidized silicon (o), and oxygen (■) vs. sputter time.

large quantities, up to 10 at%. Obviously, the interface is not sharp. The interface width, defined as the depth over which the tin atomic concentration increases from 10% to 90% of its bulk value, appears to vary with the kind of substrate. For standard Glaverbel glass/TCO subtrates the interface width is about 45 nm (figure 3a). For sputter-deposited Corning glass/TCO substrates this value is 30 nm (figure 3b). Roughness measurements yield values for the roughness of 29 nm for Glaverbel, and 8 nm for Corning substrates. The SEM photographs of figure 4 confirm this difference qualitatively.

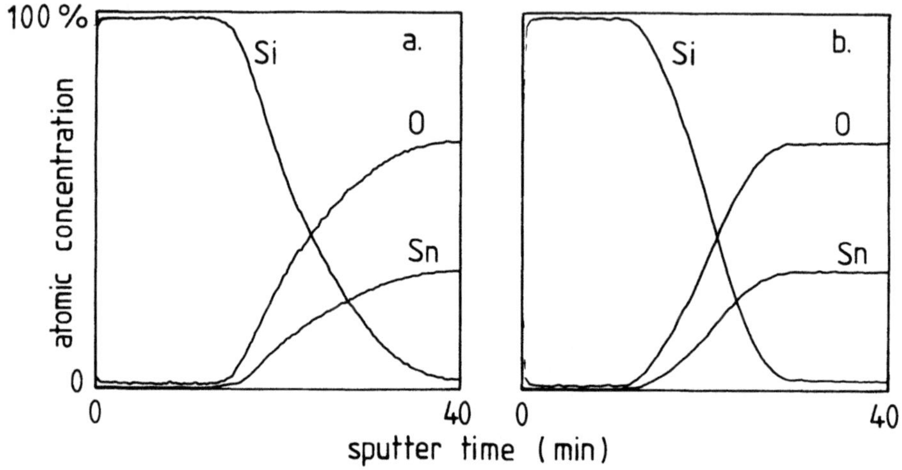

Fig.3: AES depth profiles showing the total amounts of silicon, tin, and oxygen vs. sputter time; a. Glaverbel glass/TCO substrate; b. Corning glass/TCO substrate.

Fig.4: SEM photographs of the surfaces of a. Glaverbel TCO, and b. Corning TCO.

4. CONCLUDING DISCUSSION

The width of the a-Si:H/TCO interface, as observed by AES depth profiling, depends on the TCO surface roughness. However, the interface width is 15 to 20 nm larger than can be accounted for by the surface roughness only. The difference is probably caused by diffusion and by sputtering. No distinction can as yet be made between the contributions of each of these effects. Nevertheless, we may conclude that surface roughness must be taken into account when interpreting AES depth profiles.

The relatively large, fairly constant amount of oxidized silicon observed throughout the a-Si:H layer (about 5 at%) may be ascribed to oxidation during the AES depth profiling or during the glow-discharge deposition, or both. At the interface the abundance of oxidized silicon increases to 10 at%. The extra amount of oxidized silicon might be explained by assuming a reaction between tin oxide and silicon in the first stages of the a-Si:H deposition. The observed maximum atomic concentration of 5 at% reduced tin seems realistic. It is within the range of values found in other experiments [5,6], where in general the deposition conditions are different. The chemical shift associated with the transition from metallic to oxidized tin is found to be -4.5 ± 0.5 eV, which is in agreement with literature. Therefore, we may conclude that the method used to decompose the tin AES profile spectra into reduced and oxidized contributions is reliable.

ACKNOWLEDGEMENTS

The authors wish to thank V. Van den Bergh (SCK Mol) for the deposition of the a-Si:H layers, L. Knaepen (SCK Mol) for the SEM photographs, and K. Struik (Eindhoven University of Technology, Dept. of Mechanical Engineering) for kindly putting the experimental roughness meter at our disposal. They are indebted to J.P.M. Schmitt (SOLEMS S.A.) for putting the roughness problem to our attention.

5. REFERENCES

[1] H.H. Brongersma, A.S. Verlinde, H.J. van Daal, *Proc. ISES Solar World Congress*, Hamburg (1987), contract number EN 3S/0052 (in press).
[2] O. Kuboi, *Jap. J. Appl. Phys.* **20**, (1981) L783.
[3] J. Smeets, P. Gunst, M. van Roy, M. Denayer, P. Nagels, *Proc. Conf. Poly- microcrystalline and amorphous semiconductors*, eds. P. Pinard and S. Kalbitzer (1984) 557.
[4] K.G. Struik, Fan Muchang, *Proc. 4th Int. Precision Engineering Seminar*, Cranfield (1987) 81.
[5] J.H. Thomas III, A. Catalano, *Appl. Phys. Lett.* **43**, (1983) 101.
[6] E. Eser, F. Ramos, *Proc. 7th EPSEC*, Sevilla (1986) 560.

GAS PHASE COMPOSITION RESULTING FROM THE SiH_4 PHOTODISSOCIATION WITH A PULSED ArF EXCIMER LASER

E. BOCH, C. FUCHS, E. FOGARASSY, P. SIFFERT
Nuclear Research Centre (IN2P3)
Physics and Applications of Semiconductors Laboratory (PHASE) (UA du CNRS n°292)
23, rue du LOESS
F-67037 STRASBOURG CEDEX (FRANCE)

ABSTRACT - The photodissociation of SiH_4, using the 193 nm resonance line of an ArF pulsed excimer laser has been recognized as an interesting process to deposit high quality hydrogenated silicon films at low temperature. Therefore, it is of primary importance to clarify the fundamental chemical reactions involved in the gas phase during this process. The gas composition, and specially the dissociation yield of SiH_4 has been determined as a function of the number of laser pulse and initial pressure of SiH_4, by measuring the partial pressure of different species created in the sealed chamber during the photodissociation of silane, using a condensation technique. Our results prove the existence of a stationary state and demonstrate the presence of reverse reactions in the chemical cinetic model discussed in this work.

1. INTRODUCTION

The photodissociation of SiH_4, using the 193 nm resonance line of an ArF pulsed excimer laser has been recognized as an interesting process to deposit high quality hydrogenated silicon films at low temperature. At this wavelength, silane molecules are dissociated by a two-photon excitation of molecular electronic states [1], which require high energy densities provided by the laser.

Therefore, it is of primary importance to clarify the fundamental chemical reactions involved in the gas phase during the photoassited deposition process. We propose here a simple model describing the 193 nm photochemical process. By analogy with the reactions involved in the photodissociation of SiH_4 with a Xe lamp at 147 nm [2] and the mercury sensitized dissociation of SiH_4 molecules.

The gas composition, and specially the dissociation yield of SiH_4 has also been determined as a function of the number of laser pulse and initial pressure of SiH_4, by measuring the partial pressure of different species created in the gas cell during the photodissociation of silane, using a condensation technique. These results prove the existence of a stationary state and demonstrate the presence of reverse reactions in the chemical cinetic model.

2. EXPERIMENTAL

Pure silane, sealed in a suprasil quartz cylindrical reaction chamber ($ø \approx 3$ cm, $L = 10$ cm) at pressure ranging between 1 and 100 torr, was irradiated under normal incidence with a Lambda Physik (EMG 201 MSC) ArF excimer laser providing, at 193 nm, pulses of 20 ns duration with energy densities ranging between 10 and 200 mJ/cm^2. The irradiated area (≈ 0.25 cm^2) is continuously moved along the reactor in order to prevent the deposition of silicon on the walls. We neglect in this work the diffusion of the species in the reactor during the 20 ns irradiation pulse since the SiH_4 diffusion coefficient (20 cm^2s^{-1}) is not far from those of the other species created in the reaction chamber.

The photochemical decomposition of silane leads to the formation of both stable species, H_2, Si_2H_6, SiH_4 and a-Si, and intermediate radicals like SiH_2, SiH_3 and H. These radicals do not interfere with the experimental results since a calculation modeling has shown that only a few milliseconds are necessary for them to react and form stable species. Thus, the gas composition has reached a stable state (no more radicals) in a time much shorter that ω^{-1} (pulse frequency $\omega = 10$ s^{-1}). The final SiH_4 partial pressure has been determined after different irradiation times (i.e. number of laser pulse) or different initial pressures of SiH_4, in the following manner. In the first step, SiH_4 and its higher derivates were condensed at liquid nitrogen temperature. In this case, the residual pressure may be related to the presence of hydrogen formed during the decomposition. In the second step, after pumping off the H_2 molecules, the partial pressures of SiH_4 and Si_2H_6 were determined by heating the reactor above their respective boiling temperature (T_{SiH4} = - 112°C, T_{Si2H6} = - 14°C). The partial pressures of SiH_4 and Si_2H_6 at 20°C have been deduced from the experimental measurements by using calibration curves (P = f(t)) previously established. The precision of these values has been estimated to be in order of 5 %. In this work, the small contribution of Si_3H_8 (< 1 %) formed during the photochemical reactions is neglected. The results have shown that Si_2H_6 pressure do not exceed 5% of the total pressure. Thus, we will not take it into account in the following discussion.

3. RESULTS AND DISCUSSION

The results, obtained at 254 nm (Hg lamp) and at 147 nm, show that UV excitation of SiH_4 molecules can be achieved through different ways which depend on the wavelength and intensity of the light source. With a low pressure mercury lamp, Hg atoms mixed with SiH_4 are excited by the 254 nm resonance line and transfer their energy to SiH_4 molecules by collision. With the Xe-lamp, 8.4 eV photons (147 nm) directly dissociate SiH_4 by a single photon excitation of molecular electronic states.

In the case of ArF excimer laser, at 193 nm, the photon energy (6.4 eV) is too low to decompose SiH_4 by a single photon process because the significant absorption starts only above 7.8 eV (160 nm) [4]. However, as deduced from Fig. 1, which shows the non-linear behaviour of the process as a function of energy density of the ArF laser, the decomposition of SiH_4 by a two photon excitation of molecular electronic states might occur above a laser threshold energy of about 30 mJ/cm^2.

$$SiH_4 + 2\,h\nu \rightarrow SiH_4^*$$

This behaviour confirms the experimental results of a-Si deposition rate vs laser energy density [1].

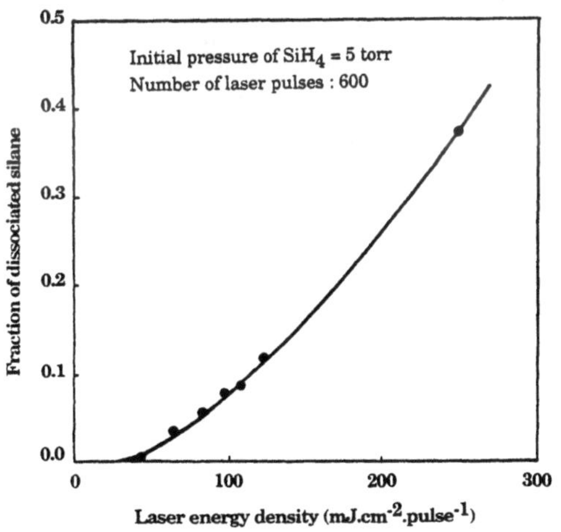

Figure 1 : Dissociation yield of silane as a function of the laser energy density.

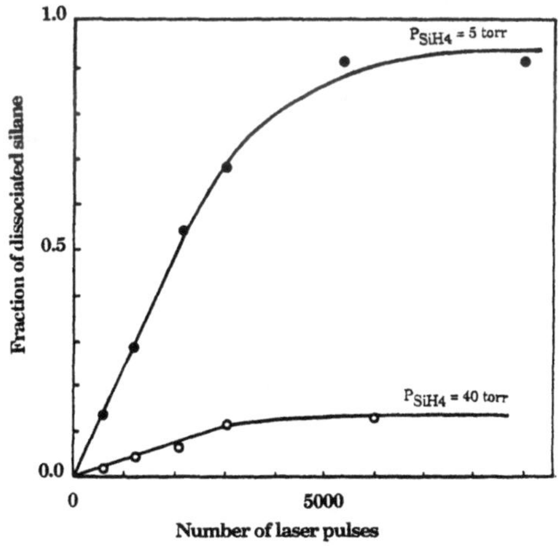

Figure 2 : Silane dissociation as a function of the number of laser pulses.

At the same time, the nature of the decomposition of photoexcited SiH_4^* molecules into reactive species and intermediate radicals depend strongly on the way of excitation, and specially on the wavelength of the light source. Two main reactions have been proposed [2, 5] :

$SiH_4{}^* \rightarrow SiH_3 + H \quad \Delta H = -91.71$ Kcal/mol (3.98 eV/mol) (1)

$SiH_4{}^* \rightarrow SiH_2 + 2H \quad \Delta H = -164.16$ Kcal/mol (7.14 eV/mol) (2)

With the 254 nm (4.88 eV) resonance line, the decomposition of SiH_4 by Hg sensitization is only possible through reaction (1) which needs 3.98 eV. Thus, only SiH_3 radicals are created. By contrast, the 147 nm light (8.4 eV) induce both reactions (1) and (2) leading to the formation in the gas of SiH_2 and SiH_3 radicals. Austin and Lamp [2] propose the quantum yield $\phi_1 = 0.17$ for reaction (1) and $\phi_2 = 0.83$ for reaction (2).

The two-photon absorption of coherent excitation light from ArF laser also permit the formation of SiH_2 and SiH_3 from both reations (1) and (2). However, the molecular electronic states of SiH_4 involved in the excitation with a 193 nm ArF laser are not those excited by the 8.4 eV photons from the 147 nm light. Therefore, the quantum yields ϕ_1 and ϕ_2 are unlikely to be exactly those calculated by Austin and Lamp.

We suppose, nevertheless, in a first step of the cinetic modeling, and before being able to determine them experimentally, that $\phi_1 = 0.17$ and $\phi_2 = 0.83$.

The radical species responsible for film deposition and growth strongly depend on secondary reactions which take place in the gas phase. The secondary reactions more generally considered are the same that in 254 nm Hg CVD and 147 nm photolysis.

$H + SiH_4 \rightarrow H_2 + SiH_3$ (3)

$SiH_3 + SiH_3 \rightarrow SiH_4 + SiH_2$ (4)

$SiH_2 + SiH_4 \rightarrow Si_2H_6$ (5)

We show in Fig. 2 the results of the determination of the dissociation yield ($1 - P_{SiH4}/P^o_{SiH4}$) as a function of the number laser pulses for two different values of the initial SiH_4 pressure.

In both cases, the yield of dissociation initially increases with the number of laser pulses and then reaches a constant value following approximately 5000 pulses of irradiation. This constant value strongly increases when the initial pressure of SiH_4 in the reaction chamber decreases.

This result may be related to the existence of a state of chemical equilibrium. The phenomenon has already been observed in Hg photo CVD. Therefore, in order to take into account the quasi equilibrium state, we included in the cinetic modeling reverse reactions leading to the regeneration of SiH_4 molecules. We only consider the reactions having the highest rate constants [6] :

$H + Si_2H_6 \rightarrow SiH_4 + SiH_3$ (6)

$H_2 + SiH_2 \rightarrow SiH_4$ (7)

This scheme must also be completed by the reactions describing the deposition on the walls

$SiH_3 + \text{wall} \rightarrow Si + 1.5\, H_2$ (8)

$SiH_2 + \text{wall} \rightarrow Si + H_2$ (9)

We have summarized in Table I the different chemical constants k corresponding the the reactions 1 to 9 found in the literature. The rate constant of the excitation reaction of SiH_4 (0) has been calculated using the value of the two-photon absorption cross section of SiH_4 ($\sigma = 6.10^{-44}$ cm^4s) measured in a previous work [3]. The relation between k_o and σ is :

$k_o = 0.5\, \sigma\, I^2$ for a biphotonic process

I in photon.$cm^{-2}\, s^{-1}$

k_o in s^{-1}

For numerical example, when I = 100 mJ/cm^2, $k_o = 7.5.10^5\ s^{-1}$.

	Rate constants at 300 K ($cm^3.molecule^{-1}.s^{-1}$)
k_0	$7.5 . 10^5$ s^{-1}
ϕ_1	0.17
ϕ_2	0.83
k_3	$(4.3 \pm 0.9) . 10^{-13}$ [7], $8.5 . 10^{-12}$ [8]
k_4	$1.5 . 10^{-10}$ [9] ; $2 . 10^{-10}$ [10, 11]
k_5	$6 . 10^{-14} < k_5 < 1.5 . 10^{-11}$ [12] ; $1.1 . 10^{-10}$ [13]
k_6	$1.1 . 10^{-12}$ [14] ; $3.7 . 10^{-12}$ [7]
k_7	$1.6 . 10^{-17} < k_6 < 2.5 . 10^{-15}$ [12]
	$1 . 10^{-13}$ [13] ; $\approx 10^{-9}$ [6]
k_8	$1.1 . 10^2$ s^{-1} [15]
k_9	$k_9 \approx 5.k_8$ [9]

TABLE I [6]

We have also determined the yield of dissociation of SiH_4 as a function of P^oSiH_4, in order to explain why all the experiments of a-Si:H deposition by ArF laser CVD of SiH_4 are performed at low initial pressure of silane (a few torr).

The results of Fig. 3 show that the yield of dissociation strongly increases when the initial pressure of SiH_4 decreases, and reaches a constant value depending on the duration of irradiation and laser energy density.

At 1 torr of initial SiH_4 pressure, it is possible to dissociate more than 60 percent of initial SiH_4 and it would certainly be better with at least 5000 pulses. These results are in total agreement with those of the Fig. 2. The dissociation reaction of SiH_4 is the more efficient as the pressure is low. This result should give interesting applications on optical and electrical properties of the deposit.

The results presented here on the decomposition of SiH_4 by ArF laser photo CVD should be confirmed soon by a mathematical simulation. Finally, it already appears that we should also consider the single-photon absorption at 193 nm of disilane which is created in the gas phase during the two-photon excitation of SiH_4.

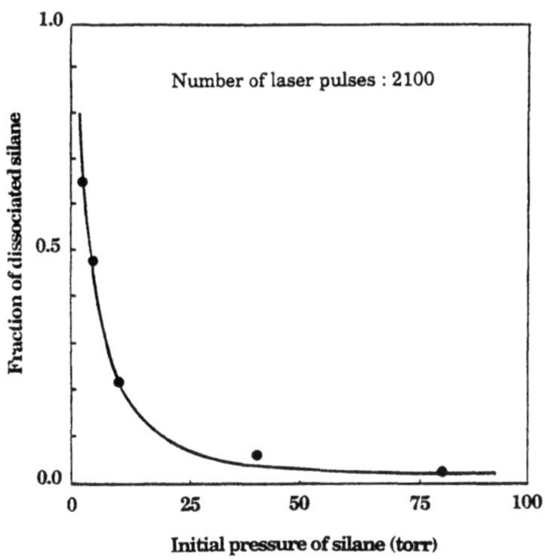

Figure 3 :
Dissociation yield of silane as a function of the initial pressure.

4. CONCLUSION

We have proposed in this work, a simple mechanism for photodissociation at 193 nm of silane with an ArF excimer laser. The rate constant of the two photon excitation reaction of the SiH_4 molecules has been calculated using the two-photon cross section of silane.

The analysis of the gas composition has also shown the existence of an equilibrium state in the reaction chamber which is explained by considering reverse reactions leading to the regeneration in the gas phase of SiH_4 molecules. Finally, we have shown that the yield of decomposition of SiH_4 increases rapidly when initial SiH_4 pressure decreases.

REFERENCES

[1] C. FUCHS, E. FOGARASSY
in *Photon, Beam and Plasma Stimulated Chemical Processes at Surfaces* edited by Donelly, Herman and Hirose, (MRS Proceedings) **75**, 195 (1986) (North Holland, New York, Amsterdam, Oxford).

[2] G.C.A. PERKINS, E.R. AUSTIN, F.W. LAMPE
Am. Chem. Soc. **28**, 1109 (1979).

[3] C. FUCHS, E. BOCH, E. FOGARASSY, B. AKA, P. SIFFERT
MRS Boston, 1987.

[4] Y. HARADA, J.N. MURRELL, H.H. SHEENA
Chem. Phys. Lett. **1**, 595 (1968).

[5] H. NIKI, G.J. MAINS
J. Phys. Chem. **68**, 304 (1964).

[6] B. AKA, C. FUCHS, E. FOGARASSY, P. SIFFERT
in *Photon beam and plasma enhenced processing* edited by Golanski, Krimmel and Nguyen, (EMRS Proceedings). p. 147 (1987) (Les Editions de Physiques, Les Ulis, France).

[7] E.R. AUSTIN and F.W. LAMPE
J. Phys. Chem. **81**, 12 (1975) 1139.

[8] J.A. COWFER, K. P. LYNCH and J.V. MICHAEL
J. Phys. Chem. **79**, 12 (1975) 1139.

[9] J. PERRIN and T. BROEKHUISEN
Appl. Phys. Lett. **50**, 8 (1987) 433.

[10] D. MIHELCIC, V. SCHUBERT, R.N. SCHINDLER and P. POTZINGER
J. Chem. Phys. **81**, 16 (1977) 1543.

[11] B. REIMANN, A. MATTERN, R. LAUPERT and P. POTZINGER
Berichte der Bunsen-Gesellschaft **81**, 5 (1977) 500.

[12] P. JOHN and J.H. PURNELL
J. Chem. Soc. Trans. Faraday I, **69** (1973) 1455.

[13] G. INOUE and M. SUZUKI
Chem. Phys. Lett. **122**, 4 (1985) 361.

[14] T.L. POLLOCK, H.S. SANDHU, A. JODHAN and O.P. STRAUSZ
J. Am. Chem. Soc. **95**, 4 (1973) 1017.

[15] J. BLAZEJOWSKI and F.W. LAMPE
J. Photochemistry **16** (1981) 105.

INFLUENCE OF THIN FILM STRUCTURE ON THE OPTOELECTRONIC PROPERTIES OF ITO LAYERS PRODUCED BY REACTIVE THERMAL EVAPORATION

C.Carvalho, *O.Figueiredo, E.Fortunato, R. Martins, A.Maçarico and L.Guimarães
Faculty of Science and Technology of New University of Lisbon, Portugal
*Crystallography Center, IICT, Portugal

SUMMARY

High quality transparent and conductive indium tin oxide (ITO) thin films are produced by reactive thermal evaporation (RTE), using either variable alloy compositions based on Sn_x/In_{1-x} or a fixed powder mixture $(SnO_2)_{0.1}/(In_2O_3)_{0.9}$ Films produced using Sn_x/In_{1-x} alloys ($0.05 < x < 0.20$) show a much higher degree of crystallinity and are clearly textured. Concerning the fixed powder mixture, the degree of crystallinity is low and increases with partial pressure of oxygen $P(O_2)$ and thickness. The properties are dependent on film microstructure, with minimum sheet resistivity (ρ_s) and optimum optical transmitance (Tg) for the maximum texture degree. The best performances are obtained for films based on $Sn_{0.1}/In_{0.9}$ at $P(O_2)=6\times10^{-4}$ mbar showing $\rho_s \leq 8\ \Omega/\square$ and Tg ≥ 85 % .

1 . INTRODUCTION

Thin films based on Indium Tin Oxide (ITO) layers consist of an In_2O_3 host lattice with a b.c.c. structure [1], presenting a n–type behaviour resulting from oxygen vacancies [2]. This behaviour is enhanced by doping it with tin, since the substitution of an Sn^{4+} for an In^{3+} cation leads to a donor level in the energy band gap [3]. Improvement on films conductivity (without loosing transmitance) can be achieved by a proper control on oxygen and tin incorporated in the matrix, depending on the deposition method used as well on the control of parameters. Under these conditions, ITO layers behave as wide band gap degenerated semiconductor with a high transmitance in the visible region [4]. These properties make it quite suitable to be used as anti-reflecting coating (ARC) in devices such as solar cells and other electronic devices where a ARC is required.

Several deposition techniques such as those ones based on sputtering, chemical vapour deposition (CVD), spraying, electron beam deposition or pyrolysis have been used [2]. However the reported results cover a wide range of particular values, depending on the specific application for which such films are produced.

As far as photovoltaic (PV) applications are concerned, the experience shows that not only low ρ_s associated to high T_g are needed but also textured surfaces to enhance carriers collection and so, the short circuit current of such devices.

In this work we refer the electro–optical properties of ITO layers produced by RTE method.

The obtained films present a high degree of crystallinity, low ρs and high T_g, as well as texture surfaces depending on deposition conditions used.

2 . EXPERIMENTAL DETAILS

ITO films are deposited by reactive thermal evaporation of Sn_x/In_{1-x} alloys and $(SnO_2)_{0.1}/(In_2O_3)_{0.9}$ powder mixture on soda-lime glass substrate. Pure oxygen is introduced in the system through a needle valve in order to settle the partial pressure used during the deposition process. Concerning the crucibles, a tungsten or tantalum one is used for evaporation of alloys and powder mixtures, respectively.

Prior to the evaporation process, the system is evacuated to pressures below 6×10^{-6} mbar and substrates heated to 200 °C (termocouple on substrate). The deposition rate (0.5–1.2 Å/sec) as well as film thickness is monitored through a Edwards – FTM4 with the quartz crystal positioned beside the substrate.

The various films properties are measured as follows. X–ray diffraction spectra are recorded through a Philips powder X–ray diffractometer with a Cu–K_α radiation monochromatized with a curved graphite crystal. The sheet resistivity (ρ_s) is determined by a Veeco four point probe FPP 5000. The Hall effect measurements are carried out by the Van der Pauw method where the maximum strength of the applied magnetic field is 10 KG. The optical transmission of substrate plus film in the visible region is determined by a Shimadzu 260 UV double beam spectrophotometer. Finally, the surface morphology is investigated using a Philips 505 electron scanning microscope (SEM).

3. RESULTS AND DISCUSSION

3.1 — Optical and electrical properties

Figure 1, shows the dependence of ρ_s and Tg on $P(O_2)$ for Sn_x / In_{1-x} alloys ($0.05 < x < 0.20$). The data show that ρ_s increases either for low or high $P(O_2)$ having its minimum located at $P(O_2) = 6\times10^{-4}$ mbar. The increase in ρ_s at low $P(O_2)$ is attributed to the saturation of oxygen vacancies existent on the structure, while at high $P(O_2)$ this is due to excess oxygen which causes a dissipation of vacancies [7]. Concerning the optical transmitance the recorded values are nearly the same (85%) for all alloy compositions and $P(O_2)$ used, indicating that a similar degree of oxidation is achieved.

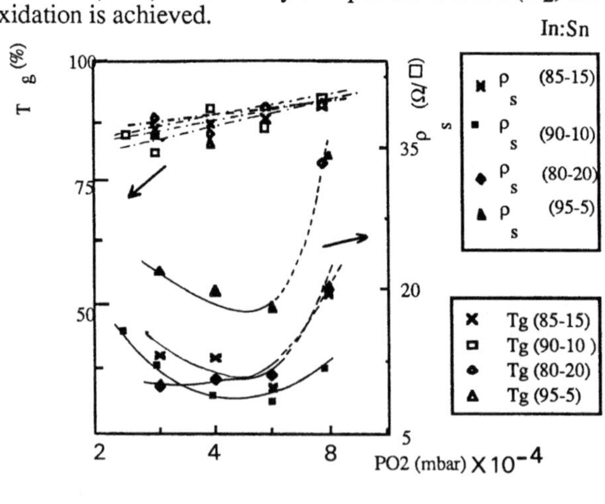

Figure 1 – Dependence of ρ_s and Tg on $P(O_2)$ for different alloy composition.

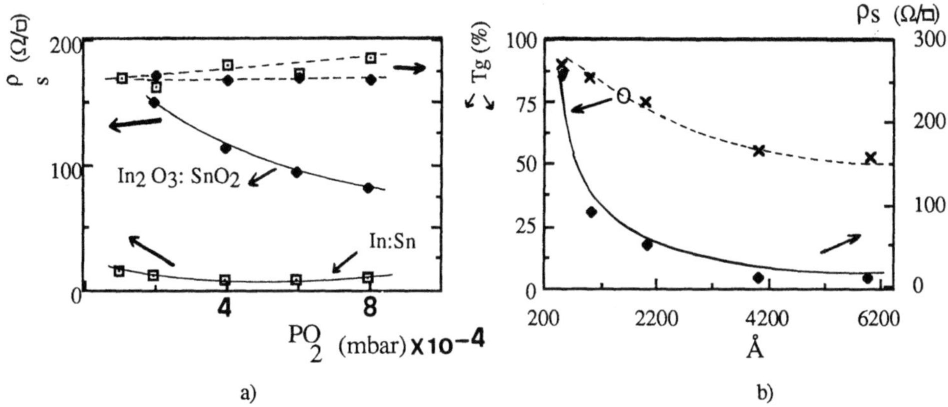

Figure 2 – Represents ρ_s and Tg as a function of $P(O_2)$ for films deposited using an alloy based on In:Sn and In_2O_3: SnO_2 produced at 90:10 a) and as a function of film thickness, for films based on powder mixtures b).

Through Hall effect measurements, carrier concentration, N_e, Hall mobility, μ_H, and carrier's sign are determined. Films with the lowest ρ_s have $Ne=4\times10^{20}$ cm^{-3} and $\mu_H=45$ cm^2V^{-1} s^{-1} and the transport mechanism is governed by electrons.

Figure 2 presents the dependence of ρ_s and T_g on P_{O2} for two kind of ITO films produced (from alloys and powder mixtures) for comparative purposes. No significant differences are detected on T_g whatever the $P(O_2)$ used. A different behaviour is observed for ρ_s which can be ascribed to the different structural composition of the obtained films. The differences observed in ρ_s are related with the degree of crystallinity induced by each one of the film sources used. The role of thickness on optoelectronic properties of ITO films based on powder mixtures is also investigated by varying the thickness between 600 to 6500 Å (see fig.3). The results show that T_g and ρ_s decrease significatively by increasing the thickness behind 1500 Å which can be ascribed to the nucleation process, as we shall discuss below.

As far as refractive index n and optical gap are concerned absorbance results show that they are not affected either by changing the evaporation source or p_{O2}.This is an indication that optical transitions on such structures are mainly governed by the amorphous tissue imbibing the crystallites,i.e., the degree of oxidation is similar in both kind of ITO thin films analysed.The obtained results reveal n values of the order of 2.1 with Eop of the order of 3.1eV, which are quite suitable for PV applications.

3.2 — Structural characterization by X-ray diffraction studies

X–ray diffraction patterns recorded under normal θ/2θ scanning mode were entirely assigned to the cubic bixbyite-type structure of In_2O_3 irrespective the starting mixture, alloy or oxide. No evidence could be formed for the presence of any extra crystalline phase, namely SnO_2. Tin is therefore incorporate in the cubic lattice of indium oxide, probably promoting a certain degree of stuffing the vacant sites in the anionic sublattice. The hypothesis of Sn acting as a substitutional impurity is supported by the decreasing of the unit cell parameter with increasing Sn content (1.0135 nm for Sn0.05/In0.95 alloy, compared to 1.0119 nm for Sn0.2/In0.8), a fact explained by the ionic crystal radii of Sn^{4+} and In^{3+} in octahedral coordination — 0.069 nm and 0.080 nm, respectively. Films obtained from oxide $(SnO_2)_x/(In_2O_3)_{1-x}$ mixtures showed a lower degree of crystallinity than those produced from Sn_x/In_{1-x} alloys.Film thickness slightly increased the crystallinity (figure 4 a)), the same holding for the oxygen partial pressure under constant thickness (figure 4b)).

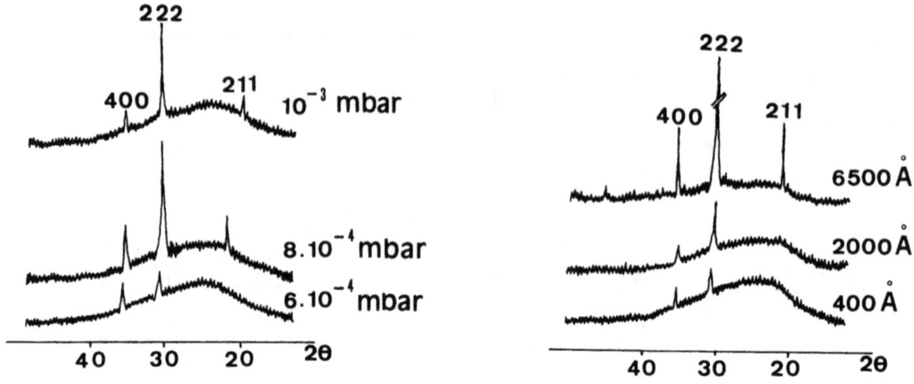

a) b)

Fig.4— X-ray diffraction patterns for ITO films obtained from: $(SnO_2)_{0.1}/(In_2O_3)_{0.9}$ powder mixture as a function of $P(O_2)$ for a constant thickness(a) and constant $P(O_2)=8\times10^{-4}$ mbar (b).

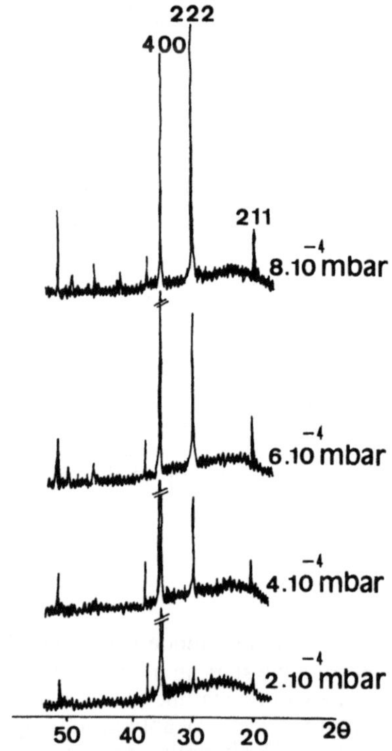

Fig.5— X-ray diffraction patterns for ITO film obtained from: a)Sn $_{0.1}$/ In $_{0.9}$ alloy as a function of $P(O_2)$

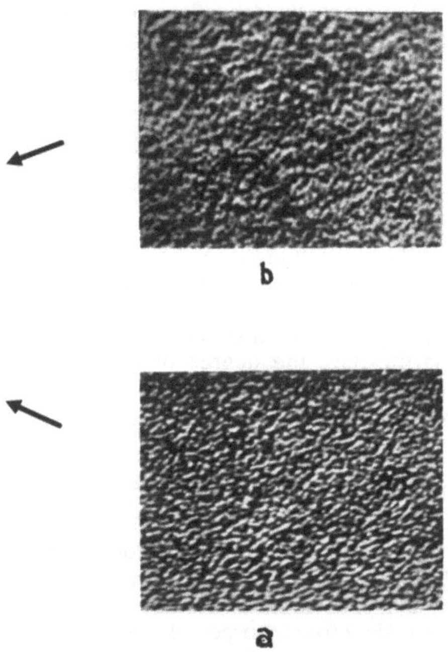

Fig.6— Scanning electron micrographs (magnification 10^4) of ITO films obtained from $Sn_{0.1}/In_{0.9}$ alloy at $P_{O_2}=6\times10^{-4}$ mbar (a) and $P(O_2)=8\times10^{-4}$ mbar (b).

Beyond being crystalline, the films obtained from Sn_x/In_{1-x} alloys also exhibited different textures (figure 4 c)) : for oxygen partial pressure up to 6×10^{-4} mbar, a <100> texture is evident, with <400> reflection more than twice the intensity of the usually more intense <222> reflections; for higher $P(O_2)$, the relative intensities of the diffraction pattern approach those of a randomly oriented powder or denote a slight <111> preferential orientation of the crystallites. Both textures were already noticed in ITO films produced by a spputering technique [1]. As a supplement to the X-ray study scanning electron microscope measurements are performed in order to determine surface morphology on ITO films. Results show a perceptible texture effect dependent on deposition conditions as can be seen in figure 5.On the other hand, no influence of the texture on the lattice parameter could be detected.

CONCLUSIONS

The studied performed on ITO thin films produced by RTE technique reveals that:

1— Films produced based on Sn_x/In_{1-x} alloy composition are superior to those ones based on $(SnO_2)_x/(In_2O_3)_{1-x}$ powder mixtures, whatever structure and electro-optical properties concerned.

2— Both kind of films have similar degree of oxidation as detected by x-ray studies and confirmed by their optical properties.

3— Electrical properties presented by these films are directly dependent on the degree of crystallinity and surface texture.

4— A preferential <100> texture is detected on films based on alloy composition,which in turn, is dependent on p_{O_2} and alloy composition used.

5— Overall, we conclude that the best films performances are obtained using $Sn_{0.1}/In0.9$ which leads to films having $\rho_s < 8\ \Omega/\square$, $Tg > 85\%$ $\mu_e=45\ cm^2V^{-1}s^{-1}$, with the higher degree of crystallization together with surface texturization. This property make them quite suitable to be used in PV applications.

ACKNOWLEDGEMENTS

The authors are grateful to J.Laheria (University of Tel-Aviv) for performing SEM measurements as well as to I.Ferreira and M. Santos for the help given during the realization of this work .This work was supported by EEC contract Nº EN 3S–C1–602–P, Junta Nacional de Investigação Cientifíca e Tecnológica (JNICT) – Project Nº 87.375 and also by Instituto de Investigação Científica (INIC) – Project B.

REFERENCES

[1]— J.C.C. Fan et al., *J.of Appl.Phys.* **48** (1977) 3524
[2]— J.C.Manifacier et al., *Thin Solid Films,* **41** (1977) 127
[3]— C.A.Vincent,*J.Electrochem. Soc.*, **119** (1972) 515
[4]— R.Pommier et al.,*Thin Solid Films* , **77** (1981) 91

A NOVEL P-I-N a-Si SOLAR CELL STRUCTURE WITH VERY LOW ACTIVE-AREA LOSS AND INTERCONNECTION POWER LOSS

J. Xi, S. Muhl, B. von Roedern and A. Madan

Glasstech Solar, Inc.
12441 W. 49th Avenue
Wheat Ridge, CO, 80033 USA (303) 425-6600

We present a radically new concept in the fabrication of complete integrated amorphous silicon solar cell panels with an active area loss of less than 4%. This is to be contrasted with 10-15% area loss which is generally exhibited by conventional methods used with fabrication of amorphous silicon solar panels. To date, unoptimized 100 cm^2 a-Si p-i-n cells have produced an efficiency of 8.5% with an active area loss of 3.7%. Fill factor achieved for a 100 cm^2 area device was 0.67 indicating that the loss associated with interconnections is also extremely low even when the short circuit current was as high as 2 amperes. We have begun to make 1 ft^2 prototype panels and the initial results are very encouraging.

INTRODUCTION

For large area a-Si solar cell, it is important to reduce the active area loss as well as to further improve active area efficiency. To date, the most popular structure of large area amorphous silicon (a-Si) solar cells is based on laser-scribed series-interconnected design [1]. Although the scribed line can be thin, each subunit requires three scribed lines and a sufficient space between them to avoid electrical shorts. This results in a 9-31% active area loss of the solar cell panel [2,3].

In this paper we report a radically new concept in the design and manufacture of monolithic series-interconnected amorphous silicon solar cell panels. It involves the utilization of point-like grid design which dramatically reduces active area loss while keeping excellent current collection with small power loss. The current collection is achieved via many small metallic points contacted to the trasnparent conducting oxide (TCO); the interconnection of these points is performed at the backside of the cell. Therefore, the area of shading and spacing due to these metallic points is greatly reduced. Below we describe the details of the device structure, and the results of calculation and experiment followed by discussion.

Device Structure and Grid-Related Performance Analysis

Figure 1 shows the top view of a typical grid pattern of a cell with a shading area loss which can be minimized to about 1% as we discuss below. Figure 2 shows a schematic of the basic structure of the large area a-Si p-i-n cell. A metallic grid consisting of many tiny metal points is deposited onto the TCO coated glass substrate. The diameter of each metal point is about 1.4 mm and the space between these points is about 17 mm. The active layers (amorphous silicon p-i-n) are deposited next. Since the metal points are much thicker than the thickness of the a-Si film, they necessarily penetrate through the a-Si films. A metallic back contact layer is deposited next onto the a-Si film. With appropriate masking, the metal points are separated from the metallic back contact. Then an insulating film is formed on top of the metal layer, leaving an opening for each metallic point. Finally, a metallic network is deposited onto the insulating film to interconnect the metal points.

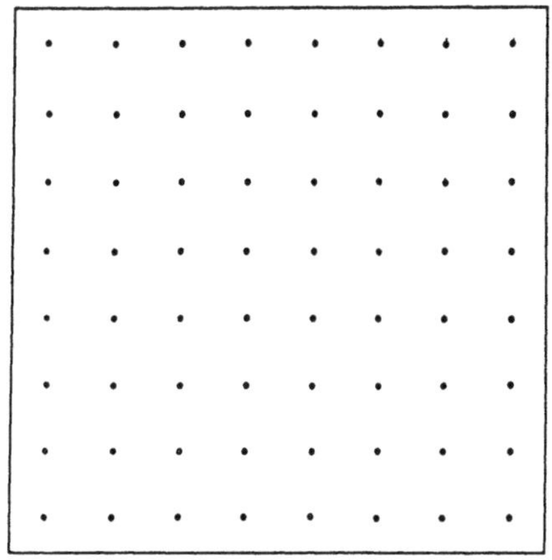

Fig. 1 A typical pattern of grid point-like structure

From Figure 2 it can be seen that the main active area loss is the shading presented by the metallic point contacts. In our prototype device, by using relative simple masking techniques, a 3.7% active area loss has been achieved, of which the shading loss of the metallic points is about 0.5%. Therefore, by applying more sophisticated masking technology to reduce the spacing loss, the total active area loss can be further reduced.

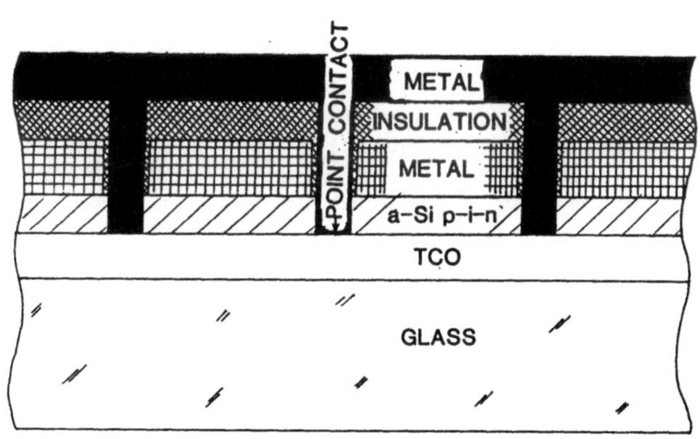

Fig. 2 A schematic of the device structure using the point-like grid concept.

Loss Considerations

We now consider the electrical losses associated with this stucture. For simplicity, we use a circular area representing active area from which the current will be collected by the center metallic point through the TCO layer, as shown in Figure 3.

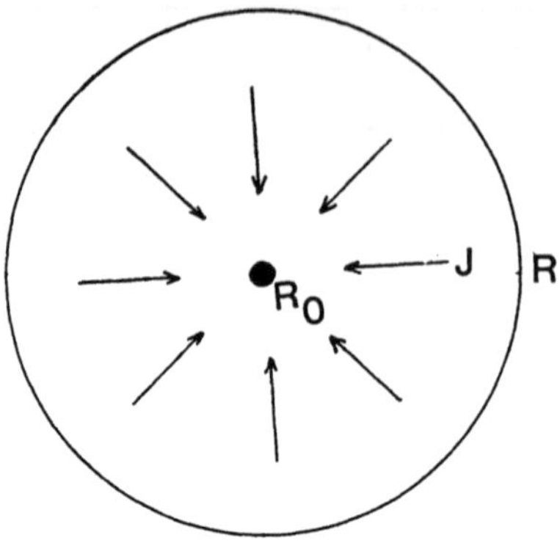

Fig. 3 Current collection pattern through TCO in point-like grid design

The power loss (ΔP) due to TCO for this configuration is given by,

$$\frac{\Delta P}{P} = \frac{\int_{R_o}^{R} J_p^2 \pi^2 (R^2 - r^2)^2 \frac{\rho_{sh} dr}{2\pi r}}{J_p \pi (R^2 - R_o^2) \quad V_p}$$

Here, J_p and V_p are current density, and voltage at the power point respectively, ρ_{sh} is sheet resistance of TCO film, R is radius of the circle area and R_o is radius of the metallic point. Table 1 lists the calculated results of the electrical power loss which can be as low as 3.3%.

Table 1 - Power loss in TCO for optimized design of point-like and line grid design a move. The calculation is based on the following assumptions: sheet resistance $\rho sh = 5\Omega$/square, current density at maximum power point $J_p = 15mA/cm^2$, voltage ouput at maximum power point $V_p = 0.63$ V.

grid geometric parameters	$R = 0.7cm$ $R_o = 0.07cm$	$R = 0.7cm$ $R_o = 0.1cm$
relative power loss $\Delta P/p$ (%)	4.5%	3.3%

RESULTS AND DISCUSSION

By using a GSI-commercially available multichamber R & D glow discharge depositon system, it is a common practice to fabricate single junction a-Si p-i-n cell with active area efficiency approaching 11% [4]. In our R & D system, we have fabricated a complete integrated panel using the point-grid structure with an area loss of only 3.7%. Table 2 lists the experimental results of thin prototype (unoptimized) panels and Figure 4 shows the I-V data under global AM1.5 spectra (It should be stressed that the intrinsic layer of the cell was made deliberately thin (5000 Å) for stability reasons).

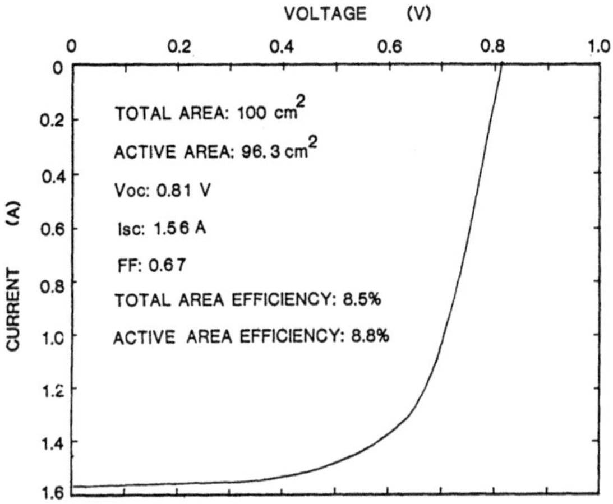

Fig. 4 I-V characteristic of GSI's 100cm^2 prototype a-Si p-i-n module using a point-like grid structure.

Recently, we began to make 1 ft^2 prototype panel using the point-like contact design, in GSI's in-house production line. We have successfully fabricated the first complete panel with total area efficiency of 5.7% with a thin i layer (~ 3500 Å) (Fig. 5.)

Fig. 5 GSI's 1 ft^2 prototype point-like TCO contact a-Si solar panel

With further optimization of processing, we anticipate a 7% total area efficiency can be achieved on the large scale production level in the near future.

From Table 2 it is readly seen that the difference in the performance between small area unit cell (2 cm^2) and the complete large area panel is indeed small. Also, the large fill factor obtained for 100 cm^2 cell indicates that the total electrical losses due to the current paths in the TCO, grid, interconnection, and interfacing contact, are extremely low even when the short circuit current is close to 2 amperes.

Table 2 - Performance of GSI's prototype panel using point-grid structure under global AM1.5 spectra.

Panel Geometry			Small Area ($2cm^2$) Cell with point-like grid TCO contact on the sample panel	Complete Panel ($100\ cm^2$)
Illuminated Area (cm^2)	100	V_{oc} (Volt)	0.81	0.81
Active Area (cm^2)	96.3	J_{sc} ($mA\ cm^{-2}$)	16.2	15.6 (total area)
Relative Active Area Loss (%)	3.7	I_{sc} ($mA\ cm^{-2}$)	32.4	1560
Relative Active Area Loss Due to Metallic Point Contact (%)	0.51	FF	0.67	0.67
		η(%)	8.8	8.5 (total area) 8.8 (active area)

SUMMARY

Our calculation and experimental results have shown that the point-like grid design of large area a-Si p-i-n cells offers an extremely low active area loss and a very low interconnection power loss. This approach also represents the least perforamnce difference between small area and large area type devices. We have shown that a 3.7% active area loss has been achieved in a 100 cm^2 completed panel whose efficiency is 8.5%. Further, by the use of more precise masking techniques, it is possible to further reduce the active area loss to only 1%. From our experience, there are no fundamental difficulties in transferring this technology to large scale production of inexpensive a-Si solar cells and we have fabricated the first 1 ft^2 prototype panels with point contact design and the initial results are encouraging. Finally, this type of structure can be applied to other thin film photovoltaic devices.

ACKNOWLEDGEMENTS

We sould like to express our thanks to Dr. P.K. Bhat, M. Miller, C. Marshall, E. Valentich, R. Sedlak, and E. Diede for their useful discussion and technical support.

REFERENCES

[1] See, for instance, J.J. Hanak, U.S. Pat 4292092 (Sept. 29, 1981).

[2] S. Yamazaki, K. Itoh, S. Watable, A. Mase, K. Urata, K. Shibata, and H. Shinohara, 17th IEEE PV Specilist Conf., 205, May 1984, Kissimmee.

[3] D.E. Carlson, A. Catalano, R.V. D'Aiello, C.R. Dickson, and R.S. Oswald, SPIE Conference, Sept. 1986, Boston.

[4] A. Madan, SPIE, VOl 706, 72, 1986; J. Xi, R.E. Hollingsworth, P.K. Bhat, and A. Madan, AIP Conf. Proceedings, Vol 157, 183, 1987; R.E. Hollingsworth, P.K. Bhat and A. Madan, Proc. 19th IEE Photovoltaic Conf, New Orleans, 1987 - to be published.

RESIDUAL DOPING EFFECTS USING DIBORANE: THE IMPORTANCE OF REACTOR DESIGN

P. ROCA i CABARROCAS, A. LABDI and B. EQUER
Laboratoire de Physique des Interfaces et des Couches Minces
(ER 258 - CNRS) Ecole Polytechnique 91128 Palaiseau, Cedex, (FRANCE)

ABSTRACT: Thermal decomposition (CVD) of diborane-silane mixtures at temperatures as low as 100°C is demonstrated by the characterization of intrinsic a-Si:H films exposed to the mixture. The CVD effects are studied by different techniques: ellipsometry, SIMS, conductivity and analyses of device properties, which allow us to determine the effect of temperature, surface reactivity and duration of CVD. These dependences are taken into account in order to discuss the importance of CVD effects in the production of a-Si:H based solar cells for different reactor configurations: monochambers, multiplasma and multichambers.

1. INTRODUCTION

In amorphous silicon solar cells production, control of **p/i** interface and **i** layer intrinsic quality are of crucial importance to obtain a high performance and good stability device /1/. Control of both parameters is difficult in a monochamber deposition system where boron, deposited on walls during the growth of **p** layers, is removed by plasma and restituted into **i** layers. We have developed a new strategy, between the two common ones: multichambers and monochambers, in order to avoid boron recycling effects and to reduce multichamber system cost. Our machine (ARCAM) is a multiplasma monochamber reactor in which three chambers are located in a hot wall single vaccuum vessel /2,3/. Each of the plasma chambers is devoted to deposition of either intrinsic, boron doped or phosphorous doped a-Si:H respectively. In this way we can effectively avoid the problem of boron recycling by the plasma which is dominant in a monochamber deposition system as shown in figure 1. As a matter of fact, even after a nine hours deposition of intrinsic a-Si:H in the "**p** -chamber" (i.e. the chamber normally use to grow **p** doped layers), the boron level in the material produced is about ten times higher than that produced in the "**i** -chamber".

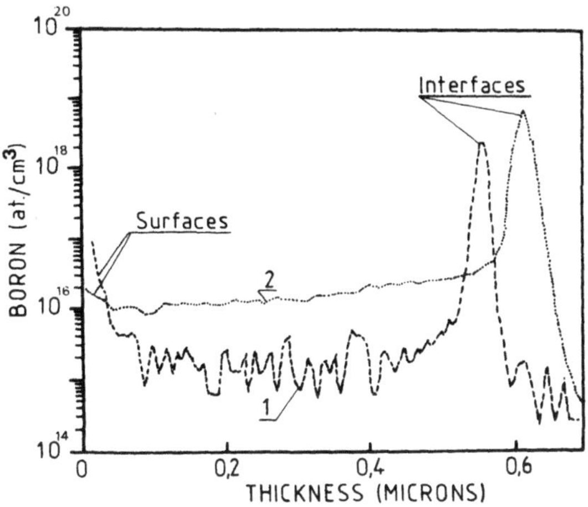

Figure 1. Boron concentration profile measured by SIMS in two intrinsic a-Si:H films deposited in a clean (1) or in a boron contaminated chamber (2).

Further, as no physical separation exists between the three chambers of our reactor, some diborane (or phosphine) can flow from the **p** (or **n**) chamber to the others where some contamination by thermal decomposition of B_2H_6 (or PH_3) can then be expected. As we did not observe any important effect associated with thermal decomposition of phosphine, we will focus on the effects of diborane.

2. RESULTS

Thermal decomposition of diborane (0.2%)-silane mixture (commonly used for the production of the **p** layer in a-Si:H based solar cells) has been studied through its effects on intrinsic a-Si:H films and characterized by various techniques. The results presented here concern:

- the optical properties as measured by spectroscopic ellipsometry,
- the boron profile and concentration measured by secondary ion mass spectrometry (SIMS),
- the dark conductivity and activation energy measured in coplanar configuration,
- the open circuit voltage of devices for which the **p** layer is obtained by CVD.

2.1. Spectroscopic ellipsometry

We have used the high sensitivity of spectroscopic ellipsometry to detect surface effects associated with CVD. For this study we first deposited 1 μm thick a-Si:H films on glass substrates. Immediatly after stopping the plasma, we passed diborane diluted at 0.2% in silane in the reactor for 45 minutes, keeping the temperature constant. Figure 2 shows the effect on a sample deposited at 200°C. The whole displacement of the imaginary part of the dielectric constant can be interpreted as an increase of the surface roughness /4/ produced by the CVD or as the deposition of a less dense a-Si:H layer.

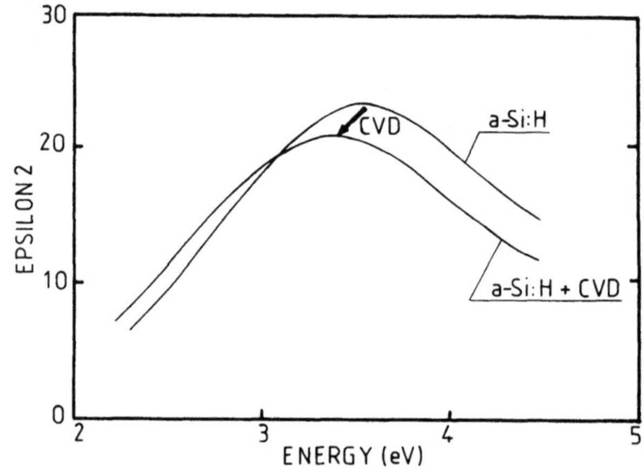

Figure 2. Modification of the imaginary part of the dielectric constant of a freshly deposited a-Si:H film, as a consequence of flowing diborane diluted at 0.2% in silane for 45 minutes (T_s=200°C).

Performing the same type of experiments at different temperatures and different substrates shows that:

- There is a very small effect when the a-Si:H is grown at 100°C which can be understood by a higher a-Si:H surface roughness when deposition is performed at lower temperature /5/.

As a consequence the experiment is less sensitive to the CVD effect. On the other hand, this temperature dependence excludes a simple gas adsorption effect which would be more pronounced at low temperature.

- The ellipsometric spectra of a crystalline silicon or glass substrates are not modified by the CVD indicating that CVD depends on the chemical reactivity of the surface.
- If the CVD is performed on oxidized a-Si:H samples, no effect is observed on ellipsometric spectra. However SIMS measurements show that a thin boron contaminated layer exists on the top of the sample.This can be explained by the fact that both CVD and oxidation produce an effect equivalent to an increase of the surface roughness /4/.

2.2. Secondary ion mass spectroscopy (SIMS)

SIMS measurements have been performed on the samples characterized by ellipsometry and clearly show the existence of a boron contaminated surface layer. However, in order to avoid surface effects in the SIMS measurements, we have produced samples where the boron contaminated layer has been sandwiched between two layers of intrinsic a-Si:H. Figure 3 shows boron concentration profile on this type of samples.

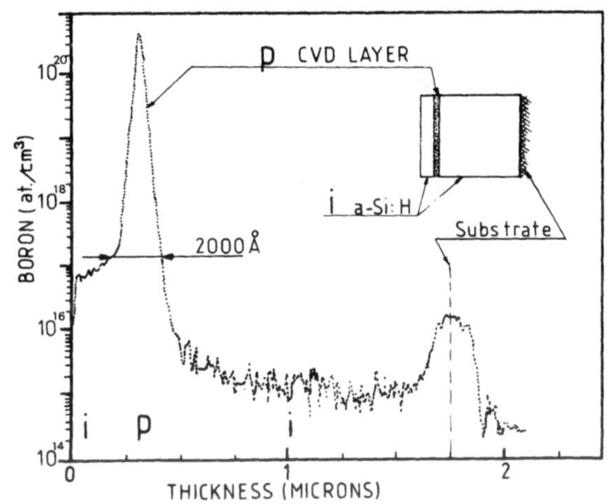

Figure 3. Boron concentration profile measured by SIMS in a sample deposited at 200°C, consisiting of one layer obtained by 45 minutes CVD of 0.2% diborane in silane, between two intrinsic a-Si:H layers grown by standard RF plasma.

We observe :

- The presence of a boron peak at the interface with the substrate. This peak is systematically observed in SIMS measurements and is associated to the substrate itself /6/.
- A very low concentration of boron in the bulk of the sample ($\approx 10^{15}$ boron atoms/cm^3).
- The presence of a boron containing layer produced by CVD during 45 minutes at 200°C. The boron concentration and profile of this layer are comparable to that obtained by plasma deposition /7/.
- The effect of plasma recycling of the boron deposited by CVD on the reactor walls is evidenced by the high concentration of boron in the **i** layer near the surface (compare to figure 1).

Thus CVD of 0.2% diborane in silane mixtures produces boron contaminated layers on surfaces covered by a-Si:H. Moreover this layer affects the electrical properties of samples as measured by coplanar conductivity.

3.3. Coplanar conductivity

The formation of boron containing layers by CVD has dramatic consequences on the electrical properties of the films. Table I shows the values of dark conductivity σ_d, pre-exponential factor σ_o and activation energy E_a for an intrinsic sample, a doped one and a sample with a thin layer deposited by CVD:

	σ_d(50°C) $(\Omega.cm)^{-1}$	σ_o $(\Omega.cm)^{-1}$	E_a (eV)
Intrinsic	5.10^{-9}	6.10^{4}	0.80
Boron doped	1.10^{-4}	166	0.40
i+CVD	1.10^{-6}	60	0.48

Table I. Effect of a **p** layer produced by CVD on a freshly deposited intrinsic a-Si:H layer. The samples are 0.7μm thick and were deposited at P=40 mTorr, RF power=2 watts and T_s= 200°C.

We note that the values of conductivity and activation energy for the contaminated sample are very close to that of a doped one and are in good agreement with an overlayer of 100 Å of boron doped material. This demonstrates that **p** layers could be simply produced by thermal decomposition of diborane-silane mixtures.

2.4. Device characterization

We have used V_{oc} measurements of Substrate/SnO_2/**n-i-p**/Al devices, in which the **p** doped layer was grown by CVD, to study the duration and temperature dependence of CVD. This configuration has been chosen in order to avoid the effects of boron recycling by plasma, which would perturbe the results in a Substrate/SnO_2/**p-i-n**/Al configuration as is shown in figure 3. Table II summarizes the results of this characterization:

	V_{oc} (mV)
Reference sample, Substrate/SnO_2/**n-i-**/Al, T_s=200°C	136
Sample with **p** layer produced CVD during 45' at 200°C	443
Sample with **p** layer produced CVD during 45' at 150°C	444
Sample with **p** layer produced CVD during 45' at 100°C	195
Sample with **p** layer produced CVD during 5' at 200°C	167

Table II. The V_{oc} under AM1/4 illumination conditions of **n-i-p** devices in which the **p** layer has been obtained by the CVD of diborane.

The main results of this characterization are:

- The effect of CVD of diborane on the open circuit voltage subsists even at 100°C. It should be mentioned that even at 100°C, intrinsic a-Si:H produced in the ARCAM reactor has a low density of states /2,3/.

- The effect is of the same order at 150°C and 200°C. This can be understood if we take into account that the diffusion coefficient of boron is greater in the lower density sample deposited at 150°C /8/.

- The effect depends on the duration of CVD: a five minutes CVD produces little change in V_{oc} as compared to the reference sample.

3. CONSEQUENCES ON REACTOR DESIGN

Three types of deposition systems (monochamber, multiplasma-monochamber and multichamber) are used for the production of a-Si:H based solar cells. The monochamber solution is less expensive but plasma recycling of boron deposited on the walls prevents the production of abrupt interfaces and a boron-free material in the **i** layer (fig.1). In order to decide between multiplasma and multichamber reactors, we must take into account the thermal decomposition of diborane in the multiplasma system and the higher cost of the multichamber system. However, the effects of CVD depend on its duration and are practically negligible after two minutes deposition of the **p** layer in a typical a-Si:H based solar cell. Therefore the multiplasma system is competitive in front of a multichamber one for the production of abrupt interfaces at a lower cost. Further, we have compared the performances of **p-i-n** solar cells produced in the ARCAM reactor (multiplasma) to that of devices deposited in the same plasma conditions but using the same chamber for the three layers (monochamber). For all the deposition conditions we obtain a higher blue response using the multiplasma reactor which indicates a better **p/i** interface.

4. CONCLUSION

Thermal decomposition of diborane mixed in silane produces a **p** doped layer on freshly deposited a-Si:H films. The parameters which control the thickness of the **p** layer are the CVD temperature and duration: small effects have been observed at 100°C or after a five minutes CVD at 200°C. Therefore thermal decomposition of diborane mixed in silane is a problem to be considered in the use of any reactor and especially with multiplasma, hot walls reactors. Nevertheless, owing to the short time nedeed for the **p** layer deposition in a-Si:H based solar cells, this effect is of little importance and the multiplasma reactor is a performant and economic solution.

ACKNOWLEDGEMENTS

This work is supported by PIRSEM/CNRS, AFME and CEC contracts. P. Roca i Cabarrocas thanks AFME for a grant.

REFERENCES

1) W. Y. Kim, S. Yamanaka, M. Hallerdt, M. Konagai and K. Takahashi, Third Photovoltaic Science and Engineering Conference, Tokyo, Japan (1987)
2) P. Roca i Cabarrocas, B. Equer, J. Huc, A. Lloret and J. P. M. Schmitt, Seventh E. C. Photovoltaic Solar Energy Conference, Sevilla , p. 533 (1986)
3) P. Roca i Cabarrocas, These de doctorat de l'Université Paris VII, Paris (1988)
4) B. Drevillon and F. Vaillant, Thin Solid Films **124**, 217 (1985)
5) A. M. Antoine, B. Drevillon and P. Roca i Cabarrocas, J. Appl. Phys. **61**, 2501 (1987)
6) G. J. Slusser and L. MacDowell, J. Vac. Sci. Technol. **A 5**, 1649 (1987)
7) C. C. Tsai, M. J. Thompson, R. A. Street, M. Stutzman and F. Ponce, J. of Non Cryst. Solids, **77&78**, 995 (1985)
8) G. H. Bauer, Private communication

DEVELOPMENT OF AN EFFICIENT HYDROGENATION METHOD FOR ATOMIC AND IONIZED CLUSTER BEAM DEPOSITION OF SOLAR GRADE AMORPHOUS SILICON*

P. Tzanetakis, Y. Franghiadakis and A. Christou

Institute of Electronic Structure and Laser
Research Center of Crete
P.O. Box 1527, 711 10 Heraklion, Crete, Greece

ABSTRACT

The role of hydrogen in reactive growth of amorphous silicon (a-Si:H) films is extremely important in determining the structure and the properties of the deposited material. An atomic hydrogen source of very simple design and operation was developed and incorporated in a deposition system based on Ionized Cluster Beam (ICB) of silicon. This type of source can provide the activated hydrogen necessary to passivate films grown at high rates.

The control of the energy per depositing atom in the beam permits to avoid the non-uniform growth (columnar structure) usually associated with reactive evaporation.

INTRODUCTION

Among the alternative to silane CVD techniques for deposition of a-Si:H thin films, the Ionized Cluster Beam (ICB) method appears to be the most promising for attaining high rate, low substrate temperature growth of device-quality films.

The fundamental advantage of ICB growth of thin films lies in the fact that the energy and the flux of the film forming species can be varied independently.

The ICB technique has successfully been implemented in a multitude of amorphous and epitaxial thin film growing experiments involving a broad range of materials : metals, semiconductors, ceramics and even organic compounds.

The applicability of the ICB technique to a-Si:H deposition was first demonstrated a few years ago by Yamada (1). According to the published literature, little progress has been made since and no report of solar cell fabrication by this process is known to the authors. One of the reasons lies certainly in practical difficulties in the cluster forming device of an ICB system. This is usually the nozzle of a crucible heated to about 2000 C, a temperature high enough for the vapor pressure of Si to be in the 1 to 10 torr range. At this elevated temperature , Si vapor is reactive and great care must be taken to avoid damage and clogging of the nozzle. The choice of the crucible material is important and is discussed in the next paragraph.

The atomic hydrogen source presented here was designed and tested in the framework of a project aimed at the development of an ICB deposition process for a-Si:H than can serve as a viable alternative to silane CVD. It is worth reminding that besides the potential for high-rate, high quality film growth the ICB deposition offers other attractive features such as a good

*Work supported by the C.E.C., "Photovoltaic Power Generation Programme" under Contract No. EN3S/0039.

utilization of source material and the possibility to deposit over a limited surface removable from the vacuum system together with the substrates. This can eliminate the time consuming, wall cleaning step necessary with CVD reactors.

EXPERIMENTAL

1) The activated Hydrogen source

It is well established that the incorporation of hydrogen in a-Si films involves activated (at least partly dissociated or ionized) hydrogen. In the ICB setups described in the literature hydrogen is admitted in the deposition system at pressures of the order of a few 10^{-5} to 10^{-4} torr. Activation and ionization of H is achieved on the heated W filaments of the ioniser or by electron impact. A detailed analysis of previous work with evaporated a-Si (2) indicates that film quality is very sensitive to the precise arrival rate of atomic hydrogen at the surface of the growing film. In order to passivate films grown at high rates the pressure of hydrogen in the deposition chamber should increase proportionally to the growth rate. The required pressure can easily become prohibitively high for practical purposes.

The activated hydrogen source is depicted on Figure 2 as it is arranged in the ICB deposition system. It is based on the catalytic dissociation of H_2 on the surface of a hot W filament placed inside a small chamber into which H_2 is fed at constant flow. The pressure inside the dissociator chamber can be higher by orders of magnitude than in the main chamber. The rate of atomic hydrogen flowing out of the aperture increases with increasing H_2 pressure inside the source. The recombination of hydrogen in the source is limited, at least at small dissociation fractions, because of a buffering effect of molecular hydrogen. H atoms evaporating from the W filament at 2200 K thermalize with the molecules through collisions with the latter before leaving the dissociator.

2) The ICB setup

Figure 2 shows the basic components of the ICB system used in this investigation. The high potential (0 to 5 kVolts) for the acceleration of the clusters is applied here on the substrate-equipotential box assembly. This configuration is less convenient than the usual method of grounding the substrate holder and raising the potential of the ICB source, but permits the use of conventional power supplies. This results in a considerably lower cost of the system.

The use of an equipotential box which can easily be removed and cleaned, serves a dual purpose of allowing a tight arrangement of the system components in the vacuum system without disturbing the ionized cluster trajectories and to prevent particulate formation in the deposition system by peeling-off Si layers already deposited on the chamber walls.

RESULTS AND DISCUSSION

The total amount of hydrogen retained in a-Si:H films under various operating aconditions of the source was calculated from the corresponding IR absorption spectra. Figure 2 shows the variation of total hydrogen content of the films as a function of the hydrogen atom arrival rate at the surface of the growing film. These data were taken with samples deposited by electron gun evaporation. The probability of retaining hydrogen is close to 1 for the low fluxes. The solid line of Figure 2 is calculated on the basis of a very simple assumption : the incorporation probability is taken proportional to the number of available (total minus occupied) H sites. A total number of .25 H sites per Si atom gave the best fit with the experimental results. This behaviour is in sharp disagreement with the results of a similar investigation where a sticking coefficient smaller by two orders of magnitude was observed (3). The hydrogen dissociator used for the latter investigation produces atoms in thermal equilibrium with H_2 gas at 2000 K. We believe this disagreement is due to the difference in energy per H atom between the two techniques. A much lower incorporation coefficient was also observed in our laboratory when plasma dissociation of H_2 was used.

A very similar behaviour was observed with ICB grown samples. In this case a complete series of samples with varying H content could not be obtained because of serious difficulties associated with graphite crucibles. It is not yet clear what the exact reaction of Si and graphite at 2000 C is and if the low pressure of H_2 in the vacuum chamber ($2x10^{-5}$ torr) plays a role in this reaction. The practical consequence is that after a few hours of operation the nozzle of graphite crucibles must be replaced.

Work is in progress to investigate the suitability of refractory metals, especially Ta, as crucible material and to find a cluster forming mechanism not involving Si vapor expansion through a nozzle.

REFERENCES

1. I. Yamada, in "Semiconductors and semimetals" Vol. **32A** edited by J. Pankove (1984).

2. Y. Franghiadakis and P. Tzanetakis, "Hydrogen Dissociation and Incorporation into Amorphous Silicon", Submitted for publication.

3. R.E. Viturro and K. Weiser, Phil. Mag. **B53(2)**, 93 (1984).

Fig.1 : The ICB set-up: 1. Nozzle, 2. Graphite crucible, 3. Heater, 4. Heat shields, 5. Water cooling lines, 6. Ionizer, 7. Ionizer filament, 8. Adjustable electrostatic lens, 9. Equipotential box, 10. Substrate holder, 11. Crystal thickness monitor, 12. Hydrogen dissociator, 13. Dissociator filament, 14. Molecular hydrogen inlet.

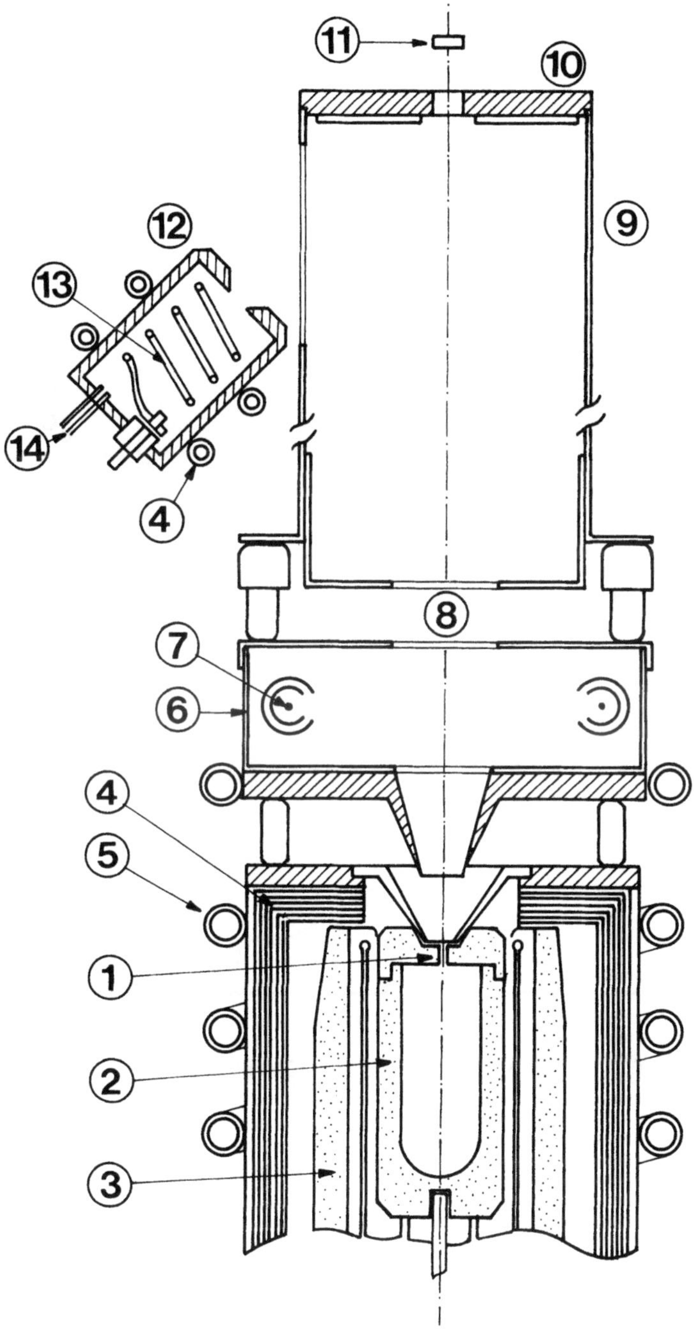

Fig.2 : Hydrogen content of evaporated films as a function of H atom arrival rate during growth.

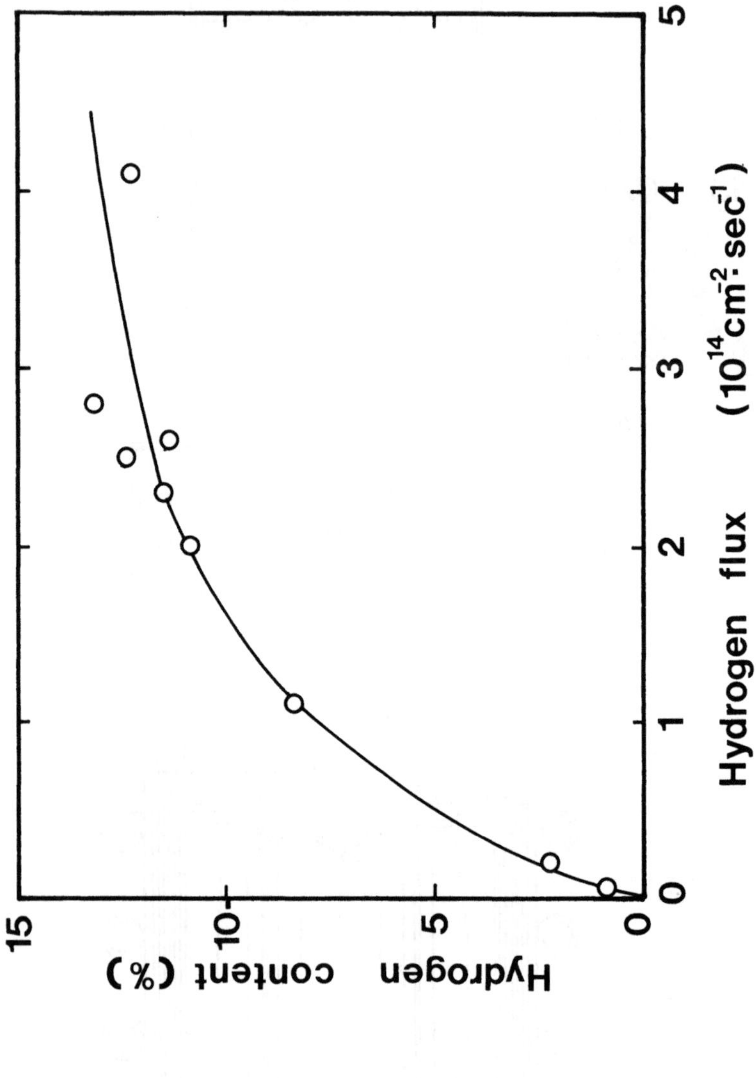

FOUR-TERMINAL TANDEM SOLAR MODULE USING a-Si:H AND c-Si

W. Krühler, R. Plättner and J. Grabmaier
Siemens Research Laboratories, D-8000 Munich 83

ABSTRACT. Amorphous silicon (a-Si:H) and crystalline silicon (c-Si) are materials with proper bandgaps (1.8 eV, resp. 1.1 eV) for tandem structures in order to match the sun spectrum optimally. We propose two types of series-connected four-terminal a-Si:H/c-Si tandem solar modules: One consists of a semitransparent thin-film a-Si:H module on top of a polycrystalline silicon module made from conventional S-Web- or Silso material; the second one is a thin-film tandem solar module consisting of a semitransparent thin-film a-Si:H module on top of a thin-film c-Si module, using the advantage that series-connected thin-film cells covering large areas can be manufactured as monolithic devices. Patterning processes for thick- and thin-film c-Si solar modules are presented for the first time.

1. INTRODUCTION

For supplementing or replacing conventional fuels the photovoltaic systems should be based on high-efficient, low-cost, large-area, stable and ecological solar cells. Up to now, there is no solar cell fullfilling all these requirements simultaneously.

A very promising strategy for enhancing photovoltaic technology is based on the development of multijunction or tandem modules, especially of thin-film tandem solar modules.

The tandem solar modules proposed in this work consist of two separated submodules: a semitransparent top module utilizing amorphous, hydrogenated silicon (a-Si:H) and a bottom module utilizing polycrystalline silicon (c-Si). Two patterning processes for the thick- as well as for the thin-film c-Si solar modules are described.

2. PHYSICAL ASPECTS

Assuming sunlight with an AM1 spectrum theoretical calculations /1,2,3/ yield, that maximum efficiencies for a tandem solar cell consisting of two cells are achieved, if the top cell is made from a semiconductor with a bandgap E_{g1} of about 1.8 eV and if the bottom cell is made from a semiconductor with a bandgap E_{g2} of about 1.2 eV. These two requirements together with the condition of non-toxic semiconductor materials limit the selection of proper materials very drastically. From this point of view only a-Si:H (E_g = 1.75 eV) remains for the top cell fabrication, and for the bottom cell a-$Si_{1-x}Ge_x$:H-alloys (1.0 eV < E_g < 1.75 eV for 1 > x > 0), c-Si (E_g = 1.1 eV) and copper indium diselenide (CIS, E_g = 1.0 eV) can be used. Up to date, the system a-Si:H on CIS has achieved the highest efficiency of 13.7 % /4/, while 13.0 % efficiency was reported /5/ from a triple solar cell consisting of fluorinated amorphous silicon materials, and 12.4 % was obtained with a two-terminal a-Si:H/c-Si solar cell /6/.

Fig. 1 shows the AM1,5 sun spectrum and the spectral response of todays a-Si:H-and c-Si solar cells, indicating the better utilization of the sunlight by the combination of both cell types in a tandem structure. The curve denoted with c-Si(f) denotes the spectral response of a c-Si solar cell illuminated with sunlight which is filtered by the a-Si:H top cell.

If two solar cells are well coupled, then the tandem cell efficiency η_M of the combination is about the efficiency η_{TC} of the top cell, which is exposed to the entire solar spectrum, plus half of the efficiency η_{BC} of the bottom cell illuminated with filtered light; i.e. $\eta_M \approx \eta_{TC} + {}^1/_2\eta_{BC}$. This rough formula contains, that a tandem cell with an total efficiency η_M in the 20 % range (Fig. 2) should ingeniously consist of a top cell having an efficiency η_{TC} of about 13 % and of a bottom cell with an efficiency η_{BC} around 7 % under filtered light (F).

The possibility of designing tandem solar cells in a two- or four-terminal arrangement (Fig. 3) requires some considerations about advantages and disadvantages of the corresponding structures.

The advantages of a four-terminal cell are: i: the separated optimation of each cell, ii: no current-matching between the both cells, iii: no interdiffusion of doping atoms, iv: the separated patterning of each module. However, the four-terminal tandem solar cell requires an optical coupler, two additional contact layers and two substrates.

The two-terminal tandem solar cell has the advantages: i: the simplest structure without an optical coupler, ii: the most inexpensiv version of the tandem structures, iii: the reflection and absorption losses are minimal. On the other hand, the two-terminal tandem cell requires current matching and a low-loss diffusion-free tunnel junction connecting both cells. In a first development step it seems to be easier to start with a four-terminal tandem solar module.

3. PATTERNING OF a-Si:H/c-Si TANDEM SOLAR MODULES

The four-terminal a-Si:H/c-Si tandem solar cell (module) approach requires a semitransparent a-Si:H top solar cell (module) which is combined with the c-Si bottom solar cell (module) via an optical coupler (Fig. 4). The pin a-Si:H top module has the series-connected structure with a front as well as a back electrical contact made from transparent conductive oxide (TCO) materials such as zinc oxide (ZnO) or tin oxide (SnO_2). The patterning of such an a-Si:H top module is carried out by the well-known methods like screen-printing/lift-off or laser-scribing /7/.

The c-Si submodule in Fig. 4 can consist of thick (~ 0.5 mm) polycrystalline silicon (Silso- or S-Web material /8/) or of thin-film (≈ 50 µm) polycrystalline silicon. The thick c-Si submodul is fabricated in the following procedure. The striped metallic back contacts are evaporated by a mask process or screen-printed on an isolating substrate (glass or ceramics). The automatically cutted striped c-Si solar cells are sticked on the back contacts by welding or pasting with conductive adhesives. The metallic contacts and the cells are separated by small distances (< 0.5 mm). A grid is deposited by an usual screen-printing process. The series-connection is made by welding of small ribbons between the grid and the back contact of the adjacent solar cell. The contact holes, which are necessary for this process, are provided by milling a small part at the corner of the striped solar cells. This can be done in a low-cost procedure before sticking. An antireflection coating (not shown in the figure) finishes the production of the c-Si submodule.

The fabrication process of a thin-film c-Si solar module is shown in Fig. 5. On an insulating substrate (a) coated with a conducting layer (b) for the back contacts a layer of p-type polycrystalline silicon (c) is deposited by a deposition process such as CVD (or liquid phase epitaxy). After the preparation of the p/n-junction (d) by diffusion or by implantation of phosphorus a patterning process (such as laser-scribing or milling) separates the silicon film and the conducting layer into single striped solar cells (e). The resulting grooves (≈ 0.5 mm wide) are refilled with an insulating organic material (f). Then a dot-shaped selectiv etch process provides the contact holes into the silicon cells only (g). In a screen-printing process the grid as well as the series-connection of the cells is applied (h). An antireflection coating (not shown in the figure) finishes the production of the integrated c-Si submodule. With that procedure thin-film silicon solar modules covering large areas can be manufactured as monolithic devices. The size of the single cell stripes can be designed as large as several cm in width and about 1 m in length depending on the fabrication process.

The optical coupler in a four-terminal tandem solar module (Fig. 4) consists of an insulating layer in form of a transparent foil. Using an organic foil such as Tedlar it was proved that the losses range in the order of about 3 %.

4. EXPERIMENTAL RESULTS

In first experiments samples of a-Si:H/c-Si tandem modules were manufactured and measured.

Polycrystalline Si solar modules (10 cm x 10 cm) each consisting of 10 solar cells (a 1 cm x 10 cm x 0.5 mm) were fabricated from Silso- as well as from S-Web-Material. The efficiencies were 8.2 % and 5.3 %, respectively. Without any coupler, the efficiencies of the c-Si modules are strongly reduced to 2.7 % and 1.5 % under AM1 illumination filtered by a semitransparent a-Si:H-filter with pin structure. The highest efficiency of semitransparent a-Si:H solar cells (3.9 cm^2) reported so far /4/ is 10.6 %. We employed a-Si:H cells with efficiencies of about 8 %.

The combination of our first results leads to a calculated tandem efficiency of about 11 % (a-Si:H/Silso) and of about 10 % (a-Si:H/S-Web). Fig. 6 shows a photo of a model of an a-Si:H/c-Si tandem solar module.

The deposition of thin-film polycrystalline Si layers on graphite foils and molybdenum foils was carried out by decomposition of $SiHCl_3/H_2$ gas mixtures in a CVD process at temperatures between 950°C and 1150°C. The deposition rate was 7 µm/min, the thickness of the Si layer was about 100 µm, the grains were in the order of about 30 µm - 40 µm in size. In order to deposit polycrystalline Si layers with grains of about 100 µm in size and with diffusion length of about 100 µm new deposition methods are under investigations.

5. CONCLUSION AND OUTLOOK

The concept of the a-Si:H/c-Si tandem solar module is based on the idea to combine materials having the optimal bandgaps for a more efficient utilization of the sunlight. A proposal was given to fabricate large-area solar tandem modules made from only silicon materials with the potential, that efficiencies in the 20 % range could be attained. First experiments with a-Si:H/c-Si tandem structures having c-Si submodules made from S-Web or

Silso material lead to efficiencies of 10 % and 11 %, resp. The feasibility of a new deposition methode to produce c-Si films with sufficient large grains and diffusion lengths has to be demonstrated in future. As reported in the literature /9, 10/ with the liquid phase epitaxy (LPE) thin-film Si solar cells were prepared having efficiencies over 12 %. The combination of this result with the efficiency of 10.6 % reported for the semitransparent a-Si:H solar cell /4/ yields a calculated efficiency of over 16 % meaning a promising outlook.

6. REFERENCES

/1/ J.C.C. Fan, B-Y. Tsaur and B.J. Palm: Proc. 16th IEEE Photovoltaic Specialists Conf., 1982, San Diego, USA, p. 692

/2/ I. Chambouleyron and F. Alvarez: Proc. 18th IEEE PHotovoltaic Specialists Conf., 1985, Las Vegas, USA, p. 533

/3/ C. Eberspacher, B. Felder, R.R. Potter, R.D. Wieting: Proc. 18th IEEE Photovoltaic Specialists Conf., 1985, Las Vegas, USA, p.1031

/4/ K. Mitchell, R. Potter, J. Ermer, R. Wieting, C. Eberspacher, D.P. Tanner, K. Knapp, R. Gay: Proc. 19th IEEE Photovoltaic Specialists Conf., 1987, New Orleans, USA, p. 13

/5/ S. Guha, J. Yang, P. Nath and M. Hack: Appl. Phys. Lett. 49 (1986) 218

/6/ K. Okuda, H. Okamoto and Y. Hamakawa: Jap. Journ. of Appl. Phys. Vol.22, 1983, L 605

/7/ W. Juergens, R. Plättner, H. Kausche, W. Peters, W. Stetter: 7th E.C. Photovoltaic Solar Energy Conference, Sevilla, Spain,1986,494

/8/ J.G. Grabmaier, G. Hoyler, R. Falckenberg: 18th IEEE Photovoltaic Specialists Conf., Las Vegas, USA, 1985, 1124

/9/ A.M. Barnett, R.B. Hall, D.H. Ford, P.G. Lasswell, J.A. Rand: Proc. 7th E.C. Photovoltaic Solar Energy Conf., 1986, Sevilla, Spain, p. 836

/10/A.M. Barnett, J.A. Rand, R.B. Hall, D.A. Fardig: Techn. Digest of the Int. PVSEC-3, Tokyo, Japan, 1987, p. 101

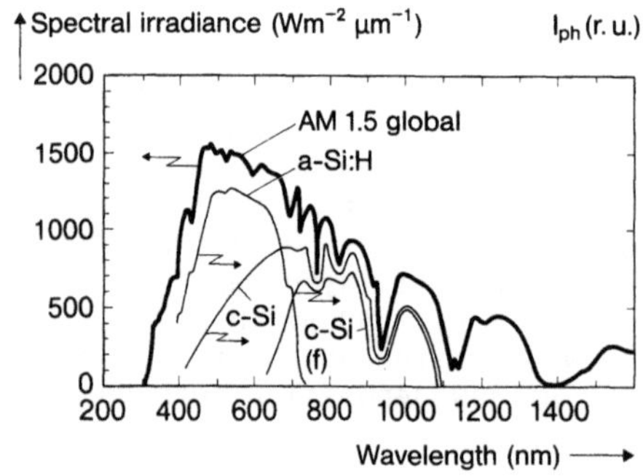

Fig.1:

AM 1,5 sun spectrum and the spectral response of a-Si:H- and c-Si solar cells; c-Si(f) denotes the spectral response of an a-Si:H-filtered c-Si solar cell

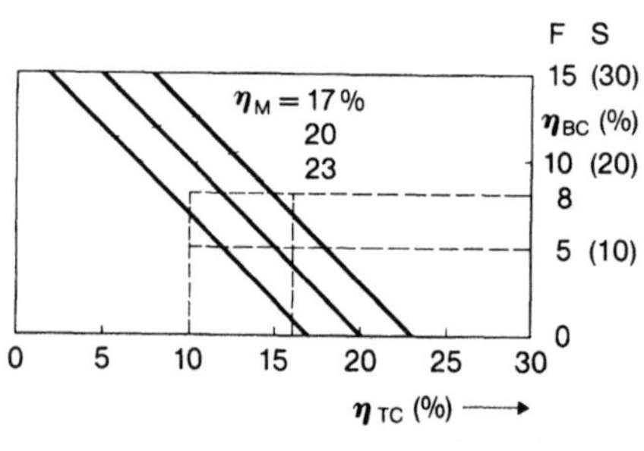

$$\eta_M = \eta_{TC} + \tfrac{1}{2}\eta_{BC}$$

Fig.2:

Total efficiency η_M of tandem solar cells consisting of top cells (TC) and bottom cells (BC); S=stand-alone, F=filtered light

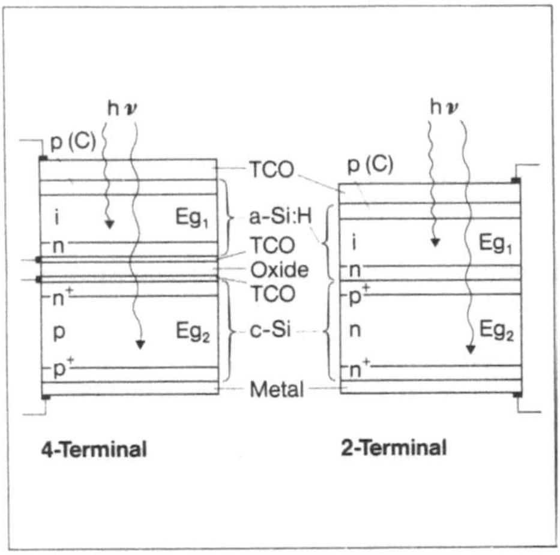

Fig.3:

Tandem solar cells in a four- and a two-terminal arrangement

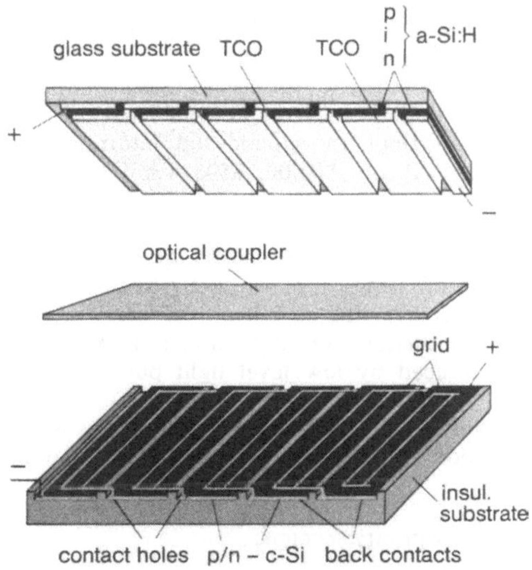

Fig.4:

A four-terminal a-Si:H/c-Si tandem solar module approach

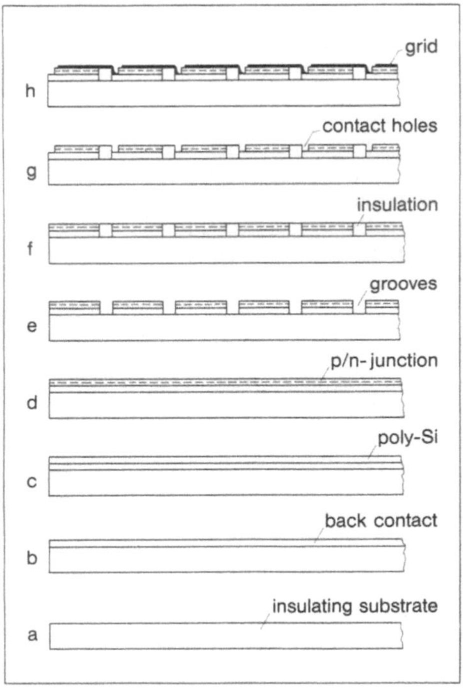

Fig.5:

Fabrication process of a thin-film c-Si solar module

Fig.6:

Photo of a model of an a-Si:H/S-Web-tandem solar module ($10x10cm^2$)

THE DENSITY OF STATES IN THE BAND GAP OF AMORPHOUS SILICON DETERMINED BY THERMOSTIMULATED CONDUCTIVITY

G. Landweer, G. van Noort, H. Simons, J. Bezemer and W.F. van der Weg,
Dept. of Atomic and Interface Physics, University of Utrecht,
P.O.B. 80.000, 3508 TA Utrecht, the Netherlands.

ABSTRACT. The density of states (DOS) between the conduction band and the Fermi level in glow-discharge a-Si:H has been determined by measuring the thermostimulated current of carriers released from traps. The $\mu\tau$ product of the carriers was determined simultaneously by measuring a photo current, induced by low level light pulses. We have made an attempt to determine μ and τ separately, by combining different models for the TSC process. We have found values of the microscopic mobility which depend strongly on assumptions made about the density of states at the mobility edge.

1. INTRODUCTION

Main parameters of the intrinsic layer of an amorphous silicon solar cell are the electron mobility and lifetime, which determine how far electrons can travel in the material between excitation and recombination. The $\mu\tau$ product is strongly related to the density of states in the band gap. Determination of the DOS of intrinsic a-Si:H is a helpful tool in optimizing the growth parameters, including reactor geometry, to produce high quality photovoltaic active material. The measurement of the DOS is also useful to study physical properties like the Staebler Wronski effect.

In order to prevent confusion about the meaning of the mobilities and the lifetimes in the multiple trapping transport model which we use, the following definitions are given. The microscopic mobility μ_m is a measure for the real velocity of carriers in the extended states under influence of an applied electric field. The drift mobility μ_d is much smaller than the microscopic mobility, since it is based on the mean velocity of a carrier, including the time that the carrier is captured in a trap.

Often the recombination lifetime τ_r is defined as the total time a carrier is present in extended states between excitation and recombination, the trapping time *not* included. We will use this definition and mention also two other lifetimes. The trapping lifetime τ_t can be defined as the time between two trapping events, the drift lifetime τ_d, which is very large in amorphous silicon, is the time between excitation and final recombination. If we further define *m* as the mean number of trapping events of a carrier during its lifetime in the extended states and the traps, the following relations for $\mu\tau$ products hold.

$$\mu_m\tau_r = \mu_d\tau_d = m.\mu_m\tau_t \qquad (1)$$

Our method of measuring the DOS is based on early work of Simmons et al. [1], who showed that the release of carriers from traps can be used to scan the density of states in an insulator or the band gap of a semiconductor. After filling traps at a low temperature, electrons are released during heating from an energy interval between the conduction band and the Fermi level with a width of about kT. Simmons et al. used a non-trapping transport model. Zhu and Fritzsche [2] were the first who applied the method to amorphous silicon. The multiple trapping during the electron transport was taken into account by introducing an *effective* attempt to escape frequency, several orders of magnitude lower than the escape frequency for a single event. They also showed that special photoconductivity experiments are needed to determine the $\mu\tau$ product of the electrons at the specific temperatures and trap occupancies that exist during the TSC measurement.

We introduced a method to perform the TSC measurements and the photoconductivity measurements simultaneously [3]. This method will be illustrated here.

Misra et al. [4] have performed TSC experiments at very low temperatures, down to 30 K, and performed model computations on the band tails. They did not use a constant attempt to escape frequency, nor a discrete energy level approximation for electron excitation. In their method a temperature determined distribution of electrons between the bottom of the conduction band and the region of the quasi Fermi level was applied. Their model needs a critical estimation of the density of states at the energy of extended state conductivity. Although different assumptions have to be made, we propose a combination of both models in order to derive interesting results concerning the values of μ and τ.

2. DESCRIPTION OF THE METHOD AND MEASUREMENTS

2.1 *The thermostimulated conductivity*

Samples are annealed at 170°C in the dark and subsequently cooled to 80 K. At that temperature, a moderate illumination of the sample is applied to fill the gap states above the Fermi level. We will assume that all states are filled, which is in fact a rather arbitrary approximation. The actual occupational function appears to reproduce very well, since a variation of the illumination by several orders of magnitude does not alter the TSC results. We only take electron processes above the Fermi level into account because of the low mobility of holes compared to the electron mobility.

After a short relaxation time an equilibrium is reached and the occupation of the states can be described by means of quasi Fermi levels. Then the sample is heated slowly at a constant heating rate β (K/s), and the thermally stimulated current is measured. Fig. 1 shows a typical TSC curve. In fig. 3 the occupation function is represented schematically at three stages of the measurement. The samples of about 1 μm thickness have coplanar electrodes at a mutual distance of 0.5 mm. A DC voltage of 100 V is applied to measure the conductivity.

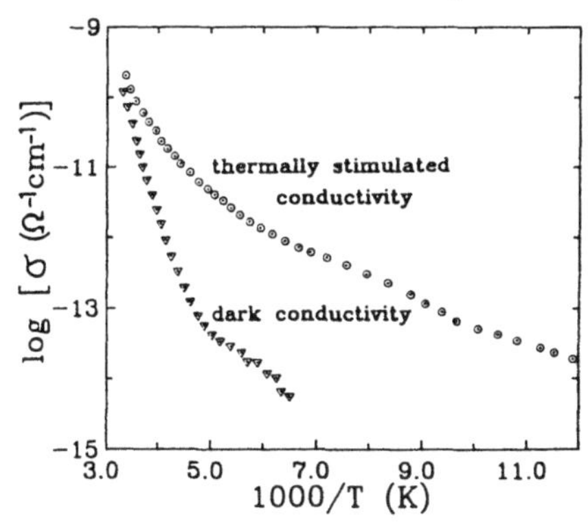

Fig. 1. Thermostimulated conductivity and dark conductivity of an amorphous silicon sample at a heating rate of 0.08 K/s.

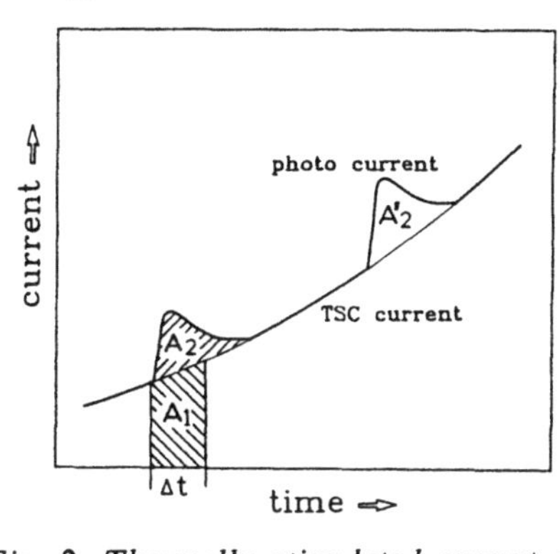

Fig. 2. Thermally stimulated current as a function of time on linear scales. The areas A_2 indicate integrated currents induced by low level light pulses of about 0.3 s. The current peaks are several seconds wide.

According to the non-trapping model, the electrons are released from a small energy interval around E_m, which is called the energy of maximum TSC. With a constant heating rate ($T = T_o + \beta t$) and an attempt to escape frequency ν_o, it follows [1] :

$$E_m = -kT \ln(41\nu_o/\beta). \qquad (2)$$

The zero point of the energy is at the mobility edge of the conduction band. To account for the multiple trapping, ν_o has to be replaced by the effective attempt to escape frequency ν_e. The value of ν_e is experimentally found by assuming that $E_m = E_F$ at a temperature where the thermostimulated current becomes small compared to the dark conductivity. The activation energy of the dark conductivity is equal to the Fermi energy E_F. The best numerical approximation of E_m depends somewhat on the starting temperature and the value of β. We used

$$E_m = -0.97\ kT \ln(52\nu_e/\beta) + 0.02, \qquad (3)$$

where E_m is expressed in eV, ν_e in s^{-1} and β in K/s. An estimate of the number of trapping events *m* of an electron before recombination can be deduced from the ratio ν_o/ν_e, which is of the order of 10^2 to 10^4.

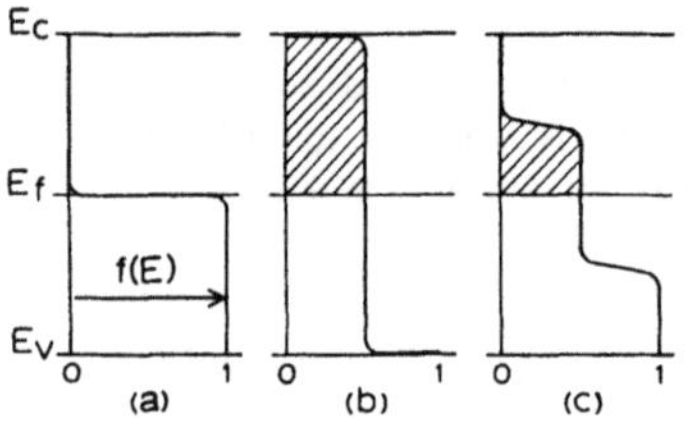

Fig. 3. Occupation of traps above the Fermi level at three stages of a TSC measurement, schematically. (a) Annealed and cooled, (b) Illuminated at 80 K, (c) During heating at about 200 K.

2.2 The $\mu\tau$ product and the DOS.

The thermostimulated current is related to the density of states by the $\mu\tau$ product. This $\mu\tau$ product is measured in our method by determination of the photo current due to low level light pulses during the TSC experiment. This is illustrated in fig. 2. In fact the number of released electrons from the traps is compared to the number of electrons introduced by the light pulse. The duration of the light pulse is about 0.3 second. The area of the photo current pulse is

$$A_2 = N\ e\ \frac{V}{l}\ \frac{\mu_m\tau_r}{l}\ , \qquad (4)$$

where N is the total number of absorbed photons, e the electron charge, V the applied voltage and l the distance between the electrodes. The last factor represents the fraction of the electrons that reaches the electrode. E_m shifts ΔE in energy during a small time Δt. The area A_1 in fig. 2 under the TSC curve is given by

$$A_1 = e\ v\ \frac{V}{l}\ \frac{\mu_m\tau_r}{l}\ g(E)\Delta E\ , \qquad (5)$$

with v the volume of the active part of the sample. From (4) and (5) it follows that

$$g(E)\Delta E = \frac{A_1}{A_2}\ \frac{N}{v}\ . \qquad (6)$$

With formulas (3), (4) and (5) the density of states g(E) and the values of $\mu_m\tau_r$ can be found experimentally.

In figs. 4 and 5 the results are shown for two sets of samples, grown at various substrate holder temperatures (a) and several glow-discharge powers (b). The starting temperature of 80 K was not low enough to show the conduction band tails, which are indicated schematically. A considerable increase of the DOS is found towards the middle of the gap.

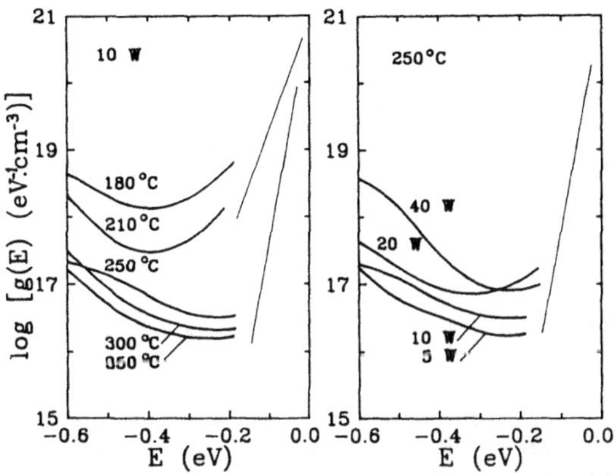

Fig. 4a,b. Density of states of a-Si:H samples produced at different growth parameters. (a) Series grown at a glow-discharge power of 10 W and various substrate holder temperatures, (b) series grown at 250°C and various discharge powers.

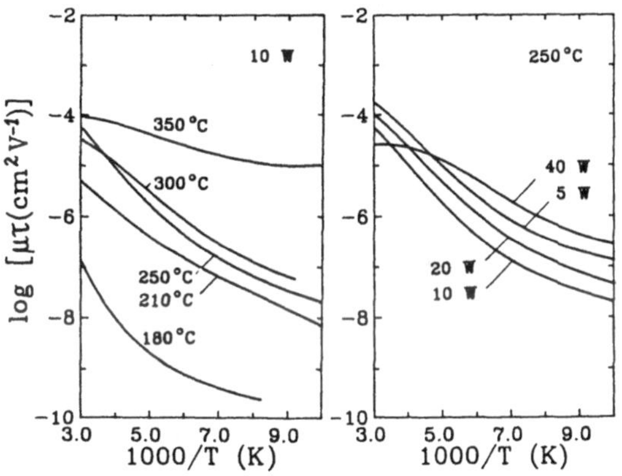

Fig. 5a,b. Electron $\mu\tau$ product of the two series of samples mentioned in Fig. 4a,b. The results have been obtained from photoconductivity peaks superimposed on TSC curves.

2.3 The values of μ and τ

According to our model, the thermally stimulated conductivity is described by

$$\sigma_{TSC} = e\ \mu_m \tau_r\ f_o(E)\ g(E_m)\ \beta \left| E_m - 0.02 \right| / T\ , \qquad (7)$$

in which $f_o(E)$ is the occupation function at the starting temperature T_o of the experiment. The model is rather simple and the density of states as well as the $\mu\tau$ product can be determined with a relatively high accuracy. The results are very insensitive to variations in the value of ν_e. An incorrect value of ν_0 would rather shift the location of traps than change their number. An error of less than a factor of two is expected due to the approximation of the initial occupational function and the neglecting of the contributions of holes. The error due to the approximation of the occupational function by a discrete step at E_m is probably not serious if the g(E) function is not too steep like at the band tails.

The model described by Misra et al. results in the following expression for the TSC.

$$\sigma_{TSC} = e\ \mu_m\ N_c \int_{E_F}^{E_c} f_o(E)\ \exp - \left[\frac{E}{kT} + \frac{N_c}{\beta \tau_r g(E)} \int_{T_o}^{T} e^{-E/kT} dT \right] dE. \quad (8)$$

N_c is the density of states at the conduction band edge, which is supposed to be between 10^{19} and 10^{22} $eV^{-1}cm^{-3}$. The essential difference of formula (8) with respect to formula (7) is that it is not based on an effective attempt to escape frequency. Relations for the temperature dependent ratio of gap state and conduction band state occupancies are used instead. If eq. 8 is simplified by the same approximations which were also used to derive eq. 7, it is found that the combination of both equations yields :

$$\nu_e = \frac{N_c}{\tau(E)\, g(E)} \ . \quad (9)$$

From this expression we obtained values for μ and τ separately as shown for one sample in fig. 6. We used the value of $N_c = 3.10^{20}$ $eV^{-1}cm^{-3}$ and the measured values of the $\mu\tau$ product. The value of ν_e was kept constant at the value of 10^{-9} s^{-1} which was determined experimentally at 300 K where $E_m \approx E_F$. It should be noted that the resulting values of μ may vary one or two orders of magnitude due to the uncertainty in the determination of N_c and ν_e.

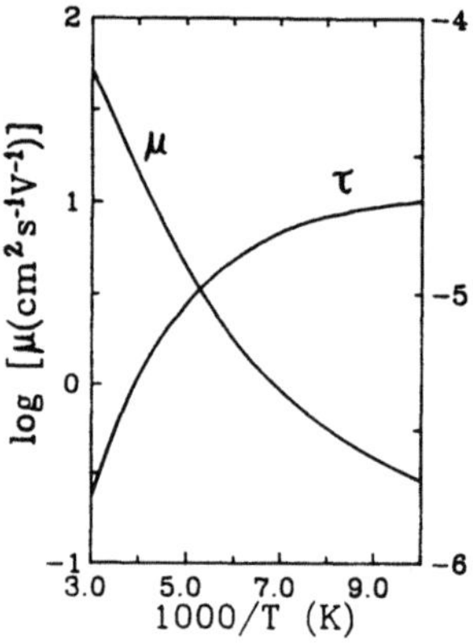

Fig.6. The values of μ and τ separately, computed for the sample grown at 250°C and 10 W.

3. CONCLUSIONS

It has been shown that TSC can be used advantageously to determine the density of states and the $\mu\tau$ product. As illustrated, useful comparisons between different samples can be made, for instance to optimize growth parameters. The method is relatively simple and in this way no serious systematic errors are expected. The first results of combining two different models are surprising. Values of the microscopic mobility are found which are strongly temperature dependent.

REFERENCES

[1] J.G.Simmons, G.W.Taylor and M.C.Tam, Phys. Rev B 7 (1973) 3714.

[2] M.Zhu and H.Fritzsche, Phil Mag. B **53** (1986) 41.

[3] G.Landweer, R.Schreutelkamp, H.Simons, J.Bezemer and W.F.van der Weg, J. of Non-Cryst. Solids **97 & 98** (1987) 711.

[4] D.S.Misra, Vijay A.Sing and S.C.Agarwal, Solid St. Commun. **55** (1985) 147.

FLUORINATED HYDROGENATED AMORPHOUS SILICON CARBIDE

G.Landweer, M.E.M.de Winkel, J.Bezemer, C.P.M.Dunselman
F.H.P.M.Habraken and W.F.van der Weg
Dept. of Atomic and Interface Physics, University of Utrecht
P.O.B. 80.000, 3508 TA Utrecht, The Netherlands.

ABSTRACT. Optical properties, chemical composition and density of states of a-SiC:F:H have been studied as a function of deposition parameters. The material was grown from SiH_4 and CF_4 by glow-discharge CVD. Some properties of this material were compared with properties of a-SiC:H grown from SiH_4 and CH_4. ERD-measurements showed that more oxygen is present in fluorinated than in hydrogenated layers. Two reasons for oxygen incorporation were considered, i.e. release of oxygen from the reactor wall during the glow-discharge and oxidation by exposure of the sample to air. Addition of CF_4 results in a larger carbon content in the layers than addition of CH_4. The optical band gap was found to depend only on the carbon concentration. The carbon incorporation rate depends strongly on the glow-discharge power.

1. INTRODUCTION

Incorporation of carbon widens the band gap of amorphous silicon. Doped amorphous silicon carbide is used as a window layer material in a-Si solar cells. Fluorination of a-Si has several interesting aspects, for instance its influence on the Staebler-Wronski effect and on solar cell stability. Therefore we started a study on a-SiC:F:H grown by glow-discharge decomposition of SiH_4 and CF_4. In previous work, by Uesegi et al.[1] who investigated fluorinated amorphous silicon carbide using SiF_4 and CF_4 or CH_4, were obtained the best photoconductive properties by growing with CF_4. Morimoto et al.[2] measured a smaller spin density in a-SiC:F:H grown from SiH_4 and CF_4. We studied the optical properties of a-SiC:F:H to investigate its suitability as a window layer in a-Si solar cells in comparison to a-SiC:H. Furthermore we examined the chemical composition of our layers by means of Elastic Recoil Detection (ERD) as a part of a study on the incorporation mechanism of carbon and fluorine. Infra red measurements are also in progress for that purpose. Thermostimulated conductivity measurements[3] have been performed to determine the density of states and to estimate the quality of fluorinated a-SiC as a photovoltaic active layer in solar cells.

2. EXPERIMENTAL

The layers were grown in a bakeable UHV deposition system. We were able to limit the oxygen and water background pressures to 1.10^{-9} and 3.10^{-9} mbar, respectively. First we have grown a-Si and a-SiC layers from SiH_4 and SiH_4 with CF_4, recpectively, at various growth parameters. The influence of deposition temperature and glow-discharge power on the optical band gap was compared and the concentration of the different elements in the layers was measured. The optical band gap has been determined by optical reflection and transmission measurements, using the method described by Klazes et al.[4].

Furthermore we have grown two series of a-SiC samples using SiH_4 and either CF_4 or CH_4 at various gas concentrations to examine the bandgap and the chemical composition at different carbon contents. To study the incorporation of oxygen we also have prepared a series of a-SiC:F:H samples with a very thin intrinsic a-Si layer on top.

Typical growth parameters are, unless mentioned otherwise, gas pressure: 0.17 mbar, glow-discharge power: 7 W (about 0.02 W/cm^2), deposition temperature: 250°C and gas flow rate: 10 sccm. The gas ratio is defined as the flow of CF_4 or CH_4 divided by the total gas flow.

3. RESULTS and DISCUSSION

3.1 Influence of deposition temperature.

Variation of the growth temperature between 180 and 350°C showed a decrease in the optical band gap of intrinsic a-Si from 1.71 to 1.59 eV with increasing temperature . The carbon layers showed a similar behaviour. The optical band gap changed from 1.77 to 1.61 eV for a CF_4 fraction of 0.24 in the reactor gas. The incorporated carbon content was independent of the growth temperature. The decrease of the optical band gap is accompanied by a decrease in the hydrogen concentration.

3.2 Influence of glow-discharge power.

We varied the glow-discharge power between 5 and 40 W. This did not lead to a change in the optical band gap of intrinsic a-Si. However the optical band gap of a-SiC layers grown with CF_4 varied from 1.71 to 1.97 eV with increasing power as shown in fig.1. ERD measurements revealed that the carbon incorporation increases at a high power, see fig.2. At a low power there is no appreciable dissociation of CF_4 and no carbon is incorporated in the layer. We conclude that the optical band gap is directly controlled by the carbon content while it only indirectly depends on the glow-discharge power.

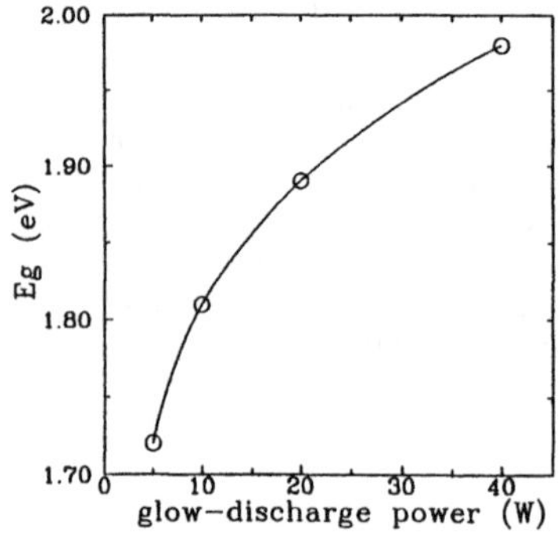

fig.1 The optical bandgap versus the glow-discharge power for a-SiC:F:H layers grown from SiH_4 and CF_4, the gas ratio was 0.34, the growth temperature was 250°C.

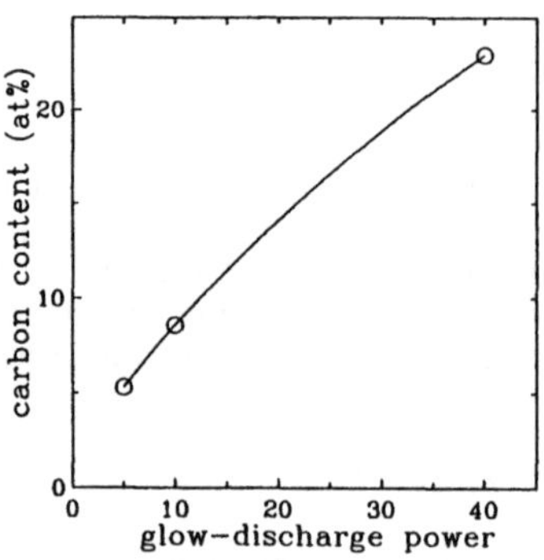

fig.2 The carbon content of a-SiC:F:H versus the glow-discharge power. The CF_4 gas ratio was 0.50, the growth temperature was 250°C.

3.3 The chemical composition.

At the production of the two series of samples, grown to compare fluorinated and hydrogenated silicon carbide, the fraction of the carbon containing gas varied between 0.18 and 0.8. The optical band gap increased from 1.68 to 2.36 and from 1.67 to 2.05 eV for CF_4 and CH_4, respectively, as shown in fig.3. Fig.4 shows the carbon content of these layers. The optical band gap as a function of the carbon content is given in fig.5. The value of the optical band gap only depends on the carbon content and not on the gas source. It appears that the carbon incorporation from CF_4 is more effective than from CH_4.

Furthermore we measured the fluorine, the hydrogen and the oxygen content, fig.6,7. For an increasing CF_4 fraction the fluorine content increases up to 21 at%. The hydrogen content decreases to a value of 10 at%. On the other hand the hydrogen content in the layers grown with CH_4 increases to 23 at%. Usually in a-Si there is about 15 at% hydrogen at a deposition temperature of 250°C. In the fluorinated layers a part of the hydrogen is replaced by fluorine. The high amount of hydrogen or fluorine content in carbon containing layers might be due to incorporation of a large amount of radicals CF_x and CH_x while SiH_x is more completely dissociated at the surface.

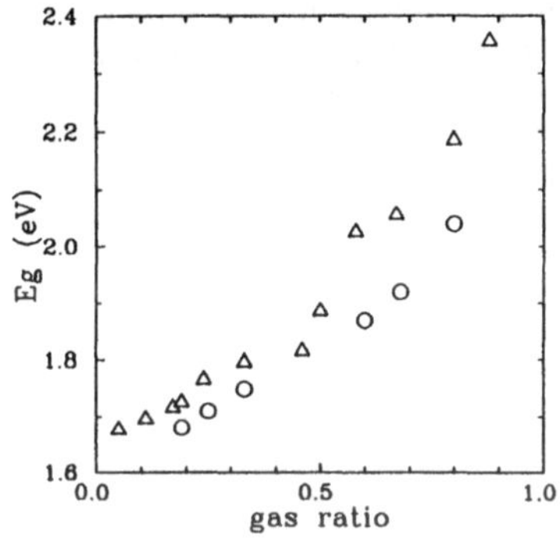

fig.3 The optical bandgap of a-SiC layers versus the gas ratio. The layers are grown from SiH_4 and CF_4 (Δ) or CH_4 (O).

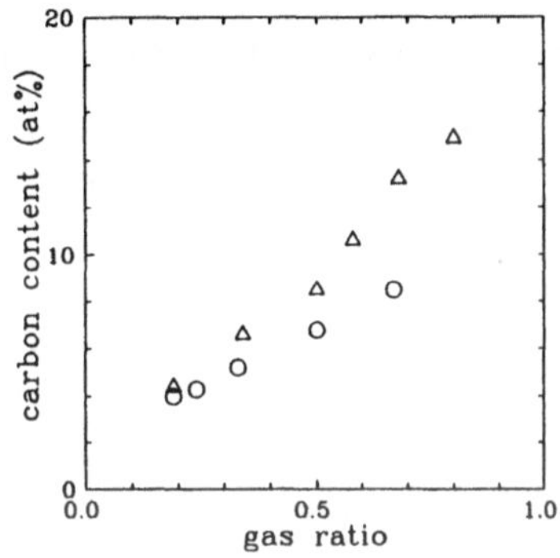

fig.4 The carbon content of a-SiC layers versus the gas ratio. The layers are grown from SiH_4 and CF_4 (Δ) or CH_4 (O).

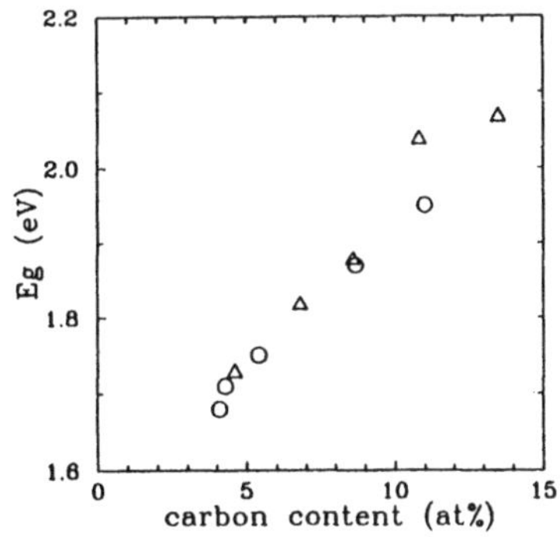

fig.5 The optical bandgap of a-SiC layers versus the carbon content. The layer are grown from SiH_4 and CF_4 (Δ) or CH_4 (O).

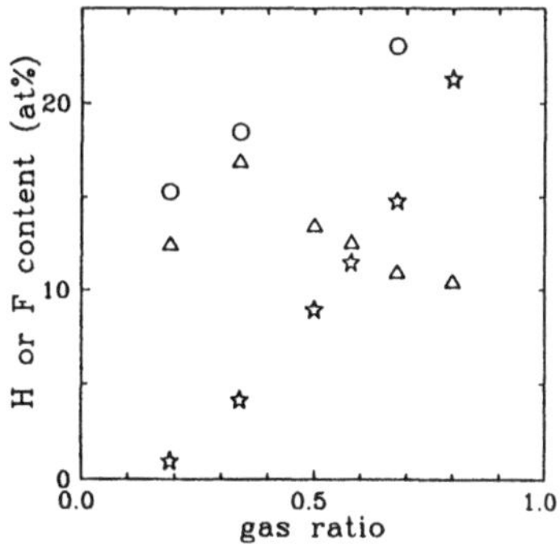

Fig.6 The fluorine (☆) and hydrogen (Δ) content of a-SiC:F:H layers grown from SiH_4 and CF_4, and the hydrogen content (O) of a-SiC:H grown from SiH_4 and CH_4, both versus the gas ratio.

The behaviour of the oxygen incorporation is very different for the two types of layers. We found a high oxygen content, up to 20 at%, proportional to the fluorine concentration in the fluorinated layers, as shown in fig.7. Several experiments have been performed to investigate the origin of this oxygen incorporation. The a-SiC:F:H layers were covered in situ by a thin intrinsic a-Si layer. Two different thicknesses of these layers were used, 6 and 25 nm. A strong reduction in the oxygen content was observed in the fluorinated layers. The reduction amounted to 25 % of the initial value when covering the sample with the thicker layer. We investigated the fluorine content of the covered layers as well and found no difference between the uncovered and covered layers. Apparently a part of the oxygen is incorporated by exposure to air. A possible explanation is a combination of filling of voids caused by incorporation of radicals of CF_4 and passivation of unsaturated bonds in the a-SiC:F:H network. Contrary to our results, Flamm et al.[5] found an outdiffusion of fluorine after exposure to air, but they used NF_3 as a doping gas and their layers contained nitrogen which diffuses out of the layers as well.

In the covered fluorinated layers there still remained an oxygen content up to 10 % which must have been incorporated during the deposition of the layer. We have grown layers at two different oxygen and water background pressures to relate these pressures to the oxygen content of the layers. After reducing the background pressures from 5.10^{-9} to less then 1.10^{-9} mbar for oxygen and from 5.10^{-8} to 2.10^{-9} mbar for water by baking the reactor chamber we found an oxygen concentration of about 1 at% in the covered layers. The uncovered layers with highest fluorine and carbon content still contained about 10 at% oxygen. For the a-SiC layers without fluorine there was no difference in the oxygen content which was already rather low.

The origin of the oxygen in the fluorinated layers, incorporated during the growth, is probably related to an etching process. Fluorine reacts with absorbed water molecules on the reactor chamber wall, forming fluorine-oxygen radicals which, we supose, are preferentially incorporated in the layer. From the ERD spectrum in fig.9 a decrease of the oxygen content during the growth of the layer is observed. This can be caused by the decrease of the water content of the wall as well as by the covering of the wall with an a-SiC layer.

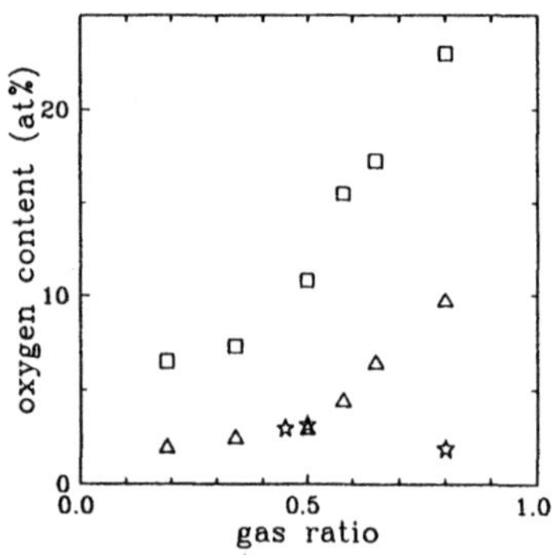

fig.7 The oxygen content in a-SiC:F:H grown from SiH_4 and CF_4 versus the gas ratio. The squares and triangles represent values for uncovered layers, the latest grown at low water background pressure. The stars represent values for covered layers grown under the same conditions.

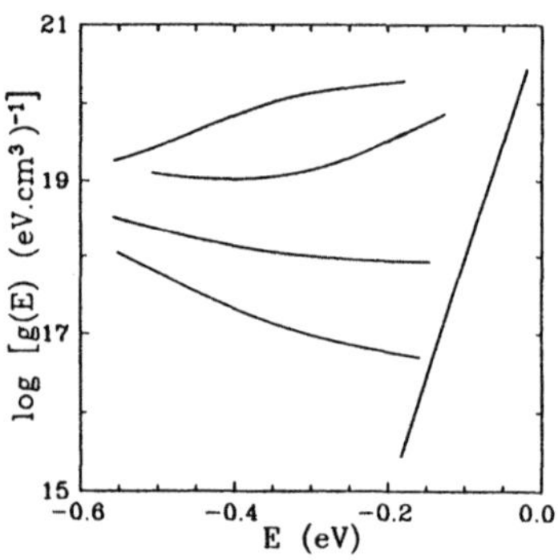

fig.8 Density of states of four samples a-SiC:F:H grown from SiH_4 and CF_4. The DOS increases for increasing CF_4 ratio, which are 0.34, 0.50, 0.60 and 0.68. The straight line symbolizes the conduction band tail, in the figure E_c is 0 eV.

3.4 The density of states

The density of states of the fluorinated a-SiC layers was determined by thermostimulated conductivity measurements (TSC). In fig.8 we show the density of gap states (DOS) in a region between the Fermi level and the conduction band edge. For increasing carbon content the DOS increases from 10^{17} to 10^{20} $(eV.cm^3)^{-1}$. It should be remarked that the TSC-measurements have been performed with the layers without covering. The rather high DOS might therefore be caused by the relatively high oxygen content (up to 10 at%) of these samples.

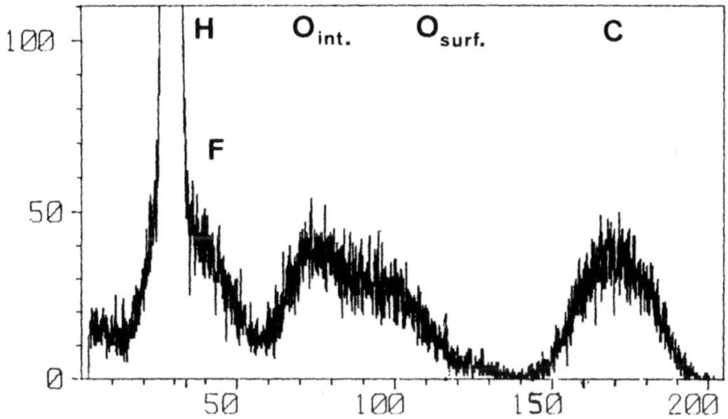

fig.9 An ERD spectrum of an a-SiC:F:H sample, the three peaks from the left to the right represent the fluorine, which peak coincidences with the hydrogen peak, the oxygen and the carbon content. In the oxygen peak a slight decrease of the concentration is visible. The horizontal scale is an energy scale which for each analyzed element corresponds to a depth scale.

4. CONCLUSIONS

With regard to solar cell applications we conclude that it is possible to grow a-SiC:F:H, using CF_4, as a window layer with a large optical band gap. The optical band gap depends directly on the carbon content. A carbon concentration of 12 at% causes an increase of the optical gap of 0.5 eV. The carbon incorporation from CF_4 is more effective than from CH_4.

It is remarkable that layers grown from a fluorine compound may contain a considerable amount of oxygen due to etching processes during the growth. A background pressure of 10^{-8} mbar of water causes an increase of 10 % of oxygen in the layers. The present TSC measurements yield a DOS which is too high for a photovoltaic active layer. We infer this to be caused by the high oxygen content.

REFERENCES

1) T.Uesegi, H.Ihara, H.Matsumura, Jap.J.of Appl. Phys. **24** (1985) 1263
2) A.Morimoto, T.Miura, M.Kumeda, T.Shimizu, Jap.J.of Appl. Phys. **22** (1983) 908
3) G.Landweer, R.Schreutelkamp, H.Simons, J.Bezemer, W.F.van der Weg, J.of Non-Cryst. Sol. **97&98** (1987) 711
4) R.H.Klazes, M.H.L.M.van den Broek, J.Bezemer and S.Radelaar, Phil.Mag.B, **45** (1982) 377
5) D.L.Flamm, C.-P.Chang, D.E.Ibbotson, J.A.Mucha, Solid State Techn. march 1987, 43.

TRANSIENT RESPONSE OF CHARGE CARRIERS IN a-Si

J.Furlan, S.Amon, University of Ljubljana,
Fakulteta za elektrotehniko, Tržaška 25, Ljubljana, Yugoslavia

F.Smole, Iskra-Tovarna polprevodnikov, Trbovlje, Yugoslavia

Abstract: The usual approach in describing generations recombinations assuming equal generation recombination rates of free electron and hole densities is not in general permissible in describing generations and recombinations in a-Si under time variable conditions. The rates of change of free and trapped charge carrier densities can be analytically derived on the basis of SRH model and using simplified zero temperature occupancy distribution. Examination of transient response of free and trapped carrier densities in an undoped and a doped a-Si shows that the initial recombination rate after ceasing the excitation is always equal for free electrons and holes. The excess trapped carrier densities in an undoped a-Si exist and expire a long time following the decay of free charge carriers. In a doped a-Si the response time of minority carriers is very short and that of majority carriers can be very long.

Introduction

The usual approach in describing analytically thermal generations and recombinations in a-Si assumes equal rates of changes of free electrons and free holes /1/,/2/,/3/

$$\frac{dn}{dt} = G_n - R_n = G_p - R_p = \frac{dp}{dt}$$

This approach being customary in evaluating generations recombinations in single crystal Si is not in general permissible in a-Si due to very large densities of localized states in the mobility gap. The assumption of equal generation recombination rates of free electrons and holes in a-Si is allowable only in steady state conditions where thermal generation recombination term of free charge carriers is compensated either by diverging current flow or by light generations and where the rate of thermal transitions of arriving charge carriers to localized states is compensated by equal rate of their departures thus maintaining the fixed occupation of localized states.

In time variable conditions any change of free carrier densities alters the occupation probability affecting the densities of trapped charge carriers. The resulting time variation of occupancy distribution within the mobility gap reflects back the rate of change of free charge carriers creating in general unequal rates dn/dt and dp/dt.

The rates of change of free and trapped charge carrier densities in a-Si depend on the localized states distribution and the density of added impurities. The time response of charge carriers clearly depends on the manner of thermal nonequilibrium excitations.

An illustrative picture about time variations of free and captured charge carriers in dynamic nonequilibrium conditions can be obtained observing transient response of light generated carriers approaching their thermal equilibrium densities. In the paper a simplified analysis governing time variations of free and trapped electron and hole densities is described. Solutions of transient response of charge carriers are then discussed for the specific cases of an undoped and a doped a-Si having a symmetrical double exponential distribution of localized gap states.

Time variations of free and trapped charge carrier densities

The rates of change of free and trapped electron and hole densities out of thermal equilibrium can be described on the basis of Fig.1, showing the basic transitions of charge carriers due to thermal generations and recombinations.

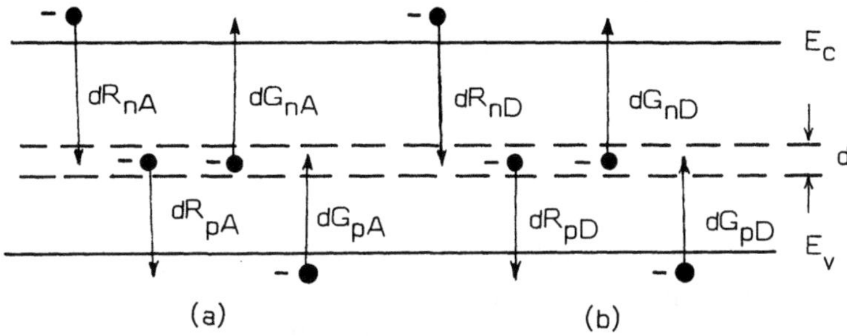

Fig.1. Four electron capture and emission transitions between the conduction or valence band states and localized acceptor-like states (a), and donor-like states (b) in the energy interval dE

Using SRH generation recombination model /4/ for a-Si having a continuous distribution of localized states in the mobility gap and discriminating between capture cross sections for electrons and holes in acceptorlike and donorlike centers, the rates of change of free and trapped carrier densities due to transitions of carriers to localized states inside the energy interval dE at an energy E can be easily described /3/ and then the contributions of all localized states are obtained by integrating over all energies inside the mobiliy gap. To enable analytical integration, the simplified approach using zero-temperature occupancy distribution was employed /2/,/3/, resulting in a simple analytical description of time variations of free and trapped charge carrier densities

$$\frac{dn}{dt} = G_n - R_n = (G_{nD}-R_{nD})+(G_{nA}-R_{nA}) = \sigma v_{th}(n_i^2 - pn) \cdot \left(\frac{1}{Rn+p} \int_{E_{1D}}^{E_{2D}} g_D dE + \frac{1}{n+Rp} \int_{E_{1A}}^{E_{2A}} g_A dE \right) + \frac{Rn}{Rn+p}\frac{dp_t}{dt} - \frac{n}{n+Rp}\frac{dn_t}{dt} \quad (1)$$

$$\frac{dp}{dt} = G_p - R_p = (G_{pD}-R_{pD})+(G_{pA}-R_{pA}) = \sigma v_{th}(n_i^2 - pn) \cdot \left(\frac{1}{Rn+p} \int_{E_{1D}}^{E_{2D}} g_D dE + \frac{1}{n+Rp} \int_{E_{1A}}^{E_{2A}} g_A dE \right) + \frac{Rp}{n+Rp}\frac{dn_t}{dt} - \frac{p}{Rn+p}\frac{dp_t}{dt} \quad (2)$$

$$\frac{dn_t}{dt} = R_{nA}+G_{pA}-G_{nA}-R_{pA} = \sigma v_{th}\frac{n+Rp}{R}\left(\frac{n}{n+Rp} \int_{E_{1A}}^{E_{2A}} g_A dE - n_t \right) \quad (3)$$

$$\frac{dp_t}{dt} = R_{pD}+G_{nD}-G_{pD}-R_{nD} = \sigma v_{th}\frac{Rn+p}{R}\left(\frac{p}{Rn+p} \int_{E_{1D}}^{E_{2D}} g_D dE - p_t \right) \quad (4)$$

with integration limits defined by electron and hole trap quasi-Fermi levels /1/, /2/,/3/

$$E_{1A} = E_i - kT \ln(p+n/R)/n_i \qquad E_{1D} = E_i - kT \ln(Rn+p)/n_i$$
$$E_{2A} = E_i + kT \ln(n+Rp)/n_i \qquad E_{2D} = E_i + kT \ln(n+p/R)/n_i \qquad (5)$$

where g_D and g_A represent donorlike and acceptorlike density distributions, and the relations between capture cross sections are given by

$$\sigma_{nD} = \sigma_{pA} = \sigma \qquad \sigma_{nD}/\sigma_{nA} = \sigma_{pA}/\sigma_{pD} = R \qquad (6)$$

Transient response of free and trapped charge carrier densities

As an application of equations describing time variations of charge carriers in a-Si two specific examples will be examined both representing the transient response of light generated carriers approaching their thermal equilibrium densities after interrupting light generations.

With the purpose to simplify analytical expressions and physical interpretation of events during transients a symmetrical double-exponential distribution of gap states expressed by

$$g_A = A \exp(E-E_c)/W \qquad g_D = A \exp(E_v-E)/W \qquad (7)$$

will be used in both cases.

A) Transient response in an undoped a-Si:

Due to symmetrical distribution of donorlike and acceptorlike states the densities of electrons in an undoped a-Si equal the densities of holes

$$n = p \qquad n_t = p_t \qquad (8)$$

This simplifies the equations (1) to (4), giving only two expressions describing the rates of $dn/dt=dp/dt$ and $dn_t/dt=dp_t/dt$. Adding these two rates and assuming that during the examined transition time $pn \gg n_i^2$, results in a very simple expression

$$\frac{dn}{dt} + \frac{dn_t}{dt} = \frac{dp}{dt} + \frac{dp_t}{dt} \doteq -2\sigma v_{th}\, n\, n_t \qquad (9)$$

which can serve as a convenient basis for the interpretation of events during transient.

During the illumination at time $t<0$, the steady state conditions, shown in Fig.2, are described by

$$\frac{dn_t}{dt}(0) = 0 \qquad G_L = -(G_n-R_n)(0) = -\frac{dn}{dt}(0) = 2\sigma v_{th}\, n(0)\, n_t(0)$$

At the moment of switching off the light source (t=0), the occupation of states is still unchanged and the thermal generation recombination rates at t=0 are therefore the same as just before light interruption.

Since variations of trapped charge carrier densities arise not before substantial changes of free charge carrier concentrations, it can be assumed for short elapsed times $t>0$ that

$$\frac{dn}{dt} = -2\,\sigma v_{th}\, n_t(0)\, n$$

indicating that free charge carriers follow approximately an exponential decay having time constant

$$\tau = 1/2\,\sigma v_{th} n_t(0)$$

and that trapped carrier densities do not change considerably from their steady state values.

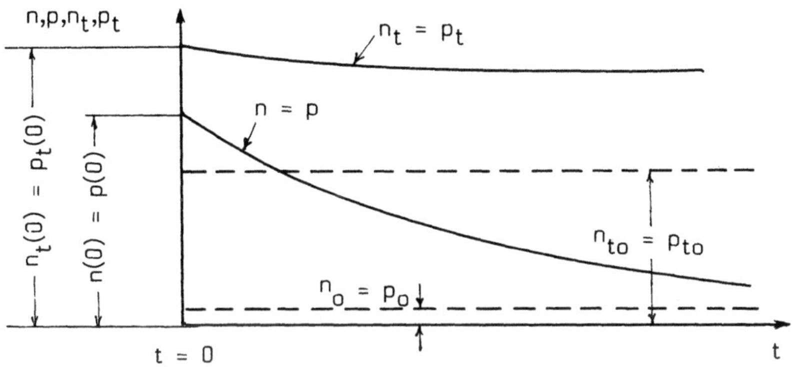

Fig.2. Transient response of charge carrier densities in an undoped a-Si with symmetrical distribution of donorlike and acceptorlike states

It can be therefore concluded that the transient time of free electron and hole densities are very short compared with transient response of trapped charge carrier densities.

B) Transient response in N-type doped a-Si:

In this example it is assumed that a-Si is intentionally doped with donor impurities introducing shallow localized states below the conduction band shifting Fermi levels towards the conduction band and not attributing to variations of localized states occupation during transient response of the free and trapped charge carriers.

Assuming again that in nonequilibrium conditions $pn \gg n_i^2$, the rearrangement of eqs.(1) to (4) results in a simple expression

$$\frac{dn}{dt} + \frac{dn_t}{dt} = \frac{dp}{dt} + \frac{dp_t}{dt} = -\,\sigma v_{th}(n\, p_t + p\, n_t) \tag{10}$$

During continuous illumination at $t < 0$, the steady state conditions are

$$\frac{dn_t}{dt}(0) = \frac{dp_t}{dt}(0) = 0$$

$$G_L = -\frac{dn}{dt}(0) = -\frac{dp}{dt}(0) = \sigma v_{th}\left[\,n(0)p_t(0) + p(0)n_t(0)\right]$$

where concentrations of free and trapped charge carriers can be determined from steady-state analysis of illuminated a-Si /5/.

At the moment the light generations are turned off, the densities of charge carriers are equal to those just before light interruption and thermal generation recombination rates at t=0 remain unchanged. It can be further concluded that during short elapsed times $t > 0$, electron and hole densities decay with an initial time constant

$$\tilde{\tau}_n = \tilde{\tau}_p \doteq 1/\sigma v_{th} n_t(0) \doteq 1/\sigma v_{th} N_D$$

associated with recombination rate of both free charge carriers, and that trapped carrier densities do not change considerably immediately after ceasing illumination.

Within a short time this situation is changed drastically. Since majority free electron density predominates over minority hole density, equal rates of dn/dt and dp/dt result in a large relative drop of hole density compared to a very small change of electron density, as shown in Fig.3. As a consequence, the eq.(10) can be in a very short elapsed time rearranged into the form

$$\frac{dn}{dt} \doteq \frac{dp_t}{dt} \doteq - \sigma v_{th} n p_t \doteq - \sigma v_{th} n(0) p_t$$

describing equal decaying rates of free electrons and trapped holes with the time constant

$$\tilde{\tau}_n = \tilde{\tau}_{p_t} = 1/\sigma v_{th} n(0)$$

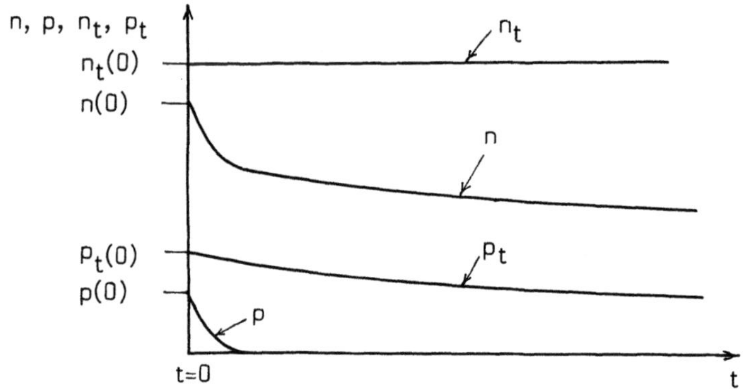

Fig.3. Transient response of charge carriers in N-type doped a-Si

Conclusions

It is shown that the initial rate of change after ceasing the excitation is always equal for free electrons and holes, both decaying with the steady-state time constant. The transient response in an undoped a-Si is very short for free electrons and holes and long for trapped charge carriers. In a doped a-Si the transient response of minority carriers is very short and that of majority carriers can be very long.

References:

1. G.W.Taylor,J.G.Simmons, J.Non-Cryst.Solids, vol.8-10, pp.940-946, 1972.
2. M.Hack, M.Shur, J.Appl.Phys., vol.54, pp.5858-5863, 1983.
3. J.Furlan, S.Amon, MRS Symp.Proc., vol.70, pp.149-154, 1986.
4. W.Shockley, W.T.Read, Phys.Rev., vol.87, pp.835-842, 1952.
5. J.Furlan, S.Amon, F.Smole, Digest Int.PVSEC-3, pp.279-282,1987.

INVESTIGATION OF p-i INTERFACES BY MEANS OF TRANSIENT PHOTOCONDUCTIVITY MEASUREMENTS

H.M. Wentinck, W. Stefan, J.W. Metselaar

Delft University of Technology, Department of Electrical Engineering, P.O.Box 5031, 2600 GA Delft, The Netherlands.

ABSTRACT The built-in electric field in two 1 μm thick amorphous silicon solar cells was determined by means of transient photoconductivity measurements. In order to improve the spatial resolution for determining the electric field, a measurement system with a high time resolution of 1 ns, and an extensive model which predicts the current in the device, were developed.

INTRODUCTION Because the built-in electric field of an amorphous silicon solar cell affects the efficiency of the cell, it is desirable to determine the field experimentally. A promising technique for the determination of the field is the transient photoconductivity experiment, as has been explored by Street[1]. In this experiment electrons are excited at the p-i interface of the cell by means of a short light pulse. After excitation, the electrons move through the electric field towards the opposite electrode, thereby generating a current in the electric circuit connected to the device. Since the current reflects the velocity with which the electrons move, and therefore also the electric field, it is in principle possible to determine the field, provided that the electron mobility is known. It is generally believed that at room temperature the electron drift mobility is well defined, and electron transport exhibits mainly Gaussian diffusion, see for example, Michiel et.al[2].

Accepting these ideas, the striking limitation of the transient photoconductivity measurement lies in the time resolution with which the current transient can be recorded. So far, the best time resolution of published experiments is about 5 ns, which is not sufficient for the determination of the strongly varying field close to the interface. In the case of solar cells the application of a retarding external field which slows down the excited electrons and which increases the spatial resolution (as has been applied by Street to Schottky diodes), may not be a suitable solution. Immediately after the retarding field is turned on, injection of holes and electrons takes place and a significant change of the built-in field can be expected. Alternatively, a new measurement system has been developed with a significantly higher time resolution. It will be shown that full advantage of such a system can be obtained if the duration of the light pulse, the excitation profile, and the response functions of the electric circuit and amplifiers which are connected to the device are incorporated in the analysis.

EXPERIMENT The solar cells which were investigated are two 1.1 μm thick ITO/ZnO-p-i-n-AL amorphous silicon solar cells. They have been deposited by a conventional glow discharge process in a one chamber system on Corning 7059 glass which has been coated with an indium tinoxide zinkoxide ITO/ZnO layer. The intrinsic layer and the 10 nm thick n^+ layer have been deposited from SiH_4 and a 1% PH_3/SiH_4 gas mixture under the following conditions: rf power density 2 mW/cm^3, pressure 0.35 mbar, temperature 250^0C, flowrate 1 sccs, leak/desorption rate of the reactor 10^{-5} sccs. Solar cell A has a 30 nm thick amorphous p layer which is deposited from a 1% B_2H_6/SiH_4 gas mixture under the previously described deposition conditions. Solar cell B has a 20 nm thick p layer which was grown under conditions favoring the formation of microcrystals. These conditions are rf power density 50 mW/cm^3, pressure 2 mbar, temperature 170^0C, 2% of 1% B_2H_6/SiH_4 in H_2. The microcrystalline phase has been identified by Raman Spectroscopy. On Corning 7059 glass these conditions have led to layers as thin as 30 nm, with a conductivity of 1 Ω^{-1} cm^{-1}. To suppress contamination of the intrinsic layer with the residual boron left in the reactor after the deposition of the p layer, the specimens were stored under a steel plate during a dummy deposition of 900 s. The top contact is formed by 1 mm^2 500 nm thick aluminium dots. The sheet resistance of the ITO/ZnO layer is about 10 $\Omega/\square$.

The measurement system resembles the conventional time-of-flight measurement system of Spear[3]. A HP 54100A (DC-1 GHz) digital oscilloscope and a Wattkins & Johnson 6202-419 (10-1500 MHz) pre-amplifier detect the current transient. A light pulse of 0.6 ns is generated by a PRA LN 103/102 nitrogen pulse/dye laser. The solar cells are mounted in a special device holder which is shown in figure 1. The Al-contact and the ZnO/ITO layer are connected to the 50 Ω impedances R_{amp} and R_s of the pre-amplifier and the voltage source. Since the distance between the ITO/ZnO layer and the device holder is only 0.1 mm, the ITO/ZnO layer and the device holder form a capacitor C_{ito} which shunts R_s. The electric circuit which is connected to the device can then be represented by the equivalent circuit in figure 1. In this figure, i(t) is the current generated in the device, and $C_{dev} = \varepsilon A/L$ is the device capacitance. Here ε is the dielectric constant of amorphous silicon, A is the area of the aluminium dot, and L is the thickness of the device.

<u>THEORY</u> The built-in electric field has been determined in the following way. First, the field is described by a suitable set of functions and parameters. Assuming that only electrons contribute to the generated current, an appropriate electron transport equation is solved and the resulting current transient is calculated. Then, the current at the entrance of the amplifier, and subsequently the voltage transient u(t), which is simulated on the screen of the digital oscilloscope, are calculated. In order to make these calculations, the response functions of the electric circuit, the pre-amplifier, and the oscilloscope amplifier were determined. Finally, the simulated voltage transient u(t) is compared with the measured voltage transient $u_m(t)$. The complete calculation is subjected to a Levenberg-Marquardt iteration scheme which adjusts the parameters describing the field until a satisfactory agreement between u(t) and $u_m(t)$ is achieved[4]. Disregarding the disturbance of the electric field by the excited electrons, the following transport equation holds:

$$\frac{\partial n}{\partial t} = \frac{1}{q}\frac{\partial j}{\partial x} + f.g - r.n \quad \text{with} \quad j = q\mu nE + qD\frac{\partial n}{\partial x} \qquad (1)$$

Here n=n(x,t) is the electron concentration as a function of the space x and the time t, q is the elementary charge, j=j(x,t) is the current density, f=f(t) is a function which describes the light pulse, g=g(x) is the excitation profile, r=r(x) is a function which describes recombination and deep trapping, μ is the electron mobility, E=E(x) is the electric field which has to be determined, and D is the diffusion constant. Surface recombination takes place at the boundaries x=0 and x=L. Equation (1) can be solved numerically by forward differencing, applying the Lax-Wendroff method and the Crank-Nicholson method for the evaluation of the field term and the diffusion term, respectively, see, for example, Press et.al[5]. From elementary electrostatic theory, it can be shown that in the quasi-static approximation the generated current i(t) is given by

$$i(t) = \frac{A}{L}\int_0^L j(x,t)\,dx \qquad (2)$$

The accurate solution of eq.(1) costs much computer time. The time step Δt should be much smaller than $\Delta x/\mu E$, with Δx being the space between two adjacent grid points. At the interface, fields can be as high as 10^5 V/cm. For $\mu = 1\,cm^2/Vs$ and $\Delta x = L/100 = 10^{-6}$ cm. $\Delta t < 10^{-11}$ s, which implies that one has to resort to fast computers.

RESULTS Figure 2 shows the current voltage (I-V) characteristics of the solar cells A and B in darkness and under AM1 illumination. The efficiency of A and B is about 2.5 and 3.5% respectively. That the efficiency is low is due partly to the incorporation of a few ppm boron in the intrinsic layer, which was concluded from SIMS analysis. The dummy growth, after the deposition of the p-layer seems insufficient to eliminate residual boron in the reactor completely. The efficiency of A is lower than B. This is mainly due to the fact that the p-layer of B absorbs less light and has a lower conductivity activation energy than the p-layer of A. Figure 3 shows the results of the transient photo conductivity experiments at zero bias, and also the simulations. Figure 4 shows the electric fields E(x) and the electrostatic potentials $\Psi(x)$ which correspond to these transients. In figure 3 the simulated current transient i(t), and the corresponding voltage transient u(t) which fits the measured voltage transient $u_m(t)$, are depicted against the time t. The transients are the result of series of 0.6 ns light pulses, wavelength 480 nm, repetition rate 3 Hz, penetrating the solar cells at the p-i interface x=0. The excited charges were about 1 pC and were determined from charge collection measurements at large reverse bias. The curves in figure 3 have been normalised to 1 pC to facilitate the comparison between A and B. The time shift of the voltage transients in respect to the current transients is due to time delays in the measurement system and is incorporated in the simulation. The sensitivity of the measurement is illustrated by the dashed lines in figures 3 and 4 showing the simulation of a somewhat different electric field in solar cell B. For the simulations the following assumptions were made: 1) the excitation profile of the light pulse is $g(x) = \exp(-\alpha x)$ with $a = 1.35\,10^5\,cm^{-1}$. 2) The electron drift mobility is 1 cm^2/Vs. This value is somewhat smaller than the drift mobility derived from a time-of-flight experiment applied to a 3 µm thick Schottky diode fabricated under the same conditions as the solar cells, but with no boron contamination, and which was 1.5 cm^2/Vs. 3) The surface recombination rate at x=0 is $S_0 = 5.10^4$ cm/s, and takes into account the strong recombination in the p-layer. The recombination function $r(x) = 10^8\,s^{-1}$ for $0<x<0.2$ µm and $r(x) = 3.10^7\,s^{-1}$ for $x<0.2$ µm. This corresponds to $10^{16}\,cm^{-3}$ and $3.10^{15}\,cm^{-3}$ recombination centers for $0<x<0.2$ µm, and for $x<0.2$ µm respectively, with a capture cross section of $10^{-15}\,cm^2$. At the p-i interface the concentration of positively charged and neutral dangling bonds which are the most active recombination centers is of this order, as was concluded from computer modelling of the steady state in amorphous silicon solar cells.

The estimation of the recombination rates is disputable, since it was not verified by other means. That the main problem lies probably here is illustrated by figure 5, which shows the simulated voltage transients for solar cell B for S_o=5.10^4 cm/s and S_o=10^4 cm/s. One sees that the simulations result in rather different transients. For future work additional spectral response measurements are suggested. Further, it seems more proper to determine the values of parameters describing the density of states instead of parameters describing the electric field, since E(x) and R(x) are coupled to the density of states, and the Levenberg-Marquardt iteration scheme allows this coupling.

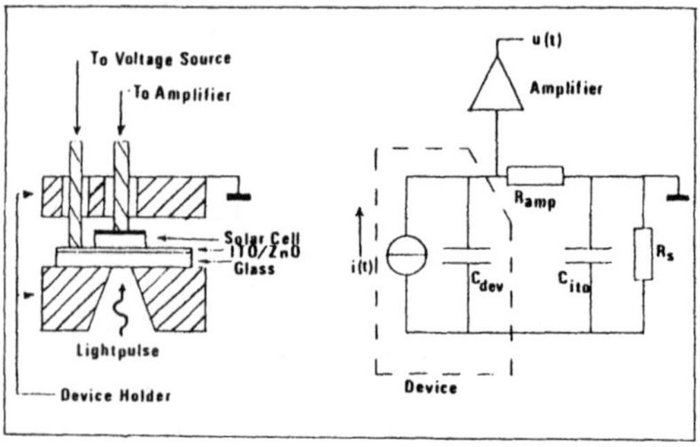

figure 1, Device holder and electric circuit.

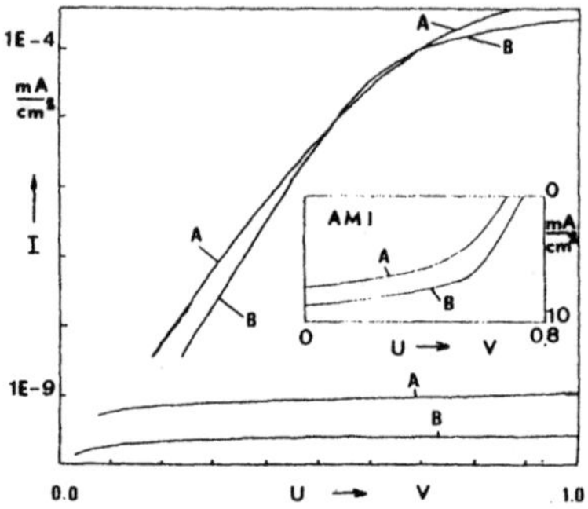

figure 2, I-V characteristics of solar cells A and B in darkness and under AM1.

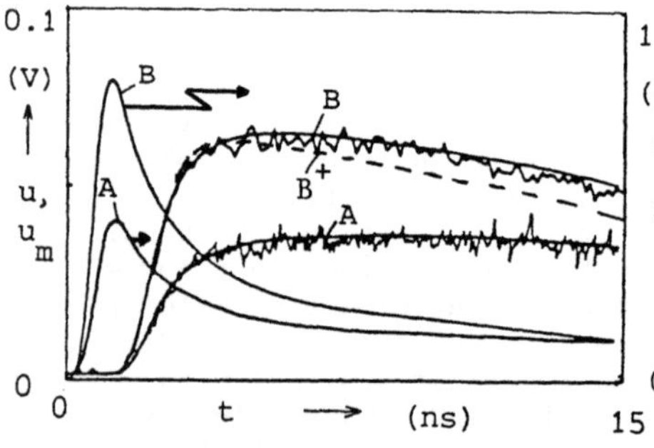

figure 3, Measured voltage transients $u_m(t)$ and simulated current and voltage transients i(t) and u(t) for solar cell A and B.

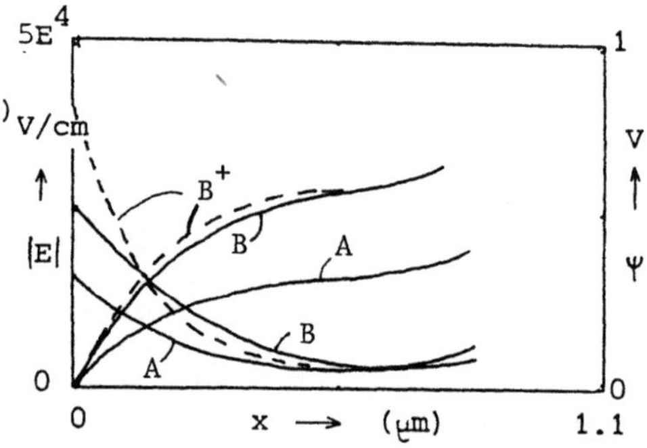

figure 4, Electric field E(x) and potential Ψ(x) derived from these transients. The dashed curve in this figure corresponds to the dashed curve in figure 3.

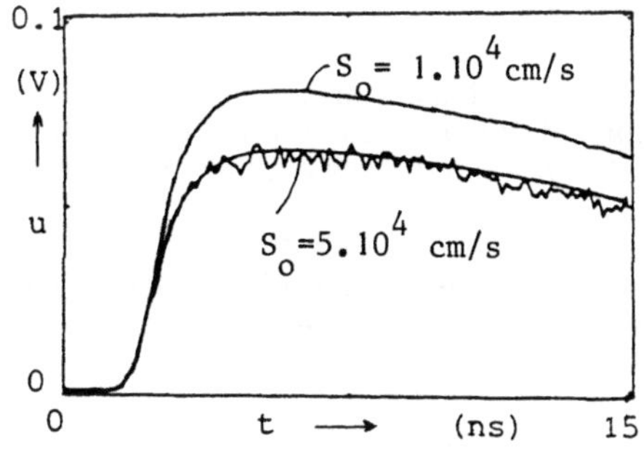

figure 5, Influence of the surface recombination S_0 on the simulated voltage transient.

CONCLUSION A sensitive method for the determination of the electric field at the p-i interface has been explored. The method can be succesful, provided that the recombination of electrons is correctly modelled.

ACKNOWLEDGEMENT We thank G. Bilger and M. Schubert of the University of Stuttgart (FRG) for the substrates with the ITO/ZnO coating, and the Raman Spectroscopy, G.J.L. Ouwerling and F. van Rijs for the Levenberg-Marquardt routine, and A. v.d. Male and J. van Staden for their contribution to the preparation of the samples.

LITERATURE

[1] Street, R.A., Phys. Rev. B 27(8), 4924 (1983).

[2] Michiel, H., G.J. Adriaenssens, and E.A. Davis, Phys. Rev. B 34(4), 2486 (1986).

[3] Spear, W.E., J. Non-Cryst. Solids 1, 197 (1968).

[4] Marquardt, D.W., J. SIAM, 11, 431 (1963).

[5] Press, W.H., B.P. Flannery, S.A. Tenkolsky, and W.T. Vetterling, Numerical Recipes, Camb. Univ. Press (1986).

GAP STATE STUDIES IN a-$Si_{1-x}Ge_x$: H ALLOYS

L. Chahed, M.L. Thèye
Laboratoire d'Optique des Solides, UA CNRS 781
Université P. et M. Curie, 4 place Jussieu, 75252 Paris Cedex 05, France

S. Basrour, J.C. Bruyère
Laboratoire des Propriétés Electroniques des Solides, CNRS,
BP 166, 38042 Grenoble Cedex, France

C. Godet, A. Lloret
Laboratoire de Physique des Interfaces et des Couches Minces, UA CNRS 258,
Ecole Polytechnique, 91128 Palaiseau Cedex, France

ABSTRACT. The changes in gap state density with increasing Ge concentration in glow-discharge a-$Si_{1-x}Ge_x$: H alloys prepared under well controlled conditions are investigated by analyzing the optical transmission and photothermal deflection measurements. The results are correlated with those of electron spin resonance experiments. A model is proposed for the defect state density.

1. INTRODUCTION

Hydrogenated amorphous SiGe alloys (a-$Si_{1-x}Ge_x$: H) have lately attracted much interest due to their potential applications in photovoltaïcs as a low band gap material. However, the photoelectronic properties of these alloys are known to deteriorate rapidly with increasing x (1). This has been correlated with an enhancement of the gap state density as well as with the development of an heterostructure (2-5). The aim of the present paper is to study the modifications of the density of states in the pseudo-gap, as deduced from combined optical and photothermal deflection spectroscopies between 0.6 and 2 eV and from electron spin resonance experiments, as a function of Ge concentration for r.f. glow-discharge alloys prepared under well-controlled conditions. The results are analysed in relation with those of exodiffusion experiments. A model is proposed for the density of defect-induced localized states in the alloys.

2. EXPERIMENT

The samples were 0.5 to 5 μm thick undoped a-$Si_{1-x}Ge_x$: H films deposited by r.f. glow-discharge decomposition of a (SiH_4, GeH_4) gas mixture, the germane flow-rate ratio R_G ranging from 0 to 17 %. Since it has been established that the electronic properties of a-SiGe : H alloys are improved by the H_2 dilution method (6), we have chosen to set a constant flow-rate ratio D (H_2)/D (GeH_4) = 9. The oven-like multiplasma monochamber ARCAM reactor allowed to control the substrates temperature accurately between 100 and 250° C, a thermal decomposition of GeH_4 being observed above 300° C. The r.f. power was explored in the 1-5 Watt range. The standard plasma pressure was set to 90 mTorr, with a 40 SCCM total flow-rate.

The alloy composition was determined by electron microprobe and Rutherford Back Scattering measurements. The total hydrogen content was measured by Elastic Recoil Detection Analysis. Information about hydrogen bonding was provided by exodiffusion experiments (7).

The photothermal deflection spectroscopy (PDS) experiments were performed on 0.5 to 4 μm thick films deposited on glass substrates, in CCl_4, at a 13 Hz chopping frequency, with a 250 W tungsten halogen lamp as exciting source. The PDS spectra were calibrated by fitting to the true film absorbance A = 1 - (R + T) calculated from the optical constants and thickness values deduced from transmission experiments, in the overlap range at high energies (for $\alpha = 1 - 5 \times 10^3$ cm^{-1} in all cases). The optical absorption coefficient spectra were then obtained by a fringe-averaging procedure based on appropriate thin film formulas.

The electron spin resonance (ESR) experiments were performed with a Brücker ESP 300 spectrometer at X-band with 100 KHz modulation at room temperature. The samples were 5 μm thick films deposited on Al foils, peeled off with HCl and collected in a tetrasil tube (2 to 9 mg material).

3. RESULTS AND DISCUSSION

Systematic optical absorption studies have shown that, for identical deposition conditions, the results depend quite strongly on the film thickness. This is illustrated in figure 1, which shows a comparison of the optical absorption edges of two films deposited at a r.f. power of 5 W and a substrate temperature of 150° C, with the same Ge concentration x = 0.16 but with different thicknesses : 0.75 and 4.1 μm. The two spectra coïncide in the high energy range and the two films have the same optical gap E_{04} = 1.83 eV. However, the inverse slope of the exponential part of the edge E_o is larger for the thin film (66 meV compared to 56 meV) and the sub-band gap absorption is higher by an order of magnitude down to 0.6 eV. The observed E_o increase in the thin film may indicate a disorder-induced broadening of the band tails. Alternatively, it may reflect the presence of an extra-contribution of defect states located close to the band edges, which partially overlap the tail states. The density of deeper defect states, especially close to mid-gap, is also much larger for the thin film. These results must be correlated with those of exodiffusion experiments on the same samples (7), which show that, not only the total hydrogen content, but also the proportion of weakly-bonded hydrogen, is larger for the thin film. This indicates that the high density of defect states throughout the pseudo-gap in thin films is linked to the presence of a microstructure, and that its significant decrease in thicker films must result from atomic rearrangements during the film deposition which reduce the microstructure and eliminate at least part of the weakly-bonded hydrogen.

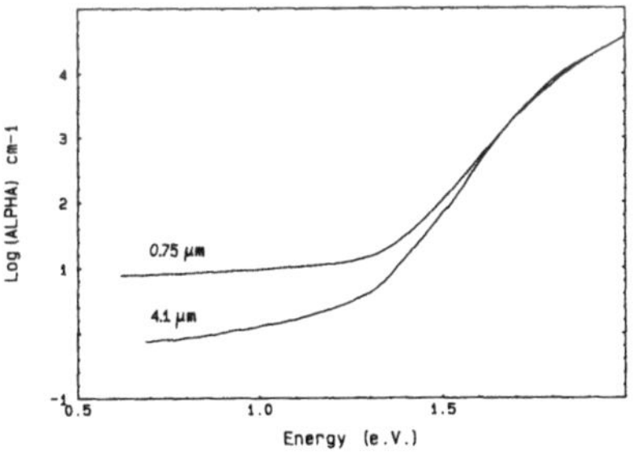

Figure 1 : Optical absorption edges for two alloy films (x = 0.16) with different thicknesses : 0.75 and 4.1 μm.

From investigations performed on 0.5 μm thick films whose properties are most sensitive to deposition conditions, it could be deduced that increasing the r.f. power from 1 to 5 Watts at a given substrate temperature had practically no effect either on the optical parameters E_{04} and E_o, or on the level and spectral variation of the sub-band gap absorption. On the contrary, increasing the substrate temperature T_s from 100 to 200° C had a large influence on the optical absorption edge. E_{04} decreased, essentially because of an important increase of the Ge concentration in the film (for the same R_G value). The E_o disorder parameter decreased significantly, which can be correlated with the drastic decrease of the proportion of weakly-bonded hydrogen as deduced from exodiffusion experiments (7) (while the total hydrogen content decreases only slightly), and therefore to the consequent reduction of the heterostructure. The sub-band gap absorption followed the same trends as the E_o parameter, indicating a concomitant decrease of the amount of defects.

We have chosen to study the influence of the Ge concentration, for x varying from 0 to 0.60, on a series of 3 to 4 m thick films deposited at a 5 W r.f. power and at a rather low substrate temperature, T_s = 150° C, because under such conditions the alloy photoelectronic properties deteriorate quite rapidly with increasing x. These samples are thus well adapted to a study of the causes of this deterioration. Their total hydrogen content increases slightly with increasing x, from 0.09 to 0.11 in the explored composition range. Figure 2 shows the modifications of the optical absorption edge upon alloying. There is both a shift to lower energies and changes in the shape of the spectra. The variations of the E_{04} and E_o parameters are summarized in figure 3 and table 1. E_{04} decreases linearly with increasing x according to : E_{04} (eV) = 1.92 - 0.63 x. The variation rate is slower than that reported in recent work (2, 3), but strikingly similar to that obtained for non-hydrogenated sputtered a-SiGe alloys (8). This suggests that the narrowing of the gap is essentially a Ge alloying effect. The E_o parameter, which can still be determined at the highest x value (0.60) because of the rather low sub-band gap absorption, increases slightly with increasing x, indicating a moderate increase in structural disorder upon alloying.

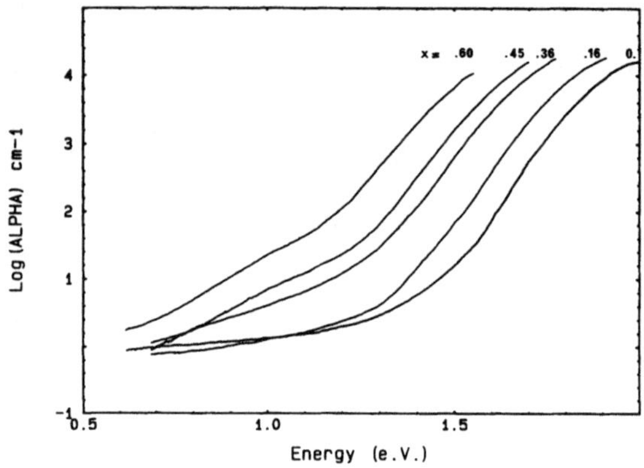

Figure 2 : Optical absorption edges for thick alloy films with increasing Ge concentrations x = 0, 0.16, 0.36, 0.45 and 0.60.

x (at %)	0	16	36	45	60
E_{04} (eV)	1.92	1.83	1.704	1.65	1.53
E_o (meV)	52	56	60	65	69
$N_D (10^{16} cm^{-3})$	1.5	1.2	2.4	3.7	4.5
N_s $(10^{16} cm^{-3})$	-	4.3	12	60	40
g	-	2.014	2.015	2.016	2.018
ΔH_{pp} (Gauss)	-	23	32.5	38.6	38.2

Table 1

It is particularly interesting to compare the optical absorption spectra in the low energy range. For x = 0.16, there is practically no difference with respect to the pure a-Si : H spectrum, apart from the shift and slight broadening of the upper part of the edge. The defect-related extra-absorption $\Delta\alpha = \alpha - \alpha_0 \exp(\hbar\omega/E_0)$ presents the same smooth decrease towards low energies, characteristic of good quality a-Si : H, with approximately the same values. In particular, no structure which could indicate the presence of a well-defined maximum in the gap density of states can be detected on the $\Delta\alpha$ spectra. On the contrary, for higher x values ($x \geqslant 0.36$), one observes both a significant overall increase of the sub-band gap absorption and the gradual emergence of a bump which can clearly be seen on figure 2 for x = 0.60. In spite of the cut-off of the data on the high energy side due to the substraction of the exponential tail, which partially overlaps the defect contribution, the $\Delta\alpha$ spectra definitely show a maximum, the intensity of which increases linearly with increasing x. The values of the defect density N_D deduced from $\Delta\alpha$ by the method proposed for a-Si : H (9), which also increase with increasing x, are reported in table 1 and in figure 4. The energy E_{max} of the $\Delta\alpha$ maximum as a function of x is reported on figure 3. One can see that its variation is practically parallel to the E_{04} one, the (E_{04} - E_{max}) difference being of the order of 0.45 eV for all compositions. Combined XPS and SXS experiments on the same samples have shown that the E_{04} decrease upon alloying is essentially due to a shift of the top of the valence band, the bottom of the conduction band remaining at about the same energy (10). It can therefore be inferred that the $\Delta\alpha$ maximum is due to transitions from extended states in the valence band to a band of empty defect states located at about 0.45 eV from the bottom of the conduction band, related to the Ge atoms. Time-of-flight experiments performed on similarly prepared alloys also indicate the apparition of additional states just below the broadened conduction band tail (11). Our interpretation is in agreement with the model proposed recently for the density of gap states in a-SiGe : H alloys (3), the defect states being then associated with Ge dangling bonds.

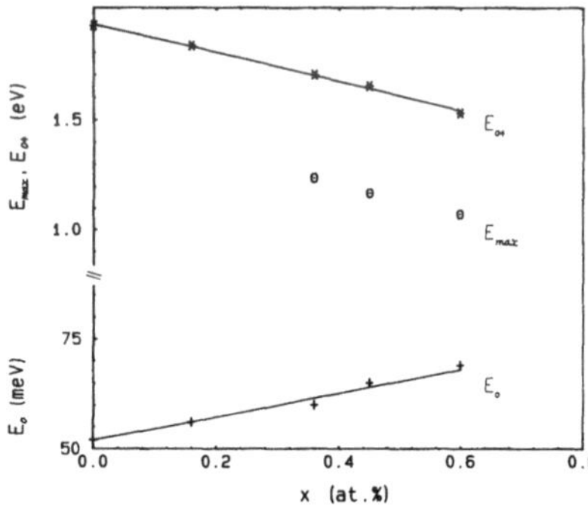

Figure 3 : Variations of the optical gap E_{04}, of the exponential edge inverse slope E_0 and of the energy of the $\Delta\alpha$ maximum E_{max} versus Ge concentration x.

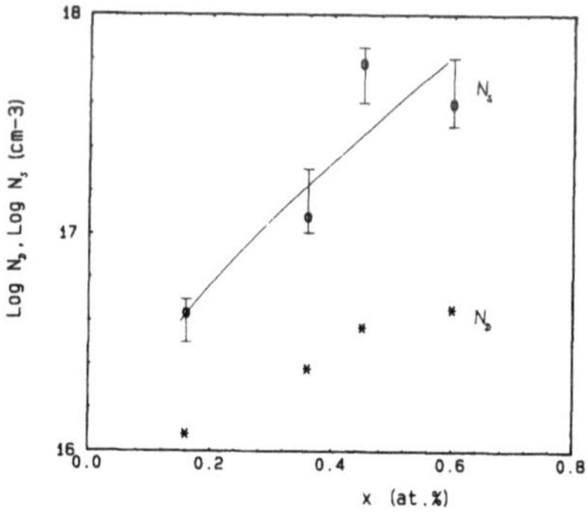

Figure 4 : Variations of the defect density deduced from optical absorption N_D and of the spin density N_S versus Ge concentration x.

The values of the spin density N_S, the g factor and the line width ΔH_{pp} deduced from ESR experiments performed on alloys prepared under identical conditions are reported in table 1. The observed signals correspond essentially to the Ge dangling bond line (g = 2.020 in a-Ge : H). The evolution of the g value with x is characteristic of the changes in the Ge environment. No signal corresponding to the Si dangling bond (g = 2.0055 in a-Si : H) could be detected even at x = 0.16, as expected in particular from the preferential attachment of hydrogen atoms to Si (12).

We focus now on the variation of the spin density N_S with Ge concentration, which is reported in figure 4 together with the N_D variation. The N_S values follow the same trends as in previous work (13, 14), but they are slightly lower. When comparing the density of paramagnetic centers given by N_S to the density of defects estimated from optical data, one can see that both increase with increasing x, the N_S variation being a little more rapid. It seems obvious that the two methods probe the same Ge-related defects. However, the N_S values are much higher than the N_D values, the discrepancy increasing with increasing x. Such an unexpected effect has already been reported (3, 14). It probably means that the optical cross section is different for transitions involving localized states related to defects on Si and Ge atoms respectively, and that the estimation of N_D from $\Delta\alpha$ according to the procedure proposed for a-Si : H must be questioned in the alloys.

One can notice that the exodiffusion experiments on the same samples show a drastic increase in the proportion of weakly-bonded hydrogen for $x \geqslant 0.36$. Such a change in the alloy microstructure, which roughly coïncides with a more rapid deterioration of their photoelectronic properties, cannot be detected in the variation of the disorder parameter E_0 but can be correlated with the apparition of a large density of Ge-related defect states in the optical data. The two phenomena are therefore linked and must strongly contribute to the poor quality of Ge-rich alloys.

ACKNOWLEDGEMENTS. This work has been supported in part by the CCE under contract EN3S-0062-F, and by the AFME and the PIRSEM-CNRS. Fruitful discussions with E. Bustarret are gratefully acknowledged.

REFERENCES

1. J. BULLOT, this conference.
2. K.D. MACKENZIE, J.R. EGGERT, D.J. LEOPOLD, Y.M. LI, S. LIN and W. PAUL, Phys. Rev. B31, 2198 (1985).
3. A. SKUMANICH, A. FROVA and N.M. AMER, Sol. St. Commun. 54, 597 (1985).
4. A.H. MAHAN, P. RABOISSON and R. TSU, Appl. Phys. Lett. 50, 335 (1987).
5. W. BEYER, J. Non-Cryst. Sol. 97-98, 1027 (1987).
6. A. MATSUDA, M. KOYAMA, N. IKUCHI, Y. IMANISHI and K. TANAKA, Jap. J. Appl. Phys. 25, L54 (1986).
7. G. SARDIN, C. ROCH, J. ANDREU, J.L. MORENZA, C. GODET, A. LLORET and J.P. STOQUERT, this conference.
8. J. DIXMIER, P. ELKAIM, M. CUNIOT, H. LABIDI, L. CHAHED, A. GHEORGHIU and M.L. THEYE, this conference.
9. W.B. JACKSON and N.M. AMER, Phys. Rev. B25, 5559 (1982).
10. C. SENEMAUD, I. ARDELEAN, L. CHAHED, A. GHEORGHIU and M.L. THEYE, this conference.
11. R. VANDERHAGHEN and C. LONGEAUD, this conference.
12. A. MORIMOTO, T. MIURA, M. KUMEDA and T. SHIMIZU, Jap. J. Appl. Phys. 20, L833 (1981).
13. F. FINGER, R. CARIUS, W. FUHS and A. SCHRIMPF, J. Non-Cryst. Sol. 77-78, 731 (1985).
14. M. KUMEDA, Y. JINNO and T. SHIMIZU, Phys. Stat. Sol. (b) 81, K71 (1977).

SURFACE–INDUCED EFFECTS ON THE STRUCTURE AND COMPOSITION OF THIN FILM a–Si p–i–n SOLAR CELLS

R.Martins,M.Santos,E.Fortunato,A.Maçarico,M.Lavado,I.Ferreira,L.Guimarães, *J.Gordon and *Y.Shapira
Faculty of Science and Technology of New University of Lisbon, Portugal.
*Faculty of Engineering, Tel–Aviv, Israel.

ABSTRACT

Surface effects induced by impurity atoms at ITO / p–i–n junction interfaces are analysed through AES and SIMS techniques and correlated with optoelectronic properties and density of states (DOS) of i-layers inferred from C–V/ field effect measurements. Through such analysis it is possible to determine deleuterious effects promoted by atoms interdiffusing towards the intrinsic layer and how to control such effects by using an a–Si:C:H interlayer to stop autodiffusion processes. The obtained results also indicate the appearance of a "dead layer" near the interface, mainly due to oxygen autodiffusion from the transparent conductive oxide layer, which can be well determined by a proper analysis of AES spectra. These results together with ways developed for improving structures performances will be presented.

1 . INTRODUCTION

The usefulness of a–Si:H/μc–Si:H and a–Si:C:H/μc–Si:C:H films for the production of solar cells depends on their capability to satisfy all cost, efficiency and stability requirements. Nevertheless, it is difficult to reconcile all these demands as far as conventional processes are concerned. For instance, thermal and light induce degradation depend on i–layer thickness, its structural, optoelectronic properties and on contamination of autodiffusion impurities [1,2]. Thermal diffusion from back side metal electrode also degradates cell performances by destroying the i/n junction. Device performances are also affected by surface states and finishing of p/i, i/n and n(p)/metal interfaces which control interface recombination processes. Reducing on surface roughness, transparency and conductivity of the transparent conductive oxide layer, TCO, after a brief exposure to a hydrogen plasma, will increase absorption and ohmic losses [3]. This also leaves free metal which conceivably can be incorporated into doped layer, degrading rapidly the diode junction [3]. All these difficulties are not only dependent on chemical properties of constituent elements but also by defects promoted by electron and ion bombardment during the deposition process.

In order to determine surface induced effects at junctions and how they are correlated with optoelectronic properties of films produced through plasma chemistry and deposition separation method (TCDDC system [4]), AES, SIMS and C–V measurements are performed on ITO/a–Si:C:H/a–Si:H and SS/n^+/a–Si:H structures. These results, together with optoelectronic properties presented by individual layers and ways developed for improving interdiffusion blocking in these structures will be discussed.

2 . EXPERIMENTAL RESULTS AND DETAILS

Films and structures under investigation have been produced by a TCDDC system as described elsewhere [4]. In order to analyse the interface behaviour on ITO/semiconductor junction as well as to determine the role of an a–Si:C:H undoped layer as a blocking layer against indium (In)or tin (Sn) interdiffusion, AES and SIMS measurements have been carried out on structures referred above. In figure 1 we represent the Auger sputtered profile obtained in a glass/ITO/a–Si:C:H/a–Si:H structure, while in figure 2 a SIMS profile obtained on glass/SnO_2/a–Si:C:H(0,B)/a–Si:H structures is shown. C–V and field effect measurements performed in similar structures are shown in figure 3 for inferring DOS. Dark conductivity (σ_d) measurements are carried out on the upper layer of structures using the gap cell configuration method to observe if electric properties of i–layer changes due to contamination by impurities autodiffusion. These results are shown in figure 4.

3 . ANALYSIS AND DISCUSSION

The analysis of AES spectrum and depth profile (fig.1) have been performed following [5] . The obtained results show three different regions concerning atoms profile. At a–Si:C:H/ITO region, we observe a fast rising up on profile distribution for In atoms, which it slows down as the ITO region is approached. Here, a constant concentration through all bulk region is obtained. Similar behaviour is observed for Sn atoms. On the other hand, the oxygen profile is shifted towards the a–Si:C:H layer. These results show a combination of a "natural" interface and diffusion of In (or other elements) into a–Si:C:H films, perturbed by surface roughness of ITO [6]. The profile shape for elements mentioned above suggest that oxygen from ITO layer diffuses towards a–Si:C:H films. In order to estimate the different contributions, we have to follow the changes in energy of In line shape [5], since In will be in its oxide state in the interface (In_1=396 eV and In_2=402 eV) whilst it will be in its metallic (elemental) state in the diffusion region (In_1=402 eV and In_2=408 eV). Thus, we estimate the real interface, the surface roughness and "diffused" width as being 200–300 Å, 500–750 Å and 40–60 Å, respectively.

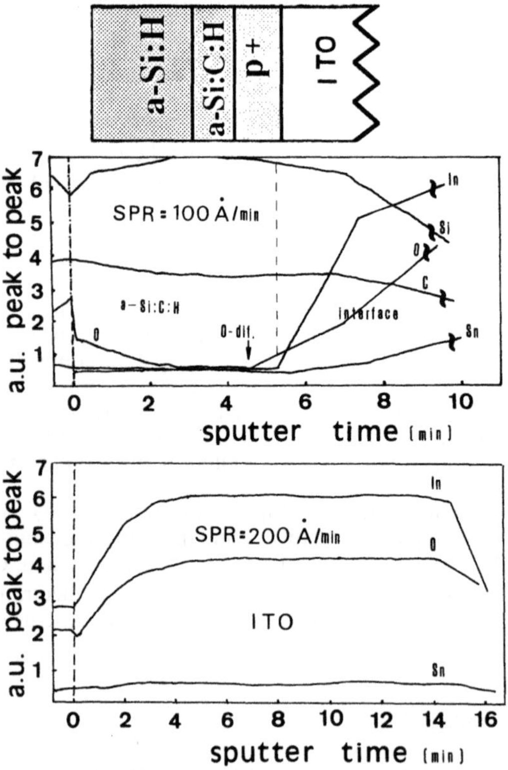

Fig.1—Auger sputtered profile of an ITO/a-Si:C:H structure.

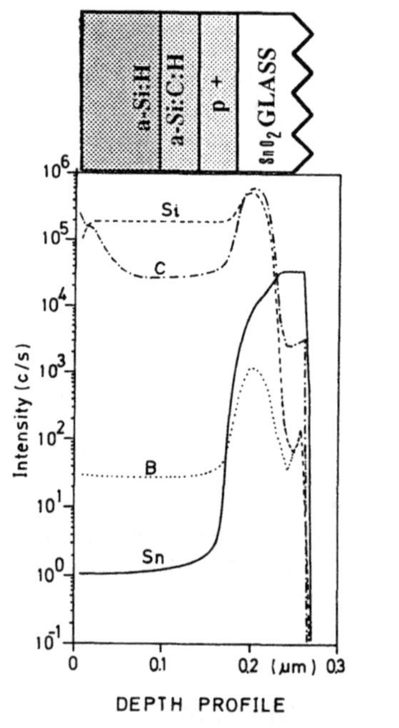

Fig.2— SIMS profile of SnO_2 / a-Si: C(0;1%B) / a-Si:H structure.

In figure 2 a typical SIMS depth profile is shown in order to observe the interdiffusion of Sn and boron atoms towards a–Si:H layer and also to observe how efficient is an a–Si:C:H interlayer for blocking impurity autodiffusion. The results show a steep Sn profile behind the a–Si:C:H interlayer whilst a light diffusion is observed in the i–layer. This means that Sn atoms are chemically trapped at this interface. Concerning boron profile, the contamination level observed within the a–Si:H layer is below 10 ppm, which proves again, that interlayer reduces impurity contaminations of i–layer.

C–V / field effect measurements performed in similar structures than those ones referred above (fig.3), were analysed following [7,8]. The results obtained reveal a discrepancy between DOS inferred through the two techniques. While the former one show $g(E_F) \sim 10^{15}$ $eV^{-1}cm^{-3}$ the second one shows $g(E_F) \sim 5.10^{16}$ $eV^{-1}cm^{-3}$. Such disagreement was expected [7] since DOS determined by field effect measurements are over estimated. Thus we can conclude that films under investigation have DOS, at most of 10^{16} $eV^{-1}cm^{-3}$.

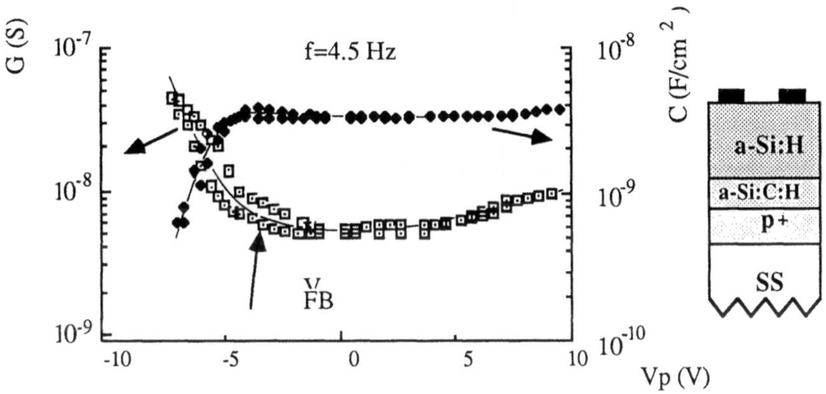

Fig.3— C–V measurements performed on metal/a–Si:C:H/a–Si:H MIS structures.

In figure 4 we show the plot of σ_D versus $10^3/T$ performed on different structures (structure 1 has an a–Si:C:H layer based on CH_4:SiH_4 mixtures in the proportion of 1:5 while in structure 2 a 2: 5 proportion was used). The analysis of obtained data show that σ_D increases, due to impurity contamination when a–Si:C:H interlayer (structure 2) has not the required composition to work as an effective barrier. The same measurements performed on a–Si:H films chemical etched at a constant rate show that σ_D is uniform through the bulk of the material, showing that homogeneous films are produced by the preparation method used.

4 . CONCLUSIONS

The experimental results indicate that surface induced effects in a–Si:H ITO[SnO_2(F)]/p–i–n junctions can be well determined through Auger techniques, whilst SIMS techniques shows the behaviour of impurity autodiffusion towards i–layer. From such results we conclude that the tails of the real interface (with a width of the order of 200–250 Å) are ascribed with outwards impurity diffusion in extensions of 40–50 Å. ITO layer surface roughness of the order of 800 Å was determined from AES depth profile, confirming the existence of a textured surface in agreement with X–ray diffraction measurements [6]. From such measurements a "dead layer" ascribed with interface is inferred which is responsible for degradation of structure composition and increase on recombination processes that takes place near interfaces.

a–Si:C:H undoped interlayers prove to be effective in stopping interdiffusion processes, since no significant changes on i–layer properties were detected (the degree of contamination of this layer by boron atoms, determined through SIMS profile is below 10 ppm, while in structures without such interlayer it is one order of magnitude higher).

Finally we would like to mention that p–i–n solar cells based on such structures and using μc–Si:C:H as doped layers, present open circuit voltages higher than 1 V [9], indicating that the proposed structure and preparation method are quite suitable to improve solar cell performances.

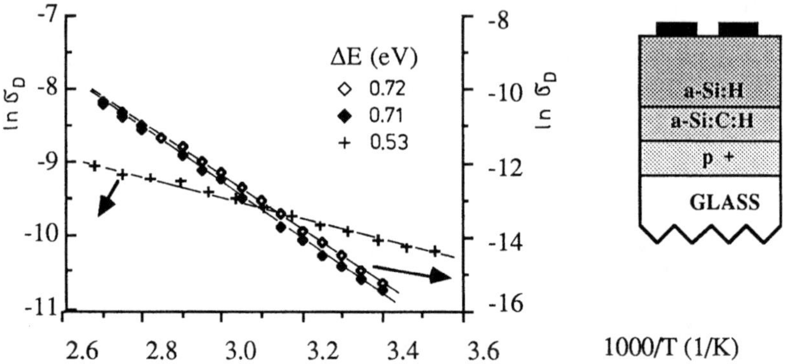

Fig.4— Represents σ_D as a function of 1000 /T performed on structure 1 (◆ up ◇down) and structure 2 (+ down).

ACKNOWLEDGEMENTS

The authors are grateful to Dr. G. Willeck (IMEC) for the help and contribution given on film's characterization. Thanks are also due to N. Carvalho, M. Vieira, L. Rodrigues and A. Branquinho for the help during the realization of this work.Thanks are also due to J. Bregman for the helpful discussions concerning the interpretation of the obtained results. This work was supported by U.S.A.I.D.–C.D.R. Program (Contract ψ7–178) and by Instituto Nacional de Investigação Científica (INIC)–Project B.

REFERENCES

[1] – Y.Tawada et al., *Proc. of Inter. Soc. for Optical Eng.Conf.*,Cambridge, USA (1986)
[2] – H.Yamagishi et al., *Proc. of 7th E.C.P.V.S.E.Conf.* , Sevilha , Spain (1986)
[3] – A.H.Mahan et al., *Appl. Phys. Lett.*, **44** (1984) 220
[4] – R.Martins et al., *Proc. of 5th E.C.P.V.S.E. Conf.*, Athens, Greece (1983)
[5] – J.Bregman et al., *J.Vac.Sci.Technol. B* **3**, 4 (1985) 959 ; *J.Vac.Sci.Technol.B* **4**, 5 (1986) 1195 ; *Appl. Phys. Lett.* **46**, 1 (1985) 48
[6] – C.Carvalho et al., *Presented at this Conference*
[7] – T.Suzuki et al., *Jap.J.of Appl. Phys.* **21** (1982) Suppl.21–2, 213
[8] – A.Madan et al., *Proc. of the 7th ICALS* , Edinburgh (1977) 377
[9] – R.Martins et al., *MRS Spring Meeting,* Reno, USA (1988)

FABRICATION AND STABILITY OF a-Si SINGLE AND TANDEM TYPE SOLAR CELLS

K.Tsuge, H.Yamagishi, K.Asaoka, K.Kobayashi, H.Nishio, W.A.Nevin and Y.Tawada
Central Research Laboratories, Kanegafuchi Chemical Industry Co. Ltd., 2-80, 1-chome, Yoshida-cho, Hyogo-ku, Kobe 652, Japan

ABSTRACT. Developments in fabrication and materials technologies for amorphous silicon solar cells have resulted in conversion efficiencies of 11.2% for 1 sq. cm single and tandem type solar cells. Thermal and light-induced degradation has also been investigated for these cells. It is found that our heat-resisting sub-module is expected to degrade by only 10% over a period of 40,000 h at 120°C. The light-induced degradation of tandem type cells is discussed. Our heat-resisting tandem cells showed no degradation during light exposure for 1,500 h at a cell temperature of 90°C.

1. INTRODUCTION

Research efforts in amorphous silicon solar cells are focused on the improvement of both conversion efficiency and stability. Concerning the former, conversion efficiencies of more than 11% for single type solar cells and more than 10% for two-stacked a-Si/a-Si cells have been reported (1),(2). With respect to the latter, light-induced degradation is thought to be the most important problem. Intensive research is therefore being continued in order to understand the mechanism of the Staebler-Wronski effect and solve its problem. In spite of this, the exact details of the light-induced degradation mechanism are not clear. However, practical methods for decreasing this degradation have been proposed, and recently, tandem cells with higher stability have been reported (3),(4).

In this paper, we first present improvements in the conversion efficiencies of single and tandem type solar cells on the basis of improvements in the fabrication technology. Secondly, thermal stability of the solar cells is discussed and progress in this area is presented. Finally, we discuss the light-induced degradation of tandem type cells having high efficiencies.

2. FABRICATION OF SINGLE TYPE SOLAR CELL

All solar cells were prepared by a three-separated chamber system with turbomolecular pumps. Basic structure of the single type cell was glass/tin oxide/p (a-SiC:H)/i (a-Si:H)/n (μc-Si:H)/metal.

Recently single chamber systems have been developed by several groups(5),(6). It appears that each system has its own method for avoiding contamination by impurities. We have found that contamination by impurities in the i layer decreased cell performance even in our separate chamber system. We have therefore developed techniques for operation and maintenance of the separate chamber system in order to decrease the contamination. Figure 1 shows some results concerning the profiles of impurities in n/i/p layers obtained by an IMA.

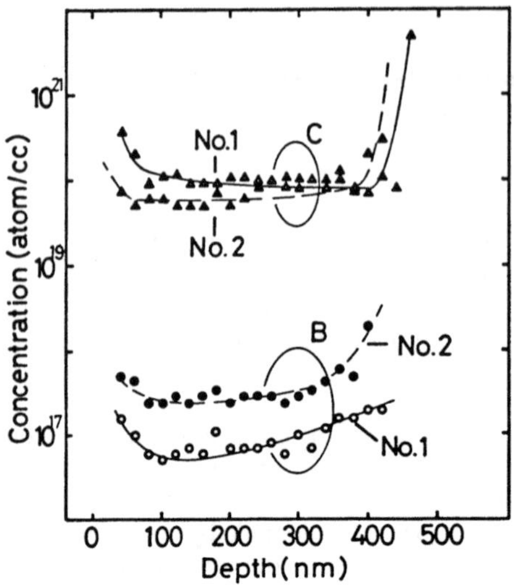

Fig.1 IMA profiles of impurities in n/i/p layers.

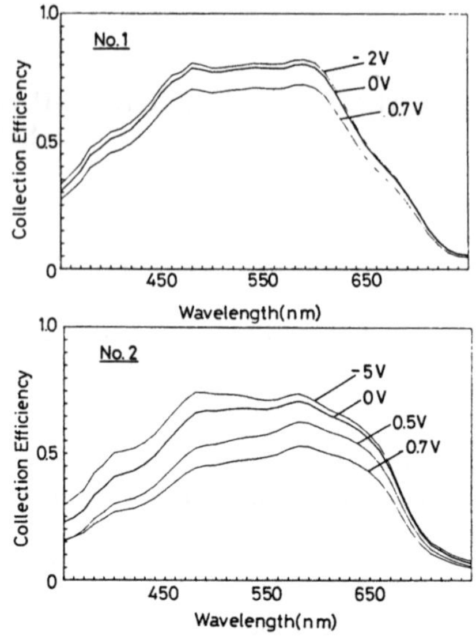

Fig.2 Spectral response of the No.1 and No.2 cells at various dc biases.

The No.1 cell was prepared using the techniques developed and the No.2 cell was prepared so that the No.1 and No.2 cells had the same concentrations of carbon and oxygen in each i layer. Concentration profiles of oxygen are not shown in Fig.1, because absolute values of oxygen concentration had some uncertainty. However, the ratios of ion intensity (O/Si) at each measurement point in these two cells were almost the same. As observed in Fig.1, the boron concentration of the No.1 cell was lower by about an order of magnitude than that of the No.2 cell.

Spectral responses of the No.1 and No.2 cells are shown for different dc biases in Fig.2. The No.2 cell has a larger dependence of spectral response on dc bias than the No.1 cell and this dependence is larger in the short wavelength region. The solar cell performance of each cell (1 cm^2) under AM-1, 100 mW/cm^2 illumination is compared in Table I.

Table I. Solar cell performance of No.1 and No.2 cells

Cell	Voc (volts)	Jsc (mA/cm^2)	FF (%)	Eff. (%)
No.1	0.892	16.9	71.0	10.7
No.2	0.872	15.0	65.0	8.50

As indicated in Table I, the No.2 cell showed lower performance especially in Jsc and FF. This result coincides with the difference in spectral response and its dependence on dc bias, shown in Fig.2.

From these data it is found that boron concentration in the i layer has a large influence on the solar cell performance. A decrease in $\mu\tau$ product of electrons by the contamination of boron seems to occur within the i layer. Contamination of oxygen and moisture is also found to affect the FF of the solar cell resulting in deterioration of its efficiency. It is therefore concluded that there remains a problem of contamination even in a separate chamber system. On the basis of this understanding, we optimized the design parameters of the single type cell, the deposition conditions of each layer and the configuration of tin oxide to obtain a highest efficiency of 11.2%.

3. FABRICATION OF TANDEM TYPE SOLAR CELL

All tandem type cells were prepared by the same system as single type cells. The basic configuration of the tandem type cell was the same as the single type cell except for the thickness of each layer, and was a two stack type in which the i layers of both top and bottom cell were composed of the same material, that is, a-Si:H. Impurities were decreased by using the same methods as for the single type cell.

In order to achieve a high efficiency, we are now optimizing the tandem type cell for the same parameters as the single type cell, stated above. Figure 3 shows the dependence of solar cell properties on the top cell i layer thickness, keeping the bottom cell i layer thickness constant at 5000Å. The dimensions of each tandem type cell were 1x1 cm^2. We have obtained an efficiency of 10.8% with a blocking layer at the n/p interface and 11.2% without it. V-I characteristics of these cells are shown in Fig.4. Textured tin oxide and graded layers of both boron and carbon concentration at the two p/i interfaces were used for these cells.

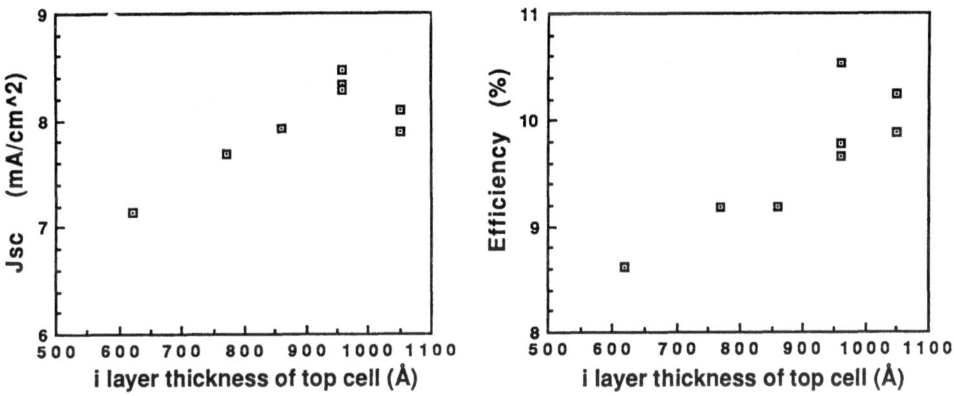

Fig.3 Dependence of solar cell properties on the top cell i layer thickness. Thickness of the bottom cell i layer was 5000Å.

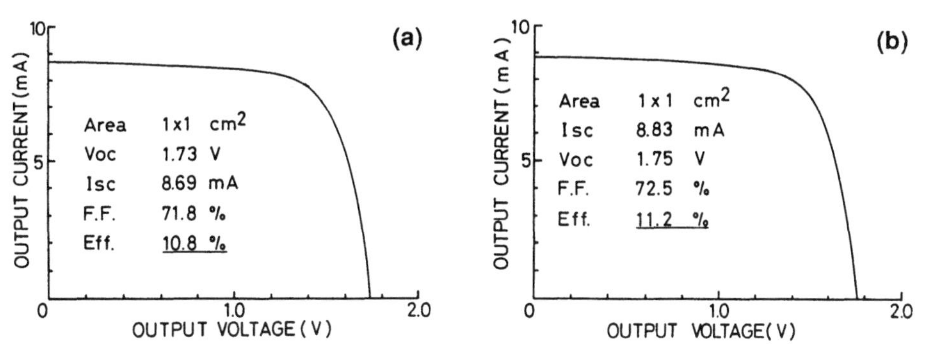

Fig.4 V-I characteristics of a-Si/a-Si tandem type cells: (a) with, and (b) without a blocking layer at the n/p interface.

4. THERMAL STABILITY

There are two problems reported so far concerning thermal degradation (7), (8). One is the diffusion of aluminum which is used as the back electrode of a conventional solar cell (7) or the diffusion of silicon into the aluminum electrode (8) with rise of cell temperature. Another is observed only in a monolithically connected solar cell. Contact resistance at the interface between the tin oxide and aluminum electrode increases as cell temperature becomes higher, which results in the degradation of the solar cell performance (8).

On the other hand, Willing et al. have reported that although these phenomena were observed at high cell temperatures, the solar cell performance was not affected at the cell temperatures experienced in real terrestrial use (9). In addition, it was also shown that silicon diffused into the aluminum electrode by heat treatment.

Concerning the problem of diffusion, Eser et al. have reported that they had developed a special aluminum deposition technique which gave a high degree of thermal stability to the solar cells (8). Alternatively, we have solved this problem by introducing a blocking layer at the n/aluminum interface (3). Recently, we have also found that the thermal stability of the solar cell depends to a large extent on the deposition system of aluminum. Figure 5 shows the difference in thermal stability between samples obtained by means of a vacuum evaporation system and a magnetron sputtering one. As indicated in Fig.5, the magnetron sputtering system gives the solar cell a high degree of thermal stability, while the solar cell fabricated by vacuum evaporation is thermally unstable. It is also found that moisture in the sputtering chamber has an influence on the degree of thermal stability, tending to decrease it.

With respect to the problem at the interconnect region, we measured the change of contact resistance (Rc) at this region after an anneal. Rates of change in Rc at the annealing temperature of 180°C (at the treatment time of 100 h) are compared in Table II.

Table II. Rate of change in Rc at the annealing temperature of 180°C

Electrode	Condition of laser-scribing	Rate (ohm/h)
Al	Multi-mode	1.2×10^{-1}
Al	Single-mode	5.6×10^{-2}
Blocking layer/Al	Single-mode	5.5×10^{-3}

From this table, it is clear that the rate of increase in Rc depends on the conditions used for laser-scribing of the amorphous layers and that the insertion of a blocking layer at the tin oxide/aluminum interface can suppress a rise of Rc even at 180°C.

The origin of the increase in Rc was also investigated. Table III shows the free energies of oxide formation to the equilibrium state, which suggest the ease of oxidizing a material by tin oxide although they give us no information on the rate of oxidation.

Table III. Free energy of oxide formation at 150°C

Electrode	Free energy (kcal/mol)
Al	-195
Ag	177
Blocking layer/Al	-68

It is suggested from this table that aluminum attached directly to tin oxide may be easily oxidized by the tin oxide. In order to confirm this experimentally, we analysed aluminum and oxygen near the interface by means of a surface analysis technique. However, differences of the oxygen content in aluminum near the interface and the chemical shift related to oxygen

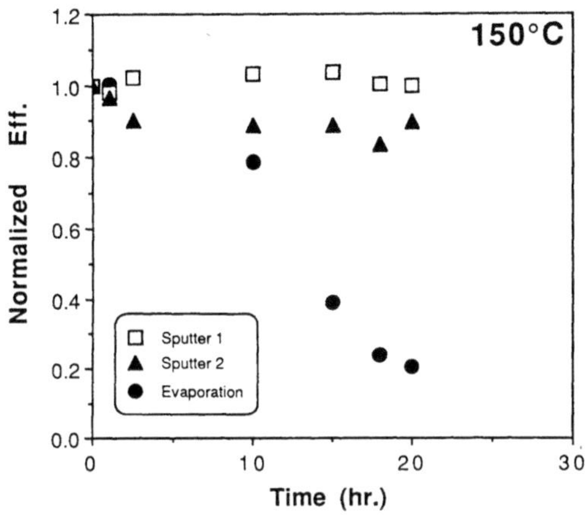

Fig.5 Variation of efficiency with time at 150°C for samples prepared by vacuum evaporation and magnetron sputtering.

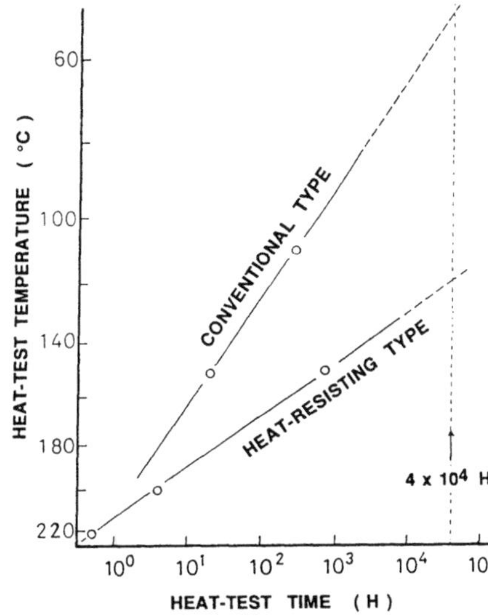

Fig.6 Overall heat-test for sub-modules of the conventional type and heat-resisting type. Heat-test time until 10% degradation of out-put at each heat-test temperature is plotted.

configuration could not be observed. We think this was caused by the oxidation of samples during their preparation and the knock-on of oxygen during sputtering. We therefore compared etching rates of aluminum in 5% NaOH solution. The sample without a blocking layer had twice the etching rate of aluminum as that with a thin blocking layer.

From these data, it is thought that oxidation of aluminum by tin oxide causes a rise of Rc, which decreases the out-put of the solar cell.

We have evaluated the overall heat-resisting temperature of sub-modules. Figure 6 shows the results of heat-tests made for the conventional type in which the back electrode is composed of only aluminum and the heat-resisting type in which a blocking layer is inserted between the tin oxide or n layer and aluminum. These two types consist of monolithically interconnected solar cells. In this figure, the heat-test time until 10% degradation of out-put at each heat-test temperature is plotted. Choosing the heat-resisting temperature as that which gives 10% degradation during 40,000 h, it is estimated as less than 60°C for the conventional type, and about 120°C for the heat-resisting type.

5. LIGHT-INDUCED DEGRADATION

It is now widely recognized that the tandem type solar cell has an advantage over the single type cell with respect to the light-induced degradation. Tandem type solar cells are expected to be promising candidates for terrestrial use. Tawada et al. had reported a stable tandem cell with a blocking layer at the n/p interface (3), while Hiroe et al. have discussed the role of its blocking layer (10).

However, the reason for its good stability is not clearly understood and there may be another possible explanation for this. It may be considered that the tandem type solar cells evaluated in these studies had efficiencies of about 8% or less and that they might not be well-refined. From this viewpoint, we made light exposure tests for a series of samples which had the basic structure mentioned in 3 and efficiencies of around 10%. A typical result is shown in Fig.7, which indicates the dependence on the blocking layer thickness. Figure 8 shows the initial stage of the light-induced degradation of tandem cells fabricated without a blocking layer at the n/p interface. In Figs.7 and 8, light-induced degradation is observed and is found to reach 8% after about 500 h exposure to 100 mW/cm^2 intensity light even in the sample having a 50Å blocking layer. It is therefore concluded that although the introduction of a blocking layer between the n and p layers is effective, it is not sufficient to solve the problem of light-induced degradation.

We are now carring out research to improve the stability of the i layer to prolonged light exposure, for example, by means of hydrogen plasma treatment (11).

It is also found that our tandem cell retains its initial efficiency even after a light exposure time of 1,500 h at the cell temperature of 90°C. Since our tandem cell also has thermal stability, it can be preferably used for automobile and roof tile applications. In these applications, the module temperature sometimes becomes high enough to recover the light-induced degradation to a considerable extent.

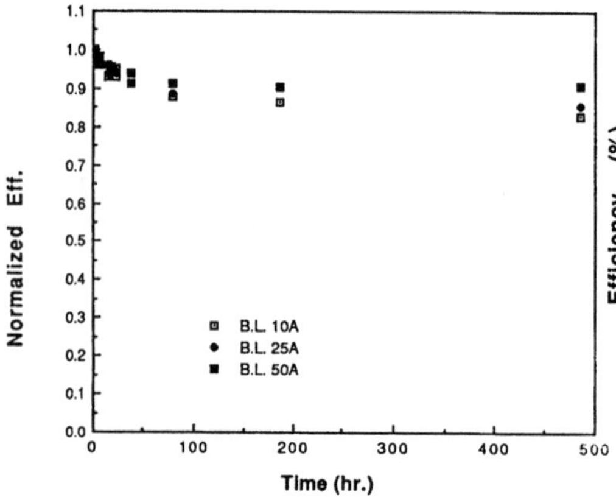

Fig.7 Dependence of light-induced degradation of a-Si/a-Si tandem type cells on the thickness of a blocking layer inserted at the n/p interface.

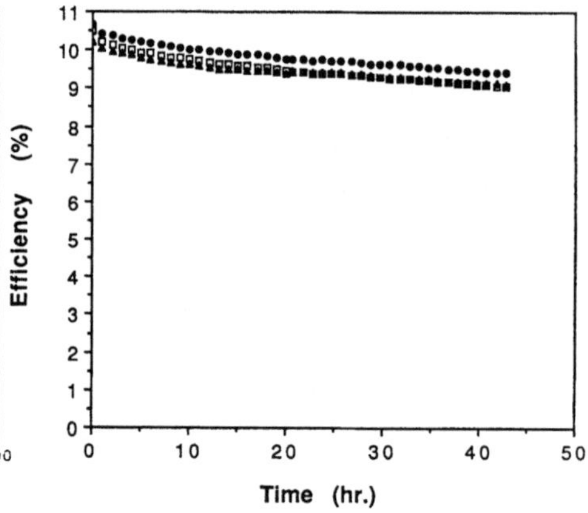

Fig.8 Initial stage of light-induced degradation of tandem type cells without a blocking layer at the n/p interface.

6. CONCLUSIONS

Fabrication and materials technologies were developed for the improvement of the conversion efficiency and stability of amorphous silicon solar cells.

We obtained 11.2% efficiency for 1cm^2 single and tandem type solar cells by decreasing the contamination and optimizing the design parameters, the deposition conditions and the tin oxide configuration.

We developed heat-resisting solar cells and sub-modules by introducing a blocking layer at the n/aluminum and tin oxide/aluminum interfaces.

It is clear that although the insertion of a blocking layer between the n and p layers is effective, it is not sufficient to solve the problem of light-induced degradation for the tandem type solar cells having high efficiency. However, at a cell temperature of 90°C, our tandem type solar cells showed no degradation during a light exposure time of 1,500 h.

ACKNOWLEDGEMENT

We are grateful to Prof. Y.Hamakawa of Osaka Univ. for his advice and to our colleagues for their helpful discussion and support.

This work was partially supported by the New Energy Development Organization as a part of the Sunshine Project.

REFERENCES

(1) J.L.Stone, *Tech. Dig. 3rd Int. Photovoltaic Science and Engineering Conf.*, 161 (1987)
(2) Y.Hamakawa, *Tech. Dig. 3rd Int. Photovoltaic Science and Engineerimg Conf.*, 147 (1987)
(3) Y.Tawada, J.Takada, N.Fukada, M.Yamaguchi, H.Yamagishi, M.Kondo, Y.Hosokawa, K.Tsuge, T.Nakayama and I.Hatano, *Proc. 18th IEEE Photovoltaic Specialists Conf.*, 1712 (1985)
(4) Y.Uchida, *Proc. 7th European Photovoltaic Solar Energy Conf.*, 395 (1986)
(5) M.Bohm, A.E.Delahoy, F.B.Ellis, E.Eser, S.C.Gau, F.J.Kampas, Z.Kiss, K.R.Ramaprasad, H.Volltrauer and H.Weakliem, *Proc. 18th IEEE Photovoltaic Specialists Conf.*, 888 (1985)
(6) C.R.Dickson, J.Pickens and A.Wilczynski, *Solar Cells*, **19**, 179 (1986-1987)
(7) N.Fukada, J.Takada, M.Yamaguchi, K.Tsuge and Y.Tawada, *Tech. Dig. 1st Int. Photovoltaic Science and Engineering Conf.*, 229 (1984)
(8) E.Eser, K.R.Ramaprasad, H.Volltrauer, F.Ramos, S.C.Gau and R.Vos, *Solar Cells*, **21**, 25 (1987)
(9) F.Willing, M. Bennett and J.Newton, *Proc. 19th IEEE Photovoltaic Specialists Conf.*, 1086 (1987)
(10) A.Hiroe, H.Yamagishi, H.Nishio, M.Kondo and Y.Tawada, *Proc. 19th IEEE Photovoltaic Specialists Conf.*, 1111 (1987)
(11) to be published

ON THE INFLUENCE OF AN EXTERNAL DC SUBSTRATE BIAS ON BORON DOPING IN AMORPHOUS SILICON CARBIDE

F. Alvarez, A. Gobbi, F.C. Marques and I. Chambouleyron
Instituto de Fisica, UNICAMP, CP 6165, CAMPINAS, SP, 13081, BRAZIL

ABSTRACT. Past works shows that a DC substrate polarization influences the impurities incorporation in hydrogenated amorphous silicon layers deposited by RF glow discharge. In the present paper these previous results are extended to the case of amorphous silicon carbide films, the final objective being the optimization of the deposition conditions of the window layers in solar cells heterostructures.

1. INTRODUCTION

In the past years considerable effort has been devoted to study the properties of amorphous silicon carbide alloys. Its band gap controllability, doping and photoconductivity properties are unique for its use as a window material in amorphous silicon solar cells heterostructures (1). It is also well documented that the optoelectronic properties of a-Si:C:H films depend strongly on the preparation conditions. However, a unified opinion on which plasma discharge conditions would produce the best material is not yet existent. The substrate temperature and the RF power density (defined as RF power/chamber pressure) are among the most important parameters controlling the film quality (2). The self polarization of the powered electrode is not generally controlled in an asymmetric capacitively glow discharge system, so far the most popular method used in the deposition of amorphous semiconductors for solar cells applications. The ion bombardment of the film, as well as the change of the plasma sheath, are then leaved as uncontrolled parameters. Nevertheless, using a suitable fixed electrode area and an external DC power supply, it is possible to increase (positive) or decrease (negative) the DC potential of the powered electrode and thus change the plasma-substrate potential (3).

In this work, the influence of several deposition parameters on the optoelectronic properties of amorphous silicon carbide were studied. Special emphasis was done on the (photo) conductivity dependence on substrate bias. The methodology follows. First, the optical gap of the material grown onto grounded substrates was investigated as a function of the $(CH_4)/(SiH_4 + CH_4 + Ar)$ gaseous ratio. The use of Ar could not be avoided because the available doping impurities were argon diluted. Second, a material having an optical band-gap of 2 eV was chosen. Its dark and (photo) conductivity was afterwards optimized as a function of the diborane content in the gaseous mixture. Finally, the best boron doped material was deposited under different DC substrate biases.

2. EXPERIMENTAL RESULTS AND DISCUSSION

Figure 1 represents the optical gap (determined from a Tauc's type plot) as a function of the gaseous mixture. The methane + silane gaseous mixture flow was kept at approximately 7 sccm during the different runs. As already mentioned, the use of argon could not be avoided when preparing doped films. In the case of undoped material a constant argon flux was

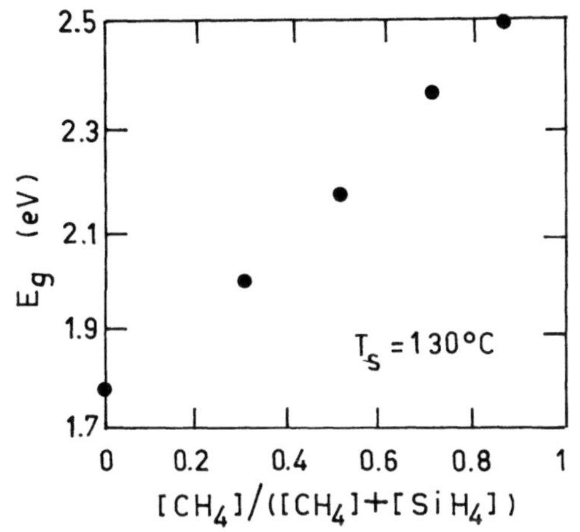

Figure 1. Optical gap vs. the ratio of methane/(methane + silane) in the deposition chamber. The (methane + silane) total flux was kept constant (7 sccm) for the different runs. A constant argon flux (5 sccm) was added to the mixture.

added to the (CH_4 + SiH_4) mixture in order to guarantee plasma conditions similar to those of the doped samples. In this way, excepting their impurity content, the materials will have close composition. The experimental dots of figure 1 correspond to samples grown onto grounded substrates (Corning glass 7059) at 130°C, total flux = 12 sccm, reactor pressure = 50 Pa and an RF power of 2.5 W. The optical gap increases motonically up to 2.5 eV. The deposition rate however, slowly decreases up to a methane to (methane + silane) ratio of 0.7. For higher ratios it drops drastically. This experimental finding is explained assuming that the growth limiting species is silicon. The above results confirm that the incorporation of carbon has the primary effect of widening the band-gap. In our samples however, infra-red spectra (not displayed) indicate that samples having higher carbon content posses a much higher hydrogen concentration. As a consequence, it becomes difficult to quantify the respective carbon and hydrogen contributions to the band-gap widening effect. In figure 2 we have represented the (photo) conductivity against the diborane/(silane + methane) gaseous ratio for an alloy having a band-gap of 2 eV. The photoconductivity was measured in a planar geometry under 100 mW/cm^2 illumination from a mercury-halogen lamp (ELH). The maximum photoconductivity was found to happen for gaseous ratios in the range 10^{-4} and 10^{-3}. This is an indication that too many boron atoms in the discharge increase the density of gap states in the deposited material. It is important to note that in amorphous silicon films, on the contrary, even larger amounts of boron do not produce this effect.

The dependence of the (photo) conductivity of films deposited with varying substrate DC biases is represented in figure 3. We remark the fact that "positive substrate bias" does not necessarily mean that the substrate-plasma potential (being more precise we should say the "potential mean value") is positive (4). For positive DC biases the substrate will only be "less negative" with respect to the plasma potential. Previous results show that, in hydrogenated amorphous silicon films, a negative (positive) DC substrate bias improves the incorporation of boron (phosphorous) (3,4). The results of figure 3 indicate, however, an opposite trend. Figure 2 may help to understand this peculiar behavious. As already mentioned large amounts of boron atoms in silicon carbide alloys increase the density of gap defects, degradating the photoconductivity. Assuming that a negative bias enhances the incorporation of boron atoms in the material (as it happens in a-Si:H), a concomitant increasing density of defects is being generated. Thus, the two opposite effects might mutually cancel giving no conductivity dependence on negative substrate biases.

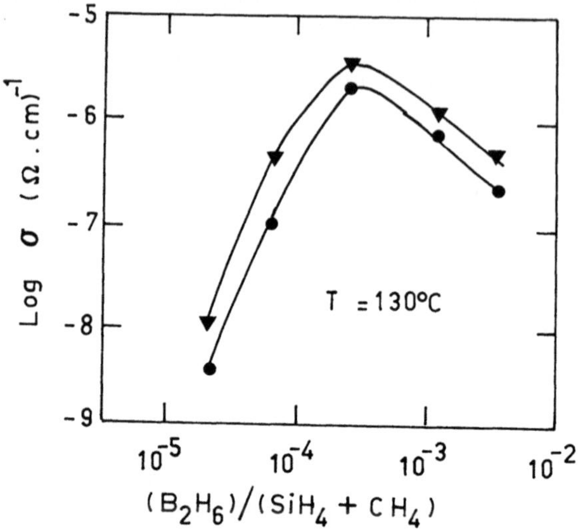

Figure 2. Dark and photo conductivity vs. the ratio of the dopant gas flux to (silane + methane).

Conversely, for positive bias substrate, the boron incorporation is prevented and a less defective material is obtained. Although smaller amounts of boron are incorporated, they act in a more effective way due to the relative low defect density.

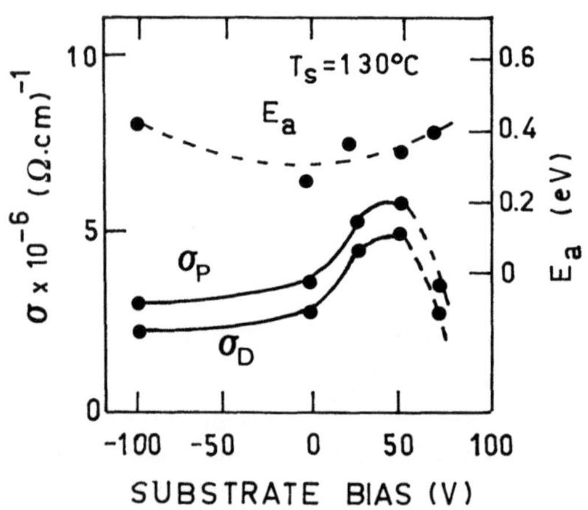

Figure 3. Conductivity and activation energy vs. DC substrate bias for boron doped amorphous silicon carbide films.

An alternative explanation could be found arguing that amorphous silicon carbide is very sensitive to ion bombardment. In this picture, a negative bias will produce severe ion damage in the material. This point of view agrees with the results of Mahan et al. (2). These authors have shown that due to ion bombardment, higher power densities (decreasing the reactor pressure) produce low quality intrinsic amorphous silicon carbide (higher density of midgap states and less photoconductivity). This is so because lowering the chamber pressure during the plasma discharge increases the self-polarization of the powered electrode, which is equivalent to increase the plasma substrate potential. For positive bias substrate, the opposite effect should be expected. However, for a substrate potential varying between 0 to 50 V the substrate plasma potential changes 5 volts aproximately. Incidentally, this is an expected result, since the plasma potential (due to the large mobility differences between ions and electrons) is always positive with respect to any potential inserted in the system. Nevertheless, for large positive substrate biases, the system becomes instable and the conditions for a controllable film deposition are no more met. In figure 3 the observed conductivity decreases for large positive substrate biased, could be attributed to such instabilities.

3. SUMMARY AND CONCLUSIONS

In agreement with other authors, the optical gap of amorphous silicon carbide films deposited onto grounded substrates, increases monotonically with the methane/silane gaseous ratio up to 2.5 eV. The incorporation of hydrogen however, is also increased (as determined by the integrated strengh of the 2000 cm^{-1} absorption band). It has proven to be very difficult in this stage to evaluate the separate contribution of carbon and hydrogen to the optical gap widening mechanism.

An appropriate positive bias on the substrate during the material growth improves the (photo) conductivity of boron doped amorphous silicon carbide films of 2 eV optical gap. A negative bias substrate does not produces changes (within the experimental errors) in the (photo) conductivity. These results are interpreted assuming, either that a negative bias substrate enhance the boron incorporation increasing the defect density, or that a strong ion bombardment damage is produced by the negative substrate bias, with a consequent degradation of the material properties. An appropriate positive substrate bias on the contrary, produce a better material because of the reduction of the substrate-plasma potential and a concomitant ion bombardment decrease. The fewer boron atoms incorporated in the network become in fact a more effective dopant. The conclusionsof the present work agree well with existing results on the influence of the deposition conditions (gaseous mixture, partial pressure and RF power) on the boron doping of amorphous silicon carbide films.

ACKNOWLEDGMENTS

The authors are indebted to R. Koropecki, M. Battioni and R. Buitrago, INTEC, Argentina, for their help with the optical measurements. This work was partially supported by TWAS, ICTP, Trieste, Italy, RG86-15, UNICAMP-FAP 74-87 and FAPESP, Proc. 87-0627-7, Brazil.

REFERENCES

1.- Y. Tawada, K. Tsuge, M. Kondo, K. Nishimura, H. Okamoto and Y. Hamakawa, 4th E.C. Photovoltaic Solar Energy Conf., Reidel, Holland, 698 (1982).
2.- A.H. Mahan, B. von Roedern, D.L. Williamson and A. Madan, J. Appl. Phys., 57, 2717 (1985).
3.- F. Alvarez, I. Chambouleyron, A. Gobbi, C. Mendonça and F.L. Castro, J. of Non-Cryst. Sol., 77&78, 527 (1985).
4.- F. Alvarez, Appl. Phys. Letters, 47, 960 (1986).

EFFECT OF LINEAR PHOSPHORUS PROFILES AT THE P-I INTERFACE OF TCO/PIN/METAL a-Si:H SOLAR CELLS

H. Kausche, W. Kusian, W. Juergens
Siemens Research Laboratories
Otto-Hahn-Ring 6, D-8000 Munich 83

ABSTRACT. Amorphous silicon solar cells were fabricated in a large-area two-chamber deposition system by RF glow discharge. The blue response of these cells was improved by incorporating a linearly decreasing phosphorus doping profile into the i layer starting at the p/i interface.This improvement is due to a reduction of the surface recombination of electrons in the p/i space-charge barrier layer. Changing the initial doping concentration from 0 to 4 ppm and the profile length from 0 up to 160 nm, we found a maximum for the short-circuit current and efficiency (under AM1 illumination) as a function of the initial phosphine concentration. The maximum shifts to lower concentration values with increasing profile length.

1. INTRODUCTION

Amorphous silicon solar cells fabricated by a plasma enhanced CVD process usually have the structure: glass/TCO/p(C)in/metal. The deposition sequence starts with the p layer. The illumination direction in these cells is through the p layer. To ensure a good performance, the p/i junction must be carefully prepared. However, residual boron atoms incorporated into the first part of the i layer deteriorate the p/i junction, which results in a poor blue response of the cell. This problem occurs in single-chamber reactors and less seriously in large-area multichamber reactors. In single-chamber reactors the boron remaining on the inner surfaces is responsible for this effect. The dominant influence in large-area multichamber reactors is the degassing of boron from previously deposited layers, especially from the moving substrate holder /1/. This boron contamination has concentrations of a few ppm and decreases the efficiency /2/. One way of reducing the amount of entraining boron is, in addition to a pump down process, to flush the reactor with a reactive gas to reduce the residual diborane in the system /3/. Another way of increasing the blue response of solar cells of this type is to dope the i layer with small amounts of phosphorus /4/. But the disadvantage of this method is that the enhancement of the blue response takes place at the expense of the red response. To avoid this, we tried in this work to improve the p/i junction by compensating the entrained boron by a linearly decreasing doping profile with phosphorus, starting at the p/i interface.

2. PREPARATION

The solar cells used for this investigation were deposited in a large-area two-chamber diode deposition system with an electrode size of 40 cm x 40 cm. The sequence of the deposition was as follows: p layer deposition in chamber 1 followed by i layer deposition in chamber 2 and then an n layer deposition again in chamber 1. More detailed deposition parameters are given in /2/. The two chambers were cleaned after approximately 100 deposition runs so that the inner walls were covered with relatively thick amorphous silicon layers. Nevertheless, excellent vacuum conditions with small amounts of impurities were obtained. In spite of this, steeply decreasing boron profiles with initial concentrations of a few ppm at the p/i interface were estimated in the i layer by SIMS measurements. The upper part of Fig. 1 shows this schematically. To compensate this unintentional boron doping, we incorporated linearly decreasing doping profiles with phosphorus as indicated in the lower part of Fig. 1. The investigated profiles start with different concentrations c in the range from 0 to 4 ppm PH_3 in SiH_4 and different profile lengths w of 0, 40, 80 and 160 nm representing 0, 7, 14 and 28 % of the entire i layer thickness.

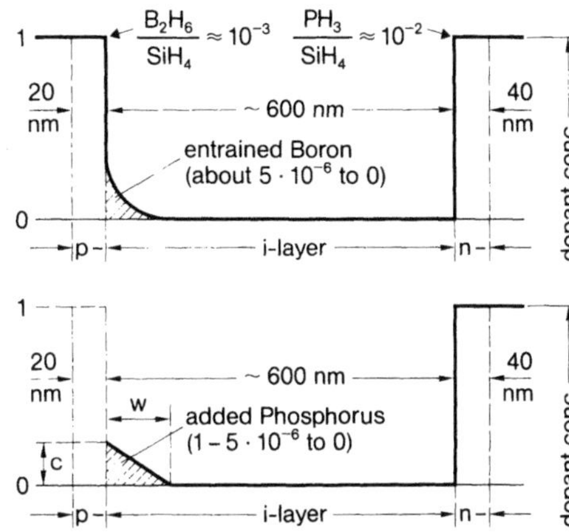

Fig.1: Dopant concentration across the pin-structure

3. EXPERIMENTAL RESULTS

At these solar cells, we measured the AM1 light characteristics and the voltage dependent spectral response as described in /5/. In Fig. 2 the parameters of the AM1 light I-V curves, i.e. efficiency η, short-circuit current I_{sc}, open-circuit voltage V_{oc} and fill factor FF are plotted versus the initial phosphorus concentration c for various profile lengths w. For a short profile length of w = 40 nm, V_{oc} and FF are independent of the initial concentration c (see Fig. 2a). With increasing w (see Fig. 2b and 2c) the open circuit voltage as well as the fill factor decrease and become functions falling linearly with c. On the other hand the short-circuit current exhibits a different dependence on c. I_{sc} shows a soft maximum behavior with initial concentration c. The maximum value shifts to lower c values as the profile length w increases. The short-circuit current dominates the efficiency η whose behavior as a function of c and w is thus quite similar.

To find out the nature of the improvement of I_{sc} by phosphorus doping, we also performed spectral response measurements. In Fig. 3 the external short-circuit (V = 0) collection efficiency Q versus wavelength λ is shown for various values of c. The profile length w was kept constant at a value of 40 nm. It is obvious that with increasing c the blue response of the solar cell is improved, while the red response remains unchanged. This is due to the compensating effect of the phosphorus, which increases the electron lifetime in the p/i space-charge barrier layer and therefore reduces the surface recombination at this junction /6/. To confirm this result, we also looked at the voltage dependence of the spectral photocurrents and the internal collection efficiency q, which is defined by $q(\lambda,V) := -Q(\lambda,V)/Q(\lambda,-2V)$. We estimated the flatband voltage V_f together with the internal flatband collection $q(V_f) = q_f$ as described in /7, 8/ and listed the values in Table I together with the cell parameters. It is seen

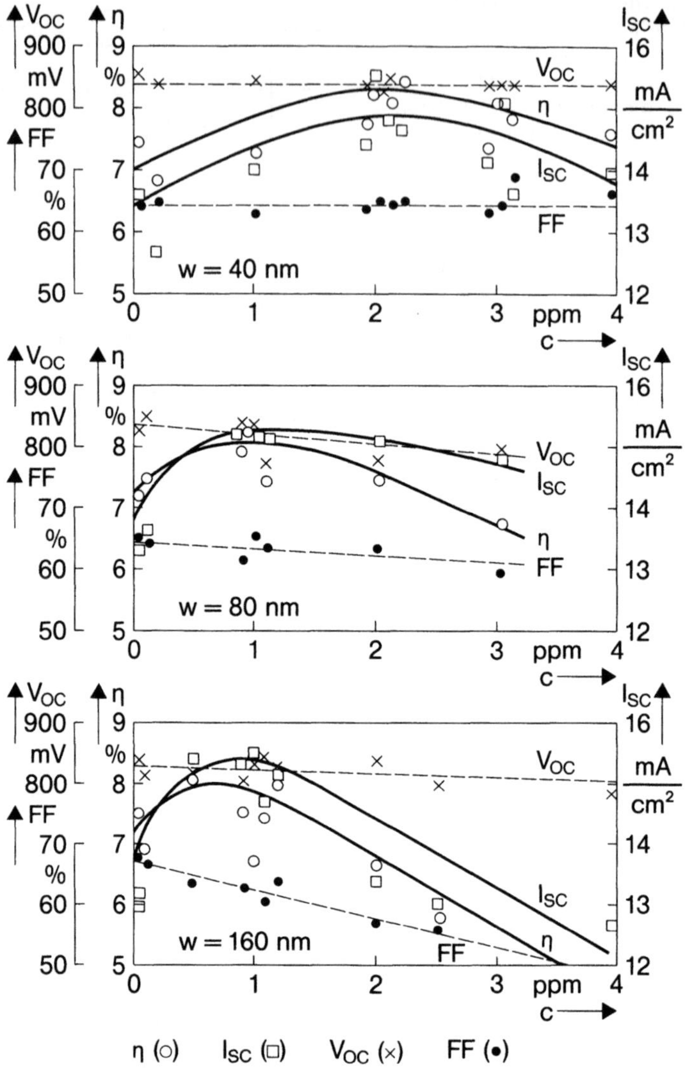

Fig. 2: Efficiency η, short-circuit current I_{sc}, open-circuit voltage V_{oc} and fill factor FF versus the initial phosphorus concentration c.

that in samples with phosphorus doping, the profile q_f is apparently higher than in the undoped sample, which indicates an improvement of the p/i junction with doping compensation.

An overcompensation of the boron entrained in the i layer leads to a decrease of the conversion efficiency. We therefore suggest, in line with /1/, that the most promising phosphorus doping profile has nearly the same shape as the unintentional boron profile.

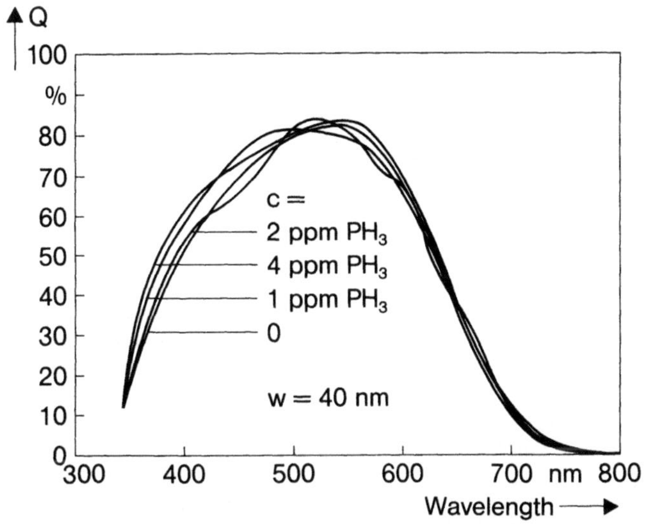

Fig. 3: External collection efficiency Q versus wavelength λ for various doping concentrations c.

Table I: Overview of cell parameters in the annealed and lightsoaked state.

	c	n [%]	I_{sc} [mA/cm²]	V_{oc} [V]	FF [%]	V_f [V]	$-q_f$
A annealed	0 ppm	7.6	13.4	0.85	66.1	0.65	0.8
	1 ppm PH_3	7.9	15.1	0.86	61.2	0.53	0.92
	2 ppm PH_3	7.5	14.5	0.82	62.8	0.55	0.92
	4 ppm PH_3	7.6	14.6	0.84	62.1	0.45	0.93
B lightsoaked	0 ppm	5.9	12.9	0.84	55.8	0.5	0.66
	1 ppm PH_3	5.5	13.9	0.83	47.6	0.47	0.47
	2 ppm PH_3	5.7	13.7	0.81	51.3	0.45	0.69
	4 ppm PH_3	5.4	13.6	0.82	48.4	0.47	0.5

We also looked at the light-degradation behavior (16h AM1 illumination under open circuit condition) of the doped cells and found no significant difference in stability between the samples with and without doping profiles.

4. SUMMARY

In conclusion, it was shown that with a well-chosen linearly decreasing phosphorus doping profile starting at the p/i interface, the blue response of an a-Si pin solar cell is improved and the red response is unchanged, thus giving higher efficiencies. The increase was interpreted in terms of a reduced surface recombination of electrons at the p/i space-charge barrier layer.

REFERENCES

/1/ A. Catalano, M. Bennett, D. Carlson, C.R. Dickson, K. Rajan
'Role of impurities in the batch processing of a-Si:H pin devices'
Proc. of the 6th E.C. Photovoltaic Solar Energy Conference, London (1985) 751

/2/ H. Kausche, W. Jüergens, E. Günzel, R. Plättner, J. Grabmaier
'Change of cell performance of a-Si:H solar cells by impurities and doping profiles in a large two chamber reactor'
Proc. of the 2nd International Photovoltaic Science and Engeneering Conference, Beijing (1986) 377

/3/ A. Catalano, G. Wood
'Short wavelength response in a-Si:H p-i-n diodes: a simple method to minimize interface recombination'
Proc. of the MRS-Spring Meeting 1988, Reno, Nevada, in press

/4/ W. Kusian, W. Krühler, B. Bullemer
'Enhanced blue response of top a-Si:H solar cells for tandem structures'
Proc. of the 19th IEEE Photovoltaic Specialists Conference, New Orleans, Lousiana (1987) 599

/5/ W. Kusian, H. Pfleiderer, B. Bullemer
'Spectral photocurrent-voltage characteristics of crystalline and amorphous silicon solar cells'
Solar Cells 22 (1987) 239

/6/ W. Kusian, H. Pfleiderer, B. Bullemer
'Spectral photocurrent-voltage characteristics of a-Si pνn and pπn diodes'
Journ. of Non-Cryst. Solids **97&98** (1987) 647

/7/ W. Kusian, H. Pfleiderer, B. Bullemer
'Elementary degradation aspects for a-Si solar cells'
Proc. of the MRS-Spring Meeting 1988, Reno, Nevada, in press

/8/ Y. Sun, H. Pfleiderer, W. Krühler
'The transition voltage of a-Si pin solar cells before and after degradation',
this conference

AMORPHOUS $Si_{1-x}Ge_x$:H ALLOYS FOR TANDEM SOLAR CELLS WITH OPTIMIZED ELECTRON AND HOLE $\mu\tau$-PRODUCTS

E. Günzel, F. Karg, R. Plättner
Research Laboratories of Siemens AG,
D-8000 Munich 83

ABSTRACT. Several deposition parameters for the fabrication of a-$Si_{0.7}Ge_{0.3}$:H alloys by glow discharge were varied using both diode and triode reactors. Investigations of their transport properties placed special emphasis on the hole $\mu\tau$ product, its dependence on the doping level and the associated red sensitivity of a-SiGe:H solar cells. Best transport parameters have been achieved in the diode system under low pressure and small electrode distances.

1. INTRODUCTION

The manufacture of amorphous hydrogenated silicon-germanium alloys a-$Si_{1-x}Ge_x$:H is the object of intensive research carried out by a number of groups e.g. /1-4/ using conventional glow discharge and other different preparation methods. The aim of these activities is to further increase the conversion efficiency of amorphous thin-film solar cells as well as their stability by the use of tandem cell structures (Fig. 1). Most suitable for the bottom cell of a tandem cell is a semiconductor material with a bandgap of about 1.3 eV or even as low as 1.2 eV if a top cell with a 1.75 eV band gap material like a-Si:H is used. For the low-gap a-SiGe:H material preparation methods such as diode and triode geometry excitation were previously examined /5/. Preferentially soft deposition methods such as triode excitation but also methods based on favourable kinetics of the reactive gases were successful in improving the electronic properties of a-SiGe:H up to a germanium content of 30 %. Large area deposition with machines like triode reactors in both cross field and mesh geometry is associated with more complicated operation compared with diode reactors.

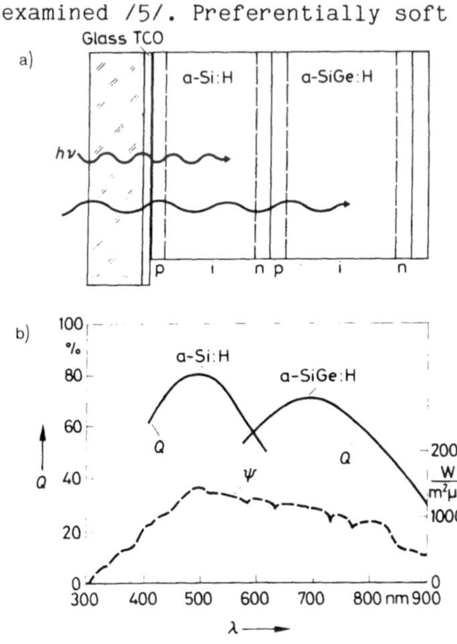

Fig.1: a) Two terminal tandem cell b) AM1 solar spectrum Ψ and required spectral response Q for tandem cells

In this work we therefore used the conventional diode glow discharge method and tried to improve the a-SiGe:H film properties to attain a quality comparable with the "soft" manufactured materials. We report on the influence of the electrode distance and the gas pressure on the transport properties of a-SiGe:H films while keeping the temperature, r.f. power, total gas flow and hydrogen dilution constant.

For the characterization of the film properties we preferentially used time of flight measurements. This method is especially suitable for the characterization of amorphous semiconductors for solar cells because it results in separate values for the $\mu\tau$ products of holes and electrons. Holes are the limiting carriers in a-SiGe:H solar cells with the p/i junction at the light entrance side in today's best structure TCO/pin. In contrast, the suitability of the photo- to dark-conductivity ratio σ_{ph}/σ_d (the photosensitivity) as an empirical quality factor in a-SiGe:H alloys has not yet been rigorously proven. Secondary photocurrent mainly yields only transport properties for the majority carriers.

2. EXPERIMENTAL

Amorphous $Si_{1-x}Ge_x$:H films were deposited by r.f. glow discharge in two types of capacitively coupled reactors: a stainless steel diode reactor and a quartz triode reactor (Fig. 2). The distance between the two heatable electrodes of the diode reator was varied between 1.6 cm und 4.0 cm.The substrates were placed on the grounded bottom electrode. The r.f. power (13.5 MHz) was about 10 W (0.07 W/cm^2). The r.f. electrode and one of the two grounded electrodes of the triode reactor were mounted outside the quartz cylinder. The substrates were placed on the bottom grounded electrode. This electrode was heatable, its distance to the r.f. electrode was about 10 cm. The r.f. power (13.5 MHz) was between 40 W and 50 W. A constant substrate temperature of 230°C was used for both reactors.

The gas mixtures used consisted of SiH_4 and GeH_4 diluted in hydrogen (1:20 and 1:10 for the diode and triode reactor respectively). The total gas flow rate was between 70 sccm and 100 sccm, the gas pressure between 1.0 mbar and 3.0 mbar. The compositions of the a-$Si_{1-x}Ge_x$:H films were determined by electron beam microanalysis.
Three methods were employed to characterize the transport properties of the a-SiGe:H material, requiring three kinds of sample structure:
i) Stationary dark- and photoconductivity (PC) was measured at 0.5 µm thick films with a gap electrode structure deposited on glass substrates. Monochromatic illumination by a HeNe laser with a photon flux of $1.6\cdot10^{15}cm^{-2}s^{-1}$ was used in conjunction with reflectivity and transmission data to determine the $\eta\mu\tau$ product for the majority carriers (electrons). In addition the photosensitivity σ_{ph}/σ_d was measured under AM1 illumination to allow our results to be compared with other published data.

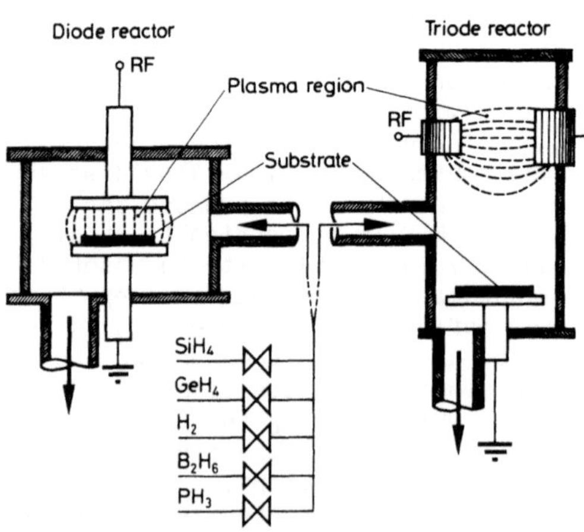

Fig.2: Diode and triode reactor set up

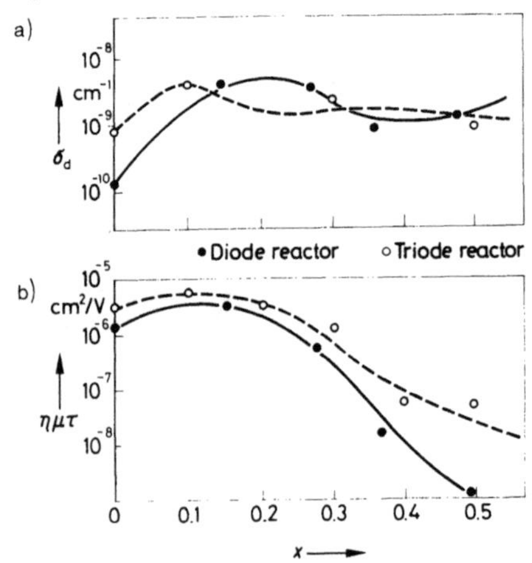

Fig.3: a) Dark conductivity and b) $\eta\mu\tau$ values as a function of Ge content x

ii) The time-of-flight measurements (TOF) were performed on a p-(SiC)/i-(SiGe)/n-(Si) structure, deposited on tin oxide covered glass substrates and a semitransparent Pd back contact. The i-layer thickness was approximately 2 µm in this series. This sample was illuminated from both sides alternatively to determine the $\mu\tau$ products for electrons and holes.
iii) The spectral response as a function of external bias U was measured at a solar cell structure p-(Si)/i-(SiGe)/n-(Si). The internal collection efficiency Q_i under short circuit displayed in Fig. 6 was determined by division of the external collection efficiency Q at U = 0 by the saturating collection efficiency under high reverse bias.

3. RESULTS

Optoelectronic transport properties for a-$Si_{1-x}Ge_x$:H alloys as a function of the Ge concentration x have been frequently reported. In general the investigated materials were deposited with the same plasma parameters for the whole alloy range x. An example of these types of investigation is given in Fig. 3. The $\eta\mu\tau$ product for the majority carriers (electrons) was determined from secondary photoconductivity. Although two widely different deposition systems were employed, - a diode and a triode reactor - both sets of data have one feature in common: the declining $\eta\mu\tau$ product for films with a Ge concentration in excess of about 30 %. Deposition parameters suitable for Si rich alloys are obviously unsuitable for Ge rich alloys. In a second stage of our experiments we exclusively investigated alloys with the critical Ge concentration of 30 %. Through rigorous variations of the deposition parameters for the diode reactor, we tried to improve the transport parameters of this alloy.
In Fig. 4 we sum up the results obtained in the course of our investigations. Two parameters were varied: the electrode distance d and the total gas pressure p. Fig. 4a shows the photosensitivity σ_{ph}/σ_d under AM1 illumination, 4b the dark conductivity and 4c the associated $\eta\mu\tau$ product of electrons. In addition we determined the $\mu\tau$ product for both electrons and holes by time-of-flight measurements.

The results from stationary photo conductivity show a distinct, but small improvement of the photosensitivity especially for small electrode distances and low pressures. Additionally to these properties of the majority carriers we observed

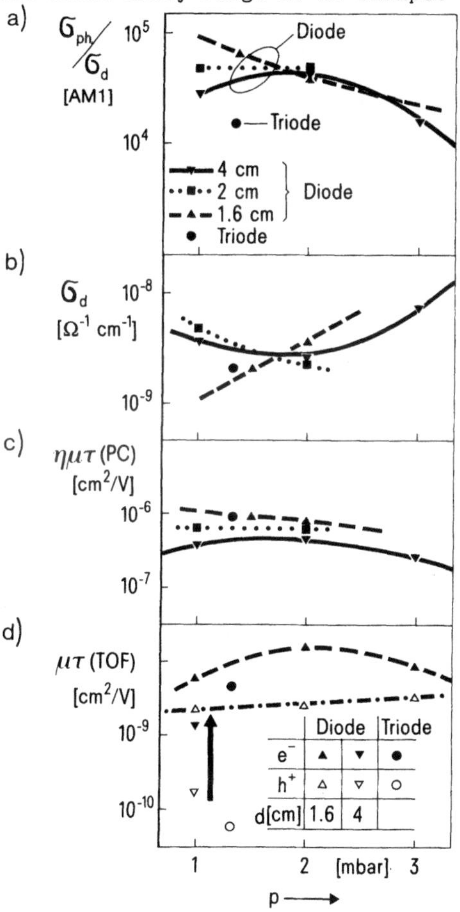

Fig.4: a) Photosensitivity b) Dark conductivity c) $\eta\mu\tau$ values (PC) d) $\mu\tau$ values (TOF) as a function of gas pressure and electrode distance

a markedly higher improvement in the minority carrier properties: the hole $\mu\tau$ product increases by about one order of magnitude when reducing the electrode distance from 4 cm to 1.6 cm (Fig.4d).

This will have important implications for charge collection in the red spectral range of SiGe solar cells. When illuminated by long wavelengths, carrier generation occurs homogeneously within the i-layer of these cells. An efficient carrier collection then requires equal $\mu\tau$ products for electrons and holes. As we noted earlier /5/, holes are the limiting carriers in a-SiGe:H solar cells as their $\mu\tau$ product decreases severely with the Ge concentration. In order to level out both $\mu\tau$ products we proposed boron fine doping as a possible method of increasing the red sensitivity of the cells. As a basis for finding out the optimum amount of boron we used the $\mu\tau$ products for electrons and holes measured earlier /5/ for variously doped a-Si:H and a-SiGe:H (Fig. 5).

The symmetrical behaviour of the $\mu\tau$ product for minority and majority carriers becomes obvious. However, at a doping level around 1 ppm B_2H_6 the $\mu\tau$ values are equal.

In consequence, we looked for improvements in the long wavelength carrier collection efficiency at doping levels in the 0 ppm to 2 ppm range. In Fig. 6 we plotted the internal collection efficiency under short circuit conditions for the undoped as well as the 1 ppm and 2 ppm boron doped "i"-layer of a p-Si/i-$Si_{0.7}Ge_{0.3}$:H/n-Si solar cell. The carrier collection increases above 650 nm as expected at a doping level of 1 ppm B_2H_6.

4. DISCUSSIONS AND CONCLUSIONS

Two salient results of our investigations will be noted and discussed.
i) A reduced electrode distance d significantly increases the hole $\mu\tau$ product.
ii) Small values of both d and gas pressure p further slightly increase the photosensitivity of the a-$Si_{0.7}Ge_{03.}$:H films. The distance of approximately

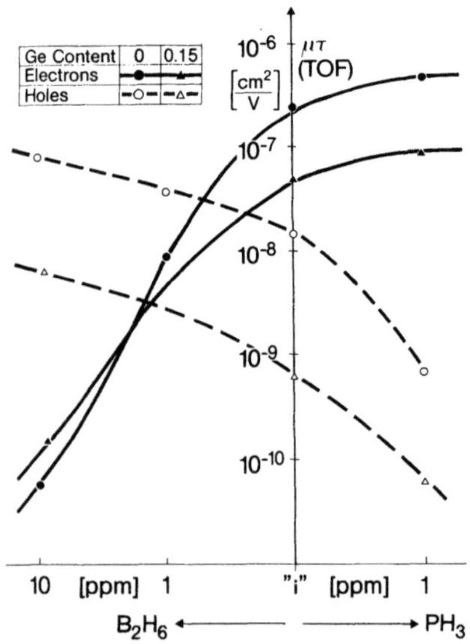

Fig.5: Influence of B_2H_6 and PH_3 fine doping on $\mu\tau_e$ and $\mu\tau_h$ values of a-Si:H and a-SiGe:H

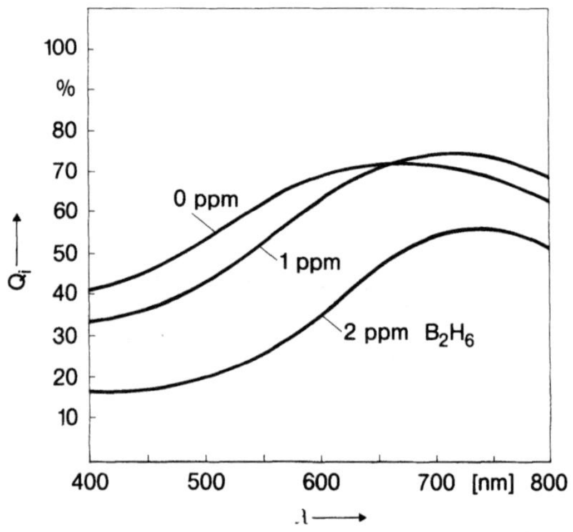

Fig.6: Internal quantum efficiency of a-SiGe:H cells with doped i layers

10 cm between generation and deposition is obviously unfavourable for a-SiGe:H deposition as is apparent from in the remarkably lower photosensitivity of these films (Fig. 4). Even the results of the triode reactor fit into this principle (i), though the electrode distance is not defined in the above mentioned sense.

The lower gas pressure is connected with a higher kinetic energy of the particles impinging upon the film surface. This might enhance the surface mobility of the heavier Ge (as compared to Si) containing radicals enabling them to find a local energy minimum on the growing surface.

These tendencies also fit in well with a recently published theory /6/, which regards the suppression of polymer formation in the plasma and on the growing surface as the key factors for high quality film growth. Polymer formation can be suppressed on the surface by a sufficiently high deposition temperature ($T_S > 200°C$), in the plasma volume by low power, low pressures and/or small electrode distances. The latter two points could be confirmed by the present work.

Regarding the influence of the r.f. power we cannot as yet give a definite answer. Our results have shown an increase in the hydrogen content in the a-SiGe:H produced in the triode reactor when working with 50 W instead of 10 W in the diode system. The origin of the inferior photoelectronic properties despite the higher hydrogen content might originate in a different bonding structure which will be investigated in the near future.

Further improvement of the transport properties of a-SiGe:H films is expected by varying of the other plasma parameters which were held constant in this work.

ACKNOWLEDGEMENT. We wish to thank Dr. W. Kusian and K. Dietrich for taking spectral response and time of flight measurements and Dr. R. Weyl for the electron beam microanalysis.

REFERENCES

/1/ Guha, S.; Payson, J.S.; Agarwal, S.C.; Ovshinsky, S.R.: 'Fluorinated amorphous silicon-germanium alloys deposited from disilan-germane mixture' Journal of Non-Cryst. Sol. 97 & 98 (1987) p. 1455

/2/ Matsuda, A.; Tanaka, K.: 'Guiding principle for preparing highly photosensitive Si-based amorphous alloys' ibid. p. 1367

/3/ MacKenzie, K.D.; Burnett, J.H.; Eggert, J.R.; Li, Y.M.; Paul, W.: 'Improvement of photoconductive response in amorphous silicon-germanium alloys produced from fluorides instead of hydrides' ibid, p. 1019

/4/ Aljishi, S.; Chu, V.; Smith, Z.E.; Shen, D.S.; Conde, J.P.; Slobodin, D.; Kolodzey, J.; Wagner, S.: 'Steady state and transient transport in a-SiGe:H,F alloys' ibid. p. 1023

/5/ Karg, F.; Dietrich, K.; Krühler, W.; Plättner, R.: 'Electronic transport properties of differently fabricated a-SiGe:H films for tandem solar cells' Technical Digest of the International PVSEC-3 Tokyo, Japan, 1987,p.661

/6/ Ross, R.C.; Chao, S.S.; Tyler, J.E.; Czubatyi, W.; Lucovsky, G.: 'Plasma phase polymerization reactions in the deposition of glow discharge deposited a-Ge:H alloy films' J.Vac.Sci.Technol.A3(3) (1985) p.958

HIGH-RATE DEPOSITION OF AMORPHOUS SILICON BY VHF GLOW DISCHARGE FOR SOLAR CELL APPLICATIONS

A.Shah, N.Wyrsch, K.Prasad, M.Brêchet,R.Tscharner, M.Favre, B.Leutz and H.Curtins.
Institut de Microtechnique, Université de Neuchâtel,Breguet 2,
CH-2000 Neuchâtel, Switzerland (+41 / 38. 24.60.00).

ABSTRACT

We have recently developed a novel high-rate deposition method for amorphous hydrogenated silicon, called the Very-High-Frequency Glow Discharge Technique (VHF-GD). Its main feature is the use of plasma frequencies in the very high frequency range, i.e. between 30 and 150 MHz. Deposition rates as high as 30 Å/sec with good material properties are obtained.
In this paper we present the effect of increasing the plasma excitation frequency on the film properties; we focus on those parameters that are important for solar cell applications (defects density, transport properties, degradation,...). Finally, we present first device measurements on both Schottky and MIS type solar cells.

1. INTRODUCTION

A high deposition rate is a key issue for reducing the cost of amorphous hydrogenated silicon (*a*-Si:H) solar cells; it reduces the processing time and therefore increases production capacity. To achieve this goal, we can basically follow up one of two different approaches : either to retain the conventional RF-Glow Discharge (GD) technique and modify certain deposition parameters (e.g. increase the RF power [4], use a higher silane [5,6]); or to employ a novel deposition method, such as Photo-CVD [7] , the plasma confinement method [8], Activated Reactive Evaporation (ARE) [9], or microwave GD (i.e. Electron Cyclotron Resonance CVD) [10,11]. Both these basic approaches have sofar not lead to satisfactory results in terms of material quality, large area depositions or processing costs as compared with the standard GD technique at 13.56 MHz.
The first approach, i.e. a modification of the deposition parameters, generally leads to higher hydrogen contents, mostly in polymeric forms, and to poor opto-electronic properties [4]. With the use of higher silane (Si_2H_6) , satisfactory material quality has been achieved at deposition rates up to 20 Å/sec and p-i-n solar cells with 7% efficiency have been produced at 10 Å/sec [6]. This is indeed one of the few successful high-rate deposition techniques; however, the price of disilane (four times the price of silane) is high and the cell stability is still unknown. With the second approach, i.e. by using a novel deposition method, a variety of problems appear such as heavy ion bombardement, difficulties in industrial implementation, problems in large-area deposition, poor uniformity, etc.
The VHF method may be placed somewhere between these two approaches. It uses a conventionally designed reactor, but takes the advantage of a higher plasma excitation frequency. In fact, most plasma systems operate at the standard industrial frequency of 13.56 MHz, for which a lot of commercial equipment is readily available. Outside the DC to 15 MHz range, no systematic studies have been carried out except at particular values such as f=2.45 GHz (microwave frequencies). Microwave GD has been successful in producing *a*-Si:H with satisfactory opto-electronic properties at rates higher than 20 Å/sec (rates over 100 Å/sec have been achieved, however with rather poor quality). This is a strong indication that plasma excitation frequency does play an important role in the plasma chemistry and deposition kinetics [12,13] and influences the deposition rate. The use of microwaves, however, is connected with limitations in large-area deposition, uniformity and with other practical restrictions.
The VHF-GD technique avoids these difficulties by achieving, in a conventionally designed reactor, good material quality at depositions rates upto 30 Å/sec, on medium-size areas with sufficient uniformity. Furthermore, layers with thicknesses of 100 µm can be easily deposited.

2. EXPERIMENTAL

The VHF-GD system used has been described previously [1-3]. Dow Corning 7059 glass substrates with a surface of 64 cm^2 were employed. Sample thicknesses were determined gravimetrically from the weight increase of the substrate before and after deposition and assuming a density of 2.2 g/cm^3 for *a*-Si:H. Dark and AM1.5 conductivities are measured in gap configuration in a vacuum system at pressure lower than 10^{-7} mbar. Dark conductivity σ_{dark} is deduced from an Arrhenius plot of conductivity versus reciprocal absolute temperature; the photoconductivity σ_{ph} is measured under 100 mW/cm^2 illumination by halogen lamps (coarse approximation for AM1.5 spectrum). The optical bandgap E_{opt} is deduced from Tauc's plot. Defect density and Urbach tail energy E_0 were calculated from photothermal deflection spectroscopy (PDS) curves using the model and conversion factor given by Jackson and Amer [14]. All samples were first annealed at 220 °C for 3 hours under vacuum, and cooled down to room temperature with a gradient of 0.5 °C/min prior to measurement. The Staebler-Wronski effect has been monitored by conductivity and PDS measurements, both during degradation and after saturation (14 hours with 500 mW/cm^2 white light illumination).

3. RESULTS and DISCUSSIONS

3.1 Effect of the plasma excitation frequency

A series of samples has been deposited with a fixed set of experimental conditions: substrate temperature T_s=280 °C, RF power density P=0.1 W/cm^3, total pressure p=0.28 mbar, SiH_4 flow F=20 sccm and a plasma excitation frequency in the range 25 - 150 MHz. Fig. 1 shows the effect of the plasma excitation frequency on the deposition rate: we remark a maximum for $f=f_{opt}\approx 70$ MHz with a deposition rate R=21 Å/sec. Below and above this optimum frequency f_{opt}, we observe a rapid decrease of R.

This effect can be explained if we consider the Electron Energy Distribution Function (EEDF) [2,3]. A change in f modifies significantly the EEDF as follows [12,13]: with increasing frequency, the tail of high energy electrons is enhanced (increase of the density of these electrons). As a consequence, we expect an increase in the ionization rate (more electrons having an energy higher than the ionization energy threshold for silane) and also an increase in the total electron density n_e. A larger n_e implies a higher silane dissociation rate and an enhanced deposition rate R. However, at somewhat higher frequencies, a second effect starts to counteract the first one; an increase in f begins to lead to a decrease in the average electron energy, reducing now the available electrons for ionization and dissociation processes. The total electron density n_e therefore shows a maximum and we can reasonably assume that the silane dissociation and also the deposition rate are correlated with n_e. This interpretation is supported by the fact that the plasma can, at the 'optimum frequency' f_{opt}=70 MHz, be sustained with a power density as low as 5 mW/cm^3; this is attributed to the low plasma impedance (large n_e) prevailing at f_{opt}. Another interesting effect we observe, is the self-confinement of the plasma between the two electrodes, which could also contribute to increase the deposition rate (as observed with mechanical confinement in [8]).

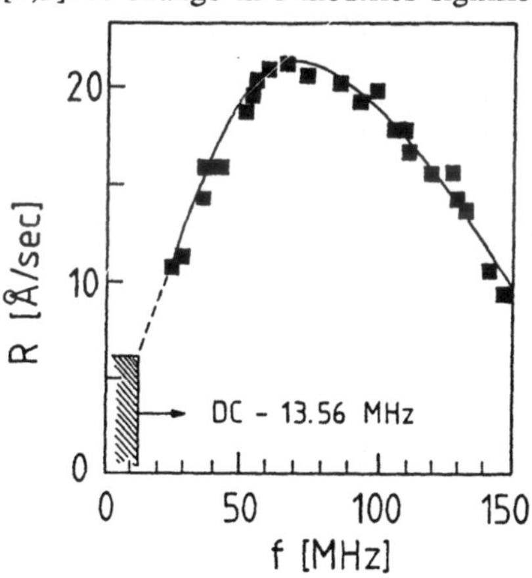

Fig. 1: Deposition rate R as function of plasma excitation frequency f.

3.2 Film properties

Film properties are well known to be thickness-dependent [15-17] and therefore comparisons must be carried out on samples with approximately the same thickness. In Fig. 2, we present the opto-electronic properties of *a*-Si:H layers having all approximately the same thickness d≈2.3 μm.

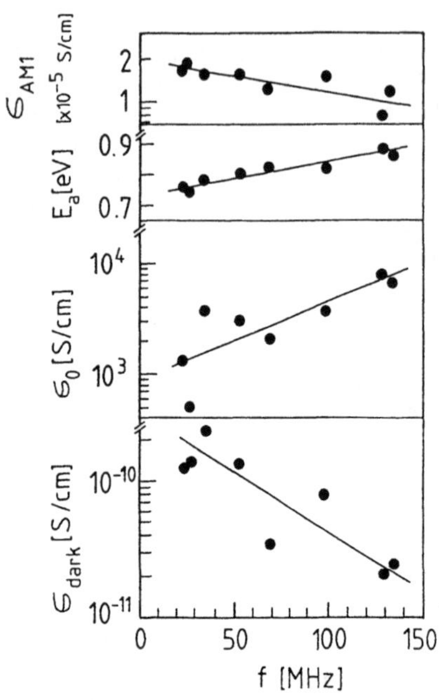

Fig. 2: Electrical conductivity data for a-Si:H as a function of plasma excitation frequency f.

The dark conductivity σ_{dark} decreases from ~10^{-10} S/cm at 25 MHz to ~10^{-11} S/cm at 150 MHz. We observe a small but significant increase of the activation energy E_a from 0.75 to 0.85 eV and also an increase of σ_0 with increasing frequency. The photo-conductivity σ_{ph} decreases slightly (from ~$2x10^{-5}$S/cm to ~$1x10^{-5}$S/cm), consistently with the increase of E_a. We suspect the change in E_a and σ_0 to be correlated with a change in the internal stress of the material. As remarked by Heintze and Spear [18], mechanical stress, in fact, does influence E_a and σ_0 (in their case the experiments were carried out using <u>externally</u> applied mechanical stress). The total defect density (surface/interface + bulk) is pratically constant at $N_D \approx 2x10^{16}$ cm^{-3} for all frequencies; the same applies to the Urbach energy at $E_0 \approx 58$ meV. As PDS is sensitive to both surface and bulk defects (N_s and N_b respectively), we have used a series of samples of different thicknesses to separate both contributions; thereby, the following formula is used: $N_D=(N_s / d) + N_b$, d being the sample thickness [17]. We thereby found constant values of $N_b \approx 3x10^{15}$ cm^{-3}, $N_s \approx 3x10^{12}$ cm^{-2} and E_0 for plasma excitation frequencies f between 25 and 150 MHz. The optical gap also remains constant at a value of ~1.75 eV.

In Table 1, we present the material properties at a fixed frequency of 70 MHz for three different deposition rates, namely for 3.6 Å/sec, 9.5 Å/sec and 15.7 Å/sec. As one can see, we observe no noticeable change in the defect densities of the films. σ_{dark} as well as σ_{ph} decrease with increasing R; however the activation energy E_a shows the opposite trend. Results of Bhat et al. for high rate deposition using disilane [6] are given as a comparison. As far as the conductivity data are concerned, they are no noticeable differences between the disilane and the VHF depositions, except for the activation energy values E_a , where the disilane deposition method yields higher values. Urbach energy E_0 values as well defect density N_D values and density of states $g(E_F)$ cannot be compared directly because of the different measurement techniques used and the lack of thickness indications for the disilane depositions. In both cases (VHF-GD technique and RF-GD with Si_2H_6), material are of reasonably good quality. Tentative measurements of hole diffusion length by the differential surface photovoltage technique have failed due to basic practical and theoretical limitations of this method when applied to a-Si:H [19].

Parameters	VHF-GD of SiH_4			RF-GD of Si_2H_6	
T_s [°C]	300	300	300	270	270
P [W/cm^3]	0.015	0.09	0.1	?	?
p [mbar]	0.14	0.28	0.28	0.34	0.34
F_{SiH4} [sccm]	6.6	9.5	18.3		
F_{Si2H6} [sccm]				24	24
f [MHz]	70	70	70	13.56	13.56
d [µm]	1.95	1.87	2.15	?	?
R [Å/sec]	3.6	9.5	15.7	4	20
c_H [at. %]	8.9	6.3	5.4	?	?
E_{opt} [eV]	~1.75	~1.75	~1.75	1.72	1.74
N_D [cm^{-3}]	$1.5x10^{16}$	$1.8x10^{16}$	$1.7x10^{16}$	-	-
$g(E_F)$ [1/cm^3eV]	-	-	-	$2x10^{16}$	$3.5x10^{16}$
E_0 [meV] (PDS)	55	56	59	-	-
E_0 [meV] (CPM)	-	-	-	40	50
σ_d [1/Ωcm]	$8.5x10^{-10}$	$1.5x10^{-10}$	$1.7x10^{-10}$	$2x10^{-10}$	$1x10^{-10}$
E_a [eV]	0.73	0.81	0.79	0.82	0.90
σ_{ph} [1/Ωcm]	$6.5x10^{-5}$	$3.5x10^{-5}$	$1.8x10^{-5}$	$4x10^{-5}$	$2x10^{-5}$

Table 1: Material properties of different a-Si:H layers thicknesses for VHF-GD of SiH_4 compared with RF-GD of Si_2H_6 [6].

These results show the versatility of the VHF-GD technique for the deposition of good quality *a*-Si:H layers at varying deposition rates. <u>The possibility of varying the activation energy and the Fermi level without doping, could be useful in shaping</u>

the internal field within the i-layer of a device. However, for solar cell applications, hole transport properties have yet to be investigated in detail for the VHF-GD material.

3.3 Degradation

Degradation (Staebler-Wronski effect) is an important limiting factor for a solar cell. In fact, one should always take into acount the material properties (such as dark and photo- conductivity, defect density, Urbach energy) after degradation. We have studied the Staebler-Wronski effect (SWE), on 1.1 and 5.1 μm thick samples deposited at 17 Å/sec. Using two different thicknesses, we were able to deduce both surface and bulk contributions of the defect densities. In Table 2, we present the measured properties of films in both state 'A' (annealed state) and state 'B' (after saturation). One important result is that we do not see any change of N_S upon light soaking (see Fig. 3); the SWE is, in our case at least, clearly a bulk effect. The bulk defect density after saturation N_b ('B') ≈ $3x10^{16}$ cm^{-3} is relatively low compared with that observed for materials of other groups [20-22]. An interesting observation is that N_b ('B') seems not dependent on the Urbach tail energy E_0 (the latter being a measure of latent weak Si-Si bond density) (see details in ref. [23,24]). These results are quite encouraging. However, to make a more complete analysis of the SWE, one should take into acount possible inhomogeneities (in defect density, transport properties and Urbach energy) throughout the sample [17,24].

	State 'A'		State 'B'	
R [Å/sec]	17	17		
d [mm]	1.1	5.1	1.1	5.1
σ_d [$10^{-12}/\Omega cm$]	55	280	2.0	4.6
E_a [eV]	0.81	0.73		
σ_{ph} [$10^{-8}/\Omega cm$]*	220	570	4.2	13
N_b [$10^{15}/cm^3$]	4-5	4-5	29	29
N_s [$10^{12}/cm^2$]	~3.2	~3.2	~3.0	~3.0
E_0 [meV]	59	51	54	49
E_g [eV]	1.76	1.76		
c_H [at. %]	5-6	5-6		
A_{sw} [$cm^{-3}s$]		~9100		

* G=$1.0x10^{19}/cm^3$ for bandpass λ=680±20 nm

Table 2 : Summary of material properties in state 'A' and 'B'.

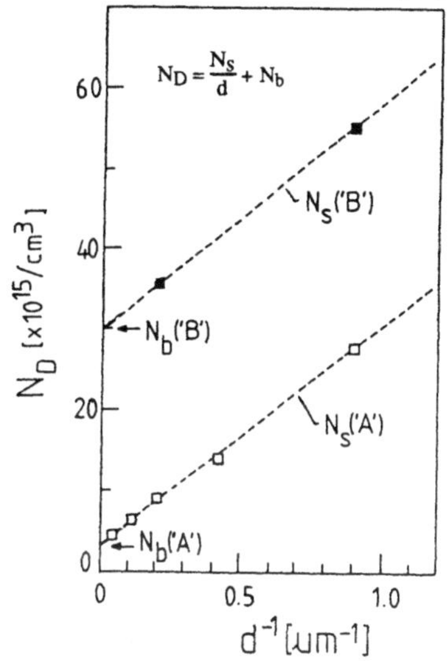

***Fig. 3:** Measured PDS total defect density N_D as function of reciprocal thickness d in annealed state 'A' and saturated state 'B'.*

3.4 Device results

We have investigated n-type doping with the VHF-GD technique and this mainly at low deposition rates. As layers of a few hundreds of Ångströms have to be deposited for the n layer of typical p-i-n solar cells, we should have low deposition rates to be able to control the deposition proccess. We achieved deposition at rates down to R=2.3 Å/sec with a very low RF power density of P=20 mW/cm^3 (at f=70 MHz) and a doping efficiency comparable with published values [25].

A set of *a*-Si:H Schottky diodes have been made with the following structure: Cr / n (500 Å) / i (1.5 μm) / Pd (50 - 100 Å), as a first test for photovoltaic applications of our VHF-GD material. Intrinsic layers of the diodes were deposited at 17.5 Å/sec using a plasma excitation frequency of 70 MHz. At 10 mW/cm^2 white light illumination, we measured V_{oc} ≈ 460 mV, a J_{sc} ≈ 1.3 -1.8 mA/cm^2 , a fill-factor FF ≈ 0.55 and a conversion efficiency under the barrier η ≈ 4.5 %. These first results for non-optimalized diodes are quite encouraging, despite the yet relatively low value of FF. Further studies should be carried out to investigate the possible limiting factors in these diodes (hole transport properties, barrier quality, contact diffusion, back contact barrier,etc).

MIS solar cells have been made with our VHF-material, using i-layers deposited at f=70 MHz and R=14 Å/sec: these MIS devices were fabricated by the Konstanz group, using a special surface passivation technique of the i-layer and Pt as a barrier material [26,27]. They show a remarkably high V_{oc} value over 900 mV under AM1.5 illumination (77 mW/cm^2). However, the short circuit value J_{sc} remains rather low and further investigations are needed to increase it.

4. CONCLUSIONS

We have shown the versatility and the scope of high-rate *a*-Si:H deposition using the VHF-GD technique. The plasma excitation frequency, with its marked influence on the electron energy distribution function (EEDF), is a one of the important parameters controlling the plasma chemistry and deposition kinetics. With the VHF-GD method, high-rate deposition can be achieved without incurring at the same time heavy ion bombardement (that generally leads to bad quality material). With this new technique, we are able to produce satisfactory material in view of photovoltaic applications, at rates up to 25 Å/sec. However, more experiments have to be executed to characterize hole transport properties and to insure performant device fabrication.

This work was supported by Swiss Federal Research Grant OFEN(REN) 86/9.

REFERENCES

[1] H.Curtins et al., Electronics Lett., **23** (1986) 228.
[2] H.Curtins et al., Plasma Chem. and Plasma Process., **7** (1987) 267.
[3] H.Curtins et al., MRS Symp., **95** (1987) 249.
[4] C.M.Fortmann et al., 7. EC-PSEC, Seville (1986) 435.
[5] H.Itoh et al., Proc. of the 1. International Photovoltaic Science and Enginneering
[6] P.K.Bhat et al., SERI Contractors Meeting, Denver, Colorado, (1987).
[7] A.Yamada et al., Jap. J. of Appl. Phys., **24** (1985) 1586.
[8] T.Hamasaki et al., Appl. Phys. Lett., **44** (1984) 600.
[9] J.C.Anderson and S.Biswas, J. of Non-Cryst. Sol., **77 & 78** (1985) 817.
[10] A.E.Delahoy, J. of Non-Cryst. Sol., **77 & 78** (1985) 833.
[11] S.Kato and T.Akoi, J. of Non-Cryt. Sol., **77 & 78** (1985) 813.
[12] M.R.Wertheimer, J. Vac. Technol., **A4** (1986) 729.
[13] D.L.Flamm, J. Vac. Technol., **A3** (1985) 2643.
[14] W.B.Jackson and N.M.Amer, Phys. Rev., **B 25** (1982) 5559.
[15] F.Boulitrop et al., J. of Appl. Phys., **58** (1985) 3494.
[16] W.B.Jackson et al., Appl. Phys. Lett., **42** (1983) 105.
[17] M.Favre et al., J. of Non-Cryst. Sol., **97 & 98** (1987) 731.
[18] M.Heintze and W.E.Spear, J. of Non-Cryst. Sol., **97 & 98** (1987) 495.
[19] R.Schwarz et al., J. of Non-Cryst. Sol., **97 & 98** (1987) 759.
[20] M.Stutzmann et al., Phys. Rev., **B 32** (1985) 23.
[21] T.L.Chu et al., Appl. Phys. Lett., **52** (1988) 807.
[22] C.R.Wronski et al., AIP Conf. Proc., **157** (1987) 70.
[23] Z E.Smith and S.Wagner, Phys. Rev. Lett., **59** (1987) 688.
[24] H.Curtins et al., presented at MRS Symp., Reno, (1988).
[25] W.E.Spear and P.LeComber, S.S.Commun., **17** (1975) 1193.
[26] J.Meier et al., Appl. Phys. A, **46** (1988) 31.
[27] J.Meier et al., presented at this conference.

LOW TEMPERATURE MICROCRYSTALLINE SILICON COMPATIBLE WITH A-SI SOLAR CELL DESIGN

G. Willeke, J. Nijs and R. Mertens
Interuniversity Microelectronics Centre
Kapeldreef 75
B-3030 Leuven
BELGIUM

ABSTRACT 300Å thick n+ μc-Si films with $\sigma \approx 20$ (Ωcm)-1 have been obtained at Ts=200°C. The influence of reactor design on the properties of n+ microcrystalline silicon films has been investigated.

1 - Introduction

In the past few years microcrystalline silicon has received a considerable amount of attention for its possible use as a highly conductive weakly absorbing contact material in a-Si solar cells and other thin film applications. Typically high quality microcrystalline phase is deposited under conditions of strong dilution of SiH4 in H2, a somewhat higher rf power density and rather high substrate temperatures, conditions where chemical equilibrium is approached at the growing substrate film interface [1]. The latter condition of high substrate temperature appears to be incompatible with a-Si solar cell design and the deposition of highly conductive thin films at low substrate temperatures is desired.

The effect of substrate temperature on material properties have been studied by Nakatani et al [2]. They found an increase in the crystalline volume fraction from 20% to about 70% when Ts increased from 150°C to 300°C. At the same time the electrical conductivity increased from 10-2 (Ωcm)-1 to about 23 (Ωcm)-1 for 0.5μm thick heavily P-doped films. In a more detailed study [3], Matsuda carefully scanned the deposition parameter space. Fig. 1 shows the crystalline volume fraction Xc and average grain size δ - determined from FWHM and integrated intensity of the measured x-ray diffraction line - as a function of substrate temperature for otherwise different dilution and rf power levels. δ increases monotonously with Ts for these approximately 1μm thick samples. A strong dependence of Xc on Ts is observed with a maximum at about 400°C and Xc tending to zero in the lower as well as the higher Ts range. This has been explained by a thermally activated surface diffusion coefficient Ds.Therefore at lower temperatures SiHx adsorbates cannot diffuse into stable sites for μc nuclei formation. At higher temperatures the decreasing surface coverage by H makes the surface more reactive, leading to a higher binding energy between SiHx molecules and adsorbing sites, thereby decreasing Ds. The deposition rate was about 30Åmin-1 for the highest Xc-material, approximately independent of substrate temperature.

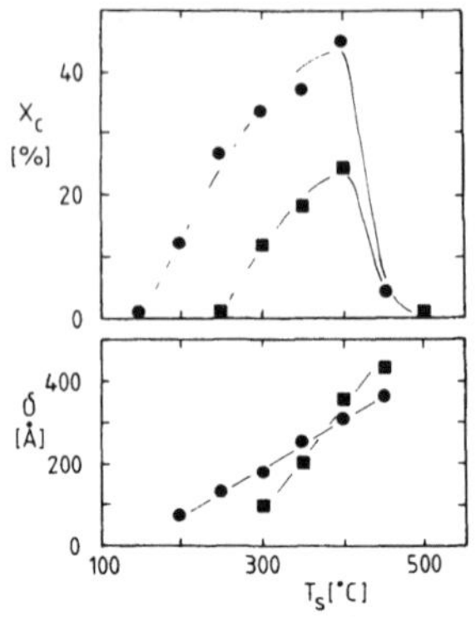

Fig.1: Crystalline volume fraction Xc and grain size δ as a function of Ts (after ref. [3]).

2 - Preparation method

In this study two different rf glow discharge systems have been used, (a) a hot wall type quartz tube reactor with external rf electrodes and (b) a parallel plate cold wall type system were films were deposited on a grounded, heated Al substrate holder. Both systems were capacitively fed from a 13.56 MHz source and type (b) was turbomolecular pumped to a base pressure of about 10-3 μbar, whereas type (a) was simply fitted with a rotary oil pump and had a base pressure of approximately 3μbar. Type (a) reactor had a shielded thermocouple for temperature monitoring during processing; the substrate surface temperature was carefully calibrated by means of a thermocouple directly connected to the substrate surface. The effect of substrate surface heating during processing due to bombardment with energetic particles -if any- was not taken into account but appears to be similar for both systems.

Thin layers of μc-Si were deposited on Corning 7059 glass substrates for transmission and conductivity measurements; the latter have been carried out by the four point probe method. The film thickness was obtained from surface profilometric studies of the film edge which was prepared by shading a small area during deposition. Optical absorption coefficients have been calculated using the method described in [4] from simple transmission data; for simplicity a refractive index n=4 independent of photon energy has been taken for all specimens. The resulting error is only a few percent even if the refractive index changes by up to 50% [5].

The deposition conditions have been varied and typical process parameters for both systems were as follows: 200°C $\leq$ Ts $\leq$ 340°C, 20W $\leq$ P $\leq$ 50W, 100μbar $\leq$ p $\leq$ 1000μbar, 0.01 $\leq$ SiH4(0.5-1% PH3) / (SiH4(0.5-1%PH3)+H2) $\leq$ 0.20.

3 - Experimental results

Fig. 2 summarizes room temperature conductivity data σRT as a function of substrate temperature Ts. Full circles indicate values obtained in type (a) reactor, whereas crosses denote σRT for type (b) specimens. Values for (a)-specimens lie consistently higher by a factor 3-5 in the higher temperature regime and by a factor 10-20 at lower temperatures. An increase with Ts is observed in both cases. For comparison data taken from literature have been added (see figure caption). Open circles and arrow indicate the increase in conductivity observed when a 40.68MHz rf-source was used instead of the conventional 13.56MHz under otherwise similar deposition conditions [8].

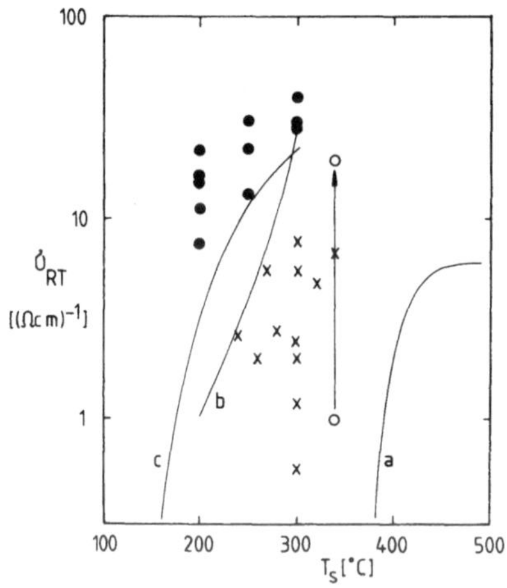

Fig.2: Room temperature conductivity σRT as a function of substrate temperature Ts (•,x: type(a),(b)-specimens; a: ref.[6],b: ref.[7],c: ref.[2],o: ref.[8]).

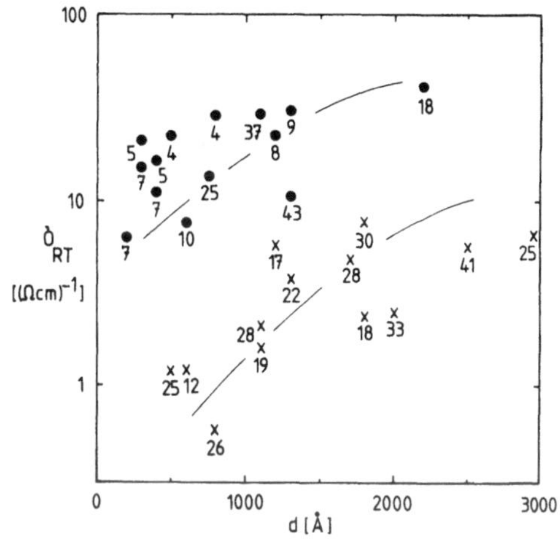

Fig.3: Room temperature conductivity σRT as a function of film thickness d. The numbers indicate the growth rate in Åmin-1 (•,x: type(a),(b)-specimens).

In fig. 3 room temperature conductivity data is shown as a function of film thickness for samples deposited under different conditions and substrate temperatures. Full circles and crosses denote (a) and (b) specimens respectively and the numbers given in the figure refer to the growth rate in Åmin-1. σ increases with d as expected and lies consistently higher by a factor 3-10 for (a)-specimens of comparable growth rates.

Fig. 4 summarizes optical absorption coefficient data for (a)-type specimens (o,•) and (b)-samples (x,+). The conductivity in (Ωcm)-1 is indicated. α decreases with increasing σ as expected, indicating a larger crystalline volume fraction and/or grain size for the type (a) specimens. For comparison absorption data of cr-Si and a-Si:H are included [9][10].

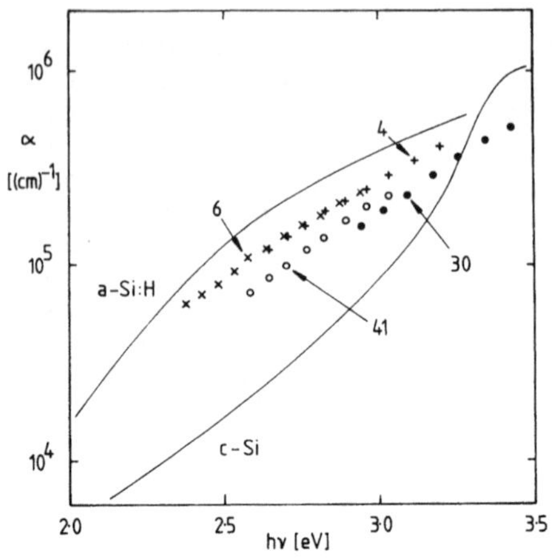

Fig.4: Optical absorption coefficient α as a function of photon energy hν (o,●: type (a) x,+: type (b) specimens; the numbers denote σRT in (Ωcm)-1).

4 - Discussion

300Å thick films of n+ μc-Si with conductivities of about 20 (Ωcm)-1 have been obtained in a hot wall low vacuum type reactor at Ts=200°C and 13.56 MHz. In this case growth rates have been rather small, but films grown at 250 °C with reasonable growth rates (20-30Åmin-1) showed only slightly inferior properties. Films deposited in a high vacuum cold wall type reactor (b) have conductivities consistently lower by a factor 3-10 for similar thicknesses and substrate temperatures. Since Ts and d have been carefully measured and similar deposition conditions have been used in both cases, the reactor design appears to be influential.

Our experimental results can be discussed in the light of Veprek's findings [11]. He has investigated the influence of DC bias on the properties of microcrystalline silicon. With increasing negative bias the grain size decreases until the formation of amorphous phase is observed at large negative bias, resulting from radiation-induced damage in the growing silicon film. In addition the deposition rate increases with increasing negative bias, indicating a departure from chemical equilibrium. In previous work a dependence of the properties on rf frequency has been found [6](see fig. 2). This is not unexpected, as a smaller excitation frequency increases the electron drift length in the plasma and since ions can be considered immobile at these frequencies, the glass substrates used are charged to a larger negative potential. Comparing the two reactors used in the present work, a major difference consists in the grounded substrate holder used in reactor (b). A larger negative potential is expected in this case as compared to the rather symmetrical electrode configuration in (a), where substrates and surrounding surfaces are at a (smaller negative) floating potential. The somewhat larger growth rates expected in case (b) are indeed observed, leading to the conclusion that in reactor (b) grain sizes and/or crystalline volume fractions are limited by ion bombardment resulting in an 'amorphization effect'.

Acknowledgements

This work was performed in the frame of the photovoltaic R&D programme of the Commission of the European Communities under contract number EN3S/0050 and co-sponsored by the Belgian Government under contract number CE/ENE/03. We would like to thank P. DeSchepper for his help in the preparation of the specimens.

References

[1] Z. Iqbal, A.P. Webb and S. Veprek, Appl. Phys. Lett., 36, 163 (1980)
[2] K. Nakatani, M. Yano, K. Suzuki and H. Okaniwa, J. Non-Cryst. Sol., 59&60, 827 (1983)
[3] A. Matsuda, J. Non-Cryst. Sol., 59&60, 767 (1983)
[4] R. Swaenepoel, J. Phys. E, 16, 1214 (1983)
[5] E. Demesmaeker and P. Gossele, M. Sc. Thesis, K.U. Leuven, (1988) unpublished
[6] D.E. Carlson et. al., Amorphous silicon solar cells, Final Report, SERI, March 1982
[7] Y. Mishima, T. Hamasaki, H. Kurata, M. Hirose and Y. Osaka, Jap. J. Appl. Phys., 20, L121 (1981)
[8] G. Willeke, Ph.D. Thesis, University of Dundee, (1983) unpublished
[9] a-Si:H data: G.D. Cody. B. Abeles, C.R. Wronski, B. Brooks and W.A. Lanford, J. Non-Cryst. Sol., 35&36, 463 (1980)
[10] cr-Si data below 3.0eV: W.C. Dash and R. Newman, Phys. Rev., 99, 1151 (1955)
cr-Si data above 3.0eV: H.R. Philipp and E.A. Taft, Phys. Rev., 120, 37 (1960)
[11] F.A. Sarott, Z. Iqbal and S. Veprek, Solid St. Commun., 42, 465 (1982)

ACCELERATED PHOTO-DEGRADATION OF UNDOPED AMORPHOUS SILICON FILMS AT LOW TEMPERATURES: AN EXPERIMENTAL STUDY

E. Eser, and E. Urbanski
Chronar Corp., P.O. Box 177, Princeton, NJ 08540, USA

ABSTRACT

An extensive experimental study of light induced degradation in hydrogenated amorphous silicon films was performed in order to test various kinetic models that were proposed in recent years. In order to distinguish clearly between various models of the time dependence of the photo-resistance, we have numerically differentiated the experimentally obtained r vs t (where r is the photo-resistance normalized to the initial value) and investigated the dependence of dr/dt on r. Under most of the experimental conditions investigated we have found that, the data fits quite well to an expression of the form $dr/dt = (D/r) - E$. The rate constant D was found to be thermally activated with an activation energy (averaged over four different samples) of 0.14 eV, and to depend linearly on light soaking intensity. These results seem to support the "Special Defect Creating Center" model. At low intensities and/or deep degradations there is evidence of a different degradation mechanism. The rate constant D was also found to depend (albeit weakly) on the hydrogen content of the films.

INTRODUCTION

Since its first observation[1], light induced degradation in a-Si:H has been the subject of many phenomanalogic[2,3,4] and kinetic[5,6,7,8,9,10] studies. As a result of these extensive studies, it is generally agreed that the degradation is the result of the breaking of the weak Si:Si bonds in the material. However, the processes responsible in transferring the photon energy to the weak bonds are still not clearly understood.

In an earlier paper[10], we presented initial experimental results supporting "Special Defect Creating Center" (SDCC) model where energy released by the recombination of the photo-generated carriers at special defects in the energy gap causes breaking of the weak Si:Si bonds. This is in contrast to a previous model where bond breaking energy is released by the band-to-band recombination of the photo-generated carriers[6].

In the present paper we present more comprehensive experimental results in support of the SDCC model. Our experimental procedure is exactly the same as the one used in our previous study[10], and therefore will not be described in detail. Briefly however, it consists of measuring the AM1 photo-resistance "R" of the material as a funtion of time "t" during a degradation under a given intensity "I" of illumination. After numerically differentiating the data, and normalizing it to the initial value of the photo-resistance, we analyze the dependence of the degradation rate "dr/dt" on "r" where "r" is the normalized photo-resistance. In order to probe a wide range of degradations and at the same time have negligible thermal recovery during the experiments, we used high illumination intensities (up to 25xAM1) and low temperatures (in the range of 150 K to 280 K).

EXPERIMENTAL RESULTS AND DISCUSSIONS

For the present work we have used three glow discharge deposited (GD #1, GD #2, GD #3) and one Photo-Chemical Vapor Deposited (PCVD) deposited a-Si:H samples. Some results on the GD #1 sample have been reported previously[10], and are duplicated here.

Fig. 1 shows the variation of the degradation rate dr/dt as a function of the inverse of the photo-resistance 1/r. The data were obtained under 16xAM1 illumination at 200 K except for sample GD #2 where the degradation was measured at 195 K. The data clearly shows a relationship of the form,

$$\frac{dr}{dt} = \frac{D}{r} - E \qquad (1)$$

The form of this equation supports a defect recombination mechanism since band-to-band recombination would have resulted in a $1/r^2$ type of dependence[8,10]. The constant term "E", however does not fit into a simple defect recombination picture, and will be discussed later.

One of the predictions of the SDCC model is that the degradation rate constant "D" should be proportional to the illumination intensity[8]. In order to test this prediction we determined the degradation rate constant (which is simply the slopes of the straight lines in plots such as figure 1) for various illumination intensities at two different temperatures for the sample GD #3. The results are given in Fig. 2. The data taken at 170 K shows a good linearity except for the point corresponding to 16xAM1 intensity. At 200 K we also see a linear relationship with some scatter for the 20xAM1 intensity but the linear relationship does not extrapolate to (0,0) point. Even with these deficiencies, we believe that the data is strongly suggestive of a linear relationship between "D" and "I". Consequently, the data of Figs. 1 and 2, taken together, lead us to conclude that the SDCC model properly describes the photo-degradation process in a-Si:H films under the experimental conditions examined here.

We have also examined the temperature dependence of the degradation rate constant for all four of the samples. The data presented in Fig. 3 shows that the rate constant is thermally activated with similar activation energies for all four samples : 0.14 ± 0.03 eV. Even though the available data is not sufficient to assign a physical meaning to this activation energy, the fact that we obtained similar values for a variety of samples suggests that the underlying mechanism is intrinsic to a-Si:H, and is independent of the sample origin.

As pointed out previously, the experimentally determined rate equation (eq. 1) contains a constant term "E" and needs to be discussed further. In our earlier paper[10] we showed that eq. 1 describes the degradation kinetics down to a certain degradation level, and that at deeper degradation levels, the rate decreased more slowly than 1/r. We ascribed this observation to the movement of the quasi-Fermi levels: as the degradation proceeds quasi-Fermi level moves deeper into the gap, and as it crosses the energy level of the SDCC, these defects are deactivated and another degradation mechanism takes over. Since the position of the quasi-Fermi levels are also controlled by the illumination intensity, at low intensities we would expect to see the transition from one degradation mechanism to another at much lower degradation levels. The data presented in Fig. 4, taken at AM1 illumination, shows that this transition takes place at x2 degradation (200 K data) while we observe no transition down to x10 degradation for 16xAM1 illumination (Fig. 1). Under these circumstances, we believe that the constant term in eq. 1 is an artifact of our analysis which does not take into account the second degradation mechanism. Assigning any physical meaning to the constant term "E" would mean that eq. 1 is valid for up to infinite degradation which is clearly not the case.

The final aspect of the photo-degradation phenomena in a-Si:H that we have examined is the dependence of the degradation rate constant on the hydrogen content of the films. The role hydrogen plays in the photo-degradation of a-Si:H has both practical and theoretical importance. However, in the present paper, we will simply present our initial observation without any detailed analysis. We have prepared a series of a-Si:H films at various deposition temperatures, measured their hydrogen content by Secondary Ion Mass Spectrscopy, and determined the degradation rate constants for the case of 16xAM1 illumination. The data is given in Fig. 5. We observe a maximum in the degradation rate in the range of 13 to 15 at % hydrogen. Below this level the degradation rate constant decreases rapidly while above this level the decrease is more gradual.

CONCLUSION

Experimental evidence presented in the present study indicates that recombination of the photo-generated carriers via "Special Defect Creating Centers" is responsible for the photo-degradation of a-Si:H films, for the case of low-to-medium degradation levels. At low intensities and/or high degradation a second degradation mechanism is also observed. In the former case, which was investigated more in detail, the degradation rate constant is found to be thermally activated with an activation energy of 0.14 ± 0.03 eV. This activation energy is intrinsic to a-Si:H and independent of the material preparation method. The degradation is also found to be dependent on the hydrogen content of the material, with a maximum around 13 to 15 at % hydrogen.

ACKNOWLEDGEMENTS

We are grateful to Dr. H. Shade in providing us with the PCVD sample and to Mr. F. Ramos for the SIMS measurements. This work was partially supported by the Solar Energy Research Institute under subcontracts No. ZB-3-03056-1 and No. ZB-7-06003-1.

REFERENCES

1- D. L. Staebler and C. R. Wronski, Appl. Phys. Lett. 31, 292 (1977)
2- D. Adler, J. Phys. (Paris) 42, C-3 (1981)
3- R. S. Crandall, Phys. Rev. B 24, 7457 (1981)
4- H. Dersch, J. Stuke, and J. Beichler, Appl. Phys. Lett. 38, 456 (1981)
5- C. Lee, W. D. Ohlsen, P. C. Taylor, H. S. Ullal, and G. P. Ceasar, Phys. Rev. B 21, 100 (1985)
6- M. Stutzmann, W. B. Jackson, and C. C. Tsai, Phys. Rev. B 32, 23 (1985)
7- E. Eser, J. Appl. Phys. 59, 3508 (1986)
8- M. Hack, S. Guha, and W. den Boer, Phys. Rev. B 33, 2512 (1986)
9- D. Redfield, Appl. Phys. Lett. 49, 1517 (1986)
10- E. Eser and E. Urbanski, Appl. Phys. Lett. 51, 2124 (1987)

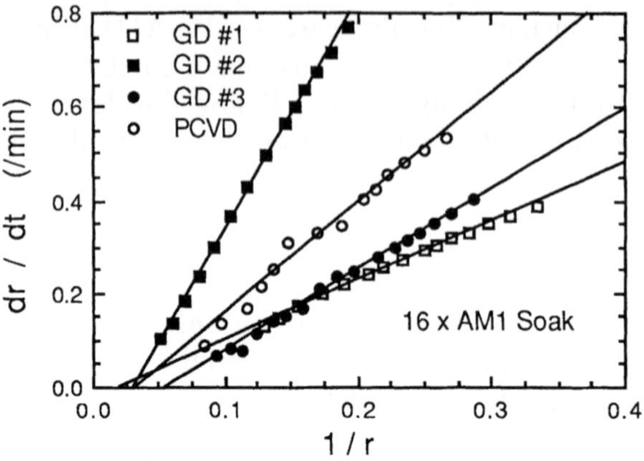

FIG. 1. Rate of change of the normalized photo-resistance as a function of inverse normalized photo-resistance for three glow discharge (GD) and one photo-CVD (PCVD) samples. T = 200 K except for GD #2 where T = 195 K.

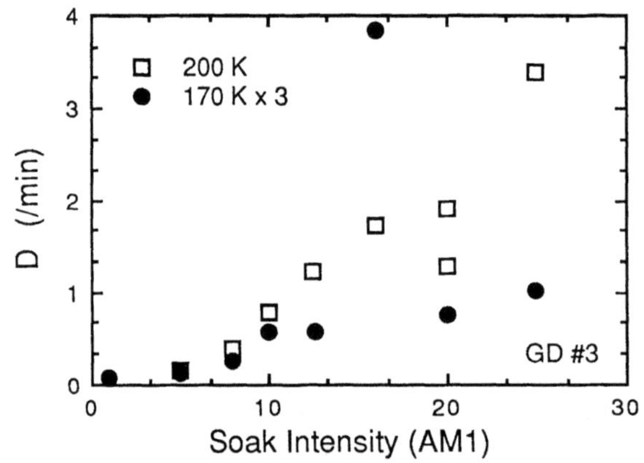

FIG. 2. Variation of the photo-degradation rate constant (D) with soak intensity at two different temperatures.

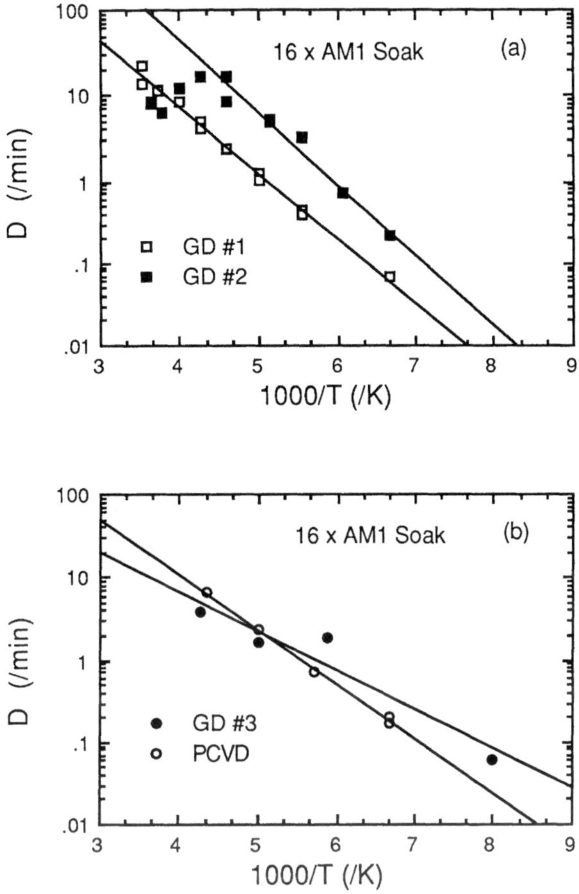

FIG. 3. Variation of the photo-degradation rate constant (D) with temperature. (a) : samples GD #1 and GD #2; (b) : samples GD #3 and PCVD.

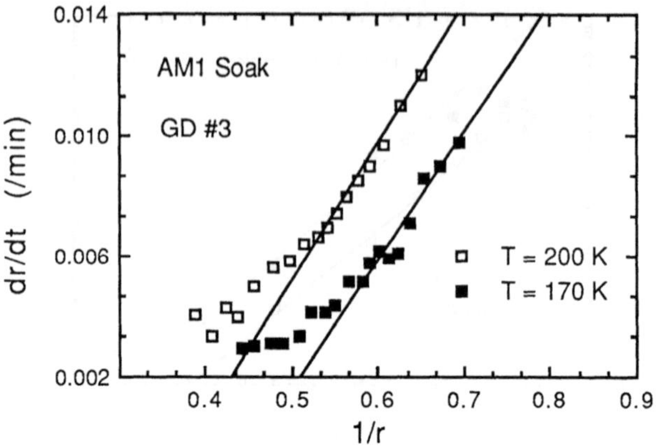

FIG. 4. Rate of change of the normalized photo-resistance as a function of inverse normalized photo-resistance for the sample GD #3.

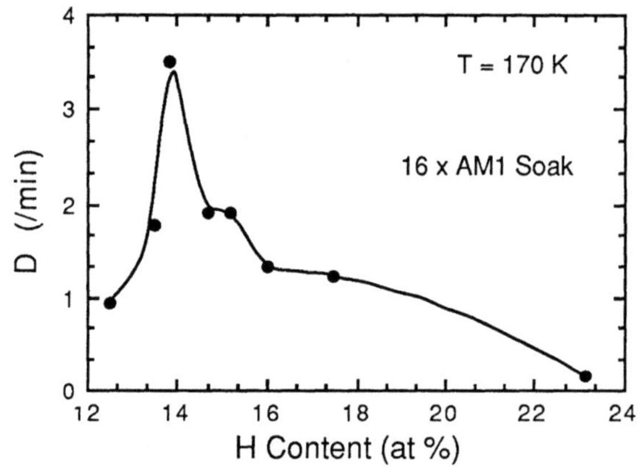

FIG. 5. Variation of the photo-degradation rate constant (D) with hydrogen content in glow discharge deposited samples.

MULTI-JUNCTION a-SiC:H/a-Si:H SOLAR CELLS

Takashi Yoshida, Takuro Ihara, Sinji Fujikake, Yukimi Ichikawa and Hiroshi Sakai
Fuji Electric Corporate Research and Development, Ltd.
2-2-1 Nagasaka, Yokosuka, Kanagawa 240-01, Japan

ABSTRACT. Double-stacked a-Si/a-Si or a-SiC/a-Si tandem solar cell is more stable than single junction a-Si solar cells. A key to improve the conversion efficiency of the tandem solar cell is open-circuit voltage. We have studied the photo and dark current-voltage (I-V) characteristics of the a-Si and a-SiC solar cells with a view to clarify the effect of the bandgap of I layer and Interface (IF) layer at the P/I junction on the open-circuit voltage (Voc). It was found that increase in the Voc by increasing the bandgap of I layer and by increasing the IF layer corresponded with a shift of the dark I-V characteristics to higher voltages. The fill factor, however, decreased with increasing bandgap of these layers because of deterioration in the film quality. On the basis ofthese results, we have optimized the design for tandem solar cells and obtainedconversion efficiencies of 10.8% and 9.1% for 1cm^2 a-Si/a-Si and a-SiC/a-Si tandem solar cells, respectively.

1. INTRODUCTION

The stability of a-Si:H based P-I-N solar cells to light soaking is the most important key to govern the applicability to practical power generation systems. In previous papers(1),(2),(3), we have reported that double-stacked a-Si/a-Si tandem solar cell is more stable than the conventional single-junction a-Si solar cell. Moreover, if the open-circuit voltage of the top cell were increased by contriving the device structure, the conversion efficiency of the tandem solar cell could exceed that of the single-junction solar cell; potentially, photons absorbed by the top cell have high energy which enables us to give a high Voc. The following two techniques are applicable to increase the Voc:

(1) Application of wide bandgap materials such as a-SiC:H and a-Si:H deposited at lower temperatures to the I layer of top cell.
(2) Insertion of the interface (IF) layer composed of a-SiC:H to the P/I junction region.

The detailed mechanism of these techniques, however, has not yet been understood very well.

In this paper, therefore, we have carried out experimental studies of these high voltage solar cells, and fabricated double-stacked tandem cells consisting of a-Si/a-Si and a-SiC/a-Si configurations.

2. EXPERIMENTAL STUDIES FOR IMPROVING Voc

We carried out a series of experiments to understand the effects of the wide bandgap I layer and the IF layer on the Voc of a-Si based solar cells. Sample cells used for the experiments were deposited with a newly developed IVE (Interdigital-Vertical-Electrode) plasma CVD apparatus(4). This consists of a load-lock chamber and two plasma CVD chambers; one is for P and N layer depositions and the other is for I layer deposition as

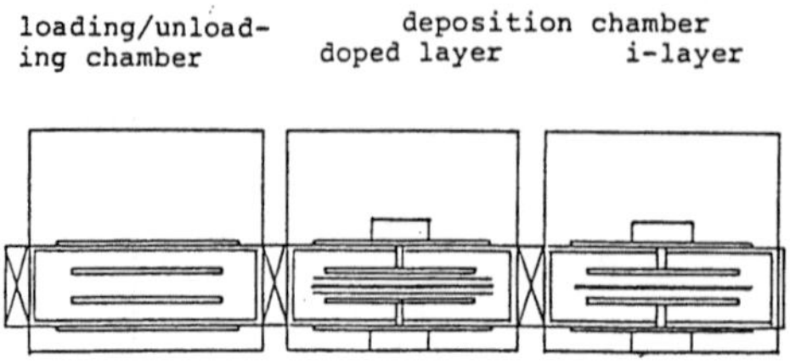

Fig.1 Schematics of IVE apparatus.

Fig.2 Voc as a function of Eg(I).

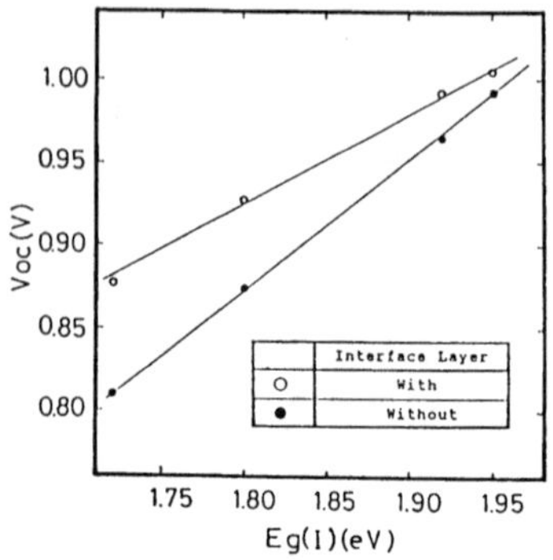

shown in Figure 1. Using this apparatus, we can deposit amorphous films on two glass substrates of up to 40cm x 120cm, simultaneously.

2.1 Effect of I layer bandgap and IF layer on Voc

First, we studied the effect of the bandgap of I layer, Eg(I) and the IF layer on the Voc. The solar cells fabricated had the configuration of Glass/SnO_2/P(a-SiC)-I(a-SiC)-N(a-Si)/Metal or Glass/SnO_2/P(a-SiC)-IF(a-SiC)-I(a-Si(C))-N(a-Si)/Metal. The bandgap Eg(I) was controlled by changing substrate temperature during a-Si:H film deposition for 1.72eV$<$ Eg(I) $<$1.83eV and by changing the content of carbon in a-SiC:H film for Eg(I) $>$ 1.85eV. In these sample cells, the bandgaps of P layer and N layer were fixed at 1.98eV and 1.70eV, respectively.

Figure 2 shows the Voc as a function of Eg(I) for cells with and without IF layer. In this experiment, the Eg(IF) was arranged to be (Eg(P)+Eg(I))/2, and the thickness of I layer and IF layer were about 1500A and 100A, respectively, where Eg(P) and Eg(IF) are the bandgaps of P layer and IF layer. As seen from this result, the Voc increases linearly for both cells with and without the IF layer as the Eg(I) increases, and the slope for the cell with the IF layer is almost one half of that for the cell without the IF layer. This means that the Voc of P-I-N a-Si based solar cells is dominated mainly by the bandgap of the intrinsic layer within an order of 100A from the P/I interface.

2.2 Photo and Dark I-V Characteristics

To understand the detailed mechanism for increase in the Voc, we next measured the photo current Iph and the dark current Id of these solar cells as a function of bias voltage. In a simplified model, the photo voltaic characteristics is given by the summation of Iph and Id as shown in Figure 3. The validity to apply the simplified model to the a-Si based solar cell was evaluated experimentally. To do that, we measured Iph and Id separately, and compared the summation of Iph and Id with the I-V curve under light illumination condition. As a result, we found that the actual I-V curve was expressed by the simplified model very well. In the experiment, the Iph was measured by using lock-in amplifier with chopped light ($100mW/cm^2$). Since the simplified model was confirmed to be useful, next we measured Iph and Id as a function of bias voltage for various sample cells used for the experiment in Figure 2 to study the effect of the bandgap of I layer and the insertion of IF layer on these characteristics.

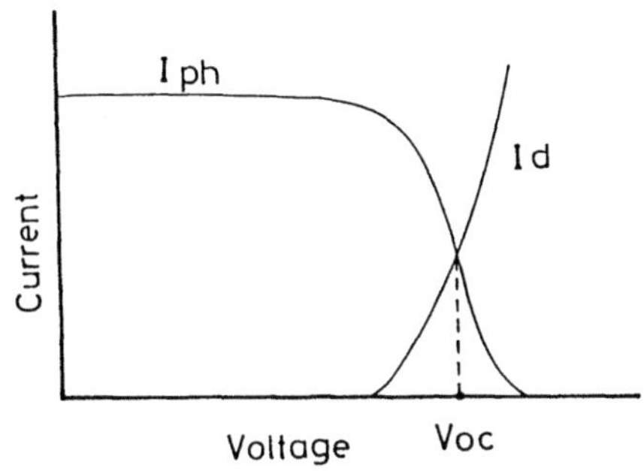

Fig.3 Photo and dark I-V curve.

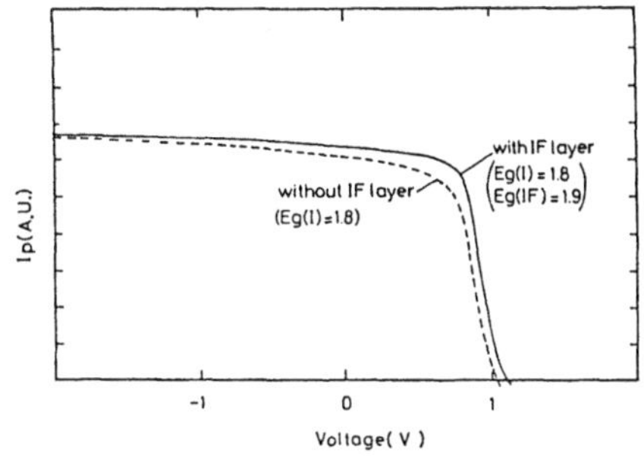

Fig.4 Photo I-V curves for cells with and without IF layer.

Figure 4 shows the voltage dependance of Iph for cells with and without IF layer (Eg(I)=1.8eV,Eg(IF)=1.9eV). As seen from this figure, flat-band voltage Vf is increased and the voltage dependance of the Iph becomes flat by inserting the IF layer. These result from prevention of electron back diffusion to the P/I interface region by generating a bilt-in fieled as reported (2).

Figure 5 shows the voltage dependance of Iph for cells with various Eg(I)(Eg(I)=1.8eV,1.9eV,1.95eV). The Vf of the cell with Eg(I)=1.95eV is increased by 0.16V, compared with that with Eg(I)=1.8eV. Thus the increase in Vf is almost equal to the difference of Eg(I). These results indicate that the built-in potential of the a-Si based P-I-N cells with the same P and N layers is dominated by Eg(I), especially near P/I interface.

Next we show the dark I-V characteristics of the cells. Figure 6 shows the dark I-V characteristics of the cell (Eg(I)=1.8eV) with and without IF layer (Eg(IF)=1.9eV). The Id-V curve is shifted towards a higher forward bias voltage by inserting the IF layer. Figure 7 shows dark I-V characteristics of cells with various Eg(I)(1.8eV, 1.9eV, 1.95eV). The Id-V curves also shifted toward higher forward bias voltages as Eg(I) increases.

To understand above results, we discuss the recombination current of P-I-N diodes. The forward dark current of P-I-N diode can be approximated by the sum of diffusion current and recombination current. The shift of Id-

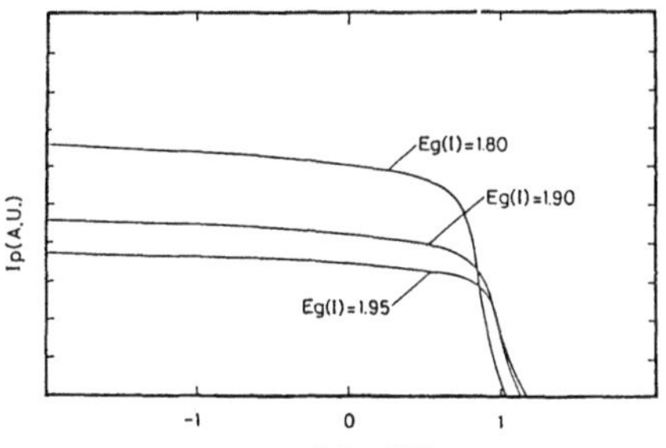

Fig.5 Photo I-V curves for Eg(I)= 1.80, 1.90 and 1.95eV.

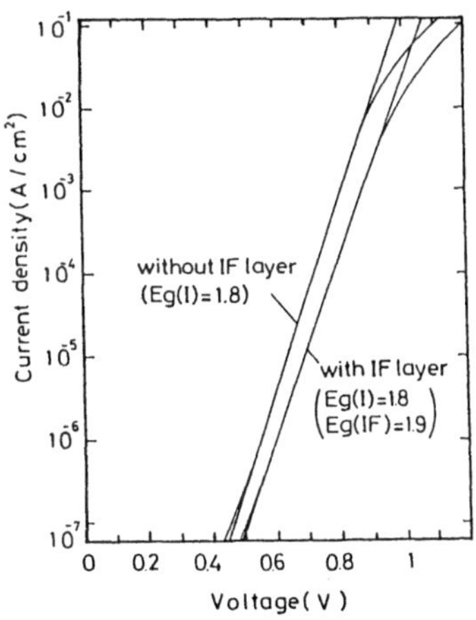

Fig.6 Park I-V curves for cells with and without IF layer.

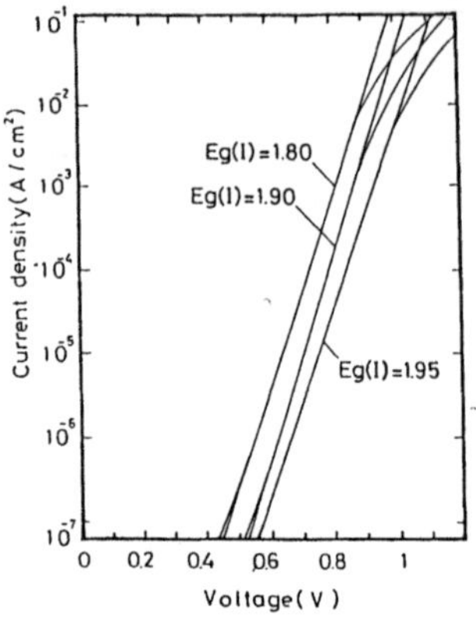

Fig.7 Dark I-V curves for Eg(I)= 1.80, 1.90 and 1,95eV.

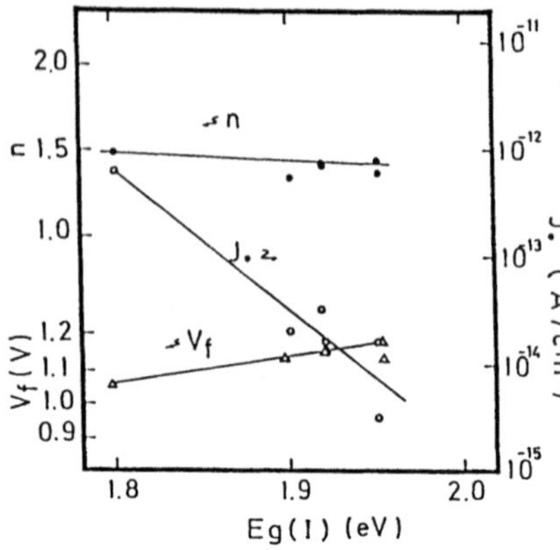

Fig.8 n, Vf and Jo as a function of Eg(I).

V curve by insertion of the IF layer means decrease in recombination current, especially near the P/I interface region. The change in recombination current, Irec(IF), may be explained by decrease in the intrinsic carrier density Ni at the P/I interface region since Irec(IF) is propotional to Ni(IF)/τ, where τ is the carrier life time at P/I interface region.

The shift of the Id-V curve to higher forward bias voltages for the cells with various Eg(I) is also explained by change in the intrinsic carrier density of I layer Ni(I). Figure 8 shows diode ideality factor n, saturation current density Jo, and Vf as a function of Eg(I); the experimental results are expressed by J=Joexp(qV/nkT). The value of n is about 1.5 for Eg(I)=1.8 to 1.95 eV. On the contrary, Jo decreases with increasing Eg(I). The other experimental results showed that Jo and n increased rapidly because of reduction of τ , when Eg(I) exceeded a certain value (e.g. 2.0eV) or low quality a-SiC:H films were used.

From the above discussion we consider that the recombination current for P(a-SiC)-I-N diode consists of Irec(IF) and others (e.g. recombination in I layer Irec(I) and recombination at I/N interface Irec(I/N)). As seen from Figures 6 and 7, I-d curves for both a-Si:H I layer cell with IF layer (Eg(I)=1.8eV, Eg(IF)=1.9eV) and a-SiC:H I layer cell (Eg(I)=1.9eV) are in good agreement in the current range of 10^{-2}-10^{-3} A. This result suggests that the main term of the recombination current is Irec(IF) for a-Si based cells with an I layer thickness of 1500A or less.

3. FABRICATION OF TANDEM CELLS

Two techniques described above have been applied to the tandem cells with the Glass/SnO_2/P(a-SiC)-IF(a-SiC)-I(a-Si or a-SiC)-N(μc-Si)/P-IF-I(a-Si)-N/Metal structure; one is to optimize the IF layer and the other is to change the Eg(I) of the top cell by using a-SiC or changing deposition temperature. The highest conversion efficiency of the a-Si/a-Si tandem cell obtained to date is 10.8% (AM1, 100mW/cm^2). The I-V curve is shown in Figure 9. The Eg(I) of the top and bottom cells used were 1.8eV and 1.73eV, respectivery.

Figure 10 shows the I-V curve of a-SiC/a-Si tandem cell under simulated sunlight. The value of Voc exceeded a-Si/a-Si tandem cell by 0.1V

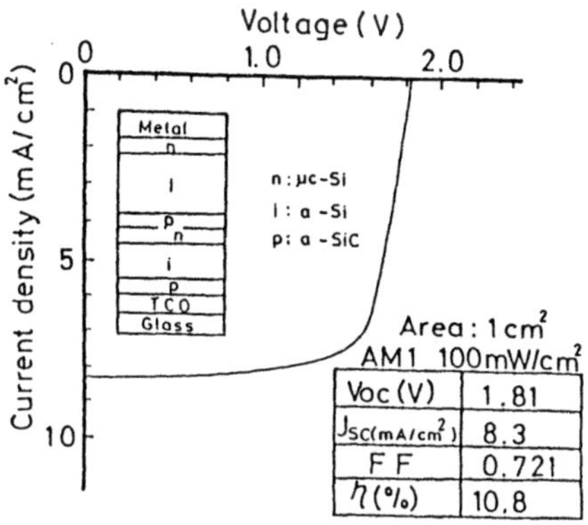

Fig.9 I-V curve of 10.8%a-Si/a-Si tandem cell.

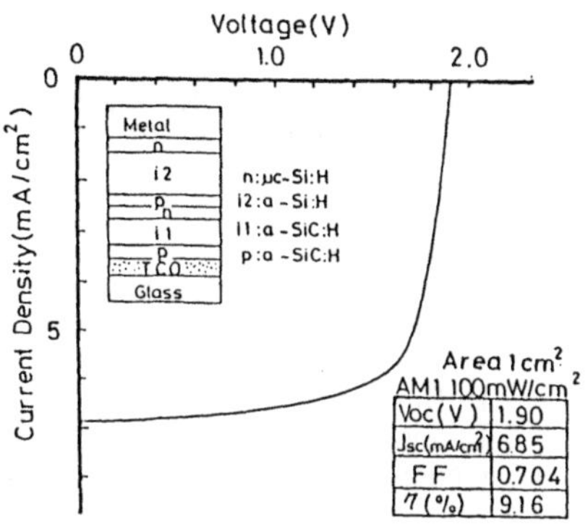

Fig.10 I-V curve of 9.1% a-SiC/a-Si tandem cell.

because of a higher Eg(I) (1.95eV) for the top cell. The short-circuit current Jsc is, however, small compared with that of a-Si/a-Si tandem cell. This is probably due to the quality of a-SiC:H film, because the FF decreases with increasing thickness of the I layer. Thus, we believe that the improvement in conversion efficiency should be attained by increasing the quality of a-SiC:H film.

4. CONCLUSIONS

We studied the effect of the bandgap of I layer and the interface layer on the open-circuit voltage experimentally, and the following results are obtained:

(1) Theopen-circuit voltage is dominated by mainly the bandgap of the intrinsic layer within an order of 100A from the P/I interface.
(2) Increase in the open-circuit voltage is attributed to a shift of dark I-V characteristics to higher bias voltages.
(3) To increase the open-circuit voltage, therefore, we should reduce the saturation current in the dark I-V characteristics, or increase the ideality factor without increasing the saturation current.
(4) Applying the obtained results to the design for double-stacked tandem cells, we have obtained conversion efficiencies of 10.8% and 9.1% for a-Si/a-Si and a-SiC/a-Si 1cm^2 tandem cells, respectively.

ACKNOWLEDGEMENT

The authors are grateful to Dr. Haruki of Fuji Electric for his advice. This work was supported by the New Energy Development Organization as a part of Sunshine Project.

REFERENCES

(1) Y.Uchida, Proc. 17th E.C. PVSEC, Sevillia, 1986, p.395.
(2) T.Yoshida, K.Maruyama, O.Nabeta, Y.Ichikawa, H.Sakai and Y.Uchida, Proc. 19th IEEE Photovoltaic Speci.conf., New Orleans, 1987, p.1095.
(3) T.Yoshida, S.Fujikake, H.Shimabukuro, Y.Ichikawa and H.Sakai, Proc. 3rd Int. PVSEC, TOKYO, p.465.
(4) H.Sakai and Y.Uchida, Proc. 3rd Int. PVSEC, TOKYO, p.377.

RELATIONS BETWEEN RAMAN SCATTERING AND OPTICAL ABSORPTION PARAMETERS IN PECVD a-Si:N:H ALLOYS

E. Morgado
Centro de Física Molecular das Universidades de Lisboa
Av. Rovisco Pais, Complexo I, IST, 1000 Lisboa, Portugal

E. Bustarret
Laboratoire d'Études des Propriétés Électroniques des Solides
CNRS, BP 166, 38042 Grenoble Cedex, France

ABSTRACT. Thin films of a-Si:N:H prepared by r.f. glow-discharge were characterized by Raman scattering and PDS spectroscopies. Estimates of the N atomic concentration in the films were obtained by IR spectroscopy. In the Raman spectrum the fullwidth at half maximum $\Delta\omega_{TO}$ and upper half width at half maximum $\Gamma/2$ of the TO peak both increase with the N content, as well as the relative intensity of the LO band, while the peak maximum ω_{TO} slightly shifts to lower frequencies. New correlations are found between the parameters $1/\Delta\omega_{TO}$ or $1/\Gamma$ and the Urbach reciprocal slope E_U in the optical absorption spectrum. In the light of previous theoretical works the present data on a-Si:N:H indicate that the characteristic energy of the valence band tail in the electronic density of states is directly related to the Si-Si bond angle distortions which are enhanced by the incorporation of Nitrogen atoms.

1. INTRODUCTION

Silicon rich a-Si:N:H alloys have been currently studied due to their potential use as a variable band gap material in spectrally selective optoelectronic devices such as photovoltaic cells and light emitting diodes. It has been generally found that the incorporation of Nitrogen atoms in the a-Si matrix not only widens the optical gap but also strongly affects the electronic properties and the gap-states density. A marked effect is the increase of the width of the valence band-tail of states with the N concentration [1,2].

The Raman spectrum reflects the phonon density of states and can show how the Si-matrix short-range order is affected by the incorporation of Nitrogen. The optical absorption spectrum provides information about the electronic density of states. Searching for relations between Raman spectrum shape parameters and optical absorption parameters may help in the understanding of the effects of microstructure on the optoelectronic properties.

2. EXPERIMENTAL DETAILS

a-Si:N:H thin films with varying Nitrogen contents were prepared in a capacitively coupled r.f. glow-discharge system from a SiH_4 (10% in Argon) + N_2 mixture in conditions described elsewhere [1]. Control of the N incorporation in the films was achieved varying the gas molar ratio N_2/SiH_4 in the gas mixture. Photothermal Deflection Spectroscopy (PDS) was used to obtain a reliable estimate of the Urbach parameter E_U deduced from the exponential region of the optical absorption spectrum. The PDS set-up has been described elsewhere [3]. Raman measurements were performed at 5145 Å in the backscattering geometry. The polarization configuration was VV and the incident power was kept around 50 mW. The spectral resolution was between 2,5 and 6 cm^{-1}.

3. RESULTS AND DISCUSSION

In the present analysis we concentrate on shape parameters associated with the TO-like peak in the Raman spectrum : the full width at half maximum FWHM $\Delta\omega_{TO}$, the half width at half maximum measured on the high frequency side $\Gamma/2$ and the frequency ω_{TO} of the optic peak maximum. The evolution of these parameters with the Nitrogen content is shown in Fig 1. The Nitrogen concentration C_N was estimated from the integrated absorption of the Si-N asymmmetric stretching band around 850 cm^{-1} in the IR spectra, using the calibration factor of Sasaki [4] :

The Γ parameter, which represents the width of the TO peak, increases with the N content (Fig. 1-b). Several theoretical models associate the width of the TO peak with the degree of angular disorder . A relation between Γ and the r.m.s. deviation $\Delta\theta_b$ of the bond angle distribution of the Silicon matrix was derived by Beeman et al. [5]:

$$\Gamma = 15 + 6\,\Delta\theta_b \tag{1}$$

Another ordinate scale in Fig. 1-b indicates the corresponding $\Delta\theta_b$ according to this formula. A similar relation proposed by Tsu et al [6] gives pratically the same values.

The parameter $\Delta\omega_{TO}$ depends not only on the TO band width but also on the relative intensity of the LO band expressed by the ratio I_{LO}/I_{TO}. The increse of $\Delta\omega_{TO}$ with C_N in Fig. 1-a probably includes these two effects.

The maximum of the TO peak, ω_{TO}, shifts to lower frequencies with increasing N content, as is shown in Fig. 1- c. This trend is not inconsistent with recent EXAFS results which indicate a small increase of the Si-Si bond length with the N/Si ratio [7].

Lannin et al. [8] observed a positive correlation between the reciprocal FWHM $1/\Delta\omega_{TO}$ and the optical gap in a set of sputttered and CVD a-Si samples, prepared at different deposition temperatures, the first reported correlation between Raman and optical parameters. An optically derived quantity which seems more closely related with disorder is the Urbach energy E_U deduced from the exponential region of the optical absorption spectrum. Thus, it appears quite appropriate to search for a correlation between $1/\Delta\omega_{TO}$ or $1/\Gamma$ and E_U , and that is indeed observed as shown in Fig. 2. If we remember that in a-Si:H an increase in the optical gap is associated with a decrease in the the Urbach energy E_U[9] we can conclude that a-Si:H material should also obey the trends depicted in Fig. 2. However, a-Si:N:H does not obey the Lannin correlation [8] because in these alloys an increase in the optical gap is generally associated with an increase in E_U [1, 2, 10].

A more simple correlation is that one we have reported in a previous paper between Γ and E_U [10], which was regarded as an indication that the width of the exponential valence band tail of states was closely related to the bond-angle distortions of the Silicon tetrahedra. Recently, Smith et al. [11] have proposed a relationship between the characteristic energy for the formation of valence band-tail states E_{form} and the r.m.s. bond-angle deviation $\Delta\theta_b$. According to Bar-Yam et al. [12] the characteristic energy of the valence band tail of states should be equal to the formation energy of these states, i.e., $E_{form} = E_U$. This allows to write :

$$E_U = (3/4)\,K_\theta\,(r_b\,\Delta\theta_b)^2 \tag{2}$$

where K_θ is the spring constant for bond-bending and r_b the Si-Si bond length.

In Fig. 3 one can see that our a-Si:N:H samples agree reasonably well with the model expressed by equation (2). The values for the constants are those of the V.F.F. model, $k_\theta = 9.2 \times 10^3$ dynes/cm and $r_b = 2.35$ Å, and the value of $\Delta\theta_b$ for each sample was calculated from the measured TO width Γ through the Beeman relation (1). A more accurate approach taking into account the H-Si-Si bond angles has also been proposed recently [13]. We point out that this is the first comparison between equation (2) and experimental values of E_U and $\Delta\theta_b$ (or Γ) measured on the same samples. The discussion made in Smith's paper [11] was based on values of those parameters measured on different samples which had in common the fact that they have been prepared at similar deposition temperatures. It seems to us that the principal meaning of the general agreement of the experimental points with the theoretical curve in Fig. 3 is that the characteristic energy of the valence band tail of states is strongly determined by the bond-angle distortions.

4. CONCLUSIONS

A study was made of the evolution of the Raman spectrum of PECVD a-Si:N:H thin films with different Nitrogen contents and its correlation with optical parameters.

The width of the TO peak increases with the N concentration which can be related mainly to an enhancement of the Si-Si bond-angle distortions induced by the incorporation of trivalent N atoms. A shift of the optic peak maximum to lower frequencies and an increase in the relative intensity of the LO region with increasing N content were also observed.

New correlations were found between the parameters of the Raman spectrum $1/\Delta\omega_{TO}$ or $1/\Gamma$ and the Urbach energy E_U deduced from the exponential region of the optical absorption spectrum . These correlations, which clarify and include those previously found in a-Si:H, show a dependence between structural and electronic disorders. Our data agree reasonably well with models identifying the Urbach energy with local elastic potential fluctuations.

In summary, the present results show that a marked effect of Nitrogen incorporation into the Silicon matrix is the enhancement of angular disorder, which accounts for the broadening of the valence band-tail of states and may, therefore, affect the photoelectrical properties.

REFERENCES

[1] E. Morgado and L. Guimarães, J. Non-Cryst. Sol. 77/78 (1985) 949
[2] E. Morgado and E. Bustarret, Proc. 7th Eur. Photovoltaic Sol. En. Conf. (D. Reidel, Dordrecht, 1987) p. 567
[3] F. Boulitrop, J. Bullot, M. Gauthier, M.P. Schmidt and Y. Catherine, Solid State Comm. 54 (1985) 107
[4] G. Sasaki, M. Kondo, S. Fujita and A. Sasaki, Jap. J. Appl. Phys. 21 (1982) 1394
[5] D. Beeman, R. Tsu and M.F. Thorpe, Phys. Rev. B 32 (1985) 874
[6] R. Tsu, J.G. Hernandez and F.H. Pollak, J. Non-Cryst. Sol. 66 (1984) 109
[7] S. Mobilio and A. Fillipponi, Proc. 12th ICALS (North-Holland, Amsterdam, 1987) p. 365
[8] J.S. Lannin, L.J. Pillione, S.T. Kchirsagar, R. Messier and R.C. Ross, Phys. Rev. B 26 (1986) 3506
[9] G.D. Cody, T. Tiedje, B. Abeles, B. Brooks and Y. Goldstein, Phys. Rev. Lett. 47 (1981) 1480
[10] E. Bustarret and E. Morgado, Solid State Comm. 63 (1987) 581
[11] Z.E. Smith, S. Aljishi and S. Wagner, Proc. 12th ICALS (North-Holland, Amsterdam, 1987) p. 775
[12] Y. Bar-Yam, D. Adler and J.D. Joannopoulos, Phys. Rev. Lett. 57 (1986) 467
[13] E. Bustarret, F. Vaillant and B. Hebb, M.R.S. Symposium E, Reno (April 1988)

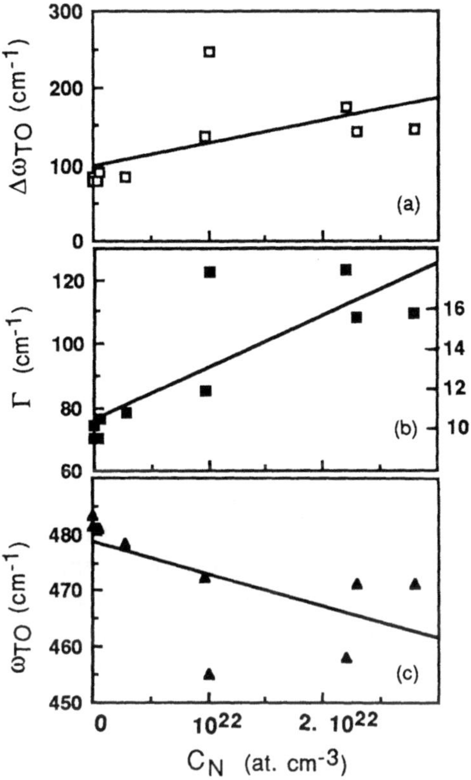

Fig. 1 - Dependence of the Raman TO peak parameters $\Delta\omega_{TO}$ (a), Γ (b) and ω_{TO} (c) on the Nitrogen concentration C_N.

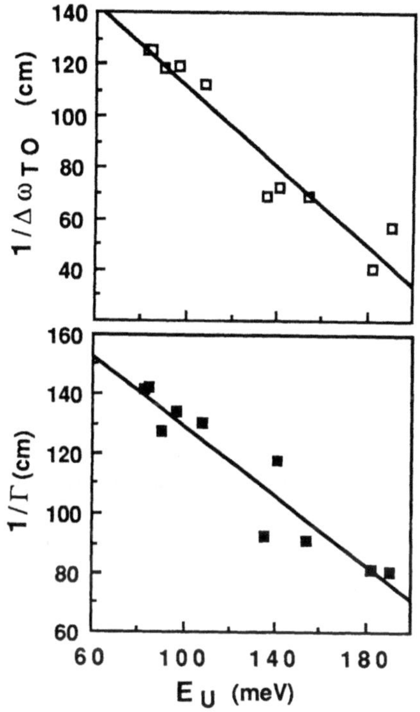

Fig. 2 - Correlations between the reciprocal widths of the Raman TO peak $1/\Delta\omega_{TO}$ or $1/\Gamma$ and the Urbach energy E_U.

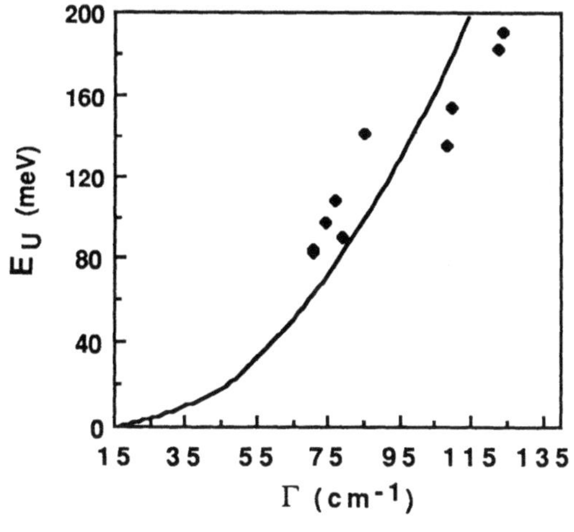

Fig. 3 - Comparison between experimental data (◆) and equation 2 (—).

HYDROGENATED AMORPHOUS SILICON BY DISILANE LPCVD (*)

C. Manfredotti, F. Fizzotti
Experimental Physics Dept., University of Torino, Via Giuria 1, Torino (Italy)
and National Group of Matter Structure of National Council for Researches (Italy)

G. Amato
National Electrotechnical Institute "G.Ferraris", Strada delle Cacce, 91 Torino (Italy)

and T. Shasmi, E. Vittone
TEAM Srl, Via Verbano 2, Ispra (Varese, Italy)

ABSTRACT - Intrinsic amorphous silicon films have been deposited from Si_2H_6 by using a standard production LPCVD apparatus hosting more than 100 3" wafers in stand-up position. A deposition temperatures around 400°C, pressures in the 5-10 Torr interval and gas holding times of 250 s turned out to be suitable for a reasonable homogeneity along the tube axis, a good stability and reproducibility of the films, together with fairly good electronic properties.

The samples have been characterized by IR absorption, SIMS, dark conductivity, AM 1.5 photoconductive gain, spectral response of photoconductivity, optical absorption and PDS measurements.

The results of the characterization indicate hydrogen contents of the order of 15%, homogeneously distributed along the depth of the samples, oxygen concentration lower than 10^{19} cm^{-3}, dark conductivities of the order of 10^{-10} $(\Omega\,cm)^{-1}$ and AM 1.5 photoconductivity up to 10^{-5} $(\Omega\,cm)^{-1}$, optical energy gaps above 1.7 eV, Urbach's tails with a logarithmic slope of 70 meV, mobility-lifetime products for electrons just below 10^{-7} cm^2 V^{-1} and defects densities of the order of 10^{16} cm^{-3}, even for samples thinner than 0.5 µm.

All these results, which presently confirm preliminary results previously reported, indicate clearly that LPCVD technique is really suitable for obtaining "solar grade" amorphous silicon and that the envisaged limits for this method could be overcome by using suitable apparatus and deposition conditions.

(*) Work partially supported by E.C. under Contract N. EN3S-0092-I(A)

INTRODUCTION

The Low Pressure Chemical Vapour Deposition (LPCVD) technique, which is largely used in electronic industry, has been tried in the past also for depositing a-Si:H films[1,2,3] with results that, at those times, were not extremely different from GD ones. However, solar cell efficiencies never larger than 5% have been reported so far, with Jsc values up to 10 $mAcm^{-2}$, but with fill factors below 0.5, which were essentially attributed to an i-Si defects density, above 10^{17} cm^{-3} and to a bad p-i interface[4]. The apparatuses used until then were very small and the short gas residence times necessarily adopted made the deposition conditions and rates strongly not-uniform along the growth tube, while the low volume-to-surface ratio created probably contamination problems.

In the present work[5] we adopted a quite different approach, by using a big deposition apparatus and adopting very long gas holding times, both in order to get uniform deposition conditions and not to vaste disilane. The pressure was kept below 10 Torr in order to avoid powder formation and deposition. By putting the substrates on a stand-up position and by spacing them by distances of the order of 1 mm or shorter, it is possible to get even more than 500 deposited samples in a single run. Unfortunately, powder problems did not allow to overcome deposition rates of 1-2 A/s^{-1}, which are nevertheless comparable to standard SiH_4 GD growth velocities. The deposition temperature (420°C) was selected in order to be compatible with a whole p-i-n cell deposition process, to avoid strongly bound hydrogen evolution and not to allow $(SiH_2)_n$ formation in the film. At temperatures above 420°C p-layer formation begins to be uncontrollable because of the high deposition rate. Moreover, at temperatures above 400°C interdiffusion of dopants and out diffusion of hydrogen in pre-deposited layers could be important.

2 - EXPERIMENTAL

The depositions were carried out in a 5" fused silica ampoule, 2.20 m long, at 420°C and, generally, at 5 Torr, with a Si_2H_6 flux of 15 SCCM and a gas residence time of 270 s. Gas supply was regulated by MFC and all the process steps were automatically controlled. Pure electronic grade Si_2H_6 from Matheson and Air Liquide (less than 5 ppm H_2O) was used together with electronic grade 99.999% N_2 for purging. The flat temperature zone (within 10°C) was 50 cm long in the center of the growth tube.
The substrates (silicon, corning glass, quartz) were etched and carefully cleaned by standard procedures; before deposition, they were kept in the hot tube under N_2 flushing for several hours. The samples charging was carried out under a laminar flow bench. The vacuum apparatus was equipped with a water trap at the end of the tube and with and alumina trap at the mechanical pump inlet. Even without using the pumping down before deposition, oxygen and carbon content in the samples were largely below

0.1% (see Fig. 1), i.e. lower than in GD samples obtained with pumping down by a standard apparatus. Therefore, the outgassing produced by the higher temperature is more than balanced by the absence of that due to the discharge. The hydrogen concentration was uniform and generally above 4% as determined by N-15 technique (see Fig. 2). This demonstrates that, even by using long gas holding times, we do not have a time-dependence of the gas-phase composition. No dependence of the hydrogen concentration on the sample position along the tube was observed.
IR absorption spectra (Fig. 3) indicate clearly that SiH bonding is largely dominant in our samples, at least for deposition temperatures above 400°C. The disagreement between IR and N-15 data concerning total hydrogen concentration will be discussed in details in a future work.

3 - RESULTS AND DISCUSSION

Optical data (see Fig. 4) indicate consistently that optical energy gaps are above 1.7 eV and reach in some cases even 1.8 eV. If hydrogen concentration of 4% is correct, this means that hydrogen is more efficient in enlarging the optical gaps in our samples. The values of B are in 700 $cm^{-1/2}\ eV^{-1/2}$ range. It has to be noticed that values of Eg larger than 1.7 eV have never been reached by LPCVD.

Two indicative PDS absorption spectra are shown in Fig. 5: from these one can evaluate the Urbach's tail logarithmic slope E_o and, by integrating the absorption coefficient from 1.25 eV down to 0.75 eV and by eliminating the absorption due to tail states, one can estimate a total defect density N_d which should be comparable to total spin density [6]. The N_d values quoted in Tab. I inversely follow the sample thickness, as reported also for GD samples [7], and indicate clearly the possibility of reaching 10^{16} cm^{-3} for samples of suitable thickness. While the values of E_o are certainly larger than those reported for GD samples, the values of Nd are comparable and are much lower than those previously reported [2,4].

Finally, a set of normalized spectral responses of photoconductivity are reported in Fig. 6. From these spectra it is possible to get the absorption coefficient in the Urbach's tail region, by a normalization with the measured absorption coefficient in the 1.8÷2.4 eV region, and also to obtain the mobility-lifetime product $\mu\tau$, evaluated at 1.8 eV for incident fluxes of about 10^{17} $cm^{-2}S^{-1}$. The results are reported in Tab. II together with E_o values determined this way and AM 1.5 photoconductive gains, AM 1.5 photoconductivities and dark conductivities. One can observe that, while E_o values are compatible with those ones reporte in Tab. I, and G, δp and δd values are comparable with the average data obtained for GD samples, $\mu\tau$ values seem to be slightly low, even if still comparable with quoted GD data [8].

TABLE I

SAMPLE	E_o(meV)	N_d(cm^{-3})	Eg(eV)	d(μm)
7/8/87 P4	83	$1.5 10^{11}$	1.73	0,3
27/8/87 P4	79.5	$2.7 10^{17}$	1.75	0,16
8/3/88 P5	77.5	$3.9 10^{16}$	1.62	0,5

Values of E_o, defects density N_d, optical energy gap Eg and thickness d as determined by PDS measurements.

TABLE II

SAMPLE	Eo(meV)	ημτ (cm^2V^{-1})	G	δp	δd
12/4/88 P1	75 meV	$6.5\ 10^{-8}$	$1.4\ 10^4$	$1.35\ 10^{-5}$	$1.25\ 10^{-9}$
22/3/88 P1	72 meV	$3.9\ 10^{-8}$	$1.1\ 10^4$	$4.8\ 10^{-6}$	$4.7\ 10^{-10}$
22/3/88 P2	74 meV	$3.5\ 10^{-8}$	10^{-4}	$4.2\ 10^{-6}$	$4.2\ 10^{-10}$
8/3/88 P5	73 meV	$1.8\ 10^{-8}$	$7.5\ 10^3$	$5.7\ 10^{-6}$	$7.7\ 10^{-10}$

Values of E_o,ημτ, AM 1.5 photoconductive gain, AM 1.5 photoconductivity δ_p and dark conductivity δ_d for the latest samples, as determined by photoconductivity methods.

4 - CONCLUSIONS

LPCVD a-Si:H seems to be now at such a stage, that it is really not unrealistic to forecast applications to the optoelectronics field in general and to the photovaltaic field in particular.

Its properties are similar to those of GD films deposited few years ago, but it is certainly more stable and reproducible. Moreover, Stabler -Wronski effect is practically absent, and the classical instabilities due to weakly bound hydrogen are not present.

We are presently planning to realize p-i-n solar cells with films entirely deposited by LPCVD in the near future: the expectation is clearly to overcome a 5% conversion efficiency, which represents the best obtained by LPCVD so far.

ACKNOWLEDGEMENTS

Thanks are due to Prof. G. Della Mea of Trento University for N-15 measurements, which were carried out at Legnaro INFN Laboratories, and to Dr. Grillo of ENI Ricerche (Monterotondo, Italy) for PDS measurements.

REFERENCES

1) Y. Ashida, Y. Mishima, M. Hirose, Y. Osaka and K. Kojima, Jap. J. Appl. Phis. 23, L 129 (1984)
2) S.S. Hegedus, R.E. Rocheleau, J.M. Cebulka and B.N. Baron, J. Appl. Phys. 60, 1046 (1986)
3) F.B. Ellis, R.G. Gordon, W. Paul and B.G. Yakobi, Appl. Phys. Lett. 47, 135 (1985)
4) S.S. Hegedus, R.E. Rocheleau, W. Buchaman and B.N. Baron, J. Appl. Phys. 61, 381 (1987)
5) C. Manfredotti, G. Amato, F. Fizzotti, T. Shamsi and E. Vittone, ISES Conference, Hambourg 1987
6) N.M. Amer and W.B. Jackson, "Semiconductors and Semimetals", Vol. 27, Part. B, Chapter 3
7) M. Farre, H. Curtins, and A.V. Shah, Int. Conf. on Amorphous Liquids and Semiconductors, Prague 1987
8) H. Okamoto, Jarect Vol. 16, Amorphous Tehnologies and Devices (1984), Ed. Y. Hamakawa, OHMSHA Ltd. p. 120.

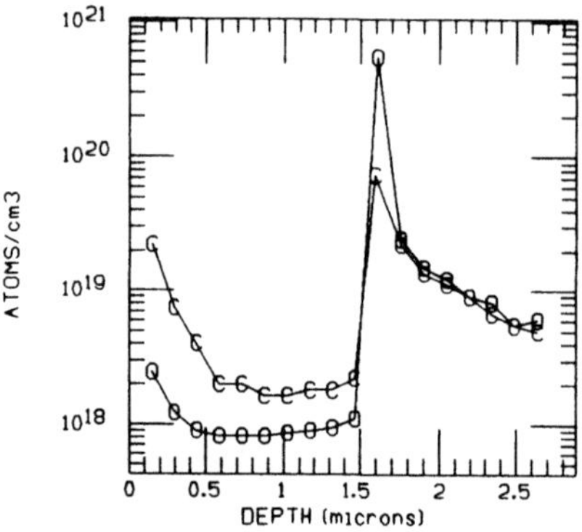

Fig. 1 - SIMS depth profile for oxygen and carbon in a typical sample.

Fig.2 - Hydrogen concentration profile in two samples as measured by N-15 technique.

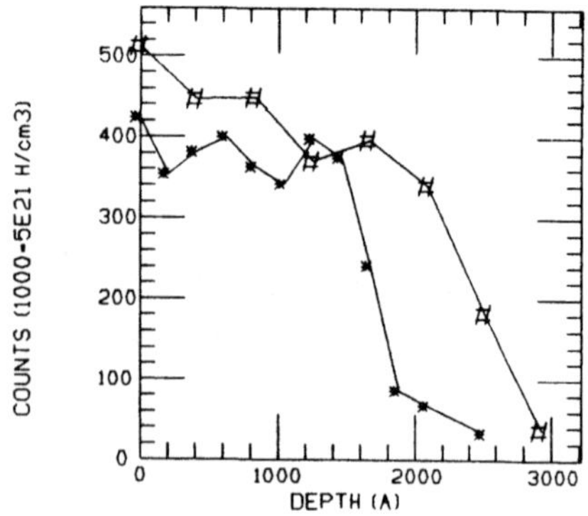

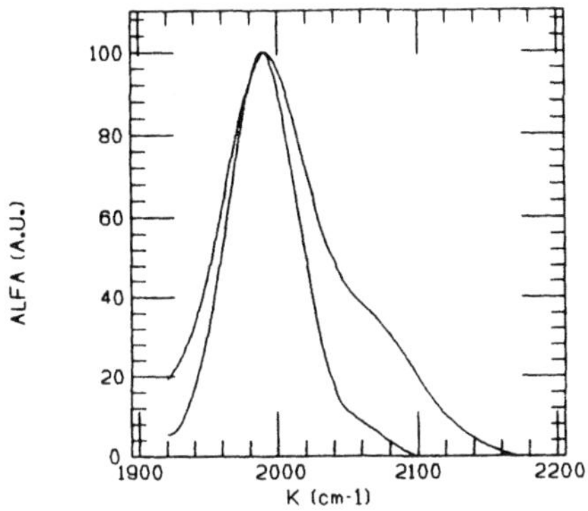

Fig.3- IR absorption coefficient versus wave number for SiH (1990 cm^{-1}) and (SiH_2) (2070 cm^{-1}) stretching modes, for two typical samples.

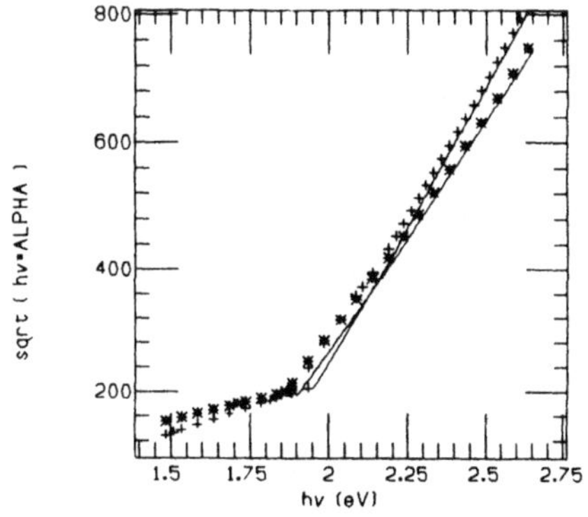

Fig. 4 - Tauc's plots for two typical samples

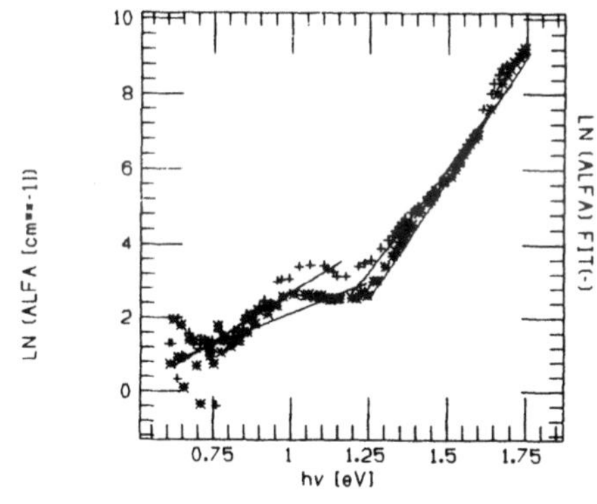

Fig. 5 - PDS absorption spectra for two indicative samples

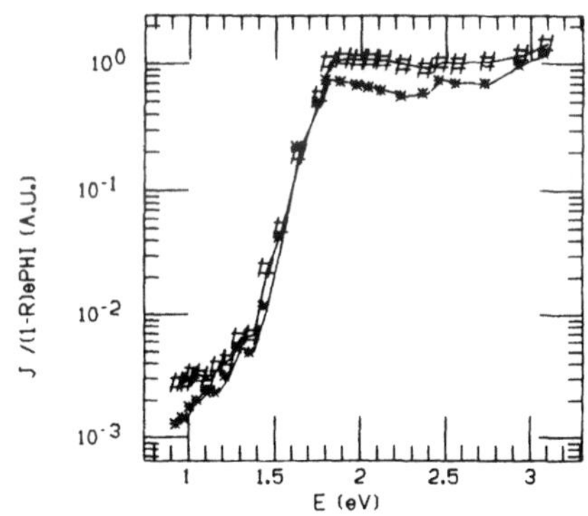

Fig.6- Normalized photoconductivity spectral response for two typical samples.

VARIABLE ENERGY EBIC ANALYSIS OF a-Si:H PIN DEVICES

B. EQUER, A. KARAR, Laboratoire de Physique des Interfaces et des Couches Minces. (ER 258 CNRS), Ecole Polytechnique, 91128 FRANCE
J. ANDREU, Depart.de Fisica Aplicada, Universitat de Barcelona, SPAIN
S. NAJAR, Dept. de Génie Electrique, ENIS, SFAX, Tunisie

ABSTRACT :

Electron beams can be used as a probe to analyse collection efficiencies in a-Si:H p-i-n devices. In this study, electrons in the 1-30 KeV range, are used to generate electron-hole pairs at different and well defined depths in p-i-n devices. In well characterized materials, analysis of the induced current can be related to collection efficiencies and to electric potential profiles, the knowledge of which is an important factor in the optimization of single gap or multigap solar cells and other thin film devices. This paper demonstrates the feasibility of the method and discuss it's possible future enhancements.

1.INTRODUCTION

There is a widely recognized need for experimental methods measuring the detailed characteristics of complete amorphous semiconductors devices. As a matter of fact, measurements of material parameters (mobilities, time of life, densities of states) are not sufficient to make accurate predictions on devices performances. Detailed computer calculations need often to be qualified by comparing their output to experimental informations. For p-i-n devices, using light beam of different wavelenghts as a probe to analyse, at least qualitatively, device behaviour is common practice and methods to extract quantitative informations have been proposed /1/. In the present work, we suggest that using electrons beams instead of light beams bring several advantages when quantitative analysis of device behaviour is considered.

The electron beam induced current (EBIC) technique has been largely used with crystalline materials /2/. In many cases, the technique makes use of the small spot size of Scanning Electron Microscope (SEM) to produce excess carriers in the vicinity of electrodes or junctions and relates the induced current to quantities such as diffusion lenght and time of life. Changing the energy of electron beam, has also been used /3/ to analyse vertical structures. In amorphous semiconductors, SEM has been used both to generate irradiation defects and to analyse their effect /5/. In this work, an electron beam of variable energy is used to analyse charge collection as a function of the depth in amorphous silicon p-i-n structures.

Compared to light beams electron beams used as probes present some interesting characteristics : 1-the absolute beam intensity can be easily and accurately measured. There is no uncertainty analoguous to those related to light confinement (internal reflections, interferences, diffusion), 2-the shape of the generation function is favourable to collection efficiency analysis as it exhibits a peak at a depth increasing with electron energy. (Figure 1)

The density of electron-hole pairs per unit of lenght, produced in a semiconductor by an incident electron beam of energy E can be described by a generation function g(z,E) :

$$g\,(z,E)\,dz = \frac{E}{\varepsilon_p}(1 - r)\,f\left({}^{z}\!/\!{}_{R(E)} \right) \frac{dz}{R(E)} \qquad (1)$$

where f(u) is a dimensionless function, defined between 0 and 1 and of unit area. The factor 1-r takes into account secondary electrons and was taken as 92%. For f(u), we use a parametrization given by Everhart and Hoff /4/ : Figure 1. The electron range R(E) is a function of electron energy which is well approximed by Gruen's formula /2/ :

$$R(E) = \frac{k}{\rho} E^{1,75} \qquad \text{with : } \quad {}^{k}/_{\rho} = 2{,}01\ 10^{-2} \quad \text{for E in KeV and R in } \mu m \qquad (2)$$

The mean energy ε_p required to produce one electron-hole pair is well known in crystallline materials (ε_p = 3.6eV in crystalline Si), but there is still questions on it's value in amorphous semiconductors /8-9/. Measurements performed on thick (4μm) devices, under high reverse polarizations (80 V) lead to an upper bound of 5.0 eV (Figure 5). Final results are calculated with the two hypothesis 3.6 eV or 5eV, but higher values are not excluded (see discussion) indicating that calculated efficiencies might have to be scaled up by ≈ 40%. Finally, it is important to notice that both parameters implied in the generation profiles, ε_p and R(E) are practically independent of material quality.

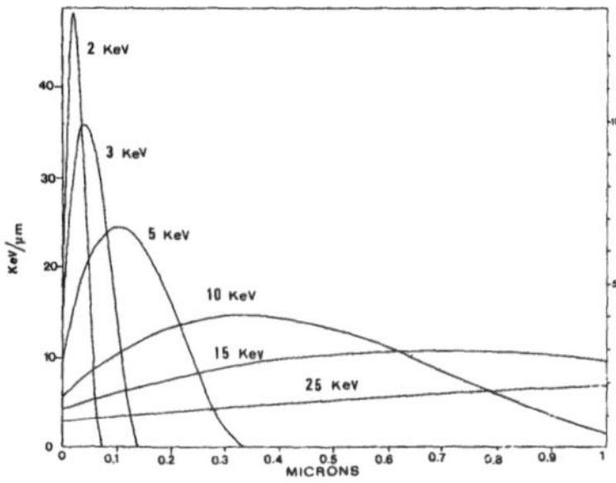

Figure 1 : the electron-hole pair generation function for electrons in silicon is sharply peaked near the entrance electrode. Analysis of collected carriers provides information on different depths depending on electron energy

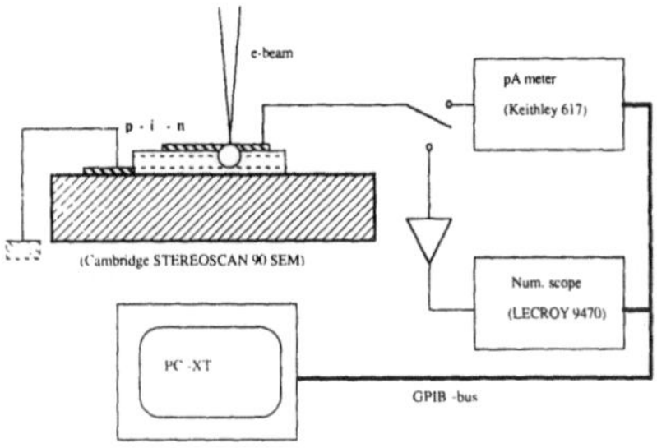

Figure 2 : The electron beam, in the 2 -25 KeV range , is used to generate pairs in a p-i-n device. The induced current (EBIC) is measured either in mean value by a pA-meter or digitized to allow further processing.

2.EXPERIMENTAL

The beam of electrons, in the 1 to 30 KeV range, is produced by a Scanning Electron Microscope. (Stereoscan 90, Cambridge Instruments). The beam current is measured in a Faraday cup with a Keithley 617 electrometer. Induced current is measured either with the electrometer in the picoammeter range for mean value measurements or with a numerical oscilloscope Lecroy 9400 in the line scanning mode (Figure 2). Both instruments readings are recorded by a PC microcomputer via the GPIB bus. The p-i-n diodes can be operated under reverse polarization from 0 to 100V (for thick devices). Due to the large amount of energy brought by the beam, the following procedure is used in order to avoid material degradation : low beam intensity (1-10 pA), unfocalized beam (spot size 10μm), all measurements made in raster TV scan. Special care is taken no to expose the diode to the beam except for the measurement period (a few seconds). No degradation is observed when the above mentionned conditions are fulfilled. It has been checked in both measurement mode that no space charge effect was present. In the mean value mode, an area of about 200 x 200 μm^2 is used.

In the line scanning mode, the current is amplified in a current to voltage amplifier and accumulation of a hundred line scanning is required to produce reduced noise output. This mode is mainly used for qualitative analysis. Non uniformity of more than 10% are observed which have been related to local changes in the potential profile.

Three series of devices have been measured. They were produced on the ARCAM RF plasma reactor /6/ using standard conditions. Detailed characterization of the a-Si:H has been performed /7/. In order to qualify the EBIC technique thick devices were used (from 1 to 4μm) instead of the usual ≈.5μm thickness of solar cells. For the same reason, and to allow reverse polarizations of the devices, some of them were deposited on chromium covered glass (instead of ITO) and the p+ and n+ layers were increased up to 25nm. The back electrode is aluminium deposited by vacuum evaporation (≈ 100nm). Notice that electrons enters the device through this back electrode. We plan to produce n-i-p devices in order to make direct comparisons with the spectral photoresponse technique.

3.RESULTS

From the measured beam current i_b and EBIC current i_c one defines the EBIC gain and the EBIC efficiency :

$$G(E) = \frac{i_c(E)}{i_b(E)} \qquad \text{and} \qquad \eta(E) = G(E)/G_{th}(E) \tag{3}$$

where the theoretical or expected gain G_{th} is the mean number of electron-hole pairs produced in the structure per incoming electron. It is simply related to the integral of g(z,E) over the p-i-n structure. This takes into account the lost of energy in the substrate and in the electrode.

Figure 3: Measured EBIC efficiencies (ratio between collected and expected number of electron-hole pairs) as a function of electron energy and for three p-i-n of different **i**-layer thicknesses. (solid lines are parabolic fit to the data)

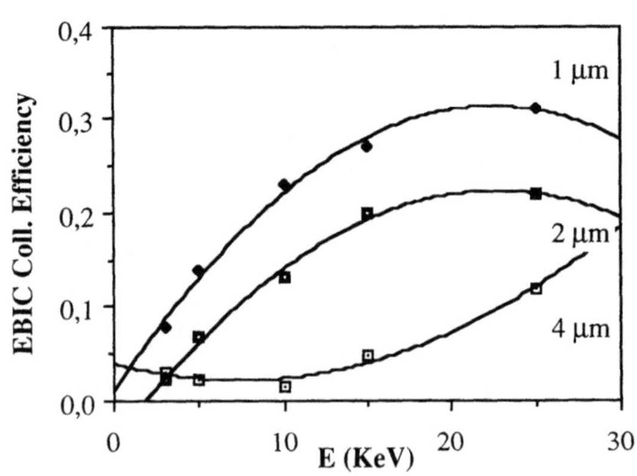

Efficiencies as a function of electron energy (and then of generation profiles) are presented on figure 3 for devices of different thicknesses and different types of devices. It is observed, as expected, that efficiencies increase for high electron energies, i.e. when more electron-hole pairs are created in the vicinity of the P^+ - i junction. In this case most of the induced current is related to electron transport, whereas, for low energy incident electrons, electron-hole pairs are generated in the n^+-i area and induced curent is related to hole transport.

The EBIC efficiency can be related to the generation function :

$$\eta\,(E) = \frac{\int_{z_1}^{z_2} T(z)\, g(z,E)\, dz}{D} \qquad \text{where } D = \int_{z_1}^{z_2} g(z,E)\, dz \tag{4}$$

where T(z) is the mean value of the induced charge for one electron-hole pair created at point z. It can also be thought of as a collection efficiency. One can first attempt to extract T(z) from measured values of η (E) by inverting the preceding integral equation. This has been done in the case of spectral photo-response by Takahama and al. /1/ (DICE method) by a numerical inversion of the integral equation. The same technique can be applied to EBIC measurements, but it is interesting to make use of the special shape of EBIC generation functions to develop the following more physical calculation : we consider a set of discrete values of energy E_i such as $R(E_i) < d$, the device's thickness, and we consider the points $z_i = R(E_i)$. Then , equation (4) can be translated in a matrix equation :

$$\eta_i = \frac{g_{ij}}{D} T_j \qquad \text{with} \quad g_{ij} = 0 \text{ for } j > i \tag{5}$$

where T_j is the mean value of T(z) over the interval (z_{j-1}, z_j). Here g_{ij} is a square matrix and T_i can be directly calculated by substitution due to the special form of g_{ij} . Choosing equidistant values of E_i leads to a higher density of points z_i near the aluminium electrode. The procedure is very sensitive to entering the right value for the thickness of this electrode. For correct values it gives positive and non oscillating results (although there is no positivity or any other constraint on T_i). Results for T(z) are given on Figure 4. They correspond to measured values of η (E) for three **p-i-n** with different **i** layer thicknesses shown on Figure 3. The technique has been applied to three different set of devices with very similar results for devices elaborated in the same conditions. Application to thinner devices was not possible with our SEM due to it's limited number of available energies, but should give better accuracies as generation functions peak more sharply at low energies.

4.DISCUSSION

Calculated collection profiles are increasing from the n-i towards the i-p interface, a feature which can be related to a better hole collection at the p-i junction. Efficiencies are higher , in mean value, in 1µm than in 4µm devices as expected, but they are almost a factor 2 below values obtained in solar cell analysis. A possible explanation migh be that we have underestimated the mean energy per electron-hole pair ε_p. A value of ≈7eV, compatible with Klein's empirical formula /10/ would bring our maximum efficiencies near 1 but would conflict with results measured under high reverse polarization (Figure 5) except if one invoke some charge carrier multiplication. To give the same multiplication factor for all electron energies, such a mechanism should be attributed to holes near the p-i junction.

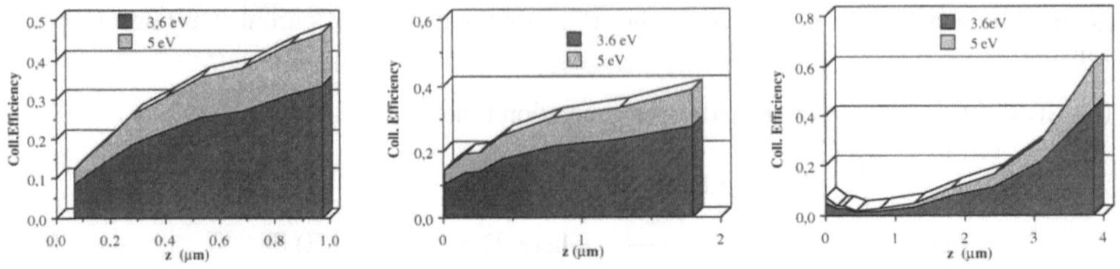

Figure 4 : Calculated EBIC collection efficiencies as a function of depth for three samples with different **i**-layer thicknesses. The electron beam enters at z = 0 on the **n-i** side through the aluminium electrode (≈100 nm).

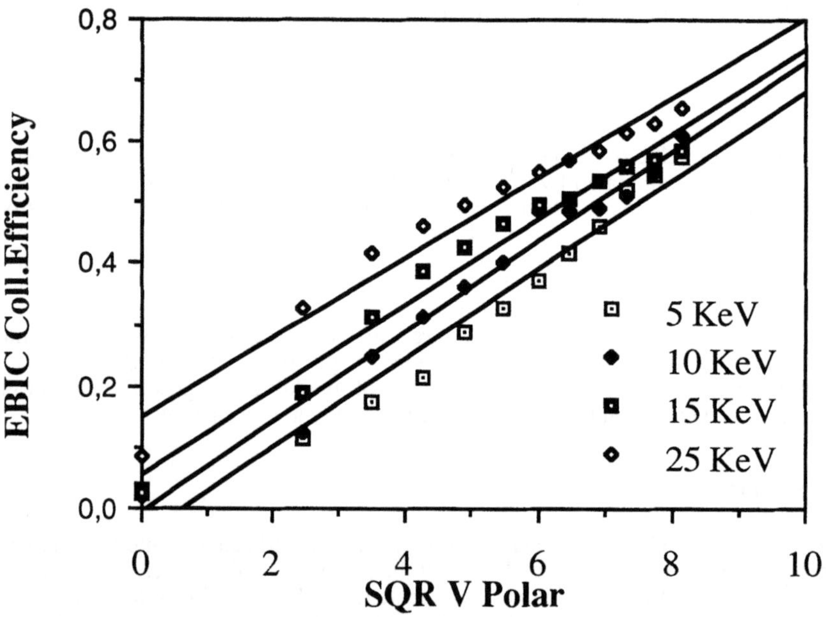

Figure 5 : EBIC collection efficiencies for a thick (4µm) **p-i-n** device under high reverse polarizations for different electron energies. A value of ε_p = 3.6 eV per electron-hole pair was assumed. Although no saturation is observed, a value of 5eV can be extrapolated. (see text)

5.CONCLUSION

The EBIC technique, at variable energy, is an efficient way to measure and calculate internal collection efficiencies. This is related 1/ to the fact that electron currents can be more easily measured than photon current for which both quantum efficiency and light distribution in a device are difficult to define accurately and 2/ to the shape of electron generation function. The accuracy of the technique should allow to extract some more physical quantities like electric field profiles, which are related to derivatives of collection efficiencies. Combined with the usual advantages of EBIC as a mean to perform a 2D-map of device quality at the micron level, this quantitative analysis should provide a new practical mean to analyze and optimize photovoltaic devices.

ACKNOWLEDGEMENTS

We are pleased to thank P. Roca i Cabarocas for preparing the samples and for many unvaluable discussions. One of the authors (B.E) want to thank A.Rocher and F.Battistella from Laboratoire d'Optique Electronique, Toulouse, for the help they gave us to start this work.

This work is supported by PIRSEM/CNRS, AFME and CEC contracts.

1 T.Takahama, M. Isomura, S.Tsuda, H.Tarui, Y.Hishikawa, N.Nakamura,Y. Nakashima, T. Matsuoka, H. Nishiwaki, M. Ohnishi, S. Nakano and Y. Kuwano, Japan. J. of Applied Physics, **25** (10), 1986 ,pp 1538
2 H.J. Leamy, J.Appl.Phys.**53** (6)1982
3 C.J Wu and D.B.Wittry, J.Appl.Phys.**49** (5), 1978
4 T.E. Everhart and P.H.Hoff, J. of Applied Physics **42** (13), (1971) pp5837
5 U.Schneider, B.Schröder, F.Finger, Proceedings of the 12th ICALS, Prague, North-Holland Pub. ,1987, p795
6 P.Roca i Cabarocas et al, Proceedings of the 7th European Solar Energy Conference, Sevilla 1986, D.Reidel Pub.
7 P.Roca i Cabarocas, Ph D Thesis, Paris 1988
8 B.Equer and A.Karar, Nucl. Instr. and Methods (to be published)
9 S.N.Kaplan, I.Fujieda, V.Perez-Mendez,S.Qureshi, W.Ward and R.A.Street, London Conference on Position-Sensitive Detectors, 1987
10 C.A. Klein, J.Appl.Phys.**39**2029, 1968

VACUUM-UV-PHOTO-CVD SETUP FOR a-Si(C):H WINDOW LAYERS

F. Kessler, H.-D. Mohring, G.H. Bauer, W.H. Bloss
Institut fuer Physikalische Elektronik, Universitaet Stuttgart,
Pfaffenwaldring 47, D-7000 Stuttgart 80, F. R. Germany

ABSTRACT. A vacuum-photo-CVD reactor for direct photo decomposition of SiH_4, Si_2H_6, B_2H_6, and of hydrocarbons has been built up with a gaseous transmission filter between deposition chamber and vacuum ultra violet light source (deuterium lamp) in order to investigate the effect of high energy photons onto generation of radicals and film properties. The filter cut-off-frequency has been continuously shifted across the relevant threshold energies for the formation of different types of radicals by use of O_2 as filter gas at various partial pressures. Growth rates mainly depend on total gas pressure, window-to-substrate distance and spatial gas flow distribution in the reactor. Variation of the substrate temperature does not affect growth rates thus indicating a pure photon-controlled decomposition process, whereas the strong substrate temperature dependence of the optical bandgap, which is much more pronounced than in glow discharge films, points towards different surface processes for H-incorporation.

1. INTRODUCTION

Recent activities for the deposition of p^+-layers in a-Si:H based solar cells by photochemical vapour deposition methods give indication for an improved electronic quality of the p/i-interface originating from reduced interaction of high energetic species with the surface of the growing film in comparison to glow discharge systems /1-3/.

Since silanes show only negligible optical cross sections in the energy range $h\nu \leq 6$ eV, enhanced film growth rates are achieved either by the use of sensitizers, which in fact brings up the possibility of additional contamination, or by direct photodissociation with vuv light. The direct decomposition method offers the selection of various reaction mechanisms with different threshold energies, but , however, is essentially coupled to a suitable high intensity vuv-source.

We present a vacuum-uv-photo-CVD system with the possibility for continuous adjustment of the maximum photon energy, describe the spatial distribution of radicals with help of simple model calculations and consider properties of photo-CVD a-Si:H films deposited under different experimental conditions.

2. EXPERIMENTAL

Our deposition system for vuv-photo-CVD consists of three chambers separated by vuv-transmissive MgF_2 windows ($\lambda \geq 115$ nm) :

i) a chamber for the low power D_2-discharge lamp (typically 15 W, Cathodeon Ltd.) flowed with H_2, He or Ar for cooling.

ii) a gaseous transmission filter system for the continuous adjustment of transmission edges versus wavelength (O_2 at different partial pressures) cutting off the high energetic range of the spectral distribution of the lamp.

iii) a deposition chamber.

Film deposition on the window was prevented either by N_2/H_2 purge or for steady state conditions by use of fomblin oil (only applicable for Si_2H_6 deposition).

In order to measure the photon flux across the filter at various gas pressures a grating monochromator with photomultiplier is attached instead of the reaction chamber.

Fermi level position of the films has been calculated from temperature dependent dark conductivity measurements with coplanar contact configuration on Corning 7059 glass substrates; optical properties are determined from spectral transmission and reflection from 200 nm - 3000 nm (Infrasil substrates).

3. RESULTS AND DISCUSSION

3.1. Film growth

In a first step the filter has been evacuated to 10^{-5} mbar in order to get the influence of window absorption. The spectrum extends to 115 nm, but the strong Lyman line at 121.6 nm (10.2 eV) of the original lamp spectrum /4/ is remarkably reduced by the two 6 mm thick MgF_2 windows (Fig. 1).

The spectral distribution of the deuterium lamp overlaps with the regime of noticeable cross sections for SiH_4- and Si_2H_6- photodissociation (Fig. 1). The number of photons contributing to dissociation of SiH_4 and Si_2H_6 can be calculated according to

$$F=\iint \{S(\lambda)/h\nu\}\{1 - \exp[-nd\sigma(\lambda)]\}d\lambda\, d\Omega$$

S -spectral radiance of the lamp

n -gas density

d -light path in the gas

σ -effective cross section

Ω -sol.angle of the lamp (0.0537 sr).

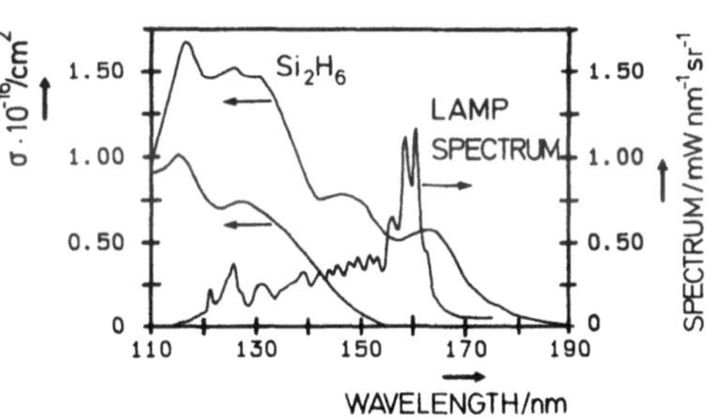

Fig. 1: Spectrum of the D_2-lamp and SiH_4/Si_2H_6 cross sections

At total absorption the number of absorbed photons (2×10^{14}/s for SiH_4 and 7×10^{14}/s for Si_2H_6) gives the theoretical upper limit for the radical generation rate with our reduced spectrum. Without MgF_2 windows the maximum rate of absorption processes would be 5×10^{14}/s and 9×10^{14}/s resp..

Under the assumptions that each absorption process generates an active species for film growth and that all these radicals could be forced within their lifetime to impinge onto a 1 cm^2 substrate and stick there with probability 1, we could expect a growth rate of 3 Å per second. In practice, however, none of these conditions is met and at best some percent of this figure can be taken as a first unromantic estimation.

In order to simulate the actual growth conditions, two model calculations have been carried out to describe the spatial distribution of radicals in the reactor as borderline cases of the real deposition chamber geometry:

1) the neutral gas inlet is a small pipe in the centre of the substrate ($r = 0$)
2) homogenous density of neutral gas.

The first case gives the density of activated species and from that the film growth rate rises as r near the gas inlet and drops as 1/r in the region where the diffusion length $L = (D\tau)^{1/2}$ is smaller than r /5/.

The good lateral homogeneity of our films demonstrates, that the second case is the more adequate description for our experimental configuration. Assuming homogenous generation rate G and the boundary conditions $dn^*/dz = 0|_{z=0}$ (window) and $n^* = 0|_{z=d}$ (sticking coefficient 1 at the substrate) the density distribution of the photogenerated radicals is

$$n^*(z) = G\tau \{1 - sh(z/L)/sh(d/L)\} \qquad \text{sh -hyperbolic sine.}$$

Using Fick's law we can calculate the current density of activated species at the substrate:

$$j|_{z=d} = GL \, cth(d/L) \qquad \text{cth -hyperbolic cotangent.}$$

It is obvious from this equation that high values of j are obtained for $d/L \ll 1$; so window-to-substrate distance should be chosen as small as possible to increase deposition rates.

Taking our data for Si_2H_6 film growth rate (5 nm per hour on 10 cm^2 substrate area) and assuming sticking coefficient 1 we calculate 7×10^{12} radicals per second and cm^2 contributing to film growth. With actual values for gas pressure (0.5 mbar), window transmission and Si_2H_6 optical absorption cross sections 4×10^{13} radicals per second hit 1 cm^2 substrate area; this leads to a real sticking coefficient of ≈ 0.2 as an average value for SiH_x species.

In Fig. 2 relative film growth rates are outlined as a function of neutral gas pressure at fixed d (10 cm^2 substrate illumination). We recognize maximum rates near 0.5 mbar; at lower pressures absorption is insufficient, whereas for high pressure radicals are generated too far from the substrate and radical concentration declines as $1/r^2$ with the distance from the place of generation.

The absolute deposition rates strongly depend on lamp-to-substrate-distance but are very low for deposition on substrate areas $\geq cm^2$. For practical applications this problem can only be solved with a high intensity vuv source.

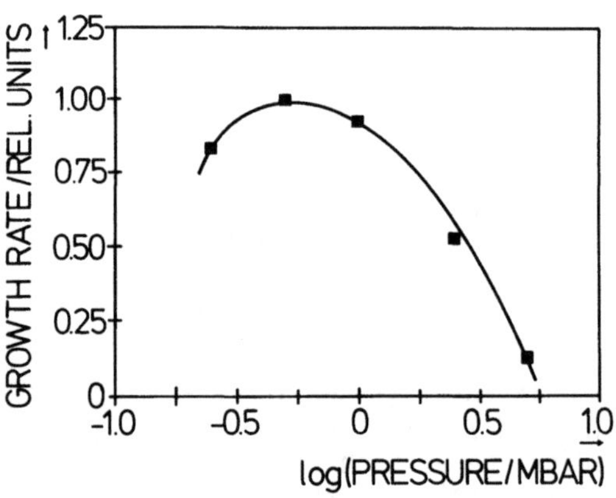

Fig. 2: Relative growth rates vs. pressure

3.2. Filter for radical selection

In Figs. 3 and 4 the number of absorbed photons per wavelength is given for SiH_4 and Si_2H_6 together with the thresholds for the formation of different types of radicals /6/. We see that different photodissociation processes and ratios of reaction products (radicals) can be controlled and adjusted by a suitable cut-off filter for the exciting vuv radiation.

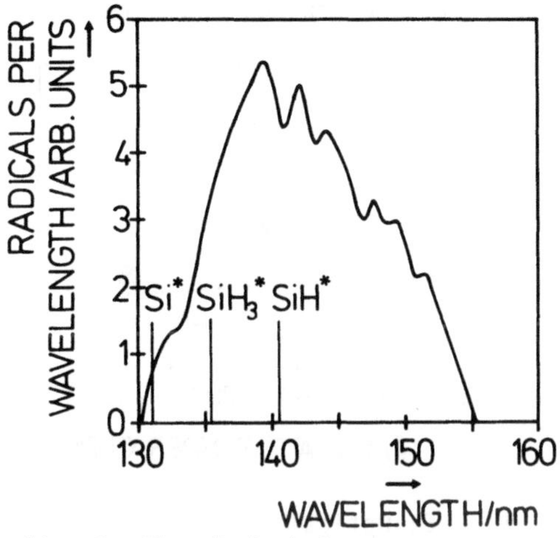

Fig. 3: Absorbed photons per wavelength in SiH_4 (D_2-lamp illum.)

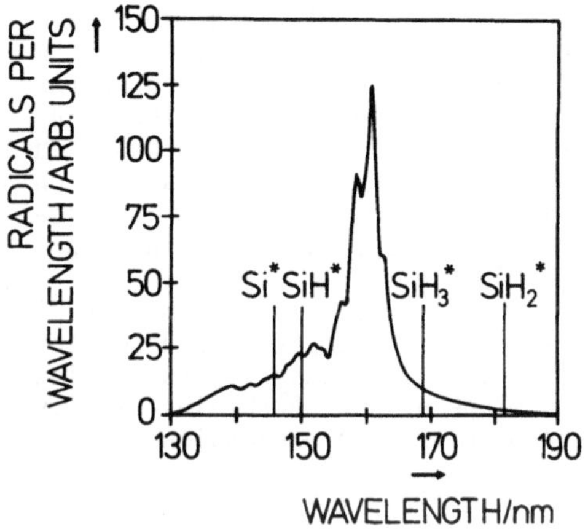

Fig. 4: Absorbed photons per wavelength in Si_2H_6 (D_2-lamp illum.)

Nitrogen and hydrogen have been found to be completely transparent above 130 nm and can't therefore be used as filter gases but the more can be used as purging gases. Very promising results are obtained with O_2, which has an absorption continuum (Schumann-Runge continuum)in the relevant wavelength region. Below 175 nm O_2 photodissociates to produce one ground state and one excited state oxygen atom /7/:

$$O_2 + h\nu \rightarrow O(^3P) + O(^1D).$$

Figure 5 shows the transmission of the O_2 gaseous filter at different partial pressures. It is clearly demonstrated, that there is a pronounced cut off energy which can be controlled by the O_2 pressure.High energy photons below 130 nm are absorbed, if CaF_2 windows are used instead of MgF_2.

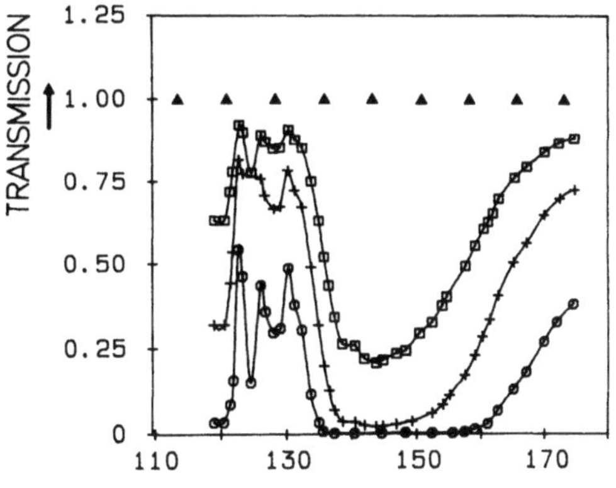

Fig. 5: Filter transmission at different O_2 partial pressures

3.3. Film properties

We have deposited intrinsic a-Si:H films by Si_2H_6 photodecomposition. In view of the observed small film growth rates (and even lower values for SiH_4 decomposition), which brings up the problem of pollution from the reactor leak rate during long time growth under steady state conditions (oxygen incorporation detected by XPS measurements), all films described in this paper have been prepared without application of the gaseous filter.

Variation of the substrate temperature strongly affects the optical bandgap determined after Tauc's relation (Fig. 6). The change in E_g is much more pronounced than in gd deposited films, which points towards different surface processes responsible for H-incorporation.

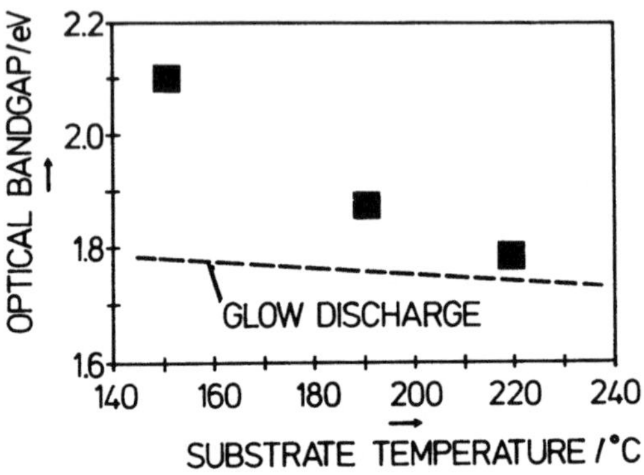

Fig. 6: Optical bandgap vs. substrate temperature

Determination of Fermi level position from temperature dependent conductivity measurements taking the slope of the Arrhenius-plot or the more precise expression

$$(E_C - E_F)_T = kT \ln[\sigma_{(T)}/\sigma_0]$$

with $\sigma_0 = 200\ 1/\Omega cm$ /8/

reveals good intrinsic films with room temperature Fermi level near midgap and ratios $\sigma_{photo}/\sigma_{dark}$ in the range of 10^4 to 10^6 (AM1 ELH illumination). For $E_g = 2.1$ eV the photoconductivity of photo-CVD a-Si:H exceeds the data of glow discharge deposited a-SiC:H and offers the possibility for a high quality window layer.

4. CONCLUSION

High quality intrinsic a-Si:H films have been prepared by direct photodissociation of Si_2H_6 using a low power D_2 lamp. The small growth rates for homogenous deposition on areas > cm^2 should be improved with a high intensity vuv source.

ACKNOWLEDGEMENT: This work has been supported by EC contract EN3S-0058-D(B)

REFERENCES

/1/ H. Takei et al., J. Appl. Phys. **58,** 3664 (1985)
/2/ T. Fuyuki et al., Conf. Rec. 19 IEEE PVSC, New York 1987, p. 569
/3/ P.A. Robertson, W.I. Milne, MRS Symp. Proc. Vol. 70, p. 31
/4/ P.J. Key et al. J. Phys. E **13**, 866 (1980)
/5/ G.H. Bauer Proc. CEC contractors meeting, Hamburg 1987
/6/ H. Stafast to be publ. in Appl. Phys B
/7/ K. Watanabe et al., Air Force Cambridge Res. Center, Bedford, Mass., Tech. Rep. No. 53-23, (1953)
/8/ J. Stuke, J. Non-Cryst. Solids **97&98**, 1 (1987)

ELECTRON TRANSPORT AND RECOMBINATION IN a-$Si_{1-x}Ge_x$:H ALLOYS

C. E. Nebel, M. Schubert, H. C. Weller, G. H. Bauer, W. H. Bloss
Institut f. Physikalische Elektronik, Universitaet Stuttgart,
Pfaffenwaldring 47, D-7000 Stuttgart 80, F. R. Germany

ABSTRACT

Steady-state-photocurrent (PC), time-of-flight (TOF), charge collection and photo-thermal-deflection-spectroscopy (PDS) measurements are evaluated to determine transport and recombination quality of a-$Si_{1-x}Ge_x$:H for $0 \leq x \leq 0.3$. Drift mobility (μ_D) data are used to calculate the tail state distribution at the conduction band. By incorporation of Ge additional localization broadens the conduction band (cb) tail. The dangling bond density, calculated from electron $\mu_D\tau_{dt}$ products (charge collection), also dramatically increases whereas the Urbach energy of the valence band tail (PDS) varies little versus Ge concentration ($x \leq 0.2$).
The increase in dangling bonds (db) quenches the lifetime of carriers and turns out to be very sensitive to preparation conditions (temp., pressure, discharge power) and methods (diode, triode reactor plus H_2 dilution). Variation of the cb-tail strongly depends on Ge concentration but not on specific preparation methods (diode/triode config., H_2 dilution). Enlarged tailing reduces μ_D of electrons considerably. Our results reveal as decisive effects an increase in db-density and localization at the conduction band which deteriorate the photoelectronic quality of a-$Si_{1-x}Ge_x$:H.

INTRODUCTION

Alloying of amorphous Si films by Ge is of considerable interest regarding thin film photovoltaics with band gaps optimized with respect to solar radiation distribution as well as for multispectral solar cells. However the photoconductive properties of alloys so far prepared are inferior to a-Si:H. In this paper we discuss results from PC, TOF, charge collection and PDS experiments intending to improve our understanding of the nature of states that control transport and recombination of photoexcited carriers.

EXPERIMENTAL

In order to determine the variation of optoelectronic quality versus Ge incorporation, samples have been prepared in a dc glow discharge of SiH_4, GeH_4, H_2 mixtures in a diode/triode reactor configuration [1]. The increase in Ge concentration (x = 0, 0.025, 0.11, 0.2, 0.3) derived from XPS data corresponds to a decrease of optical gap (E_g = 1.75eV, 1.74eV, 1.69eV, 1.62eV, 1.55eV).
A conventional TOF setup and sandwich contact configuration (substrate/Cr/ a-$Si_{1-x}Ge_x$:H/semitransparent Cr) are used to measure either the drift mobility μ_D at different temperatures (120K - 400K) and electric fields (10^4V/cm - 10^5V/cm) or the mobility-deep-trapping-lifetime product $\mu_D\tau_{dt}$ depending on charge collection Q, which is the time integral of the current pulse. The indication that carriers have completely crossed the sample is when Q is observed to be independent of the applied electric field F, and the defined transit time then gives the drift mobility.
Mobility μ_D of electrons for $0 \leq x \leq 0.2$ is shown in Fig. 1 in a log(μ_D)

versus 1/T plot for different electric field strengths and Ge contents. An increase in Ge concentration lowers μ_D from $1.5cm^2/Vs$ (x=0) to $0.13cm^2/Vs$ (x=0.2) (T=300K) and increases its activation energy. In addition mobility μ_D of electrons for samples containing 11at% Ge prepared in a diode/triode discharge configuration is shown in Fig. 2. Despite of use of different preparation methods [1] no significant differences of high temperature mobility (T = 300K, $\mu_D = 0.4cm^2/Vs$) could be deduced. Differences in electric field and sample thickness cause the variation at low temp. [2].

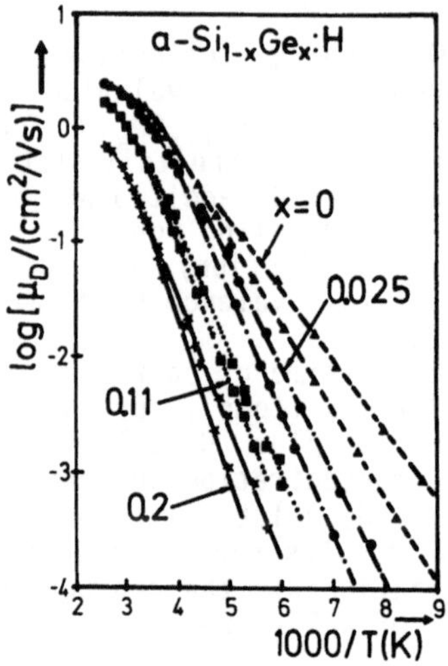

Fig.1. μ_D of electrons for various Ge content (diode discharge config.)

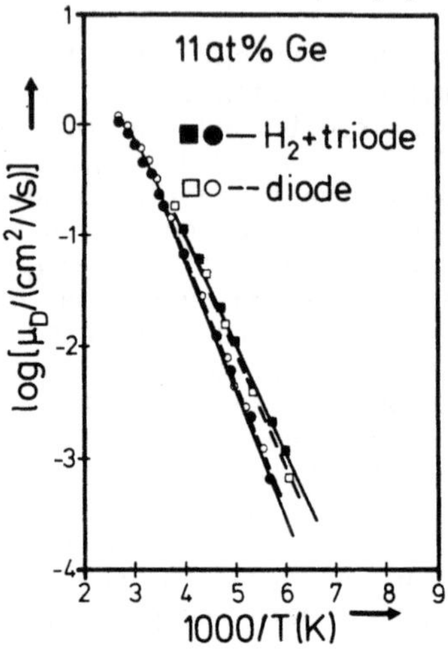

Fig.2. Comparison of μ_D of samples prepared in diode/triode+H_2 reactor config.

At sufficiently low fields deep trapping occurs before the transit is complete. Q then increases proportionally to F, and $\mu_D\tau_{dt}$ can be determined from the Hecht formula [3]. Fig. 3 shows the $\mu_D\tau_{dt}$ values and the $\eta\mu_D\tau$-products deduced from PC (coplanar contact configuration) of electrons for different alloy compositions. The observed decrease of $\mu_D\tau$ corresponds to a monotonically increasing Ge concentration from 0 to 30 at% and directly shows enlarged deep trapping and recombination of electrons.
Urbach energies (PDS) are drawn in Fig. 4, indicating that the valence band tail within 20at% Ge content is not dramatically changed for both types of samples (diode, triode plus H_2 dilution), which is in good agreement with hole mobility data [4].

DISCUSSION

The drift process of electrons at the mobility edge is characterized by multiple trapping [MT] and release from the continuous distribution of traps [5]. In order to calculate a distribution of tail states the temperature and field dependent data of μ_D have been analyzed by use of [2,8,9]:

$$\mu_D = d/\{F\ [t_0 + \sum_i t_0\ b_t\ N_i\ \nu_i^{-1}\ ]\ \} \qquad (1)$$

where t_0 is the free carrier transit time ($=d/\mu_c F$), ν_i the attempt-to-escape frequency ($\nu_i = \nu_0 \exp[-(E_c-E_i)/kT]$), b_t the trapping parameter reflecting the product of thermal carrier velocity times capture cross section of tail states ($v\sigma = 10^{-8}cm^3/s$), μ_c the extended state mobility, F the electric field strength and d the sample thickness. The continuous distribution of localized states g(E) is represented by a set of discrete levels N_i at energy E_i with densities $N_i = g(E_i)\Delta E$, ΔE being the spacing between levels [2].

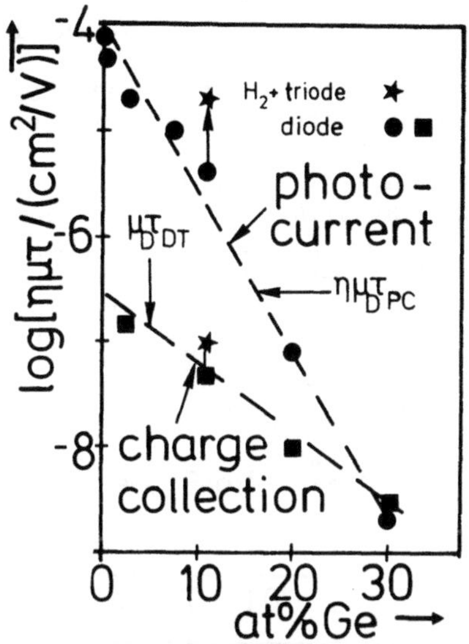

Fig.3. $\eta\mu_D\tau$(PC)/$\mu_D\tau_{dt}$(charge collection) products

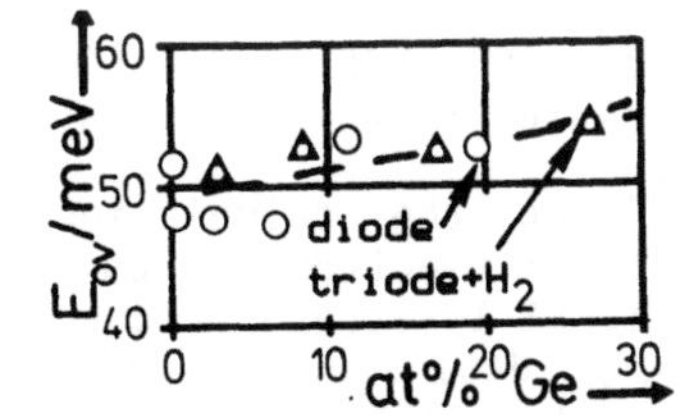

Fig.4. Urbach energies of a-SiGe:H

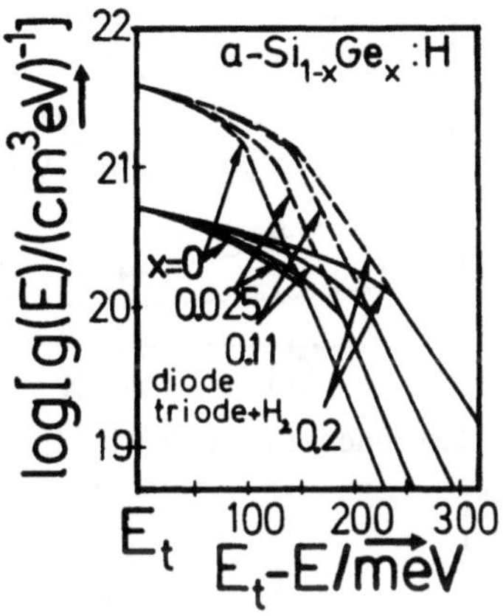

Fig.5. Conduc. band tail state distr. [12]

In order to estimate the density of states at the mobility edge $N(E_t)$ we follow the argument of Tiedje [8] and Monroe [9] who introduce a hopping/multiple-trapping transport model. For a slowly varying density of localized states at E_c (transition from extended to localized states, $N(E_c) = (2\text{-}6)\times10^{21}(\text{cm}^3\text{eV})^{-1}$ [10,11]), the level of transport can be calculated by [8,9,12]:

$$E_c - E_t = [N(E_c)/m] - \{0.73\ [\gamma^3/m\Delta]^{0.25}\ [kT]^{0.75}\} \quad (2)$$

where m is the slope of a linearly decreasing tail state distribution, γ the inverse localization length for an occupied conduction band tail state ($\sim10^7\,\text{cm}^{-1}$ [13]), and $\Delta \sim kT$. E_t shifts (150-250)meV down into the localized tail states (T=300K) in a region where the density of states $N(E_t) \approx (5 - 10)\times 10^{20}(\text{cm}^3\text{eV})^{-1}$. In consideration of this uncertainty the presented model has been calculated in the limits of $5\times10^{20}(\text{cm}^3\text{eV})^{-1} \leq N(E_t) \leq 4\times10^{21}(\text{cm}^3\text{eV})^{-1}$. The results will be discussed with respect to these values.

Mobility of carriers above E_t (μ_c) is assumed to be temperature dependent $\mu_c \sim T^{-1}$ which is adequate to either the random phase approximation [11] or a phonon assisted hopping mobility [14].
Excellent agreement between experimental data (see Fig. 1,2) and the calculated model (Eq. 1) for μ_D in the range of temperatures and fields considered here including the temperature dependence of μ_c ($\sim T^{-1}$) has been achieved, when introducing two energetic regimes for the tail state distribution: a flat one close to the transport level E_t which is approximated by a linear decrease in g(E) versus E, and a steep one, which shows an exponential decrease vs. energy. Tail state distribution depends on the assumed density of states at E_t as shown in Fig. 5. By incorporation of Ge even in low concentration (2.5at.%) an increase in density of tail states at the conduction band is observed. Additional localization introduced by chemical disorder broadens the band tail independently of preparation method (diode,

triode), and thus lowers the measured drift mobility and gives rise to the dramatic increase in activation energy elucidated from $\log(\mu_D)$ versus 1/T plots.
In addition the calculations reveal a decrease of the extended state mobility of electrons μ_c as shown in Fig. 6 coinciding with an increase of the attempt-to-escape frequency ν_0 (see Fig. 7).
The $\mu_D\tau_{dt}$ data shown in Fig. 3 are used to evaluate the dangling bond (db) density. From ESR measurements it is found that at alloy compositions above

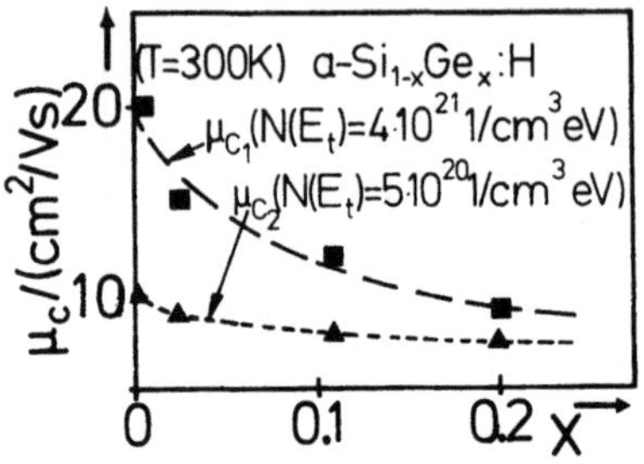

Fig.6. Extended-state-mobility

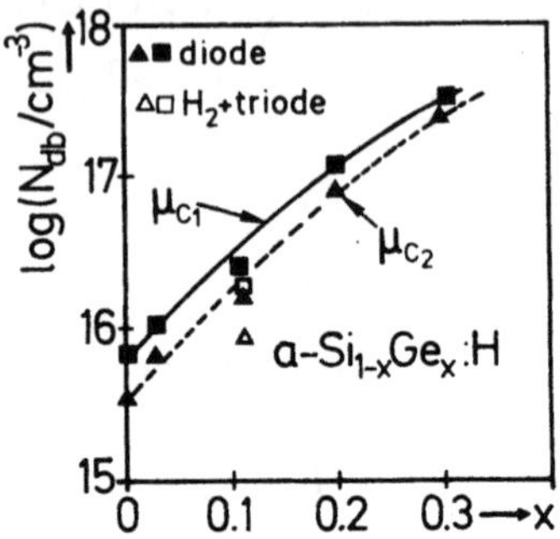

Fig.8. db-density versus Ge content

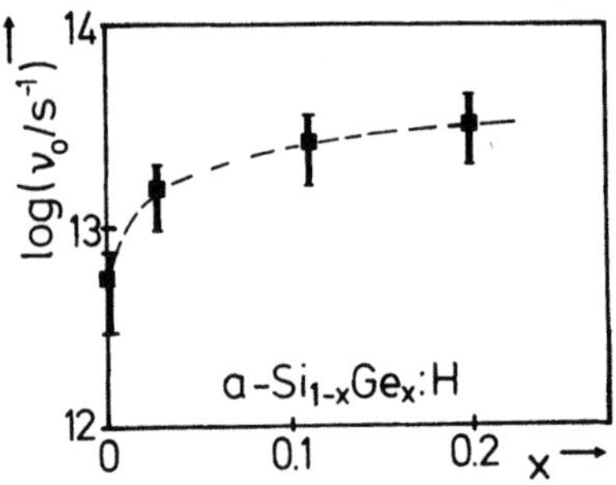

Fig.7. Attempt-to-escape frequency

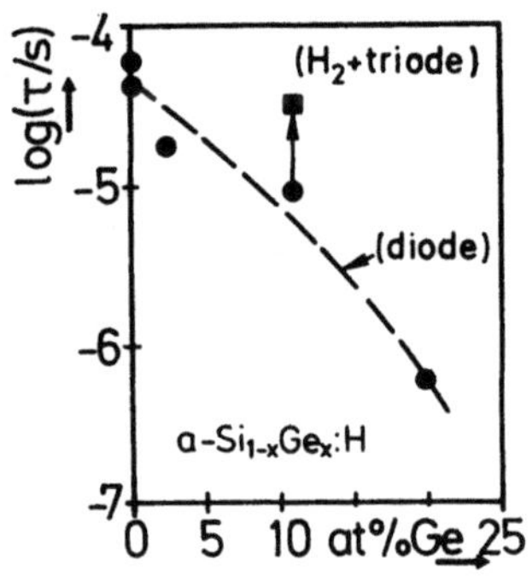

Fig.9. τ_{eff} of electrons and holes

10at% Ge dangling bonds completely dominate the defect density [15]. The data show that the predominating db changes from Si to Ge, their relative position in the gap presumably changes, however, the capture cross section for electrons appears to be independent of alloy composition [16]. The db density is related to products of $\mu_D\tau_{dt}$ by the relation

$$N_{db} = \mu_c/(v\ \sigma\ \mu_D\ \tau_{dt}) \qquad (3)$$

where σ is the capture cross section of dangling bonds $\sim 10^{-15}cm^2$.
N_{db} has been calculated from the data shown in Fig. 3,6. The defect density rises from $(3\text{-}6)\times10^{15}cm^{-3}$ (x=0) to $(2\text{-}4)\times10^{17}cm^{-3}$ (x=0.3) as a consequence of Ge alloying (Fig. 8), and dramatically quenches the effective lifetime of carriers (deduced from PC, TOF results) as shown in Fig. 9 which directly reflects the enlarged recombination due to Ge incorporation.
The defect density and therefore τ strongly depends on preparation conditions. By use of a triode reactor and H_2 dilution N_{db} decreases by a factor of two (11at% Ge) and enlarges τ by a factor of 4 (see Fig. 8,9).

CONCLUSION

Interpretation of TOF data shows that additional localization broadens the

Interpretation of TOF data shows that additional localization broadens the cb tail. Incorporation of Ge also dramatically rises dangling bond density (charge collection) but keeps the Urbach energy E_{0V} (PDS) nearly unchanged ($x \leq 0.2$). Results are summarized in Fig. 10 showing the variations schematically. The cb tail turned out to be hybrid. An increase of localization at the cb drops μ_D of electrons and depends strongly on Ge concentration but does not show significant dependence on preparation conditions. The transport level E_t shifts deeper into the tail as a consequence of enlarged tailing (see Fig. 11) and affects μ_c and ν_0 [12].
The rise of dangling bond density speeds up recombination strongly and is sensitively dependent on preparation conditions and methods.

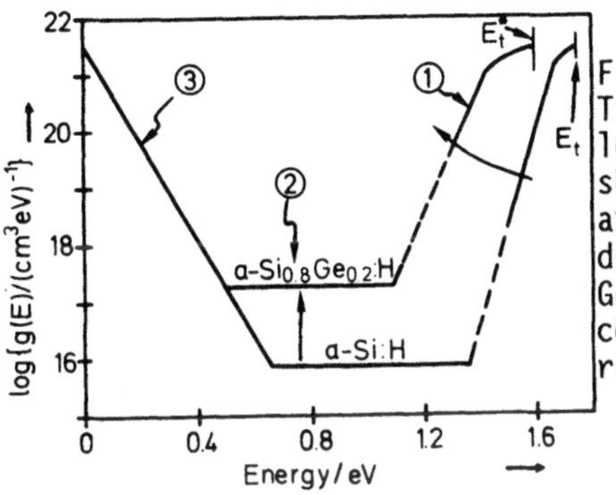

Fig.11 Transport level shift at cb due to Ge incorporation

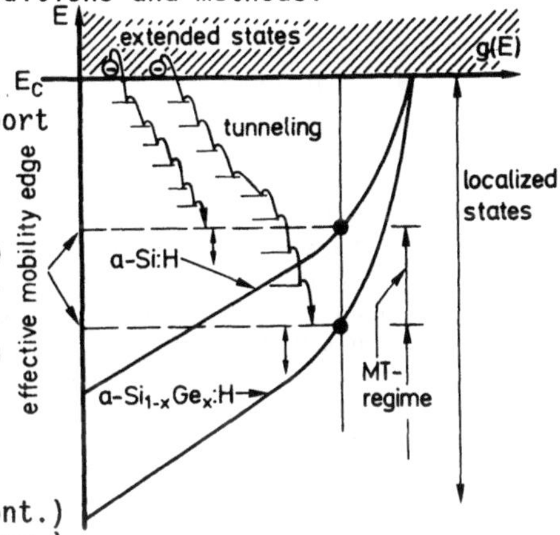

Fig.10. 1 incr. of cb-tail (dep.on Ge cont.)
2 incr. of db (dep. on Ge and prep.)
3 E_{0V} remains ~const. ($x \leq 0.2$)

In the considered limits ($0 \leq x \leq 0.3$) the Urbach energy of the valence band tail varies little revealing clearly the decisive effects of increased defect density and localization at the cb which deteriorate the photoelectronic quality of a-$Si_{1-x}Ge_x$:H alloys ($x \leq 0.3$).

ACKNOWLEDGEMENT

This work has been supported by BMFT contr. no. 03E-8327-B.

REFERENCES

1. H.C. Weller et al., J. Non-Cryst. Solids **97&98**, 1071 (1987).
2. C.E. Nebel, G.H. Bauer, to be published.
3. Z. Hecht, Z. Physik **77**, 235 (1932).
4. G.H. Bauer et al., Proc. MRS Spring Meeting 1988,Symp. E, in press.
5. D. Monroe and M.A. Kastner, Phil. Mag. B **47**, 605 (1983).
6. F.W. Schmidlin, Phys.Rev.B **16**, 2362 (1977); Phil.Mag.B **41**,535 (1980).
7. E.A. Davis, H. Michiel, G.J. Adriaenssens, Phil. Mag. B **52**, 261 (1985).
8. T. Tiedje,Proc.of Int.Workshop on a-Sem.,World Scient.Pub.Co,113(1987).
9. D. Monroe, Phys. Rev. Lett. **54**, 146 (1984).
10. N.F. Mott, J. Non-Cryst. Solids **97&98**, 531 (1987).
11. W.B. Jackson et al., J. Non-Cryst. Solids **77&78**, 2810 (1985).
12. C.E. Nebel et al., Proc. MRS Spring Meeting 1988, Symp. E, in press.
13. M. Stutzmann, J. Non-Cryst. Solids **97&98**, 105 (1987).
14. P. Nagels, Topics in Applied Physics **36**, 124 (1979).
15. F. Finger et al, J. Non-Cryst. Solids **77&78**, 731 (1985).

Barrier Enhancement by Passivation of a-Si:H Surfaces

J. Meier, H. Keppner, G. Kragler, E. Bucher, Fakultät für Physik, Universität Konstanz, D-7750 Konstanz, FRG.

H. Curtins and A.V. Shah, Institut de Microtechnique, Université de Neuchâtel, CH-2000 Neuchâtel, Switzerland.

S. Paasche and W.H. Bloss, Institut für Physikalische Elektronik, Universität Stuttgart, D-7000 Stuttgart 80, FRG.

Abstract.
By means of a new microwave plasma passivation technique we demonstrate a substantial improvement of V_{oc} in a-Si:H Schottkybarrier solar cells. Some samples showed one of the highest values of V_{oc} = 905 mV ever observed at AM1.5 illumination intensity. An excellent basematerial and a perfect interface encouraged us to optimize the cells to achieve higher photocurrents than J_{sc} = 5 mA/cm² and higher fillfactors than 60 % as previously published. As semitransparent barrier metal we used Pt and Ir continuous thin films. In this paper we are trying to find a correlation between the high values of V_{oc} and ellipsometric studies of oxide-growth at the interface. Our solar cells with a-Si:H as base-material exhibit efficiencies around 5%, with ZrO_2 as AR coating.

1. Introduction

It is well known, that a thin tunneling oxide at the interface between the absorbing semiconductor and the barrier forming metal leads to an increase of V_{oc} and thus to the efficiency of the cell. This fact is documented especially for c-silicon (1,2,3) but also when a-Si:H (4,5) is used as base material. The outdiffusion of hydrogen leads to problems for a-Si:H passivation, thus all process temperatures are limited to some 250 °C. In order to reduce the activation energy of the oxide forming process, a plasma process was developed for surface passivation.

2. Experimental

The structure of our MIS Schottky barrier solar cells is given in Fig.1. Both, polished and sand-blasted stainless steel substrates of 4 cm² were used, simultaneously forming the rear contact. First a 500 Å thick phosphorous doped layer with a dark conductivity of $\sigma_d = 2\text{-}3 \times 10^{-3}$ /Ohmcm and an activation energy of E_a = 0.20 eV was deposited onto the stainless steel as ohmic contact. On top of the ohmic contact, the intrinsic a-Si:H layer with the following properties was deposited:

Dark conductivity :	σ_d	$= 10^{-11}\text{-}10^{-10}$ /Ohmcm.
AM 1 conductivity :	σ_{ph}	$= 2.5 \times 10^{-5}$ /Ohmcm.
Bulk defect density:	N_b	$= 3 \times 10^{15}$ /cm³
Bandgap :	E_{opt}	= 1.70 eV
Bonded hydrogen content:	c_H	= 6 at. %

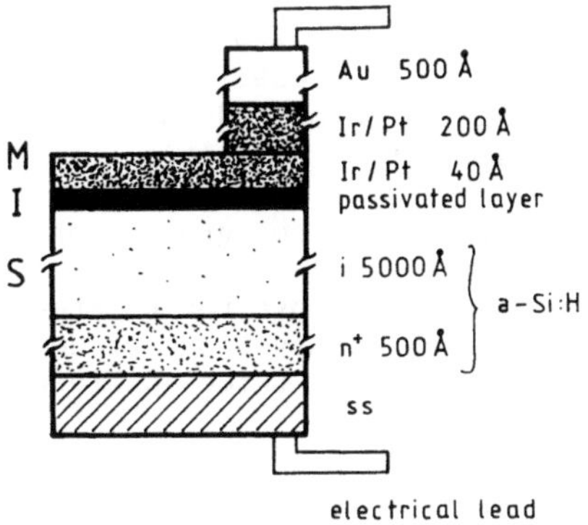

Fig.1. Structure of the a-Si:H MIS solar cell.
Fig.2. Surface passivation plasma-reactor.

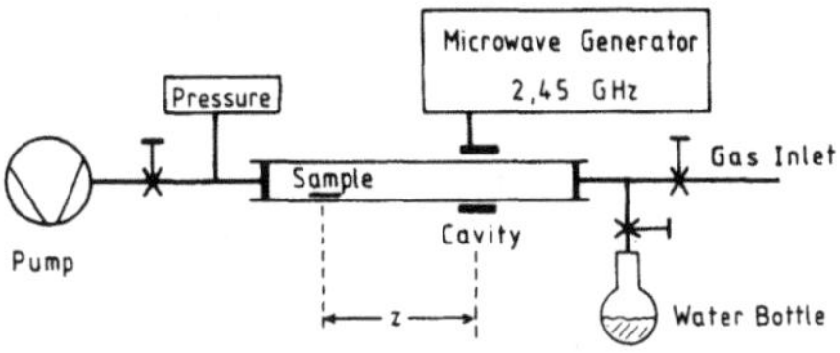

Both, n and i depositions were done in the same chamber and the same vacuum, where the high rate glow discharge process was powered by a capacitively coupled 70 MHz generator. More details of the process are described elsewhere (6). After the a-Si:H deposition, the surface treatment in a plasma reactor followed, as illustrated in Fig.2. A quartz-tube reactor with a vacuum system and gas supply was powered by a 2.45 GHz microwave generator. The treated samples were then coated by electron beam evaporation of Pt or Ir semitransparent barrier films. To achieve a good reproducibility, the coating process was controlled by a quartz-monitor.

3. Results and Discussion

i. Remarks to the open circuit voltage

In Fig. 3 it is shown, that the value of V_{OC} basically depends on the gas which is used in the passivating reactor, where water vapour tends to lead to the best results. Furthermore there is a weak dependence between V_{OC} and the gas pressure within a certain range and the distance of the sample in the reactor. In all the following experiments, the ideal position and the ideal pressure was used for the passivation process.

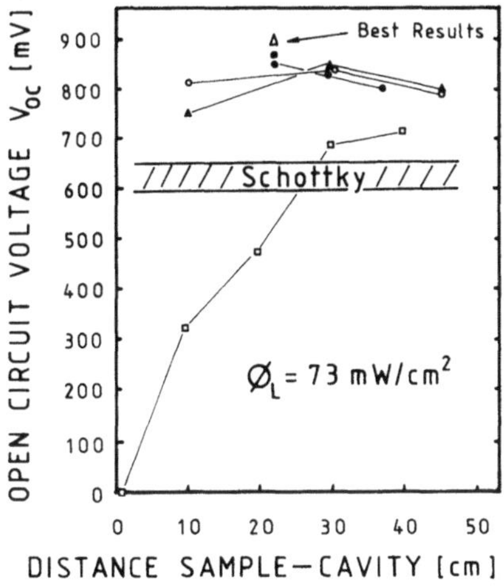

Fig.3. Open circuit voltage as a function of the sample distance from the microwave-cavity at various atmospheres and pressures: □ = 2 mbar O_2, Ir; o = 30 mbar H_2O, Ir; ▲ = 16 mbar H_2O, Ir; • = 10 mbar H_2O, Ir; Δ = 10 mbar H_2O, Pt.

Because water vapour leads to best results, we propose two competitive models to explain the barrier enhancement due to our process:

- The hydrogen depleted interfacial region may be saturated by atomic hydrogen generated by the water vapour plasma, thus resulting in a perfect Schottky barrier.
- A doping specific MIS surface passivation by the growth of a tunneling oxide (2).

The second model can be supported if it is possible to prove the existence of an oxide on the a-Si:H as a function of the passivation process. Therefore we performed the following experiment with three different samples:

- An ordinary Schottky diode (samples a)
- The barrier forming metal layer was prepared after the base material was exposed for 40 days in air (samples b)
- A water vapour passivated sample (c) and also exposed for 40 days in air

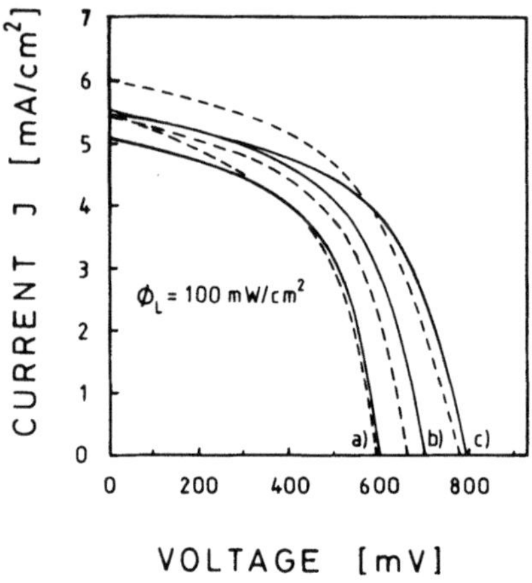

Fig.4. I-V characteristic of an a-Si:H solar cell with Schottky-diode, b: after long exposure at air, c: dashed lines indicates Pt as barrier metal, the full ones Ir barriers.

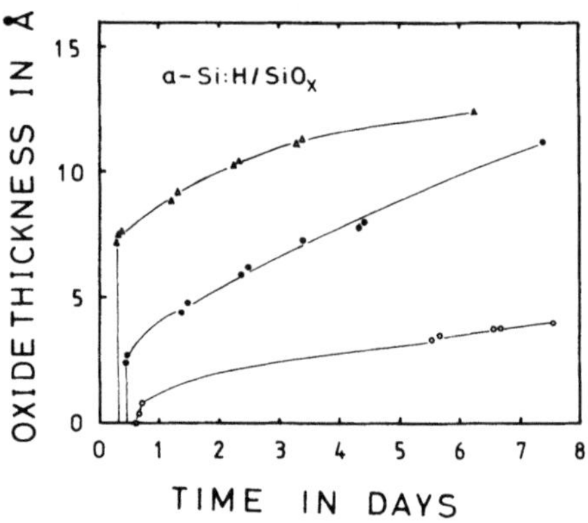

Fig.5. Ellipsometric measurement of oxide growth on intrinsic a-Si:H as a function of time for different pretreatments.

Fig.4 shows, that samples c exhibit higher V_{OC}, as expected. Samples b also show an increase of V_{OC} and it has been shown (7) that a-Si:H formes a native oxide after some time at air exposure. Fig.5 illustrates the results of our ellipsometrical measurements, where the development of samples b is plotted on the lowest curve, showing an index of refraction of approximately 1.5 which indicates an oxide growth in agreement with (7). More details on oxide growth will be presented in a different paper of this conference. The above curves were taken from samples c where an immediate film growth after the plasma-treatment can be measured, which leads during the some time period to a higher oxidation rate value than samples b. Thus the existence of a thicker oxide may indicate, that we may use model 2 to explain the phenomena. However a significant increase of the diode quality factor, indicative of stronger majority carrier suppression, could not be detected.

ii. Remarks to the photocurrent

If the thickness of the intrinsic layer is varied from 0.3 µm to 0.5 µm - 1 µm a rise in J_{SC} is observed, although the 0.3 µm layer is thick enough to absorbe the total AM1 spectrum. In Fig. 6 the i-layer thickness dependence of the photocurrent is plotted.

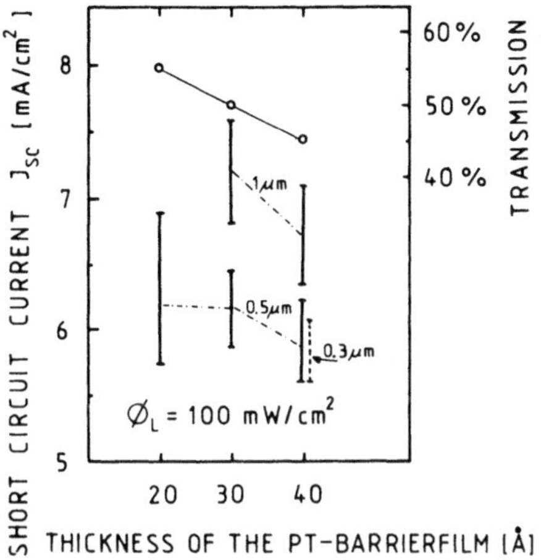

Fig.6. J_{SC} vs barrier metal thickness with different thicknesses of the base material. The above curve presents the transmission of Pt films on quartz substrates.

It also can be seen that a film-thickness around 30 Å leads to the highest J_{SC} values. As usual, an antireflective coating leads to an increase of J_{SC}. When ZrO_2 at 470 Å is used, we note an improvement of 50% at polished samples on 30 Å thick Ir films (Fig. 7a). A further improvement of J_{SC} could be measured, if the stainless steel substrate was roughened by sand-blasting illustrated in Fig 7b for the case of an MIS cell and in Fig 7c for the case of an unpassivated Schottky cell. We can conclude from 7c that additive improvement of the light coupling mechanism is possible.

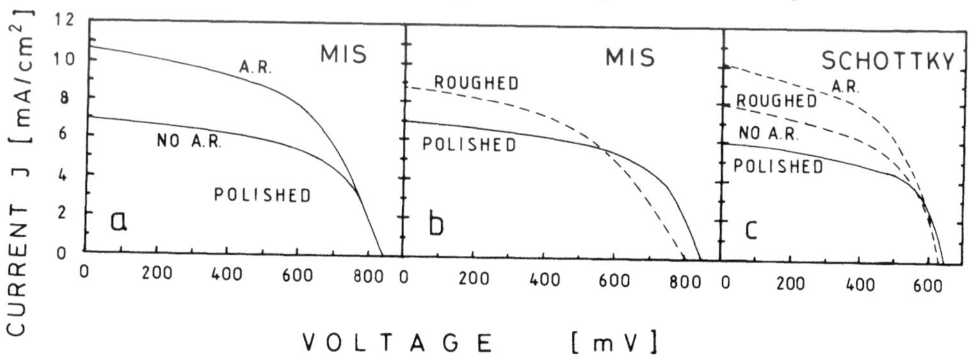

Fig.7. I-V Characteristics of MIS a-Si:H solar cells (a,b) and Schottky cells (c) with respect of photocurrent enhancement due to roughening of the substrate and the use of a AR coating.

It is remarkable, that sand-blasting of the substrate doesn't affect the perfect interface properties of the a-Si:H MIS cell.

iii. Remarks about the fillfactor.

When we used polished substrates, the fillfactor of all cells varied between 50 % and 60 %. As demonstrated in Fig. 8 the spectral response shows a bias dependent shift of its maximum.

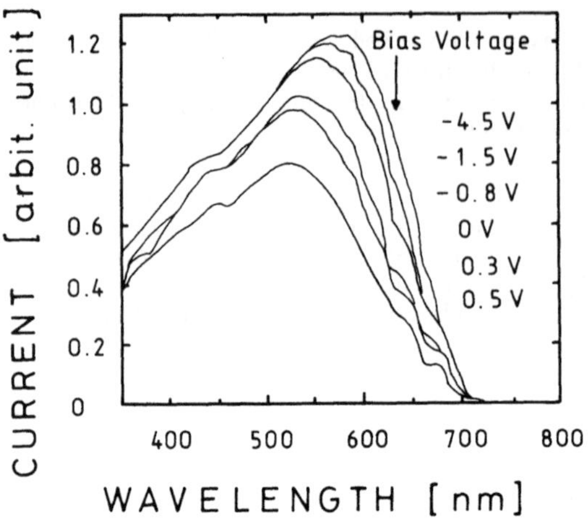

Fig.8. Sectral response at different bias voltages.

At high reverse bias voltages the spectral response curve shifts to higher sensitivity in the red, which indicates, that either the field in the intrinsic layer might not be homogenious, or that there is a dominant recombination process in the i-layer. MIS and Schottky cells on rough substrates exhibit lower fillfactors due to a higher series resistance if 30 Å to 40 Å of barrier metal is taken as optimized thickness from previous experiments with polished substrates.

4. Conclusions

With the plasma surface passivation process we achieved the highest V_{oc} values with a-Si:H material of the above mentioned properties. We conclude from ellipsometric measurements, that the growth of a thin oxidelayer leads to a barrier enhancement. The photocurrent could be improved by variation of the basematerial properties, an antireflective coating and using a roughened substrate, from 5 mA/cm^2 up to 11 mA/cm^2. The roughing procedure is still not optimized, because we only used a 50 µm grain size for sand-blasting. Also the thickness of the barrier metal and the AR coating on the rough a-Si:H surface must be newly optimized in order to obtain higher fillfactors. By extrapolation of Schottky cell data we think that photocurrents of about 15 mA/cm^2 are possible.

Acknowledgements.
This work was supported by the BMFT, project No. 03E-8453-B and 03E-8019-B1 as well as EC Project No. EN 3S-0107-D(B). In Switzerland it was supported through Swiss Federal Research grant OFEN(REN) 85/16. We are greatful for constructive discussions with H.Berner and valuable technical assistance from R.Burckhardt and J.Hufnagl.

References

1. J. Shewchun, M.A. Green, F.D. King: Solid State Electron. 17, 563 (1974)
2. D.L. Pulfrey: IEEE Trans. Electr. Devices Vol. ED-25 (11), 1308 (1978)
3. P. Munz, R. Schultheiss: Proc. 17th Photovoltaic Specialist's Conference IEEE New York (1984)
4. A. Madan, J. McGill, W. Czubatyj, J. Yang, S.R. Ovshinsky: Appl. Phys. Lett. 37(9), 826 (1980)
5. J. Meier, H. Keppner, E. Bucher, P. Munz, H. Curtins, A.V. Shah, S. Paasche, W.H. Bloss: Appl. Phys. A 46, 31 (1988)
6. H. Curtins, N. Wyrsch, A.V. Shah: Electron. Lett. 23, 228 (1987)
7. J.P. Ponpon, B. Bourdon: Solid State Electron. 25, 875 (1982)

TOWARDS A BETTER UNDERSTANDING OF THE STAEBLER-WRONSKI EFFECT IN AMORPHOUS SILICON

C.M. Fortmann
Solarex Thin Film Division
826 Newtown-Yardley Road
Newtown, PA 18940 USA

ABSTRACT. The saturated dangling bond density is explored as a function of charge carrier density and temperature. Solar cell RBA (reverse bias anneal) and FBA (forward bias anneal) are used for this study. The defects that are responsible for the SWE (Staebler-Wronski effect) [1] are shown also to be related to the RBA and FBA effects. At annealing temperatures greater than 150°C forward bias increases the density of dangling bonds in the i-layer, while reverse bias decreases the density. The temperature and cooling rate dependence of FBA and RBA are investigated. The dangling bond density is proportional to the square root of the recombination rate, rather than the linear relationship predicted by kinetic models. These results are more consistent with an equilibrium description of the SWE.

INTRODUCTION

The SWE has received intense examination in recent years as it is one of the major impediments to more wide spread commercial application of amorphous silicon solar cells. These investigations have for the most part focussed on the rate of change of either a device or film property due to light exposure. In the work presented here the attention is directed to the high temperature saturated degradation. Amorphous materials will also degrade at high temperatures (T>150°C), however unlike room temperature degradation, the competing processes of thermal annealing and light induced degradation lead quickly to a net balance of forward (degradation) and reverse processes (annealing). The temperature and charge density of the saturated degradation is explored by this work. The experimental framework provided by the examination of the saturated high temperature will be used to review some of the concepts of degradation developed through the investigation of kinetics of the SWE.

The rate of degradation due to the SWE in a-Si:H has been related to the recombination rate of carriers [2], and a balance of recombination driven degradation, thermal degradation, recombination driven annealing [3], and with thermal degradation also taken into account [4]. Recombination rate induced degradation would lead to a high temperature saturated concentration of dangling bonds that is proportional to the recombination rate. This possibility is examined in detail. In this work the dangling bond is assumed to be the defect related to the SWE, although there is not complete agreement on this indentification.

The relationship between recombination and degradation was explored at temperatures high enough (T>150°C) to insure saturation in a reasonable period of time (<1hr). Solar cells (p-i-n) were degraded at high temperatures by FBA or enhanced by RBA. FBA results in the creation of dangling bonds. FBA was chosen over light exposure to degrade the solar cells as the recombination rate, as well as carrier densities could be more carefully controlled (the recombination rate 'R' $\sim J_{injected}$). Relatively uniform recombination rates in the i-layer result from FBA, as it is reported [5] that nonuniformly absorbed illumination of solar cells under forward bias results in nearly constant recombination rates across the i-layer. Another advantage of solar cells is that they are completely encapsulated, thus increasing the resistance to reactions with the environment during high temperature annealing. The current density used for FBA was varied between 0 to 250mA/cm^2, while the temperature range examined was 150° to 250°C. RBA was used to study the effect of reduced carrier densities during annealing. Quantitative analysis was only applied to the FBA case, since the recombination rates across the i-layer of the solar cell are much more uniform. The cooling rate dependence of FBA and RBA were also investigated.

Solar cell performance characteristics (measured at room temperature) were used to determine the extent of FBA degradation. An analysis of the SWQE [6] (short wavelength quantum efficiency) was used to obtain better quantitative information concerning the density of dangling bonds (N_r) created by FBA in the solar cells. In this technique the recombination in the low field region increases resulting in reduced SWQE as:

$$\Delta SWQE = qp_o(D-W_n-W_p)(\sigma_p \Delta N_r V_{th})/J_F \qquad (1)$$

where J_F is the current that would result from 100% collection efficiency, q is the electron charge, p_o is the dark equilibrium concentration of holes in the low field regions of the i-layer, D is the i-layer thickness, W_n is the depletion width at the i/n interface, W_p is the depletion region width at the p/i interface, σ_p is the capture cross section for holes by a dangling bond, which has density N_r and V_{th} is the thermal velocity of the carriers.

The effect of degradation in solar cells is apparent through a decrease in SWQE which results from an increase in N_r. It is necessary to examine cells with thickness greater than ~0.5μm to provide a sufficiently large low field region to observe the reduced SWQE. In some cases the ratio of SWQE with no light bias is taken to that with red light bias to eliminate calibration errors. The preparation of solar cells used in this study have been described previously [7], while other experimental details are given in [8].

To further explore the temperature dependence of the degradation due to current injection on an alternate system, a limited number of film studies (i-layers deposited onto quartz substrates) were also carried out. In this case recombination currents were provided by exposure to white light with intensity of ~800mW/cm^2. The density of defects in films was deduced from below bandap optical absorption (alpha at 1eV) measured by PDS (photothermal deflection spectroscopy). Fast cooling (rates greater than ~1000°/min) were used for the illumination degraded films.

DISCUSSION

RBA increases solar cell performance [9], this could be due to an improvement in doping efficiency in the doped layers or a reduction in the dangling bond density in the i-layer. To determine if this effect is related to the i-layer rather than the doped layers, the effect of RBA as a function of i-layer thickness was measured. Solar cells were prepared with thickness ranging from 2000Å to 8500Å. The series of cells were subjected to RBA at a temperature of ~175°C. Figure 1 shows the reverse bias necessary for a performance gain (both 12% fill factor gain and a 65% fill factor) as a function of cell thickness. The voltage needed for a given amount of performance gain increases with increasing i-layer thickness, indicating

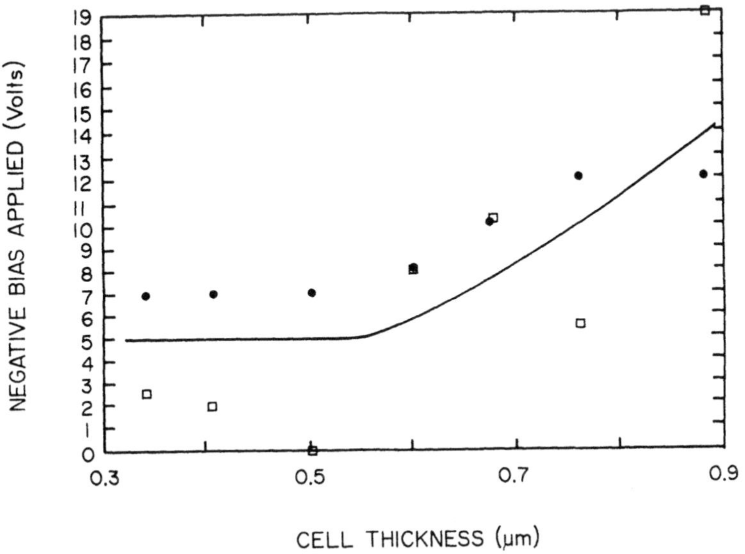

Figure 1. Voltage required for a 12% fill factor gain (•) and a 65% fill factor (□) as a function of i-layer thickness.

that the effect is associated with the i-layer. The effect of RBA completely saturates in less than one hour at this temperature. RBA is not cooling rate dependent (cooling rates from ∿15 to ∿1000°C/min) when bias is maintained during cooling. The effects of reverse bias heat treatment can be reversed through an anneal at 150°C for one hour. The SWQE of cells after RBA increases indicating again that the effect is related to SWE defects in the i-layer.

To investigate the recombination rate dependence of the FBA treatment, solar cells (∿5500Å and ∿6500Å thick) were annealed at currents ranging from 0 to 250mA/cm^2. Figure 2 shows the current dependence of the decrease in fill factor that results from current injection. The decrease in performance seen here also saturates in less than one hour and can be reversed by an anneal at 150°C in one hour.

The effect of current injection degradation using the analysis of the SWQE was carried out on the thicker (∿6500Å thick) solar cells. The degradation of SWQE due to defect density increases is shown in Figure 3 (left). The change in the SWQE ratio (no light bias to that with red light bias) versus the injection current is shown in Figure 3 (right).

The lack of a substantial cooling rate dependene suggests that neither RBA or FBA are strongly temperature dependent. To further investigate the temperature dependence, FBA studies of solar cells were carried out as a function of temperature. Current injection degradation FBA (J=150mA/cm^2) experiments were carried out on solar cells in the temperature range 150° to 250°C. Films were also degraded as a function of temperature. Control devices and films were only annealed as a function of temperature. Films were degraded by light exposure (∿800mW/cm^2) in the same temperature range. No strong temperature dependence emerged in either the annealing or the degradation of devices or films (Figure 4).

The issue of recombination induced degradation can now be addressed. The saturation value [3] for the dangling bond density is:

$$N_r = A_1 R \qquad (2)$$

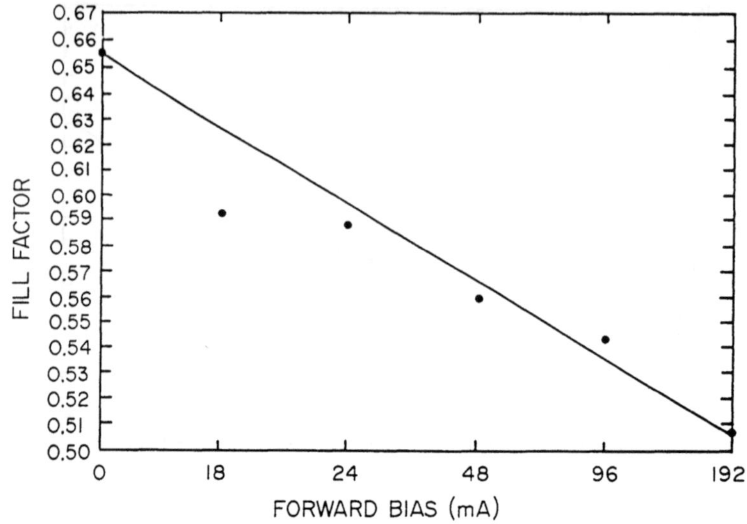

Figure 2. A plot of FF versus the injection current.

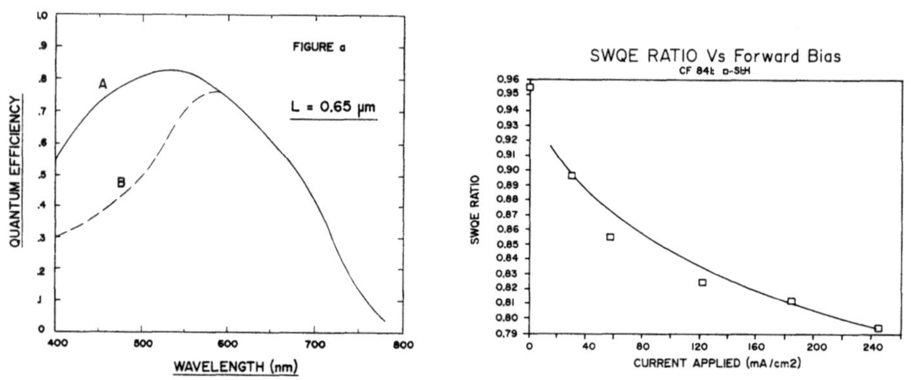

Figure 3. Left the QE of annealed (a) and degraded (b) solar cell. SWQE (ratio) versus forward injection current (right) for a p-i-n solar cell.

where R is the recombination rate and A_1 is a constant. From equation (1) ΔSWQE is proportional to ΔN_r. Therefore the expected relationship between ΔSWQE and J(aR) is given by:

$$\Delta\text{SWQE ratio} = A_3 J \quad (3)$$

The experimental values of the SWQE loss were compared to those obtained using equation (3) where A_3 was adjusted to yield the equality at J=31mA/cm . The fit to the experiment is poor. The relationship between FBA current and SWQE loss appears to be a square root dependency (solid line in Figure 3):

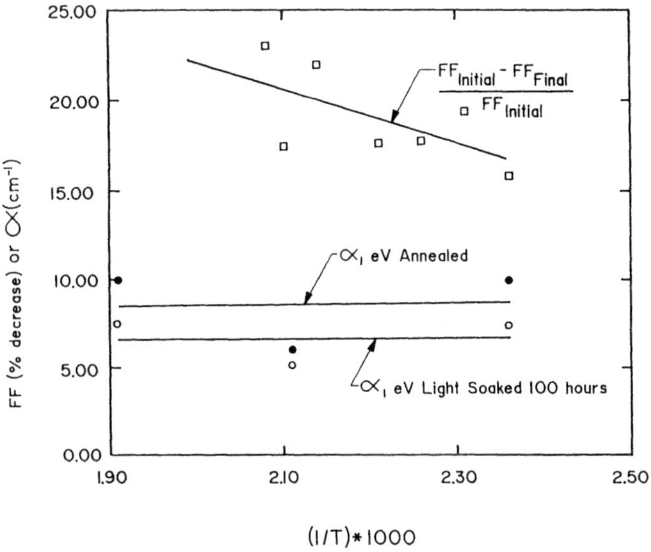

Figure 4. Fill factor versus inverse temperature used for FBA solar cells degraded at a current level of ~150mA/cm^2, also the absorption at 1eV of annealed (∘), and light soaked (•) films is also shown.

$$\Delta\text{SWQE ration} = A_4(J_{app})^{\frac{1}{2}} \tag{4}$$

This relation is not expected from any standard model of recombination induced degradation, however such a relation is easily obtained from equilibrium considerations; that is, the imposed charge density is in equilibrium with the dangling bond density, albeit with the assumption that this equilibrium exists even though the values of the carrier densities are not themselves equilibrium values. For example, an equilibrium between a species that is proportional to the injected current [i] (a charge carrier), a weak bond site 'W_r' and two dangling bonds is:

$$i + W_r \leftrightarrow 2N_r \qquad \Delta G \tag{5}$$

$$[N_r]^2/[i]*[W_r] = K\sim 1 \tag{6}$$

Either the weak bond site or one of the dangling bonds must be a charged center for charge conservation. Equation (6) and (4) are equivalent. Substitution of the expected values of N_r (~10^{16}cm^{-3}) with i (~10^{12} to 10^{13}cm^{-3}), and K~1 (consistent with the small temperature dependence) yields reasonable values of the weak bond density (10^{19} to 10^{20}cm^{-3}). The value of A_4 was adjusted to yield the equality at the experimental point SWQE ratio = 0.897 and J_{app} = 31mA/cm^{-2}. The relation that results from using this value for A_4 in equation (4) is a good fit to the experimental data shown in Table I.

CONCLUSIONS

The saturated value of the degradation of a-Si:H is examined as a function

TABLE I

J_{app}	SWQE Ratio	Eq(3)	Eq(4)
0	0.955		
31	0.897	0.897	0.897
58	0.855	0.846	0.875
123	0.824	0.725	0.839
185	0.812	0.609	0.813
246	0.794	0.495	0.791

of temperature and charge density. The results of this investigation are not consistent with recombination rate driven degradation. An equilibrium description of the SWE offers the best fit to the data presented here.

ACKNOWLEDGEMENT

This work was partially supported by SERI under contract no. ZB-7-06003-2.

REFERENCES

1. D.L. Staebler and C.R. Wronski, Appl. Phys. Lett. 31, 292 (1977)

2. M. Stutzmann, W.B. Jackson and C.C. Tsai, Phys. Rev. B 32, 23 (1985).

3. D. Redfield, AIP Conf. Proc., 157 Stability of Amorphous Silicon Alloy Materials and Devices, B.L. Stafford and E. Sabisky, p. 257, AIP, NY 1987.

4. D. Redfield, Journal of the Non-Crystalline Solids 97&98, 783, North Holland, Amsterdam (1987).

5. M. Hack and M. Shur, J. Appl. Phys. 58 (2) (1985).

6. C.M. Fortmann, S. Lange, M. Hicks and C. Wronski, to be published.

7. C.M. Fortmann, J.L. Newton, J. O'Dowd and J. Fischer, AIP Conf. Proc. 157, Stability of Amorphous Silicon Alloy Materials and Devices, B.L. Stafford and E. Sabisky, p. 103, AIP, NY (1987).

8. C.M. Fortmann, Materials Research Society, (Sym E) meeting held in Reno, NV to be published (1988).

9. G.A. Swartz, Appl. Phys. Lett. 44, 697 (1984).

HIGH EFFICIENCY SINGLE AND STACKED JUNCTION a-Si:H BASED ALLOY SOLAR CELLS

R.R. Arya, J.L. Newton, C.M. Fortmann and A. Catalano
Solarex Corporation, Thin Film Division
826 Newtown-Yardley Road, Newtown, PA 18940 USA

ABSTRACT. Amorphous silicon-carbon alloy single junction solar cells have resulted in open-circuit voltages as high as 1.05V. Single junction a-SiGe:H solar cells with conversion efficiencies above 10% are reported. Stacked junction cells employing these alloys have resulted in conversion efficiencies as high as 10.5%.

INTRODUCTION

Hydrogenated amorphous silicon (a-Si:H) has an optical bandgap of 1.7eV and its absorption co-efficient is such that the response of single junction p-i-n solar cells is restricted to about 750nm. This limits the short-circuit current density to about $19mA/cm^2$ and the conversion efficiency to about 12-13%. In order to further improve the conversion efficiency it is necessary to better utilize the solar spectrum. This requires the development of amorphous silicon based alloys which have both wider and narrower optical bandgaps than amorphous silicon and to employ them in monolithic stacked junction configurations. Amorphous silicon-carbon (a-SiC:H) alloys and amorphous silicon-germanium (a-SiGe:H) alloys have emerged as promising candidate materials for wider and for narrower bandgap materials respectively.

In this paper we shall discuss the development of a-SiC:H and a-SiGe:H single junction and two-junction stacked solar cells employing these materials.

SINGLE JUNCTION SOLAR CELLS

Silicon-Germanium Cells

We have studied amorphous silicon-germanium alloys which have optical bandgaps in the range of 1.7eV to 1.43eV. These alloys were deposited by dc glow-discharge from a gaseous mixture of silane and germane. The opto-electronic properties of these films have been discussed elsewhere [1,2].

Single junction solar cells have been fabricated with i-layer material which has optical bandgap of either 1.61eV or 1.57eV. The concept of p/i graded interface has been extended to i(a-Si:H)/i(a-SiGe:H). These cells have the same p-layer and the same p/i graded interfaces as those employed in our high efficiency single junction a Si:H cells [3]. The graded interface between the a-Si:H buffer and the bulk a-SiGe:H i-layer is employed to reduce recombination of carriers at that interface since an abrupt transition from a-Si:H (1.7eV) to a-SiGe:H (1.57eV) can increase the number of recombination centers at that interface [4]. Figure 1 shows the device structure employed for a-SiGe:H single junction cells. Even with a graded interface there is evidence that hole transport is relatively poor in these alloys as compared to a-Si:H. Figure 2 shows the forward-bias dependence of photo-generated carriers as a function of wavelength.

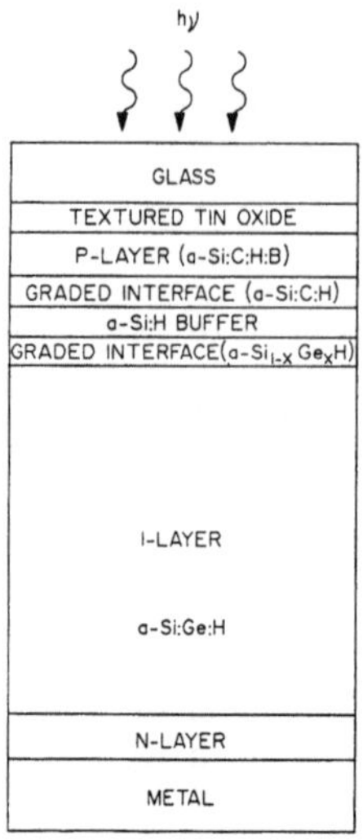

Figure 1. Device structure of a-SiGe:H single junction solar cell.

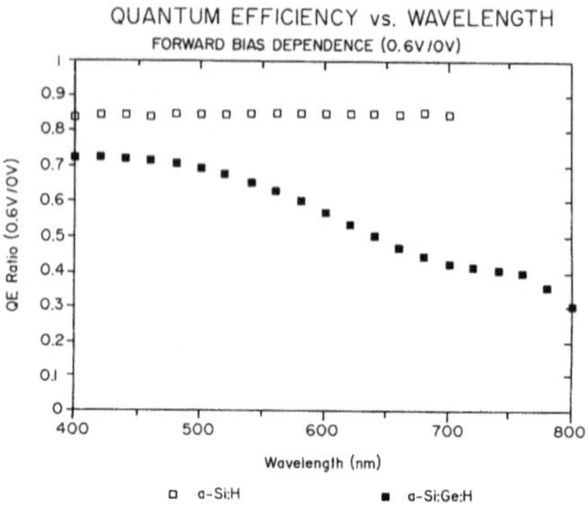

Figure 2. Ratio of quantum efficiency (0.6V/0V) as a function of wavelength.

In this figure the ratio of quantum efficiency at an applied bias of 0.6V to that at 0V is plotted for an a-Si:H cell and an a-SiGe:H cell. For an a-Si:H cell there is no wavelength dependence of the collection of carriers in forward bias while there is a strong wavelength dependence in the case of a-SiGe:H cell. Collection of carriers generated in the bulk i-layer (long wavelength) are more bias dependent indicating hole transport limitation due to the low $\mu\tau$ product of holes or a low field region in the i-layer. The best single junction cell fabricated with an i-layer of bandgap 1.61eV had a conversion efficiency of 10.77%. The best cell fabricated with an i-layer bandgap of 1.57eV had a conversion efficiency of 10.1%. Figure 3 shows the J-V characteristics of this cell under AM1.5 global conditions. Figure 4 shows the quantum efficiency versus wavelength measurement of this cell. Integrating the quantum efficiency measurement with the global AM1.5 spectrum yields a J_{sc} value of 20mA/cm^2 which is in good agreement with the measured value of J_{sc} under solar simulation.

Silicon-Carbon Cells

Amorphous silicon-carbon films were prepared by dc glow discharge from a gaseous mixture of silane and methane. The composition of films were determined by Rutherford Back Scattering (RBS). Figure 5 shows the atomic fraction of carbon in the solid as a function of gas phase concentration. The incorporation ratio is far less than unity owing to the higher energy needed to create the necessary reactive carbon based intermediates. The optical bandgap varies linearly with the composition of the solid according to:

$$E_g = 1.694 + 3.25\ (x)$$

where x is the atom fraction of the carbon in the solid. The opto-electronic properties of these films have been reported elsewhere [5].

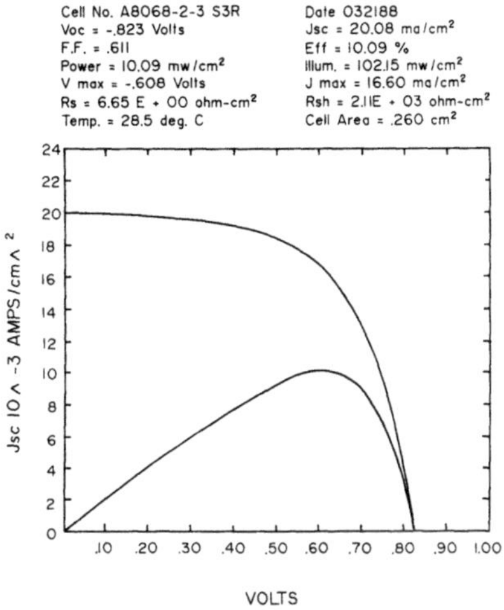

Figure 3. J-V characteristics of a-SiGe:H single junction cell under simulated AM1.5 global illumination.

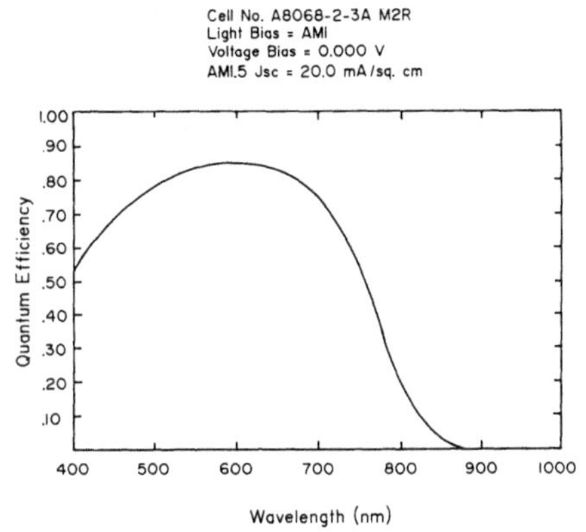

Figure 4. Quantum efficiency versus wavelength measurement of the device in Figure 3.

Single junction solar cells were fabricated with a-SiC:H i-layers. The thickness of the i-layer was restricted to 2000Å since only thin i-layers are needed for optimum two or three layer stacked cells. In these cells the p- and the n-layers were kept the same and only the bandgap of the i-layer was varied. Further, no p/i graded interface was employed in these cells. Figure 6 shows a plot of the measured V_{oc} and fill-factor of solar cells versus the methane concentration from which the i-layers were deposited. We find that V_{oc} initially increases linearly as the methane concentration is increased in the mixture of silane and methane. The V_{oc} increases from a low value of 0.86V to a high value of 1.05V for a gas composition of 50% methane. The fill-factor remains in the range of 65-70% for gas phase concentrations of up to 15% (E_g = 1.9eV) but falls off more rapidly at higher methane concentrations. The best devices were made with methane concentrations between 10-15% in the i-region. In these devices the increase in open-circuit voltage can only be associated with the increased bandgap of the i-region [5].

STACKED JUNCTION SOLAR CELLS

Monolithic stacked junction solar cells have been fabricated in which the i-layers are composed of a-SiC:H, a-Si:H, and a-SiGe:H. A two junction stacked cell employing an a-SiC:H i-layer (E_g = 1.9eV) top cell and an a-Si:H i-layer bottom cell has resulted in a conversion efficiency of 10.2%. Figure 7 shows the J-V characteristics of this cell under simulated AM1.5 global conditions. This cell has current matching at 8.2mA/cm^2. Employing a similar top cell with an a-SiGe:H i-layer (E_g = 1.57eV) bottom cell has resulted in a conversion efficiency of 10.5%. Figure 8 shows the J-V characteristics of this cell under AM1.5 global conditions. Figure 9 shows the quantum efficiency versus wavelength

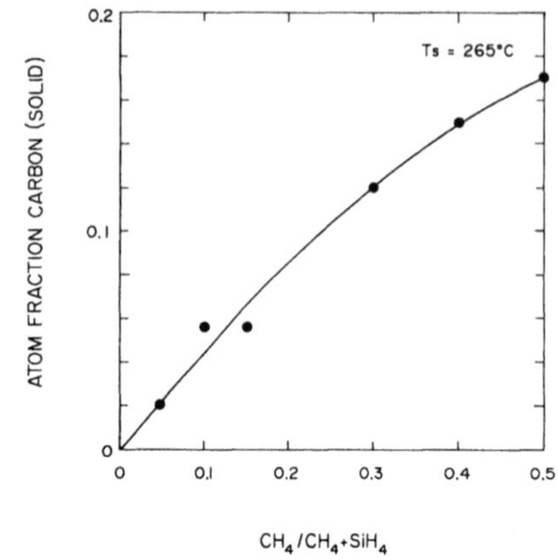

Figure 5. Atomic fraction of carbon in the solid plotted as a function of methane in the gas phase.

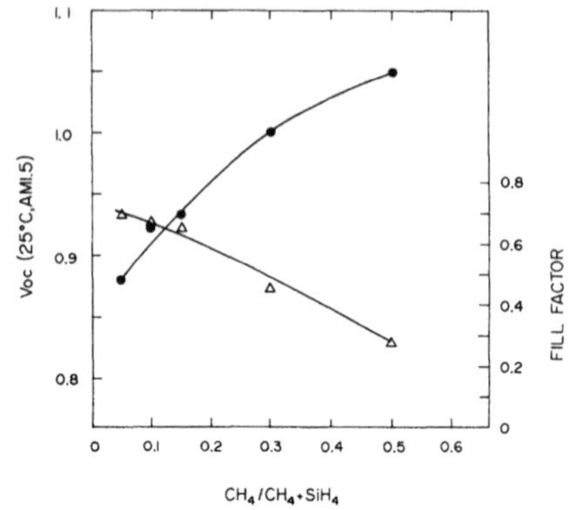

Figure 6. Open-circuit voltage and fill-factor of a-SiC:H devices as a function of optical bandgap of i-layer.

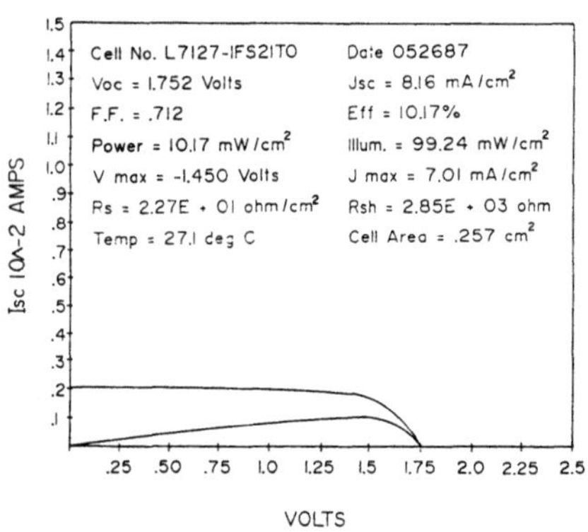

Figure 7. J-V characteristics of a-SiC:H/a-Si:H stacked junction cell under AM1.5 global illumination.

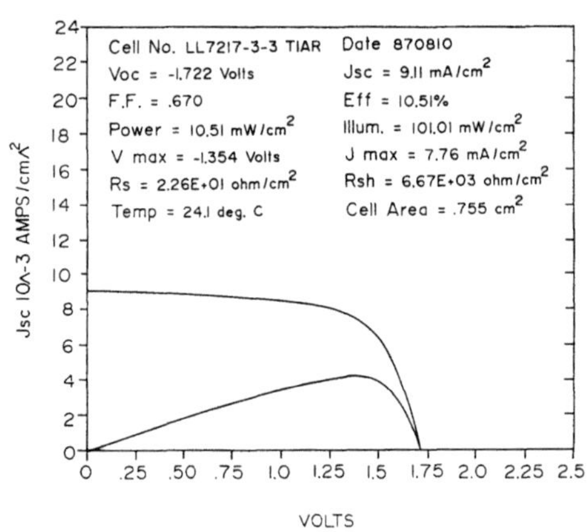

Figure 8. J-V characteristics of a-SiC:H/a-SiGe:H stacked junction cell under AM1.5 global illumination.

measurement of this device. In this device the short-circuit current density is 9.1mA/cm^2 with the front cell limiting the current. Even though the individual cells of this stacked cell are slightly mismatched, the total current generated by this cell is 19.2mA/cm^2. The photovoltaic parameters of single and stacked junction a-Si:H based alloy cells are summarized in Table I.

In summary, a-SiGe:H single junction cells have resulted in conversion efficiencies of 10.77% (E_g = 1.61eV) and 10.1% (E_g = 1.57eV). Single junction a-SiC:H cells have resulted in open-circuit voltages in excess of 1.05V. Stacked junction cells employing a-Si:H based alloys

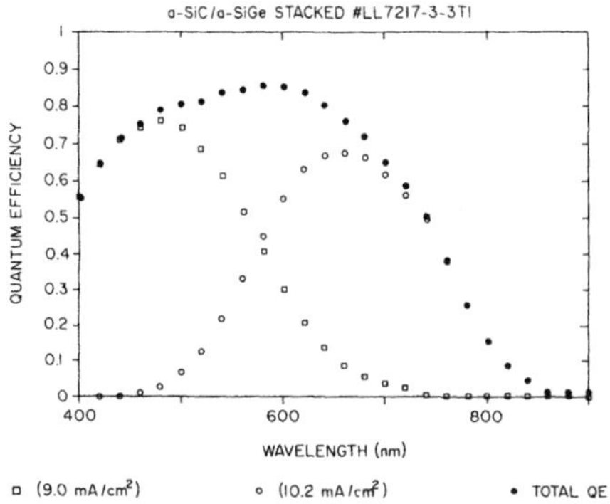

Figure 9. Quantum efficiency versus wavelength measurement of the device shown in Figure 8.

have resulted in conversion efficiencies as high as 10.5%.

TABLE I

Cell Number	Single Junction	Stacked Junction	E_g (eV)	V_{oc} (mV)	J_{sc} (mA/cm^2)	FF	Eff. (%)
LL711133S3	a-SiGe		1.61	851	19.34	0.654	10.77
A806823S3	a-SiGe		1.57	823	20.08	0.611	10.09
L71271FS2		SiC/Si:H	1.9/1.7	1752	8.16	0.712	10.17
LL712733T1		SiC/SiGe	1.85/1.57	1722	9.11	0.67	10.51

REFERENCES

1. J. Newton, G. Wood and A. Catalano, Proc. 19th IEEE PVSC, p. 862 (1987).

2. A. Catalano, J.L. Newton, R.R. Arya and S. Wiedeman, Tech. Dig. 3rd PVSEC, p. 61 (1987).

3. A. Catalano, R.R. Arya, C. Fortmann, J. Morris, J. Newton and J.G. O'Dowd, Proc. 19th IEEE PVSC, p. 1506 (1987).

4. R.R. Arya, A. Catalano, J.L. Newton and J. Morris, Tech. Dig. 3rd PVSEC, p. 725 (1987).

5. A. Catalano, C. Fortmann, J. Newton, R.R. Arya and G. Wood, Tech. Dig. 3rd PVSEC, p. 705 (1987).

ACKNOWLEDGEMENTS

The authors wish to thank G. Wood and D. Bradley for technical assistance and Ms. Karen Puskar for typing the manuscript. This work has been partially supported by SERI under contract number ZB-7-06003-2.

EFFECT OF SLIGHTLY PROFILED PHOSPHORUS OR BORON DOPING OF THE CENTRAL PIN-DIODE LAYER ON THE FILL FACTOR

M. Gorn, N. Kniffler
Messerschmitt-Bölkow-Blohm GmbH
-Energy and Process Technology-
P.Box 80 11 09
D-8000 Munich 80
West Germany

ABSTRACT. The effect of fine-doping profiles in the i-layer of a-Si:H based TCO/pin/Al solar cells on initial performance, particulary the fill factor, is studied. Except for rising phosphorus profiles, all other phosphorus and boron profiles resulted in lower fill factors. The results are discussed in terms of changes in the electric field distribution and in the mobility-lifetime products of carriers. Long term illumination of cells with rising phosphorus profiles over 10.000 h shows no accelerated degradation for large times.

1. INTRODUCTION

Fine-doping of the i-layer in a-Si:H based pin-type solar cells has been often used to influence initial efficiency, spectral response and long term stability. According to initial efficiency, constant fine-doping levels are not used any more because better results are obtained by reduction of background impurities /1/. Fine-doping profiles on the other hand are interesting in the context of long term stability. It was found that such doping profiles improved the stability of a-Si:H based solar cells, for high initial efficiencies $\geq$ 8 % at least for several hours of AM1 illumination /2/.

For the technological application of solar cells both initial efficiency and long term stability for the period between 10^3 and 10^4 h are relevant. There has been reported e.g. on an anomalous degradation time dependence, initially slow and faster for long times, in cells with constant fine-doping levels /1/.

The purpose of this paper is to present:

a) extended data (with respect to ref. 2) on the influence of several types of linearly increasing or decreasing doping profiles on the initial performance of TCO/pin/Al solar cells, particulary the fill factor
b) discussion of the fill factors considering the electric field distribution and mobility-lifetime products
c) data on cells after 10.000 h of AM1 illumination.

A new feature in these data lies also in the use of a three-chamber system, where no contribution to the profiles can come from the boron outgasing from the chamber walls due to the initial p^+-layer deposition.

2. PREPARATION OF CELLS WITH FINE-DOPING PROFILES

Structures of the type glas/TCO/pin were deposited in a three-chamber glow discharge system. During i-layer deposition, doping gases PH_3 or B_2H_6 were added with concentrations either linearly increasing or decreasing with deposition time, controlled by computer. The maximal doping at the boundary of a profile was either. 5 or 1 vppm. The extension of the profiles was either along the whole i-layer or half of the i-layer (as indicated in the insert of fig. 1).

The total cell thicknesses were around 4.000 A as measured by a profilometer ("Alphastep").

3. CELL PARAMETERS RESULTING FROM DOPING PROFILES

There were four undoped reference substrates with average fill factors of .64$\pm$ 0.02 and average Voc of around 750 mV. Efficiencies (AM1) ranged around 6 to 8 % for undoped i-layers and cells with phosphorus profiles. Cells with boron profiles had efficiencies of 2 to 6 %.

In the following, we concentrate on the fill factor, which was most sensitive to the doping profiles we used.

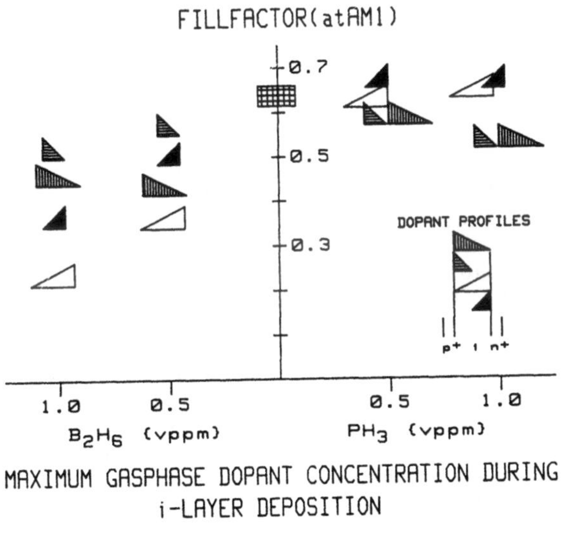

Fig. 1
fill factors of TCO/pin solar cells with various fine-doping profiles in the i-layer (center-of-mass of triangles)

As displayed in fig. 1, the rising phosphorus profiles (open and closed triangles on the right side) did not affect the fill factor, whereas phosphorus in the front did so (hatched triangles, right side). The boron profiles resulted in much lower fill factors (left side of fig. 1).

4. DISCUSSION OF THE RESULTS

Lower fill factors from boron in the front of the i-layer (hatched triangles in fig. 1, boron side) result from a suppression of the blue and green response, as can be seen in fig. 2. This can be explained by a reduction of the electric field by boron at the p/i-interface and in the front of the i-layer, where most of the carriers are created by the light entering through the p-window. By this reduction, the field should be enhanced elsewhere (if one assumes a constant built-in voltage), which is in the rear of the i-layer. This seems however to be unfavourable for light entering through the p-window.

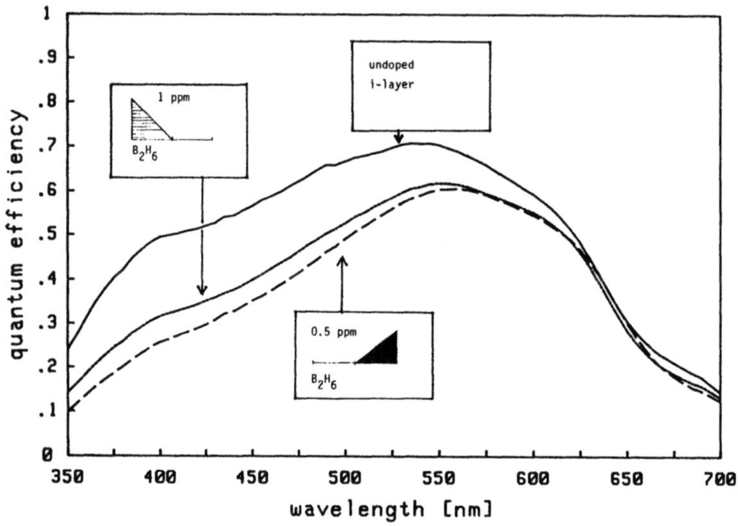

Fig. 2: Spectral response at 0 Volt bias for rising and falling boron profiles, compared with an undoped i-layer cell

Boron in the rear of the cell was observed to yield the lowest fill factors (open and closed triangles on left side of fig. 1). Already with a low boron concentration, the blue and green response is strongly suppressed, as shown in fig. 2. Boron in the rear, towards the i/n-interface, probably creates a strong field spike at the i/n-interface. Under the assumption of constant built-in voltage the field should be recuced elsewhere, which is in the front and the middle of the i-layer. This does however not really explain the large differences between boron in the rear and boron in the front (see fig. 1).

Boron is known to have also an influence on the carrier $\mu\tau$-products. After Street and Zesch /3/, the electron-$\mu\tau$ is lowered by boron, whereas the hole-$\mu\tau$ remains almost unaffected. This is deleterious towards the rear of the i-layer, where the electrons, in the case of blue and green light (strongly absorbed in the front) limit the current. With boron in the front, this effect on the electrons should also play a role. However, in our experimental data the effect of boron in the front on the fill factor is not as strong as that of boron in the rear. Maybe this is due to additional contributions to the electron-current from diffusion which change more into a drift-like current towards the rear. This electron-current to the rear would be more affected by boron in the rear than by boron in the front, which could explain the above mentioned difference between the two types of boron profiles.

The phosphorus profiles which rise towards the rear of the i-layer (open and closed triangles on right side of fig. 1) are observed to have fill factors which are similar to the cells with undoped i-layers. Considering the electric field, there will be a field reduction in the rear and consequently (assuming constant built-in voltage) a field enhancement towards the front. This may be not disadvantageous to some extent in the case of light entering through p^+. Furthermore the $\mu\tau$-product of the holes is reduced which seems to have almost no effect in the rear of the i-layer, however.

Phosphorus in the front (hatched triangles, right side in fig. 1) lowered the fill factor slightly. The field should get a stronger spike at the p/i-interface lowering the field everywhere else (assuming constant built-in voltage) which is in the middle and the rear, affecting the blue/green as well as the red performance of the cell because of current continuity. The $\mu\tau$-product of the holes which is reduced by the phosphorus in the front should however only affect the red response (in which case the holes are the limiting carrier), but not the green and blue response (limited by the electrons).

The general observation in fig. 1, that boron profiles deteriorate the fill factor more strongly than phosphorus profiles can consequently be explained by the fact that the light enters through the p-window and that green light, which is not uniformly absorbed, is more affected by either of the boron profiles (rising or falling).

5. REMARK ON DEGRADATION

The TCO/pin cells with the highest initial fill factors have been tested for short term (10 to 10^3 h) and long term degradation (10^4h) by AM1

	fill factor after				$\frac{\Delta FF\ (10^4 h)}{FF\ (0h)}$
	0h	10 h	10^3h	10^4h	
undoped+)	.64	.61	.57	.56	-13 %
1ppm PH_3 ◿	.64	.62	.59	.56	-13 %
1ppm PH_3 ◢	.66	.62	.59	.58	-12 %
0.5ppm PH_3 ◢	.66	.63	.59	.57	-14 %

+) average over two substrates

Table 1: Fill factors for short-term (10^3 h) illumination and after 10^4h, for cells with undoped and rising phosphorus doped cells (as in fig. 1) illumination at 40° C. After 10^3 h, the fill factor of cells with rising phosphorus profiles was .59 (.57 for undoped), see table 1. Also after 10^4 h there was no significant difference between undoped and rising phosphorus profile cells, see table 1. We do therefore not see (compared to ref. 2) a significant improvement of stability by the rising phosphorus profiles but we did not use as high concentrations. There is furthermore no evidence for an accelerated degradation beyond 10^3 h (the open-circuit voltage remained almost constant over 10^4 h, within a few percent). After 10^4 h, the efficiencies were between 4.5 and 5.5 %.

6. CONCLUSIONS

Rising phosphorus profiles in TCO/pin cells do not affect initial performance to some extent because a field enhancement in the front can be advantageous in the case of sunlight entering through the p^+-layer. Boron profiles in these cells lead to much lower fill factors because of a reduction in the green and blue, either by field suppression or by reduction of the mobility-lifetime product for electrons which are the limiting carrier in this case. Long term illumination of rising phosphorus profile cells resulted in a degradation dependence similar to cells with undoped i-layer, as well as for short periods as well as for 10.000 h.

7. ACKNOWLEDGEMENTS

For valuable discussions we are very grateful to R. Geyer, P. Lechner, H. Rübel and B. Scheppat.

8. REFERENCES

/1/ A. Catalano, M. Bennett, R. Arya, K. Rajan and J. Newton, Proc. of the 18th IEEE Photovoltaic Spec. Conf., Las Vegas, Nevada (1985), p.1378

/2/ M. Möller, H. Kausche, E. Günzel, W. Jürgens and W. Stettner: MRS-Symposium on Mat. Issues in Appl. of Amorphous Silicon Technology, San Francisco, Calif. (1985)

/3/ R.A. Street and J. Zesch, J. of Non-Crystalline Solids 59 & 60 (1983) 449

PROPERTIES OF a-Si:H DEPOSITED BY DC MAGNETRON SPUTTERING

D.Gracin,M.Tasevski[*] and B.Orel[**]

Ruđer Bošković Institute,41000 Zagreb,p.o.b. 1016
Yugoslavia
* Institute for electronics and vacuum technique
61115 Ljubljana, p.o.b.1016 Yugoslavia
** Boris Kidrič Institute,61115 Ljubljana ,p.o.b.1016
Yugoslavia

ABSTRACT

Amorphous thin films were deposited by means of DC magnetron source , using Ar-H mixture as a working gas.The variation of deposition parameters, e.g. distance between the cathode and growing film, cathode current density, working gas pressure and composition, different optical properties were obtained.

The variation in optical gap in our films was between 1.5 and 2.1 eV and the dependence of E_g upon the Urbach slope ,E_o,was similar with glow discharge samples.

The result of our calculations leads to a good correlation between optical gap , and in turn with type of hydrogen bonding and E_o value, on total energy transported by particles impinging upon growing film.

INTRODUCTION

In the reactive sputtering a-Si:H deposition, the film is growing mostly trough Si atoms and SiH_x radical species , with sticking coefficient near unit[1] The energy for further rearrangement is provided by energetic particle bombardment.The main problem is well adjust the energy of those particles .If it is to low , there is no beneficial efect and if it is to large , the damage of film is produced[2].In our experiments ,the energy most of particles impinging the growing surface are supstantialy less than 100 eV.In such regime , we tried to find out the influence of total amount of energy transported towards the growing film on their properties.

EXPERIMENTAL

A DC magnetron source was used for deposition of a-Si:H films with thicknes between 0.2 and 1 µm on glass or monocrystal Si at near room temperature.The magnetron discharge was established in Ar -Hydrogen (5-15%) mixture with total presure between 5 and 25 mTorr..The distance between cathode and growing film were varied from 4 to 6 cm.

The apsorption coefficient of a-Si:H films was calculated from transmitance and reflectance in the visible and IR part of spectra. The optical gap was derived using Tauc plot down to apsorption coefficient value of 5 10^3 cm-1 and the Urbach slope Eo from the exponential part of spectral dependence.

The type of hydrogen bonding were estimated from part of IR spectra corresponding to stretching mode vibration assuming the Gaussian peaks with centers at 2000 and 2090 cm-1 for monohidride and dihydride respectively.

The total energy Etot transported from the near cathode region towards the growin film was made according to the calculation proposed by Somekh [3].

RESULTS AND DISCUSSION

In the fig. 1 are shown typical apsorption IR spectra for a-Si:H samples formed at the lowest (a) and the most high (b) total energy of particle bombard m ent using in our experiment. In the first case , the SiH2 bonds are dominant and in the second SiH.

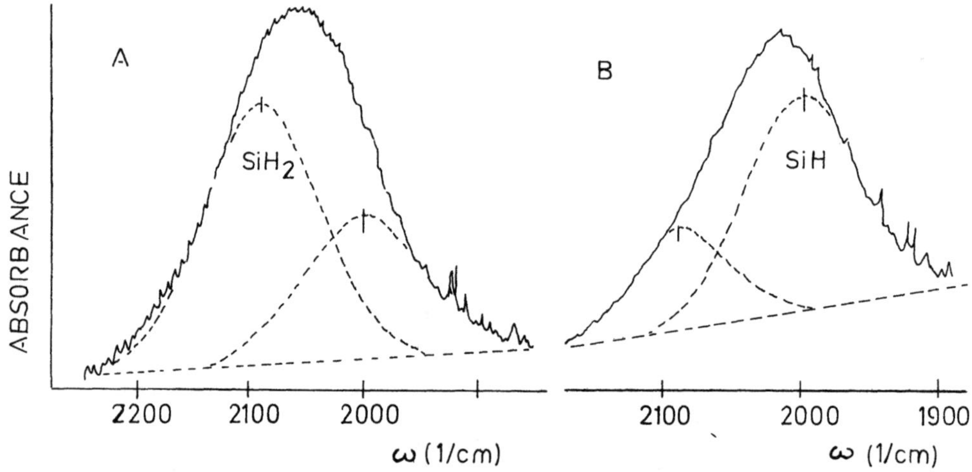

Fig.1. The IR absorpti o n spectra of two magnetron sputt. samples deposited at low (a) and high (b) total energy of particle bombardment during the film formation (the extrema in our experiment s.fig.4)

Comparing the value of optical gap and ratio of hydrogen bonds as monohydride and dihidride , the corelation shown in fig.2.was obtained.The homogeneity of film (hydrogen is mostly bonded as SiH)is accompanied with lowering the Eg opt value

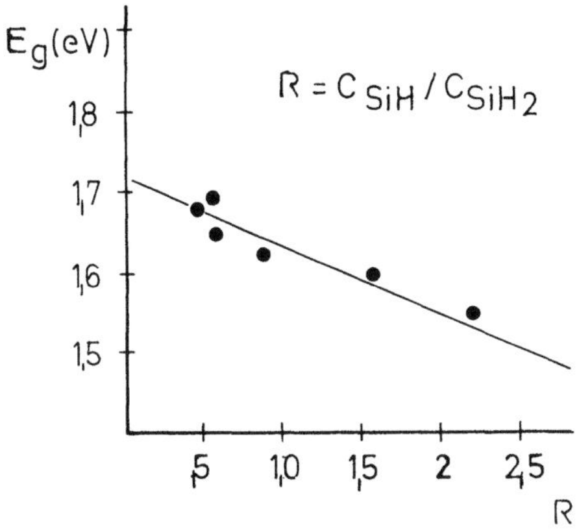

Fig.2.The corelation of the ratio of the concentration of the hydrogen bonds mono- and di hydride bonds in magnetron sputt. amorphous Si:H

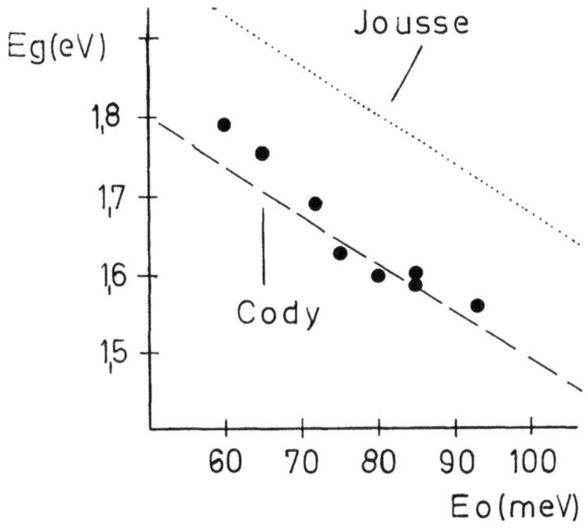

Fig.3 The correlation of the Urbach slope Eo and Eg-opt

The dependence of Eg opt on the Urbach slope Eo is shown in fig.3 . According to Cody |4| interpretation , the more density films egsibit more disorder. Our results beter fit to the Cody|4| slope (dashed lin e) from a isohronal annealings of GD a:Si:H than the D.Jousse at al.|5| slope from measurement on rf sputtered samples.(dotted line).

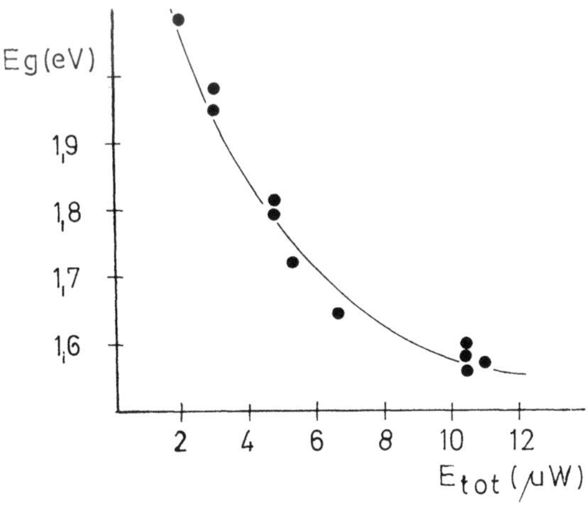

Fig.4.The corelation between optical gap and the total amount of energy transported towards growing film by energetic particles

The numerical data for dependence of Eg opt on total energy of bombarding particles in unit ti m e per squer cm is given on fig 4.This dependence , togeder with above mentioned corelation in properties , showes the posibilities and lim i tations of magnetron sputtered a-Si:H.

CONCLUSION

The increasing the total energy of particle bombardment , in the range it is used i n our experiments , leads to more homogeneous structure ,lowers the optical gap , but in turn enhances the disorder .The last effect is acco m panied with increasing the trap densitie and lowers the electrical performance of layers.

LITERATURE

1.A.Gallaher , Mat.Res.Soc.Symp.Pro c .vol. 70 (1986) 3
2.J.A.Thornton , Sol.Cells 21 (1987) 41
3.R.E.Somekh, J.Vac.Sci.Technol. A 2 (1984) 1285
4.G.D.Cody,B.G.Brooks and B.Abeles , Solar Energy Mater.,8 (1982) 231
5.D.Jousse,E.Bustarret,A.Deneuville and J.P.Stoquert , Phys.Rev. B 34 (1986) 7031

CHARACTERIZATION OF SCREENPRINTED CONTACTS ON A-SI:H SOLAR CELLS WITH A MICROC-SI TOP LAYER

K.Baert, J.Roggen and J.Nijs
Interuniversity Microelectronics Centre
Kapeldreef 75 ; B-3030 Leuven ; BELGIUM

ABSTRACT. Low temperature screenprintable pastes have been further developed. Contact resistances down to 4 $\Omega.cm^2$ have been found with Ti/Ag pastes, fired at 250 ^oC. A-Si:H solar cells with micro-crystalline back layer are fabricated, using this paste as back contact. The results (η =4.5 %) are comparable to cells with evaporated contacts. Optical studies show that most of the light reflection at this backside paste contact is governed by a silicon-air interface.

1.INTRODUCTION

The advantages of screenprinting for the backside metal contact of a-Si:H are very straightforward: higher throughput, patterning and alignation during the printing and possibilities for lower cost. Screenprintable pastes using glass frits have already been extensively used for crystalline silicon solar cells. Unfortunately the firing temperatures are too high for a-Si:H solar cells. Therefore pastes have to be developed using polymers that can be fired at temperatures below 300 ^oC. There are some requirements concerning these pastes in terms in resistances: contact resistances should be not much higher than 1 $\Omega.cm^2$ and the sheet resistance should be down to 1 Ω/sq in order to be used in integrated lay-outs.

It has already been shown in our previous publications [1,2] that (a) n^+-microcrystalline silicon is needed at the backside of the cell to ensure low contact resistance with the paste, that (b) addition of Ti or Mo to the paste is necessary for the same reason, that (c) Ag particles in the paste are essential for low sheetresistance and that (d)the contact resistance on the TCO-layer (SnO_x) can be made as low as 0.02 $\Omega.cm^2$.

2.NEW DEVELOPMENTS IN THE SCREENPRINTABLE PASTES

The same Mo(40 %)/Ag(60%) paste has been used as in the previous publication [2] ,while the Ti(40 %)/Ag(60 %) paste has been improved using smaller Ti-particles. The results on sheetresistance R_{sh} and and contact resistance R_c are shown respectively in Fig.1 and Table I. The results on contact resistance have been obtained on a n^+ microcrystalline silicon layer with a conductivity of 6 S/cm. This microcrystalline silicon layer has been directly deposited by plasma glow discharge .

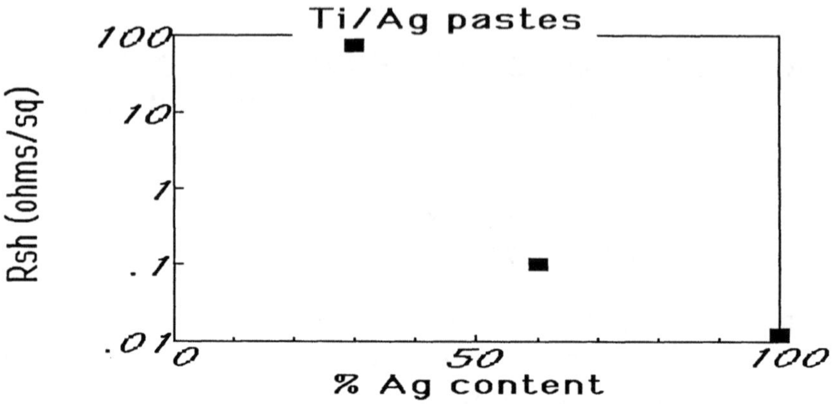

Fig.1

R_c ($\Omega.cm^2$)	new Ti(40%)/Ag(60%)	old Mo(40%)/Ag(60%)
sample #1	4.0	7.7
sample #2	4.8	7.7

Table I

Further improvements can still be reached by increasing the conductivity of the microcrystalline silicon layer and by further optimization of the pastes (relative amounts of Ag and Ti or Mo, firing temperature,temperature profile, firing time, polymer content...).

3.A-SI:H SOLAR CELLS

A-Si:H p-i-n solar cells have been made with the TiAg paste(fired at 250°C for 30 min) contacting the n^+-microcrystalline layer that was put on top of the amorphous p-i-n structure (of course the amorphous n-layer is not necessary in this case). This is compared with evaporated Ti(500Å)Al(0.5μm) contacts on top of the p-i-n structure (see Fig.2a and 2b).

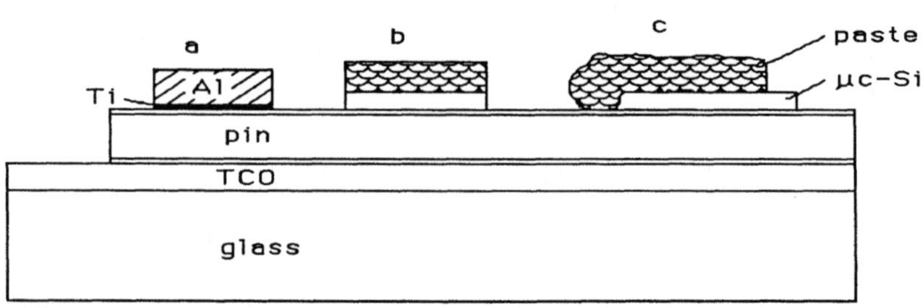

Fig.2

Before deposition of the microcrystalline(μc) silicon an HF(10%)-dip was done during 10 sec in order to remove any oxide. Nevertheless another unexpected problem was encountered, because due to the microcrystalline silicon deposition, the sheetresistance of the TCO-layer was drastically increased. This can be solved by lowering the deposition temperature of the μc-Si layer, but then it is difficult to obtain microcrystallinity. Therefore a compromise had to be defined for the deposition conditions, being T_{depo}=220 ^{o}C ,resulting in a sheetresistance for the TCO of 80 Ω/sq and in a conductivity for the μc-Si layer of 10-20 S/cm. The results for the configurations of Fig. 2a and 2b are shown in Fig.3 for AM1.5 illumination. The efficiency in both cases is about 4.5%. The area of the cells is 1 mm^2.

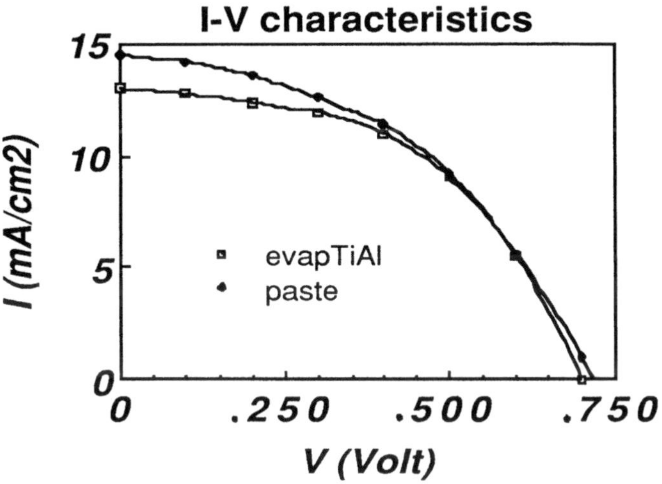

Fig.3

No V_{oc} reduction or increase in the reverse current has been observed, which proves that the screenprinting process itself does not cause damage to the cell. The typical curve as shown in Fig.3 proves that the cells with the pastes have a somewhat lower FF , but also a somewhat higher short-circuit current than for a Ti-Al back side contact. These effects can be explained by the more realistic drawing as in Fig.2c, taking into account misalignments in the screenprinting process. Also an extra contact resistance might exist between the amorphous n-layer and the μc-layer.

4.OPTICAL ENHANCEMENT

It is well known that light trapping enhances the short circuit current and therefore the efficiency of a solar cell [3] . For metal pastes this has so far only been investigated by Garozzo et al [4] for crystalline silicon. In order to investigate the reflection properties of our pastes a test vehicle was made as shown in Fig.4a. The results for light reflection are shown in Fig. 5a and 5b.

By comparison of the specular reflection on the paste (Ag-paste,S) with the specular reflection on air (air,S) (Fig.4a without paste), there does not seem to be much difference. Also the diffuse reflection (D) on the paste for larger wavelengths is quite a lot smaller than the specular reflection (S).

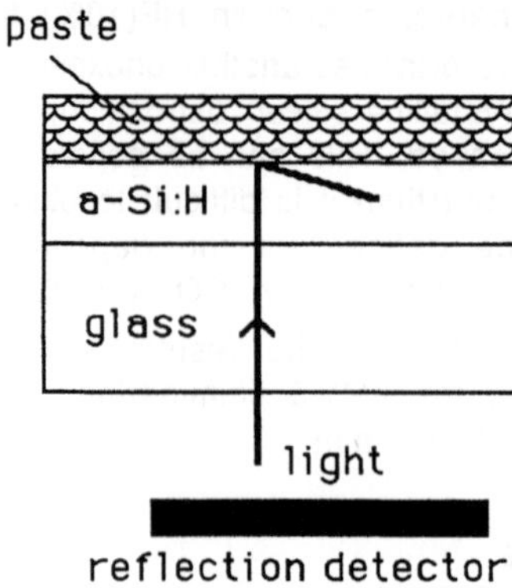

Fig.4a

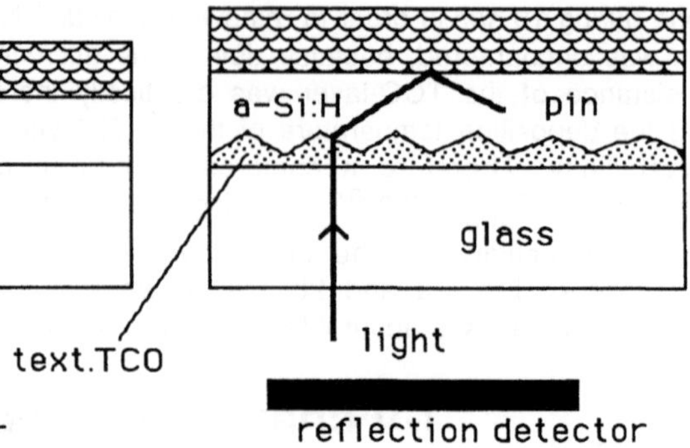

Fig. 4b

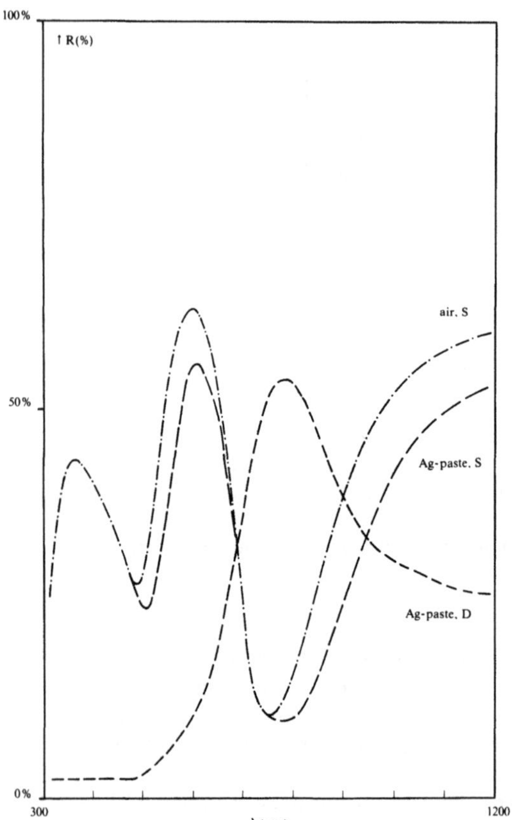

Fig. 5a

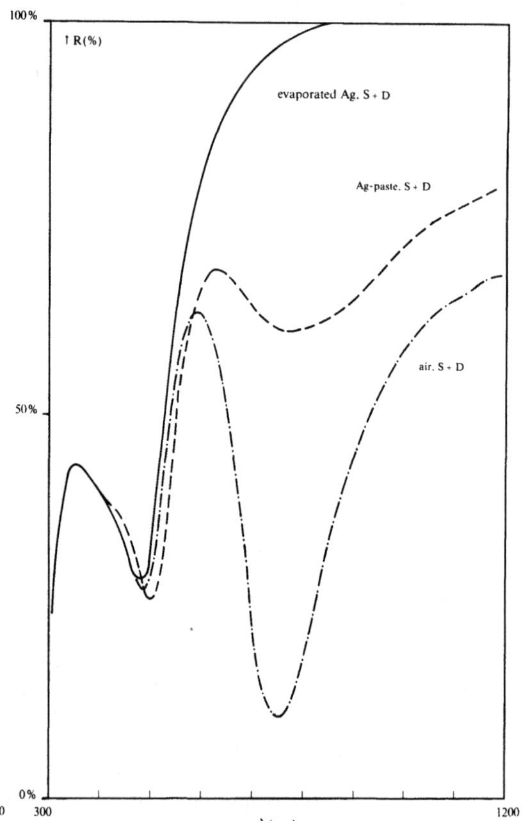

Fig. 5b

Moreover the total reflection (S+D) is quite smaller than the reflection at a mirrorlike evaporated Ag-contact,as can be seen from Fig.5b.

Conclusion is that most of the interface between a-Si:H and the paste is just air,with the metal particles point contacting the a-Si:H. Therefore the right structure for optical enhancement is the structure as shown in Fig.4b where the light diffraction occurs at the structured TCO and exceeds the critical angle at the back side silicon-air interface.

5.CONCLUSIONS

It is proven that thick film technology for the back side metal contacts of amorphous silicon solar cells can easily result in efficiencies of more than 4%. With a further optimalization of the pin structure (1 chamber system), the microcrystalline layer and the paste composition and firing, it must be possible to realize contact resistances of less than 1 $\Omega.cm^2$ and efficiencies of 7...8%. Optical enhancement can be obtained by texturing the TCO-layer.

Technical problems that remain are:stability of the contacts, printability of the pastes, feasibility of HF-dip before screenprinting in a production environment. Economical questions that remain are: the longer deposition time of the µc-Si, the cost of the Ag in the pastes and the cost trade-offs with Al-scribing.

REFERENCES

[1] K.Baert,G.Willeke,M.Honore,W.Vandervorst,J.Roggen,J.Nijs ; 'Screenprinting technology for backside metallization of a-Si:H solar cells ' ;*Proceedings of the 19th IEEE Photovoltaic Specialists Conference,* p.1081; New Orleans; May 1987.

[2] K.Baert,G.Willeke,M.Honore,J.Roggen,J.Nijs ; 'Amorphous silicon solar cells with a screenprinted back contact '; *Techn.Digest of the 3rd International Photovoltaic Science and Engineering Conference,* p.577;Tokyo; November 1987.

[3] Fujitomo et al ; *Solar cells,*11 ,p.357 ;1984

[4] M.Garozzo,L.Pirozzi,L.Fornarini,G.Martinelli and D.Palmeri; 'Optical confinement in silicon solar cells ' ; *Proceedings of the 7th European Photovoltaic solar Energy Conference* , p.962 ; Sevilla ; October 1986

ACKNOWLEDGEMENTS

This work was performed in the frame of the Photovoltaic Programme of the Commission of the European Communities under contract number EN3S-0050-B and cosponsored by the Belgian Government under contract number CE/ENE/03.

The authors also like to thank Dr. Kniffler of MBB ,B.R.D. for the delivery of amorphous pin solar cells.

The authors also like to thank M.Honore and A.Janssens respectively for the paste fabrication and the screenprinting technology and Dr.G.Willeke and Prof.R.Mertens for the solar cell physics and technology back-up.

CARRIER DIFFUSION AND DRIFT IN a-Si:H SCHOTTKY BARRIER DEVICES

R. Schwarz, A. M. Asenov[1], S. Goedecker[2], E. Chitate
Physik-Department E16, Technische Universität München, D-8046 Garching, BRD

ABSTRACT: The three basic semiconductor equations, that is the Poisson and the continuity equations, are solved by an iterative program using the Scharfetter-Gummel procedure. The resulting potential and carrier density distributions and the corresponding drift and diffusion current terms are discussed for an amorphous silicon based Schottky barrier solar cell. An analytical solution for the case of nonlinear recombination can be found under the assumption of negligible electric fields. The solution suggests a modified definition of the minority carrier diffusion length and implies a strong intensity dependence.

1. BASIC EQUATIONS

To determine the performance of solar cells based on hydrogenated amorphous silicon (a-Si:H) one needs to solve simultaneously Poisson's equation and the two continuity equations for electrons and holes, respectively, under appropriate boundary conditions. In general it is not possible to solve the three basic semiconductor equations analytically. The problem is well-known in the field of crystalline semiconductor research and development /1/ and was described over the past few years in a number of publications also for the case of amorphous materials /2-4/.

The basic equations one needs to solve simultaneously are the Poisson equation

$$\frac{\partial^2 \Psi}{\partial x^2} = \frac{-q}{\epsilon}(p - n + n_t + N_{dop}) \quad (1)$$

and the steady-state continuity equations for electrons and holes, respectively

$$\frac{\partial n}{\partial t} = G - R + \frac{1}{q}\nabla j_n = 0 \quad (2)$$

$$\frac{\partial p}{\partial t} = G - R - \frac{1}{q}\nabla j_p = 0 \quad (3)$$

For intrinsic films $N_{dop} = 0$. n_t denotes the trap density, q is the electronic charge. The electron and hole current densities are given by

$$j_n = -q\mu_n(n\frac{\partial \Psi}{\partial x} - \frac{kT}{q}\frac{\partial n}{\partial x}) \quad (4)$$

$$j_p = -q\mu_p(p\frac{\partial \Psi}{\partial x} - \frac{kT}{q}\frac{\partial p}{\partial x}) \quad (5)$$

[1]Present address: Institute of Microelectronics, Sofia, Bulgaria
[2]Present address: Ecole Polytechnique Federale de Lausanne, Lausanne, Switzerland

In eqs. (4) and (5) we have used Einstein's relation $D_{n,p} = \frac{kT}{q}\mu_{n,p}$. It is also introduced into the definition of the diffusion length L

$$L = \sqrt{D\tau} = \sqrt{\frac{kT}{q}\mu\tau} \tag{6}$$

This set of equations shall now be used in a simulation of an a-Si:H based Schottky barrier (SB) solar cell.

2. SIMULATION PROGRAM

A number of approaches could be followed to solve eqs. (1) to (3). For example, the problem can be transformed into a system of first-order differential equations and integrated by the Newton-Raphson method. To avoid problems with convergence most authors approximate the initial differential equations by either a finite-element or finite-difference method. The latter was employed in our calculations together with the Scharfetter-Gummel /5/ approximation. Here initial distributions are given for the potential distribution $\Psi(x)$ and the electron and hole density distributions n(x) and p(x), respectively. Then Poisson's equation and the two continuity equations are solved separately and iteratively until convergence is reached.

All variables are normalized appropriately. Tridiagonal matrices from the discretization scheme are solved by Gauss elimination. Large carrier concentration gradi-

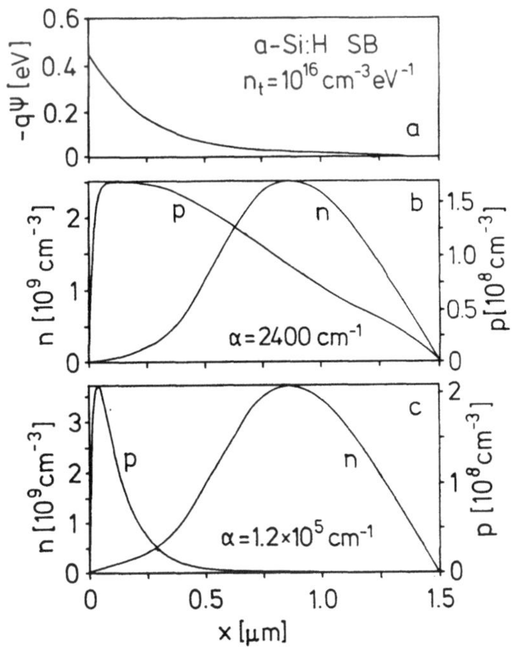

Fig. 1: Simulation of an a-Si:H Schottky barrier solar cell. (a) Potential, electron n and hole p densities for two absorption coefficients (b) 2400 cm^{-1} (c) $1.2x10^{5}$ cm^{-1}.

ents are taken into account by proper choice of a non-equidistant mesh. The program is written in Fortran language and runs on a Cyber computer.

The structure simulated is a 1.5 μm thick i-layer of a-Si:H with a front Schottky barrier and a back ohmic contact (see Fig. 1 a). Short-circuit operation is imposed by the boundary conditions n(0) = n(d) = n_i and p(0) = p(d) = p_i. Recombination is assumed to be of Shockley-Read-Hall type. To account for the particular midgap density-of-state distribution in a-Si:H a trap density of $N_t = 10^{16} cm^{-3} eV^{-1}$ is introduced into the Poisson equation via $n_t = N_t \Psi$. This leads to an expontial field distribution with a decay length of 0.26 μm, as shown in Fig. 1 (a).

Fig. 1 (b) and (c) show the resulting electron and hole density profiles for two different absorption coefficients. Electron and hole mobilities are 3 and 1 cm^2/Vs, respectively. We wanted to consider the case of relatively low free carrier concentration where the exponential field distribution is maintained. So a photon flux of 10^{14} $1/cm^2s$ is chosen in the simulated example. Photogenerated holes drift towards the front surface, their distribution is strongly peaked for large absorption coefficients. Electrons are driven towards the sample center by the electric field.

The resulting drift and diffusion currents, as defined in the first and second terms of eqs. (4) and (5), respectively, are shown in Fig. 2 for $\alpha = 2400 cm^{-1}$. It is interesting to note that the electron diffusion current changes sign at the peak location of the carrier distribution. The drift component is dominant except for $x \succ 1.1$ μm. There

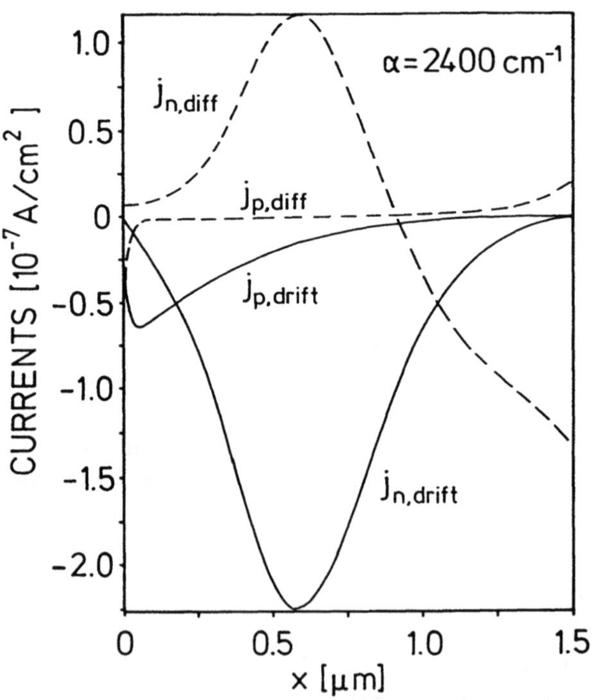

Fig. 2: Drift and diffusion currents for electrons and holes (cf. Fig. 1 b).

the electric field is already strongly reduced. For holes the drift current component dominates nearly everywhere.

It has been recently discussed in the literature whether the convential measurement of the minority carrier diffusion length by the surface photovoltage method (SPV) is still correct in amorphous semiconductors /6,7/. We have taken the collection efficiency η from the simulated Schottky barrier to plot $1/\eta$ versus $1/\alpha$ (see Fig. 3). Extrapolation of the linear part to zero ordinate yields a value of ca. 0.25 μm, which can be considered as an average collection length in the region of the surface barrier field. Considering Fig. 2 it can be expected that for higher light intensities the region of dominant drift current components will become smaller. We plan to extend this study also to the high injection case.

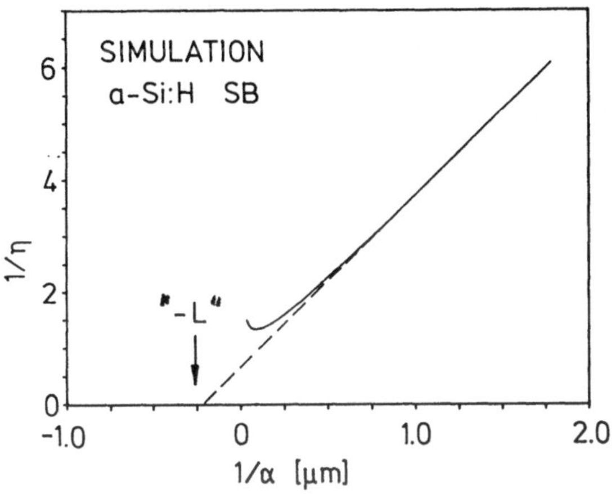

Fig. 3: Simulated SPV plot under 0.001 sun. Parameters are the same as in Fig. 1 and 2.

3. POWER LAW ANSATZ FOR RECOMBINATION

Analytical solutions of the above equations for a-Si:H solar cells are described in the literature for some special cases. For example, even the approximation that recombination is zero throughout the device leads to a surprisingly good description of the spectral dependence of the current collection /8/.

As we have discussed in a recent paper /9/ the power law dependence of the photocurrent I_{ph} on incident light flux Φ with a power γ suggest to use this recombination term:

$$R = cn^{\gamma} \tag{7}$$

Neglecting again the field terms and using the boundary conditions for the carrier density one can integrate the continuity equation once. This results in an expression for the current density and a new definition of the diffusion length L

$$L = \sqrt{\frac{2D}{c(1+\gamma)}}\, n^{s}, \qquad s = \frac{\gamma - 1}{2\gamma} \tag{8}$$

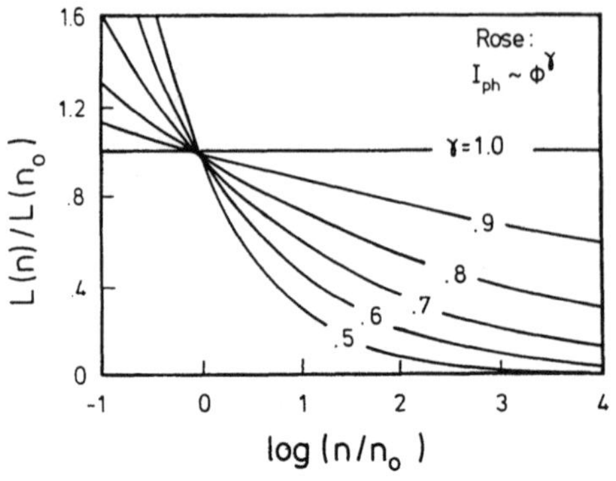

Fig. 4: Calculated density dependence of the diffusion length under a power law ansatz for recombination.

For a typical a-Si:H film with γ=0.8 /7/ the exponent s equals -.125. This is illustrated in Fig. 4. If the diffusion length is reported as a function of the incident ligth flux the exponent should be $s\gamma$ = -0.1. It would be desirable to find a reliable method to measure L and verify this power law dependence.

4. SUMMARY

An a-Si:H based Schottky barrier solar cell structure was simulated with an iterative program using the Scharfetter-Gummel approximation. Simulation of short-circuit current operation shows that drift currents dominate under low light injection (.001 sun). The theoretical solution of the continuity equation with a power law ansatz for the recombination term, while neglecting the internal electric field, leads to a new definition for the diffusion length which is strongly dependent on the carrier density, the incident light intensity, and on the position inside the device. In a next step we want to include this ansatz into the simulation program.

ACKNOWLEDGEMENTS: We thank Dr. H. Pfleiderer for stimulating discussions and T. Muschik for help with the art work.

REFERENCES

/1/ A.E. Iverson, D.L. Smith, IEEE Trans. El. Dev. ED-34 (1987) 2098.
/2/ G.A. Swartz, J. Appl. Phys. 53 (1982) 712.
/3/ M. Hack, M. Shur, J. Appl. Phys. 58 (1985) 997.
/4/ R.J. Schwartz, J.L. Gray, G.B. Turner, D. Kanani, H. Ullal, Proc. 17th IEEE PVSC (IEEE, New York, 1984), p. 369.
/5/ D.L. Scharfetter, H.K. Gummel, IEEE Trans. El. Dev. ED-16 (1969) 64.
/6/ P.J. McElheny, J.K. Arch, S.J. Fonash, Appl. Phys. Lett. 51 (1987) 1611.
/7/ R. Schwarz, S. Goedecker, T. Muschik, N. Wyrsch, A.V. Shah, H. Curtins, J. Non-Cryst. Sol. 97/98 (1987) 759.
/8/ D. Gutkowicz-Krusin, C.R. Wronski, T. Tiedje, Appl. Phys. Lett. 38 (1981) 87.
/9/ S. Goedecker, R. Schwarz, Helvetica Physica Acta 61 (1988) 144.

ALLOYING EFFECTS ON THE ELECTRONIC DENSITY OF STATES OF GLOW-DISCHARGE $a\text{-}Si_{1-x}Ge_x : H$

C. SENEMAUD, I. ARDELEAN
Laboratoire de Chimie Physique, UA CNRS 176,
Université P. et M. Curie, 11 rue P. et M. Curie,
75231 Paris Cedex 05, France

L. CHAHED, A. GHEORGHIU, M.L. THEYE
Laboratoire d'Optique des Solides, UA CNRS 781,
Université P. et M. Curie, 4 Place Jussieu,
75252 Paris Cedex 05, France

ABSTRACT. The electronic density of states in the valence and conduction bands of amorphous hydrogenated silicon-germanium alloys prepared by glow discharge in the ARCAM reactor have been investigated as a function of their composition by combining soft X-ray, photoelectron, optical and photo thermal deflection spectroscopies. It is shown that the decrease of the optical gap Eg as Ge concentration increases is mainly due to an upwards shift of the valence band edge ; the conduction band edge is practically not shifted but is spread out, due to the progressive shift of the first maximum away from E_F. The modifications of the density of states as Ge is incorporated are discussed in relation with both alloying and structural disorder effects.

1. INTRODUCTION

As part of a comprehensive study of glow-discharge amorphous hydrogenated silicon-germanium alloys ($a\text{-}Si_{1-x}Ge_x : H$) prepared under well-controlled deposition conditions (1), various complementary spectroscopic techniques have been combined in order to determine the modifications upon alloying of the electronic state distributions in both the valence and the conduction bands, especially near the pseudo-gap, as well as in the Si and Ge core levels.

The Si 2p and Ge 3d core levels have been studied by X-ray induced photoelectron spectroscopy (XPS). The valence band states have been investigated both by XPS and by soft X-ray emission spectroscopy (SXES), while the conduction band states have been probed by soft X-ray absorption spectroscopy (SXAS). The data obtained for the valence and conduction band edges have been compared with optical absorption data determined from combined optical transmission and photothermal deflection spectroscopy (PDS), which give more insight on the localized states in the pseudo-gap. All the results are discussed in terms of chemical and structural disorder effects related to Ge incorporation.

2. EXPERIMENT

The $a\text{-}Si_{1-x}Ge_x : H$ alloys were prepared by r.f. glow discharge decomposition of suitable (SiH_4, GeH_4, H_2) gas mixtures in the ARCAM reactor under well-controlled conditions (1). The series under investigation was deposited at a r.f. power of 5 W, and a substrate temperature of 150°C. The atomic Ge concentrations, as determined by electron microprobe and Rutherford backscattering measurements, were x = 0.16, 0.36 and 0.45.

The total hydrogen concentration as deduced from Elastic Recoil Detection experiments, increased slightly from 0.19 to 0.22 for x varying from 0.16 to 0.45.

The SXS experiments were performed with a bent crystal spectrometer equipped with a gypsum (020) crystal (2). The samples were 1 μm thick films deposited on Al substrates and Be foils for the emission (SXES) and absorption (SXAS) measurements respectively. The energy resolution was about 2.10^{-4}. The XPS spectra were induced, either by monochromatized Mg $K\alpha$ radiation (for the core levels), or by non-monochromatized Al $K\alpha$ radiation (for the valence band). The corresponding total instrumental broadening was respectively 0.5 and 1.0 eV. The Au $4f_{7/2}$ level at 83.8 eV was taken as a reference for the binding energy scale.

The optical absorption coefficient α was determined between 0.6 and 2 eV on samples deposited under identical conditions on glass substrates, from a combination of optical transmission and photothermal deflection measurements, as explained elsewhere (1).

3. RESULTS AND DISCUSSION

XPS investigations of the Si 2p and Ge 3d electron distributions show that no chemical shift of these two lines can be detected within experimental uncertainties when varying the Ge concentration, in agreement with previous results (2,3). This is consistent with the fact that Si and Ge have about the same electronegativity and that very little charge transfer between Si and Ge atoms is thus expected upon alloying. On the other hand, we observe a slight broadening of the two lines, more important for Ge 3d than for Si 2p, as x increases from 0.16 to 0.45. This is illustrated in figure 1, for the Ge 3d line. Although a possible influence of oxidation on the high binding energy side of the electron distributions cannot entirely be ruled out, such a broadening probably reflects an increase of the local charge fluctuations due to additional short range disorder, especially around Ge atoms, as more Ge is incorporated. This is in agreement with the concomitant increase of the inverse slope of the exponential part of the absorption edge E_o (from 56 to 65 meV), usually considered as a disorder parameter (1). It must be noticed that our experimental results are in contradiction with those of previous work (3) in which the Ge 3d line had been separated into its two spin orbit split components, which exhibited a significant decrease of the line width with increasing x in the same composition range. This discrepancy suggests differences in the changes

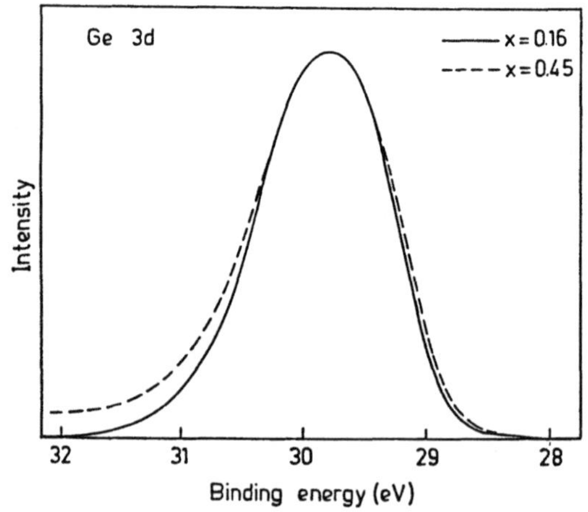

Fig 1. Ge 3d core level spectra for a-Si_{1-x} Ge_x : H with x = 0.16 and 0.45

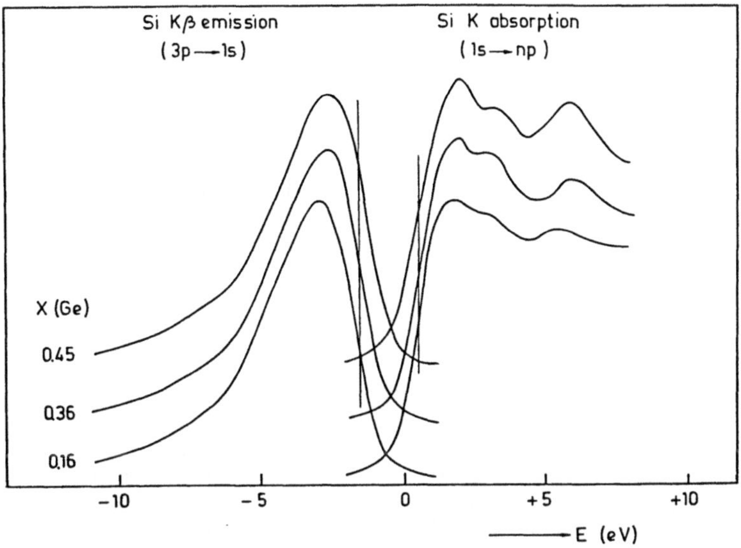

Fig 2. SiK X-ray emission and absorption spectra from a-Si_{1-x} Ge_x : H with x = 0.16, 0.36 and 0.45.

of the local structure when incorporating Ge to Si in the two types of alloys.

The experimental Kβ X-ray emission (transitions from the valence band to the Si 1s core level) and K photoabsorption (transitions from the Si 1s core level to the conduction band) spectra are reported in figure 2 as a function of energy for the three a-Si_{1-x} Ge_x : H alloys. These spectral distributions correspond respectively to 3p valence band and p conduction band states around Si atoms. Since for each alloy the same core level (Si 1s) is involved in both the emission and absorption transitions, the relative energy position of the corresponding spectra can directly be obtained from the measurements without further adjustment. The energy E(Si 1s) has been determined by combining XPS measurements of $2p_{3/2}$ core level energy and SXES determination of $2p_{3/2} \rightarrow 1s$ transition energy. Since the energy of the Si 1s core level remains practically constant upon alloying, the energy positions of the spectra for the different alloys can easily be compared. The Fermi level, located on the energy scale from the value of the Si 1s energy for each alloy, is taken as the origin of the energies in figure 2. Its position is determined with an uncertainty of about 0.2 eV.

As x increases from 0.16 to 0.45, the upper edge of the 3p valence band distribution is shifted upwards (towards low binding energies), while the lower edge of the Si p conduction band distribution remains practically at the same energy. Figure 3 shows a comparison between the variations upon alloying of the optical gap E_g, defined as the energy at which $\alpha = 10^4$ cm^{-1}, and of the distance ΔE_V between the upper edge of the valence band, measured at half maximum and the Fermi level. Both E_g and ΔE_V decrease linearly with increasing x at practically the same rate. It can be concluded that the narrowing of the optical gap in amorphous hydrogenated Si Ge alloys is essentially due to a shift of the top of the valence band towards lower binding energies, the bottom of the conduction band remaining practically at the

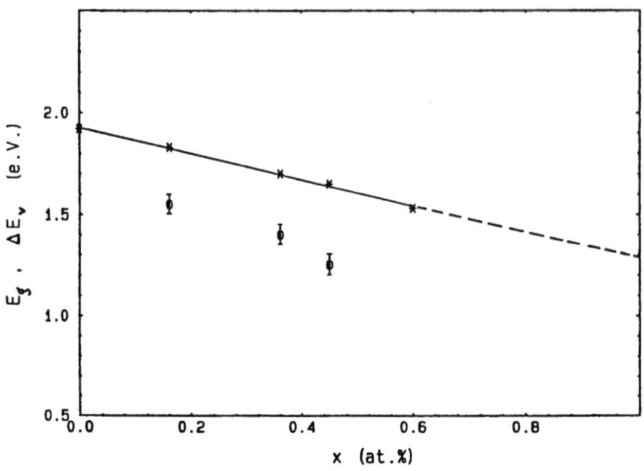

Fig 3. Variation of the optical gap Eg (*) and of the energy distance E_V measured on X-ray emission spectra between the upper edge and E_F (ɸ)

same energy . These results are confirmed by the XPS experiments, which show a shift of the upper edge of the valence band very similar to that observed on the SXES spectra.

Our experimental results are in good agreement with recent density of states calculations performed for a-Si Ge : H alloys within the CPA approximation (4).

It must be pointed out that the effects of alloying also show up in the distribution of the conduction states, the three structures observed beyond the edge being gradually shifted away from E_F with increasing x.

A slight broadening of the 3p valence band distribution can be detected as x increases from 0.16 to 0.45 (of about 5 %). This can be attributed, like for the core levels, to a gradual increase of the local disorder around Si atoms, especially of the bond angle fluctuations, as Ge is incorporated into the Si network. On the other hand, while the upper edge of the 3p valence band distribution is shifted practically parallel to itself, the lower edge of the p conduction band distribution tends to broaden as x increases. More precisely, while the energy at half maximum remains the same as described before, the maximum is definitely shifted upwards. This might be the consequence of a redistribution of antibonding Si (and Ge) states upon alloying. SXAS studies of Ge s conduction band states are under way in order to clear up this problem.

It is worth-noticing that a comparison of the SXES results obtained for two alloys with x = 0.16 deposited at 150 and 250°C respectively, show slight differences in the 3p valence band distributions. One observes essentially that, for the low substrate temperature alloy, the maximum recedes slightly and there is a faint increase in intensity at about 7 eV below E_F, i.e. in the region of mixed sp states in crystalline Si. If we refer to a previous study on a-Si : H (5), this suggests differences in the network topology, with a tendency to a better local ordering in the 150°C alloy.

Unlike previous XPS and SXES studies on differently prepared a-Si_{1-x} Ge_x : H alloys (2), the present work does not show any evidence for the existence of a shoulder on the tail of the upper edge of the valence band, growing with the Ge concentration. This is again in agreement with the results of the optical absorption studies, which show that the sub-band gap absorption is significantly lower in the present alloys, even for high Ge contents (1), and therefore that the density of defect states in the pseudogap must be too low to be detected either by XPS or by SXES.

ACKNOWLEDGEMENTS. The work presented in this paper was supported in part by the CCE under contract EN 3S-0062-F and by the AFME and the PIRSEM-CNRS.

REFERENCES.

(1) L. CHAHED, M.L. THEYE, S. BASROUR, J.C. BRUYERE, C. GODET, A. LLORET this conference.

(2) C. CARDINAUD, C. SENEMAUD, G. VILLELA
J. non cryst. solids 88, 55, 1986.

(3) F. EVANGELISTI, F. BOSCHERINI, R. CIMINO, F. PATELLA, P. PERFETTI, C. QUARESIMA
J. non cryst. solids 97,98, 407, 1987

(4) D.A. PAPACONSTANTOPOULOS, C. SENEMAUD, E. BELIN
Europhysics Letters to be published

(5) B. DREVILLON, C. SENEMAUD, C. CARDINAUD, M. DRISS-KHODJA, C. GODET
Philos. Mag. B 24, 335, 1986

"NEW REACTOR DESIGN FOR LOW CONTAMINATION AMORPHOUS SILICON DEPOSITION"

J.P.M. SCHMITT *, J. MEOT *, P. ROUBEAU **, P. PARRENS ***

* SOLEMS - 3 rue Léon Blum - F 91124 PALAISEAU Cedex, France
** PROMOPHY - 3 Place de la gare de Lozère - F 91120 PALAISEAU, France
*** NEXTRAL - ZAC Pré Millet - F 38330 SAINT-ISMIER, France

ABSTRACT

The pressurized plasma box concept is presented. This design addresses the problem of impurity control in amorphous semiconductor deposited by silane based glow discharge without requiring the heavy and expansive features of ultra high vacuum. Impurity level is predicted to decrease by a factor 100 with respect to standard technique, a factor 40 improvement is actually observed in a large plasma box machine capable of coating two 30 x 30 cm^2 substrates. Other implications of the pressurized box design, such as gas compositional transition time, are also reported.

1. INTRODUCTION

Among the specifications which are important for the deposition of device grade amorphous silicon (aSi:H), we address in this work chemical purity. It is known that the major contaminants found in standard glow discharge aSi:H are oxygen (with densities ranging from 2 x 10^{19} to 5 x 10^{20} cm^{-3}), carbon and to a smaller extend nitrogen (1,4). All these impurities are found to reduce the carrier diffusion length (1). Moreover the level of background impurities was clearly correlated to part of the light induced degradation of aSi:H and subsequent solar cell decay (5,6,7). Although various impurities were detected in silane cylinders (8) when the feedstock gases are electronic grade and the gas line thoroughly degassed, most contaminants are likely to come from the vacuum vessel. As a first indication, the major contaminants of aSi:H (O and C) are also, with hydrogen, the most common elements met in the background level of standard vacuum technology (H_2O, CO_2, CO, oil...). This is why several deposition machines were built using ultra high vacuum technology (4,9) which allowed to reduce oxygen concentration to 2-5 x 10^{18} cm^{-3}.
However, as aSi:H devices approach maturity, industry is seeking for large surface, low cost and high throughput deposition machines. Those requirements are conflicting with the high cost and poor flexibility of UHV technology. We propose and evaluate a novel reactor geometry where the convenience of standard vacuum technology is preserved while UHV standard purity is achieved in the deposition zone. The new concept is described and the corresponding benefit estimated. A reactor for the simultaneous deposition of two 30 x 30 cm^2 substrates was designed accordingly. The machine performances were evaluated in terms of impurity analysis, density of state and deposition uniformity. Not only impurity but also particulates are of great importance to the device oriented thin film specialist. It relates to machine maintenance : regular reactor cleaning prevents flaking of the accumulated silicon. We describe here preliminary results concerning an etching process for the reactor cleaning.

2. THE PRESSURIZED REACTOR CONCEPT

2.1. Classical geometry

A typical PACVD reactor is a single vacuum vessel. As shown in Fig. 1, prior to deposition, the chamber is generally evacuated by a pump of conductance C_o achieving a base pressure P_o. The base pressure mainly originates from wall and seal degassing, the corresponding flow, $f \approx P_o C_o$ consisting mainly in water, carbon oxides and hydrocarbons. When switching to deposition, a silane flow F is injected in the chamber and the pump is throttled to a conductance C_p allowing the pressure to reach a selected value P_p with $F \approx P_p C_p$. The solid relative impurity concentration after deposition remains of the order of the contaminant relative partial pressures, then

$r \approx [\ O, C\ ] / [\ Si\ ] \approx [\ H_2O, \ldots\ ] / [\ SiH_4\ ] \approx f / F \approx (P_o/P_p)\ (C_o/C_p)$.

Typical values for deposition reactors are :

$P_o \approx 10^{-4}$ Pa, $P_p \approx 20$ Pa, $C_o \approx 500$ l/s, $F \approx 60$ sccm, $C_p \approx 5$ l/s.

Hence we infer a ratio $r_A \approx 5 \times 10^{-4}$ in agreement with the $2\text{-}5 \times 10^{19}$ cm^{-3} measured for the oxygen content in standard aSi:H.

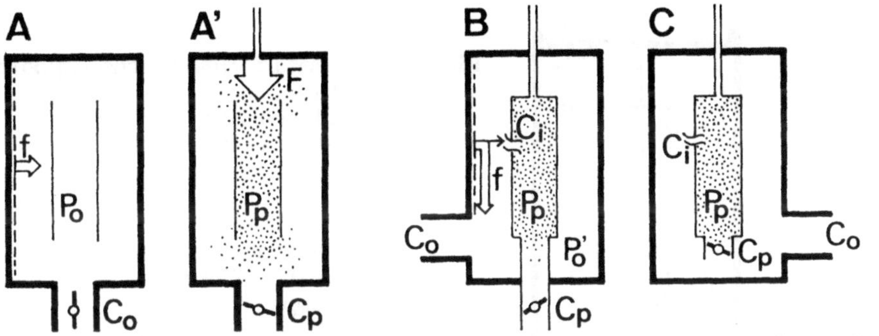

Fig. 1. Schematic views of the gas flows in a standard configuration before (A) and during the process (A'). The plasma box configuration is shown with double pumping (B) and internal throttling (C).

With the classical geometry (type A), two actions can reduce background impurities : either lower degassing flow by going UHV or increase F facing large running cost (silane waste) and capital cost (pumping and scrubbing capacity).

2.2. The double shell geometry

Here the reactor consists in two shells, the inner one is called plasma box. It is not vacuum tight, but relies on mechanical alignment to limit the leakage conductance to a low level C_i. All walls of the plasma box are at process temperature (150°C to 250°C) hence degassing can be completed prior to process. Most inner surfaces of the box are exposed to the reactive plasma. The reactive gases are injected in the box. For the gas pumping, two options are discussed : in B (double pumping), the box is connected to a pump via a separated pipe and throttle, a section which must be UHV grade ; in C (internal throttling), the box opens via a non standard throttle to the external vessel.

The outer shell is a standard vacuum vessel. It encloses the plasma box and all the hardware for heating and substrate handling.

Items such as seals, feedthroughs and mechanical gears are allowed to remain cold, the degassing flow f is still present in the outer vessel. The chamber is pumped via a large conductance C_o while the box is throttled to C_p. The contribution of f to the relative contamination r in the plasma box can be easily evaluated since $C_o >> C_p > C_i$,

in option B, $r_B \approx (f/F)\ (C_i/C_o)$; in option C, $r_C \approx (f/F)\ (C_p/C_o)$.

It appears that if $C_i \approx 1$ l/s, the ratio r is reduced by 100 in case B and 20 in case C with respect to geometry A. Here we assumed molecular regime and neglected temperature and gas mass effects. Introducing viscosity would further improve r, thanks to the collisions keeping out the counterstreaming impurities. During process, there is a large pressure differential between both sides of the plasma box, the pressure ratio being of the order of r/r_A.

3. EXPERIMENT

3.1. Description of the machine (option B)

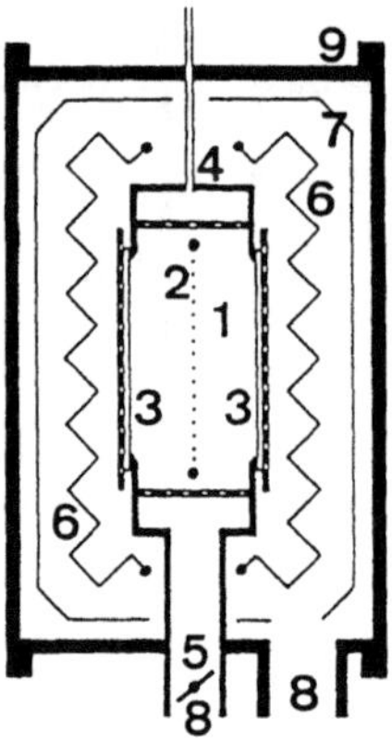

Fig. 2. Schematic view of the deposition machine.
1) Plasma box
2) RF grid electrode
3) Substrates
4) Gas inlet and distribution
5) Throttle
6) Radiant heaters
7) Heat reflectors
8) Pumps

A double pumping machine was built using two 400 l/s turbopumps. The reactive gases are flowing laminarly along the 42 x 42 x 9 cm^3 plasma box. Two 30 x 30 cm^2 substrates close the box sides, the mediant electrode is a grid. The ensemble is radiantly heated at process temperature (up to 330°C). After pumping, the base pressure is around 10^{-4} Pa.

3.2. Impurity control

Two runs were performed with similar loading, pumping and plasma conditions and configurations A and B, A was mocked up by valving off the chamber pump and substituting a grid to one substrate. The samples were deposited on crystalline Si and SIMS analyzed. In this large machine, C_i was measured to be 4 l/s for nitrogen. The expected benefit from geometry A to B should be a factor 100. As seen in Fig. 3, a factor 40 is observed for [O] and 5 for [C]. Results are convincing for oxygen, for a smaller effect on carbon (hopefully not as harmful) potential explanations are cylinder contamination, poor degassing of the throttling section and/or pump backstreaming.
Unfortunately, SIMS measurement were not absolute but, in the B sample, we infer from the substrate [O] < 10^{-19} cm^{-3}.

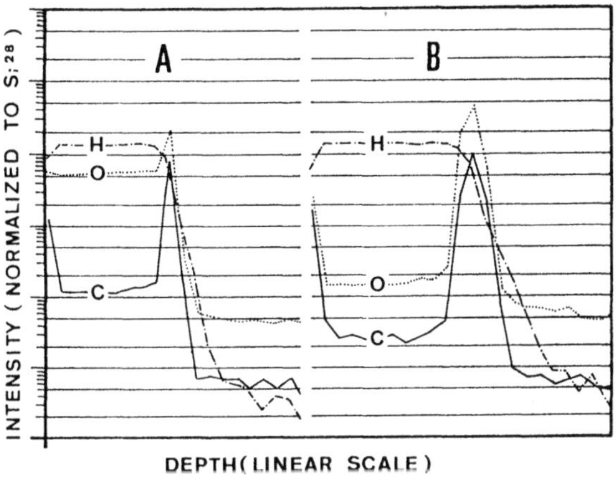

Fig. 3. Cesium SIMS of two aSi:H samples deposited in configurations A and B. Signals are normalized to mass 28 (Si).

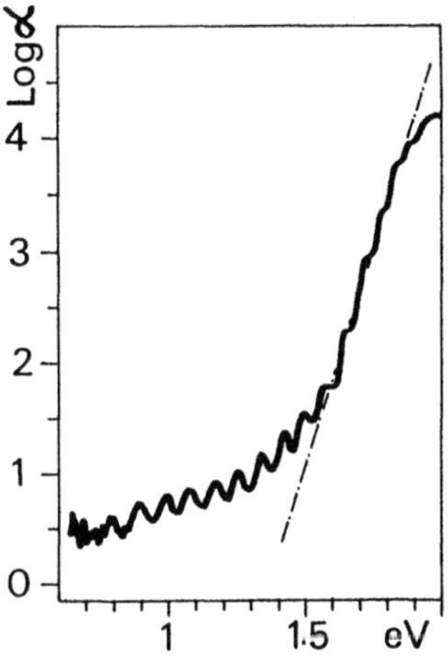

Fig. 4. Photodeflection spectrum of a sample deposited in the B configuration.

3.3. Material quality

The process conditions were explored. On 28 x 28 cm^2, a thickness uniformity better than ± 3 % is found for deposition rate up to 3 Å /s, faster rates are accessible at the expense of a poorer uniformity. PDS measurements (Fig. 4) reveals a device grade density of state around 4 x $10^{16}cm^{-3}$ and a sharp Urbach tail slope of 55 meV.

3.4. Gas compositional transition

After plasma ignition or gas compositional transition, a process transient can alter sharp interfaces in devices. Figure 5 shows that, after ignition, the silane pressure drops while hydrogen pressure builds up although in intermediate gas flow regime the total pressure is practically unaffected. The transition time depends on the gas flow and the reactor volume. This is why in configuration A, for a volume of 150l, the time is 10 sec. while for B the reactor is reduced to a minimum 20l and the transition time is below 1.5 sec. : the process reaches steady state in a few monolayers.

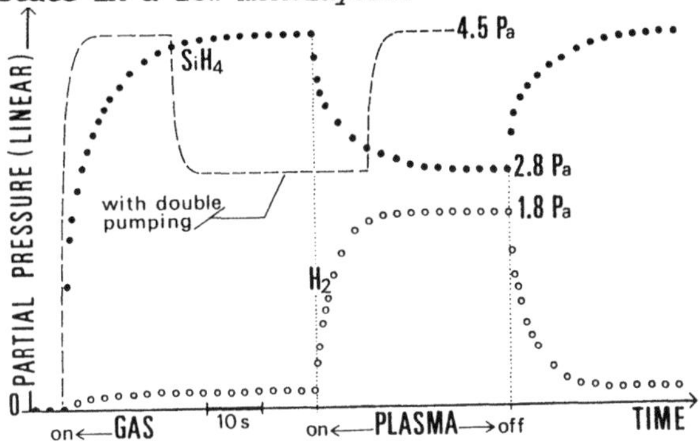

Fig. 5. Sampling mass spectrometry of the reactor gases. Mass 30 (dark circle) corresponds to silane, mass 2 (open circle) to hydrogen with a minor silane contribution. A pure silane 40 sec. plasma is shown in configuration A, the interrupted line is for configuration B.

3.5. Reactor plasma cleaning

NF_3 was selected because CF_4/O_2 was found too slow etchant and SF_6 was feared to bring sulfur contamination. The etching rate was measured by laser reflectometry. As seen in Fig. 6, the etching rate saturates at high plasma power, the saturation level varying with the NF_3 flow rate. At saturation, the conversion efficiency of fluorine into SiF_4 is nearly complete.
Note that the measured rates are for a fully coated reactor, a clear load effect is observed at the end of the cleaning process.

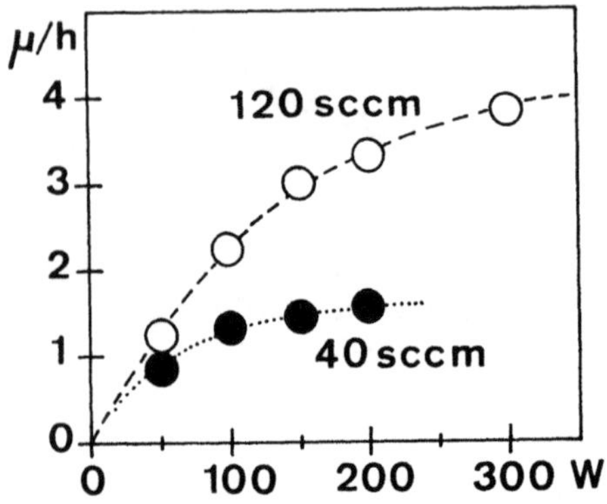

Fig. 6. Etching rate as a function of power and NF_3 flow in a fully aSi:H coated reactor at 180°C measured by laser reflectometry.

4. CONCLUDING REMARKS

We have demonstrated that an alternative exists to ultra high vacuum technology for the design of a deposition machine where amorphous silicon contamination is reduced. The pressurized box concept, by relaxing the purity constraints on the outer part of the vessel, allows easy design of in line carrier automatic transportation or any other production oriented item. The same design also offers complementary conveniences such as short transient times at interfaces and easy plasma cleaning.

REFERENCES

1. D.E. CARLSON in Tetrahedrally bounded amorphous semiconductors, ed. by D. ADLER and H. FRITZSCHE, (Plenum, New York, 1985), 165.
2. C.W. MAGEE and D.E. CARLSON, Solar Cells 2 (1980), 365.
3. D.E. CARLSON, J. Vac. Sciences & Technol. 20 (1982), 290.
4. S. TSUDA et al., Jap. J. of App. Phys. 26 (1987), 33.
5. M. STUTZMANN, W.B. JACKSON, C.C. TSAI, Phys. Rev. B 32 (1985), 23.
6. N. NAKAMURA et al., AIP Conf. Proceed. N° 120, (AIP, New York, 1984), 303.
7. V. DALAL, Proceed. 19th IEEE Photovoltaic Specialist Conf., (IEEE, New York, 1987).
8. R.R. CORDEMAN and P.E. VANIER, J. Appl. Phys. 54 (1983), 3987.
9. C.C. TSAI et al., J. Non Cryst. Solids 59-60 (1983), 731.

ELECTRICAL, OPTICAL AND STRUCTURAL PROPERTIES OF p^+ MICROCRYSTALLINE Si:C:H DEPOSITED BY RF GLOW DISCHARGE

Bernard Goldstein and C. Robert Dickson
Solarex Thin Film Division, 826 Newtown-Yardley Road
Newtown, PA 18940 USA

and

Ian H. Campbell and Philippe M. Fauchet
Department of Electrical Engineering
Princeton University
Princeton, NJ 08544 USA

The incorporation of thin p^+ films of amorphous (a) Si:C:H into a-Si:H solar cells as transparent contacts has been one of the major advances in solar cell technology. The overall performance of such films, however, remains severely limited because increasing the bandgap, through the introduction of carbon, decreases the conductivity and thereby increases the series resistance. Microcrystalline (μc) films of Si:C:H seem to offer some chance of improvement but highly doped p^+ microcrystalline material made from conventional feedstock gases, using standard glow discharge techniques has not been reported. Recently, Hattori and co-workers reported the fabrication of highly conductive p^+ μc-Si:C:H films using a unique electron cyclotron-chemical vapor deposition process in which the film is deposited at low growth rates (~ 0.05 Å /sec) from a low pressure ($\sim 10^{-4}$ torr) magnetically confined microwave plasma[1].

In this letter we report the fabrication and properties of similar p^+ μc-Si:C:H films deposited using disilylmethane (DSM) as the carbon feedstock gas in a conventional RF glow discharge process that can be readily used to fabricate both small and large area solar cells and solar cell modules. Our major findings include conversion from microcrystalline to amorphous growth as the carbon concentration increases, the critical importance of the substrate for the growth of microcrystalline material, and a significant improvement in conduction properties when films deposited at low temperatures are annealed at higher temperatures.

In the deposition process, both methane and disilylmethane (DSM) were used as the carbon feedstock gas, but using DSM permits the incorporation of about twice as much carbon into the film before its growth converts from microcrystalline to amorphous. Typical discharge conditions include: RF power=350 W, $H_2/SiH_4 \sim 50$, $B_2H_6/SiH_4 \sim 0.015$, $DSM/SiH_4 \sim 0.1$, P=1 torr, and substrate temperatures from 125 °C to 300 °C. Deposition under these conditions produced growth rates of about 0.6 Å /sec. The films were deposited on fused quartz substrates for Raman spectroscopy, optical absorption and Secondary Ion Mass Spectroscopy (SIMS) measurements; single crystal silicon substrates were used for Fourier Transform Infra-Red (FTIR) transmission measurements. The films varied in thickness from 200-7000 Å .

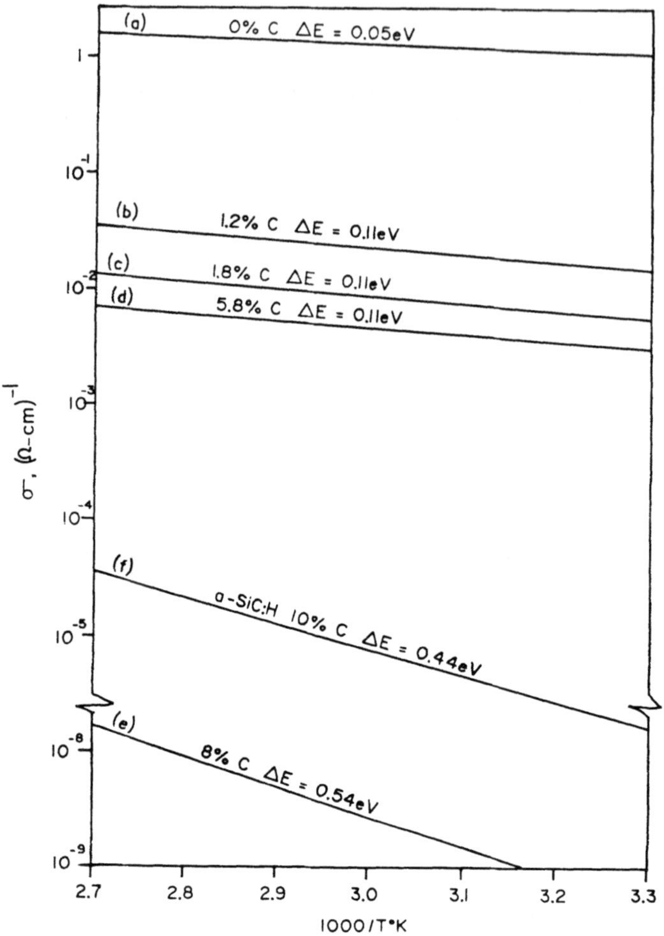

Fig. 1. Conductivity *vs* reciprocal temperature for films with increasing carbon concentration and concomitant decrease in microcrystallinity. Curve (f) is typical of films of a-Si:C:H commonly used in solar cells.

The temperature dependence of the conductivity of p^+ μc-Si:C:H films containing increasing carbon fractions is plotted in Fig. 1. The result from a typical p^+ a-Si:C:H film, commonly used in solar cells, is also included for reference. At carbon concentrations up to 6 atomic percent in the film, the conductivity, σ, and its activation energy, ΔE, are superior for solar cell applications to those of the reference p^+ amorphous layer; at higher carbon concentrations σ and ΔE rapidly become comparable or inferior to the reference film. In Fig. 2 we show Raman spectra of an equivalent series of films. The degree of microcrystallinity in the films is indicated by the ratio of the narrow, microcrystalline Si vibrational line at $\sim$ 517 cm^{-1} to the much broader amorphous Si vibrational line at $\sim$ 476 cm^{-1}. Increasing the carbon content suppresses the microcrystallinity until at $\gtrsim 6\%$ the films are completely amorphous. The microcrystalline films seem to consist of crystalline grains $\sim$ 50-80 Å in diameter, as determined by the Raman line shape[2], surrounded by amorphous material. We attribute the observed decrease in conductivity primarily to the structural changes in the films since the conductivity of the microcrystalline phase is much larger than that of the amorphous phase, and there is a strong correlation between the complete disappearance of the microcrystallinity and the large conductivity changes at 6-8% carbon. Transmission Electron Microscope (TEM) studies confirm this picture of the microcrystalline films. Fig. 3 contains cross-sectional TEM micrographs of an n^+ and a p^+ layer at a magnification of 600,000. Note the

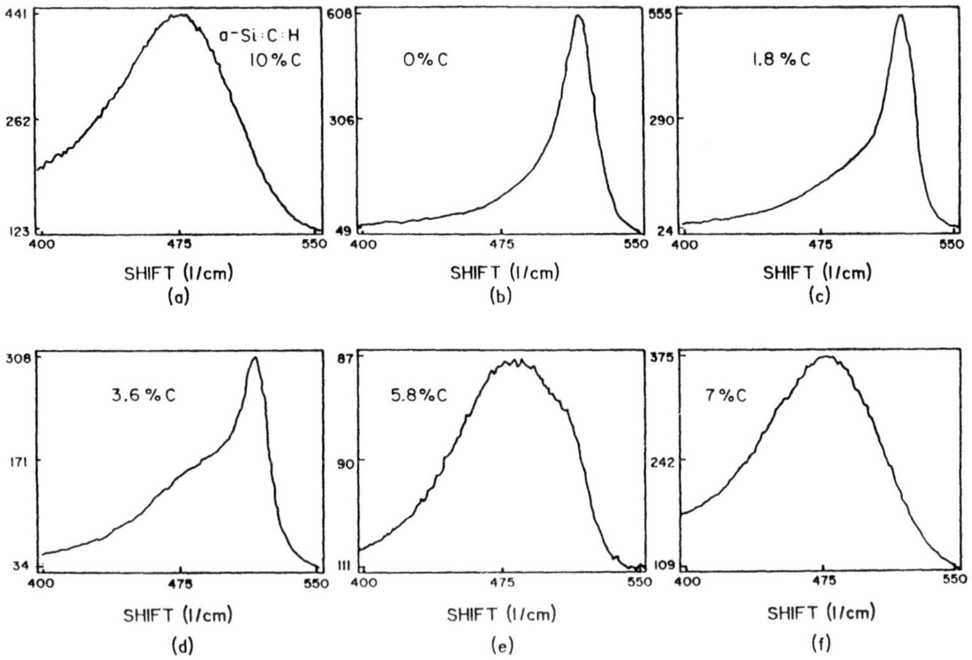

Fig. 2. Raman spectra of films similar to those shown in Fig. 1. Curve (a) is for reference.

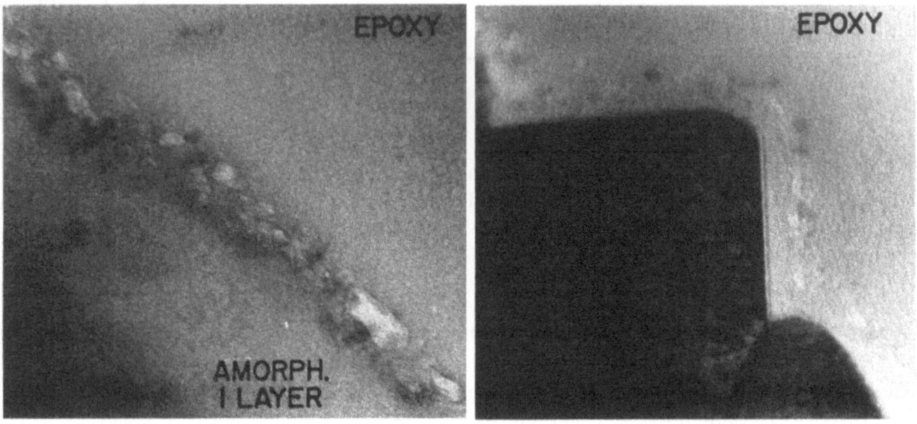

Fig. 3. Transmission electron micrographs of n^+ and p^+ microcrystalline Si:C:H layers on an a-Si:H substrate. Note the significantly poorer quality of the p^+ layer (1 cm = 400 Å).

clear indication of crystalline grains ~ 50-100 Å in size. In this case the n^+ layer is more microcrystalline due, presumably, to the action of phosphorus as a crystallization enhancer[3].

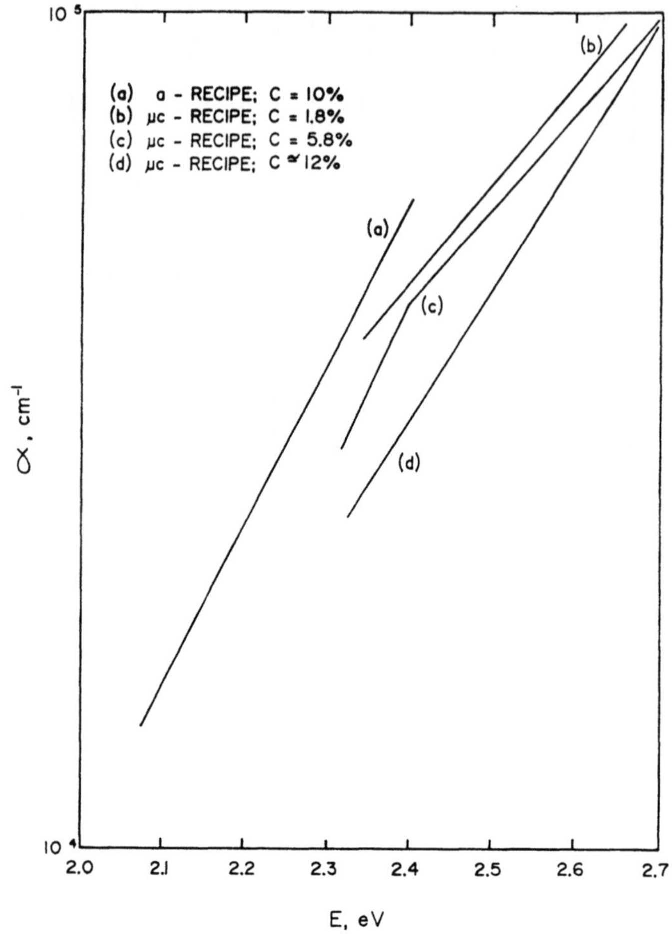

Fig. 4. Optical absorption *vs* energy of (a) typical a-Si:C:H films used in solar cells, and (b)-(d) films deposited using the microcrystalline process. Note that sample (d) is amorphous, yet has lower absorption than film (a).

The effects of microcrystallinity and carbon content on the absorption coefficient, in the visible region, of the p^+ microcrystalline films are shown in Fig. 4. The result from a typical p^+ amorphous layer is again included for reference. It is well known that the absorption coefficient, at a fixed energy, decreases as either the microcrystalline fraction[4] or the carbon content increases. For the data shown in Fig. 4 the effect of the increasing carbon concentration dominates the concomitant decrease in microcrystallinity. The appearance of two slopes in the absorption coefficient as a function of energy, as in curve (c), is often observed and may be due to the presence of both amorphous and microcrystalline phases in the film. The simultaneous presence of these two two phases is indicated by the Raman spectra in Fig. 2-(e). Of special interest is the observation that when a-Si:C:H is produced by the microcrystalline deposition process, where the high carbon content completely suppresses the microcrystallinity, the absorption coefficient is lower than that of a-Si:C:H of similar carbon content produced by the amorphous deposition process - compare curves (d) and (a) of Fig. 4. SIMS analysis shows that the hydrogen content of the curve (d) film is two and a half times that of the curve (a) film, which probably accounts for its absorption decrease.

When the thickness of the films discussed above decreases to less than ~ 1000Å their measured properties change dramatically. The conductivity decreases, and its activation energy increases as shown in Fig. 5. Using Raman spectroscopy,

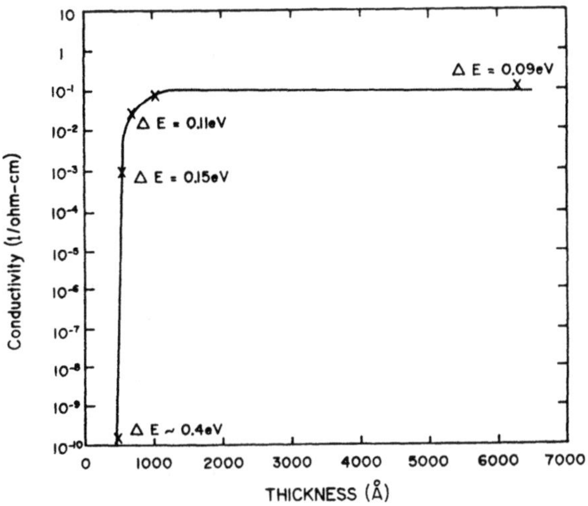

Fig. 5. Conductivity and its activation energy as a function of film thickness for samples made by the microcrystalline process.

these conductivity changes were traced to the fact the first few hundred angstroms of the film are amorphous and not microcrystalline. The initial film growth, however, can be made microcrystalline by using different substrates. Fig. 6 contains Raman spectra of 200 Å thick films deposited by the microcrystalline process onto three different substrates: conducting tin oxide (CTO), poly-crystalline germanium, and a-Si:H along with reference spectra of an amorphous and a microcrystalline film. The material on the CTO substrate is completely amorphous, that on the Ge is about half microcrystalline and half amor-

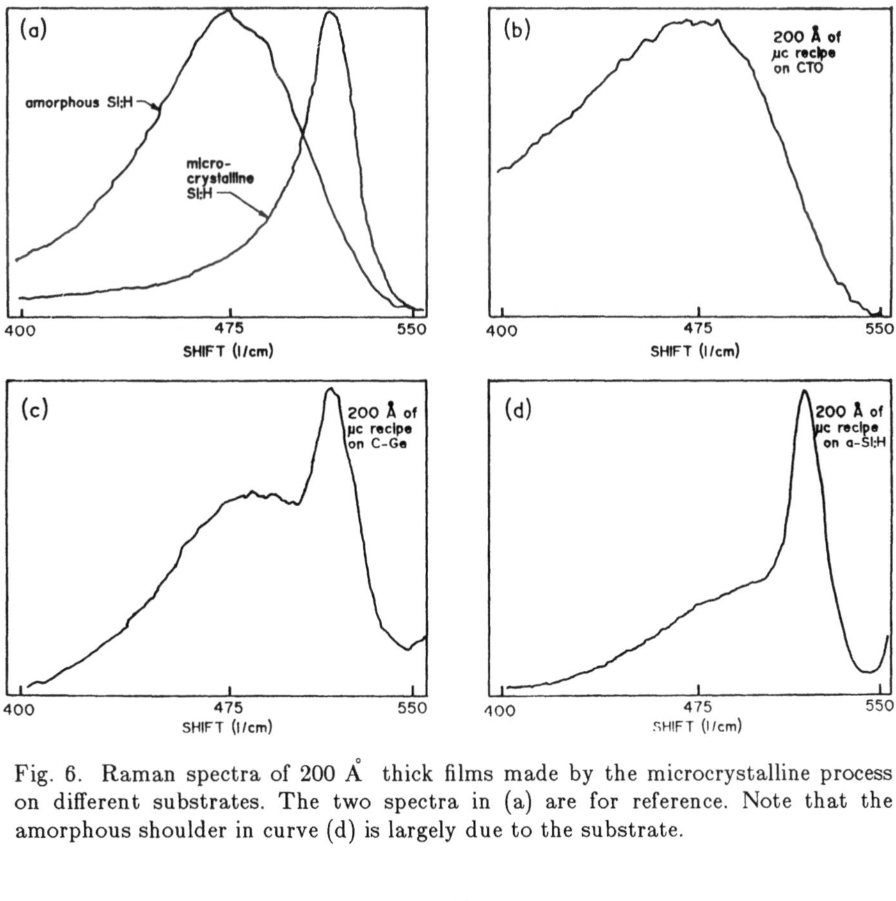

Fig. 6. Raman spectra of 200 Å thick films made by the microcrystalline process on different substrates. The two spectra in (a) are for reference. Note that the amorphous shoulder in curve (d) is largely due to the substrate.

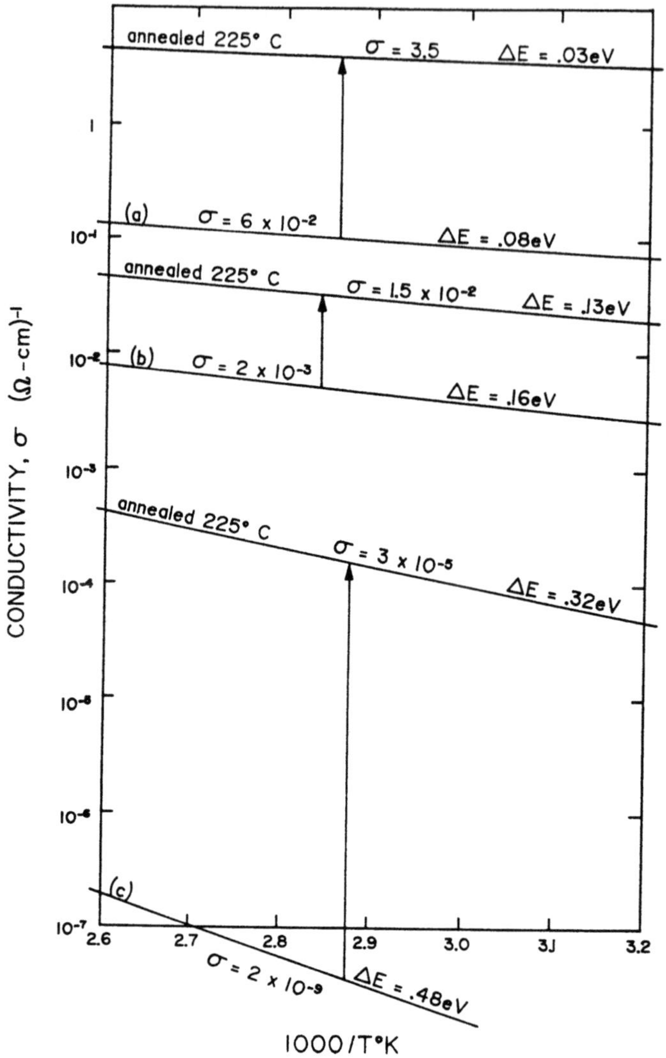

Fig. 7. Changes in conductivity and its activation energy of films made by the microcrystalline process at low substrate temperatures and annealed at 225 °C for 30 minutes.

phous, while that on a-Si:H is almost completely microcrystalline. Based on these results, the incorporation of a microcrystalline p^+ layer, typically 300 Å thick, into a CTO/p/i/n solar cell structure would require the initial deposition of, for example, an a-Si:H layer onto the CTO substrate to permit the growth of the thin microcrystalline film.

Thermal annealing dramatically affects the electrical properties of both amorphous and microcrystalline films made using the microcrystalline process at low substrate temperatures. Films deposited at substrate temperatures below 125 °C, then annealed in an inert atmosphere at a temperature higher than 150 °C, had significant improvement in both conductivity and its activation energy. Some representative changes in σ and ΔE are shown in Fig. 7. The initial conductivity of the films made below 125 °C is larger than that of the films made at 225 °C and becomes even higher when annealed at 225 °C. The effect is seen at all carbon concentrations up to 10%, the highest concentration measured. Neither phosphorus doped n^+ microcrystalline films, deposited using the microcrystalline parameters, nor amorphous material made with no hydrogen dilution of silane, relatively low power, and high growth rate, improve significantly upon annealing. In all cases the annealing temperature used was 225 °C which was near the broad peak found for maximum improvement. Temperatures higher than 250 °C were not used in order to avoid hydrogen evolution[5].

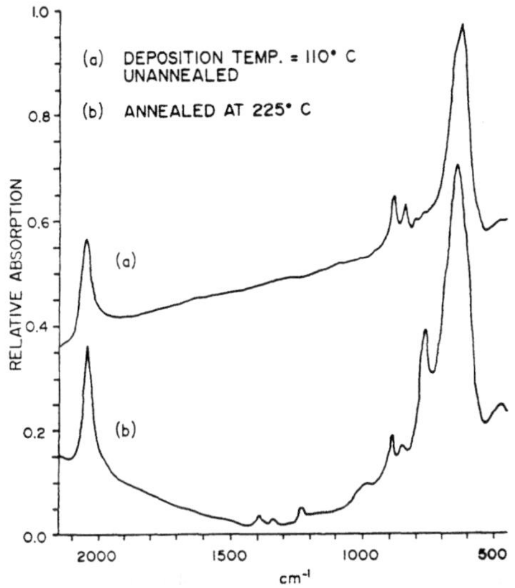

Fig. 8. Infra-red spectra of a microcrystalline film made at 110 °C, before and after annealing at 225 °C for 30 minutes. Lines at 780 and 1200-1400 cm^{-1} are due to Si-CH_3.

SIMS measurements show that the total boron concentration does not change significantly over the range of deposition temperatures. Absorption in the visible region and Raman scattering measurements made on these films do not change upon annealing, indicating that the degree of microcrystallinity remains constant. The FTIR spectrum, however, is significantly affected by annealing as illustrated by the pre- and post anneal spectra shown in Fig. 8. In the unannealed sample none of the Si-CH_3 lines at 780 cm^{-1} or at 1200-1400 cm^{-1} are observed, while these lines are present after annealing. Since, at these high volume fractions, the conductivity of the crystallite determines the conductivity of the film, we suggest that the changes reported here indicate crystallite conductivity changes. Furthermore, since the observation of Si-CH_3 centers requires the formation of C-H bonds and it is also well established hydrogen forms complexes with both deep dangling-bond states and boron directly, we suggest that annealing produces a redistribution of hydrogen within the film that results in the observed changes. A more detailed model cannot be offered at this time; further experiments such as ESR would help clarify the microscopic mechanism.

The Solarex group acknowledges partial support from the Solar Energy Research Institute under contract No. ZB-7-060003-2. The Princeton group acknowledges partial support from the National Science Foundation through the Presidential Young Investigator program.

References

1. Y. Hattori et al., 19th IEEE Photovoltaics Specialists Conference, May 1987, New Orleans.
2. I.H. Campbell and P.M. Fauchet, Solid State Commun., **58**, 739 (1986).
3. A. Matsuda et al., Jpn. J. Appl. Phys., **19**, L305 (1980).
4. T.D. Moustakas, H.P. Maruska and R. Friedman, J. Appl. Phys., **58** 983, (1985).
5. M.H. Brodsky, M.A. Frisch and J.F. Ziegler, Appl. Phys. Lett., **30**, 561 (1977).

STUDY OF DISORDER IN AMORPHOUS Si-Ge ALLOYS BY COUPLED X-RAY DIFFRACTOMETRY AND OPTICAL AND PHOTOTHERMAL SPECTROSCOPIES

J. DIXMIER, P. ELKAIM, M. CUNIOT, H. LABIDI
Laboratoire de Physique des Solides, CNRS,
1 place A. Briand, 92195 Meudon Principal Cedex, France

L. CHAHED, A. GHEORGHIU, M.L. THEYE
Laboratoire d'Optique des Solides, UA CNRS 781
Université P. et M. Curie, 4 place Jussieu,
75252 Paris Cedex 05, France

ABSTRACT. The structure changes as a function of Ge concentration and H incorporation are investigated by pseudo-grazing incidence X-ray diffractometry applied to non-hydrogenated and hydrogenated amorphous SiGe alloys prepared by different techniques. The results are correlated with those of optical absorption studies in the pseudo-gap region performed on the same samples.

1. INTRODUCTION

The observed rapid deterioration of the photoelectronic properties of amorphous hydrogenated SiGe alloys ($a\text{-}Si_{1-x}Ge_x : H$) when increasing the Ge concentration, has been attributed, at least in part, to the occurrence of heterostructure in Ge-rich alloys (1, 2). It seemed therefore interesting to perform comparative X-ray diffraction studies of the structural disorder in non-hydrogenated and hydrogenated alloys prepared by different techniques as a function of composition, and to correlate the results with those of optical absorption studies on the same samples, giving information on the electronic density of states in the pseudo-gap region. It has already been shown (3) that a comparison between the experimental interference functions directly deduced from the diffraction experiments and the ones calculated from two continuous random network (CRN) models with different topologies (4, 5) allow to characterize the structural changes upon alloying. The same approach is followed here, by concentrating on the positions k_1 and k_2 of the first and second peaks of the interference functions. The variations of these parameters with the Ge concentration for the different types of alloys are analyzed in relation with the concomitant variations of optical parameters sensitive to alloying and disorder effects, the optical gap E_g and the inverse slope of the Urbach edge E_o, as well as of defect-related optical absorption.

2. EXPERIMENT

The samples were 3-4 μm thick films deposited simultaneously on Si monocrystalline substrates for X-ray diffraction experiments and on glass substrates for optical and photothermal deflection experiments. Three series of samples were investigated : a) deposited by cathode sputtering in the triode configuration at $T_s \simeq 250°$ C (SP) ; b) deposited by reactive sputtering under the same conditions as before but with 10 % hydrogen in the argon plasma (SPH) ; c) deposited by r.f. glow-discharge decomposition of (SiH_4, GeH_4, H_2) mixtures in the ARCAM reactor at T_s = 150°C (GDH) (6). The alloy composition was determined by electron microprobe analysis.

The X-ray diffraction experiments were performed with CuKα radiation (λ = 1.54 A) at nearly grazing incidence (2θ = 2°) in order to increase the signal for

3-4 μm thick films. A high efficiency curved detector allowed to reduce the exposure time drastically (3). The diffracted intensities, which could be measured up to $k = 4\pi \sin\theta/\lambda = 7.5$ A^{-1}, were recorded in a multichannel and obtained as a function of the diffraction angle 2θ after calibration. The data were treated on line by a microcomputer. The Compton radiation was eliminated by substracting the spectrum of the bare substrate.

The optical absorption coefficient α was determined between 0.6 and 2 eV from a combination of optical transmission and photothermal deflection (PDS) measurements. The PDS spectra were calibrated by fitting to the optical data in the region of overlap at high energies, for $\alpha = 1 - 5 \times 10^3$ cm^{-1}. The α spectra were then deduced by a fringe-averaging procedure based on appropriate thin film formulas.

3. RESULTS AND DISCUSSION

Figure 1 shows, as a typical example, the experimental X-ray diffracted intensity spectra obtained for SP and GDH pure amorphous Si. The two first peaks are very well defined and the third peak with its characteristic shoulder can clearly be seen. As pointed out earlier, we are essentially interested here in the variations of the positions k_1 and k_2 of the first and second peaks with the Ge concentration for the three methods of deposition. These variations are reported in figure 2, together with the k_1 and k_2 variations calculated from the relaxed Polk (P) and Connell Temkin (CT) models.

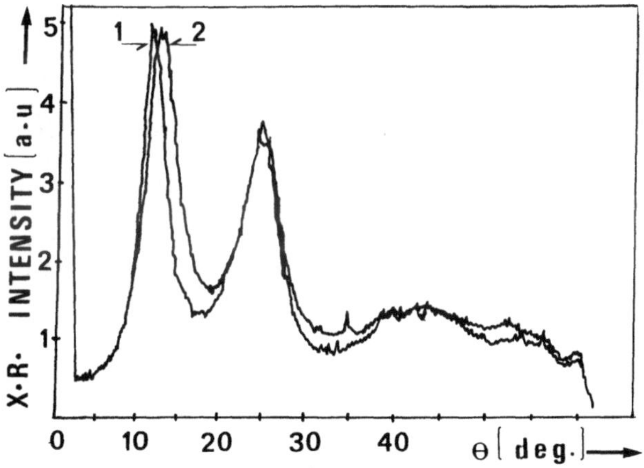

Figure 1
X-ray diffracted intensity spectra for GDH (1) and SP (2) pure amorphous Si

The experimental k_2 values are, within experimental uncertainties, independent on the method of deposition and on the incorporation of hydrogen. Besides, they are quite close to the calculated values, which are themselves independent on the type of CRN model after relaxation, and decrease linearly at the same rate with increasing x. It can thus be inferred that the position of the second peak of the diffracted intensity for amorphous SiGe alloys is essentially determined by the average interatomic distance which, for a given composition, is practically the same whatever the method of preparation, and very close to the model value. The second diffraction peak is consequently very little sensitive to the distances of more remote neighbours, which depend on the degree of structural disorder, the

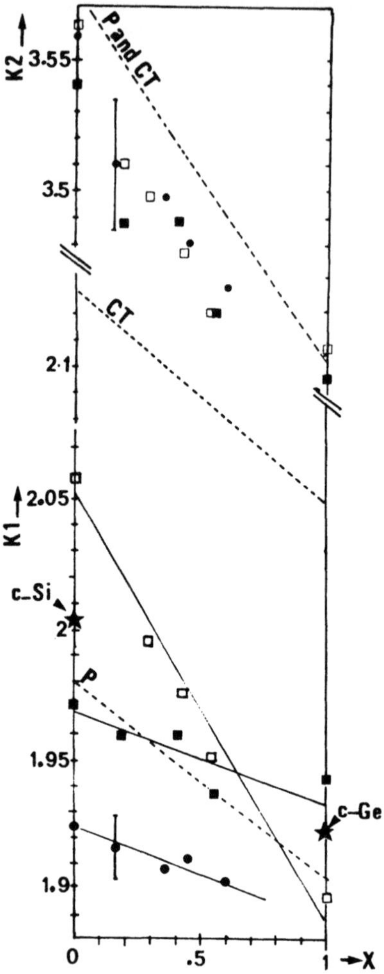

Figure 2
Positions of the first (k_1) and second (k_2) maxima for SP (□), SPH (■), GDH (•) amorphous samples, for crystalline Si and Ge (*), and for relaxed P and CT models (---)

microstructure, the hydrogen incorporation, etc...

The results concerning the positions of the first peak of the diffracted intensities are strikingly different. The experimental k_1 values depend very strongly on the preparation conditions. In particular, non-hydrogenated and hydrogenated alloys can clearly be distinguished. On the other hand, the k_1 values calculated for the two types of relaxed models are also quite different. While the P model values are rather close to the values obtained by linear interpolation between the positions of the (111) diffraction peak in crystalline Si and Ge, the CT model values are much higher. However, they decrease at the same rate with increasing x. When comparing the experimental and calculated data, it appears that the values for the hydrogenated samples are of the same order as the P model values, the GDH values being slightly smaller than the SPH values, but vary less rapidly with x. The non-hydrogenated samples have a very different behaviour. The value for pure a-Si is rather close to the CT model value, then k_1 decreases rapidly with increasing x down to a value of the order of the P model value for pure a-Ge.

The differences between the k_1 values calculated from the two types of CRN models can be understood if one remembers that, even after relaxation, the environment of a given atom (say over the four nearest neighbours) is locally more dense in the CT model than in the P model (the diamond crystalline structure being in between, but closer to the P model). This observation can be correlated with the relative positions of the first diffraction peaks for the three different structures. The local distortions of the network are also significantly larger in the CT model than in the P model, which is more flexible because of the presence of a larger variety (5-, 6- and 7-fold) of rings.

The experimental results can therefore be tentatively interpreted in terms of local structural disorder beyond the first neighbours. Due both to the method of deposition and to the absence of hydrogen, SP pure a-Si must present a locally dense and distorted structure and contain a large amount of defects of the dangling and weak bond types in order to relieve part of the resulting microscopic strains.

The addition of Ge atoms under identical deposition conditions seems to induce rather drastic changes in the local structure, which may be explained by less rigidity in the bonding. For SPH and GDH samples, the incorporation of hydrogen during deposition leads, as usually, to a relaxation of the network ; this effect is particularly important for pure Si and Si-rich alloys. The GDH samples seem to be less disordered than the SPH ones, which can be explained both by different growth process and by more efficient hydrogen bonding. The fact that the k_1 values for both types of hydrogenated alloys decrease less rapidly with increasing x than the model values may be correlated with an increase of disorder in Ge-rich alloys resulting from the preferential bonding of hydrogen to Si and from the development of an heterostructure.

The results of the optical absorption studies performed on the same samples confirm the conclusions drawn from the X-ray diffraction studies and bring additional information. Figure 3 shows, for $x = 0$ and $x \simeq 0.45$, a comparison of the optical absorption edges obtained for samples deposited by the three different methods. The $x = 0$ case is particularly instructive. The two hydrogenated samples have the

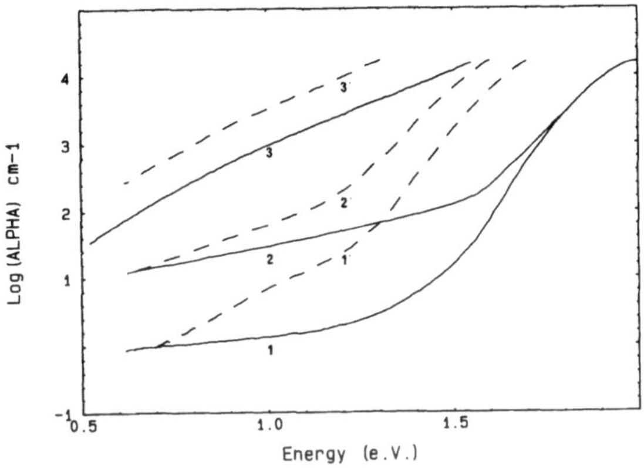

Figure 3
Optical absorption edges of GDH (1), SPH (2) and SP (3) samples with $x = 0$ (continuous lines) and $x \simeq 0.45$ (dashed lines).

same optical gap but the inverse slope E_0 of the exponential part of the edge, related to the width of the disorder-induced band tails, is larger for the SPH than for the GDH sample (81 meV compared to 52 meV). Moreover, the absorption at lower energies, which corresponds to defect states distributed in the pseudo-gap, is higher by about an order of magnitude in the SPH sample. This means that this sample is not only more disordered, as deduced from the diffraction studies, but also contains a larger amount of defects. In the non-hydrogenated sample, the increase in both the band tails and the defect state density is so important that their respective contributions completely overlap and cannot be distinguished any more in the absorption edge, which is very high and varies smoothly with energy. From the data presented in figure 3 for $x \simeq 0.45$, it can be seen that the comments made about pure Si also apply to the alloys. For hydrogenated samples, there is a clear correlation between the respective E_0 and k_1 values for the two methods of preparation at all compositions, which confirms that these two parameters are essentially determined by the disorder. For non-hydrogenated samples however, for which the optical absorption is dominated by the contribution of defect states,

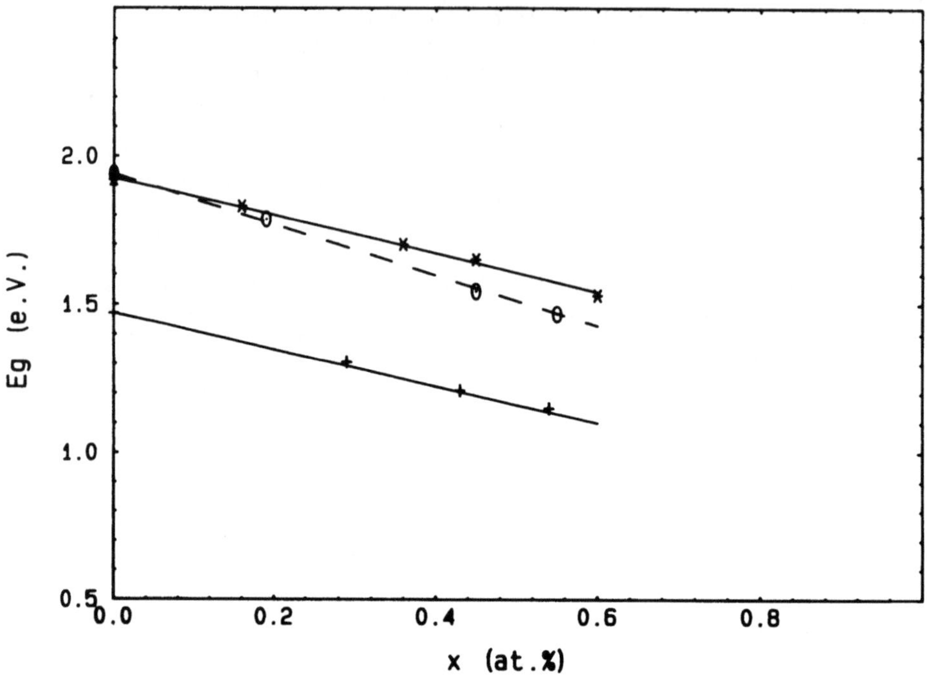

Figure 4
Variations of the optical gap Eg with Ge concentration x for SP (+), SPH (0) and GDH (x) samples.

such a correlation between structure and optical studies is lost, because the k_1 values are not directly sensitive to the presence of defects in the network.

It is worth noting in the end that, as shown by figure 4, the variation of the optical gap (defined as the energy where $\alpha = 10^4$ cm^{-1}) with Ge concentration is the same for the SP samples and for the GDH samples, which are known to contain about the same content of bonded hydrogen for all compositions (responsible for the 0.45 eV difference between the two sets of values). This suggests that the observed narrowing of the optical gap upon Ge incorporation is essentially an alloying effect (like the k_2 variation), and is independent on the proportion of defects in the samples.

ACKNOWLEDGEMENT. This work has been supported in part by the CCE under contract EN3S-0062 F and by the AFME and the PIRSEM-CNRS.

REFERENCES

1. K.D. MACKENZIE, J.R. EGGERT, D.J. LEOPOLD, Y.M. LI, S. LIN and W. PAUL, Phys. Rev. B31, 2198 (1985)
2. A.H. MAHAN, P. RABOISSON and R. TSU, Appl. Phys. Lett. 50, 335 (1987)
3. M. CUNIOT, J. DIXMIER and R. MOSSERI, J. Non-Cryst. Sol. 97-98, 171 (1987)
4. D.E. POLK, J. Non-Cryst. Sol. 5, 365 (1971) ; D.E. POLK and D.S. BOUDREAUX, Phys. Rev. Lett. 31, 92 (1973)
5. G.A.N. CONNELL and R.J. TEMKIN, Phys. Rev. B9, 5323 (1974)
6. L. CHAHED, M.L. THEYE, S. BASROUR, J.C. BRUYERE, C. GODET and T. LLORET, this conference.

a-CSiSn:H AND a-CSiGe:H ALLOYS

F. Demichelis, G. Kaniadakis, E. Tresso, A. Tagliaferro° and P. Rava[+]
Dip. Fisica, Politecnico di Torino
C.so Duca degli Abruzzi, 24 - 10129 Torino (Italy)
° IEN Galileo Ferraris - Torino
[+] Elettrorava S.p.A. - Savonera, Torino

ABSTRACT The properties of two new semiconducting tetrahedral alloys, a-CSiSn:H and a-CSiGe:H, are compared, in order to investigate the different influence of tin and germanium on film properties. Measurements of transmittance and reflectance were analyzed in order to deduce the optical gap and the optical and dielectric constants, while from resistance measurements as a function of temperature, information over conductivity are extracted. As a result, the influence of Argon flow rate, that is Ge or Sn amount, on optical and electrical properties is deduced. For both materials, optical gap increases with decreasing Ar flow rate. Moreover, the properties of two samples having similar values of optical gap, one for each material, are compared: optical and electrical properties are greatly different.

1. INTRODUCTION

a-SiSn:H [1-4] and a-SiGe:H [5-8] alloys were investigated by several authors as materials having narrower band gap than a-Si:H and very promising for photovoltaic and photoconductive devices. a-SiSn:H is an alloy whose optical gap can be varied from 0.34 eV to 1.73 eV by a mixture of tin atoms [4] . However, the introduction in a-Si:H alloy of tin atoms which do not bind very well with hydrogen atoms tends to create a large number of dangling bonds and because of their large atomic radius an increase in a-Si:H network stress. The optical gap of a-SiGe:H alloy can be varied from 1.0 eV to 1.7 eV by varying germanium content [5]. The introduction of germanium atoms in a-Si:H alloy could give rise to SiGe fourfold coordinated structure. However, since hydrogen atoms are more likely to bind silicon than germanium atoms a-SiGe:H characteristics are inferior to those of a-Si:H. On the other hand, it is known that introduction of carbon in a-Si:H produce an alloy having wider band gap than a-Si:H [9-10]. The carbon has remarkable characteristics: thermal stability and binding energy to hydrogen are higher than those of Si, Ge and Sn.

Furthermore, while the a-Si:H normally contains much less hydrogen than that required to reduce the average coordination number to anywhere near the optimal value, in a-C:H alloy the average coordination number can be reduced to the optimal value by the addition of hydrogen or of significant amounts of sp bonds present in threefold coordinated carbon. For all above considerations we have studied the new tetrahedrally carbon based alloy a-CSiSn:H [11-13] and a-CSiGe:H [14] in order to improve the optoelectronic properties of the narrower band gap a-SiSn:H and a-SiGe:H alloys. In this paper we report a comparison of the results on the optical and electrical properties of a-CSiSn:H an a-CSiGe:H sets as a function of different tin and germanium content.

2. SAMPLE PREPARATION AND ANALYSIS

Sputter Assisted Plasma Chemical Vapor Deposition (SAPCVD) which uses

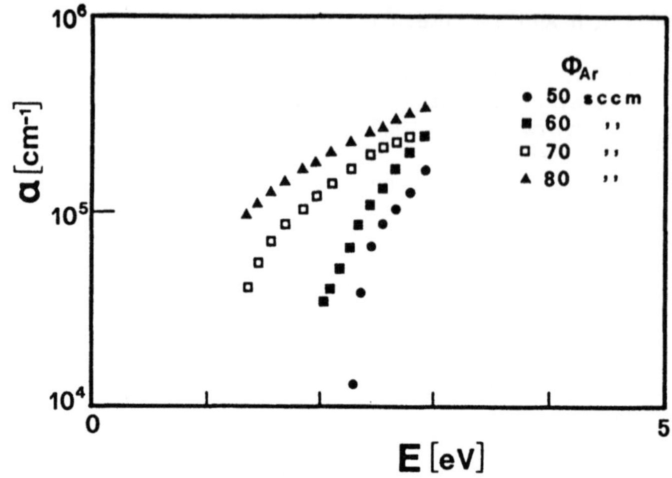

Fig. 1

α vs $h\nu$ for a-CSiSn:H

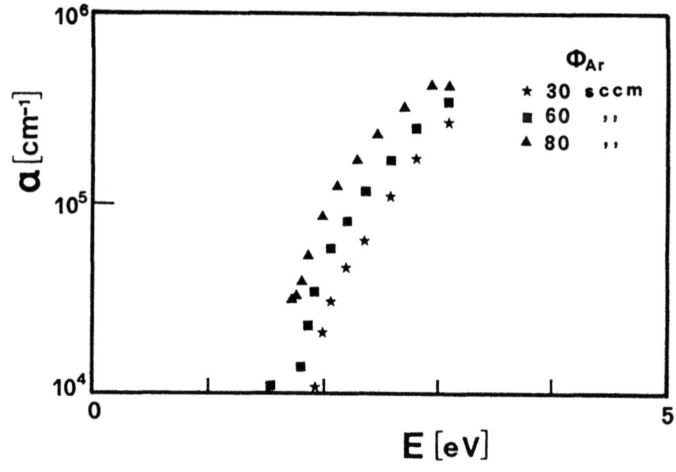

Fig. 2

α vs $h\nu$ for a-CSiGe:H

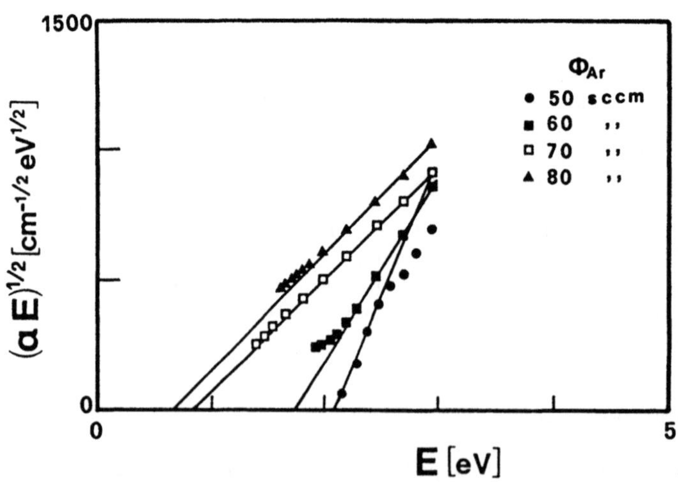

Fig. 3

a-CSiSn:H Tauc's plots

Table I
Deposition Conditions for a-CSiSn:H and a-CSiGe:H sets

			a-CSiSn:H	a-CSiGe:H
Substrate Temp.	(°C)		150	50
RF Power	(W)		500	250
Pressure	(Pa)		7.5	2.3
DC Voltage	(V)		400	800
		SiH_4	10	5
Flow rate	(sccm)	CH_4	10	5
		Ar	50-80	30-80

both the glow discharge method and the sputter method simultaneously was employed in the fabrication of the films. Magnetron cathodes of tin (purity 99.9995%) or germanium (purity 99.999%) were used as a target. A mixture of SiH_4, CH_4 and Ar was introduced into the chamber controlling separately the flow of each gas. The deposition conditions of the two sets of films were summarized in Table I.

The films were deposited onto optically polished fused silica for optical measurements and onto Corning glass for conductivity measurements. The optical measurements of transmittance and reflectance were recorded on a Perkin Elmer Lambda 9 spectrophotometer in the region 0.2 to 1.5 μm. Transmittance measurements were taken at normal incidence in the double beam mode. The reflectance was measured from both the air and substrate sides of the films and compared with the reflectance of a NBS mirror standard. The thickness of the films was measured by a Dektak IIA Stylus Profiler. The procedure described in ref. 15 was used to deduce the optical constants n and k and then the complex dielectric constant $\tilde{\varepsilon}$. The conductivity was measured as a function of the temperature in a vacuum chamber, with a Hewlett Packard high resistance meter 4329 A. DC voltage in a wide range from 10V to 500V in direct and reverse bias was applied. The ohmic behaviour of the electrodes contacts was checked.

RESULTS AND DISCUSSION

Figs. 1 and 2 show the asborption coefficient as a function of $h\nu$ for a-CSiSn:H and a-CSiGe:H respectively, at various Argon flow, that is various tin or germanium content, as a function of photon energy. Actually, from previous measurements of elemental composition [12-13] we have verified the linear dependence of tin content upon argon flow. We observe that absorption onset is shifted to lower energy with increasing Ar, that is increasing Sn or Ge content. Figs. 3 and 4 show the plots of $(\alpha E)^{1/2}$ vs $h\nu$, used to determine optical gap Eg for a-CSiSn:H and a-CSiGe:H respectively. As Ar flow increases, $(\alpha E)^{1/2}$ vs $h\nu$ plots shift to lower energy. Figs. 5a and 5b show the argon flow dependence of the optical band gaps.

We can deduce that addition of germanium to a-CSi:H makes band gap narrower monotonically, while for a-CSiSn:H a different trend is observed. For Ar flows lower than 60 sccm the samples do not contain tin; in fact the values of their optical gap correspond to the a-CSi:H values. This is also confirmed by the different slopes of $(\alpha E)^{1/2}$ vs $h\nu$ lines in Fig. 3. The addition of tin or germanium atoms in a-CSi:H network allows to realize samples with a wider range of Eg values with respect to a-Si:H.

To compare the influence of the addition in a-CSi:H of tin or germa-

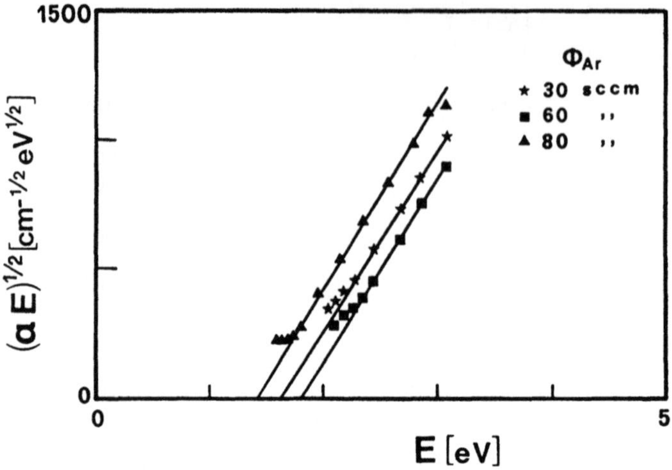

Fig. 4

a-CSiGe:H Tauc's plots

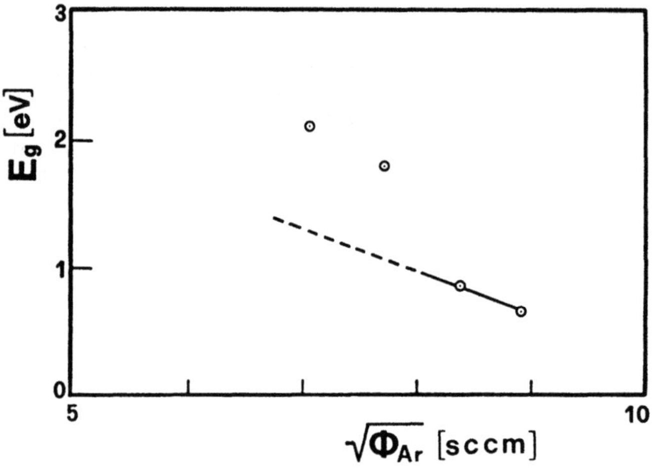

Fig. 5a

Ar flow dependence of Eg for a-CSiSn:H

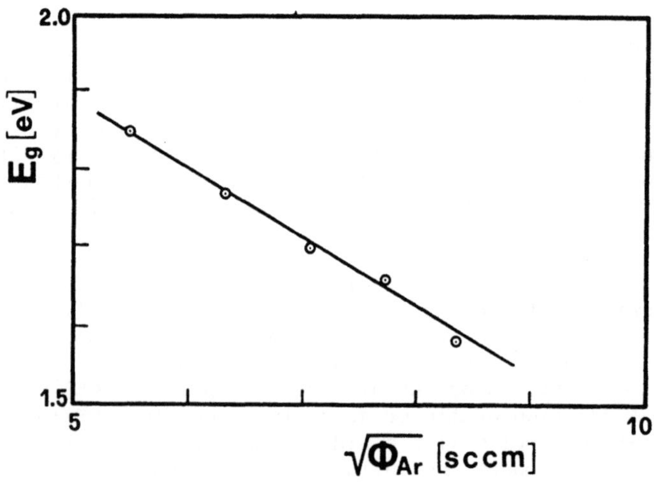

Fig. 5b

Ar flow dependence of Eg for a-CSiGe:H

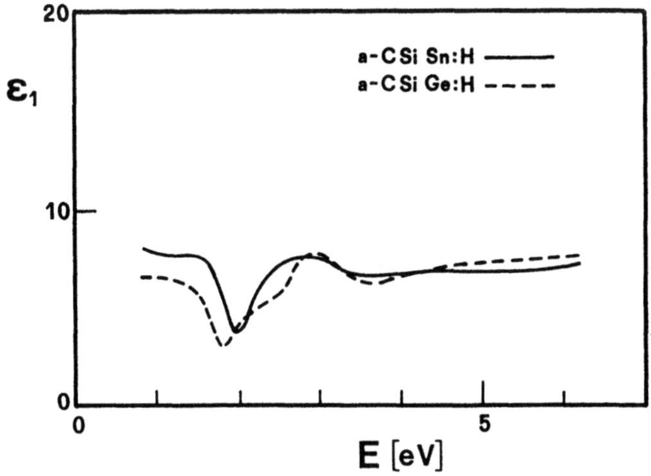

Fig. 6a

ε_1 vs $h\nu$ for a-CSiSn:H and a-CSiGe:H samples having similar Eg

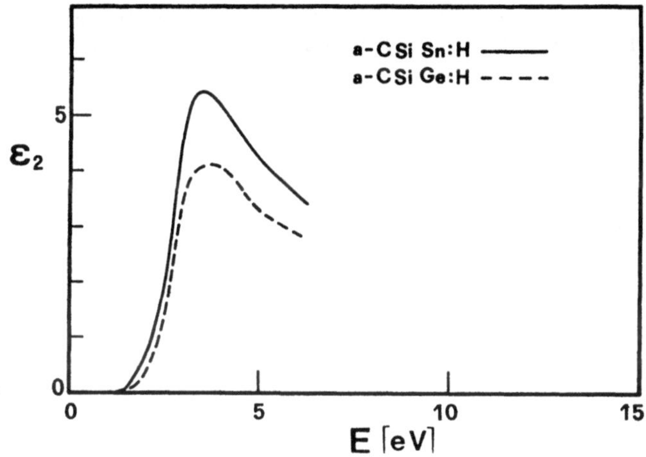

Fig. 6b

ε_2 vs $h\nu$ for fig. 6a samples

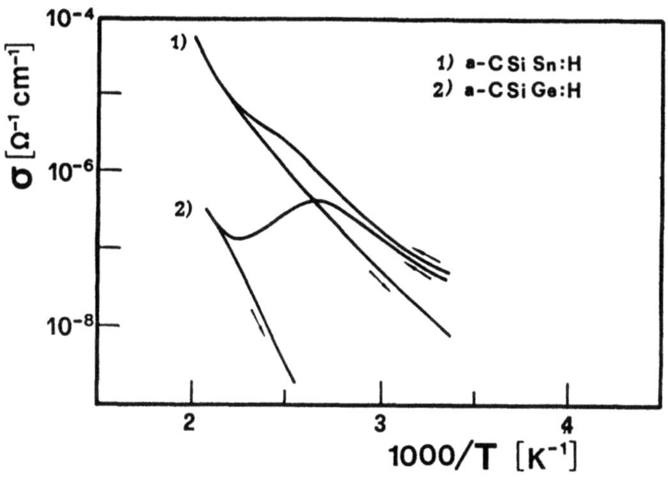

Fig. 7

Arrhenius plots for Figs. 6 samples

nium we report the optoelectronic properties of a-CSiSn:H (ϕ_{Ar} =60 sccm) and a-CSiGe:H (ϕ_{Ar} =30 sccm) samples having about the same value of the optical gap. Figs. 6 show the trend of the experimental real (ε_1) and imaginary (ε_2) part of the dielectric constant for the samples a-CSiSn:H (ϕ_{Ar} =60 sccm) and a-CSiGe:H (ϕ_{Ar} =30 sccm). All the trends indicate that the presence of carbon strongly influences the optical constants [16] and then the optical properties of the films. In Fig. 7 are shown the trends of log σ vs $10^3/T$ measured during heating and cooling processes for the same two samples. We observe in both cases a curvature in the heating process indicating probably a change in conductivity mechanism due to the presence of carbon [14] .

In conclusion we have found that the carbon present in a-CSiSn:H and a-CSiGe:H alloys strongly influences the optical and electrical properties of the samples. However introduction of convenient tin or germanium content makes optical gap narrower allowing to obtain a wide range of optical gap values.

ACKNOWLEDGEMENTS

This work was partially supported by Research Project PFE2, sponsored by the Consiglio Nazionale delle Ricerche. G. Kaniadakis wishes to thank the International Centre for Theoretical Physics, Trieste, for providing a grant.

References

1 C. Vérié, J.F. Rochette and J.P. Rebouillet - J. Phys. Paris 42, C4, 667, (1981)
2 A.H. Mahan, D.L. Williamson and A. Madan - Appl. Phys. Lett. 44(2), 220, (1984)
3 A. Morimoto, T. Kataoka and T. Shimizu - Jap. Journ. of Appl. Phys. 23, L812, (1984)
4 G.N. Parson, J.W. Cook Jr., G. Lucovsky, S.Y. Lin and M.J. Mantini - J. Vac. Sci. Technol. A4, 470, (1986)
5 G. Nakamura, K. Sato, and Y. Yukimoto - Proc. of the Sixteenth I.E.E.E. Photovoltaic Specialist Conference - San Diego, Ca, p. 1331 (1982)
6 H. Munakata and H. Kukimoto - Appl. Phys. Lett. 42, 432, (1983)
7 K. Nozawa, Y. Yamaguchi, J. Harma and T. Shimizu - J. Non Cryst. Solids 44, 713, (1983)
8 T. Ichimura, T. Ihara, T. Hama, M. Ohsawa, H. Sakai and Y. Uchida - Jap. Journ. Appl. Phys. 25, L276, (1986)
9 D.A. Anderson and W.E. Spear - Phil. Mag. 35, 1, (1977)
10 Y. Catherine and G. Turban - Thin Solid Films 70, 101, (1980)
11 F. Demichelis, G. Kaniadakis, A. Tagliaferro, E. Tresso and P. Rava - Phys. Rev. B 37, 1237, (1988)
12 F. Demichelis, G. Kaniadakis, A. Tagliaferro, E. Tresso, P. Rava and G. Della Mea - Journ. Appl. Phys. (in press)
13 F. Demichelis, G. Kaniadakis, A. Tagliaferro, E. Tresso, G. Della Mea and A. Paccagnella - Int. Journ. of Modern Physics B (in press)
14 F. Demichelis, G. Kaniadakis, A. Tagliaferro, E. Tresso and P. Rava - Proc. of Int. PVSEC-3, 313, (1987)
15 F. Demichelis, G. Kaniadakis, A. Tagliaferro and E. Tresso - Appl. Opt. 26, 1737, (1987)
16 F. Demichelis, G. Kaniadakis, A. Tagliaferro, E. Tresso and G. Della Mea - Proc. MRS Spring Meeting, Reno, (1988) (in press)

EXPERIMENTAL DETERMINATION OF ELECTRONIC TRANSPORT PARAMETERS IN HYDROGENATED AMORPHOUS SILICON GERMANIUM ALLOYS.

R. VANDERHAGHEN (*), C. LONGEAUD (**)

(*) Laboratoire de physique des interfaces et couches minces, (U.R.258), Ecole Polytechnique, 91128 Palaiseau cedex, France.

(**) Laboratoire de génie électrique de Paris, E.S.E, Plateau du moulon, 91190 Gif/Yvette, France.

Electron transport parameters (density of states shape, capture cross section, extended states mobility) are obtained from the time of flight data analysed with four methods. In one hand, we study the influence of the deposition temperature on the electronic transport parameters of high quality hydrogenated amorphous silicon. In all cases, the shape of the conduction band tail is made of an exponential distribution following a flatter distribution. The decrease of the deposition temperature produces a broadenning of the tail and an increase of the capture cross-section. We have also measured the mobility-lifetime product in a-Si:H with a new method. The order of magnitude of this product is found compatible with the mobility-lifetime product determined in steady-state photoconductivity. Alloying with Germanium results in a broadening of the conduction band tail. This broadening comes with a decrease of the extended states mobility, an increase of the capture cross-section and of the deep tail states below E_c-E=0.3 eV.

1. INTRODUCTION

It is possible, from a time of flight (T.O.F) experiment, to determine the density of tail states (D.O.S) distribution and the transport parameters (T.P) such as the capture cross-section, σ, and the extended states mobility, μ_0. However there is still a controversy about the D.O.S shape and the values of these T.P [1,2,3,4]. Therefore, we have studied the D.O.S and the T.P for a set of samples prepared in the same reactor. The samples were glow discharge hydrogenated amorphous Silicon (a-Si:H),and Silicon-Germanium alloys (a-SiGe:H).

We have studied the influence on the D.O.S shape and on the T.P of:

i) the substrate temperature during the deposition of the a-Si:H

ii) the Germanium content of the a-SiGe:H layers.

The measurement of the mobility-lifetime product, $\mu\tau$, by means of the T.O.F experiment is generally found to be two orders of magnitude lower than the $\mu\tau$ product obtained in steady state photoconductivity. This discrepancy will be analysed.

2. METHODS

The D.O.S were studied with three methods exposed elsewhere [4,5,6]:

A) a direct experimental estimate of the D.O.S shape from a Fourier transform of the derivative of the observed current I(t). This estimate is valid mainly when the transport is non (or slightly) dispersive [5].

B) an experimental estimate of the D.O.S. shape based on the estimate of the first slope $-(1-\alpha_1)$ in a Ln(I)-Ln(t) plot when the transport is dispersive [6]. Indeed, at a time t the shape of the current I(t) is linked to the trapping/release constants of the traps located at E defined by:

$$E_c-E=kTLn(\nu t),$$

where ν is the attempt to escape frequency. If we assume that, locally, the D.O.S is exponential with:

$$N(E_c-E)=N_0\exp[-(E_c-E)/kT_c],$$

then I(t) will behave, at a temperature T, as $t^{-(1-\alpha_1)}$ with $\alpha_1=T/T_c$ [7]. A variation of α_1 with the applied electric field, ξ, or with t before transit, at fixed T, is representative of a variation of the slope of the D.O.S with E_c-E. The agreement between the experimental methods A and B is excellent

C) a general analysis of the experimental data (dispersive or not), this analysis of transit times and current shapes being conforted by a numerical simulation, based on a matricial method taking the spatial position and the diffusion of the carriers into account [4]. We have assumed that ν is independent of T and E, that μ_0 is independent of T and that $N(E_c)=10^{21}$ $cm^{-3}eV^{-1}$. Consequently, σ is independent of the energy and varies as $T^{-3/2}$. The transit time is defined as $t_{0.5}$ [8], that is either as the time where I(t) equals half the extrapolation of the first linear part in a Ln(I)-Ln(t) plot [4] or as the time at which half the carriers had completed their transit identified by the corresponding fall in I(t) compared with an extraction-free curve obtained at a lower field [1]. The use of $t_{0.5}$ instead of $t_{0.8}$ explains the apparent reduction of room temperature drift mobility, μ_D, compared with previous publications.

D) Trying to give an answer to the controversy concerning the origins of the discrepancy between the $\mu\tau$ products measured in T.O.F and in steady-state photoconductivity [9], we have performed a new experiment on the a-Si:H samples. This experiment is achieved like the T.O.F but instead of applying a pulsed field until all the carriers are collected, the pulsed field is set to zero during a variable time, t_s, while carriers are still between the two electrodes. After t_s the pulsed field is applied again and we plot the collected charge, $Q(t_s)$. In this analysis, τ is assumed to be related to the deep trapping occuring during t_s. The charge loss due to the diffusion during t_s is taken into account.

3. EXPERIMENTAL RESULTS

3.1 Experimental conditions

High quality a-Si:H (a-SiGe:H alloys) was produced by glow discharge decomposition of SiH_4 (SiH_4-GeH_4-H_2 mixture) in an ARCAM reactor [10]. The substrates were held at various temperatures ranging from 150°C to 250°C. The samples studied were Pt Schottky structures whose thicknesses range from 2.4 to 5 µm. We present in table I the characteristics of the samples. The experiments were performed as exposed in reference [3]. the measurement temperatures were ranging from 120 K to 300 K.

Table I

Sample n°	1	2	3	4	5	6
% Ge	0	0	0	5	16	25
P (W)	5	5	5	5	5	5
T_{dep} (°C)	150	200	250	200	200	250
L (µm)	3.2	3.8	5	5	5	2.4

Characteristics of the samples L is the thickness, T_{dep} the substrate temperature and P the power during deposition.

3.2 a-Si:H analysis

At high temperatures (T>240 K) the transport is non-dispersive, and the transit is analysed with method A, whereas, for T<240 K the transit is analysed with method B. Experimental data from method A (dots) are presented in figure 1, where the full line represents the D.O.S obtained from method B. The D.O.S deduced from methods A and B behave like exponential distributions with characteristic temperatures T_{c2} of the order of 230 K for $0.15<E_c-E<0.3$ eV. With these experimental methods we cannot deduce the shape of the D.O.S for $0<E_c-E<0.15$ eV. Moreover, in both methods, μ_0 and ν are scaling factors. Nevertheless, these values can be estimated with method C which also demonstrates that another distribution, with a flattened slope, seems to exist above 0.15 eV. Since we cannot analyse accurately the D.O.S close to the conduction band we have chosen, in order to fit the experimental data, to parametrize the D.O.S in this region with an exponential distribution, the characteristic temperature of which is T_{c1}=370 K.

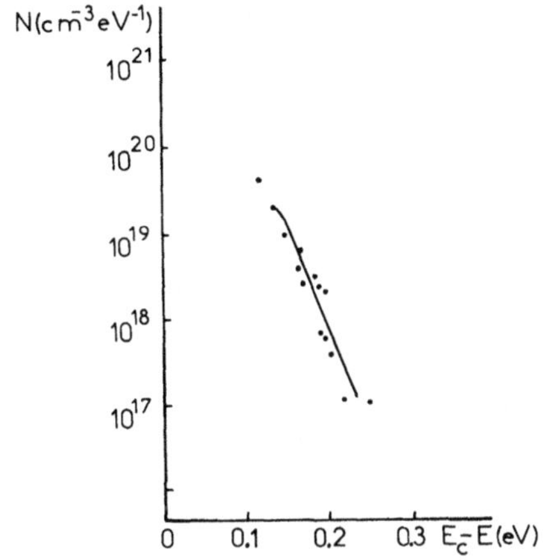

Figure 1
D.O.S of sample 2 obtained with method A (dots) and B (line)

The evolution of the slope of the D.O.S with the energy is ensured by the measurements per--formed at low temperatures with various fields. As we have said, a variation of α_1 with ξ, at fixed T, indicates a variation of the D.O.S slope with E_c-E. This behaviour was already mentionned in previous papers [3,4,6] and is well explained by the simulation. Indeed, the whole data are well represented by means of the method C.

The variations of the activation energy of μ_D must be related to the same behaviours and, as a matter of example, we present in figure 2 the the plot of μ_D versus 1/T at various fields for measured (points) and simulated (full lines) data for sample 2. We can emphasize the good agreement obtained between the experimental and simulated results.

We have applied these three methods to samples 1, 2, 3, to study the influence of the substrate temperature, T_{dep}, on the T.P. The variations of the T.P with T_{dep} can be clearly seen in table II. The characteristic temperatures of the D.O.S distributions do not seem to be affected by the variations of T_{dep}, but if we call E_{int} the energy location of the intersec--tion of the two distributions, we note that E_{int} varies with T_{dep}.

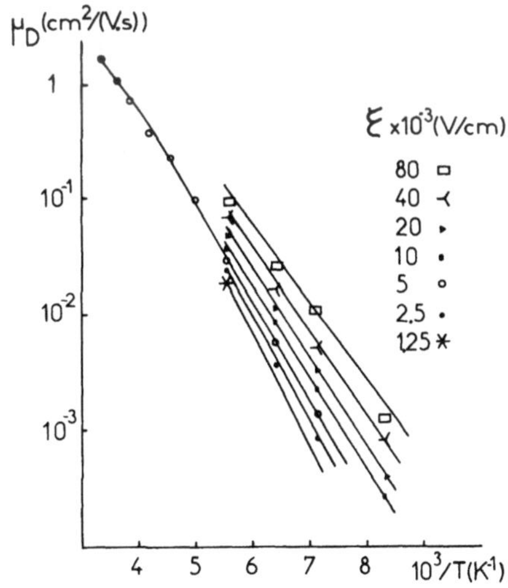

Figure 2
Activation of the drift mobility for sample 2. Experimental (points) and simulated data (lines) are in good agreement.

The D.O.S deduced from the three methods are presented in figure 3. The curves of figures

1 and 3 are plotted using the parameters listed in table II.

TABLE II

Sample n°	1	2	3
T_{dep} (°C)	150	200	250
μ_o ($cm^2/(V.s)$)	24	25	27
$\nu \times 10^{-11}$ (s^{-1})	5	2.7	2.4
T_{c1} (K)	370	370	370
T_{c2} (K)	235	220	230
E_c-E_{int} (eV)	0.16	0.15	0.15

Transport parameters deduced from methods A,B,C. T_{c1}, T_{c2} are the characteristic tem--peratures for states close to the conduction band and deeper in the gap respecti-vely. E_{int} is the energy lo--cation of the intersection of the distributions.

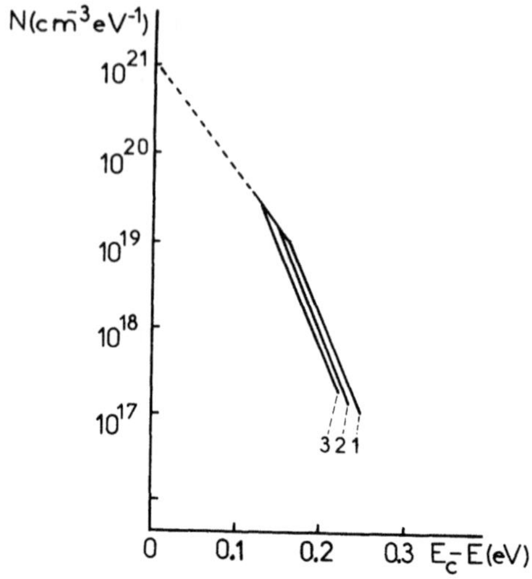

Figure 3

D.O.S of samples 1,2,3 obtained with method A,B (full lines) and method C (dashed line)

In the T.O.F experiment the deep trapping can be parametri--zed by a lifetime τ. Most of the time, τ is derived from the analy--sis of the variation of the col--lected charge with the applied field, Q(ξ), by means of the Hecht formula. Unfortunately, Q(ξ) can be modified by the interface recombination enhanced by the diffusion [4]. This was checked in changing the interface field by the application of a DC bias: Q(ξ) is strongly affected by the applied bias. On the contrary, the use of the method D suppresses the influence of the interface. The variation of $Q(t_s)$ can be related to the density of deep traps. For sample 2 we find μτ from a fit of $Q(t_s)$ as exposed in figure 4. The initial slope at small t_s and low temperature is very sensitive to τ: we find $\mu\tau=1.1\times10^{-5}$ cm^2V^{-1}. Assuming a flat distribution of deep states above E_f and according to the values of μ and ν of table II we obtain a density $N=1.2\times10^{15}$ $cm^{-3}eV^{-1}$.

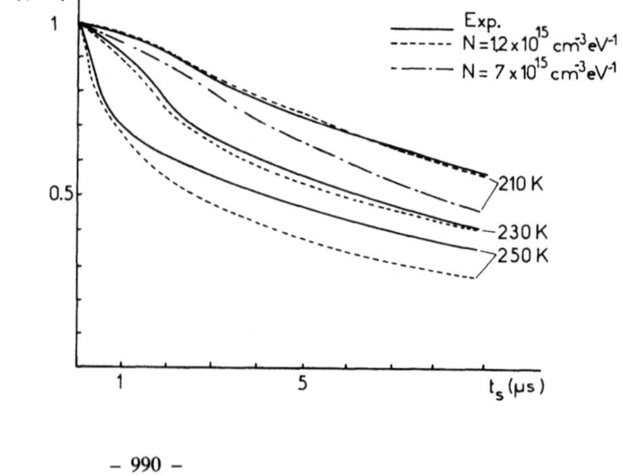

Figure 4
Variation of the collected charge $Q(t_s)$ for various temperatures. The best fit of the experimental curves (full lines) is obtained for $N=1.2\times10^{15}$ $cm^{-3}eV^{-1}$ (dashed line).

3.3 a-SiGe:H analysis

The transport in a-SiGe:H with a high Germanium content is dispersive for all temperatures. Thus, we can only apply methods B and C. We present in table III the T.P we determine from the experimental data and in figure 5 the corresponding D.O.S obtained for samples 2, 4, 5, 6.

TABLE III

Sample n°	2	4	5	6
% Ge	0	5	16	25
μ_0 ($cm^2/(V.s)$)	25	25	18	10
$\nu \times 10^{-11}$ (s^{-1})	2.7	1.5	25	50
T_{c1} (K)	370	390	390	390
T_{c2} (K)	220	260	390	390
E_c-E_{int} (eV)	0.15	0.15	--	--

Transport parameters deduced from methods B,C for samples 2,4,5,6. T_{c1} and T_{c2} are the characteristic temperatures of the D.O.S distributions E_{int} is the energy location of the intersection of the distributions.

μ_0 decreases with the content of Germanium, x, whereas ν and, hence, the capture cross-section are increasing. The characteristic temperature of the first exponential distribution is found to be slightly higher than in the a-Si:H. Actually, there is a broadening of the conduction band tail as x increases. Furthermore, the density of deep states in the tail seems to increase drastically with x. This can be seen on figure 5 where one should define a third flatter distribution for E_c-E>0.3 eV with a characteristic temperature of the order of 1000 K. The mean value of N(E) in these deep states increases by two orders of magnitude when x increases from 0 to 0.16. For E_c-E>0.3 eV the deep states tail is deduced from the va--riations of α_1. Indeed, the pre--transit slope of the Ln(I)-Ln(t) plot increases sharply leading to low values of α_1 (high T_c values).

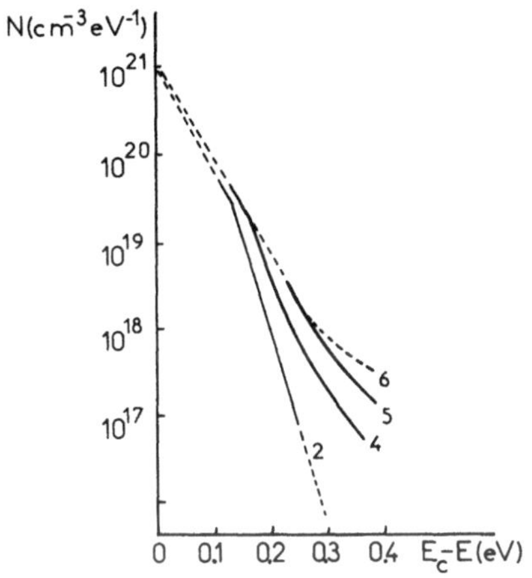

Figure 5
D.O.S of samples 2,4,5,6 obtained with methods A,B (full lines) and method C (dotted lines)

This observation is conforted by the flattening of post-transit tail, this flattening being due to the slow release of these deep traps.

4. DISCUSSION

For the a-Si:H, the D.O.S shape is made of two exponential distributions crossing near E_c-E=0.15 eV. The deepest distribution is established by means of the A, B, C methods. We have checked that the distribution between 0.1 and 0.15 eV is roughly exponential. To fit the experimental data we had to assume that the D.O.S is exponential between the conduction band and 0.15 eV. The value of μ_0 is directly related to the shape of the D.O.S close to the conduction band, whereas σ is related to the activation of μ_D. A slight error in the measurement of the sample temperature, especially for low temperatures, can lead to a simultaneous modification in the estimate of σ and μ_0. This explains the slight discrepancy with our previous measurements [3]. In the present experiment the sample temperature was measured very accurately.

With the present $\mu\tau$ measurements, we estimate that the density of states near E_f is equal to a few 10^{15} $cm^{-3}eV^{-1}$, compatible with the steady states measurements [11].

The increase of the substrate temperature results simultaneously in a decrease of σ and a shrinking of the band tail: these behaviours can be related to an improvement of the medium range order.

Alloying with Germanium increases σ and broadens the band tail: this can be related to a decrease of the order. The decrease of μ_o with the Ge content could be attributed to additional potential fluctuations at the mobility edge. The last effect related to the increase of the Ge content is the existence of a deep states tail for E_c-E>0.3 eV, also seen by other analysis methods[12]. These states, for x<0.16, do not seem to act mainly as recombination centers according to the $\mu\tau$ estimated from steady state measurements [11]: the steady states $\mu\tau$ decreases by less than a factor 10 whereas the density of states around 0.35 eV increases by two orders of magnitude for x varying from 0 to 0.16. However, these deep states can affect the potential profile in a PIN structure.

5. CONCLUSION

For the a-Si:H, we have observed exponential distribution of states 0.15 eV below E_c. Above E_c-0.15 eV the distribution is flatter. The density of deep states for $E_f<E<E_f+0.35$ eV equals a few 10^{15} $cm^{-3}eV^{-1}$. When the deposition temperature is lowered the tail states distribution broadens: this could be related to an increase of the disorder.

Alloying with Germanium results also in a broadening with a simultaneous increase of σ and of the deep states below E_c-E=0.3 eV.

AKNOWLEDGEMENTS

This work is supported by AFME-PIRSEM contracts and CEC EN3S-0062-F contracts. The samples were prepared by P. Roca y Cabarrocas, C. Godet and P. Andro.

REFERENCES

1)J.M. MARSHALL,R.A. STREET,J.M.THOMSON, Phil. Mag. B 54,(1986),51
2)F. KARG, W. KRUHLER, M. MOLLER, K. V. KLITZING, J. Appl. Phys. 60, (1986), 2016.
3)R. VANDERHAGHEN, C. LONGEAUD, Proceedings of the XII I.C.A.L.S. (1987), Journal of non-Cryst. Solids 97, p 1059
4)R. VANDERHAGHEN, Submitted to Phys. Rev. B
5)C. LONGEAUD, G.FOURNET, R. VANDERHAGHEN, Submitted to Phys. Rev. B
6)C. LONGEAUD, Submitted to Phys. Rev. B
7)T. TIEDJE and A. ROSE, Solid States Comm., 37, (1981), 49
8)G. SEYNHAEVE, G. J. ADRIAENSSENS, H. MICHIEL, H. OVERHOF, To be published
9)E. P. SCHIFF, Phil. Mag. Let. 55, (1987), 87
10)P. ROCA I CABARROCAS,B. EQUER,J. HUC,A. LLORET, J.P.M. SCHMITT, Proc. of the 7th Photovoltaic Solar Energy Conf.-Sevilla (1986), 533
11)C. GODET, Private communication
12)L. CHAHED,M.L. THEYE,S. BASROUR,J.C. BRUYERE,C. GODET,A. LLORET, This conference.

STUDY OF HYDROGEN BONDS IN AMORPHOUS GALLIUM ARSENIDE

K. SEDEEK, H. CARCHANO, J.L. SEGUIN

Université d'Aix-Marseille III
Faculté des Sciences et Techniques de Saint-Jérôme
Département de Photoélectricité
Laboratoire des Matériaux Amorphes - Service A62
F.13397 MARSEILLE Cedex 13

ABSTRACT

In the precedent report, we studied the hydrogen effects on the electrical properties of amorphous gallium arsenide. In this paper, we present results concerning the characterization of hydrogen bonds in the material by X-ray photoelectron spectroscopy (XPS) and infrared spectroscopy (IRS). XPS results show a considerable modification of the Ga 3d core level with hydrogenation, while this of As 3d is unaffected. We estimate therefore that hydrogen is preferentially bonded to gallium. The IRS spectrum presents two large absorption bands at 530 cm^{-1} and 1430 cm^{-1}. We think, in analogy with other works, that these absorption bands are induced by the three centre bonds in which hydrogen bridges two gallium atoms. For arsenic, the IRS spectrum shows three weak bands at 840 cm^{-1} and 2092 cm^{-1} assigned to As-H bonds and at 1000 cm^{-1} attributed to As-H_2 bonds. These results confirm that significant amount of hydrogen is bonded in the form of Ga-H-Ga.

1. INTRODUCTION

In a previous report /1/, we recorded the effect of hydrogenation on the electrical properties of R.F. sputtered amorphous gallium arsenide (a-GaAs). We showed that the electrical conductivity decreases by about four orders of magnitude and the activation energy increases by 0.31 eV arguing that hydrogen compensates partially dangling bonds.

With the aim of better understanding the role of hydrogen in improving the a-GaAs quality and why it is difficult to realize a Schottky barrier, we try to study the nature of hydrogen bonds in this material.

2. EXPERIMENTAL TECHNIQUES

The films were prepared by R.F. sputtering of a polycristalline GaAs target in an Ar : H_2 mixture. The total pressure (P_T) was maintained at 50 m Torr and the partial pressure of H_2 was varied between 0 and 25 % of total pressure. Samples of about 500 Å thick were deposited on Mo substrates for XPS experiments and of about 2μm thick on Si substrates for IRS experiments. Films structure was checked by X-Ray diffraction, while their composition was analysed using EDAX. The XPS spectra were induced by Al $K\alpha_{1,2}$ radiation for samples with and whithout hydrogen. The IRS measurements were carried out using a Nicolet Fourrier Transform Infrared Spectrograph (FTIR 7000), between 400 and 4000 cm^{-1} on 3 samples prepared with P_{H2} = 10, 15 and 25 %.

3. EXPERIMENTAL RESULTS

3.1. XPS measurements

Figure 1 shows the As3d core level spectrum for a-GaAs and a-GaAs:H. The shape and the width of the line are not perceptibly modified by H incorporation. (The slight shift remains within the experimental uncertainty). On the other hand, the shape of the Ga3d core level spectra (figure 2) is noticeably altered by hydrogenation. The GaAs and Ga_2O_3 peaks for a-GaAs:H are no more separated and we observe only one broad line.

3.2. IRS measurements

Figure 5 shows the IR transmission spectra for a-GaAs:H prepared with different hydrogen partial pressures (P_{H2}/P_T = 10, 15 and 25 %). On the spectrum of the most hydrogenated sample, there are two strong bands at 530 cm^{-1} and 1430 cm^{-1} which remain very sharp whatever P_{H2}/P_T is. We also observe other bands at 840, 1000, 1130, 1725 and 2010 cm^{-1} and a shoulder at 560 cm^{-1}.

4. DISCUSSION

As 3d core level spectrum suggests that a few H atoms are bonded to As since this spectrum is unaffected by hydrogenation. The modification observed on the Ga3d core level spectrum after hydrogenation shows that the chemical environnement of the Ga atoms is modified by hydrogen incorporation. To analyse the structure of these spectra, we decomposed them into a sum of gaussians. The Ga3d spectrum for a-GaAs was decomposed into a sum of two gaussians corresponding to Ga-O and Ga-As bonds contributions (figure 3). In the case of the Ga3d core level spectrum for a-GaAs:H, we used 3 gaussians (figure 4) to reproduce the experimental spectrum. The third gaussian can be attributed to Ga atoms bonded simultanously to H and O atoms. These XPS results show that H is present in the material and seems to be preferentially bonded to Ga atoms.

With the aim of trying to identify the absorbtion bands of the IRS spectra of a-GaAs:H we compared these spectra to those of Ga and As hydrides and to those obtained by other authors /2-4/ that worked on a-GaAs:H. The 840 cm^{-1} and 1000 cm^{-1} bonds can respectively be attributed to the As-H bending mode by comparison with As-H bending mode of AsH_4Br /5/ and to the As-H_2 bending mode identified by Wang /3/. The 2010 cm^{-1} peak is identified as As-H stretching mode by analogy to As-H_3 /5/. The shoulder at 530 cm^{-1} and the 1725 cm^{-1} band are respectively assigned to the Ga-H wagging and stretching modes according to Wang /3/ and to the spectra of $NaGaH_4$ /6/. The 1130 cm^{-1} peak has not been identified. Concerning the two broad bands at 530 cm^{-1} and 1430 cm^{-1}, Wang et al. /3/ argue that they arise from wagging and stretching vibrations of H atoms situated in bridging positions between two Ga atoms. This identification is supported by the theoritical work by O'Reilley and Robertson /7/ who showed that the Ga-H-Ga bridge is the most favorable binding mode for H in a-GaAs. Figure 6 shows the suggested hydrogen bonds in a-GaAs.

5. CONCLUSION

These experimental results show that in RF sputtered a-GaAs:H, hydrogen is preferentially bonded to Ga atoms, essentially in the form of Ga-H-Ga bridges. So, we think that not all the dangling bonds are saturated by hydrogen. This may explain the difficulty to realize a Schottky barrier using this material.

6. ACKNOWLEDGEMENTS

We would like to thank C. Senemaud for XPS measurements and J. Pourcin for IRS experiments.

7. REFERENCES

1. Report presented at the Hamburg Conference, Sept. 16-18, (1987).
2. D.K. Paul, J. Blake, S. Oguz and W. Paul, J. Non-Cryst. Solids, 35-36, 501, (1980).
3. Z.P. Wang, L. Ley and M. Cardona, Phys. Rev., B6 (1982).
4. B. Kussel, Ph. D. Thesis, Marbourg (G.F.R.) (1984).
5. A. Heinemann, B. Bunsenges, Phys. Chem., 68, 287 (1968).
6. A.P. Kurbakova, L.A. Leites, V.V. Garrilenko, X. Yu, N. Karuksin and L. Zakharkin, Spectrochm. Acta, 31A, 281 (1975).
7. E.P. O'Reilly and J. Robertson, Phys. Rev. B34, 8684 (1986).

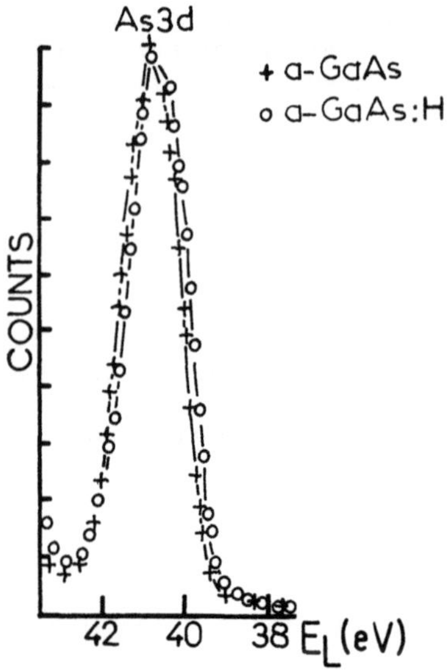

Fig.1. As3d core level spectra for a-GaAs and a-GaAs:H.

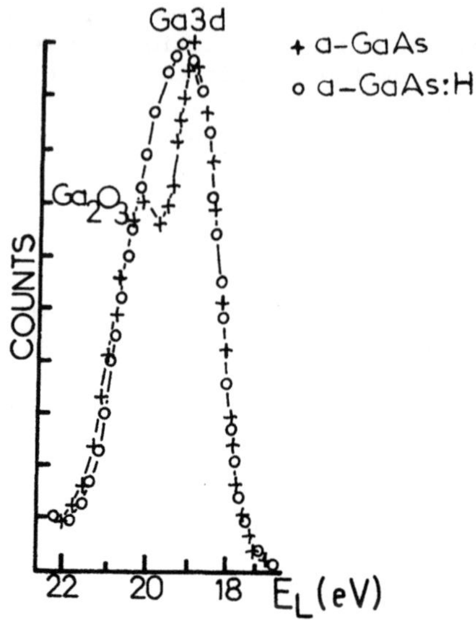

Fig.2. Ga3d core level spectra for a-GaAs and a-GaAs:H.

Fig.3. Ga3d core level spectra for a-GaAs decomposed into 2 gaussians.

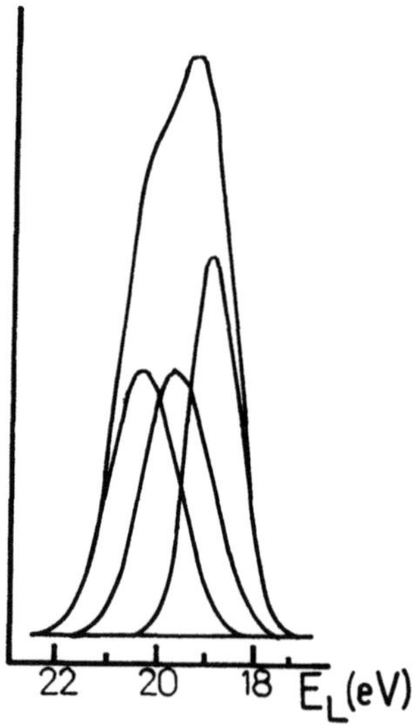

Fig.4. Ga3d core level spectra for a-GaAs:H decomposed into 3 gaussians.

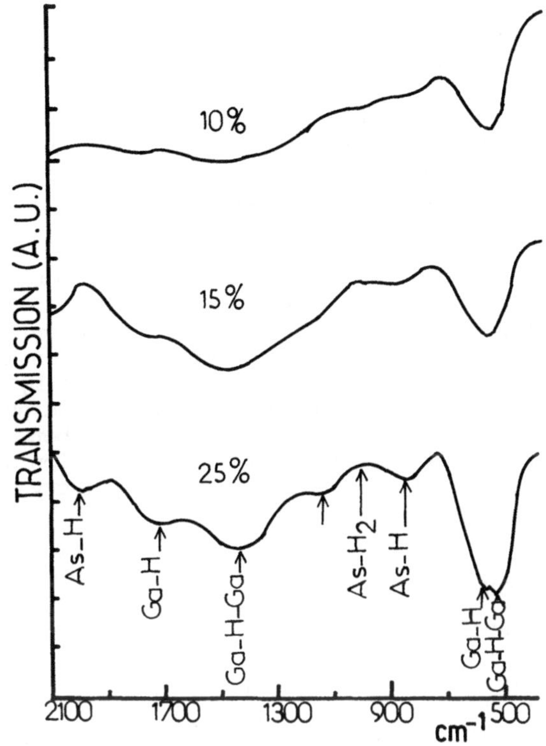

Fig.5. IR Transmission spectra for a-GaAs:H with different hydrogen partial pressures.

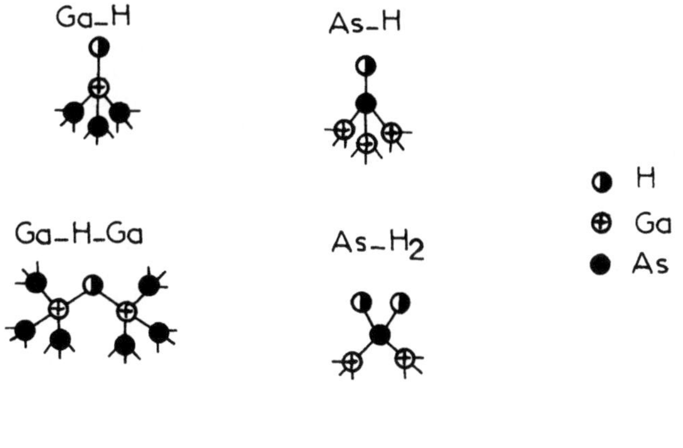

Fig.6. Suggested H bonding configurations in a-GaAs:H.

ELECTRICAL AND OPTICAL PROPERTIES OF TIN OXIDE LAYERS PREPARED BY PHYSICAL VAPOR DEPOSITION

V.Schlosser, G.Wind
Ludwig Boltzmann Institut für Festkörperphysik and
Institut für Festkörperphysik,University of Vienna
A-1060 Vienna, Kopernikusgasse 15 (AUSTRIA).

ABSTRACT. We have prepared amorphous 90nm thick tin dioxide layers by means of a physical vapor deposition method. Elemental tin is evaporated in an oxygen atmosphere onto about 300°C hot substrates cut from borosilicate glass and silicon. Free carrier concentrations as high as $3x10^{20}cm^{-3}$ are obtained by vacancy doping which was done by properly adjusting oxygen partial pressure, tin partial pressure and subsequent annealing conditions.Hall mobility of our samples varies between $5cm^2/Vsec$ and $13\ cm^2/Vsec$. Transmission spectra show that absorption due to the band gap transition arises about 0.5eV above the one in polycrystalline tin dioxide layers.

1. INTRODUCTION

Tin oxide is widely used for the preparation of conducting transparent electrodes. The content of this work is a study of electrical and optical properties of thin tin dioxide layers prepared by a reactive physical vapor deposition method using high purity tin and semiconductor grade oxygen. Amorphous and polycrystalline thin films can be prepared by this deposition technique. The influence of deposition conditions on free carrier concentration and mobility was examined. Although the sheet resistance of amorphous vacany doped tin oxide layers does not yet exceed one third of published values for spray-deposited polycrystalline films the enhanced transparency of a-SiO_2 in the ultraviolet wavelength range shows the potential for new applications.

2. EXPERIMENTAL

A modified Univex 300 vacuum unit from Leybold Hereaus equipped for high-vacuum conditions was used for the preparation of tin oxide layers. High vacuum of less then $1x10^{-6}$mbar was achieved by a turbo molecular pump. Partial pressure of argon, oxygen and nitrogen trifluoride can be adjusted and regulated by a motorized valve controlled from a Balzers RVG 040 control unit. Elemental tin was evaporated thermally using a tungsten boat or by a small electron gun using a graphite liner. Although there are no significant differences of optical and electrical properties of tinoxide prepared by using either thermal or electron gun evaporation the latter has been preferentially used since rate control and hence reproducibility is much better. A substrate heater allows to keep the substrate at elevated temperatures up to 400°C during deposition and post-treatment. A quartz crystal thin film monitor - Balzers QSG 201D - was used for rate control and thickness monitoring. In order to reduce residual water vapor the vacuum chamber and the turbo molecular pump can be preheated. Residual hydrocarbons was reduced by a glow discharge in oxygen prior to deposition. We used borosilicate glass (BSG) and pure silicon as substrates. For the preparation of transmission electron microscope studies NaCl crystals were used as substrates. BSG substrates were chemically cleaned prior to deposition. Silicon substrates were cleaned and oxidized in hot HNO_3 prior to deposition. Two electrodes pasted onto the substrates with a silver epoxy compound are used to monitor sheet resistance in-situ during deposition. After the glow discharge in oxygen for about 10 minutes - $p_{O2} \sim 10^{-2}$mbar - the deposition chamber was evacuated to its final vacuum. Then the partial pressure of oxygen was adjusted. It ranged between 1 μbar and 4.5 μbar. We used high-purity oxygen intended for semiconductor devices manufacturing. In some experiments up to four per cent nitrogen trifluoride was added to the oxygen in order to dope our tin oxide layers with fluorine. The substrate was heated to a temperature between 270 °C and 400 °C. Tin was evaporated with a partial pressure high enough to obtain growth rates between 10nm/min and 100nm/min of tin oxide on the substrate. The mean free path of evaporated tin particles in the oxygen atmosphere was about one half to one third of the distance between evaporation source and substrate. Since layers as-deposited exhibit only poor electrical conductivity a post-treatment in oxygen - $p_{O2} \sim 10$ mbar - for 1 min to 100 min at elevated substrate temperatures was carried out. In order to saturate dangling bonds in some experiments the layers were further annealed in an atmosphere containing water vapor, HCl-vapor or NF_3. Hall mobility and resistivity was determined using standard Van der Pauw geometries. Optical transmission and reflection spectra were measured in the range between 190nm and 40 μm.

3. RESULTS AND DISCUSSION

The dependence of the formation of tin oxide layers and the final free carrier concentration on the oxygen partial pressure during deposition is shown in fig.1. Below 1.5 μbar the resulting layer is composed mainly of tin monoxide and keeps its morphology and optical and electrical properties even after a post-anneal in oxygen. Since optical and electrical properties are less favorable for practical applications than the ones from tin dioxide only few experiments have been carried out using this range of partial oxygen pressure. Deposition of tin with a partial pressure of oxygen greater than 2.5 μbar and substrate temperature of more than 280 °C was always leading to crystalline tin dioxide. TEM studies show that crystals of about 30nm in diameter are formed during deposition. X-ray diffraction analyses show that reflexes correspond to the rutile lattice structure of tin dioxide with lattice constants a = 3.19 Å and c = 4.74 Å. Grain size and morphology appears to be nearly independent of oxygen partial pressure, substrate temperature in the range 280 °C to 400 °C and surface preparation. However, the as-deposited layers in the thickness range from 90nm to 300nm exhibit extremely high sheet resistance. At an oxygen partial pressure of 1.5 to 2.0 μbar tin dioxide was deposited in an amorphous state. The sheet resistance after deposition remains still very high. In the case of polycrystalline and amorphous deposition a post-treatment in oxygen - $p_{O2} \sim 10$ mbar - was leading to a reduction of resistivity to below 1×10^{-2} Ωcm in the case of c-SnO_2 and to 2.5×10^{-3} Ωcm in the case of a-SnO_2. The substrate temperature during the post-treatment was similar or identical to the temperature during deposition. Annealing time varies between 1 min and 100 min. The resistance during the post-annealing changes rapidly once the oxygen pressure is raised to 10 mbar. After several minutes it reaches its minimum value and starts to increase slowly again. Although this behavior is not fully explored we suggest that the decrease in resistivity during the first is due to a reconstruction of the surface thus leading to an improved overall mobility. No changes in morphology has been observed after a short anneal in oxygen. Furthermore there is no significant difference in the time intervall from the end of the deposition until the in-situ resistance reaches its lowest value for thin and for thick layers. The later increase of the in-situ resistance during the post heat treatment is probably due to the diffusion of oxygen into the tin oxide layers and saturating vacancies inside the layer which is leading to a reduction of free carrier concentration. For long annealing times it was observed that a-SnO_2 recrystallizes. This behavior appears to be substrate dependend since simultaneous deposition of a-SnO_2 onto silicon and BSG was leading to c-SnO_2 after an anneal for 25 min in the case of silicon only. The ratio of the mobilities obtained from Hall measurements is about a factor of two - 12.9 cm^2/Vsec for c-SnO_2 and 6.2 cm^2/Vsec for a-SnO_2. The free carrier concentration determined for the a-SnO_2 layer on BSG is 2.9×10^{20} cm^{-3} whereas the carrier concentration is 7×10^{19} cm^{-3} in the c-SnO_2 layer grown on silicon. Results of free carrier concentration and mobility are displayed in fig.2 for crystalline and amorphous tin dioxide layers. The observed difference between the use of dry and wet oxygen is not yet understood. In general it can be said that wet oxygen is leading to lower carrier concentration but higher mobilities than the deposition in dry oxygen. Furthermore samples prepared by the use of wet oxygen are less sensitive to the exposure to air than layers grown in dry oxygen. The attempt to dope our a-SnO_2 layers with F,Cl or Sb did not result in an increased carrier concentration. Therefore we conclude that the generation of free electrons in our samples is exclusively due to vacancies in the tin dioxide layers. A comparison of transmission spectra measured on 90nm thick tin dioxide layers with different composition and crystallinity is shown in fig.3. Tin monoxide starts to absorb below 3eV which is in excellent agreement with other authors [1]. The fundamental absorption of tin dioxide starts at photon energies below 4eV. For comparison a 240nm thick layer of c-SnO_2:F prepared by a German research laboratory by a pyrolytic spray-deposition method is also shown in fig.3 (dashed line). Absorption in a-SnO_2 starts about 0.5eV above that one in SnO_2. The coincidence of interference minima and maxima of c-SnO_2 and a-SnO_2 suggests that the index of refraction does not differ very much between fine polycrystalline and amorphous tin dioxide layers. It shall be remembered that thickness control during deposition is done with a crystal quartz monitor assuming a density of 6.95 gcm^{-3} in both cases. Thus we conclude that the transition from polycrystalline to amorphous SnO_2 observed in the optical experiments is less abrupt than it appears from TEM and X-ray diffraction analysis. Moreover we have observed transmission spectra of 90nm layers of tin dioxide with intermediate oxygen content which are situated in between the two displayed absorption edges (squares and triangles in fig.3).

4. CONCLUSIONS

From our results we conclude that amorphous tin dioxide layers can be prepared by adjusting the partial pressure of oxygen during the deposition of tin onto a heated substrate. The transparency of a-SnO_2 in the ultraviolet wavelength region is clearly enhanced compared to polycrystalline tin oxide layers thus suggesting that the bandgap is about 0.5eV greater than in crystalline SnO_2. Since adding of dopants does not yield an increase of free carriers vacancies are responsible for the generation of electrons in our layers. Free carrier concentrations as high as $3x10^{20}$ cm^{-3} have been observed. Further work must be done to understand the dominant mechanism for the limitation of the electron mobility in amorphous tin oxide. However, we have shown that mobilities as high as one third of the values reported for the best spray-deposited crystalline samples are already obtainable.

5. ACKNOWLEDGEMENT

We thank Dr. Pongratz from the Technical University of Vienna for carrying out transmission electron microscope and X-ray diffraction studies on our samples and for valuable discussions.

6. REFERENCES

1. J.C. Manifacier, M. DeMurcia, J.P. Fillard, and L. Vicario, Thin solid films 41, (1977) 127.

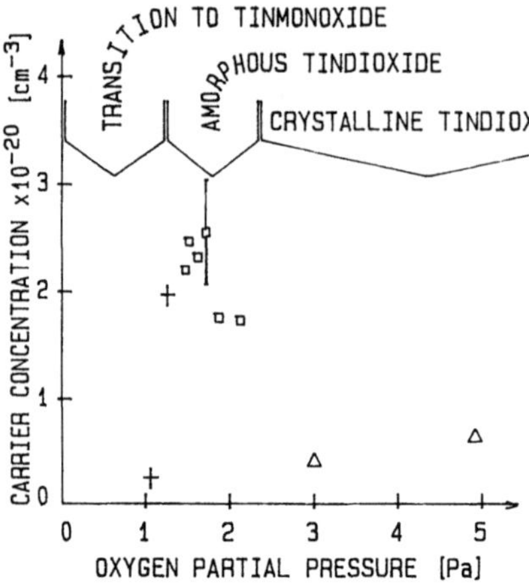

Fig.1: Dependence of the formation of tinoxide on the partial pressure of Oxygen during reactive deposition of tin. + .. SnO_X (1<X<2), □ .. $a\text{-}SnO_2$, △ .. $c\text{-}SnO_2$

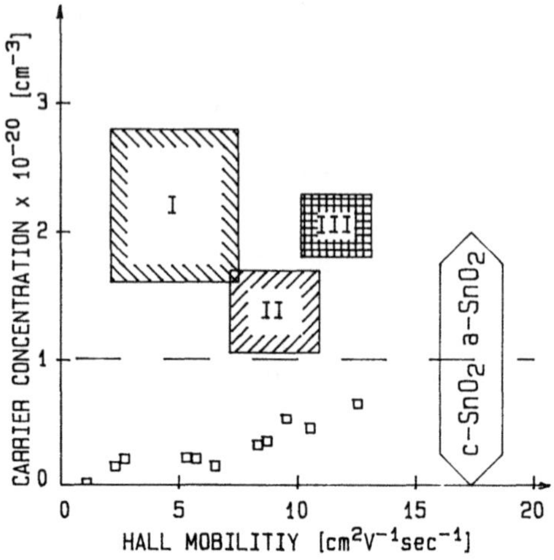

Fig.2: Dependence of free carrier concentration and mobility on different preparation conditions. □ .. crystalline SnO_2,
I .. $a\text{-}SnO_2$ annealed in dry Oxygen, t<10min,
II .. $a\text{-}SnO_2$ annealed in wet Oxygen, t>10min,
III same as II with subsequent passivation.

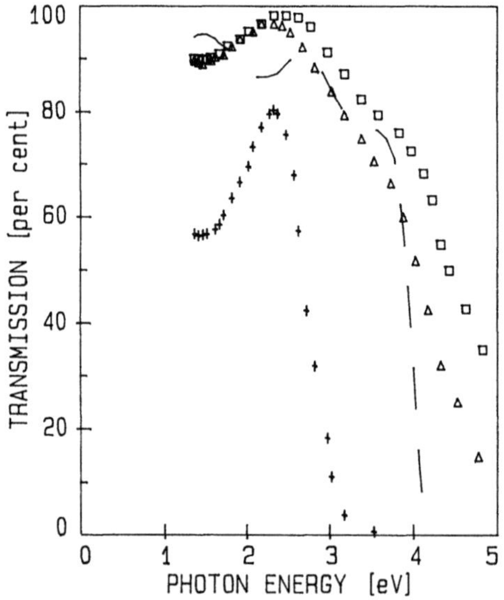

Fig.3: Transmission spectra of tinoxide layers on borosilicate-glass. □ .. 90nm $a\text{-}SnO_2$, △ .. 90nm $c\text{-}SnO_2$ + .. 90nm SnO_X (1<X<2), dashed line: 240nm $c\text{-}SnO_2$ delivered from a foreign research laboratory.

SCREEN PRINTED "ITO" THIN FILMS FOR SOLAR CONVERSION USE

B. REZIG, B. BESSAIS*

E.N.I.T. Département d'Energétique
BP 37 Le Belvedere - Tunis

TUNISIE

*I.N.R.S.T. Centre de Physique et d'Enertétique
Soliman, BP 95, 2050 Hammam-Lif

TUNISIE.

ABSTRACT

Thin transparent and conductive "ITO" layers ($In_2O_3(Sn)$) planed to be collector electrodes or barrieres in PV heterostructures have been made by screen printing technique (S.P.T.). In a first step, the mastership of the fabrication process leading to the control of the reproductibility and the stability of the thin film properties had been our main goal. In this framework, we obtained crystallized layers with high transmittance factor (90%). They showed, in a reproducible way, the expected two major lines associated to <222> and <400> hkl planes of the centered $In_2O_3(Sn)$ cubic structure. Sheet resistance controled by four probe points method had been lowered by heat treatment procedure from 5 to 10 k$\Omega/\square$ down to 500 $\Omega/\square$. In this paper we show, on one hand, how fabrication conditions affect the structural, electrical and optical properties of the as-made layers. On the other hand, why the stability of the sheet-resistance and its sensitivity to some heat treatments could be interpreted in terms of adsorption - desorption phenomena, which understanding should help to perform other efficient treatments which improve the quality factor of the film.

I. INTRODUCTION

An increasing interest had been recently focused on transparent and conductive thin oxide films, such as ZnO, SnO_2, $In_2O_3(Sn)$(ITO).... etc. These wide gap materials are considered indispensable for collection electrodes in liquid crystal displayers, backwall contacts, antireflection and barrier coatings in solar cells. They have been also used as heat mirrors in thermal isolation technology.

The most used is ITO, owing to its degenerate character and high transmittance factor relatively easily obtained by mean of some economical techniques such as Dipping [1], Spray [2,3], and more recently screen printing. Unfortunately, the reported electrooptical performances (essentially the quality factor Q ; Q = Transmittance : T%/Sheet resistance : $R_\square$) and stability of ITO layers seem to be very sensitive to the fabrication technique and conditions, and often to the layer biography itself.

In this paper, we describe the screen printing fabrication process and the as-made layer characteristics governed in their whole behaviour by the growth parameters.

We show, also, how heat treatments increase in a significant amount the quality factor Q. Stability problems are, finally, underlined.

II. FABRICATION PROCESS

II.1. Printing

The starting paste was an ESL true solution of two tin and indium based organometalic compounds (ESL # 3050) printed on a semi-industrial PRESCO machine. 325 mesh stainless steel screens are used allowing as-printed thicknesses laying in the range of 10 - 30 µm deposited on pyrex squares (20 x 20 mm2) - (fig. n°1).

II.2. Drying and firing

First, to facilitate the sample handling, layers are heated during 10 to 15 mn in dry air flux at 125°C.
Second, selected layers are submitted to an IR firing process which temperature profile is determinant to obtain good adherence and optical homogeneity, (fig. n°2). The crystallization temperature is hold at 600°C and the duration effects explored 10 by 10 mn in the 10 - 60 mn range.

One can notice, on one hand, that 400-500°C is a boiling-burning range likely related to the binder and other organic additives. The controle of this transition region determines the final mechanical aspect of the layer. On the other hand, the role of air in the crystallization process is to adjust the stoichiochiometry of the oxide (pure Nitrogen flux gave, as expected, black and bright layers).

Finally, the reflected color surface distribution is a fast and reliable diagnostic of the thickness homogeneity of the ITO film.

III. RESULTS AND DISCUSSION

III.1. As made layers

Fig. n°3 and Table (I) give the observed ITO lines and their associated reticular distances and planes. As made films are well crystallized and show that ITO is the dominant phase.

Table I : Comparison between X-Ray In_2O_3 powder (ASTM) and observed ITO film fired at 600°C during 45 mn.

hkl	d_{hkl}(Å) (ASTM)	d_{jkl}(Å) obs	I_{hkl} (ASTM)	I_{hkl} obs	a_{hkl}(Å) (ASTM)	a_{hkl}(Å) obs
222	2.921	2.928	100 %	100 %	10.118	10.14
321	2.704	2.811	2 %	44.7 %	10.118	10.51
400	2.529	2.540	30 %	28.3 %	10.117	10.16
440	1.788	1.789	35 %	22.38%	10.118	10.12
622	1.525	1.526	25 %	13.4 %	10.118	10.09

The preferred orientation (<222>) and the good agreement with standard ASTM data proved that the effective Sn/In ratio inside the grown layer is kept at a reasonable low level. Nevertheless, one can notice that the <321> line intensity seems to be highly sensitive to the firing time (t_f), (Fig.n°4). Also the mean value of the lattice constant (a = 10.2Å) is near the In_2O_3 crystalline parameter (a = 10.118Å) and the reported sprayed film values (10.0-10.05Å) [2,4], and is not seriously affected by the fabrication process detailed above.

High transmittance coefficient (≥ 90 %) is obtained with all films (Fig.n°5). Some of them show interferential fringes allowing the estimation of the thickness and the optical index (Table II). Contrast amplitude (correlated to optical index) [5,6] observed on transmittance curves could be a good indication of the volume homogeneity and surface planeity.

Table II : Optical index and transmittance (λ = 0.5 μm) dependance on firing time.

t_f(mn)	d_{opt}(μm)	n(λ = 0.5 μm)	T%(λ = 0.5 μm)	$R_\square$(K Ω/□)
30'	0.39	1.706	89.6	8
45'	0.45	1.704	93	5
80'	0.38	1.705	93.2	6.5

Short firing time produced resistive layers. Beyond 15 mn, the scattered $R_\square$ values stabilized around 5 kΩ(Fig.n°6). This rather high level could be explained by three principal reason :

i- Crystallization is made in confined air and not in dynamic flux where growth could be better involved.

ii- Cooling profile is not yet optimized. Indeed this is of a high importance because antogonism exists between low rate which avoids cracks and other damages due to thermal stresses and fast rate acting as a tempering factor which limits oxygen contamination and consequently controles layer resistivity.

iii- Ageing effect of the starting paste.

III.2. Heat treatments and stability

(Figure n°7) shows that $R_\square$ had been lowered by a significant amount (down to 500 Ω) by mean of heat treatments in air flux at 220°C during 1 hour. Longuer duration produced the reverse effect. This quite complex behaviour is compared to the reversible increasing-decreasing sheet resistance observed under various storage conditions and thought to be basically related to oxygen adsorption-desorption phenomena. Deeper investigations are being conducted in dynamic vacuum and moisture conditions. First results confirm the strong effect noticed above [7].

IV. Conclusion

Crystallized and transparent ITO thin films (0.4 - 0.5 μm) had been prepared by SPT.
Electrical conductivity of as-made and treated layers is strongly influenced

by environmental conditions through adsorption-desorption phenomena. This offers a very rich and useful opportunity to study the kinetic of these mechanisms in order to adjust the final sheet resistance at desired low values.

Acknowledgements

The authors would like to thank Prs R.BENNACEUR and M.MAALEJ, to facilitate our work conditions and for their encouragement.

REFERENCES

[1] H.Dislich and E.Hussmann, thin solid films, 77 (1981) 129.

[2] OP.Agnihotri, A.K. Sharma, BK.Gupta and R.Thangaraj, J. Phys. D. Appl. Phys. vol.II (1978).

[3] K.L. Chopra, R.C. Kainthla, D.K. Pandya and A.P.Thakour, Phys. Thin. Films 12 (1982) 167.

[4] R.Pommier. Thèse de 3ème cycle U.S.T.L. Montpellier (1981).

[5] J.C. Manifacier, J.Gasiot, J.P.Fillard, J. Phys. D. 1002 (1976).

[6] J. Szczyrbowsky J. Phys. D : appl. Phys., Vol.11 (1978)

[7] The authors, to be published.

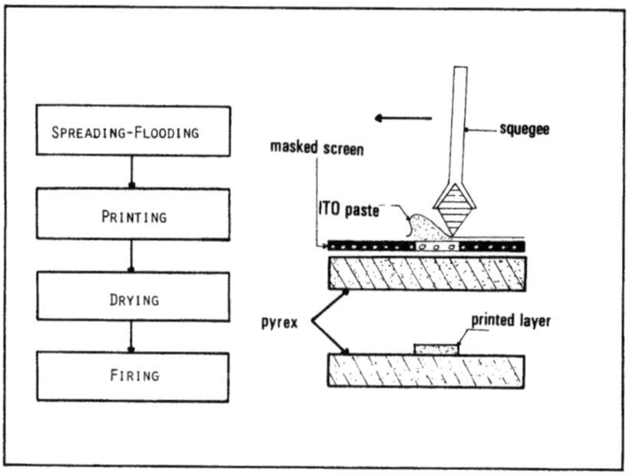

FIG N°1: SCREEN PRINTING PROCESS.

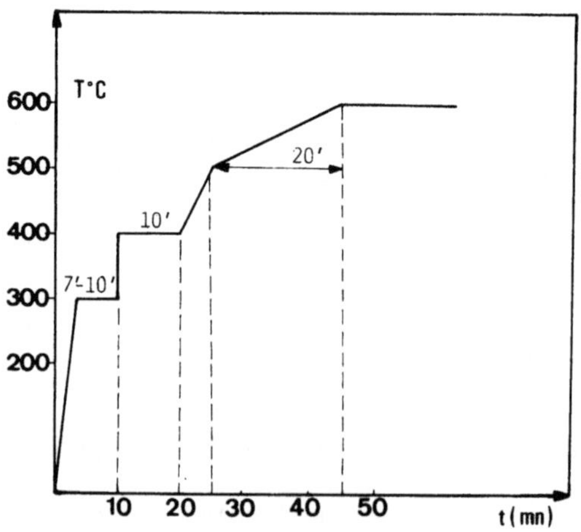

FIG N°2: FIRING PROFILE

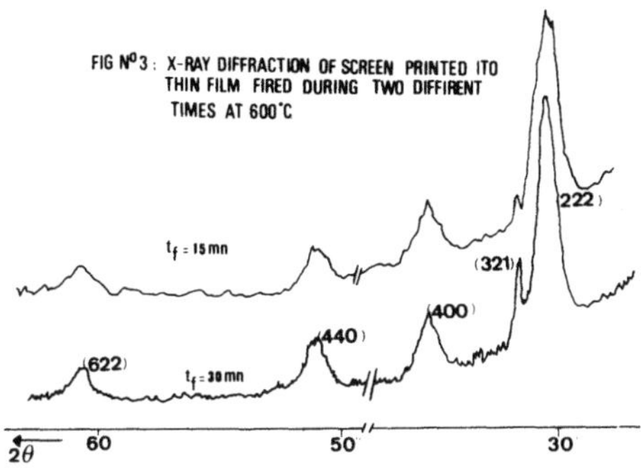

FIG N°3: X-RAY DIFFRACTION OF SCREEN PRINTED ITO THIN FILM FIRED DURING TWO DIFFIRENT TIMES AT 600°C

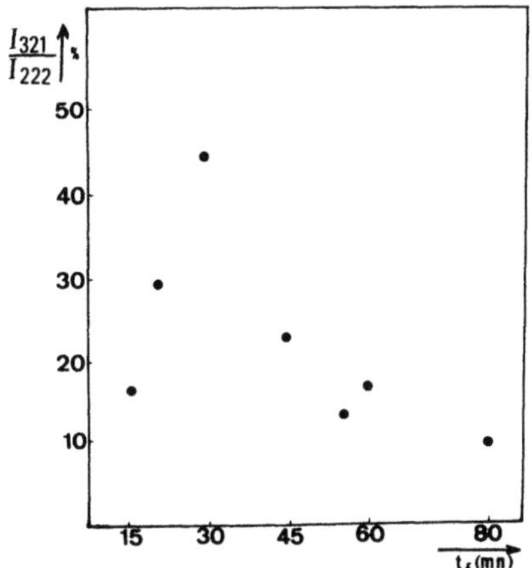

FIG N°4: (321) LINE INTENSITY EVOLUTION VERSUS FIRING TIME

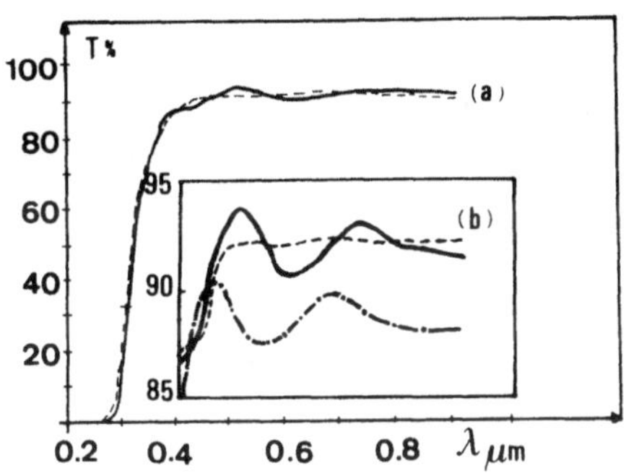

FIG N°5: TRANSMISSION SPECTRA OF ITO FILM FIRED AT 600°C (a)

(b) INTERFERENTIAL FRINGES DETAILS

FIRING TIMES 30mn (-·-·), 45mn (—), 60 mn (------)

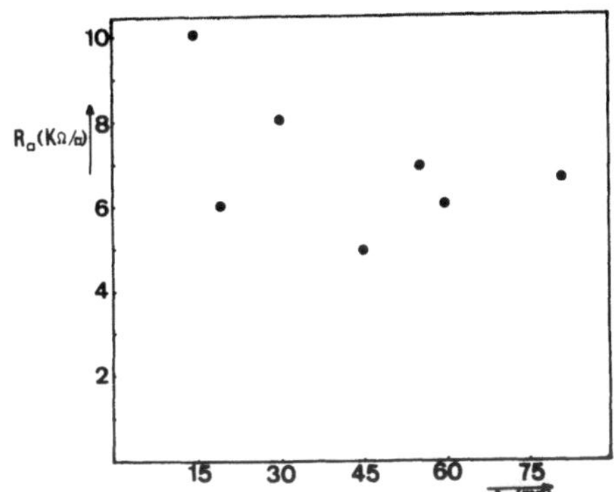

FIG N°6: SHEET RESISTANCE VERSUS FIRING TIME (for samples having the same deposited thickness)

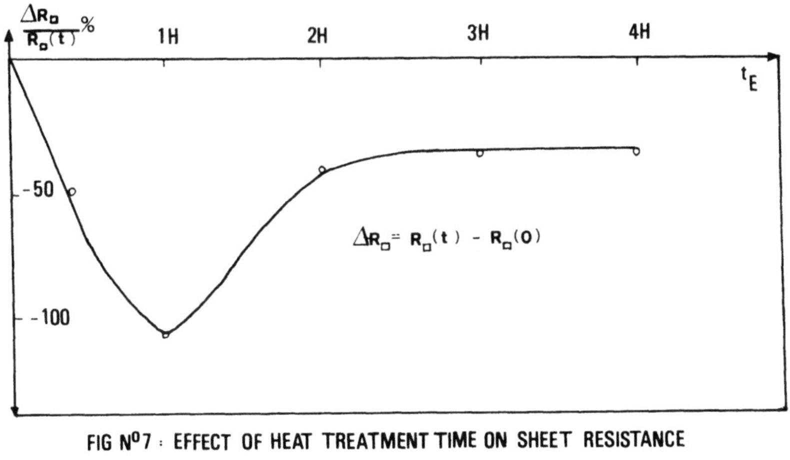

FIG N°7: EFFECT OF HEAT TREATMENT TIME ON SHEET RESISTANCE

AMORPHOUS SILICON PIN SOLAR CELLS USING INDIUM DOPED ZnO TRANSPARENT CONDUCTIVE COATINGS

A. Sánchez, A. Ortiz, E. Sansores, J. Campos and P. K. Nair
Solar Energy Laboratory IIM-UNAM
Apartado Postal 34, 62580 Temixco, Morelos, MEXICO
and
C. Falcony
Departamento de Fisica, CINVESTAV-IPN
Apartado Postal 14-740, 07000 México, D.F., MEXICO

ABSTRACT.

Indium doped zinc oxide (ZnO:In) films deposited by spray pyrolysis technique on corning 7059 glass substrate and possessing and average optical transmission of 80 to 90% and sheet resistance between 16 a 90 ohm/sq were used as a transparent conductive coating in a single junction p-i-n a-Si solar cell. Aluminium back contacts of 1mm diameter were deposited on the n-layer by thermal evaporation. The cell structure is glass/ZnO:In/pin/Al. The results show very good rectification behavior in dark. Under 690 W/m^2 simulated solar illumination through the glass, conversion efficiency of 4.5% was obtained. No significant change in the fill factor as a function of time was observed.

INTRODUCTION

In recent years, hydrogenated amorphous silicon, a-Si:H, and some of its alloys have attracted much attention because of their potential applications in thin film devices like solar cells [1], electroluminescent diodes [2], etc. However, the major effort has been concentrated on its application in solar cells. Generally, amorphous silicon solar cells are prepared by the plasma-enhanced chemical vapor deposition (PECVD) technique in a single chamber or multichamber system [3,4] and the a-Si:H film is grown on a glass substrate or a metallic substrate.

Irrespective of the substrate, the amorphous silicon solar cells require a transparent conductive coating. Usually a glass substrate coated with a layer of SnO_2, ITO (Indium Tin Oxide), or both are used [5,6] Nevertheless, studies by Auger electron spectroscopy and other techniques [6,7] on the interfaces of these material with a-Si:H prepared by PECVD have shown, in the case of ITO, the migration of In into the a-Si:H film. When both coatings are used, diffusion of Sn and In can still be seen. Zinc oxide transparent conductive films have been proposed as an alternative material instead of SnO_2 or ITO [8] The optical and electrical properties of the indium-doped zinc oxide (ZnO:In) films prepared by spray pyrolysis technique [9] make them good candidates for their use in opto-electronic devices as transparent conducting contacts. Auger studies have shown that in this case, there is no migration of the coating constituents into the a-Si:H prepared by PECVD [9]. The lack of chemical interaction therefore suggests that they can be used in a-Si:H solar cells. In this work we report results of p-i-n junction a-Si solar cells made in a single chamber using ZnO:In as the transparent conductive film (TCF).

SAMPLE PREPARATION AND MEASUREMENT

The ZnO:In TCF was prepared by the spray pyrolysis technique on corning 7059 glass substrates. In this technique a jet of small drops of 0.1M $Zn(CH_3COO)_2$ solution in isopropyl alcohol plus one part of deionized water, is driven by a carrier gas towards the heated substrate surface where the reaction occurs. The apparatus and details of the technique have been given elsewhere [9]. For the present work the substrate temperature was kept at 380°C. The gas and solution flow rates were 10 l/min and 20 cm^3/min respectively and the deposition time was 10 min. The dopping was carried out by adding 2-4 at% of $InCl_3$ to the spraying solution.

The a-Si:H pin junction was grown onto the ZnO:In films by the PECVD technique in a capacitive radio frequency single chamber system [10]. The total flow rate, chamber pressure, substrate temperature, rf power were kept constant at 50 sccm, 450 mtorr, 250°C and 5 W respectively, for all three layers. The intrinsic layer was deposited using pure silane, and the p and n layers were deposited with the addition of 1% of diborane or phosphine diluted in silane.

Optical transmission and reflectivity were measured, with references to air and aluminum respectively with a double beam SHIMADZU UV-365 sprectrophometer. Resistivity measurements were made by the four probe technique. The optical band gap, activation energy, and dark and light conductivity of the a-Si:H films were calculated by routine techniques. Efficiency measurements were carried out with a solar simulator, Oriel model 6722, at 690 W/m^2 intensity. The photocurrent response and J-V characteristics were measured with a computerised measuring system using an HP 4140B pA meter/ d.c. voltage source and an HP 21PC [11].

RESULTS AND DISCUSSION

In figure 1-a, we show the optical transmission curve of a typical ZnO:In (TCF) using a dopping concentration of 4% $InCl_3$. It could be observed that the average transmission in the interval 500-600 nm is 85%. The reflectivity of this film when it is illuminated through the glass is shown in figure 1-b, where the average value at the above interval is within 14% to 18%. Using the interference peaks occuring in the optical transmission curve and a value of 2.45 for the refractive index, the TCF thickness is found to be 0.4 μm. The sheet resistance for this is 90 ohm/sq. However under an optimun dopping concentration of 2 at% of $InCl_3$, a sheet resistance of 16 ohm/sq with nearly the same optical transmission as in figure 1-a was obtained. Thus the optical and electrical properties for the present TCF are acceptable for transparent conductive coating applications in solar cells [9].

The properties of the a-Si material obtained in our system are shown in table 1. We can see that the opto-electronic properties of the semiconducting layers are excellent for solar cells applications. Figure 2 shows the photocurrent response of an intrinsic a-Si:H layer under 1 kW/m^2 illumination versus time, provided by a tungsten-halogen lamp. It could be seen that the change in conductivity upon illumination is of five orders of magnitude. The fast decay of the photocurrent when the light is turned off is also notable. SEM pictures have shown that these films are devoid of any microstructural defects.

The device were fabricated simultaneously in the configuration glass/ZnO:In/p-i-n/Al employing both the 90 ohm/sq (TCF1) and 16 ohm/sq (TFC2). The quality of the p-i-n junctions were investigated by dark J vs V characteristics. The J vs V characteristics of the defect-free diodes are shown in figure 3. We note that they show very good rectification, $>10^5$ at 0.8 V, diode quality factor of n=1.9 and n=1.7 and a reverse saturation current Jo= 2×10^{-8} A/cm^2 and Jo= 8×10^{-9} A/cm^2 for the TCF1 and TFC2 respectively. The values for n and Jo are not optimum. However, the sheet resistance of the substrate (90 and 16 ohm/sq)is not seen to make significant change in these values. Hence we believe that the recombination problem due to the boron contamination of the i-layer is responsible for the inferior n and Jo values. It is known that the residual boro dopping of the i-layer creates extra defect centers and consequently result in an increase in the density of gap states [12]

Figure 4 shows the typical J vs V curves for the cells using TCF1 and TCF2 under 690 W/m^2 simulated solar radiation. The following characteristics are observed:

TCF1: Voc= 0.76 V, Isc= 5.6 mA/cm^2, FF= 0.61, η = 3.3%.

TCF2: Voc= 0.77 V, Isc= 6.9 mA/cm^2, FF= 0.65, η = 4.5%

We believe that the Voc is not optimum because of the lack of optimization of the p,i and n layer thicknesses -these had less than the ideal values [6]. The small values of the Jsc may be due to the recombination currents in the i-layer, resistive losses in the TCF, or due to thin i and n layers [13]. But since the fill factor is good,we consider that the incomplete light absorption due to a thin i-layer is mainly responsible for the reduction in Jsc.

The effect of storage of the cells under laboratory ambient on the cell parameters have been observed over a period of four months. No appreciable change was seen in Voc. The change in FF and Jsc were within 10%. We must mention that the cells were not encapsulated and hence the observed stability in the cell parameters is very good and are indicative of the stability of the TCF/a-Si:H interface as well.

CONCLUSIONS

We have reported in this paper the fabrication of pin a-Si:H solar cells on spray pyrolitically coated ZnO:In films. The diode characteristics as well as the cell parameter are not the best reported in the literature for pin a-Si:H solar cells, but this is due to the lack of optimization of the thickness of the p,i and n layers. Our observation show that the ZnO:In as a transparent conductive coating for the pin amorphous silicon solar cell fabrication overcomes most of the contamination problems reported for the SnO_2 and ITO transparent coatings.

ACKNOWLEDGEMENTS

The authors are grateful to Consejo Nacional de Ciencia y Tecnologia (CONACyT) and the Organization of American States (OAS) for financial support for the work. We also acknowledge D. Morales for his drawing assistance.

REFERENCES

[1] D.E. Carlson and C.R. Wronski, Appl, Phys. Lett. 28 (1976) 671.
[2] D. Kruangan, M. Deguchi, Y. Hattori, T. Toyama, H. Okamoto, and Y. Hamakawa, Spring Meeting Mat. Res. Soc., April 21-25, 1987, Anaheim, CA, USA.
[3] Y. Kuwano, M. Ohnishi, S. Tsuda, Y. Nakashima, and N. Nakamura, Jpn. J. Appl. Phys. 21 (1982) 413.
[4] A. Madan, Proccedings 1987 Amorphous Silicon Subcontractors Review Meeting, 1987, Palo Alto, CA, USA.
[5] J. I. Pankove, ed., Semiconductors and Semimetals, Vol. 21, Hydrogenated Amorphous Silicon, part D, Device Applications (Academic Press, New York, 1984).
[6] K. Takahashi and M. Konagai, Amorphous Silicon Solar Cells (North Oxford Academic Publishers Ltd.,1986).
[7] F. Sanchez-Sinecio and R. Williams, J. Appl. Phys. 54 (1983),2757.
[8] M. Shimizu, Y. Matsueda, T. Shiosaki, and A. Kawabata, J. Cryst. Growth 71 (1985) 209.
[9] A. Ortiz, A. Sanchez, C. Falcony, M. H. Farias, G. A. Hirata, and L. Cota-Araiza, J. Non Cryst. Solids (in press 1988).
[10] A. Sanchez and E. Sansores, Proccedings IX Reunion Nacional de Energia Solar, October 2-4, 1985, Merida, Yuc., Mexico, (Publisher,ANES) 273.
[11] J. Campos, E. Sansores, A. Sanchez, P. K. Nair, Proccedings X Reunion Nacional de Energia Solar, October 1-3, 1986, Guanajuato, Gto., Mexico, (Publisher,ANES) 27.
[12] Harris,A.J., R.S. Walker, and R. Sneddon, J. Appl. Phys. 51,(1980),4287.
[13] Y. Hamakawa, Amorphous Semiconductor Technologies and Devices, Ohmu-sha, 1982, p. 164.

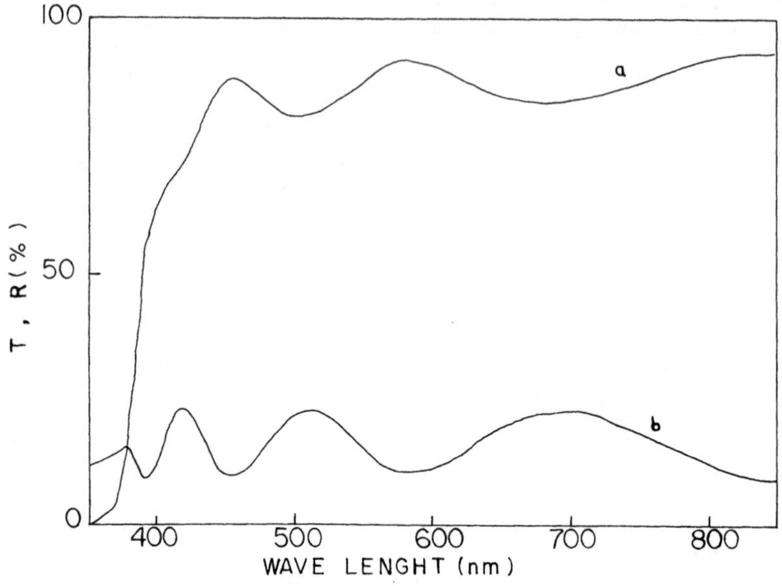

FIG. 1 (a) Optical transmission and (b) Optical reflection curves for Zn O: In transparent conductive coating

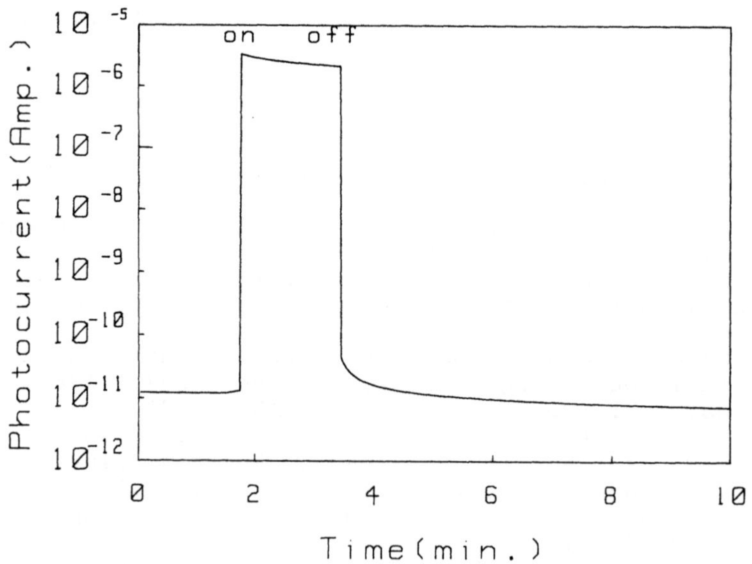

Figure 2. Dark and photocurrent response of a-Si:H intrinsic layer under 1KW/M^2

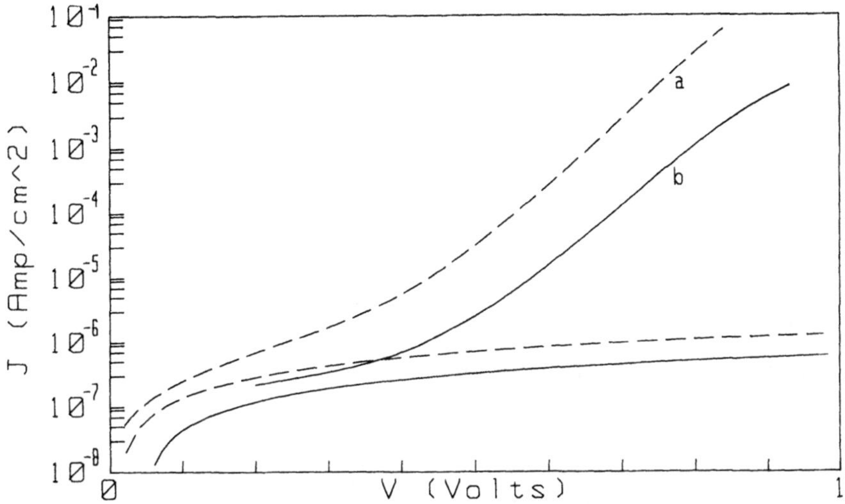

Fig. 3 Ln(J) vs V characteristics of a-Si p-i-n junctions with ZnO:In transparent conductive film. (a) TCF1 with sheet resistence of 90 ohm-sq. (b) TCF2 with sheet resistence of 16 ohm-sq.

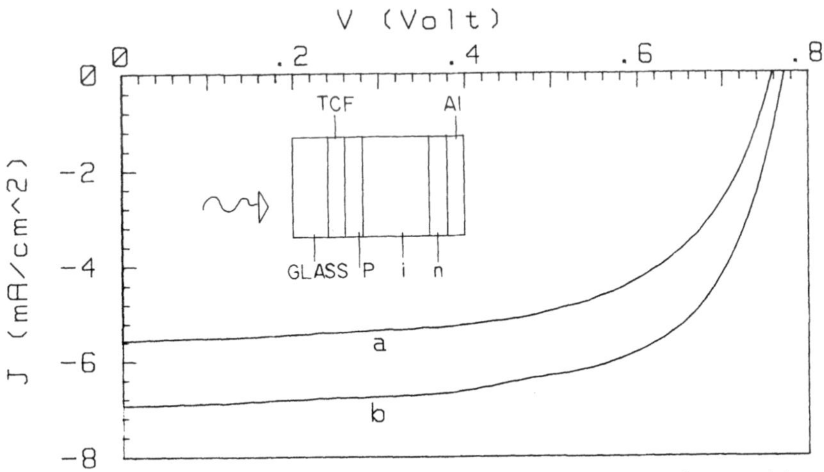

Fig. 4 J vs V characteristics and cell structure of p-i-n solar cells witn ZnO:In transparent conductive coating. (a) using TCF1 and (b) using TCF2.

Table 1. Optical and electrical properties of a-Si:H produced in a capacitively coupled PECVD with deposition rate of 3.0 Å/sec.

Layer a-Si:H	t (μm)	Conductivity light $(ohm\text{-}cm)^{-1}$	dark	Energy Activation (eV)	Optical Band Gap (eV)	Infrared Peak (cm^{-1})
Intrinsec	0.5	$2x10^{-5}$	$5x10^{-11}$	0.8-0.9	1.74-1.8	2000
n-type	0.2		$4x10^{-2}$	0.18	1.74	
p-type	0.2		$2x10^{-3}$	0.30	1.65	

LIGHT TRAPPING IN a-Si:H SOLAR CELLS : A COMPARISON OF VARIOUS TEXTURIZED TCO SUBSTRATES.

P. VILATO, O. MARIS, Saint-Gobain Recherche. 39, quai Lucien Lefranc, B.P. 135, F-93304 AUBERVILLIERS Cédex (FRANCE).

Abstract : Light scattering ITO/glass samples were prepared either by using scattering glass substrates (chemicaly etched or rolled glass) or by depositing on flat glass an ITO film containing a scattering component. Deposition of a-Si:H and Al layers was performed on these substrates. From reflectance spectra of these multilayered samples, we evaluated light trapping. Most performant sample was the etched substrate, but because of the roughness of the surface, high density of pin-holes was observed in solar cells. Rolled glass exhibited low light trapping and did not seem to be suitable for this application. Scattering ITO samples had intermediate performances. Further work is necessary on the late method to enhance reproducebility and to test solar cells made with these substrates.

1. INTRODUCTION

In solar cells, trapping light increases optical absorption [1]. This effect may be achieved by the texturization of the Transparent Conductive Oxide (TCO) front electrode. Texturization is obtained by (i) roughening the glass substrate [2, 3], (ii) depositing a scattering base-coat on glass before TCO deposition on glass (e.g. SiO_2 by dipping [4], (iii) depositing a scattering TCO layer (e.g. a SnO_2:F with large grains [5]. We have developed here approaches (i) and (iii). In case (i), we used chemically etched or rolled glass ; in case (iii), a scattering TCO film was deposited on flat glass. In both cases, TCO layer was a pyrolytic ITO thin film of ≈ 220 nm. SOLEMS used these different ITO/glass samples as substrates to deposit a-Si:H and subsequent Al film. In this paper, we mainly present spectrophotometric results on these multilayered samples.

2. EXPERIMENTAL

2.1 Texturized glass substrates

Chemical frosting experiments have been described elsewhere [3]. Briefly, etching with an ammonium bifluoride / hydrochloridric acid solution resulted in a pyramid-like structure. The pits were ≈ 10-20 μm diameter and 2-5 μm height. Characteristics are displayed in Table I. The reference was an unscattering ITO film on flat glass. Note that the haze ratio H (>80 %) was the largest of the samples studied here. The rolled glass substrate was 3 mm thick and both sides were rough. Mean pit diameter in surface was ≈ 200 μm. Haze (≈ 5%) was the lowest of those reported in Table I.

2.2 Scattering ITO thin film on flat glass

We used 1.3 mm thick extra-white glass substrates of 5 cm x 7 cm. A SiO_2 powder (Aerosil) with a mean single particle diameter of 12 nm was included in the pyrolysis powder. Relevant light scattering was seen for Aerosil concentration more than 1%. We present here results of 5 and 10%-Aerosil samples. At the surface, we have observed clusters (~10 μm) or small agregates (< 1 μm) (Fig. 1). Preliminary electron microprobe analysis suggested that these grains had similar composition to that of flat areas of the film. The samples presented here were the substrates with highest H in the 5 and 10% series. Unreproducebility was noticeable in the 10% series (H(550 nm) = 11-24%), probably due to the large difference between Aerosil and pyrolysis starting material densities. Therefore, 5% concentration was a compromise between acceptable H values (>10%) and reproducebility. Spectrophotometric measurements with liquids of refractive index 1.5 and 1.8 suggested that scattering was due to the grains at the surface.

Classical a-Si:H and Al depositions were performed on these scattering substrates and on the unscattering reference. Reflectance measurements in the 350-1000 nm range were carried out with these multilayered samples.

3. RESULTS AND DISCUSSION

Roughly, two ranges could be distinguished in the optical spectra (Fig.2, 3) : the 400-600 nm range with low variations of reflectance and the 600-800 nm range where oscillations dramatically increased. This was due to the relatively poor absorption of a-Si:H in the high-λ range. Then, interferences caused by the a-Si:H/ITO system appear to be predominant. Due to oscillations, mean parameters are better adapted than values at a given λ. In what follows, we will take as parameters: R_i (half-sum of the maximum and minimum reflectance) and Δ_i (relative decrease of reflectance of the considered sample with respect to the reference, i.e. [R_i (sample) - R_i (reference)]/R_i (reference)). Subscripts 1 and 2 stand for 400-600 nm and 600-800 nm ranges, respectively. Δ_i gave an evaluation of light trapping by the multilayered system. Results are summarized in Table II.

3.1 Texturized glass substrates

3.1.1 <u>Etched sample</u>. Reflectance dramatically decreased with respect to the reference (Fig.2), especially in the 600-800 nm range where Δ_2 was down to - 35 %. At SOLEMS, solar cells were made with some of these substrates. About 10% of the cells could be measured, the others presented too many pin-holes. At 2500 lux, in the average, J_{sc} = 190 μA/cm² and V_{oc} = 640 mV, while J_{sc} = 220 μA/cm² and V_{oc} = 668 mV for the reference. These bad performances of the etched samples could be attributed to large light absorption in the ITO/glass system (9% at 550 nm, 12% at 750 nm).

3.1.2 Rolled glass. Large variations of reflectance were seen (Fig.2), especially in the 600-800 nm range, where trapping did not seem to be very efficient (Δ_2 = -16% which is the poorest performance in this study). However, trapping was not negligeable at lower-λ (Δ_1 = -20%), but, this region is less interesting than the high-λ range.

3.2 Scattering ITO thin films on flat glass

Spectra corresponding to 5 and 10% - Aerosil concentration were similar (Fig.3). In the 400-600 nm range, maybe the 5% sample is slightly more efficient for trapping (Table II), but the small difference in R_1 values of both samples was not believed to be significative. In the 600-800 nm range the 10% sample seem to better trapp light than the 5% one. It is remarkable that it is difficult to predict the trend of final values from haze ratio before Al/a-Si:H coating : H(5% sample) > H(10% sample) at 750 nm (Table I), while the opposite trend was seen for Δ_2.

4. CONCLUSION

Optically, etched substrates exhibited the highest haze ratio and also the best trapping performance, especially in the 600-800 nm, where trapping is wanted in a-Si solar cells. Unfortunatly, cells made with these substrates showed a large density of pin-holes. The rolled glass substrate did not seem suitable for PV applications, as trapping was poor in the high-λ range. Intermediate results were obtained with ITO + Aerosil substrates. Further work is necessary to enhance reproducebility. Integration in solar cells, as well extrapolation to a larger size could be further stages.

ACKNOLEDGEMENTS
The authors wish to thank C. Papret for valuable experimental work.

REFERENCES

[1] M.Green and P.Campbell.19th PV.Spec.Conf.New Orleans, 4-8 May 1987.
[2] H.W.Deckman and J.H.Dunsmuir. Appl.Phys.Lett. **41(4)**,377 (1982).
[3] P.Vilato. CEC Photovoltaïcs Contractors' Catalogue ISES Solar World Congress. Hamburg, 13-18 Sept. 1987.
[4] M.Misonou, M.Hyodo, H.Nagayama and H.Kawahara. 18th PV Spec. Conf. Las Vegas, 21-25 Oct. 1985.
[5] H.Ida, N.Shiba, T.Mishuku, H.Karasawa, A.Ito, M.Yamanaka and Y.Hayashi. IEEE Electron. Device Lett **4**, 157 (1983).

Parameters / Samples	550 nm			750 nm			
	T(%)	H(%)	R(%)	T(%)	H(%)	R(%)	R (Ω)
Reference	86	3.5	9	85	3.5	7.5	10
Chemically etched glass	82	82	9.5	82	80.5	7.5	15
Rolled glass	80.5	5.5	13	79	4	8	10
ITO + Aerosil (5 %)	77.5	22	14	79	14	12	18
ITO + Aerosil (10 %)	75	24	15	83.5	9	12.5	17

Table I : Transmittance, haze ratio, reflectance and sheet resistance of the different ITO substrates before Al/a-Si:H deposition. Reference : unscattering ITO on flat glass.

Parameters / Samples	R_1(%)	Δ_1(%)	R_2(%)	Δ_2(%)
Reference	20	0	38	0
Chemically etched glass	18	-10	24.5	-35
Rolled glass	20	0	32	-16
ITO + Aerosil (5 %)	16	-20	27	-29
ITO + Aerosil (10 %)	17	-15	25.5	-33

Table II : Calculated values drawn from spectrophotometric spectra of multilayered samples. Range 1 : 400-600 nm, range 2 : 600-800 nm. R_i and Δ_i are respectively representative of mean reflectance andrelative decrease of reflectance with respect to the reference (unscattering sample) due to light trapping.

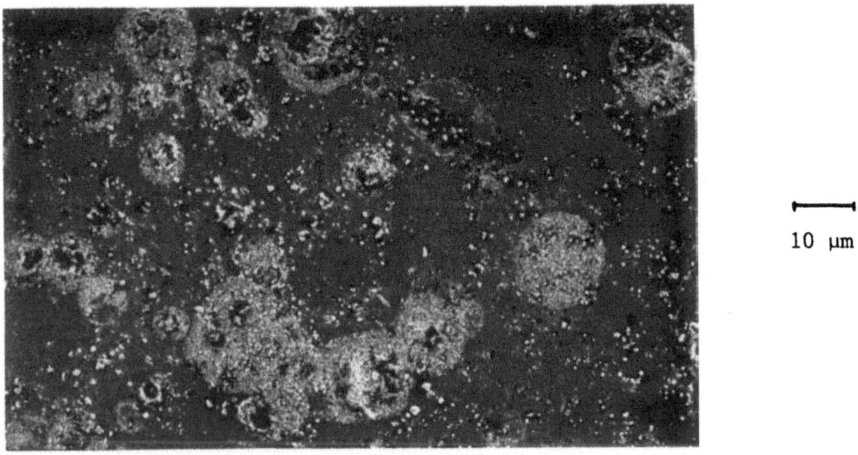

Fig. 1 : **SEM photography of the surface an ITO + Aerosil thin film deposited on flat glass. Aerosil concentration is 5 % for this sample.**

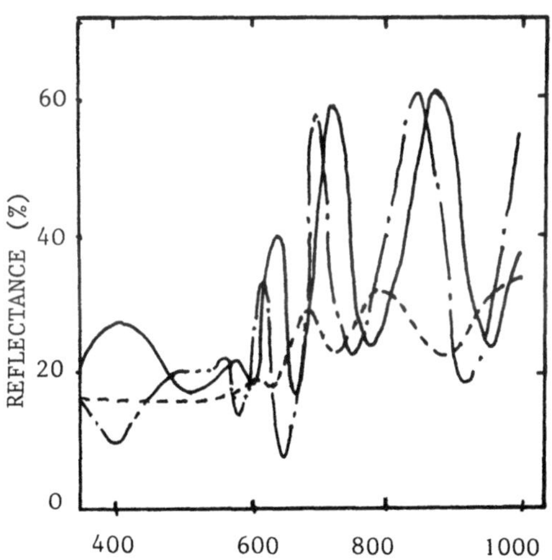

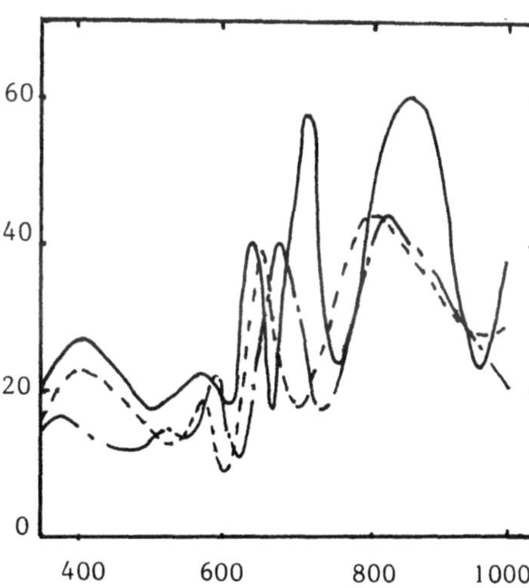

Reflectance spectra of the various scattering ITO/glass substrates. Also reported unscattering reference.

Fig. 2

———— : reference
- - - - - : chemically etched glass
—·—·— : rolled glass

Fig. 3

———— : reference
- - - - - : ITO + 5% Aerosil
—·—·— : ITO + 10% Aerosil

MIX
Papier aus verantwortungsvollen Quellen
Paper from responsible sources
FSC® C105338

If you have any concerns about our products,
you can contact us on
ProductSafety@springernature.com

In case Publisher is established outside the EU,
the EU authorized representative is:
Springer Nature Customer Service Center GmbH
Europaplatz 3, 69115 Heidelberg, Germany

Printed by Libri Plureos GmbH
in Hamburg, Germany